APPLICATION OF CONTROL VOLUME BASED FINITE ELEMENT METHOD (CVFEM) FOR NANOFLUID FLOW AND HEAT TRANSFER

APPLICATION OF CONTROL VOLUME BASED FINITE ELEMENT METHOD (CVFEM) FOR NANOFLUID FLOW AND HEAT TRANSFER

MOHSEN SHEIKHOLESLAMI
Department of Mechanical Engineering, Babol Noshirvani University of Technology, Babol, Islamic Republic of Iran

ELSEVIER

Elsevier
Radarweg 29, PO Box 211, 1000 AE Amsterdam, Netherlands
The Boulevard, Langford Lane, Kidlington, Oxford OX5 1GB, United Kingdom
50 Hampshire Street, 5th Floor, Cambridge, MA 02139, United States

British Library Cataloguing-in-Publication Data
A catalogue record for this book is available from the British Library

Library of Congress Cataloging-in-Publication Data
A catalog record for this book is available from the Library of Congress

ISBN: 978-0-12-814152-6

For Information on all Elsevier publications
visit our website at https://www.elsevier.com/books-and-journals

Publisher: Matthew Deans
Acquisition Editor: Simon Holt
Editorial Project Manager: Jeniffer Pierce/John Leonard
Production Project Manager: Sruthi Satheesh
Cover Designer: Harris Greg

Typeset by MPS Limited, Chennai, India

Contents

Biography

Mohsen Sheikholeslami Kandelousi (M. Sheikholeslami)
Department of Mechanical Engineering, Babol Noshirvani University of Technology, Babol, Islamic Republic of Iran

Dr. ***Mohsen Sheikholeslami*** works in the Department of Mechanical Engineering, Babol Noshirvani University of Technology, Iran. He is the head of renewable energy systems and nanofluid applications in the heat transfer laboratory, Babol Noshirvani University of Technology. His research interests include nanofluid, CFD, simulation, mesoscopic modeling, nonlinear science, magnetohydrodynamic, ferrohydrodynamic, electrohydrodynamic, and heat exchangers. He has written several papers and books in various fields of mechanical engineering. He is also the first author of *Applications of Nanofluid for Heat Transfer Enhancement, Application of Semi-Analytical Methods for Nanofluid Flow and Heat Transfer, Hydrothermal Analysis in Engineering Using Control Volume Finite Element Method, and External Magnetic Field Effects on Hydrothermal Treatment of Nanofluid*, published by Elsevier. According to Thomson Reuters' report (Clarivate Analytics), he was selected as the Web of Science *Highly Cited Researcher* (Top 0.01%) in 2016 and 2017.

Preface

In this book, I provide readers the new powerful numerical method, namely Control Volume-based Finite Element Method (CVFEM). This method uses the advantages of both the finite volume and finite element methods. It combines the flexibility of the finite element methods to discretize complex geometry with the conservative formulation of the finite volume methods, in which variables can be easily interpreted physically in terms of fluxes, forces, and sources. Applications of CVFEM in nanofluid simulation in the presence of various external forces are presented. This text is suitable for senior undergraduate students, postgraduate students, engineers, and scientists. In the first and second chapters, the basic idea of CVFEM is presented and then Navier–Stokes equations in the form of vorticity stream function formulation are solved. The third chapter of this book deals with the necessary fundamentals of nanotechnology. The various models for simulation of nanofluid are discussed. Homogeneous and Buongiorno models in various geometries are investigated in Chapter 4, Single-phase Model for Nanofluid Free Convection Heat Transfer by Means of CVFEM, and Chapter 5, Buongiorno Model for Nanofluid Treatment Using CVFEM. Chapter 6, Nanofluid Forced and Mixed Convection Heat Transfer by Means of CVFEM, deals with nanofluid forced and mixed convection heat transfer. Chapter 7, Effect of Uniform Lorentz Forces on Nanofluid Flow Using CVFEM, and Chapter 8, Influence of Variable Lorentz Forces on Nanofluid Free Convection Using CVFEM, give a complete account of uniform and nonuniform magnetic field effects on nanofluid natural convection heat transfer. Nanofluid forced convective heat transfer in the presence of a variable magnetic field is demonstrated in Chapter 9, Nanofluid Forced Convective Heat Transfer in Presence of Variable Magnetic Field Using CVFEM. Chapter 10, Influence of Shape Factor on Nanofluid Heat Transfer Improvement Using CVFEM, deals with the effect of nanoparticles' shape on nanofluid behavior. The influence of an electric field on natural and forced convection heat transfer of nanofluid is explained in Chapter 11, Electrohydrodynamic Nanofluid Natural Convection Using CVFEM, and Chapter 12, Forced Convection of Nanofluid in Existence of Electric Field Using CVFEM. There exist several models for the simulation of porous media: the Darcy model, the non-Darcy model, and the thermal nonequilibrium model. These various models are investigated in Chapters 13–15. Nonuniform magnetic field effect on nanofluid convective flow in a porous cavity is analyzed in Chapter 16, Nonuniform Magnetic Field Effect on Nanofluid Convective Flow in a Porous Cavity. The influence of thermal radiation on nanofluid flow in a porous medium in the presence of Coulomb forces is illustrated in Chapter 17, Thermal Radiation Influence on Nanofluid Flow in a Porous Medium in the Presence of Coulomb Forces Using CVFEM. Chapter 18, Influence of Electric Field on Forced Convection of Nanofluid in a Porous Medium by Means of CVFEM, deals with the influence of Coulomb forces on forced convection of nanofluid in a porous medium. Nanofluid heat transfer enhancement in the presence of a melting surface is analyzed in Chapter 19, Nanofluid Heat Transfer Enhancement in Presence of Melting Surface Using CVFEM. Chapter 20, Nanofluid Convective Heat Transfer Considering Magnetic Field Dependent (MFD) Viscosity by Means of CVFEM, gives the reader a full account of the theory and practice associated with nanofluid convective heat transfer considering magnetic field-dependent viscosity. A sample FORTRAN code is presented in the Appendix. The readers can extend this code for other complex geometries and physics.

Mohsen Sheikholeslami

Department of Mechanical Engineering, Babol Noshirvani University of Technology, Babol, Islamic, Republic of Iran

CHAPTER

1

Detailed Explanation of Control Volume-based Finite Element Method

1.1 INTRODUCTION

Fluid flow has several applications in engineering and nature. Mathematically, real flows are governed by a set of nonlinear partial differential equations in complex geometry. So, suitable solutions can be obtained through numerical techniques such as the finite difference method (FDM), the finite volume method (FVM), and the finite element method (FEM). In last decade, the finite element method has been developed for use in the area of computational fluid dynamics (CFD), and now this method has become a powerful method to simulate complex geometry. However, the finite volume method is applied the most to calculate fluid flows. The control volume finite element method (CVFEM) combines interesting characteristics from both the finite volume and finite element methods. The CVFEM was presented by Baliga and Patankar [1,2] using linear triangular finite elements, and by Raw and Schneider [3] using linear quadrilateral elements. Several authors have improved CVFEM from that time to now. Raw et al. [4] applied the nine-noded element to solve heat conduction problems. Banaszek [5] made a comparison of the Galerkin and CVFEM methods in diffusion problems using six-noded and nine-noded elements. Campos Silva [6] developed a computational program using nine-noded finite elements based on a control volume formulation to simulate 2-D transient, incompressible, viscous fluid flows. Campos Silva and Moura [7] and Campos Silva et al. [8] presented results for fluid flow problems. CVFEM combines the flexibility of the finite element methods to discretize complex geometry with the conservative formulation of the finite volume methods in which the variables have easy physical interpretation in terms of fluxes, forces, and sources. Saabas and Baliga [9,10] referenced a list of several works in FVMs and CVFEMs. Voller [11] presented the application of CVFEM for Fluids and Solids. Sheikholeslami et al. [12] studied the problem of natural convection between a circular enclosure and a sinusoidal cylinder. They concluded that streamlines, isotherms, and the number, size, and formation of the cells inside the enclosure strongly depend on the Rayleigh number, values of amplitude, and the number of undulations of the enclosure. Since the mid-2010s, this method has been applied for various applications [13–125].

1.2 THE DISCRETIZATION: GRID, MESH, AND CLOUD

Generally, three ways exist to place the node points into the domain [11].

1.2.1 Grid

A basic approach assigns the location of nodes using a structured grid where, in a two-dimensional domain, the location of a node is uniquely specified by a row and a column index, see Fig. 1.1A. Although such a structured approach can lead to very convenient and efficient discrete equations it lacks flexibility in accommodating complex geometries or allowing for the local concentration of nodes in solution regions of particular interest.

Application of Control Volume based Finite Element Method (CVFEM) for Nanofluid Flow and Heat Transfer.
DOI: https://doi.org/10.1016/B978-0-12-814152-6.00001-1

1.2.2 Mesh

Geometric flexibility, usually at the expense of solution efficiency, can be added by using an unstructured mesh. Fig. 1.1B shows an unstructured mesh of triangular elements. In two-dimensional domains triangular meshes are good selections because they can tessellate any planar surface. Note however, other choices of elements can be used in place of or in addition to triangular elements. The mesh can be used to determine the placement of the nodes. A common choice is to place the nodes at the vertices of the elements. In the case of triangles, this will allow for the approximation of a dependent variable, over the element, by linear interpolation between the vertex nodes. Higher order approximations can be arrived by using more nodes (e.g., placing nodes at midpoints) or alternative elements (e.g., quadrilaterals). In considering an unstructured mesh it is important to recognize the following:

1. The quality of the numerical solution obtained is critically dependent on the mesh. For example, a key quality requirement for a mesh of triangular elements is to avoid highly acute angles. The generation of appropriate meshes for a given domain is a complex topic worthy of a monograph in its own right. Fortunately, for two-dimensional problems in particular, there is a significant range of commercial and free software that can be used to generate quality meshes.
2. The term unstructured is used to indicate a lack of a global structure that relates the position of all the nodes in the domain. In practice, however, a local structure—the region of support—listing the nodes connected to a given node i, is required. Establishing, storing, and using this local data structure is one of the critical ingredients in using an unstructured mesh.

1.2.3 Cloud

The most flexible discretization is to simply populate the domain with node points that have no formal background grid or mesh connecting the nodes. Solution approaches based on this mesh-less form of discretization create "cloud" of neighboring nodes (see Fig. 1.1C [11]).

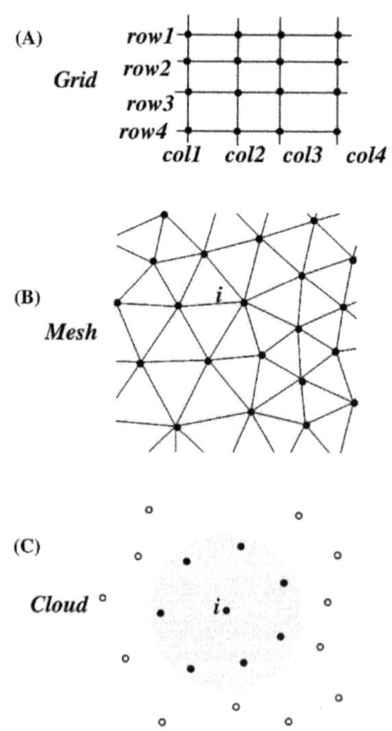

FIGURE 1.1 Different forms of discretization.

1.3 THE ELEMENT AND THE INTERPOLATION SHAPE FUNCTIONS

The building block of the discretization is the triangular element, Fig. 1.2. For linear triangular elements the node points are placed at the vertices. In Fig. 1.2 the nodes, moving in a counterclockwise direction, are labeled 1, 2, and 3. Values of the dependent variable ϕ are calculated and stored at these node points.

In this way, values at an arbitrary point (x,y) within the element can be approximated with linear interpolation

$$\phi \approx ax + by + c \tag{1.1}$$

where the constant coefficients a, b, and c satisfy the nodal relationships

$$\phi_i = ax_i + by_i + c, \qquad i = 1,2,3 \tag{1.2}$$

Eq. (1.1) can be more conveniently written in terms of the shape function, N_1, N_2, and N_3, where

$$N_i(x,y) = \begin{cases} 1 & \text{At node } i \\ 0 & \text{At all points on side opposite node } i \end{cases} \tag{1.3}$$

$$\sum_{i=1}^{3} N_i(x,y) = 1 \text{ At every point in the element} \tag{1.4}$$

such that, over the element, the continuous unknown field can be expressed as the linear combination of the values at nodes $i = 1,2,3$

$$\phi(x,y) \approx \sum_{i=1}^{3} N_i(x,y)\phi_i \tag{1.5}$$

With linear triangular elements a straightforward geometric derivation for the shape functions can be obtained. With reference to Fig. 1.2, observe that the area of the element is given by

$$A^{123} = \frac{1}{2}\begin{vmatrix} 1 & x_1 & y_1 \\ 1 & x_1 & y_1 \\ 1 & x_1 & y_1 \end{vmatrix} = \frac{1}{2}\left[(x_2y_3 - x_3y_2) - x_1(y_3 - y_2) + y_1(x_3 - x_2)\right] \tag{1.6}$$

and the area of the subelements with vertices at points $(p,2,3)$, $(p,3,1)$and $(p,1,2)$, where $p$is an arbitrary and variable point in the element, are given by

$$\begin{aligned} A^{p23} &= [(x_2y_3 - x_3y_2) - x_p(y_3 - y_2) + y_p(x_3 - x_2)] \\ A^{p31} &= [(x_3y_1 - x_1y_3) - x_p(y_1 - y_3) + y_p(x_1 - x_3)] \\ A^{p12} &= [(x_1y_2 - x_2y_1) - x_p(y_2 - y_1) + y_p(x_2 - x_1)] \end{aligned} \tag{1.7}$$

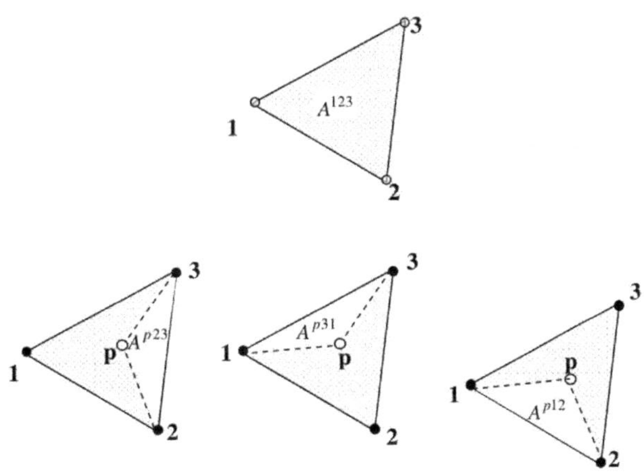

FIGURE 1.2 An element indicating the areas used in shape function definitions.

With these definitions it follows that the shape functions are given by

$$N_1 = A^{P23}/A^{123}, N_2 = A^{P31}/A^{123}, N_3 = A^{P12}/A^{123} \tag{1.8}$$

Note that, when point p coincides with node $i\,(1,2\ or\ 3)$, the shape function$N_i = 1$, and when point p is anywhere on the element side opposite node i, the associated subelement area is zero, and through Eq. (1.8), the shape function$N_i = 0$. Hence the shape functions defined by Eq. (1.8) satisfy the required condition in Eq. (1.3). Further, note that at any point p, the sum of the areas:

$$A^{P23} + A^{P31} + A^{P12} = A^{123} \tag{1.9}$$

such that the shape functions at (x_p, y_p)will sum to unity. Hence the shape functions defined by Eq. (1.8) also satisfy the condition Eq. (1.4). For future reference it is worthwhile to note that the derivatives of the shape functions in Eq. (1.8) over the element are the following constants

$$\begin{aligned}
N_{1x} &= \frac{\partial N_1}{\partial x} = \frac{(y_2 - y_3)}{2A^{123}}, \quad N_{1y} = \frac{\partial N_1}{\partial y} = \frac{(x_2 - x_3)}{2A^{123}} \\
N_{2x} &= \frac{\partial N_2}{\partial x} = \frac{(y_3 - y_2)}{2A^{123}}, \quad N_{2y} = \frac{\partial N_2}{\partial y} = \frac{(x_1 - x_3)}{2A^{123}} \\
N_{3x} &= \frac{\partial N_3}{\partial x} = \frac{(y_1 - y_2)}{2A^{123}}, \quad N_{3y} = \frac{\partial N_3}{\partial x} = \frac{(x_2 - x_1)}{2A^{123}}
\end{aligned} \tag{1.10}$$

1.4 REGION OF SUPPORT AND CONTROL VOLUME

The local structure on the mesh in Fig. 1.1B is defined in terms of the region of support—the list of nodes that share a common element with a given node i, in Fig. 1.3 [11]. In this region of support, as illustrated in Fig. 1.3, a control volume is created by joining the center of each element in the support to the midpoints of the element sides that pass through node i. This creates a closed polygonal control volume with $2m$ sides (faces); where m is the number of elements in the support. Each element contributes 1/3 of its area to the control volume area and the volumes from all the nodes tessellate the domain without overlap.

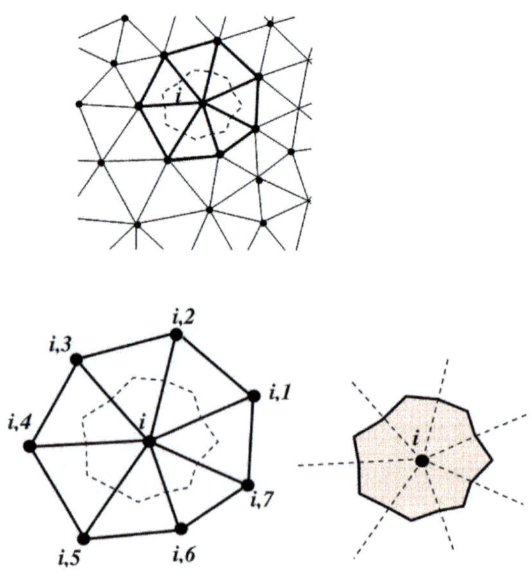

FIGURE 1.3 Region of support and control volume for node i in an unstructured mesh of Linear triangular elements.

1.5 DISCRETIZATION AND SOLUTION

1.5.1 Steady State Advection–Diffusion With Source Terms

To illustrate the solution procedure using the CVFEM, one can consider the general form of advection–diffusion equation for node i in integral form:

$$-\int_V Q\, dV - \int_A k\nabla\phi\cdot n\, dA + \int_A (v\cdot n)\phi\, dA = 0 \tag{1.11}$$

or point form

$$-\nabla\cdot(k\nabla\phi) + \nabla\cdot(v\phi) - Q = 0 \tag{1.12}$$

which can be represented by the system of CVFEM discrete equations as:

$$[a_i + Qc_i + Bc_i]\phi_i = \sum_{j=1}^{n_i} a_{i,j}\phi_{S_{i,j}} + Q_{B_i} + B_{B_i} \tag{1.13}$$

In the above, the a s are the coefficients, the index (i, j) indicates the jth node in the support of node i, the index $S_{i,\,j}$ provides the node number of the jth node in the support, the Bs account for boundary conditions, and the Qs for source terms. For the selected triangular element which is shown in Fig. 1.4 this approximation without considering the source term leads to

$$-\left(a_1^k + a_1^u\right)\phi_i + \left(a_2^k + a_2^u\right)\phi_{S_{i,3}} + \left(a_2^k + a_2^u\right)\phi_{S_{i,4}} = 0 \tag{1.14}$$

Using upwinding the advective coefficients, identified with the superscripts $(\,)^u$, are given by

$$\begin{aligned} a_1^u &= \max\left[q_{f1}, 0\right] + \max\left[q_{f2}, 0\right] \\ a_2^u &= \max\left[-q_{f1}, 0\right] \\ a_3^u &= \max\left[-q_{f2}, 0\right] \end{aligned} \tag{1.15}$$

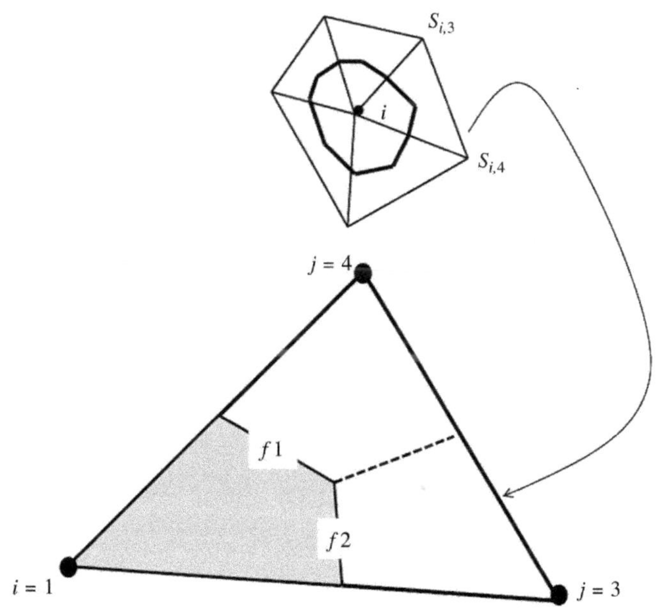

FIGURE 1.4 A sample triangular element and its corresponding control volume.

And the diffusion coefficients, identified with the superscripts $(\,)^k$, are given by

$$\begin{aligned}
a_1^k &= -k_{f1}N_{1x}\Delta\vec{y}_{f1} + k_{f1}N_{1y}\Delta\vec{x}_{f1} - k_{f2}N_{1x}\Delta\vec{y}_{f2} + k_{f2}N_{1y}\Delta\vec{x}_{f2} \\
a_2^k &= -k_{f1}N_{2x}\Delta\vec{y}_{f1} + k_{f1}N_{2y}\Delta\vec{x}_{f1} - k_{f2}N_{2x}\Delta\vec{y}_{f2} + k_{f2}N_{2y}\Delta\vec{x}_{f2} \\
a_2^k &= -k_{f1}N_{3x}\Delta\vec{y}_{f1} + k_{f1}N_{3y}\Delta\vec{x}_{f1} - k_{f2}N_{3x}\Delta\vec{y}_{f2} + k_{f2}N_{3y}\Delta\vec{x}_{f2}
\end{aligned} \tag{1.16}$$

In Eq. (1.27) the volume flow across face 1 and 2 in the direction of the outward normal, is

$$\begin{aligned}
q_{f1} &= v \cdot n\, A\Big|_{f1} = v_x^{f1}\Delta\vec{y}_{f1} - v_y^{f1}\Delta\vec{y}_{f1} \\
q_{f2} &= v \cdot n\, A\Big|_{f2} = v_x^{f2}\Delta\vec{y}_{f2} - v_y^{f2}\Delta\vec{y}_{f2}
\end{aligned} \tag{1.17}$$

The value of the diffusivity at the mid-point of face 1 can be obtained as

$$k_{f1} = [N_1k_1 + N_2k_2 + N_3k_3]_{f1} = \frac{5}{12}k_1 + \frac{5}{12}k_2 + \frac{2}{12}k_3 \tag{1.18}$$

and at the midpoint of face 2

$$k_{f2} = [N_1k_1 + N_2k_2 + N_3k_3]_{f2} = \frac{5}{12}k_1 + \frac{2}{12}k_2 + \frac{5}{12}k_3 \tag{1.19}$$

The velocity components at the midpoint of face 1 are:

$$\begin{aligned}
v_x^{f1} &= \frac{5}{12}v_{x_1} + \frac{5}{12}v_{x_2} + \frac{2}{12}v_{x_3} \\
v_y^{f1} &= \frac{5}{12}v_{y_1} + \frac{5}{12}v_{y_2} + \frac{2}{12}v_{y_3}
\end{aligned} \tag{1.20}$$

And on face 2:

$$\begin{aligned}
v_x^{f2} &= \frac{5}{12}v_{x_1} + \frac{2}{12}v_{x_2} + \frac{5}{12}v_{x_3} \\
v_y^{f2} &= \frac{5}{12}v_{y_1} + \frac{2}{12}v_{y_2} + \frac{5}{12}v_{y_3}
\end{aligned} \tag{1.21}$$

These values can be used to *update* the ith support coefficients through the following equation:

$$\begin{aligned}
a_i &= a_i + a_1^k \\
a_{i,3} &= a_{i,3} + a_2^k \\
a_{i,4} &= a_{i,4} + a_3^k
\end{aligned} \tag{1.22}$$

In Eq. (1.28), moving counterclockwise around node i, the signed distances are:

$$\begin{aligned}
\Delta\vec{x}_{f1} &= \frac{x_3}{3} - \frac{x_2}{6} - \frac{x_1}{6}, \quad \Delta\vec{x}_{f2} = -\frac{x_2}{3} + \frac{x_3}{6} + \frac{x_1}{6} \\
\Delta\vec{y}_{f1} &= \frac{y_3}{3} - \frac{y_2}{6} - \frac{y_1}{6}, \quad \Delta\vec{y}_{f2} = -\frac{y_2}{3} + \frac{y_3}{6} + \frac{y_1}{6}
\end{aligned} \tag{1.23}$$

the derivatives of the shape functions are:

$$\begin{aligned}
N_{1x} &= \frac{\partial N_1}{\partial x} = \frac{(y_2 - y_3)}{2V^{ele}}, N_{1y} = \frac{\partial N_1}{\partial y} = \frac{(x_3 - x_2)}{2V^{ele}} \\
N_{2x} &= \frac{\partial N_2}{\partial x} = \frac{(y_3 - y_1)}{2V^{ele}}, N_{2y} = \frac{\partial N_1}{\partial y} = \frac{(x_1 - x_3)}{2V^{ele}} \\
N_{3x} &= \frac{\partial N_2}{\partial x} = \frac{(y_1 - y_2)}{2V^{ele}}, N_{3y} = \frac{\partial N_3}{\partial y} = \frac{(x_2 - x_1)}{2V^{ele}}
\end{aligned} \tag{1.24}$$

and the volume of the element is

$$V^{ele} = \frac{(x_2y_3 - x_3y_2) + x_1(y_2 - y_3) + y_1(x_3 - x_2)}{2} \tag{1.25}$$

The obtained algebraic equations from the discretization procedure using CVFEM are solved by the Gauss–Seidel Method.

1.5.2 Implementation of Source Terms and Boundary Conditions

The boundary conditions for the present problem can be enforced using $B_{B_i}$and B_{C_i}as follows:

$$\text{Insulated boundary: } B_{B_i} = 0 \text{ and } B_{C_i} = 0 \tag{1.26}$$

$$\text{Fixed Value Boundary: } B_{B_i} = \phi_{value} \times 10^{16} \text{ and } B_{C_i} = 10^{16} \tag{1.27}$$

$$\text{Fixed Flux Boundary: } B_{B_i} = A_k \times q'' \text{ and } B_{C_i} = 0 \tag{1.28}$$

Where ϕ_{value} is the prescribed value on the boundary and $A_k$is the length of the control volume surface on the boundary segment.

To provide a general treatment for boundary conditions some preliminary calculation of the boundary face areas associated with each node j in a given boundary segment are required. Fig. 1.5 shows a schematic of the kth ($k = 3$) boundary, indicating the data structure. Assuming unit depth, the face area associated with any node j of the boundary segment highlighted in Fig 1.5 is given by

$$A_{k,j} = \begin{cases} Upper_1 \\ Upper_j + Lower_j, \; 2 \le j \le n_{B,k} - 1 \\ Lower_{n_{B,k}} \end{cases} \tag{1.29}$$

where

$$\begin{aligned} Upper_j &= \frac{1}{2}\sqrt{(x_j - x_{j+1})^2 + (y_j - y_{j+1})^2} \\ Lower_j &= \frac{1}{2}\sqrt{(x_j - x_{j-1})^2 + (y_j - y_{j-1})^2} \end{aligned} \tag{1.30}$$

The first and last lines on the right-hand side of Eq. (1.29) account for the first and last node on the boundary segment. This treatment assumes that there are at least two boundary segments—i.e., at least two contiguous regions of the domain boundary where different boundary conditions are applied. In cases where only one boundary condition is applied (e.g., a constant value over the whole boundary) the condition of a two-segment boundary can be artificially imposed. The volume source terms can be applied to Eq. (1.13) as:

$$\sum_{j=1}^{elements} \int_{V_j} Q \, dV \approx Q_i V_i \tag{1.31}$$

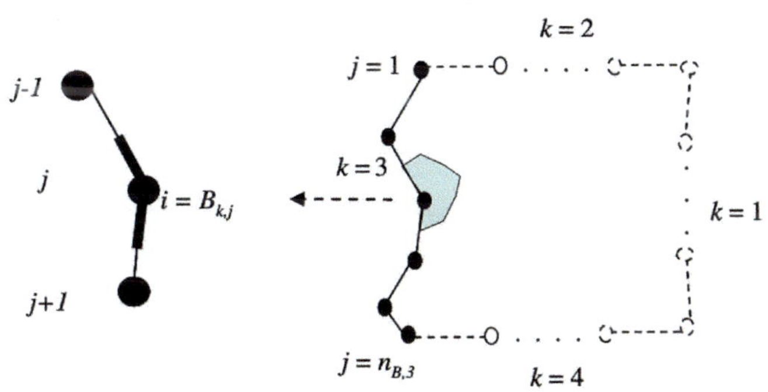

FIGURE 1.5 A domain boundary segment with three sequential points detailed.

or after linearizing as the source term

$$Q_i V_i = -Q_{C_i}\phi_i + Q_{B_i} \tag{1.32}$$

1.5.3 Unsteady Advection–Diffusion With Source Terms

From the above the net flow rate of quantity into the control volume around node i can be approximated as

$$Net_i = \int_{V_i} Q dV + \sum_{j=1}^{n_i} \int_{A^j} \kappa \nabla \phi . n dA - \int_{A^j} (\nu . n)\phi dA \approx -[a_i + Q_{C_i} + B_{C_i}]\phi_i + \sum_{j=1}^{n_i} a_{i,j}\phi_{S_{i,j}} + Q_{B_i} + B_{B_i} \tag{1.33}$$

In a steady-state problem this flow rate will be identically zero. In a transient problem, however, it will result in a change in storage of the quantity in the control volume, i.e.,

$$\frac{d}{dt}\int_{V_i} \phi dV = Net_i \tag{1.34}$$

Using nodal lumping for the volume integration and finite difference in time this equation can be used to evaluate the nodal field values at time $t + \Delta t$in terms of the nodal field values at time t,

$$V_i \phi_i^{new} = V_i \phi_i + \Delta t[(1-\theta)Net_i + \theta Net_i^{new}] \tag{1.35}$$

where $\Delta t$is a time step and the superscript $()^{new}$indicates evaluation at time $t + \Delta t$. The parameter $0 \le \theta \le 1$is a user-defined weighting factor, used to approximate the net flow into control volume i during the time interval $[t, t + \Delta t]$, in terms of the net flows at the beginning and end of the time step. Neglecting, for now, contributions from the boundaries and sources, the resulting discrete equations for three choices of θ are:

Fully Implicit $\theta = 1$:

$$V_i \phi_i^{new} = V_i \phi_i + \Delta t \left(\sum_{j=1}^{n_i} a_{i,j}\phi_{S_{i,j}}^{new} - a_i \phi_i^{new} \right) \tag{1.36}$$

The advantage of this choice is that it is unconditionally stable—i.e., for any choice of time step errors (induced or inherent) will not grow. The downside is that a system of equations needs to be solved to obtain the field values at the new time step $\phi_i^{new}(i = 1..n)$.

Crank–Nicolson $\theta = 0.5$:

$$V_i \phi_i^{new} = V_i \phi_i + \frac{\Delta t}{2}\left(\sum_{j=1}^{n_i} a_{i,j}\phi_{S_{i,j}}^{new} - a_i \phi_i^{new} \right) + \frac{\Delta t}{2}\left(\sum_{j=1}^{n_i} a_{i,j}\phi_{S_{i,j}} - a_i \phi_i \right) \tag{1.37}$$

This unconditionally stable scheme also requires a system solution. Unlike the 1st order in time, fully implicit method, however, the Crank Nicolson is 2nd order in time (i.e., time errors scale withΔt^2)

Fully Explicit $\theta = 0$:

$$V_i \phi_i^{new} = V_i \phi_i + \Delta t \left(\sum_{j=1}^{n_i} a_{i,j}\phi_{S_{i,j}} - a_i \phi_i \right) \tag{1.38}$$

This choice can be used to directly update the new time level values from the current time values without solution of a system of equations. This, however, comes at the price of a restriction on the time step size to ensure stability. The solution of Eq. (1.38) is likely to become unstable (error grow as the solution advances) if the net coefficient for $\phi_i$becomes negative. This requires that the time step is chosen such that

$$\Delta t < \min\left(\frac{V_i}{a_i}\right), \; (i = 1 \ldots n) \tag{1.39}$$

If a fine grid is used this value could be prohibitively small, i.e., the advance of time could be too slow to reach a practical value within reasonable computational resources. In many cases, however, the drawback for using a small time step is offset by the ability to solve Eq. (1.29) without iteration. Driven by its simplicity and flexibility of modeling complex nonlinear terms the explicit time integration approach is the preferred choice in this work. When sources and boundary condition treatments are added the explicit scheme Eq. (1.29) can be written as

$$(V_i + B_{C_i})\phi_i^{new} = V_i\phi_i + \Delta t\left(\sum_{j=1}^{n_i} a_{i,j}\phi_{S_{i,j}} - a_i\phi_i + Q_{B_i} - Q_{C_i}\phi_i\right) + B_{B_i} \tag{1.40}$$

The addition of the source term in Eq. (1.40) could require a further reduction in the time step to retain positive coefficients and stability, i.e.,

$$\Delta t < \min\left(\frac{V_i}{a_i + Q_{C_i}}\right),\ (i = 1\ldots n) \tag{1.41}$$

References

[1] B.R. Baliga, S.V. Patankar, A new finite element formulation for convection diffusion problems, Numer. Heat Transfer 3 (1980) 393–409.

[2] B.R. Baliga, S.V. Patankar, A control volume finite-element method for two dimensional fluid flow and heat transfer, Numer. Heat Transfer 6 (1983) 245–261.

[3] M.J. Raw, G.E. Schneider, A. Skewed, Positive influence coeffcient up-winding procedure for control volume based finite element convection-diffusion computation, Numer. Heat Transfer 9 (1986) 1–26.

[4] M.J. Raw, G.E. Schneider, V. Hassani, A nine-noded quadratic control volume based finite element for heat conduction, J. Spacecraf t 22 (1985) 523–529.

[5] J. Banaszek, Comparison of control volume and galerkin finite element methods for diffusion type problems, Numer. Heat Transfer 16 (1989) 59–78.

[6] J.B. Campos Silva, F.D.R. Vieira, L.F.M. Moura, Control volume finite element and flow visualization methods applied for unsteady viscous flow past a circular cylinder, Proc. V Congresso de Engenhari a Mecanica do Norte-Nordeste, Fortaleza, CE, Brazil vol. II (1998) 80–87.

[7] J.B. Campos Silva, L.F.M. Moura, Numerical Simulation of Fluid Flow by the Control Volume Finite Element Method, COBEM 1997, Proc. XIV Brazilian Congres s of Mechanical Engineering, Bauru, SP, Brazil, CD-ROM paper code 041, pp. 1-8, 1997.

[8] J.B. Campos Silva, Numerical Simulation of Fluid Flow by the Finite Element Method Based on Control Volumes (in Portuguese), Ph.D. thesis, State University of Campinas, Campinas, SP, Brazil, 1998.

[9] H.J. Saabas, B.R. Baliga, Co-located equal order control volume finite element method for multidimensional incompressible fluid flow-Part I: formulation, Numer. Heat Transfer 26 (1994) 381–407.

[10] H.J. Saabas, B.R. Baliga, Co-located equal order control volume finite element method for multidimensional incompressible fluid flow-Part II: verification, Numer. Heat Transfer B vol. 26 (1994) 409–424.

[11] V.R. Voller, 5 Tohccxxvc Basic Control Volume Finite Element Methods for Fluids and Solids, World Scientific Publishing Co. Pte. Ltd., 2009.

[12] M. Sheikholeslami, M. Gorji-Bandpy, I. Pop, Soheil Soleimani, Numerical study of natural convection between a circular enclosure and a sinusoidal cylinder using control volume based finite element method, Int. J. Therm. Sci. 72 (2013) 147–158.

[13] M. Sheikholeslami, M. Gorji-Bandpy, D.D. Ganji, Soheil Soleimani, Natural convection heat transfer in a cavity with sinusoidal wall filled with CuO-water nanofluid in presence of magnetic field, J. Taiwan Inst. Chem. Eng. 45 (2014) 40–49.

[14] M. Sheikholeslami, M. Gorji-Bandpy, D.D. Ganji, Soheil Soleimani, Effect of a magnetic field on natural convection in an inclined half-annulus enclosure filled with Cu–water nanofluid using CVFEM, Adv. Powder Technol. 24 (2013) 980–991.

[15] M. Sheikholeslami, I. Hashim, Soheil Soleimani, Numerical Investigation of the Effect of Magnetic Field on Natural Convection in a Curved-Shape Enclosure, Hindawi Publishing Corporation Mathematical Problems in Engineering Volume 2013, Article ID 831725, 11 pages https://doi.org/10.1155/2013/831725.

[16] M. Sheikholeslami, D.D. Ganji, M. Gorji-Bandpy, Soheil Soleimani, Magnetic field effect on nanofluid flow and heat transfer using KKL model, J. Taiwan Inst. Chem. Eng. 45 (2014) 795–807.

[17] M. Sheikholeslami, M. Gorji-Bandpy, D.D. Ganji, P. Rana, Soheil Soleimani, Magnetohydrodynamic free convection of Al2O3-water nanofluid considering Thermophoresis and Brownian motion effects, Comput. Fluids 94 (2014) 147–160.

[18] M. Sheikholeslami, D.D. Ganji, Ferrohydrodynamic and Magnetohydrodynamic effects on ferrofluid flow and convective heat transfer, Energy 75 (2014) 400–410.

[19] M. Sheikholeslami, D.D. Ganji, M.M. Rashidi, Ferrofluid flow and heat transfer in a semi annulus enclosure in the presence of magnetic source considering thermal radiation, J. Taiwan Inst. Chem. Eng. 47 (2015) 6–17.

[20] M. Sheikholeslami, R. Ellahi, M. Hassan, Soheil Soleimani, A study of natural convection heat transfer in a nanofluid filled enclosure with elliptic inner cylinder, Int. J. Numer. Methods Heat & Fluid Flow 24 (2014) 8.

[21] M.S. Kandelousi, Effect of spatially variable magnetic field on ferrofluid flow and heat transfer considering constant heat flux boundary condition, Europ. Phys. J. Plus (2014) 129–248.

[22] M. Sheikholeslami, M.M. Rashidi, Effect of space dependent magnetic field on free convection of Fe3O4-water nanofluid, J. Taiwan Inst. Chem. Eng. 56 (2015) 6–15.

[23] M. Sheikholeslami, M.M. Rashidi, D.D. Ganji, Effect of non-uniform magnetic field on forced convection heat transfer of Fe3O4- water nanofluid, Comput. Methods Appl. Mech. Engrg. 294 (2015) 299–312.

[24] M. Sheikholeslami, R. Ellahi, Electrohydrodynamic nanofluid hydrothermal treatment in an enclosure with sinusoidal upper wall, Appl. Sci. 5 (2015) 294–306. Available from: https://doi.org/10.3390/app5030294.

[25] M. Sheikholeslami, M.M. Rashidi, D.D. Ganji, Numerical investigation of magnetic nanofluid forced convective heat transfer in existence of variable magnetic field using two phase model, J. Mol. Liquids 212 (2015) 117–126.

[26] M. Sheikholeslami, K. Vajravelu, M.M. Rashidi, Forced convection heat transfer in a semi annulus under the influence of a variable magnetic field, Int. J. Heat Mass Transfer 92 (2016) 339–348.

[27] Mohsen Sheikholeslami, Ali J. Chamkha, Electrohydrodynamic free convection heat transfer of a nanofluid in a semi-annulus enclosure with a sinusoidal wall, Numer. Heat Transfer, Part A 69 (7) (2016) 781–793. Available from: http://dx.doi.org/10.1080/10407782.2015.1090819.

[28] M. Sheikholeslami, Soheil Soleimani, D.D. Ganji, Effect of electric field on hydrothermal behavior of nanofluid in a complex geometry, J. Mol. Liquids 213 (2016) 153–161.

[29] M. Sheikholeslami, M.M. Rashidi, Non-uniform magnetic field effect on nanofluid hydrothermal treatment considering Brownian motion and thermophoresis effects, J. Braz. Soc. Mech. Sci. Eng. 38 (2016) 1171–1184.

[30] M. Sheikholeslami, M.M. Rashidi, Ferrofluid heat transfer treatment in the presence of variable magnetic field, Eur. Phys. J. Plus 130 (2015) 115.

[31] M. Sheikholeslami, A.J. Chamkha, Flow and convective heat transfer of a ferro-nanofluid in a double-sided lid-driven cavity with a wavy wall in the presence of a variable magnetic field, Numer. Heat Transfer, Part A 69 (10) (2016) 1186–1200. Available from: https://doi.org/10.1080/10407782.2015.1125709.

[32] M. Sheikholeslami, M.M. Rashidi, T. Hayat, D.D. Ganji, Free convection of magnetic nanofluid considering MFD viscosity effect, J. Mol. Liquids 218 (2016) 393–399.

[33] M. Sheikholeslami, T. Hayat, A. Alsaedi, MHD free convection of Al2O3–water nanofluid considering thermal radiation: a numerical study, Int. J. Heat Mass Transfer 96 (2016) 513–524.

[34] M. Sheikholeslami, M. Nimafar, D.D. Ganji, M. Pouyandehmehr, CuO-H2O nanofluid hydrothermal analysis in a complex shaped cavity, Int. J. Hydrogen Energy 41 (2016) 17837–17845.

[35] M. Sheikholeslami, T. Hayat, A. Alsaedi, Numerical study for external magnetic source influence on water based nanofluid convective heat transfer, Int. J. Heat Mass Transfer 106 (2017) 745–755.

[36] M. Sheikholeslami, Influence of Coulomb forces on Fe3O4-H2O nanofluid thermal improvement, Int. J. Hydrogen Energy 42 (2017) (2016) 821–829.

[37] M. Sheikholeslami, S.A. Shehzad, Magnetohydrodynamic nanofluid convection in a porous enclosure considering heat flux boundary condition, Int. J. Heat Mass Transfer 106 (2017) 1261–1269.

[38] M. Sheikholeslami, T. Hayat, A. Alsaedi, S. Abelman, EHD nanofluid force convective heat transfer considering electric field dependent viscosity, Int. J. Heat Mass Transfer 108 (2017) 2558–2565. Available from: http://dx.doi.org/10.1016/j.ijheatmasstransfer.2016.10.099.

[39] M. Sheikholeslami, Houman B. Rokni, Magnetohydrodynamic CuO-water nanofluid in a porous complex shaped enclosure, ASME J. Therm. Sci. Eng. Applicat. 9 (4) (2017) 041007. Available from: https://doi.org/10.1115/1.4035973.

[40] Mohsen Sheikholeslami, CVFEM for magnetic nanofluid convective heat transfer in a porous curved enclosure, Eur. Phys. J. Plus 131 (2016) 413. Available from: https://doi.org/10.1140/epjp/i2016-16413-y.

[41] M. Sheikholeslami, D.D. Ganji, Ferrofluid convective heat transfer under the influence of external magnetic source, Alexandria Eng. J. (2016). Available from: http://dx.doi.org/10.1016/j.aej.2016.11.007.

[42] M. Sheikholeslami, H.B. Rokni, Nanofluid two phase model analysis in existence of induced magnetic field, Int. J. Heat Mass Transfer 107 (2017) 288–299.

[43] M. Sheikholeslami, D.D. Ganji, CVFEM for free convective heat transfer of CuO-water nanofluid in a tilted semi annulus, Alexandria Eng. J. (2016). Available from: http://dx.doi.org/10.1016/j.aej.2016.11.012.

[44] M. Sheikholeslami, M. Barzegar Gerdroodbary, D.D. Ganji, Numerical investigation of forced convective heat transfer of Fe3O4-water nanofluid in presence of external magnetic source, Comput. Methods Appl. Mech. Eng. 315 (2017) 831–845.

[45] M. Sheikholeslami, K. Vajravelu, Nanofluid flow and heat transfer in a cavity with variable magnetic field, Appl. Math. Comput. 298 (2017) 272–282.

[46] M. Sheikholeslami, Magnetic source impact on nanofluid heat transfer using CVFEM, Neural Comput. Applicat. (2016). Available from: https://doi.org/10.1007/s00521-016-2740-7.

[47] M. Sheikholeslami, Numerical simulation of magnetic nanofluid natural convection in porous media, Phys. Lett. A 381 (2017) 494–503.

[48] M. Sheikholeslami, Magnetic field influence on nanofluid thermal radiation in a cavity with tilted elliptic inner cylinder, J. Mol. Liquids J. Mol.r Liquids 229 (2017) 137–147.

[49] M. Sheikholeslami, D.D. Ganji, Transportation of MHD nanofluid free convection in a porous semi annulus using numerical approach, Chem. Phys. Lett. 669 (2017) 202–210.

[50] M. Sheikholeslami, Houman B. Rokni, Numerical modeling of nanofluid natural convection in a semi annulus in existence of Lorentz force, Comput. Methods Appl. Mech. Eng. 317 (2017) 419–430.

[51] M. Sheikholeslami, Numerical investigation of MHD nanofluid free convective heat transfer in a porous tilted enclosure, Eng. Comput. 34 (6) (2017) 1939–1955.

[52] M. Sheikholeslami, D.D. Ganji, Free convection of Fe3O4-water nanofluid under the influence of an external magnetic source, J. Mol. Liq. 229 (2017) 530–540. Available from: http://dx.doi.org/10.1016/j.molliq.2016.12.101.

[53] M. Sheikholeslami, D.D. Ganji, Impact of electric field on nanofluid forced convection heat transfer with considering variable properties, J. Mol. Liq. 229 (2017) 566–573. Available from: https://doi.org/10.1016/j.molliq.2016.12.107.

[54] M. Sheikholeslami, CuO-water nanofluid free convection in a porous cavity considering Darcy law, Eur. Phys. J. Plus 132 (2017) 55. Available from: https://doi.org/10.1140/epjp/i2017-11330-3.

[55] M. Sheikholeslami, P. Rana, Soheil Soleimani, Numerical study of mhd natural convection liquid metal flow and heat transfer in a wavy enclosure using CVFEM, Heat Transfer Res. 48 (2) (2017) 121–138.

[56] M. Sheikholeslami, R. Ellahi, K. Vafai, Study of Fe3O4-water nanofluid with convective heat transfer in the presence of magnetic source, Alexandria Eng. J. (2017). Available from: http://dx.doi.org/10.1016/j.aej.2017.01.027.
[57] M. Sheikholeslami, Z. Ziabakhsh, D.D. Ganji, Transport of Magnetohydrodynamic nanofluid in a porous media, Colloids Surf., A 520 (2017) 201–212.
[58] M. Sheikholeslami, S.A. Shehzad, Thermal radiation of ferrofluid in existence of Lorentz forces considering variable viscosity, Int. J. Heat Mass Transfer 109 (2017) 82–92.
[59] M. Sheikholeslami, M.M. Bhatti, Active method for nanofluid heat transfer enhancement by means of EHD, Int. J. Heat Mass Transfer 109 (2017) 115–122.
[60] M. Sheikholeslami, M. Shamlooei, Fe3O4- H2O nanofluid natural convection in presence of thermal radiation, Int. J. Hydrogen Energy 42 (9) (2017) 5708–5718.
[61] M. Sheikholeslami, Houman B. Rokni, Influence of EFD viscosity on nanofluid forced convection in a cavity with sinusoidal wall, J. Mol. Liq. 232 (2017) 390–395.
[62] M. Sheikholeslami, D.D. Ganji, Numerical approach for magnetic nanofluid flow in a porous cavity using CuO nanoparticles, Mater. Design 120 (2017) 382–393.
[63] M. Sheikholeslami, K. Vajravelu, Forced convection heat transfer in Fe3O4- ethylene glycol nanofluid under the influence of Coulomb force, J. Mol. Liq. 233 (2017) 203–210.
[64] M. Sheikholeslami, D.D. Ganji, Numerical analysis of nanofluid transportation in porous media under the influence of external magnetic source, J. Mol. Liq. 233 (2017) 499–507.
[65] M. Sheikholeslami, A. Zeeshan, Analysis of flow and heat transfer in water based nanofluid due to magnetic field in a porous enclosure with constant heat flux using CVFEM, Comput. Methods Appl. Mech. Eng. 320 (2017) 68–81.
[66] M. Sheikholeslami, Houman B. Rokni, Magnetic nanofluid natural convection in presence of thermal radiation considering variable viscosity, Europ. Phys. J. Plus (2017).
[67] M. Sheikholeslami, M.M. Bhatti, Forced convection of nanofluid in presence of constant magnetic field considering shape effects of nanoparticles, Int. J. Heat Mass Transfer 111 (2017) 1039–1049.
[68] M. Sheikholeslami, Numerical simulation for external magnetic field influence on Fe3O4-water nanofluid forced convection, Engineering Computations.
[69] M. Sheikholeslami, S. Abelman, Numerical study of the effect of magnetic field on Fe3O4-water ferrofluid convection with thermal radiation, Engineering Computations.
[70] M. Sheikholeslami, Ali J. Chamkha, P. Rana, R. Moradi, Combined thermophoresis and Brownian motion effects on nanofluid free convection heat transfer in an L-shaped enclosure, Chin. J. Phys. 55 (2017) 2356–2370.
[71] M. Sheikholeslami, H.B. Rokni, Melting heat transfer influence on nanofluid flow inside a cavity in existence of magnetic field, Int. J. Heat Mass Transfer 114 (2017) 517–526.
[72] Mohsen Sheikholeslami, M.K. Sadoughi, Simulation of CuO- water nanofluid heat transfer enhancement in presence of melting surface, Int. J. Heat Mass Transfer 116 (2018) 909–919.
[73] M. Sheikholeslami, R. Ellahi, C. Fetecau, CuO-water nanofluid magnetohydrodynamic natural convection inside a sinusoidal annulus in presence of melting heat transfer, Hindawi, Mathematical Problems in Engineering, Volume 2017, Article ID5830279, 9 pages https://doi.org/10.1155/2017/5830279.
[74] M. Sheikholeslami, H.B. Rokni, Simulation of nanofluid heat transfer in presence of magnetic field: a review, Int. J. Heat Mass Transfer 115 (2017) 1203–1233.
[75] M. Sheikholeslami, H. Oztop, MHD free convection of nanofluid in a cavity with sinusoidal walls by using CVFEM, Chin. J. Phys. 55 (6) (2017) 2291–2304.
[76] M. Sheikholeslami, D.D. Ganji, R. Moradi, Heat transfer of Fe3O4 -water nanofluid in a permeable medium with thermal radiation in existence of constant heat flux, Chem. Eng. Sci. 174 (2017) 326–336.
[77] M. Sheikholeslami, T. Hayat, A. Alsaedi, On simulation of nanofluid radiation and natural convection in an enclosure with elliptical cylinders, Int. J. Heat Mass Transfer 115 (2017) 981–991.
[78] M. Sheikholeslami, M.K. Sadoughi, Numerical modeling for Fe3O4 -water nanofluid flow in porous medium considering MFD viscosity, J. Mol. Liq. 242 (2017) 255–264.
[79] M. Sheikholeslami, H.B. Rokni, Nanofluid convective heat transfer intensification in a porous circular cylinder, Chem. Eng. Proces. Process Intensif. 120 (2017) 93–104.
[80] M. Sheikholeslami, M. Shamlooei, Convective flow of nanofluid inside a lid driven porous cavity using CVFEM, Phys. B: Condensed Matter 521 (2017) 239–250.
[81] M. Sheikholeslami, M. Seyednezhad, Nanofluid heat transfer in a permeable enclosure in presence of variable magnetic field by means of CVFEM, Int. J. Heat Mass Transfer 114 (2017) 1169–1180.
[82] M. Sheikholeslami, S.A. Shehzad, CVFEM for influence of external magnetic source on Fe3O4 – H2O nanofluid behavior in a permeable cavity considering shape effect, Int. J. Heat Mass Transfer 115 (2017) 180–191.
[83] M. Sheikholeslami, M. Shamlooei, Magnetic source influence on nanofluid flow in porous medium considering shape factor effect, Phys. Lett. A 381 (2017) 3071–3078.
[84] M. Sheikholeslami, Influence of Lorentz forces on nanofluid flow in a porous cavity by means of Non- Darcy model, Engineering Computations, Vol. 34 Issue: 8, pp. 2651–2667, https://doi.org/10.1108/EC-01-2017-0008.
[85] M. Sheikholeslami, R. Ellahi, Three dimensional mesoscopic simulation of magnetic field effect on natural convection of nanofluid, Int. J. Heat Mass Transfer 89 (2015) 799–808.
[86] M. Sheikholeslami, D.D. Ganji, Numerical modeling of magnetohydrodynamic CuO -water transportation inside a porous cavity considering shape factor effect, Colloids Surfaces A: Physicochem. Eng. Aspects Colloids Surfaces A 529 (2017) 705–714.
[87] M. Sheikholeslami, A. Zeeshan, Numerical simulation of Fe3O4 -water nanofluid flow in a non-Darcy porous media, International Journal of Numerical Methods for Heat and Fluid Flow.

[88] M. Sheikholeslami, Magnetic field influence on CuO -H2O nanofluid convective flow in a permeable cavity considering various shapes for nanoparticles, Int. J. Hydrogen Energy, Int. J. Hydrogen Energy 42 (2017) 19611–19621.
[89] M. Sheikholeslami, H.B. Rokni, Numerical simulation for impact of Coulomb force on nanofluid heat transfer in a porous enclosure in presence of thermal radiation, International Journal of Heat and Mass Transfer, https://doi.org/10.1016/j.ijheatmasstransfer.2017.11.041.
[90] M. Sheikholeslami, Magnetohydrodynamic nanofluid forced convection in a porous lid driven cubic cavity using Lattice Boltzmann Method, J. Mol. Liq. 231 (2017) 555–565.
[91] M.S. Kandelousi, KKL correlation for simulation of nanofluid flow and heat transfer in a permeable channel, Phys. Lett. A 378 (45) (2014) 3331–3339.
[92] M. Sheikholeslami, Lattice Boltzmann Method simulation of MHD non-Darcy nanofluid free convection, Physica B 516 (2017) 55–71.
[93] M. Sheikholeslami, Influence of magnetic field on nanofluid free convection in an open porous cavity by means of Lattice Boltzmann Method, J. Mol. Liq. 234 (2017) 364–374.
[94] M. Sheikholeslami, H.B. Rokni, Influence of melting surface on MHD nanofluid flow by means of two phase model, Chin. J. Phys. 55 (2017) 1352–1360.
[95] M. Sheikholeslami, H.B. Rokni, Magnetic nanofluid natural convection in the presence of thermal radiation considering variable viscosity, Europ. Phys. J. Plus 132 (2017) 238. Available from: https://doi.org/10.1140/epjp/i2017-11498-4.
[96] M. Sheikholeslami, M. Shamlooei, R. Moradi, Fe3O4- Ethylene glycol nanofluid forced convection inside a porous enclosure in existence of Coulomb force, Journal of Molecular Liquids, https://doi.org/10.1016/j.molliq.2017.11.048.
[97] M. Sheikholeslami, S.A. Shehzad, Numerical analysis of Fe3O4 –H2O nanofluid flow in permeable media under the effect of external magnetic source, International Journal of Heat and Mass Transfer, In press.
[98] M. Sheikholeslami, T. Hayat, A. Alsaedi, S. Abelman, Numerical analysis of EHD nanofluid force convective heat transfer considering electric field dependent viscosity, Int. J. Heat Mass Transfer 108 (2017) 2558–2565.
[99] M. Sheikholeslami, Numerical simulation for solidification in a LHTESS by means of nano-enhanced PCM, J. Taiwan Inst. Chem. Eng. 86 (2018) 25–41.
[100] M. Sheikholeslami, Numerical modeling of nano enhanced PCM solidification in an enclosure with metallic fin, J. Mol. Liq. 259 (2018) 424–438.
[101] M. Sheikholeslami, A. Ghasemi, Solidification heat transfer of nanofluid in existence of thermal radiation by means of FEM, Int. J. Heat Mass Transf. 123 (2018) 418–431.
[102] M. Sheikholeslami, S.A. Shehzad, CVFEM simulation for nanofluid migration in a porous medium using Darcy model, Int. J. Heat Mass Transf. 122 (2018) 1264–1271.
[103] M. Sheikholeslami, M. Darzi, M.K. Sadoughi, Heat transfer improvement and pressure drop during condensation of refrigerant-based nanofluid: an experimental procedure, Int. J. Heat Mass Transf. 122 (2018) 643–650.
[104] M. Sheikholeslami, H.B. Rokni, CVFEM for effect of Lorentz forces on nanofluid flow in a porous complex shaped enclosure by means of non-equilibrium model, J. Mol. Liq. 254 (2018) 446–462.
[105] M. Sheikholeslami, S.A. Shehzad, Zhixiong Li, Water based nanofluid free convection heat transfer in a three dimensional porous cavity with hot sphere obstacle in existence of Lorenz forces, International Journal of Heat and Mass Transfer 125 (2018) 375–386.
[106] M. Sheikholeslami, M. Jafaryar, D.D. Ganji, Zhixiong Li, Exergy loss analysis for nanofluid forced convection heat transfer in a pipe with modified turbulators, Journal of Molecular Liquids 262 (2018) 104–110.
[107] M. Sheikholeslami, M. Jafaryar, Z. Li, Nanofluid turbulent convective flow in a circular duct with helical turbulators considering CuO nanoparticles, Int. J. Heat Mass Transf. 124 (2018) 980–989.
[108] M. Sheikholeslami, Z. Li, M. Shamlooei, Nanofluid, MHD natural convection through a porous complex shaped cavity considering thermal radiation, Phys. Lett. A 382 (2018) 1615–1632.
[109] M. Sheikholeslami, H.B. Rokni, Magnetic nanofluid flow and convective heat transfer in a porous cavity considering Brownian motion effects, Phys. Fluids, 30 (1), https://doi.org/10.1063/1.5012517.
[110] M. Sheikholeslami, S.A. Shehzad, Simulation of water based nanofluid convective flow inside a porous enclosure via non-equilibrium model, Int. J. Heat Mass Transf 120 (2018) 1200–1212.
[111] M. Sheikholeslami, S.A. Shehzad, Non-Darcy free convection of Fe_3O_4-water nanoliquid in a complex shaped enclosure under impact of uniform Lorentz force, Chin. J. Phys. 56 (2018) 270–281.
[112] M. Sheikholeslami, M. Seyednezhad, Simulation of nanofluid flow and natural convection in a porous media under the influence of electric field using CVFEM, Int. J. Heat Mass Transf. 120 (2018) 772–781.
[113] M. Sheikholeslami, M. Shamlooei, R. Moradi, Numerical simulation for heat transfer intensification of nanofluid in a permeable curved enclosure considering shape effect of Fe_3O_4 nanoparticles, Chem. Eng. Process.: Process Intensif. 124 (2018) 71–82.
[114] M. Sheikholeslami, T. Hayat, T. Muhammad, A. Alsaedi, MHD forced convection flow of nanofluid in a porous cavity with hot elliptic obstacle by means of Lattice Boltzmann method, Int. J. Mech. Sci. 135 (2018) 532–540.
[115] M. Sheikholeslami, Numerical investigation of nanofluid free convection under the influence of electric field in a porous enclosure, J. Mol. Liq. 249 (2018) 1212–1221.
[116] M. Sheikholeslami, CuO-water nanofluid flow due to magnetic field inside a porous media considering Brownian motion, J. Mol. Liq. 249 (2018) 921–929.
[117] M. Sheikholeslami, H.B. Rokni, Numerical simulation for impact of Coulomb force on nanofluid heat transfer in a porous enclosure in presence of thermal radiation, Int. J. Heat Mass Transf 118 (2018) 823–831.
[118] M. Sheikholeslami, Numerical investigation for CuO-H_2O nanofluid flow in a porous channel with magnetic field using mesoscopic method, J. Mol. Liq. 249 (2018) 739–746.
[119] M. Sheikholeslami, M. Shamlooei, R. Moradi, Fe_3O_4-ethylene glycol nanofluid forced convection inside a porous enclosure in existence of Coulomb force, J. Mol. Liq. 249 (2018) 429–437.
[120] M. Sheikholeslami, S.A. Shehzad, Numerical analysis of Fe_3O_4–H_2O nanofluid flow in permeable media under the effect of external magnetic source, Int. J. Heat Mass Transf 118 (2018) 182–192.

[121] M. Sheikholeslami, T. Hayat, A. Alsaedi, Numerical simulation for forced convection flow of MHD CuO-H_2O nanofluid inside a cavity by means of LBM, J. Mol. Liq. 249 (2018) 941–948.
[122] M. Sheikholeslami, H.R. Kataria, A.S. Mittal, Effect of thermal diffusion and heat-generation on MHD nanofluid flow past an oscillating vertical plate through porous medium, J. Mol. Liq. 257 (2018) 12–25.
[123] M. Sheikholeslami, S.A. Shehzad, Z. Li, Nanofluid heat transfer intensification in a permeable channel due to magnetic field using Lattice Boltzmann method, Phys. B (2018). Available from: https://doi.org/10.1016/j.physb.2018.03.036.
[124] M. Sheikholeslami, D.D. Ganji, Numerical investigation of nanofluid transportation in a curved cavity in existence of magnetic source, Chem. Phys. Lett. 667 (2017) 307–316.
[125] M. Sheikholeslami, Influence of Lorentz forces on nanofluid flow in a porous cylinder considering Darcy model, J. Mol. Liq. 225 (2017) 903–912.

CHAPTER

2

Simulation of Vorticity Stream Function Formulation by Means of CVFEM

2.1 CVFEM STREAM FUNCTION-VORTICITY SOLUTION FOR A LID DRIVEN CAVITY FLOW

2.1.1 Definition of the Problem and Governing Equation

A classic test problem in computational fluid mechanics is a lid driven cavity flow. In this two-dimensional problem, the flow in a square conduit is induced by sliding the upper surface (the lid) at a constant × velocity *U*, see Fig. 2.1 [1].

The nature of the flow in the cavity is controlled by a Reynolds Number defined as:

$$Re = \frac{UL}{\upsilon} \tag{2.1}$$

where L is the dimension of the cavity and υ is the kinematic viscosity.

In terms of the stream function Ψ and vorticity ω the governing equations, in integral form

$$\int_S \omega \nu . n - \upsilon \nabla \omega . n dS \tag{2.2}$$

$$-\int_A \omega dA = \int_S \nabla \Psi . n dS \tag{2.3}$$

In terms of the velocity field, v, the vorticity and stream function are defined by

$$\omega = \frac{\partial \nu_y}{\partial x} - \frac{\partial \nu_x}{\partial y} \tag{2.4}$$

$$\nu_x = \frac{\partial \Psi}{\partial y}, \quad \nu_y = -\frac{\partial \Psi}{\partial x} \tag{2.5}$$

The boundary condition on Eq. (2.3) is

$$\Psi = 0 \text{ on all solid boundaries} \tag{2.6}$$

There is not an explicit boundary condition for the vorticity equation Eq. (2.2). In setting up the numerical solution of Eqs. (2.2) and (2.3) a boundary condition for the solution of the discrete form of Eq. (2.2) is established by using the discrete form of Eq. (2.3) at boundary nodes coupled with the known velocity conditions, i.e.,

$$\begin{aligned} &u_x = U \text{ on lid}, \ u_x = 0 \text{ on all other boundaries} \\ &u_y = 0 \text{ on all boundaries} \end{aligned} \tag{2.7}$$

Application of Control Volume based Finite Element Method (CVFEM) for Nanofluid Flow and Heat Transfer.
DOI: https://doi.org/10.1016/B978-0-12-814152-6.00002-3

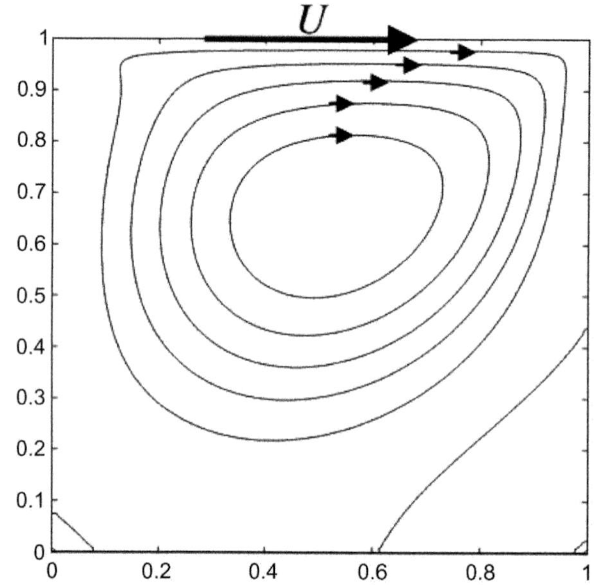

FIGURE 2.1 Lid driven cavity flow with streamline indicating flow.

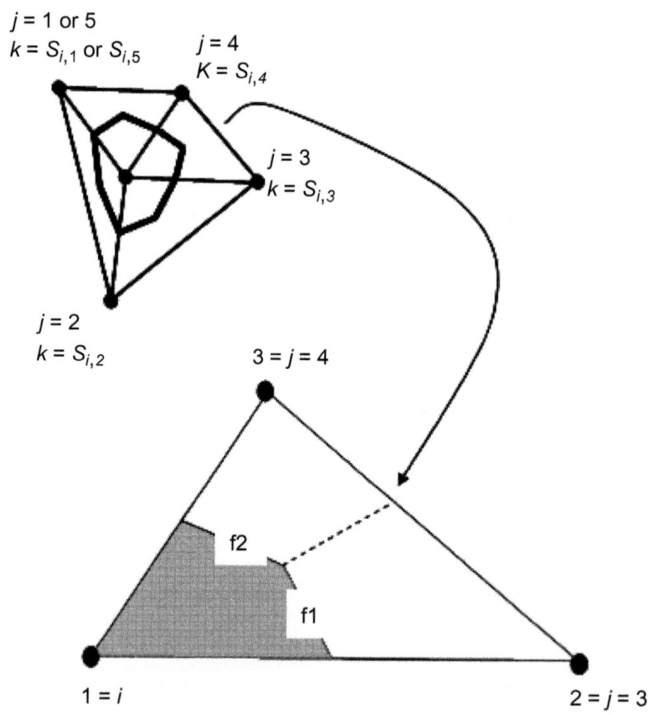

FIGURE 2.2 Arrangement of control volumes and elements.

2.1.2 The CVFEM Discretization of the Stream Function Equation

In keeping with the rest of this book the improvement of the discrete equations is based on linear triangular elements. The key components of this discretization, the support, control volume, the element, and the control volume faces ($f1$, $f2$) in an element are illustrated in Fig. 2.2 [1]. It is convenient to first consider the discretization of the stream function equation Eq. (2.3). The form of this equation can be identified as steady-state diffusion with a volume source. The general form for the discrete equation, for node i in the support shown in Fig. 2.2, is

$$\left[a_i^{\Psi} + Q_{C_i}^{\Psi} + B_{C_i}^{\Psi}\right]\Psi_i = \sum_{j=1}^{n_i} a_{i,j}^{\Psi}\Psi_{S_{i,j}} + Q_{B_i}^{\Psi} + B_{B_i}^{\Psi} \tag{2.8}$$

where the as are the coefficients, the index (i,j) indicates the jth node $(j = 1,2,..,n_i)$ in the support of node i, the index $S_{i,j}$ provides the global node number $(i = 1,2,..,n)$ of the jth node in the support, the Bs account for boundary conditions, and the Qs for source terms.

2.1.2.1 *Diffusion Contributions*

Following the presentation in the previous chapter the contribution to the coefficients as in Eq. (2.8), obtained from consideration of the diffusion flux across the control volume faces ($f1$, $f2$) for the particular element in Fig. 2.2, are

$$\begin{aligned} a_1{}^k &= -N_{1x}\Delta\vec{y}_{f1} + N_{1y}\Delta\vec{x}_{f1} - N_{1x}\Delta\vec{y}_{f2} + N_{1y}\Delta\vec{x}_{f2} \\ a_2{}^k &= N_{2x}\Delta\vec{y}_{f1} - N_{2y}\Delta\vec{x}_{f1} + N_{2x}\Delta\vec{y}_{f2} - N_{2y}\Delta\vec{x}_{f2} \\ a_3{}^k &= N_{3x}\Delta\vec{y}_{f1} - N_{3y}\Delta\vec{x}_{f1} + N_{3x}\Delta\vec{y}_{f2} - N_{3y}\Delta\vec{x}_{f2} \end{aligned} \tag{2.9}$$

These values can be used to update the ith support coefficients through

$$\begin{aligned} a_i^{\Psi} &= a_i^{\Psi} + a_1^{\kappa} \\ a_{i,3}^{\Psi} &= a_{i,3}^{\Psi} + a_2^{\kappa} \\ a_{i,4}^{\Psi} &= a_{i,4}^{\Psi} + a_4^{\kappa} \end{aligned} \tag{2.10}$$

In Eq. (2.9), moving counterclockwise around node i, the signed distances

$$\begin{aligned} \Delta\vec{x}_{f1} &= \frac{x_3}{3} - \frac{x_2}{6} - \frac{x_1}{6}, \quad \Delta\vec{x}_{f2} = -\frac{x_2}{3} + \frac{x_3}{6} + \frac{x_1}{6} \\ \Delta\vec{y}_{f1} &= \frac{y_3}{3} - \frac{y_2}{6} - \frac{y_1}{6}, \quad \Delta\vec{y}_{f1} = -\frac{y_2}{3} + \frac{y_3}{6} + \frac{y_1}{6} \end{aligned} \tag{2.11}$$

the derivatives of the shape functions

$$\begin{aligned} N_{1x} &= \frac{\partial N_1}{\partial x} = \frac{(y_2 - y_3)}{2V^{ele}}, \quad N_{1y} = \frac{\partial N_1}{\partial y} = \frac{(x_3 - x_2)}{2V^{ele}} \\ N_{2x} &= \frac{\partial N_2}{\partial x} = \frac{(y_3 - y_1)}{2V^{ele}}, \quad N_{2y} = \frac{\partial N_2}{\partial y} = \frac{(x_1 - x_3)}{2V^{ele}} \\ N_{3x} &= \frac{\partial N_3}{\partial x} = \frac{(y_1 - y_2)}{2V^{ele}}, \quad N_{3y} = \frac{\partial N_3}{\partial y} = \frac{(x_2 - x_1)}{2V^{ele}} \end{aligned} \tag{2.12}$$

and the volume of the element

$$V^{ele} = \frac{(x_2y_3 - x_3y_2) + x_1(y_2 - y_3) + y_1(x_3 - x_2)}{2} \tag{2.13}$$

2.1.2.2 *Source Terms*

The left-hand side of Eq. (2.3) is identified as a volume source term which can be represented in Eq. (2.8) by the setting

$$Q_{Bi}^{\Psi} = \omega_i V_i, \quad Q_{Ci}^{\Psi} = 0 \tag{2.14}$$

where V_i is the volume of the control volume, made up of contributions of $V^{ele}/3$ from each element in the support of node i. The coupling with the nodal vorticity field will require an iterative solution of Eq. (2.8). For any iteration, the source term in Eq. (2.14) will need to be reevaluated to reflect the updating of the iterative values of the nodal vorticity field.

2.1.2.3 Boundary Conditions

The boundary conditions of $\Psi = 0$ on all boundaries are set in Eq. (2.8) by setting, for each node i lying on the domain boundary

$$B^{\Psi}_{Bi} = 0, \quad B^{\Psi}_{Ci} = 10^{15} \tag{2.15}$$

2.1.3 The CVFEM Discretization of the Vorticity Equation

The vorticity equation Eq. (2.2) is seen to have the form of a steady state advection–diffusion equation for a scalar. The general form for this equation for node i in the support shown in Fig. 2.2 is

$$[a^{\omega}_i + B^{\omega}_{Ci}]\omega_i = \sum_{j=1}^{n_i} a^{\omega}_{i,j}\omega_{S_{i,j}} + B^{\omega}_{Bi} \tag{2.16}$$

2.1.3.1 Diffusion Contributions

The contributions to the coefficients a in Eq. (2.16), from consideration of the diffusion flux across the control volume faces ($f1, f2$) in Fig. 2.2, are

$$\begin{aligned}
a_1{}^{\kappa} &= -\nu N_{1x}\Delta\vec{y}_{f1} + \nu N_{1y}\Delta\vec{x}_{f1} - \nu N_{1x}\Delta\vec{y}_{f2} + \nu N_{1y}\Delta\vec{x}_{f2} \\
a_2{}^{\kappa} &= \nu N_{2x}\Delta\vec{y}_{f1} - \nu N_{2y}\Delta\vec{x}_{f1} + \nu N_{2x}\Delta\vec{y}_{f2} - \nu N_{2y}\Delta\vec{x}_{f2} \\
a_3{}^{\kappa} &= \nu N_{3x}\Delta\vec{y}_{f1} - \nu N_{3y}\Delta\vec{x}_{f1} + \nu N_{3x}\Delta\vec{y}_{f2} - \nu N_{3y}\Delta\vec{x}_{f2}
\end{aligned} \tag{2.17}$$

These values can be used to update the ith support coefficients through

$$\begin{aligned}
a^{\omega}_i &= a^{\omega}_i + a_1{}^{\kappa} \\
a^{\omega}_{i,3} &= a^{\omega}_{i,3} + a_2{}^{\kappa} \\
a^{\omega}_{i,4} &= a^{\omega}_{i,4} + a_4{}^{\kappa}
\end{aligned} \tag{2.18}$$

2.1.3.2 The Advection Coefficients

The contributions to the coefficients a in Eq. (2.8), from consideration of the advection flux across the control volume faces ($f1, f2$) in Fig. 2.2, are, using upwinding,

$$\begin{aligned}
a^u_1 &= \max[q_{f1}, 0] + \max[q_{f2}, 0] \\
a^u_2 &= \max[-q_{f1}, 0] \\
a^u_3 &= \max[-q_{f2}, 0]
\end{aligned} \tag{2.19}$$

These values can be used to update the ith support coefficients through

$$\begin{aligned}
a^{\omega}_i &= a^{\omega}_i + a^k_1 + a^u_1 \\
a^{\omega}_{i,3} &= a^{\omega}_{i,j} + a^k_2 + a^u_2 \\
a^{\omega}_{i,4} &= a^{\omega}_{i,j} + a^k_3 + a^u_3
\end{aligned} \tag{2.20}$$

In Eq. (2.19) the volume flow rates out across the control volume faces are calculated by

$$\begin{aligned}
q_{f1} &= \nu.nA|_{f1} = \nu_x{}^e\Delta\vec{y}_{f1} - \nu_y{}^e\Delta\vec{x}_{f1} \\
q_{f2} &= \nu.nA|_{f2} = \nu_x{}^e\Delta\vec{y}_{f2} - \nu_y{}^e\Delta\vec{x}_{f2}
\end{aligned} \tag{2.21}$$

where the velocities at the midpoint of the faces are approximated in terms of the constant element velocity, $\nu^e = (\nu_x{}^e, \nu_y{}^e)$. This value, in turn is calculated directly from the nodal stream function field through the approximation of Eq. (2.5), e.g., for the element in Fig. 2.2

$$\nu_x{}^e \approx N_{1y}\Psi_i + N_{2y}\Psi_{S_{i,3}} + N_{3y}\Psi_{S_{i,4}}$$
$$\nu_y{}^e \approx -N_{1x}\Psi_i - N_{2x}\Psi_{S_{i,3}} - N_{3x}\Psi_{S_{i,4}} \tag{2.22}$$

This coupling with the nodal stream-function field will require an iterative solution of Eq. (2.16). At each iteration, the advection coefficients in Eq. (2.19) will need to be evaluated to reflect the updating of the iterative values of the nodal stream-function field.

2.1.3.3 *Boundary Conditions*

Before Eq. (2.16) can be solved boundary conditions need to be prescribed. Fixed value boundary conditions are used. At each node i on a domain boundary the discrete form of the stream function Eq. (2.3) can be used to prescribe a value for the nodal vorticity. Use of this equation is allowed since, due to the fact that the stream function takes the known fixed value $\Psi = 0$ on all boundaries, it is not needed in the stream function solution. At a node i on the boundary the control volume finite element discretization of Eq. (2.3), informed by Eq. (2.8), can be written as

$$\omega_i = \frac{1}{V_i}\left[a_i^{\Psi}\Psi_i - \sum_{j=1}^{n_i} a_{i,j}^{\Psi}\Psi_{S_{i,j}} - \sum_{boundary} A\nabla\Psi.n\right] \tag{2.23}$$

where the coefficients a^{Ψ} are given by Eq. (2.10) and the definition of the source term in Eq. (2.14) has been used to isolate ω_i. The last term in Eq. (2.23) represents contributions from control volume faces that coincide with a boundary segment—see the double line in Fig 2.3 [1]. In the sliding lid problem of Fig. 1.1 the only nonzero contribution from this term will be for nodes on the sliding lid and will have the form

$$\left[\sum_{boundary} A\nabla\Psi.n\right] = \Delta U \tag{2.24}$$

where Δ is the length of the control volume surface on the boundary segment, see Fig. 2.3 [1].

Hence, with current iterative values of the stream function known, Eqs. (2.23) and (2.24) can be used to estimate fixed nodal values of vorticity for boundary nodes i

$$\omega_i = \frac{1}{V_i}\left[a_i^{\psi}\psi_i - \sum_{j=1}^{n_i} a_{i,j}^{\psi}\psi_{S_{i,j}} - \begin{cases}\Delta U & i\epsilon \text{ lid}\\ 0 & \text{otherwise}\end{cases}\right] \tag{2.25}$$

These values are forced into the solution of Eq. (2.16) by setting

$$B_{B_i}^{\omega} = \omega_i \times 10^{15}, \quad B_{C_i}^{\omega} = 10^{15} \tag{2.26}$$

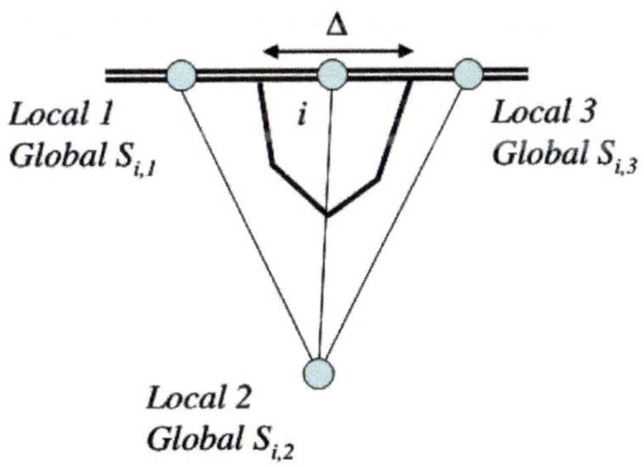

FIGURE 2.3 Node point on sliding lid boundary.

2.1.4 Calculating the Nodal Velocity Field

Following the calculation of the converged Ψ field, the nodal velocity field v can be calculated. Integrating over the i^{th}control volume results in

$$\int_{V_i} \nu_x dV = \int_{V_i} \frac{\partial \Psi}{\partial y} dV, \quad \int_{V_i} \nu_y dV = -\int_{V_i} \frac{\partial \Psi}{\partial x} dV, \tag{2.27}$$

or using the divergence theorem

$$\int_{V_i} \nu_x dV = \oint_S \Psi n_y dS, \quad \int_{V_i} \nu_y dV = -\oint_S \Psi n_x \tag{2.28}$$

where the surface integral is over the control volume faces, and $n = (n_x.n_y)$ is the outward pointing normal on the control volume face. Using midpoint integration approximations

$$\nu_{xi} = \frac{1}{V_i} \sum_{faces} \Psi A n_y, \quad \nu_{yi} = -\frac{1}{V_i} \sum_{faces} \Psi A n_x \tag{2.29}$$

where A is the area (length $\times$ unit depth) of a given control volume face, and, since boundary values of $\Psi = 0$, the summations are restricted to control faces that do not coincide with the domain boundaries. On recalling that the outward normal area product on a volume face is

$$A n_x = \Delta \vec{y}, \quad A n_y = -\Delta \vec{x} \tag{2.30}$$

According to definitions of $\Delta \vec{y}$ and $\Delta \vec{x}$, the contributions to control volume i from the right-hand side of Eq. (2.29), for the selected element from the control volume in Fig. 2.2, can be calculated as

$$\begin{aligned}
\sum_{f1,f2} \Psi A n_y &= -\left(\frac{5}{12}\Psi_i + \frac{5}{12}\Psi_{S_{i,3}} + \frac{2}{12}\Psi_{S_{i,4}} \right) \Delta \vec{x}_{f1} - \left(\frac{5}{12}\Psi_i + \frac{2}{12}\Psi_{S_{i,3}} + \frac{5}{12}\Psi_{S_{i,4}} \right) \Delta \vec{x}_{f2} \\
\sum_{f1,f2} \Psi A n_x &= -\left(\frac{5}{12}\Psi_i + \frac{5}{12}\Psi_{S_{i,3}} + \frac{2}{12}\Psi_{S_{i,4}} \right) \Delta \vec{y}_{f1} - \left(\frac{5}{12}\Psi_i + \frac{2}{12}\Psi_{S_{i,3}} + \frac{5}{12}\Psi_{S_{i,4}} \right) \Delta \vec{y}_{f2}
\end{aligned} \tag{2.31}$$

With the nodal velocity field calculated from Eq. (2.28) the velocity at given point $x = (x, y)$can be determine by determining the triangular element (local nodes 1, 2, and 3) which contains the point x and then using the shape function interpolations

$$\begin{aligned}
\nu_x(x) &= N_1 \nu_{x1} + N_2 \nu_{x2} + N_3 \nu_{x3} \\
\nu_y(x) &= N_1 \nu_{y1} + N_2 \nu_{y2} + N_3 \nu_{y3}
\end{aligned} \tag{2.32}$$

where the shape functions $N_i (i = 1, 2, 3)$ are defined in Eqs. (2.7) and (2.8).

2.1.5 Results

Fig. 2.4 shows the streamline for different Reynolds number. In *Appendix* a FORTRAN code for the CVFEM solution of a steady-state advection–diffusion equation is provided.

2.2 CVFEM STREAM FUNCTION-VORTICITY SOLUTION FOR NATURAL CONVECTION

2.2.1 Definition of the Problem and Governing Equation

The physical model and the corresponding triangular elements used in the present CVFEM program are shown in Fig. 2.5 [2], where r_{out} is the radius of the outer cylinder. The surfaces of the inner and outer cylinders

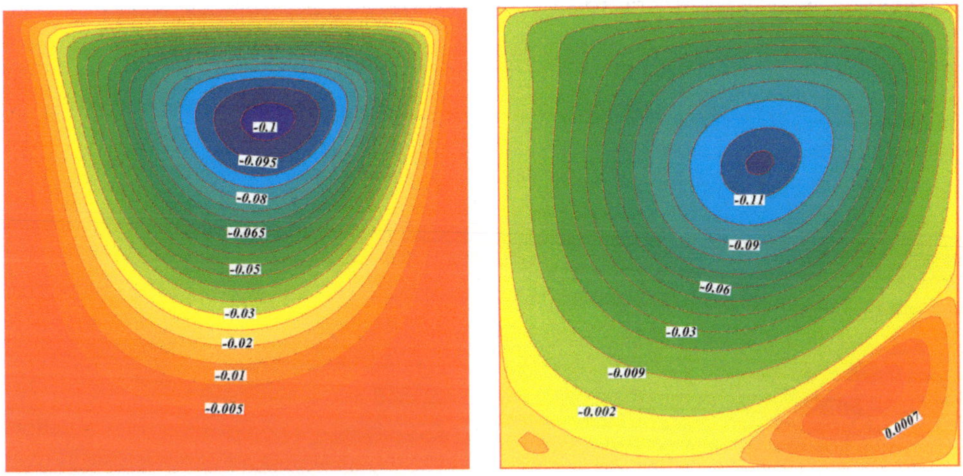

$Re = 50$ $Re = 500$

FIGURE 2.4 Streamlines for different Reynolds number.

are maintained at constant temperatures $T_h$and T_c, respectively, under the condition that $T_h > T_c$. The profile of the inner cylinder follows the pattern:

$$r = r_{in} + A\cos\left(N\left(\zeta - \zeta_0\right)\right) \tag{2.33}$$

where r_{in} is the base circle radius ($r_{in} = 1$), A and N are amplitude and number of undulations, respectively, and $\zeta$is the rotation angle.

Under the Boussinesq approximation, the continuity, momentum, and energy equations for the laminar and steady-state natural convection in the cylindrical enclosure can be written as follows:

$$\frac{\partial u}{\partial x} + \frac{\partial v}{\partial y} = 0 \tag{2.34}$$

$$u\frac{\partial u}{\partial x} + v\frac{\partial u}{\partial y} = -\frac{1}{\rho}\frac{\partial P}{\partial x} + \upsilon\left(\frac{\partial^2 u}{\partial x^2} + \frac{\partial^2 u}{\partial y^2}\right) \tag{2.35}$$

$$u\frac{\partial v}{\partial x} + v\frac{\partial v}{\partial y} = -\frac{1}{\rho}\frac{\partial P}{\partial y} + \upsilon\left(\frac{\partial^2 v}{\partial x^2} + \frac{\partial^2 v}{\partial y^2}\right) + g\,\beta\,(T - T_c) \tag{2.36}$$

$$u\frac{\partial T}{\partial x} + v\frac{\partial T}{\partial y} = \alpha\left(\frac{\partial^2 T}{\partial x^2} + \frac{\partial^2 T}{\partial y^2}\right) \tag{2.37}$$

Where u and v are the velocity components along the axes x and y, $T$is the fluid temperature, P is the pressure, and the physical meaning of the other quantities is mentioned in the Nomenclature.

We introduce the following dimensionless variables defined as

$$X = \frac{x}{L}, \quad Y = \frac{y}{L}, \quad \Omega = \frac{\omega L^2}{\alpha}, \quad \Psi = \frac{\psi}{\alpha}, \quad \Theta = \frac{T - T_c}{T_h - T_c} \tag{2.38}$$

where $L = r_{out} - r_{in} = r_{in} = 1$. Using these variables, Eqs. (2.34)–(2.37) can be written in the dimensionless form as

$$\frac{\partial \Psi}{\partial Y}\frac{\partial \Omega}{\partial X} - \frac{\partial \Psi}{\partial X}\frac{\partial \Omega}{\partial Y} = \Pr\left(\frac{\partial^2 \Omega}{\partial X^2} + \frac{\partial^2 \Omega}{\partial Y^2}\right) + Ra\Pr\left(\frac{\partial \Theta}{\partial X}\right) \tag{2.39}$$

$$\frac{\partial \Psi}{\partial Y}\frac{\partial \Theta}{\partial X} - \frac{\partial \Psi}{\partial X}\frac{\partial \Theta}{\partial Y} = \left(\frac{\partial^2 \Theta}{\partial X^2} + \frac{\partial^2 \Theta}{\partial Y^2}\right) \tag{2.40}$$

$$\frac{\partial^2 \Psi}{\partial X^2} + \frac{\partial^2 \Psi}{\partial Y^2} = -\Omega \tag{2.41}$$

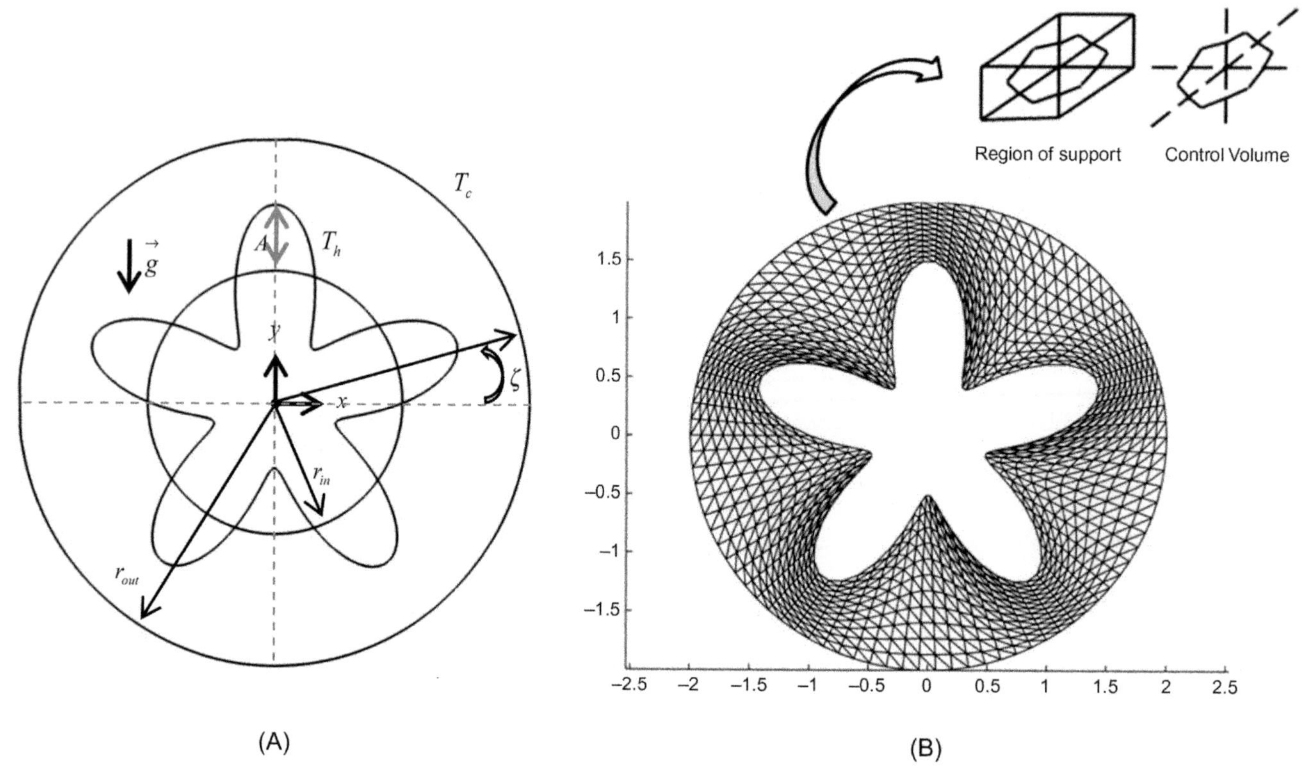

FIGURE 2.5 (A) Geometry and boundary conditions. (B) Computational mesh.

where $Ra = g\,\beta\,(T_h - T_c)\,L^3/(\alpha\upsilon)$ is the Rayleigh number and $\Pr = \upsilon/\alpha$ is the Prandtl number. The boundary conditions of Eqs. (2.39)–(2.41) are, as shown in Fig. 2.5,

$$\begin{aligned} &\Theta = 1.0 \quad \text{on the inner sinusoidal circular cylinder} \\ &\Theta = 0.0 \quad \text{on the outer circular enclosure} \\ &\Psi = 0.0 \quad \text{on all solid boundaries} \end{aligned} \tag{2.42}$$

The values of the vorticity Ω on the boundaries of the enclosure can be obtained using the definition of the stream function and the known velocity conditions during the iterative solution procedure. The local Nusselt number on the cold wall Nu_{loc} can be expressed as

$$Nu_{loc} = \frac{\partial \Theta}{\partial n} \tag{2.43}$$

where n is the distance normal to the surface of the outer cylinder. The average Nusselt number on the cold circular wall Nu_{ave} is evaluated as

$$Nu_{ave} = \frac{1}{2\pi} \int_0^{360^\circ} Nu_{loc}(\zeta)\, d\zeta \tag{2.44}$$

2.2.2 Effect of Active Parameters

Calculations are performed for various values of Rayleigh numbers $Ra = 10^3,\ 10^4, 10^5$, and 10^6, amplitude parameter $A = 0.1,\ 0.3$ and 0.5, and the number of undulations $N = 2,\ 3,\ 5$ and 6 at constant Prandtl number, $Pr = 0.71$ (air). Streamlines and isotherm obtained by the present code are compared with those reported by Kim et al. [3] for different values of the Rayleigh numbers (Fig. 2.6). The shape size and location of vortexes are the same in the result obtained by Kim et al. [3] and the present code. These comparisons illustrate an excellent agreement between the present calculations and the previously published results. We are, therefore, confident that the present results are correct and very accurate using the CVFEM code. Recently, this code was developed for various applications [4–93].

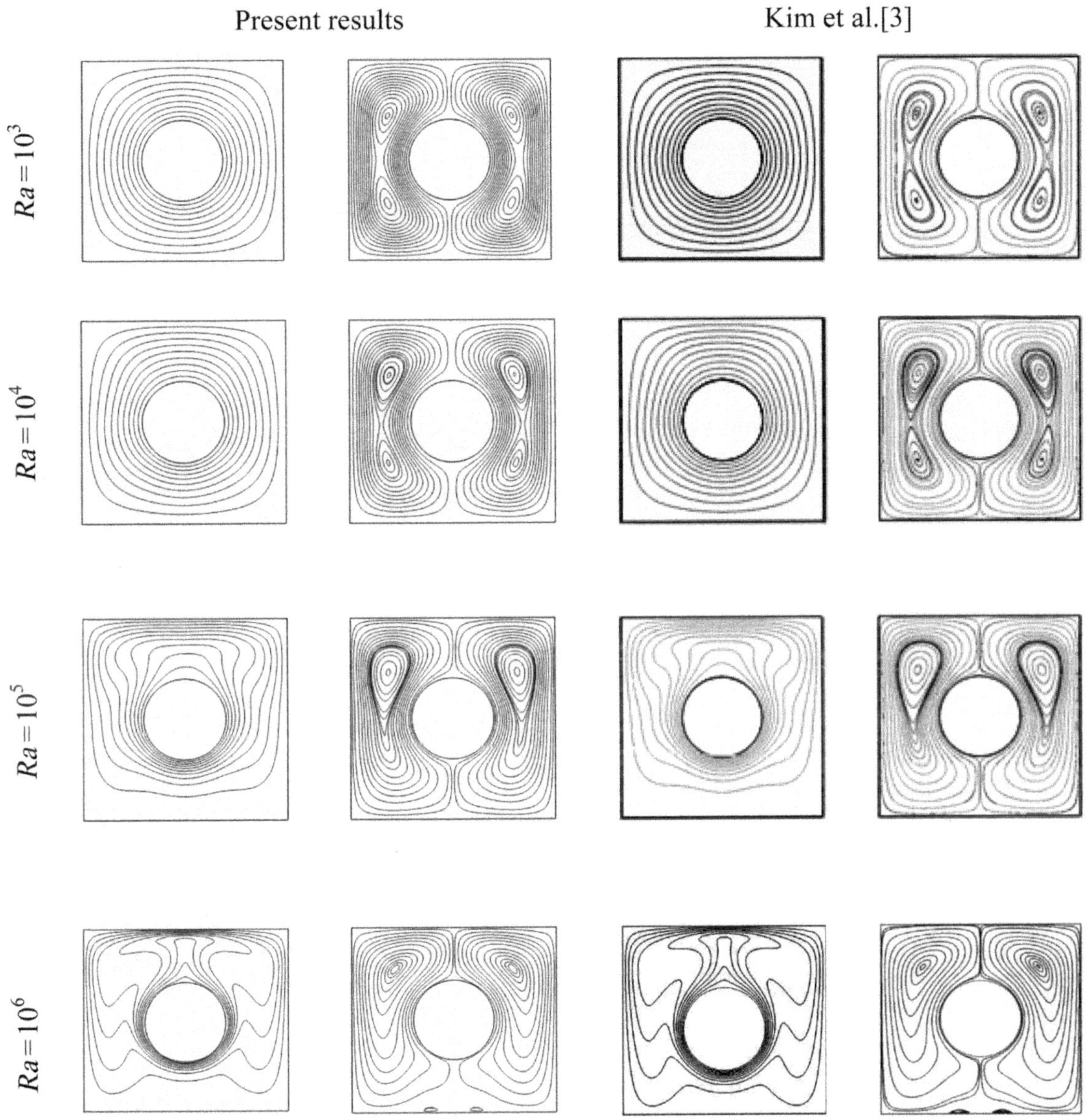

FIGURE 2.6 Comparison of the present results with those reported by Kim et al. [3] for different Rayleigh numbers when $Pr = 0.7$.

Fig. 2.7 shows the isotherms and streamlines for different values of amplitude parameter and number of undulations at $Ra = 10^3$. For all cases at this Rayleigh number the isotherms follow the shape of inner and outer walls. Moreover for lower Rayleigh numbers the absolute values of stream function are nearly small, which indicates the domination of conduction heat transfer mechanism. For $N = 3$ increasing the amplitude of undulation parameter A, enhances the absolute value of stream function $|\Psi_{max}|$. It is due to the fact that with the increase of A the gap between the hot and cold walls decreases which results in a higher heat transfer.

Fig. 2.7 also shows that as the amplitude of undulation increases from $A = 0.1$ to $A = 0.5$ the single vortex inside the enclosure divides into two vortices with different strength located at the top and bottom of the enclosure. Moreover, the streamlines indicate that the upper vortex is stronger than the lower one. The reason could be the more available space for circulation of fluid at the top of the enclosure and existence of the hot surface beneath the cold outer wall in this region, which helps the flow circulation. As seen at the bottom of the enclosure a very small secondary vortex appears near the vertical centerline, which appears due to the existence of the crest in that region. A similar flow pattern is seen at this region for $N = 5$. The values of streamlines show that for $N = 4$ increasing A from 0.1 to 0.3 enhances the values of $|\Psi_{max}|$, but as A increases furthermore its value decreases. It could be related to the effect of less available space for fluid circulation in this area.

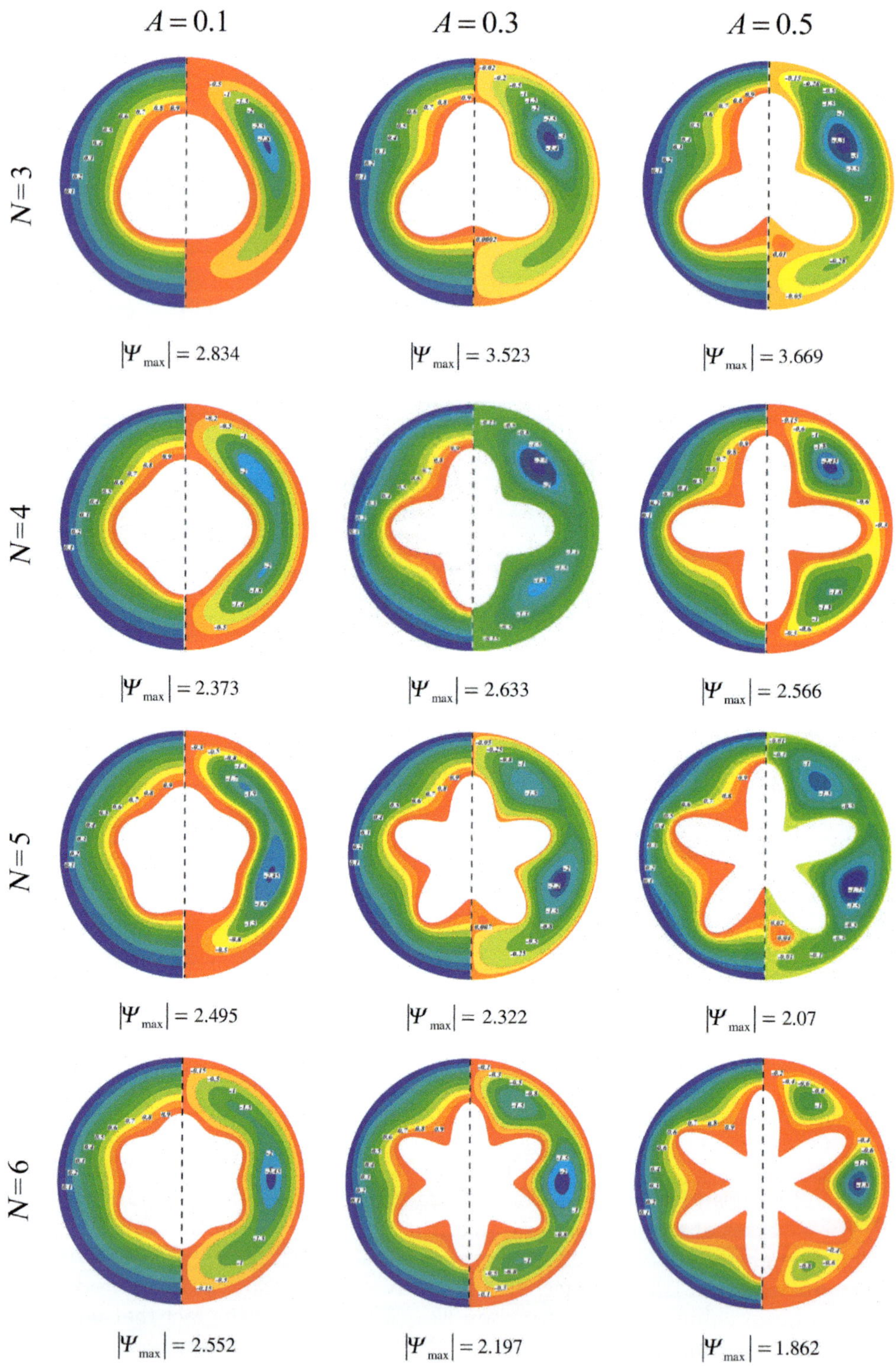

FIGURE 2.7 Isotherms (left) and streamlines (right) for different values of amplitude A and number of undulations N at $Ra = 10^3$.

The streamlines also show that there are two main vortices between the inner and outer cylinders for all values of A; an upper stronger vortex and a lower weaker one. Since the available spaces for circulation are equal for both vortexes, it can be concluded that the existence of the hot surface beneath the cold one assists the flow circulation at the upper half of the enclosure, while the existence of the hot surface over the cold one in the lower half of the enclosure has a reverse effect on the flow circulation. At $N = 4$, there is no small secondary vortex at the bottom half of the enclosure because the lower crest is parallel to the flow path. A similar pattern exists for $N = 6$. Also the flow pattern, size, and formation of vortices are nearly similar for $N = 5$ and 6. As seen, increasing A for these values of the undulation number divides the space between the inner and outer cylinders into three circulation regions. Increase of A decreases $|\Psi_{max}|$ considerably at $N=5$ and 6. Although an increase of A expands the effective surface area of hot surface, the fluid flow will be damped, especially in the gaps between undulated surfaces so these areas are almost stagnant. The streamlines and isotherms are shown for different values of N and A at $Ra=10^4$ in Fig. 2.8. As seen, the isotherms are disturbed at this Ra and thermal plumes gradually appear on the hot surface of the inner cylinder, which indicates that the convective heat transfer became comparable with the conduction mode of heat transfer at this Rayleigh number. In addition, the thickness of the thermal boundary layer between the crests of inner the cylinder becomes thinner in comparison with that of $Ra=10^3$, which indicates that the flow ventilates the region between each of the two crests; hence the thermal boundary layer is thinner at these areas.

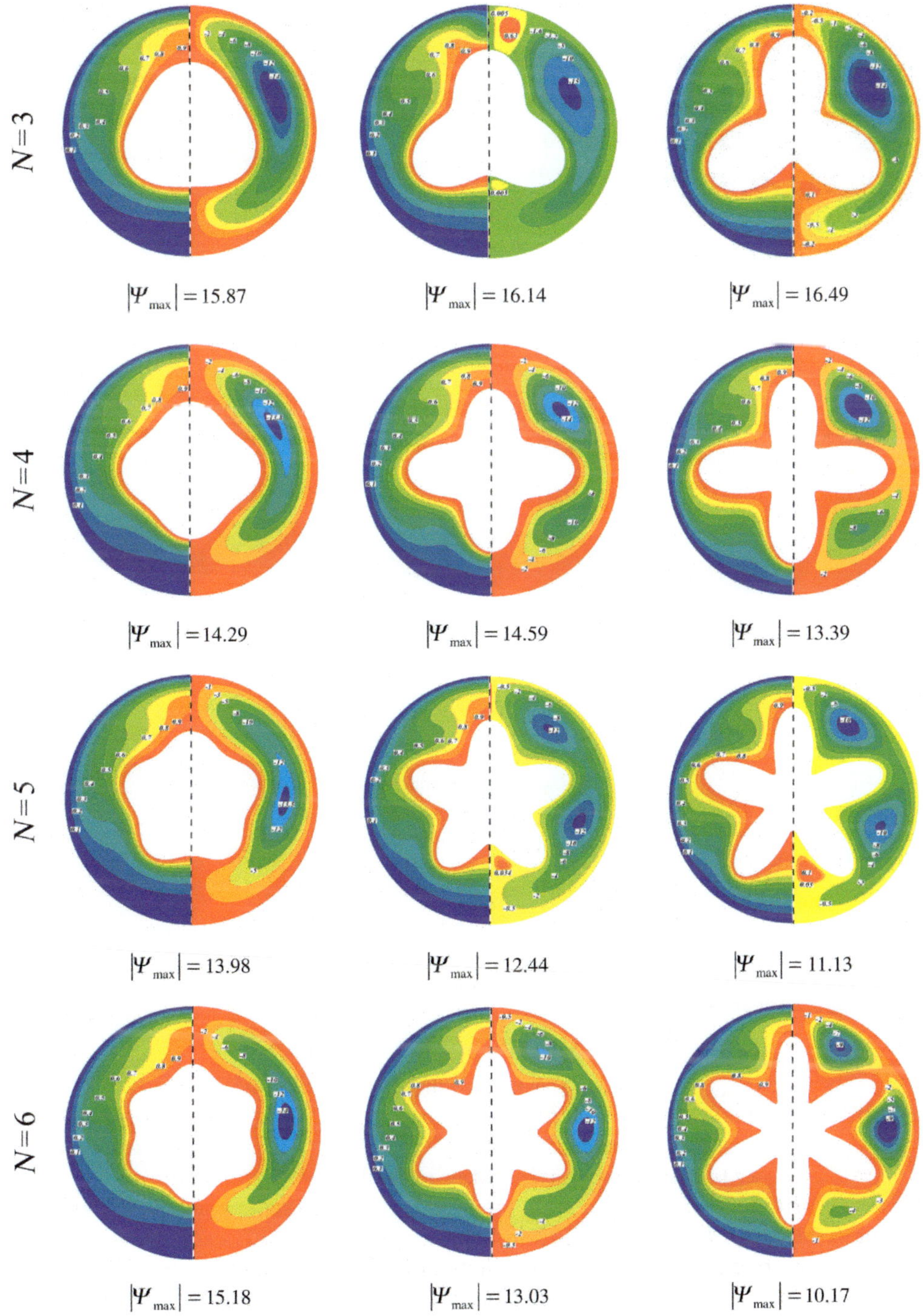

FIGURE 2.8 Isotherms (left) and streamlines (right) for different values of the amplitude A and the number of undulations N at $Ra=10^4$.

FIGURE 2.9 Isotherms (left) and streamlines (right) for different values of the amplitude A and number of undulations N at $Ra = 10^5$.

At $Ra = 10^4$, the variation of $|\Psi_{max}|$ with N and A is similar to that of $Ra = 10^3$. For $N = 3$ with the increase of A from 0.1 to 0.3, a thermal plume is formed over the upper crest at $\varsigma = 90$ degrees with reverse direction owing to the two secondary vortices which are generated newly at this area. With more increase in A these secondary vortices disappear because the available space between the inner hot wall and outer cold cylinder decreases. For $N = 4$, the primary vortex divides into two smaller vortices; an upper strong and a lower weak one. The flow pattern and number of vortices for $N = 5$ and 6 is similar to those of $Ra = 10^3$ except the value of $|\Psi_{max}|$ is higher at $Ra = 10^4$ because the convective heat transfer mechanism is more pronounced.

Figs. 2.9 and 2.10 show the effects of A and N on the streamlines and isotherms for $Ra = 10^5$ and 10^6, respectively. As a general observation the isotherms are crowed near the crests which demonstrate the diminution in

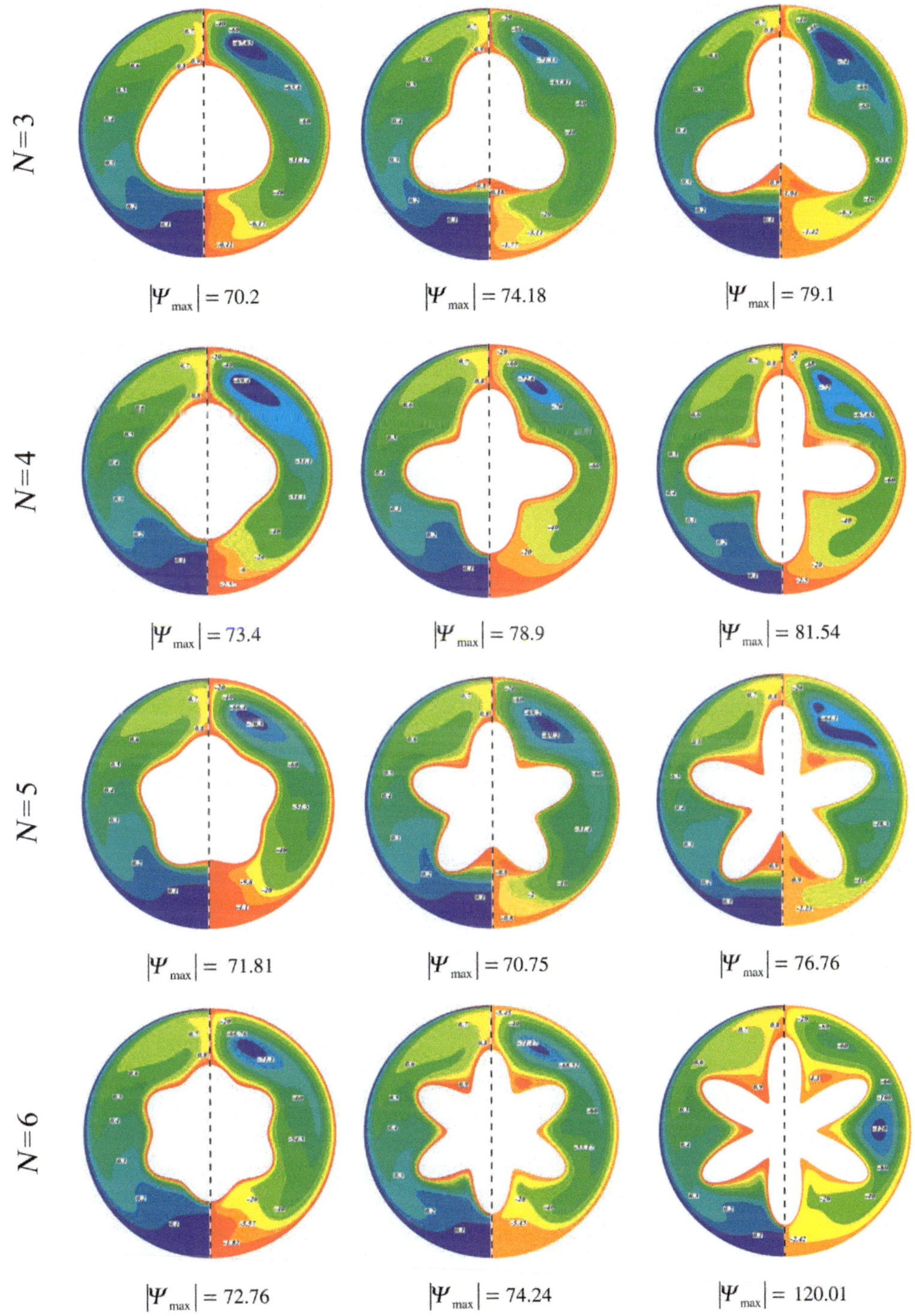

FIGURE 2.10 Isotherms (left) and streamlines (right) for different values of the amplitude A and number of undulations N at $Ra = 10^6$.

thermal boundary layer thickness at the bottom of the enclosure. At the top of the inner cylinder ($\varsigma = 90$ degrees), a strong plume appears which strongly impinges upon the outer wall of the enclosure for all values of A and N.

The maximum value of the stream function has a direct relation to the value of A for $N = 6$ in which it enhances up to the value $A = 0.3$ and then decreases. For $N = 3$ and $A = 0.1$, a primary vortex exists at the top of the inner cylinder. When A increases, this main vortex divides into two vortices with small cores which are located at the upper half of the cylinder. Finally for $A = 0.5$ these two vortices merge together and form a single

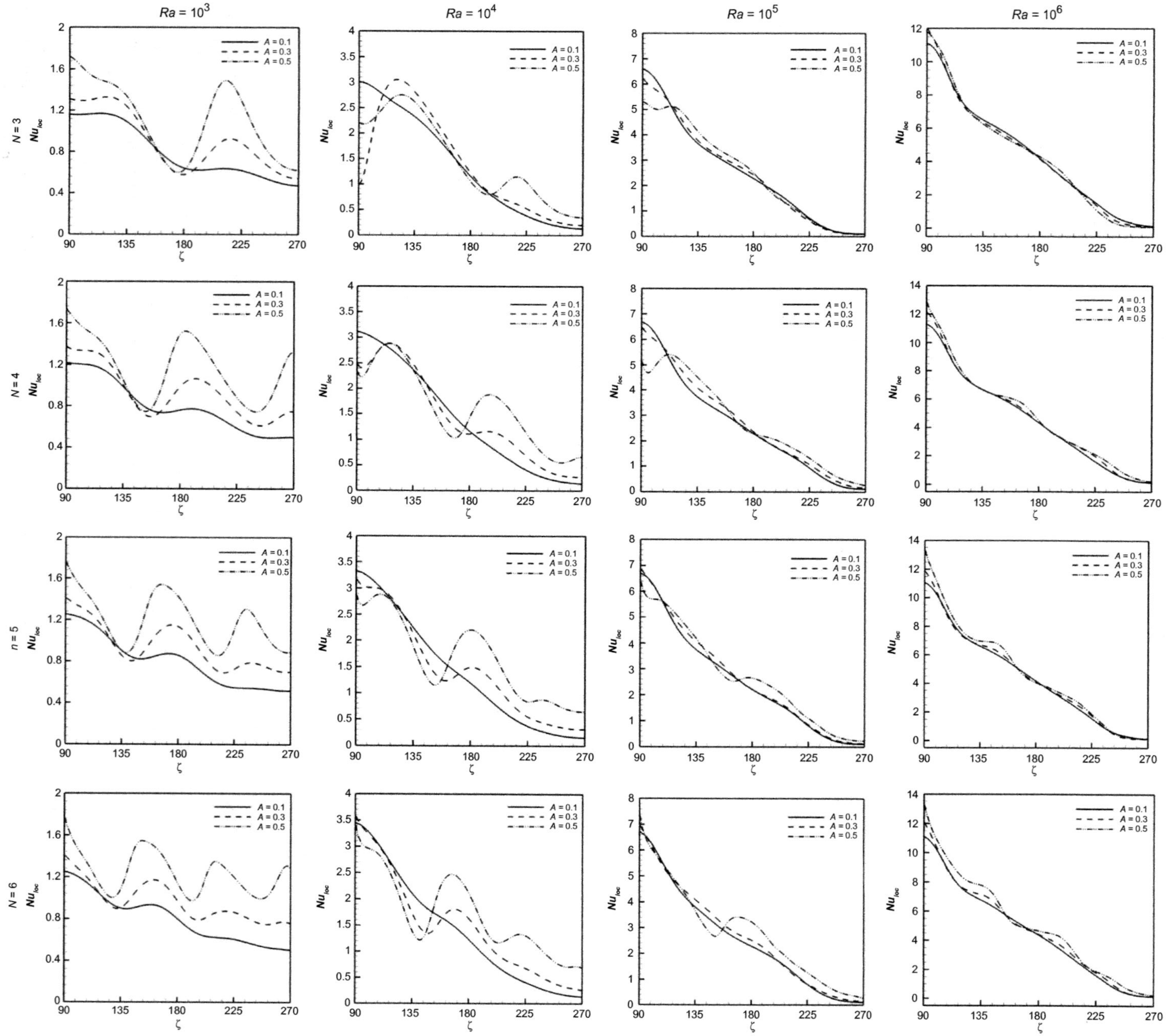

FIGURE 2.11 Effects of the amplitude A, number of undulations N, and Rayleigh number Ra on the local Nusselt number Nu_{loc}.

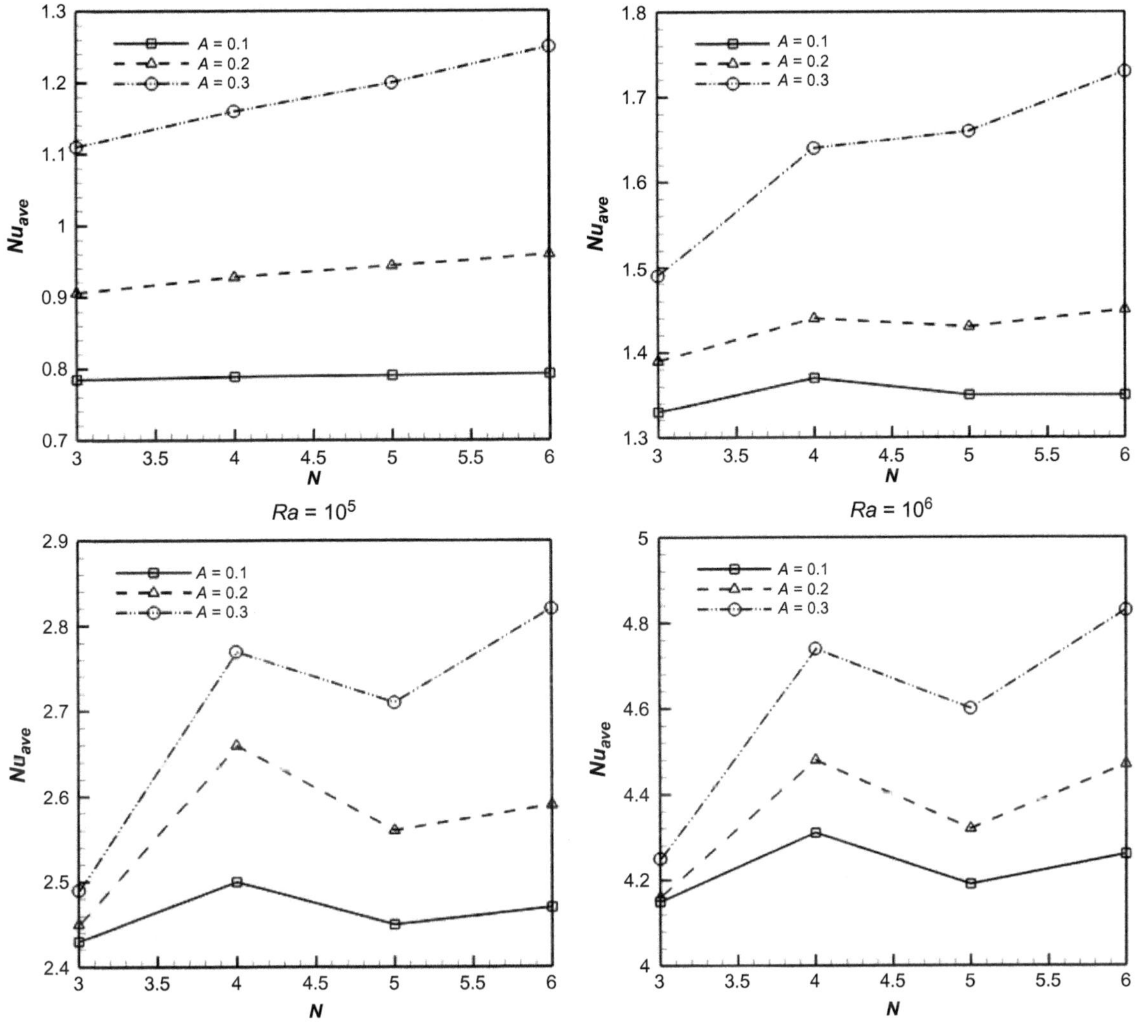

FIGURE 2.12 Effects of the amplitude A, number of undulations N, and Rayleigh number Ra on the average Nusselt number Nu_{ave}.

vortex which is stronger than the first one. For $N=4$, increasing A from 0.1 to 0.3 and then to 0.5 divides the main cell into two cells with different strength. For $N=5$, 6 and $A=0.1$ a single vortex exists inside the enclosure. The increase of A up to 0.3 and then to 0.5 pushes this single cell upwards and then divides it into two cells. The flow pattern and number of vortices at this Rayleigh number is similar to those of $Ra=10^5$ except the value of $|\Psi_{max}|$ is greater at $Ra=10^6$. It should be noted that all of the computations in the case of $Ra=10^6$ are obtained by using an unsteady solution and time averaging for a period of time.

Variation of the local Nusselt number Nu_{loc} over the cold wall and between $\varsigma=90$ and 270 degrees is depicted in Fig. 2.11. In general, the effects of increasing A on Nu_{loc} are less pronounced at higher values of the Rayleigh number. In addition, the variation of Nu_{loc} is greater for higher values of the undulation amplitude, while this variation decreases when the gap between the hot inner and cold outer cylinders increases. The number of extremum in the local Nusselt number profiles corresponds to the number of undulations of the inner cylinder. For $N=3$ there exist two local maximum, three maximum for $N=4$ and 5, and four local maximum for $N=6$.

Fig. 2.11 also shows that Nu_{loc} decreases as the angle ς increases from $\varsigma=90$ to 270 degrees; at the top of the inner cylinder ($\varsigma=90$ degrees), the flow accompanying cold fluid impinges upon the upper cold surface, which results in a thicker thermal boundary layer on the enclosure's wall. Whereas in the case of $\varsigma=270$ degrees, the hot surface locates at the top of the cold wall and the conduction mechanism is more pronounced. Thus, Nu_{loc} decreases noticeably. At $Ra=10^4, 10^5$, and 10^6, the Nu_{loc} profile uniformly reaches from its maximum value at $\varsigma=90$ degrees to its minimum value near $\varsigma=270$ degrees when $A=0.1$. For $A=0.3$ and 0.5, the profiles of the Nu_{loc} profiles have local extremes, which are related to the thermal plumes and crests over the inner cylinder.

Fig. 2.12 shows the effects of A, N, and Ra on the average Nusselt number Nu_{ave}. As the Ra increases, the value of Nu_{ave} enhances due to the convective heat transfer mechanism at higher values of Ra. At $Ra=10^3$, the average Nusselt number is approximately constant at $A=0.1$ for various numbers of undulations, whereas it increases uniformly for $A=0.3$ and 0.5 due to decreasing the space between the hot and cold surfaces. At $Ra=10^4$, the maximum value for Nu_{ave} is obtained for $N=4$ and $A=0.1$ meaning more space is available for fluid flow, while it is obtained for $N=6$ at $A=0.3$ and 0.5. The Nu_{ave} profiles for $Ra=10^5$ indicate that the minimum values of Nu_{ave} are obtained for $N=3$, whereas the maximum values of Nu_{ave} occur at $N=4$ for $A=0.1$, 0.3 and at $N=6$ for $A=0.5$. It is worth pointing out that for higher values of the Rayleigh number the profiles of the average Nusselt number have a local minimum at $N=5$ for all values of$A$.

References

[1] V.R. Voller, 5 Tohccxxvc Basic Control Volume Finite Element Methods for Fluids and Solids, World Scientific Publishing Co. Pte. Ltd., 2009.

[2] M. Sheikholeslami, M. Gorji-Bandpy, I. Pop, Soheil Soleimani, Numerical study of natural convection between a circular enclosure and a sinusoidal cylinder using control volume based finite element method, Int. J. Therm. Sci. 72 (2013) 147–158.

[3] B.S. Kim, D.S. Lee, M.Y. Ha, H.S. Yoon, A numerical study of natural convection in a square enclosure with a circular cylinder at different vertical locations, Int. J. Heat Mass Transf. 51 (2008) 1888–1906.

[4] M. Sheikholeslami, Influence of Lorentz forces on nanofluid flow in a porous cylinder considering Darcy model, J. Mol. Liq. 225 (2017) 903–912.

[5] M. Sheikholeslami, CVFEM for magnetic nanofluid convective heat transfer in a porous curved enclosure, Eur. Phys. J. Plus 131 (2016) 413. Available from: https://doi.org/10.1140/epjp/i2016-16413-y.

[6] M. Sheikholeslami, Influence of Coulomb forces on Fe3O4-H2O nanofluid thermal improvement, Int. J. Hydrogen Energy 42 (2017) 821–829.

[7] M. Sheikholeslami, Magnetic field influence on nanofluid thermal radiation in a cavity with tilted elliptic inner cylinder, J. Mol. Liq. 229 (2017) 137–147.

[8] M. Sheikholeslami, Numerical simulation of magnetic nanofluid natural convection in porous media, Phys. Lett. A 381 (2017) 494–503.

[9] M. Sheikholeslami, Magnetic source impact on nanofluid heat transfer using CVFEM, Neural Comput. Applicat. (2016). Available from: https://doi.org/10.1007/s00521-016-2740-7.

[10] M. Sheikholeslami, CuO-water nanofluid free convection in a porous cavity considering Darcy law, Europ. Phys. J. Plus 132 (2017) 55. Available from: https://doi.org/10.1140/epjp/i2017-11330-3.

[11] M. Sheikholeslami, Numerical investigation of MHD nanofluid free convective heat transfer in a porous tilted enclosure, Eng. Comput. 34 (6) (2017) 1939–1955.

[12] M.S. Kandelousi, Effect of spatially variable magnetic field on ferrofluid flow and heat transfer considering constant heat flux boundary condition, Europ. Phys. J. Plus (2014) 129–248. Available from: https://doi.org/10.1140/epjp/i2014-14248-2.

[13] M. Sheikholeslami, S. Abelman, Two phase simulation of nanofluid flow and heat transfer in an annulus in the presence of an axial magnetic field, IEEE Trans. Nanotechnol. 14 (3) (2015) 561–569.

[14] M. Sheikholeslami, S.A. Shehzad, CVFEM for influence of external magnetic source on Fe3O4 – H2O nanofluid behavior in a permeable cavity considering shape effect, Int. J. Heat Mass Transfer 115 (2017) 180–191.

[15] M. Sheikholeslami, M. Seyednezhad, Nanofluid heat transfer in a permeable enclosure in presence of variable magnetic field by means of CVFEM, Int. J. Heat Mass Transfer 114 (2017) 1169–1180.

[16] M. Sheikholeslami, Numerical simulation for external magnetic field influence on Fe3O4-water nanofluid forced convection, Engineering Computations.

[17] M. Sheikholeslami, Influence of Lorentz forces on nanofluid flow in a porous cavity by means of Non- Darcy model, Engineering Computations, Vol. 34 Issue: 8, pp.2651-2667, https://doi.org/10.1108/EC-01-2017-0008.

[18] M. Sheikholeslami, H.B. Rokni, Nanofluid convective heat transfer intensification in a porous circular cylinder, Chem. Eng. Proc. Process Intensif. 120 (2017) 93–104.

[19] M. Sheikholeslami, M.K. Sadoughi, Numerical modeling for Fe3O4 -water nanofluid flow in porous medium considering MFD viscosity, J. Mol. Liq. 242 (2017) 255–264.

[20] M. Sheikholeslami, M. Shamlooei, Magnetic source influence on nanofluid flow in porous medium considering shape factor effect, Phys. Lett. A 381 (2017) 3071–3078.

[21] M. Sheikholeslami, M. Shamlooei, Convective flow of nanofluid inside a lid driven porous cavity using CVFEM, Phys. B: Condensed Matter 521 (2017) 239–250.

[22] M. Sheikholeslami, K. Vajravelu, Forced convection heat transfer in Fe3O4- ethylene glycol nanofluid under the influence of Coulomb force, J. Mol. Liq. 233 (2017) 203–210.

[23] M. Sheikholeslami, H.B. Rokni, Influence of EFD viscosity on nanofluid forced convection in a cavity with sinusoidal wall, J. Mol. Liq. 232 (2017) 390–395.

[24] M. Sheikholeslami, H.B. Rokni, Magnetic nanofluid natural convection in the presence of thermal radiation considering variable viscosity, Europ. Phys. J. Plus 132 (2017) 238. Available from: https://doi.org/10.1140/epjp/i2017-11498-4.

[25] M. Sheikholeslami, H.B. Rokni, Numerical simulation for impact of Coulomb force on nanofluid heat transfer in a porous enclosure in presence of thermal radiation, International Journal of Heat and Mass Transfer, https://doi.org/10.1016/j.ijheatmasstransfer.2017.11.041.

[26] M. Sheikholeslami, M. Shamlooei, R. Moradi, Fe3O4- Ethylene glycol nanofluid forced convection inside a porous enclosure in existence of Coulomb force, Journal of Molecular Liquids, https://doi.org/10.1016/j.molliq.2017.11.048.

[27] M. Sheikholeslami, S.A. Shehzad, Numerical analysis of Fe3O4 –H2O nanofluid flow in permeable media under the effect of external magnetic source, International Journal of Heat and Mass Transfer, In press.

[28] M. Sheikholeslami, S.A. Shehzad, Magnetohydrodynamic nanofluid convection in a porous enclosure considering heat flux boundary condition, Int. J. Heat Mass Transfer 106 (2017) 1261–1269.

[29] M. Sheikholeslami, S.A. Shehzad, Thermal radiation of ferrofluid in existence of Lorentz forces considering variable viscosity, Int. J. Heat Mass Transfer 109 (2017) 82–92.
[30] M. Sheikholeslami, H.B. Rokni, Numerical modeling of nanofluid natural convection in a semi annulus in existence of Lorentz force, Comput. Methods Appl. Mech. Eng. 317 (2017) 419–430.
[31] M. Sheikholeslami, K. Vajravelu, Nanofluid flow and heat transfer in a cavity with variable magnetic field, Appl. Math. Comput. 298 (2017) 272–282.
[32] M. Sheikholeslami, H.B. Rokni, Magnetohydrodynamic CuO-water nanofluid in a porous complex shaped enclosure, ASME, J. Therm. Sci. Eng. Applicat. 9 (4) (2017) 041007. Available from: https://doi.org/10.1115/1.4035973.
[33] M. Sheikholeslami, H.B. Rokni, Nanofluid two phase model analysis in existence of induced magnetic field, Int. J. Heat Mass Transfer 107 (2017) 288–299.
[34] M. Sheikholeslami, T. Hayat, A. Alsaedi, S. Abelman, Numerical analysis of EHD nanofluid force convective heat transfer considering electric field dependent viscosity, Int. J. Heat Mass Transfer 108 (2017) 2558–2565.
[35] M. Sheikholeslami, Magnetic field influence on CuO -H2O nanofluid convective flow in a permeable cavity considering various shapes for nanoparticles, Int. J. Hydrogen Energy 42 (2017) 19611–19621.
[36] M. Sheikholeslami, M.M. Bhatti, Forced convection of nanofluid in presence of constant magnetic field considering shape effects of nanoparticles, Int. J. Heat Mass Transfer 111 (2017) 1039–1049.
[37] M. Sheikholeslami, A. Zeeshan, Analysis of flow and heat transfer in water based nanofluid due to magnetic field in a porous enclosure with constant heat flux using CVFEM, Comput. Methods Appl. Mech. Eng. 320 (2017) 68–81.
[38] M. Sheikholeslami, M. Shamlooei, Fe3O4- H2O nanofluid natural convection in presence of thermal radiation, Int. J. Hydrogen Energy 42 (9) (2017) 5708–5718.
[39] M. Sheikholeslami, M.M. Bhatti, Active method for nanofluid heat transfer enhancement by means of EHD, Int. J. Heat Mass Transfer 109 (2017) 115–122.
[40] M. Sheikholeslami, T. Hayat, A. Alsaedi, Numerical study for external magnetic source influence on water based nanofluid convective heat transfer, Int. J. Heat Mass Transfer 106 (2017) 745–755.
[41] M. Sheikholeslami, Ali J. Chamkha, Flow and convective heat transfer of a ferro-nanofluid in a double-sided lid-driven cavity with a wavy wall in the presence of a variable magnetic field, Numer. Heat Transfer, Part A VOL. 69 (NO. 10) (2016) 1186–1200. Available from: https://doi.org/10.1080/10407782.2015.1125709.
[42] A.J.C. Mohsen Sheikholeslami, Electrohydrodynamic free convection heat transfer of a nanofluid in a semi-annulus enclosure with a sinusoidal wall, Numer. Heat Transfer, Part A VOL. 69 (NO. 7) (2016) 781–793. Available from: https://doi.org/10.1080/10407782.2015.1090819.
[43] M. Sheikholeslami, M.K. Sadoughi, Simulation of CuO- water nanofluid heat transfer enhancement in presence of melting surface, Int. J. Heat Mass Transfer 116 (2018) 909–919.
[44] M. Sheikholeslami, H.B. Rokni, Simulation of nanofluid heat transfer in presence of magnetic field: a review, Int. J. Heat Mass Transfer 115 (2017) 1203–1233.
[45] M. Sheikholeslami, H.B. Rokni, Melting heat transfer influence on nanofluid flow inside a cavity in existence of magnetic field, Int. J. Heat Mass Transfer 114 (2017) 517–526.
[46] M. Sheikholeslami, S.A. Shehzad, Magnetohydrodynamic nanofluid convective flow in a porous enclosure by means of LBM, Int. J. Heat Mass Transfer 113 (2017) 796–805.
[47] M. Sheikholeslami, Magnetohydrodynamic nanofluid forced convection in a porous lid driven cubic cavity using Lattice Boltzmann Method, J. Mol. Liq. 231 (2017) 555–565.
[48] M.S. Kandelousi, KKL correlation for simulation of nanofluid flow and heat transfer in a permeable channel, Phys. Lett. A 378 (45) (2014) 3331–3339.
[49] M. Sheikholeslami, Effect of uniform suction on nanofluid flow and heat transfer over a cylinder, J Braz. Soc. Mech. Sci. Eng. 37 (2015) 1623–1633.
[50] M. Sheikholeslami, D.D. Ganji, Ferrohydrodynamic and Magnetohydrodynamic effects on ferrofluid flow and convective heat transfer, Energy 75 (2014) 400–410.
[51] M. Sheikholeslami, D.D. Ganji, M.M. Rashidi, Ferrofluid flow and heat transfer in a semi annulus enclosure in the presence of magnetic source considering thermal radiation, J. Taiwan Inst. Chem. Eng. 47 (2015) 6–17.
[52] M. Sheikholeslami, R. Ellahi, M. Hassan, Soheil Soleimani, A study of natural convection heat transfer in a nanofluid filled enclosure with elliptic inner cylinder, Int. J. Numer. Methods Heat Fluid Flow 24 (2014) 8.
[53] M. Sheikholeslami, M.M. Rashidi, D.D. Ganji, Effect of non-uniform magnetic field on forced convection heat transfer of Fe3O4- water nanofluid, Comput. Methods Appl. Mech. Engrg. 294 (2015) 299–312.
[54] M. Sheikholeslami, K. Vajravelu, M.M. Rashidi, Forced convection heat transfer in a semi annulus under the influence of a variable magnetic field, Int. J. Heat Mass Transfer 92 (2016) 339–348.
[55] M. Sheikholeslami, Soheil Soleimani, D.D. Ganji, Effect of electric field on hydrothermal behavior of nanofluid in a complex geometry, J. Mol. Liq. 213 (2016) 153–161.
[56] M. Sheikholeslami, M.M. Rashidi, Non-uniform magnetic field effect on nanofluid hydrothermal treatment considering Brownian motion and thermophoresis effects, J Braz. Soc. Mech. Sci. Eng. 38 (2016) 1171–1184.
[57] M. Sheikholeslami, M.M. Rashidi, Ferrofluid heat transfer treatment in the presence of variable magnetic field, Eur. Phys. J. Plus 130 (2015) 115.
[58] M. Sheikholeslami, M.M. Rashidi, T. Hayat, D.D. Ganji, Free convection of magnetic nanofluid considering MFD viscosity effect, J. Mol. Liq. 218 (2016) 393–399.
[59] M. Sheikholeslami, T. Hayat, A. Alsaedi, MHD free convection of Al2O3–water nanofluid considering thermal radiation: a numerical study, Int. J. Heat Mass Transfer 96 (2016) 513–524.
[60] M. Sheikholeslami, M. Nimafar, D.D. Ganji, M. Pouyandehmehr, CuO-H2O nanofluid hydrothermal analysis in a complex shaped cavity, Int. J. Hydrogen Energy 41 (2016) 17837–17845.
[61] M. Sheikholeslami, D.D. Ganji, Numerical investigation of nanofluid transportation in a curved cavity in existence of magnetic source, Chem. Phys. Lett. 667 (2017) 307–316.

[62] M. Sheikholeslami, D.D. Ganji, Transportation of MHD nanofluid free convection in a porous semi annulus using numerical approach, Chem. Phys. Lett. 669 (2017) 202–210.
[63] M. Sheikholeslami, D.D. Ganji, Free convection of Fe3O4-water nanofluid under the influence of an external magnetic source, J. Mol. Liq. 229 (2017) 530–540.
[64] M. Sheikholeslami, D.D. Ganji, Impact of electric field on nanofluid forced convection heat transfer with considering variable properties, J. Mol. Liq. 229 (2017) 566–573.
[65] M. Sheikholeslami, Z. Ziabakhsh, D.D. Ganji, Transport of Magnetohydrodynamic nanofluid in a porous media, Colloids Surf., A 520 (2017) 201–212.
[66] M. Sheikholeslami, D.D. Ganji, Numerical approach for magnetic nanofluid flow in a porous cavity using CuO nanoparticles, Mater. Design 120 (2017) 382–393.
[67] M. Sheikholeslami, D.D. Ganji, Numerical analysis of nanofluid transportation in porous media under the influence of external magnetic source, J. Mol. Liq. 233 (2017) 499–507.
[68] M. Sheikholeslami, D.D. Ganji, R. Moradi, Heat transfer of Fe3O4 -water nanofluid in a permeable medium with thermal radiation in existence of constant heat flux, Chem. Eng. Sci. 174 (2017) 326–336.
[69] M. Sheikholeslami, D.D. Ganji, Numerical modeling of magnetohydrodynamic CuO -water transportation inside a porous cavity considering shape factor effect, Colloids Surfaces A: Physicochem. Eng. Aspects Colloids Surfaces A 529 (2017) 705–714.
[70] M. Sheikholeslami, P. Rana, Soheil Soleimani, Numerical study of mhd natural convection liquid metal flow and heat transfer in a wavy enclosure using CVFEM, Heat Transfer Res. 48 (2) (2017) 121–138.
[71] M. Sheikholeslami, Numerical simulation for solidification in a LHTESS by means of Nano-enhanced PCM, J. Taiwan Inst. Chem. Eng. 86 (2018) 25–41.
[72] M. Sheikholeslami, Numerical modeling of Nano enhanced PCM solidification in an enclosure with metallic fin, J. Mol. Liq. 259 (2018) 424–438.
[73] M. Sheikholeslami, A. Ghasemi, Solidification heat transfer of nanofluid in existence of thermal radiation by means of FEM, Int. J. Heat Mass Transfer 123 (2018) 418–431.
[74] M. Sheikholeslami, S.A. Shehzad, CVFEM simulation for nanofluid migration in a porous medium using Darcy model, Int. J. Heat Mass Transfer 122 (2018) 1264–1271.
[75] M. Sheikholeslami, M. Darzi, M.K. Sadoughi, Heat transfer improvement and pressure drop during condensation of refrigerant-based Nanofluid; an experimental procedure, Int. J. Heat Mass Transfer 122 (2018) 643–650.
[76] M. Sheikholeslami, H.B. Rokni, CVFEM for effect of Lorentz forces on nanofluid flow in a porous complex shaped enclosure by means of Non-equilibrium model, J. Mol. Liq. 254 (2018) 446–462.
[77] M. Sheikholeslami, S.A. Shehzad, Z. Li, Water based nanofluid free convection heat transfer in a three dimensional porous cavity with hot sphere obstacle in existence of Lorenz forces, Int. J. Heat Mass Transfer 125 (2018) 375–386.
[78] M. Sheikholeslami, M. Jafaryar, D.D. Ganji, Z. Li, Exergy loss analysis for nanofluid forced convection heat transfer in a pipe with modified turbulators, J. Mol. Liq. 262 (2018) 104–110.
[79] M. Sheikholeslami, M. Jafaryar, Z. Li, Nanofluid turbulent convective flow in a circular duct with helical turbulators considering CuO nanoparticles, Int. J. Heat Mass Transfer 124 (2018) 980–989.
[80] M. Sheikholeslami, Z. Li, M. Shamlooei, Nanofluid MHD natural convection through a porous complex shaped cavity considering thermal radiation, Phys. Lett. A 382 (2018) 1615–1632.
[81] M. Sheikholeslami, H.B. Rokni, Magnetic nanofluid flow and convective heat transfer in a porous cavity considering Brownian motion effects, Phys. Fluids, 30, 1, 10.1063/1.5012517, https://doi.org/10.1063/1.5012517.
[82] M. Sheikholeslami, S.A. Shehzad, Simulation of water based nanofluid convective flow inside a porous enclosure via Non-equilibrium model, Int. J. Heat Mass Transfer 120 (2018) 1200–1212.
[83] M. Sheikholeslami, S.A. Shehzad, Non-Darcy free convection of Fe_3O_4-water nanoliquid in a complex shaped enclosure under impact of uniform Lorentz force, Chin. J. Phys. 56 (2018) 270–281.
[84] M. Sheikholeslami, M. Seyednezhad, Simulation of nanofluid flow and natural convection in a porous media under the influence of electric field using CVFEM, Int. J. Heat Mass Transfer 120 (2018) 772–781.
[85] M. Sheikholeslami, M. Shamlooei, R. Moradi, Numerical simulation for heat transfer intensification of nanofluid in a permeable curved enclosure considering shape effect of Fe_3O_4 nanoparticles, Chem. Eng. Process.: Process Intensif. 124 (2018) 71–82.
[86] M. Sheikholeslami, T. Hayat, T. Muhammad, A. Alsaedi, MHD forced convection flow of nanofluid in a porous cavity with hot elliptic obstacle by means of Lattice Boltzmann method, Int. J. Mech. Sci. 135 (2018) 532–540.
[87] M. Sheikholeslami, Numerical investigation of nanofluid free convection under the influence of electric field in a porous enclosure, J. Mol. Liq. 249 (2018) 1212–1221.
[88] M. Sheikholeslami, CuO-water nanofluid flow due to magnetic field inside a porous media considering Brownian motion, J. Mol. Liq. 249 (2018) 921–929.
[89] M. Sheikholeslami, H.B. Rokni, Numerical simulation for impact of Coulomb force on nanofluid heat transfer in a porous enclosure in presence of thermal radiation, Int. J. Heat Mass Transfer 118 (2018) 823–831.
[90] M. Sheikholeslami, Numerical investigation for CuO-H_2O nanofluid flow in a porous channel with magnetic field using mesoscopic method, J. Mol. Liq. 249 (2018) 739–746.
[91] M. Sheikholeslami, T. Hayat, A. Alsaedi, Numerical simulation for forced convection flow of MHD CuO-H_2O nanofluid inside a cavity by means of LBM, J. Mol. Liq. 249 (2018) 941–948.
[92] M. Sheikholeslami, S.A. Shehzad, Numerical analysis of Fe_3O_4 –H_2O nanofluid flow in permeable media under the effect of external magnetic source, Int. J. Heat Mass Transfer 118 (2018) 182–192.
[93] M. Sheikholeslami, M. Shamlooei, R. Moradi, Fe3O4- Ethylene glycol nanofluid forced convection inside a porous enclosure in existence of Coulomb force, J. Mol. Liq. 249 (2018) 429–437.

CHAPTER

3

Various Application of Nanofluid for Heat Transfer Augmentation

3.1 INTRODUCTION

Nanofluids are produced by dispersing the nanometer-scale solid particles into base liquids with low thermal conductivity such as water, ethylene glycol (EG), oils, etc. Control of heat transfer in many energy systems is crucial due to the increase in energy prices. In recent years, nanofluids technology has been proposed and studied by some researchers experimentally or numerically to control heat transfer in a process. The nanofluid can be applied to engineering problems, such as heat exchangers, cooling of electronic equipment, and chemical processes. There are two ways for simulation of a nanofluid: single-phase and two-phase. In the first method, researchers assumed that nanofluids were to be treated as the common pure fluid and conventional equations of mass, momentum, and energy are used and the only effect of nanofluid is its thermal conductivity and viscosity which are obtained from the theoretical models or experimental data. These researchers assumed that nanoparticles are in thermal equilibrium and there aren't any slip velocities between the nanoparticles and fluid molecules, thus they have a uniform mixture of nanoparticles. In the second method, researchers assumed that there are slip velocities between nanoparticles and fluid molecules. So the volume fraction of nanofluids may not be uniform anymore and there would be a variable concentration of nanoparticles in a mixture. There are several numerical and semianalytical methods which have been used by several authors in order to simulate nanofluid flow and heat transfer.

3.1.1 Definition of Nanofluid

Low thermal conductivity of conventional heat transfer fluids such as water, oil, and EG mixture is a serious limitation in improving the performance and compactness of much engineering equipment, such as heat exchangers and electronic devices. To overcome this disadvantage, there is a strong motivation to develop advanced heat transfer fluids with substantially higher conductivity. An innovative way of improving the thermal conductivities of fluids is to suspend small solid particles in the fluid. Various types of powders such as metallic, nonmetallic, and polymeric particles can be added into fluids to form slurries. The thermal conductivities of fluids with suspended particles are expected to be higher than that of common fluids. Nanofluids are a new kind of heat transfer fluid containing a small quantity of nanosized particles (usually less than 100 nm) that are uniformly and stably suspended in a liquid. The dispersion of a small amount of solid nanoparticles in conventional fluids changes their thermal conductivity remarkably. Compared to the existing techniques for enhancing heat transfer, the nanofluids show a superior potential for increasing heat transfer rates in a variety of cases [1].

3.1.2 Model Description

In the literature, convective heat transfer with nanofluids can be modeled using mainly the two-phase or single approach. In the two-phase approach, the velocity between the fluid and particles might not be zero [2] due to several factors such as gravity, friction between the fluid and solid particles, Brownian forces, Brownian diffusion, sedimentation, and dispersion. In the second approach, the nanoparticles can be easily fluidized and therefore, one may assume that the motion slip between the phases, if any, would be considered negligible [3]. The latter approach is simpler and more computationally efficient.

Application of Control Volume based Finite Element Method (CVFEM) for Nanofluid Flow and Heat Transfer.
DOI: https://doi.org/10.1016/B978-0-12-814152-6.00003-5

3.1.3 Conservation Equations

3.1.3.1 *Single-phase Model*

Although nanofluids are solid–liquid mixtures, the approach conventionally used in most studies of natural convection handles the nanofluid as a single-phase (homogenous) fluid. In fact, due to the extreme size and low concentration of the suspended nanoparticles, the particles are assumed to move with the same velocity as the fluid. Also, by considering the local thermal equilibrium, the solid particle–liquid mixture may then be approximately considered to behave as a conventional single-phase fluid with properties that are to be evaluated as functions of those of the constituents. The governing equations for a homogenous analysis of natural convection are continuity, momentum, and energy equations with their density, specific heat, thermal conductivity, and viscosity modified for nanofluid application. The specific governing equations for various studied enclosures are not shown here and they can be found in different references [4]. It should be mentioned that sometimes this assumption is not correct. For example, Ding and Wen [5] found that this assumption may not always remain true for a nanofluid. They investigated the particle migration in a nanofluid for a pipe flow and stated that at Peclet numbers exceeding 10 the particle distribution is significantly nonuniform. Nevertheless, many studies have performed the numerical simulation using single-phase assumption and reported acceptable results for the heat transfer and hydrodynamic properties of the flow.

3.1.3.2 *Two-phase Model*

Several authors have tried to establish convective transport models for nanofluids [6]. Nanofluid is a two-phase mixture in which the solid phase consists of nanosized particles. In view of the nanoscale size of the particles, it may be questionable whether the theory of conventional two-phase flow can be applied in describing the flow characteristics of nanofluid. On the other hand, several factors such as gravity, friction between the fluid and solid particles and Brownian forces, the phenomena of Brownian diffusion, sedimentation, and dispersion may affect a nanofluid flow. Consequently, the slip velocity between the fluid and particles cannot be neglected for simulating nanofluid flows. Since the two-phase approach considers the movement between the solid and fluid molecule, it may have better prediction in nanofluid study. To fully describe and predict the flow and behavior of complex flows, different multiphase theories have been proposed and used. The large number of published articles concerning multiphase flows typically employed the Mixture Theory to predict the behavior of nanofluids [7]. A comprehensive survey of convective transport in nanofluids was made by Buongiorno [8], using a model in which Brownian motion and thermophoresis are accounted for. Buongiorno developed a two-component four-equation nonhomogeneous equilibrium model for mass, momentum, and heat transfer in nanofluids. The nanofluid is treated as a two-component mixture (base fluid and nanoparticles) with the following assumptions: No chemical reactions; Negligible external forces; Dilute mixture ($\phi = 1$); Negligible viscous dissipation; Negligible radiative heat transfer; Nanoparticle and base fluid locally in thermal equilibrium. Invoking the above assumptions, the following equations represent the mathematical formulation of the nonhomogenous single-phase model for the governing equations as formulated by Buongiorno [8]:

3.1.3.2.1 Continuity Equation

$$\nabla.v = 0 \tag{3.1}$$

where v is the velocity

3.1.3.2.2 Nanoparticle Continuity Equation

$$\frac{\partial\phi}{\partial t} + v.\nabla\phi = \nabla.\left(D_B\nabla\phi + D_T\frac{\nabla T}{T}\right) \tag{3.2}$$

Here ϕ is nanoparticle volume fraction, D_B is the Brownian diffusion coefficient given by the Einstein–Stokes's equation:

$$D_B = \frac{k_B T}{3\pi\mu d_p} \tag{3.3}$$

where μ is the viscosity of the fluid, d_p is the nanoparticle diameter, $k_B = 1.385 \times 10^{-23}$ is Boltzmann constant and D_T is the thermophoretic diffusion coefficient, which is defined as

$$D_T = \left(\frac{\mu}{\rho}\right)\left(0.26\frac{k}{k+k_p}\right) \tag{3.4}$$

In Eq. (3.4), k and k_p are the thermal conductivity of the fluid and particle materials, respectively.

3.1.3.2.3 Momentum Equation

$$v.\nabla v = -\frac{1}{\rho_{nf}}\nabla p + \nabla.\tau + g \tag{3.5}$$

where

$$\tau = -\mu_{nf}\left(\nabla v + (\nabla v)^t\right) \tag{3.6}$$

where the superscript "t" indicates the transpose of ∇v. Also p is pressure.

3.1.3.2.4 Energy Equation

$$v.\nabla T = \nabla\left(\alpha_{nf}\nabla T\right) + \frac{\rho_p c_p}{\rho_{nf} c_{nf}}\left(D_B \nabla\phi.\nabla T + D_T\frac{\nabla T.\nabla T}{T}\right) \tag{3.7}$$

where ϕ and T are nanoparticle concentration and temperature of nanofluid, respectively.

This nanofluid model can be characterized as a "two-fluid" (nanoparticles + base fluid), four-equation (mass, momentum, energy), nonhomogeneous (nanoparticle/fluid slip velocity allowed) equilibrium (nanoparticle/fluid temperature differences not allowed) model. Note that the conservation equations are strongly coupled. That is, v depends on ϕ via viscosity; ϕ depends on T mostly because of thermophoresis; T depends on ϕ via thermal conductivity and also via the Brownian and thermophoretic terms in the energy equation: ϕ and T obviously depends on v because of the convection terms in the nanoparticle continuity and energy equations, respectively.

In a numerical study by Behzadmehr et al. [9] for the first time a two-phase mixture model was implemented to investigate the behavior of Cu–water nanofluid in a tube and the results were also compared with previous works using a single-phase approach. The authors claimed that the simulation done by assuming that base fluid and particles behave separately possessed results that are more precise compared to the previous computational modeling. They implemented the mixture theory for their work. It was suggested that the continuity, momentum, and energy equations be written for a mixture of fluid and a solid phase. Some assumptions were also stated for the model such as a strong coupling between two phases and the fluid being closely followed by the particles with each phase owning a different velocity leading to a term called slip velocity of nanoparticles as in Eq. (3.8):

$$V_{pf} = V_P - V_f = \frac{\rho_p d_p^2}{18\mu_f f_{drag}}\frac{\left(\rho_p - \rho_m\right)}{\rho_p}a, \quad a = g - (V_m.\nabla)V_m, f_{drag} = \begin{cases} 1 + 0.15\text{Re}_p^{0.687} \\ 0.0183\text{Re}_p \end{cases} \tag{3.8}$$

The conservation equations (continuity, momentum, and energy, respectively) will be written for the mixture as follows:

$$\nabla.\left(\rho_m V_m\right) = 0 \tag{3.9}$$

$$\nabla.\left(\rho_m V_m V_m\right) = -\nabla P_m + \nabla.[\tau - \tau_t] + \rho_m g + \nabla.\left(\sum_{k=1}^{n}\phi_k\rho_k V_{dr,k}V_{dr,k}\right) \tag{3.10}$$

$$\nabla.\left(\phi_p V_k\left(\rho_k h_k + p\right)\right) = \nabla.\left(k_{eff}\nabla T - C_p\rho_m vt\right) \tag{3.11}$$

where $V_{dr,p}$ is the particle draft velocity that is related to the slip velocity and is defined as:

$$V_{dr,p} = V_P - V_f = V_{pf} - \sum_{k=1}^{n}\frac{\phi_k\rho_k}{\rho_m}V_{fk} \tag{3.12}$$

3.1.4 Physical Properties of the Nanofluids for Single-phase Model

Base nanofluid properties have been published over the past few years in literature. However, only recently have some data on temperature-dependent properties been provided, even though they are only for nanofluid effective thermal conductivity and effective absolute viscosity.

3.1.4.1 *Density*

In the absence of experimental data for nanofluid densities, constant-value temperature independent values, based on nanoparticle volume fraction, are used:

$$\rho_{nf} = \rho_f(1-\phi) + \rho_p\phi \tag{3.13}$$

3.1.4.2 *Specific Heat Capacity*

It has been suggested that the effective specific heat can be calculated using the following equation as reported in [10]:

$$(C_p)_{nf} = (C_p)_f(1-\phi) + (C_p)_p\phi \tag{3.14}$$

Other authors suggest an alternative approach based on a heat capacity concept [11]:

$$(\rho C_p)_{nf} = (\rho C_p)_f(1-\phi) + (\rho C_p)_p\phi \tag{3.15}$$

These two formulations may of course lead to different results for specific heat. Due to the lack of experimental data, both formulations are considered equivalent in estimating nanofluid specific heat capacity [12].

3.1.4.3 *Thermal Expansion Coefficient*

Thermal expansion coefficient of nanofluid can be obtained as follows [1]:

$$(\rho\beta)_{nf} = (\rho\beta)_f(1-\phi) + (\rho\beta)_p\phi \tag{3.16}$$

3.1.4.4 *The Electrical Conductivity*

The effective electrical conductivity of nanofluid was presented by Maxwell [13] as below:

$$\sigma_{nf}/\sigma_f = 1 + 3(\sigma_P/\sigma_f - 1)\phi/((\sigma_P/\sigma_f + 2) - (\sigma_P/\sigma_f - 1)\phi) \tag{3.17}$$

3.1.4.5 *Dynamic Viscosity*

Various models have been suggested to model the viscosity of a nanofluid mixture that take into account the percentage of nanoparticles suspended in the base fluid. The classic Brinkman model [14] seems to be a proper one which has been extensively used in the studies on numerical simulation concerning nanofluids. Eq. (3.1) shows the relation between the nanofluid viscosity, base fluid viscosity, as well as the nanoparticle concentration in this model.

$$\mu_{nf} = \mu_f/(1-\phi)^{2.5} \tag{3.18}$$

However, in some recent computational studies, other models have been selected to be used in the numerical process, like the work done by Abu-Nada and Chamkha [15] to investigate the convection of CuO–EG–water nanofluid in an enclosure where Namburu correlation for viscosity [16] was applied:

$$\log(\mu_{nf}) = Ae^{-BT} \tag{3.19}$$

where

$$\begin{aligned} A &= 1.837\phi^2 - 29.643\phi + 165.65 \\ B &= 4\times10^{-6}\phi^2 - 0.001\phi + 0.0186 \end{aligned} \tag{3.20}$$

In their study, the results where compared to that of viscosity modeled by Brinkman. It was outlined that as far as a value for normalized average Nusselt number for the fluid is concerned, for various values of Rayleigh number, the Brinkman model owns a prediction of higher value compared to that for the Namburu model showing the notable role of viscosity model used in the calculations. The authors also state that a combination of different models might also be implemented that will show different dependence on volume concentration as well as the geometry aspect ratio, along with the limitation that the models include only the ones mentioned in the

TABLE 3.1 Different Models for Viscosity of Nanofluids Used in Simulation

Model	Equation
Einstein model [17]	$\mu_{nf} = (2.5\phi + 1)\mu_f, \quad \phi < 0.05$
Pak and Cho's Correlation [18]	$\mu_{nf} = \mu_f(1 + 39.11\phi + 533.9\phi^2)$
Jang et al. model [19]	$\mu_{nf} = (2.5\phi + 1)\mu_f\left[1 + \eta(d_p/H)^{-2\varepsilon}\phi^{2/3}(\varepsilon + 1)\right]$
Koo and Kleinstreuer [20]	$\mu_{nf} = 5 \times 10^4 \beta\phi\rho_f\sqrt{\frac{k_B T}{d_p\rho_p}}f(T,\phi), \quad \begin{cases} \beta = 0.0137(100\phi)^{-0.8229} \text{ for } \phi < 1\% \\ \beta = 0.0011(100\phi)^{-0.7272} \text{ for } \phi > 1\% \end{cases}$
Maiga model [21]	$\mu_{nf} = \mu_f(1 + 7.3\phi + 123\phi^2)$
Brownian model [22]	$\mu_{nf} = \mu_f(1 + 2.5\phi + 6.17\phi^2)$
Nguyen model [23]	$\mu_{nf} = \mu_f(1 + 0.025\phi + 0.015\phi^2)$
Masoumi et al. [24]	$\mu_{nf} = \mu_f + \rho_p V_B d_p^2/(72C\delta)$
Gherasim et al. [25]	$\mu_{nf} = \mu_f 0.904 e^{14.8\phi}$

study. Other studies have also shown that different models might lead to different results; a number of suggested relations for viscosity models used in numerical studies are presented in Table 3.1.

3.1.4.6 *Thermal Conductivity*

Different nanofluid models based on a combination of the different formulas for the thermal conductivity adopted in the studies of natural convection are summarized in Table 3.2. Also Table 3.3 demonstrates the values of thermo-physical properties for different materials used as suspended particles in nanofluids.

3.2 SIMULATION OF NANOFLUID FLOW AND HEAT TRANSFER

Several semianalytical and numerical methods have been applied successfully in order to simulate nanofluid flow and heat transfer. In the following sections we present these works.

3.2.1 Semianalytical Methods

Forced convective heat transfer to Sisko nanofluid past a stretching cylinder in the presence of variable thermal conductivity was presented by Khan and Malik [34]. They used Homotopy Analysis Method (HAM) to solve the governing equations. They found that the curvature parameter assisted the temperature as well as concentration profiles. Momentum and heat transfer characteristics from heated spheroids in water-based nanofluids has been investigated by Sasmal and Nirmalkar [35]. They showed that the smaller the nanoparticles' size the better the heat transfer at low Reynolds number and volume fraction. Hayat et al. [36] studied the effects of homogeneous–heterogeneous reactions in flow of magnetite-Fe_3O_4 nanoparticles by a rotating disk. They showed that the axial, radial, and azimuthal velocity profiles are a decreasing function of the Hartman number. Sheikholeslami et al. [37] utilized Least Square and Galerkin Methods to investigate MHD nanofluid flow in a semiporous channel. They indicate that velocity boundary layer thickness decreases with an increase of Reynolds number and it increases as the Hartmann number increases. Sheikholeslami et al. [38] studied the squeezing unsteady nanofluid flow using Adomian Decomposition Method (ADM). They showed that the Nusselt number increases with an increase of nanoparticle volume fraction and Eckert number. Sheikholeslami and Ganji [39] applied Homotopy perturbation method (HPM) to analysis heat transfer of Cu-water nanofluid flow between parallel plates. They indicated that the Nusselt number has a direct relationship with nanoparticle volume fraction, the squeeze number, and Eckert number when two plates are separated. Application of ADM for nanofluid Jeffery–Hamel flow with high magnetic field has been presented by Sheikholeslami et al. [40]. They proved that in greater angles or Reynolds numbers a high Hartmann number is needed to reduce backflow.

TABLE 3.2 Different Models for Thermal Conductivity of Nanofluids Used in Simulation

Model	Equation
Koo and Kleinstreuer [20]	$k_{nf} = k_f \left[\frac{k_p + 2k_f - 2\phi(k_f - k_p)}{k_p + 2k_f + \phi(k_f - k_p)}\right] + 5\times10^4 \beta\phi\rho_f (C_p)_f \sqrt{\frac{k_B T}{d_p \rho_p}} f(T,\phi)$
Jang and Choi model [26]	$\frac{k_{nf}}{k_f} = (1-\phi) + Bk_p\phi + 18\times10^6 \frac{3d_f}{d_p} k_f \mathrm{Re}_{dp}^2 \mathrm{Pr}\,\phi$
Bruggeman model [27]	$k_{nf} = 0.25k_f(3\phi - 1)k_p/k_f + [3(1-\phi) - 1] + \sqrt{\Delta_B}$ $\Delta_B = \left[(3\phi-1)k_p/k_f + (3(1-\phi)-1)\right]^2 + 8k_p/k_f$
Chon et al. model [28]	$k_{nf} = k_f\left[1 + 64.7\phi^{0.7640}(d_f/d_p)^{0.369}(k_f/k_p)^{0.7476}\mathrm{Pr}_T^{0.9955}\mathrm{Re}^{1.2321}\right]$
Charuyakorn et al. [29]	$\frac{k_{nf}}{k_f} = \left[\frac{k_p + 2k_f - 2\phi(k_f - k_p)}{k_p + 2k_f + \phi(k_f - k_p)}\right]\left(1 + b\phi Pe_p^m\right)$
Staionary model [30]	$k_{nf} = k_f\left[1 + k_p\phi d_f/(k_f(1-\phi)d_p)\right]$
Yu and Choi [31]	$\frac{k_{nf}}{k_f} = \left[\frac{k_p + 2k_f - 2\phi(k_f - k_p)(1+\eta)^3}{k_p + 2k_f + \phi(k_f - k_p)(1+\eta)^3}\right]$
Patel et al. [32]	$\frac{k_{nf}}{k_f} = 1 + \frac{k_p d_f \phi}{k_f d_p (1-\phi)}\left[1 + c\frac{2k_B T d_p}{\pi\alpha_f \mu_f d_p^2}\right]$
Mintsa et al. [33]	$k_{nf} = k_f(1.72\phi + 1.0)$

TABLE 3.3 The Thermo-physical Properties of the Nanofluid

	ρ (kg/m^3)	$C_p(j/kgk)$	$k(W/m.k)$	$\beta(K^{-1})$	$\sigma(\Omega\cdot m)^{-1}$
Pure water	997.1	4179	0.613	21×10^{-5}	0.05
Copper (Cu)	8933	385	401	1.67×10^{-5}	5.96×10^7
Silver (Ag)	10 500	235	429	1.89×10^{-5}	3.60×10^7
Alumina (Al_2O_3)	3970	765	40	0.85×10^{-5}	1×10^{-10}
Titanium Oxide (TiO_2)	4250	686.2	8.9538	0.9×10^{-5}	1×10^{-12}

Flow and heat transfer of Cu-water nanofluid between a stretching sheet and a porous surface in a rotating system was studied by Sheikholeslami et al. [41]. They showed that for both suction and injection, the heat transfer rate at the surface increases with increasing nanoparticle volume fraction, Reynolds number, and injection/suction parameter and it decreases with the power of rotation parameter. Sheikholeslami et al. [42] used HAM to describe nanofluid flow over a permeable stretching wall in a porous medium. They found that an increase in the nanoparticle volume fraction will decrease the momentum boundary layer thickness and entropy generation rate while this increases the thermal boundary layer thickness. Sheikholeslami and Ganji [43] utilized Galerkin's optimal homotopy asymptotic method to investigate magnetohydrodynamic nanofluid flow in a permeable channel. They showed that velocity boundary layer thickness decreases with the increase of Reynolds number and nanoparticle volume fraction, and it increases as the Hartmann number increases. Sheikholeslami et al. [44] presented an application of HPM for simulation of two-phase unsteady nanofluid flow and heat transfer between parallel plates in the presence of a time-dependent magnetic field. Nanofluid flow and heat transfer between parallel plates considering Brownian motion has been investigated by Sheikholeslami and Ganji [45]. They used Differential Transformation Method (DTM) to solve the governing equations. They showed that skin friction coefficient increases with the increase of the squeeze number and Hartmann number. Sheikholeslami et al. [46] studied the steady nanofluid flow between parallel plates. They indicated that the Nusselt number augments with an increase of viscosity parameters but it decreases with the augment of the magnetic parameter, thermophoretic parameter, and Brownian parameter. DTM has been applied by Domairry et al. [47] to solve the problem of free convection heat transfer of non-Newtonian nanofluid between two vertical flat plates. They showed that as the nanoparticle volume fraction increases, the momentum boundary layer thickness increases. Table 3.4 shows the summary of the semianalytical method studies on nanofluid.

TABLE 3.4 Summary of the Semianalytical Method Studies on Nanofluid

Authors	Geometry of study	Method	Type of nanofluid	Properties	Remarks
Sheikholeslami et al. [37]	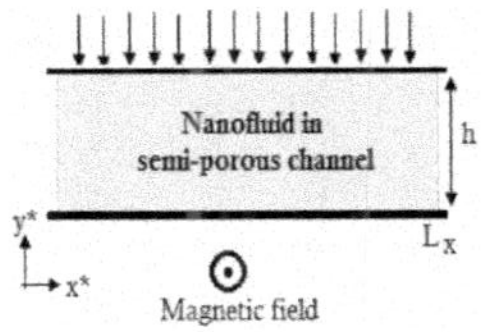	Least Square and Galerkin Methods	Cu-water	$1 \le Re \le 10$ $0 \le \phi \le 0.06$ $0 \le Ha \le 10$ $Pr = 6.8$	Velocity boundary layer thickness decreases with increase of Reynolds number and it increases as Hartmann number increases
Sheikholeslami et al. [38]	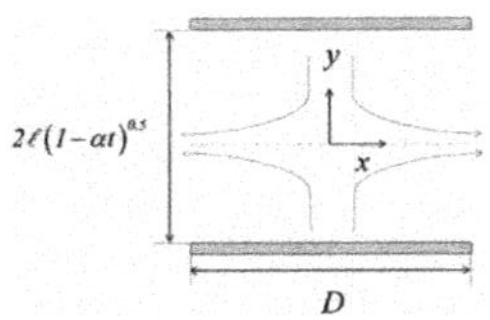	Adomian Decomposition Method	Cu-water Ag-water Al_2O_3-water TiO_2-water	$-1 \le S \le 1$ $0 \le \phi \le 0.06$ $0.01 \le Ec \le 2$ $Pr = 6.2$	Nusselt number increases with increase of nanoparticle volume fraction and Eckert number but it decreases with increase of the squeeze number
Sheikholeslami and Ganji [39]	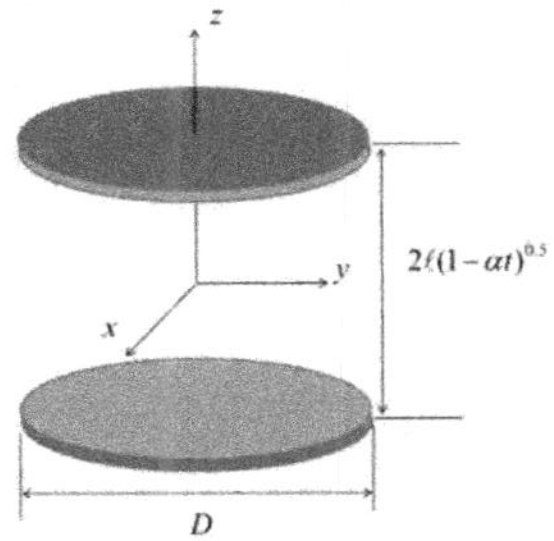	Homotopy perturbation method	Cu-water	$-1 \le S \le 1$ $0 \le \phi \le 0.06$ $0 \le Ec \le 1$ $0 \le \delta \le 1$ $Pr = 6.2$	Nusselt number has direct relationship with nanoparticle volume fraction, δ, the squeeze number, and Eckert number when two plates are separated but it has reverse relationship with the squeeze number when two plates are squeezed
Sheikholeslami et al. [40]	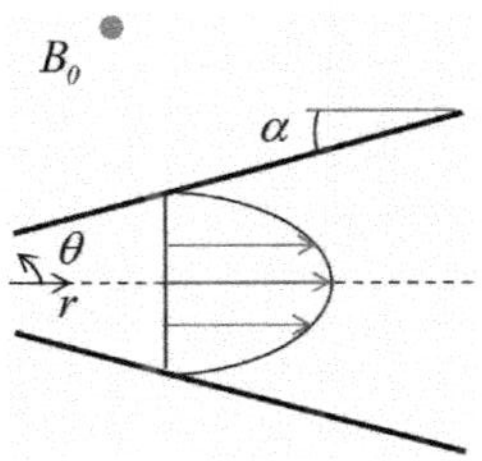	Adomian decomposition method	Cu-water	$100 \le Re \le 300$ $2.5^\circ \le \alpha \le 5^\circ$ $0 \le \phi \le 0.06$ $0 \le Ha \le 2000$ $Pr = 6.2$	In greater angles or Reynolds numbers high Hartmann number is needed to reduce backflow
Sheikholeslami et al. [41]	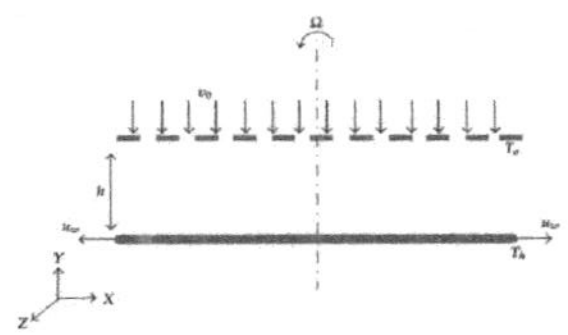	Homotopy Analysis Method	Cu-water	$5 \le R \le 20$ $0.5 \le Kr \le 6$ $-1 \le \lambda \le 1$ $0 \le \phi \le 0.2$ $0 \le Ha \le 2000$ $Pr = 6.2$	For both suction and injection, the heat transfer rate at the surface increases with increasing nanoparticle volume fraction, Reynolds number, and injection/suction parameter and it decreases with power of rotation parameter
Sheikholeslami et al. [42]	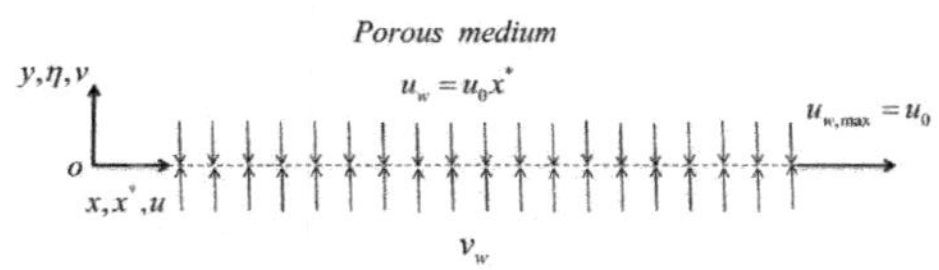	Homotopy Analysis Method	Cu-water Ag-water Al_2O_3-water TiO_2-water	$0.1 \le Re \le 1.5$ $-0.1 \le f_w \le 1$ $0 \le n \le 5$ $0 \le \phi \le 0.2$ $0 \le Ha \le 2000$ $Pr = 6.2$	Increase in the nanoparticle volume fraction will decrease momentum boundary layer thickness and entropy generation rate while this increases the thermal boundary layer thickness. Such effects are found to be more noticeable in the Ag-water solution than in the other solutions

(Continued)

TABLE 3.4 *(Continued)*

Authors	Geometry of study	Method	Type of nanofluid	Properties	Remarks
Sheikholeslami and Ganji [43]	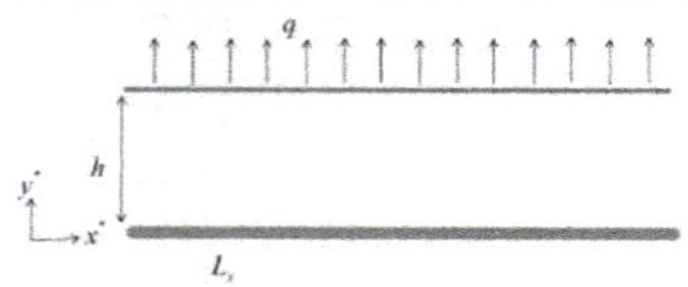	Galerkin Optimal Homotopy Asymptotic Method	Cu-water	$0 \le Ha \le 20$ $1 \le Re \le 20$ $0 \le \phi \le 0.06$ $Pr = 6.2$	Velocity boundary layer thickness decreases with increase of Reynolds number and nanoparticle volume fraction and it increases as the Hartmann number increases
Sheikholeslami et al. [44]	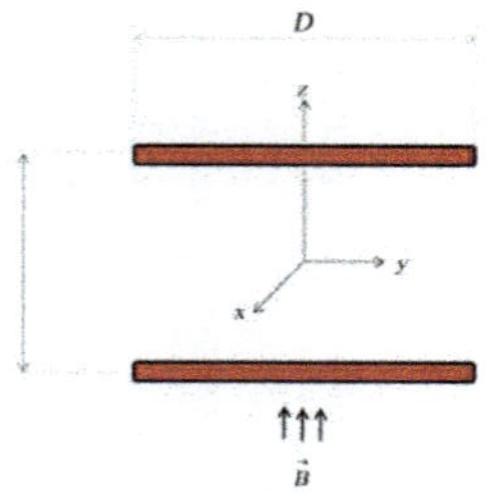	Homotopy perturbation method	Al_2O_3-water	$0 \le Ha \le 12$ $0 \le S \le 12$ $0.01 \le Ec \le 0.4$ $0.5 \le Sc \le 12$ $Pr = 10$	Nusselt number is an increasing function of the Hartmann number, Eckert number, and Schmidt number but it is a decreasing function of the squeeze number
Sheikholeslami and Ganji [45]	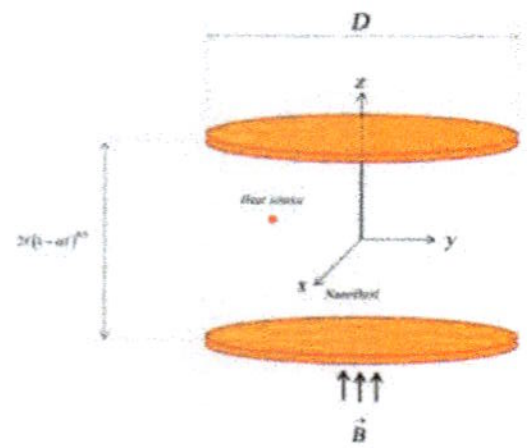	Differential Transformation Method	CuO-water, Al_2O_3-water	$0 \le S \le 10$ $0 \le Ha \le 20$ $-1 \le Hs \le -10$ $0.01 \le Ec \le 0.4$ $0.5 \le Sc \le 12$ $Pr = 10$	Skin friction coefficient increases with an increase of the squeeze number and Hartmann number but it decreases with an increase of nanofluid volume fraction. Nusselt number increases with augment of nanoparticle volume fraction, and Hartmann number while it decreases with increase of the squeeze number
Sheikholeslami et al. [46]	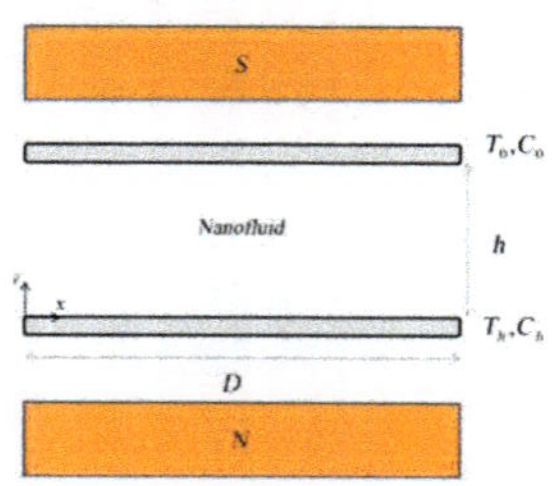	Differential Transformation Method	Al_2O_3-water	$0 \le S \le 10$ $0 \le Ha \le 20$ $-1 \le Hs \le -10$ $0.01 \le Ec \le 0.4$ $0.5 \le Sc \le 12$ $Pr = 10$	Nusselt number augments with increase of viscosity parameters but it decreases with augment of Magnetic parameter, thermophoretic parameter, and Brownian parameter
Domairry et al. [47]	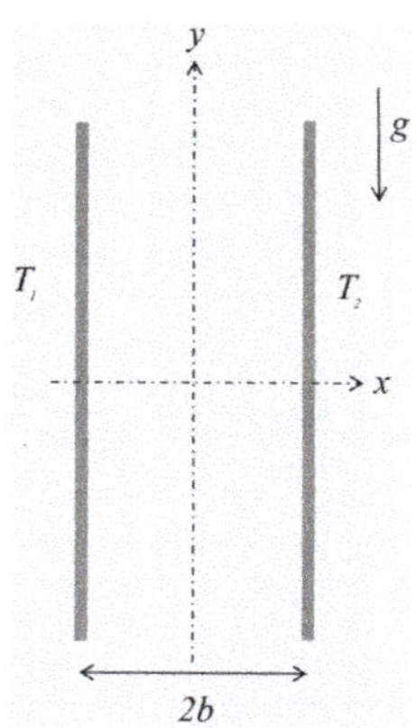	Differential Transformation Method	Cu-water	$0 \le Ha \le 20$ $0 \le \delta \le 100$ $0 \le \phi \le 0.1$ $0 \le Ec \le 6$ $Pr = 6.2$	As the nanoparticle volume fraction increases, the momentum boundary layer thickness increases, whereas the thermal boundary layer thickness decreases

3.2.2 Runge–Kutta Method

Slip effects on unsteady stagnation point flow of a nanofluid over a stretching sheet has been investigated by Malvandi et al. [48]. They showed that Cu-water nanofluids exhibit a better thermal performance than the other considered nanofluids. Malvandi [49] investigated the unsteady flow of a nanofluid in the stagnation point region of a time-dependent rotating sphere. Ashorynejad et al. [50] studied nanofluid flow and heat transfer due to a stretching cylinder in the presence of magnetic field. They showed that choosing copper (for a small magnetic parameter) and alumina (for large values of magnetic parameter) leads to the highest cooling performance for this problem. Heating a permeable stretching surface in a porous medium was studied by Sheikholeslami and Ganji [51]. Three-dimensional nanofluid flow, heat and mass transfer in a rotating system has been presented by Sheikholeslami and Ganji [52]. They showed that the Nusselt number has a direct relationship with Reynolds number while it has a reverse relationship with rotation parameter and magnetic parameter.

Sheikholeslami et al. [53] studied the nanofluid flow and heat transfer in a rotating system in the presence of a magnetic field. Sheikholeslami and Ganji [54] studied two-phase modeling of nanofluid in a rotating system with a permeable sheet. Unsteady nanofluid flow and heat transfer in the presence of a magnetic field considering thermal radiation has been investigated by Sheikholeslami and Ganji [55]. Sheikholeslami et al. [56] studied MHD nanofluid flow and heat transfer considering viscous dissipation. They showed that the magnitude of the skin friction coefficient is an increasing function of the magnetic parameter, rotation parameter, and Reynolds number and it is a decreasing function of the nanoparticle volume fraction. Sheikholeslami et al. [57] studied the effect of thermal radiation on magnetohydrodynamics nanofluid flow and heat transfer by means of a two-phase model. Sheikholeslami [58] used KKL model for simulating nanofluid flow and heat transfer in a permeable channel. The effect of uniform suction on nanofluid flow and heat transfer over a cylinder has been studied by Sheikholeslami [59]. Sheikholeslami and Abelman [60] studied two-phase simulation of nanofluid flow and heat transfer in an annulus in the presence of an axial magnetic field. Nanofluid spraying on an inclined rotating disk for cooling process has been investigated by Sheikholeslami et al. [61]. Sheikholeslami et al. [62] investigated nanofluid flow and heat transfer over a stretching porous cylinder considering thermal radiation. They showed that skin friction coefficient increases with the increase of Reynolds number and suction parameter but it decreases with the increase of nanoparticle volume fraction. Table 3.5 shows the summary of the Runge–Kutta method studies on nanofluid. Chamkha and Aly [63] have studied the boundary layer flow of a nanofluid past a vertical flat plate. They have considered the Brownian motion and the thermophoresis effect. They have transformed the governing equations to a nonsimilar form and used numerical techniques to solve the same. They have reported that the local skin friction coefficient increased as any of the suction, injection parameter, thermophoresis parameter, Lewis number, or heat generation or absorption parameters increased, while it decreased as any of the buoyancy ratio, Brownian motion parameter, or the magnetic field parameter increased.

3.2.3 Finite Difference Method

Chamkha and Rashad [64] have studied the flow of a nanofluid around a nonisothermal wedge. They have considered the Brownian movement and the thermophoresis effects. They have concluded that the local skin friction coefficient, local Nusselt number, and the local Sherwood number reduced as either of the magnetic parameter or the pressure gradient parameter was increased. The presence of the Brownian motion and the thermophoresis effects caused the local Nusselt number to decrease and the Sherwood number to increase. Sheremet and Pop [65] used Buongiorno's mathematical model for conjugate natural convection in a square porous cavity filled with nanofluid. They showed that high thermophoresis parameter, low Brownian motion parameter, low Lewis and Rayleigh numbers, and high thermal conductivity ratio reflect essential nonhomogeneous distribution of the nanoparticles inside the porous cavity. Sheremet et al. [66] studied the three-dimensional natural convection in a porous enclosure filled with a nanofluid using Buongiorno's mathematical model. Sheremet et al. [67] investigated the effect of thermal stratification on free convection in a square porous cavity filled with a nanofluid using Tiwari and Das' nanofluid model.

Ghalambaz et al. [68] studied the free convection heat transfer in a porous cavity filled with a nanofluid using Tiwari and Das' nanofluid model. Double-diffusive mixed convection in a porous open cavity filled with a nanofluid using Buongiorno's model has been studied by Sheremet et al. [69]. Sheremet and Pop [70] studied nanofluid free convection in a triangular porous cavity porous. Natural convection in a horizontal cylindrical annulus filled with a porous medium saturated by a nanofluid has been investigated by Sheremet and Pop [71]. Magnetic field effect on the unsteady natural convection in a wavy-walled cavity filled with a nanofluid has been studied by Sheremet et al. [72].

TABLE 3.5 Summary of the Runge–Kutta Method Studies on Nanofluid

Authors	Geometry of study	Method	Type of nanofluid	Properties	Remarks
Malvandi et al. [48]	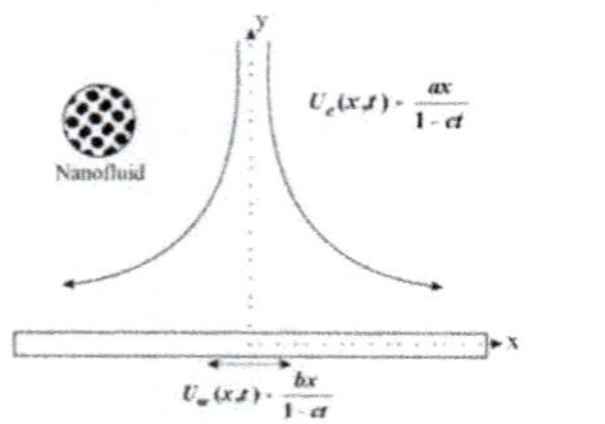	Runge–Kutta–Fehlberg method	Cu-water, Al_2O_3-water and TiO_2-water nanofluids	$1 \le d_p \le 100$ nm $0 \le \phi \le 0.2$ $Pr = 6.2$	Cu-water nanofluids exhibit a better thermal performance than the other considered nanofluids. Slip velocity at the walls increase the heat transfer rate
Malvandi [49]	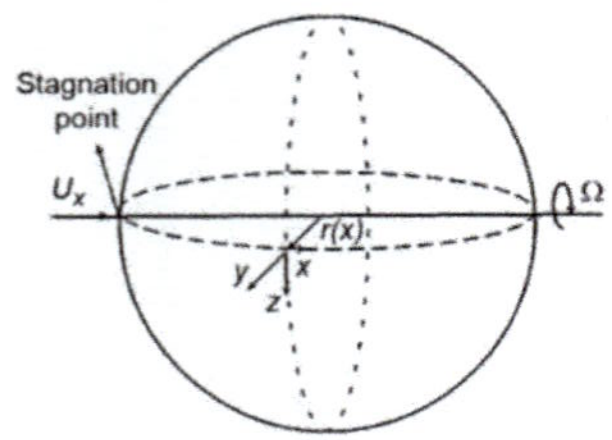	Runge–Kutta–Fehlberg method	Nonhomogeneous mixtures	$1 \le d_p \le 100$ nm $0 \le \phi \le 0.2$	Increasing the thermophoresis is found to decrease heat transfer and concentration rates. This decrease is suppressed for higher thermophoresis numbers. In addition, it was observed that unlike the heat transfer rate, a rise in Brownian motion leads to an increase in concentration rate
Ashorynejad et al. [50]	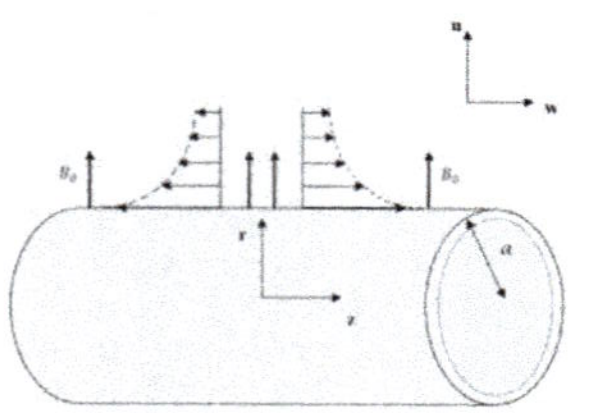	Fourth order Runge–Kutta method	Cu-water Ag-water Al_2O_3-water TiO_2-water	$1 \le Re \le 5$ $0 \le M \le 10$ $0 \le \phi \le 0.2$ $Pr = 6.2$	Choosing copper (for small of magnetic parameter) and alumina (for large values of magnetic parameter) leads to the highest cooling performance for this problem
Sheikholeslami and Ganji [51]	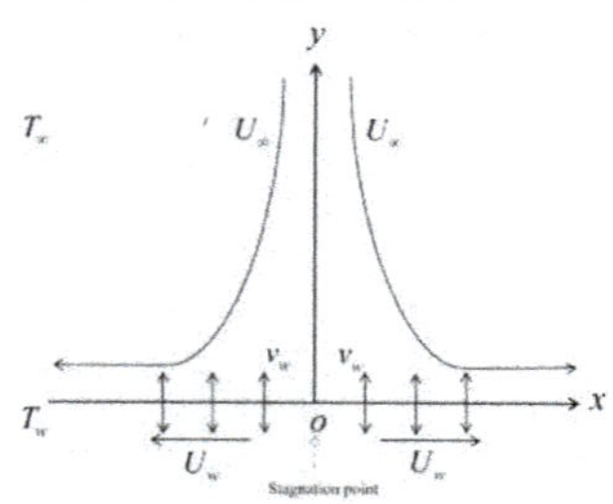	Fourth order Runge–Kutta method	Cu-water Ag-water Al_2O_3-water TiO_2-water Cu- Ethylene glycol	$1 \le Re \le 5$ $-0.5 \le \gamma \le 1$ $0 \le \varepsilon \le 1$ $0 \le \phi \le 0.2$ $Pr = 6.2$	Choosing titanium oxide as the nanoparticle and ethylene glycol as base fluid proved to have the highest cooling performance for this problem
Sheikholeslami and Ganji [52]	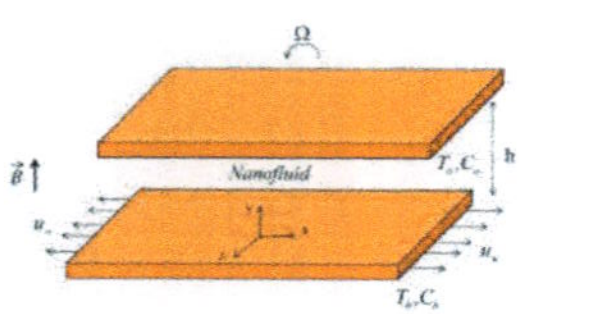	Fourth order Runge–Kutta method	Al_2O_3-water	$1 \le R \le 10$ $0 \le M \le 12$ $-0.5 \le \gamma \le 1$ $K_r = 0.5, Nt = 0.1$ $Nb = 0.1, Sc = 0.1$ $Pr = 10$	Nusselt number has a direct relationship with Reynolds number while it has a reverse relationship with rotation parameter, magnetic parameter, Schmidt number, thermophoretic parameter and Brownian parameter

Sheikholeslami et al. [53]	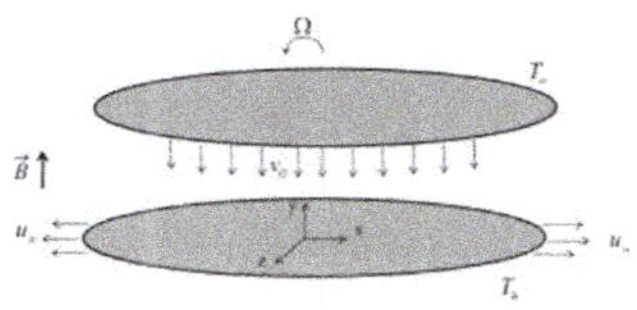	Fourth order Runge–Kutta method	Cu-water Ag-water Al_2O_3-water TiO_2-water	$5 \le R \le 20$ $0 \le M \le 50$ $0.5 \le Kr \le 6$ $0 \le \phi \le 0.2$ $-0.5 \le \lambda \le 1$ $Pr = 6.2$	The highest values are obtained when titanium oxide is used as nanoparticle. Also it can be found that the Nusselt number decreases with an increase of magnetic parameter due to the presence of Lorentz forces
Sheikholeslami and Ganji [54]	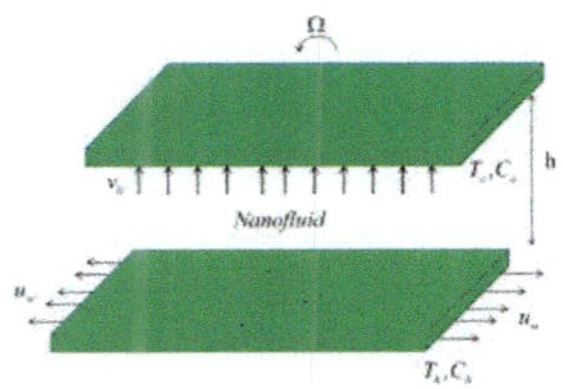	Fourth order Runge–Kutta method	Al_2O_3-water	$1 \le R \le 10$ $0.5 \le Kr \le 6$ $0 \le \lambda \le 3$ $Nt = Nb = 0.1$ $Pr = 10$	The Nusselt number has a direct relationship with the Reynolds number and injection parameter, while it has a reverse relationship with rotation parameter, Schmidt number, thermophoretic parameter and Brownian parameter
Sheikholeslami and Ganji [55]	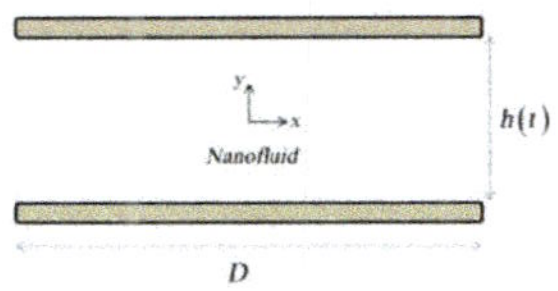	Fourth order Runge–Kutta method	Al_2O_3-water	$0.5 \le S \le 12$ $0.01 \le Ec \le 0.4$ $0.01 \le Nt \le 0.4$ $0.5 \le Sc \le 20$ $0 \le Rd \le 16$ $Pr = 10$	Concentration boundary layer thickness increases with increase of radiation parameter
Sheikholeslami et al. [56]	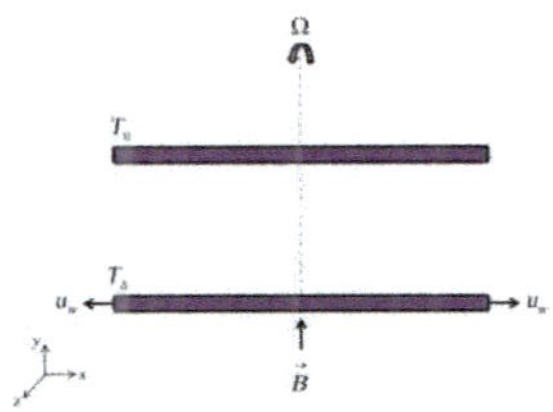	Fourth order Runge–Kutta method	CuO-water, Al_2O_3-water	$0.1 \le R \le 2$ $0 \le M \le 9$ $0 \le \phi \le 0.04$ $1 \le Kr \le 6$ $0.01 \le Ec \le 0.04$ $Pr = 10$	The magnitude of the skin friction coefficient is an increasing function of the magnetic parameter, rotation parameter, and Reynolds number and it is a decreasing function of the nanoparticle volume fraction
Sheikholeslami et al. [57]	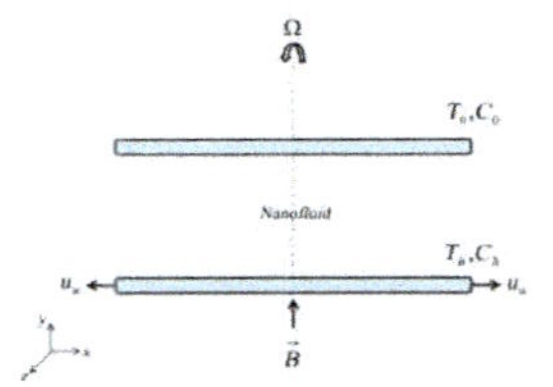	Fourth order Runge–Kutta method	Al_2O_3-water	$0.1 \le R \le 20$ $0 \le Rd \le 12$ $0.01 \le Sc \le 6$ $0 \le M \le 16$ $0 \le \phi \le 0.04$ $0.1 \le Kr \le 12$ $Pr = 10$	Nusselt number has a direct relationship with the radiation parameter and Reynolds number while it has a reverse relationship with other active parameters. Also it can be found that concentration boundary layer thickness decreases with the increase of the radiation parameter
Sheikholeslami [58]	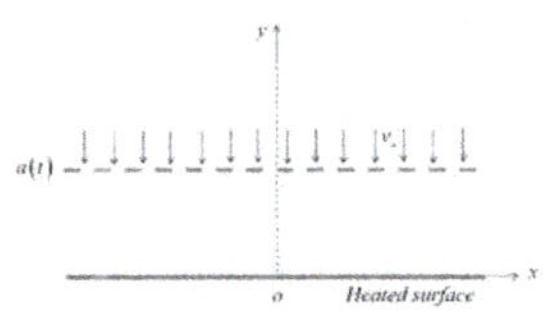	Fourth order Runge–Kutta method	Al_2O_3-water	$-1 \le \alpha \le 2$ $1 \le R \le 4$ $0 \le \phi \le 0.04$ $0 \le m \le 3$ $Pr = 6.8$	Heat transfer enhancement has a direct relationship with the Reynolds number when the power law index is equal to zero but an opposite trend is observed for other values of the power law index

(*Continued*)

TABLE 3.5 (Continued)

Authors	Geometry of study	Method	Type of nanofluid	Properties	Remarks
Sheikholeslami [59]	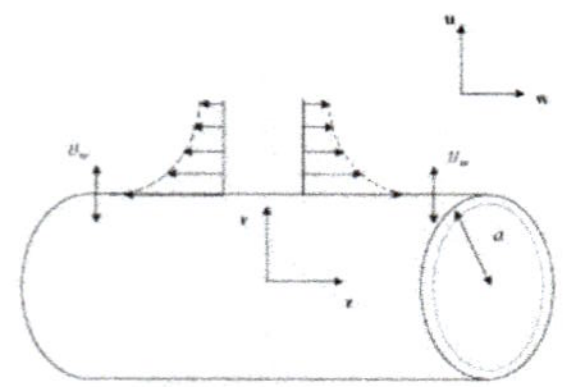	Fourth order Runge–Kutta method	CuO-water, Al_2O_3-water	$0.1 \le Re \le 2.7$ $0 \le \gamma \le 2$ $0 \le \phi \le 0.04$ $Pr = 6.2$	Adding nanoparticle into the base fluid of this problem is capable of changing the flow pattern. It is found that the Nusselt number is an increasing function of nanoparticle volume fraction, suction parameter, and Reynolds number
Sheikholeslami and Abelman [60]	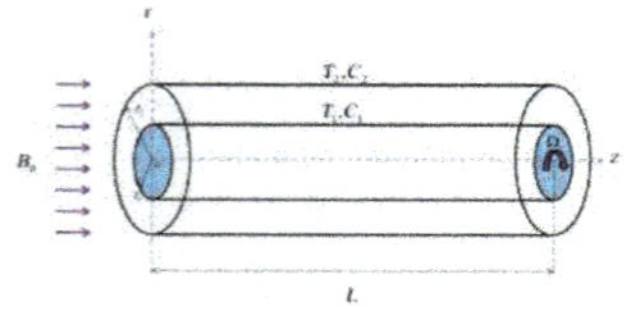	Fourth order Runge–Kutta method	Al_2O_3-water	$1 \le R \le 4$ $1 \le M \le 4$ $0.01 \le Ec \le 0.1$ $0.01 \le Nt \le 0.4$ $0.01 \le Nt \le 0.6$ $0.001 \le Ec \le 0.02$ $Pr = 10$	The Nusselt number has a direct relationship with the aspect ratio and Hartmann number but it has a reverse relationship with the Reynolds number, Schmidt number, Brownian parameter, thermophoresis parameter. and Eckert number
Sheikholeslami et al. [61]	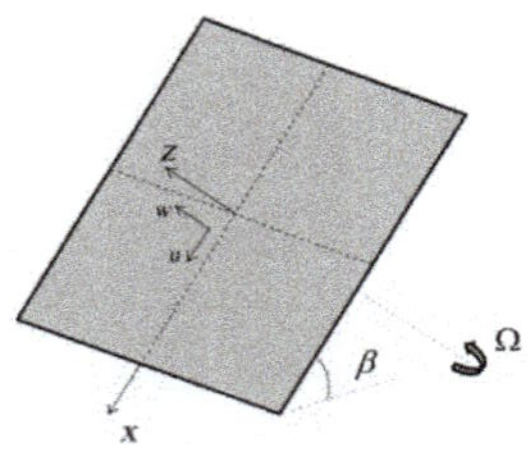	Fourth order Runge–Kutta method	Al_2O_3-water	$0.2 \le \delta \le 1$ $0.1 \le Nt = Nb \le 0.4$ $0.1 \le Sc \le 8$ $Pr = 10$	The Nusselt number is an increasing function of each active parameter. Latent heat has a direct relationship with the Schmidt number, Brownian parameter, and thermophoretic parameter but it has a reverse relationship with normalized thickness.
Sheikholeslami et al. [62]	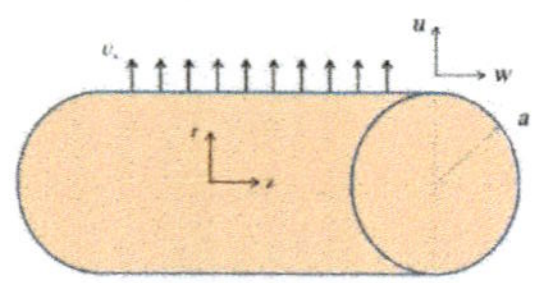	Fourth order Runge–Kutta method	Cu-water	$0.5 \le \gamma \le 2$ $0.5 \le Rd \le 2$ $0 \le \phi \le 0.04$ $Pr = 6.8$	Skin friction coefficient increases with increase of Reynolds number and suction parameter but it decreases with increase of nanoparticle volume fraction

Khan et al. [73] studied the three-dimensional flow of nanofluid induced by an exponentially stretching sheet. They showed that existence of interesting Sparrow–Gregg-type hills for temperature distribution corresponding to some ranges of parametric values. Nanofluid flow with multimedia physical features for conjugate mixed convection and radiation has been studied by Hsiao [74]. Table 3.6 shows the summary of the finite difference method studies on nanofluid.

3.2.4 Finite Volume Method

Garoosi and Hoseininejad [75] investigated the natural and mixed convection heat transfer between differentially heated cylinders in an adiabatic enclosure filled with nanofluid. Garoosi et al. [76] applied Buongiorno model for mixed convection of the nanofluid in heat exchangers. Two-phase mixture modeling of mixed convection of nanofluids in a square cavity with internal and external heating has been studied by Garoosi et al. [77]. Teamah et al. [78] studied the augmentation of natural convective heat transfer in a square cavity by utilizing nanofluids in the presence of magnetic field. They showed that for a weak magnetic field the addition of nanoparticles is necessary to enhance the heat transfer but for a strong magnetic field there is no need for nanoparticles because the heat transfer will decrease. Santra et al. [79] studied the heat transfer augmentation in a differentially heated square cavity using copper–water nanofluid. Das and Ohal [80] investigated Natural convection heat transfer augmentation in a partially heated and partially cooled square cavity utilizing nanofluids. Oztop et al. [81] analyzed the nonisothermal temperature distribution on natural convection in nanofluid filled enclosures. They showed that an enhancement in heat transfer rate was registered for the whole range of Rayleigh numbers. Table 3.7 shows the summary of the finite volume method studies on nanofluid.

3.2.5 Finite Element Method

MHD mixed convection of nanofluid filled partially heated triangular enclosure with a rotating adiabatic has been investigated by Selimefendigil and Oztop [82]. They showed that local and average heat transfer and total entropy generation are enhanced as the solid volume fraction of nanoparticle and angular rotational speed of the cylinder increases and the Hartmann number decreases. Heat transfer enhancements around 30% are achieved for the highest volume fraction compared to base fluid. Selimefendigil and Oztop [83] studied the natural convection and entropy generation of nanofluid filled cavity having different shaped obstacles under the influence of magnetic field and internal heat generation. Selimefendigil and Oztop [84] studied pulsating nanofluids jet impingement cooling of a heated horizontal surface. They showed that the combined effect of pulsation and inclusion of nanoparticles is not favorable for the stagnation point heat transfer enhancement for some combinations of Reynolds number and nanoparticle volume fraction. Selimefendigil and Oztop [85] studied MHD mixed convection in a nanofluid filled lid driven square enclosure with a rotating cylinder. Selimefendigil and Oztop [86] investigated numerical investigation and reduced order model of mixed convection at a backward facing step with a rotating cylinder subjected to nanofluid. Effect of nanoparticle shape on mixed convection due to rotating cylinder in an internally heated and flexible walled cavity filled with SiO_2-water nanofluids has been investigated by Selimefendigil et al. [87]. They indicated that Nusselt number enhances with external Rayleigh number and nanoparticle volume fraction while the opposite behavior is seen as the value of internal Rayleigh number and flexibility of the wall increase. Conjugate natural convection in a cavity with a conductive partition and filled with different nanofluids on different sides of the partition has been studied by Selimefendigil and Oztop [88]. They proved that as the value of the Grashof number, thermal conductivity ratio (Kr), and nanoparticle volume fraction increase, average Nusselt number increases. Table 3.8 shows the summary of the finite element method studies on nanofluid.

3.2.6 Control Volume-based Finite Element Method

Heatline analysis has been used by Sheikholeslami et al. [89] to investigate two-phase simulation of nanofluid flow and heat transfer. They found that Nusselt number decreases as buoyancy ratio number increases until it reaches a minimum value and then starts increasing. As Lewis number increases, this minimum value occurs at higher buoyancy ratio number. Natural convection heat transfer in a cavity with a sinusoidal wall filled with CuO-water nanofluid in presence of magnetic field has been studied by Sheikholeslami et al. [90]. Effects of a magnetic field on natural convection in different enclosures filled with naofluids have been examined by Sheikholeslami et al. [91–93]. Soleimani et al. [94] studied the natural convection heat transfer in a nanofluid filled semiannulus enclosure. They found that there is an optimum angle of turn in which the average Nusselt

TABLE 3.6 Summary of the finite Difference Method Studies on Nanofluid

Authors	Geometry of study	Method	Type of nanofluid	Properties	Remarks
Sheremet and Pop [65]	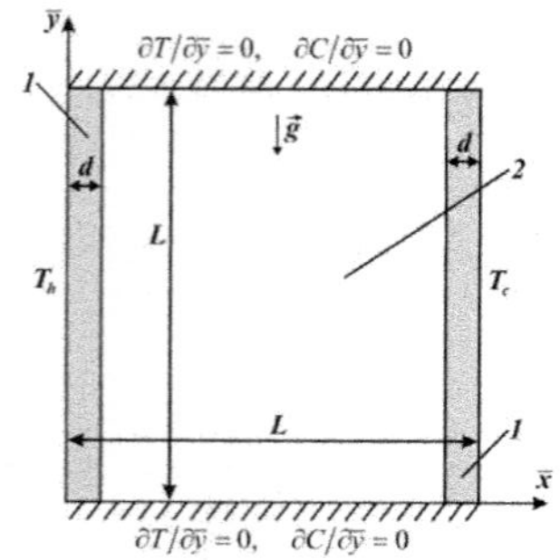	Finite difference method	Porous cavity with water-based nanofluid	$Ra=100-500$; $Le=1-10$; $Nb=0.1-0.4$; $Nt=0.1-0.4$; $Nr=0.1-0.4$; $K_r=0.1-10.0$; $D=0.1-0.3$.	High thermophoresis parameter, low Brownian motion parameter, low Lewis and Rayleigh numbers, and high thermal conductivity ratio reflect essential nonhomogeneous distribution of the nanoparticles inside the porous cavity
Sheremet et al. [66]	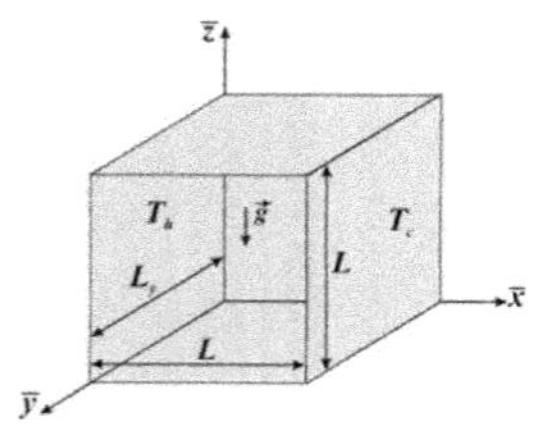	Finite difference method	Porous cavity with water-based nanofluid	$Ra=30-500$, $Le=1-10$, $Nr=0.1-0.4$, $Nb=0.1-0.4$, $Nt=0.1-0.4$, aspect ratio ($A=0.2-5.0$).	Average Nusselt number at the hot surface is an increasing function of the Rayleigh number and Brownian motion parameter, and a decreasing function of the Lewis number, buoyancy-ratio, and thermophoresis parameters. The average Sherwood number at the hot surface is an increasing function of the Rayleigh number and thermophoresis parameter, and a decreasing function of the Lewis number, buoyancy-ratio, and Brownian motion parameters.
Sheremet et al. [67]	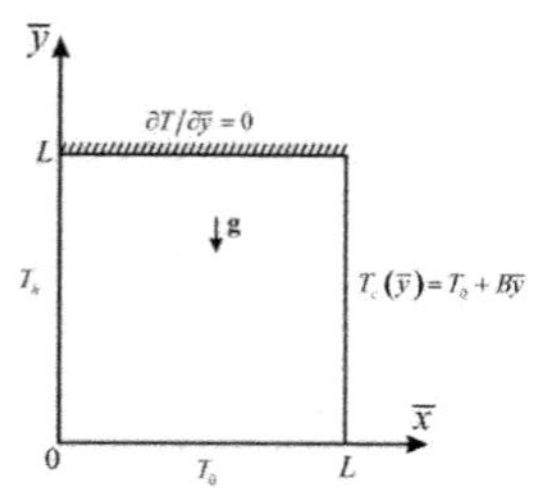	Finite difference method	Cu-water	$Ra=10-500$, $\varphi=0.0-0.05$, porosity of porous medium ($\varepsilon=0.1-0.8$), thermal stratification parameter ($b=0-1.0$) solid matrix of the porous medium (aluminum foam and glass balls)	An increase in the thermal stratification parameter leads to an essential reduction of the convective core sizes, displacement of this core close to the lower left corner and rotation of this vortex along the horizontal axis. More intensive attenuation of convective flow with ε and b occurs for low thermal conductivity material for the solid matrix of the porous medium
Ghalambaz et al. [68]	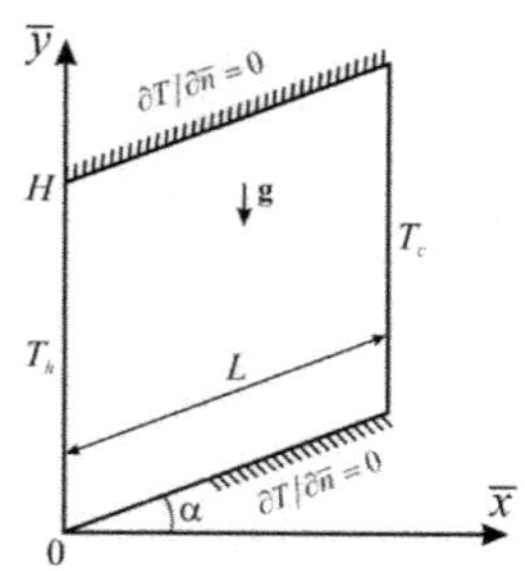	Finite difference method	Cu-water	$-60° < \alpha < 60°$, aspect ratio $0.1 < A < 10$, $10 < Ra < 1000$, $0 < \varphi < 0.1$, $\varepsilon=0.3, 0.5, 0.7$, $k_s=2k_{bf}$	Presence of nanoparticles deteriorates the heat transfer in all studied cases. The decrease of the porosity increases the porous matrix thermal conductivity while the decrease of the inclination angle and of the aspect ratio would boost the deterioration of heat transfer

Sheremet et al. [69]	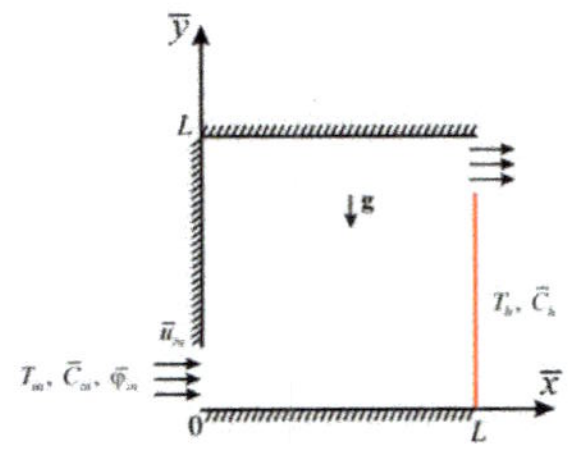	Finite difference method	Porous cavity with water-based nanofluid	$Re = 10-100$, $Pr = 0.2$, $Ra = 50-500$, usual Lewis number ($Le = 1-50$), Dufour-contaminant Lewis number ($Ld = 1-50$), nanofluid Lewis number ($Ln = 1-50$), $Nr = 0.1-0.4$, regular double-diffusive buoyancy ratio ($Nc = 0.1-0.4$), $Nb = 0.1-0.4$, $Nt = 0.1-0.4$, modified Dufour parameter ($Nd = 0.1-0.4$)	Average Nusselt number at hot vertical wall is an increasing function of the Rayleigh and Reynolds numbers, and a decreasing function of the usual Lewis number. While the average Sherwood number at this vertical wall is an increasing function of the usual Lewis. Effects of the Rayleigh and Reynolds numbers on $\lvert \overline{Sh} \rvert$ and the thermophoresis parameter on the average Nusselt and Sherwood numbers are nonmonotonic. It has been shown that in the present porous problem the Richardson number does not define the prevailing of the forced or natural convection modes
Sheremet and Pop [70]	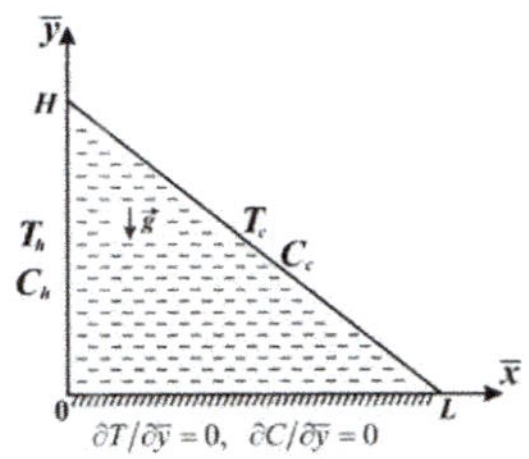	Finite difference method	Porous cavity with water-based nanofluid	$Ra = 100-500$; $Le = 1-10$; $Nb = 0.1-0.4$; $Nt = 0.1-0.4$; $Nr = 0.1-0.4$; $A = H/L = 1.0$.	Average Nusselt number is an increasing function of *Ra*, *Le*, and a decreasing function of *Nr*, *Nb*, *Nt*. At the same time the average Sherwood number is an increasing function of *Ra*, *Le*, *Nb*, *Nt* and a decreasing function of *Nr*.
Sheremet and Pop [71]	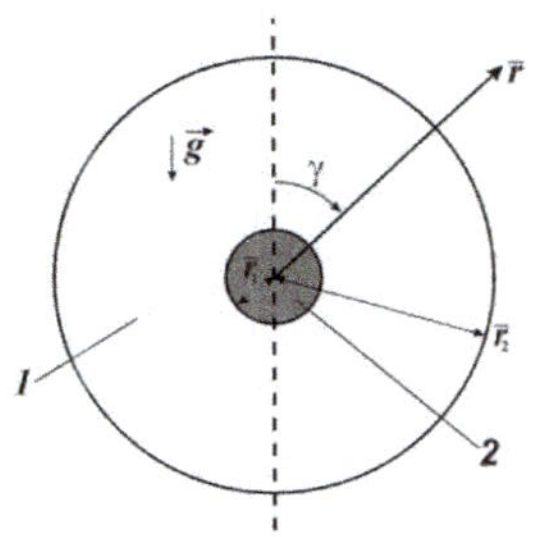	Finite difference method	Cu-water	$Ra = 200-700$, $\varepsilon = 0.1-0.8$, $\varphi = 0.0-0.05$, the annulus radius ratio ($R = 1.5, 2.0, 2.5$), solid matrix of porous medium (aluminum foam and glass balls)	Average Nusselt number is an increasing function of the Rayleigh number and annulus radius ratio and a decreasing function of the porosity of porous medium regardless of the solid matrix material. A decrease in the thermal conductivity of the solid matrix material leads to an attenuation of the convective heat transfer inside the annulus for high values of the porosity and solid volume fraction of nanoparticles
Sheremet et al. [72]	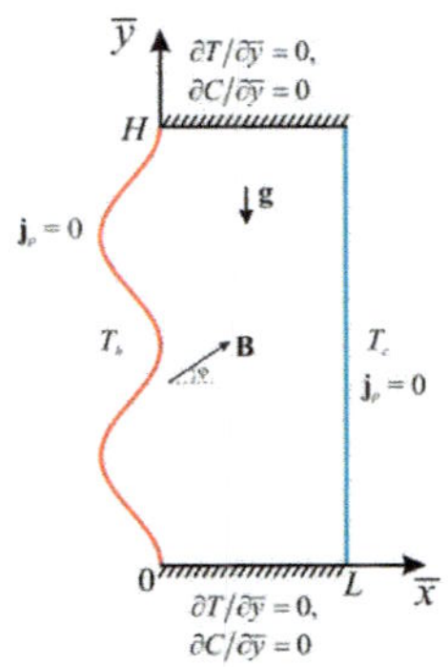	Finite difference method	Water-based nanofluid	$Ra = 10^5$, $Le = 10$, $Pr = 6.26$, $Nr = 0.1$, $Nb = 0.1$, $Nt = 0.1$, $A = 1$, $Ha = 0-100$, undulation number ($\kappa = 1-3$), wavy contraction ratio ($b = 0.1-0.3$), $\varphi = 0-\pi$, $\tau = 0-0.13$.	An increase in the Hartmann number leads to an attenuation of convective flow and heat transfer and a formation of a double-core convective cell for high values of *Ha*. An increase in the wavy contraction ratio leads to an increase in the wave amplitude and an attenuation of the convective flow with more intensive heating of the wavy troughs

(*Continued*)

TABLE 3.6 (*Continued*)

Authors	Geometry of study	Method	Type of nanofluid	Properties	Remarks
Khan et al. [73]	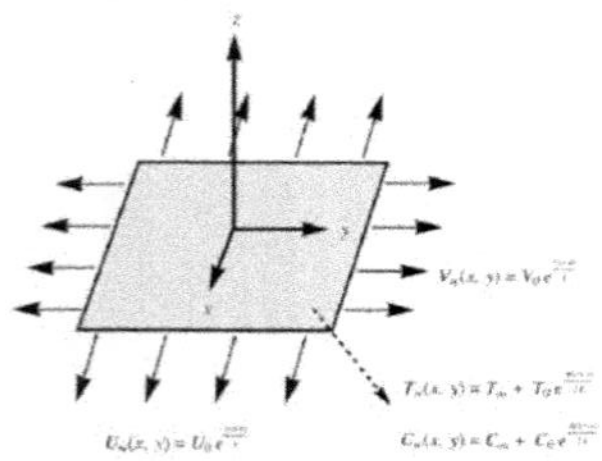	Finite difference scheme known as Keller-box method	Water-based nanofluid	$0 \le \lambda \le 1$ $0.1 \le Nb \le 0.9$ $0.3 \le Nt \le 0.5$ $5 \le Pr \le 10$ $-1 \le A \le 3$	The existence of interesting Sparrow–Gregg-type hills for temperature distribution corresponding to some range of parametric values
Kai-Long Hsiao [74]	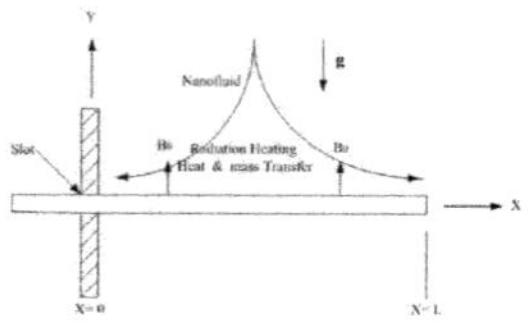	An improved finite-difference with box method	Cu-water	$K = 400$ (W/mK): $\rho = 8940\ (kg/m^3)$ $C_p = 385 (J/kgK)$ $10^1 \le Re \le 10^6$ $0 \le Pr \le 1000$	The results show that dimensionless heat transfer effects increase with increasing values of Pr, R0, Nb, or G but decrease with increasing M, Nt, or Ec. Those, parameters Pr, R0, Nb, or G are important factors in this study for increasing the heat transfer effects. The fourth, for mass transfer of the values Sc and Nb can be obtained a good mass diffusion effect, but the parameter Nt has not been processed for this kind of function

TABLE 3.7 Summary of the Finite Volume Method Studies on Nanofluid

Authors	Geometry of study	Method	Type of nanofluid	Properties	Remarks
Garoosi and Hoseininejad [75]	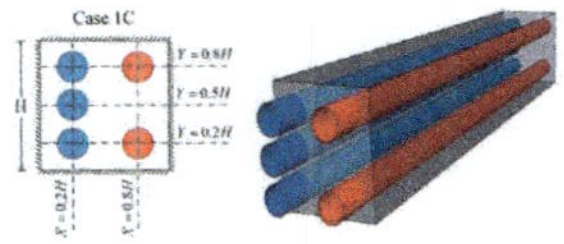	Finite volume method	Cu-water TiO_2-water Al_2O_3-water	$10^4 \le Ra \le 10^7$ $0 \le \phi \le 0.05$ $25 \le d_p \le 145$ $Pr = 6.8$	Changing the location of the heat source/sink from bottom–top to top–bottom configuration, the heat transfer rate decreases significantly
Garoosi et al. [76]	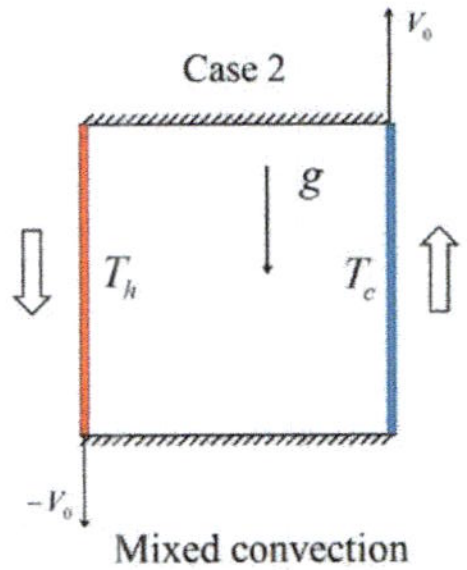	Finite volume method	Al_2O_3-water	$10^4 \le Ra \le 10^7$ $0.01 \le Ri \le 1000$ $10^2 \le Gr \le 10^4$ $0 \le \phi \le 0.05$ $25 \le d_p \le 145$	At low Rayleigh numbers and high Richardson numbers, the particle distribution is fairly nonuniform while at high Ra and low Ri values particle distribution remains almost uniform for free and mixed convection cases, respectively
Garoosi et al. [77]	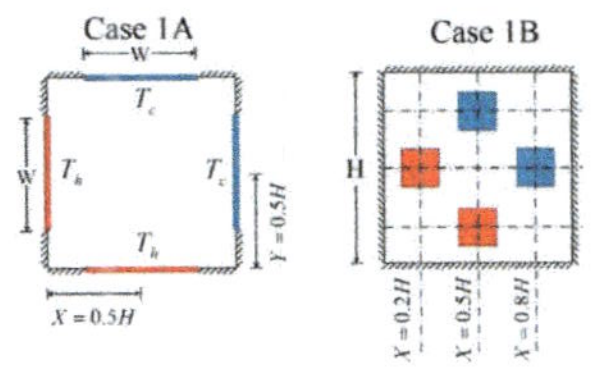	Finite volume method	Cu-water TiO_2-water Al_2O_3-water	$10^4 \le Ra \le 10^7$ $0.01 \le Ri \le 1000$ $0 \le \phi \le 0.05$ $25 \le d_p \le 145$	Thermophoretic effects are negligible for nanoparticles with high thermal conductivity. As a result, in such conditions the use of homogeneous and single-phase models is valid at any Ra and Ri
Teamah et al. [78]	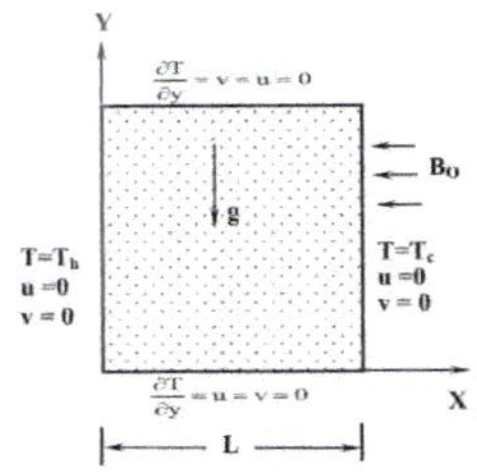	Finite volume method	Water-based nanofluid	$10^4 \le Ra \le 10^7$ $0 \le \phi \le 0.05$ $0 \le Ha \le 60$	At a weak magnetic field the addition of nanoparticles is necessary to enhance the heat transfer but for a strong magnetic field there is no need for nanoparticles because the heat transfer will decrease
Santra et al. [79]	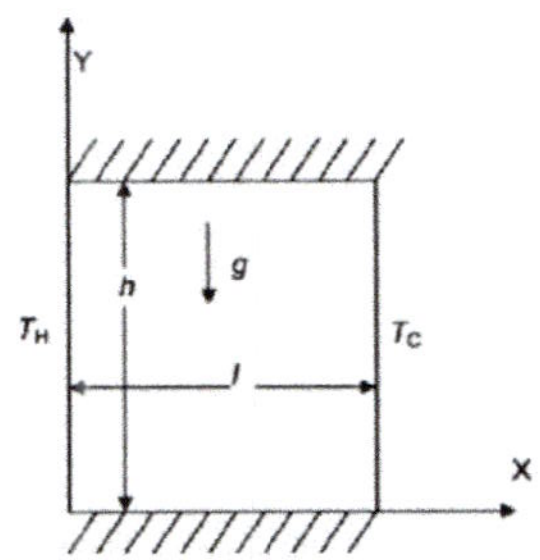	Finite volume method	Cu-water	$10^4 \le Ra \le 10^7$ $0 \le \phi \le 0.05$ $A = 1$	The heat transfer rate decreases with increase in solid volume fraction for a particular Rayleigh number. However, it increases with Rayleigh number for a particular solid volume fraction

(*Continued*)

TABLE 3.7 (Continued)

Authors	Geometry of study	Method	Type of nanofluid	Properties	Remarks
Das and Ohal [80]	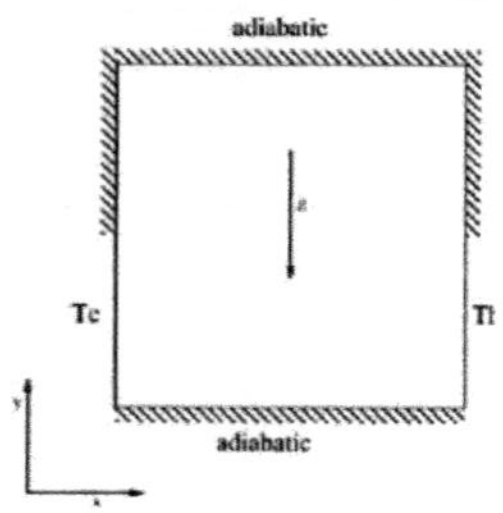	Finite volume method	Cu-water	$10^4 \leq Gr \leq 10^7$ $0 \leq \phi \leq 0.02$ $A = 1$	The rate of heat transfer increases with an increase in the nanoparticles volume fraction
Oztop et al. [81]	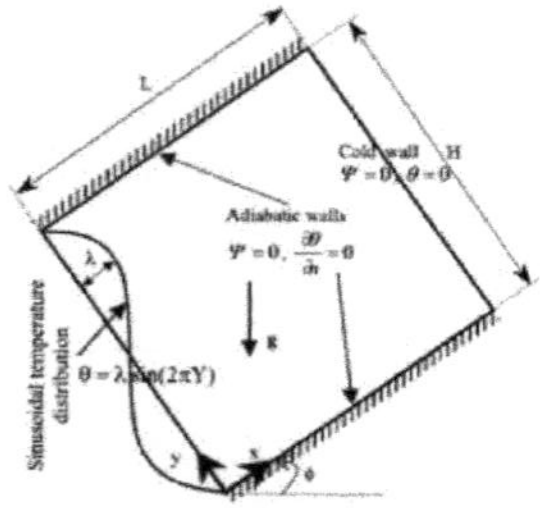	Finite volume method	Cu-water TiO_2-water Al_2O_3-water	$10^4 \leq Gr \leq 10^7$ $0 \leq \phi \leq 0.01$ $0 \leq \varphi \leq 90^{\circ}$ $A = 1$	An enhancement in heat transfer rate was registered for the whole range of Rayleigh numbers

TABLE 3.8 Summary of the Finite Element Method Studies on Nanofluid

Authors	Geometry of study	Method	Type of nanofluid	Properties	Remarks
Selimefendigil and Oztop [82]	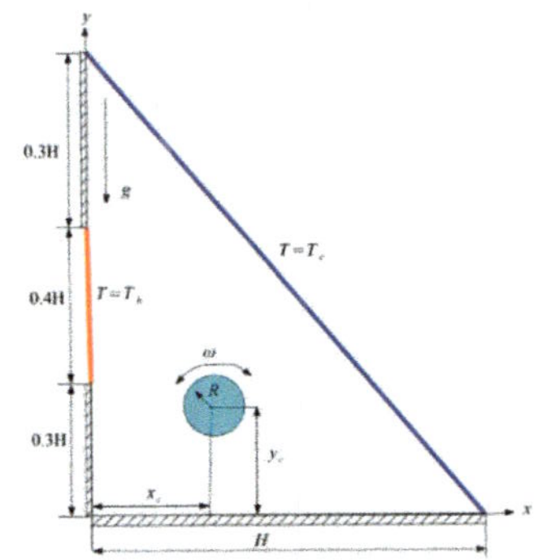	Finite element method	Cu-water	$10^4 \leq Gr \leq 10^6$ $0 \leq \phi \leq 0.05$ $0 \leq Ha \leq 50$ $-20 \leq \omega \leq 20$ $Pr = 7.1$	Local and average heat transfer and total entropy generation enhance as the solid volume fraction of nanoparticle and angular rotational speed of the cylinder increases and Hartmann number decreases. Heat transfer enhancements around 30% are achieved for the highest volume fraction compared to base fluid
Selimefendigil and Oztop [83]	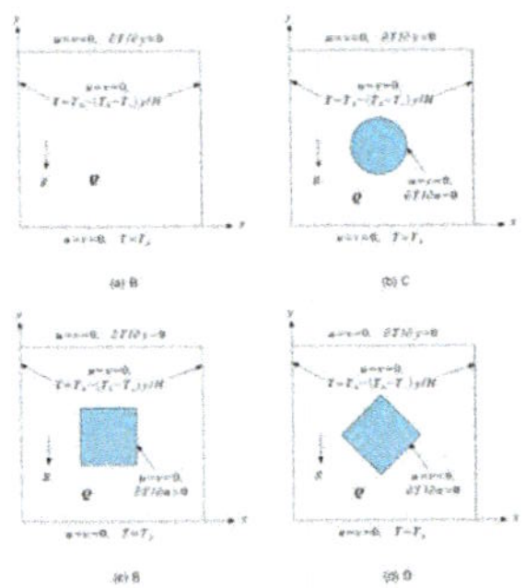	Finite element method	Cu-water	$10^4 \leq Ra_E \leq 10^6$ $10^4 \leq Ra_I \leq 10^6$ $0 \leq \phi \leq 0.05$ $0 \leq Ha \leq 50$ $Pr = 6.9$	As the value of the external Rayleigh number decreases, internal Rayleigh number and Hartmann number increase, average heat transfer enhance. Average Nusselt number enhances by about 40%–60% at the highest solid volume fraction when compared to base fluid with different obstacles installed within the cavity.
Selimefendigil and Oztop [84]	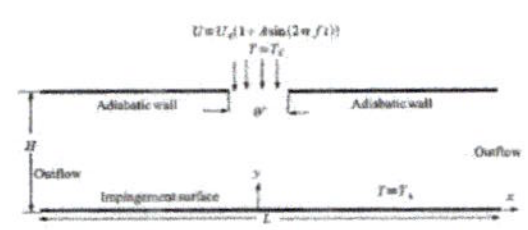	Finite volume method	Al_2O_3-water	$100 \leq Re \leq 400$ $1 \leq f \leq 10$ $0 \leq \phi \leq 0.06$ $Pr = 7.1$	The combined effect of pulsation and inclusion of nanoparticles is not favorable for the stagnation point heat transfer enhancement for some combinations of Reynolds number and nanoparticle volume fraction
Selimefendigil and Oztop [85]	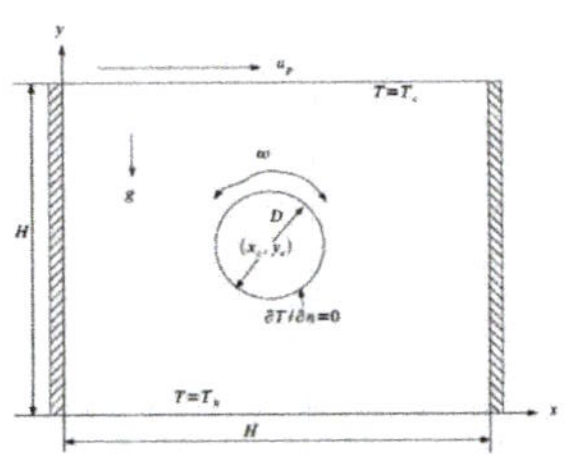	Finite element method	Cu-water	$0.001 \leq Ri \leq 10$ $0 \leq Ha \leq 50$ $0 \leq \phi \leq 0.05$ $-10 \leq \Omega \leq 10$ $Pr = 6.2$	Average heat transfer is enhanced with Richardson number, nanoparticle volume fraction, and cylinder rotation, while magnetic field reduces the convection. The combined effect of magnetic field and cylinder rotation acts in a way to enhance local heat transfer for some locations along the heated wall

(Continued)

TABLE 3.8 (Continued)

Authors	Geometry of study	Method	Type of nanofluid	Properties	Remarks
Selimefendigil and Oztop [86]	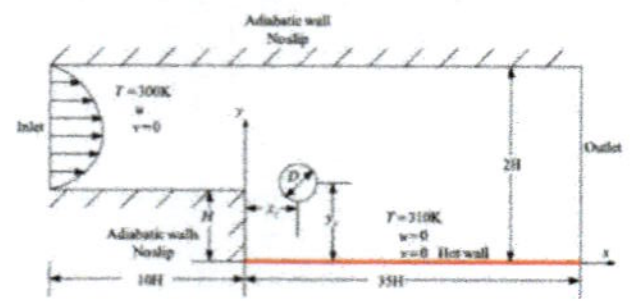	Finite element method	Al_2O_3-water	$50 \le Re \le 200$ $Gr = 10^4$ $0 \le \phi \le 0.05$ $-4 \le \Omega \le 4$ $Pr = 6.2$	Cylinder rotation affects the flow and thermal patterns behind the step. The addition of the nanoparticles enhances the averaged heat transfer along the bottom wall downstream of the step. There is almost a linear relation between heat transfer enhancement and nanoparticle volume fraction
Selimefendigil et al. [87]	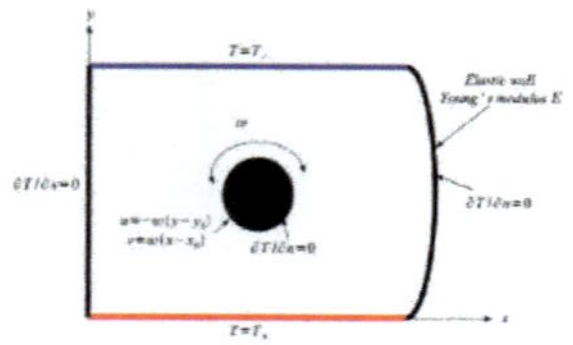	Finite element method	SiO_2-water	$10^3 \le Ra_E \le 5 \times 10^5$ $10^4 \le Ra_I \le 10^6$ $5 \times 10^2 \le E \le 10^6$ $0 \le \phi \le 0.05$ $-2000 \le \Omega \le 2000$	Average Nusselt number enhances with external Rayleigh number and nanoparticle volume fraction while the opposite behavior is seen as the value of internal Rayleigh number and flexibility of the wall increases. Among different nanoparticle types cylindrical ones show the best performance in terms of heat transfer enhancement
Selimefendigil and Oztop [88]	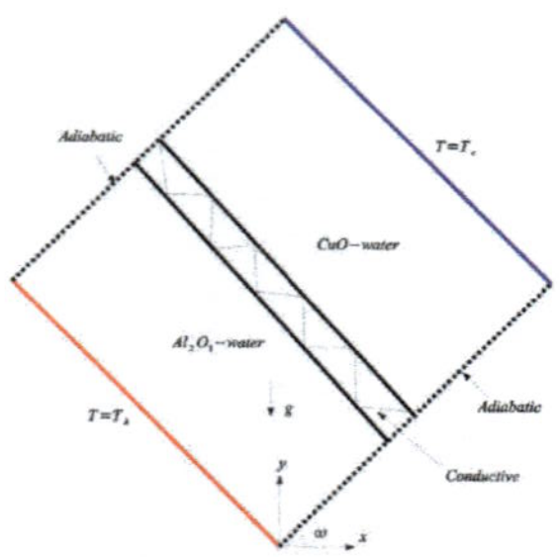	Finite element method	Al_2O_3-water and CuO-water	$10^3 \le Gr \le 10^6$ $0^o \le \theta \le 270^o$ $0.01 \le Kr \le 10$ $0 \le \phi \le 0.04$	As the value of the Grashof number, thermal conductivity ratio (Kr), and nanoparticle volume fraction increase, the average Nusselt number increases. When nanoparticles with low thermal conductivity on the right cavity are added it is more effective for the heat transfer enhancement compared to adding nanoparticles with high thermal conductivity

number is maximum for each Rayleigh number. Moreover, the angle of turn has an important effect on the streamlines, isotherms, and maximum or minimum values of local Nusselt number. Effects of MHD on Cu-water nanofluid flow and heat transfer has been studied by Sheikholeslami et al. [95]. Constant temperature and heat flux boundary conditions for an Al_2O_3-water nanofluid filled enclosure have been examined by Sheikholeslami et al. [96–99]. Sheikholeslami et al. [100] studied free convection heat transfer in a nanofluid filled inclined L-shaped enclosure. Ferrohydrodynamic and Magnetohydrodynamic effects on ferrofluid flow and convective heat transfer has been investigated by Sheikholeslami and Ganji [101]. They found that Nusselt number increases with augment of Rayleigh number and nanoparticle volume fraction but it decreases with an increase of Hartmann number. Magnetic number has a different effect on Nusselt number corresponding to the Rayleigh number.

Sheikholeslami et al. [102] considered the effect of thermal radiation on ferrofluid flow and heat transfer in a semiannulus enclosure in the presence of a magnetic source. Sheikholeslami Kandelousi [103] studied the effect of spatially variable magnetic field on ferrofluid flow and heat transfer considering constant heat flux boundary condition. Free convection heat transfer in a nanofluid filled enclosure with elliptic inner cylinder has been presented by Sheikholeslami et al. [104]. Sheikholeslami and Rashidi [105] studied the effect of space dependent magnetic field on free convection of Fe_3O_4-water nanofluid. Effect of nonuniform magnetic field on forced convection heat transfer of nanofluid has been studied by Sheikholeslami et al. [106]. Electrohydrodynamic nanofluid hydrothermal treatment in an enclosure with sinusoidal upper wall has been investigated by Sheikholeslami and Ellahi [107]. Sheikholeslami et al. [108] applied a two-phase model for magnetic nanofluid forced convective heat transfer in the existence of variable magnetic field. Sheikholeslami et al. [109] investigated forced convection heat transfer in a semiannulus under the influence of a variable magnetic field. Sheikholeslami et al. [110] studied the effect of electric field on hydrothermal behavior of nanofluid in a complex geometry. They found that the effect of an electric field on heat transfer is more pronounced at low Reynolds number. Sheikholeslami et al. [111] investigated nonuniform magnetic field effect on nanofluid hydrothermal treatment considering Brownian motion and thermophoresis effects. Sheikholeslami and Rashidi [112] studied ferrofluid heat transfer treatment in the presence of variable magnetic field. They found that the Nusselt number has direct relationship with Richardson number, nanoparticle volume fraction while it has an inverse relationship with Hartmann number and magnetic number. Table 3.9 shows the summary of the control volume-based finite element method studies on nanofluid.

3.2.7 Lattice Boltzmann Method

Investigation of nanofluid flow and heat transfer in presence of magnetic field using KKL model has been studied by Sheikholeslami et al. [113]. Nanofluid hydrothermal behaviors in an enclosure with curve boundaries have been studied by Sheikholeslami et al. [114–117]. Ashorynejad et al. [118] studied magnetic field effects on natural convection flow of a nanofluid in a horizontal cylindrical annulus. They found that flow oscillations can be suppressed effectively by imposing an external radial magnetic field. MHD effects on nanofluid flow and heat transfer in a semiannulus enclosure has been studied by Sheikholeslami et al. [119,120]. They showed that the enhancement in heat transfer increases as Hartmann number increases but it decreases with increase of Rayleigh number. Free convection of ferrofluid in a cavity heated from below in the presence of an external magnetic field has been investigated by Sheikholeslami and Gorji [121]. They found that particles with a smaller size have better ability to dissipate heat, and a larger volume fraction would provide a stronger driving force which leads to an increase in temperature profile.

Sheikholeslami et al. [122] studied MHD free convection in an eccentric semiannulus filled with nanofluid. Sheikholeslami et al. [123] simulated MHD CuO-water nanofluid flow and convective heat transfer considering Lorentz forces. Entropy generation of nanofluid in the presence of magnetic field was studied by Sheikholeslami and Ganji [124]. Sheikholeslami et al. [125] simulated magnetohydrodynamic natural convection heat transfer of Al_2O_3-water nanofluid in a horizontal cylindrical enclosure with an inner triangular cylinder. Sheikholeslami and Ellahi [126] studied ferrofluid flow for magnetic drug targeting. They showed that back flow occurs near the region where the magnetic source is located. Sheikholeslami et al. [127] simulated the magnetic field effect on hydrothermal behavior of nanofluid in a cubic cavity. Three-dimensional mesoscopic simulation of magnetic field effect on natural convection of nanofluid has been studied by Sheikholeslami and Ellahi [128]. Nanofluid heat transfer enhancement and entropy generation has been studied by Sheikholeslami et al. [129]. Effect of a magnetic source on free convection in a cavity subjugated to nanofluid has been investigated [130,131]. Table 3.10 shows the summary of the lattice Boltzmann Method studies on nanofluid. Recently several authors have applied new methods for simulation of hydrothermal behavior [132–268].

TABLE 3.9 Summary of the Control Volume-based Finite Element Method Studies on Nanofluid

Authors	Geometry of study	Method	Type of nanofluid	Properties	Remarks
Sheikholeslami et al. [89]	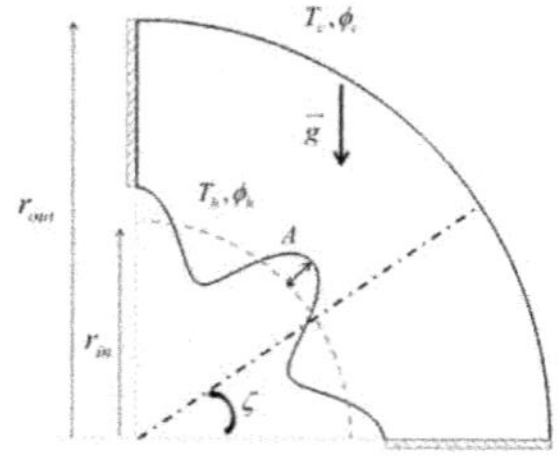	Control volume-based finite element method	Al_2O_3-water	$10^3 \le Ra \le 10^5$ $0.1 \le Nr \le 4$ $1 \le Le \le 8$ $Nb = Nt = 0.5$ $Pr = 10$	Average Nusselt number decreases as buoyancy ratio number increases until it reaches a minimum value and then starts increasing. As Lewis number increases, this minimum value occurs at higher buoyancy ratio number
Sheikholeslami et al. [90]	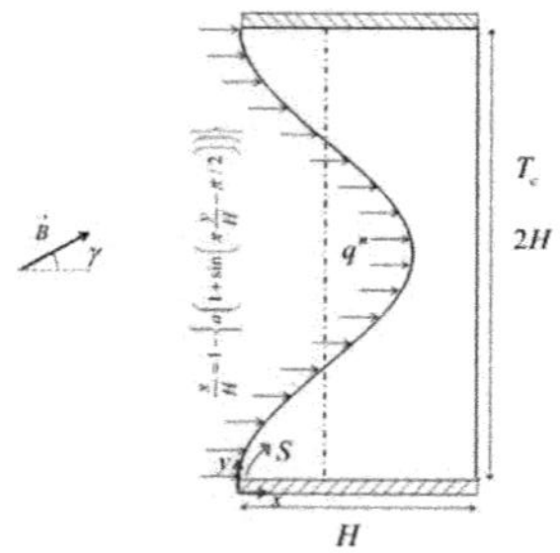	Control volume-based finite element method	CuO-water	$10^3 \le Ra \le 10^5$ $0.1 \le a \le 0.3$ $0 \le Ha \le 100$ $0 \le \phi \le 0.04$ $Pr = 6.2$	Nusselt number is an increasing function of nanoparticles volume fraction, dimensionless amplitude of the sinusoidal wall, and Rayleigh number while it is a decreasing function of Hartmann number
Sheikholeslami et al. [91]	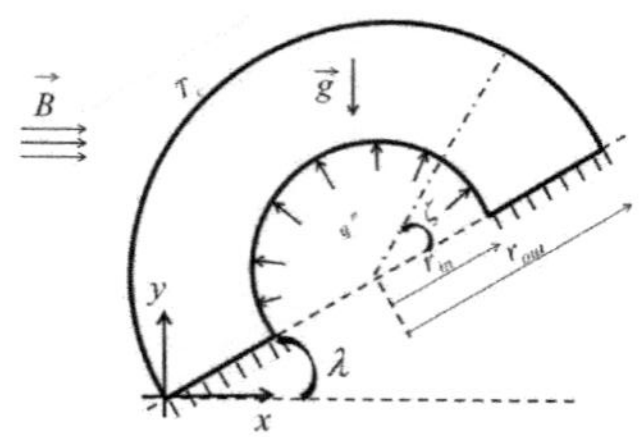	Control volume-based finite element method	Cu-water	$10^3 \le Ra \le 10^5$ $0^\circ \le \lambda \le 90^\circ$ $0 \le \phi \le 0.06$ $0 \le Ha \le 100$ $Pr = 6.2$	Hartmann number and the inclination angle of the enclosure can be control parameters at different Rayleigh numbers. In the presence of magnetic field the velocity field is retarded and hence convection and Nusselt number decreases
Sheikholeslami et al. [92]	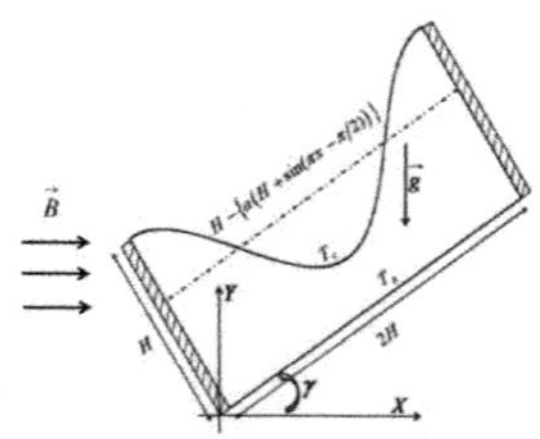	Control volume-based finite element method	Cu-water	$10^3 \le Ra \le 10^5$ $0^\circ \le \gamma \le 90^\circ$ $0 \le Ha \le 100$ $0 \le \phi \le 0.06$ $Pr = 6.2$	At $Ra = 10^3$, maximum value of enhancement for low Hartmann number is obtained at $\gamma = 0^\circ$, but for higher values of Hartmann number, maximum values of E occurs at $\gamma = 90^\circ$
Sheikholeslami et al. [93]	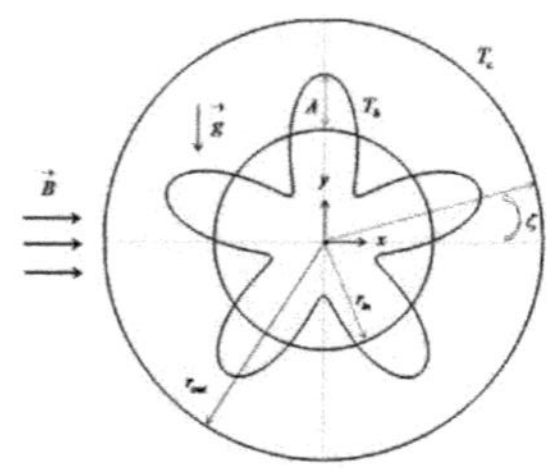	Control volume-based finite element method	Cu-water	$10^3 \le Ra \le 10^5$ $3 \le N \le 6$ $0 \le Ha \le 100$ $0 \le \phi \le 0.06$ $A = 0.5, Pr = 6.2$	Nusselt number is an increasing function of nanoparticle volume fraction, the number of undulations, and Rayleigh numbers while it is a decreasing function of Hartmann number.

Soleimani et al. [94]	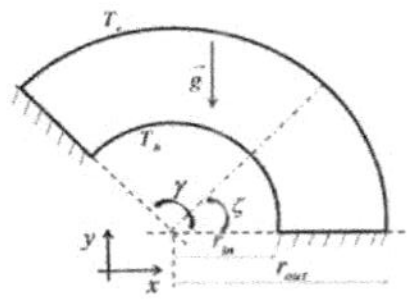	Control volume-based finite element method	Cu-water	$10^3 \leq Ra \leq 10^5$ $45^\circ \leq \gamma \leq 180^\circ$ $0 \leq \phi \leq 0.06$ $Pr = 6.2$	There is an optimum angle of turn in which the average Nusselt number is maximum for each Rayleigh number. Moreover, the angle of turn has an important effect on the streamlines, isotherms, and maximum or minimum values of local Nusselt number
Sheikholeslami et al. [95]	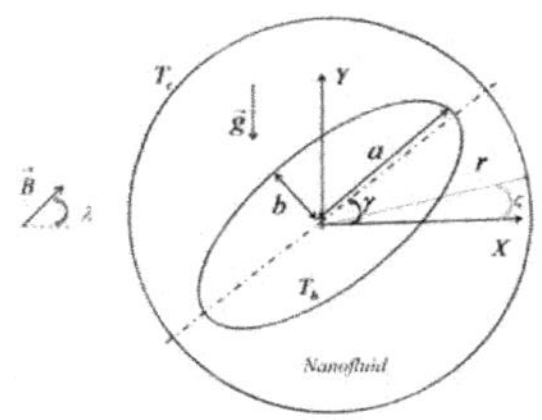	Control volume-based finite element method	Cu-water	$10^3 \leq Ra \leq 10^5$ $0^\circ \leq \gamma \leq 90^\circ$ $0 \leq \phi \leq 0.06$ $\varepsilon = 0.9, a = 0.8L$ $Pr = 6.2$	Increasing Rayleigh number leads to decreased heat transfer enhancement while the opposite trend is observed with augment of Hartmann number
Sheikholeslami et al. [96]	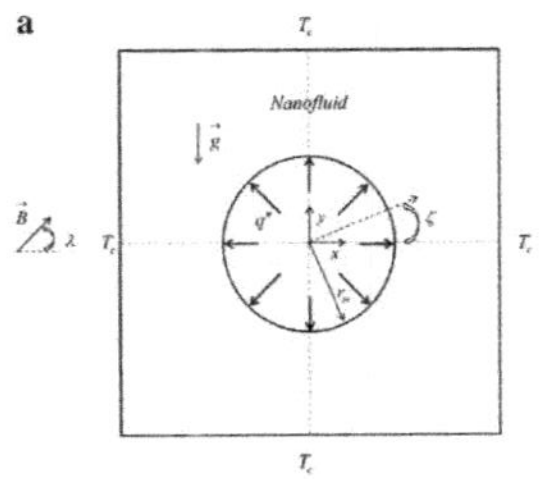	Control volume-based finite element method	Al_2O_3-water	$10^3 \leq Ra \leq 10^5$ $0^\circ \leq Ha \leq 100$ $0 \leq \phi \leq 0.04$ $0.2 \leq r_{in}/L \leq 0.4$ $Pr = 6.2$	Domination of conduction mechanism causes heat transfer enhancement to increase. So enhancement in heat transfer increases with increase of Hartmann number and aspect ratio while it decreases with augment of Rayleigh number
Sheikholeslami et al. [97]	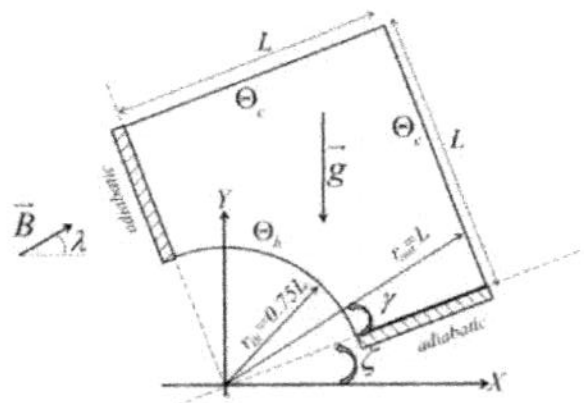	Control volume-based finite element method	Al_2O_3-water	$10^3 \leq Ra \leq 10^5$ $-90^\circ \leq \gamma \leq 0^\circ$ $0 \leq \phi \leq 0.04$ $0 \leq Ha \leq 100$ $Pr = 6.2$	The heat transfer between cold and hot regions of the enclosure cannot be well understood by using isotherm patterns so heatline visualization technique is used to find the direction and intensity of heat transfer in a domain
Sheikholeslami et al. [98]	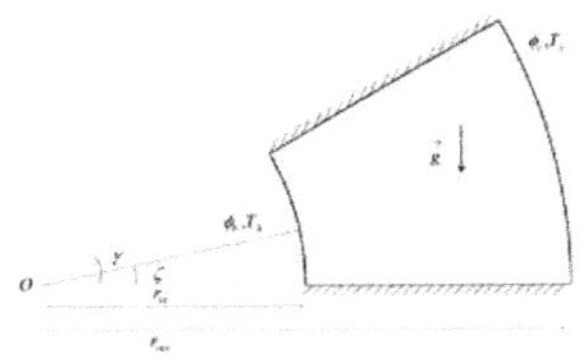	Control volume-based finite element method	Al_2O_3-water	$30^\circ \leq \gamma \leq 90^\circ$ $0.1 \leq Nr \leq 4$ $2 \leq Le \leq 8$ $Ra = 10^5$ $Nb = Nt = 0.5$ $Pr = 10$	Lewis number has no significant effect on Nusselt number at low values of buoyancy ratio number
Sheikholeslami et al. [99]	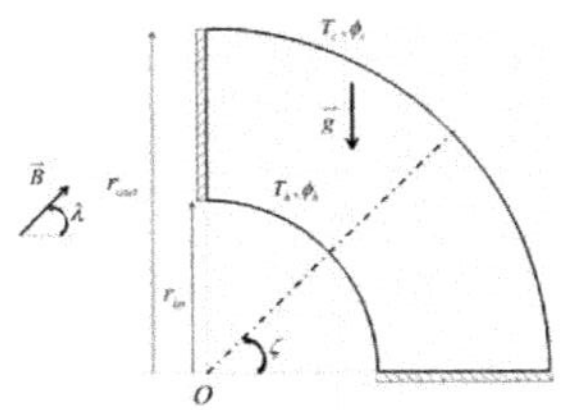	Control volume-based finite element method	Al_2O_3-water	$10^3 \leq Ra \leq 10^5$ $0 \leq Ha \leq 100$ $30^\circ \leq \gamma \leq 90^\circ$ $0.1 \leq Nr \leq 4$ $2 \leq Le \leq 8$ $Nb = Nt = 0.5$ $Pr = 10$	As buoyancy ratio number increases the effects of other active parameters are more pronounced

(*Continued*)

TABLE 3.9 (*Continued*)

Authors	Geometry of study	Method	Type of nanofluid	Properties	Remarks
Sheikholeslami et al. [100]	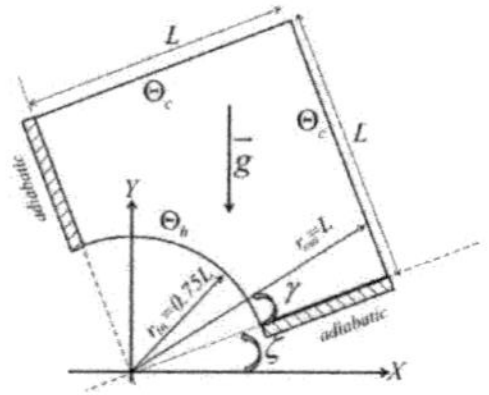	Control volume-based finite element method	Cu-water	$10^3 \le Ra \le 10^5$ $-45^\circ \le \zeta \le 45^\circ$ $0 \le \phi \le 0.06$ $Pr = 6.2$	The results show that for $Ra = 10^4$ the maximum and minimum average Nusselt number are corresponding to $\zeta = -45^\circ$ and 45° respectively, whereas the opposite trend is observed for $Ra = 10^5$
Sheikholeslami and Ganji [101]	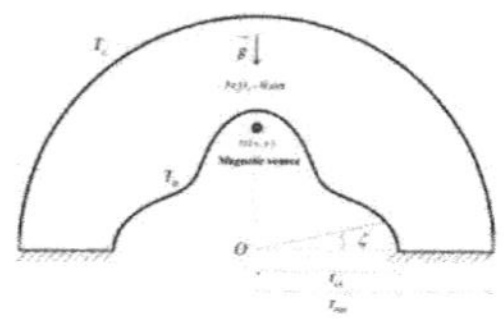	Control volume-based finite element method	Fe_3O_4-water	$10^3 \le Ra \le 10^5$ $0 \le Ha \le 5$ $0 \le Mn_F \le 500$ $0 \le \phi \le 0.04$ $Pr = 6.8$ $Ec = 10^{-5}$	Nusselt number increases with augment of Rayleigh number and nanoparticle volume fraction but it decreases with increase of Hartmann number. Magnetic number has different effect on Nusselt number corresponding to Rayleigh number
Sheikholeslami et al. [102]	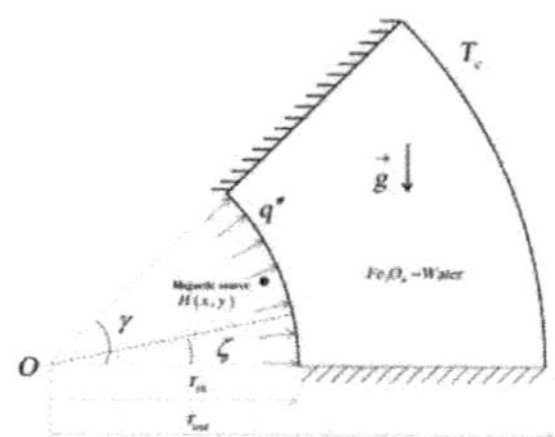	Control volume-based finite element method	Fe_3O_4-water	$10^3 \le Ra \le 10^5$ $0 \le Ha \le 10$ $0 \le Mn_F \le 100$ $0 \le Nr \le 0.027$ $0 \le \phi \le 0.04$ $Pr = 6.8$ $Ec = 10^{-5}$	Nusselt number is an increasing function of Rayleigh number, nanoparticle volume fraction, and magnetic number while it is a decreasing function of with Hartmann number and Radiation parameter
Sheikholeslami Kandelousi [103]	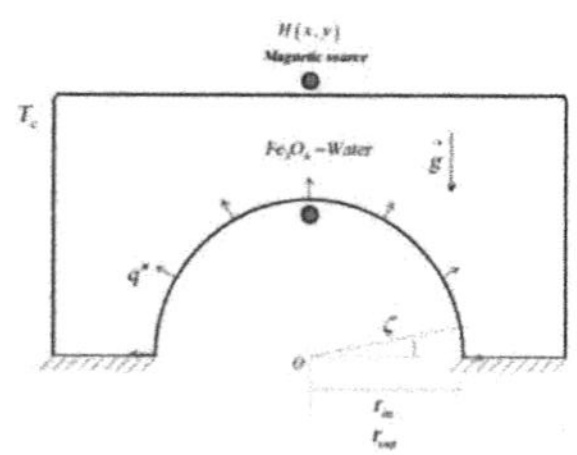	Control volume-based finite element method	Fe_3O_4- water	$10^3 \le Ra \le 10^5$ $0 \le Mn_F \le 100$ $0 \le Ha \le 10$ $0 \le \phi \le 0.04$ $Pr = 6.2$	Enhancement in heat transfer decreases with increase of Rayleigh number and magnetic number but it increases with increase of Hartmann number
Sheikholeslami et al. [104]	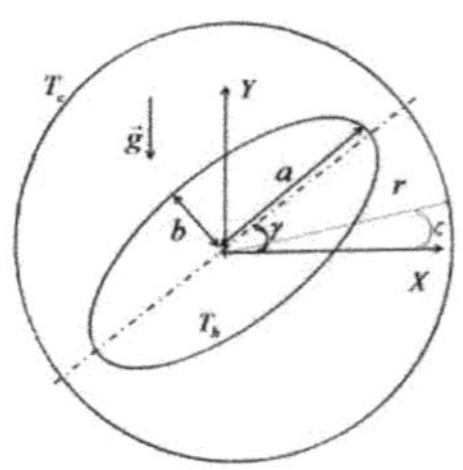	Control volume-based finite element method	CuO-water	$10^3 \le Ra \le 10^5$ $0^\circ \le \gamma \le 90^\circ$ $0 \le \phi \le 0.04$ $Pr = 6.2$	Increasing Rayleigh number leads to the decrease in ratio of heat transfer enhancement. For high Rayleigh number the minimum heat transfer enhancement ratio occurs at $\gamma = 90^\circ$

Sheikholeslami and Rashidi [105]	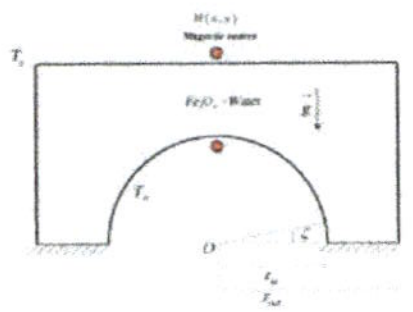	Control volume-based finite element method	Fe_3O_4- water	$10^3 \le Ra \le 10^5$ $0 \le Mn_F \le 100$ $0 \le Ha \le 10$ $0 \le \phi \le 0.04$ $Pr = 6.2$	Nusselt number is an increasing function of magnetic number, Rayleigh number, and nanoparticle volume fraction while it is a decreasing function of Hartmann number
Sheikholeslami et al. [106]	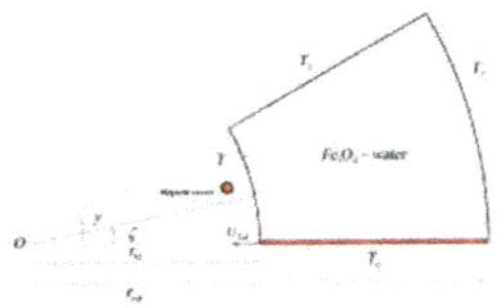	Control volume-based finite element method	Fe_3O_4- water	$10 \le Re \le 10^3$ $0 \le Ha \le 20$ $0 \le \phi \le 0.04$ $Pr = 6.2$	Nusselt number has a direct relationship with Reynolds number and nanoparticle volume fraction while it has reverse relationship with Hartmann number
Sheikholeslami and Ellahi [107]	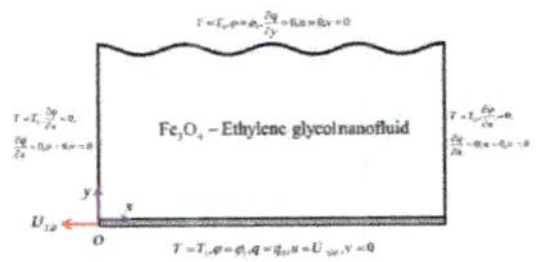	Control volume-based finite element method	Fe_3O_4-Ethylene glycol	$3000 \le Re \le 6000$ $0 \le \Delta\varphi \le 10$ $0 \le \phi \le 0.04$ $Pr = 149.54$	Heat transfer rises with augment of supplied voltage and Reynolds number
Sheikholeslami et al. [108]	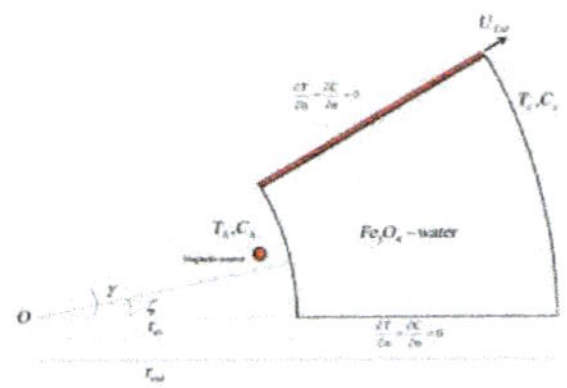	Control volume-based finite element method	Fe_3O_4- water	$10 \le Re \le 500$ $2 \le Le \le 8$ $0 \le Ha \le 20$	Nusselt number has direct relationship with Reynolds number while it has reverse relationship with Hartmann number and Lewis number
Sheikholeslami et al. [109]	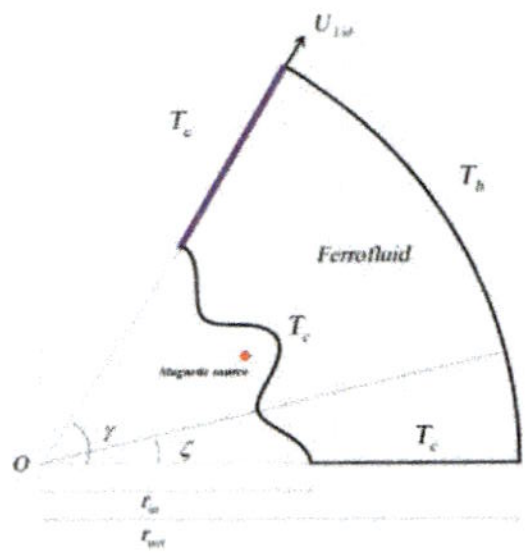	Control volume-based finite element method	Fe_3O_4- water	$10 \le Re \le 600$ $0 \le Mn_F \le 10$ $0 \le Ha \le 10$ $0 \le \phi \le 0.04$ $Pr = 6.8$	Effects of Kelvin forces are more pronounced for high Reynolds number. Heat transfer enhancement has direct relationship with the Reynolds number and the magnetic number; while it has inverse relationship with the Hartmann number
Sheikholeslami et al. [110]	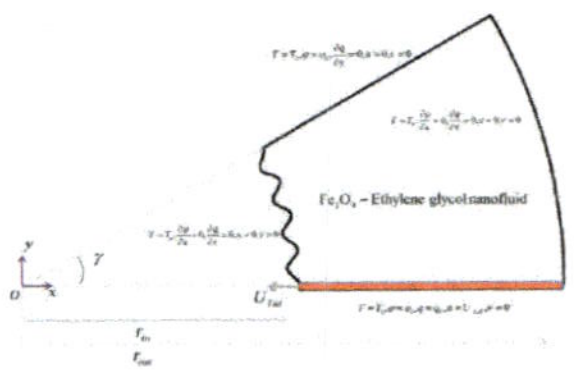	Control volume-based finite element method	Fe_3O_4-Ethylene glycol	$3000 \le Re \le 6000$ $0 \le \Delta\varphi \le 10$ $0 \le \phi \le 0.04$ $Pr = 149.54$	Effect of electric field on heat transfer is more pronounced at low Reynolds number

(*Continued*)

TABLE 3.9 (*Continued*)

Authors	Geometry of study	Method	Type of nanofluid	Properties	Remarks
Sheikholeslami et al. [111]	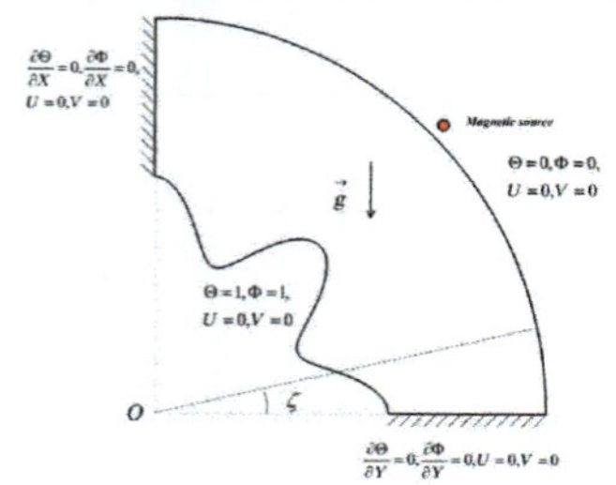	Control volume-based finite element method	Fe_3O_4- water	$10^3 \le Ra \le 10^5$ $0.1 \le Nr \le 4$ $0 \le Ha \le 10$ $2 \le Le \le 4$ $Pr = 6.85$	Nusselt number has a direct relationship with Rayleigh number, buoyancy ratio number, and Lewis number while it has a reverse relationship with Hartmann number
Sheikholeslami, and Rashidi [112]	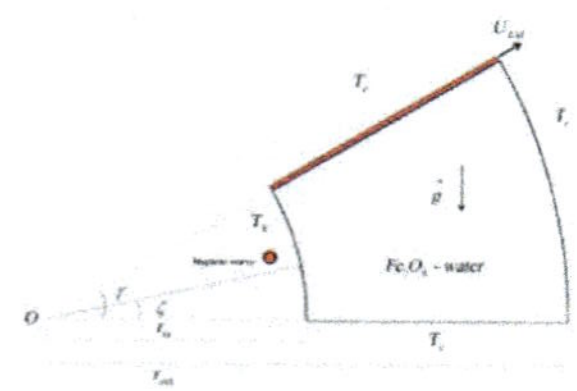	Control volume-based finite element method	Fe_3O_4- water	$0.001 \le Ri \le 10$ $0 \le \delta^* \le 0.6$ $0 \le Mn_F \le 10$ $0 \le Ha \le 10$ $Re = 100$	Nusselt number has a direct relationship with Richardson number and nanoparticle volume fraction while it has reverse relationship with Hartmann number and Magnetic number

TABLE 3.10 Summary of the Lattice Boltzmann Method Studies on Nanofluid

Authors	Geometry of study	Method	Type of nanofluid	Properties	Remarks
Sheikholeslami et al. [113]	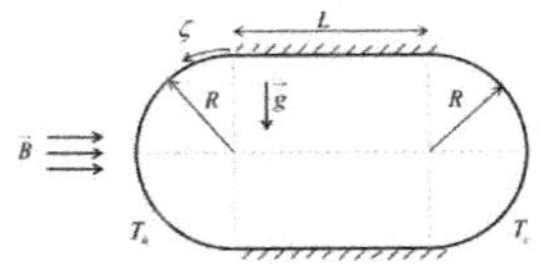	Lattice Boltzmann Method	Al_2O_3-water	$10^4 \leq Ra \leq 10^6$ $0 \leq \phi \leq 0.04$ $0 \leq Ha \leq 60$ $R/L = 0.5$ $Pr = 6.8$	Enhancement in heat transfer increases with increase of Hartmann number except for $Ra = 10^4$ in which $Ha = 40$ roles as a critical Hartmann number
Sheikholeslami et al. [114]	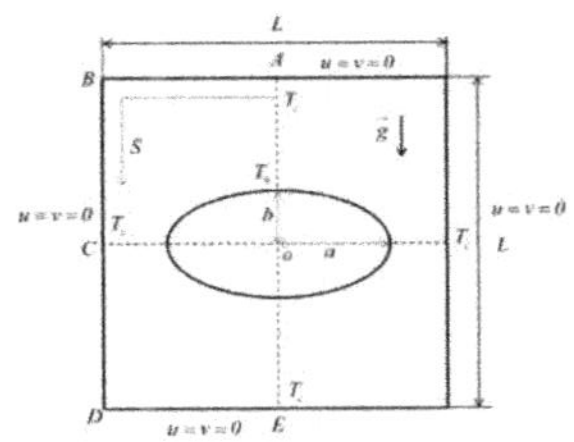	Lattice Boltzmann Method	Cu-water	$10^3 \leq Ra \leq 10^6$ $0 \leq \phi \leq 0.06$ $0.65 \leq \varepsilon \leq 0.95$ $a = 0.4$ $Pr = 6.8$	The minimum value of enhancement of heat transfer occurs at $\varepsilon = 0.95$ for $Ra = 10^5$ but for other values of Rayleigh number it is obtained at $\varepsilon = 0.65$
Sheikholeslami et al. [115]	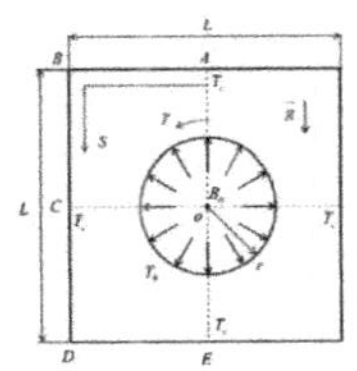	Lattice Boltzmann Method	Cu-water	$10^4 \leq Ra \leq 10^6$ $0 \leq \phi \leq 0.06$ $0 \leq Ha \leq 60$ $Pr = 6.8$	As the nanoparticle volume fraction and Rayleigh number increase, average Nusselt number increases but opposite trends are observed when the Hartmann number increases
Sheikholeslami et al. [116]	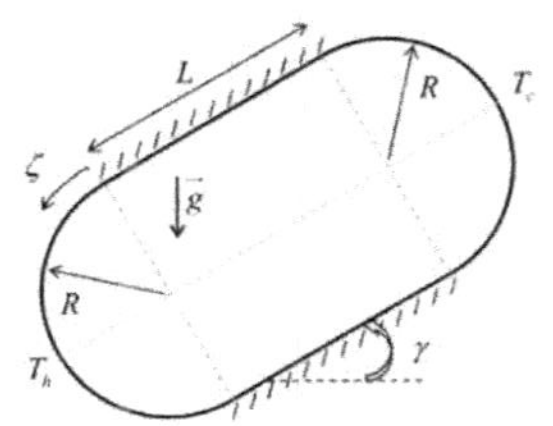	Lattice Boltzmann Method	Cu-water	$10^4 \leq Ra \leq 10^6$ $0 \leq \phi \leq 0.04$ $-60^\circ \leq \gamma \leq 60^\circ$ $R/L = 0.5$ $Pr = 6.8$	The change of inclination angle has a significant impact on the thermal and hydrodynamic flow field
Sheikholeslami et al. [117]	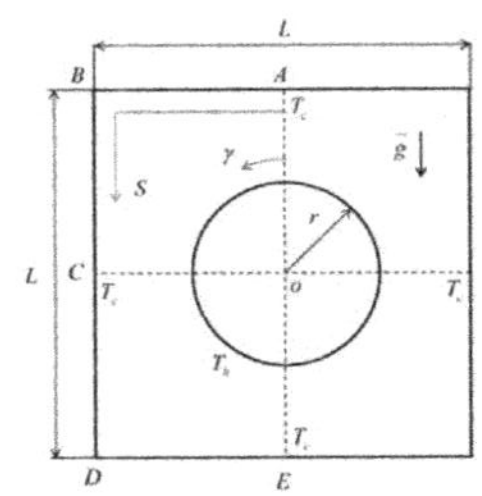	Lattice Boltzmann Method	Cu-water Ag-water Al_2O_3-water TiO_2-water	$10^3 \leq Ra \leq 10^6$ $0 \leq \phi \leq 0.04$ $1.5 \leq \lambda \leq 4.5$ $Pr = 6.8$	Choosing copper as the nanoparticle leads to obtaining the highest enhancement for this problem

(Continued)

TABLE 3.10 (*Continued*)

Authors	Geometry of study	Method	Type of nanofluid	Properties	Remarks
Ashorynejad et al. [118]	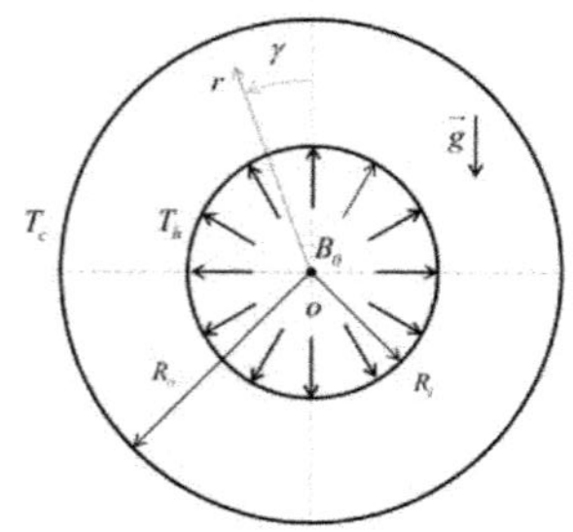	Lattice Boltzmann Method	Ag-water	$10^3 \le Ra \le 10^6$ $0 \le \phi \le 0.04$ $0 \le Ha \le 60$ $Pr = 6.8$	Flow oscillations can be suppressed effectively by imposing an external radial magnetic field
Sheikholeslami et al. [119]	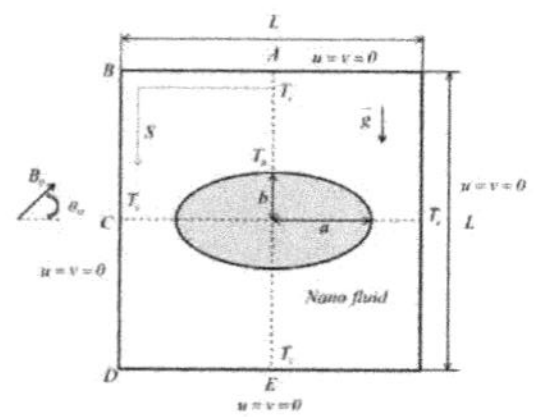	Lattice Boltzmann Method	Al_2O_3-water	$10^4 \le Ra \le 10^6$ $0 \le \phi \le 0.04$ $0 \le Ha \le 100$ $Pr = 6.8$	The enhancement in heat transfer increases as Hartmann number increases but it decreases with increase of Rayleigh number
		Lattice Boltzmann Method			
Sheikholeslami et al. [120]	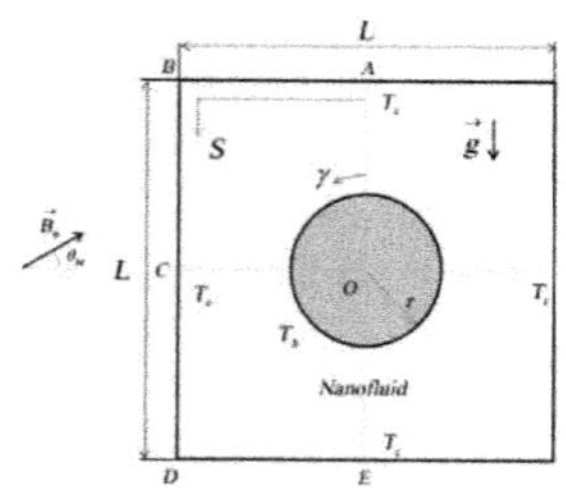	Lattice Boltzmann Method	Cu-water	$10^4 \le Ra \le 10^6$ $0 \le \phi \le 0.06$ $0 \le Ha \le 40$ $1.5 \le \lambda \le 4.5$ $Pr = 6.8$	Enhancement ratio increases with decrease of Rayleigh number and it increases with augment of Hartmann number
Sheikholeslami and Gorji [121]	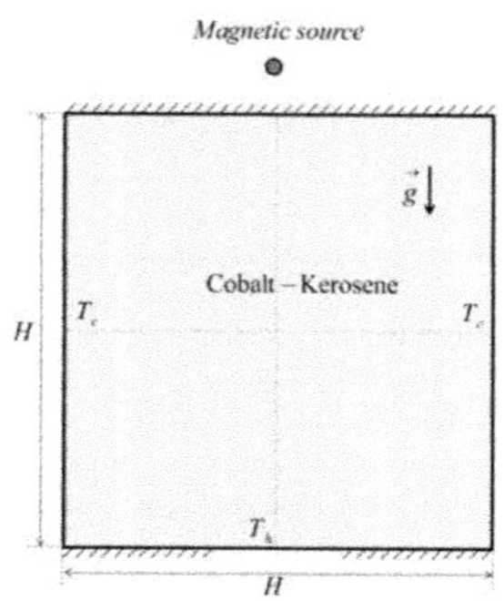	Lattice Boltzmann Method	Cobalt-kerosene	$10^4 \le Ra \le 10^6$ $0 \le \phi \le 0.04$ $0.675 \le A \le 1.325$ $0.2 \le \varepsilon \le 0.8$ $Pr = 6.8$	Particles with a smaller size have better ability to dissipate heat, and a larger volume fraction would provide a stronger driving force which leads to increase in temperature profile

Sheikholeslami et al. [122]	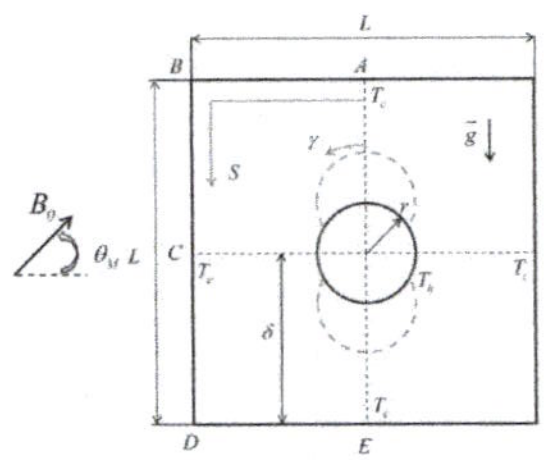	Lattice Boltzmann Method	Cu-water	$10^4 \le Ra \le 10^6$ $0 \le \phi \le 0.06$ $0 \le Ha \le 40$ $0.2 \le \frac{\delta}{L} \le 0.8$ $\lambda = 3.5$ $Pr = 6.8$	Nusselt number has direct relationship with nanoparticle volume fraction and Rayleigh number but it has inverse relationship with Hartmann number and position of inner cylinder at high Rayleigh number
Sheikholeslami et al. [123]	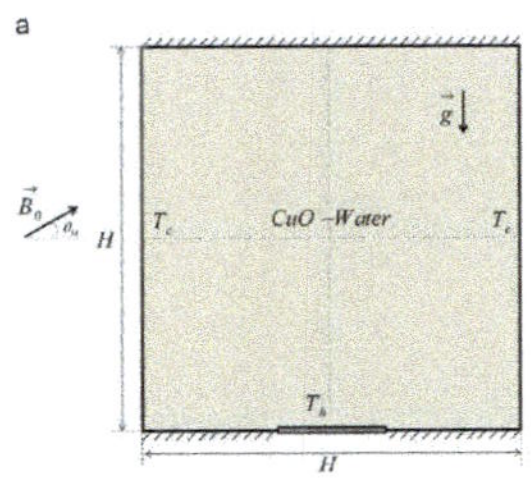	Lattice Boltzmann Method	CuO-water	$10^3 \le Ra \le 10^5$ $0 \le \phi \le 0.04$ $0 \le Ha \le 100$ $0.4 \le \varepsilon \le 0.8$ $Pr = 6.8$	Enhancement in heat transfer increases as Hartmann number and heat source length increase but it decreases with increase of Rayleigh number. Also it can be found that effect of Hartmann number and heat source length is more pronounced at high Rayleigh number
Sheikholeslami and Ganji [124]	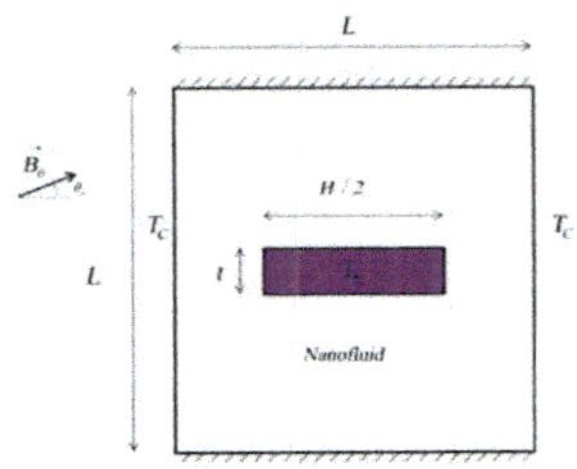	Lattice Boltzmann Method	CuO-water	$10^3 \le Ra \le 10^5$ $0 \le \phi \le 0.04$ $0 \le Ha \le 100$ $Pr = 6.8$	Heat transfer rate and Dimensionless entropy generation number increase with increase of the Rayleigh number and nanoparticle volume fraction but it decreases with increase of the Hartmann number
Sheikholeslami et al. [125]	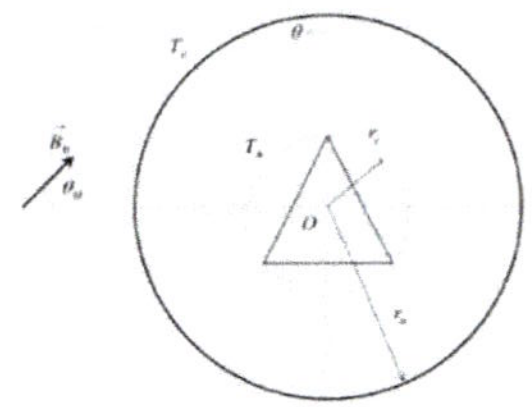	Lattice Boltzmann Method	Al_2O_3-water	$10^3 \le Ra \le 10^5$ $0 \le \phi \le 0.04$ $0 \le Ha \le 100$ $Pr = 6.8$	Lattice Boltzmann method with double-population is a powerful approach for the simulation of natural convection heat transfer in nanofluids in regions with curved boundaries
Sheikholeslami and Ellahi [126]	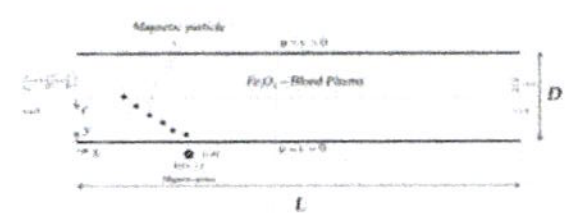	Lattice Boltzmann Method	Fe_3O_4-Plasma	$50 \le Re \le 400$ $0 \le Mn_F \le 100$ $\phi = 0.04$ $Ha = 20$	Back flow occurs near the region where the magnetic source is located. Also it can be found that skin friction coefficient is a decreasing function of Reynolds number and magnetic number
Sheikholeslami et al. [127]	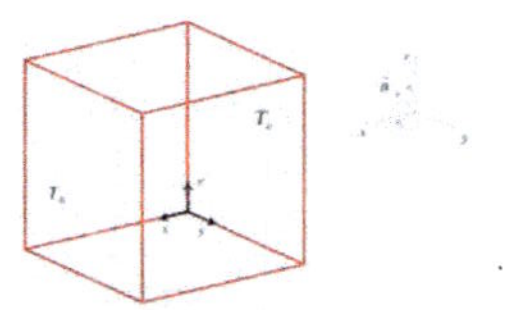	Lattice Boltzmann Method	Al_2O_3-water	$10^3 \le Ra \le 10^5$ $0 \le \phi \le 0.04$ $0 \le Ha \le 100$ $Pr = 6.8$	Nusselt number increases with increase of nanoparticle volume fraction and Rayleigh number while it decreases with increase of Hartmann number

(*Continued*)

TABLE 3.10 (Continued)

Authors	Geometry of study	Method	Type of nanofluid	Properties	Remarks
Sheikholeslami and Ellahi [128]	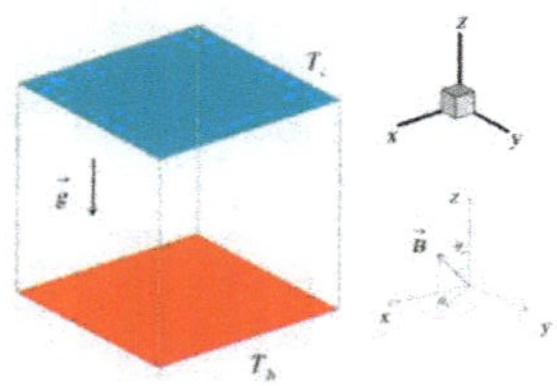	Lattice Boltzmann Method	Al_2O_3-water	$10^3 \le Ra \le 10^5$ $0 \le \phi \le 0.04$ $0 \le Ha \le 100$ $Pr = 6.8$	Applying magnetic field results in a force opposite to the flow direction that leads to drag on the flow and then reduces the convection currents by reducing the velocities
Sheikholeslami et al. [129]	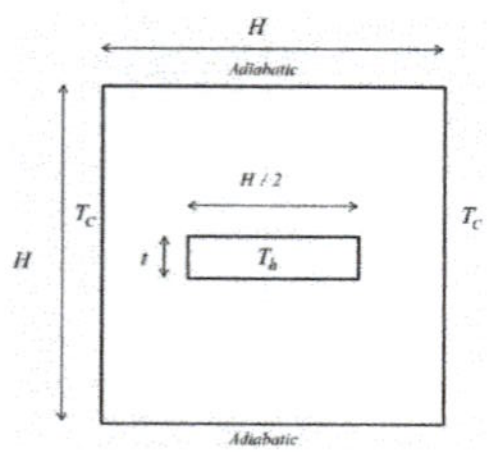	Lattice Boltzmann Method	Cu-water Ag-water Al_2O_3-water TiO_2-water	$10^3 \le Ra \le 10^5$ $0 \le \phi \le 0.04$ $2 \le H/t \le 9$ $Pr = 6.8$	The effect of nanoparticle volume fraction is found to be more pronounced for low Rayleigh number as compared to high Rayleigh number
Kefayati [130]	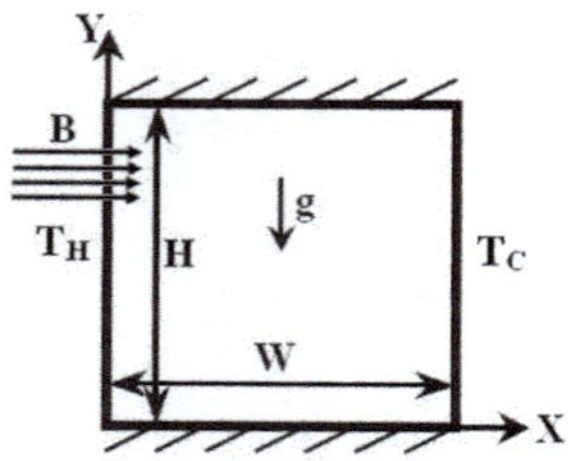	lattice Boltzmann method	Cu-water	$10^3 \le Ra \le 10^5$ $0 \le \phi \le 0.06$ $0 \le Ha \le 90$ $0.5 \le A \le 2$	The heat transfer decreases by increment of Hartmann number. Heat transfer decreases with growth of the aspect ratio but this growth causes the effect of nanoparticles to increase. Magnetic field augments the effect of nanoparticles at High Rayleigh numbers ($Ra = 10^5$). The effect of nanoparticles rises for high Hartmann numbers when the aspect ratio increases
Mejri et al. [131]	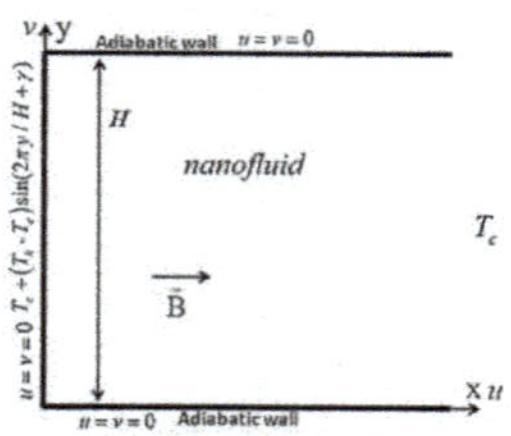	lattice Boltzmann method	Al_2O_3 -water	$10^3 \le Ra \le 10^6$ $0 \le \phi \le 0.06$ $0 \le Ha \le 60$ $\gamma = 0, \pi/4, \pi/2, 3\pi/4, \pi$	The heat transfer rate decreases with the increase of Hartmann number and increases with the rise of Rayleigh number

References

[1] K. Khanafer, K. Vafai, M. Lightstone, Buoyancy-driven heat transfer enhancement in a two-dimensional enclosure utilizing nanofluids, Int. J. Heat Mass Transfer 46 (2003) 3639–3653.

[2] Y. Xuan, Q. Li, Heat transfer enhancement of nanofluids, Int. J. Heat Fluid Flow 21 (2000) 58–64.

[3] Y. Xuan, W. Roetzel, Conceptions for heat transfer correlations of nanofluids, Int. J. Heat Mass Transfer 43 (2000) 3701–3707.

[4] E. Abu-Nada, H.F. Oztop, Effects of inclination angle on natural convection in enclosures filled with Cu–water nanofluid, Int. J. Heat Fluid Flow 30 (2009) 669–678.

[5] Y. Ding, D. Wen, Particle migration in a flow of nanoparticle suspensions, Powder Technol. 149 (2–3) (2005) 84–92.

[6] A. Akbarinia, R. Laur, Investigation the diameter of solid particles effects on a laminar nanofluid flow in a curved tube using a two phase approach, Int. J. Heat Fluid Flow 30 (2009) 706–714.

[7] S. Mirmasoumi, A. Behzadmehr, Numerical study of laminar mixed convection of a nanofluid in a horizontal tube using two-phase mixture model, Appl. Therm. Eng. 28 (2008) 717–727.

[8] J. Buongiorno, Convective transport in nanofluids, ASME J. Heat Transfer 128 (2006) 240–250.

[9] A. Behzadmehr, M. Saffar-Avval, N. Galanis, Prediction of turbulent forced convection of a nanofluid in a tube with uniform heat flux using a two phase approach, Int. J. Heat Fluid Flow 28 (2) (2007) 211–219.

[10] B.C. Pak, Y.I. Cho, Hydrodynamic and heat transfer study of dispersed fluids with submicron metallic oxide particles, Exp. Heat Transfer 11 (1998) 151–170.

[11] J.A. Eastman, S.U.S. Choi, S. Li, G. Soyez, L.J. Thompson, R.J. Di Melfi, Novel thermal properties of nanostructured materials, Mater. Sci. Forum 312–314 (1999) 629–634.

[12] S.J. Palm, G. Roy, C.T. Nguyen, Heat transfer enhancement with the use of nanofluids in radial flow cooling systems considering temperature dependent properties, Appl. Therm. Eng. 26 (2006) 2209–2218.

[13] J.C. Maxwell, A Treatise on Electricity and Magnetism, second ed, Oxford University Press, Cambridge, 1904, pp. 435–441.

[14] H. Brinkman, The viscosity of concentrated suspensions and solutions, J. Chem. Phys. 20 (1952) 571.

[15] E. Abu-Nada, A.J. Chamkha, Effect of nanofluid variable properties on natural convection in enclosures filled with a CuO–EG–Water nanofluid, Int. J. Therm. Sci. 49 (12) (2010) 2339–2352.

[16] P.K. Namburu, D.P. Kulkarni, D. Misra, D.K. Das, Viscosity of copper oxide nanoparticles dispersed in ethylene glycol and water mixture, Exp. Thermal Fluid Sci. 32 (2) (2007) 397–402.

[17] A. Einstein, Investigation on the Theory of Brownian Motion, Dover, New York, 1956.

[18] B.C. Pak, Y. Cho, Hydrodynamic and heat transfer study of dispersed fluids with submicron metallic oxide particle, Exp. Heat Transfer 11 (1998) 151–170.

[19] S.P. Jang, J.H. Lee, K.S. Hwang, S.U.S. Choi, Particle concentration and tube size dependence of viscosities of Al_2O_3-water nanofluids flowing through microand minitubes, Appl. Phys. Lett. 91 (2007) 243112.

[20] J. Koo, C. Kleinstreuer, A new thermal conductivity model for nanofluids, J. Nanopart. Res. 6 (2004) 577–588.

[21] S.E.B. Maïga, S.M. Palm, C.T. Nguyen, G. Roy, N. Galanis, Heat transfer enhancement by using nanofluids in forced convection flows, Int. J. Heat Fluid Flow 26 (2005) 530–546.

[22] D. Orozco, Hydrodynamic behavior of suspension of polar particles, Encyclopedia Surface Colloid Science 4 (2005) 2375–2396.

[23] C.T. Nguyen, F. Desgranges, N. Galanis, G. Roy, T. Mare, S. Boucher, et al., Viscosity data for Al_2O_3-water nanofluid-hysteresis: is heat transfer enhancement using nanofluids reliable? International, J. Therm. Sci. 47 (2008) 103–111.

[24] N. Masoumi, N. Sohrabi, A. Behzadmehr, A new model for calculating the effective viscosity of nanofluids, J. Phys. D: Appl. Phys. 42 (2009) 055501.

[25] I. Gherasim, G. Roy, C.T. Nguyen, D. Vo-Ngoc, Experimental investigation of nanofluids in confined laminar radial flows, Int. J. Therm. Sci. 48 (2009) 1486–1493.

[26] S.P. Jang, S.U.S. Choi, The role of Brownian motion in the enhanced thermal conductivity of nanofluids, Appl. Phys. Lett. 84 (2004) 4316–4318.

[27] B.X. Wang, L.R. Zhou, X.F. Peng, A fractal model for predicting the effective thermal conductivity of liquid with suspension of nanoparticles, Int. J. Heat Mass Transfer 46 (2003) 2665–2672.

[28] C.H. Chon, K.D. Kihm, S.P. Lee, S.U.S. Choi, Empirical correlation finding the role of temperature and particle size for nanofluid (Al_2O_3) thermal conductivity enhancement, Appl. Phys. Lett. 87 (2005). Available from: https://doi.org/10.1063/1.2093936. Article ID153107.

[29] P. Charuyakorn, S. Sengupta, S.K. Roy, Forced convection heat transfer in micro encapsulated phase change material slurries, Int. J. Heat Mass Transfer 34 (1991) 819–833.

[30] J.A. Eastman, S.R. Phillpot, U.S. Choi, P. Keblinski, Thermal transport in nanofluids, Annu. Rev. Mater. Res. 34 (2000) 219–246.

[31] W. Yu, S.U.S. Choi, The role of interfacial layers in the enhanced thermal conductivity of nanofluids: a renovated Maxwell model, J. Nanopart. Res. 5 (2003) 167–171.

[32] H.E. Patel, T. Sundarrajan, T. Pradeep, A. Dasgupta, N. Dasgupta, S.K. Das, A microconvection model for thermal conductivity of nanofluid, Pram. J. Phys. 65 (2005) 863–869.

[33] H. Angue Mintsa, G. Roy, C.T. Nguyen, D. Doucet, New temperature dependent thermal conductivity data for water-based nanofluids, Int. J. Therm. Sci. 48 (2009) 363–371.

[34] M. Khan, R. Malik, Forced convective heat transfer to Sisko nanofluid past a stretching cylinder in the presence of variable thermal conductivity, J. Mol. Liq. 218 (2016) 1–7.

[35] C. Sasmal, N. Nirmalkar, Momentum and heat transfer characteristics from heated spheroids in water based nanofluids, Int. J. Heat Mass Transfer 96 (2016) 582–601.

[36] T. Hayat, M. Imtiaz, A. Alsaedi, F. Alzahrani, Effects of homogeneous–heterogeneous reactions in flow of magnetite-Fe_3O_4 nanoparticles by a rotating disk, J. Mol. Liq. 216 (2016) 845–855.

[37] M. Sheikholeslami, M. Hatami, D.D. Ganji, Analytical investigation of MHD nanofluid flow in a Semi-Porous Channel, Powder Technol. 246 (2013) 327–336.

[38] M. Sheikholeslami, D.D. Ganji, H.R. Ashorynejad, Investigation of squeezing unsteady nanofluid flow using ADM, Powder Technol. 239 (2013) 259–265.
[39] M. Sheikholeslami, D.D. Ganji, Heat transfer of Cu-water nanofluid flow between parallel plates, Powder Technol. 235 (2013) 873–879.
[40] M. Sheikholeslami, D.D. Ganji, H.R. Ashorynejad, H.B. Rokni, Analytical investigation of Jeffery-Hamel flow with high magnetic field and nano particle by Adomian decomposition method, Appl. Math. Mech.-Engl. Ed. 33 (1) (2012) 1553–1564.
[41] M. Sheikholeslami, H.R. Ashorynejad, G. Domairry, I. Hashim, Flow and Heat Transfer of Cu-Water Nanofluid between a Stretching Sheet and a Porous Surface in a Rotating System, Hindawi Publishing Corporation J. Appl. Math. Volume (2012) 19. Available from: https://doi.org/10.1155/2012/421320. Article ID 421320.
[42] M. Sheikholeslami, R. Ellahi, H.R. Ashorynejad, G. Domairry, T. Hayat, Effects of Heat transfer in flow of nanofluids over a permeable stretching wall in a porous medium, J. Comput. Theor. Nanosci. 11 (2014) 1–11.
[43] M. Sheikholeslami, D.D. Ganji, Magnetohydrodynamic flow in a permeable channel filled with nanofluid, Sci. IranicaB 21 (1) (2014) 203–212.
[44] M. Sheikholeslami, M. Hatami, G. Domairry, Numerical simulation of two phase unsteady nanofluid flow and heat transfer between parallel plates in presence of time dependent magnetic field, J. Taiwan Inst. Chem. Eng. 46 (2015) 43–50.
[45] M. Sheikholeslami, D.D. Ganji, Nanofluid flow and heat transfer between parallel plates considering Brownian motion using DTM, Comput. Methods Appl. Mech. Engrg. 283 (2015) 651–663.
[46] M. Sheikholeslami, M.M. Rashidi, D.M. Al Saad, F. Firouzi, H.B. Rokni, G. Domairry, Steady nanofluid flow between parallel plates considering Thermophoresis and Brownian effects, J. King Saud Univ. Sci. 28 (4) (2016) 380–389.
[47] D. Domairry, M. Sheikholeslami, H.R. Ashorynejad, R.S.R. Gorla, M. Khani, Natural convection flow of a non-Newtonian nanofluid between two vertical flat plates, Proc. IMechE Part N:J. Nanoeng. Nanosyst. 225 (3) (2012) 115–122. Available from: https://doi.org/10.1177/1740349911433468. ©IMechE.
[48] A. Malvandi, F. Hedayati, D.D. Ganji, Slip effects on unsteady stagnation point flow of a nanofluid over a stretching sheet, Powder Technol. 253 (2014) 377–384.
[49] A. Malvandi, The unsteady flow of a nanofluid in the stagnation point region of a time-dependent rotating sphere, Therm. Sci. 19 (5) (2015) 1603–1612.
[50] H.R. Ashorynejad, M. Sheikholeslami, I. Pop, D.D. Ganji, Nanofluid flow and heat transfer due to a stretching cylinder in the presence of magnetic field, Heat Mass Transfer 49 (2013) 427–436.
[51] M. Sheikholeslami, D.D. Ganji, Heated permeable stretching surface in a porous medium using Nanofluids, J. Appl. Fluid Mech. 7 (3) (2014) 535–542.
[52] M. Sheikholeslami, D.D. Ganji, Three dimensional heat and mass transfer in a rotating system using nanofluid, Powder Technol. 253 (2014) 789–796.
[53] M. Sheikholeslami, M. Hatami, D.D. Ganji, Nanofluid flow and heat transfer in a rotating system in the presence of a magnetic field, J. Mol. Liq. 190 (2014) 112–120.
[54] M. Sheikholeslami, D.D. Ganji, Numerical investigation for two phase modeling of nanofluid in a rotating system with permeable sheet, J. Mol. Liq. 194 (2014) 13–19.
[55] M. Sheikholeslami, D.D. Ganji, Unsteady nanofluid flow and heat transfer in presence of magnetic field considering thermal radiation, J Braz. Soc. Mech. Sci. Eng. 37 (2015) 895–902.
[56] M. Sheikholeslami, S. Abelman, D.D. Ganji, Numerical simulation of MHD nanofluid flow and heat transfer considering viscous dissipation, Int. J. Heat Mass Transfer 79 (2014) 212–222.
[57] M. Sheikholeslami, D.D. Ganji, M.Y. Javed, R. Ellahi, Effect of thermal radiation on magnetohydrodynamics nanofluid flow and heat transfer by means of two phase model, J. Magn. Magn. Mater. 374 (2015) 36–43.
[58] M.S. Kandelousi, KKL correlation for simulation of nanofluid flow and heat transfer in a permeable channel, Phys. Lett. A 378 (2014) 3331–3339.
[59] M. Sheikholeslami, Effect of uniform suction on nanofluid flow and heat transfer over a cylinder, J Braz. Soc. Mech. Sci. Eng. 37 (2015) 1623–1633.
[60] M. Sheikholeslami, S. Abelman, Two phase simulation of nanofluid flow and heat transfer in an annulus in the presence of an axial magnetic field, IEEE Trans. Nanotechnol. 14 (3) (2015) 561–569.
[61] M. Sheikholeslami, M. Hatami, D.D. Ganji, Numerical investigation of nanofluid spraying on an inclined rotating disk for cooling process, J. Mol. Liq. 211 (2015) 577–583.
[62] M. Sheikholeslami, M.T. Mustafa, D.D. Ganji, Nanofluid flow and heat transfer over a stretching porous cylinder considering thermal radiation, Iranian J. Sci. Technol. 39 (Issue 3.1) (2015) 433–440. Autumn.
[63] A.J. Chamkha, A.M. Aly, MHD free convection flow of a nanofluid past a vertical plate in the presence of heat generation or absorption effects, Chem. Eng. Commun. 198 (2011) 425–441.
[64] A.J. Chamkha, A.M. Rashad, MHD forced convection flow of a nanofluid adjacent to a non-isothermal wedge, Comput. Therm. Sci. 6 (2014) 27–39.
[65] M.A. Sheremet, I. Pop, Conjugate natural convection in a square porous cavity filled by a nanofluid using Buongiorno's mathematical model, Int. J. Heat Mass Transfer 79 (2014) 137–145.
[66] M.A. Sheremet, I. Pop, M.M. Rahman, Three-dimensional natural convection in a porous enclosure filled with a nanofluid using Buongiorno's mathematical model, Int. J. Heat Mass Transfer 82 (2015) 396–405.
[67] M.A. Sheremet, S. Dinarvand, I. Pop, Effect of thermal stratification on free convection in a square porous cavity filled with a nanofluid using Tiwari and Das' nanofluid model, Physica E 69 (2015) 332–341.
[68] M. Ghalambaz, M.A. Sheremet, I. Pop, Free convection in a parallelogrammic porous cavity filled with a nanofluid using Tiwari and Das' nanofluid model, PLoS One 10 (2015) e0126486.
[69] M.A. Sheremet, I. Pop, A. Ishak, Double-diffusive mixed convection in a porous open cavity filled with a nanofluid using Buongiorno's model, Trans. Porous Media 109 (2015) 131–145.

[70] M.A. Sheremet, I. Pop, Free convection in a triangular cavity filled with a porous medium saturated by a nanofluid: Buongiorno's mathematical model, Int. J. Numer. Methods Heat Fluid Flow 25 (2015) 1138–1161.
[71] M.A. Sheremet, I. Pop, Natural convection in a horizontal cylindrical annulus filled with a porous medium saturated by a nanofluid using Tiwari and Das' nanofluid model, Europ. Phys. J. Plus 130 (2015) 107.
[72] M.A. Sheremet, I. Pop, N.C. Rosca, Magnetic field effect on the unsteady natural convection in a wavy-walled cavity filled with a nanofluid: Buongiorno's mathematical model, J. Taiwan Inst. Chemical Engineers 61 (2016) 211–222.
[73] J.A. Khan, M. Mustafa, T. Hayat, M. Sheikholeslami, A. Alsaedi, Three-dimensional flow of nanofluid induced by an exponentially stretching sheet: an application to solar energy, PLoS One 10 (3) (2015) e0116603. Available from: https://doi.org/10.1371/journal.pone.0116603.
[74] K.-L. Hsiao, Nanofluid flow with multimedia physical features for conjugate mixed convection and radiation, Comput. Fluids 104 (2014) 1–8.
[75] F. Garoosi, F. Hoseininejad, Numerical study of natural and mixed convection heat transfer between differentially heated cylinders in an adiabatic enclosure filled with nanofluid, J. Mol. Liq. 215 (2016) 1–17.
[76] F. Garoosi, S. Garoosi, K. Hooman, Numerical simulation of natural convection and mixed convection of the nanofluid in a square cavity using Buongiorno model, Powder Technol. 268 (2014) 279–292.
[77] F. Garoosi, B. Rohani, M.M. Rashidi, Two-phase mixture modeling of mixed convection of nanofluids in a square cavity with internal and external heating, Powder Technol. 275 (2015) 304–321.
[78] M.A. Teamah, W.M. El-Maghlany, Augmentation of natural convective heat transfer in square cavity by utilizing nanofluids in the presence of magnetic field and uniform heat generation/absorption, Int. J. Therm. Sci. 58 (2012) 130.
[79] A.K. Santra, S. Sen, N. Chakraborty, Study of heat transfer augmentation in a differentially heated square cavity using copper-water nanofluid, Int. J. Therm. Sci. 47 (2008) 1113–1122.
[80] M.K. Das, P.S. Ohal, Natural convection heat transfer augmentation in a partially heated and partially cooled square cavity utilizing nanofluids, Int. J. Numer. Methods Heat Fluid Flow 19 (2009) 411–431.
[81] H.F. Oztop, E. Abu-Nada, Y. Varol, K. Al-Salem, Computational analysis of non- isothermal temperature distribution on natural convection in nanofluid filled enclosures, Superlatt. Microstruct. 49 (2011) 453–467.
[82] F. Selimefendigil, H.F. Öztop, MHD mixed convection of nanofluid filled partially heated triangular enclosure with a rotating adiabatic, J. Taiwan Inst. Chem. Eng. 45 (2014) 2150–2162.
[83] F. Selimefendigil, H.F. Öztop, Natural convection and entropy generation of nanofluid filled cavity having different shaped obstacles under the influence of magnetic field and internal heat generation, J. Taiwan Inst. Chem. Eng. 56 (2015) 42–56.
[84] F. Selimefendigil, H.F. Öztop, Pulsating nanofluids jet impingement cooling of a heated horizontal surface, Int. J. Heat Mass Transfer 69 (2014) 54–65.
[85] F. Selimefendigil, H.F. Öztop, Numerical study of MHD mixed convection in a nanofluid filled lid driven square enclosure with a rotating cylinder, Int. J. Heat Mass Transfer 78 (2014) 741–754.
[86] F. Selimefendigil, H.F. Öztop, Numerical investigation and reduced order model of mixed convection at a backward facing step with a rotating cylinder subjected to nanofluid, Comput. Fluids 109 (2015) 27–37.
[87] F. Selimefendigil, H.F. Öztop, Mixed convection due to rotating cylinder in an internally heated and flexible walled cavity filled with SiO_2–water nanofluids: effect of nanoparticle shape, Int. Commun. Heat Mass Transfer 71 (2016) 9–19.
[88] F. Selimefendigil, H.F. Öztop, Conjugate natural convection in a cavity with a conductive partition and filled with different nanofluids on different sides of the partition, J. Mol. Liq. 216 (2016) 67–77.
[89] M. Sheikholeslami, M. Gorji-Bandpy, Soheil Soleimani, Two phase simulation of nanofluid flow and heat transfer using heatline analysis, Int. Commun. Heat Mass Transfer 47 (2013) 73–81.
[90] M. Sheikholeslami, M. Gorji-Bandpy, D.D. Ganji, Soheil Soleimani, Natural convection heat transfer in a cavity with sinusoidal wall filled with CuO-water nanofluid in presence of magnetic field, J. Taiwan Inst. Chem. Eng. 45 (2014) 40–49.
[91] M. Sheikholeslami, M. Gorji-Bandpy, D.D. Ganji, Soheil Soleimani, Effect of a magnetic field on natural convection in an inclined half-annulus enclosure filled with Cu–water nanofluid using CVFEM, Adv. Powder Technol. 24 (2013) 980–991.
[92] M. Sheikholeslami, M. Gorji-Bandpy, D.D. Ganji, Soheil Soleimani, MHD natural convection in a nanofluid filled inclined enclosure with sinusoidal wall using CVFEM, Neural Comput Applic 24 (2014) 873–882.
[93] M. Sheikholeslami, M. Gorji-Bandpy, D.D. Ganji, Soheil Soleimani, S.M. Seyyedi, Natural convection of nanofluids in an enclosure between a circular and a sinusoidal cylinder in the presence of magnetic field, Int. Commun. Heat Mass Transfer 39 (2012) 1435–1443.
[94] Soheil Soleimani, M. Sheikholeslami, D.D. Ganji, M. Gorji-Bandpay, Natural convection heat transfer in a nanofluid filled semi-annulus enclosure, Int. Commun. Heat Mass Transfer 39 (2012) 565–574.
[95] M. Sheikholeslami, M. Gorji Bandpy, R. Ellahi, M. Hassan, Soheil Soleimani, Effects of MHD on Cu-water nanofluid flow and heat transfer by means of CVFEM, J. Magn. Magn. Mater. 349 (2014) 188–200.
[96] M. Sheikholeslami, M. Gorji-Bandpy, D.D. Ganji, Soheil Soleimani, Heat flux boundary condition for nanofluid filled enclosure in presence of magnetic field, J. Mol. Liq. 193 (2014) 174–184.
[97] M. Sheikholeslami, D.D. Ganji, M. Gorji-Bandpy, Soheil Soleimani, Magnetic field effect on nanofluid flow and heat transfer using KKL model, J. Taiwan Inst. Chem. Eng. 45 (2014) 795–807.
[98] M. Sheikholeslami, M. Gorji-Bandpy, D.D. Ganji, Soheil Soleimani, Thermal management for free convection of nanofluid using two phase model, J. Mol. Liq. 194 (2014) 179–187.
[99] M. Sheikholeslami, M. Gorji-Bandpy, D.D. Ganji, P. Rana, Soheil Soleimani, Magnetohydrodynamic free convection of Al_2O_3-water nanofluid considering Thermophoresis and Brownian motion effects, Comput Fluids 94 (2014) 147–160.
[100] M. Sheikholeslami, M. Gorji-Bandpy, D.D. Ganji, Soheil Soleimani, Natural convection heat transfer in a nanofluid filled inclined L-shaped enclosure, IJST, Trans. Mech. Eng. 38 (2014) 217–226.
[101] M. Sheikholeslami, D.D. Ganji, Ferrohydrodynamic and Magnetohydrodynamic effects on ferrofluid flow and convective heat transfer, Energy 75 (2014) 400–410.

[102] M. Sheikholeslami, D.D. Ganji, M.M. Rashidi, Ferrofluid flow and heat transfer in a semi annulus enclosure in the presence of magnetic source considering thermal radiation, J. Taiwan Inst. Chem. Eng. 47 (2015) 6–17.

[103] M.S. Kandelousi, Effect of spatially variable magnetic field on ferrofluid flow and heat transfer considering constant heat flux boundary condition, Europ. Phys. J. Plus (2014) 129–248.

[104] M. Sheikholeslami, R. Ellahi, Mohsan Hassan, Soheil Soleimani, A study of natural convection heat transfer in a nanofluid filled enclosure with elliptic inner cylinder, Int. J. Numer. Methods Heat Fluid Flow 24 (2014) 8.

[105] M. Sheikholeslami, M.M. Rashidi, Effect of space dependent magnetic field on free convection of Fe_3O_4-water nanofluid, J. Taiwan Inst. Chem. Eng. 56 (2015) 6–15.

[106] M. Sheikholeslami, M.M. Rashidi, D.D. Ganji, Effect of non-uniform magnetic field on forced convection heat transfer of Fe_3O_4-water nanofluid, Comput. Methods Appl. Mech. Engrg. 294 (2015) 299–312.

[107] M. Sheikholeslami, Rahmat Ellahi, Electrohydrodynamic nanofluid hydrothermal treatment in an enclosure with sinusoidal upper wall, Appl. Sci. 5 (2015) 294–306. Available from: https://doi.org/10.3390/app5030294.

[108] M. Sheikholeslami, M.M. Rashidi, D.D. Ganji, Numerical investigation of magnetic nanofluid forced convective heat transfer in existence of variable magnetic field using two phase model, J. Mol. Liq. 212 (2015) 117–126.

[109] M. Sheikholeslami, K. Vajravelu, M.M. Rashidi, Forced convection heat transfer in a semi annulus under the influence of a variable magnetic field, Int. J. Heat Mass Transfer 92 (2016) 339–348.

[110] M. Sheikholeslami, Soheil Soleimani, D.D. Ganji, Effect of electric field on hydrothermal behavior of nanofluid in a complex geometry, J. Mol. Liq. 213 (2016) 153–161.

[111] M. Sheikholeslami, M.M. Rashidi, Non-uniform magnetic field effect on nanofluid hydrothermal treatment considering Brownian motion and thermophoresis effects, J Braz. Soc. Mech. Sci. Eng. 38 (2016) 1171–1184. Available from: https://doi.org/10.1007/s40430-015-0459-5.

[112] M. Sheikholeslami, M.M. Rashidi, Ferrofluid heat transfer treatment in the presence of variable magnetic field, Eur. Phys. J. Plus 130 (2015) 115.

[113] M. Sheikholeslami, M. Gorji-Bandpay, D.D. Ganji, Investigation of nanofluid flow and heat transfer in presence of magnetic field using KKL model, Arabian J. Sci. Eng. 39 (6) (2014) 5007–5016.

[114] M. Sheikholeslami, M. Gorji-Bandpy, D.D. Ganji, Natural convection in a nanofluid filled concentric annulus between an outer square cylinder and an inner elliptic cylinder, Sci. Iranica, Trans. B: Mech. Eng. 20 (4) (2013) 1–13.

[115] M. Sheikholeslami, M.G. Bandpay, D.D. Ganji, Magnetic field effects on natural convection around a horizontal circular cylinder inside a square enclosure filled with nanofluid, Int. Commun. Heat Mass Transfer 39 (2012) 978.

[116] M. Sheikholeslami, M. Gorji-Bandpy, S.M. Seyyedi, D.D. Ganji, H.B. Rokni, Soheil Soleimani, Application of LBM in simulation of natural convection in a nanofluid filled square cavity with curve boundaries, Powder Technol. 247 (2013) 87–94.

[117] M. Sheikholeslami, M. Gorji-Bandpy, G. Domairry, Free convection of nanofluid filled enclosure using lattice Boltzmann method (LBM), Appl. Math. Mech. -Engl. Ed. 34 (7) (2013) 1–15.

[118] H.R. Ashorynejad, A.A. Mohamad, M. Sheikholeslami, Magnetic field effects on natural convection flow of a nanofluid in a horizontal cylindrical annulus using Lattice Boltzmann method, Int. J. Therm. Sci. 64 (2013) 240–250.

[119] M. Sheikholeslami, M. Gorji-Bandpy, D.D. Ganji, Numerical investigation of MHD effects on Al_2O_3-water nanofluid flow and heat transfer in a semi-annulus enclosure using LBM, Energy 60 (2013) 501–510.

[120] M. Sheikholeslami, M. Gorji-Bandpy, D.D. Ganji, Lattice Boltzmann method for MHD natural convection heat transfer using nanofluid, Powder Technol. 254 (2014) 82–93.

[121] M. Sheikholeslami, M. Gorji-Bandpy, Free convection of ferrofluid in a cavity heated from below in the presence of an external magnetic field, Powder Technol. 256 (2014) 490–498.

[122] M. Sheikholeslami, M. Gorji-Bandpy, D.D. Ganji, MHD free convection in an eccentric semi-annulus filled with nanofluid, J. Taiwan Inst. Chem. Eng. 45 (2014) 1204–1216.

[123] M. Sheikholeslami, M.G. Bandpy, R. Ellahi, A. Zeeshan, Simulation of MHD CuO–water nanofluid flow and convective heat transfer considering Lorentz forces, J. Magn. Magn. Mater. 369 (2014) 69–80.

[124] M. Sheikholeslami, D.D. Ganji, Entropy generation of nanofluid in presence of magnetic field using Lattice Boltzmann Method, Physica A 417 (2015) 273–286.

[125] M. Sheikholeslami, M. Gorji-Bandpy, K. Vajravelu, Lattice Boltzmann Simulation of Magnetohydrodynamic Natural Convection Heat Transfer of Al_2O_3-water Nanofluid in a Horizontal Cylindrical Enclosure with an Inner Triangular Cylinder, Int. J. Heat Mass Transfer 80 (2015) 16–25.

[126] M. Sheikholeslami, R. Ellahi, Simulation of ferrofluid flow for magnetic drug targeting using Lattice Boltzmann method, J. Zeitschrift Fur Naturforschung A, Verlag der Zeitschrift für Naturforschung 70 (2) (2015) 115–124.

[127] M. Sheikholeslami, M.G. Bandpy, H.R. Ashorynejad, Lattice Boltzmann Method for simulation of magnetic field effect on hydrothermal behavior of nanofluid in a cubic cavity, Phys. A: Statist. Mech. Applicat. 432 (2015) 58–70.

[128] M. Sheikholeslami, R. Ellahi, Three dimensional mesoscopic simulation of magnetic field effect on natural convection of nanofluid, Int. J. Heat Mass Transfer 89 (2015) 799–808.

[129] M. Sheikholeslami, H.R. Ashorynejad, P. Rana, Lattice Boltzmann simulation of nanofluid heat transfer enhancement and entropy generation, J. Mol. Liq. 214 (2016) 86–95.

[130] G.H.R. Kefayati, Lattice Boltzmann simulation of natural convection in nanofluid-filled 2D long enclosures at presence of magnetic source, Theor. Comput. Fluid Dynam. 27 (2013) 865–883.

[131] G.H.R. Kefayati, Lattice Boltzmann simulation of natural convection in a nanouid-filled inclined square cavity at presence of magnetic field, Sci. Iranica 20 (5) (2013) 1517–1527.

[132] M. Sheikholeslami, H.R. Ashorynejad, D.D. Ganji, A. Yıldırım, Homotopy perturbation method for three-dimensional problem of condensation film on inclined rotating disk, Sci.Iranica B 19 (3) (2012) 437–442.

[133] R. Kumar, S. Sood, S.A. Shehzad, M. Sheikholeslami, Radiative heat transfer study for flow of non-Newtonian nanofluid past a Riga plate with variable thickness, J. Mol. Liq. 248 (2017) 143–152.
[134] M. Sheikholeslami, M. Azimi, D.D. Ganji, Application of differential transformation method for nanofluid flow in a semi-permeable channel considering magnetic field effect, Int. J. Comput. Methods Eng. Sci. Mech. 16 (4) (2015) 246–255.
[135] M. Sheikholeslami, Ali J. Chamkha, P. Rana, R. Moradi, Combined thermophoresis and Brownian motion effects on nanofluid free convection heat transfer in an L-shaped enclosure, Chin. J. Phys. 55 (2017) 2356–2370.
[136] M. Sheikholeslami, H.R. Kataria, A.S. Mittal, Radiation effects on heat transfer of three dimensional nanofluid flow considering thermal interfacial resistance and micro mixing in suspensions, Chin. J. Phys. (2017). Available from: https://doi.org/10.1016/j.cjph.2017.09.010.
[137] M. Sheikholeslami, F. Bani Sheykholeslami, S. Khoshhal, H. Mola-Abasia, D.D. Ganji, H.B. Rokni, Effect of magnetic field on Cu–water nanofluid heat transfer using GMDH-type neural network, Neur. Comput Applic 25 (2014) 171–178.
[138] M. Sheikholeslami, M.T. Mustafa, D.D. Ganji, Effect of Lorentz forces on forced-convection nanofluid flow over a stretched surface, Particuology 26 (2016) 108–113.
[139] M. Sheikholeslami, D.D. Ganji, Numerical modeling of magnetohydrodynamic CuO -water transportation inside a porous cavity considering shape factor effect, Colloids Surfaces A 529 (2017) 705–714.
[140] M. Sheikholeslami, M. Nimafar, D.D. Ganji, Analytical approach for effect of melting heat transfer on nanofluid heat transfer, Eur. Phys. J. Plus 132 (2017) 385. Available from: https://doi.org/10.1140/epjp/i2017-11669-3.
[141] M.M. Bhatti, M. Sheikholeslami, A. Zeeshan, Entropy analysis on Electro-kinetically Modulated peristaltic propulsion on magnetized nanofluid flow through a microchannel, Entropy 19 (2017) 481. Available from: https://doi.org/10.3390/e19090481.
[142] H. Xu, I. Pop, Fully developed mixed convection flow in a vertical channel filled with nanouids, Int. Commun. Heat Mass Transfer 39 (2012) 1086–1092.
[143] M. Sheikholeslami, D.D. Ganji, R. Moradi, Heat transfer of Fe_3O_4-water nanofluid in a permeable medium with thermal radiation in existence of constant heat flux, Chem. Eng. Sci. 174 (2017) 326–336.
[144] M. Sheikholeslami, D.D. Ganji, Print Book External Magnetic Field Effects on Hydrothermal Treatment of Nanofluid, William Andrew, Elsevier, 2016, ISBN: 9780323431385, pp. 1–354.
[145] M. Sheikholeslami, D.D. Ganji, R. Moradi, Forced convection in existence of Lorentz forces in a porous cavity with hot circular obstacle using nanofluid via Lattice Boltzmann method, J. Mol. Liq. 246 (2017) 103–111.
[146] M. Sheikholeslami, M.M. Rashidi, T. Hayat, D.D. Ganji, Free convection of magnetic nanofluid considering MFD viscosity effect, J. Mol. Liq. 218 (2016) 393–399.
[147] M. Sheikholeslami, T. Hayat, A. Alsaedi, MHD free convection of Al_2O_3–water nanofluid considering thermal radiation: a numerical study, Int. J. Heat Mass Transfer 96 (2016) 513–524.
[148] H. Safarnia, M. Sheikholeslami, D.D. Ganji, Electrohydrodynamic nanofluid flow and forced convective heat transfer in a channel, Eur. Phys. J. Plus 131 (2016) 96. Available from: https://doi.org/10.1140/epjp/i2016-16096-4.
[149] M. Sheikholeslami, H.B. Rokni, Magnetic nanofluid natural convection in the presence of thermal radiation considering variable viscosity, Europ. Phys. J. Plus 132 (2017) 238. Available from: https://doi.org/10.1140/epjp/i2017-11498-4.
[150] M. Sheikholeslami, Ali J. Chamkha, Electrohydrodynamic free convection heat transfer of a nanofluid in a semi-annulus enclosure with a sinusoidal wall, Numer. Heat Transfer, Part A 69 (7) (2016) 781–793. Available from: https://doi.org/10.1080/10407782.2015.1090819.
[151] M. Sheikholeslami, Mofid Gorji-Bandpy, D.D. Ganji, Review of heat transfer enhancement methods: focus on passive methods using swirl flow devices, Renew. Sustain. Energy Rev. 49 (2015) 444–469.
[152] M. Sheikholeslami, D.D. Ganji, Print Book Hydrothermal Analysis in Engineering Using Control Volume Finite Element Method, Academic Press, 2015, ISBN: 9780128029503, pp. 1–226.
[153] M. Sheikholeslami, M. Shamlooei, Convective flow of nanofluid inside a lid driven porous cavity using CVFEM, Phys. B: Condensed Matter 521 (2017) 239–250.
[154] M. Sheikholeslami, M. Shamlooei, Magnetic source influence on nanofluid flow in porous medium considering shape factor effect, Phys. Lett. A 381 (2017) 3071–3078.
[155] M. Sheikholeslami, M.K. Sadoughi, Numerical modeling for Fe_3O_4-water nanofluid flow in porous medium considering MFD viscosity, J. Mol. Liq. 242 (2017) 255–264.
[156] M. Sheikholeslami, D.D. Ganji, Nanofluid convective heat transfer using semi analytical and Numerical approaches: a review, J. Taiwan Inst. Chem. Eng. 65 (2016) 43–77.
[157] M. Sheikholeslami, H.B. Rokni, Nanofluid convective heat transfer intensification in a porous circular cylinder, Chem. Eng. Processing: Process Intensif. 120 (2017) 93–104.
[158] M. Sheikholeslami, A. Zeeshan, Mesoscopic simulation of CuO-H_2O nanofluid in a porous enclosure with elliptic heat source, Int. J. Hydrogen Energy (2017).
[159] M. Sheikholeslami, Lattice Boltzmann Method simulation of MHD non-Darcy nanofluid free convection, Physica B 516 (2017) 55–71.
[160] M. Sheikholeslami, H.B. Rokni, Melting heat transfer of nanofluid in existence of magnetic field considering Buongiorno Model, Chin. J. Phys. (2017).
[161] M. Sheikholeslami, H.B. Rokni, Free convection of CuO-H_2O nanofluid in a curved porous enclosure using mesoscopic approach, Int. J. Hydrogen Energy (2017).
[162] M. Sheikholeslami, M.M. Bhatti, Forced convection of nanofluid in presence of constant magnetic field considering shape effects of nanoparticles, Int. J. Heat Mass Transfer 111 (2017) 1039–1049.
[163] M. Sheikholeslami, D.D. Ganji, Analytical investigation for Lorentz forces effect on nanofluid Marangoni boundary layer hydrothermal behavior using HAM, Indian J. Phys. 91 (12) (2017) 1581–1587.
[164] M. Sheikholeslami, H.B. Rokni, Magnetic nanofluid natural convection in presence of thermal radiation considering variable viscosity, Europ. Phys. J. Plus (2017).
[165] M. Sheikholeslami, D.D. Ganji, Numerical investigation of nanofluid melting heat transfer between two pipes, Alexandria Eng. J. (2017). Available from: http://doi.org/10.1016/j.aej.2017.03.028.

[166] M. Sheikholeslami, Influence of magnetic field on nanofluid free convection in an open porous cavity by means of Lattice Boltzmann Method, J. Mol. Liq. 234 (2017) 364–374.

[167] M. Sheikholeslami, A. Zeeshan, Analysis of flow and heat transfer in water based nanofluid due to magnetic field in a porous enclosure with constant heat flux using CVFEM, Comput. Methods Appl. Mech. Eng. 320 (2017) 68–81.

[168] M. Sheikholeslami, D.D. Ganji, Numerical analysis of nanofluid transportation in porous media under the influence of external magnetic source, J. Mol. Liq. 233 (2017) 499–507.

[169] M. Sheikholeslami, K. Vajravelu, Forced convection heat transfer in Fe_3O_4-ethylene glycol nanofluid under the influence of Coulomb force, J. Mol. Liq. 233 (2017) 203–210.

[170] M. Sheikholeslami, D.D. Ganji, Numerical approach for magnetic nanofluid flow in a porous cavity using CuO nanoparticles, Mater. Design 120 (2017) 382–393.

[171] M. Sheikholeslami, D.D. Ganji, Applications of Nanofluid for Heat Transfer Enhancement, William Andrew, Elsevier, 2017, ISBN: 9780081021729, .

[172] M. Sheikholeslami, P. Jalili, D.D. Ganji, Magnetic field effect on nanofluid flow between two circular cylinders using AGM, Alexandria Eng. J. (2017).

[173] M. Sheikholeslami, H.B. Rokni, Influence of EFD viscosity on nanofluid forced convection in a cavity with sinusoidal wall, J. Mol. Liq. 232 (2017) 390–395.

[174] M. Sheikholeslami, Magnetohydrodynamic nanofluid forced convection in a porous lid driven cubic cavity using Lattice Boltzmann Method, J. Mol. Liq. 231 (2017) 555–565. Available from: https://doi.org/10.1016/j.molliq.2017.02.020.

[175] M. Sheikholeslami, M. Shamlooei, Fe_3O_4-H_2O nanofluid natural convection in presence of thermal radiation, Int. J. Hydrogen Energy 42 (9) (2017) 5708–5718.

[176] M. Sheikholeslami, M.M. Bhatti, Active method for nanofluid heat transfer enhancement by means of EHD, Int. J. Heat Mass Transfer 109 (2017) 115–122.

[177] M. Sheikholeslami, S.A. Shehzad, Thermal radiation of ferrofluid in existence of Lorentz forces considering variable viscosity, Int. J. Heat Mass Transfer 109 (2017) 82–92.

[178] M. Sheikholeslami, Z. Ziabakhsh, D.D. Ganji, Transport of Magnetohydrodynamic nanofluid in a porous media, Colloids Surf., A 520 (2017) 201–212.

[179] M. Sheikholeslami, R. Ellahi, K. Vafai, Study of Fe_3O_4-water nanofluid with convective heat transfer in the presence of magnetic source, Alexandria Eng. J. (2017). Available from: http://dx.doi.org/10.1016/j.aej.2017.01.027.

[180] M. Sheikholeslami, M. Nimafar, D.D. Ganji, Nanofluid heat transfer between two pipes considering Brownian motion using AGM, Alexandria Eng. J. (2017).

[181] M. Sheikholeslami, T. Hayat, A. Alsaedi, Numerical simulation of nanofluid forced convection heat transfer improvement in existence of magnetic field using Lattice Boltzmann Method, Int. J. Heat Mass Transfer 108 (2017) 1870–1883.

[182] M. Sheikholeslami, P. Rana, Soheil Soleimani, Numerical study of mhd natural convection liquid metal flow and heat transfer in a wavy enclosure using CVFEM, Heat Transfer Res. 48 (2) (2017) 121–138.

[183] M. Sheikholeslami, CuO-water nanofluid free convection in a porous cavity considering Darcy law, Europ. Phys. J. Plus 132 (2017) 55. Available from: https://doi.org/10.1140/epjp/i2017-11330-3.

[184] M. Sheikholeslami, D.D. Ganji, Impact of electric field on nanofluid forced convection heat transfer with considering variable properties, J. Mol. Liq. 229 (2017) 566–573.

[185] M. Sheikholeslami, D.D. Ganji, Free convection of Fe_3O_4-water nanofluid under the influence of an external magnetic source, J. Mol. Liq. 229 (2017) 530–540.

[186] M. Sheikholeslami, Influence of Lorentz forces on nanofluid flow in a porous cavity by means of Non- Darcy model, Eng. Comput. 34 (8) (2017) 2651–2667. Available from: https://doi.org/10.1108/EC-01-2017-0008.

[187] M. Sheikholeslami, H.B. Rokni, Numerical modeling of nanofluid natural convection in a semi annulus in existence of Lorentz force, Comput. Methods Appl. Mech. Eng. 317 (2017) 419–430.

[189] M. Sheikholeslami, D.D. Ganji, Transportation of MHD nanofluid free convection in a porous semi annulus using numerical approach, Chem. Phys. Lett. 669 (2017) 202–210.

[190] M. Sheikholeslami, Magnetic field influence on nanofluid thermal radiation in a cavity with tilted elliptic inner cylinder, J. Mol. Liq. 229 (2017) 137–147.

[191] M. Sheikholeslami, Numerical simulation of magnetic nanofluid natural convection in porous media, Phys. Lett. A 381 (2017) 494–503.

[192] M. Sheikholeslami, Magnetic source impact on nanofluid heat transfer using CVFEM, Neural Comput. Applicat. (2016). Available from: https://doi.org/10.1007/s00521-016-2740-7.

[193] M. Sheikholeslami, K. Vajravelu, Nanofluid flow and heat transfer in a cavity with variable magnetic field, Appl. Math. Comput. 298 (2017) 272–282.

[194] M. Sheikholeslami, M. Barzegar Gerdroodbary, D.D. Ganji, Numerical investigation of forced convective heat transfer of Fe_3O_4-water nanofluid in presence of external magnetic source, Comput. Methods Appl. Mech. Eng. 315 (2017) 831–845.

[195] M. Sheikholeslami, Influence of Lorentz forces on nanofluid flow in a porous cylinder considering Darcy model, J. Mol. Liq. 225 (2017) 903–912.

[196] M. Sheikholeslami, D.D. Ganji, CVFEM for free convective heat transfer of CuO-water nanofluid in a tilted semi annulus, Alexandria Eng. J. (2016). Available from: http://dx.doi.org/10.1016/j.aej.2016.11.012.

[197] M. Sheikholeslami, D.D. Ganji, Ferrofluid convective heat transfer under the influence of external magnetic source, Alexandria Eng. J. (2016). Available from: http://dx.doi.org/10.1016/j.aej.2016.11.007.

[198] M. Sheikholeslami, D.D. Ganji, Numerical investigation of nanofluid transportation in a curved cavity in existence of magnetic source, Chem. Phys. Lett. 667 (2017) 307–316.

[199] M. Sheikholeslami, CVFEM for magnetic nanofluid convective heat transfer in a porous curved enclosure, Eur. Phys. J. Plus 131 (2016) 413. Available from: https://doi.org/10.1140/epjp/i2016-16413-y.

[200] M. Sheikholeslami, H.B. Rokni, Magnetohydrodynamic CuO-water nanofluid in a porous complex shaped enclosure, ASME, J. Therm. Sci. Eng. Applicat. 9 (4) (2017) 041007. Available from: https://doi.org/10.1115/1.4035973.
[201] M. Sheikholeslami, H.B. Rokni, Nanofluid two phase model analysis in existence of induced magnetic field, Int. J. Heat Mass Transfer 107 (2017) 288–299.
[202] M. Sheikholeslami, Ali J. Chamkha, Influence of Lorentz forces on nanofluid forced convection considering Marangoni convection, J. Mol. Liq. 225 (2017) 750–757.
[203] M. Sheikholeslami, T. Hayat, A. Alsaedi, S. Abelman, EHD nanofluid force convective heat transfer considering electric field dependent viscosity, Int. J. Heat Mass Transfer 108 (2017) 2558–2565.
[204] M. Sheikholeslami, S.A. Shehzad, Magnetohydrodynamic nanofluid convection in a porous enclosure considering heat flux boundary condition, Int. J. Heat Mass Transfer 106 (2017) 1261–1269.
[205] M. Sheikholeslami, Q.M. Zaigham Zia, R. Ellahi, Influence of Induced Magnetic Field on Free Convection of Nanofluid Considering Koo-Kleinstreuer-Li (KKL) Correlation, Appl. Sci. 6 (2016) 324. Available from: https://doi.org/10.3390/app6110324.
[206] M. Sheikholeslami, Influence of Coulomb forces on Fe_3O_4-H_2O nanofluid thermal improvement, Int. J. Hydrogen Energy 42 (2017) 821–829.
[207] M. Sheikholeslami, T. Hayat, A. Alsaedi, Numerical study for external magnetic source influence on water based nanofluid convective heat transfer, Int. J. Heat Mass Transfer 106 (2017) 745–755.
[208] M. Sheikholeslami, D.D. Ganji, Nanofluid hydrothermal behavior in existence of Lorentz forces considering Joule heating effect, J. Mol. Liq. 224 (2016) 526–537.
[209] M. Sheikholeslami, D.D. Ganji, Influence of magnetic field on CuO-H_2O nanofluid flow considering Marangoni boundary layer, Int. J. Hydrogen Energy 42 (5) (2017) 2748–2755.
[210] M. Sheikholeslami, Numerical simulation for external magnetic field influence on Fe_3O_4-water nanofluid forced convection, Engineering Computations, In press.
[211] M. Sheikholeslami, D.D. Ganji, Heat transfer enhancement in an air to water heat exchanger with discontinuous helical turbulators; experimental and numerical studies, Energy 116 (2016) 341–352.
[212] M. Sheikholeslami, D.D. Ganji, Heat transfer improvement in a double pipe heat exchanger by means of perforated turbulators, Energy Convers. Manage. 127 (2016) 112–123.
[213] M. Sheikholeslami, D.D. Ganji, Turbulent heat transfer enhancement in an air-to-water heat exchanger, Proc IMechE Part E: J Process Mech. Eng. 0 (0) (2016) 1–14. Available from: https://doi.org/10.1177/0954408916664758.
[214] M. Sheikholeslami, M. Nimafar, D.D. Ganji, M. Pouyandehmehr, CuO-H_2O nanofluid hydrothermal analysis in a complex shaped cavity, Int. J. Hydrogen Energy 41 (2016) 17837–17845.
[215] M. Sheikholeslami, Sina Lohrasbi, D.D. Ganji, Numerical analysis of discharging process acceleration in LHTESS by immersing innovative fin configuration using finite element method, Appl. Therm. Eng. 107 (2016) 154–166.
[216] M. Sheikholeslami, D.D. Ganji, M.M. Rashidi, Magnetic field effect on unsteady nanofluid flow and heat transfer using Buongiorno model, J. Magn. Magn. Mater. 416 (2016) 164–173.
[217] M. Sheikholeslami, Numerical investigation of MHD nanofluid free convective heat transfer in a porous tilted enclosure, Eng. Computat. 34 (6) (2017) 1939–1955.
[218] M. Sheikholeslami, H.B. Rokni, Numerical simulation for impact of Coulomb force on nanofluid heat transfer in a porous enclosure in presence of thermal radiation, Int. J. Heat Mass Transfer (2018). Available from: https://doi.org/10.1016/j.ijheatmasstransfer.2017.11.041.
[219] M. Sheikholeslami, M. Shamlooei, R. Moradi, Fe_3O_4-Ethylene glycol nanofluid forced convection inside a porous enclosure in existence of Coulomb force, J. Mol. Liq. (2018). Available from: https://doi.org/10.1016/j.molliq.2017.11.048.
[220] M. Sheikholeslami, S.A. Shehzad, Numerical analysis of Fe_3O_4–H2O nanofluid flow in permeable media under the effect of external magnetic source, International Journal of Heat and Mass Transfer, Inpress.
[221] M. Sheikholeslami, T. Hayat, A. Alsaedi, S. Abelman, Numerical analysis of EHD nanofluid force convective heat transfer considering electric field dependent viscosity, Int. J. Heat Mass Transfer 108 (2017) 2558–2565.
[222] M. Sheikholeslami, Magnetic field influence on CuO-H_2O nanofluid convective flow in a permeable cavity considering various shapes for nanoparticles, Int. J. Hydrogen Energy 42 (2017) 19611–19621.
[223] M. Sheikholeslami, M.K. Sadoughi, Simulation of CuO- water nanofluid heat transfer enhancement in presence of melting surface, Int. J. Heat Mass Transfer 116 (2018) 909–919.
[224] M. Sheikholeslami, H.B. Rokni, Simulation of nanofluid heat transfer in presence of magnetic field: a review, Int. J. Heat Mass Transfer 115 (2017) 1203–1233.
[225] M. Sheikholeslami, H.B. Rokni, Melting heat transfer influence on nanofluid flow inside a cavity in existence of magnetic field, Int. J. Heat Mass Transfer 114 (2017) 517–526.
[226] M. Sheikholeslami, S.A. Shehzad, Magnetohydrodynamic nanofluid convective flow in a porous enclosure by means of LBM, Int. J. Heat Mass Transfer 113 (2017) 796–805.
[227] M. Sheikholeslami, M. Seyednezhad, Lattice Boltzmann Method simulation for CuO-water nanofluid flow in a porous enclosure with hot obstacle, J. Mol. Liq. 243 (2017) 249–256.
[228] M. Sheikholeslami, M. Sadoughi, Mesoscopic method for MHD nanofluid flow inside a porous cavity considering various shapes of nanoparticles, Int. J. Heat Mass Transfer 113 (2017) 106–114.
[229] M. Sheikholeslami, S.A. Shehzad, CVFEM for influence of external magnetic source on Fe_3O_4–H_2O nanofluid behavior in a permeable cavity considering shape effect, Int. J. Heat Mass Transfer 115 (2017) 180–191.
[230] M. Sheikholeslami, M. Seyednezhad, Nanofluid heat transfer in a permeable enclosure in presence of variable magnetic field by means of CVFEM, Int. J. Heat Mass Transfer 114 (2017) 1169–1180.
[231] M. Sheikholeslami, Numerical investigation for CuO-H_2O nanofluid flow in a porous channel with magnetic field using mesoscopic method, Journal of Molecular Liquids, Inpress.

[232] M. Sheikholeslami, H.B. Rokni, Influence of melting surface on MHD nanofluid flow by means of two phase model, Chin. J. Phys. 55 (2017) 1352–1360.

[233] M. Sheikholeslami, H.B. Rokni, Effect of melting heat transfer on nanofluid flow in existence of magnetic field considering Buongiorno Model, Chin. J. Phys. 55 (2017) 1115–1126.

[234] S.R. Hosseini, M. Sheikholeslami, M. Ghasemian, D.D. Ganji, Nanofluid heat transfer analysis in a microchannel heat sink (MCHS) under the effect of magnetic field by means of KKL model, Powder Technol. 324 (2018) 36–47.

[235] R. Kumar, S. Sood, M. Sheikholeslami, S.A. Shehzad, Nonlinear thermal radiation and cubic autocatalysis chemical reaction effects on the flow of stretched nanofluid under rotational oscillations, J. Colloid Interface Sci. 505 (2017) 253–265.

[236] M. Sheikholeslami, D.D. Ganji, Analytical investigation for Lorentz forces effect on nanofluid Marangoni boundary layer hydrothermal behavior using HAM, Indian J. Phys. 91 (12) (2017) 1581–1587.

[237] M. Sheikholeslami, Ali J. Chamkha, Flow and convective heat transfer of a ferro-nanofluid in a double-sided lid-driven cavity with a wavy wall in the presence of a variable magnetic field, Numer. Heat Transfer, Part A 69 (10) (2016) 1186–1200. Available from: https://doi.org/10.1080/10407782.2015.1125709.

[238] M. Sheikholeslami, H. Oztop, MHD free convection of nanofluid in a cavity with sinusoidal walls by using CVFEM, Chin. J. Phys. 55 (6) (2017) 2291–2304.

[239] R. Kumar, S. Sood, S.A. Shehzad, M. Sheikholeslami, Numerical modeling of time-dependent bio-convective stagnation flow of a nanofluid in slip regime, Results Phys. 7 (2017) 3325–3332. Available from: https://doi.org/10.1016/j.rinp.2017.08.059.

[240] M. Sheikholeslami, T. Hayat, A. Alsaedi, On simulation of nanofluid radiation and natural convection in an enclosure with elliptical cylinders, Int. J. Heat Mass Transfer 115 (2017) 981–991.

[241] M. Sheikholeslami, Numerical simulation for solidification in a LHTESS by means of nano-enhanced PCM, J. Taiwan Inst. Chem. Eng. 86 (2018) 25–41.

[242] M. Sheikholeslami, Numerical modeling of nano enhanced PCM solidification in an enclosure with metallic fin, J. Mol. Liq. 259 (2018) 424–438.

[243] M. Sheikholeslami, A. Ghasemi, Solidification heat transfer of nanofluid in existence of thermal radiation by means of FEM, Int. J. Heat Mass Transf. 123 (2018) 418–431.

[244] M. Sheikholeslami, S.A. Shehzad, CVFEM simulation for nanofluid migration in a porous medium using Darcy model, Int. J. Heat Mass Transf. 122 (2018) 1264–1271.

[245] M. Sheikholeslami, M. Darzi, M.K. Sadoughi, Heat transfer improvement and pressure drop during condensation of refrigerant-based nanofluid: an experimental procedure, Int. J. Heat Mass Transf. 122 (2018) 643–650.

[246] M. Sheikholeslami, H.B. Rokni, CVFEM for effect of Lorentz forces on nanofluid flow in a porous complex shaped enclosure by means of non-equilibrium model, J. Mol. Liq. 254 (2018) 446–462.

[247] M. Sheikholeslami, S.A. Shehzad, Zhixiong Li, Water based nanofluid free convection heat transfer in a three dimensional porous cavity with hot sphere obstacle in existence of Lorenz forces, International Journal of Heat and Mass Transfer 125 (2018) 375–386.

[248] M. Sheikholeslami, M. Jafaryar, D.D. Ganji, Zhixiong Li, Exergy loss analysis for nanofluid forced convection heat transfer in a pipe with modified turbulators, Journal of Molecular Liquids 262 (2018) 104–110.

[249] M. Sheikholeslami, M. Jafaryar, Z. Li, Nanofluid turbulent convective flow in a circular duct with helical turbulators considering CuO nanoparticles, Int. J. Heat Mass Transf. 124 (2018) 980–989.

[250] M. Sheikholeslami, Z. Li, M. Shamlooei, Nanofluid, MHD natural convection through a porous complex shaped cavity considering thermal radiation, Phys. Lett. A 382 (2018) 1615–1632.

[251] M. Sheikholeslami, H.B. Rokni, Magnetic nanofluid flow and convective heat transfer in a porous cavity considering Brownian motion effects, Phys. Fluids, 30 (1), https://doi.org/10.1063/1.5012517.

[252] M. Sheikholeslami, S.A. Shehzad, Simulation of water based nanofluid convective flow inside a porous enclosure via non-equilibrium model, Int. J. Heat Mass Transf 120 (2018) 1200–1212.

[253] M. Sheikholeslami, S.A. Shehzad, Non-Darcy free convection of Fe_3O_4-water nanoliquid in a complex shaped enclosure under impact of uniform Lorentz force, Chin. J. Phys. 56 (2018) 270–281.

[254] M. Sheikholeslami, M. Seyednezhad, Simulation of nanofluid flow and natural convection in a porous media under the influence of electric field using CVFEM, Int. J. Heat Mass Transf. 120 (2018) 772–781.

[255] M. Sheikholeslami, M. Shamlooei, R. Moradi, Numerical simulation for heat transfer intensification of nanofluid in a permeable curved enclosure considering shape effect of Fe_3O_4 nanoparticles, Chem. Eng. Process.: Process Intensif. 124 (2018) 71–82.

[256] M. Sheikholeslami, T. Hayat, T. Muhammad, A. Alsaedi, MHD forced convection flow of nanofluid in a porous cavity with hot elliptic obstacle by means of Lattice Boltzmann method, Int. J. Mech. Sci. 135 (2018) 532–540.

[257] M. Sheikholeslami, Numerical investigation of nanofluid free convection under the influence of electric field in a porous enclosure, J. Mol. Liq. 249 (2018) 1212–1221.

[258] M. Sheikholeslami, CuO-water nanofluid flow due to magnetic field inside a porous media considering Brownian motion, J. Mol. Liq. 249 (2018) 921–929.

[259] M. Sheikholeslami, H.B. Rokni, Numerical simulation for impact of Coulomb force on nanofluid heat transfer in a porous enclosure in presence of thermal radiation, Int. J. Heat Mass Transf 118 (2018) 823–831.

[260] M. Sheikholeslami, Numerical investigation for CuO-H_2O nanofluid flow in a porous channel with magnetic field using mesoscopic method, J. Mol. Liq. 249 (2018) 739–746.

[261] M. Sheikholeslami, M. Shamlooei, R. Moradi, Fe_3O_4-ethylene glycol nanofluid forced convection inside a porous enclosure in existence of Coulomb force, J. Mol. Liq. 249 (2018) 429–437.

[262] M. Sheikholeslami, S.A. Shehzad, Numerical analysis of Fe_3O_4–H_2O nanofluid flow in permeable media under the effect of external magnetic source, Int. J. Heat Mass Transf 118 (2018) 182–192.

[263] M. Sheikholeslami, T. Hayat, A. Alsaedi, Numerical simulation for forced convection flow of MHD CuO-H_2O nanofluid inside a cavity by means of LBM, J. Mol. Liq. 249 (2018) 941–948.

[264] M. Sheikholeslami, H.R. Kataria, A.S. Mittal, Effect of thermal diffusion and heat-generation on MHD nanofluid flow past an oscillating vertical plate through porous medium, J. Mol. Liq. 257 (2018) 12–25.

[265] M. Sheikholeslami, S.A. Shehzad, Z. Li, Nanofluid heat transfer intensification in a permeable channel due to magnetic field using Lattice Boltzmann method, Phys. B (2018). Available from: https://doi.org/10.1016/j.physb.2018.03.036.

[266] M. Sheikholeslami, A. Zeeshan, Numerical simulation of Fe_3O_4-water nanofluid flow in a non-Darcy porous media, Int. J. Num. Methods Heat Fluid Flow 28 (3) (2018) 641–660. Available from: https://doi.org/10.1108/HFF-04-2017-0160.

[267] M. Sheikholeslami, D.D. Ganji, Influence of electric field on Fe_3O_4-water nanofluid radiative and convective heat transfer in a permeable enclosure, J. Mol. Liq. 250 (2018) 404–412.

[268] M. Sheikholeslami, M. Barzegar Gerdroodbary, S. Valiallah Mousavi, D.D. Ganji, R. Moradi, Heat transfer enhancement of ferrofluid inside an 90° elbow channel by non-uniform magnetic field, J. Magnet. Magnet. Mater 460 (2018) 302–311.

CHAPTER

4

Single-phase Model for Nanofluid Free Convection Heat Transfer by Means of CVFEM

4.1 INTRODUCTION

Common heat transfer fluids such as water, ethylene glycol, and engine oil have limited heat transfer capabilities due to their low heat transfer properties. In contrast, metals thermal conductivities are up to three times higher than the fluids, so it is naturally desirable to combine the two substances to produce a heat transfer medium that behaves like a fluid, but has the thermal conductivity of a metal. The term "nanofluid" was first proposed by Choi [1] to indicate engineered colloids composed of nanoparticles dispersed in a base fluid. Khanafer et al. [2] firstly conducted a numerical investigation on the heat transfer enhancement due to adding nanoparticles in a differentially heated enclosure. They found that the suspended nanoparticles substantially increase the heat transfer rate at any given Grashof number. Influence of axial magnetic field on nanofluid thermal management has been analyzed by Sheikholeslami and Abelman [3]. Sheikholeslami and Rashidi [4] applied a single-phase model for Fe_3O_4-water nanofluid. They indicated that Lorentz forces reduce the velocity. The influence of single magnetic source on Fe_3O_4-H_2O flow style has been reported by Sheikholeslami and Ganji [5]. Sheikholeslami and Shehzad [6] employed lattice Boltzmann method for nanofluid flow in a porous cavity in the presence of a magnetic field. Sheikholeslami [7] presented a mesoscopic approach for nanofluid convective flow in a porous lid driven cubic cavity. Sheikholeslami Kandelousi [8] presented application of a Koo–Kleinstreuer–Li (KKL) model for simulation of nanofluid in a semiporous channel. Influence of uniform suction on nanofluid flow and heat transfer over a cylinder was investigated by Sheikholeslami [9]. Sheikholeslami [10] studied nanofluid single-phase model flow in a porous enclosure in the presence of a constant magnetic field. Sheikholeslami [11] simulated nanofluid flow and natural convection heat transfer in an open porous cavity. Sheikholeslami and Chamkha [12] demonstrated the impact of Marangoni convection on nanofluid forced convection in the presence of Lorentz forces. Sheikholeslami et al. [13] presented a numerical simulation of nanofluid forced convection heat transfer augmentation in the presence of a magnetic field using lattice Boltzmann method. Several numerical simulations have been presented for the simulation of a nanofluid [14–93].

4.2 NANOFLUID HYDROTHERMAL ANALYSIS IN A COMPLEX SHAPED CAVITY

4.2.1 Problem Definition

Fig. 4.1 depicts the geometry and boundary conditions. The inner cylinder is hot and follows the formula:

$$r = \cos((\zeta - \zeta_0)N)A + r_{in} \tag{4.1}$$

each parameter has been shown in Fig. 4.1.

Application of Control Volume based Finite Element Method (CVFEM) for Nanofluid Flow and Heat Transfer.
DOI: https://doi.org/10.1016/B978-0-12-814152-6.00004-7

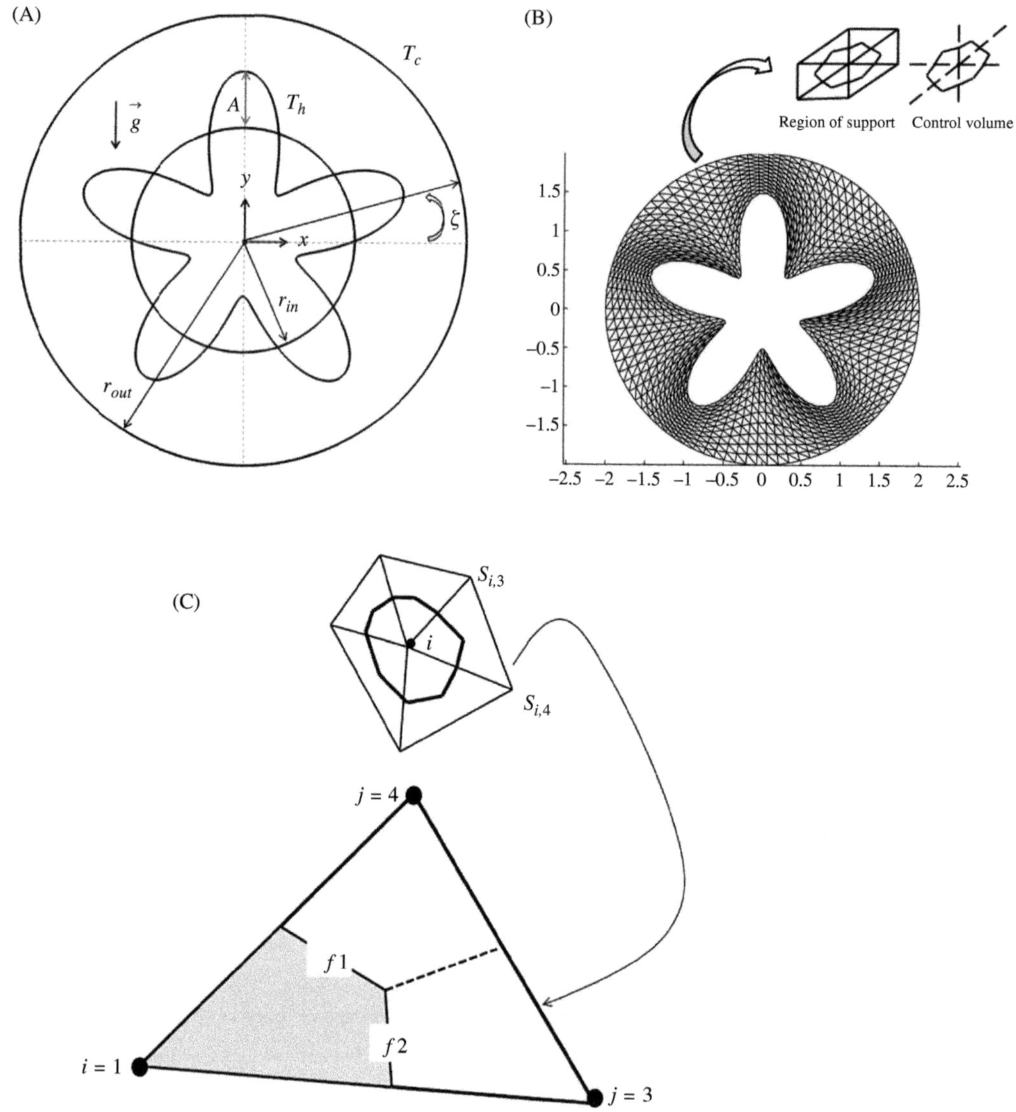

FIGURE 4.1 (A) Geometry and the boundary conditions with (B) the mesh of Geometry considered in this work; (C) A sample triangular element and its corresponding control volume.

4.2.2 Governing Equation

The governing equations for laminar CuO-water free convective heat transfer must be presented as:

$$\frac{\partial v}{\partial y} + \frac{\partial u}{\partial x} = 0 \tag{4.2}$$

$$\frac{\mu_{nf}}{\rho_{nf}}\left(\frac{\partial^2 u}{\partial x^2} + \frac{\partial^2 u}{\partial y^2}\right) - \frac{1}{\rho_{nf}}\frac{\partial P}{\partial x} = v\frac{\partial u}{\partial y} + u\frac{\partial u}{\partial x} \tag{4.3}$$

$$\frac{\mu_{nf}}{\rho_{nf}}\left(\frac{\partial^2 v}{\partial x^2} + \frac{\partial^2 v}{\partial y^2}\right) - (T_c - T)g\beta_{nf} - \frac{\partial P}{\partial y}\frac{1}{\rho_{nf}} = v\frac{\partial v}{\partial y} + u\frac{\partial v}{\partial x} \tag{4.4}$$

$$k_{nf}\left(\frac{\partial^2 T}{\partial x^2} + \frac{\partial^2 T}{\partial y^2}\right) = (\rho C_p)_{nf}\left(v\frac{\partial T}{\partial y} + u\frac{\partial T}{\partial x}\right) \tag{4.5}$$

(ρ_{nf}), $(\rho C_p)_{nf}$, and β_{nf} are defined as:

$$\rho_{nf} = \rho_f(1-\phi) + \rho_s\phi \tag{4.6}$$

$$(\rho C_p)_{nf} = (\rho C_p)_f(1-\phi) + (\rho C_p)_s\phi \tag{4.7}$$

$$\beta_{nf} = \beta_f(1-\phi) + \beta_s\phi \tag{4.8}$$

(k_{nf}) and (μ_{nf}) are obtained according to KKL model :

$$\begin{aligned} k_{nf} &= 1 + \frac{3\left(\frac{k_p}{k_f}-1\right)\phi}{\left(\frac{k_p}{k_f}+2\right)-\left(\frac{k_p}{k_f}-1\right)\phi} + 5\times10^4 g'(\phi,T,d_p)\phi\rho_f c_{p,f}\sqrt{\frac{\kappa_b T}{\rho_p d_p}} \\ g'(\phi,T,d_p) &= \left(a_6 + a_7 Ln(d_p) + a_8 Ln(\phi) + a_9 Ln(\phi)\ln(d_p) + a_{10} Ln(d_p)^2\right) \\ &+ Ln(T)\left(a_1 + a_2 Ln(d_p) + a_3 Ln(\phi) + a_4 Ln(\phi)\ln(d_p) + a_5 Ln(d_p)^2\right) \\ R_f &+ d_p/k_p = d_p/k_{p,eff},\ R_f = 4\times10^{-8} km^2/W \end{aligned} \tag{4.9}$$

$$\mu_{nf} = \frac{\mu_f}{(1-\phi)^{2.5}} + \frac{k_{Brownian}}{k_f}\times\frac{\mu_f}{\Pr} \tag{4.10}$$

All needed coefficients and properties are illustrated in Tables 4.1 and 4.2.
Vorticity and stream function should be defined as:

$$\omega + \frac{\partial u}{\partial y} - \frac{\partial v}{\partial x} = 0,\quad \frac{\partial\psi}{\partial x} = -v,\quad \frac{\partial\psi}{\partial y} = u \tag{4.11}$$

TABLE 4.1 The Coefficient Values of CuO-H_2O Nanofluid

Coefficient values	CuO-H_2O
a_1	– 26.593310846
a_2	– 0.403818333
a_3	– 33.3516805
a_4	– 1.915825591
a_5	6.4218584E-02
a_6	48.40336955
a_7	– 9.787756683
a_8	190.245610009
a_9	10.9285386565
a_{10}	– 0.72009983664

TABLE 4.2 Thermophysical Properties of Water and Nanoparticles

	$\rho(kg/m^3)$	$C_p(j/kgk)$	$k(W/m.k)$	$\beta\times10^5(K^{-1})$	$d_p(nm)$	$\sigma(\Omega\cdot m)^{-1}$
H_2O	997.1	4179	0.613	21	-	0.05
CuO	6500	540	18	29	10^{-10}	6500

By removing pressure gradient source terms from Eqs. (4.3) and (4.4), the final form of equations can be obtained as:

$$\upsilon_{nf}\left(\frac{\partial^2\omega}{\partial x^2}+\frac{\partial^2\omega}{\partial y^2}\right)-\beta_{nf}g\left(\frac{\partial T}{\partial x}\right)=\frac{\partial\omega}{\partial x}\frac{\partial\psi}{\partial y}-\frac{\partial\omega}{\partial y}\frac{\partial\psi}{\partial x} \tag{4.12}$$

$$k_{nf}\left(\frac{\partial^2 T}{\partial y^2}+\frac{\partial^2 T}{\partial x^2}\right)=\left(\frac{\partial T}{\partial x}\frac{\partial\psi}{\partial y}-\frac{\partial T}{\partial y}\frac{\partial\psi}{\partial x}\right)(\rho C_p)_{nf} \tag{4.13}$$

$$\frac{\partial^2\psi}{\partial y^2}+\frac{\partial^2\psi}{\partial x^2}+\omega=0 \tag{4.14}$$

Dimensionless parameters are as follows:

$$\Omega=\frac{\omega L^2}{\alpha_f},\Psi=\frac{\psi}{\alpha_f},\Theta=\frac{T-T_c}{T_h-T_c},Y=\frac{y}{L},\ X=\frac{x}{L} \tag{4.15}$$

Using the above formulae the governing equations change to:

$$\begin{aligned}&+Ra_f\Pr_f\left(\frac{\partial\Theta}{\partial X}\right)\left[\phi\frac{\beta_s}{\beta_f}+(1-\phi)\right]+\Pr_f/\left[(1-\phi)^{2.5}\left(\phi\frac{\rho_s}{\rho_f}+(1-\phi)\right)\right]\left(\frac{\partial^2\Omega}{\partial Y^2}+\frac{\partial^2\Omega}{\partial X^2}\right)\\&=-\frac{\partial\Psi}{\partial X}\frac{\partial\Omega}{\partial Y}+\frac{\partial\Psi}{\partial Y}\frac{\partial\Omega}{\partial X}\end{aligned} \tag{4.16}$$

$$\left(\frac{k_{nf}}{k_f}\right)/\left[(1-\phi)+\left((1-\phi)+\phi\frac{(\rho Cp)_s}{(\rho Cp)_f}\right)\right]\left(\frac{\partial^2\Theta}{\partial X^2}+\frac{\partial^2\Theta}{\partial Y^2}\right)=-\frac{\partial\Theta}{\partial Y}\frac{\partial\Psi}{\partial X}+\frac{\partial\Theta}{\partial X}\frac{\partial\Psi}{\partial Y} \tag{4.17}$$

$$\Omega+\frac{\partial^2\Psi}{\partial X^2}+\frac{\partial^2\Psi}{\partial Y^2}=0 \tag{4.18}$$

Rayleigh and Prandtl numbers are introduced as: $Ra_f=g\beta_f(T_h-T_c)L^3/(\upsilon_f\alpha_f)$, $Pr_f=\upsilon_f/\alpha_f$, respectively. Nu_{loc} and Nu_{ave} over the cold wall should be calculated as:

$$Nu_{loc}=\left(\frac{k_{nf}}{k_f}\right)\frac{\partial\Theta}{\partial r} \tag{4.19}$$

$$Nu_{ave}=\frac{1}{\gamma}\int_0^{\gamma}Nu_{loc}(\zeta)\,d\zeta \tag{4.20}$$

4.2.3 Effects of Active Parameters

CuO-water hydrothermal analysis in a cavity with hot inner sinusoidal cylinder is studied via Control Volume-based Finite Element Method (CVFEM). Different values for number of undulations ($N=2,3,5$, and 6), nanoparticles volume fraction ($\phi=0\%, 2\%$, and 4%), and Rayleigh number ($Ra=10^3$, 10^4, and 10^5) has been examined.

The influence of nanoparticles on the velocity and temperature contours is illustrated in Fig. 4.2B. The strength of flow is enhanced by adding nanoparticles in H_2O. Furthermore, augmenting the volume fraction of nanofluid causes the temperature to enhance. Influences of Rayleigh number and N on hydrothermal behavior are depicted in Figs. 4.3 and 4.4. $|\psi_{max}|$ enhances with rise of buoyancy force. At low Reynolds number, augmenting the number of undulations causes $|\psi_{max}|$ to enhance but at high Reynolds number this trend is not observed. Conduction heat transfer mode is dominant at low Reynolds number. So the isotherms are parallel to each other in those cases. Temperature gradient at the bottom of the hot wall is higher than other places. For an odd number of undulations, very small vortexes generate near the vertical centerline. The lower eddies are weaker than the upper ones because of smaller space for flow motion. As buoyancy forces augment, thermal plumes are generated on the hot wall. These plumes make isotherms to be denser and in turn the temperature gradient enhances.

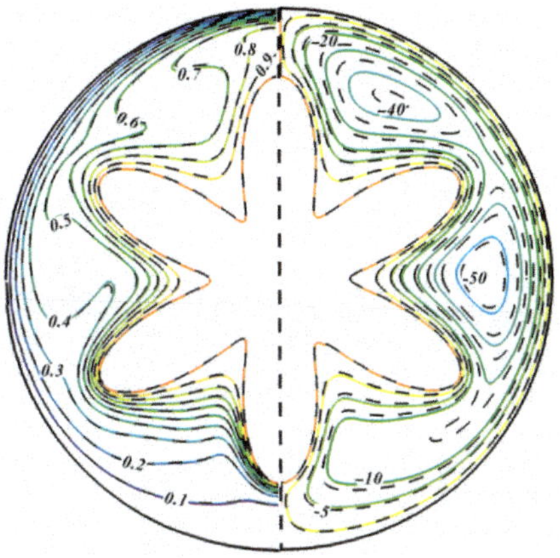

$|\Psi_{max}|_{nf} = 54.88$, $|\Psi_{max}|_{f} = 45.53$

FIGURE 4.2 Comparison of the streamlines (left) and isotherms (right) contours between nanofluid ($\phi = 0.04$) (–) and pure fluid ($\phi = 0$) (- - -) for $N = 6, A = 0.5$ at $Ra = 10^5$ and $\Pr = 6.2$.

$|\Psi_{\max}|_{nf} = 54.88$, $|\Psi_{\max}|_{f} = 45.53$

Effects of ϕ, Ra and N on Nu_{loc} and Nu_{ave} are depicted in Figs. 4.5 and 4.6, respectively. The average Nusselt number can be obtained according to the following formula:

$$\begin{aligned} Nu_{ave} &= 4.56414 - 2.58664 \log(Ra) + 0.19N - 8.075\phi + 0.021 \log(Ra)\, N + 3.791 \log(Ra)\, \phi + 0.287\, N\, \phi \\ &\quad + 0.417 \log(Ra)^2 - 0.023N^2 - 0.3074\phi^2 \end{aligned} \tag{4.21}$$

The numbers of extremum in Nu_{loc} are dependent on the number of undulations and thermal plumes. As ζ augments the conduction mechanism becomes dominant. So Nu_{loc} reduces with rise of ζ. The Nusselt number is enhanced with the rise of ϕ, Ra, and N.

4.3 NATURAL CONVECTION HEAT TRANSFER IN A NANOFLUID FILLED ENCLOSURE WITH ELLIPTIC INNER CYLINDER

4.3.1 Problem Definition

The schematic diagram and the mesh of the semiannulus enclosure used in the CVFEM program are shown in Fig. 4.7. The system consists of a circular enclosure with radius of L, within which an inclined elliptic cylinder is located and rotates from $\gamma = 0^\circ$ to 90°. T_h and T_c are the constant temperatures of the inner and outer cylinders, respectively and $T_h > T_c$. Setting a as the major axis and b as the minor axis of elliptic cylinder, the eccentricity (ε) for the inner cylinder is defined as

$$\varepsilon = \sqrt{a^2 - b^2}/a \text{ or } b = \sqrt{1 - \varepsilon^2}.\, a \tag{4.22}$$

In this study, for the inner ellipse, the eccentricity and the major axis are 0.9 and $0.8L$, respectively.

4.3.2 Governing Equation

The flow is steady, two-dimensional, laminar, and incompressible. The radiation, viscous dissipation, induced electric current, and Joule heating are neglected. The magnetic Reynolds number is assumed to be small so that the induced magnetic field can be neglected compared to the applied magnetic field. Neglecting displacement currents, induced magnetic field and using the Boussinesq approximation, the governing equations of heat transfer and fluid flow for nanofluid can be obtained as follows:

FIGURE 4.3 Comparison of the isotherms contours for different values of Rayleigh number (Ra) and number of undulations (N) at $A = 0.5$ and $\phi = 0.04$.

$$\frac{\partial u}{\partial x} + \frac{\partial v}{\partial y} = 0 \tag{4.23}$$

$$u\frac{\partial u}{\partial x} + v\frac{\partial u}{\partial y} = -\frac{1}{\rho_{nf}}\frac{\partial P}{\partial x} + \upsilon_{nf}\left(\frac{\partial^2 u}{\partial x^2} + \frac{\partial^2 u}{\partial y^2}\right) \tag{4.24}$$

$$u\frac{\partial v}{\partial x} + v\frac{\partial v}{\partial y} = -\frac{1}{\rho_{nf}}\frac{\partial P}{\partial y} + \upsilon_{nf}\left(\frac{\partial^2 v}{\partial x^2} + \frac{\partial^2 v}{\partial y^2}\right) + \beta_{nf}g(T - T_c) \tag{4.25}$$

$$u\frac{\partial T}{\partial x} + v\frac{\partial T}{\partial y} = \alpha_{nf}\left(\frac{\partial^2 T}{\partial x^2} + \frac{\partial^2 T}{\partial y^2}\right) \tag{4.26}$$

FIGURE 4.4 Comparison of the streamlines contours for different values of Rayleigh number (Ra) and number of undulations (N) at $A = 0.5$ and $\phi = 0.04$.

where the effective density (ρ_{nf}), the thermal expansion coefficient (β_{nf}), and heat capacitance $\left(\rho C_p\right)_{nf}$ of the nanofluid are defined as:

$$\begin{aligned} \rho_{nf} &= \rho_f(1-\phi) + \rho_s\phi \\ \beta_{nf} &= \beta_f(1-\phi) + \beta_s\phi \\ \left(\rho C_p\right)_{nf} &= \left(\rho C_p\right)_f(1-\phi) + \left(\rho C_p\right)_s\phi \end{aligned} \tag{4.27}$$

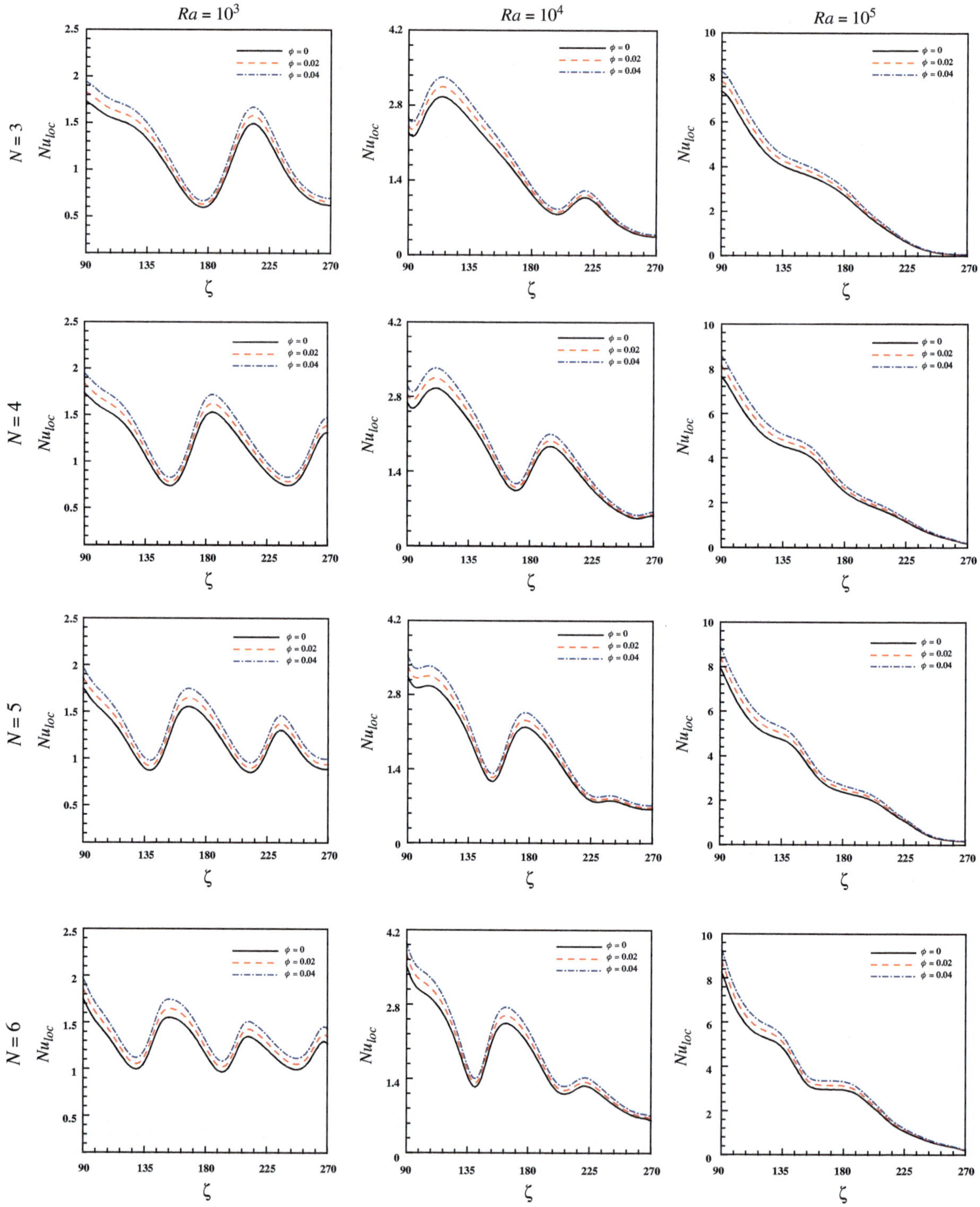

FIGURE 4.5 Effects of the nanoparticle volume fraction, number of undulations and Rayleigh number on Local Nusselt number.

The Brownian motion has a significant impact on the effective thermal conductivity.

$$k_{eff} = k_{static} + k_{Brownian} \tag{4.28}$$

$$\frac{k_{static}}{k_f} = 1 + \frac{3\left(\frac{k_p}{k_f} - 1\right)\phi}{\left(\frac{k_p}{k_f} + 2\right) - \left(\frac{k_p}{k_f} - 1\right)\phi} \tag{4.29}$$

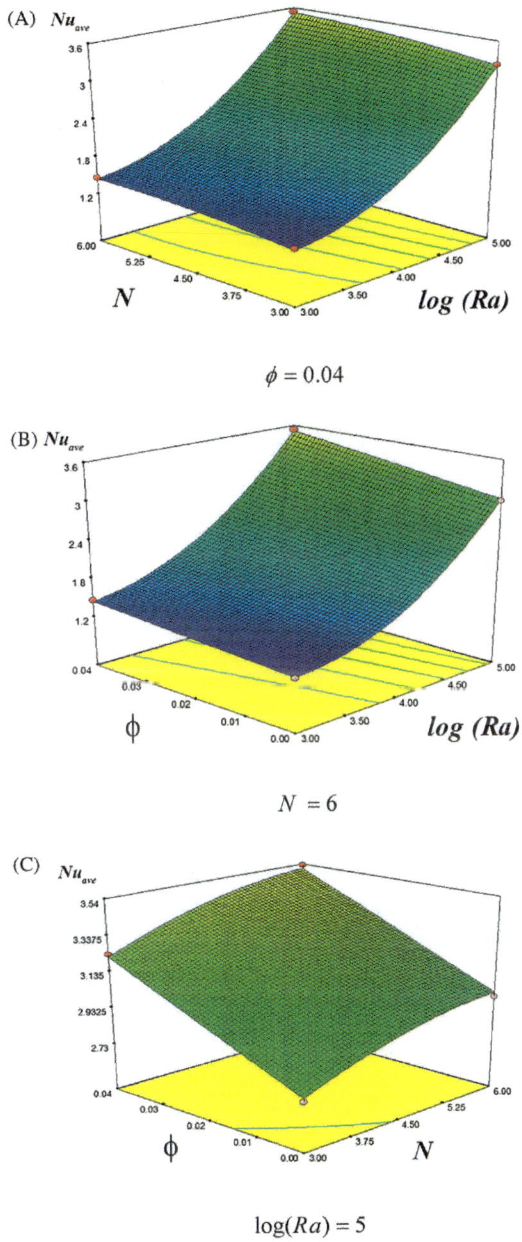

FIGURE 4.6 Effects of the nanoparticle volume fraction, number of undulations and Rayleigh number on Average Nusselt number.

where, k_{static} is the static thermal conductivity based on Maxwell classical correlation. The enhanced thermal conductivity component generated by microscale convective heat transfer of a particle's Brownian motion and affected by ambient fluid motion is obtained via simulating Stokes' flow around a sphere (nanoparticle).

$$k_{Brownian} = 5 \times 10^4 \beta \phi \rho_f c_{p,f} \sqrt{\frac{\kappa_b T}{\rho_p d_p}} f(T, \phi) \tag{4.30}$$

In recent years, there has been an increasing trend to emphasize the importance of the interfacial thermal resistance between nanoparticles and base fluids.

The KKL model combined β and f functions to develop a new g' function which captures the influences of particle diameter, temperature, and volume fraction. The empirical g'-function depends on the type of nanofluid.

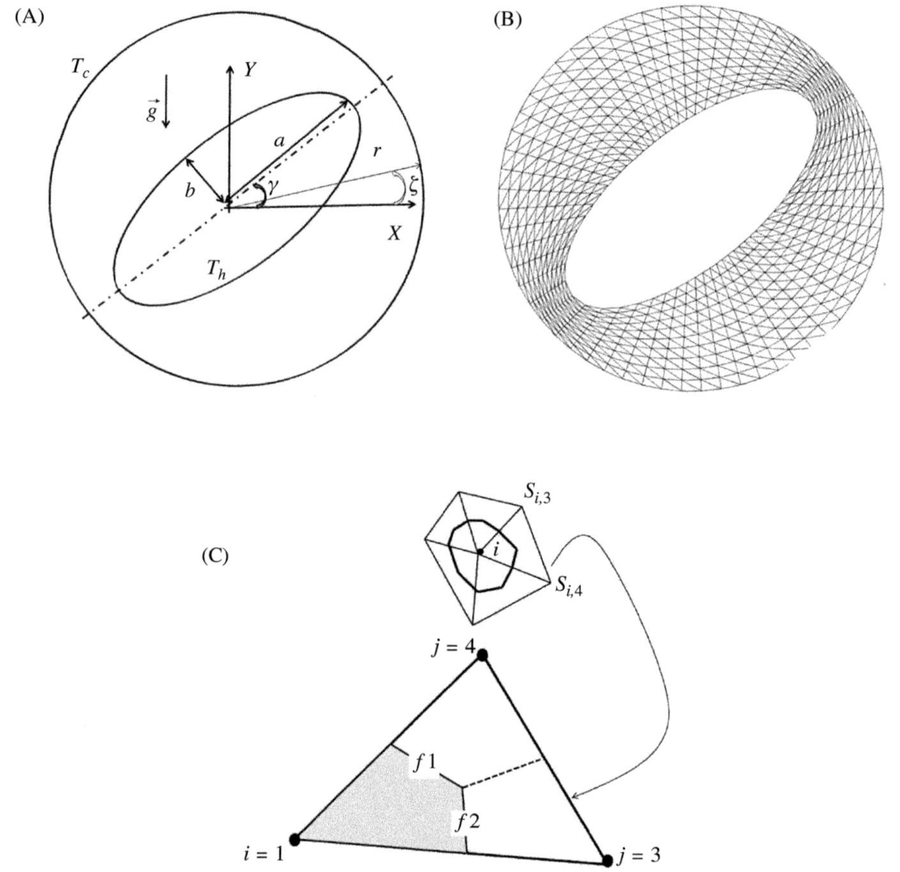

FIGURE 4.7 (A)Geometry and the boundary conditions with (B) the mesh of enclosure considered in this work; (C) A sample triangular element and its corresponding control volume.

Also, by introducing a thermal interfacial resistance $R_f = 4 \times 10^{-8} km^2/W$ the original k_p in Eq. (4.30) was replaced by a new $k_{p,eff}$ in the form:

$$R_f + \frac{d_p}{k_p} = \frac{d_p}{k_{p,eff}} \tag{4.31}$$

For different based fluids and different nanoparticles, the function should be different. Only water-based nanofluids are considered in the current study. For CuO-water nanofluids, this function follows the format:

$$\begin{aligned} g'(T, \phi, d_p) = & \left(a_1 + a_2 \ln(d_p) + a_3 \ln(\phi) + a_4 \ln(\phi) \ln(d_p) + a_5 \ln(d_p)^2\right) \ln(T) \\ & + \left(a_6 + a_7 \ln(d_p) + a_8 \ln(\phi) + a_9 \ln(\phi) \ln(d_p) + a_{10} \ln(d_p)^2\right) \end{aligned} \tag{4.32}$$

where the coefficients a_i $(i = 0.10)$ are based on the type of nanopartices, CuO-water nanofluids has an R^2 of 96% and 98%, respectively (Table 4.1). Finally, the KKL correlation is written as:

$$k_{Brownian} = 5 \times 10^4 \phi \rho_f c_{p,f} \sqrt{\frac{\kappa_b T}{\rho_p d_p}} g'(T, \phi, d_p) \tag{4.33}$$

For the effective viscosity due to micromixing in suspensions, they proposed:

$$\mu_{eff} = \mu_{static} + \mu_{Brownian} = \mu_{static} + \frac{k_{Brownian}}{k_f} \times \frac{\mu_f}{\Pr_f} \tag{4.34}$$

where $\mu_{static} = \frac{\mu_f}{(1-\phi)^{2.5}}$ is viscosity of the nanofluid, as given originally by Brinkman.

The stream function and vorticity are defined as:

$$u = \frac{\partial \psi}{\partial y}, \quad v = -\frac{\partial \psi}{\partial x}, \quad \omega = \frac{\partial v}{\partial x} - \frac{\partial u}{\partial y} \tag{4.35}$$

The stream function satisfies the continuity Eq. (4.23). The vorticity equation is obtained by eliminating the pressure between the two momentum equations, i.e., by taking the y-derivative of Eq. (4.24) and subtracting from it the x-derivative of Eq. (4.25). This gives:

$$\frac{\partial \psi}{\partial y}\frac{\partial \omega}{\partial x} - \frac{\partial \psi}{\partial x}\frac{\partial \omega}{\partial y} = \upsilon_{nf}\left(\frac{\partial^2 \omega}{\partial x^2} + \frac{\partial^2 \omega}{\partial y^2}\right) + \beta_{nf} g \left(\frac{\partial T}{\partial x}\right) \tag{4.36}$$

$$\frac{\partial \psi}{\partial y}\frac{\partial T}{\partial x} - \frac{\partial \psi}{\partial x}\frac{\partial T}{\partial y} = \alpha_{nf}\left(\frac{\partial^2 T}{\partial x^2} + \frac{\partial^2 T}{\partial y^2}\right) \tag{4.37}$$

$$\frac{\partial^2 \psi}{\partial x^2} + \frac{\partial^2 \psi}{\partial y^2} = -\omega \tag{4.38}$$

By introducing the following nondimensional variables:

$$X = \frac{x}{L}, \quad Y = \frac{y}{L}, \quad \Omega = \frac{\omega L^2}{\alpha_f}, \Psi = \frac{\psi}{\alpha_f}, \Theta = \frac{T - T_c}{T_h - T_c} \tag{4.39}$$

where in Eq. (4.39) $L = r_{out}$. Using the dimensionless parameters, the equations now become:

$$\frac{\partial \Psi}{\partial Y}\frac{\partial \Omega}{\partial X} - \frac{\partial \Psi}{\partial X}\frac{\partial \Omega}{\partial Y} = \left[\frac{\Pr_f}{(1-\phi)^{2.5}\left((1-\phi) + \phi\frac{\rho_s}{\rho_f}\right)}\right]\left(\frac{\partial^2 \Omega}{\partial X^2} + \frac{\partial^2 \Omega}{\partial Y^2}\right) + Ra_f \Pr_f \left[(1-\phi) + \phi\frac{\beta_s}{\beta_f}\right]\left(\frac{\partial \Theta}{\partial X}\right) \tag{4.40}$$

$$\frac{\partial \Psi}{\partial Y}\frac{\partial \Theta}{\partial X} - \frac{\partial \Psi}{\partial X}\frac{\partial \Theta}{\partial Y} = \left[\frac{k_{nf}/k_f}{\left((1-\phi) + \phi\frac{(\rho Cp)_s}{(\rho Cp)_f}\right)}\right]\left(\frac{\partial^2 \Theta}{\partial X^2} + \frac{\partial^2 \Theta}{\partial Y^2}\right) \tag{4.41}$$

$$\frac{\partial^2 \Psi}{\partial X^2} + \frac{\partial^2 \Psi}{\partial Y^2} = -\Omega \tag{4.42}$$

where $Ra_f = g\beta_f L^3 (T_h - T_c)/(\alpha_f \upsilon_f)$ is the Rayleigh number for the base fluid and $Pr_f = \upsilon_f/\alpha_f$ is the Prandtl number for the base fluid. The thermophysical properties of the nanofluid are given in Table 4.2. The boundary conditions as shown in Fig. 4.7 are:

$$\begin{aligned} \Theta &= 1.0 \quad \text{on the inner elliptic boundary} \\ \Theta &= 0.0 \quad \text{on the outer circular boundary} \\ \Psi &= 0.0 \quad \text{on all solid boundaries} \end{aligned} \tag{4.43}$$

The values of vorticity on the boundary of the enclosure can be obtained using the stream function formulation and the known velocity conditions during the iterative solution procedure.

The local Nusselt number of the nanofluid along the cold wall can be expressed as:

$$Nu_{loc} = \left(\frac{k_{nf}}{k_f}\right)\frac{\partial \Theta}{\partial r} \tag{4.44}$$

where r is the radial direction. The average Nusselt number on the cold circular wall is evaluated as:

$$Nu_{ave} = \frac{1}{2\pi}\int_0^{2\pi} Nu_{loc}(\zeta)\, d\zeta \tag{4.45}$$

To estimate the enhancement of heat transfer between the case of $\phi = 0.04$ and the pure fluid (base fluid) case, the enhancement is defined as:

$$E = \frac{Nu(\phi = 0.04) - Nu(basefluid)}{Nu(basefluid)} \times 100 \tag{4.46}$$

4.3.3 Effects of Active Parameters

In this section, natural convection heat transfer between a circular enclosure and an elliptic cylinder filled with nanofluid is investigated numerically using the control volume-based finite element method (CVFEM). The fluid in the enclosure is CuO-water nanofluid. Calculations are carried out for constant eccentricity ($\varepsilon = 0.9$), major axis ($a = 0.8L$), and Prandtl ($Pr = 6.2$) at different values of Rayleigh numbers ($Ra = 10^3, 10^4, 10^5$) and inclined angle of elliptic inner cylinder ($\gamma = 0^\circ, 30^\circ, 60^\circ, 90^\circ$) and volume fraction of nanoparticles ($\phi = 0\%, 2\%$, and 4%).

The effects of adding nanoparticles on the streamlines and isotherms are shown in Fig. 4.8. As can be seen, the absolute values of stream functions indicate that the strength of flow increases with an increase in the volume fraction of nanofluid. Also it is obvious to say that adding nanoparticle leads to an increase in Nusselt number due to increment in thermal conductivity. Since the reduction in temperature gradient due to the presence of nanoparticles is much smaller than the thermal conductivity ratio, therefore an enhancement in the Nusselt number occurs by adding nanoparticle.

Fig. 4.9 shows the isotherms and streamlines for different values of Ra and γ at $\phi = 0.04$. At $Ra = 10^3$ the isotherms are parallel to each other and take the form of the inner and outer wall and the stream function magnitude is relatively small which indicates the domination of the conduction heat transfer mechanism. Increasing the inclination angle leads to an increase in the absolute value of the maximum stream function ($|\Psi_{\max}|$) at this Rayleigh number. At $\gamma = 0$ degrees the streamlines and isotherms are symmetric with respect to the vertical centerline of the enclosure. Each pair of cells has two cells. The top vortex is stronger because at this area the hot surface is located beneath the cold one which helps the flow circulation, whereas the arrangement of the presence of the cold wall under the hot cylinder at the bottom of the enclosure resists the flow circulation. As $\gamma = 0$ degree increases these two pairs of cells merge together and form two single cells at different locations inside the enclosure. At $\gamma = 90$ degrees again the streamlines patterns become symmetric with respect to the vertical centerline of the enclosure. At $Ra = 10^4$ the thermal plumes start to appear over the hot elliptic cylinder. Besides, the stream function values start to grow which shows that the convection heat transfer mechanism has become comparable with conduction. At $\lambda = 0$ degree the temperature contour becomes stratified beneath the hot cylinder. With the increase of γ to 30 degrees a secondary vortex appears at the top of the enclosure and the thermal plume slants to the left because of more available space at this area. As the inclination angle of the inner cylinder increases further this secondary vortex disappears and the streamlines show two main vortices in the enclosure. Also it can be found that effect of increasing inclination angle on $|\Psi_{\max}|$ becomes less pronounced at $\gamma > 30$ degrees. When the Rayleigh number increases up to $Ra = 10^5$ isotherms are totally distorted at the top of the enclosure while it is stratified at the bottom of the enclosure, which shows the heat transfer mechanism is dominated by convection. The thermal plume is completely formed which impinges the hot fluid to the cold wall of the enclosure. As seen,

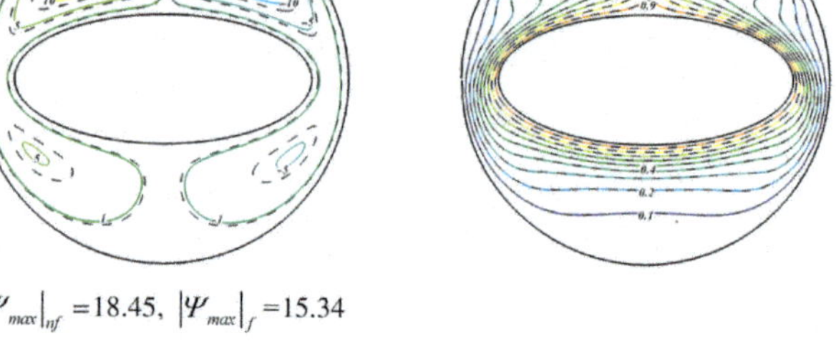

$|\Psi_{max}|_{nf} = 18.45$, $|\Psi_{max}|_{f} = 15.34$

FIGURE 4.8 Comparison of the streamlines (left) and isotherms (right) contours between nanofluid ($\phi = 0.04$) (—) and pure fluid ($\phi = 0$) (- - -) for $\varepsilon = 0.9$, $a = 0.8L$ at $Ra = 10^5$ and $\mathrm{Pr} = 6.2$.

the secondary vortex exists at $\gamma > 30$ and 60 degrees $\gamma > 30$ *degrees* and 60° which can slant the thermal plume to the left. Also, as seen in Fig. 4.9 the maximum value of $|\Psi_{\max}|$ occurs at $\gamma > 60$ degrees.

Fig. 4.10 shows the distribution of local Nusselt numbers along the surface of the outer circular wall for different inclination angle and Rayleigh number. As the Rayleigh number increases the local Nusselt number increases

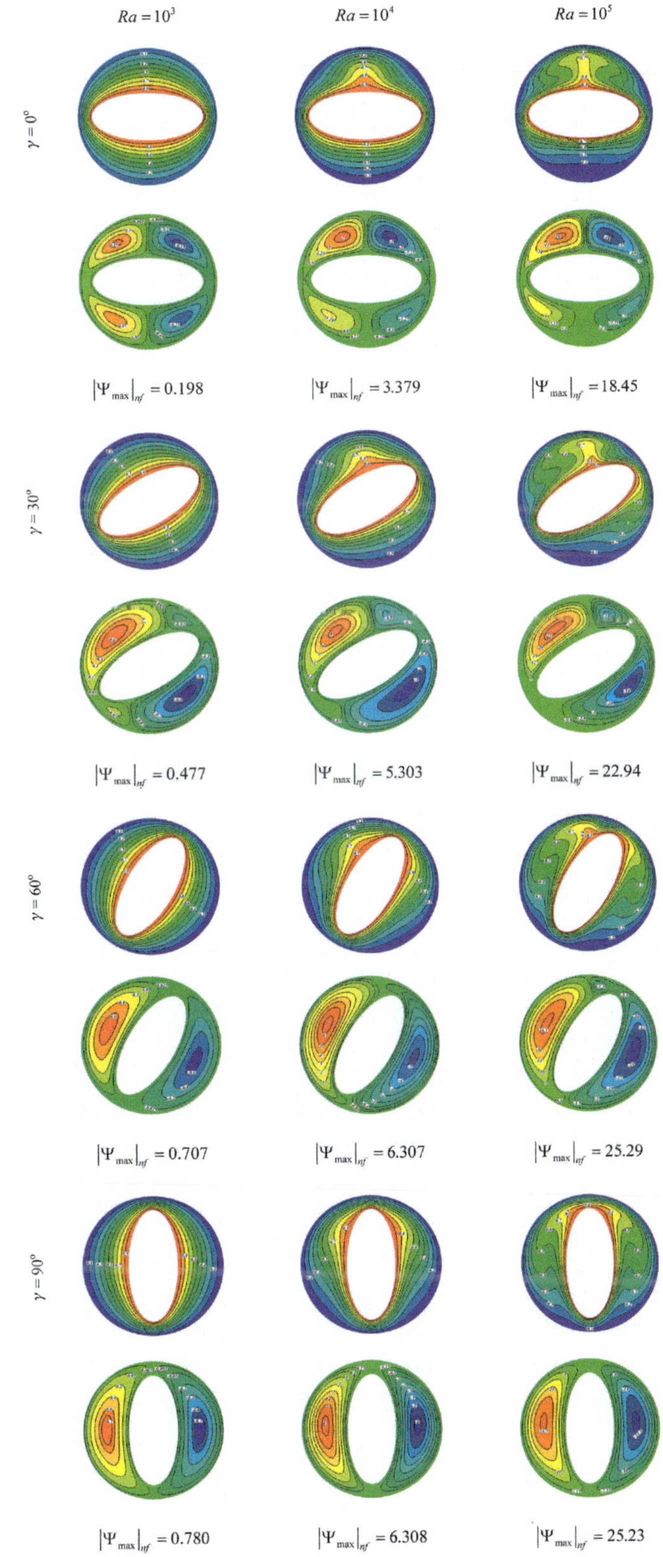

FIGURE 4.9 Isotherms (up) and streamlines (down) contours for different values of *Ra* and γ at $\phi = 0.04$

due to increasing convection effect. At $Ra = 10^3$ the Nu_{loc} profile is nearly symmetrical with respect to the horizontal centerline. As the Rayleigh number enhances (e.g., $Ra = 10^4$ and $Ra = 10^5$), the Nu_{loc} profile is no longer symmetrical and the local Nusselt number is considerably small over the bottom wall of the enclosure. These local Nusselt number profiles are more complex due to the presence of the thermal plume at the vicinity of the top wall of the enclosure.

Fig. 4.11 depicts the effects of the inclination angle and Rayleigh number on the average Nusselt number. For all values of γ, the average Nusselt number increases with the increase of Rayleigh number. As seen, the inclination angle has no significant effect on the average Nusselt number at low Rayleigh number. While for higher

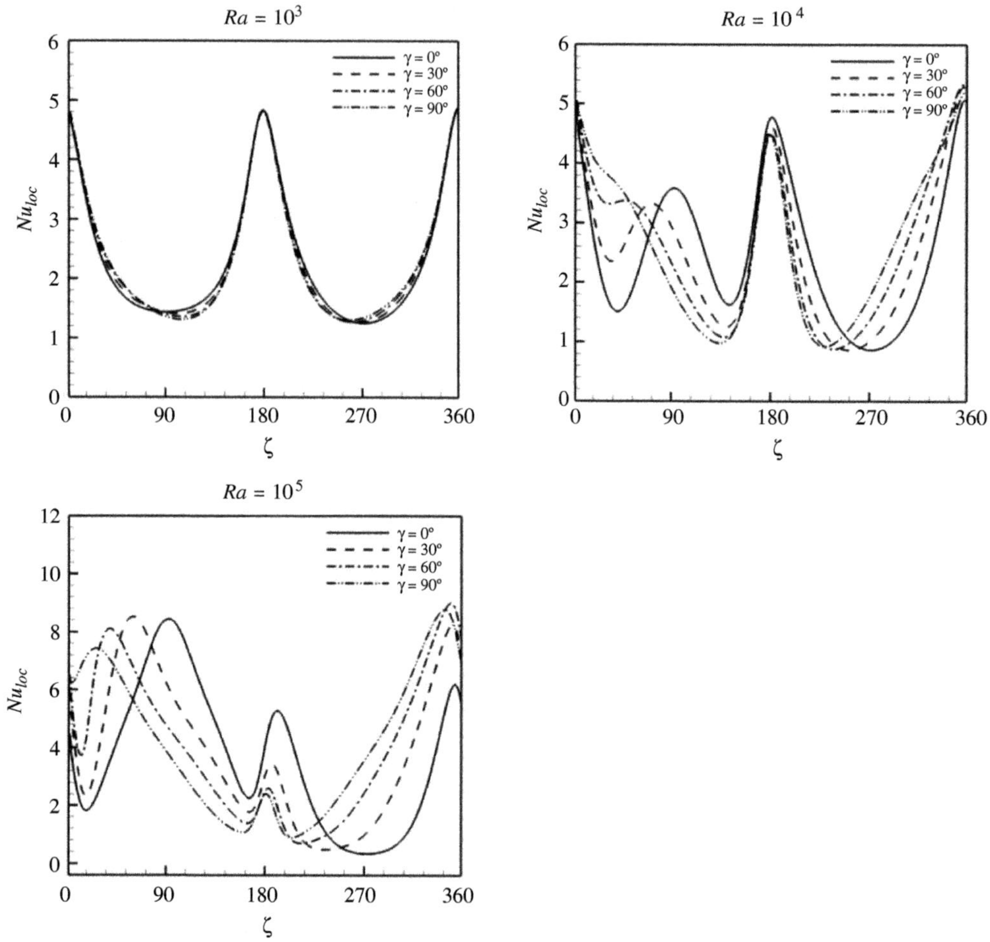

FIGURE 4.10 Effects of the inclination angle and Rayleigh number on Local Nusselt number when $\phi = 0.04$.

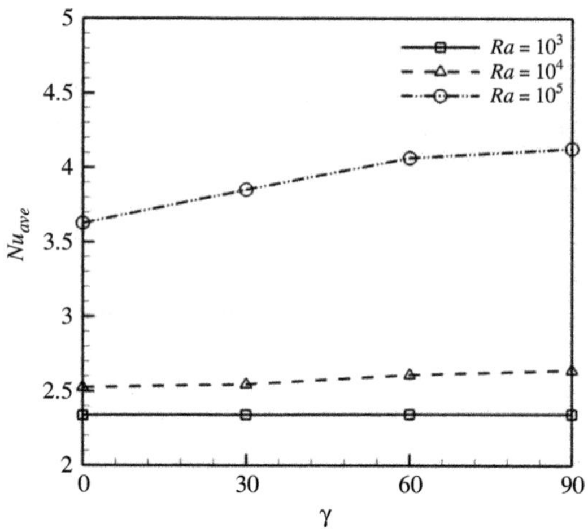

FIGURE 4.11 Effects of the inclination angle and Rayleigh number on average Nusselt number when $\phi = 0.04$.

TABLE 4.3 Constant Coefficient for Using Eq. (4.47)

a_{ij}	$i=1$	$i=2$	$i=3$	$i=4$	$i=5$	$i=6$
j = 1	7.564525	− 3.38154	− 0.25235	0.506444	0.019843	2.008494
j = 2	0.484538	− 0.31956	0.711765	− 0.04265	0.023808	0.1894

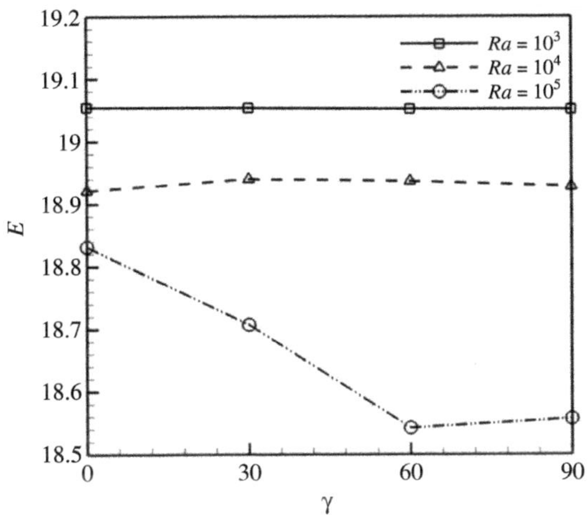

FIGURE 4.12 Effects of the inclination angle and Rayleigh number on the ratio of heat transfer enhancement due to addition of nanoparticles when Pr = 6.2.

values of the Rayleigh number, an increase in inclination angle leads to an increase in the average Nusselt number. Also, the corresponding polynomial representation of such a model for the Nusselt number is as follows:

$$\begin{aligned} Nu &= a_{12} + a_{22}\gamma + a_{32}Y_1 + a_{42}\gamma^2 + a_{52}{Y_1}^2 + a_{62}\gamma Y_1 \\ Y_1 &= a_{11} + a_{21}Ra^* + a_{31}\phi + a_{41}Ra^{*2} + a_{51}\phi^2 + a_{61}\phi Ra^* \end{aligned} \tag{4.47}$$

Also a_{ij} can be found in Table 4.3, e.g., a_{21} is equal to (−1.52352).

The heat transfer enhancement ratio due to the addition of nanoparticles for different values of inclination angle and Rayleigh number is shown in Fig. 4.12. It can be found that the effect of nanoparticles is more pronounced at low Rayleigh number than at high Rayleigh number because of the greater enhancement rate. This observation can be explained by noting that at low Rayleigh number the heat transfer is dominated by conduction. Therefore, the addition of high thermal conductivity nanoparticles will increase the conduction and make the enhancement more effective. It is an interesting observation that the enhancement in heat transfer for $Ra = 10^3$ is the same for all inclination angles. At $Ra = 10^4$ the maximum heat transfer enhancement ratio is obtained at $\gamma = 30°$. At high Rayleigh number ($Ra = 10^5$), at first the percentage of the heat transfer enhancement decreases and then increases. At this Rayleigh number the minimum heat transfer enhancement ratio occurs at $\gamma = 90°$.

4.4 NANOFLUID FREE CONVECTION HEAT TRANSFER IN A TILTED CAVITY

4.4.1 Problem Definition

Fig. 4.13 illustrates the important geometric parameters of current geometry. Also a sample mesh is presented. Constant heat flux is applied on the inner surface and the other conditions are clear in Fig. 4.13.

4.4.2 Governing Equation

The flow is laminar and steady and two-dimensional. Boussinesq approximation has been considered for momentum equations. Nanofluid is assume to be an homogenous fluid. According to these assumptions the governing equations can be presented as:

$$\frac{\partial v}{\partial y} + \frac{\partial u}{\partial x} = 0 \tag{4.48}$$

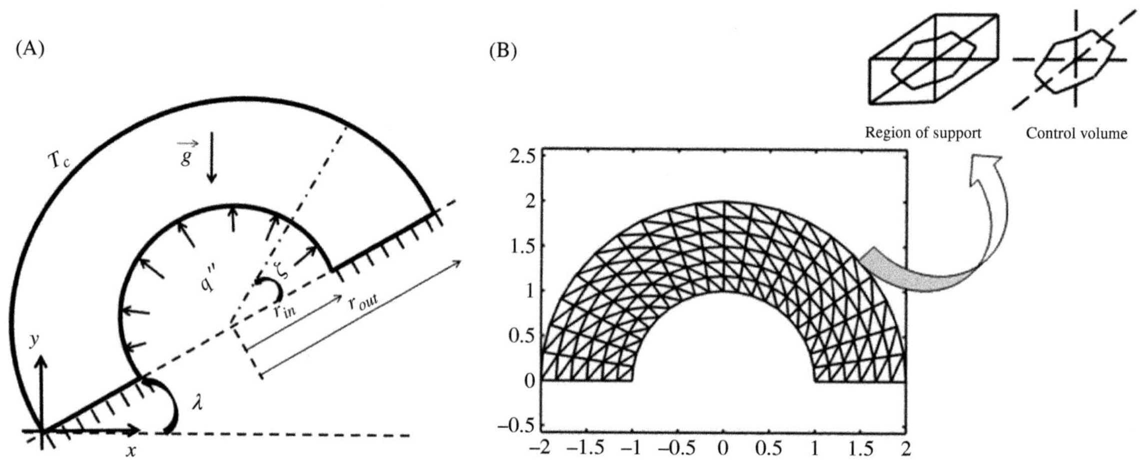

FIGURE 4.13 (A) Geometry and the boundary conditions with (B) the mesh of half-annulus enclosure considered in this work.

$$v\frac{\partial u}{\partial y}+u\frac{\partial u}{\partial x}+\frac{1}{\rho_{nf}}\frac{\partial P}{\partial x}=\frac{\mu_{nf}}{\rho_{nf}}\left(\frac{\partial^2 u}{\partial y^2}+\frac{\partial^2 u}{\partial x^2}\right) \tag{4.49}$$

$$u\frac{\partial v}{\partial x}+v\frac{\partial v}{\partial y}+\frac{\partial P}{\partial y}\frac{1}{\rho_{nf}}=\frac{\mu_{nf}}{\rho_{nf}}\left(\frac{\partial^2 v}{\partial x^2}+\frac{\partial^2 v}{\partial y^2}\right)-(T_c-T)g\beta_{nf} \tag{4.50}$$

$$\frac{k_{nf}}{(\rho C_p)_{nf}}\left(\frac{\partial^2 T}{\partial x^2}+\frac{\partial^2 T}{\partial y^2}\right)=\left(v\frac{\partial T}{\partial y}+u\frac{\partial T}{\partial x}\right) \tag{4.51}$$

β_{nf}, $(\rho C_p)_{nf}$, and ρ_{nf} are defined as:

$$\beta_{nf}=\beta_f(1-\phi)+\beta_s\phi \tag{4.52}$$

$$(\rho C_p)_{nf}=(\rho C_p)_f(1-\phi)+(\rho C_p)_s\phi \tag{4.53}$$

$$\rho_{nf}=\rho_f(1-\phi)+\rho_s\phi \tag{4.54}$$

(k_{nf}) and (μ_{nf}) are obtained according to the KKL model:

$$\begin{aligned}
k_{nf}&=1+\frac{3\left(\frac{k_p}{k_f}-1\right)\phi}{\left(\frac{k_p}{k_f}+2\right)-\left(\frac{k_p}{k_f}-1\right)\phi}+5\times10^4 g'(\phi,T,d_p)\phi\rho_f c_{p,f}\sqrt{\frac{\kappa_b T}{\rho_p d_p}}\\
g'(\phi,T,d_p)&=\left(a_1+a_2Ln(d_p)+a_3Ln(\phi)+a_4Ln(\phi)\ln(d_p)+a_5Ln(d_p)^2\right)Ln(T)\\
&+\left(a_6+a_7Ln(d_p)+a_8Ln(\phi)+a_9\ln(d_p)Ln(\phi)+a_{10}Ln(d_p)^2\right)\\
R_f&=d_p/k_{p,eff}-d_p/k_p,\ R_f=4\times10^{-8}km^2/W
\end{aligned} \tag{4.55}$$

$$\mu_{nf}=\frac{\mu_f}{(1-\phi)^{2.5}}+\frac{k_{Brownian}}{k_f}\times\frac{\mu_f}{\Pr} \tag{4.56}$$

All necessary coefficients and properties are illustrated in Tables 4.1 and 4.2.
Vorticity and stream function should be defined as:

$$\omega+\frac{\partial u}{\partial y}-\frac{\partial v}{\partial x}=0,\quad \frac{\partial\psi}{\partial x}=-v,\quad \frac{\partial\psi}{\partial y}=u \tag{4.57}$$

By removing pressure gradient source terms from Eqs. (4.49) and (4.50), the final form of the equations can be obtained as:

$$\upsilon_{nf}\left(\frac{\partial^2\omega}{\partial x^2}+\frac{\partial^2\omega}{\partial y^2}\right)-\beta_{nf}g\left(\frac{\partial T}{\partial x}\right)=\frac{\partial\omega}{\partial x}\frac{\partial\psi}{\partial y}-\frac{\partial\omega}{\partial y}\frac{\partial\psi}{\partial x} \tag{4.58}$$

$$k_{nf}\left(\frac{\partial^2 T}{\partial y^2}+\frac{\partial^2 T}{\partial x^2}\right)=\left(\frac{\partial T}{\partial x}\frac{\partial\psi}{\partial y}-\frac{\partial T}{\partial y}\frac{\partial\psi}{\partial x}\right)(\rho C_p)_{nf} \tag{4.59}$$

$$\frac{\partial^2\psi}{\partial y^2}+\frac{\partial^2\psi}{\partial x^2}+\omega=0 \tag{4.60}$$

Dimensionless parameters are as follows:

$$\Theta=\frac{T-T_c}{(q''L/k_f)},(Y,X)=\frac{(y,x)}{L},\quad(\Psi,\ \Omega)=\frac{(\psi,\ \omega L^2)}{\alpha_f} \tag{4.61}$$

Using the above formulae the governing equations change to:

$$+Ra_f\Pr_f\left(\frac{\partial\Theta}{\partial X}\right)\left[\phi\frac{\beta_s}{\beta_f}+(1-\phi)\right]+\Pr_f/\left[(1-\phi)^{2.5}\left(\phi\frac{\rho_s}{\rho_f}+(1-\phi)\right)\right]\left(\frac{\partial^2\Omega}{\partial Y^2}+\frac{\partial^2\Omega}{\partial X^2}\right)$$
$$=-\frac{\partial\Psi}{\partial X}\frac{\partial\Omega}{\partial Y}+\frac{\partial\Psi}{\partial Y}\frac{\partial\Omega}{\partial X} \tag{4.62}$$

$$\left(\frac{\partial^2\Theta}{\partial X^2}+\frac{\partial^2\Theta}{\partial Y^2}\right)\left(\frac{k_{nf}}{k_f}\right)/\left[(1-\phi)+\left((1-\phi)+\phi\frac{(\rho Cp)_s}{(\rho Cp)_f}\right)\right]=\frac{\partial\Psi}{\partial Y}\frac{\partial\Theta}{\partial X}-\frac{\partial\Psi}{\partial X}\frac{\partial\Theta}{\partial Y} \tag{4.63}$$

$$+\frac{\partial^2\Psi}{\partial Y^2}+\frac{\partial^2\Psi}{\partial X^2}=-\Omega \tag{4.64}$$

Prandtl and Rayleigh numbers are introduced as: $Pr_f=\upsilon_f/\alpha_f$, $Ra_f=g\beta_f L^4 q''/(k_f\,\alpha_f\upsilon_f)$, respectively. Nu_{loc} and Nu_{ave} over the hot wall should be calculated as:

$$Nu_{loc}=\left(\frac{k_{nf}}{k_f}\right)\frac{1}{\Theta} \tag{4.65}$$

$$Nu_{ave}=\frac{1}{\pi}\int_0^{\pi}Nu_{loc}(\zeta)\,d\zeta \tag{4.66}$$

4.4.3 Effects of Active Parameters

In this work, the influence of CuO nanoparticle on the hydrothermal behavior in an inclined half-annulus is studied using CVFEM. Simulations are made for different values of Rayleigh number ($Ra=10^3, 10^4$, and 10^5), inclination angle ($\lambda=0$, 45, 90, 135, and 180 degrees), and volume fraction of CuO ($\phi=0\%$ and 4%).

Fig. 4.14 depicts the effect of CuO nanoparticles on hydrothermal characteristics. The nanofluid velocity augments due to enhancing the solid movements. The temperature gradient augments with the rise of the volume fraction of CuO. Figs. 4.15 and 4.16 illustrate the impact of inclination angle, Rayleigh number, and CuO volume fraction. Conduction mode dominates at low Rayleigh number. Two rotating vortices exist in streamlines. As buoyancy forces grow, the temperature gradient is enhanced and a thermal plume appears near the vertical centerline. The core of vortices moves upward. At $\lambda=45$ degrees, the counterclockwise vortex is stronger than the other one due to more space for circulation. At $\lambda=90$ degrees, the two main eddies merge into one counterclockwise eddy. The streamlines and isotherms at $\lambda=135$ and 180 degrees are depicted in Fig. 4.16. As seen, the impact of the Rayleigh number on the size of eddies for $\lambda=135$ degrees is opposite to that of $\lambda=45$ *degrees*. By increasing buoyancy forces, the size of the secondary eddy decreases and finally at $Ra=10^5$ a small vortex appears on the top of the secondary eddy, which turns in a clockwise direction.

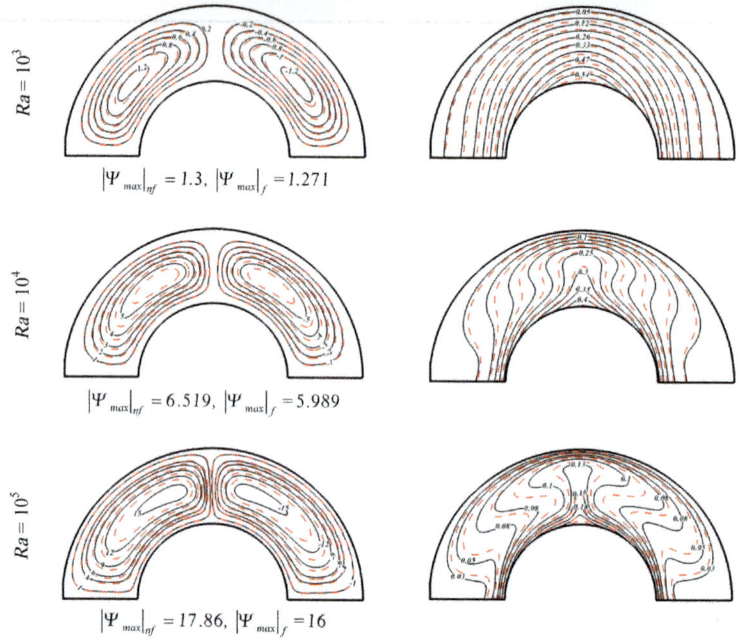

FIGURE 4.14 Comparison of the streamlines (left) and isotherms (right) contours between nanofluid ($\phi = 0.04$) (- - -) and pure fluid ($\phi = 0$) (—)for different values of Ra at $\lambda = 0^{\circ}$ and $\text{Pr} = 6.2$.

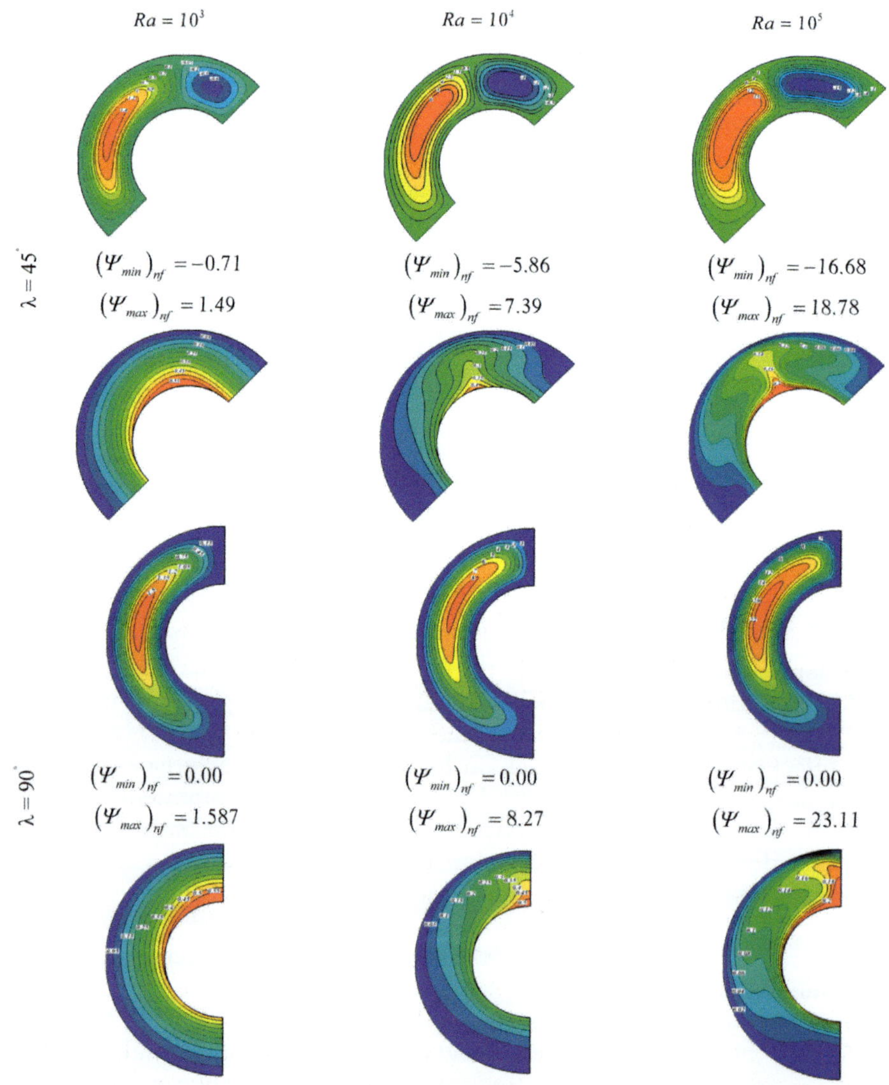

FIGURE 4.15 Isotherms (down) and streamlines (up) contours for different values of Rayleigh number when $\lambda = 45^{\circ}$ and 90° for CuO-water nanofluid ($\phi = 0.04$).

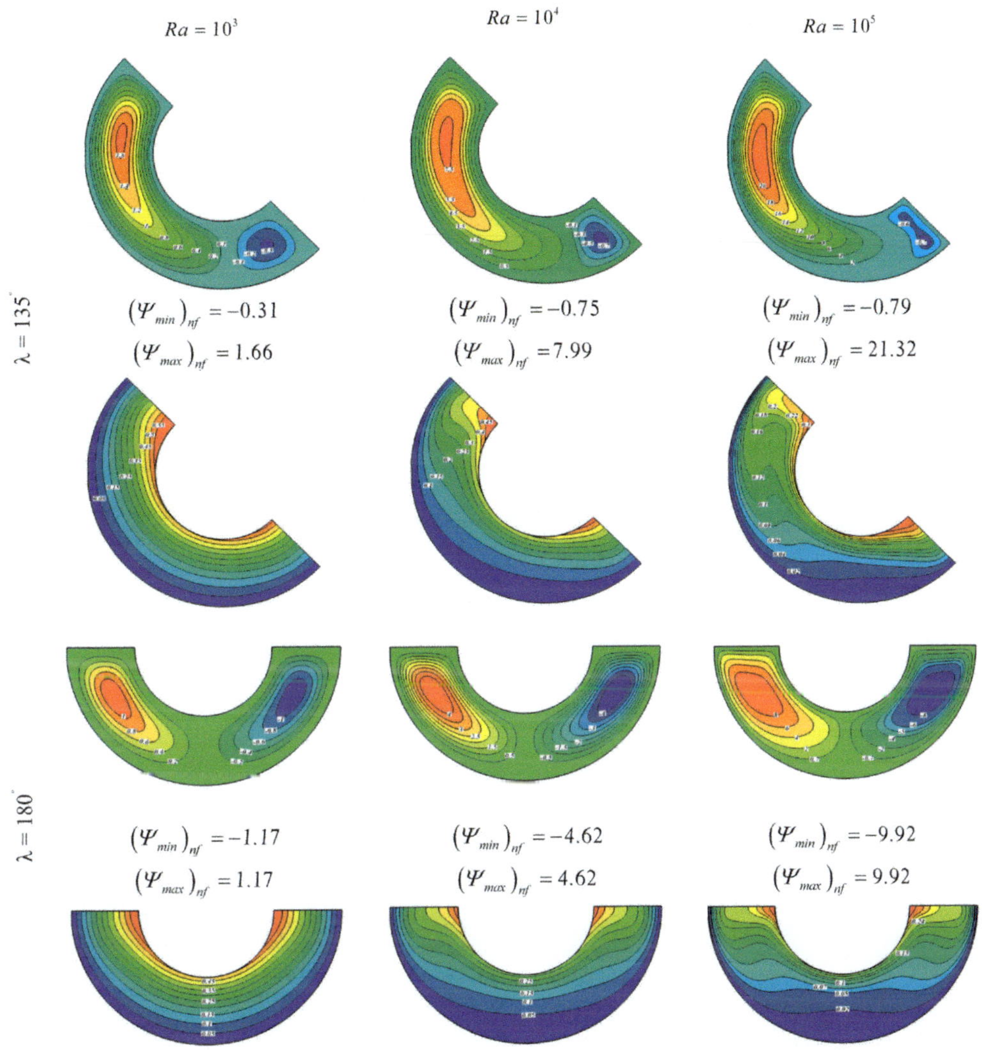

FIGURE 4.16 Isotherms (down) and streamlines (up) contours for different values of Rayleigh number when $\lambda = 135^\circ$ and 180° for CuO-water nanofluid ($\phi = 0.04$).

This behavior of the thermal boundary layer may be due to the stronger flow circulation at higher Rayleigh numbers at these points. The isotherms and streamlines for $\lambda = 180$ degrees are almost symmetric to the vertical centerline. However, enhancing the buoyancy forces has no significant influence on the location of the two main eddies' core. In fact, when the hot circular wall locates above the cold one, the impact of convection on the velocity and temperature is less pronounced. Similar to the inclination angle of 135 degrees, the temperature gradient becomes thinner near the inner wall except at the two inner corners.

The influence of important parameters on Nu_{loc} and Nu_{ave} are depicted in Figs. 4.17–4.19. The correlation for this parameter is as follows:

$$\begin{aligned} Nu_{ave} &= 2.12 - 1.2\log(Ra) + 0.01\lambda - 11.81\phi - 4.88\,\lambda\log(Ra) \\ &+ 7.94\,\phi\log(Ra) - 0.027\lambda\phi + 0.38\log(Ra)^2 + 2.19\lambda^2 + 0.07\phi^2 \end{aligned} \tag{4.67}$$

Response surface methodology (RSM) is utilized to find this correlation. This method presented polynomials formulation according to input data. As volume fraction of CuO augments, Nu_{loc} augments. Nusselt number augments with rise of Ra. Nu_{loc} profiles are symmetric to $\zeta = 90^\circ$ when $\lambda = 0^\circ$ and 180°. At $\lambda = 0^\circ$ and 45° minimum amount of Nu_{loc} is located at $\zeta = 90^\circ$ and 55° respectively. Also Fig. 4.19 shows that for $\lambda = 135^\circ$ and 180° maximum amount of Nu_{loc} is located at $\zeta = 100^\circ$ and 90° respectively. The Nusselt number decreases with

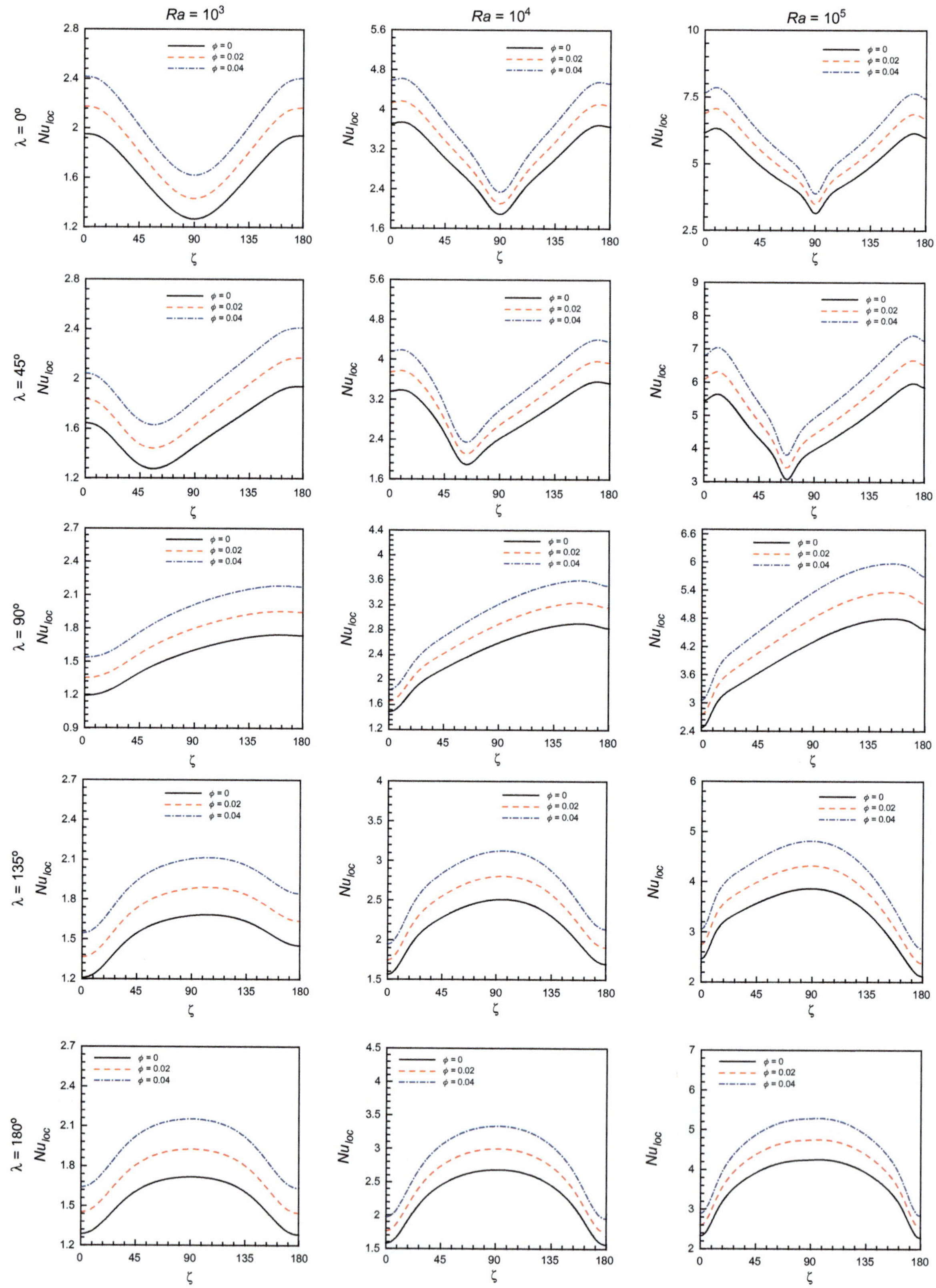

FIGURE 4.17 Effects of the nanoparticle volume fraction, Rayleigh number and inclination angle for CuO-water nanofluids on Local Nusselt number.

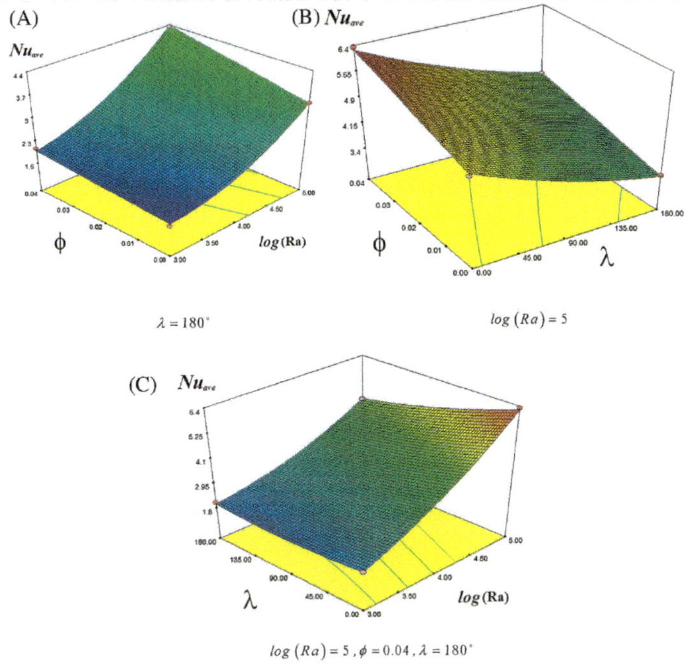

FIGURE 4.18 Effects of the nanoparticle volume fraction, Rayleigh number and inclination angle for CuO-water nanofluids on average Nusselt number.

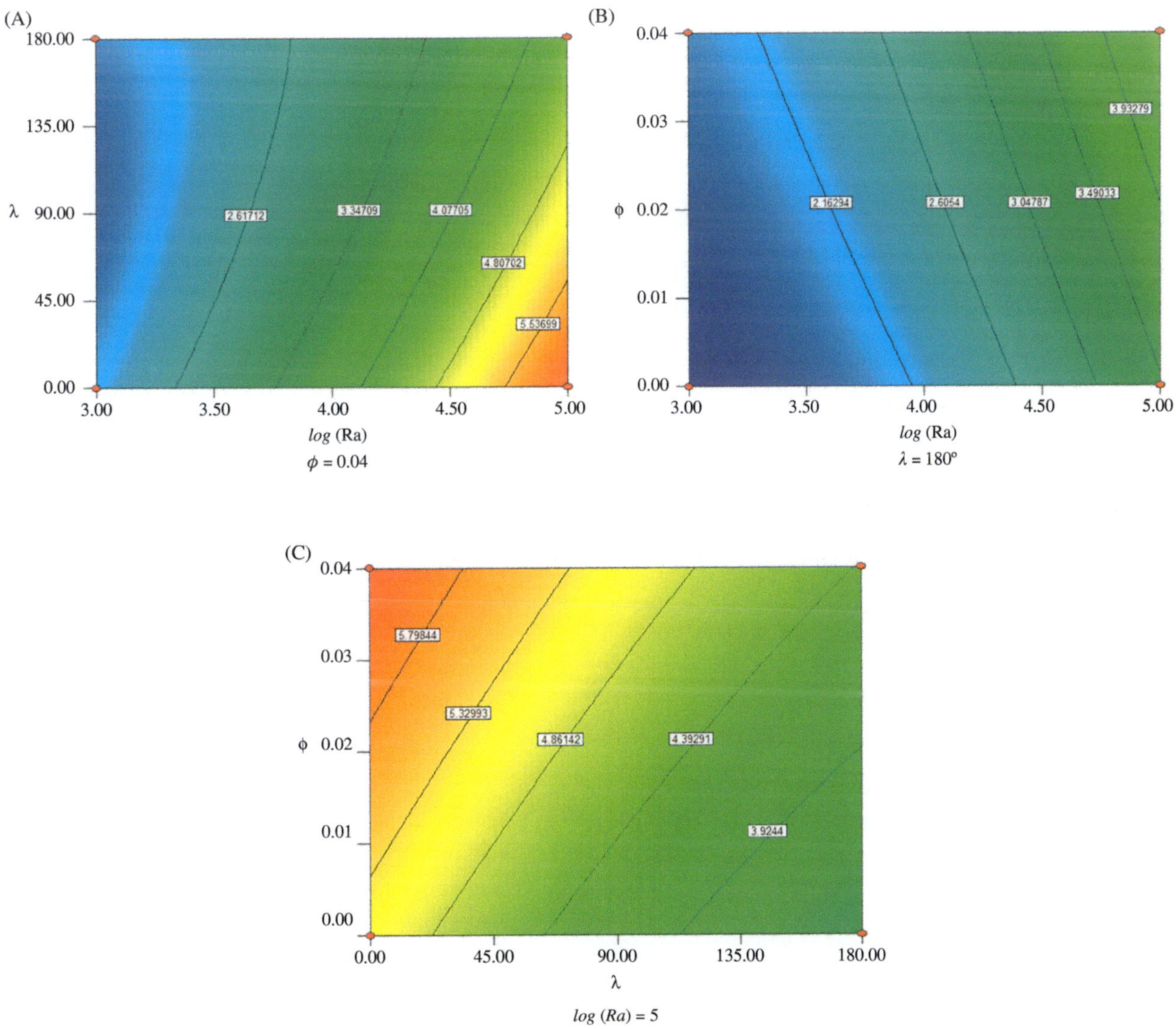

FIGURE 4.19 Contour plots of average Nusselt number.

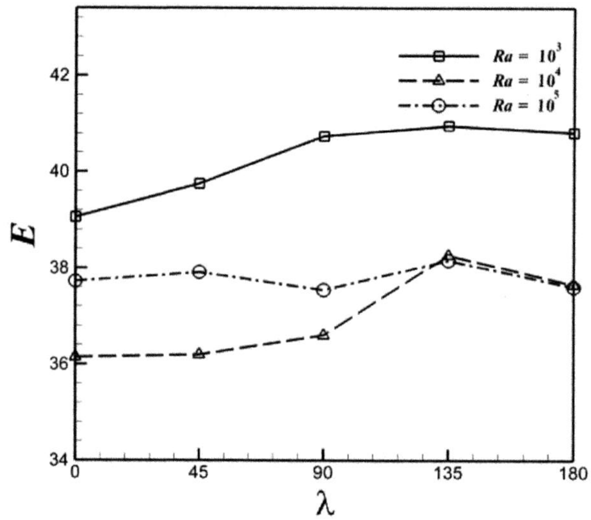

FIGURE 4.20 Effects of λ and Ra on the heat transfer enhancement due to addition of nanoparticles when $Pr = 6.2$.

enhancement of the inclination angle. Fig. 4.20 shows the influence of λ and Ra on the heat transfer enhancement. This parameter can be calculated as:

$$E = \frac{Nu(\phi = 0.04) - Nu(basefluid)}{Nu(basefluid)} \times 100 \tag{4.68}$$

According to this figure, heat transfer enhancement increases with the increase of the inclination angle. Higher Rayleigh number leads to lower values of E. At low Rayleigh number and $\lambda = 180$ *degrees*, the dominant mechanism is conduction and in this way adding nanoparticles has a significant impact on thermal conductivity.

References

[1] S. Choi, in: D.A. Siginer, H.P. Wang (Eds.), Enhancing Thermal Conductivity of Fluids With Nanoparticles in Developments and Applications of Non-Newtonian Flows, 66, ASME, 1995, pp. 99–105.

[2] K. Khanafer, K. Vafai, M. Lightstone, Buoyancy-driven heat transfer enhancement in a two-dimensional enclosure utilizing nanofluids, Int. J. Heat Mass Transfer 46 (2003) 3639–3653.

[3] M. Sheikholeslami, S. Abelman, Two phase simulation of nanofluid flow and heat transfer in an annulus in the presence of an axial magnetic field, IEEE Trans. Nanotechnol. 14 (3) (2015) 561–569.

[4] M. Sheikholeslami, M.M. Rashidi, Effect of space dependent magnetic field on free convection of Fe_3O_4-water nanofluid, J. Taiwan Inst. Chem. Eng. 56 (2015) 6–15.

[5] M. Sheikholeslami, D.D. Ganji, Ferrohydrodynamic and Magnetohydrodynamic effects on ferrofluid flow and convective heat transfer, Energy 75 (2014) 400–410.

[6] M. Sheikholeslami, S.A. Shehzad, Magnetohydrodynamic nanofluid convective flow in a porous enclosure by means of LBM, Int. J. Heat Mass Transfer 113 (2017) 796–805.

[7] M. Sheikholeslami, Magnetohydrodynamic nanofluid forced convection in a porous lid driven cubic cavity using Lattice Boltzmann Method, J. Mol. Liquids 231 (2017) 555–565.

[8] M.S. Kandelousi, KKL correlation for simulation of nanofluid flow and heat transfer in a permeable channel, Phys. Lett. A 378 (45) (2014) 3331–3339.

[9] M. Sheikholeslami, Effect of uniform suction on nanofluid flow and heat transfer over a cylinder, J Braz. Soc. Mech. Sci. Eng. 37 (2015) 1623–1633.

[10] M. Sheikholeslami, Lattice Boltzmann Method simulation of MHD non-Darcy nanofluid free convection, Phys. B 516 (2017) 55–71.

[11] M. Sheikholeslami, Influence of magnetic field on nanofluid free convection in an open porous cavity by means of Lattice Boltzmann Method, J. Mol. Liquids 234 (2017) 364–374.

[12] M. Sheikholeslami, A.J. Chamkha, Influence of Lorentz forces on nanofluid forced convection considering Marangoni convection, J. Mol. Liquids 225 (2017) 750–757.

[13] M. Sheikholeslami, T. Hayat, A. Alsaedi, Numerical simulation of nanofluid forced convection heat transfer improvement in existence of magnetic field using Lattice Boltzmann Method, Int. J. Heat Mass Transfer 108 (2017) 1870–1883.

[14] M. Sheikholeslami, R. Ellahi, Three dimensional mesoscopic simulation of magnetic field effect on natural convection of nanofluid, Int. J. Heat Mass Transfer 89 (2015) 799–808.

[15] M. Sheikholeslami, R. Ellahi, Simulation of ferrofluid flow for magnetic drug targeting using Lattice Boltzmann method, J. Zeitschrift Fur Naturforschung A 70 (2) (2015) 115–124.

[16] M. Sheikholeslami, M. Seyednezhad, Lattice Boltzmann Method simulation for CuO-water nanofluid flow in a porous enclosure with hot obstacle, J. Mol. Liquids 243 (2017) 249–256.

[17] M. Sheikholeslami, M. Sadoughi, Mesoscopic method for MHD nanofluid flow inside a porous cavity considering various shapes of nanoparticles, Int. J. Heat Mass Transfer 113 (2017) 106–114.

[18] M. Sheikholeslami, A. Zeeshan, Mesoscopic simulation of CuO-H2O nanofluid in a porous enclosure with elliptic heat source, Int. J. Hydrogen Energy 42 (22) (2017) 15393–15402.

[19] M. Sheikholeslami, S.A. Shehzad, CVFEM for influence of external magnetic source on $Fe_3O_4 - H_2O$ nanofluid behavior in a permeable cavity considering shape effect, Int. J. Heat Mass Transfer 115 (2017) 180–191.

[20] M. Sheikholeslami, M. Seyednezhad, Nanofluid heat transfer in a permeable enclosure in presence of variable magnetic field by means of CVFEM, Int. J. Heat Mass Transfer 114 (2017) 1169–1180.

[21] M. Sheikholeslami, H.B. Rokni, Free convection of CuO-H_2O nanofluid in a curved porous enclosure using mesoscopic approach, Int. J. Hydrogen Energy 42 (22) (2017) 14942–14949.

[22] M. Sheikholeslami, H.B. Rokni, Influence of melting surface on MHD nanofluid flow by means of two phase model, Chin. J. Phys. 55 (2017) 1352–1360.

[23] M. Sheikholeslami, H.B. Rokni, Influence of melting surface on MHD nanofluid flow by means of two phase model, Chinese J. Phys. 55 (2017) 1352–1360.

[24] M. Sheikholeslami, H.B. Rokni, Effect of melting heat transfer on nanofluid flow in existence of magnetic field considering Buongiorno Model, Chinese J. Phys. 55 (2017) 1115–1126.

[25] M. Sheikholeslami, Influence of Lorentz forces on nanofluid flow in a porous cylinder considering Darcy model, J. Mol. Liquids 225 (2017) 903–912.

[26] M. Sheikholeslami, CVFEM for magnetic nanofluid convective heat transfer in a porous curved enclosure, Eur. Phys. J. Plus 131 (2016) 413. Available from: https://doi.org/10.1140/epjp/i2016-16413-y.

[27] M. Sheikholeslami, Influence of Coulomb forces on Fe_3O_4-H_2O nanofluid thermal improvement, Int. J. Hydrogen Energy 42 (2017) 821–829.

[28] M. Sheikholeslami, Magnetic field influence on nanofluid thermal radiation in a cavity with tilted elliptic inner cylinder, J. Mol. Liquids 229 (2017) 137–147.

[29] M. Sheikholeslami, Numerical simulation of magnetic nanofluid natural convection in porous media, Phys. Lett. A 381 (2017) 494–503.

[30] M. Sheikholeslami, Magnetic source impact on nanofluid heat transfer using CVFEM, Neural Comput. Applicat. (2016). Available from: https://doi.org/10.1007/s00521-016-2740-7.

[31] M. Sheikholeslami, CuO-water nanofluid free convection in a porous cavity considering Darcy law, Europ. Phys. J. Plus 132 (2017) 55. Available from: https://doi.org/10.1140/epjp/i2017-11330-3.

[32] M. Sheikholeslami, Numerical investigation of MHD nanofluid free convective heat transfer in a porous tilted enclosure, Eng. Comput. 34 (6) (2017) 1939–1955.

[33] M.S. Kandelousi, Effect of spatially variable magnetic field on ferrofluid flow and heat transfer considering constant heat flux boundary condition, Europ. Phys. J. Plus (2014) 129–248. Available from: https://doi.org/10.1140/epjp/i2014-14248-2.

[34] M. Sheikholeslami, Numerical simulation for external magnetic field influence on Fe_3O_4-water nanofluid forced convection, Engineering Computations, In press.

[35] M. Sheikholeslami, Influence of Lorentz forces on nanofluid flow in a porous cavity by means of Non- Darcy model, Eng. Comput. 34 (8) (2017) 2651–2667. Available from: https://doi.org/10.1108/EC-01-2017-0008.

[36] M. Sheikholeslami, H.B. Rokni, Nanofluid convective heat transfer intensification in a porous circular cylinder, Chem. Eng. Processing: Process Intensif. 120 (2017) 93–104.

[37] M. Sheikholeslami, M.K. Sadoughi, Numerical modeling for Fe_3O_4 -water nanofluid flow in porous medium considering MFD viscosity, J. Mol. Liquids 242 (2017) 255–264.

[38] M. Sheikholeslami, M. Shamlooei, Magnetic source influence on nanofluid flow in porous medium considering shape factor effect, Phys. Lett. A 381 (2017) 3071–3078.

[39] M. Sheikholeslami, M. Shamlooei, Convective flow of nanofluid inside a lid driven porous cavity using CVFEM, Phys. B: Condensed Matter 521 (2017) 239–250.

[40] M. Sheikholeslami, K. Vajravelu, Forced convection heat transfer in Fe_3O_4- ethylene glycol nanofluid under the influence of Coulomb force, J. Mol. Liquids 233 (2017) 203–210.

[41] M. Sheikholeslami, H.B. Rokni, Influence of EFD viscosity on nanofluid forced convection in a cavity with sinusoidal wall, J. Mol. Liquids 232 (2017) 390–395.

[42] M. Sheikholeslami, H.B. Rokni, Magnetic nanofluid natural convection in the presence of thermal radiation considering variable viscosity, Europ. Phys. J. Plus 132 (2017) 238. Available from: https://doi.org/10.1140/epjp/i2017-11498-4.

[43] M. Sheikholeslami, H.B. Rokni, Numerical simulation for impact of Coulomb force on nanofluid heat transfer in a porous enclosure in presence of thermal radiation, Int. J. Heat Mass Transfer 118 (2018) 823–831.

[44] M. Sheikholeslami, M. Shamlooei, R. Moradi, Fe_3O_4- Ethylene glycol nanofluid forced convection inside a porous enclosure in existence of Coulomb force, J. Mol. Liquids 249 (2018) 429–437.

[45] M. Sheikholeslami, S.A. Shehzad, Numerical analysis of $Fe_3O_4 - H_2O$ nanofluid flow in permeable media under the effect of external magnetic source, Int. J. Heat Mass Transfer 118 (2018) 182–192.

[46] M. Sheikholeslami, S.A. Shehzad, Magnetohydrodynamic nanofluid convection in a porous enclosure considering heat flux boundary condition, Int. J. Heat Mass Transfer 106 (2017) 1261–1269.

[47] M. Sheikholeslami, S.A. Shehzad, Thermal radiation of ferrofluid in existence of Lorentz forces considering variable viscosity, Int. J. Heat Mass Transfer 109 (2017) 82–92.

[48] M. Sheikholeslami, H.B. Rokni, Numerical modeling of nanofluid natural convection in a semi annulus in existence of Lorentz force, Comput. Methods Appl. Mech. Eng. 317 (2017) 419–430.
[49] M. Sheikholeslami, K. Vajravelu, Nanofluid flow and heat transfer in a cavity with variable magnetic field, Appl. Math. Comput. 298 (2017) 272–282.
[50] M. Sheikholeslami, H.B. Rokni, Magnetohydrodynamic CuO-water nanofluid in a porous complex shaped enclosure, ASME J. Therm. Sci. Eng. Applicat. 9 (4) (2017) 041007. Available from: https://doi.org/10.1115/1.4035973.
[51] M. Sheikholeslami, H.B. Rokni, Nanofluid two phase model analysis in existence of induced magnetic field, Int. J. Heat Mass Transfer 107 (2017) 288–299.
[52] M. Sheikholeslami, T. Hayat, A. Alsaedi, S. Abelman, Numerical analysis of EHD nanofluid force convective heat transfer considering electric field dependent viscosity, Int. J. Heat Mass Transfer 108 (2017) 2558–2565.
[53] M. Sheikholeslami, Magnetic field influence on CuO -H_2O nanofluid convective flow in a permeable cavity considering various shapes for nanoparticles, Int. J. Hydrogen Energy 42 (2017) 19611–19621.
[54] M. Sheikholeslami, M.M. Bhatti, Forced convection of nanofluid in presence of constant magnetic field considering shape effects of nanoparticles, Int. J. Heat Mass Transfer 111 (2017) 1039–1049.
[55] M. Sheikholeslami, A. Zeeshan, Analysis of flow and heat transfer in water based nanofluid due to magnetic field in a porous enclosure with constant heat flux using CVFEM, Comput. Methods Appl. Mech. Eng. 320 (2017) 68–81.
[56] M. Sheikholeslami, M. Shamlooei, Fe_3O_4- H_2O nanofluid natural convection in presence of thermal radiation, Int. J. Hydrogen Energy 42 (9) (2017) 5708–5718.
[57] M. Sheikholeslami, M.M. Bhatti, Active method for nanofluid heat transfer enhancement by means of EHD, Int. J. Heat Mass Transfer 109 (2017) 115–122.
[58] M. Sheikholeslami, T. Hayat, A. Alsaedi, Numerical study for external magnetic source influence on water based nanofluid convective heat transfer, Int. J. Heat Mass Transfer 106 (2017) 745–755.
[59] M. Sheikholeslami, A.J. Chamkha, Flow and convective heat transfer of a ferro-nanofluid in a double-sided lid-driven cavity with a wavy wall in the presence of a variable magnetic field, Numer. Heat Transfer, Part A 69 (10) (2016) 1186–1200. Available from: https://doi.org/10.1080/10407782.2015.1125709.
[60] M. Sheikholeslami, A.J. Chamkha, Electrohydrodynamic free convection heat transfer of a nanofluid in a semi-annulus enclosure with a sinusoidal wall, Numer. Heat Transfer, Part A 69 (7) (2016) 781–793. Available from: http://dx.doi.org/10.1080/10407782.2015.1090819.
[61] M. Sheikholeslami, M.K. Sadoughi, Simulation of CuO- water nanofluid heat transfer enhancement in presence of melting surface, Int. J. Heat Mass Transfer 116 (2018) 909–919.
[62] M. Sheikholeslami, H.B. Rokni, Simulation of nanofluid heat transfer in presence of magnetic field: a review, Int. J. Heat Mass Transfer 115 (2017) 1203–1233.
[63] M. Sheikholeslami, H.B. Rokni, Melting heat transfer influence on nanofluid flow inside a cavity in existence of magnetic field, Int. J. Heat Mass Transfer 114 (2017) 517–526.
[64] M. Sheikholeslami, M. Nimafar, D.D. Ganji, M. Pouyandehmehr, CuO-H_2O nanofluid hydrothermal analysis in a complex shaped cavity, Int. J. Hydrogen Energy 41 (2016) 17837–17845.
[65] M. Sheikholeslami, R. Ellahi, M. Hassan, Soheil Soleimani, A study of natural convection heat transfer in a nanofluid filled enclosure with elliptic inner cylinder, Int. J. Numer. Methods Heat Fluid Flow 24 (2014) 8.
[66] M. Sheikholeslami, T. Hayat, A. Alsaedi, MHD free convection of Al_2O_3–water nanofluid considering thermal radiation: a numerical study, Int. J. Heat Mass Transfer 96 (2016) 513–524.
[67] M. Sheikholeslami, K. Vajravelu, M.M. Rashidi, Forced convection heat transfer in a semi annulus under the influence of a variable magnetic field, Int. J. Heat Mass Transfer 92 (2016) 339–348.
[68] M. Sheikholeslami, Numerical simulation for solidification in a LHTESS by means of Nano-enhanced PCM, J. Taiwan Inst. Chem. Eng. 86 (2018) 25–41.
[69] M. Sheikholeslami, Numerical modeling of Nano enhanced PCM solidification in an enclosure with metallic fin, J. Mol. Liq. 259 (2018) 424–438.
[70] M. Sheikholeslami, A. Ghasemi, Solidification heat transfer of nanofluid in existence of thermal radiation by means of FEM, Int. J. Heat Mass Transfer 123 (2018) 418–431.
[71] M. Sheikholeslami, S.A. Shehzad, CVFEM simulation for nanofluid migration in a porous medium using Darcy model, Int. J. Heat Mass Transfer 122 (2018) 1264–1271.
[72] M. Sheikholeslami, M. Darzi, M.K. Sadoughi, Heat transfer improvement and Pressure Drop during condensation of refrigerant-based Nanofluid; An Experimental Procedure, Int. J. Heat Mass Transfer 122 (2018) 643–650.
[73] M. Sheikholeslami, H.B. Rokni, CVFEM for effect of Lorentz forces on nanofluid flow in a porous complex shaped enclosure by means of Non-equilibrium model, J. Mol. Liq. 254 (2018) 446–462.
[74] M. Sheikholeslami, S.A. Shehzad, Zhixiong Li, Water based nanofluid free convection heat transfer in a three dimensional porous cavity with hot sphere obstacle in existence of Lorenz forces, Int. J. Heat Mass Transfer 125 (2018) 375–386.
[75] M. Sheikholeslami, M. Jafaryar, D.D. Ganji, Z. Li, Exergy loss analysis for nanofluid forced convection heat transfer in a pipe with modified turbulators, J. Mol. Liq. 262 (2018) 104–110.
[76] M. Sheikholeslami, M. Jafaryar, Z. Li, Nanofluid turbulent convective flow in a circular duct with helical turbulators considering CuO nanoparticles, Int. J. Heat Mass Transfer 124 (2018) 980–989.
[77] M. Sheikholeslami, Z. Li, M. Shamlooei, Nanofluid MHD natural convection through a porous complex shaped cavity considering thermal radiation, Phys. Lett. A 382 (2018) 1615–1632.
[78] M. Sheikholeslami, H.B. Rokni, Magnetic nanofluid flow and convective heat transfer in a porous cavity considering Brownian motion effects, Phys. Fluids, Volume 30, Issue 1, 10.1063/1.5012517, https://doi.org/10.1063/1.5012517.
[79] M. Sheikholeslami, S.A. Shehzad, Simulation of water based nanofluid convective flow inside a porous enclosure via Non-equilibrium model, Int. J. Heat Mass Transfer 120 (2018) 1200–1212.

[80] M. Sheikholeslami, S.A. Shehzad, Non-Darcy free convection of Fe_3O_4-water nanoliquid in a complex shaped enclosure under impact of uniform Lorentz force, Chin. J. Phys. 56 (2018) 270–281.
[81] M. Sheikholeslami, M. Seyednezhad, Simulation of nanofluid flow and natural convection in a porous media under the influence of electric field using CVFEM, Int. J. Heat Mass Transfer 120 (2018) 772–781.
[82] M. Sheikholeslami, M. Shamlooei, R. Moradi, Numerical simulation for heat transfer intensification of nanofluid in a permeable curved enclosure considering shape effect of Fe_3O_4 nanoparticles, Chem. Eng. Process. Process Intensif. 124 (2018) 71–82.
[83] M. Sheikholeslami, T. Hayat, T. Muhammad, A. Alsaedi, MHD forced convection flow of nanofluid in a porous cavity with hot elliptic obstacle by means of Lattice Boltzmann method, Int. J. Mech. Sci. 135 (2018) 532–540.
[84] M. Sheikholeslami, Numerical investigation of nanofluid free convection under the influence of electric field in a porous enclosure, J. Mol. Liq. 249 (2018) 1212–1221.
[85] M. Sheikholeslami, CuO-water nanofluid flow due to magnetic field inside a porous media considering Brownian motion, J. Mol. Liq. 249 (2018) 921–929.
[86] M. Sheikholeslami, Numerical investigation for CuO-H_2O nanofluid flow in a porous channel with magnetic field using mesoscopic method, J. Mol. Liq. 249 (2018) 739–746.
[87] M. Sheikholeslami, M. Shamlooei, R. Moradi, Fe_3O_4-ethylene glycol nanofluid forced convection inside a porous enclosure in existence of Coulomb force, J. Mol. Liq. 249 (2018) 429–437.
[88] M. Sheikholeslami, T. Hayat, A. Alsaedi, Numerical simulation for forced convection flow of MHD CuO-H_2O nanofluid inside a cavity by means of LBM, J. Mol. Liq. 249 (2018) 941–948.
[89] M. Sheikholeslami, H.R. Kataria, A.S. Mittal, Effect of thermal diffusion and heat-generation on MHD nanofluid flow past an oscillating vertical plate through porous medium, J. Mol. Liq. 257 (2018) 12–25.
[90] M. Sheikholeslami, S.A. Shehzad, Z. Li, Nanofluid heat transfer intensification in a permeable channel due to magnetic field using Lattice Boltzmann method, Phys. B (2018). Available from: https://doi.org/10.1016/j.physb.2018.03.036.
[91] M. Sheikholeslami, A. Zeeshan, Numerical simulation of Fe_3O_4-water nanofluid flow in a non-Darcy porous media, Int. J. Numer. Methods Heat Fluid Flow 28 (3) (2018) 641–660. Available from: https://doi.org/10.1108/HFF-04-2017-0160.
[92] M. Sheikholeslami, D.D. Ganji, Influence of electric field on Fe_3O_4-water nanofluid radiative and convective heat transfer in a permeable enclosure, J. Mol. Liq. 250 (2018) 404–412.
[93] M. Sheikholeslami, M.B. Gerdroodbary, S.V. Mousavi, D.D. Ganji, R. Moradi, Heat transfer enhancement of ferrofluid inside an 90o elbow channel by non-uniform magnetic field, J. Magnet. Magnet. Mater. 460 (2018) 302–311.

CHAPTER

5

Buongiorno Model for Nanofluid Treatment Using CVFEM

5.1 INTRODUCTION

Control of heat transfer in many energy systems is crucial due to the increase in energy prices. In recent years, nanofluids technology has been proposed and studied by some researchers experimentally or numerically to control heat transfer in a process. The nanofluid can be applied to engineering problems, such as heat exchangers, cooling of electronic equipments, and chemical processes. Almost all of the researchers assumed that nanofluids are to be treated as the common pure fluid and conventional equations of mass, momentum, and energy are used and the only effect of the nanofluid is its thermal conductivity and viscosity, which are obtained from the theoretical models or experimental data. These researchers assumed that nanoparticles are in thermal equilibrium and there aren't any slip velocities between the nanoparticles and fluid molecules, thus they have a uniform mixture of nanoparticles

Single-phase models assumed that there aren't any slip velocities between nanoparticles and fluid molecules and assumed that the nanoparticle concentration is uniform. It is believed that in natural convection of nanofluids, the nanoparticles could not accompany fluid molecules due to some slip mechanisms such as Brownian motion and thermophoresis, so the volume fraction of nanofluids may not be uniform anymore and there would be a variable concentration of nanoparticles in a mixture. Nield and Kuznetsov [1] studied the natural convection in a horizontal layer of a porous medium saturated by a nanofluid. Their analysis revealed that for a typical nanofluid (with large Lewis number) the prime effect of the nanofluids is via a buoyancy effect coupled with the conservation of nanoparticles, the contribution of nanoparticles to the thermal energy equation being a second-order effect. Khan and Pop [2] published a paper on boundary-layer flow of a nanofluid past a stretching sheet. They indicated that the reduced Nusselt number is a decreasing function of each dimensionless number. Cheng [3] studied the natural convection boundary layer flow over a truncated cone embedded in a porous medium saturated by a nanofluid with constant wall temperature. He showed that an increase in the thermophoresis parameter or the Brownian parameter tends to decrease the local Nusselt number. Sheikholeslami et al. [4] presented the effects of Thermophoresis and Brownian motion on nanofluid treatment in the presence of a magnetic field. Sheikholeslami et al. [5] studied the two-phase modeling of nanofluid free convection heat transfer in an L-shaped enclosure. Sheikholeslami and Abelman [6] simulated nanofluid flow and heat transfer in an annulus in the presence of an axial magnetic field by means of two-phase model. Sheikholeslami and Rokni [7] reported a numerical simulation for melting heat transfer effect on nanofluid flow by means of Buongiorno model. Sheikholeslami and Rokni [8] utilized the two-phase model for effect of Lorentz forces on nanofluid flow in the existence of melting heat transfer. Recently, numerical simulations about the application of nanofluid have been presented in various papers [9–91].

5.2 BUONGIORNO MODEL FOR NANOFLUID FLOW AND HEAT TRANSFER USING HEATLINE ANALYSIS

5.2.1 Problem Definition

The physical model and the corresponding triangular elements used in the present CVFEM program are shown in Fig. 5.1. The inner and outer walls are maintained at constant temperatures T_h and T_c, respectively. The shape of the inner cylinder profile is assumed to mimic the following pattern

Application of Control Volume based Finite Element Method (CVFEM) for Nanofluid Flow and Heat Transfer.
DOI: https://doi.org/10.1016/B978-0-12-814152-6.00005-9

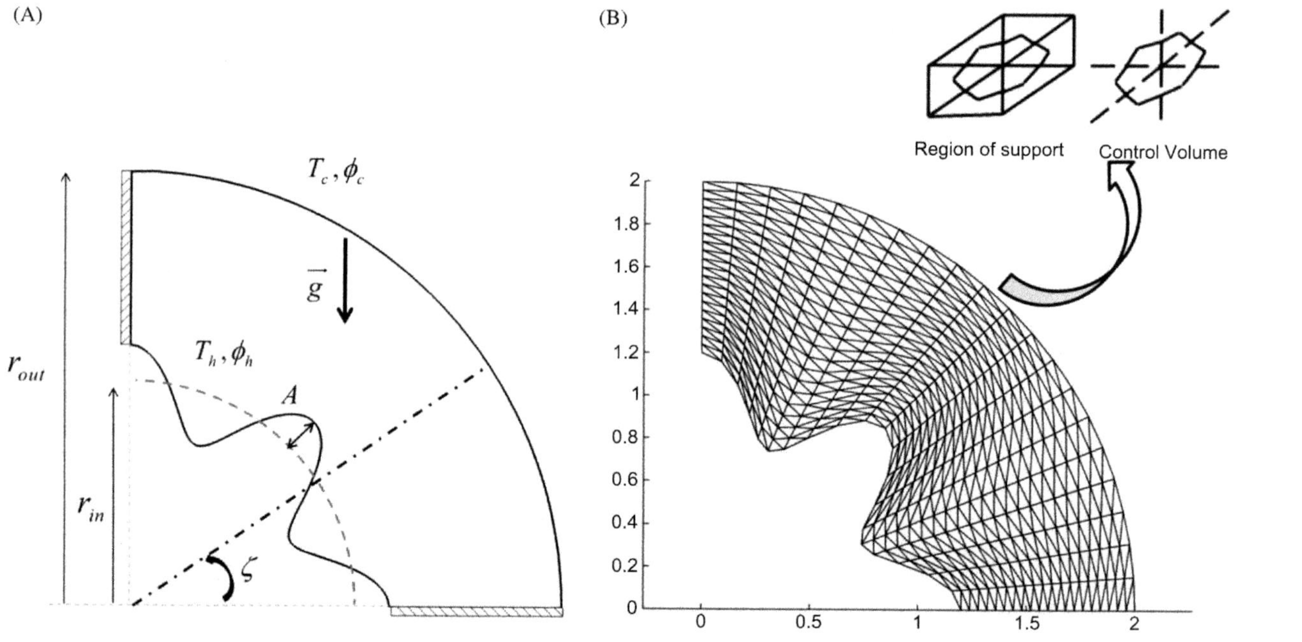

FIGURE 5.1 (A) Geometry and the boundary conditions; (B) the mesh of enclosure considered in this work.

$$r = r_{in} + A\cos(N(\zeta)) \tag{5.1}$$

in which r_{in} is the base circle radius, r_{out} is the radius of outer cylinder, A and N are amplitude and number of undulations, respectively. ζ is the rotation angle. In this study A and N are equal to 0.2 and 8, respectively.

5.2.2 Governing Equation

The nanofluid's density, ρ is

$$\begin{aligned} \rho &= \phi\rho_p + (1-\phi)\rho_f \\ &\cong \phi\rho_p + (1-\phi)\left\{\rho_{f_0}(1-\beta(T-T_c)\right\} \end{aligned} \tag{5.2}$$

where ρ_f, is the base fluid's density, T_c, is a reference temperature, ρ_{f_0} is the base fluid's density at the reference temperature, andβ is the volumetric coefficient of expansion. Taking the density of base fluid as that of the nanofluid, the density ρ in Eq. (5.2), thus becomes

$$\rho \cong \phi\rho_p + (1-\phi)\{\rho_0(1-\beta(T-T_c)\} \tag{5.3}$$

ρ_0 is the nanofluid's density at the reference temperature.

The continuity momentum under Boussinesq approximation and energy equations for the laminar and steady-state natural convection in a two-dimensional enclosure can be written in dimensional form as follows:

$$\frac{\partial u}{\partial x} + \frac{\partial v}{\partial y} = 0 \tag{5.4}$$

$$\rho_f\left\{u\frac{\partial u}{\partial x} + v\frac{\partial u}{\partial y}\right\} = -\frac{\partial P}{\partial x} + \mu\left(\frac{\partial^2 u}{\partial x^2} + \frac{\partial^2 u}{\partial y^2}\right) \tag{5.5}$$

$$\rho_f\left\{u\frac{\partial v}{\partial x} + v\frac{\partial v}{\partial y}\right\} = -\frac{\partial P}{\partial y} + \mu\left(\frac{\partial^2 v}{\partial x^2} + \frac{\partial^2 v}{\partial y^2}\right) - (\phi-\phi_c)\left(\rho_p - \rho_{f_0}\right)g + (1-\phi_c)\rho_{f_0}(T-T_c)g \tag{5.6}$$

$$u\frac{\partial T}{\partial x}+v\frac{\partial T}{\partial y}=\alpha\left(\frac{\partial^2 T}{\partial x^2}+\frac{\partial^2 T}{\partial y^2}\right)+\frac{(\rho c)_p}{(\rho c)_f}\left[\begin{array}{c} D_B\left\{\frac{\partial \phi}{\partial x}.\frac{\partial T}{\partial x}+\frac{\partial \phi}{\partial y}.\frac{\partial T}{\partial y}\right\} \\ +(D_T/T_c)\left\{\left(\frac{\partial T}{\partial x}\right)^2+\left(\frac{\partial T}{\partial y}\right)^2\right\}\end{array}\right] \tag{5.7}$$

$$u\frac{\partial \phi}{\partial x}+v\frac{\partial \phi}{\partial y}=D_B\left\{\frac{\partial^2 \phi}{\partial x^2}+\frac{\partial^2 \phi}{\partial y^2}\right\}+\left(\frac{D_T}{T_c}\right)\left\{\frac{\partial^2 T}{\partial x^2}+\frac{\partial^2 T}{\partial y^2}\right\} \tag{5.8}$$

Boundary conditions are

$$\begin{array}{ll} T=T_h, \quad \phi=\phi_h & \text{on the inner sinusoidal boundary} \\ T=T_c, \quad \phi=\phi_c & \text{on the outer circular boundary} \\ \partial T/\partial n=\partial\phi/\partial n=0.0 & \text{on two other insulation boundaries} \\ \psi=0, & \text{all the solid boundaries} \end{array} \tag{5.9}$$

The stream function and vorticity are defined as follows:

$$u=\frac{\partial \psi}{\partial y}, \quad v=-\frac{\partial \psi}{\partial x}, \quad \omega=\frac{\partial v}{\partial x}-\frac{\partial u}{\partial y} \tag{5.10}$$

The stream function satisfies the continuity Eq. (5.4). The vorticity equation is obtained by eliminating the pressure between the two momentum equations, i.e., by taking the y-derivative of Eq. (5.5) and subtracting from it the x-derivative of Eq. (5.6). Also the following nondimensional variables should be introduced:

$$X=\frac{x}{L}, \quad Y=\frac{y}{L}, \quad \Omega=\frac{\omega L^2}{\alpha}, \Psi=\frac{\psi}{\alpha}, \Theta=\frac{T-T_c}{T_h-T_c}, \quad \Phi=\frac{\phi-\phi_c}{\phi_h-\phi_c} \tag{5.11}$$

where in Eq. (5.11) $L=r_{out}-r_{in}=r_{in}$.

By using these dimensionless parameters the equations become:

$$\left[\frac{\partial \Psi}{\partial Y}\frac{\partial \Omega}{\partial X}-\frac{\partial \Psi}{\partial X}\frac{\partial \Omega}{\partial Y}\right]=\Pr\left(\frac{\partial^2 \Omega}{\partial X^2}+\frac{\partial^2 \Omega}{\partial Y^2}\right)+\Pr Ra\left(\frac{\partial \Theta}{\partial X}-Nr\frac{\partial \Phi}{\partial X}\right) \tag{5.12}$$

$$\frac{\partial \Psi}{\partial Y}\frac{\partial \Theta}{\partial X}-\frac{\partial \Psi}{\partial X}\frac{\partial \Theta}{\partial Y}=\left(\frac{\partial^2 \Theta}{\partial X^2}+\frac{\partial^2 \Theta}{\partial Y^2}\right)+Nb\left(\frac{\partial \Phi}{\partial X}\frac{\partial \Theta}{\partial X}+\frac{\partial \Phi}{\partial Y}\frac{\partial \Theta}{\partial Y}\right)+Nt\left(\left(\frac{\partial \Theta}{\partial X}\right)^2+\left(\frac{\partial \Theta}{\partial Y}\right)^2\right) \tag{5.13}$$

$$\frac{\partial \Psi}{\partial Y}\frac{\partial \Phi}{\partial X}-\frac{\partial \Psi}{\partial X}\frac{\partial \Phi}{\partial Y}=\frac{1}{Le}\left(\frac{\partial^2 \Phi}{\partial X^2}+\frac{\partial^2 \Phi}{\partial Y^2}\right)+\frac{Nt}{Nb\ Le}\left(\frac{\partial^2 \Theta}{\partial X^2}+\frac{\partial^2 \Theta}{\partial Y^2}\right) \tag{5.14}$$

$$\frac{\partial^2 \Psi}{\partial X^2}+\frac{\partial^2 \Psi}{\partial Y^2}=-\Omega \tag{5.15}$$

where the thermal Rayleigh number, the buoyancy ratio number, Prandtl number, the Brownian motion parameter, the thermophoretic parameter, and Lewis number of nanofluid are defined as: $Ra=(1-\phi_c)\rho_{f_0}g\beta L^3(T_h-T_c)/(\mu\alpha)$, $Nr=(\rho_p-\rho_0)(\phi_h-\phi_c)/\left[(1-\phi_c)\rho_{f_0}\beta L(T_h-T_c)\right]$, $\Pr=\mu/\rho_f\alpha$, $Nb=(\rho c)_p D_B(\phi_h-\phi_c)/((\rho c)_f\alpha)$, $Nt=(\rho c)_p D_T(T_h-T_c)/[(\rho c)_f\alpha T_c]$ and $Le=\alpha/D_B$, respectively.

The Eq. (5.12) has been obtained using a small temperature gradient in a dilute suspension of nanoparticles. The boundary conditions as shown in Fig. 5.1 are:

$$\begin{array}{ll} \Theta=1.0, \ \Phi=1.0 & \text{on the inner sinusoidal boundary} \\ \Theta=0.0, \ \Phi=0.0 & \text{on the outer circular boundary} \\ \partial\Theta/\partial n=\partial\Phi/\partial n=0.0 & \text{on two other insulation boundaries} \\ \Psi=0.0 & \text{on all solid boundaries} \end{array} \tag{5.16}$$

The values of vorticity on the boundary of the enclosure can be obtained using the stream function formulation and the known velocity conditions during the iterative solution procedure.

The local Nusselt number on the cold circular wall can be expressed as:

$$Nu_{loc} = -\frac{\partial \Theta}{\partial n} \tag{5.17}$$

where n is the direction normal to the outer cylinder surface. The average number on the cold circular wall is evaluated as:

$$Nu_{ave} = \frac{1}{0.5\pi} \int_{0}^{0.5\pi} Nu_{loc}(\zeta)\, d\zeta \tag{5.18}$$

The heatlines are adequate tools for visualization and analysis of 2D convection heat transfer, through an extension of the heat flux line concept to include the advection terms. Heat function (*H*) are defined in terms of the energy equation as

$$\frac{\partial H}{\partial Y} = U\Theta - \frac{\partial \Theta}{\partial X}, \quad -\frac{\partial H}{\partial X} = V\Theta - \frac{\partial \Theta}{\partial Y} \tag{5.19}$$

5.2.3 Effects of Active Parameters

In this section, CVFEM is applied to investigate natural convection heat transfer in an enclosure filled with nanofluid. Effects of thermal Rayleigh number ($Ra = 10^3, 10^4$, and 10^5), buoyancy ratio number ($Nr = 0.1$ to 4) and Lewis number ($Le = 1, 2, 4, 6$, and 8) on flow and heat transfer characteristics are examined. Brownian motion parameter of nanofluids ($Nb = 0.5$), thermophoretic parameter of nanofluids ($Nt = 0.5$), and Prandtl number ($\Pr = 10$) are fixed.

The comparison of the isotherms, streamlines, isoconcentration, and heatline contours for different values of thermal Rayleigh number is shown in Fig. 5.2. At $Ra = 10^3$ the conduction heat transfer mechanism is more pronounced. For this reason the isotherms are parallel to each other. As *Ra* increases, the distribution of the isotherm contours increases. The streamline has two eddies within the enclosure of which the lower one is stronger. With the increase of *Ra*, the role of convection in heat transfer becomes more significant. Also it can be seen that as *Ra* increases the distribution of isoconcentration contours increases. The heat flow within the enclosure is displayed using the heat function obtained from conductive heat fluxes ($\partial\Theta/\partial X, \partial\Theta/\partial Y$) as well as convective heat fluxes ($V\Theta, U\Theta$). Heatlines emanate from hot regimes and end on cold regimes illustrating the path of heat flow. The domination of conduction heat transfer in low Rayleigh number can be observed from the heatline patterns since no passive area exists. The increase of *Ra* causes the clustering of heatlines from hot to the cold wall and generates passive heat transfer area in which heat is rotated without having a significant effect on heat transfer between walls.

Comparison of the isotherms, streamlines, isoconcentration, and heatline contours for different values of buoyancy ratio number is shown in Fig. 5.3. It should be mentioned that negative *Nr* values (opposing buoyancy forces) showed more complex and interesting flow patterns, such as multicells, which is worthy of presentation and discussion. So in this study results for positive *Nr* (aiding buoyancy forces), where the temperature and species induced buoyancy forces aides each other, are not considered. For $Nr = 0$, the species induced buoyancy force has no effect on flow; the flow is solely driven by the thermal buoyancy force. However, the effect of species-induced buoyancy increases as *Nr* value increases, and reaches a certain value, where the effect of thermally induced buoyancy becomes negligible in comparison to the solutal one. For a small *Nr* value, the flow is mainly driven by the thermal buoyancy force. When $Nr = 0.1$, two upwelling plumes appear on the crests of the inner cylinder. A third plume appears above the outer circular cylinder with reverse direction owing to a counterclockwise eddy between two other clockwise eddies at this region. For $Nr > 1$ the flow is mainly driven by species concentration induced buoyancy force. So as *Nr* increases the thermal plumes disappear. The number of passive areas is increased with the increase of solutal forces.

The comparison of the isotherms, streamlines, isoconcentration, and heatline contours for different values of Lewis number is shown in Fig. 5.4. When $Le = 2$, two equal eddies with the same direction are observed. As the Lewis number increases up to 8, these eddies are converted to four eddies. The mass flow is given by $\psi_{max} \sim \delta_s\, v$, where the solutal boundary layer thickness is given by $\delta_s \sim (Ra\, Le\, Nr)^{-1/4}$ and $v \sim (Ra\, Le\, Nr)^{1/2}$ so $\psi_{max} \sim (Ra\, Le\, Nr)^{1/4}$. The solutal boundary layer, δ_s, becomes thinner by the increase of either *Ra* or *Le* or both. Heatlines are found to be more distorted as *Le* augments.

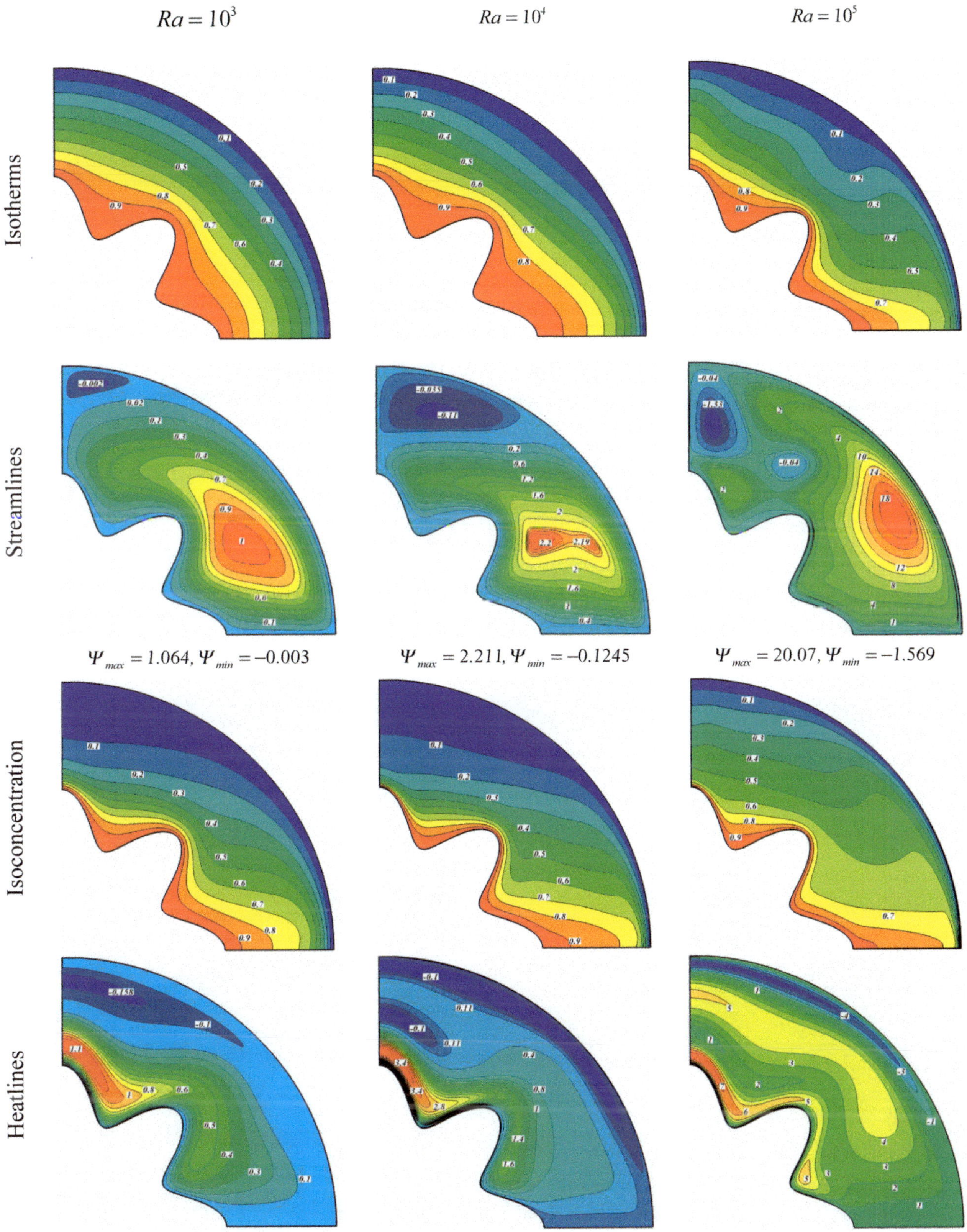

FIGURE 5.2 Comparison of the isotherms, streamlines, isoconcentration, and heatline contours for different values of thermal Rayleigh number when $Nr = 4, Le = 8, Nt = Nb = 0.5$, and Pr = 10.

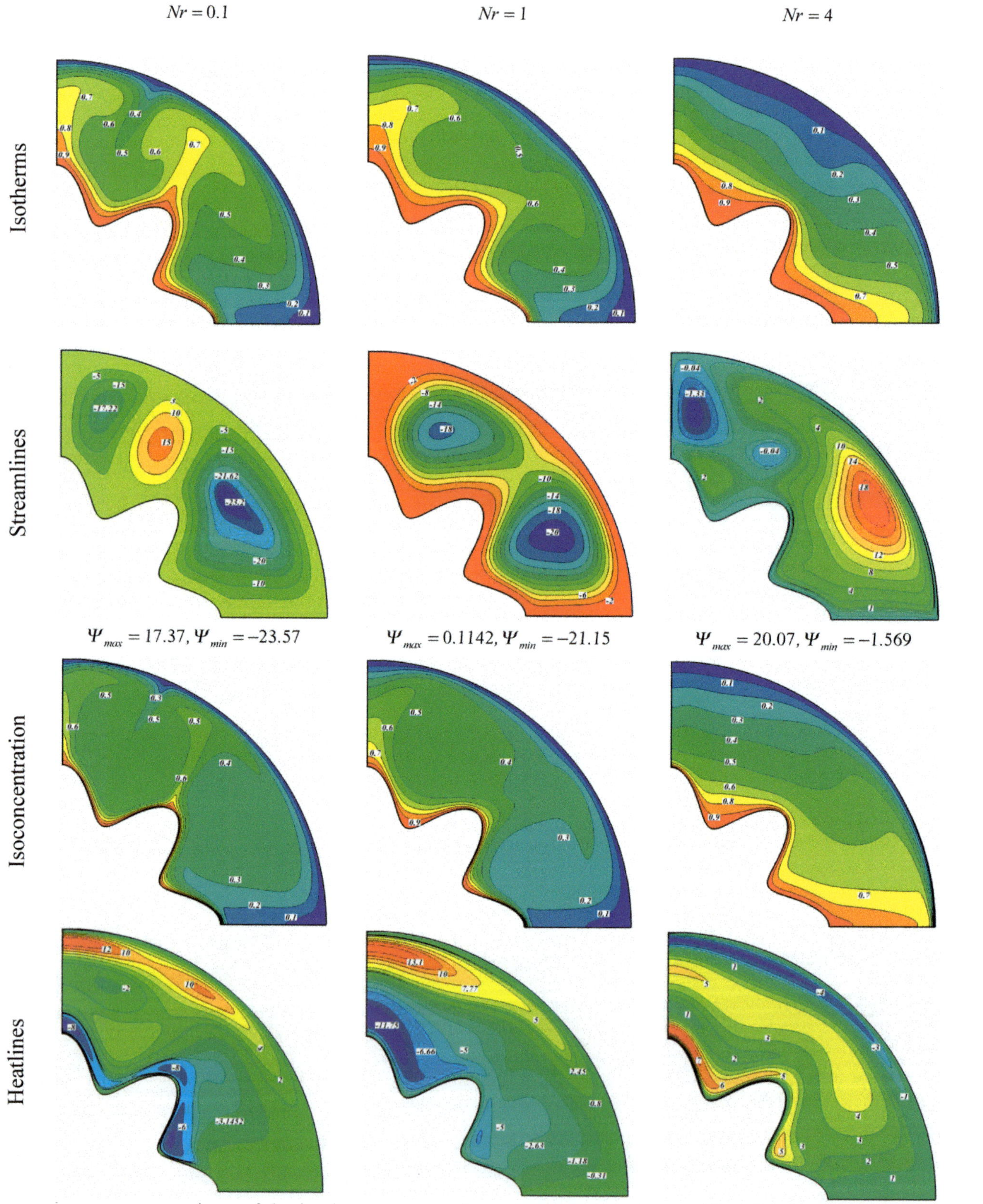

FIGURE 5.3 Comparison of the isotherms, streamlines, isoconcentration, and heatline contours for different values of buoyancy ratio number when $Ra = 10^5, Le = 8, Nt = Nb = 0.5,$ and $\Pr = 10$.

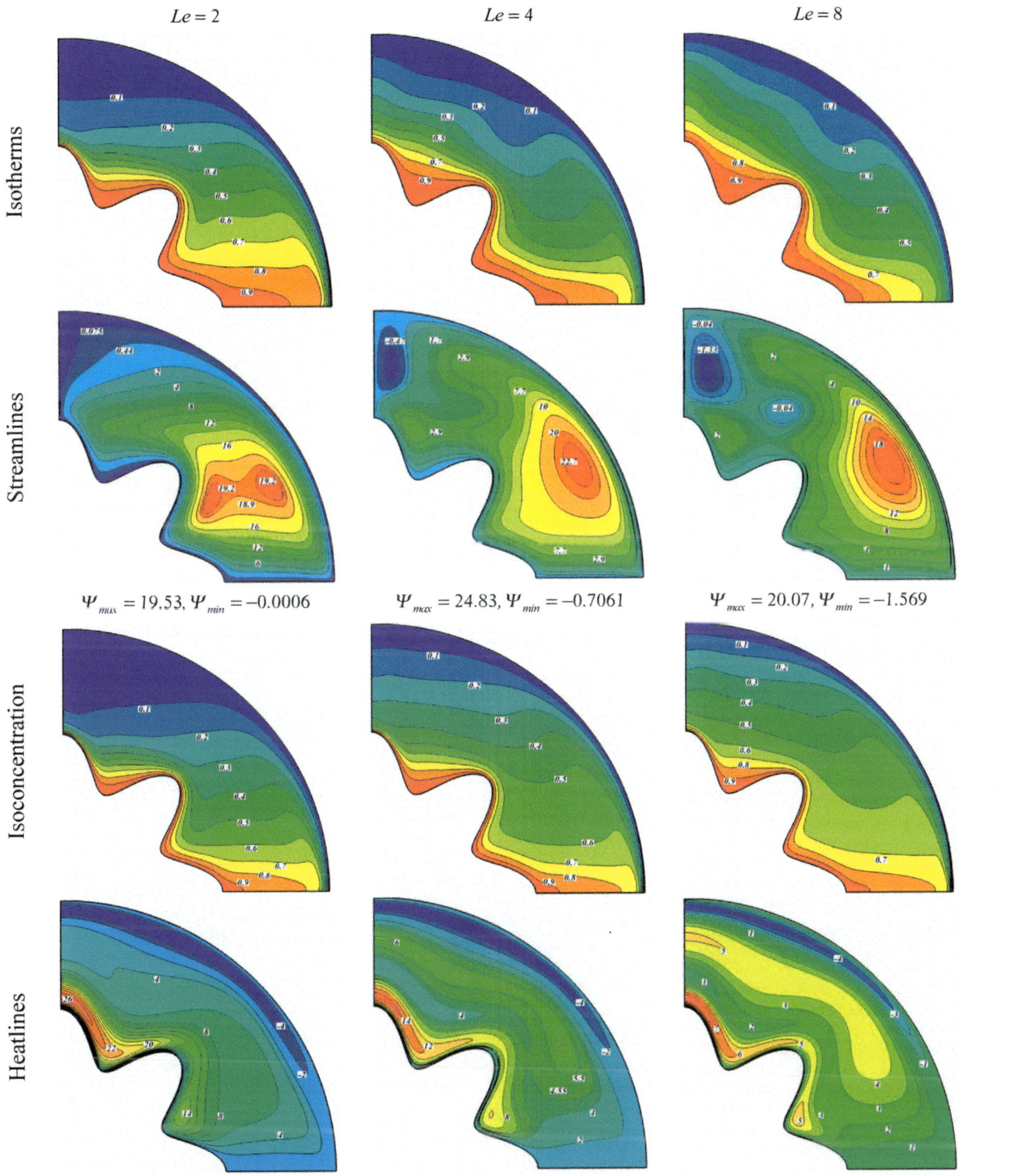

FIGURE 5.4 Comparison of the isotherms, streamlines, isoconcentration, and heatline contours for different values of Lewis number when $Ra = 10^5, Nr = 4, Nt = Nb = 0.5$, and $\Pr = 10$.

Fig. 5.5 depicts the effects of thermal Rayleigh number, buoyancy ratio number, and Lewis number on local Nusselt number. A change in a flow structure has a significant effect on the local Nusselt number. The number of extremum in the local Nusselt number profile corresponds to the number of thermal plumes of the cold wall. The occurrence of maxima for Nu_{loc} is due to dense heatlines based on conductive heat transport occurring at

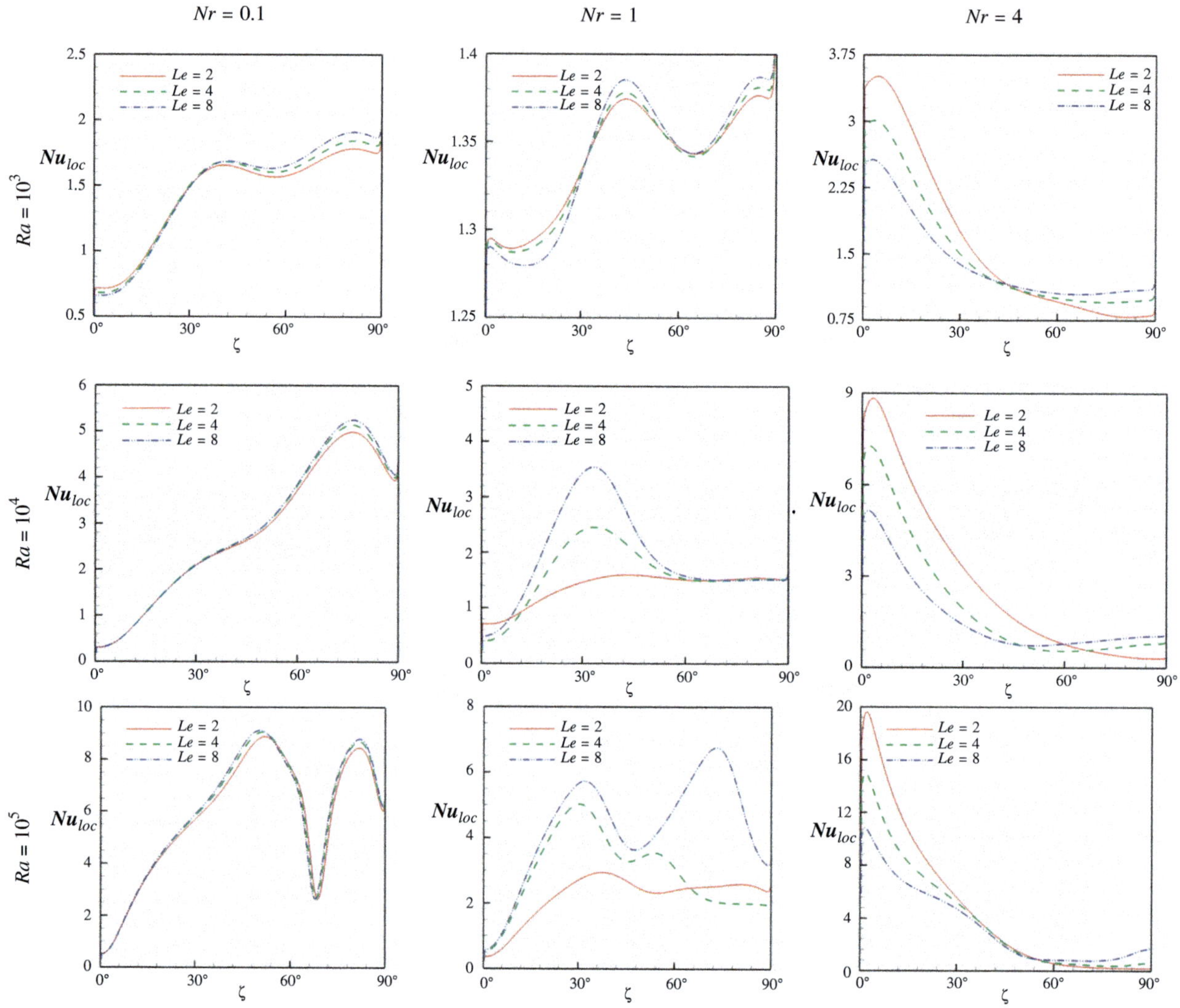

FIGURE 5.5 Effects of thermal Rayleigh number, buoyancy ratio number, and Lewis number on local Nusselt number at $Nt = Nb = 0.5$ and $\Pr = 10$.

those respective portions. For low values of Nr ($Nr < 1$) the maximum rate of heat transfer takes place at $\zeta = 90^\circ$, where the thermal boundary layer is thin. However, for

$Nr > 1$, the reverse is true, i.e., the rate of heat transfer is high at $\zeta = 0^\circ$. In fact, the trend reflects the physics of flow reversal as a function of Nr.

Fig. 5.6 shows the effects of thermal Rayleigh number, buoyancy ratio number, and Lewis number on average Nusselt number. As a general trend, the average Nusselt number decreases as Nr increases until it reaches a minimum value and then starts increasing. As Nr increases the effect of opposing buoyancy (species) increases, hence the net fluid flow driving force decreases. The Nu_{ave} reaches a minimum value when the strength of thermal buoyancy is equal to the species-induced buoyancy. The thermal boundary layer thickness differs from the species boundary layer thickness because Le is not at unity. The minimum Nu_{ave} is obtained for buoyancy ratio number values (Nr_{cr}) shifting slightly to higher Nr as Le increases. As the Le increases the species concentration boundary layer decreases in thickness compared with the thermal boundary layer thickness. Hence, the resulting species induced force decreases as the Le increases and higher Nr is needed to compensate the decrease in the species concentration boundary layer. As expected the flow is dominated by species-induced buoyancy for $Nr > Nr_{cr}$ and rate of transport increases as Nr increases. The heat transfer increases with Le for $Nr > Nr_{cr}$ and decreases for $Nr < Nr_{cr}$, for fixed Nr. Such a trend is a consequence of a separate scale on which the thermal and solutal effect acts and such a separate scale increases as the Le value increases. It is interesting to note that the rate of decrease in Nu_{ave} is more prominent for Nr values between 0 and Nr_{cr} compared to Nr greater than Nr_{cr}.

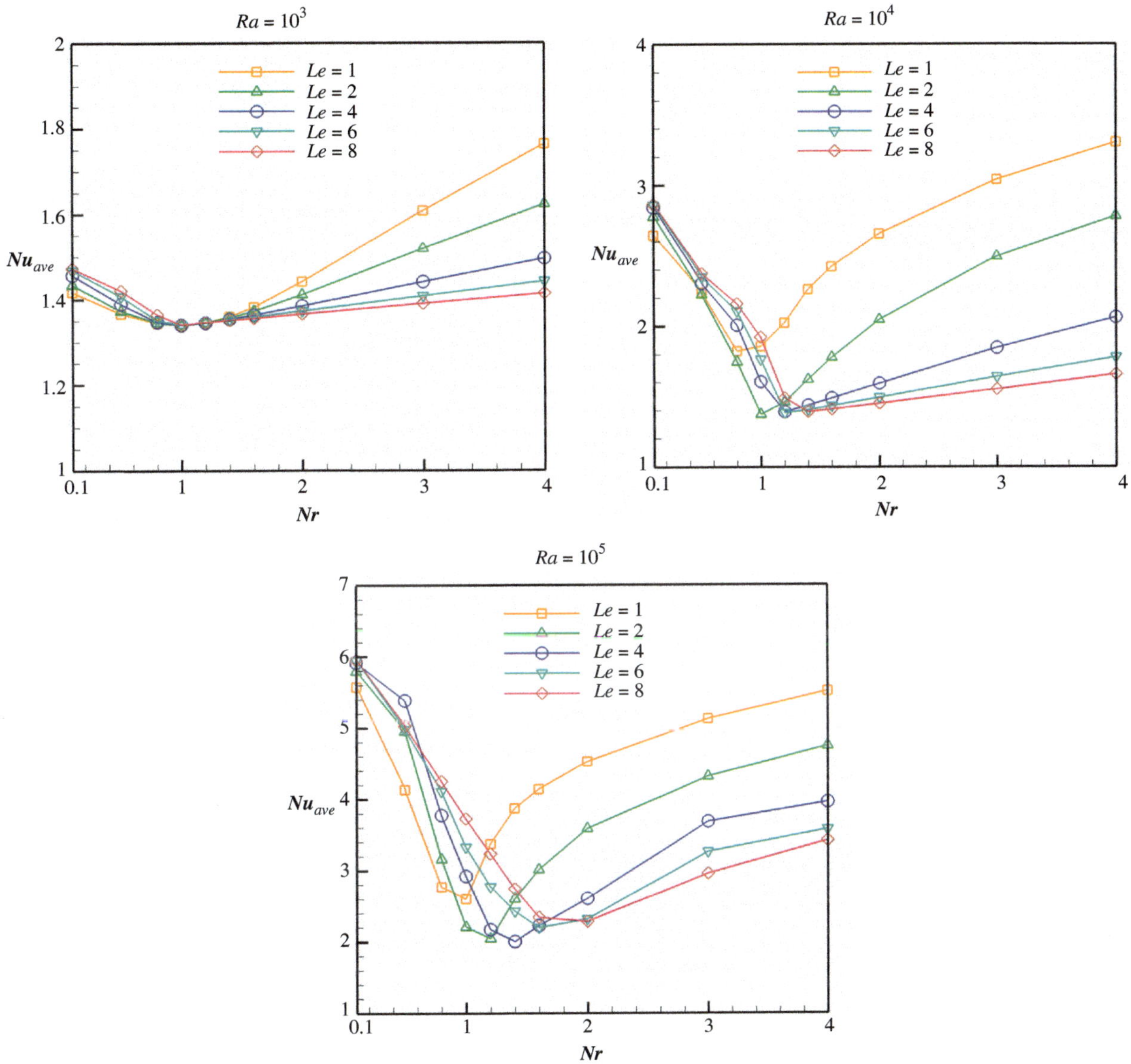

FIGURE 5.6 Effects of thermal Rayleigh number, buoyancy ratio number, and Lewis number on average Nusselt number at $Nt = Nb = 0.5$ and $\mathrm{Pr} = 10$.

This can be explained by the fact that the results are obtained for $Le \geq 2$, hence the thermal boundary layer is thicker than the species concentration boundary layer.

5.3 TWO-PHASE MODEL FOR NANOFLUID NATURAL CONVECTION HEAT TRANSFER

5.3.1 Problem Definition

The schematic diagram and the mesh of the semiannulus enclosure used in the present CVFEM program are shown in Fig. 5.7. The inner and outer walls are maintained at constant temperatures T_h and T_c, respectively, while the two other walls are thermally insulated. Also the boundary conditions of concentration are similar to temperature.

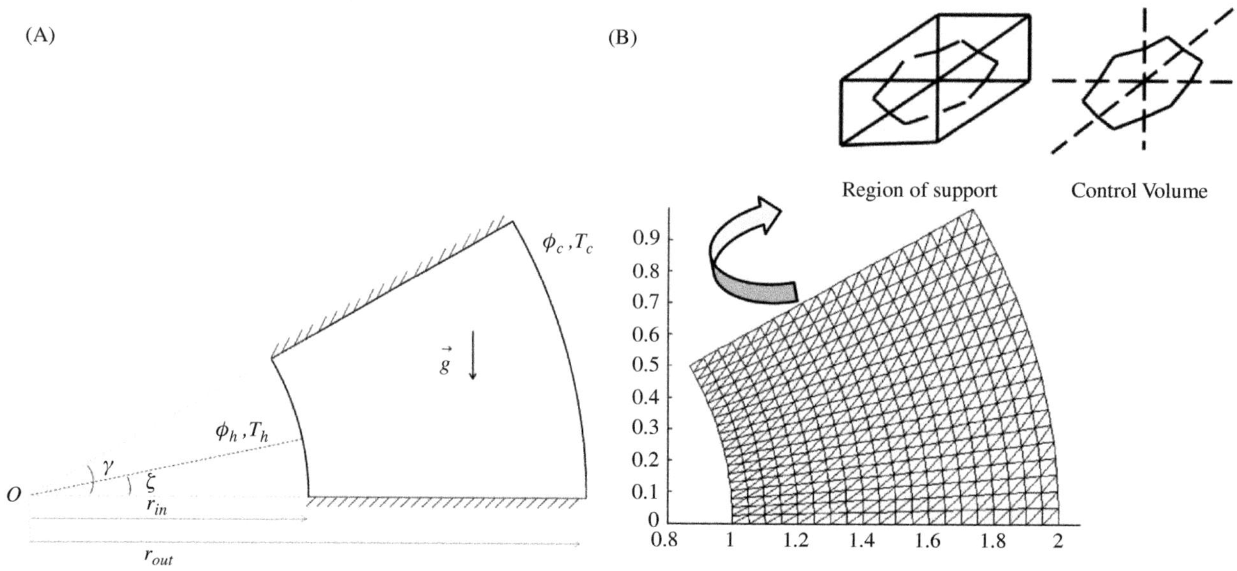

FIGURE 5.7 (A) Geometry and the boundary conditions; (B) the mesh of enclosure considered in this work.

5.3.2 Governing Equation

The nanofluid's density, ρ is

$$\begin{aligned}\rho &= \phi\rho_p + (1-\phi)\rho_f \\ &\cong \phi\rho_p + (1-\phi)\left\{\rho_{f_0}(1-\beta(T-T_c)\right\}\end{aligned} \tag{5.20}$$

where ρ_f, is the base fluid's density, T_c, is a reference temperature, ρ_{f_0} is the base fluid's density at the reference temperature, β is the volumetric coefficient of expansion. Taking the density of base fluid as that of the nanofluid, the density ρ in Eq. (5.20), thus becomes

$$\rho \cong \phi\rho_p + (1-\phi)\{\rho_0(1-\beta(T-T_c)\} \tag{5.21}$$

ρ_0 is the nanofluid's density at the reference temperature.

The continuity momentum under Boussinesq approximation and energy equations for the laminar and steady-state natural convection in a two-dimensional enclosure can be written in dimensional form as follows:

$$\frac{\partial u}{\partial x} + \frac{\partial v}{\partial y} = 0 \tag{5.22}$$

$$\rho_f\left\{u\frac{\partial u}{\partial x} + v\frac{\partial u}{\partial y}\right\} = -\frac{\partial P}{\partial x} + \mu\left(\frac{\partial^2 u}{\partial x^2} + \frac{\partial^2 u}{\partial y^2}\right) \tag{5.23}$$

$$\rho_f\left\{u\frac{\partial v}{\partial x} + v\frac{\partial v}{\partial y}\right\} = -\frac{\partial P}{\partial y} + \mu\left(\frac{\partial^2 v}{\partial x^2} + \frac{\partial^2 v}{\partial y^2}\right) - (\phi-\phi_c)\left(\rho_p - \rho_{f_0}\right)g + (1-\phi_c)\rho_{f_0}(T-T_c)g \tag{5.24}$$

$$u\frac{\partial T}{\partial x} + v\frac{\partial T}{\partial y} = \alpha\left(\frac{\partial^2 T}{\partial x^2} + \frac{\partial^2 T}{\partial y^2}\right) + \frac{(\rho c)_p}{(\rho c)_f}\left[\begin{array}{l} D_B\left\{\dfrac{\partial \phi}{\partial x}\cdot\dfrac{\partial T}{\partial x} + \dfrac{\partial \phi}{\partial y}\cdot\dfrac{\partial T}{\partial y}\right\} \\ +(D_T/T_c)\left\{\left(\dfrac{\partial T}{\partial x}\right)^2 + \left(\dfrac{\partial T}{\partial y}\right)^2\right\}\end{array}\right] \tag{5.25}$$

$$u\frac{\partial \phi}{\partial x} + v\frac{\partial \phi}{\partial y} = D_B\left\{\frac{\partial^2 \phi}{\partial x^2} + \frac{\partial^2 \phi}{\partial y^2}\right\} + \left(\frac{D_T}{T_c}\right)\left\{\frac{\partial^2 T}{\partial x^2} + \frac{\partial^2 T}{\partial y^2}\right\} \tag{5.26}$$

Boundary conditions are

$$\begin{aligned} &T = T_h, \quad \phi = \phi_h && \text{on the inner circular boundary} \\ &T = T_c, \quad \phi = \phi_c && \text{on the outer circular boundary} \\ &\partial T/\partial n = \partial \phi/\partial n = 0.0 && \text{on two other insulation boundaries} \\ &\psi = 0, && \text{all the solid boundaries} \end{aligned} \tag{5.27}$$

The stream function and vorticity are defined as follows:

$$u = \frac{\partial \psi}{\partial y}, \quad v = -\frac{\partial \psi}{\partial x}, \quad \omega = \frac{\partial v}{\partial x} - \frac{\partial u}{\partial y} \tag{5.28}$$

The stream function satisfies the continuity Eq. (5.22). The vorticity equation is obtained by eliminating the pressure between the two momentum equations, i.e., by taking the y-derivative of Eq. (5.23) and subtracting from it the x-derivative of Eq. (5.24). Also the following nondimensional variables should be introduced:

$$X = \frac{x}{L}, \quad Y = \frac{y}{L}, \quad \Omega = \frac{\omega L^2}{\alpha}, \Psi = \frac{\psi}{\alpha}, \Theta = \frac{T - T_c}{T_h - T_c}, \quad \Phi = \frac{\phi - \phi_c}{\phi_h - \phi_c}, U = \frac{uL}{\alpha}, V = \frac{vL}{\alpha} \tag{5.29}$$

where in Eq. (5.29) $L = r_{out} - r_{in} = r_{in}$.

By using these dimensionless parameters the equations become:

$$\left[\frac{\partial \Psi}{\partial Y}\frac{\partial \Omega}{\partial X} - \frac{\partial \Psi}{\partial X}\frac{\partial \Omega}{\partial Y}\right] = \Pr\left(\frac{\partial^2 \Omega}{\partial X^2} + \frac{\partial^2 \Omega}{\partial Y^2}\right) + \Pr Ra\left(\frac{\partial \Theta}{\partial X} - Nr\,\frac{\partial \Theta}{\partial X}\right) \tag{5.30}$$

$$\frac{\partial \Psi}{\partial Y}\frac{\partial \Theta}{\partial X} - \frac{\partial \Psi}{\partial X}\frac{\partial \Theta}{\partial Y} = \left(\frac{\partial^2 \Theta}{\partial X^2} + \frac{\partial^2 \Theta}{\partial Y^2}\right) + Nb\left(\frac{\partial \Phi}{\partial X}\frac{\partial \Theta}{\partial X} + \frac{\partial \Phi}{\partial Y}\frac{\partial \Theta}{\partial Y}\right) + Nt\left(\left(\frac{\partial \Theta}{\partial X}\right)^2 + \left(\frac{\partial \Theta}{\partial Y}\right)^2\right) \tag{5.31}$$

$$\frac{\partial \Psi}{\partial Y}\frac{\partial \Phi}{\partial X} - \frac{\partial \Psi}{\partial X}\frac{\partial \Phi}{\partial Y} = \frac{1}{Le}\left(\frac{\partial^2 \Phi}{\partial X^2} + \frac{\partial^2 \Phi}{\partial Y^2}\right) + \frac{Nt}{Nb\ Le}\left(\frac{\partial^2 \Theta}{\partial X^2} + \frac{\partial^2 \Theta}{\partial Y^2}\right) \tag{5.32}$$

$$\frac{\partial^2 \Psi}{\partial X^2} + \frac{\partial^2 \Psi}{\partial Y^2} = -\Omega \tag{5.33}$$

where thermal Rayleigh number, the buoyancy ratio number, Prandtl number, the Brownian motion parameter, the thermophoretic parameter, and Lewis number of nanofluid are defined as: $Ra = (1 - \phi_c)\rho_{f_0} g\beta L^3 (T_h - T_c)/(\mu\alpha)$, $Nr = (\rho_p - \rho_0)(\phi_h - \phi_c)/\left[(1 - \phi_c)\rho_{f_0}\beta L(T_h - T_c)\right]$, $\Pr = \mu/\rho_f\alpha$, $Nb = (\rho c)_p D_B(\phi_h - \phi_c)/((\rho c)_f \alpha)$, $Nt = (\rho c)_p D_T(T_h - T_c)/[(\rho c)_f \alpha T_c]$ and $Le = \alpha/D_B$, respectively.

The boundary conditions as shown in Fig. 5.7 are:

$$\begin{aligned} &\Theta = 1.0, \ \Phi = 1.0 && \text{on the inner circular boundary} \\ &\Theta = 0.0, \ \Phi = 0.0 && \text{on the outer circular boundary} \\ &\partial \Theta/\partial n = \partial \Phi/\partial n = 0.0 && \text{on two other insulation boundaries} \\ &\Psi = 0.0 && \text{on all solid boundaries} \end{aligned} \tag{5.34}$$

The values of vorticity on the boundary of the enclosure can be obtained using the stream function formulation and the known velocity conditions during the iterative solution procedure.

The local Nusselt number on the hot circular wall can be expressed as:

$$Nu_{loc} = -\frac{\partial \Theta}{\partial n} \tag{5.35}$$

where n is the direction normal to the outer cylinder surface. The average number on the hot circular wall is evaluated as:

$$Nu_{ave} = \frac{1}{0.5\pi}\int_0^{0.5\pi} Nu_{loc}(\zeta)\, d\zeta \tag{5.36}$$

5.3.3 Effects of Active Parameters

Natural convection heat transfer in an enclosure filled with nanofluid is investigated numerically using CVFEM. Effects of buoyancy ratio number ($Nr = 0.1$ to 4), Lewis number ($Le = 2, 4, 6$ and 8), and angle of turn for the enclosure ($\gamma = 30^\circ, 60^\circ$ to 90°) on flow and heat transfer characteristics are examined. Brownian motion parameter of nanofluid ($Nb = 0.5$), thermophoretic parameter of nanofluid ($Nt = 0.5$), thermal Rayleigh number ($Ra = 10^5$), and Prandtl number ($\Pr = 10$) are fixed.

The effects of angle of turn, Lewis number, and buoyancy ratio number on isotherms, streamlines, and isoconcentration are shown in Figs. 5.8 and 5.9. It should be mentioned that positive Nr values (opposing buoyancy forces) showed more complex and interesting flow patterns, such as multicells, which is worthy of presentation and discussion. So in this study, results for negative Nr (aiding buoyancy forces), where the temperature and species induced buoyancy forces aides each other, are not considered. For $Nr = 0$, the species induced buoyancy force has no effect on flow; the flow is solely driven by the thermal buoyancy force. However, the effect of species-induced buoyancy increases as Nr value increases, and reaches a certain value, where the effect of thermally induced buoyancy becomes negligible in comparison to the solutal one. For small Nr value, the flow is mainly driven by the thermal buoyancy force. When Nr increases a reverse thermal plume appears at $\zeta = 90^\circ$ when $\gamma = 90^\circ$. This phenomena is due to the existing counterclockwise eddy at this region. The isoconcentration lines are more distorted with an increase of solutal forces.

The mass flow is given by $\psi_{\max} \approx \delta_s\, v$, where the solutal boundary layer thickness is given by $\delta_s \approx (Ra\ Le\ Nr)^{-1/4}$ and $v \approx (Ra\ Le\ Nr)^{1/2}$ so $\psi_{\max} \approx (Ra\ Le\ Nr)^{1/4}$. The solutal boundary layer, δ_s, becomes thinner by increasing Le. At low Lewis number, the iscoconcentration lines are similar to isotherms. By increasing Lewis number difference between these counters becomes more visible. The eddie's core moves upward with an increase of angle of turn. Thermal boundary layer thickness increases with an increase of angle of turn. It is an interesting observation that one counterclockwise eddy is generated at $\gamma = 90^\circ$.

Effects of angle of turn, buoyancy ratio number, and Lewis number on local Nusselt number are shown in Fig. 5.10. As seen local Nusselt number increases as buoyancy ratio number increases but it decreases with increase of Lewis number. Generally, as ζ increases the local Nusselt number decreases due to the increment of the thermal boundary layer thickness. The local Nusselt number profile has a minimum point at $\zeta = 90^\circ$ because of the existence of a thermal plume at this region. Also it can be seen that the local Nusselt number increases with augmenting angle of turn.

Effects of buoyancy ratio number, angle of turn, and Lewis number on average Nusselt number are shown in Fig. 5.11. As Lewis number increases the thermal boundary layer thickness increases and in turn the Nusselt number decreases. Also it can be concluded that the Nusselt number increases with an increase of buoyancy ratio number but it decreases with increase of angle of turn. The corresponding polynomial representation of such a model for the Nusselt number is as follows (see Fig. 5.12):

$$
\begin{aligned}
Nu &= a_{12} + a_{22}Y_1 + a_{32}Le + a_{42}{Y_1}^{*2} + a_{52}Le^2 + a_{62}Y_1\,Le \\
Y_1 &= a_{11} + a_{21}\gamma + a_{31}Nr + a_{41}\gamma^2 + a_{51}Nr^2 + a_{61}\gamma\,Nr
\end{aligned}
\tag{5.37}
$$

Also a_{ij} can be found in Table 5.1 for example a_{21} equals to: (2.065756).

5.4 MHD NATURAL CONVECTION OF AL_2O_3-WATER NANOFLUID CONSIDERING THERMOPHORESIS AND BROWNIAN MOTION EFFECTS

5.4.1 Problem Definition

The schematic diagram and the mesh of the semiannulus enclosure used in the present CVFEM program are shown in Fig. 5.13. The inner and outer walls are maintained at constant temperatures T_h and T_c, respectively, while the two other walls are thermally insulated. Also the boundary conditions of concentration are similar to temperature. It is also assumed that the uniform magnetic field ($\vec{B} = B_x\vec{e_x} + B_y\vec{e_y}$) of constant magnitude $B = \sqrt{B_x^2 + B_y^2}$ is applied, where $\vec{e_x}$ and $\vec{e_y}$ are unit vectors in the Cartesian coordinate system. The orientation of the magnetic field forms an angle λ with the horizontal axis such that $\lambda = B_x/B_y$. The electric current J and the electromagnetic force F are defined by $J = \sigma\left(\vec{V} \times \vec{B}\right)$ and $F = \sigma\left(\vec{V} \times \vec{B}\right) \times \vec{B}$, respectively.

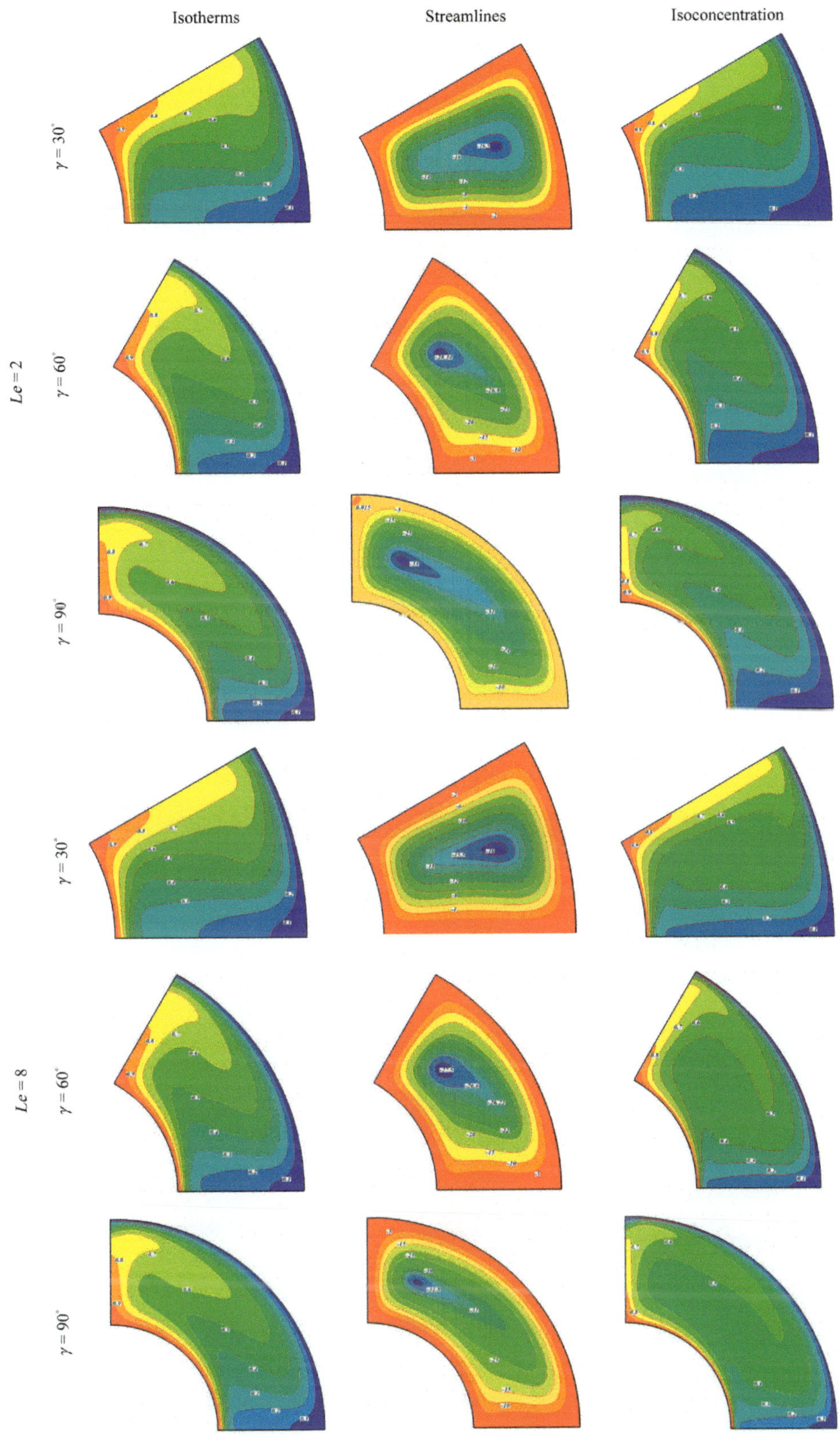

FIGURE 5.8 Comparison of the isotherms, streamlines, and isoconcentration for different values of angle of turn and Lewis number when $Nr = 0.1, Nt = Nb = 0.5, Ra = 10^5$, and $\Pr = 10$.

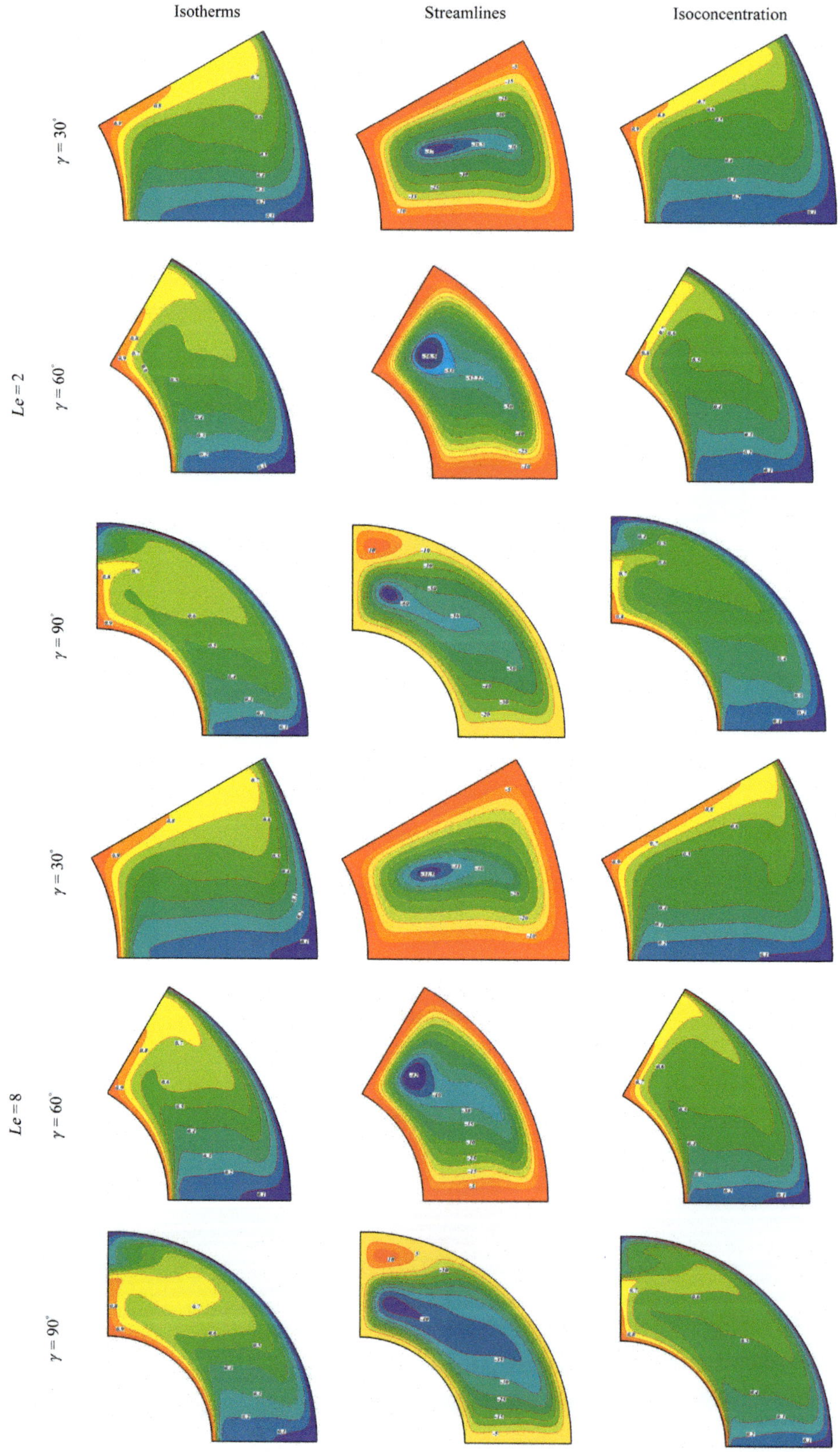

FIGURE 5.9 Comparison of the isotherms, streamlines, and isoconcentration for different values of angle of turn and Lewis number when $Nr = 4, Nt = Nb = 0.5, Ra = 10^5$, and $\Pr = 10$.

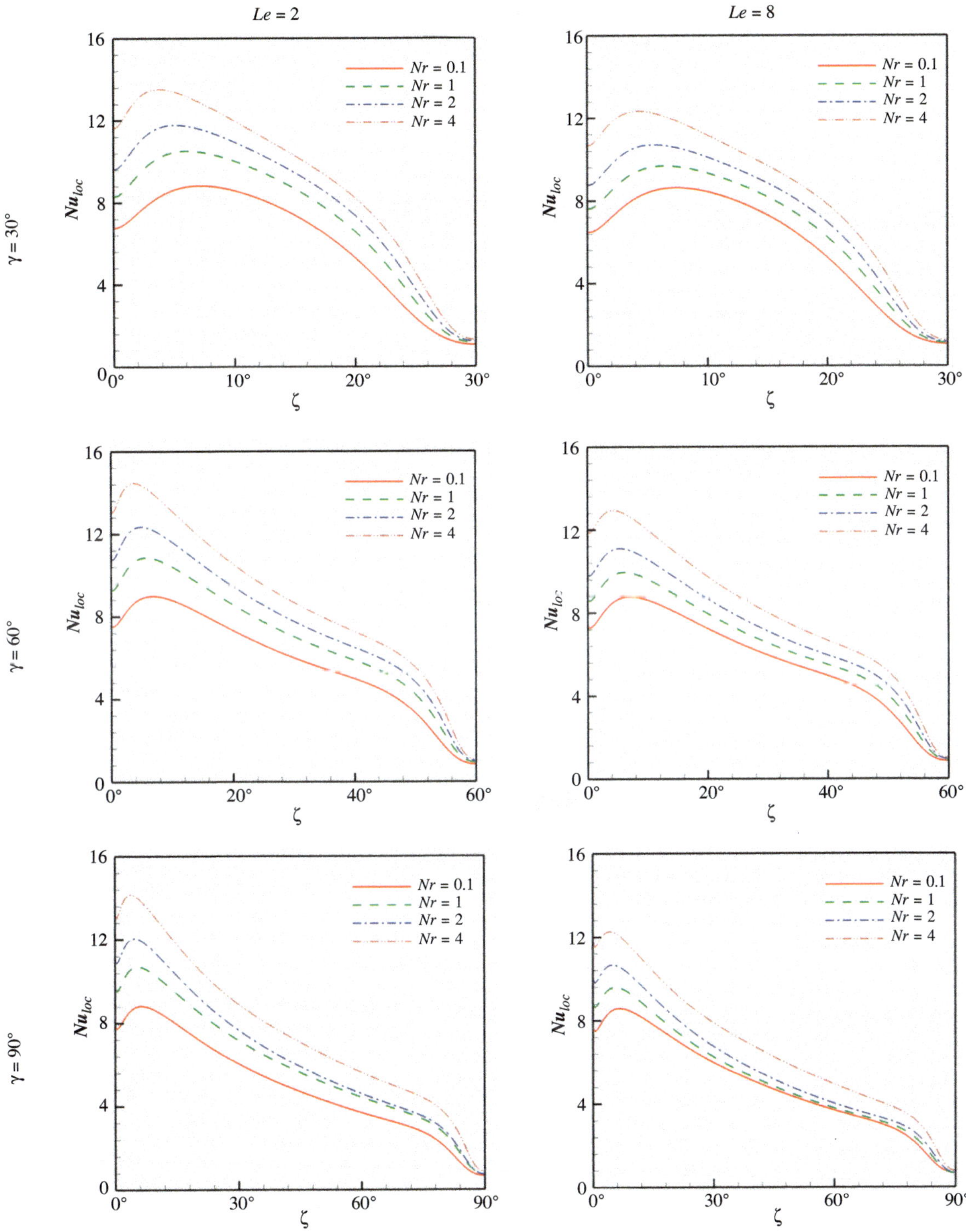

FIGURE 5.10 Effects of angle of turn for the enclosure, Hartmann number, buoyancy ratio number, and Lewis number on the local Nusselt number at $Nt = Nb = 0.5$, $Ra = 10^5$, and $\Pr = 10$.

5.4.2 Governing Equation

The nanofluid's density, ρ is

$$\begin{aligned}\rho &= \phi\rho_p + (1-\phi)\rho_f \\ &\cong \phi\rho_p + (1-\phi)\left\{\rho_{f_0}(1-\beta(T-T_c)\right\}\end{aligned} \tag{5.38}$$

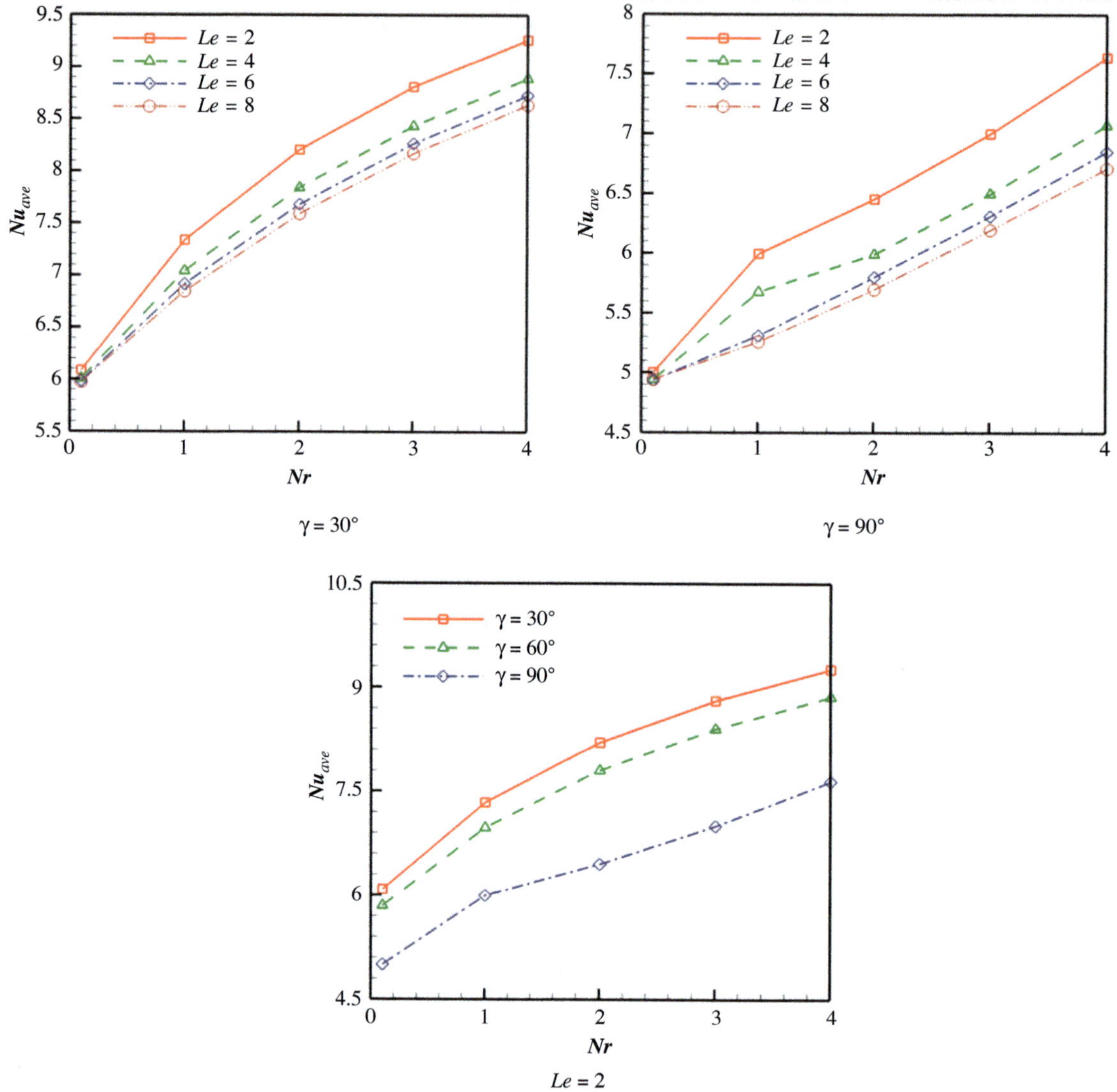

FIGURE 5.11 Effects of the angle of turn for the enclosure, buoyancy ratio number, and Lewis number on average Nusselt number at $Nt = Nb = 0.5$, $Ra = 10^5$, and $\Pr = 10$.

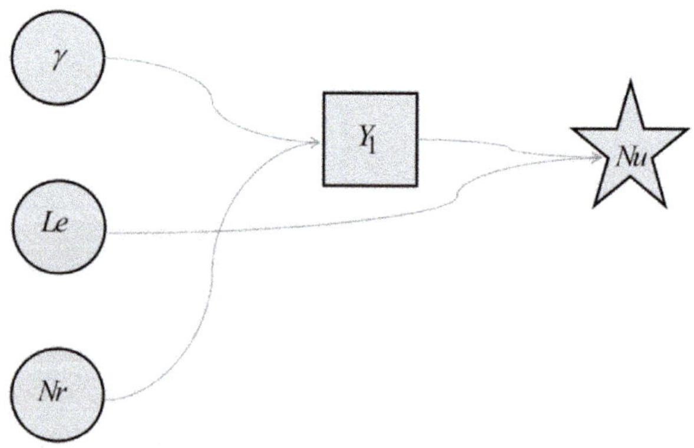

FIGURE 5.12 Evolved structure of correlation for Nusselt number.

TABLE 5.1 Constant coefficient for using Eq. (5.37)

a_{ij}	i = 1	i = 2	i = 3	i = 4	i = 5	i = 6
j = 1	5.334872	2.065756	1.111228	− 1.5454	− 0.06732	− 0.18785
j = 2	0.191916	1.077506	− 0.12466	0.001021	0.015991	− 0.01835

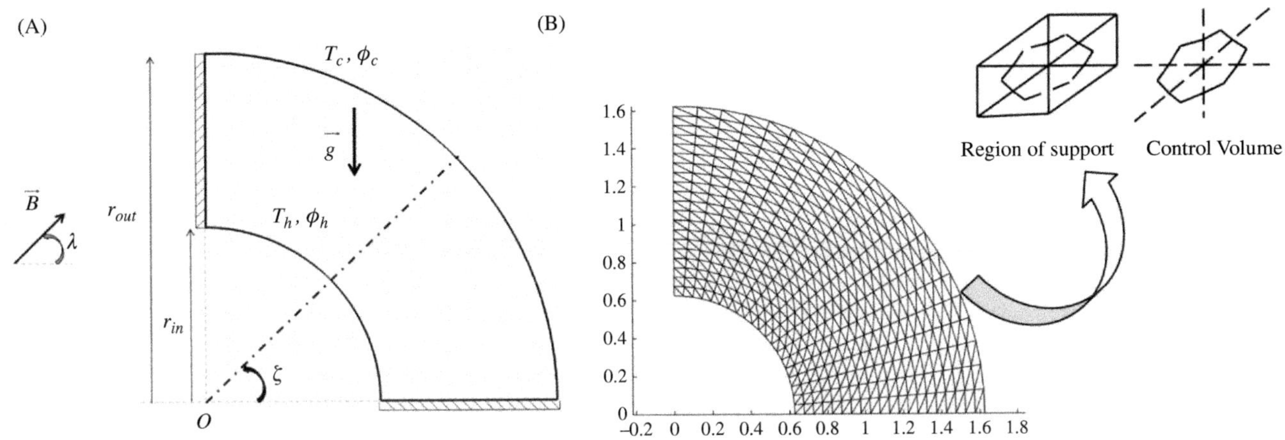

FIGURE 5.13 (A) Geometry and the boundary conditions; (B) the mesh of enclosure considered in this work.

where ρ_f, is the base fluid's density, T_c, is a reference temperature, ρ_{f_0} is the base fluid's density at the reference temperature, and β is the volumetric coefficient of expansion. Taking the density of base fluid as that of the nanofluid, the density ρ in Eq. 5.2, thus becomes

$$\rho \cong \phi\rho_p + (1-\phi)\{\rho_0(1-\beta(T-T_c)\} \tag{5.39}$$

ρ_0 is the nanofluid's density at the reference temperature.

The continuity momentum under Boussinesq approximation and energy equations for the laminar and steady-state natural convection in a two-dimensional enclosure can be written in dimensional form as follows:

$$\frac{\partial u}{\partial x} + \frac{\partial v}{\partial y} = 0 \tag{5.40}$$

$$\rho_f\left\{u\frac{\partial u}{\partial x} + v\frac{\partial u}{\partial y}\right\} = -\frac{\partial P}{\partial x} + \mu\left(\frac{\partial^2 u}{\partial x^2} + \frac{\partial^2 u}{\partial y^2}\right) + \sigma B^2(v\sin\lambda\cos\lambda - u\sin^2\lambda) \tag{5.41}$$

$$\begin{aligned}\rho_f\left\{u\frac{\partial v}{\partial x} + v\frac{\partial v}{\partial y}\right\} = {} & -\frac{\partial P}{\partial y} + \mu\left(\frac{\partial^2 v}{\partial x^2} + \frac{\partial^2 v}{\partial y^2}\right) - (\phi-\phi_c)\left(\rho_p - \rho_{f_0}\right)g + (1-\phi_c)\rho_{f_0}(T-T_c)g \\ & + \sigma B^2(u\sin\lambda\cos\lambda - v\cos^2\lambda)\end{aligned} \tag{5.42}$$

$$u\frac{\partial T}{\partial x} + v\frac{\partial T}{\partial y} = \alpha\left(\frac{\partial^2 T}{\partial x^2} + \frac{\partial^2 T}{\partial y^2}\right) + \frac{(\rho c)_p}{(\rho c)_f}\left[\begin{array}{l} D_B\left\{\dfrac{\partial \phi}{\partial x}\cdot\dfrac{\partial T}{\partial x} + \dfrac{\partial \phi}{\partial y}\cdot\dfrac{\partial T}{\partial y}\right\} \\ +(D_T/T_c)\left\{\left(\dfrac{\partial T}{\partial x}\right)^2 + \left(\dfrac{\partial T}{\partial y}\right)^2\right\}\end{array}\right] \tag{5.43}$$

$$u\frac{\partial \phi}{\partial x} + v\frac{\partial \phi}{\partial y} = D_B\left\{\frac{\partial^2 \phi}{\partial x^2} + \frac{\partial^2 \phi}{\partial y^2}\right\} + \left(\frac{D_T}{T_c}\right)\left\{\frac{\partial^2 T}{\partial x^2} + \frac{\partial^2 T}{\partial y^2}\right\} \tag{5.44}$$

Boundary conditions are

$$\begin{array}{ll} T = T_h, \quad \phi = \phi_h & \text{on the inner circular boundary} \\ T = T_c, \quad \phi = \phi_c & \text{on the outer circular boundary} \\ \partial T/\partial n = \partial\phi/\partial n = 0.0 & \text{on two other insulation boundaries} \\ \psi = 0, & \text{all the solid boundaries} \end{array} \tag{5.45}$$

The stream function and vorticity are defined as follows:

$$u = \frac{\partial \psi}{\partial y}, \quad v = -\frac{\partial \psi}{\partial x}, \quad \omega = \frac{\partial v}{\partial x} - \frac{\partial u}{\partial y} \tag{5.46}$$

The stream function satisfies the continuity Eq. (5.40). The vorticity equation is obtained by eliminating the pressure between the two momentum equations, i.e., by taking the y-derivative of Eq. (5.41) and subtracting from it the x-derivative of Eq. (5.42). Also the following nondimensional variables should be introduced:

$$X=\frac{x}{L},\quad Y=\frac{y}{L},\quad \Omega=\frac{\omega L^2}{\alpha},\quad \Psi=\frac{\psi}{\alpha},\quad \Theta=\frac{T-T_c}{T_h-T_c},\quad \Phi=\frac{\phi-\phi_c}{\phi_h-\phi_c},\quad U=\frac{uL}{\alpha},\quad V=\frac{vL}{\alpha} \tag{5.47}$$

where in Eq. (5.47) $L=r_{out}-r_{in}=r_{in}$.

By using these dimensionless parameters the equations become:

$$\begin{aligned}\left[\frac{\partial\Psi}{\partial Y}\frac{\partial\Omega}{\partial X}-\frac{\partial\Psi}{\partial X}\frac{\partial\Omega}{\partial Y}\right]&=\Pr\left(\frac{\partial^2\Omega}{\partial X^2}+\frac{\partial^2\Omega}{\partial Y^2}\right)+\Pr Ra\left(\frac{\partial\Theta}{\partial X}-Nr\frac{\partial\Phi}{\partial X}\right)\\&+Ha^2\Pr\left(-\frac{\delta V}{\delta Y}\tan\lambda+\frac{\delta U}{\delta Y}\tan^2\lambda+\frac{\delta U}{\delta X}\tan\lambda-\frac{\delta V}{\delta X}\right)\end{aligned} \tag{5.48}$$

$$\frac{\partial\Psi}{\partial Y}\frac{\partial\Theta}{\partial X}-\frac{\partial\Psi}{\partial X}\frac{\partial\Theta}{\partial Y}=\left(\frac{\partial^2\Theta}{\partial X^2}+\frac{\partial^2\Theta}{\partial Y^2}\right)+Nb\left(\frac{\partial\Phi}{\partial X}\frac{\partial\Theta}{\partial X}+\frac{\partial\Phi}{\partial Y}\frac{\partial\Theta}{\partial Y}\right)+Nt\left(\left(\frac{\partial\Theta}{\partial X}\right)^2+\left(\frac{\partial\Theta}{\partial Y}\right)^2\right) \tag{5.49}$$

$$\frac{\partial\Psi}{\partial Y}\frac{\partial\Phi}{\partial X}-\frac{\partial\Psi}{\partial X}\frac{\partial\Phi}{\partial Y}=\frac{1}{Le}\left(\frac{\partial^2\Phi}{\partial X^2}+\frac{\partial^2\Phi}{\partial Y^2}\right)+\frac{Nt}{Nb\ Le}\left(\frac{\partial^2\Theta}{\partial X^2}+\frac{\partial^2\Theta}{\partial Y^2}\right) \tag{5.50}$$

$$\frac{\partial^2\Psi}{\partial X^2}+\frac{\partial^2\Psi}{\partial Y^2}=-\Omega \tag{5.51}$$

where thermal Rayleigh number, the buoyancy ratio number, Prandtl number, the Brownian motion parameter, the thermophoretic parameter, Lewis number, and Hartmann number of nanofluid are defined as: $Ra=(1-\phi_c)\rho_{f_0}g\beta L^3(T_h-T_c)/(\mu\alpha)$, $Nr=(\rho_p-\rho_0)(\phi_h-\phi_c)/\left[(1-\phi_c)\rho_{f_0}\beta L(T_h-T_c)\right]$, $\Pr=\mu/\rho_f\alpha$, $Nb=(\rho c)_p D_B(\phi_h-\phi_c)/((\rho c)_f\alpha)$, $Nt=(\rho c)_p D_T(T_h-T_c)/[(\rho c)_f\alpha T_c]$, $Le=\alpha/D_B$ and $Ha=LB_x\sqrt{\sigma/\mu}$, respectively.

The boundary conditions as shown in Fig. 5.13 are:

$$\begin{aligned}&\Theta=1.0,\ \Phi=1.0 && \text{on the inner circular boundary}\\&\Theta=0.0,\Phi=0.0 && \text{on the outer circular boundary}\\&\partial\Theta/\partial n=\partial\Phi/\partial n=0.0 && \text{on two other insulation boundaries}\\&\Psi=0.0 && \text{on all solid boundaries}\end{aligned} \tag{5.52}$$

The values of vorticity on the boundary of the enclosure can be obtained using the stream function formulation and the known velocity conditions during the iterative solution procedure.

The local Nusselt number on the hot circular wall can be expressed as:

$$Nu_{loc}=-\frac{\partial\Theta}{\partial n} \tag{5.53}$$

where n is the direction normal to the inner cylinder surface. The average number on the hot circular wall is evaluated as:

$$Nu_{ave}=\frac{1}{0.5\pi}\int_0^{0.5\pi}Nu_{loc}(\zeta)\,d\zeta \tag{5.54}$$

5.4.3 Effects of Active Parameters

MHD effect on natural convection heat transfer in an enclosure filled with nanofluid is investigated numerically using CVFEM. Effects of Hartmann number ($Ha=0,30,60$, and 100), buoyancy ratio number ($Nr=0.1$ to 4), and Lewis number ($Le=2,4,6$, and 8) on flow and heat transfer characteristics are examined. Brownian motion parameter of nanofluids ($Nb=0.5$), thermophoretic parameter of nanofluids ($Nt=0.5$), thermal Rayleigh number ($Ra=10^5$), and Prandtl number ($\Pr=10$) are fixed.

Effects of Hartmann number, Lewis number, and buoyancy ratio number on isotherms, streamlines, isoconcentration, and heatline contours are shown in Figs. 5.14-5.17. When the magnetic field is imposed on the enclosure, the velocity field is suppressed owing to the retarding effect of the Lorenz force. So intensity of convection weakens significantly. The braking effect of the magnetic field is observed from the maximum stream function value.

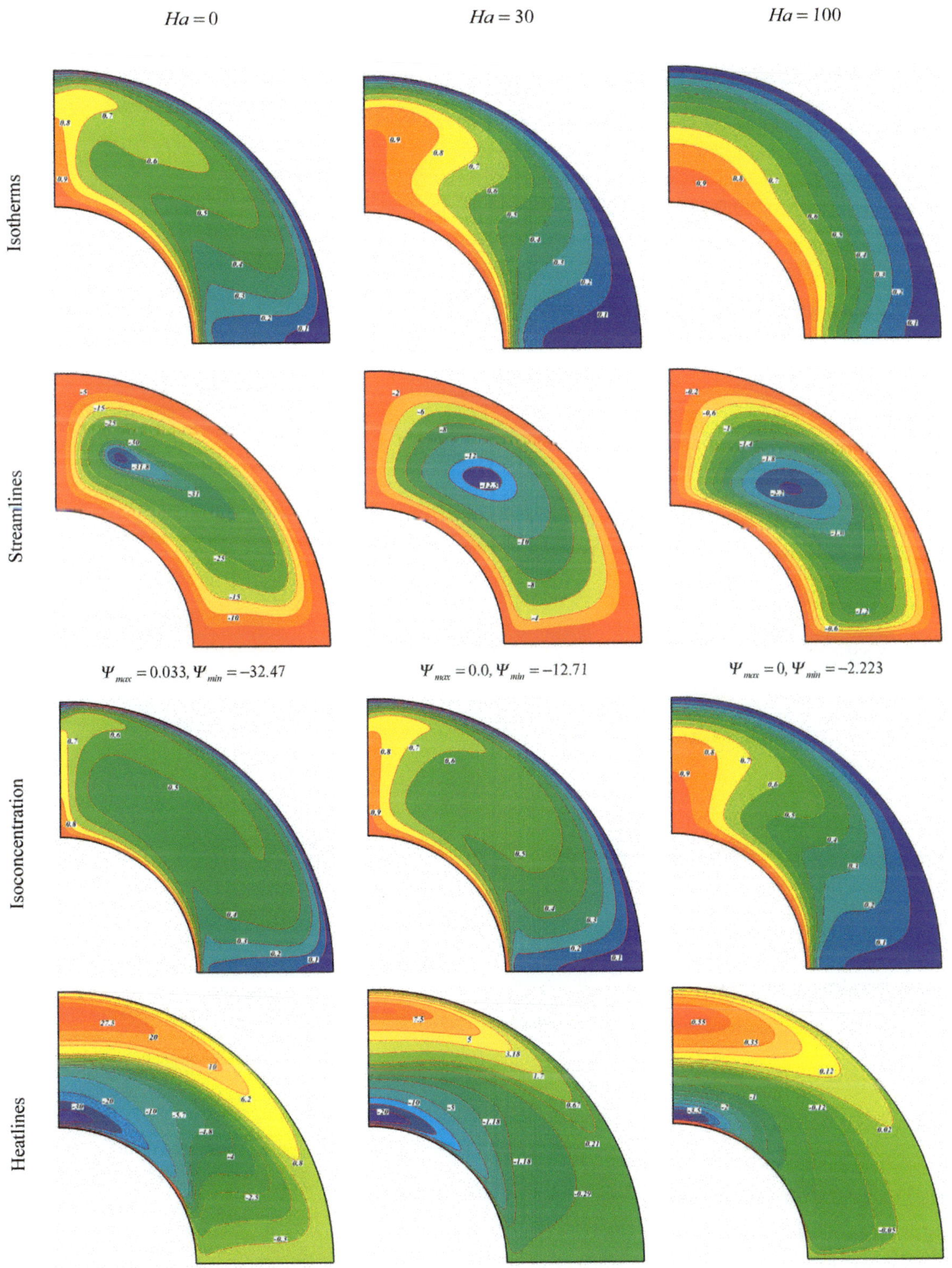

FIGURE 5.14 Comparison of the isotherms, streamlines, isoconcentration, and heatline contours for different values of thermal Hartmann number when $Nr = 0.1, Le = 8, Nt = Nb = 0.5, Ra = 10^5$, and $\Pr = 10$.

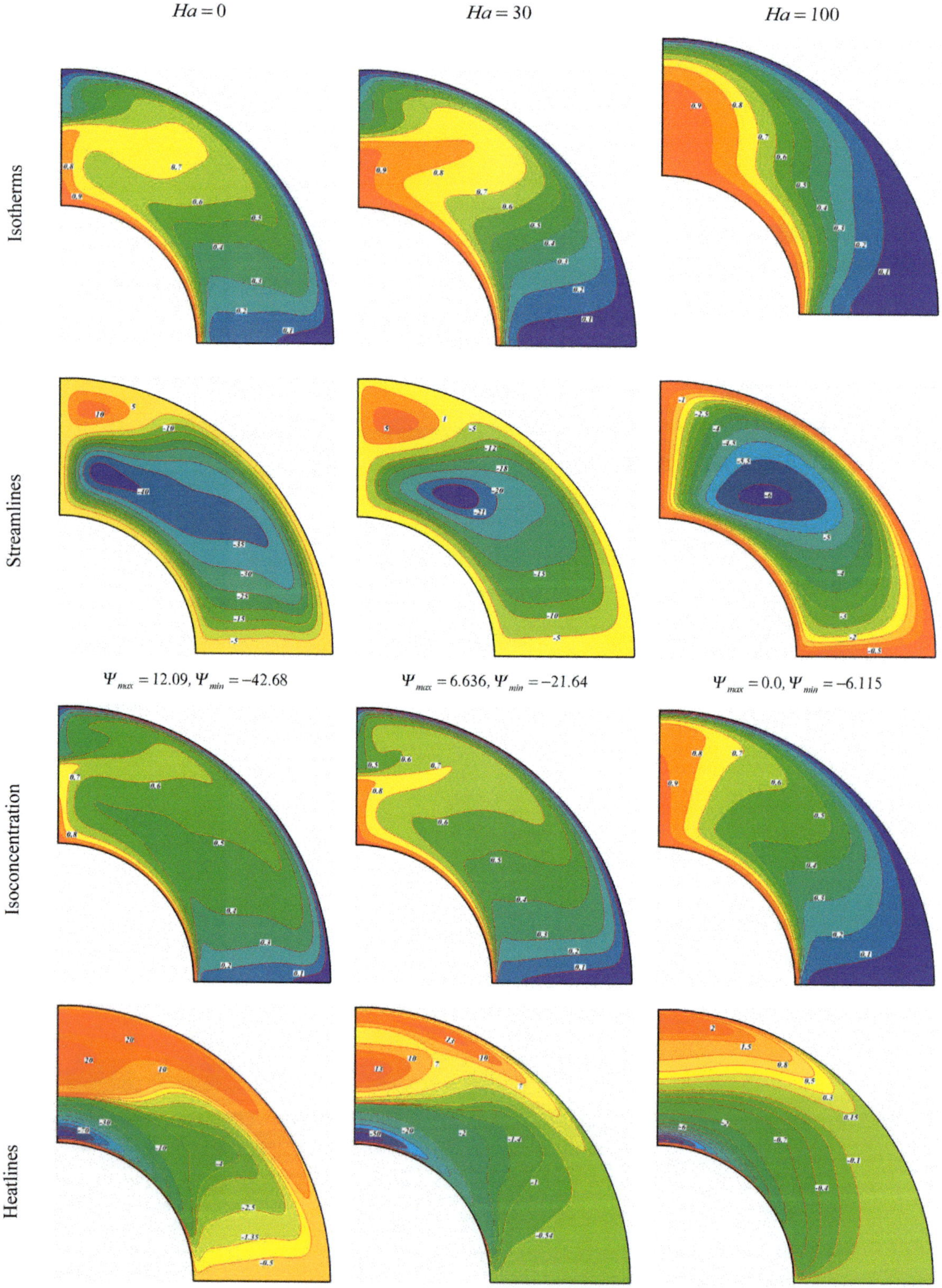

FIGURE 5.15 Comparison of the isotherms, streamlines, isoconcentration, and heatline contours for different values of thermal Hartmann number when $Nr = 4, Le = 8, Nt = Nb = 0.5, Ra = 10^5$, and $\Pr = 10$.

The core of the vortex is shift downward vertically as the Hartmann number increases. Also imposing the magnetic field leads to omitting the thermal plume over the inner wall. At high Hartmann number, the conduction heat transfer mechanism is more pronounced. For this reason the isotherms are parallel to each other. Increasing Hartmann number causes the concentration boundary layer thickness near inner wall to increase.

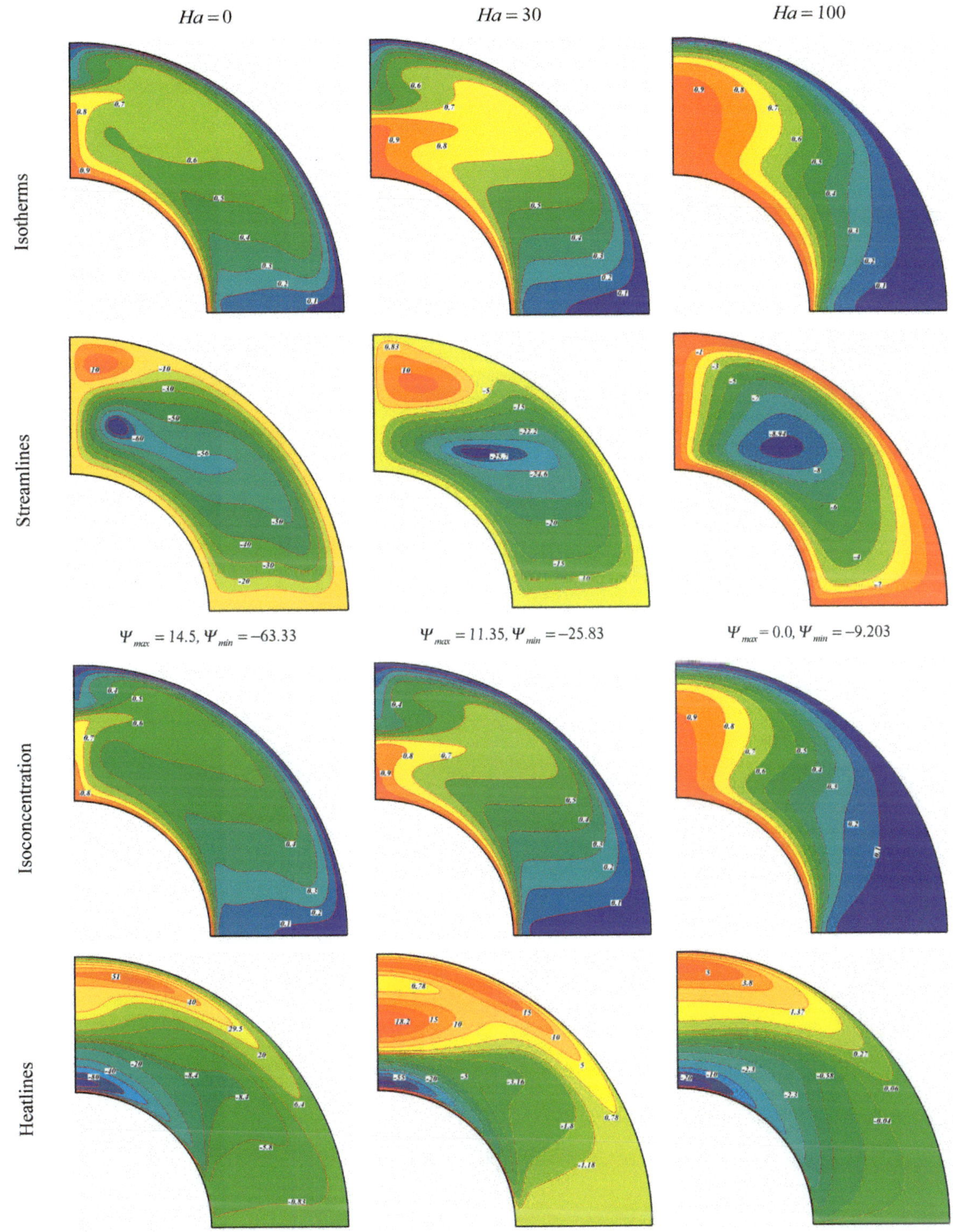

FIGURE 5.16 Comparison of the isotherms, streamlines, isoconcentration, and heatline contours for different values of thermal Hartmann number when $Nr = 4, Le = 2, Nt = Nb = 0.5, Ra = 10^5$, and $\Pr = 10$.

The heat flow within the enclosure is displayed using the heat function obtained from conductive heat fluxes $(\partial\Theta/\partial X,\ \partial\Theta/\partial Y)$ as well as convective heat fluxes $(V\Theta, U\Theta)$. Heatlines emanate from hot regimes and end on cold regimes illustrating the path of heat flow. As seen, the heatline has two regions rotating in different direction. The lower one is greater than the other which means that more heat transfer occurs in this region. As Hartmann number increases the heatlines become weaker because of the reduction of the heat transfer rate by

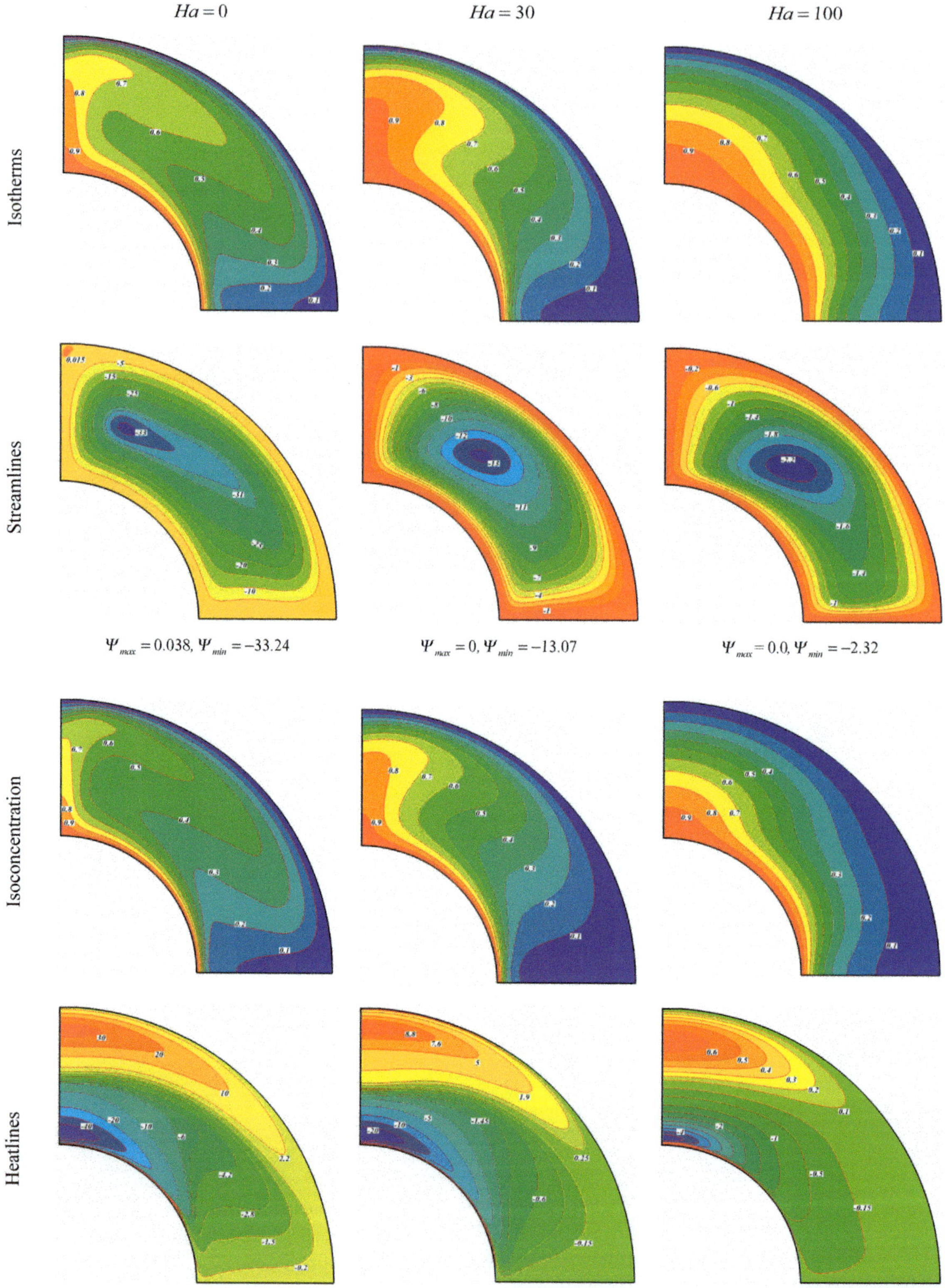

FIGURE 5.17 Comparison of the isotherms, streamlines, isoconcentration, and heatline contours for different values of thermal Hartmann number when $Nr = 0.1, Le = 2, Nt = Nb = 0.5, Ra = 10^5$, and $\Pr = 10$.

applying a magnetic field. The domination of conduction heat transfer in a high Hartmann number can be observed from the heatline patterns since no passive area exists.

It should be mentioned that negative Nr values (opposing buoyancy forces) showed more complex and interesting flow patterns, such as multicells, which is worthy of presentation and discussion. So in this study, the

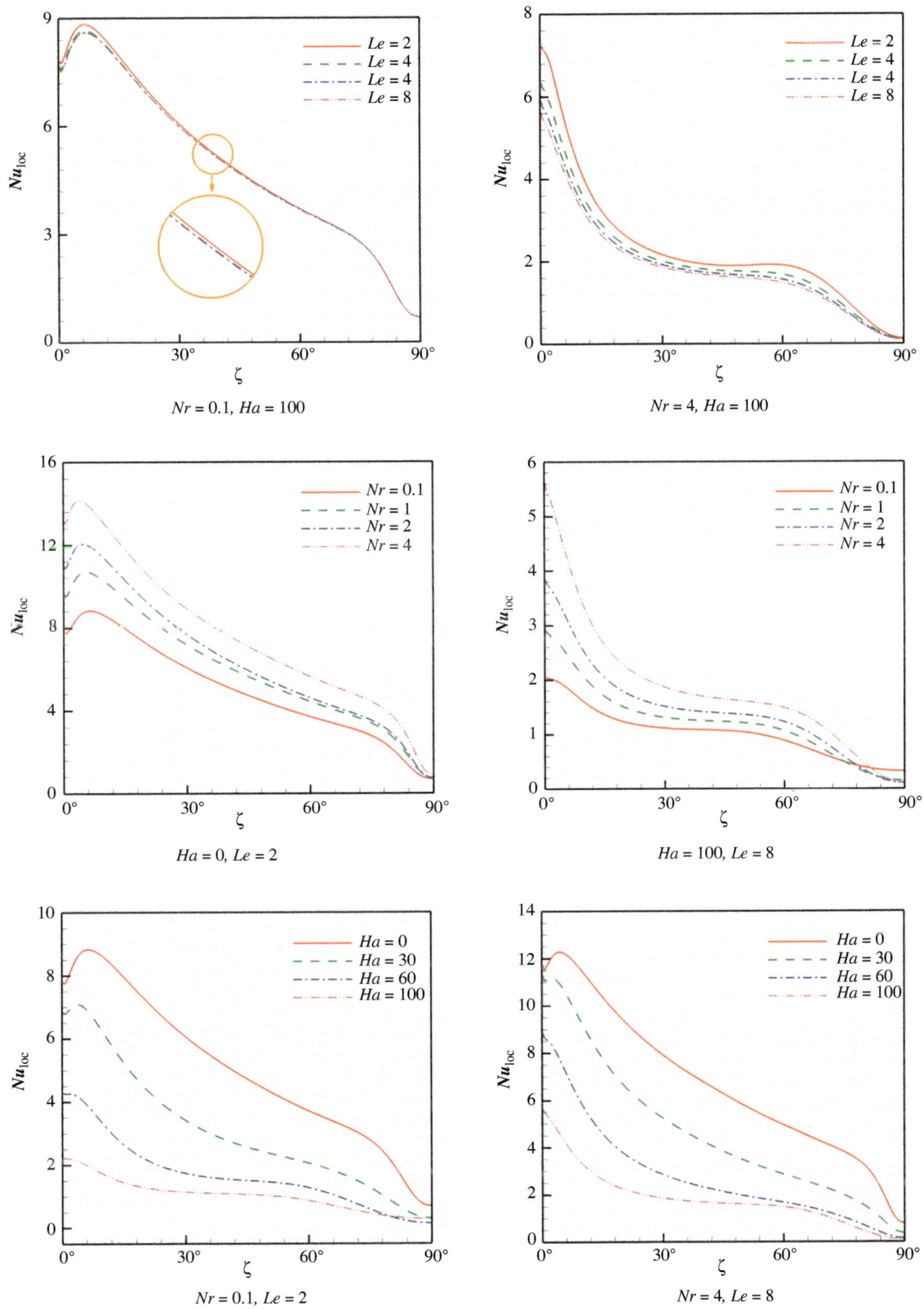

FIGURE 5.18 The effects of Hartmann number, buoyancy ratio number, and Lewis number on Local Nusselt number at $Nt = Nb = 0.5$, $Ra = 10^5$ and $\Pr = 10$.

results for positive Nr (aiding buoyancy forces), where the temperature and species induced buoyancy forces aides each other, are not considered. For $Nr = 0$, the species-induced buoyancy force has no effect on flow; the flow is solely driven by the thermal buoyancy force. However, the effect of species-induced buoyancy increases as the Nr value increases, and reaches a certain value, where the effect of thermally induced buoyancy becomes negligible in comparison to the solutal one. For small Nr value, the flow is mainly driven by the thermal buoyancy force. When Nr increases, a reverse thermal plume appears at $\zeta = 90^{\circ}$. This phenomena is due to an existing counterclockwise eddy at this region. By increasing the Hartmann number, the two main eddies merged into one counterclockwise eddy. The isoconcentration lines are more distorted with the increase of solutal forces. As Nr increases the upper region of the heatline contours are divided into two smaller ones and this new region disappears with the increase of the Hartmann number.

The mass flow is given by $\psi_{max} \approx \delta_s\, v$, where the solutal boundary layer thickness is given by $\delta_s \approx (Ra\, Le\, Nr)^{-1/4}$ and $v \approx (Ra\, Le\, Nr)^{1/2}$ so $\psi_{max} \approx (Ra\, Le\, Nr)^{1/4}$. The solutal boundary layer, δ_s, becomes thinner by increasing Le. Heatlines are found to be more distorted as Le augments.

Effects of Hartmann number, buoyancy ratio number, and Lewis number on the local Nusselt number are shown in Fig. 5.18. As seen, the local Nusselt number increases as buoyancy ratio number increases but it decreases with an increase of the Hartmann number and Lewis number. Generally as ζ increases the local Nusselt number decreases due to the increment of thermal boundary layer thickness. In the absence of a magnetic field Nu_{loc} has one local maximum near the bottom wall. The occurrence of a maxima for Nu_{loc} is due to

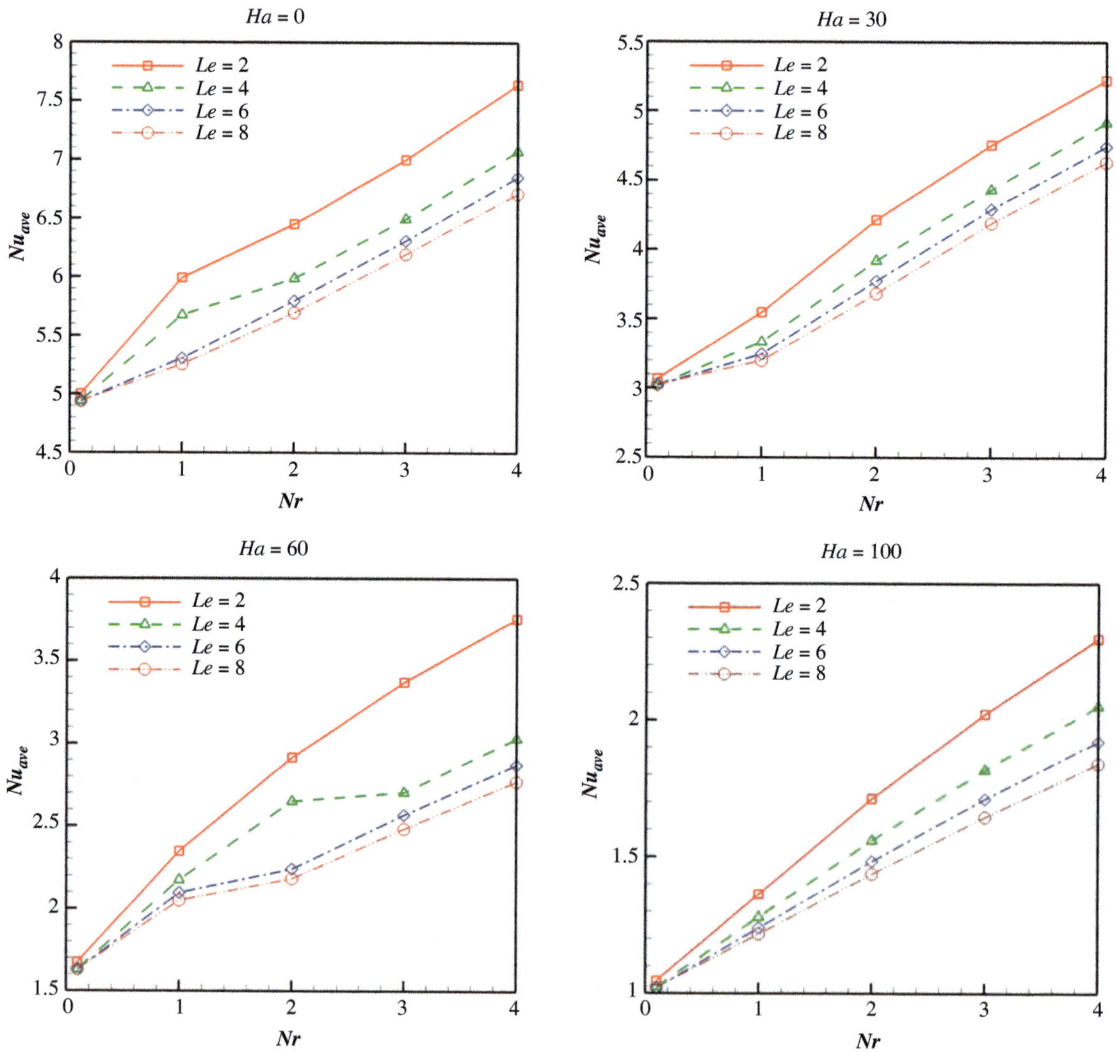

FIGURE 5.19 The effects of Hartmann number, buoyancy ratio number, and Lewis number on average Nusselt number at $Nt = Nb = 0.5, Ra = 10^5$ and $Pr = 10$.

dense heatlines based on conductive heat transport occurring at this portion. This point disappears at a high Hartmann number. The local Nusselt number profile has a minimum point at $\zeta = 90^{\circ}$ because of the existence of a thermal plume at this region. Also this figure shows that the effect of Lewis number on Nu_{loc} is negligible at low buoyancy ratio number.

The effects of Hartmann number, buoyancy ratio number, and Lewis number on average Nusselt number are shown in Fig. 5.19. The presence of a magnetic field leads to the disappearance of the thermal plume over the inner wall and makes the isotherms parallel to each other due to the domination of the conduction mode of heat transfer. Therefore the average Nusselt number decreases with the increase of the Hartmann number. As the Lewis number increases, the thermal boundary layer thickness increases and in turn the Nusselt number decreases. The effect of the buoyancy ratio number on Nu_{ave} is in contrast with Ha and Le. Also it can be found that effects of the Hartmann number and Lewis number are more pronounced at higher values of buoyancy ratio number.

References

[1] D.A. Nield, A.V. Kuznetsov, Thermal instability in a porous medium layer saturated by a nanofluid, International Journal of Heat and Mass Transfer 52 (2009) 5796–5801.

[2] W.A. Khan, I. Pop, Boundary-layer flow of a nanofluid past a stretching sheet, Int. J. Heat Mass Transfer 53 (2010) 2477–2483.

[3] C.-Y. Cheng, Natural convection boundary layer flow over a truncated cone in a porous medium saturated by a nanofluid, International Communications in Heat and Mass Transfer 39 (2012) 231–235.

[4] M. Sheikholeslami, M. Gorji-Bandpy, D.D. Ganji, P. Rana, Soheil Soleimani, Magnetohydrodynamic free convection of Al2O3-water nanofluid considering Thermophoresis and Brownian motion effects, Computers & Fluids 94 (2014) 147–160.

[5] M. Sheikholeslami, Ali J. Chamkha, P. Rana, R. Moradi, Combined thermophoresis and Brownian motion effects on nanofluid free convection heat transfer in an L-shaped enclosure, Chinese Journal of Physics 55 (2017) 2356–2370.

[6] M. Sheikholeslami, S. Abelman, Two phase simulation of nanofluid flow and heat transfer in an annulus in the presence of an axial magnetic field, IEEE Transactions on Nanotechnology 14 (3) (2015) 561–569.

[7] M. Sheikholeslami, H.B. Rokni, Effect of melting heat transfer on nanofluid flow in existence of magnetic field considering Buongiorno Model, Chinese Journal of Physics 55 (2017) 1115–1126.

[8] M. Sheikholeslami, H.B. Rokni, Influence of melting surface on MHD nanofluid flow by means of two phase model, Chinese Journal of Physics 55 (2017) 1352–1360.

[9] M. Sheikholeslami, Influence of Lorentz forces on nanofluid flow in a porous cylinder considering Darcy model, Journal of Molecular Liquids 225 (2017) 903–912.

[10] M. Sheikholeslami, CVFEM for magnetic nanofluid convective heat transfer in a porous curved enclosure, Eur. Phys. J. Plus 131 (2016) 413. Available from: https://doi.org/10.1140/epjp/i2016-16413-y.

[11] M. Sheikholeslami, Influence of Coulomb forces on Fe3O4-H2O nanofluid thermal improvement, International Journal of Hydrogen Energy 42 (2017) 821–829.

[12] M. Sheikholeslami, Magnetic field influence on nanofluid thermal radiation in a cavity with tilted elliptic inner cylinder, Journal of Molecular Liquids 229 (2017) 137–147.

[13] M. Sheikholeslami, Numerical simulation of magnetic nanofluid natural convection in porous media, Physics Letters A 381 (2017) 494–503.

[14] M. Sheikholeslami, Magnetic source impact on nanofluid heat transfer using CVFEM, Neural Computing and Applications (2016). Available from: https://doi.org/10.1007/s00521-016-2740-7.

[15] M. Sheikholeslami, CuO-water nanofluid free convection in a porous cavity considering Darcy law, The European Physical Journal Plus 132 (2017) 55. Available from: https://doi.org/10.1140/epjp/i2017-11330-3.

[16] M. Sheikholeslami, Numerical investigation of MHD nanofluid free convective heat transfer in a porous tilted enclosure, Engineering Computations 34 (6) (2017) 1939–1955.

[17] M.S. Kandelousi, Effect of spatially variable magnetic field on ferrofluid flow and heat transfer considering constant heat flux boundary condition, The European Physical Journal Plus (2014) 129–248. Available from: https://doi.org/10.1140/epjp/i2014-14248-2.

[18] M. Sheikholeslami, Numerical simulation for external magnetic field influence on Fe_3O_4-water nanofluid forced convection, Engineering Computations.

[19] M. Sheikholeslami, Influence of Lorentz forces on nanofluid flow in a porous cavity by means of Non- Darcy model, Engineering Computations 34 (8) (2017) 2651–2667. https://doi.org/10.1108/EC-01-2017-0008.

[20] M. Sheikholeslami, H.B. Rokni, Nanofluid convective heat transfer intensification in a porous circular cylinder, Chemical Engineering & Processing: Process Intensification 120 (2017) 93–104.

[21] M. Sheikholeslami, M.K. Sadoughi, Numerical modeling for Fe_3O_4 -water nanofluid flow in porous medium considering MFD viscosity, Journal of Molecular Liquids 242 (2017) 255–264.

[22] M. Sheikholeslami, M. Shamlooei, Magnetic source influence on nanofluid flow in porous medium considering shape factor effect, Physics Letters A 381 (2017) 3071–3078.

[23] M. Sheikholeslami, M. Shamlooei, Convective flow of nanofluid inside a lid driven porous cavity using CVFEM, Physica B: Condensed Matter 521 (2017) 239–250.

[24] M. Sheikholeslami, K. Vajravelu, Forced convection heat transfer in Fe_3O_4- ethylene glycol nanofluid under the influence of Coulomb force, Journal of Molecular Liquids 233 (2017) 203–210.

[25] M. Sheikholeslami, H.B. Rokni, Influence of EFD viscosity on nanofluid forced convection in a cavity with sinusoidal wall, Journal of Molecular Liquids 232 (2017) 390–395.
[26] M. Sheikholeslami, H.B. Rokni, Magnetic nanofluid natural convection in the presence of thermal radiation considering variable viscosity, The European Physical Journal Plus 132 (2017) 238. Available from: https://doi.org/10.1140/epjp/i2017-11498-4.
[27] M. Sheikholeslami, Houman B. Rokni, Numerical simulation for impact of Coulomb force on nanofluid heat transfer in a porous enclosure in presence of thermal radiation, Int. J. Heat Mass Transfer 118 (2018) 823–831.
[28] M. Sheikholeslami, M. Shamlooei, R. Moradi, Fe_3O_4- Ethylene glycol nanofluid forced convection inside a porous enclosure in existence of Coulomb force, Journal of Molecular Liquids 249 (2018) 429–437.
[29] M. Sheikholeslami, S.A. Shehzad, Numerical analysis of Fe_3O_4 $-H_2O$ nanofluid flow in permeable media under the effect of external magnetic source, Int. J. Heat Mass Transfer 118 (2018) 182–192.
[30] M. Sheikholeslami, S.A. Shehzad, Magnetohydrodynamic nanofluid convection in a porous enclosure considering heat flux boundary condition, International Journal of Heat and Mass Transfer 106 (2017) 1261–1269.
[31] M. Sheikholeslami, S.A. Shehzad, Thermal radiation of ferrofluid in existence of Lorentz forces considering variable viscosity, International Journal of Heat and Mass Transfer 109 (2017) 82–92.
[32] M. Sheikholeslami, H.B. Rokni, Numerical modeling of nanofluid natural convection in a semi annulus in existence of Lorentz force, Computer Methods in Applied Mechanics and Engineering 317 (2017) 419–430.
[33] M. Sheikholeslami, K. Vajravelu, Nanofluid flow and heat transfer in a cavity with variable magnetic field, Applied Mathematics and Computation 298 (2017) 272–282.
[34] M. Sheikholeslami, H.B. Rokni, Magnetohydrodynamic CuO-water nanofluid in a porous complex shaped enclosure, ASME, Journal of Thermal Science and Engineering Applications 9 (4) (2017) 041007. Available from: https://doi.org/10.1115/1.4035973.
[35] M. Sheikholeslami, H.B. Rokni, Nanofluid two phase model analysis in existence of induced magnetic field, International Journal of Heat and Mass Transfer 107 (2017) 288–299.
[36] M. Sheikholeslami, T. Hayat, A. Alsaedi, S. Abelman, Numerical analysis of EHD nanofluid force convective heat transfer considering electric field dependent viscosity, International Journal of Heat and Mass Transfer 108 (2017) 2558–2565.
[37] M. Sheikholeslami, Magnetic field influence on CuO $-H_2O$ nanofluid convective flow in a permeable cavity considering various shapes for nanoparticles, International Journal of Hydrogen Energy 42 (2017) 19611–19621.
[38] M. Sheikholeslami, M.M. Bhatti, Forced convection of nanofluid in presence of constant magnetic field considering shape effects of nanoparticles, International Journal of Heat and Mass Transfer 111 (2017) 1039–1049.
[39] M. Sheikholeslami, A. Zeeshan, Analysis of flow and heat transfer in water based nanofluid due to magnetic field in a porous enclosure with constant heat flux using CVFEM, Computer Methods in Applied Mechanics and Engineering 320 (2017) 68–81.
[40] M. Sheikholeslami, M. Shamlooei, Fe_3O_4- H_2O nanofluid natural convection in presence of thermal radiation, International Journal of Hydrogen Energy 42 (9) (2017) 5708–5718.
[41] M. Sheikholeslami, M.M. Bhatti, Active method for nanofluid heat transfer enhancement by means of EHD, International Journal of Heat and Mass Transfer 109 (2017) 115–122.
[42] M. Sheikholeslami, T. Hayat, A. Alsaedi, Numerical study for external magnetic source influence on water based nanofluid convective heat transfer, International Journal of Heat and Mass Transfer 106 (2017) 745–755.
[43] M. Sheikholeslami, A.J. Chamkha, Flow and convective heat transfer of a ferro-nanofluid in a double-sided lid-driven cavity with a wavy wall in the presence of a variable magnetic field, Numerical Heat Transfer, Part A 69 (10) (2016) 1186–1200. Available from: https://doi.org/10.1080/10407782.2015.1125709.
[44] M. Sheikholeslami, Ali J. Chamkha, Electrohydrodynamic free convection heat transfer of a nanofluid in a semi-annulus enclosure with a sinusoidal wall, Numerical Heat Transfer, Part A 69 (7) (2016) 781–793. Available from: https://doi.org/10.1080/10407782.2015.1090819.
[45] M. Sheikholeslami, M.K. Sadoughi, Simulation of CuO- water nanofluid heat transfer enhancement in presence of melting surface, International Journal of Heat and Mass Transfer 116 (2018) 909–919.
[46] M. Sheikholeslami, H.B. Rokni, Simulation of nanofluid heat transfer in presence of magnetic field: A review, International Journal of Heat and Mass Transfer 115 (2017) 1203–1233.
[47] M. Sheikholeslami, H.B. Rokni, Melting heat transfer influence on nanofluid flow inside a cavity in existence of magnetic field, International Journal of Heat and Mass Transfer 114 (2017) 517–526.
[48] M. Sheikholeslami, S.A. Shehzad, Magnetohydrodynamic nanofluid convective flow in a porous enclosure by means of LBM, International Journal of Heat and Mass Transfer 113 (2017) 796–805.
[49] M. Sheikholeslami, Magnetohydrodynamic nanofluid forced convection in a porous lid driven cubic cavity using Lattice Boltzmann Method, Journal of Molecular Liquids 231 (2017) 555–565.
[50] M.S. Kandelousi, KKL correlation for simulation of nanofluid flow and heat transfer in a permeable channel, Physics Letters A 378 (45) (2014) 3331–3339.
[51] M. Sheikholeslami, Effect of uniform suction on nanofluid flow and heat transfer over a cylinder, J Braz. Soc. Mech. Sci. Eng. 37 (2015) 1623–1633.
[52] M. Sheikholeslami, Lattice Boltzmann Method simulation of MHD non-Darcy nanofluid free convection, Physica B 516 (2017) 55–71.
[53] M. Sheikholeslami, Influence of magnetic field on nanofluid free convection in an open porous cavity by means of Lattice Boltzmann Method, Journal of Molecular Liquids 234 (2017) 364–374.
[54] M. Sheikholeslami, A.J. Chamkha, Influence of Lorentz forces on nanofluid forced convection considering Marangoni convection, Journal of Molecular Liquids 225 (2017) 750–757.
[55] M. Sheikholeslami, T. Hayat, A. Alsaedi, Numerical simulation of nanofluid forced convection heat transfer improvement in existence of magnetic field using Lattice Boltzmann Method, International Journal of Heat and Mass Transfer 108 (2017) 1870–1883.
[56] M. Sheikholeslami, R. Ellahi, Three dimensional mesoscopic simulation of magnetic field effect on natural convection of nanofluid, International Journal of Heat and Mass Transfer 89 (2015) 799–808.

[57] M. Sheikholeslami, R. Ellahi, Simulation of ferrofluid flow for magnetic drug targeting using Lattice Boltzmann method, Journal of Zeitschrift Fur Naturforschung A 70 (2) (2015) 115–124.
[58] M. Sheikholeslami, M. Seyednezhad, Lattice Boltzmann Method simulation for CuO-water nanofluid flow in a porous enclosure with hot obstacle, Journal of Molecular Liquids 243 (2017) 249–256.
[59] M. Sheikholeslami, M. Sadoughi, Mesoscopic method for MHD nanofluid flow inside a porous cavity considering various shapes of nanoparticles, International Journal of Heat and Mass Transfer 113 (2017) 106–114.
[60] M. Sheikholeslami, A. Zeeshan, Mesoscopic simulation of CuO-H_2O nanofluid in a porous enclosure with elliptic heat source, International Journal of Hydrogen Energy 42 (22) (2017) 15393–15402.
[61] M. Sheikholeslami, S.A. Shehzad, CVFEM for influence of external magnetic source on $Fe_3O_4 - H_2O$ nanofluid behavior in a permeable cavity considering shape effect, International Journal of Heat and Mass Transfer 115 (2017) 180–191.
[62] M. Sheikholeslami, M. Seyednezhad, Nanofluid heat transfer in a permeable enclosure in presence of variable magnetic field by means of CVFEM, International Journal of Heat and Mass Transfer 114 (2017) 1169–1180.
[63] M. Sheikholeslami, T. Hayat, A. Alsaedi, MHD free convection of Al2O3–water nanofluid considering thermal radiation: A numerical study, International Journal of Heat and Mass Transfer 96 (2016) 513–524.
[64] M. Sheikholeslami, K. Vajravelu, M.M. Rashidi, Forced convection heat transfer in a semi annulus under the influence of a variable magnetic field, International Journal of Heat and Mass Transfer 92 (2016) 339–348.
[65] M. Sheikholeslami, H.B. Rokni, Free convection of CuO-H_2O nanofluid in a curved porous enclosure using mesoscopic approach, International Journal of Hydrogen Energy 42 (22) (2017) 14942–14949.
[66] M. Sheikholeslami, Numerical investigation for CuO-H_2O nanofluid flow in a porous channel with magnetic field using mesoscopic method, J. Mol. Liq. 249 (2018) 739–746.
[67] M. Sheikholeslami, M. Barzegar Gerdroodbary, S. Valiallah Mousavi, D.D. Ganji, Rasoul Moradi, Heat transfer enhancement of ferrofluid inside an 90o elbow channel by non-uniform magnetic field, J. Magnet. Magnet. Mater. 460 (2018) 302–311.
[68] M. Sheikholeslami, Numerical simulation for solidification in a LHTESS by means of nano-enhanced PCM, J. Taiwan Inst. Chem. Eng. 86 (2018) 25–41.
[69] M. Sheikholeslami, Numerical modeling of Nano enhanced PCM solidification in an enclosure with metallic fin, J. Mol. Liq. 259 (2018) 424–438.
[70] M. Sheikholeslami, A. Ghasemi, Solidification heat transfer of nanofluid in existence of thermal radiation by means of FEM, Int. J. Heat Mass Transfer 123 (2018) 418–431.
[71] M. Sheikholeslami, S.A. Shehzad, CVFEM simulation for nanofluid migration in a porous medium using Darcy model, Int. J. Heat Mass Transfer 122 (2018) 1264–1271.
[72] M. Sheikholeslami, Milad Darzi, M.K. Sadoughi, Heat transfer improvement and pressure drop during condensation of refrigerant-based nanofluid; an experimental procedure, Int. J. Heat Mass Transfer 122 (2018) 643–650.
[73] M. Sheikholeslami, Houman B. Rokni, CVFEM for effect of Lorentz forces on nanofluid flow in a porous complex shaped enclosure by means of non-equilibrium model, J. Mol. Liq. 254 (2018) 446–462.
[74] M. Sheikholeslami, S.A. Shehzad, Zhixiong Li, Water based nanofluid free convection heat transfer in a three dimensional porous cavity with hot sphere obstacle in existence of Lorenz forces, Int. J. Heat Mass Transfer 125 (2018) 375–386.
[75] M. Sheikholeslami, M. Jafaryar, D.D. Ganji, Zhixiong Li, Exergy loss analysis for nanofluid forced convection heat transfer in a pipe with modified turbulators, J. Mol. Liq. 262 (2018) 104–110.
[76] M. Sheikholeslami, M. Jafaryar, Zhixiong Li, Nanofluid turbulent convective flow in a circular duct with helical turbulators considering CuO nanoparticles, Int. J. Heat Mass Transfer 124 (2018) 980–989.
[77] M. Sheikholeslami, Z. Li, M. Shamlooei, Nanofluid, MHD natural convection through a porous complex shaped cavity considering thermal radiation, Phys. Lett. A 382 (2018) 1615–1632.
[78] M. Sheikholeslami, H.B. Rokni, Magnetic nanofluid flow and convective heat transfer in a porous cavity considering Brownian motion effects, Phys. Fluids 30 (1) (2018). Available from: https://doi.org/10.1063/1.5012517.
[79] M. Sheikholeslami, S.A. Shehzad, Simulation of water based nanofluid convective flow inside a porous enclosure via non-equilibrium model, Int. J. Heat Mass Transfer 120 (2018) 1200–1212.
[80] M. Sheikholeslami, S.A. Shehzad, Non-Darcy free convection of Fe_3O_4-water nanoliquid in a complex shaped enclosure under impact of uniform Lorentz force, Chin. J. Phys. 56 (2018) 270–281.
[81] M. Sheikholeslami, M. Seyednezhad, Simulation of nanofluid flow and natural convection in a porous media under the influence of electric field using CVFEM, Int. J. Heat Mass Transfer 120 (2018) 772–781.
[82] M. Sheikholeslami, M. Shamlooei, R. Moradi, Numerical simulation for heat transfer intensification of nanofluid in a permeable curved enclosure considering shape effect of Fe_3O_4 nanoparticles, Chem. Eng. Process. Process Intensif. 124 (2018) 71–82.
[83] M. Sheikholeslami, T. Hayat, T. Muhammad, A. Alsaedi, MHD forced convection flow of nanofluid in a porous cavity with hot elliptic obstacle by means of Lattice Boltzmann method, Int. J. Mech. Sci. 135 (2018) 532–540.
[84] M. Sheikholeslami, Numerical investigation of nanofluid free convection under the influence of electric field in a porous enclosure, J. Mol. Liq. 249 (2018) 1212–1221.
[85] Mohsen Sheikholeslami, CuO-water nanofluid flow due to magnetic field inside a porous media considering Brownian motion, J. Mol. Liq. 249 (2018) 921–929.
[86] M. Sheikholeslami, D.D. Ganji, Influence of electric field on Fe_3O_4- water nanofluid radiative and convective heat transfer in a permeable enclosure, J. Mol. Liq. 250 (2018) 404–412.
[87] M. Sheikholeslami, M. Shamlooei, R. Moradi, Fe_3O_4- Ethylene glycol nanofluid forced convection inside a porous enclosure in existence of Coulomb force, J. Mol. Liq. 249 (2018) 429–437.
[88] M. Sheikholeslami, T. Hayat, A. Alsaedi, Numerical simulation for forced convection flow of MHD CuO-H_2O nanofluid inside a cavity by means of LBM, J. Mol. Liq. 249 (2018) 941–948.

[89] M. Sheikholeslami, H.R. Kataria, A.S. Mittal, Effect of thermal diffusion and heat-generation on MHD nanofluid flow past an oscillating vertical plate through porous medium, J. Mol. Liq. 257 (2018) 12–25.
[90] M. Sheikholeslami, S.A. Shehzad, Z. Li, Nanofluid heat transfer intensification in a permeable channel due to magnetic field using Lattice Boltzmann method, Phys. B (2018). Available from: https://doi.org/10.1016/j.physb.2018.03.036.
[91] M. Sheikholeslami, A. Zeeshan, Numerical simulation of Fe_3O_4 -water nanofluid flow in a non-Darcy porous media, Int. J. Numer. Methods Heat Fluid Flow 28 (3) (2018) 641–660. Available from: https://doi.org/10.1108/HFF-04-2017-0160.

CHAPTER

6

Nanofluid Forced and Mixed Convection Heat Transfer by Means of CVFEM

6.1 INTRODUCTION

Mixed convective flows are commonly encountered in many environmental and scientific developments, for instance, in aeronautics, fire research, heating and cooling of channels, etc. A vast amount of literature is dedicated to the studies related to forced or mixed convection in various systems which are simplified models of some of the important engineering applications [1]. Lid velocity was assumed to be constant in most of the studies related to the mixed convection in lid driven cavity problems. Many researchers studied the oscillating lid parameters and found that the flow field and heat transfer characteristics in cavities are affected by the variation of the oscillating lid parameters [2,3]. Nowadays, the energy efficiency is an extremely important topic in view of thermal conductivity enhancement amongst the researchers. For this purpose the researchers considered the involvement of nanoparticles in the base fluid. Sheikholeslami and Rashidi [4] investigated ferrofluid mixed convection heat transfer in the presence of variable magnetic field. Several publications exist about nanofluid flow in the presence of external forces [5–90].

6.2 MAGNETIC NANOFLUID MIXED CONVECTION HEAT TRANSFER TREATMENT IN THE PRESENCE OF VARIABLE MAGNETIC FIELD

6.2.1 Problem Definition

The schematic diagram and the mesh of the semiannulus enclosure used in the present CVFEM program are shown in Fig. 6.1. The inner wall is maintained at constant temperatures T_h and the other walls are maintained at constant temperature T_c ($T_h > T_c$). For the expression of the magnetic field strength it can be considered that the magnetic source represents a magnetic wire placed vertically to the x−y plane at the point $(\bar{a}, \bar{b})$. The components of the magnetic field intensity $(\overline{H_x}, \overline{H_y})$ and the magnetic field strength $(\overline{H})$ can be considered as:

$$\overline{H_x} = \frac{\gamma'}{2\pi}\frac{1}{(x-\bar{a})^2 + (y-\bar{b})^2}(y - \bar{b}) \tag{6.1}$$

$$\overline{H_y} = -\frac{\gamma'}{2\pi}\frac{1}{(x-\bar{a})^2 + (y-\bar{b})^2}(x - \bar{a}) \tag{6.2}$$

$$\overline{H} = \sqrt{\overline{H}_x^2 + \overline{H}_y^2} = \frac{\gamma'}{2\pi}\frac{1}{\sqrt{(x-\bar{a})^2 + (y-\bar{b})^2}} \tag{6.3}$$

where γ' is the magnetic field strength at the source (of the wire) and $(\bar{a}, \bar{b})$ is the position where the source is located. The contours of the magnetic field strength are shown in Fig. 6.2. In this study magnetic source is located at $(-0.01cols, 0.5rows)$. The upper wall is Lid driven with velocity of U_{Lid}. In this study γ is equal to 45°.

Application of Control Volume based Finite Element Method (CVFEM) for Nanofluid Flow and Heat Transfer.
DOI: https://doi.org/10.1016/B978-0-12-814152-6.00006-0

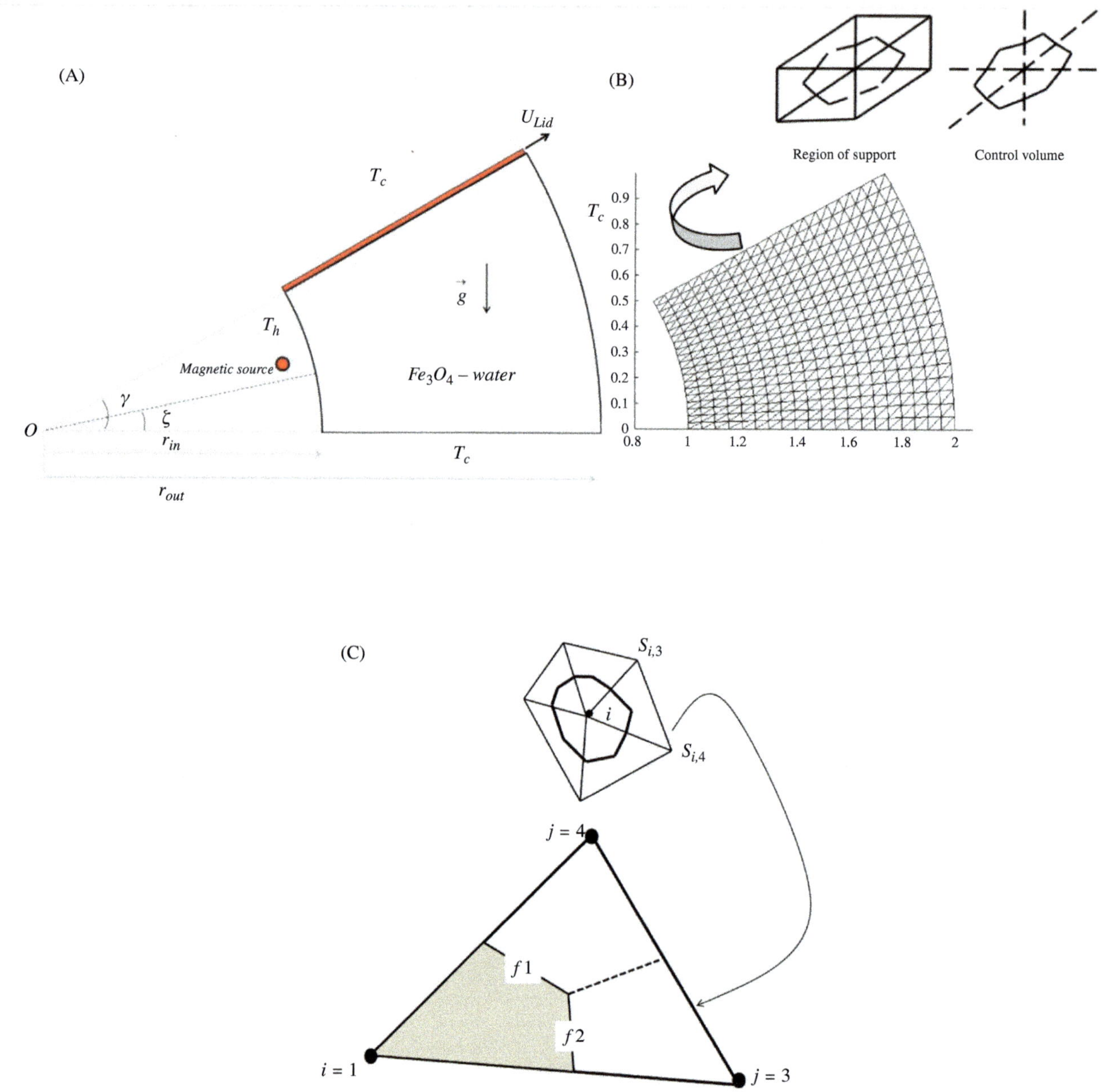

FIGURE 6.1 (A) Geometry and the boundary conditions; (B) the mesh of enclosure considered in this work; (C) A sample triangular element and its corresponding control volume.

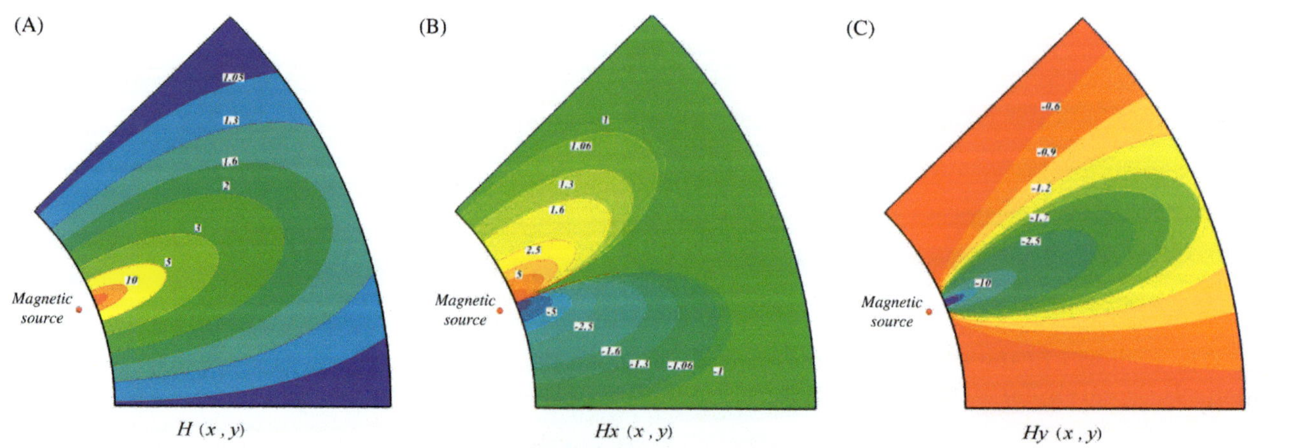

FIGURE 6.2 Contours of the (A) magnetic field strength H; (B) magnetic field intensity component in x direction Hx; (C) magnetic field intensity component in y direction Hy.

6.2.2 Governing Equation

The flow is two-dimensional, laminar, and incompressible. The magnetic Reynolds number is assumed to be small so that the induced magnetic field can be neglected compared to the applied magnetic field. The flow is considered to be steady, two-dimensional, and laminar. Using the Boussinesq approximation, the governing equations of heat transfer and fluid flow for nanofluid can be obtained as follows:

$$\frac{\partial u}{\partial x}+\frac{\partial v}{\partial y}=0 \tag{6.4}$$

$$\rho_{nf}\left(u\frac{\partial u}{\partial x}+v\frac{\partial u}{\partial y}\right)=-\frac{\partial P}{\partial x}+\eta\left(\frac{\partial^2 u}{\partial x^2}+\frac{\partial^2 u}{\partial y^2}\right)+\mu_0 M\frac{\partial \overline{H}}{\partial x}-\sigma_{nf}B_y^2u+\sigma_{nf}B_xB_yv \tag{6.5}$$

$$\rho_{nf}\left(u\frac{\partial v}{\partial x}+v\frac{\partial v}{\partial y}\right)=-\frac{\partial P}{\partial y}+\eta\left(\frac{\partial^2 v}{\partial x^2}+\frac{\partial^2 v}{\partial y^2}\right)+\mu_0 M\frac{\partial \overline{H}}{\partial y}-\sigma_{nf}B_x^2v+\sigma_{nf}B_xB_yu+\rho_{nf}\beta_{nf}g(T-T_c) \tag{6.6}$$

$$\begin{aligned}(\rho C_p)_{nf}\left(u\frac{\partial T}{\partial x}+v\frac{\partial T}{\partial y}\right)&=k_{nf}\left(\frac{\partial^2 T}{\partial x^2}+\frac{\partial^2 T}{\partial y^2}\right)+\sigma_{nf}\left(uB_y-vB_x\right)^2\\&-\mu_0 T\frac{\partial M}{\partial T}\left(u\frac{\partial \overline{H}}{\partial x}+v\frac{\partial \overline{H}}{\partial y}\right)+\eta\left\{2\left(\frac{\partial u}{\partial x}\right)^2+2\left(\frac{\partial v}{\partial x}\right)^2+\left(\frac{\partial u}{\partial x}+\frac{\partial v}{\partial y}\right)^2\right\}\end{aligned} \tag{6.7}$$

where $\eta=(1+\vec{\delta}.\vec{B})\mu_{nf}$, the variation of MFD viscosity (δ) has been taken to be isotropic, $\delta_1=\delta_2=\delta_3=\delta$. The terms $\mu_0 M\frac{\partial \overline{H}}{\partial x}$ and $\mu_0 M\frac{\partial \overline{H}}{\partial y}$ in (6.5) and (6.6), respectively, represent the components of magnetic force, per unit volume, and depend on the existence of the magnetic gradient on the corresponding x and y directions. These two terms are well known from FHD which is the so-called the Kelvin force. The terms $-\sigma_{nf}B_y^2u+\sigma_{nf}B_xB_yv$ and $-\sigma_{nf}B_x^2v+\sigma_{nf}B_xB_yu$ appearing in (6.5) and (6.6), respectively, represent the Lorentz force per unit volume towards the x and y directions and arise due to the electrical conductivity of the fluid. These two terms are known in MHD. The principles of MHD and FHD are combined in the mathematical model presented in [4] and the abovementioned terms arise together in the governing Eqs. (6.5) and (6.6). The term $\mu_0 T\frac{\partial M}{\partial T}\left(u\frac{\partial \overline{H}}{\partial x}+v\frac{\partial \overline{H}}{\partial y}\right)$ in Eq. (6.7) represents the thermal power per unit volume due to the magneto caloric effect. Also the term $\sigma_{nf}\left(uB_y-vB_x\right)^2$ in (6.7) represents the Joule heating. For the variation of the magnetization M, with the magnetic field intensity $\overline{H}$ and temperature T, the following relation, derived experimentally in [4], is considered:

$$M=K'\overline{H}(T_c'-T) \tag{6.8}$$

where K' is a constant and T_c' is the Curie temperature.

In the above equations, μ_0 is the magnetic permeability of vacuum $(4\pi\times10^{-7}Tm/A)$, $\overline{H}$ is the magnetic field strength, $\overline{B}$ is the magnetic induction $(\overline{B}=\mu_0\overline{H})$ and the bar above the quantities denotes that they are dimensional. The effective density (ρ_{nf}) and heat capacitance $(\rho C_p)_{nf}$ of the nanofluid are defined as:

$$\rho_{nf}=\rho_f(1-\phi)+\rho_s\phi \tag{6.9}$$

$$(\rho C_p)_{nf}=(\rho C_p)_f(1-\phi)+(\rho C_p)_s\phi \tag{6.10}$$

where ϕ is the solid volume fraction of nanoparticles. Thermal diffusivity of the nanofluid is

$$\alpha_{nf}=k_{nf}/(\rho C_p)_{nf} \tag{6.11}$$

and the thermal expansion coefficient of the nanofluid can be determined as

$$\beta_{nf}=\beta_f(1-\phi)+\beta_s\phi \tag{6.12}$$

The dynamic viscosity of the nanofluid given by Brinkman is

$$\mu_{nf}=\mu_f(1-\phi)^{-2.5} \tag{6.13}$$

The effective thermal conductivity of the nanofluid can be approximated by the Maxwell–Garnetts (MG) model as:

$$\frac{k_{nf}}{k_f} = \frac{k_s + 2k_f - 2\phi(k_f - k_s)}{k_s + 2k_f + \phi(k_f - k_s)} \tag{6.14}$$

and the effective electrical conductivity of nanofluid was presented by Maxwell as below:

$$\frac{\sigma_{nf}}{\sigma_f} = 1 + 3(\sigma_s/\sigma_f - 1)\phi/[(\sigma_s/\sigma_f + 2) - (\sigma_s/\sigma_f - 1)\phi] \tag{6.15}$$

By introducing the following nondimensional variables:

$$X = \frac{x}{L},\quad Y = \frac{y}{L},\quad \Theta = \frac{T - T_c}{T_h - T_c}, U = \frac{u}{U_{Lid}}, V = \frac{v}{U_{Lid}}, H = \frac{\overline{H}}{\overline{H_0}}, H_x = \frac{\overline{H_x}}{\overline{H_0}}, H_y = \frac{\overline{H_y}}{\overline{H_0}} \tag{6.16}$$

where in Eq. (6.16) $\overline{H_0} = \overline{H}(\overline{a}, 0) = \frac{\gamma}{2\pi|b|}$ and $L = r_{out} - r_{in} = r_{in}$.

Using the dimensionless parameters, the equations now become:

$$\frac{\partial U}{\partial X} + \frac{\partial V}{\partial Y} = 0 \tag{6.17}$$

$$\begin{aligned} U\frac{\partial U}{\partial X} + V\frac{\partial U}{\partial Y} &= -\frac{\partial P}{\partial X} + \frac{1}{\mathrm{Re}}\left[\frac{\mu_{nf}/\mu_f}{\rho_{nf}/\rho_f}\right](1 + \delta^*(H_x + H_y))\left(\frac{\partial^2 U}{\partial X^2} + \frac{\partial^2 U}{\partial Y^2}\right) \\ &+ Mn_F\left(\frac{\rho_f}{\rho_{nf}}\right)(\varepsilon_2 - \varepsilon_1 - \Theta)H\frac{\partial H}{\partial X} - \frac{Ha^2}{\mathrm{Re}}\left[\frac{\sigma_{nf}/\sigma_f}{\rho_{nf}/\rho_f}\right]\left(H_y^2 U - H_x H_y V\right) \end{aligned} \tag{6.18}$$

$$\begin{aligned} U\frac{\partial V}{\partial X} + V\frac{\partial V}{\partial Y} &= -\frac{\partial P}{\partial X} + \frac{1}{\mathrm{Re}}\left[\frac{\mu_{nf}/\mu_f}{\rho_{nf}/\rho_f}\right](1 + \delta^*(H_x + H_y))\left(\frac{\partial^2 V}{\partial X^2} + \frac{\partial^2 V}{\partial Y^2}\right) \\ &+ Mn_F\left(\frac{\rho_f}{\rho_{nf}}\right)(\varepsilon_2 - \varepsilon_1 - \Theta)H\frac{\partial H}{\partial Y} - \frac{Ha^2}{\mathrm{Re}}\left[\frac{\sigma_{nf}/\sigma_f}{\rho_{nf}/\rho_f}\right](H_x^2 V - H_x H_y U) + \frac{Gr}{\mathrm{Re}^2}\left[\frac{\beta_{nf}}{\beta_f}\right]\Theta \end{aligned} \tag{6.19}$$

$$\begin{aligned} U\frac{\partial \Theta}{\partial X} + V\frac{\partial \Theta}{\partial Y} &= \frac{1}{\mathrm{Pr\,Re}}\left[\frac{\frac{k_{nf}}{k_f}}{\frac{(\rho C_P)_{nf}}{(\rho C_P)_f}}\right]\left(\frac{\partial^2 \Theta}{\partial X^2} + \frac{\partial^2 \Theta}{\partial Y^2}\right) \\ &+ Ha^2\frac{Ec}{\mathrm{Re}}\left[\frac{\frac{\sigma_{nf}}{\sigma_f}}{\frac{(\rho C_P)_{nf}}{(\rho C_P)_f}}\right]\{U H_y - V H_x\}^2 + Mn_F Ec\frac{(\rho C_P)_f}{(\rho C_P)_{nf}}\left\{U\frac{\partial H}{\partial X} + V\frac{\partial H}{\partial Y}\right\}H(\varepsilon_1 + \Theta) \\ &+ \frac{Ec}{\mathrm{Re}}\left[\frac{\frac{\mu_{nf}}{\mu_f}}{\frac{(\rho C_P)_{nf}}{(\rho C_P)_f}}\right](1 + \delta^*(H_x + H_y))\left\{2\left(\frac{\partial U}{\partial X}\right)^2 + 2\left(\frac{\partial V}{\partial Y}\right)^2 + \left(\frac{\partial U}{\partial Y} + \frac{\partial V}{\partial X}\right)^2\right\} \end{aligned} \tag{6.20}$$

where $\mathrm{Re} = \frac{\rho_f L\, U_{Lid}}{\mu_f}$, $Gr = g\beta\Delta T L^3/v^2$, $Ha = LH_0\mu_0\sqrt{\sigma_f/\mu_f}$, $\varepsilon_1 = T_1/\Delta T$, $Ec = \left(\rho_f U_{Lid}^2\right)/\left[(\rho C_P)_f \Delta T\right]$, $Ri = Gr/\mathrm{Re}^2$, $\delta^* = \delta\mu_0 H_0$, and $Mn_F = \mu_0 H_0^2 K'(T_h - T_c)/\left(\rho_f\, U_{Lid}^2\right)$ are the Reynolds number, Grashof number, Hartmann number arising from MHD, temperature number, Eckert number, Richardson number, viscosity parameter, and Magnetic number arising from FHD the for the base fluid, respectively. The thermophysical properties of the nanofluid are given in Table 6.1.

TABLE 6.1 Thermophysical Properties of Water and Nanoparticles

	$\rho(kg/m^3)$	$C_p(j/kgk)$	$k(W/m.k)$	$\beta \times 10^5(K^{-1})$	$d_p(nm)$	$\sigma(\Omega \cdot m)^{-1}$
Pure water	997.1	4179	0.613	21	-	0.05
Fe_3O_4	5200	670	6	1.3	47	25000

The stream function and vorticity are defined as:

$$u = \frac{\partial \psi}{\partial y}, v = -\frac{\partial \psi}{\partial x}, \omega = \frac{\partial v}{\partial x} - \frac{\partial u}{\partial y}$$
$$\Omega = \frac{\omega L}{U_{Lid}}, \Psi = \frac{\psi}{LU_{Lid}} \tag{6.21}$$

The stream function satisfies the continuity Eq. (6.17). The vorticity equation is obtained by eliminating the pressure between the two momentum equations, i.e., by taking the y-derivative of Eq. (6.18) and subtracting from it the x-derivative of Eq. (6.19).

The boundary conditions as shown in Fig. 6.1 are:

$$\begin{aligned} &\Theta = 1.0 \quad \text{on the inner circular boundary} \\ &\Theta = 0.0 \quad \text{on the other walls} \end{aligned} \tag{6.22}$$

The values of vorticity on the boundary of the enclosure can be obtained using the stream function formulation and the known velocity conditions during the iterative solution procedure.

The local Nusselt number of the nanofluid along the hot wall can be expressed as:

$$Nu_{loc} = \left(\frac{k_{nf}}{k_f}\right)\frac{\partial \Theta}{\partial r} \tag{6.23}$$

where r is the radial direction. The average Nusselt number on the hot circular wall is evaluated as:

$$Nu_{ave} = \frac{1}{\gamma}\int_0^{\gamma} Nu_{loc}(\zeta)\, d\zeta \tag{6.24}$$

6.2.3 Effects of Active Parameters

In this study, mixed convection heat transfer of ferrofluid in the presence of variable magnetic field is investigated using CVFEM. Also the effect of magnetic field dependent (MFD) viscosity on hydrothermal behavior is considered. The mathematical model used for the formulation of the problem is consistent with the principles of FHD and MHD. Calculations are made for various values of volume fraction of nanoparticles ($\phi = 0\%$ and 4%), Richardson numbers ($Ri = 0.001, 1$ and 10), Magnetic number ($Mn_F = 0, 2, 4, 6$ and 10), Hartmann number ($Ha = 0,\ 5$ and 10), and viscosity parameter ($\delta^* = 0, 0.2, 0.4$ and 0.6). In all calculations, the Prandtl number (Pr), temperature number (ε_1), Eckert number (Ec), and Reynolds number (Re) are set to 6.8, 0.0, 10^{-5}, and 100, respectively.

Comparison of the streamlines between nanofluid and pure fluid is shown in Fig. 6.3. The velocity components of nanofluid are increased because of an increase in the energy transport in the fluid with the increase of volume fraction. Thermal boundary layer thickness decreases with increase of nanofluid volume fraction. Isotherm and streamline contours for different values of viscosity parameter, Richardson, Hartmann, and magnetic numbers are shown in Figs. 6.4, 6.5, and 6.6. When $Ri = 0.01$, the heat transfer in the enclosure is mainly dominated by the conduction mode. At $Mn_F = 0, Ha = 0$ the streamlines show one rotating eddy. When Hartmann number increases a small vortex generates near the magnetic source, so a thermal plume appears in this region. As magnetic number increases, the main vortex turns into two smaller vortexes and in turn a reverse

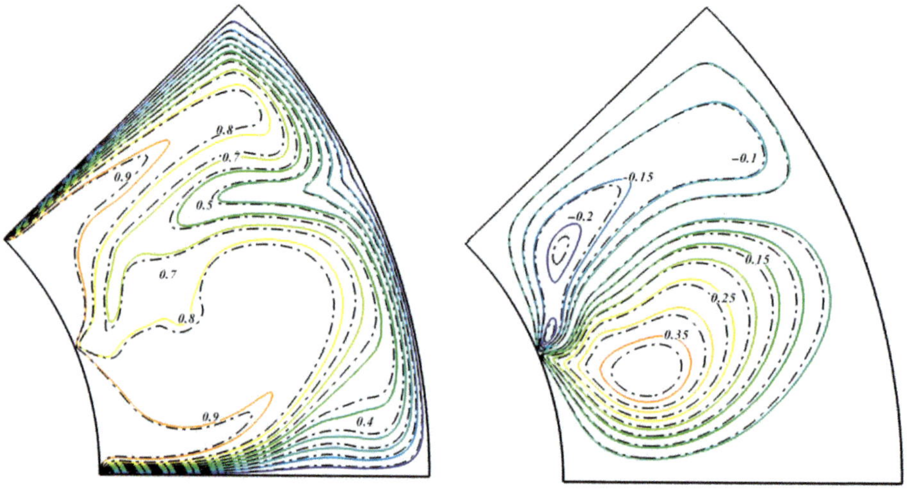

FIGURE 6.3 Comparison of the streamlines between nanofluid ($\phi = 0.04$) (— · — · —) and pure fluid ($\phi = 0$) (—) when $Ri = 10, Re = 100, Mn_F = 10, Ha = 10, Ec = 10^{-5}, \varepsilon_1 = 0, \delta^* = 0$, Pr = 6.8.

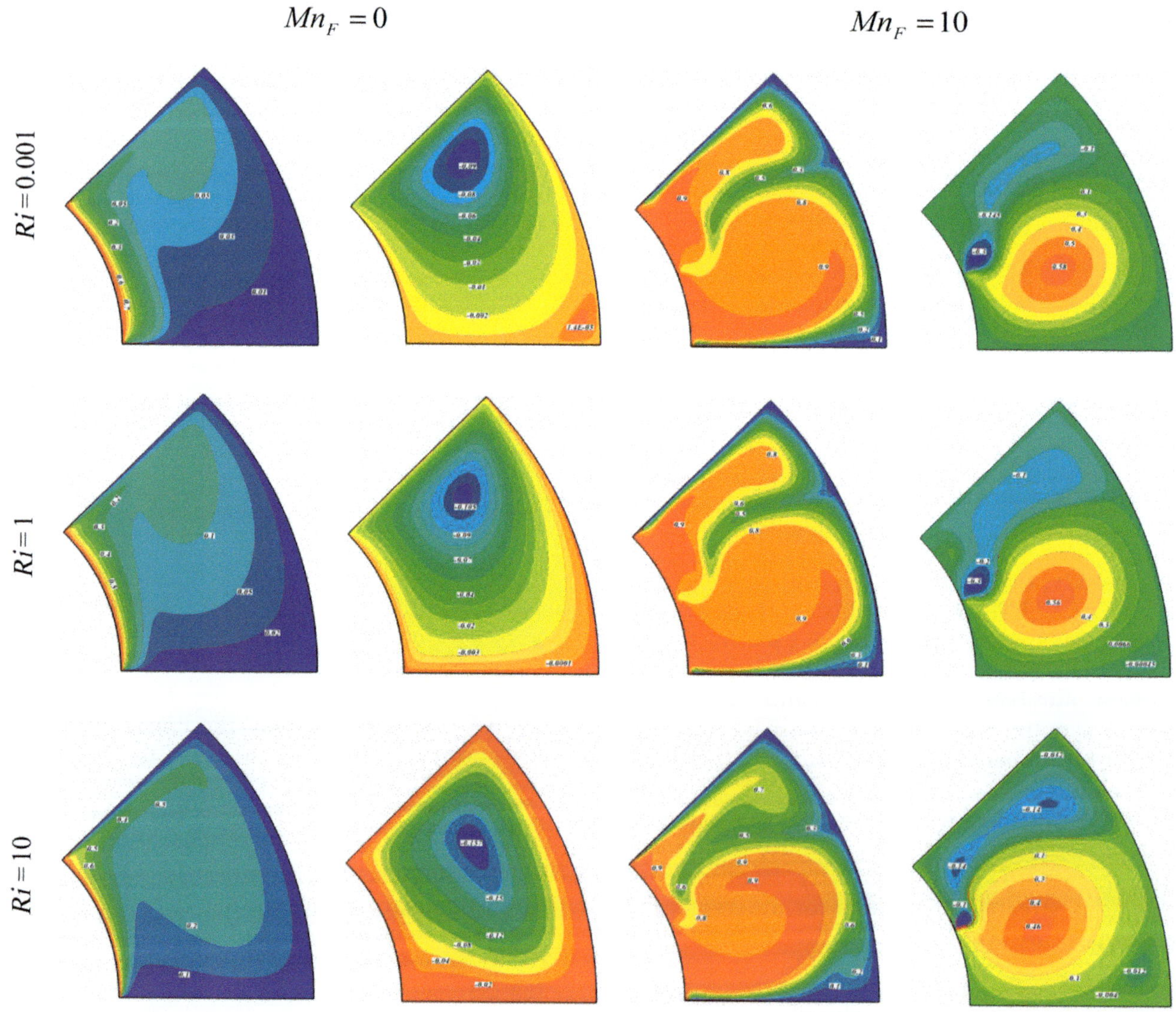

FIGURE 6.4 Isotherm (left) and streamline (right) contours for different values of Richardson number and magnetic number when $Ha = 0, \delta^* = 0$.

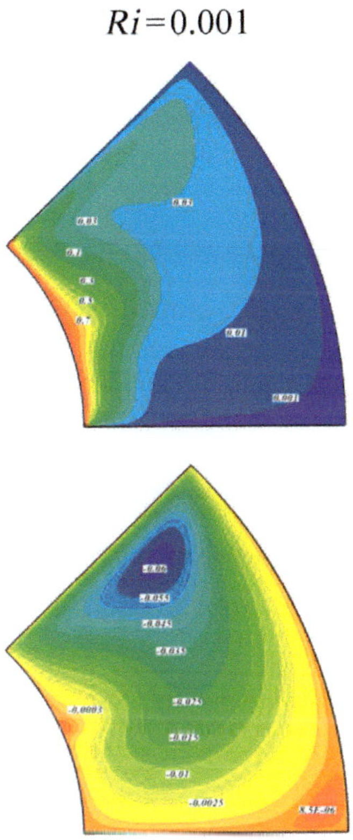

FIGURE 6.5 Isotherm (up) and streamline (down) contours for different values of Richardson number when $Mn_F = 0, Ha = 10, \delta^* = 0$.

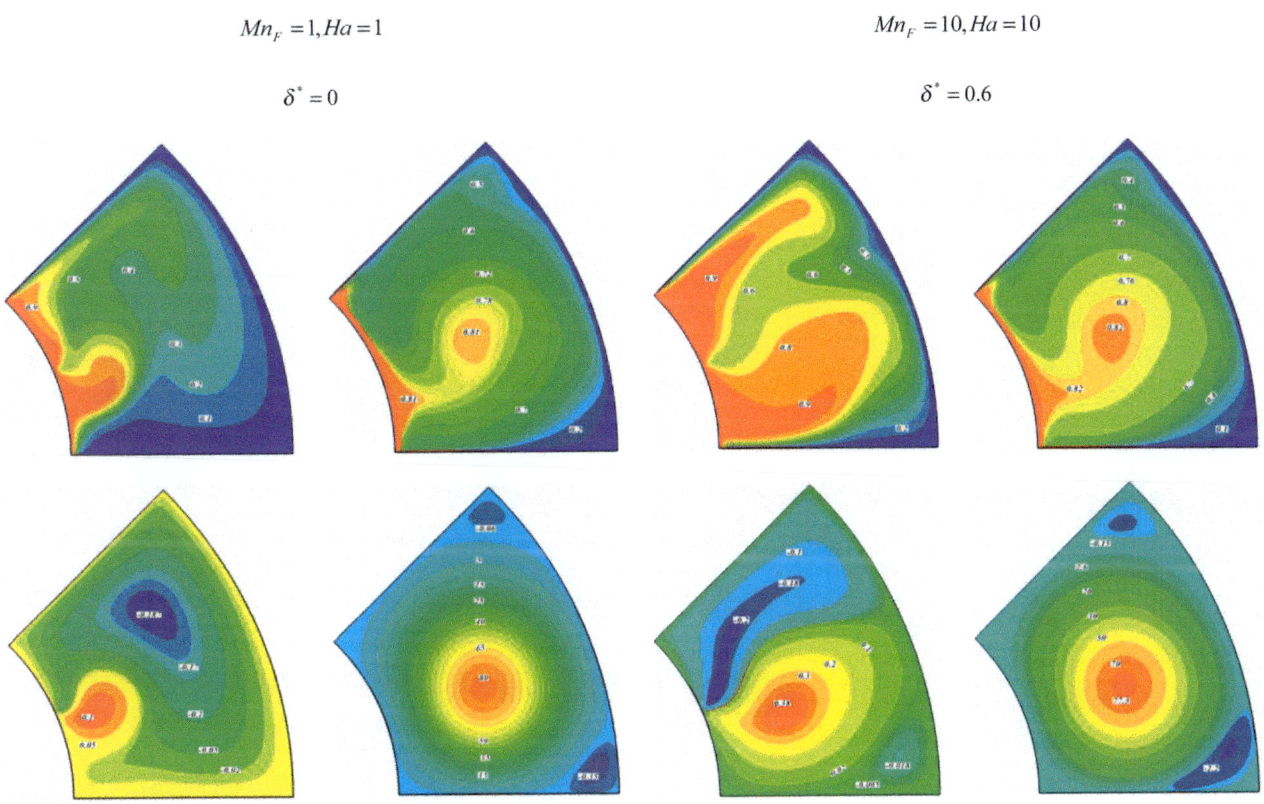

FIGURE 6.6 Isotherm (left) and streamline (right) contours for different values of δ^*, Ha, and Mn_F when $Ri = 10$.

thermal plume appears near the location of the magnetic source. As the Richardson number increases, the role of convection in heat transfer becomes more significant. At $Ri = 10$, the center of the main vortex moves downward and the thermal boundary layer thickness near the inner wall becomes thinner. When Hartmann and magnetic numbers increase up to 10, the main vortex turns into three eddies. Three thermal plumes generate over the inner wall due to the existence of eddies which are rotated in different directions. Also it can be seen that as the viscosity parameter increases the thermal boundary layer thickness decreases and one counterclockwise eddy generates at center of the enclosure.

Fig. 6.7 depicts that the effects of magnetic number, Hartmann number, viscosity parameter, r and Richardson number on average Nusselt number. The thermal boundary layer thickness increases with the increase of magnetic number and Hartmann number, while it decreases with the increase of Richardson number. So, the average Nusselt number increases with the increase of the Richardson number while it decreases with the increase of the Hartmann number and magnetic number. Also it can be concluded that by considering the effect of the magnetic field on viscosity of the fluid, the rate of heat transfer increases.

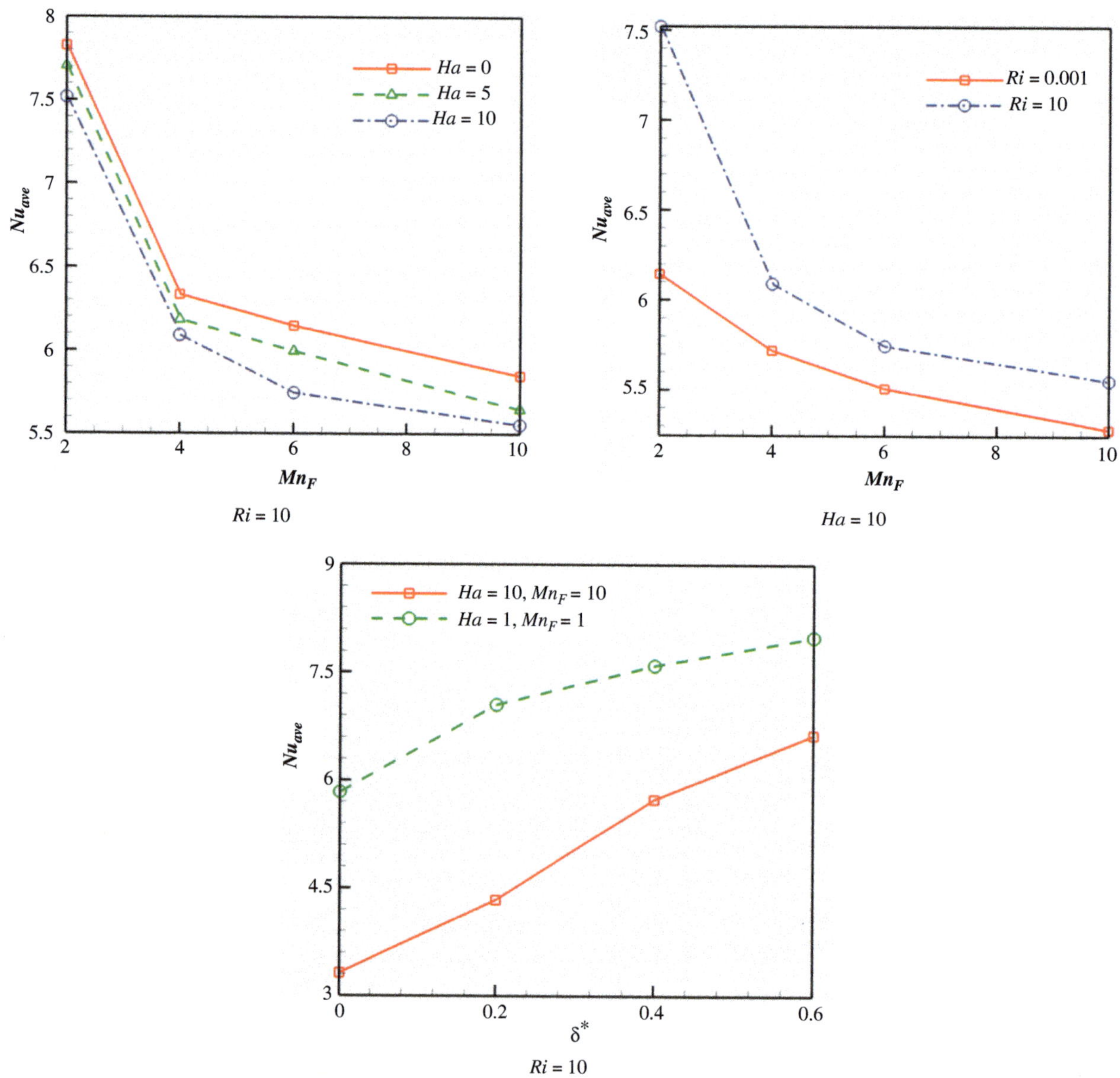

FIGURE 6.7 Effects of Ha, Mn_F, Ri, and δ^* on average Nusselt number Nu_{ave} along hot wall.

6.3 FORCED CONVECTION OF NANOFLUID IN A POROUS LID DRIVEN ENCLOSURE IN THE PRESENCE OF LORENTZ FORCES

6.3.1 Problem Definition

Fig. 6.8 depicts the geometry, boundary condition, and sample element. The formula of the inner cylinder is:

$$r = r_{in} + A\cos(N(\zeta - \zeta_0)) \tag{6.25}$$

in which r_{in} is the radius of the base circle. The inner cylinder has constant heat flux condition.

6.3.2 Governing Equation

2D steady convective flow of nanofluid in a porous media is considered in the presence of a constant magnetic field. The PDEs equations are:

$$\frac{\partial u}{\partial x} + \frac{\partial v}{\partial y} = 0 \tag{6.26}$$

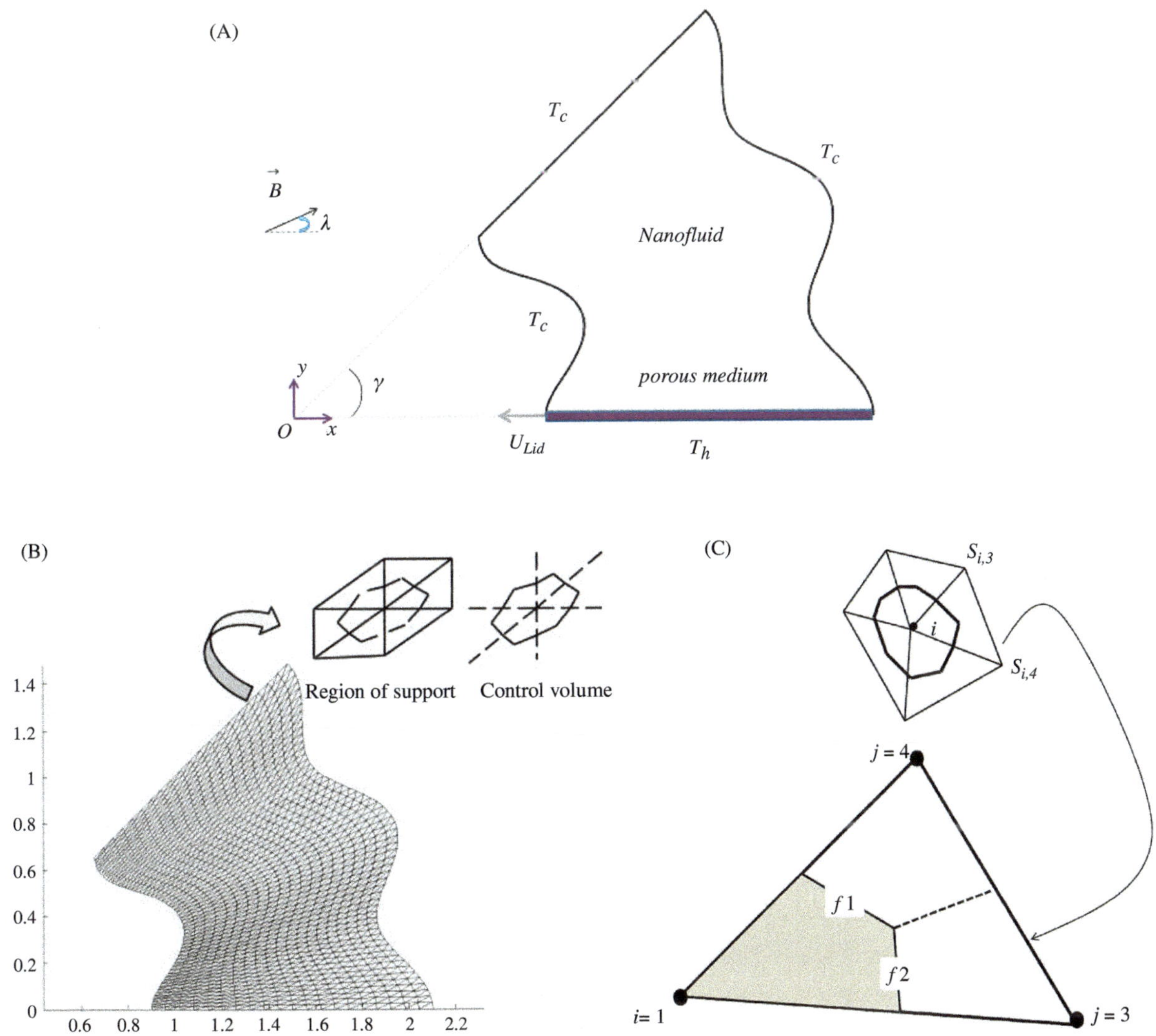

FIGURE 6.8 (A) Geometry and the boundary conditions with (B) the mesh of Geometry considered in this work. (C) A sample triangular element and its corresponding control volume.

$$\rho_{nf}\left(v\frac{\partial u}{\partial y}+\frac{\partial u}{\partial x}u\right)=\left[-\sigma_{nf}B_y^2u+\sigma_{nf}B_xB_yv+\left(\frac{\partial^2 u}{\partial y^2}+\frac{\partial^2 u}{\partial x^2}\right)\mu_{nf}-\frac{\partial P}{\partial x}\right]-\frac{\mu}{K}u \tag{6.27}$$

$$\rho_{nf}\left(\frac{\partial v}{\partial x}u+\frac{\partial v}{\partial y}v\right)=+\mu_{nf}\left(\frac{\partial^2 v}{\partial x^2}+\frac{\partial^2 v}{\partial y^2}\right)-\frac{\partial P}{\partial y}+B_y\sigma_{nf}B_xu-B_x\sigma_{nf}B_xv-\frac{\mu}{K}v, \tag{6.28}$$

$$B_x=B_o\cos\lambda,\ \ B_y=B_o\sin\lambda$$

$$\left(\rho C_p\right)_{nf}\left(v\frac{\partial T}{\partial y}+u\frac{\partial T}{\partial x}\right)=k_{nf}\left(\frac{\partial^2 T}{\partial x^2}+\frac{\partial^2 T}{\partial y^2}\right), \tag{6.29}$$

$\left(\rho C_p\right)_{nf}, (\rho\beta)_{nf}, \rho_{nf}, k_{nf}$, and σ_{nf} are defined as:

$$\left(\rho C_p\right)_{nf}=\left(\rho C_p\right)_f(1-\phi)+\left(\rho C_p\right)_s\phi \tag{6.30}$$

$$\rho_{nf}=\rho_f(1-\phi)+\rho_s\phi \tag{6.31}$$

$$\frac{\sigma_{nf}}{\sigma_f}=1+\frac{3\left(\frac{\sigma_s}{\sigma_f}-1\right)\phi}{\left(\frac{\sigma_s}{\sigma_f}+2\right)-\left(\frac{\sigma_s}{\sigma_f}-1\right)\phi} \tag{6.32}$$

k_{nf}, μ_{nf} are obtained according to Koo–Kleinstreuer–Li (KKL) model:

$$k_{nf}=k_{static}+k_{Brownian}$$

$$\frac{k_{static}}{k_f}=1+\frac{3\left(\frac{k_p}{k_f}-1\right)\phi}{\left(\frac{k_p}{k_f}+2\right)-\left(\frac{k_p}{k_f}-1\right)\phi}$$

$$\frac{k_{Brownian}}{k_f}=5\times10^4 g'(\phi,T,d_p)\phi\rho_f c_{p,f}\sqrt{\frac{\kappa_b T}{\rho_p d_p}} \tag{6.33}$$

$$g'(\phi,T,d_p)=\left(a_1+a_2Ln(d_p)+a_3Ln(\phi)+a_4Ln(\phi)\ln(d_p)+a_5Ln(d_p)^2\right)Ln(T)$$

$$+\left(a_6+a_7Ln(d_p)+a_8Ln(\phi)+a_9\ln(d_p)Ln(\phi)+a_{10}Ln(d_p)^2\right)$$

$$R_f=d_p/k_{p,eff}-d_p/k_p,\ R_f=4\times10^{-8}km^2/W$$

$$\mu_{nf}=\frac{\mu_f}{(1-\phi)^{2.5}}+\frac{k_{Brownian}}{k_f}\times\frac{\mu_f}{\Pr} \tag{6.34}$$

All required coefficients and properties are illustrated in Tables 6.2 and 6.3.
Vorticity and stream function should be used to eliminate pressure source terms:

$$\omega+\frac{\partial u}{\partial y}-\frac{\partial v}{\partial x}=0,\ \ \frac{\partial\psi}{\partial x}=-v,\ \ \frac{\partial\psi}{\partial y}=u \tag{6.35}$$

Introducing dimensionless quantities:

$$U=\frac{u}{U_{Lid}},V=\frac{v}{U_{Lid}},\ \theta=\frac{T-T_c}{\Delta T},\Delta T=T_h-T_c,(X,Y)=\frac{(x,y)}{L},$$
$$\Psi=\frac{\psi L}{U_{Lid}},\Omega=\frac{\omega}{LU_{Lid}} \tag{6.36}$$

TABLE 6.2 The Coefficient Values of $CuO-$ Water Nanofluid

Coefficient values	$CuO-$ Water
a_1	-26.5933108
a_2	-0.403818333
a_3	-33.3516805
a_4	-1.915825591
a_5	6.421858E-02
a_6	48.40336955
a_7	-9.787756683
a_8	190.245610009
a_9	10.9285386565
a_{10}	-0.72009983664

TABLE 6.3 Thermophysical Properties of Water and Nanoparticles

	$\rho(kg/m^3)$	$C_p(j/kgk)$	$k(W/m.k)$	$\beta\times10^5(K^{-1})$	$d_p(nm)$	$\sigma(\Omega\cdot\mathrm{m})^{-1}$
Water	997.1	4179	0.613	21	-	0.05
CuO	6500	540	18	29	45	10^{-10}

The final formulae are:

$$\frac{\partial^2\Psi}{\partial Y^2}+\frac{\partial^2\Psi}{\partial X^2}=-\Omega, \tag{6.37}$$

$$\begin{aligned}U\frac{\partial\Omega}{\partial X}+\frac{\partial\Omega}{\partial Y}V=&\frac{1}{\mathrm{Re}}\frac{A_5}{A_1}\left(\frac{\partial^2\Omega}{\partial Y^2}+\frac{\partial^2\Omega}{\partial X^2}\right)\\&+\frac{Ha^2}{\mathrm{Re}}\frac{A_6}{A_1}\left(\frac{\partial U}{\partial X}B_xB_y-\frac{\partial V}{\partial X}B_x^2+\frac{\partial U}{\partial Y}B_y^2-\frac{\partial V}{\partial Y}B_xB_y\right)\\&-\frac{1}{\mathrm{Re}\,Da}\frac{A_5}{A_1}\Omega,\end{aligned} \tag{6.38}$$

$$\frac{\partial\theta}{\partial X}U+\frac{\partial\theta}{\partial Y}V=\frac{1}{\mathrm{Re}\,\mathrm{Pr}}\frac{A_4}{A_2}\left(\frac{\partial^2\theta}{\partial X^2}+\frac{\partial^2\theta}{\partial Y^2}\right) \tag{6.39}$$

where dimensionless and constants parameters are illustrated as:

$$\begin{aligned}&\mathrm{Re}=\frac{\rho_f U_{Lid}L}{\mu_f},\ Ha=LB_0\sqrt{\sigma_f/\mu_f}\\&A_1=\frac{\rho_{nf}}{\rho_f},A_2=\frac{(\rho C_p)_{nf}}{(\rho C_p)_f},A_3=\frac{(\rho\beta)_{nf}}{(\rho\beta)_f},A_4=\frac{k_{nf}}{k_f},A_5=\frac{\mu_{nf}}{\mu_f},A_6=\frac{\sigma_{nf}}{\sigma_f}\end{aligned} \tag{6.40}$$

and boundary conditions are:

$$\begin{aligned}&\theta=1.0\quad\text{on bottom wall}\\&\theta=0.0\quad\text{on other walls}\\&\Psi=0.0\quad\text{on all walls}\end{aligned} \tag{6.41}$$

Local and average Nusselt over the hot cylinder can be calculated as:

$$Nu_{loc} = \frac{\partial \theta}{\partial n}\left(\frac{k_{nf}}{k_f}\right) \tag{6.42}$$

$$Nu_{ave} = \frac{1}{L}\int_{c}^{c+L} Nu_{loc}\,dy \tag{6.43}$$

6.3.3 Effects of Active Parameters

In this section, the effect of magnetic field on nanofluid forced convection in a porous complex shaped lid driven cavity is examined. Control volume-based finite element method is utilized to find the effects of Darcy number ($Da = 0.01$ to 100), Reynolds number ($\mathrm{Re} = 100$ to 1000), Hartmann number ($Ha = 0$ to 40), and volume fraction of CuO ($\phi = 0\%$ to 4%).

The influence of adding CuO nanoparticles on hydrothermal behavior is depicted in Fig. 6.9. The nanofluid velocity increases due to enhancing the solid movements. The thermal boundary layer thickness decreases with the increase in nanofluid volume fraction.

Figs. 6.10, 6.11, and 6.12 demonstrate the influences of Darcy, Hartmann, and Reynolds numbers on isotherms and stream lines. At $Da = 0.01, \mathrm{Re} = 100, Ha = 0$, only one eddy exists in streamline. By applying a magnetic field,

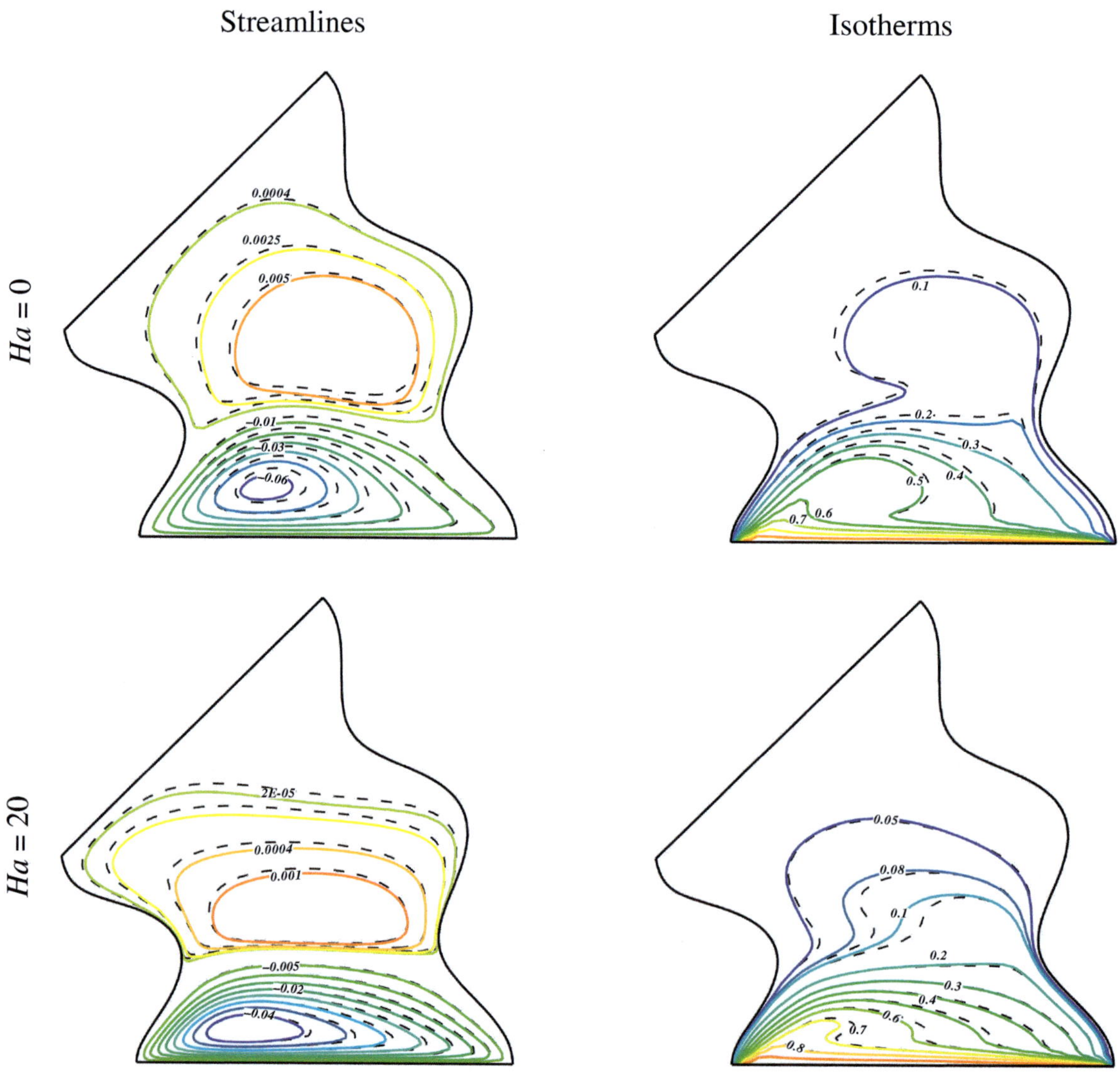

FIGURE 6.9 Impact of nanofluid volume fraction on streamline (left) and isotherm (right) contours (nanofluid ($\phi = 0.04$)(–) and pure fluid ($\phi = 0$) (- - -)) when $\mathrm{Re} = 600$.

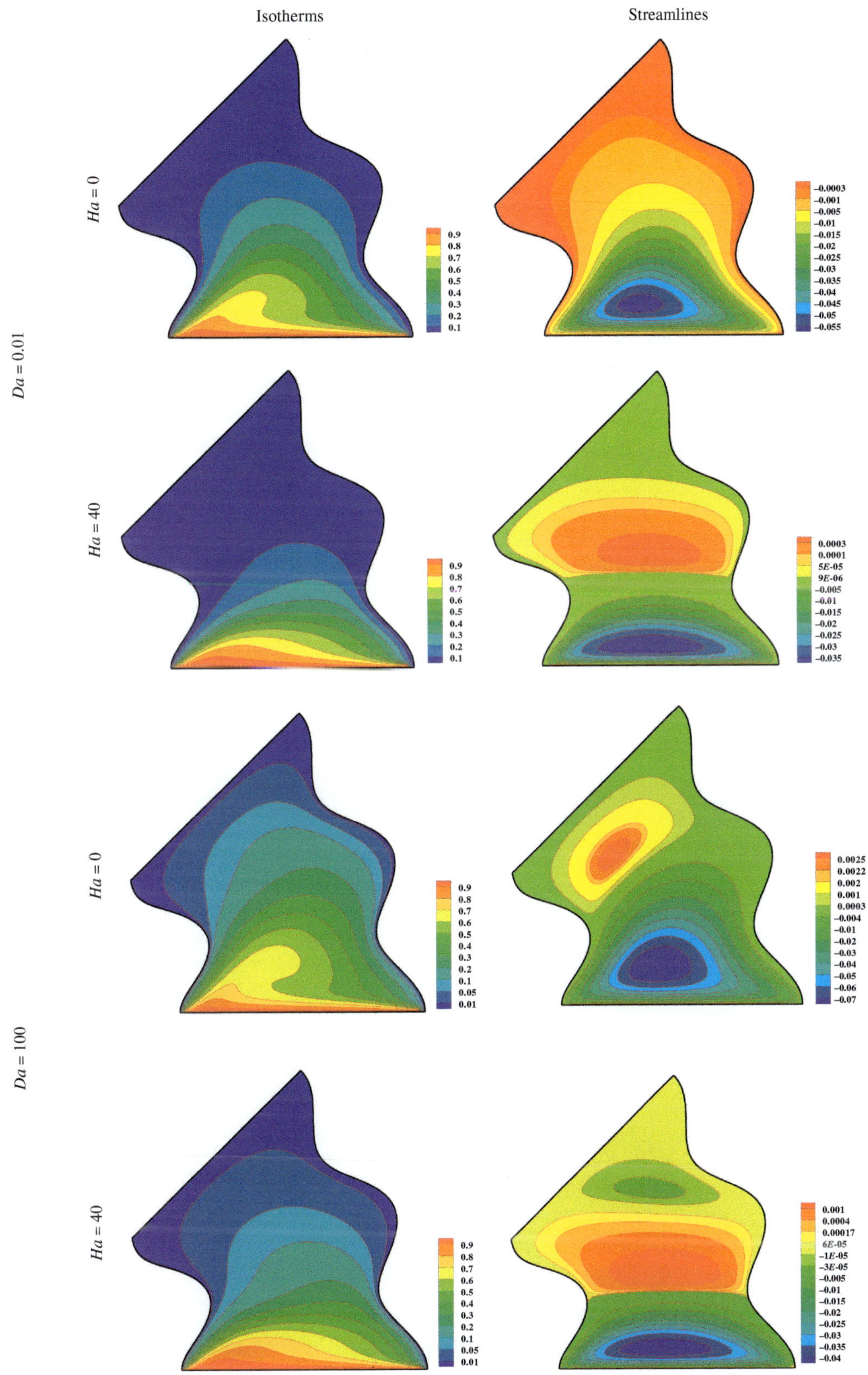

FIGURE 6.10 Isotherm (left) and streamline (right) contours for different values of Darcy and Hartmann numbers when Re = 100, $\phi = 0.04$.

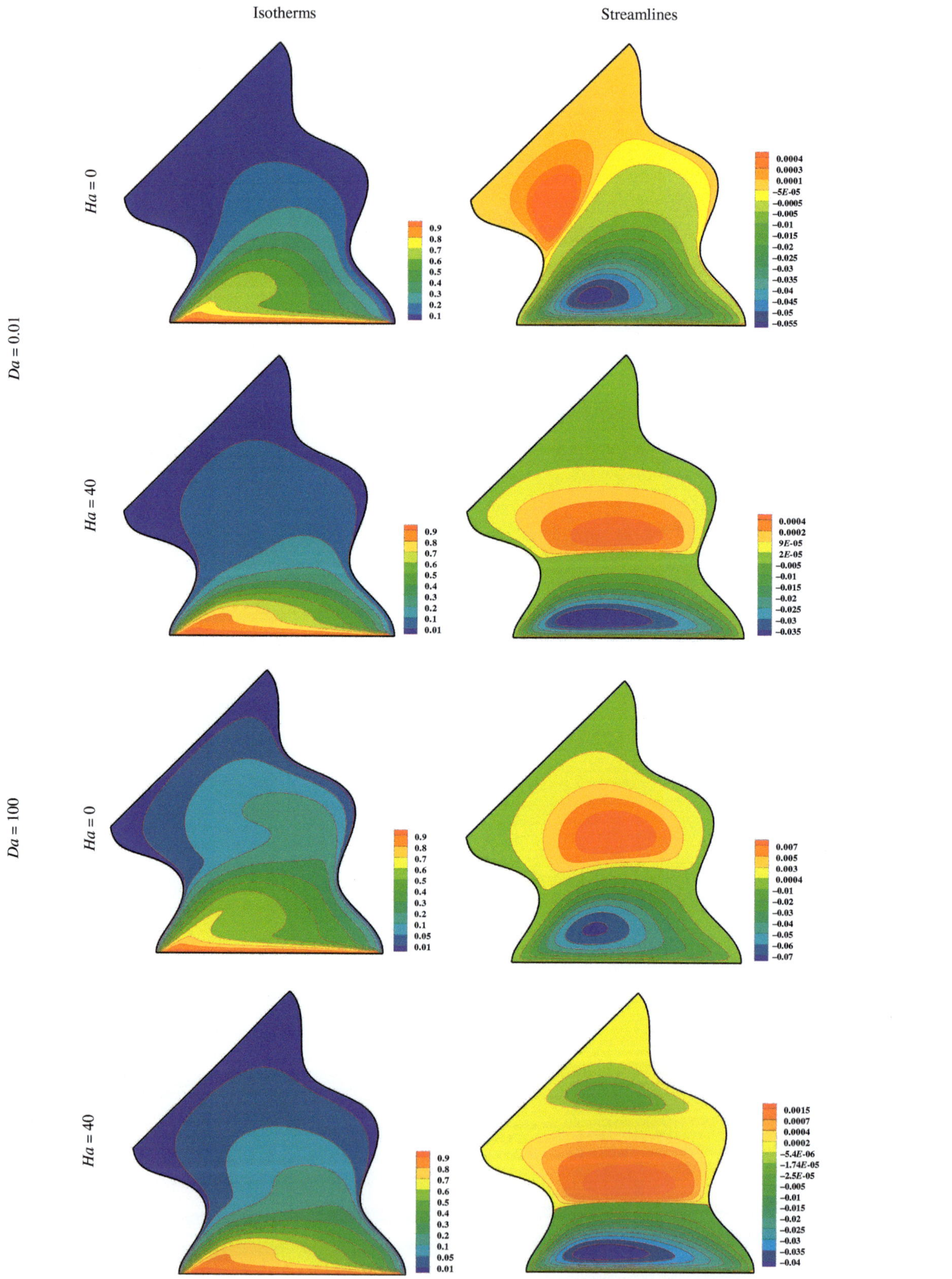

FIGURE 6.11 Isotherm (left) and streamline (right) contours for different values of Darcy and Hartmann numbers when Re = 250, $\phi = 0.04$.

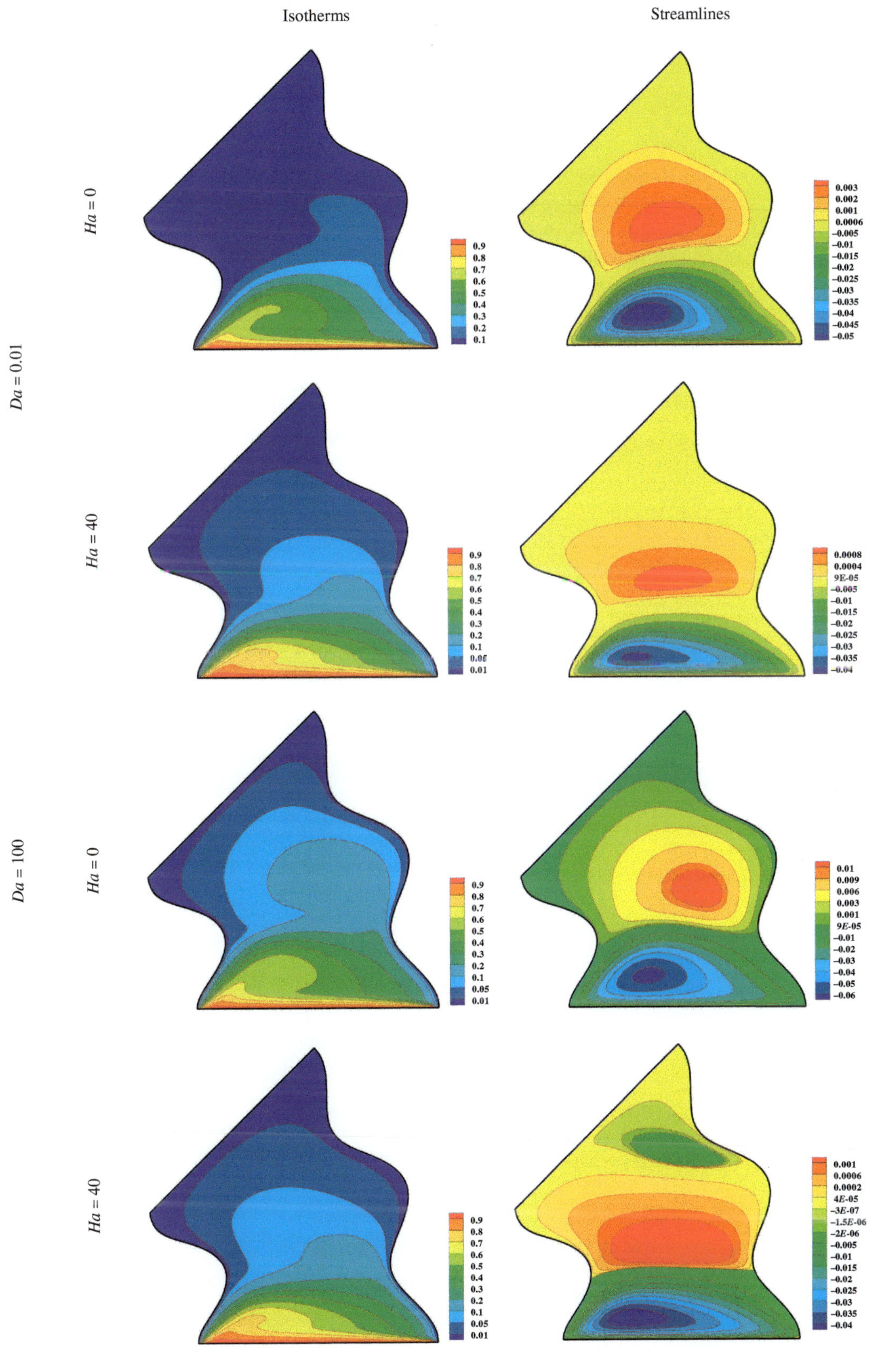

FIGURE 6.12 Isotherm (left) and streamline (right) contours for different values of Darcy and Hartmann numbers when Re = 1000, $\phi = 0.04$.

the main eddy converts to two rotating eddies in which the stronger one is situated near the lid wall and rotates counterclockwise. The distortion of isotherms becomes less than before. As the Darcy number increases, the convective mode becomes stronger and the primary eddy converts to two eddies. Increasing the Hartmann number leads to the generation of the third eddy. As the Reynolds number increases at a low Darcy number, a counterclockwise eddy generates at the left side of the enclosure. The presence of the magnetic field changes the position of the new eddy. As the Darcy number increases, the distortion of isotherms increases. The effects of the Darcy and Hartmann numbers on flow and heat transfer at $Re = 1000$ are similar to those at $Re = 250$ except for the stream function values. It means that the absolute stream function values increase with the rise of Reynolds number.

The influences of significant parameters on average Nusselt number are illustrated in Fig. 6.13. The correlation for average Nusselt number is as follows:

$$\begin{aligned} Nu_{ave} &= 7.89 + 0.7Da^* + 1.5\mathrm{Re}^* - 0.033Ha^* \\ &- 0.06Da^*\,\mathrm{Re}^* - 0.02Da^*\,Ha^* + 0.01\mathrm{Re}^*\,Ha^* \\ &+ 0.63Da^{*2} - 0.13\mathrm{Re}^{*2} - 0.1Ha^{*2} \end{aligned} \tag{6.44}$$

where $Ra^* = 0.001Ra,\;\; Ha^* = 0.1\,Ha$. Increasing Reynolds number can enhance the rate of heat transfer due to the decrease in thermal boundary layer thickness. A similar trend is observed for the Darcy number. Lorentz forces make the thermal boundary layer thickness increase. So, a temperature gradient reduces with the rise of the Hartmann number. Table 6.4 illustrates the effects of Darcy, Hartmann, and Reynolds numbers on heat transfer augmentation. Heat transfer enhancement increases with the increase of Hartmann and Reynolds numbers. This is due to the fact that this parameter is relevant to changes in thermal conductivity. The influence of Darcy number on E is the opposite to that of the Hartmann number.

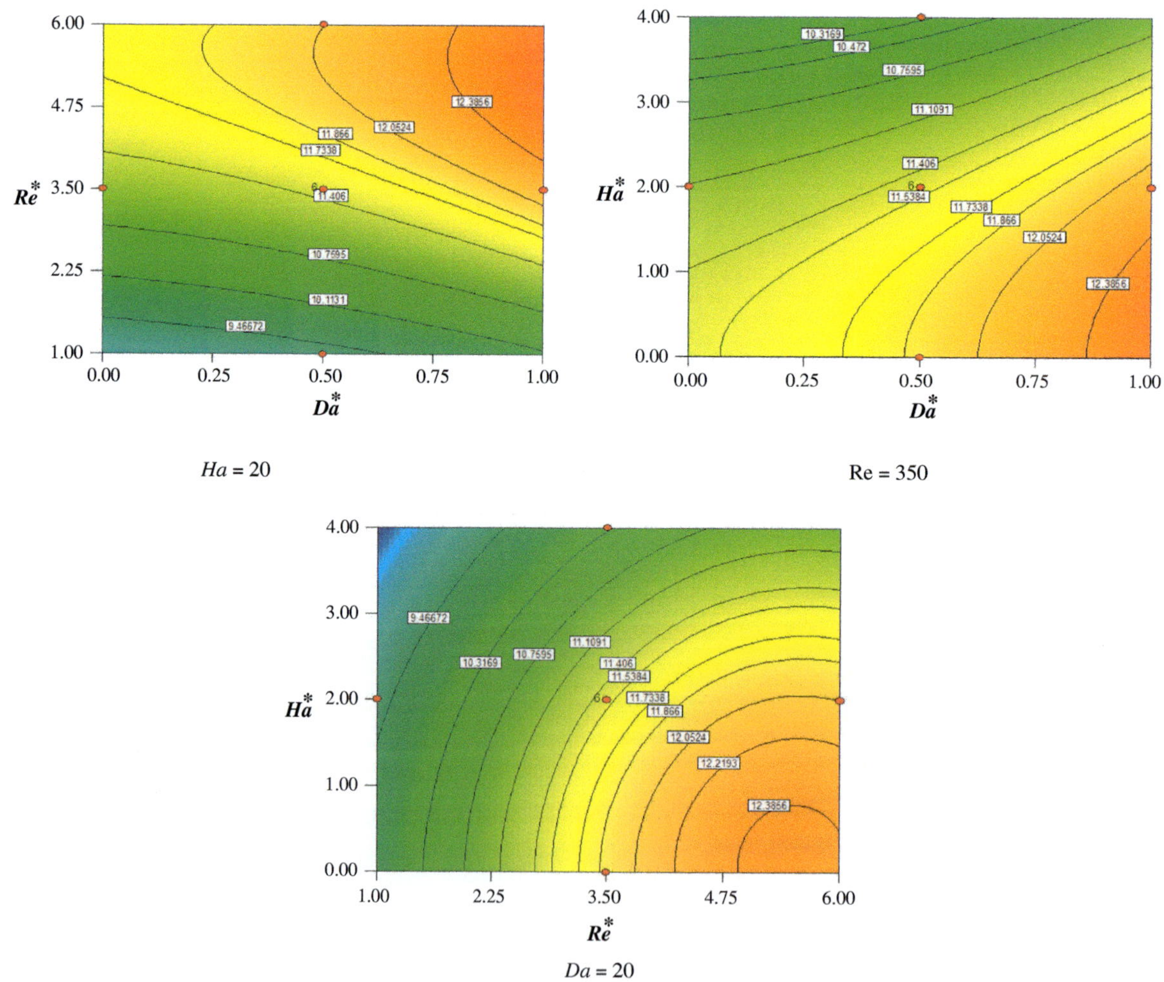

FIGURE 6.13 Effects of the Darcy, Reynolds number, and Hartmann numbers on average Nusselt number.

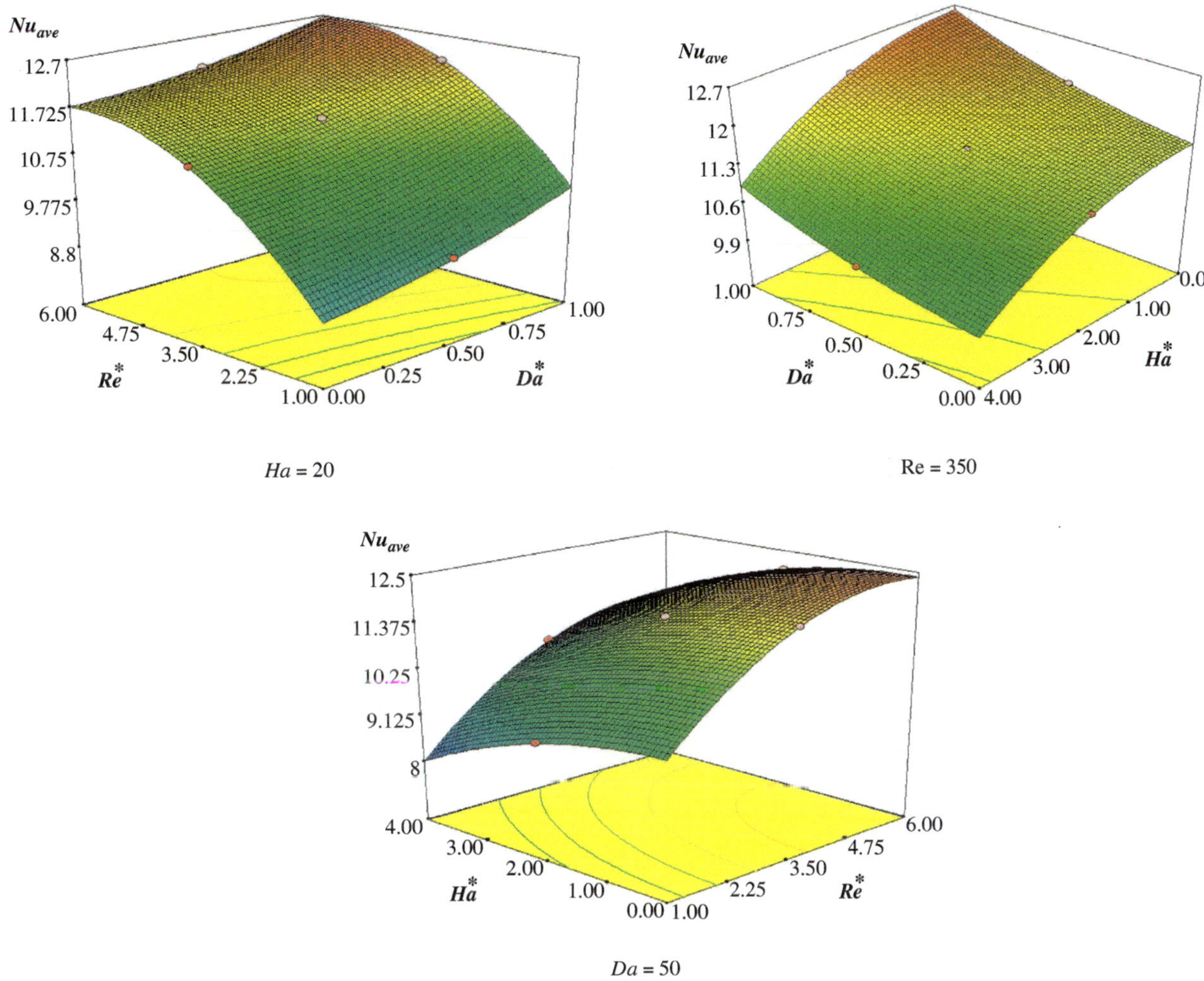

FIGURE 6.13 (Continued).

TABLE 6.4 Effects of *Da, Ha,* and *Re* on Heat Transfer Enhancement

Da	Re	Ha	E
0.01	100	0	9.347762
0.01	100	40	9.740013
0.01	300	0	10.31189
0.01	300	40	11.21865
0.01	600	0	10.89783
0.01	600	40	11.97797
100	100	0	8.812048
100	100	40	9.483864
100	300	0	9.863677
100	300	40	11.10445
100	600	0	10.06556
100	600	40	11.61814

6.4 INFLUENCE OF LORENTZ FORCES ON NANOFLUID FLOW INSIDE A POROUS ENCLOSURE WITH MOVING WALL

6.4.1 Problem Definition

Geometry, boundary condition, and sample element are depicted in Fig. 6.14. The bottom wall is hot and the other walls are cold. Also the bottom wall can move horizontally. A porous cavity is filled with nanofluid and affected by horizontal magnetic field.

6.4.2 Governing Equation

2D steady convective non-Darcy flow of nanofluid is considered in the presence of a constant magnetic field. The PDEs equations are:

$$\frac{\partial v}{\partial y}+\frac{\partial u}{\partial x}=0 \tag{6.45}$$

$$\left[-\sigma_{nf}B_y^2u+\sigma_{nf}B_xB_yv+\left(\frac{\partial^2u}{\partial y^2}+\frac{\partial^2u}{\partial x^2}\right)\mu_{nf}-\frac{\partial P}{\partial x}\right]-\frac{\mu}{K}u=\rho_{nf}\left(v\frac{\partial u}{\partial y}+\frac{\partial u}{\partial x}u\right) \tag{6.46}$$

$$\mu_{nf}\left(\frac{\partial^2v}{\partial y^2}+\frac{\partial^2v}{\partial x^2}\right)-\frac{\partial P}{\partial y}-B_x\sigma_{nf}B_xv+B_y\sigma_{nf}B_xu-\frac{\mu}{K}v=\rho_{nf}\left(\frac{\partial v}{\partial x}u+\frac{\partial v}{\partial y}v\right), \tag{6.47}$$

$$B_x=B_o\cos\lambda,\ \ B_y=B_o\sin\lambda$$

$$(\rho C_p)_{nf}\left(v\frac{\partial T}{\partial y}+u\frac{\partial T}{\partial x}\right)=k_{nf}\left(\frac{\partial^2T}{\partial x^2}+\frac{\partial^2T}{\partial y^2}\right), \tag{6.48}$$

$(\rho C_p)_{nf}$, $(\rho\beta)_{nf}$, ρ_{nf}, k_{nf}, and σ_{nf} are defined as:

$$(\rho C_p)_{nf}=(\rho C_p)_f(1-\phi)+(\rho C_p)_s\phi \tag{6.49}$$

$$\rho_{nf}=\rho_f(1-\phi)+\rho_s\phi \tag{6.50}$$

$$\frac{\sigma_{nf}}{\sigma_f}=1+\frac{3\left(\frac{\sigma_s}{\sigma_f}-1\right)\phi}{\left(\frac{\sigma_s}{\sigma_f}+2\right)-\left(\frac{\sigma_s}{\sigma_f}-1\right)\phi} \tag{6.51}$$

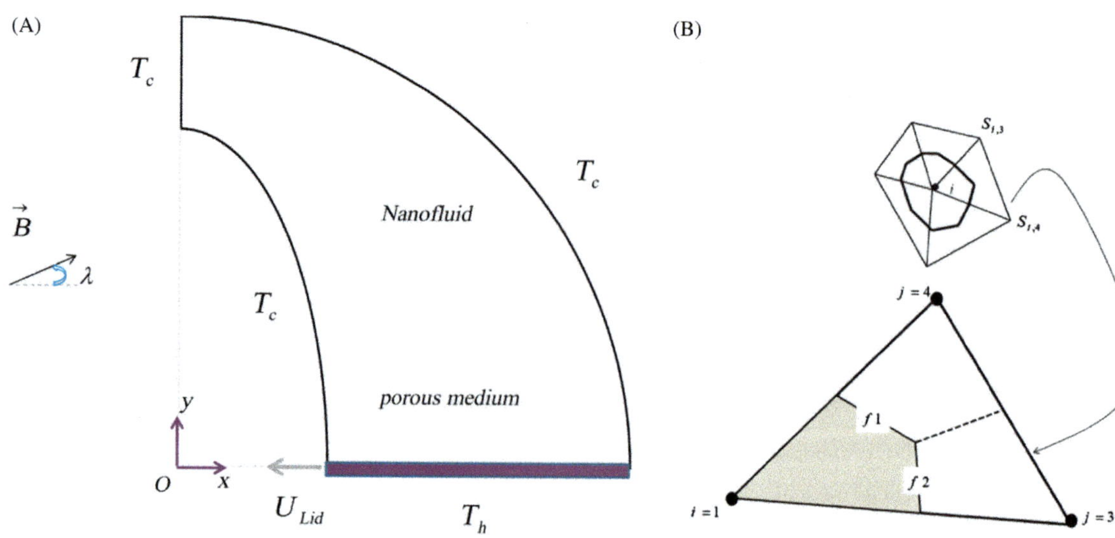

FIGURE 6.14 (A) Geometry and the boundary conditions with (B) a sample triangular element and its corresponding control volume.

μ_{nf} are calculated via KKL model:

$$\mu_{eff} = \mu_{static} + \mu_{Brownian} = \mu_{static} + \frac{k_{Brownian}}{k_f} \times \frac{\mu_f}{\Pr_f}$$

$$k_{Brownian} = 5 \times 10^4 g'(\phi, T, d_p)\phi \rho_f c_{p,f} \sqrt{\frac{\kappa_b T}{d_p \rho_p}} \tag{6.52}$$

$$g'(\phi, d_p, T) = Ln(T)\left(a_1 + a_3 Ln(\phi) + a_2 Ln(d_p) + a_5 Ln(d_p)^2 + a_4 \ln(d_p) Ln(\phi)\right)$$
$$+ \left(a_6 + a_8 Ln(\phi) + a_7 Ln(d_p) + a_{10} Ln(d_p)^2 + a_9 Ln(\phi) \ln(d_p)\right)$$

Properties and needed parameters are provided in Tables 6.2 and 6.3.
k_{nf} can be expressed as [24]:

$$\frac{k_{nf}}{k_f} = \frac{-m(k_f - k_p)\phi + (k_p - k_f)\phi + mk_f + k_p + k_f}{mk_f + (k_f - k_p)\phi + k_f + k_p} \tag{6.53}$$

Different values of shape factors for various shapes of nanoparticles are illustrated in Table 6.5.
Vorticity and stream function should be used to eliminate pressure source terms:

$$\omega + \frac{\partial u}{\partial y} - \frac{\partial v}{\partial x} = 0, \quad \frac{\partial \psi}{\partial x} = -v, \quad \frac{\partial \psi}{\partial y} = u \tag{6.54}$$

Introducing dimensionless quantities:

$$U = \frac{u}{U_{Lid}}, V = \frac{v}{U_{Lid}}, \quad \theta = \frac{T - T_c}{\Delta T}, \Delta T = T_h - T_c, (X, Y) = \frac{(x, y)}{L},$$
$$\Psi = \frac{\psi L}{U_{Lid}}, \Omega = \frac{\omega}{L U_{Lid}} \tag{6.55}$$

The final formulae are:

$$\frac{\partial^2 \Psi}{\partial X^2} + \Omega + \frac{\partial^2 \Psi}{\partial Y^2} = 0, \tag{6.56}$$

TABLE 6.5 The values of shape factor of different shapes of nanoparticles

m	Spherical	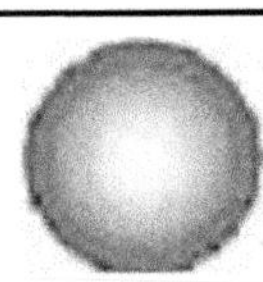	3
	Platelet		5.7
	Cylinder		4.8
	Brick	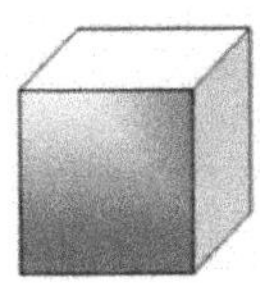	3.7

$$\frac{\partial \Omega}{\partial Y}V + \frac{\partial \Omega}{\partial X}U = \frac{1}{\mathrm{Re}}\frac{A_5}{A_1}\left(\frac{\partial^2 \Omega}{\partial Y^2} + \frac{\partial^2 \Omega}{\partial X^2}\right)$$
$$+ \frac{Ha^2}{\mathrm{Re}}\frac{A_6}{A_1}\left(\frac{\partial U}{\partial X}B_x B_y - \frac{\partial V}{\partial X}{B_x}^2 + \frac{\partial U}{\partial Y}B_y^2 - \frac{\partial V}{\partial Y}B_x B_y\right) \tag{6.57}$$
$$- \frac{1}{\mathrm{Re}\,Da}\frac{A_5}{A_1}\Omega,$$

$$\frac{A_4}{A_2}\left(\frac{\partial^2 \theta}{\partial Y^2} + \frac{\partial^2 \theta}{\partial X^2}\right) = \mathrm{PrRe}\left(\frac{\partial \theta}{\partial X}U + \frac{\partial \theta}{\partial Y}V\right) \tag{6.58}$$

where dimensionless and constants parameters are illustrated as:

$$\mathrm{Re} = \frac{\rho_f U_{Lid} L}{\mu_f},\quad Ha = LB_0\sqrt{\sigma_f/\mu_f},$$
$$A_1 = \frac{\rho_{nf}}{\rho_f}, A_3 = \frac{(\rho\beta)_{nf}}{(\rho\beta)_f}, A_5 = \frac{\mu_{nf}}{\mu_f}, \tag{6.59}$$
$$A_2 = \frac{(\rho C_p)_{nf}}{(\rho C_p)_f}, A_4 = \frac{k_{nf}}{k_f}, A_6 = \frac{\sigma_{nf}}{\sigma_f}$$

and boundary conditions are:

$$\begin{aligned} &\theta = 1.0 \quad \text{on bottom wall}\\ &\theta = 0.0 \quad \text{on other walls}\\ &\Psi = 0.0 \quad \text{on all walls}\end{aligned} \tag{6.60}$$

Nu_{loc} and Nu_{ave} over the hot wall can be calculated as:

$$Nu_{loc} = \frac{\partial \theta}{\partial y}\left(\frac{k_{nf}}{k_f}\right) \tag{6.61}$$

$$Nu_{ave} = \frac{1}{L}\int_{r_{in}}^{r_{out}} Nu_{loc}\,dx \tag{6.62}$$

6.4.3 Effects of Active Parameters

In this research, the effect of Lorentz forces on forced convection of nanofluid in a lid driven porous cavity is illustrated. μ_{nf} of CuO-water nanofluid are estimated by means of KKL model. CVFEM is utilized to find the effects of Darcy number ($Da = 0.01$ to 100), Reynolds number ($\mathrm{Re} = 100$ to 500), volume fraction of CuO ($\phi = 0\%$ to 4%), shape of nanoparticle, and Hartmann number ($Ha = 0$ to 40).

The influences of shape of the nanoparticles on Nu_{ave} are discussed in Tables 6.6. The maximum Nu is caused by Platelet, followed by Cylinder, Brick and Spherical. Therefore, Platelet nanoparticle has been selected for further investigation.

The impact of adding nanoparticles to the base fluid on velocity and temperature contours is depicted in Fig. 6.15. Temperature gradient decreases with the increase of ϕ. $|\Psi_{\max}|$ augments with the addition of nanoparticles because of the increase in the solid movements.

Figs. 6.16, 6.17, and 6.18 demonstrate the impact of Da, Re, Ha on isotherms and stream lines. At $Da = 0.01, \mathrm{Re} = 100, Ha = 0$, only one eddy exists in streamline. By applying a magnetic field, two new eddies with opposite directions generate above the main eddy. The distortion of the isotherms becomes less than before. As the Darcy number increases, the convective mode becomes stronger and the primary eddy converts to two eddies. Increasing the Hartmann number leads to generation of the third eddy. As Re augments, the strength of

TABLE 6.6 Effect of shape of nanoparticles on Nusselt number when $Da = 100, \mathrm{Re} = 600, \phi = 0.04$

	Ha	
	0	40
Spherical	15.60301	12.79417
Brick	15.95598	13.08149
Cylinder	16.50756	13.53032
Platelet	16.95605	13.89515

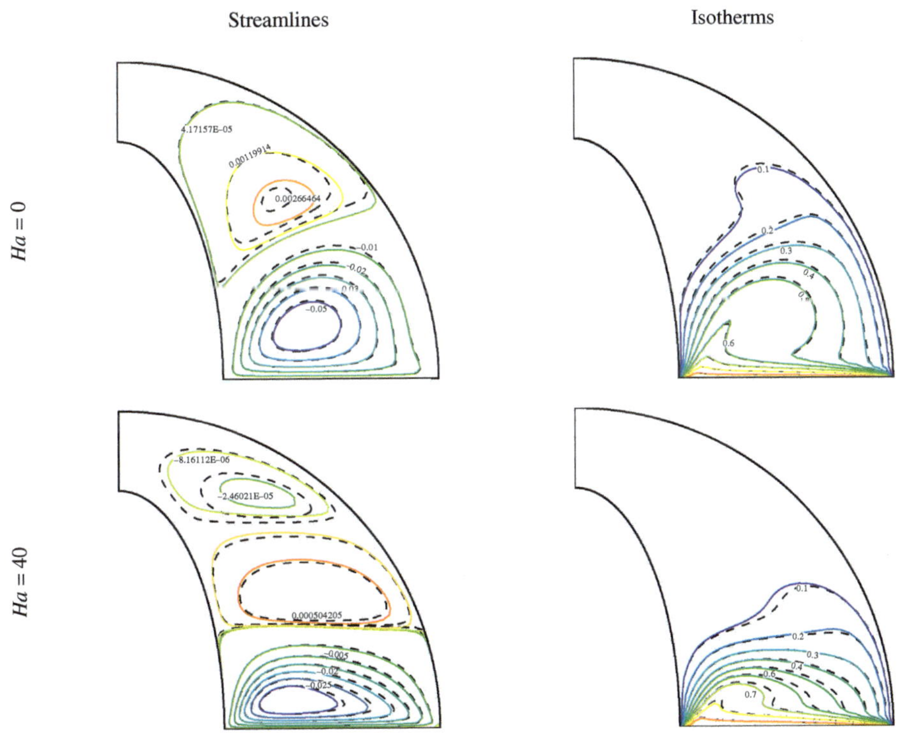

FIGURE 6.15 Influence of nanofluid volume fraction on streamline (left) and isotherm (right) contours (nanofluid ($\phi = 0.04$) (–) and pure fluid($\phi = 0$) (- - -)) when $\mathrm{Re} = 500, Da = 100$.

the secondary eddy, which is generated due to the increase of Darcy number, increases. In the absence of the magnetic field at Re = 500, a counterclockwise eddy appears above the main eddy. Impacts of Darcy and Hartmann number on isotherms and streamlines at Re = 500 are similar to those at Re = 100 except for the stream function values. It means that the absolute stream function values increase with the rise of the Reynolds number.

The influence of significant parameters on Nu_{ave} is illustrated in Fig. 6.19. The correlation for Nu_{ave} is as follows:

$$\begin{aligned} Nu_{ave} &= 12.14 + 1.46\mathrm{Re}^* + 0.54Da^* - 0.38Ha^* \\ &+ 0.01Da^*\,\mathrm{Re}^* + 0.0029\mathrm{Re}^*\,Ha^* - 0.27Da^*\,Ha^* \\ &- 0.14\mathrm{Re}^{*2} + 0.29Da^{*2} - 0.05Ha^{*2} \end{aligned} \tag{6.63}$$

where $\mathrm{Re}^* = 0.01\mathrm{Re}, Ha* = 0.1Ha, Da^* = 0.01Da$. Increasing the Reynolds number can enhance the heat transfer rate because of the reduction in thermal boundary layer thickness. A similar trend is seen for the Darcy number. The temperature gradient decreases with the increase of Lorentz forces. So, Nu_{ave} decreases with the rise of the Hartmann number.

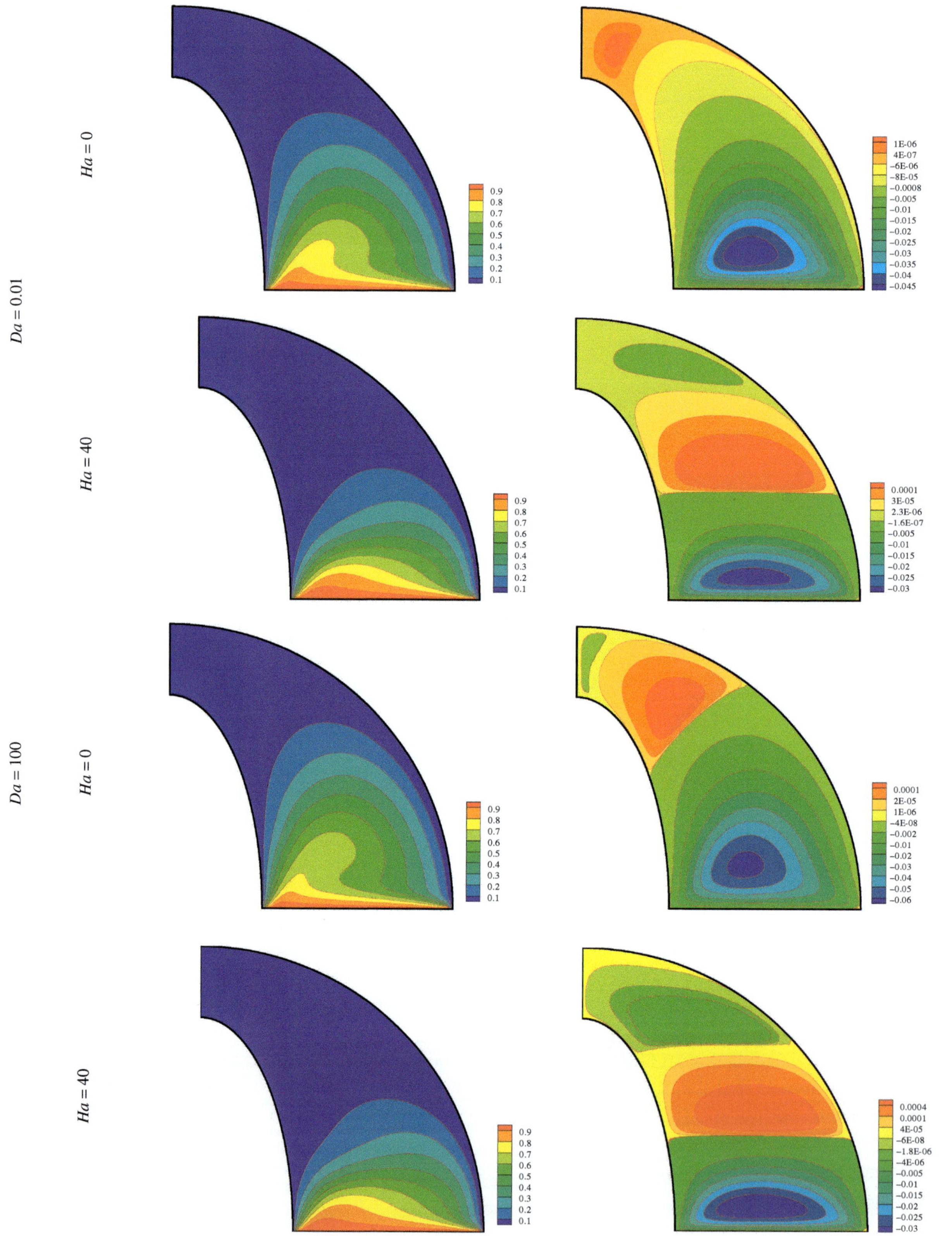

FIGURE 6.16 Effect of Darcy and Hartmann numbers on isotherm (left) and streamline (right) contours when $\phi = 0.04, \mathrm{Re} = 100$.

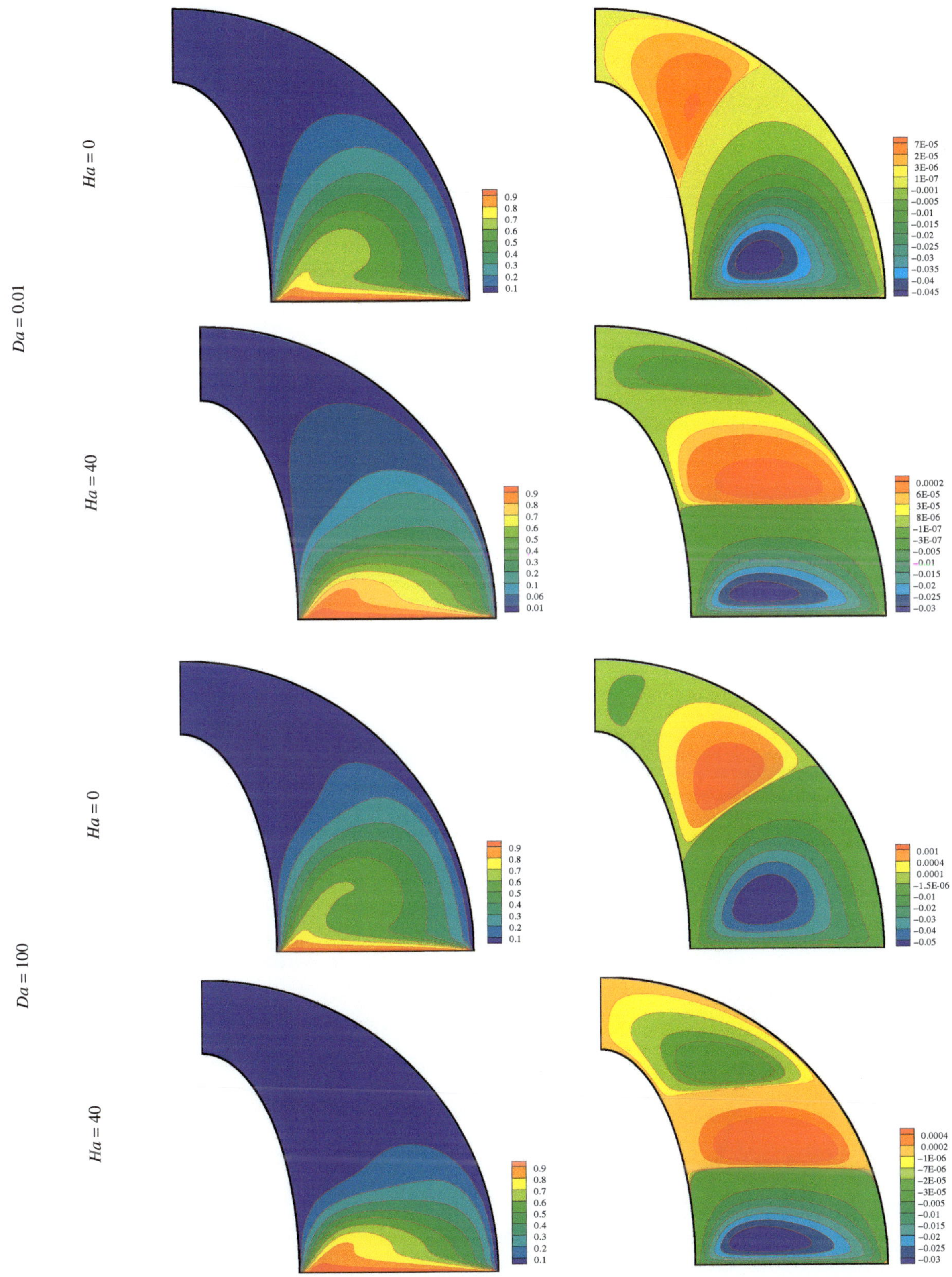

FIGURE 6.17 Effect of Darcy and Hartmann numbers on isotherm (left) and streamline (right) contours when $\phi = 0.04$, $\mathrm{Re} = 300$.

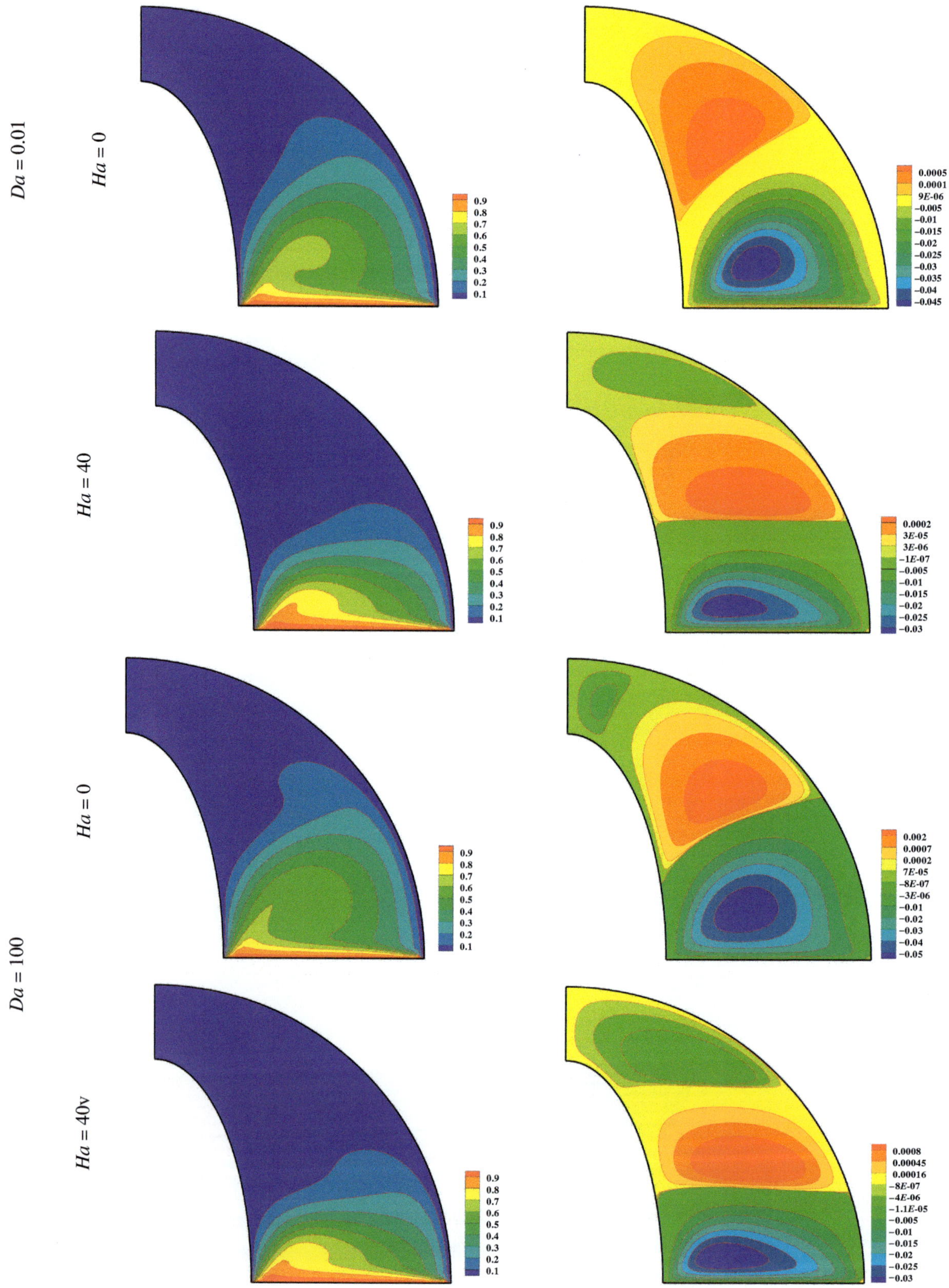

FIGURE 6.18 Effect of Darcy and Hartmann numbers on isotherm (left) and streamline (right) contours when $\phi = 0.04, \mathrm{Re} = 500$.

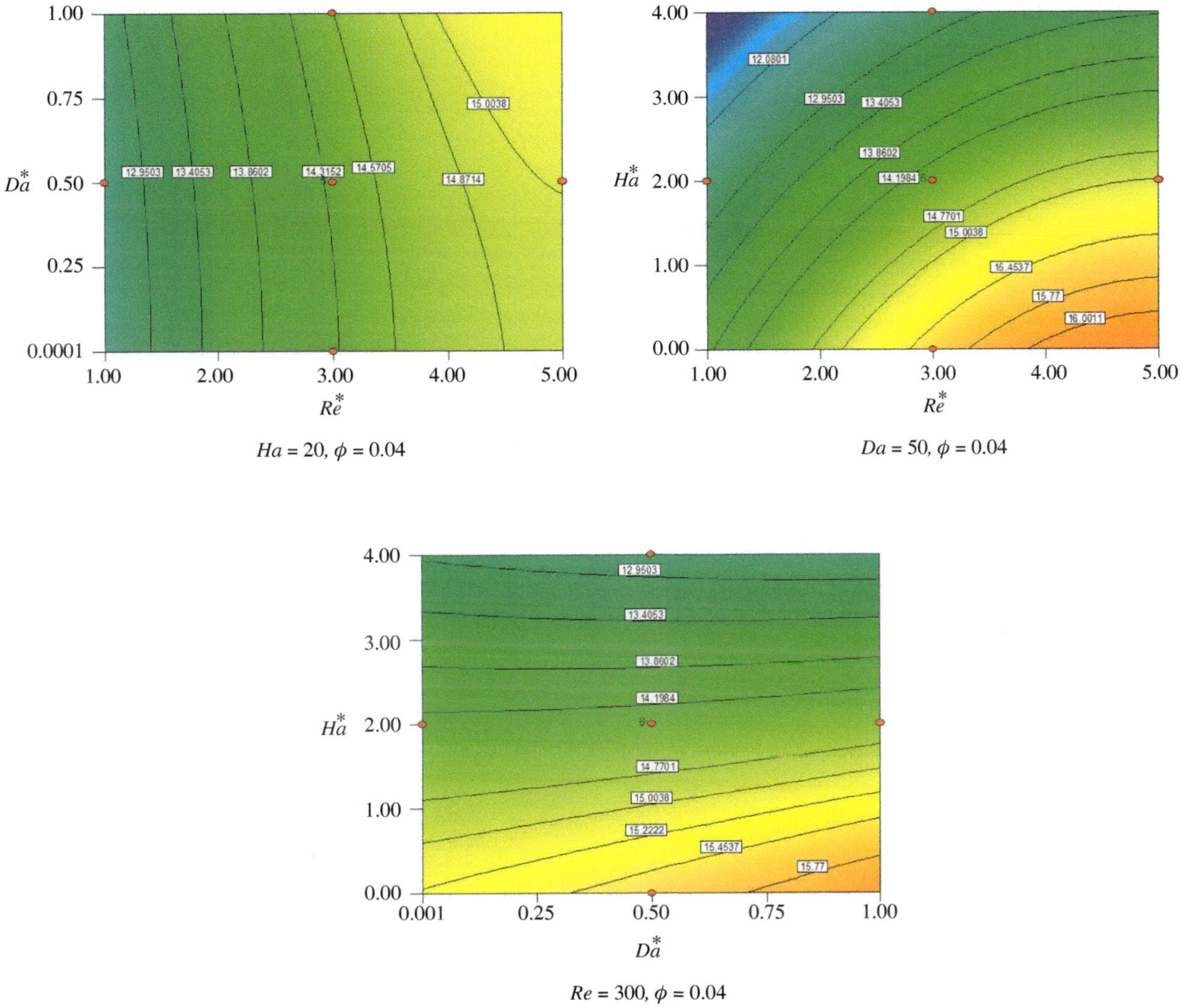

FIGURE 6.19 Influences of the Darcy, Reynolds number, and Hartmann numbers on average Nusselt number.

6.5 SINGLE-PHASE MODEL SIMULATION OF NANOFLUID FORCED CONVECTION INSIDE A PERMEABLE ENCLOSURE WITH SINUSOIDAL WALL

6.5.1 Problem Definition

Geometry, boundary condition, and sample element are depicted (see Fig. 6.20). The bottom wall is hot and other walls are cold. Also the bottom wall can move horizontally. A porous cavity is filled with nanofluid and affected by a horizontal magnetic field.

6.5.2 Governing Equation

2D steady convective non-Darcy flow of nanofluid is considered in the presence of a constant magnetic field. The PDEs equations are:

$$\frac{\partial v}{\partial y} + \frac{\partial u}{\partial x} = 0 \tag{6.64}$$

$$\left[-\sigma_{nf} B_y^2 u + \sigma_{nf} B_x B_y v + \left(\frac{\partial^2 u}{\partial y^2} + \frac{\partial^2 u}{\partial x^2}\right)\mu_{nf} - \frac{\partial P}{\partial x}\right] - \frac{\mu}{K} u = \rho_{nf}\left(v\frac{\partial u}{\partial y} + \frac{\partial u}{\partial x} u\right) \tag{6.65}$$

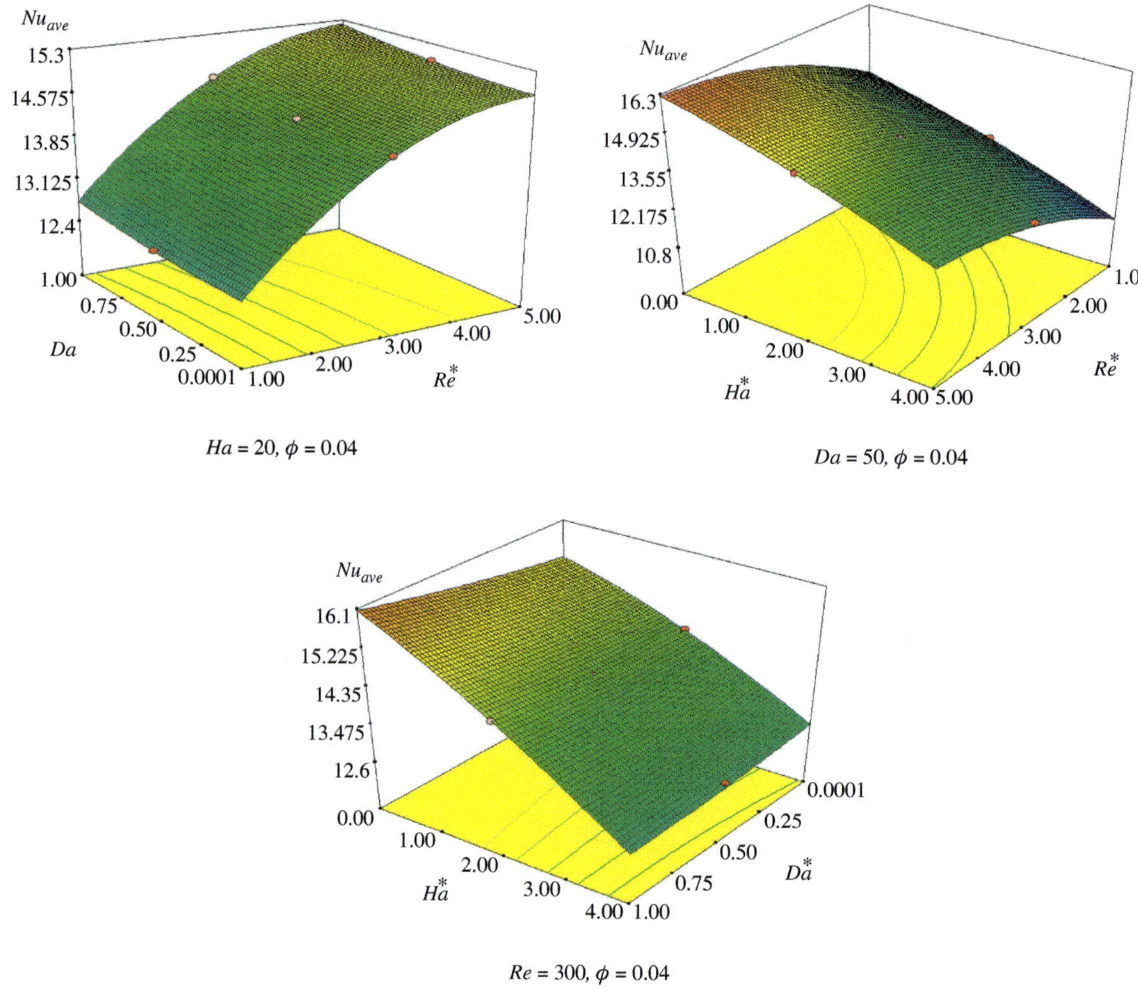

FIGURE 6.19 (Continued).

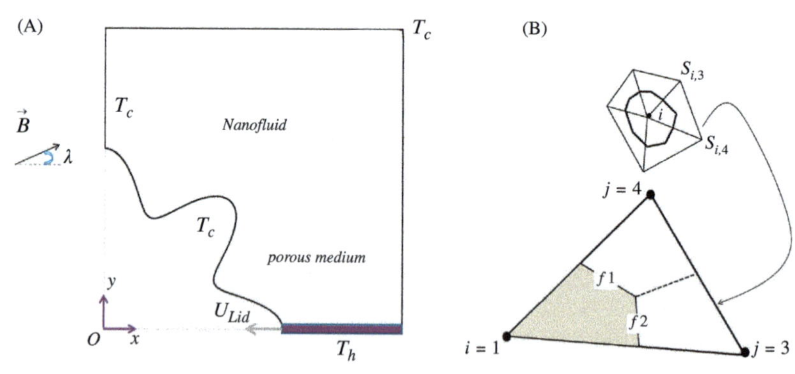

FIGURE 6.20 (a) Geometry and the boundary conditions with (b) a sample triangular element and its corresponding control volume.

$$\mu_{nf}\left(\frac{\partial^2 v}{\partial y^2}+\frac{\partial^2 v}{\partial x^2}\right)-\frac{\partial P}{\partial y}-B_x\sigma_{nf}B_x v+B_y\sigma_{nf}B_x u-\frac{\mu}{K}v=\rho_{nf}\left(\frac{\partial v}{\partial x}u+\frac{\partial v}{\partial y}v\right), \tag{6.66}$$
$$B_x=B_o\cos\lambda,\quad B_y=B_o\sin\lambda$$

$$(\rho C_p)_{nf}\left(v\frac{\partial T}{\partial y}+u\frac{\partial T}{\partial x}\right)=k_{nf}\left(\frac{\partial^2 T}{\partial x^2}+\frac{\partial^2 T}{\partial y^2}\right), \tag{6.67}$$

$(\rho C_p)_{nf}$, $(\rho\beta)_{nf}$, ρ_{nf}, k_{nf}, and σ_{nf} are defined as:

$$(\rho C_p)_{nf} = (\rho C_p)_f(1-\phi) + (\rho C_p)_s\phi \tag{6.68}$$

$$\rho_{nf} = \rho_f(1-\phi) + \rho_s\phi \tag{6.69}$$

$$\frac{\sigma_{nf}}{\sigma_f} = 1 + \frac{3\left(\frac{\sigma_s}{\sigma_f}-1\right)\phi}{\left(\frac{\sigma_s}{\sigma_f}+2\right) - \left(\frac{\sigma_s}{\sigma_f}-1\right)\phi} \tag{6.70}$$

μ_{nf} are calculated via KKL model:

$$\begin{gathered}\mu_{eff} = \mu_{static} + \mu_{Brownian} = \mu_{static} + \frac{k_{Brownian}}{k_f} \times \frac{\mu_f}{\Pr_f}\\ k_{Brownian} = 5\times10^4 g'(\phi, T, d_p)\phi\rho_f c_{p,f}\sqrt{\frac{\kappa_b T}{d_p\rho_p}}\\ g'(\phi, d_p, T) = Ln(T)\left(a_1 + a_3Ln(\phi) + a_2Ln(d_p) + a_5Ln(d_p)^2 + a_4\ln(d_p)Ln(\phi)\right)\\ +\left(a_6 + a_8Ln(\phi) + a_7Ln(d_p) + a_{10}Ln(d_p)^2 + a_9Ln(\phi)\ln(d_p)\right)\end{gathered} \tag{6.71}$$

k_{nf} can be expressed as:

$$\frac{k_{nf}}{k_f} = \frac{-m(k_f-k_p)\phi + (k_p-k_f)\phi + mk_f + k_p + k_f}{mk_f + (k_f-k_p)\phi + k_f + k_p} \tag{6.72}$$

Vorticity and stream function should be used to eliminate pressure source terms:

$$\omega + \frac{\partial u}{\partial y} - \frac{\partial v}{\partial x} = 0, \quad \frac{\partial\psi}{\partial x} = -v, \quad \frac{\partial\psi}{\partial y} = u \tag{6.73}$$

Introducing dimensionless quantities:

$$\begin{gathered}U = \frac{u}{U_{Lid}}, V = \frac{v}{U_{Lid}}, \quad \theta = \frac{T-T_c}{\Delta T}, \Delta T = T_h - T_c, (X,Y) = \frac{(x,y)}{L},\\ \Psi = \frac{\psi L}{U_{Lid}}, \Omega = \frac{\omega}{LU_{Lid}}\end{gathered} \tag{6.74}$$

The final formulae are:

$$\frac{\partial^2\Psi}{\partial X^2} + \Omega + \frac{\partial^2\Psi}{\partial Y^2} = 0, \tag{6.75}$$

$$\begin{gathered}\frac{\partial\Omega}{\partial Y}V + \frac{\partial\Omega}{\partial X}U = \frac{1}{\mathrm{Re}}\frac{A_5}{A_1}\left(\frac{\partial^2\Omega}{\partial Y^2} + \frac{\partial^2\Omega}{\partial X^2}\right)\\ + \frac{Ha^2}{\mathrm{Re}}\frac{A_6}{A_1}\left(\frac{\partial U}{\partial X}B_xB_y - \frac{\partial V}{\partial X}B_x^2 + \frac{\partial U}{\partial Y}B_y^2 - \frac{\partial V}{\partial Y}B_xB_y\right)\\ - \frac{1}{\mathrm{Re}\,Da}\frac{A_5}{A_1}\Omega,\end{gathered} \tag{6.76}$$

$$\frac{A_4}{A_2}\left(\frac{\partial^2\theta}{\partial Y^2} + \frac{\partial^2\theta}{\partial X^2}\right) = \Pr\mathrm{Re}\left(\frac{\partial\theta}{\partial X}U + \frac{\partial\theta}{\partial Y}V\right) \tag{6.77}$$

where dimensionless and constants parameters are illustrated as:

$$
\begin{aligned}
&\mathrm{Re}=\frac{\rho_f U_{Lid} L}{\mu_f},\ Ha=LB_0\sqrt{\sigma_f/\mu_f},\\
&A_1=\frac{\rho_{nf}}{\rho_f},A_3=\frac{(\rho\beta)_{nf}}{(\rho\beta)_f},A_5=\frac{\mu_{nf}}{\mu_f},\\
&A_2=\frac{(\rho C_p)_{nf}}{(\rho C_p)_f},A_4=\frac{k_{nf}}{k_f},A_6=\frac{\sigma_{nf}}{\sigma_f}
\end{aligned}
\tag{6.78}
$$

and boundary conditions are:

$$
\begin{aligned}
&\theta=1.0 \quad \text{on bottom wall}\\
&\theta=0.0 \quad \text{on other walls}\\
&\Psi=0.0 \quad \text{on all walls}
\end{aligned}
\tag{6.79}
$$

TABLE 6.7 Effect of shape of nanoparticles on Nusselt number when $Da=100, \mathrm{Re}=600, \phi=0.04$

	Ha	
	0	40
Spherical	5.795633	3.969774
Brick	5.941453	4.068879
Cylinder	6.170481	4.22454
Platelet	6.35775	4.351825

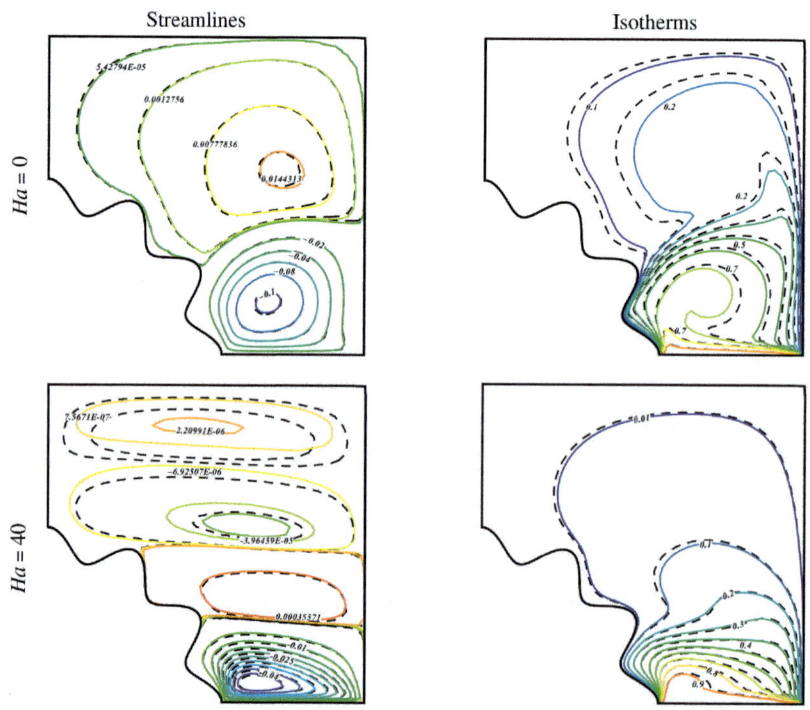

FIGURE 6.21 Influence of nanofluid volume fraction on streamline (left) and isotherm (right) contours (nanofluid ($\phi=0.04$)(–) and pure fluid($\phi=0$) (- - -)) when $\mathrm{Re}=500, Da=100$.

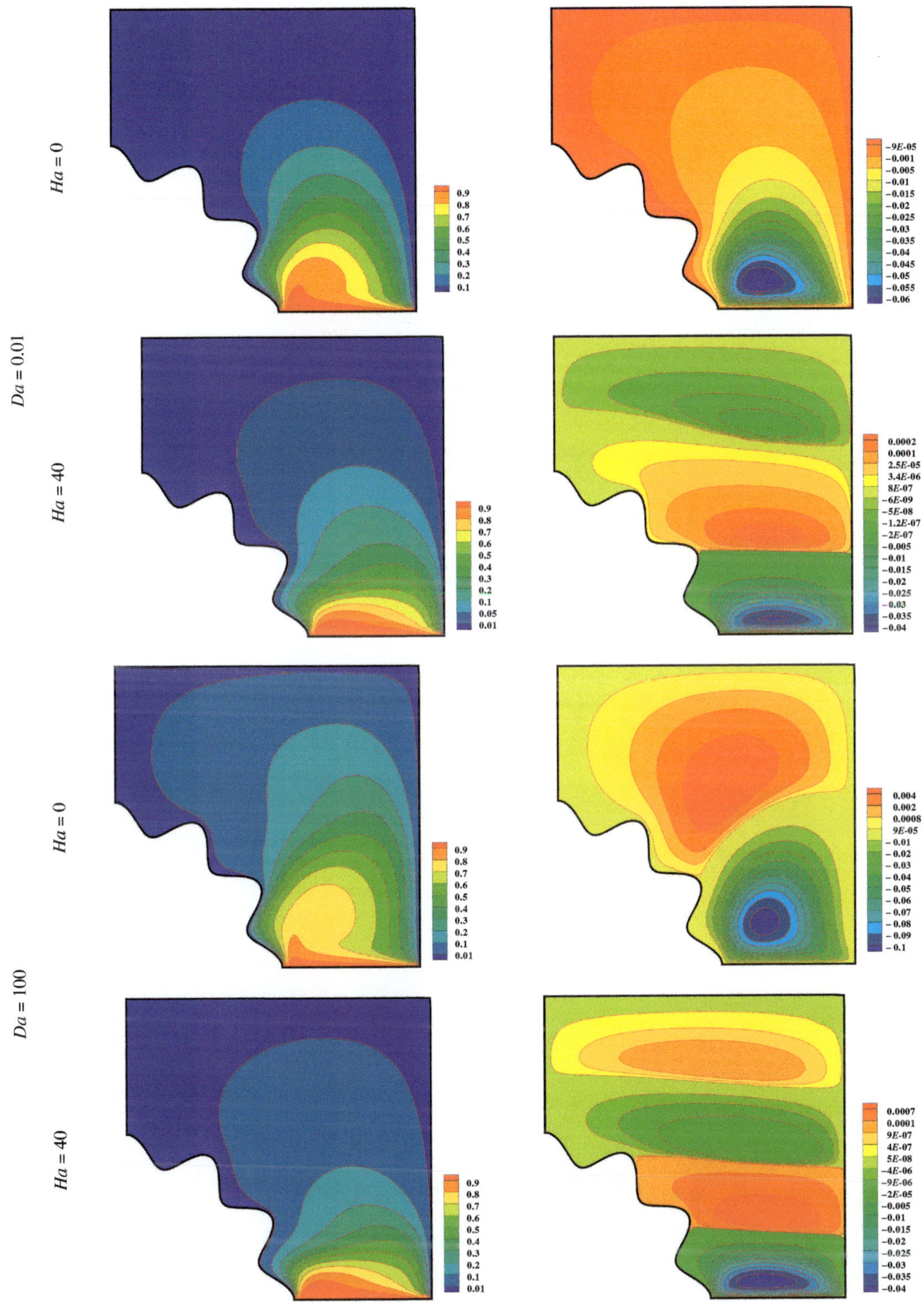

FIGURE 6.22 Effect of Darcy and Hartmann numbers on isotherm (left) and streamline (right) contours when $\phi = 0.04$, Re = 100.

Nu_{loc} and Nu_{ave} over the hot wall can be calculated as:

$$Nu_{loc} = \frac{\partial \theta}{\partial y}\left(\frac{k_{nf}}{k_f}\right) \tag{6.80}$$

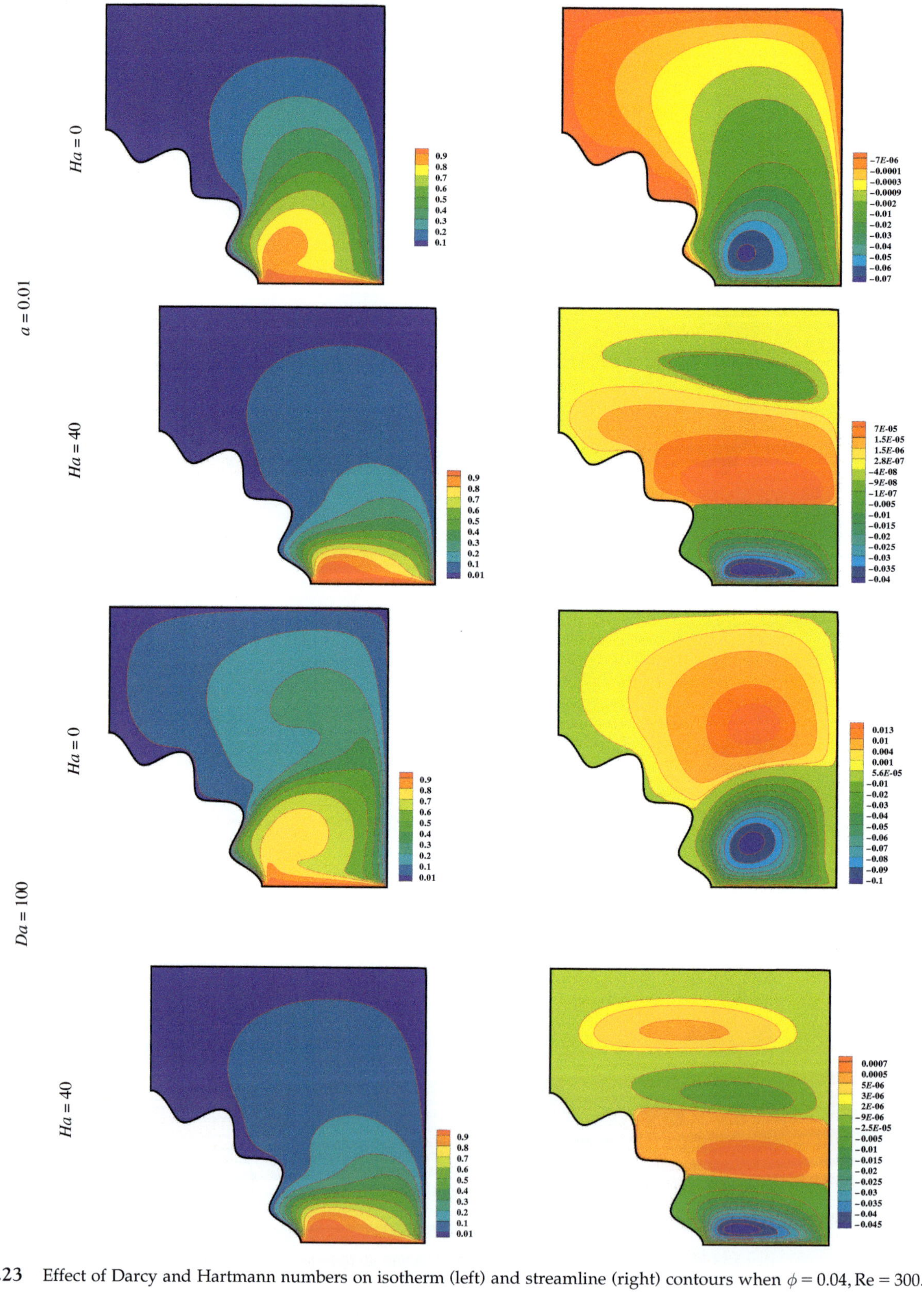

FIGURE 6.23 Effect of Darcy and Hartmann numbers on isotherm (left) and streamline (right) contours when $\phi = 0.04$, Re = 300.

$$Nu_{ave} = \frac{1}{L}\int_{r_{in}}^{r_{out}} Nu_{loc} dx \tag{6.81}$$

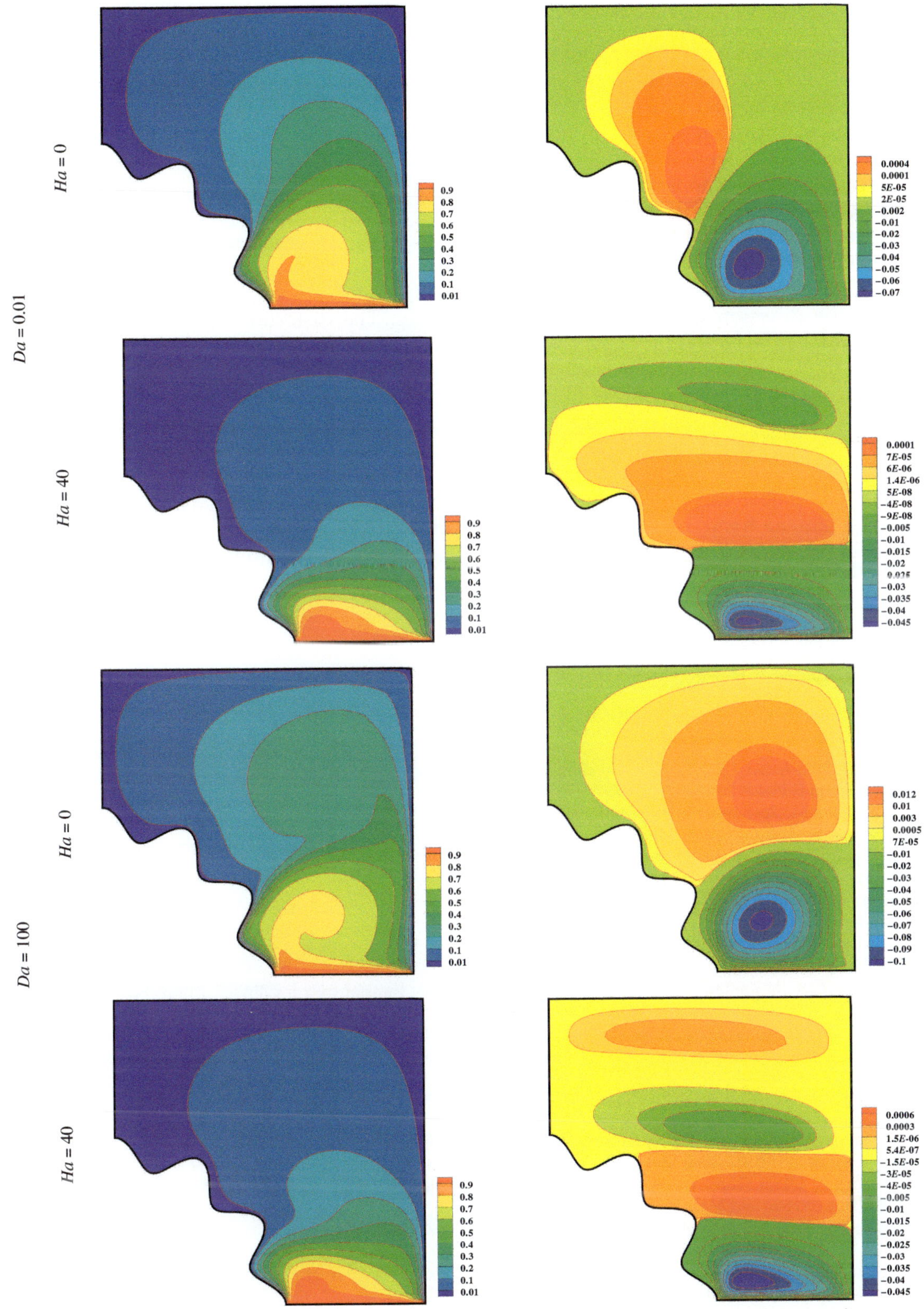

FIGURE 6.24 Effect of Darcy and Hartmann numbers on isotherm (left) and streamline (right) contours when $\phi = 0.04$, Re = 500.

6.5.3 Effects of Active Parameters

In this section, the influence of Lorentz forces on forced convection of nanofluid in a lid driven porous cavity is illustrated. μ_{nf} of CuO-water nanofluid are estimated by means of KKL model. CVFEM is utilized to find the effects of Darcy number ($Da = 0.01$ to 100), Reynolds number ($\mathrm{Re} = 100$ to 500), volume fraction of CuO ($\phi = 0\%$ to 4%), shape of nanoparticle, and Hartmann number ($Ha = 0$ to 40). The influences of shape of the nanoparticles on Nu_{ave} are discussed in Table 6.7. The maximum Nu is caused by Platelet, followed by Cylinder, Brick and Spherical. Therefore, Platelet nanoparticle has been selected for further investigation. The impact of adding nanoparticles in the base fluid on velocity and temperature contours is depicted in Fig. 6.21. The temperature gradient decreases with the increase of ϕ. $|\Psi_{\max}|$ augments with the addition of nanoparticles because of the increase in the solid movements. Figs. 6.22, 6.23, and 6.24 demonstrate the impact of Da, Re, Ha on isotherms and stream lines. At $Da = 0.01, \mathrm{Re} = 100, Ha = 0$, only one eddy exists in streamline. By applying the magnetic field, two new eddies with opposite direction generate above the main eddy. The distortion of isotherms becomes less than before. As Darcy number increases, the convective mode becomes stronger and the primary eddy converts to two eddies. Increasing the Hartmann number leads to the generation of the third eddy. As Re increases, the strength of secondary eddy, which is generated due to increase of Darcy number, enhances. In the absence of magnetic field at $\mathrm{Re} = 500$, a counterclockwise eddy appears above the main eddy. The impacts of Darcy and Hartmann number on isotherms and streamlines at $\mathrm{Re} = 500$ are similar to those at $\mathrm{Re} = 100$ except for the stream function values. It means that the absolute stream function values augment with the increase of the Reynolds number. Increasing Reynolds number can enhance the heat transfer rate because of the reduction in thermal boundary layer thickness. A similar trend is seen for the Darcy number. The temperature gradient decreases with the increase of Lorentz forces. So, Nu_{ave} decreases with the rise of the Hartmann number.

References

[1] T. Ko, K. Ting, Optimal Reynolds number for the fully developed laminar forced convection in a helical coiled tube, Energy 31 (2006) 2142–2152.

[2] C.L. Chen, C.H. Cheng, Numerical simulation of periodic mixed convective heat transfer in a rectangular cavity with a vibrating lid, Appl Therm Eng 29 (2009) 2855–2862.

[3] K.M. Khanafer, A.M. Al-Amiri, I. Pop, Numerical simulation of unsteady mixed convection in a driven cavity using an externally excited sliding lid, Eur J Mech B-Fluids 26 (2007) 669–687.

[4] M. Sheikholeslami, M.M. Rashidi, Ferrofluid heat transfer treatment in the presence of variable magnetic field, Eur. Phys. J. Plus 130 (2015) 115.

[5] M. Sheikholeslami, S.A. Shehzad, Magnetohydrodynamic nanofluid convective flow in a porous enclosure by means of LBM, International Journal of Heat and Mass Transfer 113 (2017) 796–805.

[6] M. Sheikholeslami, Magnetohydrodynamic nanofluid forced convection in a porous lid driven cubic cavity using Lattice Boltzmann Method, Journal of Molecular Liquids 231 (2017) 555–565.

[7] M.S. Kandelousi, KKL correlation for simulation of nanofluid flow and heat transfer in a permeable channel, Physics Letters A 378 (45) (2014) 3331–3339.

[8] M. Sheikholeslami, Effect of uniform suction on nanofluid flow and heat transfer over a cylinder, J Braz. Soc. Mech. Sci. Eng. 37 (2015) 1623–1633.

[9] M. Sheikholeslami, Lattice Boltzmann Method simulation of MHD non-Darcy nanofluid free convection, Physica B 516 (2017) 55–71.

[10] M. Sheikholeslami, Influence of magnetic field on nanofluid free convection in an open porous cavity by means of Lattice Boltzmann Method, Journal of Molecular Liquids 234 (2017) 364–374.

[11] M. Sheikholeslami, A.J. Chamkha, Influence of Lorentz forces on nanofluid forced convection considering Marangoni convection, Journal of Molecular Liquids 225 (2017) 750–757.

[12] M. Sheikholeslami, T. Hayat, A. Alsaedi, Numerical simulation of nanofluid forced convection heat transfer improvement in existence of magnetic field using Lattice Boltzmann Method, International Journal of Heat and Mass Transfer 108 (2017) 1870–1883.

[13] M. Sheikholeslami, R. Ellahi, Three dimensional mesoscopic simulation of magnetic field effect on natural convection of nanofluid, International Journal of Heat and Mass Transfer 89 (2015) 799–808.

[14] M. Sheikholeslami, S. Abelman, Two phase simulation of nanofluid flow and heat transfer in an annulus in the presence of an axial magnetic field, IEEE Transactions on Nanotechnology 14 (3) (2015) 561–569.

[15] M. Sheikholeslami, R. Ellahi, Simulation of ferrofluid flow for magnetic drug targeting using Lattice Boltzmann method, Journal of Zeitschrift Fur Naturforschung A 70 (2) (2015) 115–124.

[16] M. Sheikholeslami, M. Seyednezhad, Lattice Boltzmann Method simulation for CuO-water nanofluid flow in a porous enclosure with hot obstacle, Journal of Molecular Liquids 243 (2017) 249–256.

[17] M. Sheikholeslami, M. Sadoughi, Mesoscopic method for MHD nanofluid flow inside a porous cavity considering various shapes of nanoparticles, International Journal of Heat and Mass Transfer 113 (2017) 106–114.

[18] M. Sheikholeslami, A. Zeeshan, Mesoscopic simulation of CuO-H_2O nanofluid in a porous enclosure with elliptic heat source, International Journal of Hydrogen Energy 42 (22) (2017) 15393–15402.

[19] M. Sheikholeslami, S.A. Shehzad, CVFEM for influence of external magnetic source on $Fe_3O_4 - H_2O$ nanofluid behavior in a permeable cavity considering shape effect, International Journal of Heat and Mass Transfer 115 (2017) 180–191.

[20] M. Sheikholeslami, M. Seyednezhad, Nanofluid heat transfer in a permeable enclosure in presence of variable magnetic field by means of CVFEM, International Journal of Heat and Mass Transfer 114 (2017) 1169–1180.

[21] M. Sheikholeslami, H.B. Rokni, Free convection of CuO-H_2O nanofluid in a curved porous enclosure using mesoscopic approach, International Journal of Hydrogen Energy 42 (22) (2017) 14942–14949.

[22] M. Sheikholeslami, Numerical investigation for CuO-H_2O nanofluid flow in a porous channel with magnetic field using mesoscopic method, J. Mol. Liq. 249 (2018) 739–746.

[23] M. Sheikholeslami, H.B. Rokni, Influence of melting surface on MHD nanofluid flow by means of two phase model, Chinese Journal of Physics 55 (2017) 1352–1360.

[24] M. Sheikholeslami, H.B. Rokni, Effect of melting heat transfer on nanofluid flow in existence of magnetic field considering Buongiorno Model, Chinese Journal of Physics 55 (2017) 1115–1126.

[25] M. Sheikholeslami, Influence of Lorentz forces on nanofluid flow in a porous cylinder considering Darcy model, Journal of Molecular Liquids 225 (2017) 903–912.

[26] M. Sheikholeslami, CVFEM for magnetic nanofluid convective heat transfer in a porous curved enclosure, Eur. Phys. J. Plus 131 (2016) 413. Available from: https://doi.org/10.1140/epjp/i2016-16413-y.

[27] M. Sheikholeslami, Influence of Coulomb forces on Fe3O4-H2O nanofluid thermal improvement, International Journal of Hydrogen Energy 42 (2017) 821–829.

[28] M. Sheikholeslami, Magnetic field influence on nanofluid thermal radiation in a cavity with tilted elliptic inner cylinder, Journal of Molecular Liquids 229 (2017) 137–147.

[29] M. Sheikholeslami, Numerical simulation of magnetic nanofluid natural convection in porous media, Physics Letters A 381 (2017) 494–503.

[30] M. Sheikholeslami, Magnetic source impact on nanofluid heat transfer using CVFEM, Neural Computing and Applications (2016). Available from: https://doi.org/10.1007/s00521-016-2740-7.

[31] M. Sheikholeslami, CuO-water nanofluid free convection in a porous cavity considering Darcy law, The European Physical Journal Plus 132 (2017) 55. Available from: https://doi.org/10.1140/epjp/i2017-11330-3.

[32] M. Sheikholeslami, Numerical investigation of MHD nanofluid free convective heat transfer in a porous tilted enclosure, Engineering Computations 34 (6) (2017) 1939–1955.

[33] M.S. Kandelousi, Effect of spatially variable magnetic field on ferrofluid flow and heat transfer considering constant heat flux boundary condition, The European Physical Journal Plus (2014) 129–248. Available from: https://doi.org/10.1140/epjp/i2014-14248-2.

[34] M. Sheikholeslami, Numerical simulation for external magnetic field influence on Fe3O4-water nanofluid forced convection, Engineering Computations, Inpress.

[35] M. Sheikholeslami, Influence of Lorentz forces on nanofluid flow in a porous cavity by means of Non- Darcy model, Engineering Computations 34 (8) (2017) 2651–2667. Available from: https://doi.org/10.1108/EC-01-2017-0008.

[36] M. Sheikholeslami, H.B. Rokni, Nanofluid convective heat transfer intensification in a porous circular cylinder, Chemical Engineering & Processing: Process Intensification 120 (2017) 93–104.

[37] M. Sheikholeslami, M.K. Sadoughi, Numerical modeling for Fe3O4 -water nanofluid flow in porous medium considering MFD viscosity, Journal of Molecular Liquids 242 (2017) 255–264.

[38] M. Sheikholeslami, M. Shamlooei, Magnetic source influence on nanofluid flow in porous medium considering shape factor effect, Physics Letters A 381 (2017) 3071–3078.

[39] M. Sheikholeslami, M. Shamlooei, Convective flow of nanofluid inside a lid driven porous cavity using CVFEM, Physica B: Condensed Matter 521 (2017) 239–250.

[40] M. Sheikholeslami, K. Vajravelu, Forced convection heat transfer in Fe_3O_4- ethylene glycol nanofluid under the influence of Coulomb force, Journal of Molecular Liquids 233 (2017) 203–210.

[41] M. Sheikholeslami, H.B. Rokni, Influence of EFD viscosity on nanofluid forced convection in a cavity with sinusoidal wall, Journal of Molecular Liquids 232 (2017) 390–395.

[42] M. Sheikholeslami, H.B. Rokni, Magnetic nanofluid natural convection in the presence of thermal radiation considering variable viscosity, The European Physical Journal Plus 132 (2017) 238. Available from: https://doi.org/10.1140/epjp/i2017-11498-4.

[43] M. Sheikholeslami, Houman B. Rokni, Numerical simulation for impact of Coulomb force on nanofluid heat transfer in a porous enclosure in presence of thermal radiation, Int. J. Heat Mass Transfer 118 (2018) 823–831.

[44] M. Sheikholeslami, M. Shamlooei, R. Moradi, Fe3O4- Ethylene glycol nanofluid forced convection inside a porous enclosure in existence of Coulomb force, Journal of Molecular Liquids 249 (2018) 429–437.

[45] M. Sheikholeslami, S.A. Shehzad, Numerical analysis of $Fe_3O_4 - H_2O$ nanofluid flow in permeable media under the effect of external magnetic source, Int. J. Heat Mass Transfer 118 (2018) 182–192.

[46] M. Sheikholeslami, S.A. Shehzad, Magnetohydrodynamic nanofluid convection in a porous enclosure considering heat flux boundary condition, International Journal of Heat and Mass Transfer 106 (2017) 1261–1269.

[47] M. Sheikholeslami, S.A. Shehzad, Thermal radiation of ferrofluid in existence of Lorentz forces considering variable viscosity, International Journal of Heat and Mass Transfer 109 (2017) 82–92.

[48] M. Sheikholeslami, H.B. Rokni, Numerical modeling of nanofluid natural convection in a semi annulus in existence of Lorentz force, Computer Methods in Applied Mechanics and Engineering 317 (2017) 419–430.

[49] M. Sheikholeslami, K. Vajravelu, Nanofluid flow and heat transfer in a cavity with variable magnetic field, Applied Mathematics and Computation 298 (2017) 272–282.

[50] M. Sheikholeslami, H.B. Rokni, Magnetohydrodynamic CuO-water nanofluid in a porous complex shaped enclosure, ASME, Journal of Thermal Science and Engineering Applications 9 (4) (2017) 041007. Available from: https://doi.org/10.1115/1.4035973.

[51] M. Sheikholeslami, H.B. Rokni, Nanofluid two phase model analysis in existence of induced magnetic field, International Journal of Heat and Mass Transfer 107 (2017) 288–299.

[52] M. Sheikholeslami, T. Hayat, A. Alsaedi, S. Abelman, Numerical analysis of EHD nanofluid force convective heat transfer considering electric field dependent viscosity, International Journal of Heat and Mass Transfer 108 (2017) 2558–2565.

[53] M. Sheikholeslami, Magnetic field influence on CuO -H_2O nanofluid convective flow in a permeable cavity considering various shapes for nanoparticles, International Journal of Hydrogen Energy 42 (2017) 19611–19621.

[54] M. Sheikholeslami, M.M. Bhatti, Forced convection of nanofluid in presence of constant magnetic field considering shape effects of nanoparticles, International Journal of Heat and Mass Transfer 111 (2017) 1039–1049.

[55] M. Sheikholeslami, A. Zeeshan, Analysis of flow and heat transfer in water based nanofluid due to magnetic field in a porous enclosure with constant heat flux using CVFEM, Computer Methods in Applied Mechanics and Engineering 320 (2017) 68–81.

[56] M. Sheikholeslami, M. Shamlooei, Fe_3O_4- H_2O nanofluid natural convection in presence of thermal radiation, International Journal of Hydrogen Energy 42 (9) (2017) 5708–5718.

[57] M. Sheikholeslami, M.M. Bhatti, Active method for nanofluid heat transfer enhancement by means of EHD, International Journal of Heat and Mass Transfer 109 (2017) 115–122.

[58] M. Sheikholeslami, T. Hayat, A. Alsaedi, Numerical study for external magnetic source influence on water based nanofluid convective heat transfer, International Journal of Heat and Mass Transfer 106 (2017) 745–755.

[59] M. Sheikholeslami, A.J. Chamkha, Flow and convective heat transfer of a ferro-nanofluid in a double-sided lid-driven cavity with a wavy wall in the presence of a variable magnetic field, Numerical Heat Transfer, Part A 69 (10) (2016) 1186–1200. Available from: https://doi.org/10.1080/10407782.2015.1125709.

[60] M. Sheikholeslami, Ali J. Chamkha, Electrohydrodynamic free convection heat transfer of a nanofluid in a semi-annulus enclosure with a sinusoidal wall, Numerical Heat Transfer, Part A 69 (7) (2016) 781–793. Available from: http://dx.doi.org/10.1080/10407782.2015.1090819.

[61] M. Sheikholeslami, M.K. Sadoughi, Simulation of CuO- water nanofluid heat transfer enhancement in presence of melting surface, International Journal of Heat and Mass Transfer 116 (2018) 909–919.

[62] M. Sheikholeslami, H.B. Rokni, Simulation of nanofluid heat transfer in presence of magnetic field: A review, International Journal of Heat and Mass Transfer 115 (2017) 1203–1233.

[63] M. Sheikholeslami, T. Hayat, A. Alsaedi, MHD free convection of Al_2O_3–water nanofluid considering thermal radiation: A numerical study, International Journal of Heat and Mass Transfer 96 (2016) 513–524.

[64] M. Sheikholeslami, K. Vajravelu, M.M. Rashidi, Forced convection heat transfer in a semi annulus under the influence of a variable magnetic field, International Journal of Heat and Mass Transfer 92 (2016) 339–348.

[65] M. Sheikholeslami, H.B. Rokni, Melting heat transfer influence on nanofluid flow inside a cavity in existence of magnetic field, International Journal of Heat and Mass Transfer 114 (2017) 517–526.

[66] M. Sheikholeslami, A. Zeeshan, Numerical simulation of Fe_3O_4 -water nanofluid flow in a non-Darcy porous media, Int. J. Numer. Methods Heat Fluid Flow 28 (3) (2018) 641–660. Available from: https://doi.org/10.1108/HFF-04-2017-0160.

[67] M. Sheikholeslami, M. Barzegar Gerdroodbary, S. Valiallah Mousavi, D.D. Ganji, R. Moradi, Heat transfer enhancement of ferrofluid inside an 90o elbow channel by non-uniform magnetic field, J. Magnet. Magnet. Mater. 460 (2018) 302–311.

[68] M. Sheikholeslami, Numerical simulation for solidification in a LHTESS by means of nano-enhanced PCM, J. Taiwan Inst. Chem. Eng. 86 (2018) 25–41.

[69] M. Sheikholeslami, Numerical modeling of nano enhanced PCM solidification in an enclosure with metallic fin, J. Mol. Liq. 259 (2018) 424–438.

[70] M. Sheikholeslami, A. Ghasemi, Solidification heat transfer of nanofluid in existence of thermal radiation by means of FEM, Int. J. Heat Mass Transfer 123 (2018) 418–431.

[71] M. Sheikholeslami, S.A. Shehzad, CVFEM simulation for nanofluid migration in a porous medium using Darcy model, Int. J. Heat Mass Transfer 122 (2018) 1264–1271.

[72] M. Sheikholeslami, M. Darzi, M.K. Sadoughi, Heat transfer improvement and pressure drop during condensation of refrigerant-based nanofluid; an experimental procedure, Int. J. Heat Mass Transfer 122 (2018) 643–650.

[73] M. Sheikholeslami, H.B. Rokni, CVFEM for effect of Lorentz forces on nanofluid flow in a porous complex shaped enclosure by means of non-equilibrium model, J. Mol. Liq. 254 (2018) 446–462.

[74] M. Sheikholeslami, S.A. Shehzad, Z. Li, Water based nanofluid free convection heat transfer in a three dimensional porous cavity with hot sphere obstacle in existence of Lorenz forces, Int. J. Heat Mass Transfer 125 (2018) 375–386.

[75] M. Sheikholeslami, M. Jafaryar, D.D. Ganji, Z. Li, Exergy loss analysis for nanofluid forced convection heat transfer in a pipe with modified turbulators, J. Mol. Liq. 262 (2018) 104–110.

[76] M. Sheikholeslami, M. Jafaryar, Z. Li, Nanofluid turbulent convective flow in a circular duct with helical turbulators considering CuO nanoparticles, Int. J. Heat Mass Transfer 124 (2018) 980–989.

[77] M. Sheikholeslami, Z. Li, M. Shamlooei, Nanofluid, MHD natural convection through a porous complex shaped cavity considering thermal radiation, Phys. Lett. A 382 (2018) 1615–1632.

[78] M. Sheikholeslami, H.B. Rokni, Magnetic nanofluid flow and convective heat transfer in a porous cavity considering Brownian motion effects, Phys. Fluids 30 (1) (2018). Available from: https://doi.org/10.1063/1.5012517.

[79] M. Sheikholeslami, S.A. Shehzad, Simulation of water based nanofluid convective flow inside a porous enclosure via non-equilibrium model, Int. J. Heat Mass Transfer 120 (2018) 1200–1212.

[80] M. Sheikholeslami, S.A. Shehzad, Non-Darcy free convection of Fe_3O_4-water nanoliquid in a complex shaped enclosure under impact of uniform Lorentz force, Chin. J. Phys. 56 (2018) 270–281.

[81] M. Sheikholeslami, M. Seyednezhad, Simulation of nanofluid flow and natural convection in a porous media under the influence of electric field using CVFEM, Int. J. Heat Mass Transfer 120 (2018) 772–781.

[82] M. Sheikholeslami, M. Shamlooei, R. Moradi, Numerical simulation for heat transfer intensification of nanofluid in a permeable curved enclosure considering shape effect of Fe_3O_4 nanoparticles, Chem. Eng. Process. Process Intensif. 124 (2018) 71–82.
[83] M. Sheikholeslami, T. Hayat, T. Muhammad, A. Alsaedi, MHD forced convection flow of nanofluid in a porous cavity with hot elliptic obstacle by means of Lattice Boltzmann method, Int. J. Mech. Sci 135 (2018) 532–540.
[84] M. Sheikholeslami, Numerical investigation of nanofluid free convection under the influence of electric field in a porous enclosure, J. Mol. Liq. 249 (2018) 1212–1221.
[85] M. Sheikholeslami, CuO-water nanofluid flow due to magnetic field inside a porous media considering Brownian motion, J. Mol. Liq. 249 (2018) 921–929.
[86] M. Sheikholeslami, D.D. Ganji, Influence of electric field on Fe_3O_4- water nanofluid radiative and convective heat transfer in a permeable enclosure, J. Mol. Liq. 250 (2018) 404–412.
[87] M. Sheikholeslami, M. Shamlooei, R. Moradi, Fe_3O_4- Ethylene glycol nanofluid forced convection inside a porous enclosure in existence of Coulomb force, J. Mol. Liq. 249 (2018) 429–437.
[88] M. Sheikholeslami, T. Hayat, A. Alsaedi, Numerical simulation for forced convection flow of MHD CuO-H2O nanofluid inside a cavity by means of LBM, J. Mol. Liq. 249 (2018) 941–948.
[89] M. Sheikholeslami, H.R. Kataria, A.S. Mittal, Effect of thermal diffusion and heat-generation on MHD nanofluid flow past an oscillating vertical plate through porous medium, J. Mol. Liq. 257 (2018) 12–25.
[90] M. Sheikholeslami, S.A. Shehzad, Z. Li, Nanofluid heat transfer intensification in a permeable channel due to magnetic field using Lattice Boltzmann method, Phys. B (2018) 2018. Available from: https://doi.org/10.1016/j.physb.2018.03.036.

CHAPTER

7

Effect of Uniform Lorentz Forces on Nanofluid Flow Using CVFEM

7.1 INTRODUCTION

Natural convection heat transfer in the annulus between two horizontal concentric cylinders has been a subject of intensive research during the past decades due to its wide applications, such as in nuclear reactor design, cooling of electronic equipment, aircraft cabin insulation, cooling of electronic equipment, and heating and ventilation control in building design. Natural convection under the influence of a magnetic field is of great importance in many industrial applications, such as crystal growth, metal casting, and liquid metal cooling blankets for fusion reactors. Sarris et al. [1] carried out a numerical study of unsteady two-dimensional natural convection of an electrically conducting fluid in a laterally and volumetrically heated square cavity under the influence of a magnetic field. The results showed that the increasing Hartmann number is not straightforwardly connected with the resulting flow patterns in the flow configuration. Oztop et al. [2] studied the MHD natural convection in an enclosure from two semicircular heaters on the bottom wall. Their results showed that flow strength decreases with the increase of Hartmann number; this means that a magnetic field can be used as a control parameter for heat and fluid flow.

The above literature review reveals that the existence of a magnetic field has a noticeable effect on heat transfer reduction under natural convection, while in many engineering applications increasing heat transfer from solid surfaces is a goal. In this circumstance, the use of nanofluids with higher thermal conductivity can be considered as a promising solution. Sheikholeslami et al. [3] studied the natural convection in a concentric annulus between a cold outer square and heated inner circular cylinders in the presence of a static radial magnetic field. They reported that the average Nusselt number is an increasing function of nanoparticle volume fraction as well as Rayleigh number, whereas it is a decreasing function of Hartmann number. Sheikholeslami et al. [4] investigated the nanofluid flow and radiative heat transfer in the presence of constant Lorentz forces. Sheikholeslami et al. [5] simulated ferrofluid flow under the effect of Lorentz forces. Nanofluid magnetohydrodynamic flow and heat transfer have been presented by several authors [6–88].

7.2 NANOFLUID FREE CONVECTION HEAT TRANSFER IN AN ENCLOSURE BETWEEN A CIRCULAR AND A SINUSOIDAL CYLINDER IN THE PRESENCE OF MAGNETIC FIELD

7.2.1 Problem Definition

The physical model and the corresponding triangular elements used in the present CVFEM program are shown in Fig. 7.1. The inner and outer walls are maintained at constant temperatures T_h and T_c respectively. The shape of inner cylinder profile is assumed to mimic the following pattern

$$r = r_{in} + A\cos\left(N\left(\zeta - \zeta_0\right)\right) \tag{7.1}$$

Application of Control Volume based Finite Element Method (CVFEM) for Nanofluid Flow and Heat Transfer.
DOI: https://doi.org/10.1016/B978-0-12-814152-6.00007-2

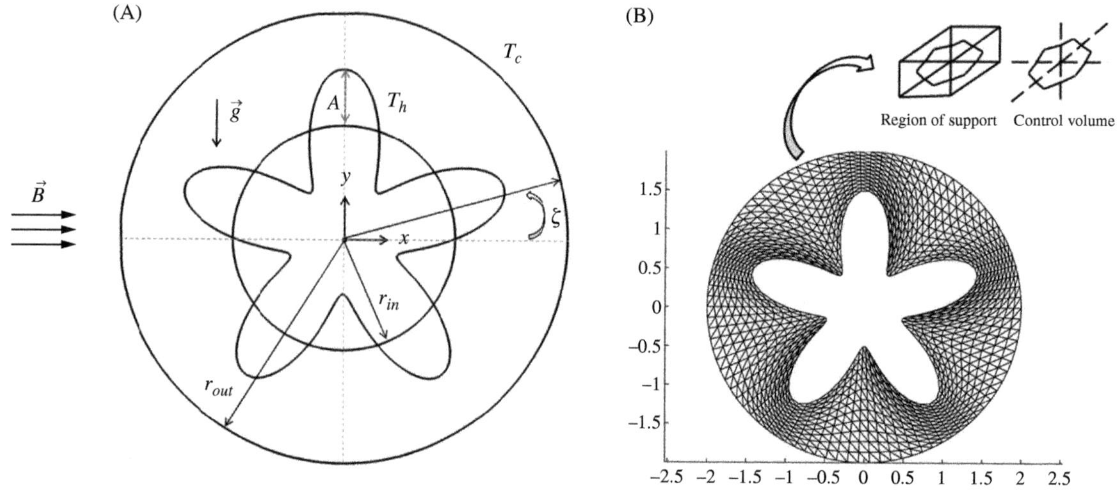

FIGURE 7.1 (A) Geometry and the boundary conditions with (B) the mesh of geometry considered in this work.

where r_{in} is the base circle radius, r_{out} is the radius of outer cylinder, A and N are the amplitude and number of undulations respectively, and ζ is the rotation angle. It is also assumed that the uniform magnetic field $\vec{B} = B_x\vec{e_x} + B_y\vec{e_y}$ of constant magnitude $B = \sqrt{B_x^2 + B_y^2}$ is applied horizontally. The electric current J and the electromagnetic force F are defined by $J = \sigma\left(\vec{V} \times \vec{B}\right)$ and $F = \sigma\left(\vec{V} \times \vec{B}\right) \times \vec{B}$, respectively.

7.2.2 Governing Equation

The flow is two-dimensional, laminar, and incompressible. The radiation, viscous dissipation, induced electric current, and Joule heating are neglected. The magnetic Reynolds number is assumed to be small so that the induced magnetic field can be neglected compared to the applied magnetic field. The flow is considered to be steady, two-dimensional, and laminar. Neglecting displacement currents, induced magnetic field, and using the Boussinesq approximation, the governing equations of heat transfer and fluid flow for nanofluid can be obtained as follows:

$$\frac{\partial u}{\partial x} + \frac{\partial v}{\partial y} = 0 \tag{7.2}$$

$$u\frac{\partial u}{\partial x} + v\frac{\partial u}{\partial y} = -\frac{1}{\rho_{nf}}\frac{\partial P}{\partial x} + \upsilon_{nf}\left(\frac{\partial^2 u}{\partial x^2} + \frac{\partial^2 u}{\partial y^2}\right) + \frac{\sigma_{nf}B^2}{\rho_{nf}}\left(v \sin\lambda \cos\lambda - u\sin^2\lambda\right) \tag{7.3}$$

$$\begin{aligned} u\frac{\partial v}{\partial x} + v\frac{\partial v}{\partial y} &= -\frac{1}{\rho_{nf}}\frac{\partial P}{\partial y} + \upsilon_{nf}\left(\frac{\partial^2 v}{\partial x^2} + \frac{\partial^2 v}{\partial y^2}\right) + \beta_{nf}g(T - T_c) \\ &+ \frac{\sigma_{nf}B^2}{\rho_{nf}}\left(u \sin\lambda \cos\lambda - v\cos^2\lambda\right) \end{aligned} \tag{7.4}$$

$$u\frac{\partial T}{\partial x} + v\frac{\partial T}{\partial y} = \alpha_{nf}\left(\frac{\partial^2 T}{\partial x^2} + \frac{\partial^2 T}{\partial y^2}\right) \tag{7.5}$$

where the effective density (ρ_{nf}) and heat capacitance $\left(\rho C_p\right)_{nf}$ of the nanofluid are defined as:

$$\rho_{nf} = \rho_f(1 - \phi) + \rho_s\phi \tag{7.6}$$

$$\left(\rho C_p\right)_{nf} = \left(\rho C_p\right)_f (1-\phi) + \left(\rho C_p\right)_s \phi \tag{7.7}$$

where ϕ is the solid volume fraction of nanoparticles. Thermal diffusivity of the nanofluids is

$$\alpha_{nf} = \frac{k_{nf}}{\left(\rho C_p\right)_{nf}} \tag{7.8}$$

and the thermal expansion coefficient of the nanofluids can be determined as

$$\beta_{nf} = \beta_f (1-\phi) + \beta_s \phi \tag{7.9}$$

The dynamic viscosity of the nanofluids given by Brinkman is

$$\mu_{nf} = \frac{\mu_f}{(1-\phi)^{2.5}} \tag{7.10}$$

The effective thermal conductivity of the nanofluid can be approximated by the Maxwell–Garnetts (MG) model as:

$$\frac{k_{nf}}{k_f} = \frac{k_s + 2k_f - 2\phi(k_f - k_s)}{k_s + 2k_f + \phi(k_f - k_s)} \tag{7.11}$$

and the effective electrical conductivity of nanofluid was presented by Maxwell as below:

$$\frac{\sigma_{nf}}{\sigma_f} = 1 + \frac{3\left(\dfrac{\sigma_s}{\sigma_f} - 1\right)\phi}{\left(\dfrac{\sigma_s}{\sigma_f} + 2\right) - \left(\dfrac{\sigma_s}{\sigma_f} - 1\right)\phi} \tag{7.12}$$

The stream function and vorticity are defined as:

$$u = \frac{\partial \psi}{\partial y}, \quad v = -\frac{\partial \psi}{\partial x}, \quad \omega = \frac{\partial v}{\partial x} - \frac{\partial u}{\partial y} \tag{7.13}$$

The stream function satisfies the continuity Eq. (7.2). The vorticity equation is obtained by eliminating the pressure between the two momentum equations, i.e., by taking the y-derivative of Eq. (7.3) and subtracting from it the x-derivative of Eq. (7.4). This gives:

$$\begin{aligned} \frac{\partial \psi}{\partial y}\frac{\partial \omega}{\partial x} - \frac{\partial \psi}{\partial x}\frac{\partial \omega}{\partial y} &= \upsilon_{nf}\left(\frac{\partial^2 \omega}{\partial x^2} + \frac{\partial^2 \omega}{\partial y^2}\right) + \beta_{nf} g \left(\frac{\partial T}{\partial x}\right) \\ &+ \frac{\sigma_{nf} B^2}{\rho_{nf}}\left(-\frac{\delta v}{\delta y}\sin\lambda\cos\lambda + \frac{\delta u}{\delta y}\sin^2\lambda + \frac{\delta u}{\delta x}\sin\lambda\cos\lambda - \frac{\delta v}{\delta x}\cos^2\lambda\right) \end{aligned} \tag{7.14}$$

$$\frac{\partial \psi}{\partial y}\frac{\partial T}{\partial x} - \frac{\partial \psi}{\partial x}\frac{\partial T}{\partial y} = \alpha_{nf}\left(\frac{\partial^2 T}{\partial x^2} + \frac{\partial^2 T}{\partial y^2}\right) \tag{7.15}$$

$$\frac{\partial^2 \psi}{\partial x^2} + \frac{\partial^2 \psi}{\partial y^2} = -\omega \tag{7.16}$$

By introducing the following nondimensional variables:

$$X = \frac{x}{L}, \; Y = \frac{y}{L}, \; \Omega = \frac{\omega L^2}{\alpha_f}, \; \Psi = \frac{\psi}{\alpha_f}, \; \Theta = \frac{T - T_c}{T_h - T_c}, \; U = \frac{uL}{\alpha_f}, \; V = \frac{vL}{\alpha_f} \tag{7.17}$$

where in Eq. (7.17) $L = r_{out} - r_{in} = r_{in}$. Using the dimensionless parameters, the equations now become:

$$\frac{\partial \Psi}{\partial Y}\frac{\partial \Omega}{\partial X} - \frac{\partial \Psi}{\partial X}\frac{\partial \Omega}{\partial Y} = \left[\frac{\Pr_f}{(1-\phi)^{2.5}\left((1-\phi)+\phi\frac{\rho_s}{\rho_f}\right)}\right]\left(\frac{\partial^2 \Omega}{\partial X^2} + \frac{\partial^2 \Omega}{\partial Y^2}\right)$$
$$+ Ra_f \Pr_f \left[(1-\phi)+\phi\frac{\beta_s}{\beta_f}\right]\left(\frac{\partial \Theta}{\partial X}\right)$$
$$+ Ha^2 \Pr_f \left(1 + \frac{3\left(\frac{\sigma_s}{\sigma_f}-1\right)\phi}{\left(\frac{\sigma_s}{\sigma_f}+2\right)-\left(\frac{\sigma_s}{\sigma_f}-1\right)\phi}\right)\left(\frac{1}{(1-\phi)+\phi\frac{\rho_s}{\rho_f}}\right)\left(-\frac{\delta V}{\delta Y}\tan\lambda + \frac{\delta U}{\delta Y}\tan^2\lambda + \frac{\delta U}{\delta X}\tan\lambda - \frac{\delta V}{\delta X}\right) \quad (7.18)$$

$$\frac{\partial \Psi}{\partial Y}\frac{\partial \Theta}{\partial X} - \frac{\partial \Psi}{\partial X}\frac{\partial \Theta}{\partial Y} = \left[\frac{\frac{k_{nf}}{k_f}}{\left((1-\phi)+\phi\frac{(\rho Cp)_s}{(\rho Cp)_f}\right)}\right]\left(\frac{\partial^2 \Theta}{\partial X^2} + \frac{\partial^2 \Theta}{\partial Y^2}\right) \quad (7.19)$$

$$\frac{\partial^2 \Psi}{\partial X^2} + \frac{\partial^2 \Psi}{\partial Y^2} = -\Omega \quad (7.20)$$

where $Ra_f = g\beta_f L^3 (T_h - T_c)/(\alpha_f \upsilon_f)$ is the Rayleigh number for the base fluid, $Ha = LB_x\sqrt{\sigma_f/\mu_f}$ is the Hartmann number, and $\Pr_f = \upsilon_f/\alpha_f$ is the Prandtl number for the base fluid. The thermophysical properties of the nanofluid are given in Table 7.1. The boundary conditions as shown in Fig. 7.1 are:

$$\begin{aligned} \Theta &= 1.0 \quad \text{on the inner sinusoidal boundary} \\ \Theta &= 0.0 \quad \text{on the outer circular boundary} \\ \Psi &= 0.0 \quad \text{on all solid boundaries} \end{aligned} \quad (7.21)$$

The values of vorticity on the boundary of the enclosure can be obtained using the stream function formulation and the known velocity conditions during the iterative solution procedure.

The local Nusselt number of the nanofluid along the cold wall can be expressed as:

$$Nu_{loc} = \left(\frac{k_{nf}}{k_f}\right)\frac{\partial \Theta}{\partial r} \quad (7.22)$$

where r is the radial direction. The average Nusselt number on the cold circular wall is evaluated as:

$$Nu_{ave} = \frac{1}{\gamma}\int_0^{\gamma} Nu_{loc}(\zeta)\, d\zeta \quad (7.23)$$

TABLE 7.1 Thermophysical Properties of Water and Nanoparticles

	$\rho(kg/m^3)$	$C_p(j/kgk)$	$k(W/m.k)$	$\beta(K^{-1})$	$\sigma(\Omega\cdot m)^{-1}$
Pure water	997.1	4179	0.613	21×10^{-5}	0.05
Copper(*Cu*)	8933	385	401	1.67×10^{-5}	5.96×10^7

7.2.3 Effects of Active Parameters

The present calculations are made for various values of Hartmann number ($Ha = 0, 20, 60$ and 100), Rayleigh number ($Ra = 10^3, 10^4$ and 10^5), volume fraction of nanoparticles ($\phi = 0\%, 2\%, 4\%$ and 6%), and number of undulations ($N = 3$ and 6) at constant amplitude ($A = 0.5$) and Prandtl number ($\Pr = 6.2$).

Comparisons of the isotherm (left) and streamline (right) contours for different values of Hartmann number and Raleigh number for $N = 3$ and $N = 6$ are shown in Figs. 7.2 and 7.3, respectively. As seen, the absolute value of stream function increases as Rayleigh number increases and it decreases as Hartmann number increases. For both cases at $Ra = 10^3$, the isotherms are uniform and follow the shape of the inner and outer walls and the values of stream function are relatively small which indicates the domination of the conduction heat transfer

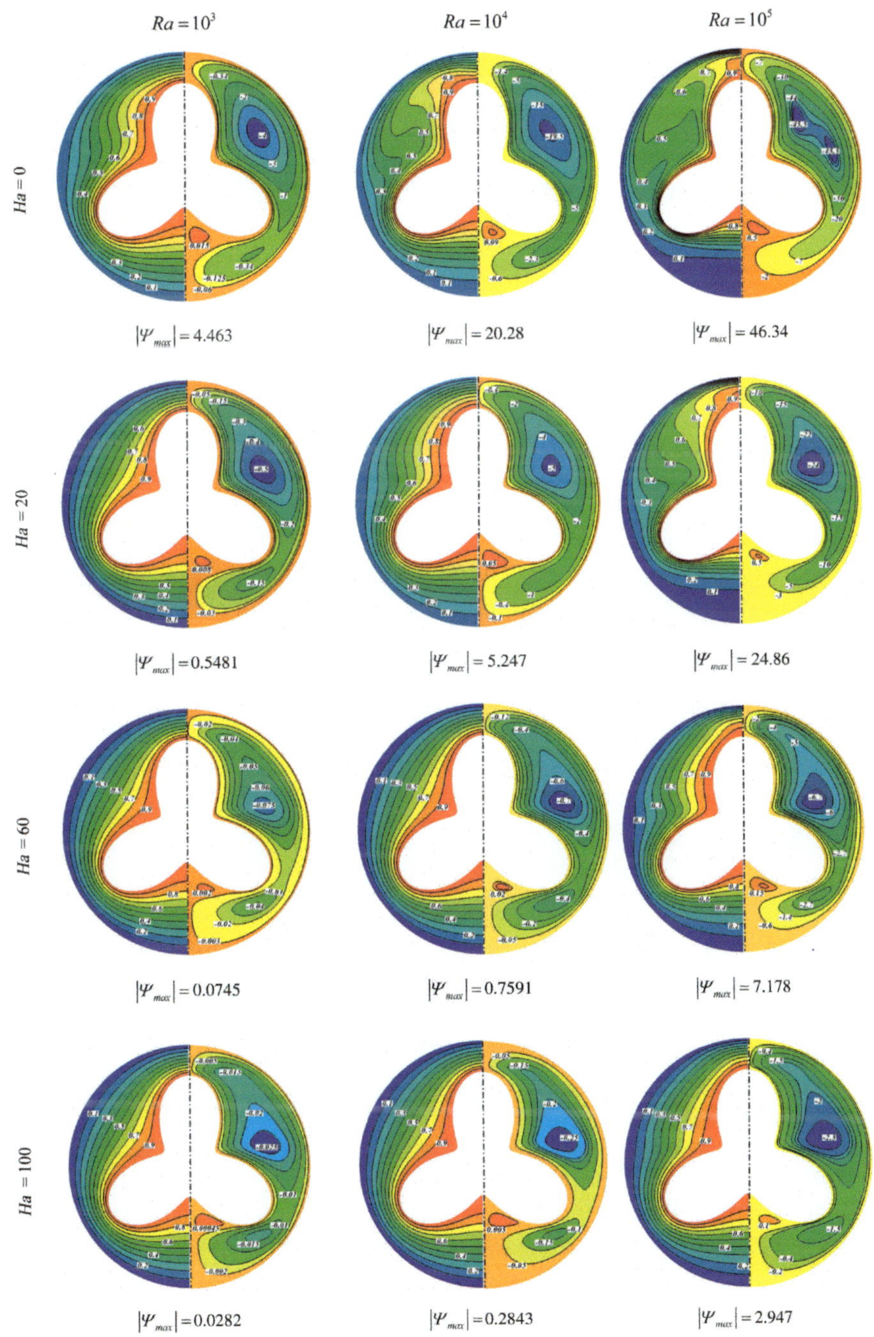

FIGURE 7.2 Comparison of the isotherm (left) and streamline (right) contours for different values of Hartmann number and Raleigh number at $A = 0.5, N = 3$.

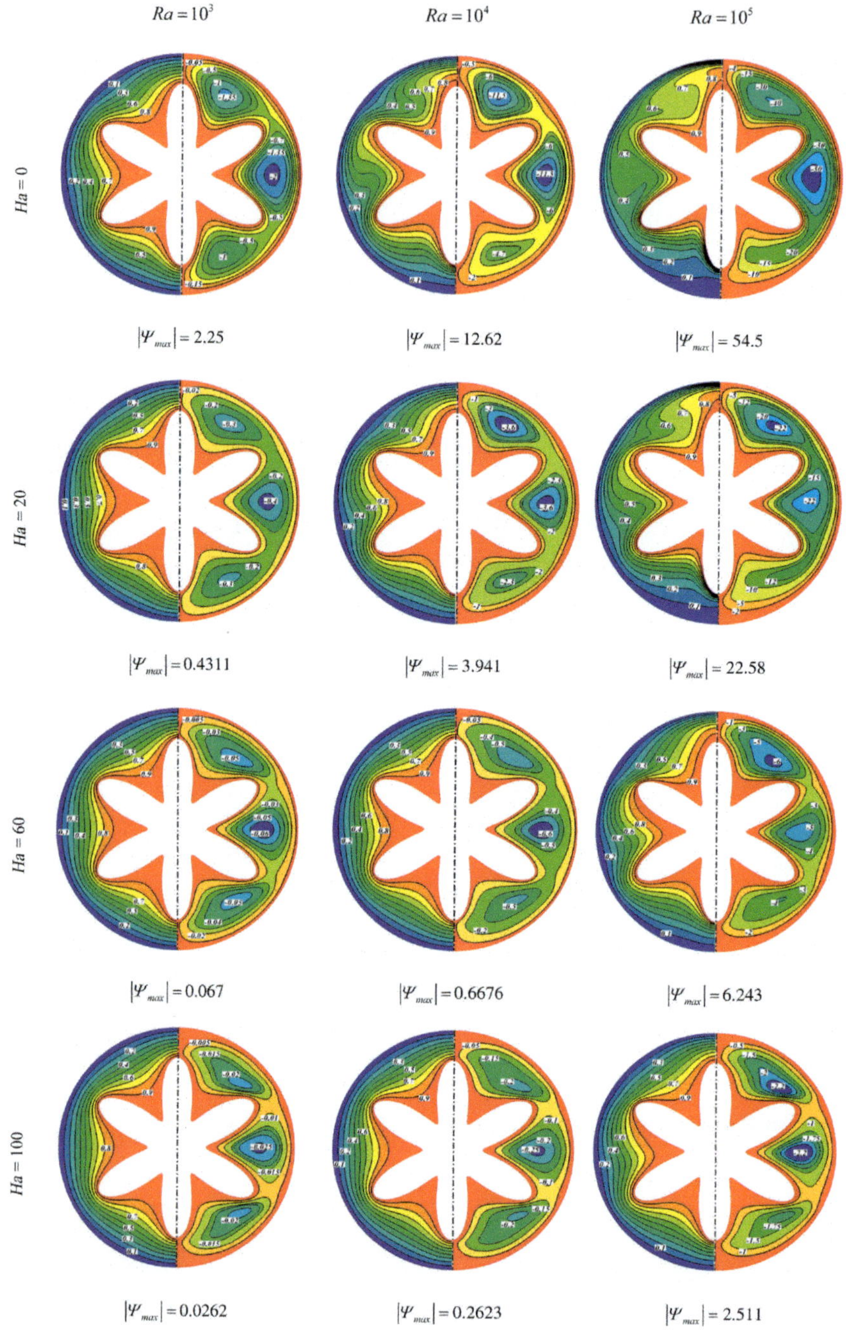

FIGURE 7.3 . Comparison of the isotherm (left) and streamline (right) contours for different values of Hartmann number and Raleigh number at $A = 0.5, N = 6$.

mechanism. For $N = 3$, under the hot surface a very small secondary eddy appears near the vertical centerline; this is due to the resistance of the crest against the flow circulation between the inner and outer cylinders. For $N = 6$, no small secondary eddy at the bottom half of the enclosure is observed because the lower crest of the inner cylinder is parallel to the flow direction. As seen, the isotherms start to be disturbed at $Ra = 10^4$ and thermal plumes gradually appear on the hot surface of the inner cylinder, especially at the upper half of the enclosure, which indicates that the convective heat transfer becomes comparable with conduction at this Rayleigh number. In addition the thermal boundary layer between the crests of the inner cylinder become thinner in comparison with those of $Ra = 10^3$ which indicates that the unventilated region between the crests diminishes; hence the thermal boundary layer is suppressed toward the inner hot cylinder at these areas. At $Ra = 10^5$, the isotherms are crowded near the crests and depict the diminution of the thermal boundary layer thickness at the bottom of the enclosure. At the top of the inner cylinder ($\zeta = 90^\circ$), a strong plume drives the flow to strongly impinge on the outer wall of the enclosure. It is worthwhile mentioning that the effect of the magnetic field is to decrease the value of the velocity magnitude throughout the enclosure because the presence of the magnetic field introduces a force called the Lorentz force, which acts against the flow if the magnetic field is applied in the normal direction. This type of resisting force slows down the fluid velocity. The increase of Hartmann number makes the core of the vortices move toward the horizontal centerline. Also the magnetic field causes the thermal plume to disappear and makes the isotherms parallel to each other due to the domination of the conduction mode of heat transfer.

The effects of the number of undulations, Rayleigh number, and Hartmann number on local Nusselt number over the cold wall and between $\zeta = 90^\circ$ and 270° are depicted in Fig. 7.4. The number of extremum in the local Nusselt number profile corresponds to the number of undulations of the inner hot surface. For $N = 3$ there is one local maximum and one local minimum while for $N = 6$, there are two maximum and three minimum. In the presence of a magnetic field, the extremum points of the local Nusselt number profiles occur at smaller ζ because of the change in the position of eddies with Hartmann number. At $Ra = 10^5$ and $Ha = 0$ the local Nusselt number decreases as ζ increases from 90° and 270°; at the top of the inner cylinder ($\zeta = 90^\circ$), the flow accompanying the cold fluid impinges on the upper cold surface which results in a thicker thermal boundary layer over the enclosure wall, whereas when we approach to $\zeta = 270^\circ$ the hot surface locates at the top of the cold wall and the conduction mechanism becomes more pronounced, hence the local Nusselt number decreases noticeably.

Fig. 7.5 shows the effects of the number of undulations, nanoparticle volume fraction, Rayleigh number, and Hartmann number on the average Nusselt number. As a whole, the average Nusselt number increases with the increase of Rayleigh number and nanoparticle volume fraction, while it decreases as Hartmann number increases. Adding high thermal conductivity nanoparticles leads to an increase in conduction. So because of the domination of conduction at low Rayleigh number, the effect of the nanoparticle volume fraction on average Nusselt number is more pronounced at low Rayleigh number than at high Rayleigh number. Also this figure depicts that increasing the number of undulations causes the average Nusselt number to increase.

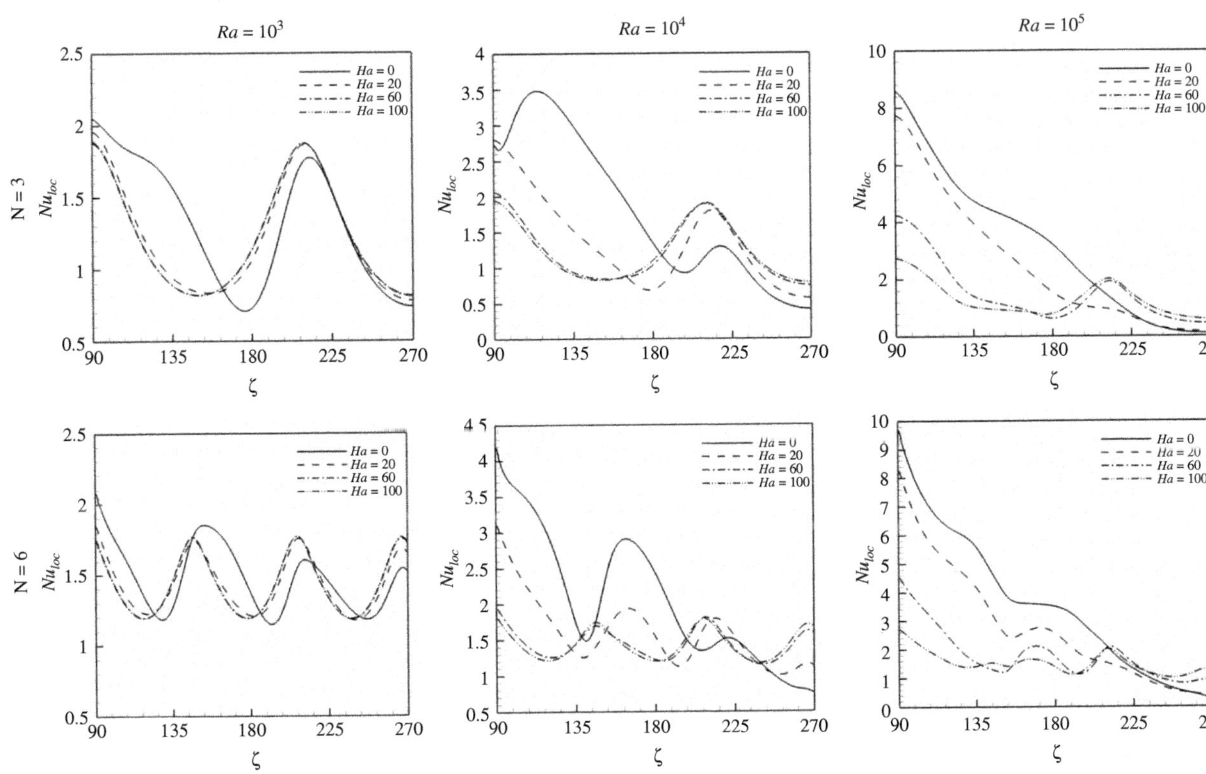

FIGURE 7.4 Effects of the number of undulations, Rayleigh number, and Hartmann number on local Nusselt number at $\phi = 0.06, A = 0.5$.

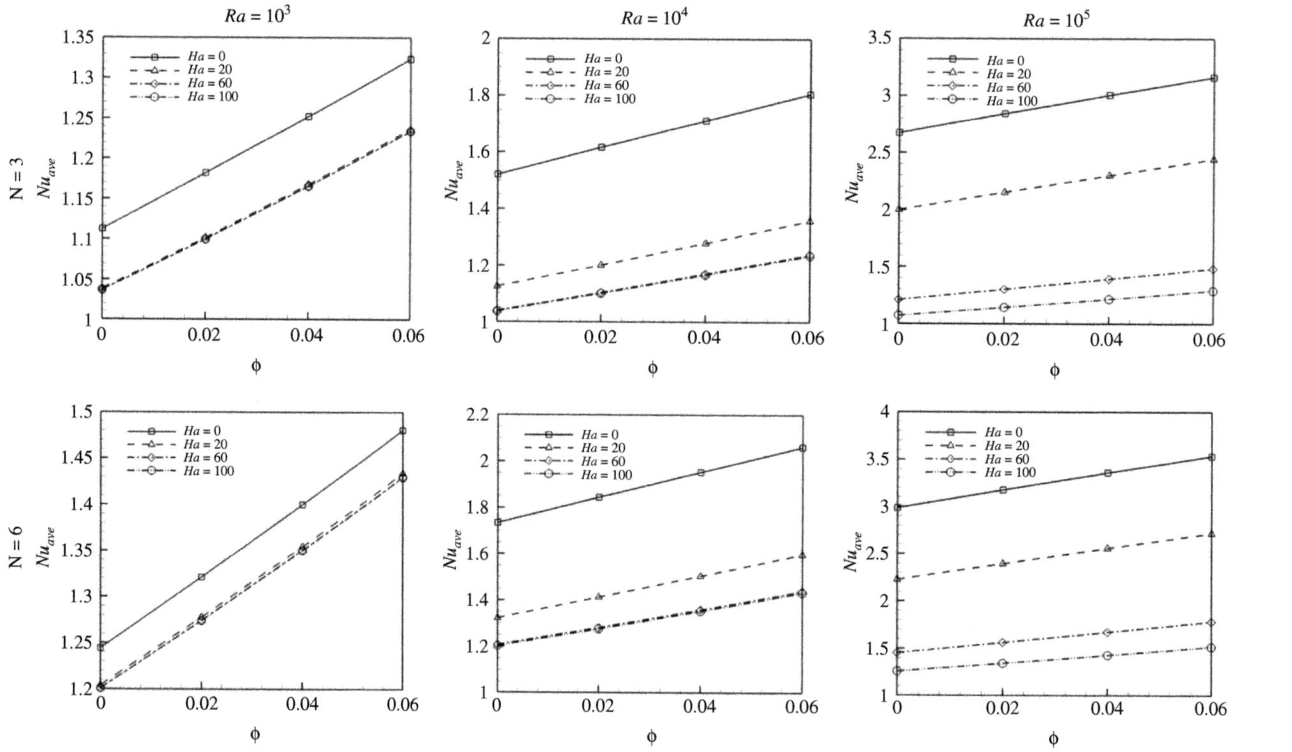

FIGURE 7.5 Effects of the number of undulations, nanoparticle volume fraction, Rayleigh number, and Hartmann number on average Nusselt number at $A = 0.5$.

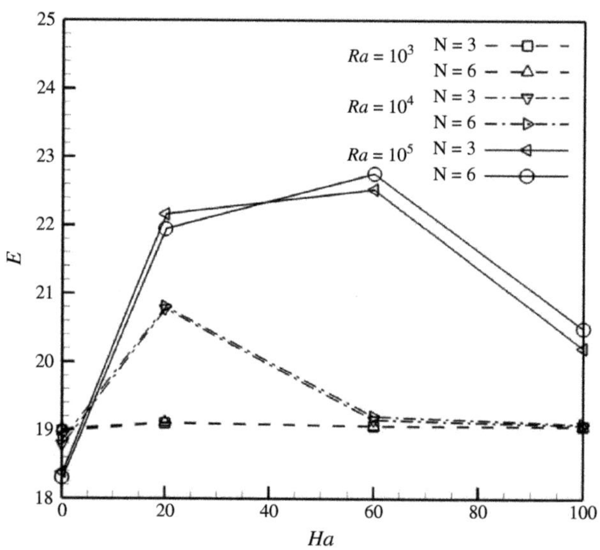

FIGURE 7.6 Effects of the number of undulations, Hartmann number, and Rayleigh number on the ratio of heat transfer enhancement due to the addition of nanoparticles.

To estimate the enhancement of heat transfer between the case of $\phi = 0.06$ and the pure fluid (base fluid) case, the enhancement is defined as:

$$E = \frac{Nu(\phi = 0.06) - Nu(basefluid)}{Nu(basefluid)} \times 100 \tag{7.24}$$

Fig. 7.6 depicts the effect of the number of undulations, Hartmann number, and Rayleigh number on heat transfer enhancement ratio. There is an insignificant change in enhancement at $Ra = 10^3$. In the absence of a magnetic field, the enhancement ratio decreases as the Rayleigh number increases while the opposite behavior is observed in the presence of a magnetic field. The maximum values of enhancement for $Ra = 10^4$ and 10^5 are observed at $Ha = 20$ and 60, respectively. It is an interesting observation that at $Ra = 10^5$ the enhancement in heat transfer for the case of $N = 6$ is greater than that obtained for the case of $N = 3$ when $Ha > 40$, while for other values of Rayleigh number the number of undulations has no effect on enhancement.

7.3 INFLUENCE OF A MAGNETIC FIELD ON FREE CONVECTION IN AN INCLINED HALF-ANNULUS ENCLOSURE FILLED WITH CU-WATER NANOFLUID

7.3.1 Problem Definition

The physical model along with the important geometric parameters and the mesh of the half-annulus enclosure used in the present CVFEM program are shown in Fig. 7.7A. The inner wall is under constant heat flux (q''), the outer wall is maintained at constant temperature (T_c) and the two other walls are thermally insulated. It is

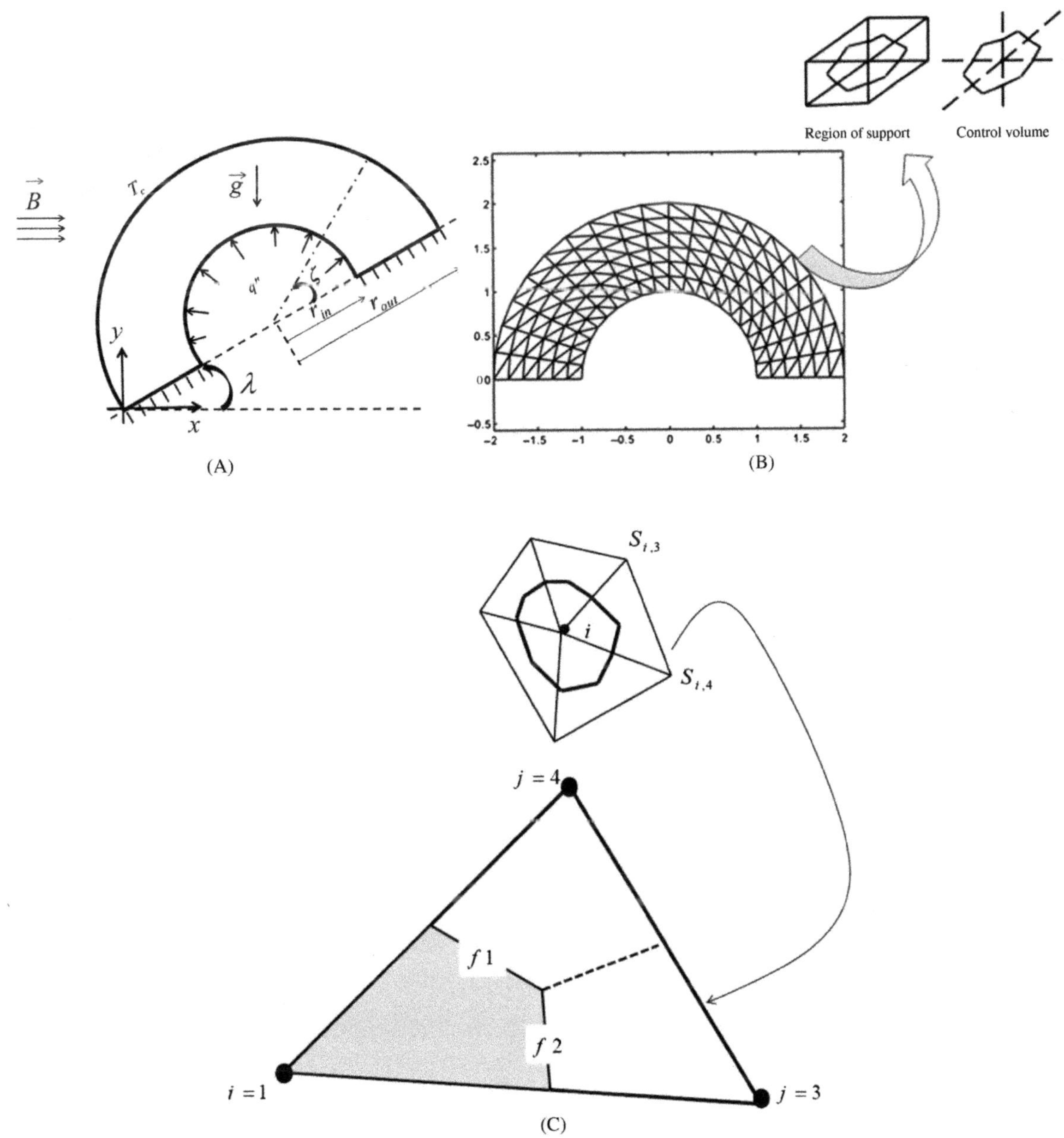

FIGURE 7.7 (A) Geometry and the boundary conditions with (B) the mesh of half-annulus enclosure considered in this work; (C) a sample triangular element and its corresponding control volume.

also assumed that the uniform magnetic field ($\vec{B} = B_x\vec{e_x} + B_y\vec{e_y}$) of constant magnitude $B = \sqrt{B_x^2 + B_y^2}$ is applied, where $\vec{e_x}$ and $\vec{e_y}$ are unit vectors in the Cartesian coordinate system. The orientation of the magnetic field forms an angle γ with the horizontal axis such that $\gamma = B_x/B_y$. The electric current J and the electromagnetic force F are defined by $J = \sigma\left(\vec{V} \times \vec{B}\right)$ and $F = \sigma\left(\vec{V} \times \vec{B}\right) \times \vec{B}$, respectively.

7.3.2 Governing Equation

The flow is two-dimensional, laminar, and incompressible. The radiation, viscous dissipation, induced electric current, and Joule heating are neglected. The magnetic Reynolds number is assumed to be small so that the induced magnetic field can be neglected compared to the applied magnetic field. The flow is considered to be steady, two-dimensional, and laminar. Neglecting displacement currents, induced magnetic field, and using the Boussinesq approximation, the governing equations of heat transfer and fluid flow for nanofluid can be obtained as follows:

$$\frac{\partial u}{\partial x} + \frac{\partial v}{\partial y} = 0 \tag{7.25}$$

$$u\frac{\partial u}{\partial x} + v\frac{\partial u}{\partial y} = -\frac{1}{\rho_{nf}}\frac{\partial P}{\partial x} + \upsilon_{nf}\left(\frac{\partial^2 u}{\partial x^2} + \frac{\partial^2 u}{\partial y^2}\right) + \frac{\sigma_{nf}B^2}{\rho_{nf}}\left(v\,\sin\lambda\,\cos\lambda - u\sin^2\lambda\right) \tag{7.26}$$

$$u\frac{\partial v}{\partial x} + v\frac{\partial v}{\partial y} = -\frac{1}{\rho_{nf}}\frac{\partial P}{\partial y} + \upsilon_{nf}\left(\frac{\partial^2 v}{\partial x^2} + \frac{\partial^2 v}{\partial y^2}\right) + \beta_{nf}g(T - T_c) + \frac{\sigma_{nf}B^2}{\rho_{nf}}\left(u\,\sin\lambda\,\cos\lambda - v\,\cos^2\lambda\right) \tag{7.27}$$

$$u\frac{\partial T}{\partial x} + v\frac{\partial T}{\partial y} = \alpha_{nf}\left(\frac{\partial^2 T}{\partial x^2} + \frac{\partial^2 T}{\partial y^2}\right) \tag{7.28}$$

where the effective density (ρ_{nf}) and heat capacitance $(\rho C_p)_{nf}$ of the nanofluid are defined as:

$$\rho_{nf} = \rho_f(1-\phi) + \rho_s\phi \tag{7.29}$$

$$(\rho C_p)_{nf} = (\rho C_p)_f(1-\phi) + (\rho C_p)_s\phi \tag{7.30}$$

where ϕ is the solid volume fraction of nanoparticles. Thermal diffusivity of the nanofluids is

$$\alpha_{nf} = \frac{k_{nf}}{(\rho C_p)_{nf}} \tag{7.31}$$

and the thermal expansion coefficient of the nanofluids can be determined as

$$\beta_{nf} = \beta_f(1-\phi) + \beta_s\phi \tag{7.32}$$

The dynamic viscosity of the nanofluids given by Brinkman is

$$\mu_{nf} = \frac{\mu_f}{(1-\phi)^{2.5}} \tag{7.33}$$

The effective thermal conductivity of the nanofluid can be approximated by the Maxwell–Garnetts (MG) model as:

$$\frac{k_{nf}}{k_f} = \frac{k_s + 2k_f - 2\phi(k_f - k_s)}{k_s + 2k_f + \phi(k_f - k_s)} \tag{7.34}$$

and the effective electrical conductivity of nanofluid was presented by Maxwell as below:

$$\frac{\sigma_{nf}}{\sigma_f}=1+\frac{3(\sigma_s/\sigma_f-1)\phi}{(\sigma_s/\sigma_f+2)-(\sigma_s/\sigma_f-1)\phi} \tag{7.35}$$

The stream function and vorticity are defined as:

$$u=\frac{\partial\psi}{\partial y},\quad v=-\frac{\partial\psi}{\partial x},\quad \omega=\frac{\partial v}{\partial x}-\frac{\partial u}{\partial y} \tag{7.36}$$

The stream function satisfies the continuity Eq. (7.25). The vorticity equation is obtained by eliminating the pressure between the two momentum equations, i.e., by taking the y-derivative of Eq. (7.26) and subtracting from it the x-derivative of Eq. (7.27). This gives:

$$\frac{\partial\psi}{\partial y}\frac{\partial\omega}{\partial x}-\frac{\partial\psi}{\partial x}\frac{\partial\omega}{\partial y}=\upsilon_{nf}\left(\frac{\partial^2\omega}{\partial x^2}+\frac{\partial^2\omega}{\partial y^2}\right)+\beta_{nf}g\left(\frac{\partial T}{\partial x}\right)$$
$$+\frac{\sigma_{nf}B^2}{\rho_{nf}}\left(-\frac{\delta v}{\delta y}\sin\lambda\cos\lambda+\frac{\delta u}{\delta y}\sin^2\lambda+\frac{\delta u}{\delta x}\sin\lambda\cos\lambda-\frac{\delta v}{\delta x}\cos^2\lambda\right) \tag{7.37}$$

$$\frac{\partial\psi}{\partial y}\frac{\partial T}{\partial x}-\frac{\partial\psi}{\partial x}\frac{\partial T}{\partial y}=\alpha_{nf}\left(\frac{\partial^2 T}{\partial x^2}+\frac{\partial^2 T}{\partial y^2}\right) \tag{7.38}$$

$$\frac{\partial^2\psi}{\partial x^2}+\frac{\partial^2\psi}{\partial y^2}=-\omega \tag{7.39}$$

By introducing the following nondimensional variables:

$$X=\frac{x}{L},\quad Y=\frac{y}{L},\quad \Omega=\frac{\omega L^2}{\alpha_f},\ \Psi=\frac{\psi}{\alpha_f},\ \Theta=\frac{T-T_c}{(q''L/k_f)},\ U=\frac{uL}{\alpha_f},\ V=\frac{vL}{\alpha_f} \tag{7.40}$$

where in Eq. (7.17) $L=r_{out}-r_{in}=r_{in}$. Using the dimensionless parameters, the equations now become:

$$\frac{\partial\Psi}{\partial Y}\frac{\partial\Omega}{\partial X}-\frac{\partial\Psi}{\partial X}\frac{\partial\Omega}{\partial Y}=\left[\frac{\Pr_f}{(1-\phi)^{2.5}\left((1-\phi)+\phi\frac{\rho_s}{\rho_f}\right)}\right]\left(\frac{\partial^2\Omega}{\partial X^2}+\frac{\partial^2\Omega}{\partial Y^2}\right)$$
$$+Ra_f\Pr_f\left[(1-\phi)+\phi\frac{\beta_s}{\beta_f}\right]\left(\frac{\partial\Theta}{\partial X}\right)$$
$$+Ha^2\Pr_f\left(1+\frac{3\left(\frac{\sigma_s}{\sigma_f}-1\right)\phi}{\left(\frac{\sigma_s}{\sigma_f}+2\right)-\left(\frac{\sigma_s}{\sigma_f}-1\right)\phi}\right)\left(\frac{1}{(1-\phi)+\phi\frac{\rho_s}{\rho_f}}\right)\left(-\frac{\delta V}{\delta Y}\tan\lambda+\frac{\delta U}{\delta Y}\tan^2\lambda+\frac{\delta U}{\delta X}\tan\lambda-\frac{\delta V}{\delta X}\right) \tag{7.41}$$

$$\frac{\partial\Psi}{\partial Y}\frac{\partial\Theta}{\partial X}-\frac{\partial\Psi}{\partial X}\frac{\partial\Theta}{\partial Y}=\left[\frac{\frac{k_{nf}}{k_f}}{\left((1-\phi)+\phi\frac{(\rho Cp)_s}{(\rho Cp)_f}\right)}\right]\left(\frac{\partial^2\Theta}{\partial X^2}+\frac{\partial^2\Theta}{\partial Y^2}\right) \tag{7.42}$$

$$\frac{\partial^2 \Psi}{\partial X^2} + \frac{\partial^2 \Psi}{\partial Y^2} = -\Omega \tag{7.43}$$

where $Ra_f = g\beta_f L^4 q''/(k_f\, \alpha_f \nu_f)$ is the Rayleigh number for the base fluid, $\Pr_f = \nu_f/\alpha_f$ is the Prandtl number, and $Ha = LB_x\sqrt{\sigma_f/\mu_f}$ is the Hartmann number for the base fluid. The boundary conditions as shown in Fig. 7.7 are:

$$\begin{aligned} &\Theta = 1.0 && \text{on the inner circular boundary} \\ &\frac{\partial \Theta}{\partial n} = -1 && \text{on the outer circular boundary} \\ &\partial\Theta/\partial n = 0.0 && \text{on the two other insulation boundaries} \\ &\Psi = 0.0 && \text{on all solid boundaries} \end{aligned} \tag{7.44}$$

The values of vorticity on the boundary of the enclosure can be obtained using the stream function formulation and the known velocity conditions during the iterative solution procedure.

The local Nusselt number of the nanofluid along the hot wall can be expressed as:

$$Nu_{loc} = \left(\frac{k_{nf}}{k_f}\right)\frac{1}{\Theta} \tag{7.45}$$

The average Nusselt number on the hot circular wall is evaluated as:

$$Nu_{ave} = \frac{1}{\pi}\int_0^{\pi} Nu_{loc}(\zeta)\, d\zeta \tag{7.46}$$

7.3.3 Effects of Active Parameters

In this paper, numerical simulations of natural convection of nanofluid flow in a half-annulus enclosure with one wall under constant heat flux in the presence of a magnetic field were performed. Calculations are made for various values of Hartmann number ($Ha = 0, 20, 60$ and 100), Rayleigh number ($Ra = 10^3, 10^4$ and 10^5), volume fraction of nanoparticles ($\phi = 0\%, 2\%, 4\%$ and 6%), and inclination angle ($\lambda = 0^{\circ}, 45^{\circ}$ and 90°) at constant Prandtl number ($\Pr = 6.2$). The fluid in the enclosure is a water-based nanofluid containing Cu nanoparticle. The thermophysical properties of Cu nanoparticles and base fluid are shown in Table 7.1.

Isotherm (down) and streamline (up) contours for different values of Rayleigh number and Hartmann number when $\lambda = 0^{\circ}, 45^{\circ}$ and 90° are shown in Figs. 7.8–7.10, respectively. Generally $|\Psi_{\max}|_{nf}$ increases with the increase of Rayleigh number and inclination angle but it decreases with the increase of Hartmann number. At $\lambda = 0^{\circ}$ and $Ra = 10^3$, the heat transfer in the enclosure is mainly dominated by the conduction mode. The streamlines show two overall rotating symmetric eddies. As the Rayleigh number increases up to 10^4, the role of convection in heat transfer becomes more significant and consequently the thermal boundary layer on the surface of the inner wall becomes thinner. In addition, a plume starts to appear on the top of the inner circular wall. At $Ra = 10^5$, a strong plume causes the flow to strongly impinge on the top of the enclosure. In addition, it can be seen that the core of the primary eddies keeps moving upward due to the increasing convection velocity. At $\lambda = 45^{\circ}$, two cells were formed, and they turn in opposite directions but they are not symmetric and the main cell, which turns in counterclockwise direction, is dominant because of more available space for circulation. It is an interesting observation that for $Ra = 10^5$, as the Hartman number increases up to 100, the secondary eddy which is turning clockwise is divided into two smaller eddies. At $\lambda = 90^{\circ}$, the two main eddies merge into one counterclockwise eddy. The isotherms are distorted at the top of the enclosure at higher Rayleigh numbers due to the stronger convection effects. When the magnetic field is imposed on the enclosure, the velocity field is suppressed owing to the retarding effect of the Lorentz force. So the intensity of convection weakens significantly. The braking effect of the magnetic field is observed from the maximum stream function value. The core vortex is shifted toward vertical as the Hartmann number increases. Also imposing the magnetic field leads to omitting the thermal plume over the inner wall.

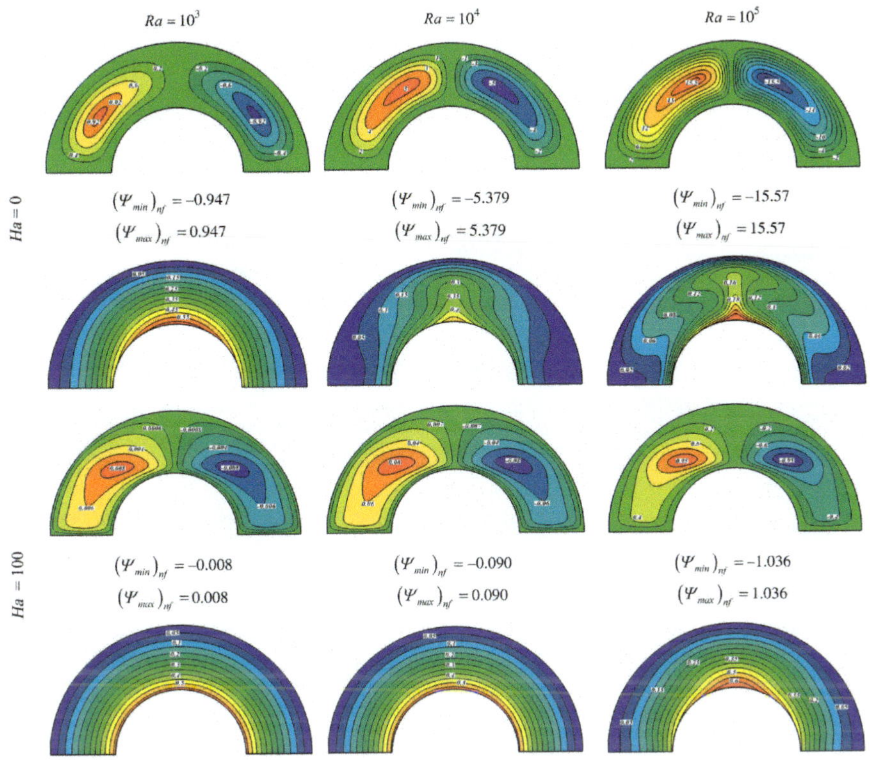

FIGURE 7.8 Isotherm (down) and streamline (up) contours for different values of Rayleigh number and Hartmann number when $\lambda = 0^{\circ}$ for Cu-water nanofluid ($\phi = 0.06$).

Fig. 7.11 shows the effects of Hartmann number, Rayleigh number, and inclination angle for Cu-water ($\phi = 0.06$) nanofluids on the Local Nusselt number. At $\gamma = 0^{\circ}$ the local Nusselt number profile is symmetric with respect to the vertical centerline of the enclosure because of the existence of symmetric geometry and boundary conditions. But for other values of inclination angle it is not symmetric. in the presence of the magnetic field, because of the domination of the conduction mechanism, the maximum value of local Nusselt number occurs at the vertical centerline. At $\gamma = 90^{\circ}$ the local Nusselt number decreases with the increase of S and the increasing Hartmann number leads to a decrease in Nusselt number. When $Ha = 0$, the number of extremum in the local Nusselt number profile corresponds to exist of thermal plume. The presence of a magnetic field leads to the thermal plume disappearing over the inner wall and makes the isotherms parallel to each other due to the domination of the conduction mode of heat transfer. So the extremum points of the local Nusselt number profiles are omitted when the Hartmann number increases.

The effects of the Hartmann number, Rayleigh number, and inclination angle on the average Nusselt number are shown in Fig. 7.12. For all values of Rayleigh number, in absence of magnetic field, the maximum value of Nu_{ave} occurs at $\lambda = 0^{\circ}$ but for high values of Hartmann number the maximum value occurs at $\lambda = 90^{\circ}$. The average Nusselt number increases with the increase of the Rayleigh number while it decreases as the Hartmann number increases. At $Ra = 10^5$ in the absence of a magnetic field, the maximum value of average Nusselt number is obtained at $\gamma = 0^{\circ}$ but for higher values of the Hartmann number the maximum value of Nu_{ave} occurs at $\gamma = 90^{\circ}$.

To estimate the enhancement of heat transfer between the case of $\phi = 0.06$ and the pure fluid (base fluid) case, the enhancement is defined as:

$$E = \frac{Nu(\phi = 0.06) - Nu(basefluid)}{Nu(basefluid)} \times 100 \tag{7.47}$$

The effects of Hartmann number, Rayleigh number, and inclination angle on heat transfer enhancement ratio are shown in Fig. 7.13. Generally, as the Rayleigh number increases the maximum values of enhancement decreases. At $Ra = 10^3$ and 10^4, it can be seen than maximum values of enhancement increases as Hartmann number increases, but for $Ra = 10^5$ minimum values of it occurred at $Ha = 40$. The increase of heat transfer

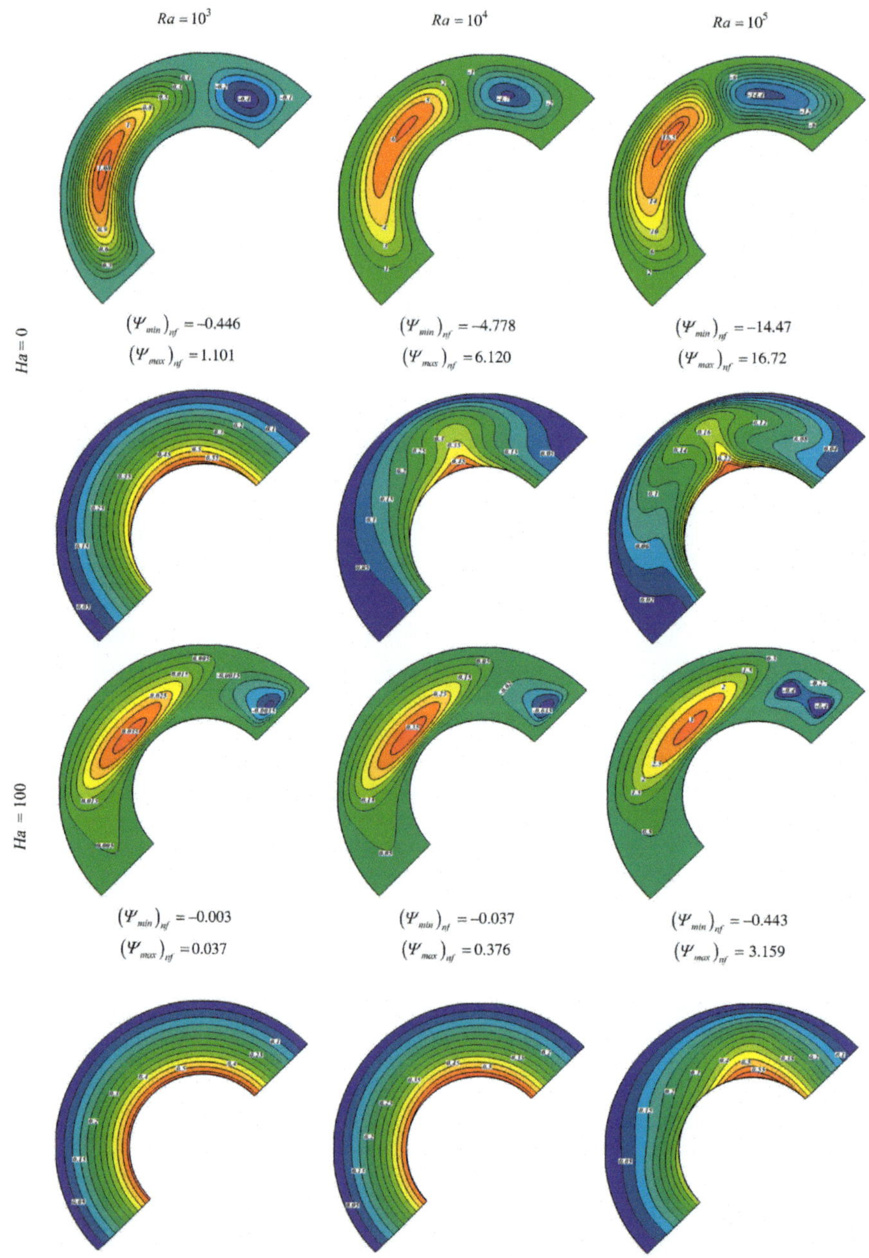

FIGURE 7.9 Isotherm (down) and streamline (up) contours for different values of Rayleigh number and Hartmann number when $\lambda = 45^{\circ}$ for Cu-water nanofluid ($\phi = 0.06$).

enhancement is related to the effect of nanoparticles on thermal conductivity of the fluid. At smaller Rayleigh numbers the dominant mechanism is conduction and an increase of the Hartmann number enhances the conduction heat transfer by slowing down the fluid circulation. For $Ra = 10^5$ the maximum proportion of heat transfer belongs to the convective heat transfer mechanism. Hence the increase of the Hartman number up to $Ha = 40$ decreases the fluid velocity and convective heat transfer. For higher Hartman numbers the magnetic field slows down the fluid flow and again conduction becomes the predominant mechanism and the value of enhancement increases.

At low Rayleigh numbers ($Ra = 10^3$ and 10^4), maximum values of enhancement for a low Hartmann number are obtained at $\lambda = 90^{\circ}$ while for high values of the Hartmann number the maximum values occur at $\lambda = 0^{\circ}$. It is an interesting observation that at $Ra = 10^5$ the enhancement in heat transfer for the case of $\lambda = 90^{\circ}$ is greater than two other cases for all values of Hartmann number.

FIGURE 7.10 Isotherm (down) and streamline (up) contours for different values of Rayleigh number and Hartmann number when $\lambda = 90^{\circ}$ for Cu-water nanofluid ($\phi = 0.06$).

7.4 MHD NANOFLUID CONVECTIVE FLOW IN AN INCLINED ENCLOSURE WITH SINUSOIDAL WALL

7.4.1 Problem Definition

The schematic of the problem and the related boundary conditions as well as the mesh of enclosure which is used in the present CVFEM program are shown in Fig. 7.14. The enclosure has a width/height aspect

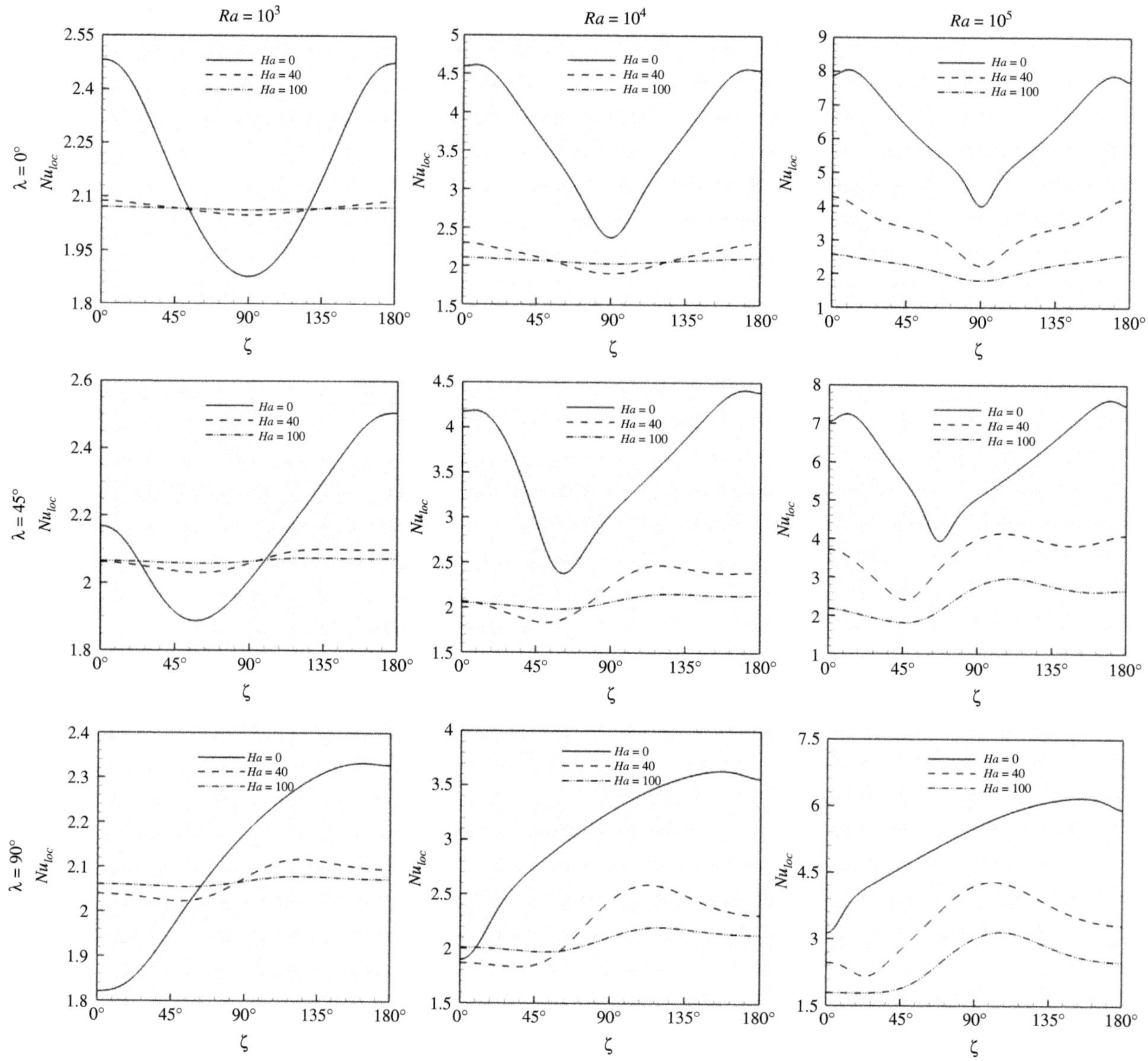

FIGURE 7.11 Effects of Hartmann number, Rayleigh number, and inclination angle for Cu-water ($\phi = 0.06$) nanofluids on the Local Nusselt number.

ratio of two. The two sidewalls with length H are thermally insulated, whereas the lower flat and upper sinusoidal walls are maintained at constant temperatures T_h and T_c, respectively. Under all circumstances $T_h > T_c$ condition is maintained. The shape of the upper sinusoidal wall profile is assumed to mimic the following pattern

$$Y = H - \{a(H + \sin(\pi x - \pi/2))\} \tag{7.48}$$

where a is the dimensionless amplitude of the sinusoidal wall. It is also assumed that the uniform magnetic field ($\vec{B} = B_x\vec{e_x} + B_y\vec{e_y}$) of constant magnitude $B = \sqrt{B_x^2 + B_y^2}$ is applied, where $\vec{e_x}$ and $\vec{e_y}$ are unit vectors in the Cartesian coordinate system. The orientation of the magnetic field form an angle γ with horizontal axis such that $\gamma = B_x/B_y$. The electric current J and the electromagnetic force F are defined by $J = \sigma\left(\vec{V} \times \vec{B}\right)$ and $F = \sigma\left(\vec{V} \times \vec{B}\right) \times \vec{B}$ respectively.

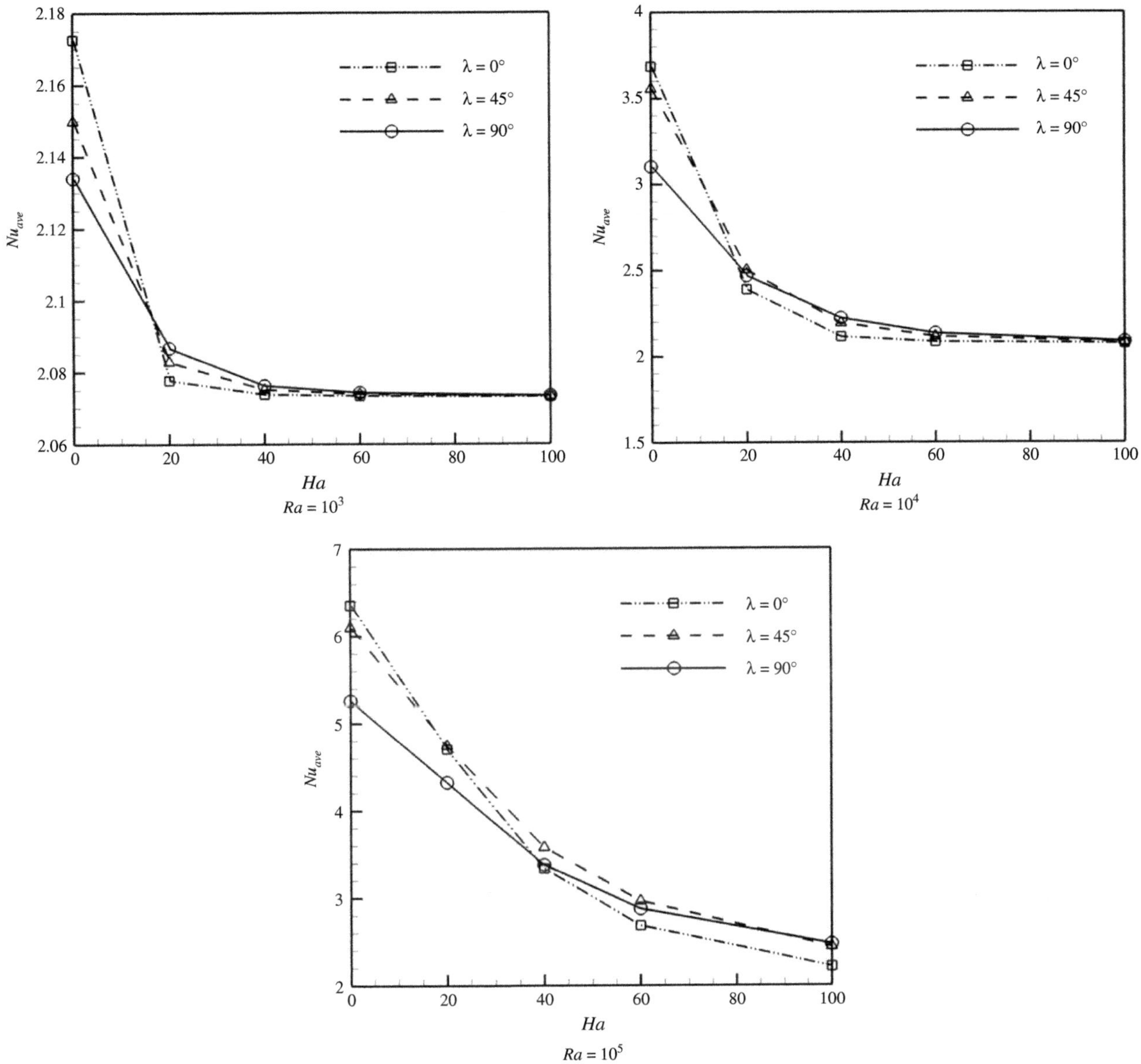

FIGURE 7.12 Effects of the Hartmann number, Rayleigh number, and inclination angle for Cu-water ($\phi = 0.06$) nanofluids on the average Nusselt number.

7.4.2 Governing Equation

The flow is two-dimensional, laminar, and incompressible. The radiation, viscous dissipation, induced electric current, and Joule heating are neglected. The magnetic Reynolds number is assumed to be small so that the induced magnetic field can be neglected compared to the applied magnetic field. The flow is considered to be steady, two-dimensional, and laminar. Neglecting displacement currents, induced magnetic field, and using the Boussinesq approximation, the governing equations of heat transfer and fluid flow for nanofluid can be obtained as follows:

$$\frac{\partial u}{\partial x} + \frac{\partial v}{\partial y} = 0 \tag{7.49}$$

$$u\frac{\partial u}{\partial x} + v\frac{\partial u}{\partial y} = -\frac{1}{\rho_{nf}}\frac{\partial P}{\partial x} + \upsilon_{nf}\left(\frac{\partial^2 u}{\partial x^2} + \frac{\partial^2 u}{\partial y^2}\right) + \frac{\sigma_{nf}B^2}{\rho_{nf}}\left(v\,\sin\lambda\,\cos\lambda - u\sin^2\lambda\right) \tag{7.50}$$

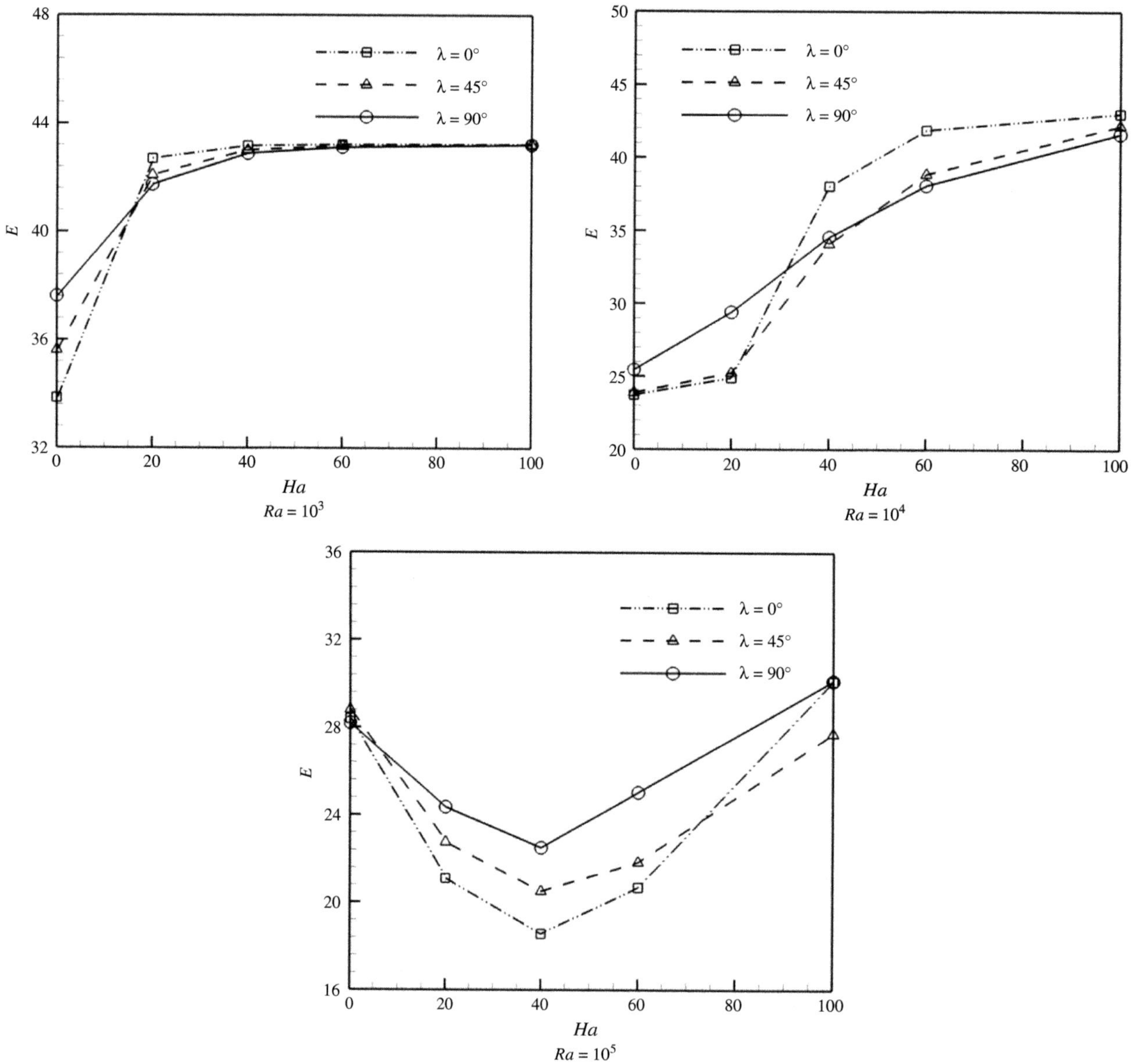

FIGURE 7.13 Effects of *Ha*, λ, and *Ra* on the heat transfer enhancement due to addition of nanoparticles when $Pr = 6.2$ (Cu-Water case).

$$u\frac{\partial v}{\partial x} + v\frac{\partial v}{\partial y} = -\frac{1}{\rho_{nf}}\frac{\partial P}{\partial y} + \upsilon_{nf}\left(\frac{\partial^2 v}{\partial x^2} + \frac{\partial^2 v}{\partial y^2}\right) + \beta_{nf} g(T - T_c) + \frac{\sigma_{nf}B^2}{\rho_{nf}}\left(u \sin\lambda \cos\lambda - v \cos^2\lambda\right) \quad (7.51)$$

$$u\frac{\partial T}{\partial x} + v\frac{\partial T}{\partial y} = \alpha_{nf}\left(\frac{\partial^2 T}{\partial x^2} + \frac{\partial^2 T}{\partial y^2}\right) \quad (7.52)$$

where the effective density (ρ_{nf}) and heat capacitance $(\rho C_p)_{nf}$ of the nanofluid are defined as:

$$\rho_{nf} = \rho_f(1-\phi) + \rho_s\phi \quad (7.53)$$

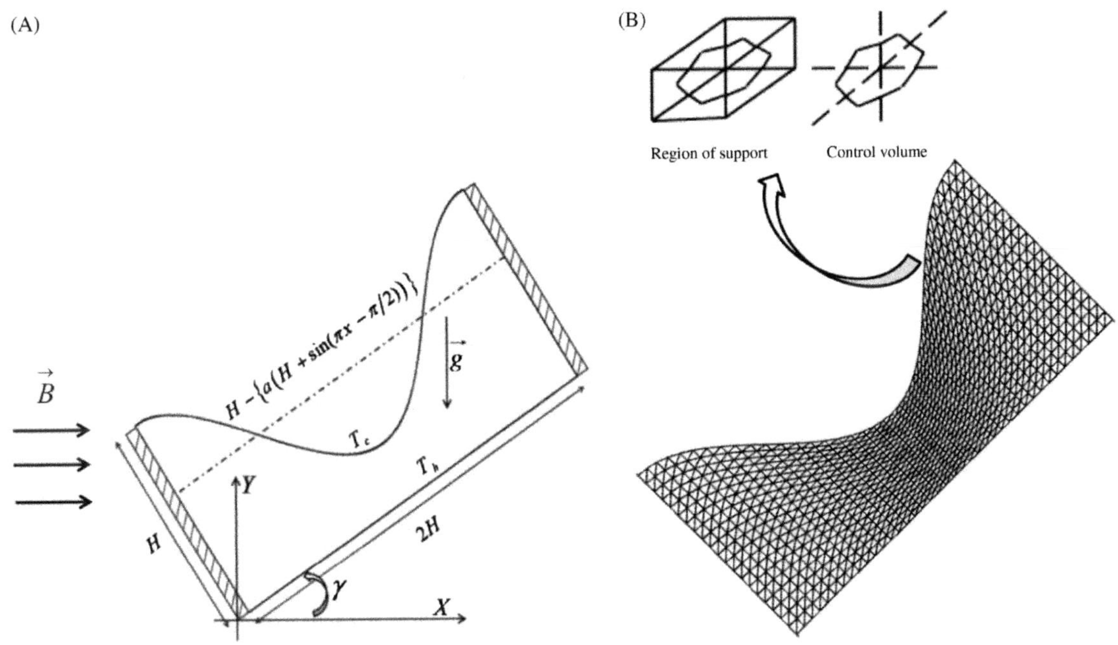

FIGURE 7.14 (A) Geometry and the boundary conditions with (B) the mesh of enclosure considered in this work.

$$(\rho C_p)_{nf} = (\rho C_p)_f(1-\phi) + (\rho C_p)_s\phi \tag{7.54}$$

where ϕ is the solid volume fraction of nanoparticles. Thermal diffusivity of the nanofluids is

$$\alpha_{nf} = \frac{k_{nf}}{(\rho C_p)_{nf}} \tag{7.55}$$

and the thermal expansion coefficient of the nanofluids can be determined as:

$$\beta_{nf} = \beta_f(1-\phi) + \beta_s\phi \tag{7.56}$$

The dynamic viscosity of the nanofluids given by Brinkman is

$$\mu_{nf} = \frac{\mu_f}{(1-\phi)^{2.5}} \tag{7.57}$$

The effective thermal conductivity of the nanofluid can be approximated by the Maxwell–Garnetts (MG) model as [14]:

$$\frac{k_{nf}}{k_f} = \frac{k_s + 2k_f - 2\phi(k_f - k_s)}{k_s + 2k_f + \phi(k_f - k_s)} \tag{7.58}$$

and the effective electrical conductivity of nanofluid was presented by Maxwell as below:

$$\frac{\sigma_{nf}}{\sigma_f} = 1 + \frac{3(\sigma_s/\sigma_f - 1)\phi}{(\sigma_s/\sigma_f + 2) - (\sigma_s/\sigma_f - 1)\phi} \tag{7.59}$$

The stream function and vorticity are defined as:

$$u = \frac{\partial\psi}{\partial y}, \quad v = -\frac{\partial\psi}{\partial x}, \quad \omega = \frac{\partial v}{\partial x} - \frac{\partial u}{\partial y} \tag{7.60}$$

The stream function satisfies the continuity Eq. (7.48). The vorticity equation is obtained by eliminating the pressure between the two momentum equations, i.e., by taking the y-derivative of Eq. (7.49) and subtracting from it the x-derivative of Eq. (7.50). This gives:

$$\frac{\partial\psi}{\partial y}\frac{\partial\omega}{\partial x}-\frac{\partial\psi}{\partial x}\frac{\partial\omega}{\partial y}=\upsilon_{nf}\left(\frac{\partial^2\omega}{\partial x^2}+\frac{\partial^2\omega}{\partial y^2}\right)+\beta_{nf}g\left(\frac{\partial T}{\partial x}\right)$$
$$+\frac{\sigma_{nf}B^2}{\rho_{nf}}\left(-\frac{\delta v}{\delta y}\sin\lambda\,\cos\lambda+\frac{\delta u}{\delta y}\sin^2\lambda+\frac{\delta u}{\delta x}\sin\lambda\,\cos\lambda-\frac{\delta v}{\delta x}\cos^2\lambda\right) \tag{7.61}$$

$$\frac{\partial\psi}{\partial y}\frac{\partial T}{\partial x}-\frac{\partial\psi}{\partial x}\frac{\partial T}{\partial y}=\alpha_{nf}\left(\frac{\partial^2 T}{\partial x^2}+\frac{\partial^2 T}{\partial y^2}\right) \tag{7.62}$$

$$\frac{\partial^2\psi}{\partial x^2}+\frac{\partial^2\psi}{\partial y^2}=-\omega \tag{7.63}$$

By introducing the following nondimensional variables:

$$X=\frac{x}{L},\quad Y=\frac{y}{L},\quad \Omega=\frac{\omega L^2}{\alpha_f},\quad \Psi=\frac{\psi}{\alpha_f},\quad \Theta=\frac{T-T_c}{T_h-T_c},\quad U=\frac{uL}{\alpha_f},\quad V=\frac{vL}{\alpha_f} \tag{7.64}$$

where in Eq. (7.64) $L=r_{out}-r_{in}=r_{in}$. Using the dimensionless parameters, the equations now become:

$$\frac{\partial\Psi}{\partial Y}\frac{\partial\Omega}{\partial X}-\frac{\partial\Psi}{\partial X}\frac{\partial\Omega}{\partial Y}=\left[\frac{\mathrm{Pr}_f}{(1-\phi)^{2.5}\left((1-\phi)+\phi\frac{\rho_s}{\rho_f}\right)}\right]\left(\frac{\partial^2\Omega}{\partial X^2}+\frac{\partial^2\Omega}{\partial Y^2}\right)$$
$$+Ra_f\mathrm{Pr}_f\left[(1-\phi)+\phi\frac{\beta_s}{\beta_f}\right]\left(\frac{\partial\Theta}{\partial X}\right) \tag{7.65}$$
$$+Ha^2\mathrm{Pr}_f\left(1+\frac{3\left(\frac{\sigma_s}{\sigma_f}-1\right)\phi}{\left(\frac{\sigma_s}{\sigma_f}+2\right)-\left(\frac{\sigma_s}{\sigma_f}-1\right)\phi}\right)\left(\frac{1}{(1-\phi)+\phi\frac{\rho_s}{\rho_f}}\right)\left(-\frac{\delta V}{\delta Y}\tan\lambda+\frac{\delta U}{\delta Y}\tan^2\lambda+\frac{\delta U}{\delta X}\tan\lambda-\frac{\delta V}{\delta X}\right)$$

$$\frac{\partial\Psi}{\partial Y}\frac{\partial\Theta}{\partial X}-\frac{\partial\Psi}{\partial X}\frac{\partial\Theta}{\partial Y}=\left[\frac{\frac{k_{nf}}{k_f}}{\left((1-\phi)+\phi\frac{(\rho Cp)_s}{(\rho Cp)_f}\right)}\right]\left(\frac{\partial^2\Theta}{\partial X^2}+\frac{\partial^2\Theta}{\partial Y^2}\right) \tag{7.66}$$

$$\frac{\partial^2\Psi}{\partial X^2}+\frac{\partial^2\Psi}{\partial Y^2}=-\Omega \tag{7.67}$$

where $Ra_f=g\beta_f L^3(T_h-T_c)/(\alpha_f\upsilon_f)$ is the Rayleigh number for the base fluid, $Ha=LB_x\sqrt{\sigma_f/\mu_f}$ is the Hartmann number, and $\mathrm{Pr}_f=\upsilon_f/\alpha_f$ is the Prandtl number for the base fluid. The boundary conditions as shown in Fig. 7.14 are:

$$\begin{aligned}&\Theta=1.0 && \text{on the hot wall}\\&\Theta=0.0 && \text{on the cold wall}\\&\partial\Theta/\partial n=0.0 && \text{on the two other insulation boundaries}\\&\Psi=0.0 && \text{on all solid boundaries}\end{aligned} \tag{7.68}$$

The values of vorticity on the boundary of the enclosure can be obtained using the stream function formulation and the known velocity conditions during the iterative solution procedure.

The local Nusselt number of the nanofluid along the hot wall can be expressed as:

$$Nu_{loc} = \frac{k_{nf}}{k_f} \frac{\partial \Theta}{\partial n} \bigg|_{hot\ wall} \tag{7.69}$$

Where n is normal to surface. The average Nusselt number on the hot wall is evaluated as:

$$Nu_{ave} = \frac{1}{2} \int_0^2 Nu_{loc}\, dS \tag{7.70}$$

7.4.3 Effects of Active Parameters

Numerical simulations of natural convection nanofluid flow in an enclosure with one sinusoidal wall in presence of magnetic field were performed. Calculations are made for various values of Hartmann number ($Ha = 0, 20, 60$ and 100), Rayleigh number ($Ra = 10^3, 10^4$ and 10^5), volume fraction of nanoparticles ($\phi = 0\%, 2\%, 4\%$ and 6%), and inclination angle ($\gamma = 0^\circ, 30^\circ, 60^\circ$ and 90°) at constant dimensionless amplitude of the sinusoidal wall ($a = 0.3$) and Prandtl number ($\Pr = 6.2$).

Figs. 7.15 and 7.16 show isotherm (up) and streamline (down) contours for different values of Rayleigh number, Hartmann number, and inclination angle. The figures show that the absolute value of stream function increases as the Rayleigh number increases and it decreases as the Hartmann number increases. Also it can be seen that the maximum values of $|\Psi_{\max}|$ are observed at $\gamma = 90^\circ$ for $Ra = 10^3, 10^4$ while it is obtained at $\gamma = 60^\circ$ for $Ra = 10^5$. At $Ra = 10^3$, for all inclination angles, the isotherms are nearly smooth curves and nearly parallel to each other, following the geometry of the sinusoidal surfaces; this pattern is the characteristic of the dominant conduction mechanism of heat transfer at low Rayleigh numbers. At $\gamma = 0^\circ$ two counterrotating vortices cores are observed. This bicellular flow pattern divides the cavity into two symmetric parts with respect to the vertical centerline of the enclosure. By increasing inclination angle, at first the upper vortex becomes stronger and then at $\gamma = 90^\circ$, streamlines become symmetric with respect to the horizontal centerline of the enclosure. In general, as the Rayleigh number increases, the buoyancy-driven circulations inside the enclosure become stronger as it is clear from greater magnitudes of stream function that more distortion appears in the isotherms. When the magnetic field is imposed on the enclosure, the velocity field is suppressed owing to the retarding effect of the Lorentz force. So the intensity of convection weakens significantly. The braking effect of the magnetic field is observed from the maximum stream function value. An increase of the Hartmann number merges two vortices into one except for $\gamma = 0^\circ$. Also magnetic field makes the thermal plume disappear over the hot wall and makes the isotherms parallel to each other due to the domination of the conduction mode of heat transfer.

Fig. 7.17 depicts the effects of the nanoparticle volume fraction, Hartmann number, inclination angle, and Rayleigh number on local Nusselt number. Generally, increasing the nanoparticles volume fraction and Rayleigh number leads to an increase in local Nusselt number. In the absence of magnetic field, at $\gamma = 0^\circ$ the local Nusselt profile is symmetric with respect to the vertical centerline of the enclosure. But in the presence of magnetic field, because of the domination of the conduction mechanism, the maximum value of the local Nusselt number occurs at the vertical centerline. At $\gamma = 90^\circ$ the local Nusselt number profile decreases with increase of S and increasing Hartmann number leads to a decrease in Nusselt number. When $Ha = 0$, the number of extremum in the local Nusselt number profile corresponds to the number of existing thermal plumes.

The effects of the Hartmann number, Rayleigh number and inclination angle on the average Nusselt number are shown in Fig. 7. 18 (A, B). Generally, the average Nusselt number increases with the increase of Rayleigh number while it decreases as the Hartmann number increases. At $Ra = 10^5$ in the absence of the magnetic field the maximum value of the average Nusselt number is obtained at $\gamma = 0^\circ$ but for higher values of the Hartmann number the maximum value of Nu_{ave} occurs at $\gamma = 90^\circ$.

To estimate the enhancement of heat transfer between the case of $\phi = 0.06$ and the pure fluid (base fluid) case, the enhancement is defined as:

$$E = \frac{Nu(\phi = 0.06) - Nu(basefluid)}{Nu(basefluid)} \times 100 \tag{7.71}$$

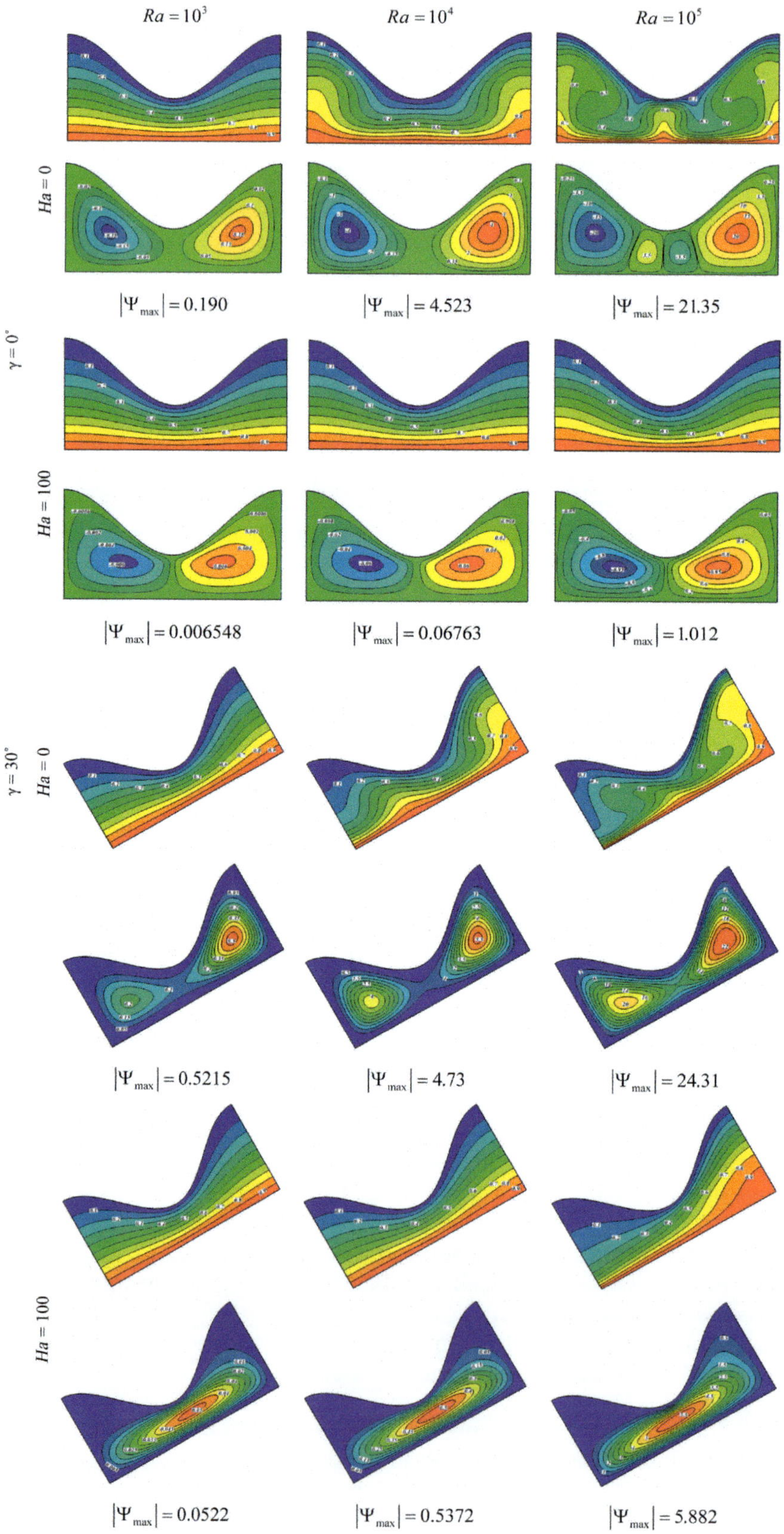

FIGURE 7.15 Isotherm (up) and streamline (down) contours for different values of Rayleigh number, Hartmann number and inclination angle at $a = 0.3$, $\phi = 0.06$ and $\mathrm{Pr} = 6.2$.

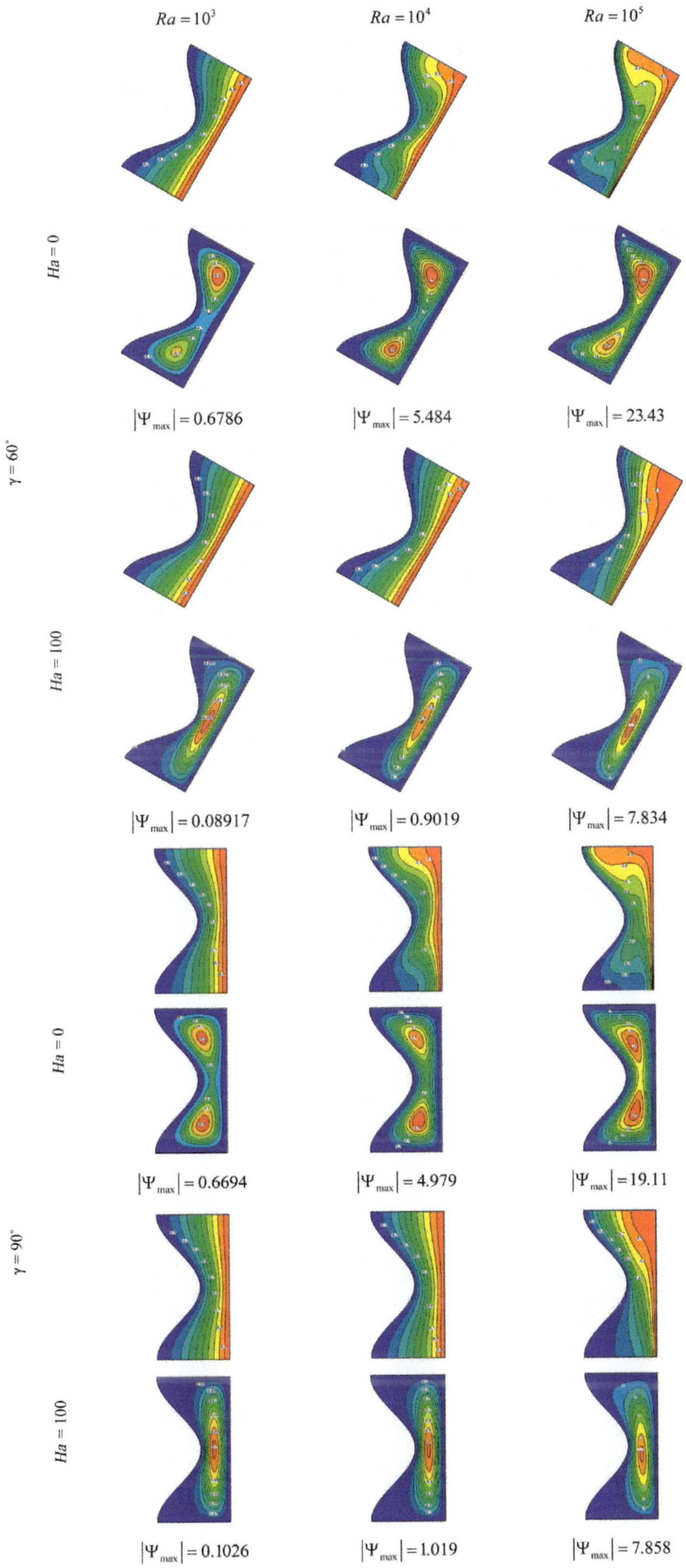

FIGURE 7.16 Isotherm (up) and streamline (down) contours for different values of Rayleigh number, inclination angle and inclination angle at $a = 0.3$, $\phi = 0.06$ and $\Pr = 6.2$.

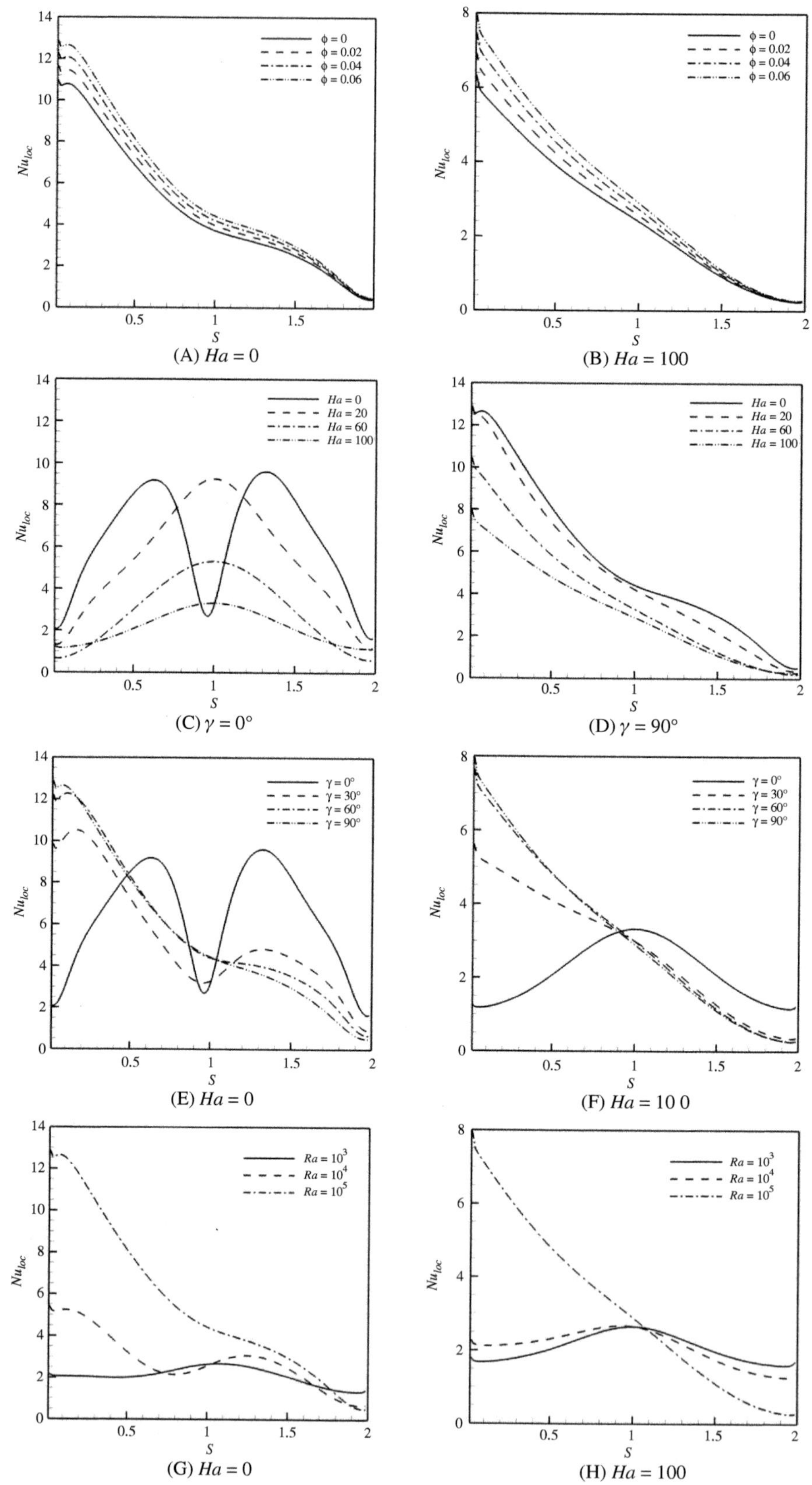

FIGURE 7.17 Effects of the nanoparticle volume fraction, Hartmann number, inclination angle, and Rayleigh number on local Nusselt number when (a), (b) $Ra = 10^5, \gamma = 90^\circ$; (c), (d) $Ra = 10^5, \phi = 0.06$; (e), (f) $Ra = 10^5, \phi = 0.06$; (g), (h) $\phi = 0.06, \gamma = 90^\circ$.

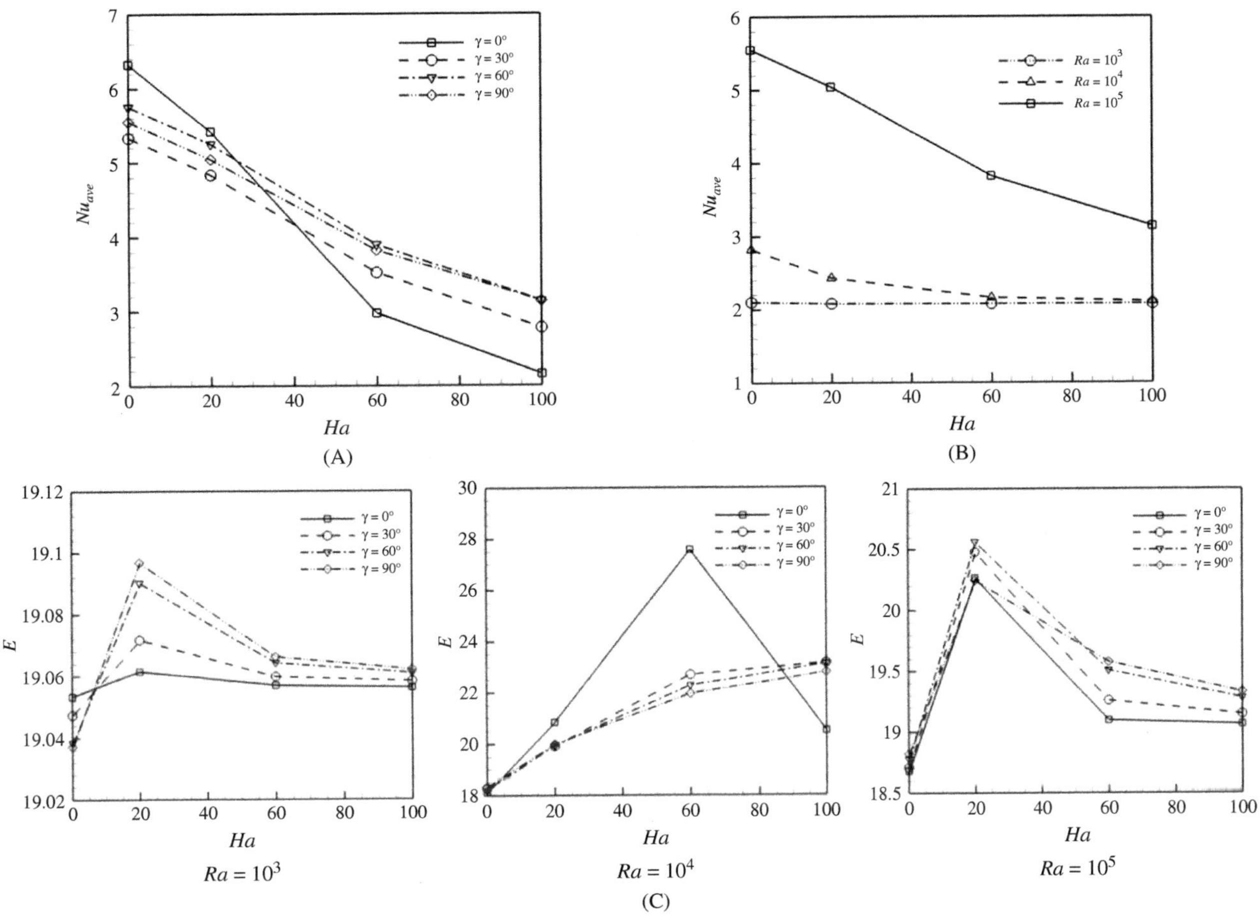

FIGURE 7.18 Effects of the Hartmann number, Rayleigh number, and inclination angle on the average Nusselt number when (A) $Ra = 10^5$; (B) $\gamma = 90^\circ$ at $a = 0.3$ and $\phi = 0.06$. (C) Effects of Hartmann number, Rayleigh number, and inclination angle on the ratio of heat transfer enhancement due to addition of nanoparticles $a = 0.3$.

The effects of Hartmann number, Rayleigh number, and inclination angle on heat transfer enhancement ratio is shown in Fig. 7. 18 (C). At $Ra = 10^3$, maximum value of enhancement for a low Hartmann number is observed at $\gamma = 0^\circ$, but for $Ha > 20$ the maximum values of it occur for $\gamma = 90^\circ$. Also it can be seen that for $Ra = 10^4$ and 10^5 the maximum value of E are obtained for $\gamma = 60^\circ$ and $\gamma = 0^\circ$, respectively. It is an interesting observation that at $Ra = 10^5$ the enhancement in heat transfer for the case of $\gamma = 0^\circ$ increases with an increase of Hartmann number when $Ha < 60$, while the opposite trend is observed for $Ha > 60$. For other values of inclination angles, enhancement in heat transfer is an increasing function of Hartmann number.

7.5 MHD NANOFLUID FLOW IN A CAVITY WITH HEAT FLUX BOUNDARY CONDITION

7.5.1 Problem Definition

The physical model along with the important geometrical parameters is as shown in Fig. 7.19 (A). The width and height of the enclosure is L. The outer cylinder is maintained at constant cold temperature T_c, whereas the inner circular wall is under constant heat flux. To assess the shape of the inner circular and outer rectangular boundary, which consists of the right and top walls, a supper elliptic function can be used as follows

$$\left(\frac{X}{a}\right)^{2\hat{n}} + \left(\frac{Y}{b}\right)^{2\hat{n}} = 1 \tag{7.72}$$

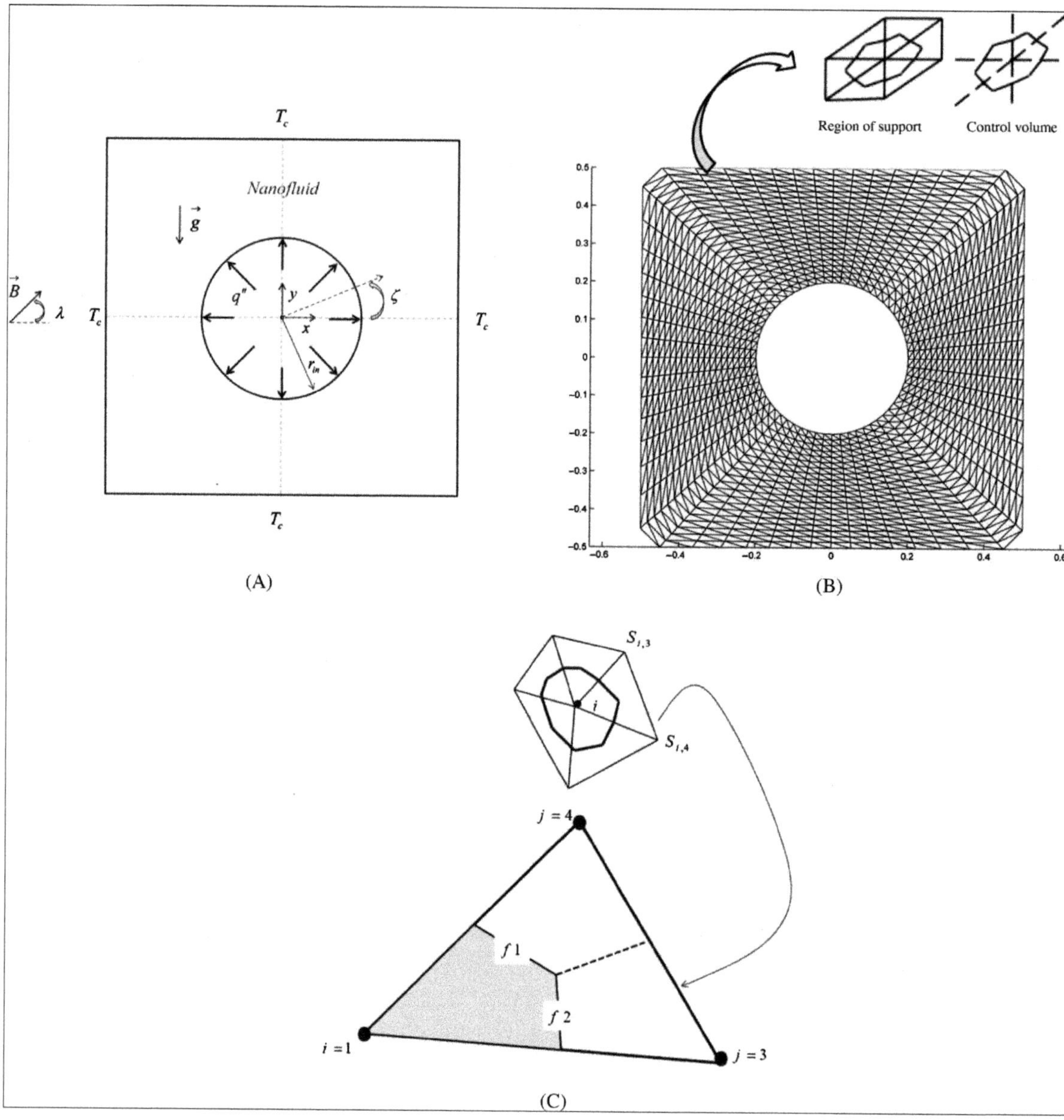

FIGURE 7.19 (A) Geometry and the boundary conditions with (B) the mesh of Geometry considered in this work; (C) a sample triangular element and its corresponding control volume.

When $a = b$ and $\hat{n} = 1$ the geometry becomes a circle. As $\hat{n}$ increases from 1 the geometry would approach a rectangle for $a \neq b$ and square for $a = b$. It is also assumed that the uniform magnetic field ($\vec{B} = B_x\vec{e_x} + B_y\vec{e_y}$) of constant magnitude $B = \sqrt{B_x^2 + B_y^2}$ is applied, where $\vec{e_x}$ and $\vec{e_y}$ are unit vectors in the Cartesian coordinate system. The orientation of the magnetic field forms an angle λ with a horizontal axis such that $\lambda = B_x/B_y$. In this study, λ equals to zero. The electric current J and the electromagnetic force F are defined by $J = \sigma\left(\vec{V} \times \vec{B}\right)$ and $F = \sigma\left(\vec{V} \times \vec{B}\right) \times \vec{B}$, respectively.

7.5.2 Governing Equation

The flow is steady, two-dimensional, laminar, and incompressible. The radiation, viscous dissipation, induced electric current, and Joule heating are neglected. The magnetic Reynolds number is assumed to be small so that the induced magnetic field can be neglected compared to the applied magnetic field. Neglecting displacement currents, induced magnetic field, and using the Boussinesq approximation, the governing equations of heat transfer and fluid flow for nanofluid can be obtained as follows:

$$\frac{\partial u}{\partial x} + \frac{\partial v}{\partial y} = 0 \tag{7.73}$$

$$u\frac{\partial u}{\partial x} + v\frac{\partial u}{\partial y} = -\frac{1}{\rho_{nf}}\frac{\partial P}{\partial x} + v_{nf}\left(\frac{\partial^2 u}{\partial x^2} + \frac{\partial^2 u}{\partial y^2}\right) + \frac{\sigma_{nf}B^2}{\rho_{nf}}\left(v \sin\lambda \cos\lambda - u \sin^2\lambda\right) \tag{7.74}$$

$$\begin{aligned} u\frac{\partial v}{\partial x} + v\frac{\partial v}{\partial y} &= -\frac{1}{\rho_{nf}}\frac{\partial P}{\partial y} + v_{nf}\left(\frac{\partial^2 v}{\partial x^2} + \frac{\partial^2 v}{\partial y^2}\right) + \beta_{nf}g(T - T_c) \\ &+ \frac{\sigma_{nf}B^2}{\rho_{nf}}\left(u \sin\lambda \cos\lambda - v \cos^2\lambda\right) \end{aligned} \tag{7.75}$$

$$u\frac{\partial T}{\partial x} + v\frac{\partial T}{\partial y} = \alpha_{nf}\left(\frac{\partial^2 T}{\partial x^2} + \frac{\partial^2 T}{\partial y^2}\right) \tag{7.76}$$

where the effective density (ρ_{nf}), the thermal expansion coefficient (β_{nf}), and heat capacitance $(\rho C_p)_{nf}$ of the nanofluid are defined as:

$$\begin{aligned} \rho_{nf} &= \rho_f(1-\phi) + \rho_s\phi \\ \beta_{nf} &= \beta_f(1-\phi) + \beta_s\phi \\ (\rho C_p)_{nf} &= (\rho C_p)_f(1-\phi) + (\rho C_p)_s\phi \end{aligned} \tag{7.77}$$

Also effective electrical conductivity of nanofluid was presented by Maxwell as below:

$$\frac{\sigma_{nf}}{\sigma_f} = 1 + \frac{3(\sigma_s/\sigma_f - 1)\phi}{(\sigma_s/\sigma_f + 2) - (\sigma_s/\sigma_f - 1)\phi} \tag{7.78}$$

The Brownian motion has a significant impact on the effective thermal conductivity.

$$k_{eff} = k_{static} + k_{Brownian} \tag{7.79}$$

$$\frac{k_{static}}{k_f} = 1 + \frac{3(k_p/k_f - 1)\phi}{(k_p/k_f + 2) - (k_p/k_f - 1)\phi} \tag{7.80}$$

$$k_{Brownian} = 5 \times 10^4 \beta\phi\rho_f c_{p,f}\sqrt{\frac{\kappa_b T}{\rho_p d_p}} f(T,\phi) \tag{7.81}$$

$$R_f + \frac{d_p}{k_p} = \frac{d_p}{k_{p,eff}} \tag{7.82}$$

For different base fluids and different nanoparticles, the function should be different. Only water-based nanofluids are considered in the current study. For Al_2O_3-water nanofluids, this function follows the format:

$$\begin{aligned} g'(T,\phi,d_p) &= \left(a_1 + a_2\ln(d_p) + a_3\ln(\phi) + a_4\ln(\phi)\ln(d_p) + a_5\ln(d_p)^2\right)\ln(T) \\ &+ \left(a_6 + a_7\ln(d_p) + a_8\ln(\phi) + a_9\ln(\phi)\ln(d_p) + a_{10}\ln(d_p)^2\right) \end{aligned} \tag{7.83}$$

with the coefficients a_i ($i = 0.10$) are based on the type of nanopartices, Al_2O_3-water nanofluids have an R^2 of 96% and 98%, respectively (Table 7.2). Finally, the KKL (Koo-Kleinstreuer-Li) correlation is written as:

$$k_{Brownian} = 5 \times 10^4 \phi\rho_f c_{p,f}\sqrt{\frac{\kappa_b T}{\rho_p d_p}} g'(T,\phi,d_p) \tag{7.84}$$

TABLE 7.2 The Coefficient Values of $Al_2O_3-Water$ Nanofluid

Coefficient values	$Al_2O_3-Water$
a_1	52.813488759
a_2	6.115637295
a_3	0.6955745084
a_4	4.17455552786E-02
a_5	0.176919300241
a_6	-298.19819084
a_7	-34.532716906
a_8	-3.9225289283
a_9	-0.2354329626
a_{10}	-0.999063481

For the *effective viscosity* due to micromixing in suspension, they proposed:

$$\mu_{eff}=\mu_{static}+\mu_{Brownian}=\mu_{static}+\frac{k_{Brownian}}{k_f}\times\frac{\mu_f}{\mathrm{Pr}_f} \tag{7.85}$$

where $\mu_{static}=\frac{\mu_f}{(1-\phi)^{2.5}}$ is viscosity of the nanofluid, as given originally by Brinkman.

The stream function and vorticity are defined as:

$$u=\frac{\partial\psi}{\partial y},\quad v=-\frac{\partial\psi}{\partial x},\quad \omega=\frac{\partial v}{\partial x}-\frac{\partial u}{\partial y} \tag{7.86}$$

The stream function satisfies the continuity Eq. (7.73). The vorticity equation is obtained by eliminating the pressure between the two momentum equations, i.e., by taking the y-derivative of Eq. (7.74) and subtracting from it the x-derivative of Eq. (7.75). This gives:

$$\begin{aligned}\frac{\partial\psi}{\partial y}\frac{\partial\omega}{\partial x}-\frac{\partial\psi}{\partial x}\frac{\partial\omega}{\partial y}&=\upsilon_{nf}\left(\frac{\partial^2\omega}{\partial x^2}+\frac{\partial^2\omega}{\partial y^2}\right)+\beta_{nf}g\left(\frac{\partial T}{\partial x}\right)\\&+\frac{\sigma_{nf}B^2}{\rho_{nf}}\left(-\frac{\delta v}{\delta y}\sin\lambda\cos\lambda+\frac{\delta u}{\delta y}\sin^2\lambda+\frac{\delta u}{\delta x}\sin\lambda\cos\lambda-\frac{\delta v}{\delta x}\cos^2\lambda\right)\end{aligned} \tag{7.87}$$

$$\frac{\partial\psi}{\partial y}\frac{\partial T}{\partial x}-\frac{\partial\psi}{\partial x}\frac{\partial T}{\partial y}=\alpha_{nf}\left(\frac{\partial^2 T}{\partial x^2}+\frac{\partial^2 T}{\partial y^2}\right) \tag{7.88}$$

$$\frac{\partial^2\psi}{\partial x^2}+\frac{\partial^2\psi}{\partial y^2}=-\omega \tag{7.89}$$

By introducing the following nondimensional variables:

$$X=\frac{x}{L},\quad Y=\frac{y}{L},\quad \Omega=\frac{\omega L^2}{\alpha_f},\Psi=\frac{\psi}{\alpha_f},\Theta=\frac{T-T_c}{(q''L/k_f)},U=\frac{uL}{\alpha_f},V=\frac{vL}{\alpha_f} \tag{7.90}$$

Using the dimensionless parameters, the equations now become:

$$\frac{\partial \Psi}{\partial Y}\frac{\partial \Omega}{\partial X} - \frac{\partial \Psi}{\partial X}\frac{\partial \Omega}{\partial Y} = \mathrm{Pr}_f / \left((1-\phi)^{2.5}\left((1-\phi) + \phi\frac{\rho_s}{\rho_f} \right) \right) \left(\frac{\partial^2 \Omega}{\partial X^2} + \frac{\partial^2 \Omega}{\partial Y^2} \right)$$
$$+ Ra_f \mathrm{Pr}_f \left[(1-\phi) + \phi\frac{\beta_s}{\beta_f} \right] \left(\frac{\partial \Theta}{\partial X} \right) \tag{7.91}$$
$$+ Ha^2 \mathrm{Pr}_f (1 + 3(\sigma_s/\sigma_f - 1)\phi / ((\sigma_s/\sigma_f + 2) - (\sigma_s/\sigma_f - 1)\phi))$$
$$\times 1/\left((1-\phi) + \phi\frac{\rho_s}{\rho_f} \right) \left(-\frac{\delta V}{\delta Y}\tan\lambda + \frac{\delta U}{\delta Y}\tan^2\lambda + \frac{\delta U}{\delta X}\tan\lambda - \frac{\delta V}{\delta X} \right)$$

$$\frac{\partial \Psi}{\partial Y}\frac{\partial \Theta}{\partial X} - \frac{\partial \Psi}{\partial X}\frac{\partial \Theta}{\partial Y} = \frac{k_{nf}}{k_f} / \left((1-\phi) + \phi\frac{(\rho Cp)_s}{(\rho Cp)_f} \right) \left(\frac{\partial^2 \Theta}{\partial X^2} + \frac{\partial^2 \Theta}{\partial Y^2} \right) \tag{7.92}$$

$$\frac{\partial^2 \Psi}{\partial X^2} + \frac{\partial^2 \Psi}{\partial Y^2} = -\Omega \tag{7.93}$$

where $Ra_f = g\beta_f L^4 q''/(k_f\, \alpha_f \upsilon_f)$ is the Rayleigh number for the base fluid, $Ha = LB_x\sqrt{\sigma_f/\mu_f}$ is the Hartmann number, and $\mathrm{Pr}_f = \upsilon_f/\alpha_f$ is the Prandtl number for the base fluid. The thermophysical properties of the nanofluid are given in Table 7.3. The boundary conditions as shown in Fig. 7.19 are:

$$\begin{aligned} &\frac{\partial \Theta}{\partial n} = -1 && \text{on the inner circular boundary} \\ &\Theta = 0.0 && \text{on the outer walls boundary} \\ &\Psi = 0.0 && \text{on all solid boundaries} \end{aligned} \tag{7.94}$$

The values of vorticity on the boundary of the enclosure can be obtained using the stream function formulation and the known velocity conditions during the iterative solution procedure. The local Nusselt number of the nanofluid along the hot wall can be expressed as:

$$Nu_{local} = \left(\frac{k_{nf}}{k_f} \right) \frac{1}{\theta} \bigg|_{inner\ wall} \tag{7.95}$$

where r is the radial direction. The average Nusselt number on hot circular wall is evaluated as:

$$Nu_{ave} = \frac{1}{2\pi}\int_0^{2\pi} Nu_{loc}\, d\zeta \tag{7.96}$$

To estimate the enhancement of heat transfer between the case of $\phi = 0.04$ and the pure fluid (base fluid) case, the enhancement is defined as:

$$E = \frac{Nu(\phi = 0.04) - Nu(basefluid)}{Nu(basefluid)} \times 100 \tag{7.97}$$

TABLE 7.3 Thermophysical Properties of Water and Nanoparticles

	$\rho(kg/m^3)$	$C_p(j/kgk)$	$k(W/m.k)$	$\beta \times 10^5 (K^{-1})$	$d_p(nm)$	$\sigma(\Omega \cdot m)^{-1}$
Pure water	997.1	4179	0.613	21	-	0.05
Al_2O_3	3970	765	25	0.85	47	1×10^{-10}

The heatlines are adequate tools for visualization and analysis of 2D convection heat transfer, through an extension of the heat flux line concept to include the advection terms. Heat function (H) are defined in terms of the energy equation as

$$\frac{\partial H}{\partial Y} = U\Theta - \frac{\partial \Theta}{\partial X}, \quad -\frac{\partial H}{\partial X} = V\Theta - \frac{\partial \Theta}{\partial Y} \tag{7.98}$$

7.5.3 Effects of Active Parameters

CVFEM is applied to solve the problem of natural convection in an enclosure filled with Al_2O_3-water nanofluid in presence of magnetic field. The effective thermal conductivity and viscosity of nanofluid are calculated by KKL correlation. Calculations are made for various values of volume fraction of nanoparticles ($\phi = 0$ to 4%), Rayleigh number ($Ra = 10^3, 10^4$ and 10^5), aspect ratio ($r_{in}/L = 0.2, 0.3$ and 0.4), and Hartmann number ($Ha = 0$ to 100) at constant Prandtl number ($\mathrm{Pr} = 6.2$).

Comparisons of the isotherms, streamlines, and heatlines for different values of Hartmann number, aspect ratio, and Rayleigh number are shown in Figs. 7.20 and 7.21. By increasing Rayleigh number the prominent heat transfer mechanism is turned from conduction to convection. Also it can be seen that increasing the aspect ratio leads to a decrease in the thermal boundary layer thickness and intensity of convection because of the

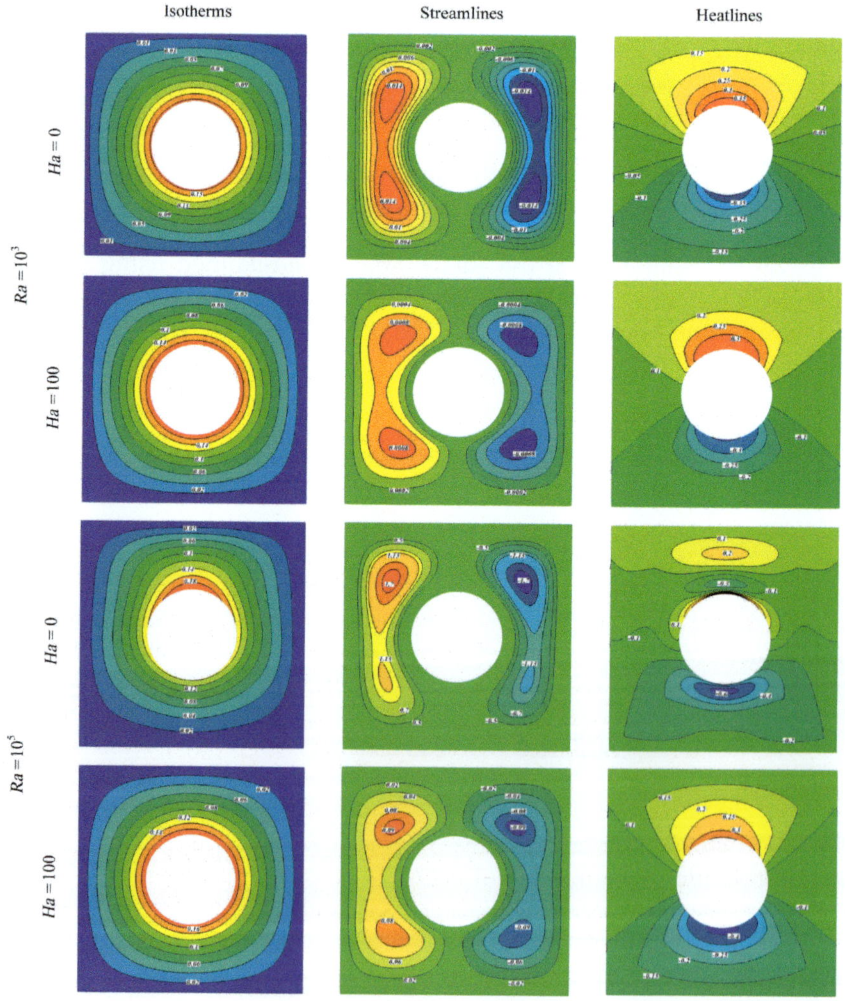

FIGURE 7.20 Comparison of the isotherms, streamlines, and heatlines for different values of Hartmann number and Raleigh number at $r_{in}/L = 0.2, \phi = 0.04$ and $\mathrm{Pr} = 6.2$.

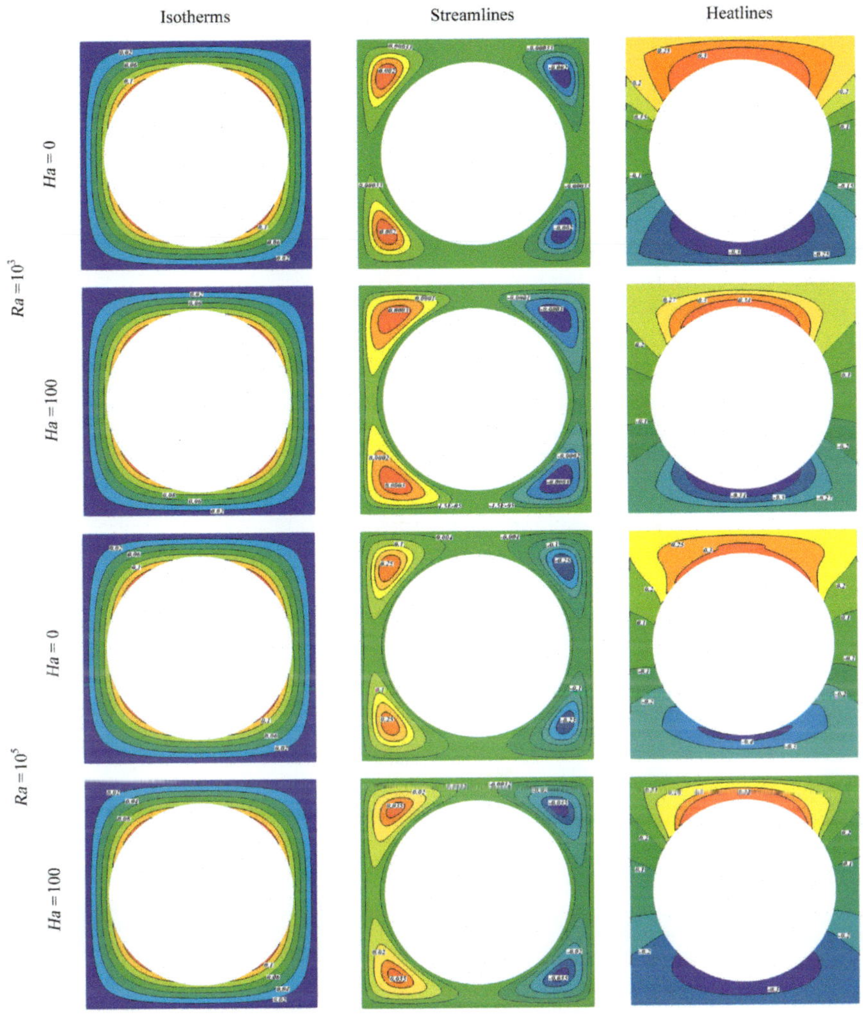

FIGURE 7.21 Comparison of the isotherms, streamlines, and heatlines for different values of Hartmann number and Raleigh number at $r_{in}/L=0.4, \phi=0.04$ and $\Pr=6.2$.

domination of the conduction heat transfer. When the magnetic field is imposed on the enclosure, the velocity field is suppressed owing to the retarding effect of the Lorentz force. So the intensity of convection weakens significantly. The braking effect of the magnetic field is observed from the maximum stream function value. The core vortex is shift upward vertically as the Hartmann number increases. Also imposing the magnetic field leads to omitting the thermal plume over the inner wall. At a high Hartmann number the conduction heat transfer mechanism is more pronounced. For this reason the isotherms are parallel to each other.

The heat flow within the enclosure is displayed using the heat function obtained from conductive heat fluxes $(\partial\Theta/\partial X, \partial\Theta/\partial Y)$ as well as convective heat fluxes $(V\Theta, U\Theta)$. Heatlines emanate from hot regimes and end on cold regimes illustrating the path of heat flow. The domination of conduction heat transfer at low Rayleigh number and high Hartmann number can be observed from the heatline patterns since no passive area exists. The increase of Ra causes the clustering of heatlines from the hot to the cold wall and generates a passive heat transfer area in which heat is rotated without having a significant effect on heat transfer between walls.

The distributions of local Nusselt numbers along the surface of the inner circular wall for different aspect ratio, Rayleigh number, and Hartmann number are shown in Fig. 7.22. Increasing Rayleigh number and aspect ratio lead to an increase in local Nusselt number but increasing Hartmann number causes local Nusselt number to decrease.

Fig. 7.23 shows the effects of aspect ratio, Rayleigh number, and Hartmann number on average Nusselt number. Increasing Hartmann number causes Lorentz force to increase and leads to a substantial suppression

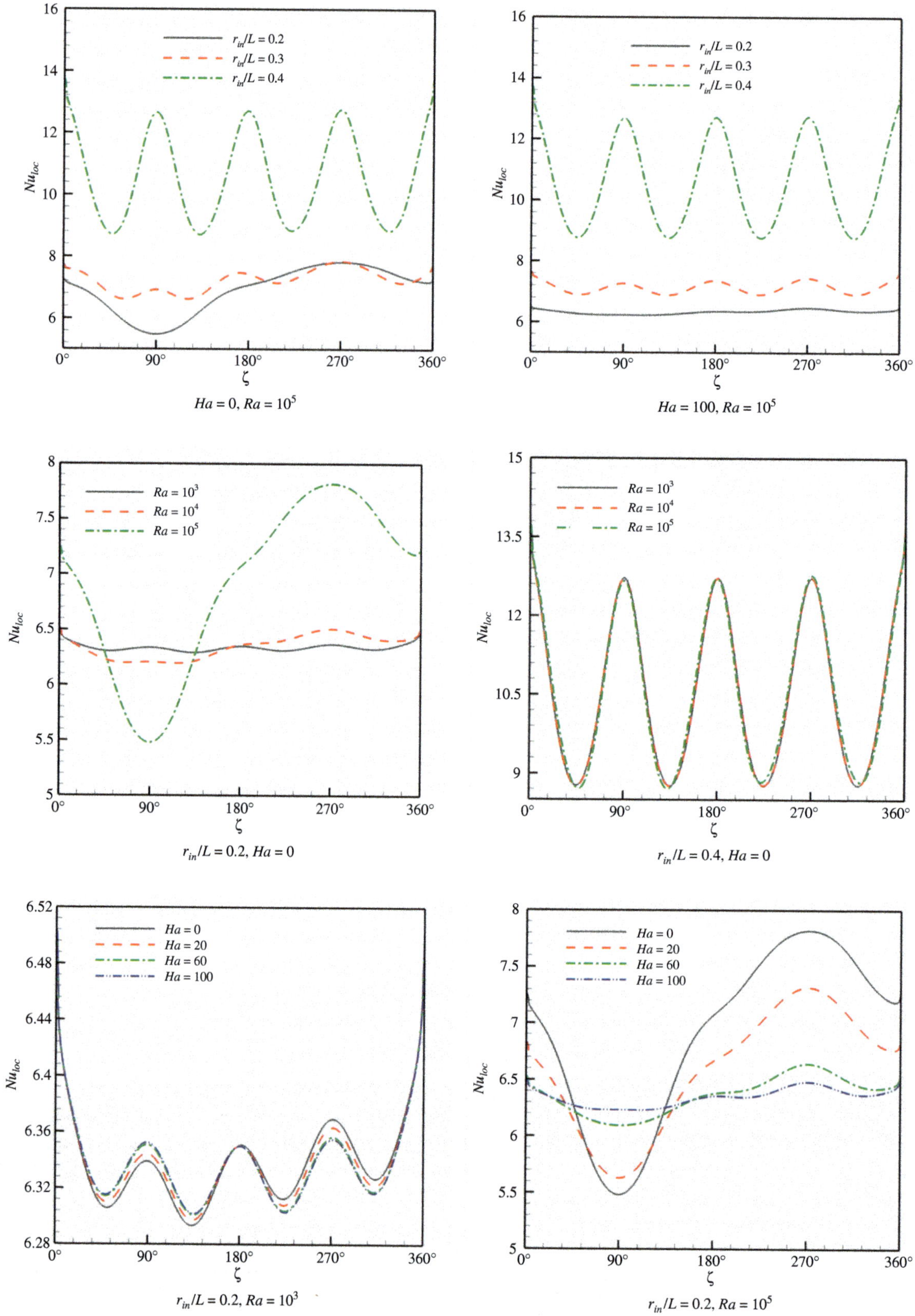

FIGURE 7.22 Effects of aspect ratio, Rayleigh number, and Hartmann number on local Nusselt number.

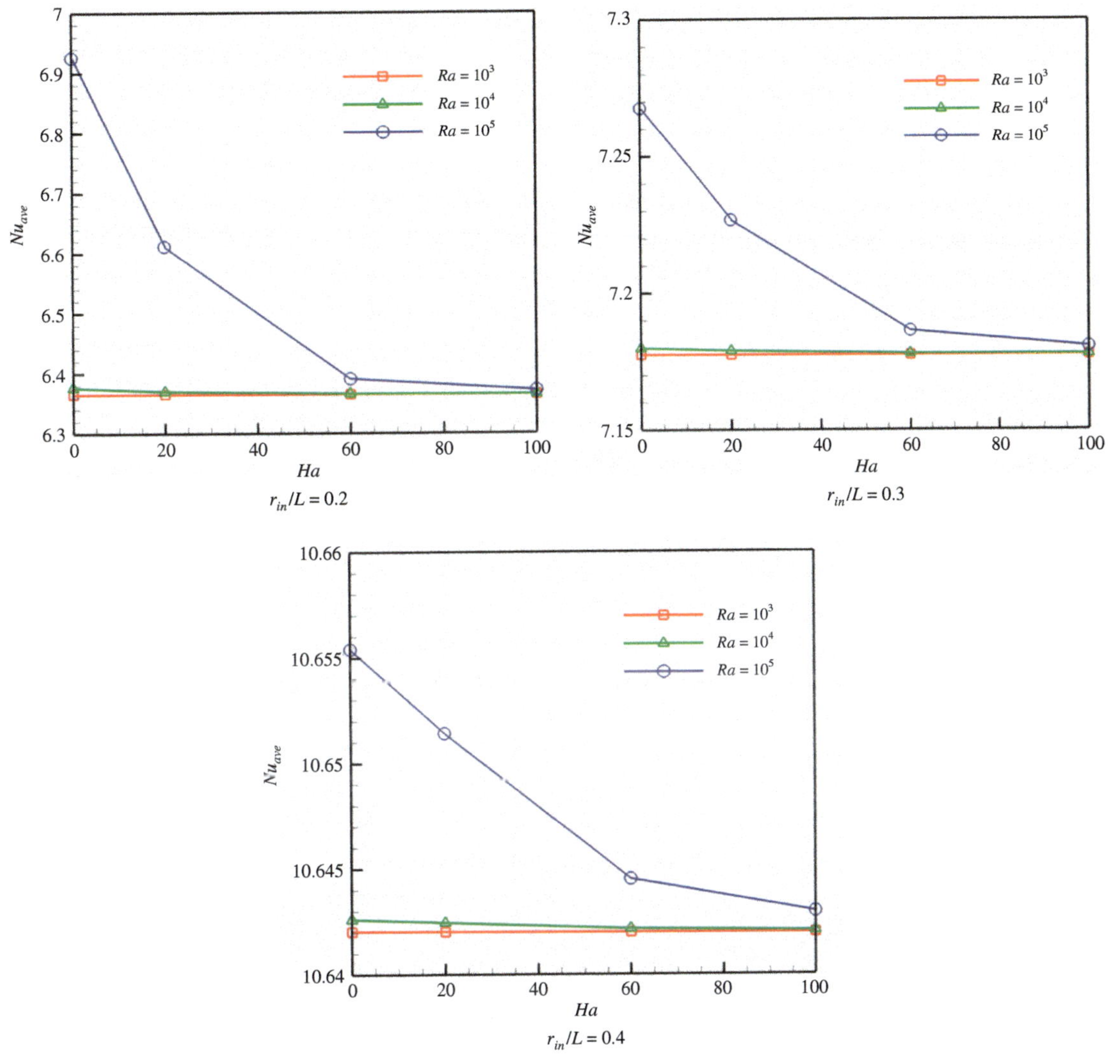

FIGURE 7.23 Effects of aspect ratio, Rayleigh number, and Hartmann number on average Nusselt number.

of the convection. So the Nusselt number has a reverse relationship with the Hartmann number. As the aspect ratio increases, space to accelerate the flow inside the cavity decreases. So the thermal boundary layer thickness decreases and in turn $1/\theta$ increases. As the volume fraction of nanoparticles increases, thermal diffusivity increases. So the high values of thermal diffusivity cause the boundary thickness to increase and accordingly decrease $1/\theta$. The Nusselt number is a function of $1/\theta$ and k_{nf}/k_f. Because the reduction in $1/\theta$ due to the presence of nanoparticles is much smaller than the thermal conductivity ratio, therefore an increase in the Nusselt number takes place by increasing the volume fraction of nanoparticles. The distance between cold and hot walls decreases with the increase of aspect ratio. So the Nusselt number decreases with the increase of r_{in}/L.

The heat transfer enhancement ratio due to the addition of nanoparticles for different values of $r_{in}/L, Ha$ and Ra is shown in Fig. 7.24. Heat transfer enhancement ratio has a direct relationship with the Hartmann number and aspect ratio but has a reverse relationship with the Rayleigh number. This observation is due to the domination of conduction heat transfer at low Rayleigh number and high Hartmann number or aspect ratio. Therefore, the addition of high thermal conductivity nanoparticles will increase the conduction and make the enhancement more effective.

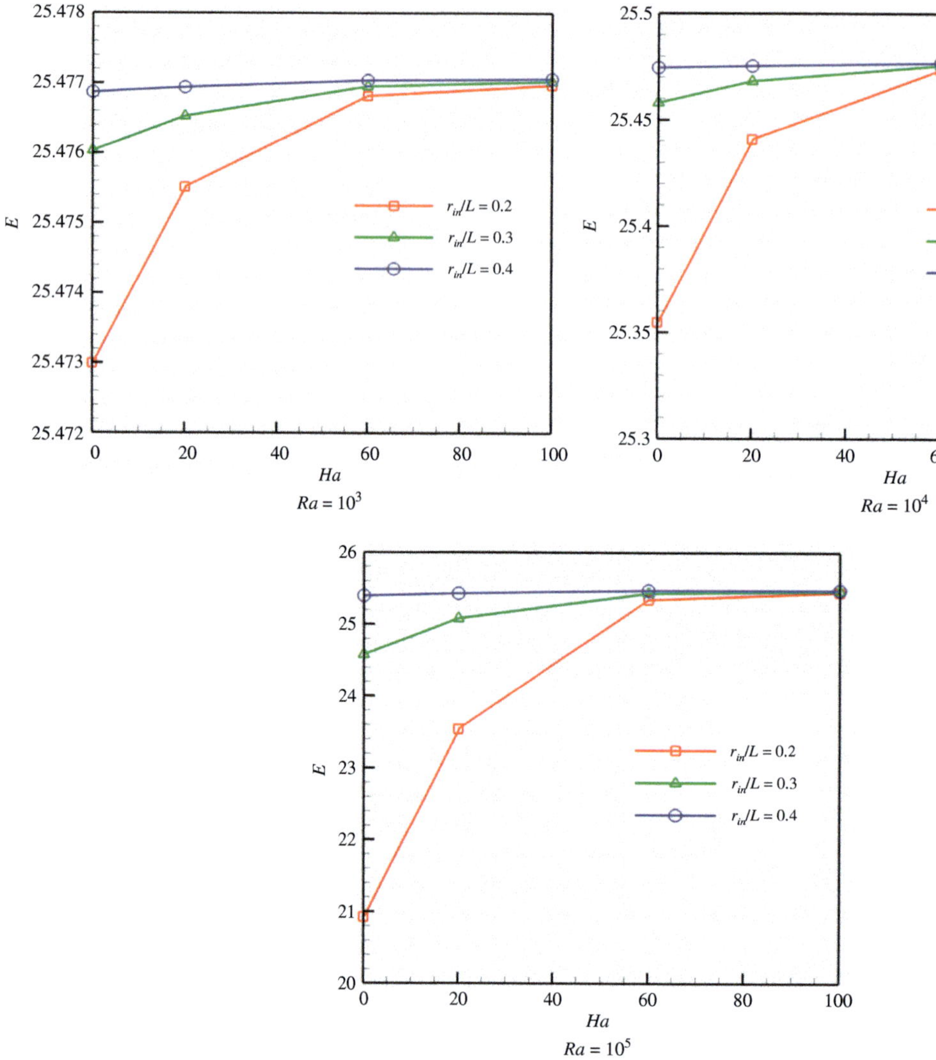

FIGURE 7.24 Effects of aspect ratio, Rayleigh number, and Hartmann number on heat transfer enhancement.

References

[1] I.E. Sarris, S.C. Kakarantzas, A.P. Grecos, N.S. Vlachos, MHD natural convection in a laterally and volumetrically heated square cavity, International Journal of Heat and Mass Transfer 48 (2005) 3443–3453.

[2] H.F. Oztop, M.M. Rahman, A. Ahsan, M. Hasanuzzaman, R. Saidur, Khaled Al-Salem, et al., MHD natural convection in an enclosure from two semi-circular heaters on the bottom wall, International Journal of Heat and Mass Transfer 55 (2012) 1844–1854.

[3] M. Sheikholeslami, M. Gorji-Bandpay, D.D. Ganji, Magnetic field effects on natural convection around a horizontal circular cylinder inside a square enclosure filled with nanofluid, International Communications in Heat and Mass Transfer 39 (2012) 978–986.

[4] M. Sheikholeslami, T. Hayat, A. Alsaedi, MHD free convection of Al_2O_3–water nanofluid considering thermal radiation: A numerical study, International Journal of Heat and Mass Transfer 96 (2016) 513–524.

[5] M. Sheikholeslami, K. Vajravelu, M.M. Rashidi, Forced convection heat transfer in a semi annulus under the influence of a variable magnetic field, International Journal of Heat and Mass Transfer 92 (2016) 339–348.

[6] M. Sheikholeslami, Influence of Lorentz forces on nanofluid flow in a porous cylinder considering Darcy model, Journal of Molecular Liquids 225 (2017) 903–912.

[7] M. Sheikholeslami, CVFEM for magnetic nanofluid convective heat transfer in a porous curved enclosure, Eur. Phys. J. Plus 131 (2016) 413. Available from: https://doi.org/10.1140/epjp/i2016-16413-y.

[8] M. Sheikholeslami, Influence of Coulomb forces on Fe_3O_4-H_2O nanofluid thermal improvement, International Journal of Hydrogen Energy 42 (2017) 821–829.

[9] M. Sheikholeslami, Magnetic field influence on nanofluid thermal radiation in a cavity with tilted elliptic inner cylinder, Journal of Molecular Liquids 229 (2017) 137–147.

[10] M. Sheikholeslami, Numerical simulation of magnetic nanofluid natural convection in porous media, Physics Letters A 381 (2017) 494–503.

[11] M. Sheikholeslami, Magnetic source impact on nanofluid heat transfer using CVFEM, Neural Computing and Applications (2016). Available from: https://doi.org/10.1007/s00521-016-2740-7.

[12] M. Sheikholeslami, CuO-water nanofluid free convection in a porous cavity considering Darcy law, The European Physical Journal Plus 132 (2017) 55. Available from: https://doi.org/10.1140/epjp/i2017-11330-3.

[13] M. Sheikholeslami, Numerical investigation of MHD nanofluid free convective heat transfer in a porous tilted enclosure, Engineering Computations 34 (6) (2017) 1939–1955.

[14] M.S. Kandelousi, Effect of spatially variable magnetic field on ferrofluid flow and heat transfer considering constant heat flux boundary condition, The European Physical Journal Plus (2014) 129–248. Available from: https://doi.org/10.1140/epjp/i2014-14248-2.

[15] M. Sheikholeslami, Numerical simulation for external magnetic field influence on Fe_3O_4-water nanofluid forced convection, Engineering Computations, Inpress.

[16] M. Sheikholeslami, Influence of Lorentz forces on nanofluid flow in a porous cavity by means of Non- Darcy model, Engineering Computations 34 (8) (2017) 2651–2667. Available from: https://doi.org/10.1108/EC-01-2017-0008.

[17] M. Sheikholeslami, H.B. Rokni, Nanofluid convective heat transfer intensification in a porous circular cylinder, Chemical Engineering & Processing: Process Intensification 120 (2017) 93–104.

[18] M. Sheikholeslami, M.K. Sadoughi, Numerical modeling for Fe_3O_4 -water nanofluid flow in porous medium considering MFD viscosity, Journal of Molecular Liquids 242 (2017) 255–264.

[19] M. Sheikholeslami, M. Shamlooei, Magnetic source influence on nanofluid flow in porous medium considering shape factor effect, Physics Letters A 381 (2017) 3071–3078.

[20] M. Sheikholeslami, M. Shamlooei, Convective flow of nanofluid inside a lid driven porous cavity using CVFEM, Physica B: Condensed Matter 521 (2017) 239–250.

[21] M. Sheikholeslami, K. Vajravelu, Forced convection heat transfer in Fe_3O_4- ethylene glycol nanofluid under the influence of Coulomb force, Journal of Molecular Liquids 233 (2017) 203–210.

[22] M. Sheikholeslami, H.B. Rokni, Influence of EFD viscosity on nanofluid forced convection in a cavity with sinusoidal wall, Journal of Molecular Liquids 232 (2017) 390–395.

[23] M. Sheikholeslami, H.B. Rokni, Magnetic nanofluid natural convection in the presence of thermal radiation considering variable viscosity, The European Physical Journal Plus 132 (2017) 238. Available from: https://doi.org/10.1140/epjp/i2017-11498-4.

[24] M. Sheikholeslami, Houman B. Rokni, Numerical simulation for impact of Coulomb force on nanofluid heat transfer in a porous enclosure in presence of thermal radiation, Int. J. Heat Mass Transfer 118 (2018) 823–831.

[25] M. Sheikholeslami, M. Shamlooei, R. Moradi, Fe_3O_4- Ethylene glycol nanofluid forced convection inside a porous enclosure in existence of Coulomb force, Journal of Molecular Liquids 249 (2018) 429–437.

[26] M. Sheikholeslami, S.A. Shehzad, Numerical analysis of Fe_3O_4 $-H_2O$ nanofluid flow in permeable media under the effect of external magnetic source, Int. J. Heat Mass Transfer 118 (2018) 182–192.

[27] M. Sheikholeslami, S.A. Shehzad, Magnetohydrodynamic nanofluid convection in a porous enclosure considering heat flux boundary condition, International Journal of Heat and Mass Transfer 106 (2017) 1261–1269.

[28] M. Sheikholeslami, S.A. Shehzad, Thermal radiation of ferrofluid in existence of Lorentz forces considering variable viscosity, International Journal of Heat and Mass Transfer 109 (2017) 82–92.

[29] M. Sheikholeslami, H.B. Rokni, Numerical modeling of nanofluid natural convection in a semi annulus in existence of Lorentz force, Computer Methods in Applied Mechanics and Engineering 317 (2017) 419–430.

[30] M. Sheikholeslami, K. Vajravelu, Nanofluid flow and heat transfer in a cavity with variable magnetic field, Applied Mathematics and Computation 298 (2017) 272–282.

[31] M. Sheikholeslami, H.B. Rokni, Magnetohydrodynamic CuO-water nanofluid in a porous complex shaped enclosure, ASME, Journal of Thermal Science and Engineering Applications 9 (4) (2017) 041007. Available from: https://doi.org/10.1115/1.4035973.

[32] M. Sheikholeslami, H.B. Rokni, Nanofluid two phase model analysis in existence of induced magnetic field, International Journal of Heat and Mass Transfer 107 (2017) 288–299.

[33] M. Sheikholeslami, T. Hayat, A. Alsaedi, S. Abelman, Numerical analysis of EHD nanofluid force convective heat transfer considering electric field dependent viscosity, International Journal of Heat and Mass Transfer 108 (2017) 2558–2565.

[34] M. Sheikholeslami, Magnetic field influence on CuO $-H_2O$ nanofluid convective flow in a permeable cavity considering various shapes for nanoparticles, International Journal of Hydrogen Energy 42 (2017) 19611–19621.

[35] M. Sheikholeslami, M.M. Bhatti, Forced convection of nanofluid in presence of constant magnetic field considering shape effects of nanoparticles, International Journal of Heat and Mass Transfer 111 (2017) 1039–1049.

[36] M. Sheikholeslami, A. Zeeshan, Analysis of flow and heat transfer in water based nanofluid due to magnetic field in a porous enclosure with constant heat flux using CVFEM, Computer Methods in Applied Mechanics and Engineering 320 (2017) 68–81.

[37] M. Sheikholeslami, M. Shamlooei, Fe_3O_4- H_2O nanofluid natural convection in presence of thermal radiation, International Journal of Hydrogen Energy 42 (9) (2017) 5708–5718.

[38] M. Sheikholeslami, M.M. Bhatti, Active method for nanofluid heat transfer enhancement by means of EHD, International Journal of Heat and Mass Transfer 109 (2017) 115–122.

[39] M. Sheikholeslami, T. Hayat, A. Alsaedi, Numerical study for external magnetic source influence on water based nanofluid convective heat transfer, International Journal of Heat and Mass Transfer 106 (2017) 745–755.

[40] M. Sheikholeslami, A.J. Chamkha, Flow and convective heat transfer of a ferro-nanofluid in a double-sided lid-driven cavity with a wavy wall in the presence of a variable magnetic field, Numerical Heat Transfer, Part A 69 (10) (2016) 1186–1200. Available from: https://doi.org/10.1080/10407782.2015.1125709.

[41] M. Sheikholeslami, A.J. Chamkha, Electrohydrodynamic free convection heat transfer of a nanofluid in a semi-annulus enclosure with a sinusoidal wall, Numerical Heat Transfer, Part A 69 (7) (2016) 781–793. Available from: http://dx.doi.org/10.1080/10407782.2015.1090819.

[42] M. Sheikholeslami, M.K. Sadoughi, Simulation of CuO- water nanofluid heat transfer enhancement in presence of melting surface, International Journal of Heat and Mass Transfer 116 (2018) 909–919.
[43] M. Sheikholeslami, H.B. Rokni, Simulation of nanofluid heat transfer in presence of magnetic field: A review, International Journal of Heat and Mass Transfer 115 (2017) 1203–1233.
[44] M. Sheikholeslami, H.B. Rokni, Melting heat transfer influence on nanofluid flow inside a cavity in existence of magnetic field, International Journal of Heat and Mass Transfer 114 (2017) 517–526.
[45] M. Sheikholeslami, S.A. Shehzad, Magnetohydrodynamic nanofluid convective flow in a porous enclosure by means of LBM, International Journal of Heat and Mass Transfer 113 (2017) 796–805.
[46] M. Sheikholeslami, Magnetohydrodynamic nanofluid forced convection in a porous lid driven cubic cavity using Lattice Boltzmann Method, Journal of Molecular Liquids 231 (2017) 555–565.
[47] M.S. Kandelousi, KKL correlation for simulation of nanofluid flow and heat transfer in a permeable channel, Physics Letters A 378 (45) (2014) 3331–3339.
[48] M. Sheikholeslami, Effect of uniform suction on nanofluid flow and heat transfer over a cylinder, J Braz. Soc. Mech. Sci. Eng. 37 (2015) 1623–1633.
[49] M. Sheikholeslami, Lattice Boltzmann Method simulation of MHD non-Darcy nanofluid free convection, Physica B 516 (2017) 55–71.
[50] M. Sheikholeslami, Influence of magnetic field on nanofluid free convection in an open porous cavity by means of Lattice Boltzmann Method, Journal of Molecular Liquids 234 (2017) 364–374.
[51] M. Sheikholeslami, A.J. Chamkha, Influence of Lorentz forces on nanofluid forced convection considering Marangoni convection, Journal of Molecular Liquids 225 (2017) 750–757.
[52] M. Sheikholeslami, T. Hayat, A. Alsaedi, Numerical simulation of nanofluid forced convection heat transfer improvement in existence of magnetic field using Lattice Boltzmann Method, International Journal of Heat and Mass Transfer 108 (2017) 1870–1883.
[53] M. Sheikholeslami, R. Ellahi, Three dimensional mesoscopic simulation of magnetic field effect on natural convection of nanofluid, International Journal of Heat and Mass Transfer 89 (2015) 799–808.
[54] M. Sheikholeslami, S. Abelman, Two phase simulation of nanofluid flow and heat transfer in an annulus in the presence of an axial magnetic field, IEEE Transactions on Nanotechnology 14 (3) (2015) 561–569.
[55] M. Sheikholeslami, R. Ellahi, Simulation of ferrofluid flow for magnetic drug targeting using Lattice Boltzmann method, Journal of Zeitschrift Fur Naturforschung A 70 (2) (2015) 115–124.
[56] M. Sheikholeslami, M. Seyednezhad, Lattice Boltzmann Method simulation for CuO-water nanofluid flow in a porous enclosure with hot obstacle, Journal of Molecular Liquids 243 (2017) 249–256.
[57] M. Sheikholeslami, M. Sadoughi, Mesoscopic method for MHD nanofluid flow inside a porous cavity considering various shapes of nanoparticles, International Journal of Heat and Mass Transfer 113 (2017) 106–114.
[58] M. Sheikholeslami, A. Zeeshan, Mesoscopic simulation of CuO-H_2O nanofluid in a porous enclosure with elliptic heat source, International Journal of Hydrogen Energy 42 (22) (2017) 15393–15402.
[59] M. Sheikholeslami, S.A. Shehzad, CVFEM for influence of external magnetic source on Fe_3O_4 – H_2O nanofluid behavior in a permeable cavity considering shape effect, International Journal of Heat and Mass Transfer 115 (2017) 180–191.
[60] M. Sheikholeslami, M. Seyednezhad, Nanofluid heat transfer in a permeable enclosure in presence of variable magnetic field by means of CVFEM, International Journal of Heat and Mass Transfer 114 (2017) 1169–1180.
[61] M. Sheikholeslami, H.B. Rokni, Free convection of CuO-H_2O nanofluid in a curved porous enclosure using mesoscopic approach, International Journal of Hydrogen Energy 42 (22) (2017) 14942–14949.
[62] M. Sheikholeslami, Numerical investigation for CuO-H_2O nanofluid flow in a porous channel with magnetic field using mesoscopic method, J. Mol. Liq. 249 (2018) 739–746.
[63] M. Sheikholeslami, H.B. Rokni, Influence of melting surface on MHD nanofluid flow by means of two phase model, Chinese Journal of Physics 55 (2017) 1352–1360.
[64] M. Sheikholeslami, H.B. Rokni, Effect of melting heat transfer on nanofluid flow in existence of magnetic field considering Buongiorno Model, Chinese Journal of Physics 55 (2017) 1115–1126.
[65] M. Sheikholeslami, T. Hayat, A. Alsaedi, Numerical simulation for forced convection flow of MHD CuO-H_2O nanofluid inside a cavity by means of LBM, J. Mol. Liq. 249 (2018) 941–948.
[66] M. Sheikholeslami, A. Zeeshan, Numerical simulation of Fe_3O_4 -water nanofluid flow in a non-Darcy porous media, Int. J. Numer. Methods Heat Fluid Flow 28 (3) (2018) 641–660. Available from: https://doi.org/10.1108/HFF-04-2017-0160.
[67] M. Sheikholeslami, M. Barzegar Gerdroodbary, S. Valiallah Mousavi, D.D. Ganji, R. Moradi, Heat transfer enhancement of ferrofluid inside an 90o elbow channel by non-uniform magnetic field, J. Magnet. Magnet. Mater. 460 (2018) 302–311.
[68] M. Sheikholeslami, Numerical simulation for solidification in a LHTESS by means of nano-enhanced PCM, J. Taiwan Inst. Chem. Eng. 86 (2018) 25–41.
[69] M. Sheikholeslami, Numerical modeling of nano enhanced PCM solidification in an enclosure with metallic fin, J. Mol. Liq. 259 (2018) 424–438.
[70] M. Sheikholeslami, A. Ghasemi, Solidification heat transfer of nanofluid in existence of thermal radiation by means of FEM, Int. J. Heat Mass Transfer 123 (2018) 418–431.
[71] M. Sheikholeslami, S.A. Shehzad, CVFEM simulation for nanofluid migration in a porous medium using Darcy model, Int. J. Heat Mass Transfer 122 (2018) 1264–1271.
[72] M. Sheikholeslami, M. Darzi, M.K. Sadoughi, Heat transfer improvement and pressure drop during condensation of refrigerant-based nanofluid; an experimental procedure, Int. J. Heat Mass Transfer 122 (2018) 643–650.
[73] M. Sheikholeslami, H.B. Rokni, CVFEM for effect of Lorentz forces on nanofluid flow in a porous complex shaped enclosure by means of non-equilibrium model, J. Mol. Liq. 254 (2018) 446–462.
[74] M. Sheikholeslami, S.A. Shehzad, Z. Li, Water based nanofluid free convection heat transfer in a three dimensional porous cavity with hot sphere obstacle in existence of Lorenz forces, Int. J. Heat Mass Transfer 125 (2018) 375–386.

[75] M. Sheikholeslami, M. Jafaryar, D.D. Ganji, Z. Li, Exergy loss analysis for nanofluid forced convection heat transfer in a pipe with modified turbulators, J. Mol. Liq. 262 (2018) 104–110.

[76] M. Sheikholeslami, M. Jafaryar, Z. Li, Nanofluid turbulent convective flow in a circular duct with helical turbulators considering CuO nanoparticles, Int. J. Heat Mass Transfer 124 (2018) 980–989.

[77] M. Sheikholeslami, Z. Li, M. Shamlooei, Nanofluid, MHD natural convection through a porous complex shaped cavity considering thermal radiation, Phys. Lett. A 382 (2018) 1615–1632.

[78] M. Sheikholeslami, H.B. Rokni, Magnetic nanofluid flow and convective heat transfer in a porous cavity considering Brownian motion effects, Phys. Fluids 30 (1) (2018). Available from: https://doi.org/10.1063/1.5012517.

[79] M. Sheikholeslami, S.A. Shehzad, Simulation of water based nanofluid convective flow inside a porous enclosure via non-equilibrium model, Int. J. Heat Mass Transfer 120 (2018) 1200–1212.

[80] M. Sheikholeslami, S.A. Shehzad, Non-Darcy free convection of Fe_3O_4-water nanoliquid in a complex shaped enclosure under impact of uniform Lorentz force, Chin. J. Phys. 56 (2018) 270–281.

[81] M. Sheikholeslami, M. Seyednezhad, Simulation of nanofluid flow and natural convection in a porous media under the influence of electric field using CVFEM, Int. J. Heat Mass Transfer 120 (2018) 772–781.

[82] M. Sheikholeslami, M. Shamlooei, R. Moradi, Numerical simulation for heat transfer intensification of nanofluid in a permeable curved enclosure considering shape effect of Fe_3O_4 nanoparticles, Chem. Eng. Process. Process Intensif. 124 (2018) 71–82.

[83] M. Sheikholeslami, T. Hayat, T. Muhammad, A. Alsaedi, MHD forced convection flow of nanofluid in a porous cavity with hot elliptic obstacle by means of Lattice Boltzmann method, Int. J. Mech. Sci. 135 (2018) 532–540.

[84] M. Sheikholeslami, Numerical investigation of nanofluid free convection under the influence of electric field in a porous enclosure, J. Mol. Liq. 249 (2018) 1212–1221.

[85] M. Sheikholeslami, CuO-water nanofluid flow due to magnetic field inside a porous media considering Brownian motion, J. Mol. Liq. 249 (2018) 921–929.

[86] M. Sheikholeslami, D.D. Ganji, Influence of electric field on Fe_3O_4- water nanofluid radiative and convective heat transfer in a permeable enclosure, J. Mol. Liq. 250 (2018) 404–412.

[87] M. Sheikholeslami, H.R. Kataria, A.S. Mittal, Effect of thermal diffusion and heat-generation on MHD nanofluid flow past an oscillating vertical plate through porous medium, J. Mol. Liq. 257 (2018) 12–25.

[88] M. Sheikholeslami, S.A. Shehzad, Z. Li, Nanofluid heat transfer intensification in a permeable channel due to magnetic field using Lattice Boltzmann method, Phys, B (2018). Available from: https://doi.org/10.1016/j.physb.2018.03.036.

CHAPTER

8

Influence of Variable Lorentz Forces on Nanofluid Free Convection Using CVFEM

8.1 INTRODUCTION

Cooling capabilities of heat transfer equipments have been constrained because of the low thermal conductivity of conventional heat transfer fluids. However, it is well known that, metals in solid form have orders-of-magnitude higher thermal conductivities than those of fluids. Therefore, the thermal conductivities of fluids that contain suspended solid metallic particles could be expected to be significantly higher than those of conventional heat transfer fluids. The goal of nanofluid is to achieve the highest possible thermal properties at the smallest possible concentrations by uniform dispersion and stable suspension of nanoparticles (preferably <10 nm) in host fluids. Nanotechnology provides a new area of research to process and produce materials with average crystallite sizes below 100 nm called nanomaterials. Sheikholeslami and Ganji [1] investigated two-phase modeling of nanofluid in a rotating system with permeable sheet. They found that the Nusselt number has a direct relationship with the Reynolds number and injection parameter while it has a reverse relationship with the Rotation parameter, Schmidt number, Thermophoretic parameter, and Brownian parameter. Magnetic nanofluid (ferrofluid) is a magnetic colloidal suspension consisting of base liquid and magnetic nanoparticles with a size range of 5–15 nm in diameter coated with a surfactant layer. Jue [2] used the semi-implicit finite element method in order to simulate magnetic gradient and thermal buoyancy-induced cavity ferrofluid flow. Their results showed the flow strength increases with the strengthening magnetic field. Numerical analysis of the heat transfer enhancement and fluid flow characteristics of a rotating cylinder under the influence of magnetic dipole in the backward facing step geometry was conducted by Selimefendigil and Oztop [3]. They found that the effect of cylinder rotation on the local Nusselt number distribution is more pronounced at low Reynolds number. Nanjundappa et al. [4] studied the effect of magnetic field dependent viscosity on the onset of ferroconvection in a ferrofluid saturated horizontal porous layer. They showed the nonlinearity of fluid magnetization has no influence on the stability of the system. The vortex dynamics behind various magnetic obstacles and characteristics of heat transfer were investigated by Zhang and Huang [5]. They found that the pressure drop penalty was not increasingly dependent on interaction parameter. FHD and MHD have been presented in various publications in recent years [6–104].

8.2 INFLUENCE OF EXTERNAL VARIABLE MAGNETIC FIELD ON FERROFLUID FLOW AND CONVECTIVE HEAT TRANSFER

8.2.1 Problem Definition

The physical model along with the important geometrical parameters and the mesh of the enclosure used in the present CVFEM program are shown in Fig. 8.1. The inner and outer walls are maintained at constant temperatures T_h and T_c, respectively. The shape of inner cylinder profile is assumed to mimic the following pattern

$$r = r_{in} + A\cos(N(\zeta)) \tag{8.1}$$

in which r_{in} is the base circle radius, r_{out} is the radius of outer cylinder, A and N are amplitude and number of undulations, respectively, ζ is the rotation angle. In this study A and N are equal to 0.2 and 4, respectively. For

Application of Control Volume based Finite Element Method (CVFEM) for Nanofluid Flow and Heat Transfer.
DOI: https://doi.org/10.1016/B978-0-12-814152-6.00008-4

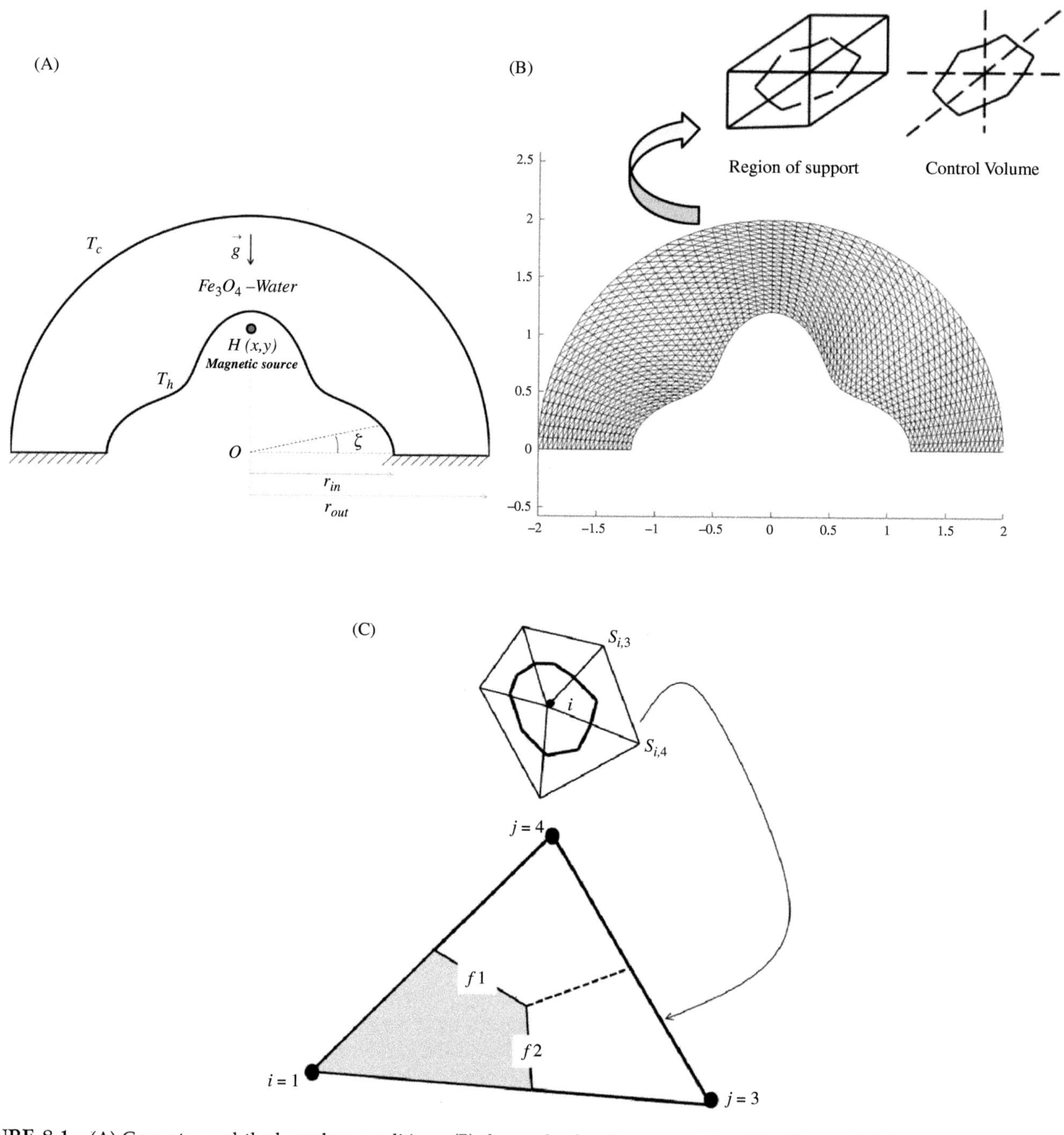

FIGURE 8.1 (A) Geometry and the boundary conditions; (B) the mesh of enclosure considered in this work; (C) a sample triangular element and its corresponding control volume.

the expression of the magnetic field strength it can be considered that the magnetic source represents a magnetic wire placed vertically to the x-y plane at the point $(\bar{a}, \bar{b})$. The components of the magnetic field intensity $(\overline{H_x}, \overline{H_y})$ and the magnetic field strength $(\overline{H})$ can be considered as:

$$\overline{H_x} = \frac{\gamma}{2\pi} \frac{1}{(x-\bar{a})^2 + (y-\bar{b})^2} (y - \bar{b}) \tag{8.2}$$

$$\overline{H_y} = -\frac{\gamma}{2\pi} \frac{1}{(x-\bar{a})^2 + (y-\bar{b})^2} (x - \bar{a}) \tag{8.3}$$

$$\overline{H} = \sqrt{\overline{H}_x^2 + \overline{H}_y^2} = \frac{\gamma}{2\pi} \frac{1}{\sqrt{(x-\overline{a})^2 + (y-\overline{b})^2}} \tag{8.4}$$

where γ the magnetic field strength at the source (of the wire) and $(\overline{a}, \overline{b})$ is the position where the source is located. The contours of the magnetic field strength are shown in Fig. 8.2. In this section magnetic source is located at $(-0.05\ cols, 0.5\ rows)$.

8.2.2 Governing Equation

The flow is steady, two-dimensional, laminar, and incompressible. The magnetic Reynolds number is assumed to be small so that the induced magnetic field can be neglected compared to the applied magnetic field. Using the Boussinesq approximation, the governing equations of heat transfer and fluid flow for nanofluid can be obtained as follows:

$$\frac{\partial u}{\partial x} + \frac{\partial v}{\partial y} = 0 \tag{8.5}$$

$$\rho_{nf}\left(u\frac{\partial u}{\partial x} + v\frac{\partial u}{\partial y}\right) = -\frac{\partial p}{\partial x} + \mu_{nf}\left(\frac{\partial^2 u}{\partial x^2} + \frac{\partial^2 u}{\partial y^2}\right) + \mu_0 M\frac{\partial \overline{H}}{\partial x} - \sigma_{nf}B_y^2 u + \sigma_{nf}B_x B_y v \tag{8.6}$$

$$\rho_{nf}\left(u\frac{\partial v}{\partial x} + v\frac{\partial v}{\partial y}\right) = -\frac{\partial p}{\partial y} + \mu_{nf}\left(\frac{\partial^2 v}{\partial x^2} + \frac{\partial^2 v}{\partial y^2}\right) + \mu_0 M\frac{\partial \overline{H}}{\partial y} - \sigma_{nf}B_x^2 v + \sigma_{nf}B_x B_y u + \rho_{nf}\beta_{nf}g(T - T_c) \tag{8.7}$$

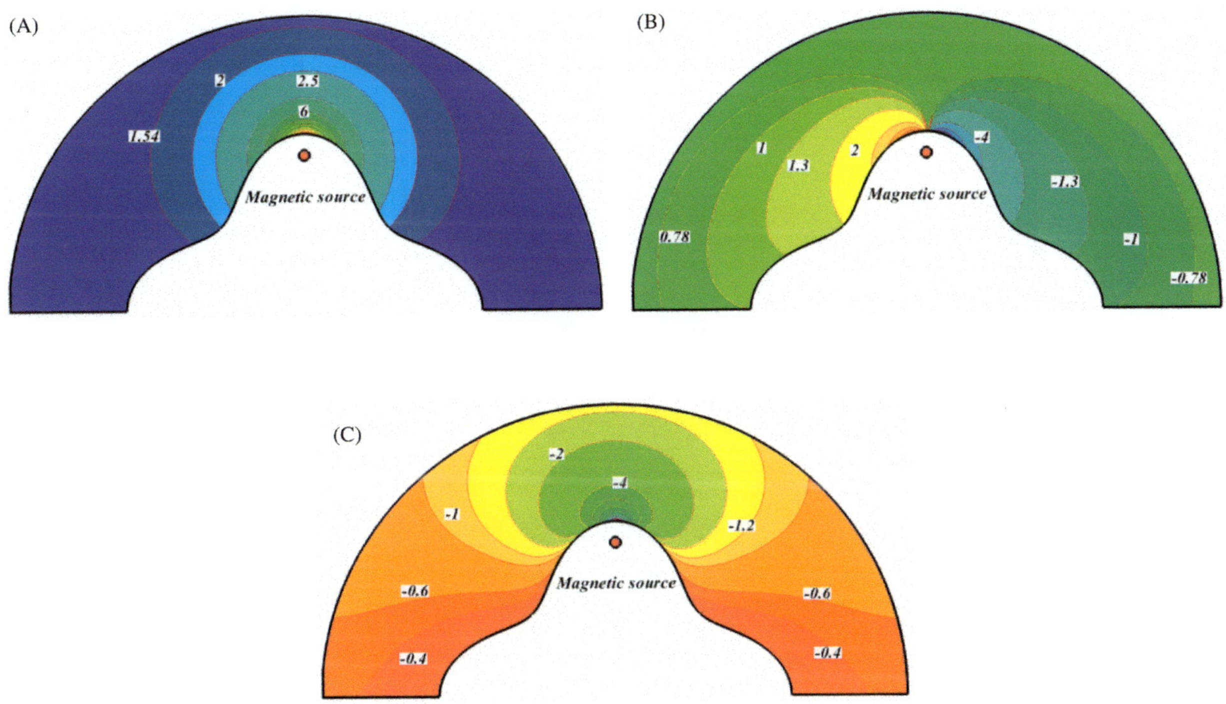

FIGURE 8.2 Contours of the (A) magnetic field strength H; (B) magnetic field intensity component in x direction Hx; (C) magnetic field intensity component in y direction Hy. (A) $H(x,y)$, (B) $Hx(x,y)$, (C) $Hy(x,y)$.

$$(\rho C_p)_{nf}\left(u\frac{\partial T}{\partial x}+v\frac{\partial T}{\partial y}\right)=k_{nf}\left(\frac{\partial^2 T}{\partial x^2}+\frac{\partial^2 T}{\partial y^2}\right)+\sigma_{nf}(uB_y-vB_x)^2$$
$$-\mu_0 T\frac{\partial M}{\partial T}\left(u\frac{\partial \overline{H}}{\partial x}+v\frac{\partial \overline{H}}{\partial y}\right)+\mu_{nf}\left\{2\left(\frac{\partial u}{\partial x}\right)^2+2\left(\frac{\partial v}{\partial y}\right)^2+\left(\frac{\partial u}{\partial y}+\frac{\partial v}{\partial x}\right)^2\right\} \tag{8.8}$$

The terms $\mu_0 M\frac{\partial \overline{H}}{\partial x}$ and $\mu_0 M\frac{\partial \overline{H}}{\partial y}$ in (8.6) and (8.7), respectively, represent the components of magnetic force, per unit volume, and depend on the existence of the magnetic gradient on the corresponding x and y directions. These two terms are well known from FHD which is the so-called Kelvin force. The terms $-\sigma_{nf}B_y^2u+\sigma_{nf}B_xB_yv$ and $-\sigma_{nf}B_x^2v+\sigma_{nf}B_xB_yu$ appearing in (8.6) and (8.7), respectively, represent the Lorentz force per unit volume towards the x and y directions and arise due to the electrical conductivity of the fluid. These two terms are known in MHD. The principles of MHD and FHD are combined in the mathematical model and the above mentioned terms arise together in the governing Eqs. (8.6) and (8.7). The term $\mu_0 T\frac{\partial M}{\partial T}\left(u\frac{\partial \overline{H}}{\partial x}+v\frac{\partial \overline{H}}{\partial y}\right)$ in Eq. (8.8) represents the thermal power per unit volume due to the magneto caloric effect. Also the term $\sigma_{nf}(uB_y-vB_x)^2$ in (8.8) represents the Joule heating. For the variation of the magnetization M, with the magnetic field intensity $\overline{H}$ and temperature T is considered:

$$M=K'\overline{H}(T_c'-T) \tag{8.9}$$

where K' is a constant and T_c' is the Curie temperature.

In the above equations, μ_0 is the magnetic permeability of vacuum $(4\pi\times10^{-7}Tm/A)$, $\overline{H}$ is the magnetic field strength, $\overline{B}$ is the magnetic induction $(\overline{B}=\mu_0\overline{H})$ and the bar above the quantities denotes that they are dimensional. The effective density (ρ_{nf}) and heat capacitance $(\rho C_p)_{nf}$ of the nanofluid are defined as:

$$\rho_{nf}=\rho_f(1-\phi)+\rho_s\phi \tag{8.10}$$

$$(\rho C_p)_{nf}=(\rho C_p)_f(1-\phi)+(\rho C_p)_s\phi \tag{8.11}$$

where ϕ is the solid volume fraction of nanoparticles. Thermal diffusivity of the nanofluids is

$$\alpha_{nf}=\frac{k_{nf}}{(\rho C_p)_{nf}} \tag{8.12}$$

and the thermal expansion coefficient of the nanofluid can be determined as

$$\beta_{nf}=\beta_f(1-\phi)+\beta_s\phi \tag{8.13}$$

The dynamic viscosity of the nanofluids given by Brinkman is

$$\mu_{nf}=\frac{\mu_f}{(1-\phi)^{2.5}} \tag{8.14}$$

The effective thermal conductivity of the nanofluid can be approximated by the Maxwell–Garnetts (MG) model as:

$$\frac{k_{nf}}{k_f}=\frac{k_s+2k_f-2\phi(k_f-k_s)}{k_s+2k_f+\phi(k_f-k_s)} \tag{8.15}$$

and the effective electrical conductivity of nanofluid was presented by Maxwell as below:

$$\frac{\sigma_{nf}}{\sigma_f}=1+\frac{3\left(\frac{\sigma_s}{\sigma_f}-1\right)\phi}{\left(\frac{\sigma_s}{\sigma_f}+2\right)-\left(\frac{\sigma_s}{\sigma_f}-1\right)\phi} \tag{8.16}$$

By introducing the following nondimensional variables:

$$a=\frac{\bar{a}}{L},b=\frac{\bar{b}}{L},X=\frac{x}{L},\quad Y=\frac{y}{L}\quad,\Theta=\frac{T-T_c}{T_h-T_c},U=\frac{uL}{\alpha_f},V=\frac{vL}{\alpha_f},$$
$$H=\frac{\overline{H}}{\overline{H_0}},H_x=\frac{\overline{H_x}}{\overline{H_0}},H_y=\frac{\overline{H_y}}{\overline{H_0}},P=\frac{p}{\rho_f(\alpha_f/L)^2} \tag{8.17}$$

where in Eq. (8.17) $\overline{H_0}=\overline{H}(\bar{a},0)=\frac{\gamma}{2\pi|b|}$ and $L=r_{out}-r_{in}$. Using the dimensionless parameters, the equations now become:

$$\frac{\partial U}{\partial X}+\frac{\partial V}{\partial Y}=0 \tag{8.18}$$

$$U\frac{\partial U}{\partial X}+V\frac{\partial U}{\partial Y}=-\frac{\partial P}{\partial X}+\Pr\left[\frac{\mu_{nf}/\mu_f}{\rho_{nf}/\rho_f}\right]\left(\frac{\partial^2 U}{\partial X^2}+\frac{\partial^2 U}{\partial Y^2}\right)$$
$$+Mn_F\Pr\left(\frac{\rho_f}{\rho_{nf}}\right)(\varepsilon_2-\varepsilon_1-\Theta)H\frac{\partial H}{\partial X}-Ha^2\Pr\left[\frac{\sigma_{nf}/\sigma_f}{\rho_{nf}/\rho_f}\right]\left(H_y^2U-H_xH_yV\right) \tag{8.19}$$

$$U\frac{\partial V}{\partial X}+V\frac{\partial V}{\partial Y}=-\frac{\partial P}{\partial X}+\Pr\left[\frac{\mu_{nf}/\mu_f}{\rho_{nf}/\rho_f}\right]\left(\frac{\partial^2 V}{\partial X^2}+\frac{\partial^2 V}{\partial Y^2}\right)$$
$$+Mn_F\Pr\left(\frac{\rho_f}{\rho_{nf}}\right)(\varepsilon_2-\varepsilon_1-\Theta)H\frac{\partial H}{\partial Y}-Ha^2\Pr\left[\frac{\sigma_{nf}/\sigma_f}{\rho_{nf}/\rho_f}\right]\left(H_x^2V-H_xH_yU\right)+Ra\Pr\left[\frac{\beta_{nf}}{\beta_f}\right]\Theta \tag{8.20}$$

$$U\frac{\partial \Theta}{\partial X}+V\frac{\partial \Theta}{\partial Y}=\left[\frac{\frac{k_{nf}}{k_f}}{\frac{(\rho C_P)_{nf}}{(\rho C_P)_f}}\right]\left(\frac{\partial^2\Theta}{\partial X^2}+\frac{\partial^2\Theta}{\partial Y^2}\right)+Ha^2\,Ec\left[\frac{\frac{\sigma_{nf}}{\sigma_f}}{\frac{(\rho C_P)_{nf}}{(\rho C_P)_f}}\right]\{U\,H_y-V\,H_x\}^2$$
$$+Mn_FEc\frac{(\rho C_P)_f}{(\rho C_P)_{nf}}\left\{U\frac{\partial H}{\partial X}+V\frac{\partial H}{\partial Y}\right\}H(\varepsilon_1+\Theta)+Ec\left[\frac{\frac{\mu_{nf}}{\mu_f}}{\frac{(\rho C_P)_{nf}}{(\rho C_P)_f}}\right]\left\{2\left(\frac{\partial U}{\partial X}\right)^2+2\left(\frac{\partial V}{\partial Y}\right)^2+\left(\frac{\partial U}{\partial Y}+\frac{\partial V}{\partial X}\right)^2\right\} \tag{8.21}$$

where $Ra_f=g\beta_fL^3(T_h-T_c)/(\alpha_f\upsilon_f)$, $\Pr_f=\upsilon_f/\alpha_f$, $Ha=L\mu_0H_0\sqrt{\sigma_f/\mu_f}$, $\varepsilon_1=T_1/\Delta T$, $\varepsilon_2=T_c'/\Delta T$, $Ec=\left(\mu_f\alpha_f\right)/\left[(\rho C_P)_f\Delta T\,L^2\right]$ and $Mn_F=\mu_0H_0^2K'(T_h-T_c)L^2/\left(\mu_f\alpha_f\right)$ are the Rayleigh number, Prandtl number, Hartmann number arising from MHD, temperature number, Curie temperature number, Eckert number, and Magnetic number arising from FHD for the base fluid, respectively. The thermophysical properties of the nanofluid are given in Table 8.1.

The stream function and vorticity are defined as:

$$u=\frac{\partial\psi}{\partial y},v=-\frac{\partial\psi}{\partial x},\omega=\frac{\partial v}{\partial x}-\frac{\partial u}{\partial y}\Omega=\frac{\omega L^2}{\alpha_f},\Psi=\frac{\psi}{\alpha_f} \tag{8.22}$$

TABLE 8.1 Thermophysical Properties of Water and Nanoparticles

	$\rho(kg/m^3)$	$C_p(j/kgk)$	$k(W/m\cdot k)$	$\beta\times10^5(K^{-1})$	$d_p(nm)$	$\sigma(\Omega\cdot m)^{-1}$
Pure water	997.1	4179	0.613	21	-	0.05
Fe_3O_4	5200	670	6	1.3	47	25000

The stream function satisfies the continuity Eq. (8.18). The vorticity equation is obtained by eliminating the pressure between the two momentum equations, i.e., by taking the y-derivative of Eq. (8.20) and subtracting from it the x-derivative of Eq. (8.19).

$$\begin{aligned} &\Theta = 1.0 \quad \text{on the inner circular boundary} \\ &\Theta = 0.0 \quad \text{on the outer circular boundary} \\ &\Psi = 0.0 \quad \text{on all solid boundaries} \end{aligned} \tag{8.23}$$

The values of vorticity on the boundary of the enclosure can be obtained using the stream function formulation and the known velocity conditions during the iterative solution procedure.

The local Nusselt number of the nanofluid along the cold wall can be expressed as:

$$Nu_{loc} = \left(\frac{k_{nf}}{k_f}\right) \frac{\partial \Theta}{\partial r} \tag{8.24}$$

where r is the radial direction. The average Nusselt number on the hot circular wall is evaluated as:

$$Nu_{ave} = \frac{1}{\pi} \int_0^{\pi} Nu_{loc}(\zeta)\, d\zeta \tag{8.25}$$

To estimate the enhancement of heat transfer between the case of $\phi = 0.04$ and the pure fluid (base fluid) case, the heat transfer enhancement is defined as:

$$En = \frac{Nu\,(\phi = 0.04) - Nu\,(basefluid)}{Nu\,(basefluid)} \times 100 \tag{8.26}$$

8.2.3 Effects of Active Parameters

The effect of an external magnetic field on ferrofluid flow and heat transfer in a semi-annulus enclosure with sinusoidal hot wall is studied. The mathematical model used for the formulation of the problem is consistent with the principles of Ferrohydrodynamics (FHD) and Magnetohydrodynamics (MHD). The thermophysical properties of Fe_3O_4 nanoparticles and base fluid (water) are shown in Table 8.1. Calculations are made for various values of volume fraction of nanoparticles ($\phi = 0\%$ and 4%), Rayleigh numbers ($Ra = 10^3$, 10^4 and 10^5), Magnetic number arising from FHD ($Mn_F = 0, 50$ and 500), and Hartmann number arising from MHD ($Ha = 0, 1, 3$ and 5). In all calculations, the Prandtl number (Pr), temperature number (ε_1), and Eckert number (Ec) are set to 6.8, 0.0, and 1×10^{-5}, respectively.

The comparisons of the isotherms and streamlines between nanofluid and pure fluid are shown in Fig. 8.3. The velocity components of nanofluid are increased because of an augment in the energy transport in the fluid. Thus, the absolute values of stream functions increase with the augment of the volume fraction of nanofluid. Also it can be seen that there is no sensible change on isotherm.

Effects of Rayleigh number, Hartmann number, and Magnetic number on isotherms and streamlines are shown in Figs. 8.4–8.6. At low Rayleigh number, conduction heat transfer mechanism is dominant in the absence of magnetic field. As the Hartmann number increases the absolute value of maximum stream function decreases. As Magnetic number increases three thermal plume appears over the hot wall. Also the primary vortex in each side turns into two smaller vortexes. As the Rayleigh number increases, the role of convection in heat transfer becomes more significant and consequently the thermal boundary layer on the surface of the inner wall becomes thinner. In addition, a plume starts to appear on the top of the inner circular wall at $\zeta = 90^{\circ}$. As Lorentz forces increases the thermal plume converts to three plumes. Also it can be seen that at high Hartmann number two secondary eddies exist near the vertical center line. At $Ra = 10^5$, the effect of Kelvin forces is not sensible while increasing Lorentz forces lead to the appearance of two equals eddies near the vertical centerline.

Effects of Magnetic number, Hartmann number and Rayleigh number on local and average Nusselt number are depicted in Fig. 8.7 and Table 8.2. Nusselt number increases with increase of Rayleigh number while it decreases with increase of Hartmann number. Increasing Magnetic number causes Nusselt number to increase except for $Ra = 10^5$. Table 8.3 illustrates the effects of Magnetic number, Hartmann number, and Rayleigh number on heat transfer enhancement. For low Rayleigh number, enhancement increases with increase of Hartmann

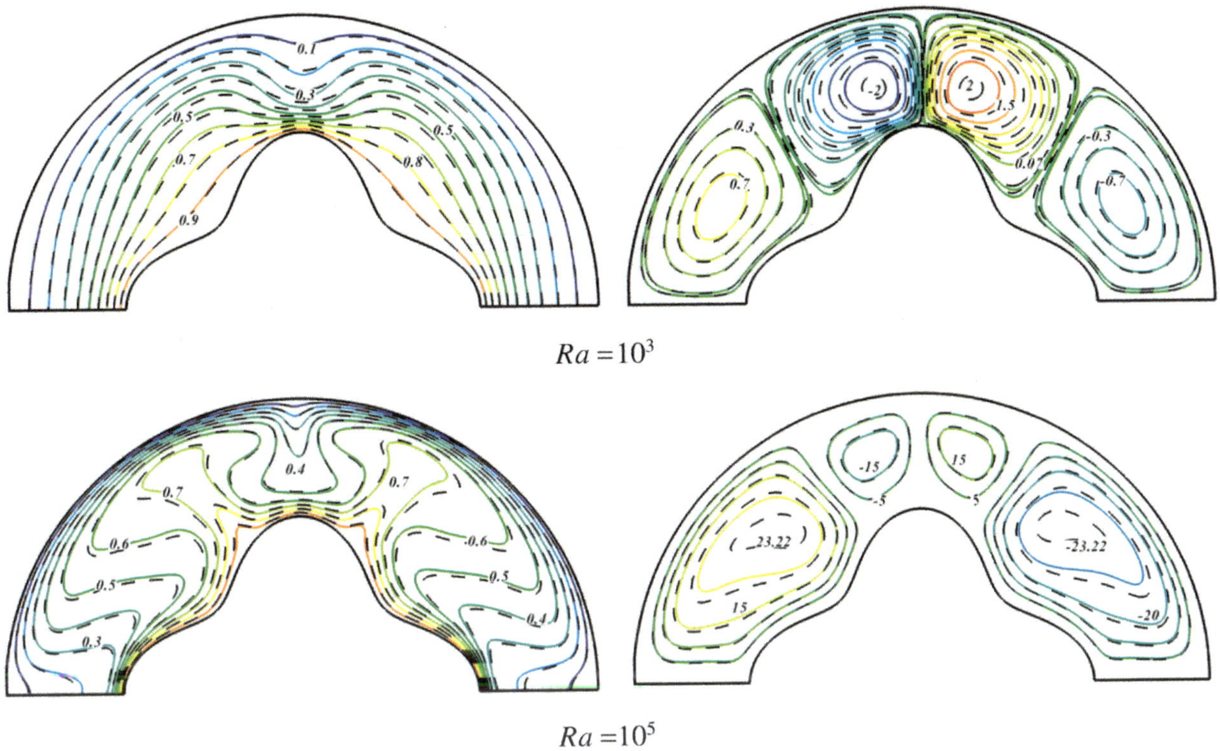

FIGURE 8.3 Comparison of the isotherms and streamlines between nanofluid ($\phi = 0.04$) (– – –) and pure fluid ($\phi = 0$) (—) when $Mn_F = 500, Ha = 5$, and $\Pr = 6.8$.

number while it decrease with increase of Magnetic number. The opposite trend is observed for the highest Rayleigh number.

8.3 FERROFLUID FLOW AND HEAT TRANSFER IN A SEMIANNULUS ENCLOSURE IN THE PRESENCE OF THERMAL RADIATION

8.3.1 Problem Definition

The schematic diagram and the mesh of the semi-annulus enclosure used in the present CVFEM program are shown in Fig. 8.8. The inner wall is under constant heat flux (q'') and the outer wall is maintained at constant temperature (T_c), while the two other walls are thermally insulated. In this study, γ is equal to 45°. For the expression of the magnetic field strength it can be considered that the magnetic source represents a magnetic wire placed vertically to the x-y plane at the point $(\bar{a}, \bar{b})$. The components of the magnetic field intensity $(\overline{H}_x, \overline{H}_y)$ and the magnetic field strength $(\overline{H})$ can be considered as:

$$\overline{H_x} = \frac{\gamma'}{2\pi} \frac{1}{(x-\bar{a})^2 + (y-\bar{b})^2}(y - \bar{b}) \tag{8.27}$$

$$\overline{H_y} = -\frac{\gamma'}{2\pi} \frac{1}{(x-\bar{a})^2 + (y-\bar{b})^2}(x - \bar{a}) \tag{8.28}$$

$$\overline{H} = \sqrt{\overline{H}_x^2 + \overline{H}_y^2} = \frac{\gamma'}{2\pi} \frac{1}{\sqrt{(x-\bar{a})^2 + (y-\bar{b})^2}} \tag{8.29}$$

where γ' is the magnetic field strength at the source (of the wire) and (a, b) is the position where the source is located. The contours of the magnetic field strength are shown in Fig. 8.9. In this study magnetic source is located at $(-0.05, 0.5)$.

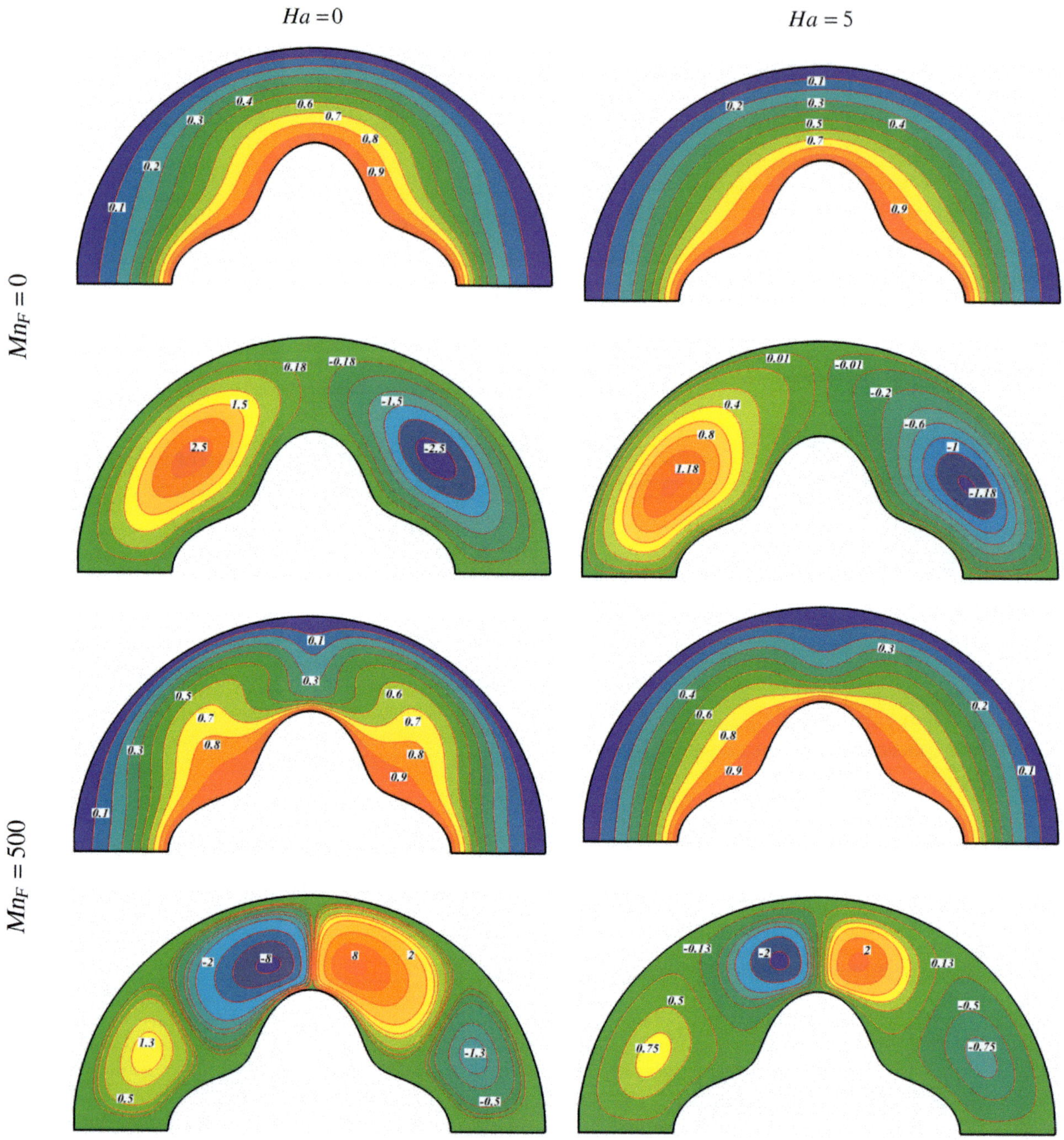

FIGURE 8.4 Isotherm (up) and streamline (down) contours for different values of Hartmann number and Magnetic number when $Ra = 10^3$, $\phi = 0.04$, $\Pr = 6.8$.

8.3.2 Governing Equation

The flow is considered to be steady, two-dimensional, incompressible, and laminar. Using the Boussinesq approximation, the governing equations of heat transfer and fluid flow for nanofluid can be obtained as follows:

$$\frac{\partial u}{\partial x} + \frac{\partial v}{\partial y} = 0 \tag{8.30}$$

$$\rho_{nf}\left(u\frac{\partial u}{\partial x} + v\frac{\partial u}{\partial y}\right) = -\frac{\partial P}{\partial x} + \mu_{nf}\left(\frac{\partial^2 u}{\partial x^2} + \frac{\partial^2 u}{\partial y^2}\right) + \mu_0 M\frac{\partial \overline{H}}{\partial x} - \sigma_{nf}B_y^2 u + \sigma_{nf}B_x B_y v \tag{8.31}$$

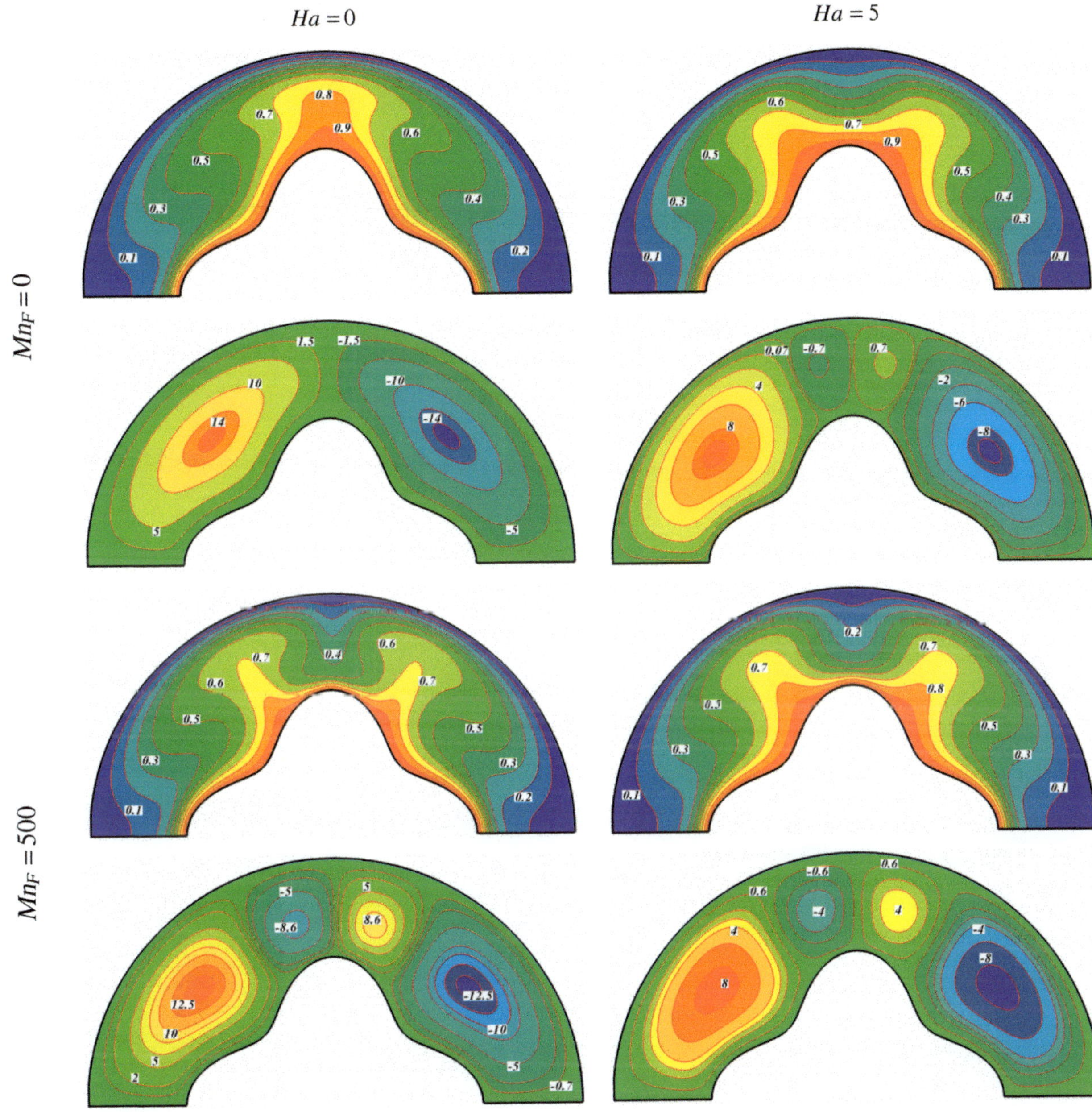

FIGURE 8.5 Isotherm (up) and streamline (down) contours for different values of Hartmann number and Magnetic number when $Ra = 10^4$, $\phi = 0.04$, $\Pr = 6.8$.

$$\rho_{nf}\left(u\frac{\partial v}{\partial x}+v\frac{\partial v}{\partial y}\right)=-\frac{\partial P}{\partial y}+\mu_{nf}\left(\frac{\partial^2 v}{\partial x^2}+\frac{\partial^2 v}{\partial y^2}\right)+\mu_0 M\frac{\partial \overline{H}}{\partial y}-\sigma_{nf}B_x^2 v+\sigma_{nf}B_xB_yu+\rho_{nf}\beta_{nf}g(T-T_c) \tag{8.32}$$

$$\begin{aligned}(\rho C_p)_{nf}\left(u\frac{\partial T}{\partial x}+v\frac{\partial T}{\partial y}\right)&=k_{nf}\left(\frac{\partial^2 T}{\partial x^2}+\frac{\partial^2 T}{\partial y^2}\right)+\sigma_{nf}(uB_y-vB_x)^2-\frac{1}{(\rho C_p)_{nf}}\frac{\partial q_r}{\partial y}\\&-\mu_0 T\frac{\partial M}{\partial T}\left(u\frac{\partial \overline{H}}{\partial x}+v\frac{\partial \overline{H}}{\partial y}\right)+\mu_{nf}\left\{2\left(\frac{\partial u}{\partial x}\right)^2+2\left(\frac{\partial v}{\partial x}\right)^2+\left(\frac{\partial u}{\partial x}+\frac{\partial v}{\partial y}\right)^2\right\}\end{aligned} \tag{8.33}$$

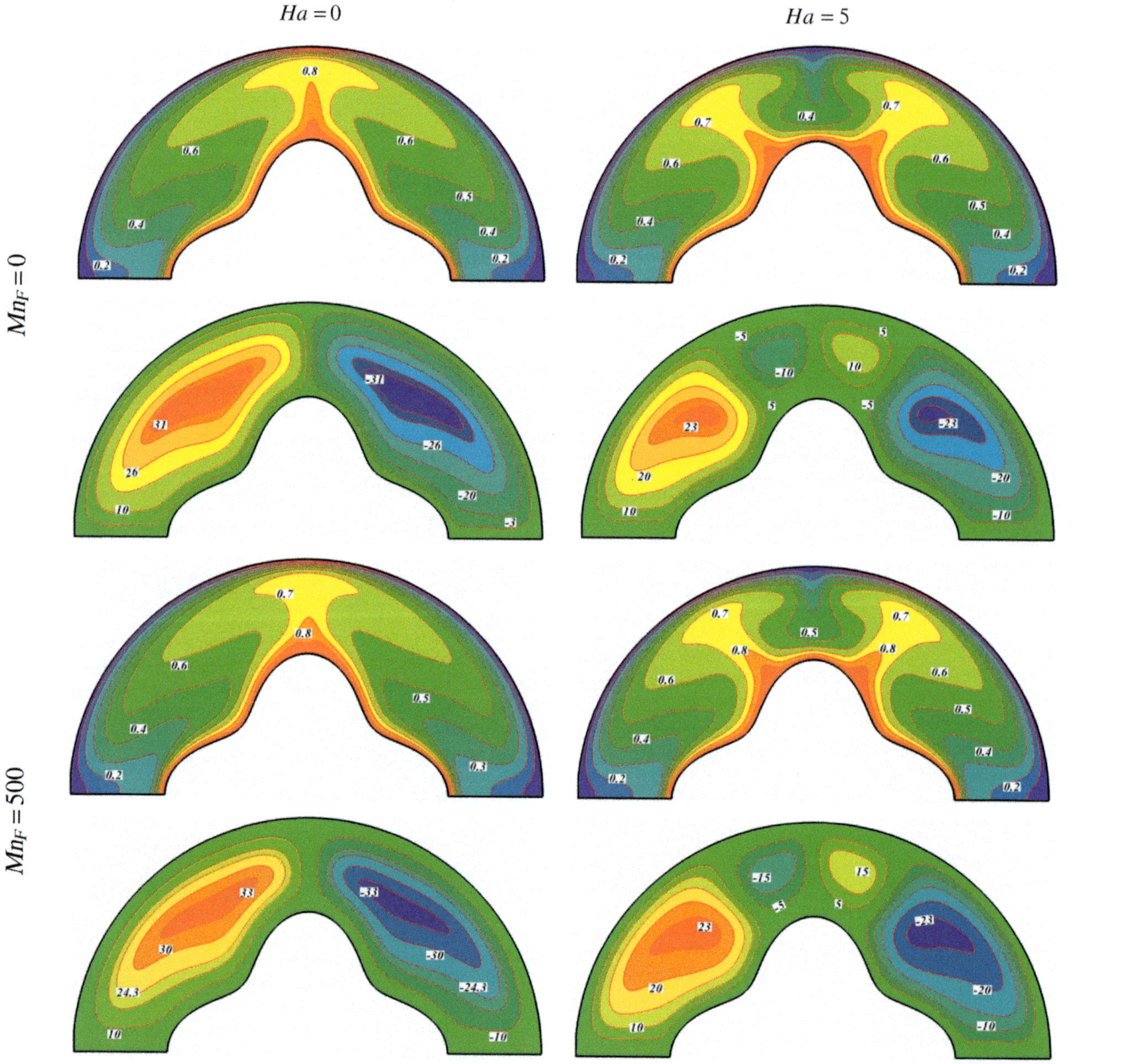

FIGURE 8.6 Isotherm (up) and streamline (down) contours for different values of Hartmann number and Magnetic number when $Ra = 10^5$, $\phi = 0.04$, $\Pr = 6.8$.

where the radiation heat flux q_r is considered according to Rosseland approximation such that $q_r = -\frac{4\sigma_e}{3\beta_R}\frac{\partial T^4}{\partial y}$ where σ_e, β_R are the Stefan–Boltzmann constant and the mean absorption coefficient, respectively. The fluid-phase temperature differences within the flow are assumed to be sufficiently small so that T^4 may be expressed as a linear function of temperature. This is done by expanding T^4 in a Taylor series about the temperature T_c and neglecting higher order terms to yield, $T^4 \cong 4T_c^3 T - 3T_c^4$.

For the variation of the magnetization M, with the magnetic field intensity $\overline{H}$ and temperature T, the following relation is considered:

$$M = K'\overline{H}\left(T_c^{'} - T\right) \tag{8.34}$$

where K' is a constant and $T_c^{'}$ is the Curie temperature.

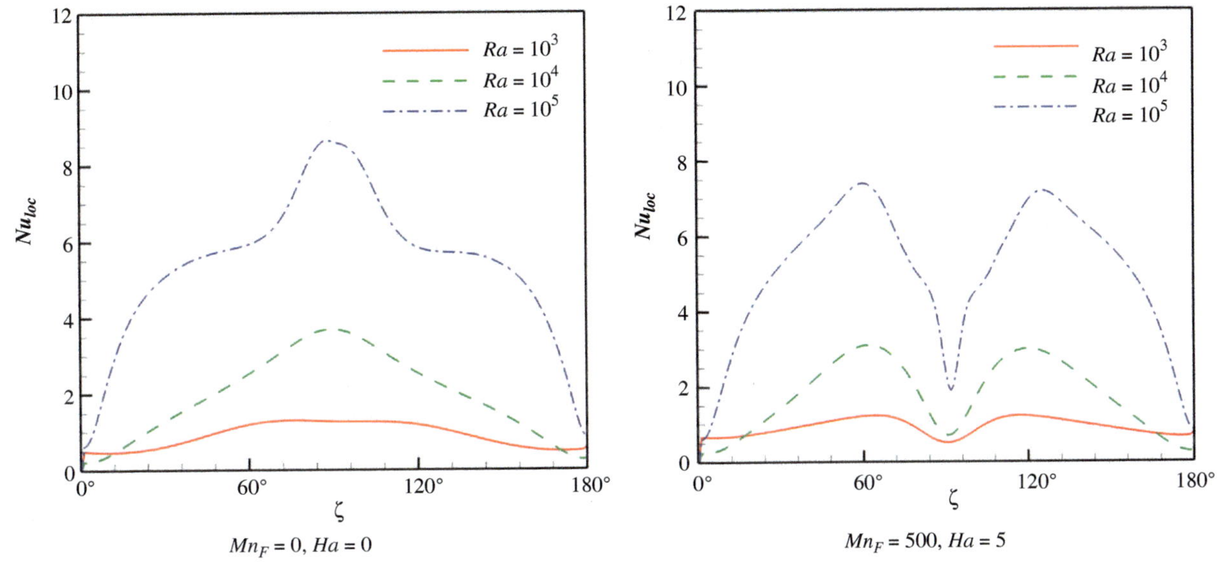

FIGURE 8.7 Effects of Magnetic number, Hartmann number, and Rayleigh number on local Nusselt number Nu_{loc} along hot wall when $\phi = 0.04$.

TABLE 8.2 Effects of Rayleigh Number, Hartmann Number, and Magnetic Number on Average Nusselt Number When $\phi = 0.04$

Ra	Mn_F	*Ha*	Nu_{ave}
10^3	0	0	0.984445
10^3	0	5	0.899518
10^3	500	0	1.478063
10^3	500	5	0.999565
10^4	0	0	2.135067
10^4	0	5	1.702675
10^4	500	0	2.365715
10^4	500	5	1.926001
10^5	0	0	5.926064
10^5	0	5	5.490067
10^5	500	0	5.843355
10^5	500	5	5.495555

In the above equations, μ_0 is the magnetic permeability of vacuum $(4\pi \times 10^{-7} Tm/A)$, $\overline{H}$ is the magnetic field strength, $\overline{B}$ is the magnetic induction $(\overline{B} = \mu_0 \overline{H})$, and the bar above the quantities denotes that they are dimensional. The effective density (ρ_{nf}) and heat capacitance $(\rho C_p)_{nf}$ of the nanofluid are defined as:

$$\rho_{nf} = \rho_f(1-\phi) + \rho_s\phi \tag{8.35}$$

$$(\rho C_p)_{nf} = (\rho C_p)_f(1-\phi) + (\rho C_p)_s\phi \tag{8.36}$$

TABLE 8.3 Effects of Rayleigh Number, Hartmann Number, and Magnetic Number on Heat Transfer Enhancement

Ra	Mn_F	*Ha*	*En*
10^3	0	0	7.461432
10^3	0	5	8.552658
10^3	500	0	4.123763
10^3	500	5	2.677329
10^4	0	0	7.135186
10^4	0	5	5.86332
10^4	500	0	6.471199
10^4	500	5	4.675113
10^5	0	0	11.38515
10^5	0	5	11.32785
10^5	500	0	11.60219
10^5	500	5	10.95681

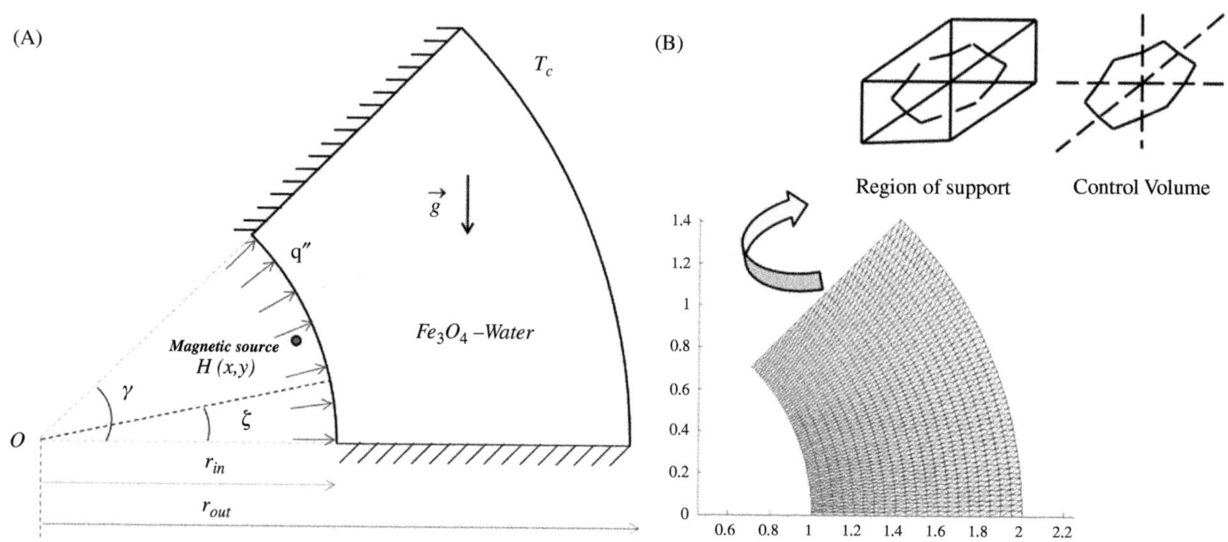

FIGURE 8.8 (A) Geometry and the boundary conditions; (B) the mesh of enclosure considered in this work.

where ϕ is the solid volume fraction of nanoparticles. Thermal diffusivity of the nanofluids is

$$\alpha_{nf} = \frac{k_{nf}}{(\rho C_p)_{nf}} \tag{8.37}$$

and the thermal expansion coefficient of the nanofluid can be determined as

$$\beta_{nf} = \beta_f(1-\phi) + \beta_s\phi \tag{8.38}$$

The dynamic viscosity of the nanofluids given by Brinkman is

$$\mu_{nf} = \frac{\mu_f}{(1-\phi)^{2.5}} \tag{8.39}$$

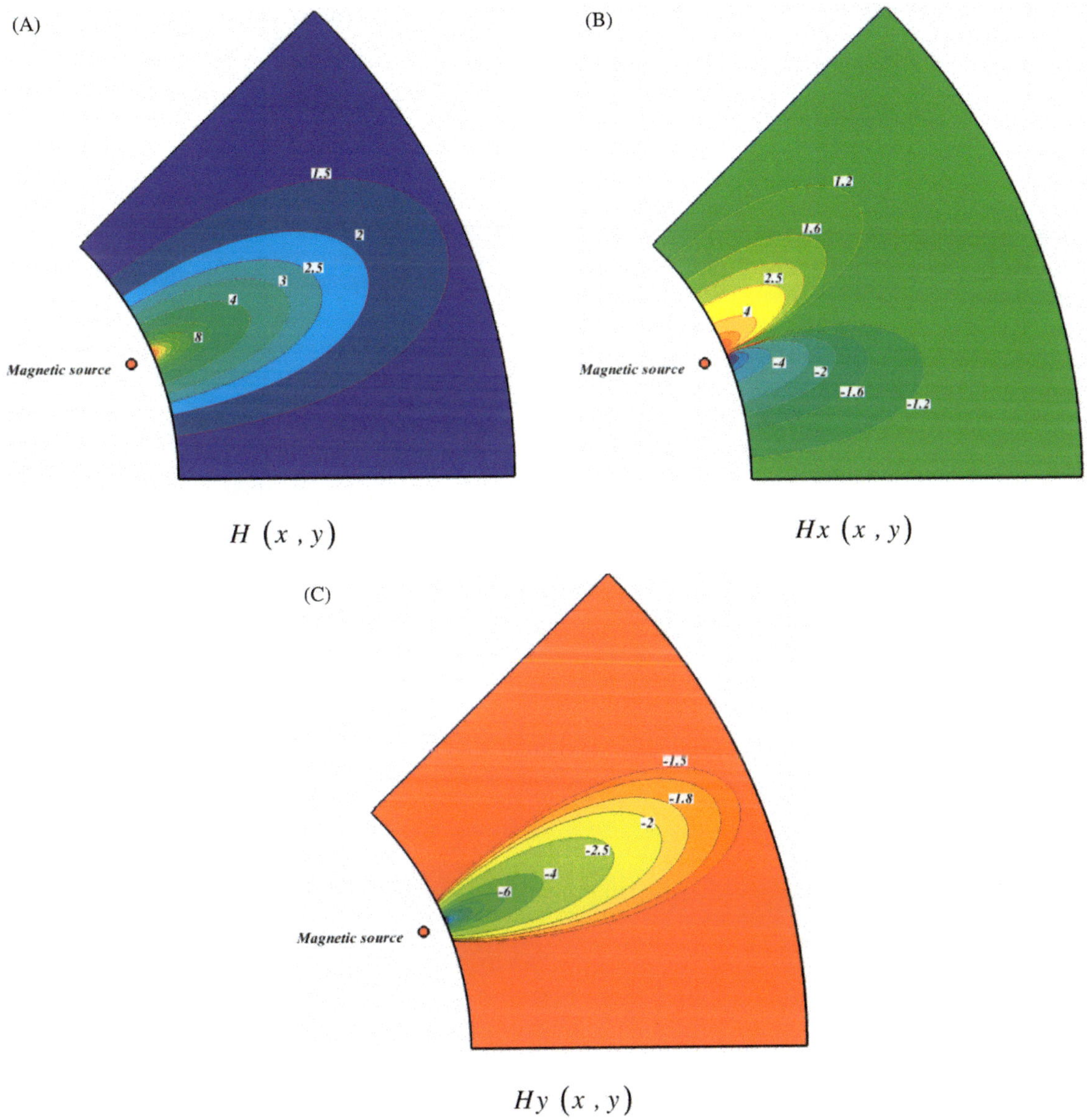

FIGURE 8.9 Contours of the (A) magnetic field strength H; (B) magnetic field intensity component in x direction Hx; (C) magnetic field intensity component in y direction Hy.

The effective thermal conductivity of the nanofluid can be approximated by the Maxwell–Garnetts (MG) model as:

$$\frac{k_{nf}}{k_f} = \frac{k_s + 2k_f - 2\phi(k_f - k_s)}{k_s + 2k_f + \phi(k_f - k_s)} \tag{8.40}$$

and the effective electrical conductivity of nanofluid was presented by Maxwell as below:

$$\frac{\sigma_{nf}}{\sigma_f} = 1 + \frac{3\left(\dfrac{\sigma_s}{\sigma_f} - 1\right)\phi}{\left(\dfrac{\sigma_s}{\sigma_f} + 2\right) - \left(\dfrac{\sigma_s}{\sigma_f} - 1\right)\phi} \tag{8.41}$$

The stream function satisfies the continuity Eq. (8.30). By introducing the following nondimensional variables:

$$X=\frac{x}{L},\quad Y=\frac{y}{L},\quad \Omega=\frac{\omega L^2}{\alpha_f},\Psi=\frac{\psi}{\alpha_f},\Theta=\frac{T-T_c}{(q''L/k_f)},U=\frac{uL}{\alpha_f},V=\frac{vL}{\alpha_f},$$
$$H=\frac{\overline{H}}{\overline{H_0}},H_x=\frac{\overline{H_x}}{\overline{H_0}},H_y=\frac{\overline{H_y}}{\overline{H_0}} \tag{8.42}$$

where in Eq. (8.42) $\overline{H_0}=\overline{H}(\overline{a},0)=\frac{\gamma}{2\pi|b|}$ and $L=r_{out}-r_{in}=r_{in}$. Using the dimensionless parameters, the equations now become:

$$\frac{\partial U}{\partial X}+\frac{\partial V}{\partial Y}=0 \tag{8.43}$$

$$U\frac{\partial U}{\partial X}+V\frac{\partial U}{\partial Y}=-\frac{\partial P}{\partial X}+\Pr\left[\frac{\mu_{nf}/\mu_f}{\rho_{nf}/\rho_f}\right]\left(\frac{\partial^2 U}{\partial X^2}+\frac{\partial^2 U}{\partial Y^2}\right)$$
$$+Mn_F\Pr\left(\frac{\rho_f}{\rho_{nf}}\right)(\varepsilon_2-\varepsilon_1-\Theta)H\frac{\partial H}{\partial X}-Ha^2\Pr\left[\frac{\sigma_{nf}/\sigma_f}{\rho_{nf}/\rho_f}\right]\left(H_y^2U-H_xH_yV\right) \tag{8.44}$$

$$U\frac{\partial V}{\partial X}+V\frac{\partial V}{\partial Y}=-\frac{\partial P}{\partial X}+\Pr\left[\frac{\mu_{nf}/\mu_f}{\rho_{nf}/\rho_f}\right]\left(\frac{\partial^2 V}{\partial X^2}+\frac{\partial^2 V}{\partial Y^2}\right)$$
$$+Mn_F\Pr\left(\frac{\rho_f}{\rho_{nf}}\right)(\varepsilon_2-\varepsilon_1-\Theta)H\frac{\partial H}{\partial Y}-Ha^2\Pr\left[\frac{\sigma_{nf}/\sigma_f}{\rho_{nf}/\rho_f}\right]\left(H_x^2V-H_xH_yU\right)+Ra\Pr\left[\frac{\beta_{nf}}{\beta_f}\right]\Theta \tag{8.45}$$

$$\frac{\partial\Psi}{\partial Y}\frac{\partial\Theta}{\partial X}-\frac{\partial\Psi}{\partial X}\frac{\partial\Theta}{\partial Y}=\left[\frac{\frac{k_{nf}}{k_f}}{\frac{(\rho C_P)_{nf}}{(\rho C_P)_f}}\right]\left(\frac{\partial^2\Theta}{\partial X^2}\right)+Ha^2\,Ec\left[\frac{\frac{\sigma_{nf}}{\sigma_f}}{\frac{(\rho C_P)_{nf}}{(\rho C_P)_f}}\right]\left\{U\,H_y-V\,H_x\right\}^2$$
$$+Mn_FEc\frac{(\rho C_P)_f}{(\rho C_P)_{nf}}\left\{U\frac{\partial H}{\partial X}+V\frac{\partial H}{\partial Y}\right\}H(\varepsilon_1+\Theta)$$
$$+Ec\left[\frac{\frac{\mu_{nf}}{\mu_f}}{\frac{(\rho C_P)_{nf}}{(\rho C_P)_f}}\right]\left\{2\left(\frac{\partial U}{\partial X}\right)^2+2\left(\frac{\partial V}{\partial X}\right)^2+\left(\frac{\partial U}{\partial X}+\frac{\partial V}{\partial Y}\right)^2\right\}+\left(\frac{k_{nf}}{k_f}\frac{(\rho C_p)_f}{(\rho C_p)_{nf}}+Nr\left(\frac{k_{nf}}{k_f}\right)^{-1}\right)\frac{\partial^2\Theta}{\partial Y^2} \tag{8.46}$$

where $Ra=g\beta_fL^4q''/(k_f\,\alpha_f\nu_f)$, $\Pr=\upsilon_f/\alpha_f$, $Ha=LB_x\sqrt{\sigma_f/\mu_f}$, $\varepsilon_1=T_1/(q''L/k_f)$, $Ec=\left(\mu_f\alpha_f\right)/\left[(\rho C_P)_f(q''L/k_f)\,L^2\right]$, $Mn_F=\mu_0H_0^2K'(q''L/k_f)L^2/\left(\mu_f\alpha_f\right)$, and $Nr=16\sigma_eT_c^3/(3\beta_Rk_f)$ are the Rayleigh number, Prandtl number, Hartmann number arising from MHD, temperature number, Eckert number, Magnetic number arising from FHD the for the base fluid, and Radiation parameter, respectively. The thermophysical properties of the nanofluid are given in Table 8.1.

The stream function and vorticity are defined as:

$$u=\frac{\partial\psi}{\partial y},v=-\frac{\partial\psi}{\partial x},\omega=\frac{\partial v}{\partial x}-\frac{\partial u}{\partial y}$$
$$\Omega=\frac{\omega L^2}{\alpha_f},\Psi=\frac{\psi}{\alpha_f} \tag{8.47}$$

The stream function satisfies the continuity Eq. (8.43). The vorticity equation is obtained by eliminating the pressure between the two momentum equations, i.e., by taking the y-derivative of Eq. (8.44) and subtracting from it the x-derivative of Eq. (8.45).

$$\begin{aligned} &\Theta = 1.0 \quad \text{on the inner circular boundary} \\ &\Theta = 0.0 \quad \text{on the outer circular boundary} \\ &\Psi = 0.0 \quad \text{on all solid boundaries} \end{aligned} \tag{8.48}$$

The values of vorticity on the boundary of the enclosure can be obtained using the stream function formulation and the known velocity conditions during the iterative solution procedure.

The local Nusselt number of the nanofluid along the inner wall can be expressed as:

$$Nu_{loc} = \left(\frac{k_{nf}}{k_f}\right)\frac{1}{\theta}\bigg|_{r=r_{in}} \tag{8.49}$$

where r is the radial direction. The average Nusselt number on the hot circular wall is evaluated as:

$$Nu_{ave} = \frac{1}{\gamma}\int_0^{\gamma} Nu_{loc}(\zeta)\, d\zeta \tag{8.50}$$

8.3.3 Effects of Active Parameters

Ferrofluid (Fe_3O_4-water) flow and heat transfer in a semi-annulus enclosure in presence of an external magnetic field is investigated. CVFEM is used to solve the governing equations. The mathematical model used for the formulation of the problem is consistent with the principles of FHD and MHD. Calculations are made for various values of volume fraction of nanoparticles ($\phi = 0\%$ and 4%), Rayleigh numbers ($Ra = 10^3$, 10^4 and 10^5), Magnetic number arising from FHD ($Mn_F = 0, 20, 60$ and 100), Radiation parameter ($Nr = 0.0$ and 0.027), and Hartmann number arising from MHD ($Ha = 0, 2, 6$ and 10). In all calculations, the Prandtl number (Pr), temperature number (ε_1), and Eckert number (Ec) are set to 6.8, 0.0, and 1×10^{-5}, respectively.

Fig. 8.10 shows the comparison of the isotherms and streamlines between nanofluid and pure fluid. The velocity components of nanofluid are increased because of an increase in the energy transport in the fluid with the increasing of volume fraction. Thus, the absolute values of stream functions indicate that the strength of flow increases with increasing the volume fraction of nanofluid. Also it can be found that thermal boundary layer thickness increases with the increase of volume fraction of nanofluid. Fig. 8.11 shows the effect of Hartmann number on isotherm and streamlines. As Hartmann number increases the absolute value of maximum stream function decreases. Also this figure shows that streamlines become more disturbed near the magnetic source. Isotherm and streamline contours for different values of Rayleigh number, Hartmann number, Radiation parameter, and Magnetic number are shown in Figs. 8.12 and 8.13.

When $Ra = 10^3$, the heat transfer in the enclosure is mainly dominated by the conduction mode. At $Mn_F = 0, Ha = 0$ the streamlines show one rotating eddy. In the presence of a magnetic field, the vortex center moves upward. As the Rayleigh number increases, the role of convection in heat transfer becomes more significant and consequently the thermal boundary layer thickness on the surface of the inner wall becomes thinner. In addition, a plume starts to appear on the top of the inner circular wall at $\zeta = 45^\circ$. The isotherms and streamlines become denser near the magnetic source $(\zeta = 22.5^\circ)$. Also Fig. 8.13 shows that absolute value of maximum stream function increases with augment of radiation parameter. The center of the vortex move downward left by increasing radiation parameter.

Fig. 8.14 depicts that the effects of Magnetic number, Hartmann number, Radiation parameter, and Rayleigh number on local Nusselt number. Local Nusselt number increases with the increase of Rayleigh number while it decreases with the increase of Hartmann number and Radiation parameter. Local minimum point at $\zeta = 22.5^\circ$ in these profiles is related to the existence of magnetic source at this region. Effects of volume fraction of nanofluid, Magnetic number, Hartmann number, Radiation parameter, and Rayleigh number on average Nusselt number are shown in Fig. 8.15 and Table 8.4. Average Nusselt number increases with increase of volume fraction of nanofluid, Magnetic number, and Rayleigh number while it decreases with increase of Hartmann number and Radiation parameter.

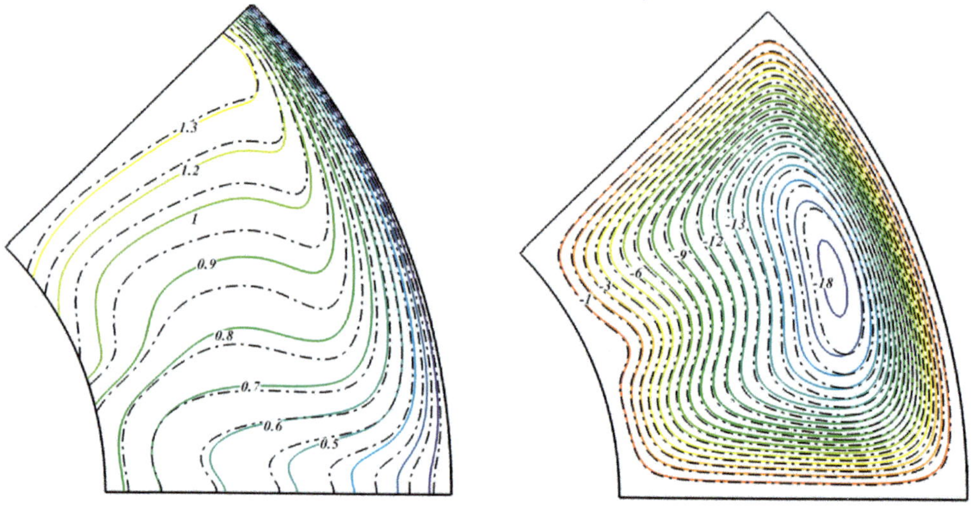

FIGURE 8.10 Comparison of the isotherms and streamlines between nanofluid ($\phi = 0.04$) (—) and pure fluid ($\phi = 0$) (– · – · –) when $Nr = 0.027, Ra = 10^5, Mn_F = 100, Ha = 10$, and $\mathrm{Pr} = 6.8$.

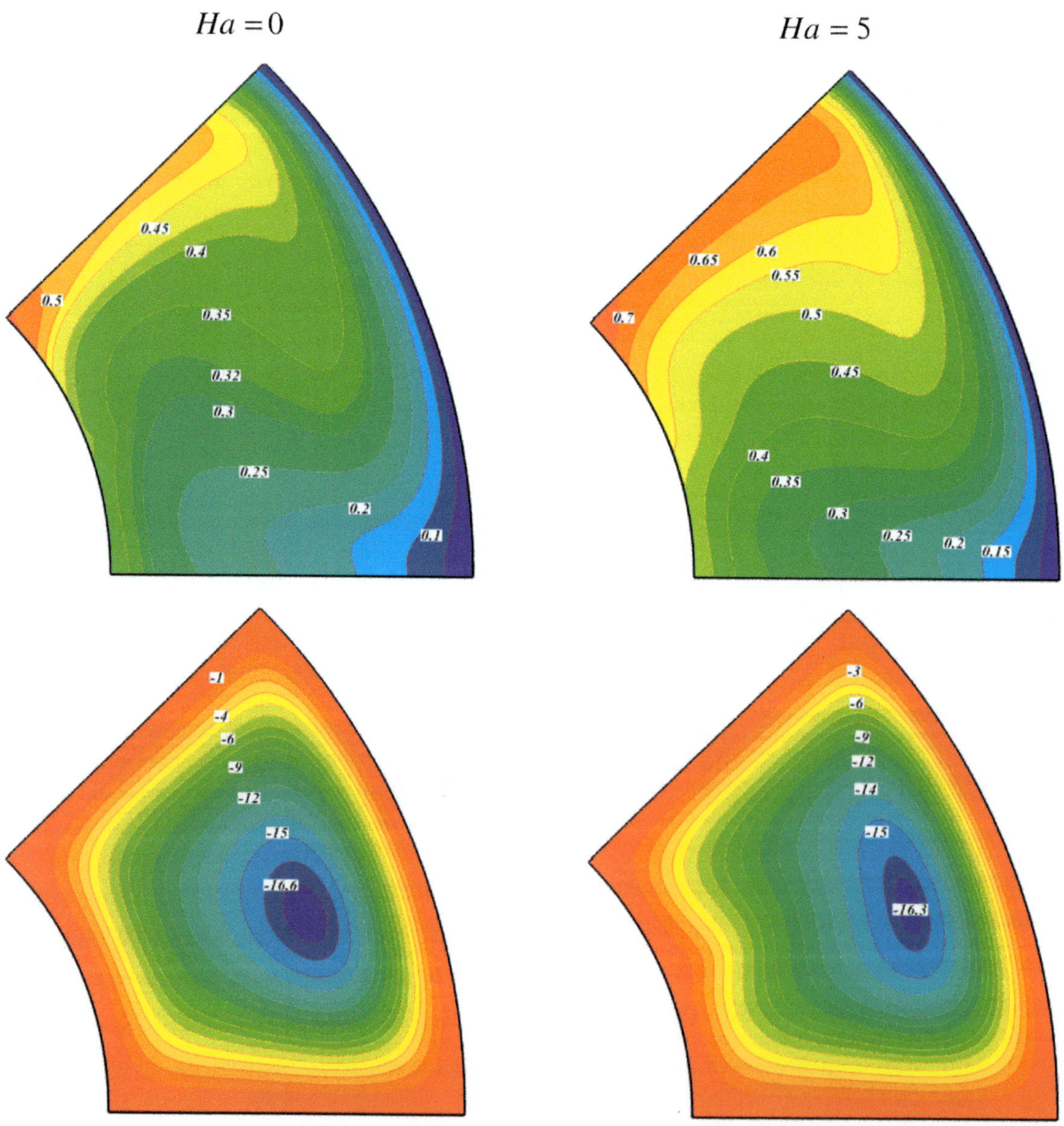

FIGURE 8.11 Effect of Hartmann number on isotherms and streamlines when $Nr = 0.027, Ra = 10^5, Mn = 100, \phi = 0.04$.

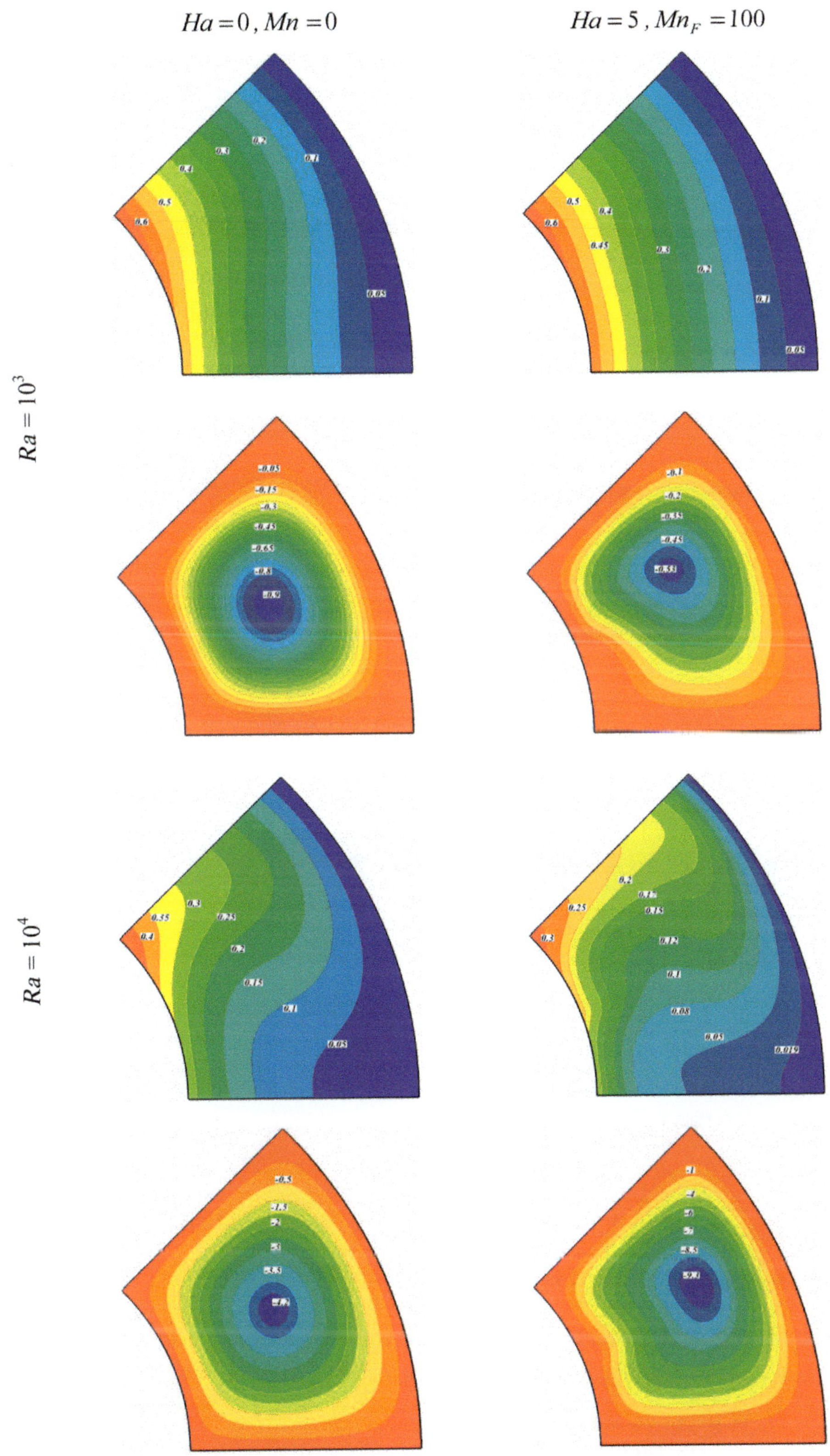

FIGURE 8.12 Isotherms (up) and streamlines (down) contours for different values of Rayleigh number, Hartmann number, and Magnetic number when $Nr=0.0$, $\phi=0.04$, $\Pr=6.8$.

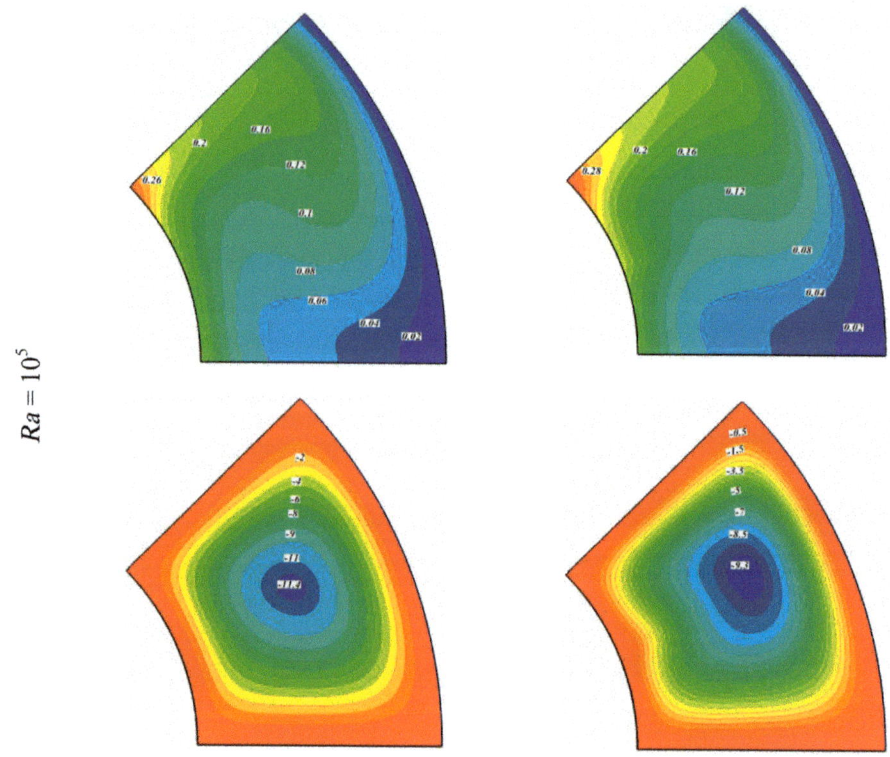

FIGURE 8.12 (Continued).

8.4 INFLUENCE OF SPATIALLY VARIABLE MAGNETIC FIELD ON FERROFLUID FLOW AND HEAT TRANSFER CONSIDERING CONSTANT HEAT FLUX BOUNDARY CONDITION

8.4.1 Problem Definition

The physical model along with the important geometrical parameters and the mesh of the enclosure used in the present CVFEM program are shown in Fig. 8.16. The two horizontal walls are thermally isolated; the outer cylinder is maintained at constant cold temperature T_c, whereas the inner circular wall is under constant heat flux. To assess the shape of inner circular and outer rectangular boundary which consists of the right and top walls, a super elliptic function can be used as follows;

$$\left(\frac{X}{a}\right)^{2n} + \left(\frac{Y}{b}\right)^{2n} = 1 \tag{8.51}$$

When $a = b$ and $n = 1$ the geometry becomes a circle. As n increases from 1 the geometry would approach a rectangle for $a \neq b$ and square for $a = b$. In this study, $r_{in}/r_{out} = 0.6$.

For the expression of the magnetic field strength it can be considered that two magnetic sources at the points $(\bar{a}_1, \bar{b}_1)$ and $(\bar{a}_2, \bar{b}_2)$. The components of the magnetic field intensity $(\overline{H_x}, \overline{H_y})$ and the magnetic field strength $(\overline{H})$ can be considered as:

$$\overline{H_x} = \frac{\gamma}{2\pi}\frac{1}{(x-\overline{a_1})^2 + (y-\overline{b_1})^2}(y - \overline{b_1}) - \frac{\gamma}{2\pi}\frac{1}{(x-\overline{a_2})^2 + (y-\overline{b_2})^2}(y - \overline{b_2}) \tag{8.52}$$

$$\overline{H_y} = -\frac{\gamma}{2\pi}\frac{1}{(x-\overline{a_1})^2 + (y-\overline{b_1})^2}(x - \overline{a_1}) - \frac{\gamma}{2\pi}\frac{1}{(x-\overline{a_2})^2 + (y-\overline{b_2})^2}(x - \overline{a_2}) \tag{8.53}$$

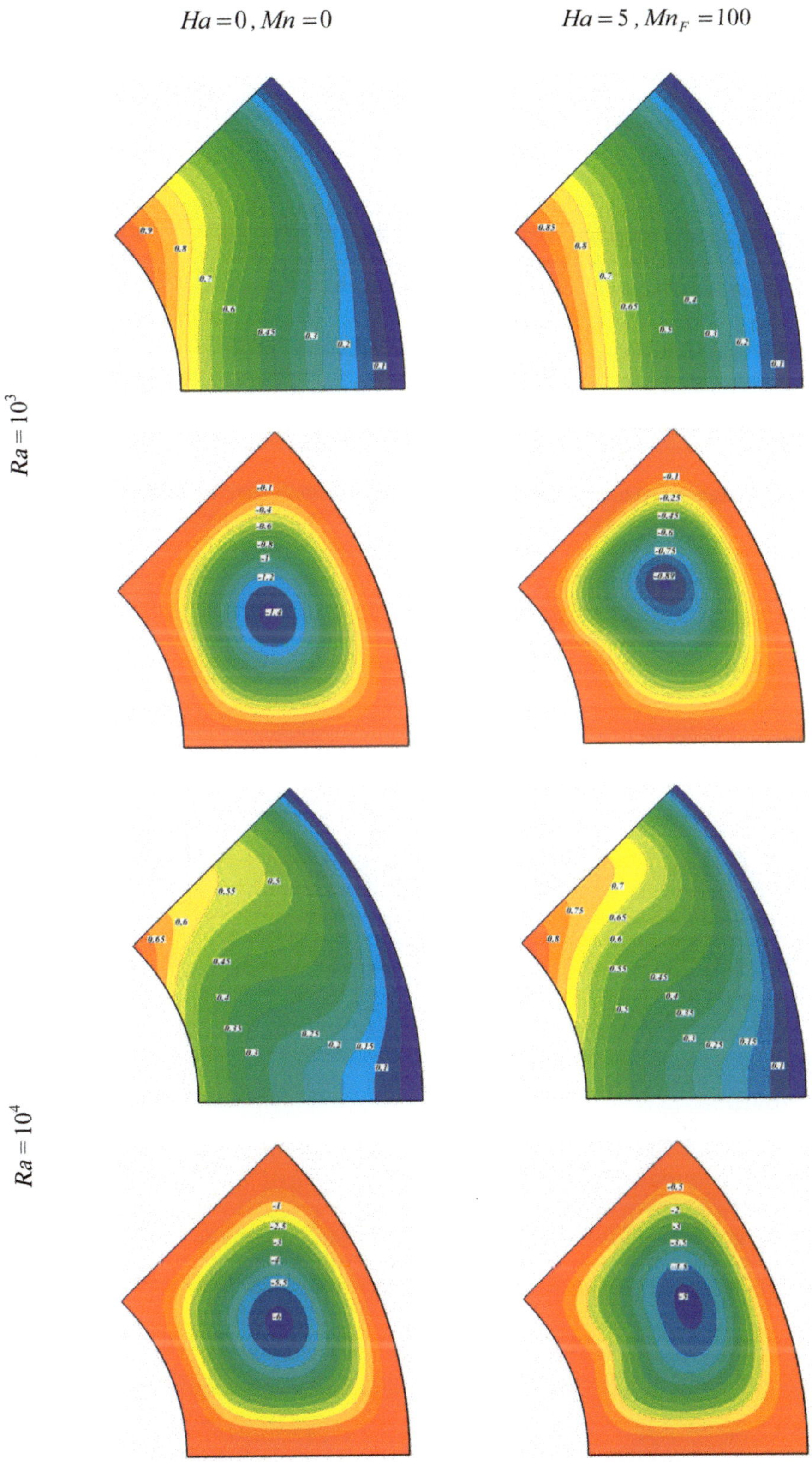

FIGURE 8.13 Isotherms (up) and streamlines (down) contours for different values of Rayleigh number, Hartmann number, and Magnetic number when $Nr = 0.027$, $\phi = 0.04$, $\Pr = 6.8$.

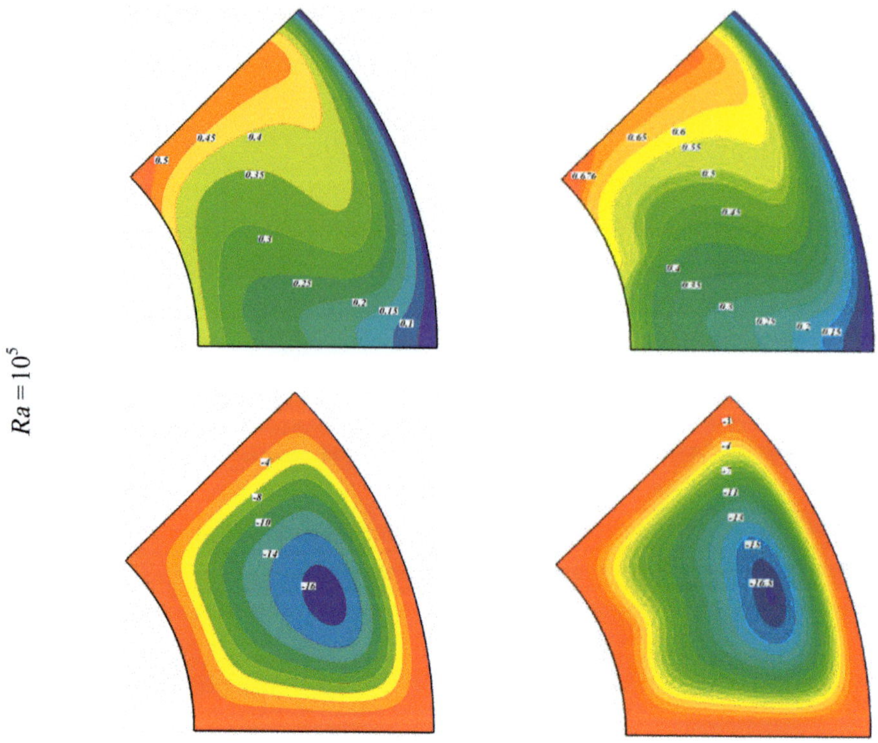

FIGURE 8.13 (Continued).

$$\overline{H} = \sqrt{\overline{H}_x^2 + \overline{H}_y^2} \tag{8.54}$$

where γ the magnetic field strength at the source (of the wire). The contours of the magnetic field strength are shown in Fig. 8.17. In this study magnetic source is located at $\left(\overline{a}_1 = 1.05\, cols,\ \overline{b}_1 = 0.5\, rows\right)$ and $\left(\overline{a}_2 = -0.05\, cols, \overline{b}_2 = 0.5\, rows\right)$.

8.4.2 Governing Equation

The flow is two-dimensional, laminar, and incompressible. The magnetic Reynolds number is assumed to be small so that the induced magnetic field can be neglected compared to the applied magnetic field. The flow is considered to be steady, two-dimensional, and laminar. Using the Boussinesq approximation, the governing equations of heat transfer and fluid flow for nanofluid can be obtained as follows:

$$\frac{\partial u}{\partial x} + \frac{\partial v}{\partial y} = 0 \tag{8.55}$$

$$\rho_{nf}\left(u\frac{\partial u}{\partial x} + v\frac{\partial u}{\partial y}\right) = -\frac{\partial P}{\partial x} + \mu_{nf}\left(\frac{\partial^2 u}{\partial x^2} + \frac{\partial^2 u}{\partial y^2}\right) + \mu_0 M\frac{\partial \overline{H}}{\partial x} - \sigma_{nf}B_y^2 u + \sigma_{nf}B_x B_y v \tag{8.56}$$

$$\rho_{nf}\left(u\frac{\partial v}{\partial x} + v\frac{\partial v}{\partial y}\right) = -\frac{\partial P}{\partial y} + \mu_{nf}\left(\frac{\partial^2 v}{\partial x^2} + \frac{\partial^2 v}{\partial y^2}\right) + \mu_0 M\frac{\partial \overline{H}}{\partial y} - \sigma_{nf}B_x^2 v + \sigma_{nf}B_x B_y u + \rho_{nf}\beta_{nf}g(T - T_c) \tag{8.57}$$

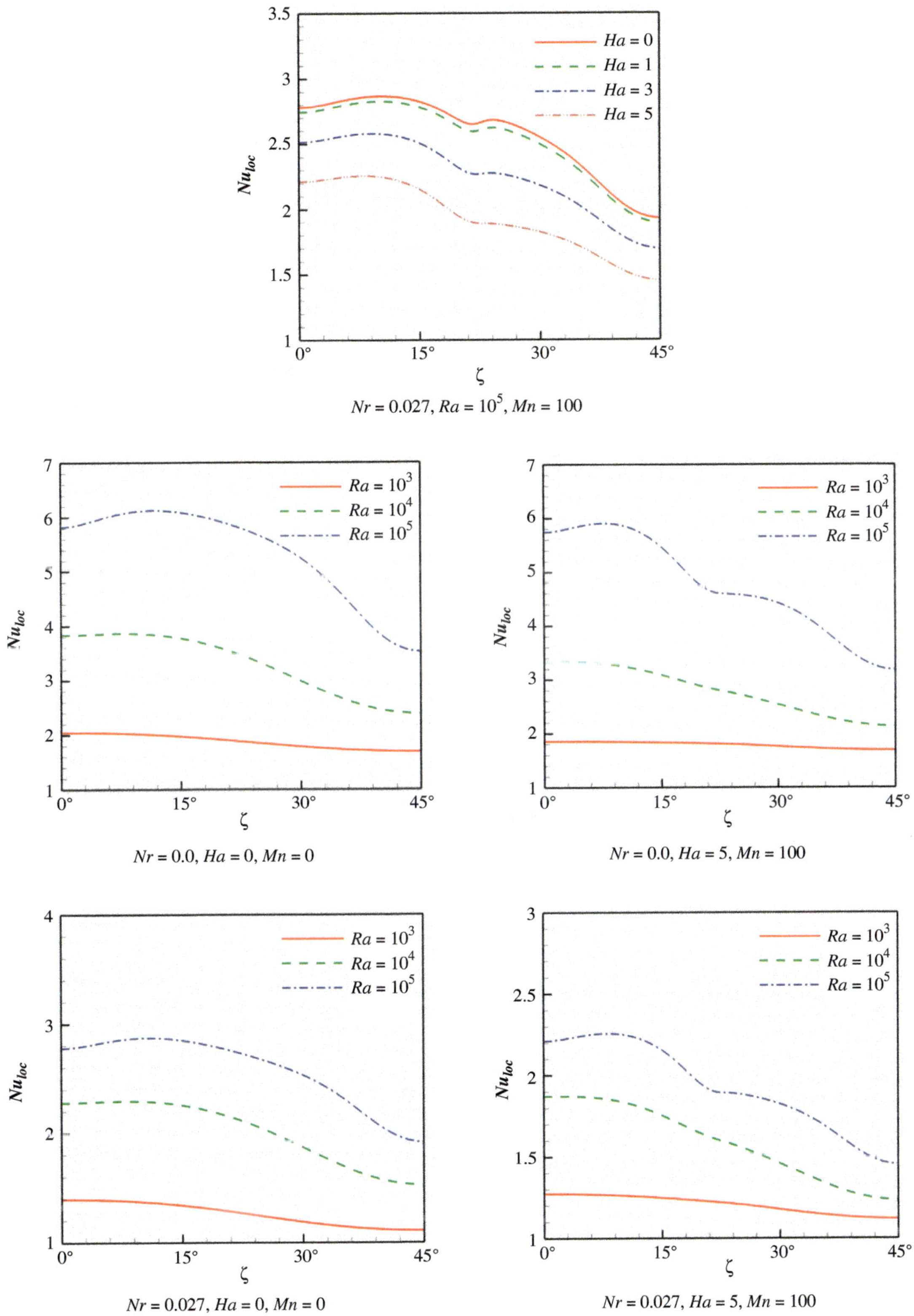

FIGURE 8.14 Effects of Magnetic number, Hartmann number, Radiation parameter, and Rayleigh number on local Nusselt number Nu_{loc} along hot wall when $\phi = 0.04$.

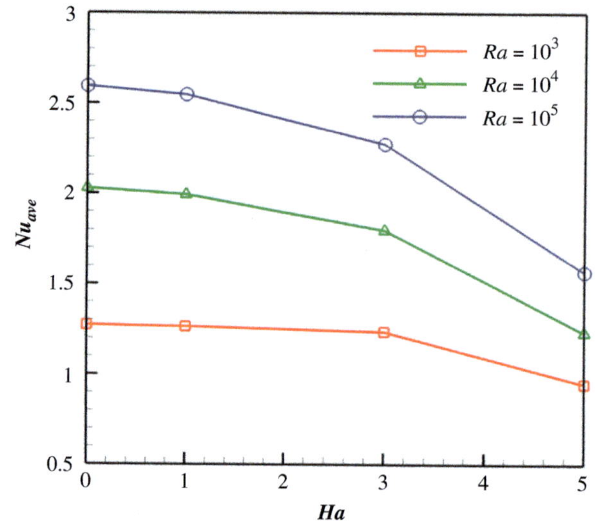

FIGURE 8.15 Effects of Hartmann number and Rayleigh number on average Nusselt number Nu_{ave} along hot wall when $Nr = 0.027, \phi = 0.04$.

TABLE 8.4 Effects of Hartmann Number, Radiation Parameter, Nanoparticle Volume Fraction, and Magnetic Number on Average Nusselt Number

		$Mn_F = 0, Ha = 0$		$Mn_F = 100, Ha = 5$		$Mn_F = 100, Ha = 0$		$Mn_F = 0, Ha = 5$	
		Nr		*Nr*		*Nr*		*Nr*	
Ra	ϕ	**0**	**0.027**	**0**	**0.027**	**0**	**0.027**	**0**	**0.027**
10^3	0	1.611611	0.995565	1.519592	0.928174	1.711999	1.010865	1.510564	0.953924
10^3	0.04	1.890241	1.270623	1.801949	1.215039	1.972767	1.285426	1.796937	1.234266
10^4	0	2.906977	1.609061	2.477273	1.244099	2.934773	1.636442	2.459938	1.226934
10^4	0.04	3.312842	2.022216	2.802561	1.596926	3.34569	2.048861	2.784897	1.582475
10^5	0	4.799285	2.103907	4.316693	1.570677	4.775208	2.106432	4.331907	1.564888
10^5	0.04	5.372904	2.595023	4.789146	1.942158	5.345175	2.588536	4.807063	1.940414

$$(\rho C_p)_{nf}\left(u\frac{\partial T}{\partial x}+v\frac{\partial T}{\partial y}\right)=k_{nf}\left(\frac{\partial^2 T}{\partial x^2}+\frac{\partial^2 T}{\partial y^2}\right)+\sigma_{nf}(uB_y-vB_x)^2$$
$$-\mu_0 T\frac{\partial M}{\partial T}\left(u\frac{\partial \overline{H}}{\partial x}+v\frac{\partial \overline{H}}{\partial y}\right)+\mu_{nf}\left\{2\left(\frac{\partial u}{\partial x}\right)^2+2\left(\frac{\partial v}{\partial x}\right)^2+\left(\frac{\partial u}{\partial x}+\frac{\partial v}{\partial y}\right)^2\right\} \tag{8.58}$$

The terms $\mu_0 M\frac{\partial \overline{H}}{\partial x}$ and $\mu_0 M\frac{\partial \overline{H}}{\partial y}$ in (8.56) and (8.57), respectively, represent the components of magnetic force, per unit volume, and depend on the existence of the magnetic gradient on the corresponding x and y directions. These two terms are well known from FHD which is the so-called Kelvin force. The terms $-\sigma_{nf}B_y^2u+\sigma_{nf}B_xB_yv$ and $-\sigma_{nf}B_x^2v+\sigma_{nf}B_xB_yu$ appearing in (8.56) and (8.57), respectively, represent the Lorentz force per unit volume towards the x and y directions and arise due to the electrical conductivity of the fluid. These two terms are known in MHD. The principles of MHD and FHD are combined in the mathematical model and the abovementioned terms arise together in the governing Eqs. (8.56) and (8.57). The term $\mu_0 T\frac{\partial M}{\partial T}\left(u\frac{\partial \overline{H}}{\partial x}+v\frac{\partial \overline{H}}{\partial y}\right)$ in Eq. (8.58) represents the thermal power per unit volume due to the magneto caloric effect. Also the term $\sigma_{nf}(uB_y-vB_x)^2$ in

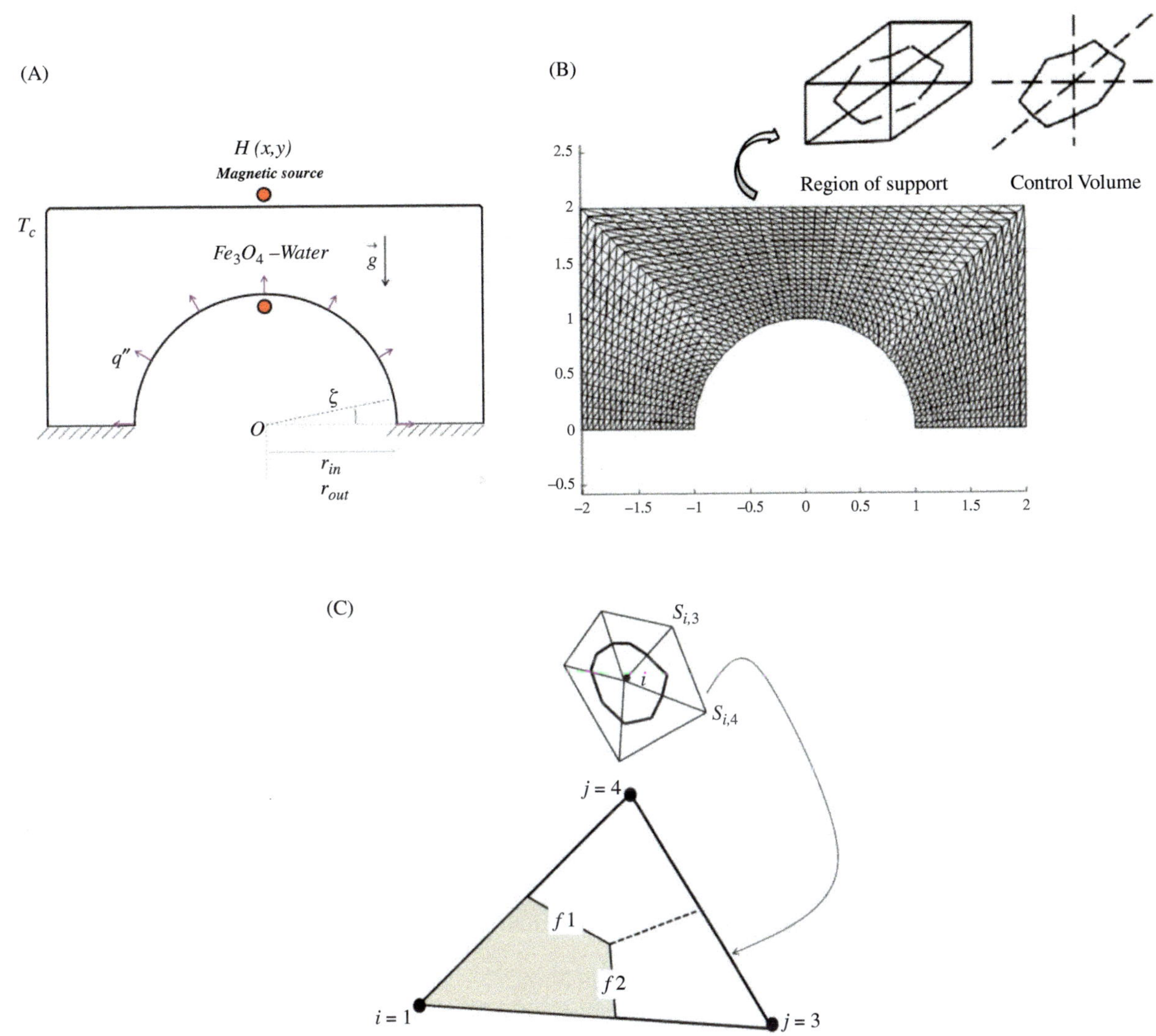

FIGURE 8.16 (A) Geometry and the boundary conditions; (B) the mesh of enclosure considered in this work; (C) a sample triangular element and its corresponding control volume.

(8.58) represents the Joule heating. For the variation of the magnetization M, with the magnetic field intensity $\overline{H}$ and temperature is considered:

$$M = K'\overline{H}(T_c' - T) \tag{8.59}$$

where K' is a constant and T_c' is the Curie temperature.

In the above equations, μ_0 is the magnetic permeability of vacuum $(4\pi \times 10^{-7} Tm/A)$, $\overline{H}$ is the magnetic field strength, $\overline{B}$ is the magnetic induction $(\overline{B} = \mu_0 \overline{H})$, and the bar above the quantities denotes that they are dimensional. The effective density (ρ_{nf}) and heat capacitance $(\rho C_p)_{nf}$ of the nanofluid are defined as:

$$\rho_{nf} = \rho_f(1-\phi) + \rho_s\phi \tag{8.60}$$

$$(\rho C_p)_{nf} = (\rho C_p)_f(1-\phi) + (\rho C_p)_s\phi \tag{8.61}$$

where ϕ is the solid volume fraction of nanoparticles. Thermal diffusivity of the nanofluid is

$$\alpha_{nf} = \frac{k_{nf}}{(\rho C_p)_{nf}} \tag{8.62}$$

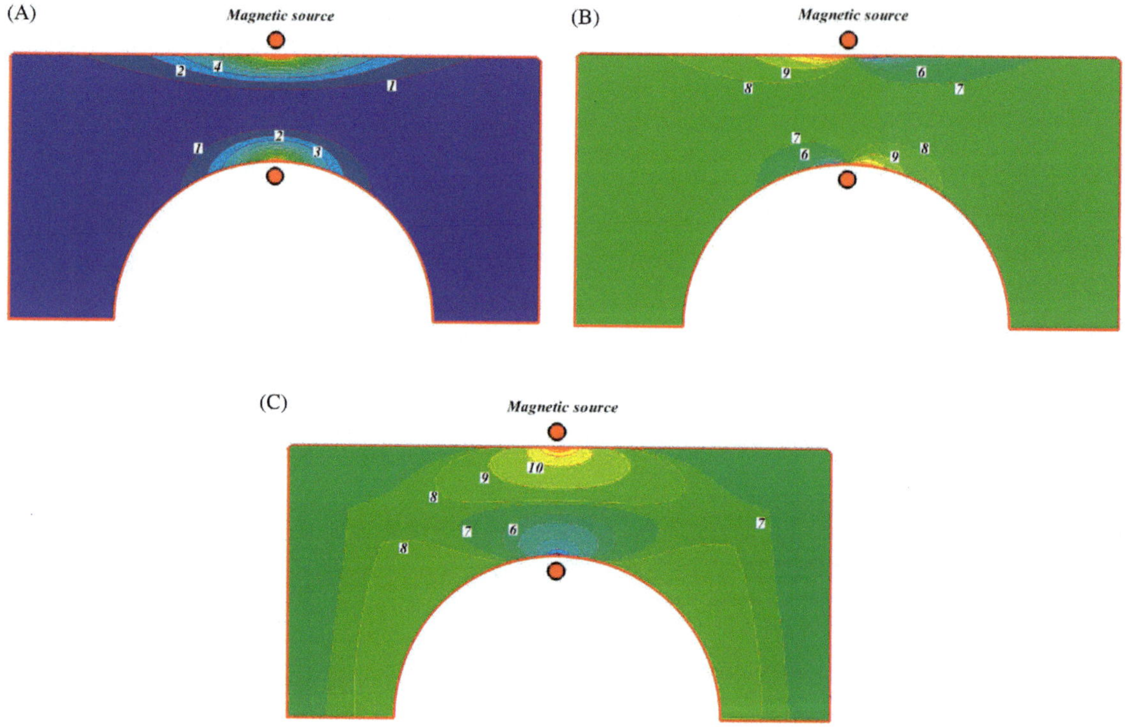

FIGURE 8.17 Contours of the (A) magnetic field strength H; (B) magnetic field intensity component in x direction Hx; (C) magnetic field intensity component in y direction Hy. (A) $H(x,y)$, (B) $Hx(x,y)$, (C) $Hy(x,y)$.

and the thermal expansion coefficient of the nanofluid can be determined as

$$\beta_{nf} = \beta_f(1-\phi) + \beta_s\phi \tag{8.63}$$

The dynamic viscosity of the nanofluid given by Brinkman is

$$\mu_{nf} = \frac{\mu_f}{(1-\phi)^{2.5}} \tag{8.64}$$

The effective thermal conductivity of the nanofluid can be approximated by the Maxwell–Garnetts (MG) model as:

$$\frac{k_{nf}}{k_f} = \frac{k_s + 2k_f - 2\phi(k_f - k_s)}{k_s + 2k_f + \phi(k_f - k_s)} \tag{8.65}$$

and the effective electrical conductivity of nanofluid was presented by Maxwell as below:

$$\frac{\sigma_{nf}}{\sigma_f} = 1 + \frac{3\left(\frac{\sigma_s}{\sigma_f} - 1\right)\phi}{\left(\frac{\sigma_s}{\sigma_f} + 2\right) - \left(\frac{\sigma_s}{\sigma_f} - 1\right)\phi} \tag{8.66}$$

The stream function and vorticity are defined as:

$$u = \frac{\partial\psi}{\partial y}, v = -\frac{\partial\psi}{\partial x}, \omega = \frac{\partial v}{\partial x} - \frac{\partial u}{\partial y} \tag{8.67}$$

The stream function satisfies the continuity Eq. (8.54). The vorticity equation is obtained by eliminating the pressure between the two momentum equations, i.e., by taking the y-derivative of Eq. (8.56) and subtracting from it the x-derivative of Eq. (8.55). By introducing the following nondimensional variables:

$$X=\frac{x}{L},\quad Y=\frac{y}{L},\quad \Omega=\frac{\omega L^2}{\alpha_f},\Psi=\frac{\psi}{\alpha_f},\Theta=\frac{T-T_c}{(q''L/k_f)},U=\frac{uL}{\alpha_f},V=\frac{vL}{\alpha_f},$$
$$H=\frac{\overline{H}}{\overline{H_0}},H_x=\frac{\overline{H_x}}{\overline{H_0}},H_y=\frac{\overline{H_y}}{\overline{H_0}} \tag{8.68}$$

where in Eq. (8.68) $\overline{H_0}=\overline{H}(\overline{a},0)=\frac{\gamma}{2\pi|b|}$ and $L=r_{out}-r_{in}=r_{in}$. Using the dimensionless parameters, the equations now become:

$$\begin{aligned}\frac{\partial\Psi}{\partial Y}\frac{\partial\Omega}{\partial X}-\frac{\partial\Psi}{\partial X}\frac{\partial\Omega}{\partial Y}&=\Pr\left[\frac{\mu_{nf}/\mu_f}{\rho_{nf}/\rho_f}\right]\left(\frac{\partial^2\Omega}{\partial X^2}+\frac{\partial^2\Omega}{\partial Y^2}\right)+Ra\Pr\left[\frac{\beta_{nf}}{\beta_f}\right]\left(\frac{\partial\Theta}{\partial X}\right)\\&+Mn_F\Pr\left(\frac{\rho_f}{\rho_{nf}}\right)\left\{\frac{\partial H}{\partial X}\frac{\partial\Theta}{\partial Y}-\frac{\partial H}{\partial Y}\frac{\partial\Theta}{\partial X}\right\}-Ha^2\Pr\left[\frac{\sigma_{nf}/\sigma_f}{\rho_{nf}/\rho_f}\right]\\&\times\left\{\frac{\partial V}{\partial X}H_x^2+V\left(2H_x\frac{\partial H_x}{\partial X}\right)-\frac{\partial U}{\partial X}H_xH_y-U\frac{\partial H_x}{\partial X}H_y-U\frac{\partial H_y}{\partial X}H_x\right.\\&\left.-\frac{\partial U}{\partial Y}H_y^2-U\left(2H_y\frac{\partial H_y}{\partial Y}\right)+\frac{\partial V}{\partial Y}H_xH_y+V\frac{\partial H_x}{\partial Y}H_y+V\frac{\partial H_y}{\partial Y}H_x\right\}\end{aligned} \tag{8.69}$$

$$\begin{aligned}\frac{\partial\Psi}{\partial Y}\frac{\partial\Theta}{\partial X}-\frac{\partial\Psi}{\partial X}\frac{\partial\Theta}{\partial Y}&=\left[\frac{\frac{k_{nf}}{k_f}}{\frac{(\rho C_P)_{nf}}{(\rho C_P)_f}}\right]\left(\frac{\partial^2\Theta}{\partial X^2}+\frac{\partial^2\Theta}{\partial Y^2}\right)+Ha^2\,Ec\left[\frac{\frac{\sigma_{nf}}{\sigma_f}}{\frac{(\rho C_P)_{nf}}{(\rho C_P)_f}}\right]\{U\,H_y-V\,H_x\}^2\\&+Mn_FEc\frac{(\rho C_P)_f}{(\rho C_P)_{nf}}\left\{U\frac{\partial H}{\partial X}+V\frac{\partial H}{\partial Y}\right\}H(\varepsilon_1+\Theta)+Ec\left[\frac{\frac{\mu_{nf}}{\mu_f}}{\frac{(\rho C_P)_{nf}}{(\rho C_P)_f}}\right]\left\{2\left(\frac{\partial U}{\partial X}\right)^2+2\left(\frac{\partial V}{\partial X}\right)^2+\left(\frac{\partial U}{\partial X}+\frac{\partial V}{\partial Y}\right)^2\right\}\end{aligned} \tag{8.70}$$

$$\frac{\partial^2\Psi}{\partial X^2}+\frac{\partial^2\Psi}{\partial Y^2}=-\Omega \tag{8.71}$$

where $Ra_f=g\beta_fL^4q''/(k_f\,\alpha_f\nu_f)$, $\Pr_f=\upsilon_f/\alpha_f$, $Ha=LB_x\sqrt{\sigma_f/\mu_f}$, $\varepsilon_1=T_1/\Delta T$, $\Delta T=(q''L/k_f)$, $Ec=(\mu_f\alpha_f)/\left[(\rho C_P)_f\Delta T\,L^2\right]$ and $Mn_F=\mu_0H_0^2K'\Delta TL^2/(\mu_f\alpha_f)$ are the Rayleigh number, Prandtl number, Hartmann number arising from MHD, temperature number, Eckert number, and Magnetic number arising from FHD the for the base fluid, respectively. The thermophysical properties of the nanofluid are given in Table 8.1. The boundary conditions as shown in Fig. 8.16 are:

$$\begin{aligned}&\frac{\partial\Theta}{\partial n}=-1 &&\text{on the inner circular boundary}\\&\Theta=0.0 &&\text{on the outer circular boundary}\\&\Psi=0.0 &&\text{on all solid boundaries}\end{aligned} \tag{8.72}$$

The values of vorticity on the boundary of the enclosure can be obtained using the stream function formulation and the known velocity conditions during the iterative solution procedure.

The local Nusselt number of the nanofluid along the inner wall can be expressed as:

$$Nu_{local} = \left(\frac{k_{nf}}{k_f}\right)\frac{1}{\theta}\bigg|_{inner\ wall} \tag{8.73}$$

where r is the radial direction. The average Nusselt number on the hot circular wall is evaluated as:

$$Nu_{ave} = \frac{1}{\gamma}\int_0^{\gamma} Nu_{loc}(\zeta)\, d\zeta \tag{8.74}$$

To estimate the enhancement of heat transfer between the case of $\phi = 0.04$ and the pure fluid (base fluid) case, the heat transfer enhancement is defined as:

$$En = \frac{Nu\,(\phi = 0.04) - Nu\,(basefluid)}{Nu\,(basefluid)} \times 100 \tag{8.75}$$

8.4.3 Effects of Active Parameters

Fe_3O_4-water flow and heat transfer in the presence of variable external magnetic field is studied numerically using the Control Volume-based Finite Element Method (CVFEM). Constant heat flux boundary condition is considered for the inner wall. Calculations are made for various values of volume fraction of nanoparticles ($\phi = 0\%$ and 4%), Rayleigh numbers ($Ra = 10^3$, 10^4, and 10^5), Magnetic number arising from FHD ($Mn_F = 0, 20, 60$, and 100) and Hartmann number arising from MHD ($Ha = 0, 5$ and 10). In all calculations, the Prandtl number (Pr), temperature number (ε_1), and Eckert number (Ec) are set to 6.8, 0.0, and 10^{-6}, respectively.

Fig. 8.18 shows the comparison of the isotherms and streamlines between nanofluid and pure fluid. The velocity components of nanofluid are increased because of an increase in the energy transport in the fluid with the

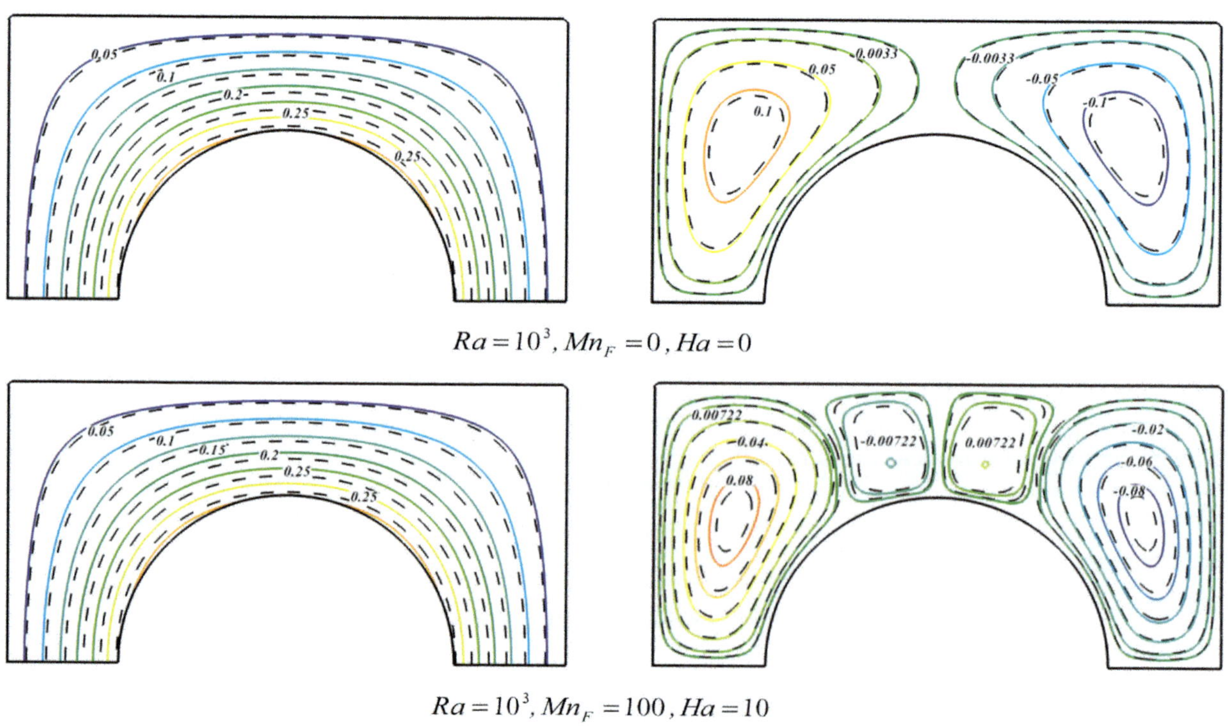

FIGURE 8.18 Comparison of the isotherms and streamlines between nanofluid ($\phi = 0.04$) (---) and pure fluid ($\phi = 0$) (—) when Pr = 6.8.

increasing of volume fraction. Thus, the absolute values of stream functions indicate that the strength of flow increases with increasing the volume fraction of nanofluid. Also it can be seen that thermal boundary layer thickness around the hot wall decreases with the increase of nanoparticle volume fraction.

Fig. 8.19 depicts the effects of Rayleigh number, Hartmann number, and Magnetic number on isotherms and streamlines. Absolute value of stream function increases with increase of Rayleigh number while it decreases with increase of Hartmann number and Magnetic number. When $Ra = 10^3$, the heat transfer in the enclosure is mainly dominated by the conduction mode. At $Mn_F = 0, Ha = 0$ the streamlines show one rotating eddy. When Magnetic parameter increases, another vortex appears near the centerline which rotates in the reverse direction in comparison to the primary vortex. Lorentz force has no significant effect on isotherm and streamline in this Rayleigh number. As the Rayleigh number increases up to 10^5, the role of convection in heat transfer becomes

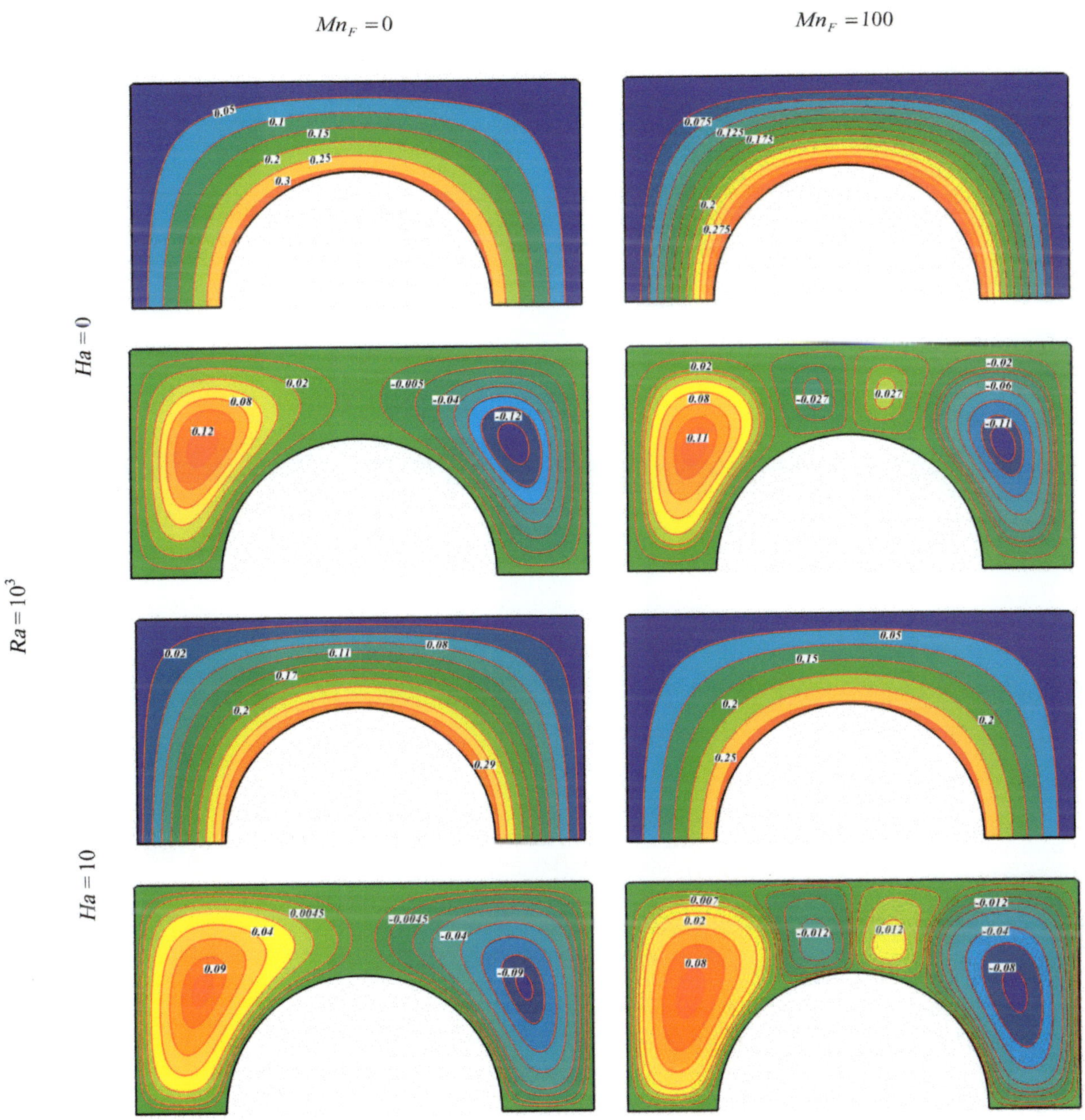

FIGURE 8.19 Isotherm (left) and streamline (right) contours for different values of Rayleigh number, Hartmann number, and Magnetic number when $Pr = 6.8, \phi = 0.04$.

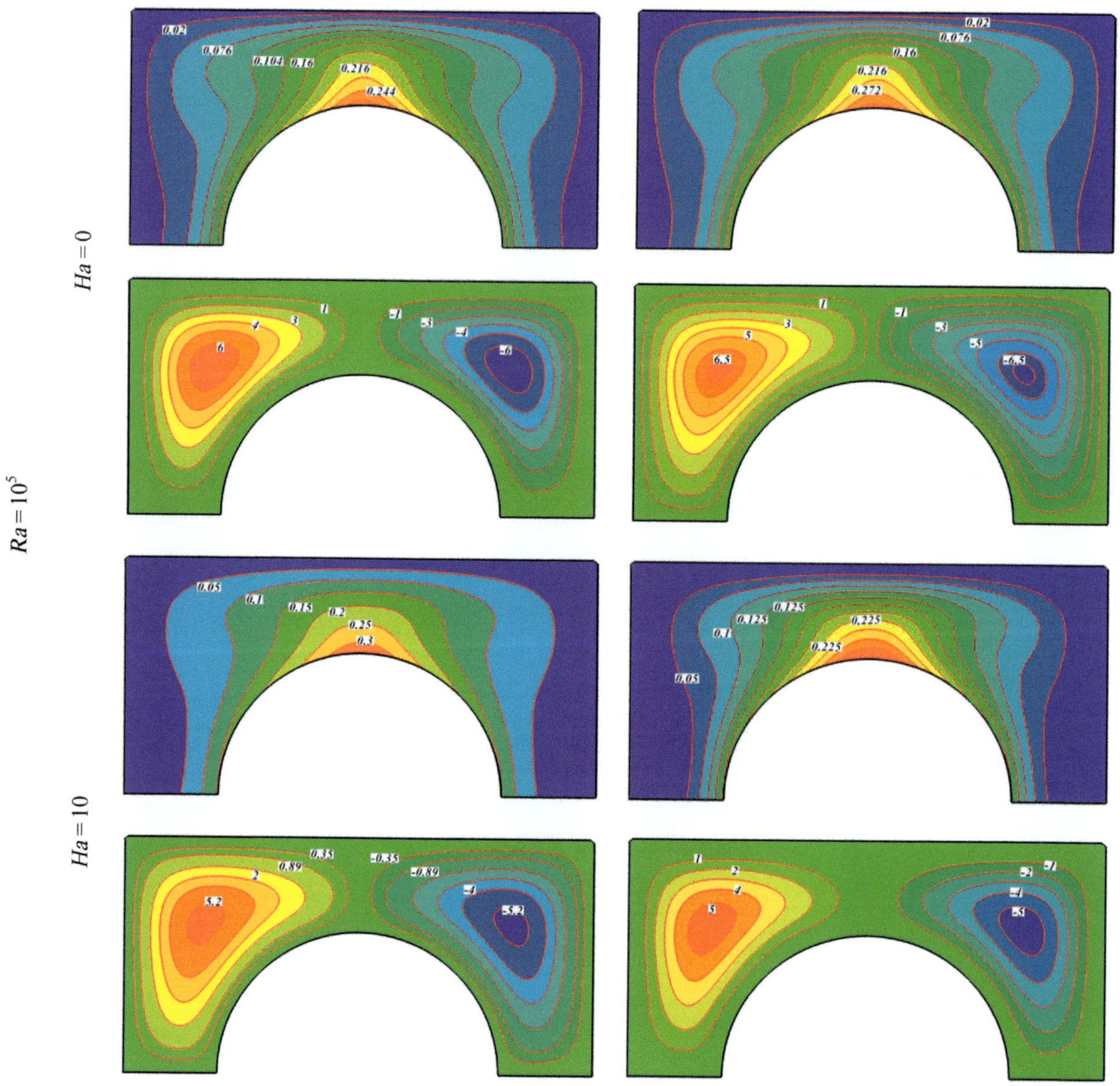

FIGURE 8.19 (Continued).

more significant and consequently the thermal boundary layer on the surface of the inner wall becomes thinner. In addition, a plume starts to appear on the top of the inner circular wall. The effect of magnetic field on hydrothermal behavior is more pronounced for low Rayleigh number.

Fig. 8.20 shows the effects of Magnetic parameter, Hartmann number, and Rayleigh number on average Nusselt number along inner wall. Nusselt number increases with increase of Rayleigh number and Magnetic parameter while it decreases with increase of Hartmann number. Effects of Magnetic number, Rayleigh number, and Hartmann number on heat transfer enhancement are shown in Fig. 8.21. Heat transfer enhancement is an increasing function of Hartmann number while it is a decreasing function of Magnetic number and Rayleigh number. The effect of nanoparticles is more pronounced at low Rayleigh number than at high Rayleigh number because of the greater enhancement rate. This observation can be explained by noting that at low Rayleigh number the heat transfer is dominated by conduction.

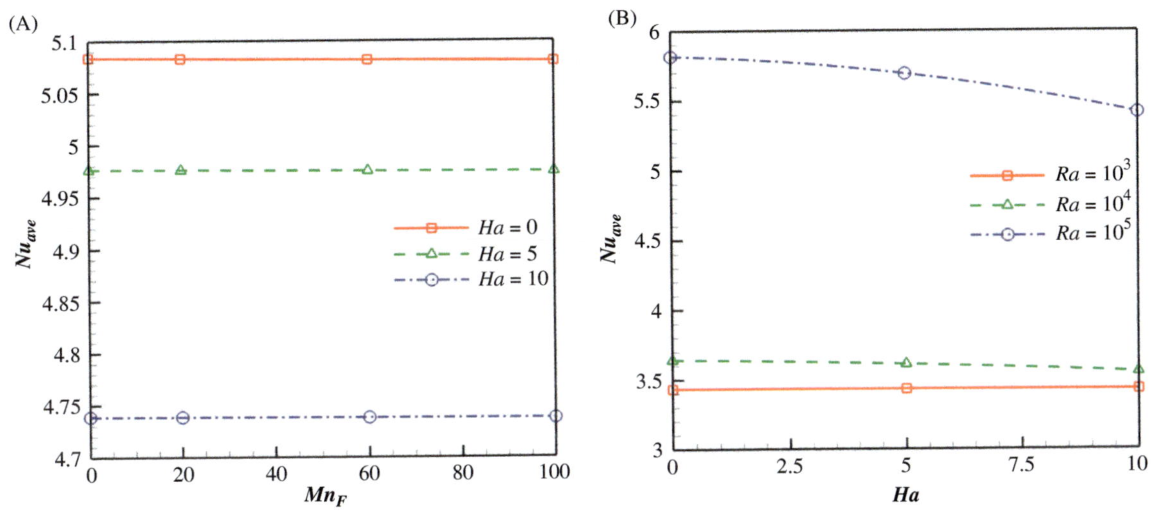

FIGURE 8.20 Effects of Magnetic parameter, Hartmann number, and Rayleigh number on average Nusselt number Nu_{ave} along hot wall when $Pr = 6.8$, $\phi = 0.04$. (A) $Ra = 10^5$, (B) $Mn_F = 100$.

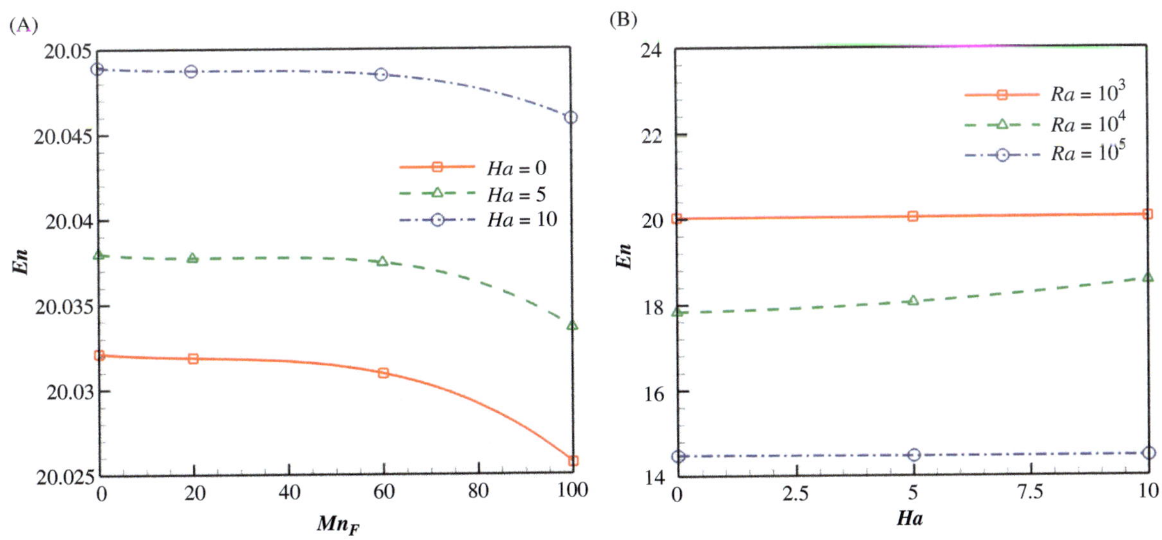

FIGURE 8.21 Effects of Magnetic number, Rayleigh number, and Hartmann number on heat transfer enhancement when $Pr = 6.8$. (A) $Ra = 10^3$, (B) $Mn_F = 0$.

8.5 EFFECT OF SPACE DEPENDENT MAGNETIC FIELD ON FREE CONVECTION OF FE_3O_4-WATER NANOFLUID

8.5.1 Problem Definition

The physical model along with the important geometrical parameters and the mesh of the enclosure used in the present CVFEM program are shown in Fig. 8.22. The inner circular wall is maintained at constant hot temperature T_h, the two horizontal walls are thermally isolated and other walls are maintained at constant cold temperature T_c ($T_h > T_c$). To reach the shape of inner circular and outer rectangular boundary which consists of the right and top walls, a supper elliptic function can be used as follows:

$$\left(\frac{X}{a}\right)^{2n} + \left(\frac{Y}{b}\right)^{2n} = 1 \tag{8.76}$$

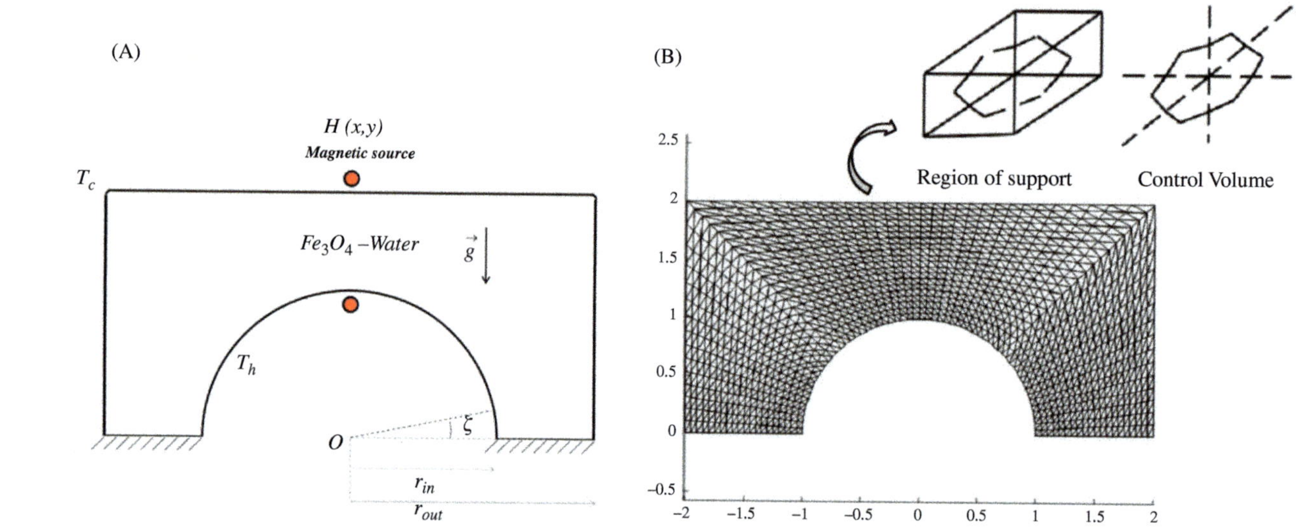

FIGURE 8.22 (A) Geometry and the boundary conditions; (B) the mesh of enclosure considered in this work.

When $a = b$ and $n = 1$ the geometry becomes a circle. As n increases from 1 the geometry would approach a rectangle for $a \neq b$ and square for $a = b$. In this study we assumed that $r_{in}/r_{out} = 0.6$.

For the expression of the magnetic field strength it can be considered that two magnetic sources are at points $(\bar{a}_1, \bar{b}_1)$ and $(\bar{a}_2, \bar{b}_2)$. The components of the magnetic field intensity $(\overline{H}_x, \overline{H}_y)$ and the magnetic field strength $(\overline{H})$ can be considered as:

$$\overline{H_x} = \frac{\gamma}{2\pi}\frac{1}{(x-\overline{a_1})^2 + (y-\overline{b_1})^2}(y-\overline{b_1}) - \frac{\gamma}{2\pi}\frac{1}{(x-\overline{a_2})^2 + (y-\overline{b_2})^2}(y-\overline{b_2}) \tag{8.77}$$

$$\overline{H_y} = -\frac{\gamma}{2\pi}\frac{1}{(x-\overline{a_1})^2 + (y-\overline{b_1})^2}(x-\overline{a_1}) - \frac{\gamma}{2\pi}\frac{1}{(x-\overline{a_2})^2 + (y-\overline{b_2})^2}(x-\overline{a_2}) \tag{8.78}$$

$$\overline{H} = \sqrt{\overline{H}_x^2 + \overline{H}_y^2} \tag{8.79}$$

where γ the magnetic field strength at the source (of the wire). The contours of the magnetic field strength are shown in Fig. 8.23. In this study a magnetic source is located at $(\overline{a_1} = 1.05\, cols, \overline{b_1} = 0.5\, rows)$ and $(\bar{a}_2 = -0.05\, cols, \overline{b_2} = 0.5\, rows)$.

8.5.2 Governing Equation

The flow is two-dimensional, laminar, and incompressible. The magnetic Reynolds number is assumed to be small so that the induced magnetic field can be neglected compared to the applied magnetic field. Using the Boussinesq approximation, the governing equations of heat transfer and fluid flow for nanofluid can be obtained as follows:

$$\frac{\partial u}{\partial x} + \frac{\partial v}{\partial y} = 0 \tag{8.80}$$

$$\rho_{nf}\left(u\frac{\partial u}{\partial x} + v\frac{\partial u}{\partial y}\right) = -\frac{\partial P}{\partial x} + \mu_{nf}\left(\frac{\partial^2 u}{\partial x^2} + \frac{\partial^2 u}{\partial y^2}\right) + \mu_0 M\frac{\partial \overline{H}}{\partial x} - \sigma_{nf}B_y^2 u + \sigma_{nf}B_x B_y v \tag{8.81}$$

$$\rho_{nf}\left(u\frac{\partial v}{\partial x} + v\frac{\partial v}{\partial y}\right) = -\frac{\partial P}{\partial y} + \mu_{nf}\left(\frac{\partial^2 v}{\partial x^2} + \frac{\partial^2 v}{\partial y^2}\right) + \mu_0 M\frac{\partial \overline{H}}{\partial y} - \sigma_{nf}B_x^2 v + \sigma_{nf}B_x B_y u + \rho_{nf}\beta_{nf}g(T - T_c) \tag{8.82}$$

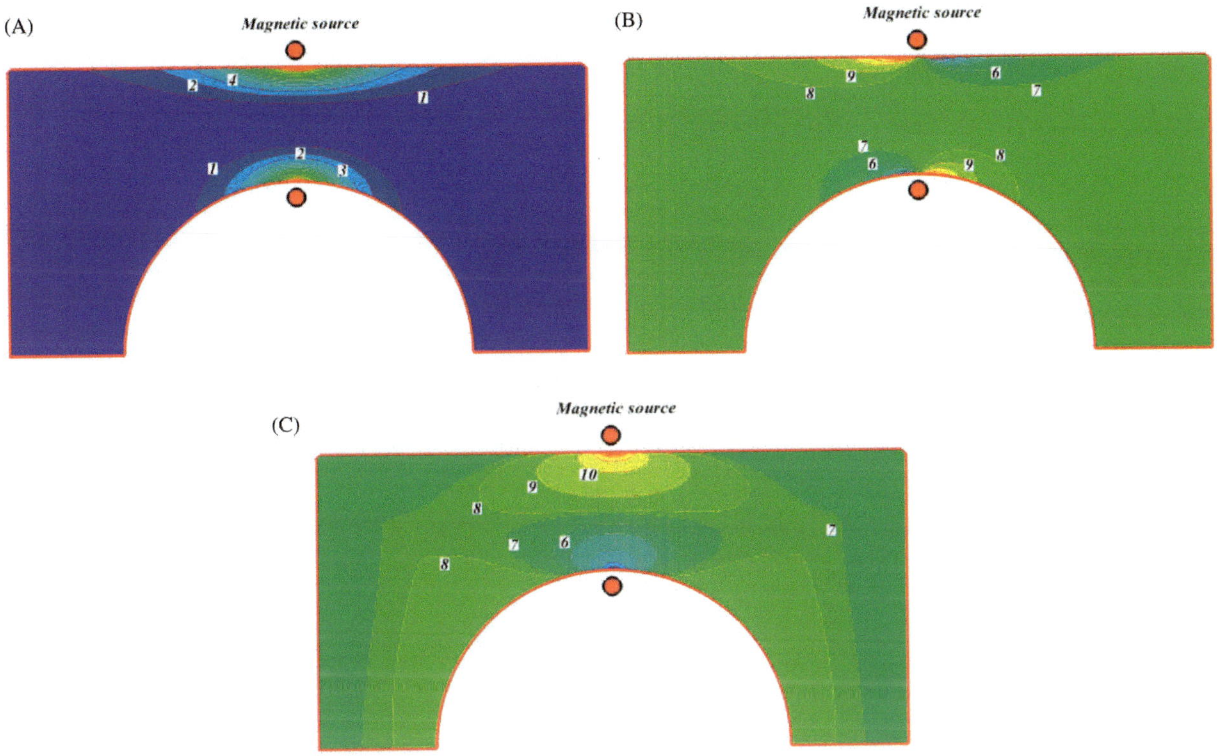

FIGURE 8.23 Contours of the (A) magnetic field strength H; (B) magnetic field intensity component in x direction Hx; (C) magnetic field intensity component in y direction Hy. (A) $H(x, y)$, (B) $Hx\ (x, y)$, (C) $Hy\ (x, y)$

$$(\rho C_p)_{nf}\left(u\frac{\partial T}{\partial x}+v\frac{\partial T}{\partial y}\right)=k_{nf}\left(\frac{\partial^2 T}{\partial x^2}+\frac{\partial^2 T}{\partial y^2}\right)+\sigma_{nf}(uB_y-vB_x)^2$$
$$-\mu_0 T\frac{\partial M}{\partial T}\left(u\frac{\partial \overline{H}}{\partial x}+v\frac{\partial \overline{H}}{\partial y}\right)+\mu_{nf}\left\{2\left(\frac{\partial u}{\partial x}\right)^2+2\left(\frac{\partial v}{\partial x}\right)^2+\left(\frac{\partial u}{\partial x}+\frac{\partial v}{\partial y}\right)^2\right\} \quad (8.83)$$

$$M=K'\overline{H}(T_c'-T) \quad (8.84)$$

In the above equations, μ_0 is the magnetic permeability of vacuum $(4\pi\times10^{-7}Tm/A)$, $\overline{H}$ is the magnetic field strength, $\overline{B}$ is the magnetic induction $(\overline{B}=\mu_0\overline{H})$, and the bar above the quantities denotes that they are dimensional. The effective density (ρ_{nf}) and heat capacitance $(\rho C_p)_{nf}$ of the nanofluid are defined as:

$$\rho_{nf}=\rho_f(1-\phi)+\rho_s\phi \quad (8.85)$$

$$(\rho C_p)_{nf}=(\rho C_p)_f(1-\phi)+(\rho C_p)_s\phi \quad (8.86)$$

where ϕ is the solid volume fraction of nanoparticles. Thermal diffusivity of the nanofluid is

$$\alpha_{nf}=\frac{k_{nf}}{(\rho C_p)_{nf}} \quad (8.87)$$

and the thermal expansion coefficient of the nanofluid can be determined as

$$\beta_{nf}=\beta_f(1-\phi)+\beta_s\phi \quad (8.88)$$

The dynamic viscosity of the nanofluids given by Brinkman is

$$\mu_{nf}=\frac{\mu_f}{(1-\phi)^{2.5}} \quad (8.89)$$

The effective thermal conductivity of the nanofluid can be approximated by the Maxwell–Garnetts (MG) model as:

$$\frac{k_{nf}}{k_f} = \frac{k_s + 2k_f - 2\phi(k_f - k_s)}{k_s + 2k_f + \phi(k_f - k_s)} \tag{8.90}$$

and the effective electrical conductivity of nanofluid was presented by Maxwell as below:

$$\frac{\sigma_{nf}}{\sigma_f} = 1 + \frac{3\left(\frac{\sigma_s}{\sigma_f} - 1\right)\phi}{\left(\frac{\sigma_s}{\sigma_f} + 2\right) - \left(\frac{\sigma_s}{\sigma_f} - 1\right)\phi} \tag{8.91}$$

By introducing the following nondimensional variables:

$$\begin{gathered} a = \frac{\bar{a}}{L}, b = \frac{\bar{b}}{L}, X = \frac{x}{L}, Y = \frac{y}{L}, \Theta = \frac{T - T_c}{T_h - T_c}, U = \frac{uL}{\alpha_f}, V = \frac{vL}{\alpha_f}, \\ H = \frac{\overline{H}}{\overline{H_0}}, H_x = \frac{\overline{H_x}}{\overline{H_0}}, H_y = \frac{\overline{H_y}}{\overline{H_0}}, P = \frac{p}{\rho_f (\alpha_f / L)^2} \end{gathered} \tag{8.92}$$

where in Eq. (8.92) $\overline{H_0} = \overline{H}(\bar{a}, 0) = \frac{\gamma}{2\pi|b|}$ and $L = r_{out} - r_{in}$. Using the dimensionless parameters, the equations now become:

$$\frac{\partial U}{\partial X} + \frac{\partial V}{\partial Y} = 0 \tag{8.93}$$

$$\begin{aligned} U\frac{\partial U}{\partial X} + V\frac{\partial U}{\partial Y} &= -\frac{\partial P}{\partial X} + \Pr\left[\frac{\mu_{nf}/\mu_f}{\rho_{nf}/\rho_f}\right]\left(\frac{\partial^2 U}{\partial X^2} + \frac{\partial^2 U}{\partial Y^2}\right) \\ &+ Mn_F \Pr\left(\frac{\rho_f}{\rho_{nf}}\right)(\varepsilon_2 - \varepsilon_1 - \Theta)H\frac{\partial H}{\partial X} - Ha^2 \Pr\left[\frac{\sigma_{nf}/\sigma_f}{\rho_{nf}/\rho_f}\right]\left(H_y^2 U - H_x H_y V\right) \end{aligned} \tag{8.94}$$

$$\begin{aligned} U\frac{\partial V}{\partial X} + V\frac{\partial V}{\partial Y} &= -\frac{\partial P}{\partial X} + \Pr\left[\frac{\mu_{nf}/\mu_f}{\rho_{nf}/\rho_f}\right]\left(\frac{\partial^2 V}{\partial X^2} + \frac{\partial^2 V}{\partial Y^2}\right) \\ &+ Mn_F \Pr\left(\frac{\rho_f}{\rho_{nf}}\right)(\varepsilon_2 - \varepsilon_1 - \Theta)H\frac{\partial H}{\partial Y} - Ha^2 \Pr\left[\frac{\sigma_{nf}/\sigma_f}{\rho_{nf}/\rho_f}\right](H_x^2 V - H_x H_y U) + Ra\Pr\left[\frac{\beta_{nf}}{\beta_f}\right]\Theta \end{aligned} \tag{8.95}$$

$$\begin{aligned} U\frac{\partial \Theta}{\partial X} + \frac{\partial \Theta}{\partial Y} &= \left[\frac{\frac{k_{nf}}{k_f}}{\frac{(\rho C_P)_{nf}}{(\rho C_P)_f}}\right]\left(\frac{\partial^2 \Theta}{\partial X^2} + \frac{\partial^2 \Theta}{\partial Y^2}\right) + Ha^2 Ec\left[\frac{\frac{\sigma_{nf}}{\sigma_f}}{\frac{(\rho C_P)_{nf}}{(\rho C_P)_f}}\right]\{U H_y - V H_x\}^2 \\ &+ Mn_F Ec\frac{(\rho C_P)_f}{(\rho C_P)_{nf}}\left\{U\frac{\partial H}{\partial X} + V\frac{\partial H}{\partial Y}\right\}H(\varepsilon_1 + \Theta) + Ec\left[\frac{\frac{\mu_{nf}}{\mu_f}}{\frac{(\rho C_P)_{nf}}{(\rho C_P)_f}}\right]\left\{2\left(\frac{\partial U}{\partial X}\right)^2 + 2\left(\frac{\partial V}{\partial Y}\right)^2 + \left(\frac{\partial U}{\partial Y} + \frac{\partial V}{\partial X}\right)^2\right\} \end{aligned} \tag{8.96}$$

where $Ra_f = g\beta_f L^3(T_h - T_c)/(\alpha_f \upsilon_f)$, $\Pr_f = \upsilon_f/\alpha_f$, $Ha = L\mu_0 H_0\sqrt{\sigma_f/\mu_f}$, $\varepsilon_1 = T_1/\Delta T$, $\varepsilon_2 = T_c'/\Delta T$, $Ec = (\mu_f \alpha_f)/\left[(\rho C_P)_f \Delta T\, L^2\right]$, and $Mn_F = \mu_0 H_0^2 K'(T_h - T_c)L^2/(\mu_f \alpha_f)$ are the Rayleigh number, Prandtl number, Hartmann number arising from MHD, temperature number, Curie temperature number, Eckert number, and Magnetic number arising from FHD the for the base fluid, respectively. The thermophysical properties of the nanofluid are given in Table 8.1.

The stream function and vorticity are defined as:

$$u = \frac{\partial \psi}{\partial y}, v = -\frac{\partial \psi}{\partial x}, \omega = \frac{\partial v}{\partial x} - \frac{\partial u}{\partial y}$$
$$\Omega = \frac{\omega L^2}{\alpha_f}, \Psi = \frac{\psi}{\alpha_f} \quad (8.97)$$

The stream function satisfies the continuity Eq. (8.93). The vorticity equation is obtained by eliminating the pressure between the two momentum equations, i.e., by taking the y-derivative of Eq. (8.94) and subtracting from it the x-derivative of Eq. (8.95).

The boundary conditions as shown in Fig. 8.22 are:

$$\begin{aligned} &\Theta = 1.0 && \text{on the inner circular boundary} \\ &\Theta = 0.0 && \text{on the outer wall boundary} \\ &\Psi = 0.0 && \text{on all solid boundaries} \end{aligned} \quad (8.98)$$

The values of vorticity on the boundary of the enclosure can be obtained using the stream function formulation and the known velocity conditions during the iterative solution procedure.

The local Nusselt number of the nanofluid along the hot wall can be expressed as:

$$Nu_{loc} = \left(\frac{k_{nf}}{k_f}\right)\frac{\partial \Theta}{\partial r} \quad (8.99)$$

where r is the radial direction. The average Nusselt number on the hot circular wall is evaluated as:

$$Nu_{ave} = \frac{1}{\gamma}\int_0^{\gamma} Nu_{loc}(\zeta)\, d\zeta \quad (8.100)$$

To estimate the enhancement of heat transfer between the case of $\phi = 0.04$ and the pure fluid (base fluid) case, the heat transfer enhancement is defined as:

$$En = \frac{Nu\,(\phi = 0.04) - Nu\,(basefluid)}{Nu\,(basefluid)} \times 100 \quad (8.101)$$

8.5.3 Effects of Active Parameters

In this work, effects of FHD and MHD on Fe_3O_4- water flow and heat transfer is investigated numerically using the CVFEM. The thermophysical properties of Fe_3O_4 nanoparticles and base fluid (water) are shown in Table 8.1 [66]. Calculations are made for various values of volume fraction of nanoparticles ($\phi = 0.0$ and 0.04), Rayleigh numbers ($Ra = 10^3$, 10^4, and 10^5), Magnetic number arising from FHD ($Mn_F = 0$, 20, 60, and 100) and Hartmann number arising from MHD ($Ha = 0, 5$ and 10). In all calculations, the Prandtl number (Pr), temperature number (ε_1), and Eckert number (Ec) are set to 6.8, 0.0, and 10^{-6}, respectively.

Fig. 8. 24A and B shows the comparison of the isotherms and streamlines between nanofluid and pure fluid. The velocity components of nanofluid are enhanced because of an augment in the energy transport in the fluid with the increasing of volume fraction. Thus, the absolute values of stream functions increase with increasing the volume fraction of nanofluid. Also it can be seen that thermal boundary layer thickness around the hot wall increases with the increase of nanoparticle volume fraction.

Fig. 8.25 depicts the effects of Rayleigh number, Hartmann number, and Magnetic number on isotherms and streamlines. $|\Psi_{max}|$ increases with the increase of Rayleigh number while it decreases with augment of Hartmann number and Magnetic number. When $Ra = 10^3$, the heat transfer in the enclosure is mainly dominated by the conduction mode. At absence of magnetic field the streamlines show one rotating eddy. When the Magnetic parameter increases, another vortex appears near the centerline which rotates in the reverse direction in comparison with the primary vortex. Lorentz force has no significant effect on isotherm and streamline at low Rayleigh number. As the Rayleigh number increases up to 10^4, the role of convection in heat transfer becomes more important and consequently the thermal boundary layer on the surface of the inner wall becomes thinner. In addition, a

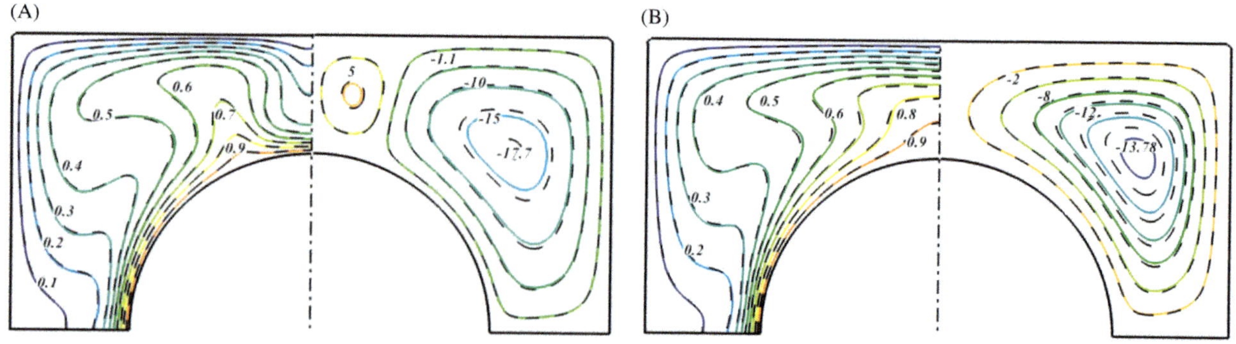

FIGURE 8.24 (A, B) Comparison of the isotherms and streamlines between nanofluid ($\phi = 0.04$) (– – –) and pure fluid ($\phi = 0$) (—) when $Pr = 6.8$. (A) $Ra = 10^5, Mn_F = 0, Ha = 0$, (B) $Ra = 10^5, Mn_F = 100, Ha = 10$.

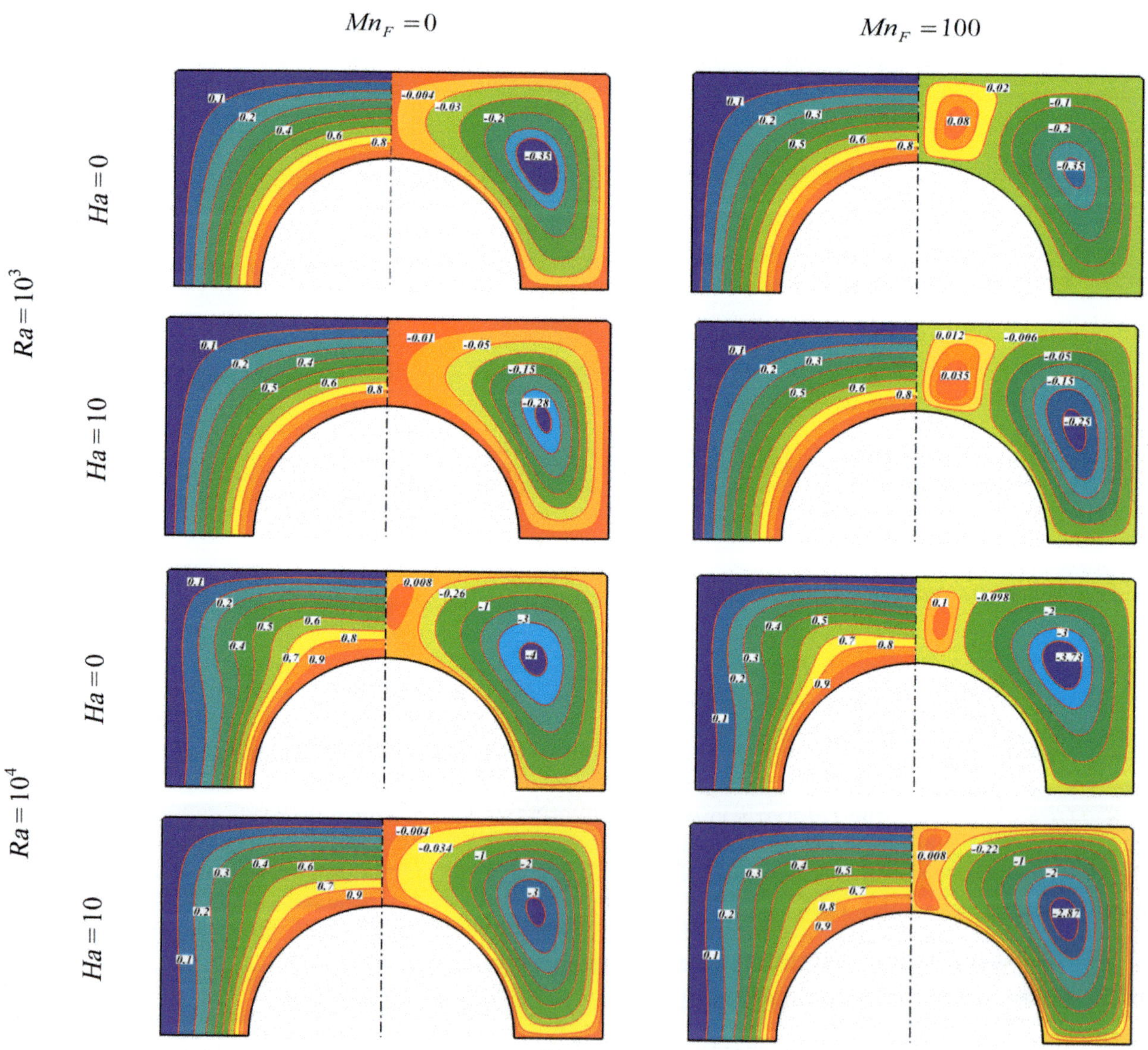

FIGURE 8.25 Isotherm (left) and streamline (right) contours for different values of Rayleigh number, Hartmann number, and Magnetic number when $Pr = 6.8$, $\phi = 0.04$.

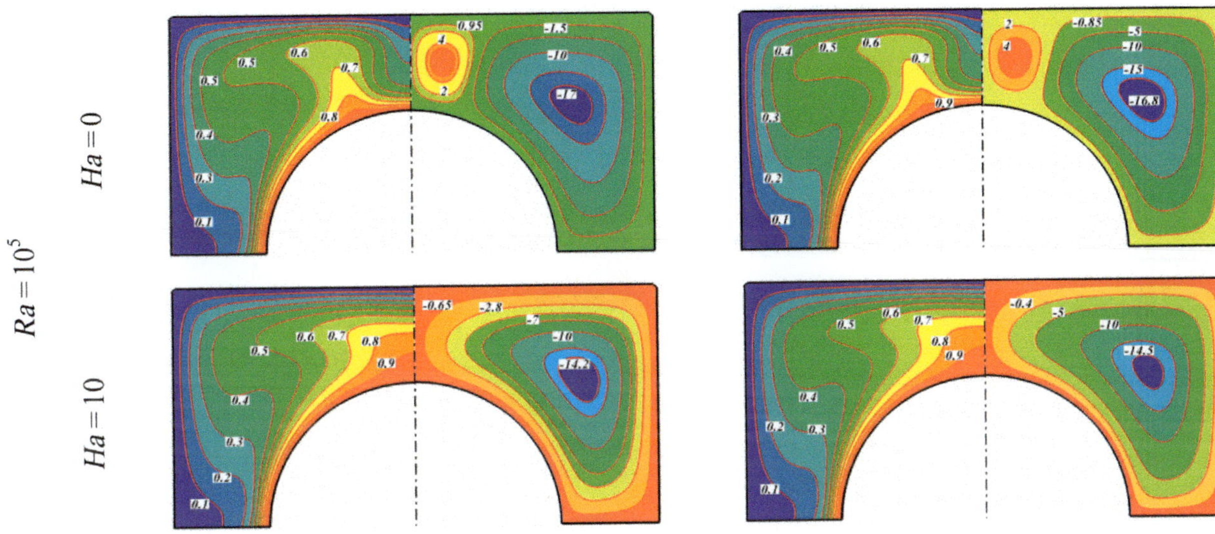

FIGURE 8.25 (Continued).

plume starts to appear on the top of the inner circular wall. It is interesting remark that at $Mn_F = 100, Ha = 10$ the smaller primary vortex turns to two smaller eddies near the magnetic sources. At $Ra = 10^5$ two reverse plumes are observed in the absence of FHD and MHD effects. As the Hartmann number increases one of the plumes disappears due to Lorentz forces.

Fig. 8.26 shows the effects of Magnetic number, Hartmann number, and Rayleigh number on local Nusselt number along the hot wall. As the Hartmann number increases the Nusselt number decreases but an opposite trend is observed for the Magnetic number and Rayleigh number. The existence of maximum and minimum points in the local Nusselt number profile corresponds to the presence of thermal plumes. Fig. 8.27A–C shows the effects of Hartmann number and Rayleigh number on average Nusselt number along a hot wall. The effects of Magnetic number, Rayleigh number, and Hartmann number on heat transfer enhancement are shown in Fig. 8.27D–F. The corresponding polynomial representations for Nusselt number and Enhancement in heat transfer are as the following:

$$\begin{aligned}
Nu_{ave} &= a_{15} + a_{25}Y_3 + a_{35}Y_4 + a_{45}{Y_3}^2 + a_{55}{Y_4}^2 + a_{65}Y_4Y_3 \\
Y_1 &= a_{11} + a_{21}Ra^* + a_{31}Mn_F{}^* + a_{41}Ra^{*2} + a_{51}Mn_F^{*2} + a_{61}Ra^*\,Mn_F^* \\
Y_2 &= a_{12} + a_{22}\phi + a_{32}Ra^* + a_{42}\phi^2 + a_{52}Ra^{*2} + a_{62}\phi Ra^* \\
Y_3 &= a_{13} + a_{23}Y_1 + a_{33}Ha^* + a_{43}{Y_1}^2 + a_{53}Ha^{*2} + a_{63}Y_1Ha^* \\
Y_4 &= a_{14} + a_{24}Ha^* + a_{34}Y_2 + a_{44}Ha^{*2} + a_{54}{Y_2}^2 + a_{64}Ha^*\,Y_2
\end{aligned} \tag{28}$$

$$\begin{aligned}
En &= b_{13} + b_{23}Y_1 + b_{33}Y_2 + b_{43}{Y_1}^2 + b_{53}{Y_2}^2 + b_{63}Y_1Y_2 \\
Y_1 &= b_{11} + b_{21}Ra^* + b_{31}Ha^* + b_{41}Ra^{*2} + b_{51}Ha^{*2} + b_{61}Ra^*\,Ha^* \\
Y_2 &= b_{12} + b_{22}Ra^* + b_{32}Mn_F^* + b_{42}Ra^{*2} + b_{52}Mn_F^{*2} + b_{62}\mathrm{Re}Mn_F^*
\end{aligned} \tag{29}$$

where $Ra^* = \log(Ra), Ha^* = Ha/10, Mn_F{}^* = Mn_F/100$. Also a_{ij}, b_{ij} can be found in Tables 8.5 and 8.6.

As buoyancy force (due to temperature gradient) and Kelvin force (due to external nonuniform magnetic field) increase, the rate of heat transfer increases while the opposite trend is observed for Lorentz forces. It means that increasing Lorentz forces lead to a decrease of Nusselt number. Also it can be concluded that the effect of Magnetic parameter on Nu_{ave} is more pronounced at low Rayleigh number. At low Rayleigh number, heat transfer enhancement increases with the increase of the Hartmann number while the opposite trend is observed for the Magnetic parameter. At high Rayleigh number the Magnetic parameter has no significant effect on enhancement but increasing Hartmann number causes the heat transfer enhancement to decrease. The effect of nanoparticles is more pronounced at low Rayleigh number than at high Rayleigh number due to the domination of conduction heat transfer mechanism.

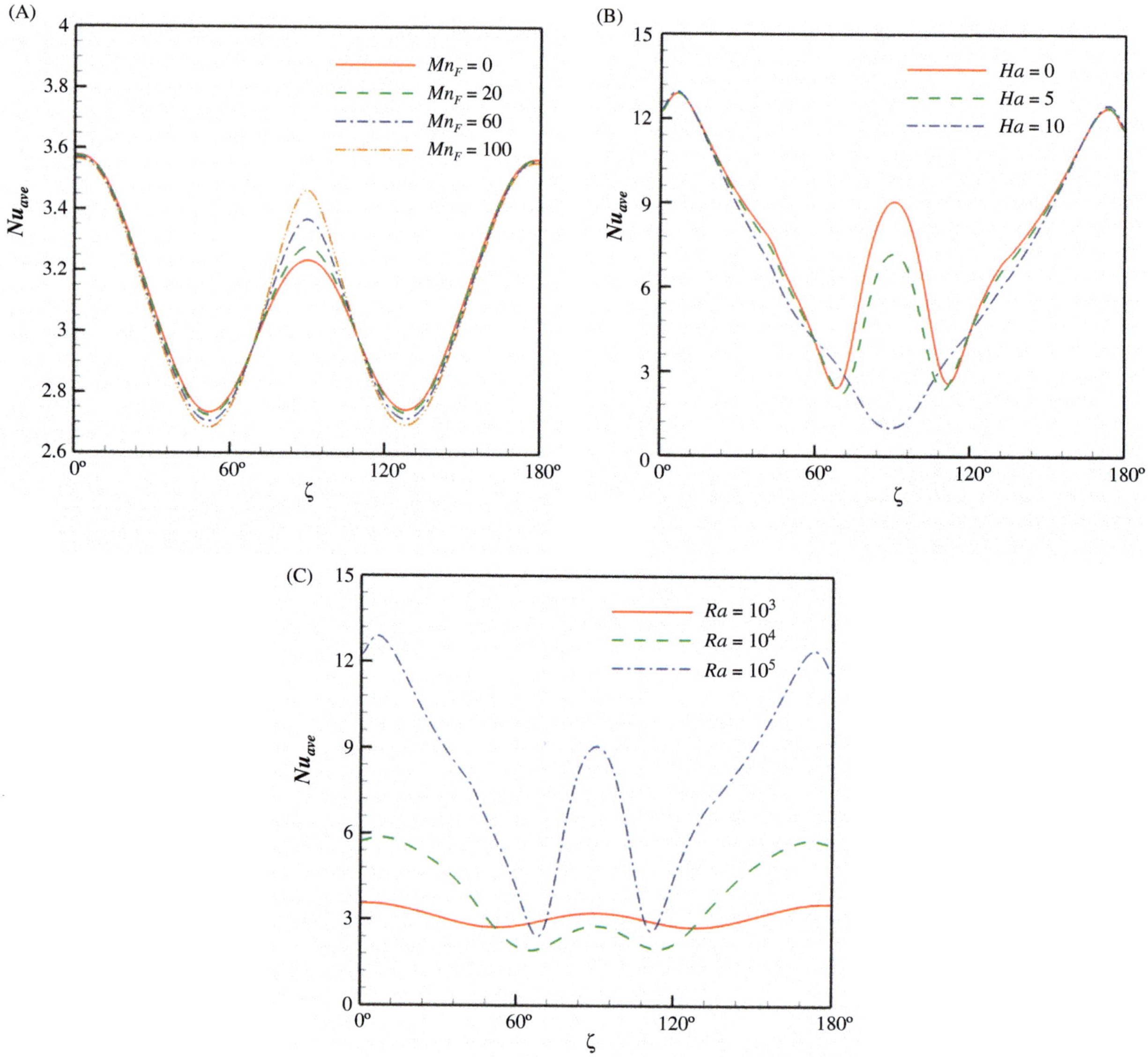

FIGURE 8.26 Effects of Magnetic number, Hartmann number, and Rayleigh number on local Nusselt number Nu_{loc} along hot wall when $Pr = 6.8$, $\phi = 0.04$. (A) $Ra = 10^3, Ha = 0$, (B) $Ra = 10^5, Mn_F = 0$, (C) $Mn_F = 0, Ha = 0$.

8.6 NONUNIFORM MAGNETIC FIELD EFFECT ON NANOFLUID HYDROTHERMAL TREATMENT CONSIDERING BROWNIAN MOTION AND THERMOPHORESIS EFFECTS

8.6.1 Problem Definition

The physical model along with the important geometrical parameters and the mesh of the enclosure used in the present CVFEM program are shown in Fig. 8.28. The inner wall is maintained at constant heat flux. The outer wall is maintained at constant temperature T_h. The inner and outer walls are maintained at constant concentration C_h and C_c, respectively. The shape of the inner cylinder profile is assumed to mimic the following pattern

$$r = r_{in} + A\cos(N(\zeta)) \tag{8.102}$$

in which r_{in} is the base circle radius, r_{out} is the radius of outer cylinder, A and N are amplitude and number of undulations, respectively, and ζ is the rotation angle. In this study A and N are equal to 0.2 and 4, respectively.

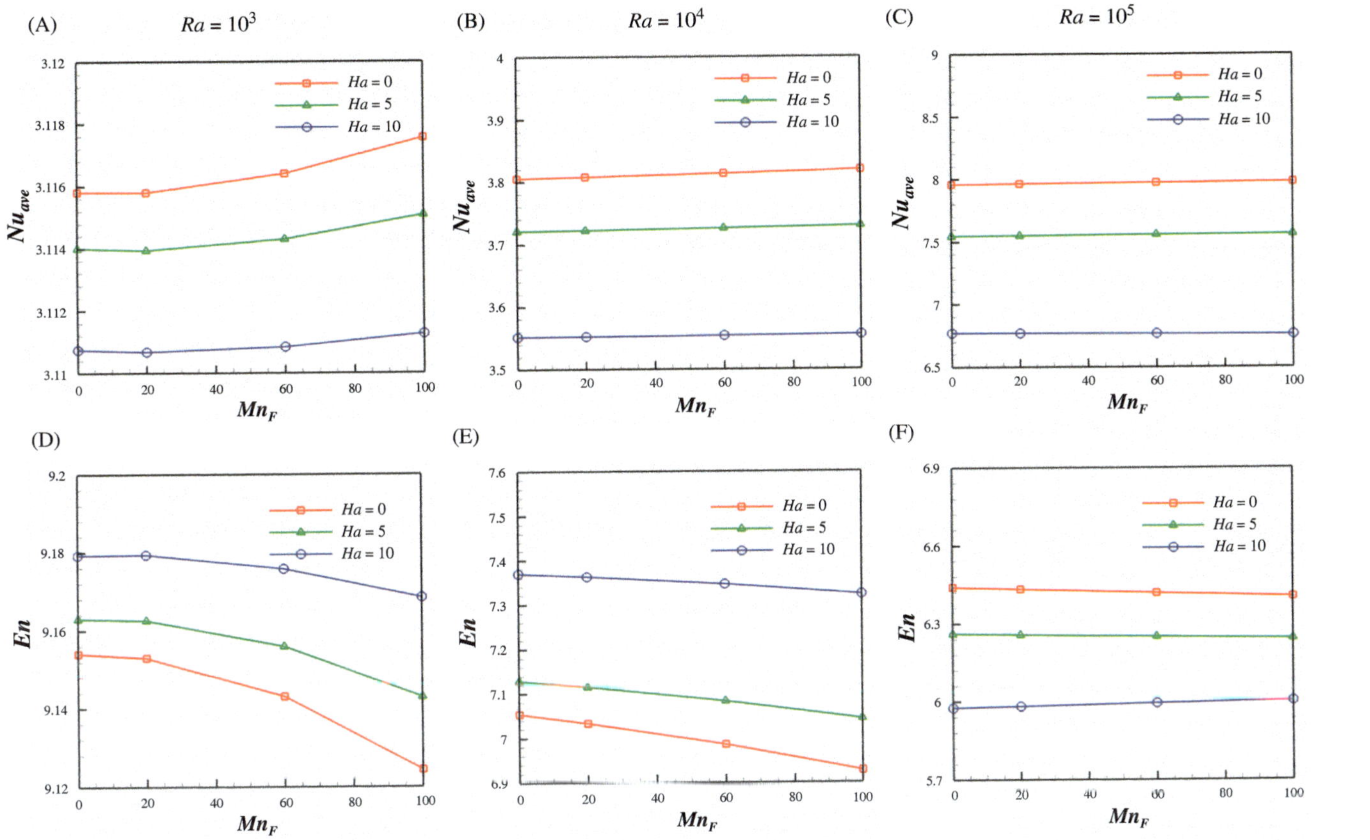

FIGURE 8.27 (A–C) Effects of Magnetic parameter, Hartmann number, and Rayleigh number on average Nusselt number Nu_{ave} along hot wall when $Pr = 6.8$, $\phi = 0.04$; (D–F) Effects of Magnetic number, Rayleigh number, and Hartmann number on heat transfer enhancement when $Pr = 6.8$.

TABLE 8.5 Constant Coefficient for Using Eq. (8.100)

a_{ij}	$i=1$	$i=2$	$i=3$	$i=4$	$i=5$	$i=6$
$j=1$	19.51647	−10.0868	−0.00804	1.525112	0.001086	0.003349
$j=2$	19.53085	−0.87952	−10.1291	−0.03518	1.525112	2.186806
$j=3$	−0.29704	1.107821	0.654435	−0.00017	0.041509	−0.24967
$j=4$	−0.15202	0.647229	1.043281	0.044365	0.005999	−0.24883
$j=5$	−0.07156	0.161803	0.861233	1.818237	1.846604	−3.66821

TABLE 8.6 Constant Coefficient for Using Eq. (8.101)

b_{ij}	$i=1$	$i=2$	$i=3$	$i=4$	$i=5$	$i=6$
$j=1$	21.33634	−5.71464	0.798586	0.542492	0.064253	−0.21888
$j=2$	21.72597	−5.81042	−0.0488	0.542492	−0.01392	0.006107
$j=3$	0.061649	4.86387	−3.87977	−0.38679	0.13164	0.25619

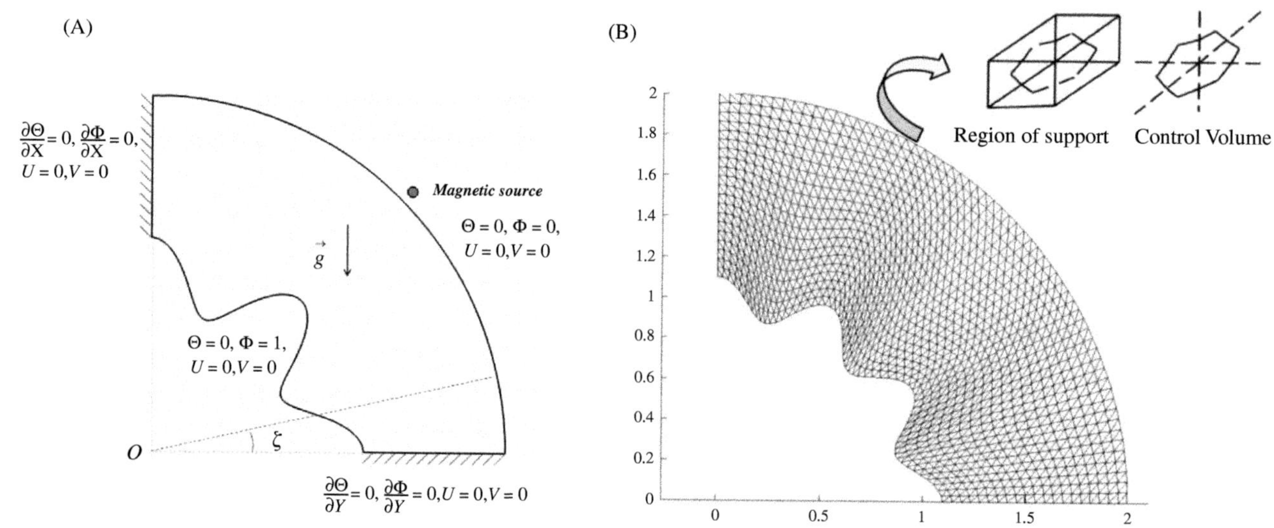

FIGURE 8.28 (A) Geometry and the boundary conditions; (B) the mesh of enclosure considered in this work.

For the expression of the magnetic field strength it can be considered that the magnetic source represents a magnetic wire placed vertically to the x-y plane at the point $(\bar{a}, \bar{b})$. The components of the magnetic field intensity $(\overline{H_x}, \overline{H_y})$ and the magnetic field strength $(\overline{H})$ can be considered as:

$$\overline{H_x} = \frac{\gamma}{2\pi} \frac{1}{(x-\bar{a})^2 + (y-\bar{b})^2} (y - \bar{b}) \tag{8.103}$$

$$\overline{H_y} = -\frac{\gamma}{2\pi} \frac{1}{(x-\bar{a})^2 + (y-\bar{b})^2} (x - \bar{a}) \tag{8.104}$$

$$\overline{H} = \sqrt{\overline{H}_x^2 + \overline{H}_y^2} = \frac{\gamma}{2\pi} \frac{1}{\sqrt{(x-\bar{a})^2 + (y-\bar{b})^2}} \tag{8.105}$$

where γ the magnetic field strength at the source (of the wire) and $(\bar{a}, \bar{b})$ is the position where the source is located. The contours of the magnetic field strength are shown in Fig. 8.29. In this study magnetic source is located at $(-0.05\ cols, 0.5\ rows)$.

8.6.2 Governing Equation

The nanofluid's density, ρ is

$$\begin{aligned} \rho &= \phi\rho_p + (1-\phi)\rho_f \\ &\cong \phi\rho_p + (1-\phi)\left\{\rho_{f_0}(1-\beta(T-T_c)\right\} \end{aligned} \tag{8.106}$$

Where ρ_f, is the base fluid's density, T_c, is a reference temperature, ρ_{f_0} is the base fluid's density at the reference temperature, β is the volumetric coefficient of expansion. Taking the density of base fluid as that of the nanofluid, the density ρ in Eq. (8.106), thus becomes

$$\rho \cong \phi\rho_p + (1-\phi)\{\rho_0(1-\beta(T-T_c)\} \tag{8.107}$$

ρ_0 is the nanofluid's density at the reference temperature.

The continuity, momentum under Boussinesq approximation, and energy equations for the laminar and steady-state natural convection in a two-dimensional enclosure can be written in dimensional form as follows:

$$\frac{\partial u}{\partial x} + \frac{\partial v}{\partial y} = 0 \tag{8.108}$$

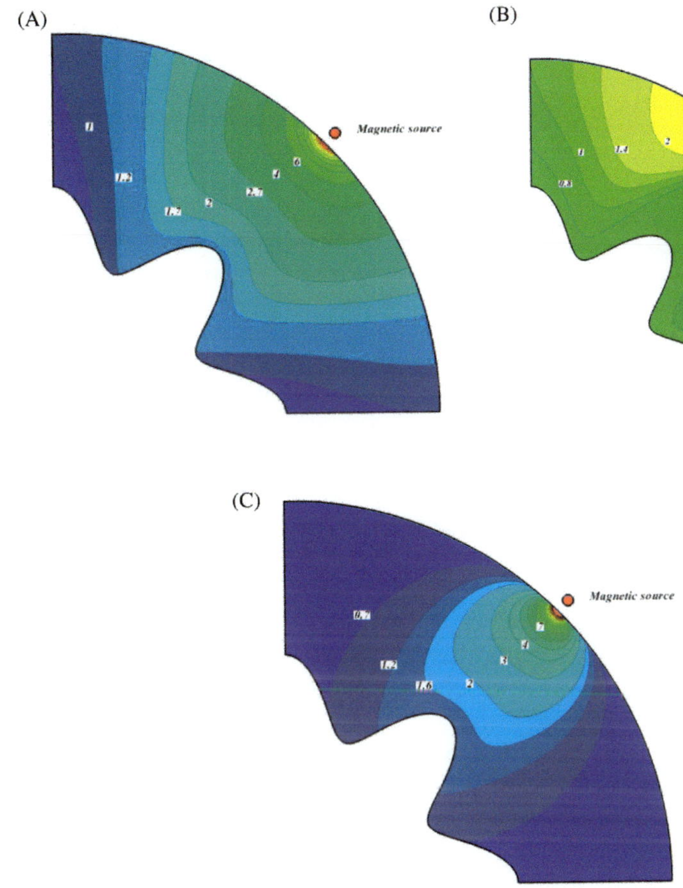

FIGURE 8.29 Contours of the (A) magnetic field strength H; (B) magnetic field intensity component in x direction Hx; (C) magnetic field intensity component in y direction Hy. (A) $H(x,y)$, (B) $Hx(x,y)$, (C) $Hy(x,y)$.

$$\rho_f\left\{u\frac{\partial u}{\partial x}+v\frac{\partial u}{\partial y}\right\}=-\frac{\partial P}{\partial x}+\mu\left(\frac{\partial^2 u}{\partial x^2}+\frac{\partial^2 u}{\partial y^2}\right)+\mu_0 M\frac{\partial \overline{H}}{\partial x}-\sigma B_y^2 u+\sigma B_x B_y v \tag{8.109}$$

$$\begin{aligned}\rho_f\left\{u\frac{\partial v}{\partial x}+v\frac{\partial v}{\partial y}\right\}=&-\frac{\partial P}{\partial y}+\mu\left(\frac{\partial^2 v}{\partial x^2}+\frac{\partial^2 v}{\partial y^2}\right)-(\phi-\phi_c)\left(\rho_p-\rho_{f0}\right)g+(1-\phi_c)\rho_{f_0}(T-T_c)g\\&+\mu_0 M\frac{\partial \overline{H}}{\partial y}-\sigma B_x^2 v+\sigma B_x B_y u\end{aligned} \tag{8.110}$$

$$\begin{aligned}u\frac{\partial T}{\partial x}+v\frac{\partial T}{\partial y}=\alpha\left(\frac{\partial^2 T}{\partial x^2}+\frac{\partial^2 T}{\partial y^2}\right)+\frac{(\rho c)_p}{(\rho c)_f}&\left[\begin{array}{c}D_B\left\{\frac{\partial \phi}{\partial x}\cdot\frac{\partial T}{\partial x}+\frac{\partial \phi}{\partial y}\cdot\frac{\partial T}{\partial y}\right\}\\+(D_T/T_c)\left\{\left(\frac{\partial T}{\partial x}\right)^2+\left(\frac{\partial T}{\partial y}\right)^2\right\}\end{array}\right]+\left(uB_y-vB_x\right)^2\\-\mu_0 T\frac{\partial M}{\partial T}\left(u\frac{\partial \overline{H}}{\partial x}+v\frac{\partial \overline{H}}{\partial y}\right)+\mu&\left\{2\left(\frac{\partial u}{\partial x}\right)^2+2\left(\frac{\partial v}{\partial x}\right)^2+\left(\frac{\partial u}{\partial x}+\frac{\partial v}{\partial y}\right)^2\right\}\end{aligned} \tag{8.111}$$

$$u\frac{\partial \phi}{\partial x}+v\frac{\partial \phi}{\partial y}=D_B\left\{\frac{\partial^2 \phi}{\partial x^2}+\frac{\partial^2 \phi}{\partial y^2}\right\}+\left(\frac{D_T}{T_c}\right)\left\{\frac{\partial^2 T}{\partial x^2}+\frac{\partial^2 T}{\partial y^2}\right\} \tag{8.112}$$

$$M = K'\overline{H}\left(T_c' - T\right) \tag{8.113}$$

where K' is a constant and T_c' is the Curie temperature.

In the above equations, μ_0 is the magnetic permeability of vacuum $(4\pi \times 10^{-7} Tm/A)$, $\overline{H}$ is the magnetic field strength, $\overline{B}$ is the magnetic induction $(\overline{B} = \mu_0 \overline{H})$ and the bar above the quantities denotes that they are dimensional.

The stream function and vorticity are defined as follows:

$$u = \frac{\partial \psi}{\partial y}, v = -\frac{\partial \psi}{\partial x}, \omega = \frac{\partial v}{\partial x} - \frac{\partial u}{\partial y} \tag{8.114}$$

The stream function satisfies the continuity Eq. (8.108). The vorticity equation is obtained by eliminating the pressure between the two momentum equations, i.e., by taking the y-derivative of Eq. (8.109) and subtracting from it the x-derivative of Eq. (8.110). Also the following nondimensional variables should be introduced:

$$X = \frac{x}{L}, \quad Y = \frac{y}{L}, \quad \Omega = \frac{\omega L^2}{\alpha}, \Psi = \frac{\psi}{\alpha}, \Theta = \frac{T - T_c}{T_h - T_c}, \quad \Phi = \frac{\phi - \phi_c}{\phi_h - \phi_c}, U = \frac{uL}{\alpha}, V = \frac{vL}{\alpha}$$
$$H = \frac{\overline{H}}{\overline{H_0}}, H_x = \frac{\overline{H_x}}{\overline{H_0}}, H_y = \frac{\overline{H_y}}{\overline{H_0}} \tag{8.115}$$

By using these dimensionless parameters the equations become:

$$\begin{aligned}
&\left[\frac{\partial \Psi}{\partial Y}\frac{\partial \Omega}{\partial X} - \frac{\partial \Psi}{\partial X}\frac{\partial \Omega}{\partial Y}\right] = \Pr\left(\frac{\partial^2 \Omega}{\partial X^2} + \frac{\partial^2 \Omega}{\partial Y^2}\right) + \Pr Ra\left(\frac{\partial \Theta}{\partial X} - Nr\frac{\partial \Phi}{\partial X}\right) + Mn_F \Pr\left\{\frac{\partial H}{\partial X}\frac{\partial \Theta}{\partial Y} - \frac{\partial H}{\partial Y}\frac{\partial \Theta}{\partial X}\right\}H \\
&- Ha^2 \Pr \times \left\{\frac{\partial V}{\partial X}H_x^2 + V\left(2H_x\frac{\partial H_x}{\partial X}\right) - \frac{\partial U}{\partial X}H_x H_y - U\frac{\partial H_x}{\partial X}H_y - U\frac{\partial H_y}{\partial X}H_x\right. \\
&\left. - \frac{\partial U}{\partial Y}H_y^2 - U\left(2H_y\frac{\partial H_y}{\partial Y}\right) + \frac{\partial V}{\partial Y}H_x H_y + V\frac{\partial H_x}{\partial Y}H_y + V\frac{\partial H_y}{\partial Y}H_x\right\}
\end{aligned} \tag{8.116}$$

$$\begin{aligned}
&\frac{\partial \Psi}{\partial Y}\frac{\partial \Theta}{\partial X} - \frac{\partial \Psi}{\partial X}\frac{\partial \Theta}{\partial Y} = \left(\frac{\partial^2 \Theta}{\partial X^2} + \frac{\partial^2 \Theta}{\partial Y^2}\right) + Nb\left(\frac{\partial \Phi}{\partial X}\frac{\partial \Theta}{\partial X} + \frac{\partial \Phi}{\partial Y}\frac{\partial \Theta}{\partial Y}\right) + Nt\left(\left(\frac{\partial \Theta}{\partial X}\right)^2 + \left(\frac{\partial \Theta}{\partial Y}\right)^2\right) \\
&+ Ha^2\, Ec\{U\, H_y - V\, H_x\}^2 + Mn_F Ec\left\{U\frac{\partial H}{\partial X} + V\frac{\partial H}{\partial Y}\right\}H(\varepsilon_1 + \Theta) + Ec\left\{2\left(\frac{\partial U}{\partial X}\right)^2 + 2\left(\frac{\partial V}{\partial X}\right)^2 + \left(\frac{\partial U}{\partial X} + \frac{\partial V}{\partial Y}\right)^2\right\}
\end{aligned} \tag{8.117}$$

$$\frac{\partial \Psi}{\partial Y}\frac{\partial \Phi}{\partial X} - \frac{\partial \Psi}{\partial X}\frac{\partial \Phi}{\partial Y} = \frac{1}{Le}\left(\frac{\partial^2 \Phi}{\partial X^2} + \frac{\partial^2 \Phi}{\partial Y^2}\right) + \frac{Nt}{Nb\ Le}\left(\frac{\partial^2 \Theta}{\partial X^2} + \frac{\partial^2 \Theta}{\partial Y^2}\right) \tag{8.118}$$

$$\frac{\partial^2 \Psi}{\partial X^2} + \frac{\partial^2 \Psi}{\partial Y^2} = -\Omega \tag{8.119}$$

where thermal Rayleigh number, the buoyancy ratio number, Prandtl number, the Brownian motion parameter, the thermophoretic parameter, Lewis number, Hartmann number, Eckert number, and Magnetic number arising from FHD of nanofluid are defined as: $Ra = (1 - \phi_c)\rho_{f_0} g\beta L^3 (T_h - T_c)/(\mu\alpha)$, $Nr = \left(\rho_p - \rho_0\right)(\phi_h - \phi_c)/[(1 - \phi_c)\rho_{f_0}\beta L(T_h - T_c)]$, $\Pr = \mu/\rho_f \alpha$, $Nb = (\rho c)_p D_B(\phi_h - \phi_c)/\left((\rho c)_f \alpha\right)$, $Nt = (\rho c)_p D_T(T_h - T_c)/[(\rho c)_f \alpha T_c]$, $Le = \alpha/D_B$, $Ha = LH_0\mu_0\sqrt{\sigma/\mu}$, $Ec = (\alpha\mu)/[(\rho C_P)\Delta T L^2]$ and $Mn_F = \mu_0 H_0^2 K'(T_h - T_c)L^2/(\mu\alpha)$, respectively.

The local Nusselt number of the nanofluid along the inner wall can be expressed as:

$$Nu_{local} = \left(\frac{k_{nf}}{k_f}\right)\frac{1}{\theta}\bigg|_{inner\ wall} \tag{8.120}$$

The average Nusselt number on the hot wall is evaluated as:

$$Nu_{ave} = \frac{1}{0.5\pi} \int_0^{0.5\pi} Nu_{loc}\, d\gamma \tag{8.121}$$

8.6.3 Effects of Active Parameters

Natural convection heat transfer in an enclosure filled with nanofluid external magnetic field is investigated numerically. Effects of Rayleigh number ($Ra = 10^3, 10^4, 10^5$), buoyancy ratio number ($Nr = 0.1$ and 4), Hartmann number ($Ha = 0, 2, 6$ and 10), and Lewis number ($Le = 2$ and 4) on flow and heat transfer characteristics are examined. In all calculations, the Prandtl number (Pr), temperature number (ε_1), Eckert number (Ec), Brownian motion parameter (Nb), and thermophoretic parameter of nanofluid (Nt) are set to $6.85, 0.0, 10^{-6}, 0.5$, and 0.5, respectively.

Effects of Hartmann number, Rayleigh number, buoyancy ratio number, and Lewis number on isotherms, streamlines, isoconcentration, and heatline contours are shown in Figs. 8.30–8.33. At $Ra = 10^3$ the conduction heat transfer mechanism is more pronounced. For this reason the isotherms are parallel to each other. As Ra increases, the distribution of isotherm contours increases. At Ra $= 10^3$, two equal eddies exist which are symmetric respect to $\zeta = 45^\circ$. The strength of upper eddy increases with increase of Hartmann number.

With increase of Ra, the role of convection in heat transfer becomes more significant. Also it can be seen that as Ra increases the distribution of isoconcentration contours increases. The heat flow within the enclosure is displayed using the heat function obtained from conductive heat fluxes ($\partial\Theta/\partial X, \partial\Theta/\partial Y$) as well as convective heat fluxes ($V\Theta, U\Theta$). Heatlines emanate from hot regimes and end on cold regimes illustrating the path of heat flow. The domination of conduction heat transfer in low Rayleigh number can be observed from the heatline patterns since no passive area exists. The increase of Ra causes the clustering of heatlines from hot to the cold wall and generates passive heat transfer area in which heat is rotated without having significant effect on heat transfer between walls. By increasing buoyancy ratio, a small eddy which rotates clockwise appears near the vertical wall. This eddy disappears by increasing the Hartmann number. As the Hartmann number increases, the Lorentz forces increase and in turn the nanofluid flow is suppressed. So, the thermal boundary layer thickness increases with the increase of Lorentz forces. As the Rayleigh number increases, the buoyancy forces increase and in turn thermal boundary layer thickness near the hot wall decreases. A similar trend is observed for buoyancy ratio number and Lewis number.

The effects of Hartmann number, Rayleigh number, buoyancy ratio number, and Lewis number on local Nusselt number are shown in Fig. 8.34. The profiles of the Nu_{loc} profiles have local extremes, which are related to the thermal plumes and crests over the inner cylinder. Also it can be found that local Nusselt number decreases with increase of Hartmann number. Fig. 8.35 shows the effect of Hartmann number, Rayleigh number, buoyancy ratio number and Lewis number on average Nusselt number. Average Nusselt number increases with increase of Rayleigh number, buoyancy ratio number, and Lewis number while it decreases with increase of Hartmann number.

8.7 EXTERNAL MAGNETIC SOURCE EFFECT ON WATER BASED NANOFLUID CONVECTIVE HEAT TRANSFER

8.7.1 Problem Definition

Fig. 8.36 depicts the sample mesh and diagram of present study. Boundary conditions are shown in this figure. Magnetic external source has been applied to this geometry. The formula of H can be present as follows:

$$\overline{H_x} = \gamma \frac{(y - \overline{b})}{2\pi} \left((\overline{b} - y)^2 + (\overline{a} - x)^2 \right)^{-2} \tag{8.122}$$

$$\overline{H_y} = \frac{(\overline{a} - x)}{2\pi} (\gamma) \left((\overline{b} - y)^2 + (\overline{a} - x)^2 \right)^{-2} \tag{8.123}$$

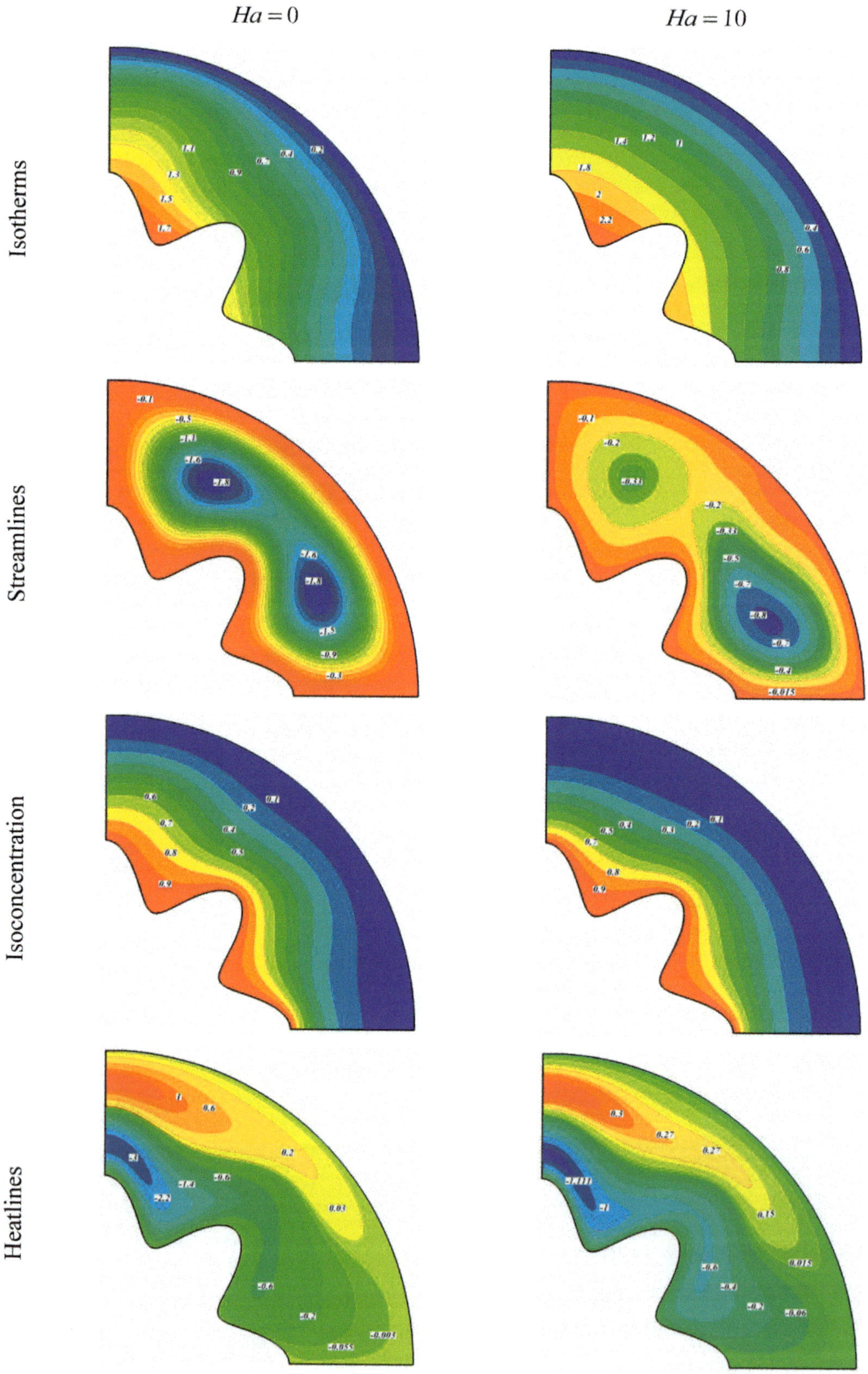

FIGURE 8.30 Comparison of the isotherm, streamline, isoconcentration, and heatline contours for different values of Hartmann number when $Nr = 0.1, Le = 2, Nt = Nb = 0.5, Ra = 10^3, Mn_F = 5$ and $\Pr = 6.85$.

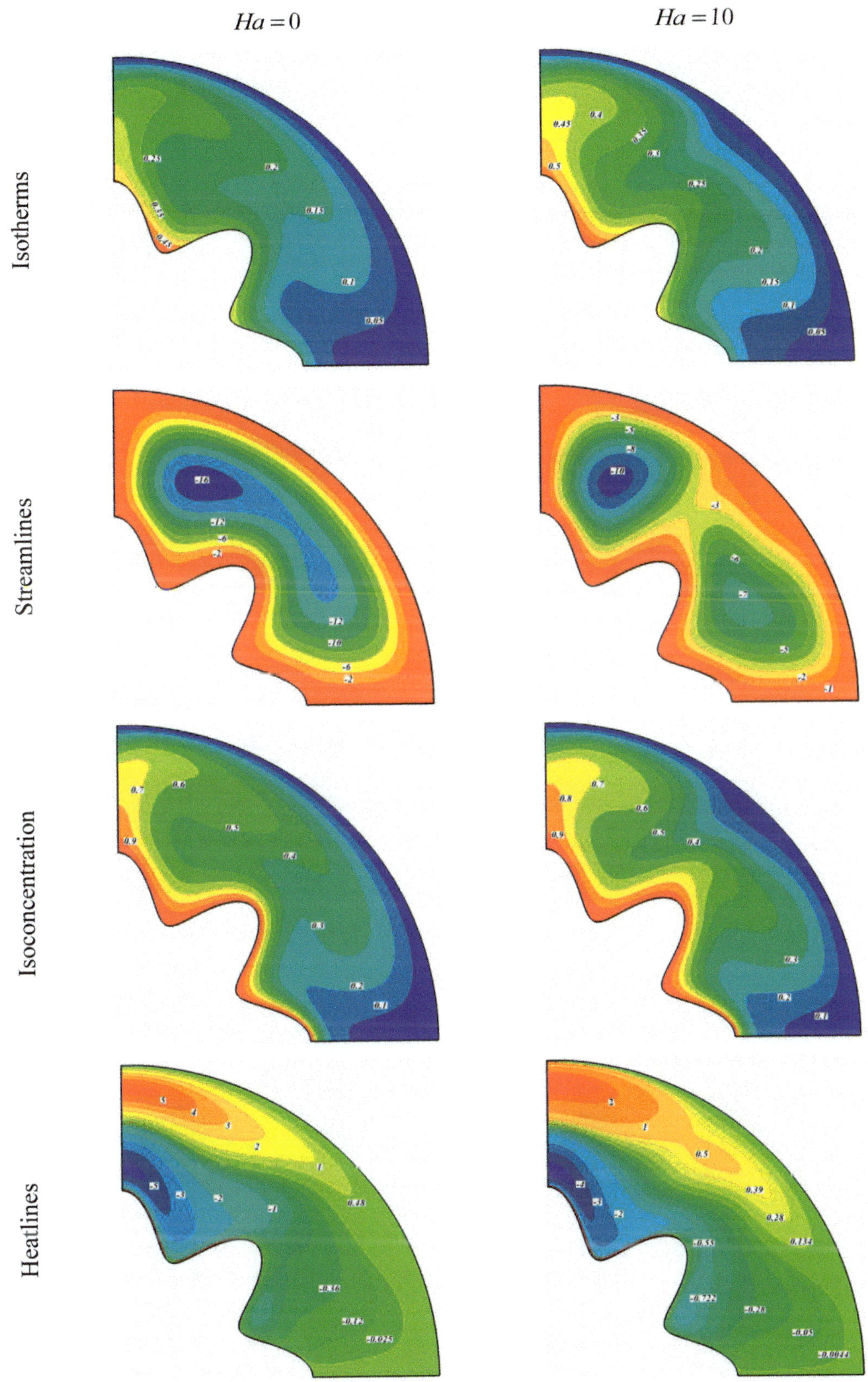

FIGURE 8.31 Comparison of the isotherm, streamline, isoconcentration, and heatline contours for different values of Hartmann number when $Nr = 0.1, Le = 2, Nt = Nb = 0.5, Ra = 10^5, Mn_F = 5$ and $\Pr = 6.85$.

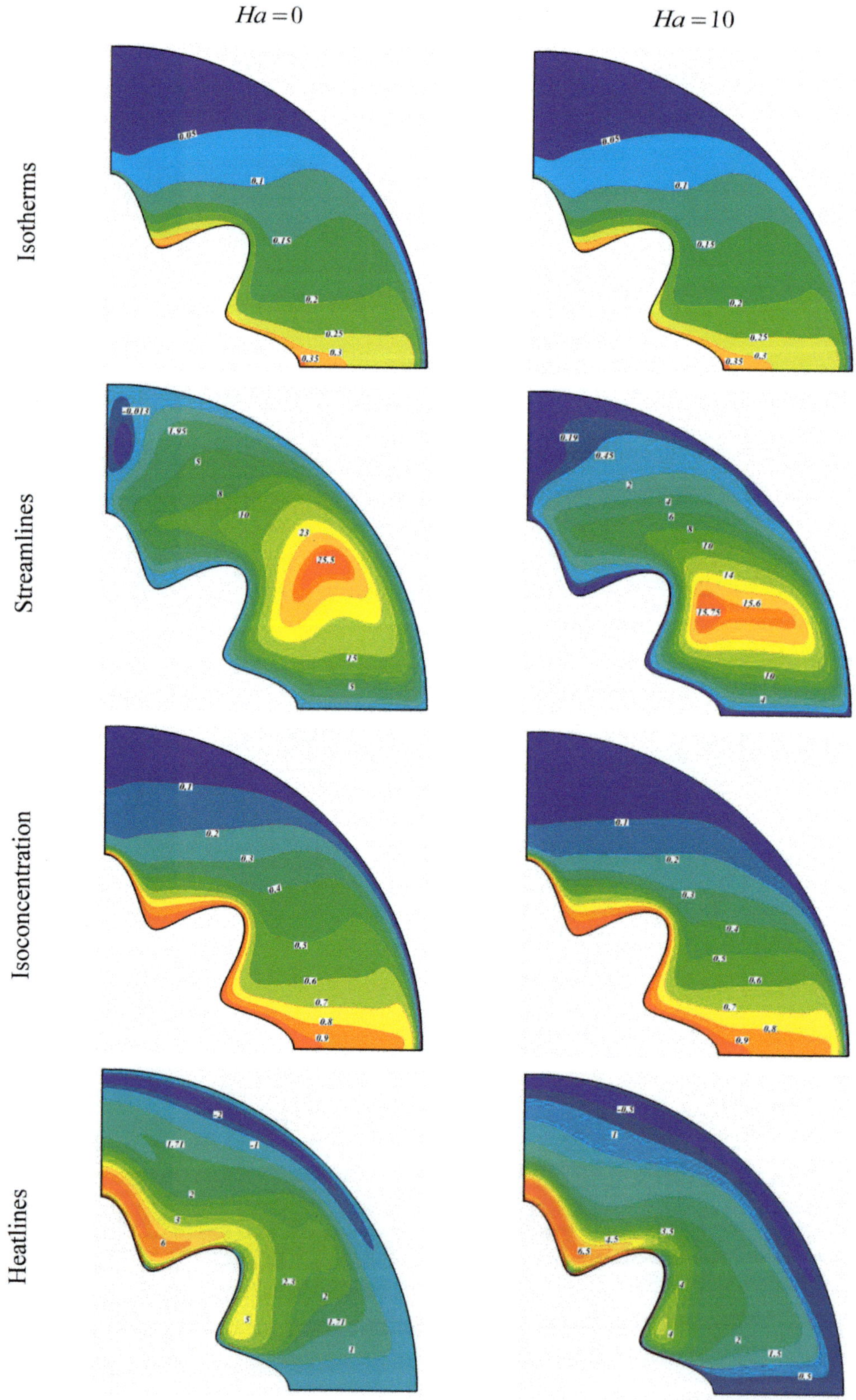

FIGURE 8.32 Comparison of the isotherm, streamline, isoconcentration, and heatline contours for different values of Hartmann number when $Nr = 4, Le = 2, Nt = Nb = 0.5, Ra = 10^5, Mn_F = 5$ and $\Pr = 6.85$.

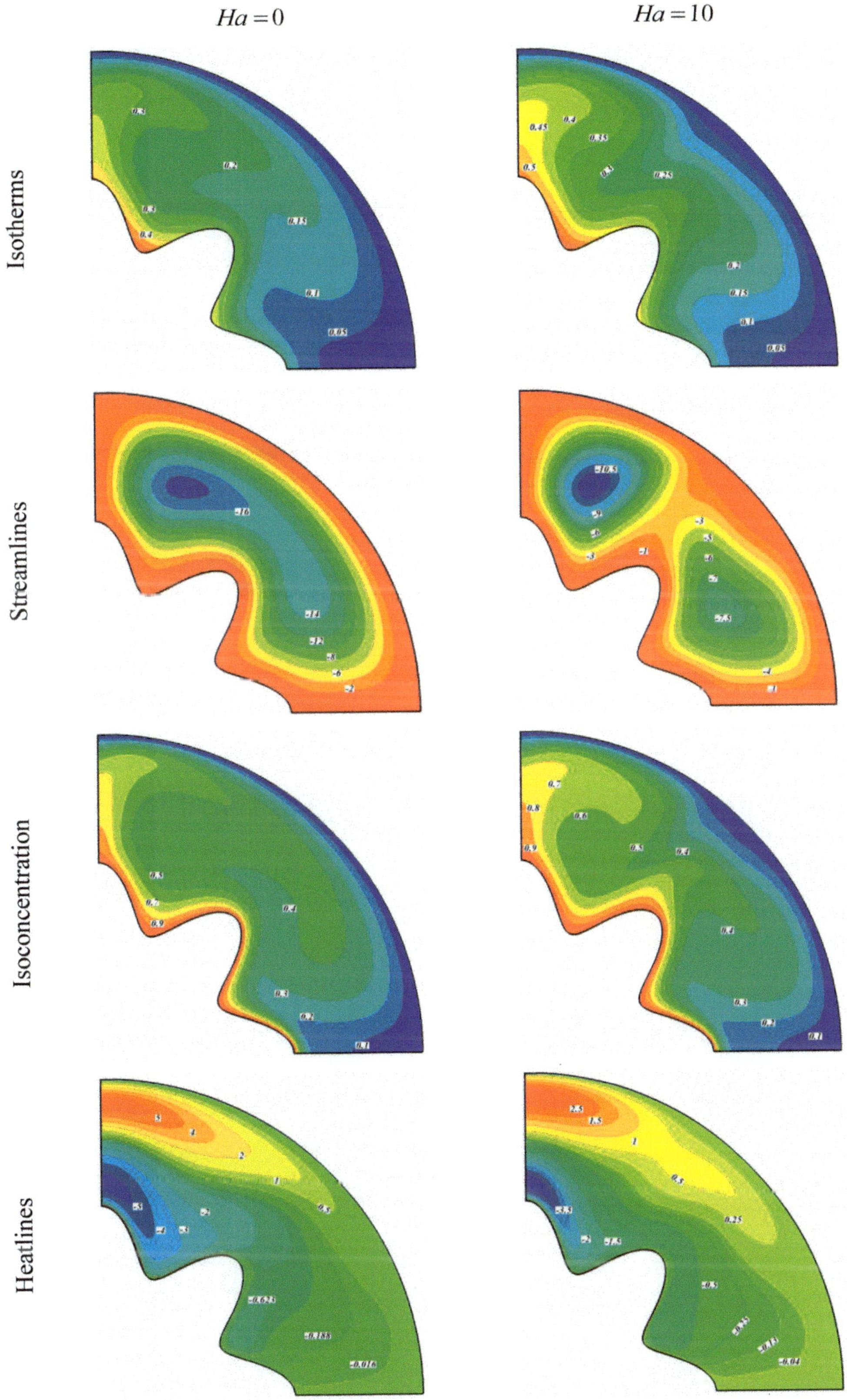

FIGURE 8.33 Comparison of the isotherm, streamline, isoconcentration, and heatline contours for different values of Hartmann number when $Nr = 0.1, Le = 4, Nt = Nb = 0.5, Ra = 10^5, Mn_F = 5$ and $\Pr = 6.85$.

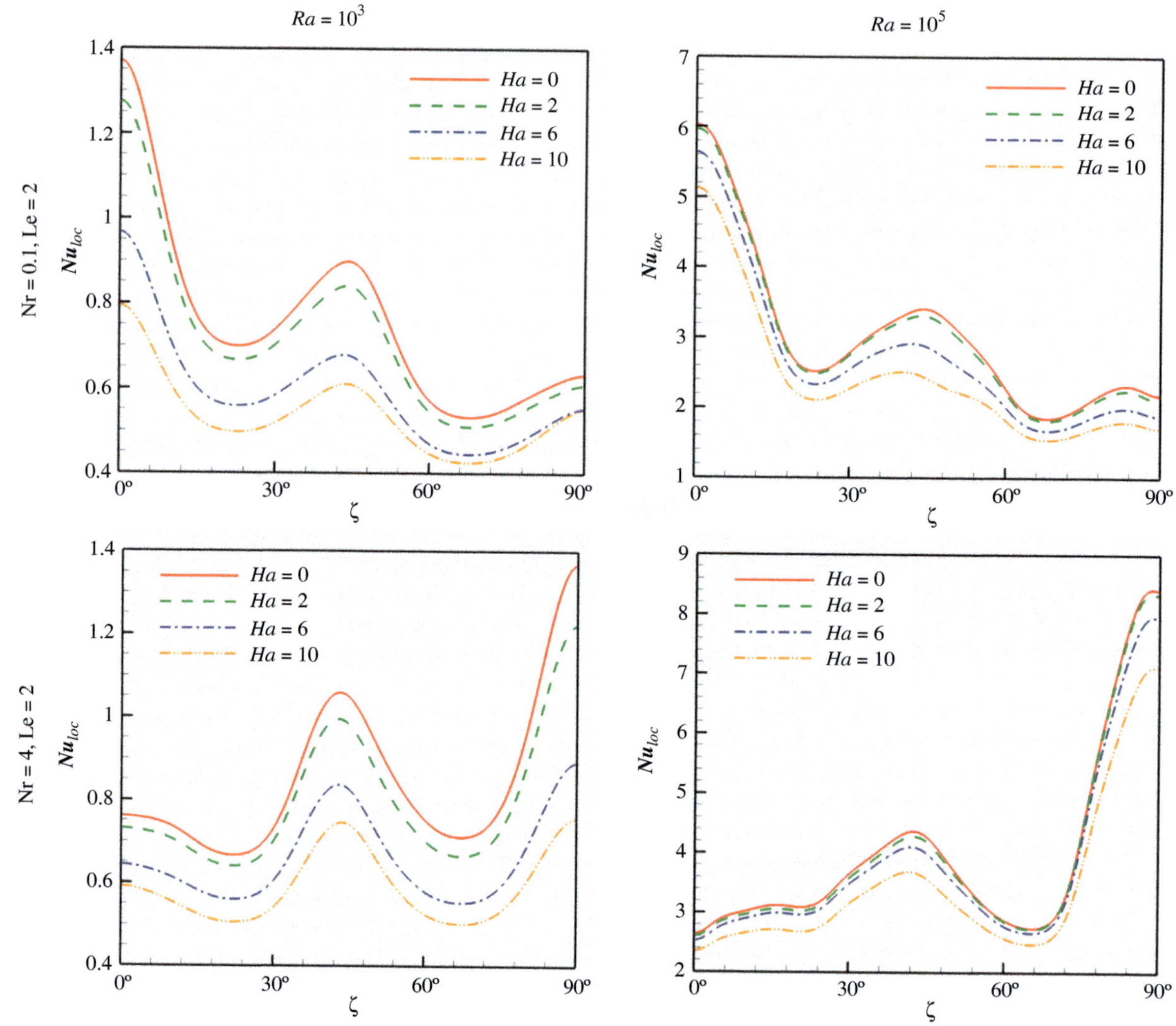

FIGURE 8.34 Effects of Hartmann number, Rayleigh number, buoyancy ratio number, and Lewis number on local Nusselt number Nu_{loc} along cold wall.

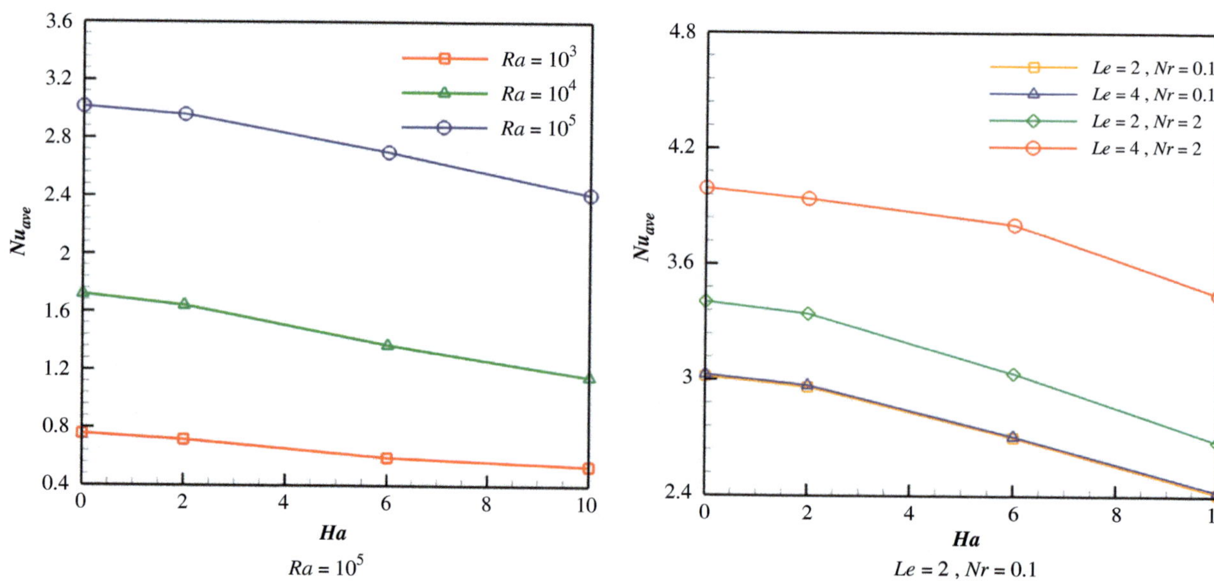

FIGURE 8.35 Effects of Hartmann number, Rayleigh number, buoyancy ratio number, and Lewis number on average Nusselt number.

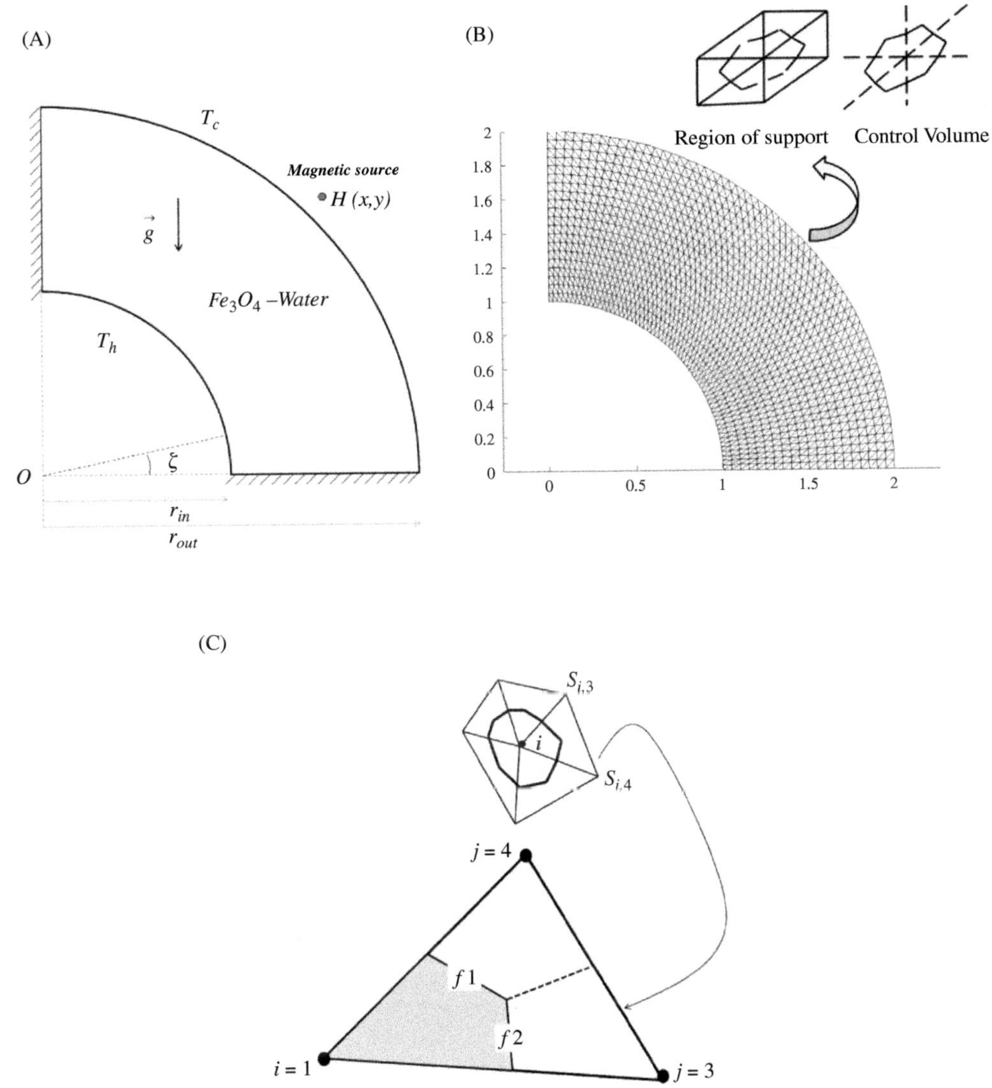

FIGURE 8.36 (A) Geometry and the boundary conditions; (B) the mesh of enclosure considered in this work; (C) a sample triangular element and its corresponding control volume.

$$\overline{H} = \sqrt{\overline{H}_y^2 + \overline{H}_x^2} = \left[(\overline{b}-y)^2 + (\overline{a}-x)^2\right]^{-0.5} \frac{\gamma}{2\pi} \tag{8.124}$$

Fig. 8.37 illustrates the strength of magnetic field due to magnetic source.

8.7.2 Governing Equation

Two-dimensional steady incompressible flow is taken into account. Governing equations are defined as:

$$\frac{\partial u}{\partial x} = -\frac{\partial v}{\partial y} \tag{8.125}$$

$$\left(\frac{\partial u}{\partial y}v + \frac{\partial u}{\partial x}u\right) = \left(\rho_{nf}\right)^{-1}\left[\mu_{nf}\left(\frac{\partial^2 u}{\partial x^2} + \frac{\partial^2 u}{\partial y^2}\right) - \frac{\partial P}{\partial x} + \mu_0 M\frac{\partial \overline{H}}{\partial x} - \sigma_{nf}B_y^2 u + \sigma_{nf}B_x B_y v\right] \tag{8.126}$$

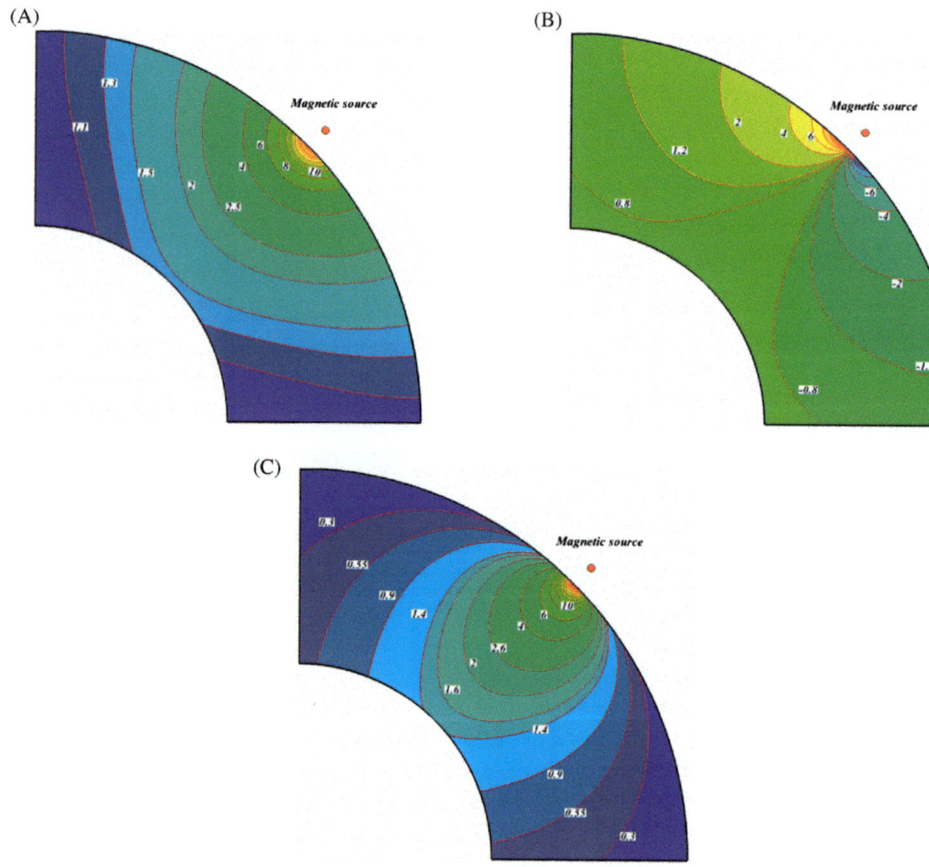

FIGURE 8.37 Contours of the (A) magnetic field strength H; (B) magnetic field intensity component in x direction Hx; (C) magnetic field intensity component in y direction Hy. (A) $H(x, y)$, (B) $Hx(x, y)$, (C) $Hy(x, y)$.

$$\rho_{nf}\left(u\frac{\partial v}{\partial x}+v\frac{\partial v}{\partial y}\right)=-\frac{\partial P}{\partial y}+\mu_{nf}\left(\frac{\partial^2 v}{\partial x^2}+\frac{\partial^2 v}{\partial y^2}\right)+\mu_0 M\frac{\partial \overline{H}}{\partial y}+B_y\sigma_{nf}B_x u-B_x\sigma_{nf}B_x v+(T-T_c)\beta_{nf}g\rho_{nf} \tag{8.127}$$

$$\begin{aligned}\left(u\frac{\partial T}{\partial x}+v\frac{\partial T}{\partial y}\right)&=\frac{k_{nf}}{(\rho C_p)_{nf}}\left(\frac{\partial^2 T}{\partial x^2}+\frac{\partial^2 T}{\partial y^2}\right)+\frac{\sigma_{nf}}{(\rho C_p)_{nf}}(uB_y-vB_x)^2\\&-\frac{\mu_0}{(\rho C_p)_{nf}}T\frac{\partial M}{\partial T}\left(u\frac{\partial \overline{H}}{\partial x}+v\frac{\partial \overline{H}}{\partial y}\right)+\frac{\mu_{nf}}{(\rho C_p)_{nf}}\left\{2\left(\frac{\partial u}{\partial x}\right)^2+2\left(\frac{\partial v}{\partial x}\right)^2+\left(\frac{\partial u}{\partial x}+\frac{\partial v}{\partial y}\right)^2\right\}\end{aligned} \tag{8.128}$$

The terms $\mu_0 M\frac{\partial \overline{H}}{\partial x}$ and $\mu_0 M\frac{\partial \overline{H}}{\partial y}$ are Kelvin forces and the terms $-\sigma_{nf}B_y^2 u+\sigma_{nf}B_x B_y v$ and $-\sigma_{nf}B_x^2 v+\sigma_{nf}B_x B_y u$ are Lorentz forces. Joule heating and magneto caloric effect are considered as second and third terms of the energy equation, respectively. The magnetization M is defined as follows:

$$M=\overline{H}K'(T_c'-T) \tag{8.129}$$

where T_c' is the Curie temperature and K' is a constant.

Magnetic induction definition is $\overline{B}=\mu_0\overline{H}$. ρ_{nf}, $(\rho C_p)_{nf}$, α_{nf}, β_{nf}, μ_{nf}, k_{nf}, and σ_{nf} are as follows:

$$\rho_{nf}=\rho_f(1-\phi)+\rho_s\phi \tag{8.130}$$

$$(\rho C_p)_{nf}=(\rho C_p)_f(1-\phi)+(\rho C_p)_s\phi \tag{8.131}$$

$$\alpha_{nf} = \frac{k_{nf}}{(\rho C_p)_{nf}} \tag{8.132}$$

$$\beta_{nf} = \beta_f(1-\phi) + \beta_s\phi \tag{8.133}$$

$$\mu_{nf} = \frac{\mu_f}{(1-\phi)^{2.5}} \tag{8.134}$$

$$k_{nf} = \frac{-2\phi(k_f - k_s) + 2k_f + k_s}{\phi(k_f - k_s) + 2k_f + k_s} k_f \tag{8.135}$$

$$\frac{\sigma_{nf}}{\sigma_f} = 1 + \frac{3(-1+\sigma_s/\sigma_f)\phi}{(+2+\sigma_s/\sigma_f) - (-1+\sigma_s/\sigma_f)\phi} \tag{8.136}$$

In order to eliminate pressure source terms, the vorticity-stream function formulation has been utilized:

$$\omega = \frac{\partial v}{\partial x} - \frac{\partial u}{\partial y}, \quad u = \frac{\partial \psi}{\partial y}, v = -\frac{\partial \psi}{\partial x} \tag{8.137}$$

Nondimensional parameters which are used in this study are:

$$Y = y/L\ X = x/L, \quad \Omega = \frac{\omega L^2}{\alpha_f}, \Psi = \frac{\psi}{\alpha_f}, (H, H_x, H_y) = \frac{(\overline{H}, \overline{H_x}, \overline{H_y})}{\left(\frac{\gamma}{2\pi|b|}\right)}$$
$$V = \frac{vL}{\alpha_f}, \Theta = \frac{T - T_c}{T_h - T_c}, U = \frac{uL}{\alpha_f} \tag{8.138}$$

The final dimensionless equations are:

$$\begin{aligned}
&-\frac{\partial\Omega}{\partial Y}\frac{\partial\Psi}{\partial X} + \frac{\partial\Omega}{\partial X}\frac{\partial\Psi}{\partial Y} = \Pr\left[\frac{\mu_{nf}/\mu_f}{\rho_{nf}/\rho_f}\right]\left(\frac{\partial^2\Omega}{\partial X^2} + \frac{\partial^2\Omega}{\partial Y^2}\right) + Ra\Pr\left[\frac{\beta_{nf}}{\beta_f}\right]\left(\frac{\partial\Theta}{\partial X}\right) \\
&+ Mn_F\Pr\left(\frac{\rho_f}{\rho_{nf}}\right)\left\{\frac{\partial H}{\partial X}\frac{\partial\Theta}{\partial Y} - \frac{\partial H}{\partial Y}\frac{\partial\Theta}{\partial X}\right\}H - Ha^2\Pr\left[\frac{\sigma_{nf}/\sigma_f}{\rho_{nf}/\rho_f}\right] \\
&\times\left\{\frac{\partial V}{\partial X}H_x^2 + V\left(2H_x\frac{\partial H_x}{\partial X}\right) - H_yH_x\frac{\partial U}{\partial X} - UH_y\frac{\partial H_x}{\partial X} - H_x\frac{\partial H_y}{\partial X}U\right. \\
&\left. - H_y\frac{\partial U}{\partial Y}H_y - 2H_y\frac{\partial H_y}{\partial Y}U + \frac{\partial V}{\partial Y}H_yH_x + VH_y\frac{\partial H_x}{\partial Y} + H_xV\frac{\partial H_y}{\partial Y}\right\}
\end{aligned} \tag{8.139}$$

$$\begin{aligned}
&-\frac{\partial\Theta}{\partial Y}\frac{\partial\Psi}{\partial X} + \frac{\partial\Theta}{\partial X}\frac{\partial\Psi}{\partial Y} = (k_{nf}/k_f)\left(\frac{\partial^2\Theta}{\partial Y^2} + \frac{\partial^2\Theta}{\partial X^2}\right)\left((\rho C_P)_{nf}/(\rho C_P)_f\right)^{-1} + \frac{\sigma_{nf}}{\sigma_f}Ha^2\,Ec\{V\,H_x - U\,H_y\}^2\left((\rho C_P)_{nf}/(\rho C_P)_f\right)^{-1} \\
&+ Mn_F Ec\frac{(\rho C_P)_f}{(\rho C_P)_{nf}}\left\{\frac{\partial H}{\partial Y}V + \frac{\partial H}{\partial X}U\right\}H(\varepsilon_1 + \Theta) + \frac{\mu_{nf}}{\mu_f}Ec\left\{2\left(\frac{\partial U}{\partial X}\right)^2 + 2\left(\frac{\partial V}{\partial X}\right)^2 + \left(\frac{\partial U}{\partial X} + \frac{\partial V}{\partial Y}\right)^2\right\}\left((\rho C_P)_{nf}/(\rho C_P)_f\right)^{-1}
\end{aligned} \tag{8.140}$$

$$\Omega = -\left(\frac{\partial^2\Psi}{\partial Y^2} + \frac{\partial^2\Psi}{\partial X^2}\right) \tag{8.141}$$

The nondimensional parameters are defined as:

$$Ra_f = g\beta_f L^3 (T_h - T_c)/(\alpha_f \upsilon_f), Ha = LB_x\sqrt{\sigma_f/\mu_f}, \Pr_f = \upsilon_f/\alpha_f,$$
$$Ec = (\mu_f \alpha_f)/\left[(\rho C_P)_f \Delta T\, L^2\right], Mn_F = \mu_0 H_0^2 K'(T_h - T_c)L^2/(\mu_f \alpha_f) \tag{8.142}$$

Table 8.1 shows the thermophysical properties of the nanofluid. Local, average Nusselt numbers and heat transfer improvement are defined as:

$$Nu_{loc} = \left(\frac{k_{nf}}{k_f}\right)\frac{\partial \Theta}{\partial r} \tag{8.143}$$

$$Nu_{ave} = \frac{1}{0.5\pi}\int_0^{0.5\pi} Nu_{loc}(\zeta)\, d\zeta \tag{8.144}$$

$$En = \frac{Nu(\phi = 0.04) - Nu(basefluid)}{Nu(basefluid)} \times 100 \tag{8.145}$$

8.7.3 Effects of Active Parameters

The influences of applying an outside magnetic field on Fe_3O_4-water hydrothermal treatment in a cavity are studied. Combined influences of Lorentz and Kelvin forces are taken into account. Influences of the Hartmann number ($Ha = 0$, 2, 6, and 10), nanoparticle volume fraction ($\phi = 0\%$ and 4%), Rayleigh number ($Ra = 10^3$, 10^4, and 10^5) and Magnetic number ($Mn_F = 0$, 20, 60, and 100) on velocity and temperature distribution are examined.

Fig. 8.38 depicts the influence of adding nanoparticles in water on streamlines. By adding nanoparticles, the energy transport of the fluid augments and in turn fluid velocity enhances. So, the power of flow enhancing according to changes occurs for values of stream functions.

Streamline and isotherm contours for various active parameters are demonstrated in Figs. 8.39–8.41. Heat conduction is the dominating mode when buoyancy forces are weak. In the absence of a magnetic source, the streamlines illustrate one rotating eddy. As Lorentz forces augment the primary eddy turns into three smaller eddies. One of them which is located near the magnetic source rotates counterclockwise. As Kelvin forces augment, the primary vortex turns into two eddies in which the clockwise eddy is greater than the other. By increasing buoyancy forces, the temperature gradient near the hot wall enhances due to the dominant convective mode. Moreover, a thermal plume generates near the vertical axis due to the existence of eddies in this region.

As Lorentz forces increase the thermal plume changes to two plumes which are located at $\zeta = 45^\circ$ and $\zeta = 90^\circ$. For high Rayleigh number ($Ra = 10^5$), Lorentz forces have no noteworthy impact on flow style in the absence of

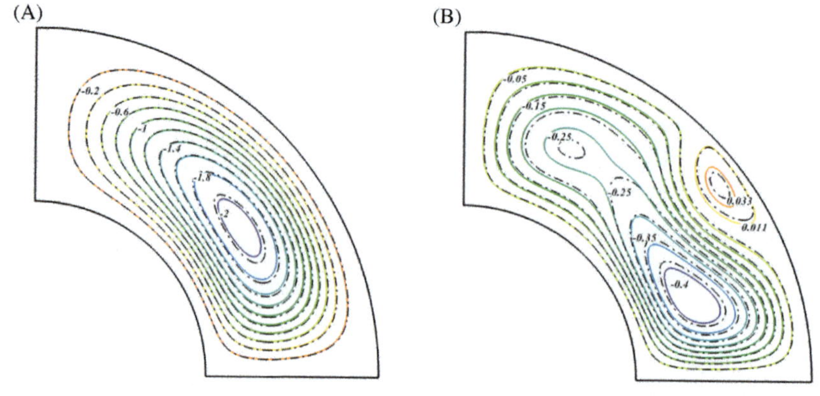

FIGURE 8.38 Comparison of the streamlines between nanofluid ($\phi = 0.04$) (–·–·–) and pure fluid ($\phi = 0$) (—) when $Pr = 6.8$. (A) $Ra = 10^3, Mn_F = 0, Ha = 0$, (B) $Ra = 10^3, Mn_F = 100, Ha = 10$.

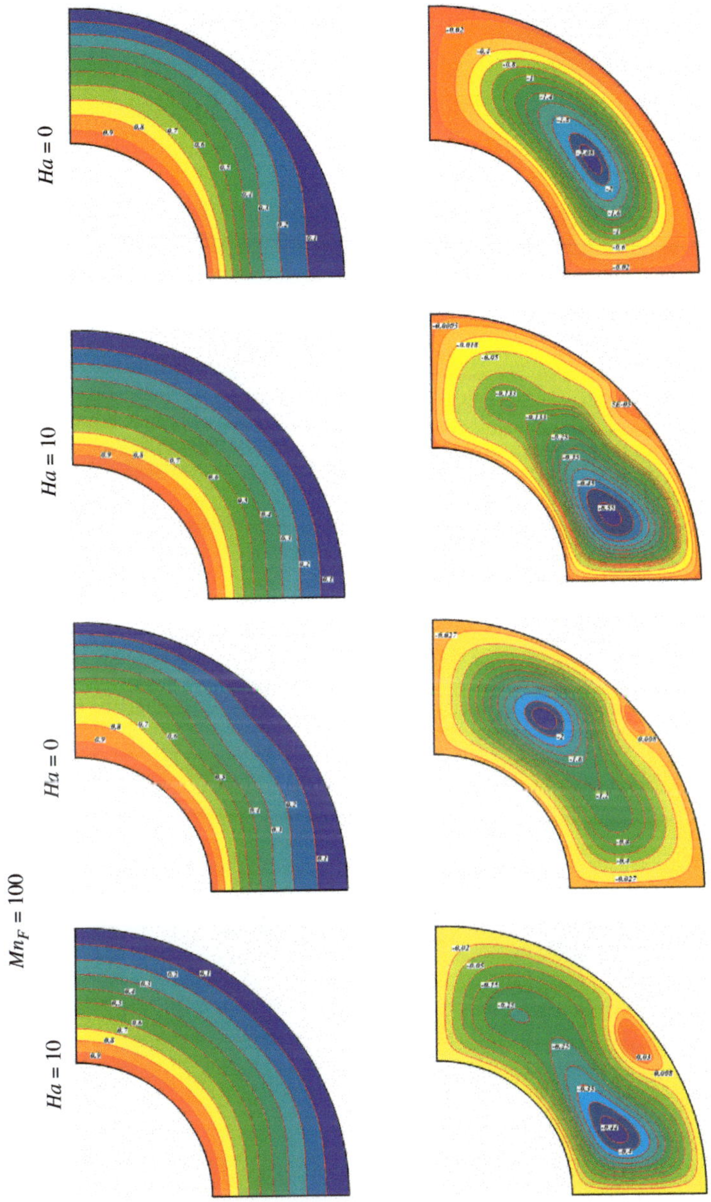

FIGURE 8.39 Isotherm (left) and streamline (right) contours for different values of Hartmann number and Magnetic number when $Ra = 10^3$.

Kelvin forces. When $Mn_F = 100$, the thermal plume slants to $\zeta = 45°$ with augment of Lorentz forces. Also another reverse plume is observed near the vertical axis. This observation is due to the existence of two eddies which rotate in different directions.

Fig. 8.42 depicts the impacts of Magnetic, Hartmann, and Raleigh numbers on local Nusselt number. This profile increases with the increase of Magnetic and Rayleigh numbers, however it reduces with the rise of the Hartmann number. Local maximum and minimum points in these profiles are related to existence thermal plumes. Fig. 8.43 and Table 8.7 show the influences of Hartmann, Magnetic, and Rayleigh numbers on Nu_{ave}. Nu_{ave} is an augmenting function of Mn_F, Ra, and ϕ but it is a reducing function of Ha.

Fig. 8.44 depicts the influences of Rayleigh, Magnetic, and Hartmann numbers on heat transfer improvement. Heat transfer improvement enhances with the augment of the Lorentz force but it reduces with the rise of buoyancy force. It is because of this point that adding nanoparticles is more practical when conduction is the dominant mode. This figure indicates that in the absence of Lorentz forces, at first En increases then decreases with increase of Mn_F.

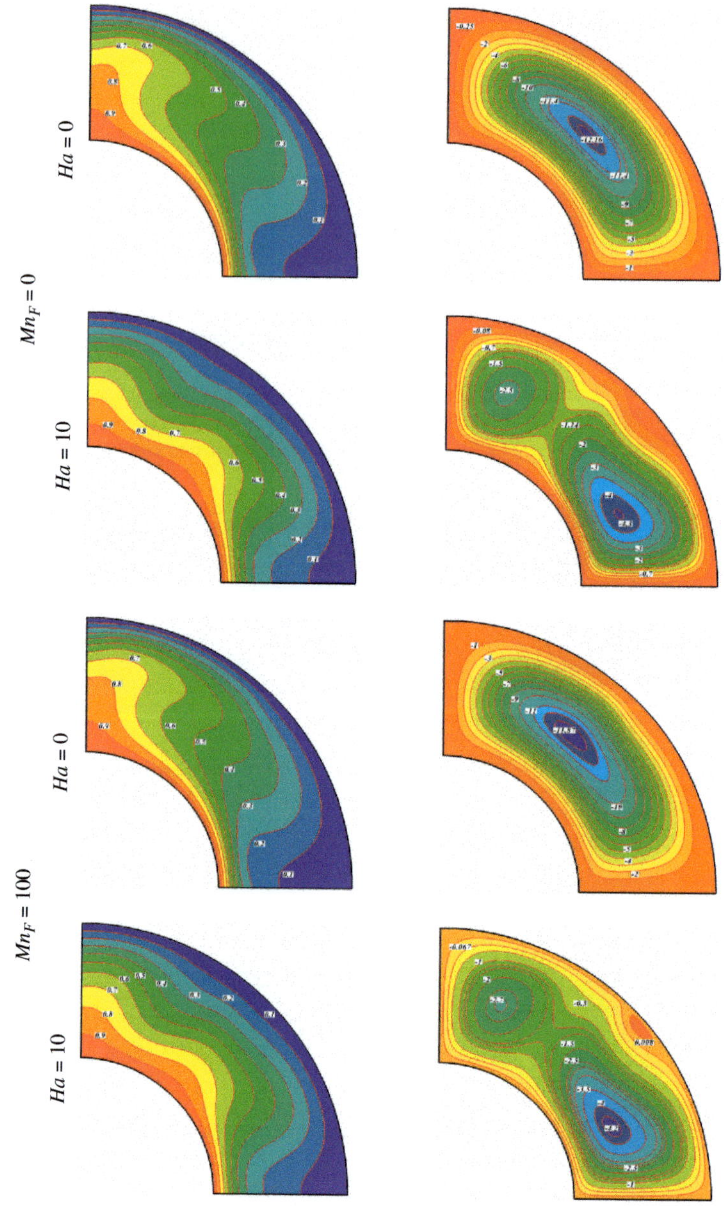

FIGURE 8.40 Isotherm (left) and streamline (right) contours for different values of Hartmann number and Magnetic number when $Ra = 10^4$.

8.8 NANOFLUID TRANSPORTATION IN A CURVED CAVITY IN THE PRESENCE OF MAGNETIC SOURCE

8.8.1 Problem Definition

A curved cavity with hot left wall is considered. Boundary conditions are depicted in Fig. 8.45. A magnetic source exists near the left bottom corner. A magnetic source has been considered as shown in Fig. 8.46. $\overline{H_x}, \overline{H_y}, \overline{H}$ can be calculated as follows:

$$\overline{H_y} = \left[(\overline{b}-y)^2 + (\overline{a}-x)^2\right]^{-1} \frac{\gamma}{2\pi}(\overline{a}-x), \tag{8.146}$$

$$\overline{H_x} = \left[(\overline{b}-y)^2 + (\overline{a}-x)^2\right]^{-1} \frac{\gamma}{2\pi}(y-\overline{b}), \tag{8.147}$$

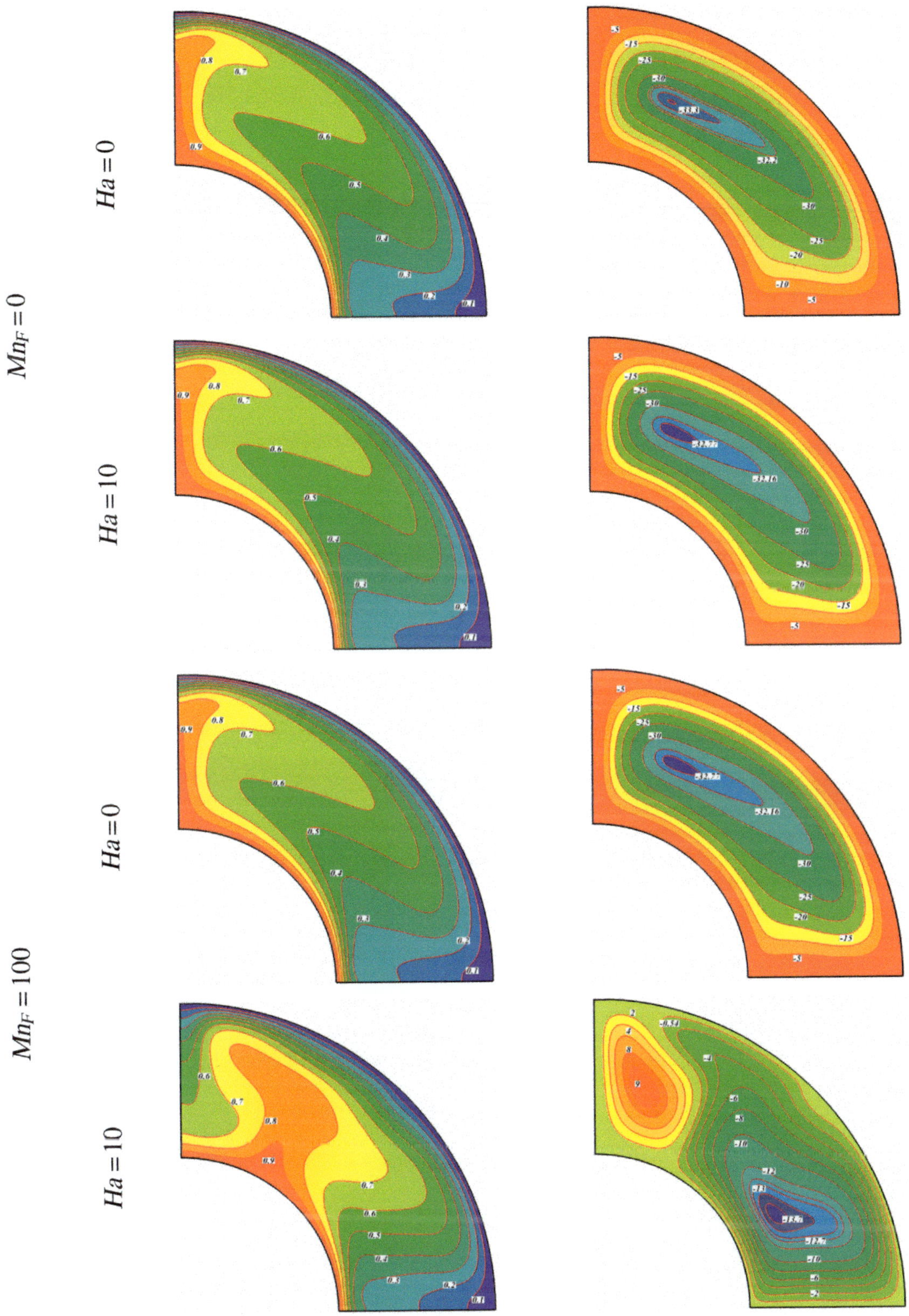

FIGURE 8.41 Isotherm (left) and streamline (right) contours for different values of Hartmann number and Magnetic number when $Ra = 10^5$.

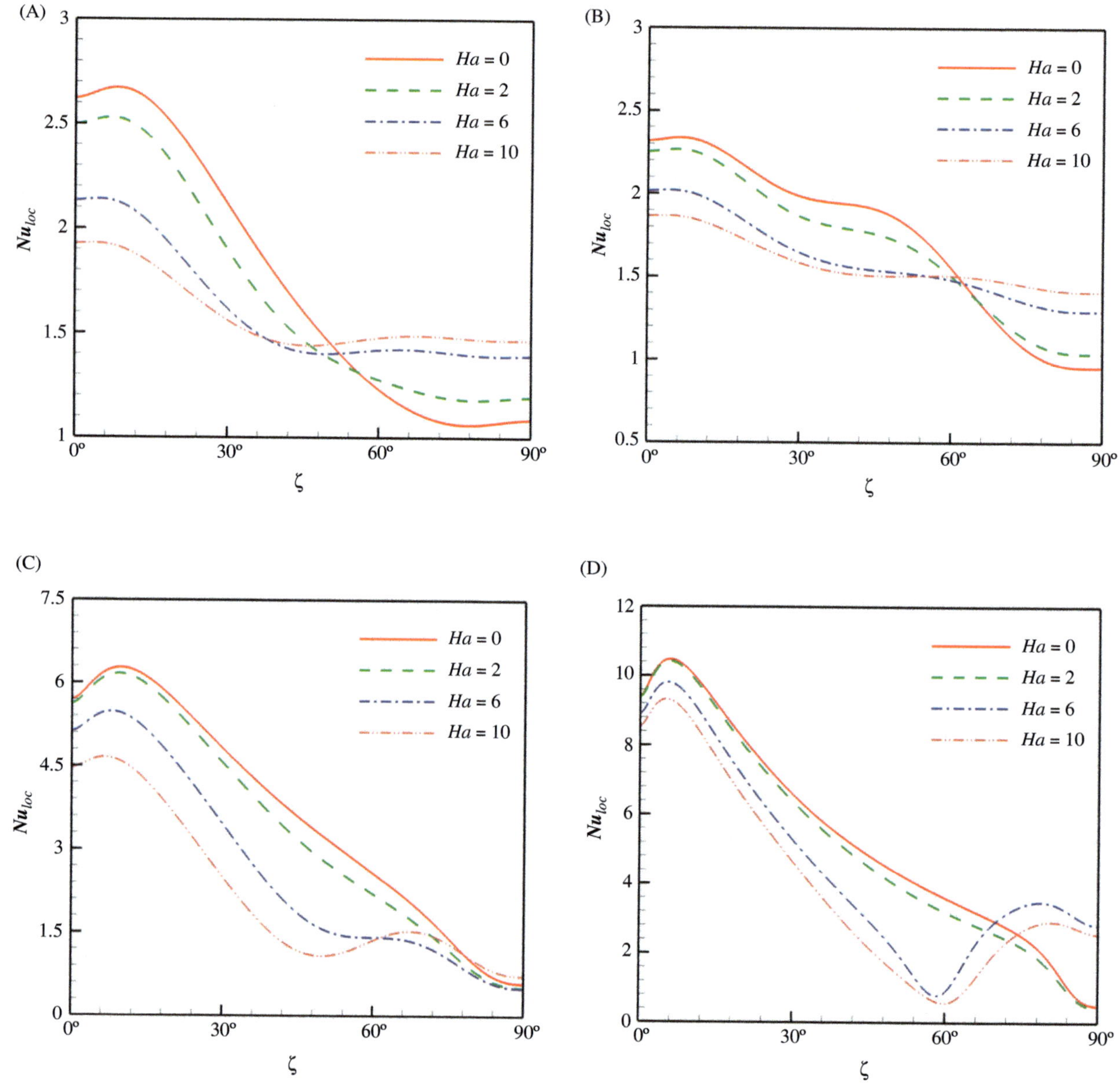

FIGURE 8.42 Effects of Magnetic number, Hartmann number, and Rayleigh number on local Nusselt number Nu_{ave} along hot wall. (A) $Ra = 10^3, Mn_F = 0$, (B) $Ra = 10^3, Mn_F = 100$, (C) $Ra = 10^4, Mn_F = 100$, (D) $Ra = 10^5, Mn_F = 100$.

$$\overline{H} = \sqrt{\overline{H}_x^2 + \overline{H}_y^2} \tag{8.148}$$

8.8.2 Governing Equation

2D laminar nanofluid flow and forced convective heat transfer are taken into account. The governing PDEs are:

$$\frac{\partial v}{\partial y} + \frac{\partial u}{\partial x} = 0, \tag{8.149}$$

$$\left(u\frac{\partial u}{\partial x} + \frac{\partial u}{\partial y}v\right) = \left(\rho_{nf}\right)^{-1}\left[\left(\frac{\partial^2 u}{\partial y^2} + \frac{\partial^2 u}{\partial x^2}\right)\mu_{nf} - \frac{\partial P}{\partial x} - \sigma_{nf}\mu_0^2 H_y^2 u + \sigma_{nf}\mu_0^2 H_x H_y v\right] \tag{8.150}$$

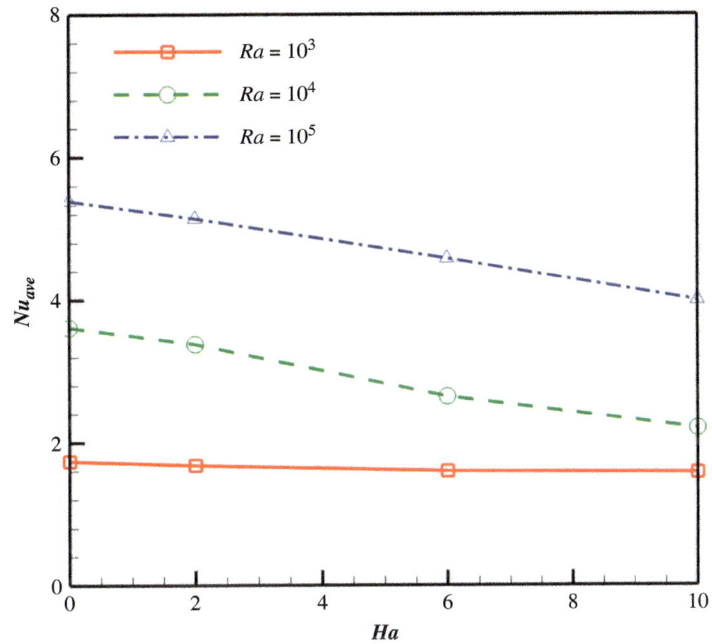

FIGURE 8.43 Effects of Hartmann number and Rayleigh number on average Nusselt number Nu_{ave} along hot wall when $Mn_F = 100$.

TABLE 8.7 Effect of Hartmann Number and Magnetic Number on Average Nusselt Number Nu_{ave} Along Hot Wall When $Ra = 10^5$

	Ha		
Mn_F	0	2	10
0	5.380431	5.136342	3.997727
20	5.382286	5.138101	3.997764
60	5.385799	5.141522	3.99787
100	5.389023	5.144803	3.99802

$$\rho_{nf}\left(\frac{\partial v}{\partial x}u + \frac{\partial v}{\partial y}v\right) = +\mu_{nf}\left(\frac{\partial^2 v}{\partial x^2} + \frac{\partial^2 v}{\partial y^2}\right) - \frac{\partial P}{\partial y} + \mu_0^2 H_y \sigma_{nf} H_x u - \mu_0^2 H_x \sigma_{nf} H_x v + (T - T_c)\beta_{nf} g \rho_{nf} \tag{8.151}$$

$$\begin{aligned}(\rho C_p)_{nf}\left(v\frac{\partial T}{\partial y} + u\frac{\partial T}{\partial x}\right) &= \sigma_{nf}\mu_0^2(H_x v - H_y u)^2 + k_{nf}\left(\frac{\partial^2 T}{\partial x^2} + \frac{\partial^2 T}{\partial y^2}\right)\\ &+ \mu_{nf}\left\{2\left(\frac{\partial u}{\partial x}\right)^2 + 2\left(\frac{\partial v}{\partial y}\right)^2 + \left(\frac{\partial u}{\partial y} + \frac{\partial v}{\partial x}\right)^2\right\} - \frac{1}{(\rho C_p)_{nf}}\frac{\partial q_r}{\partial y}, \left[q_r = -\frac{4\sigma_e}{3\beta_R}\frac{\partial T^4}{\partial y}, \quad T^4 \cong 4T_c^3 T - 3T_c^4\right]\end{aligned} \tag{8.152}$$

$\rho_{nf}, (\rho C_p)_{nf}, \alpha_{nf}, \beta_{nf}, \mu_{nf}, k_{nf}$, and σ_{nf} are calculated as

$$\rho_{nf} = \rho_f(1-\phi) + \rho_s\phi, \tag{8.153}$$

$$(\rho C_p)_{nf} = (\rho C_p)_f(1-\phi) + (\rho C_p)_s\phi, \tag{8.154}$$

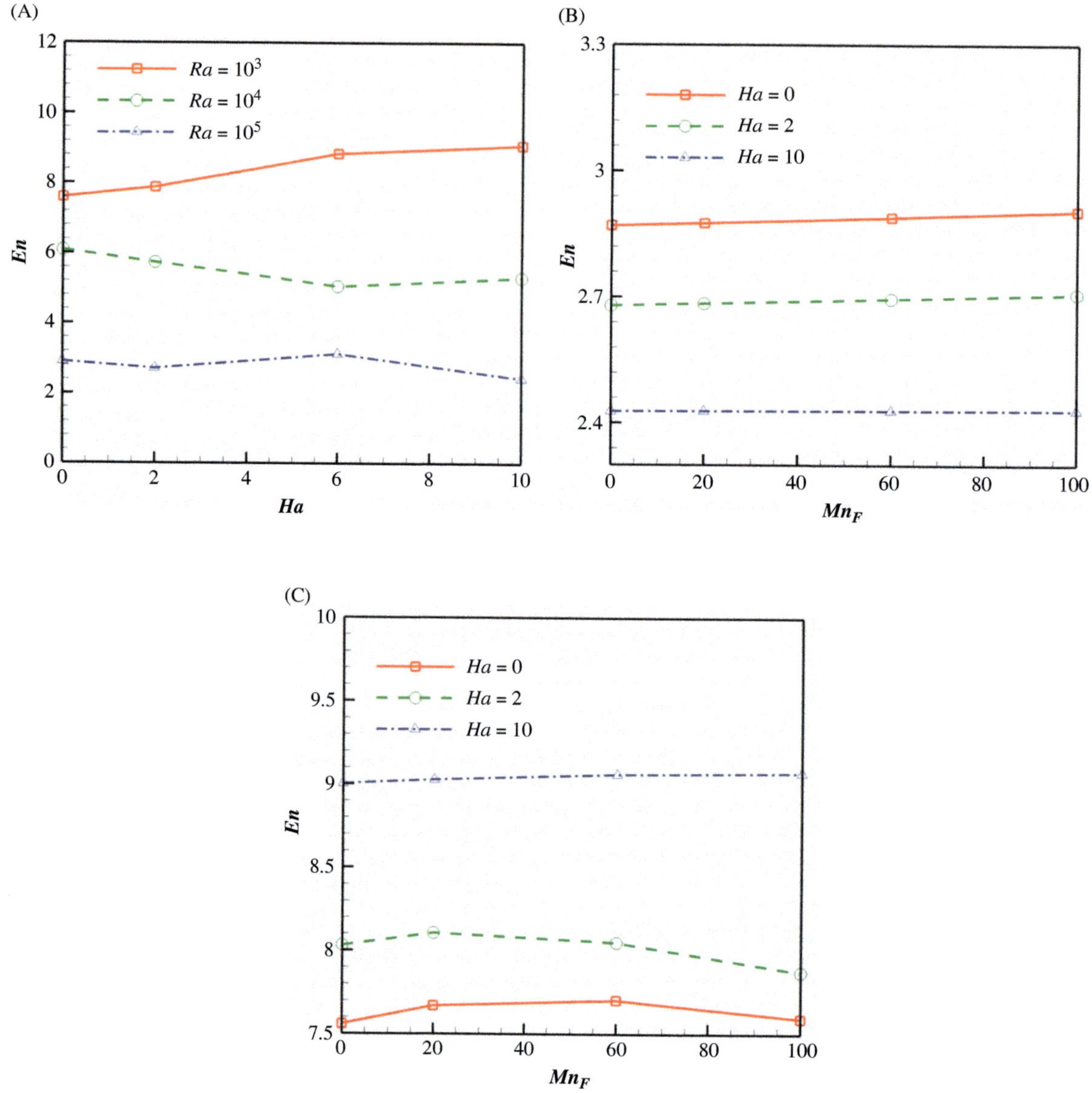

FIGURE 8.44 Effects of Magnetic number, Rayleigh number and Hartmann number on heat transfer enhancement. (A) $Mn_F = 100$, (B) $Ra = 10^5$, (C) $Ra = 10^3$.

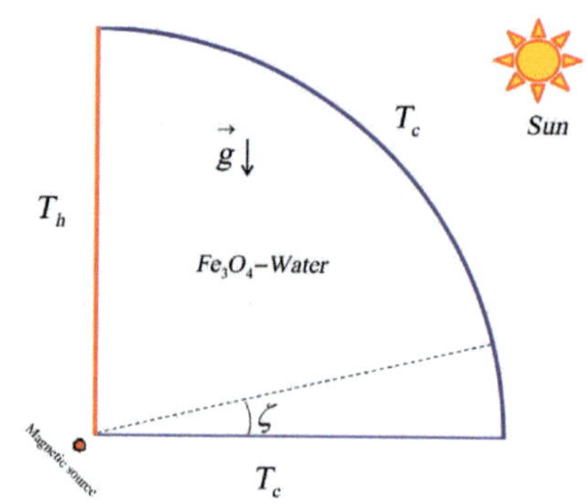

FIGURE 8.45 Geometry and the boundary conditions.

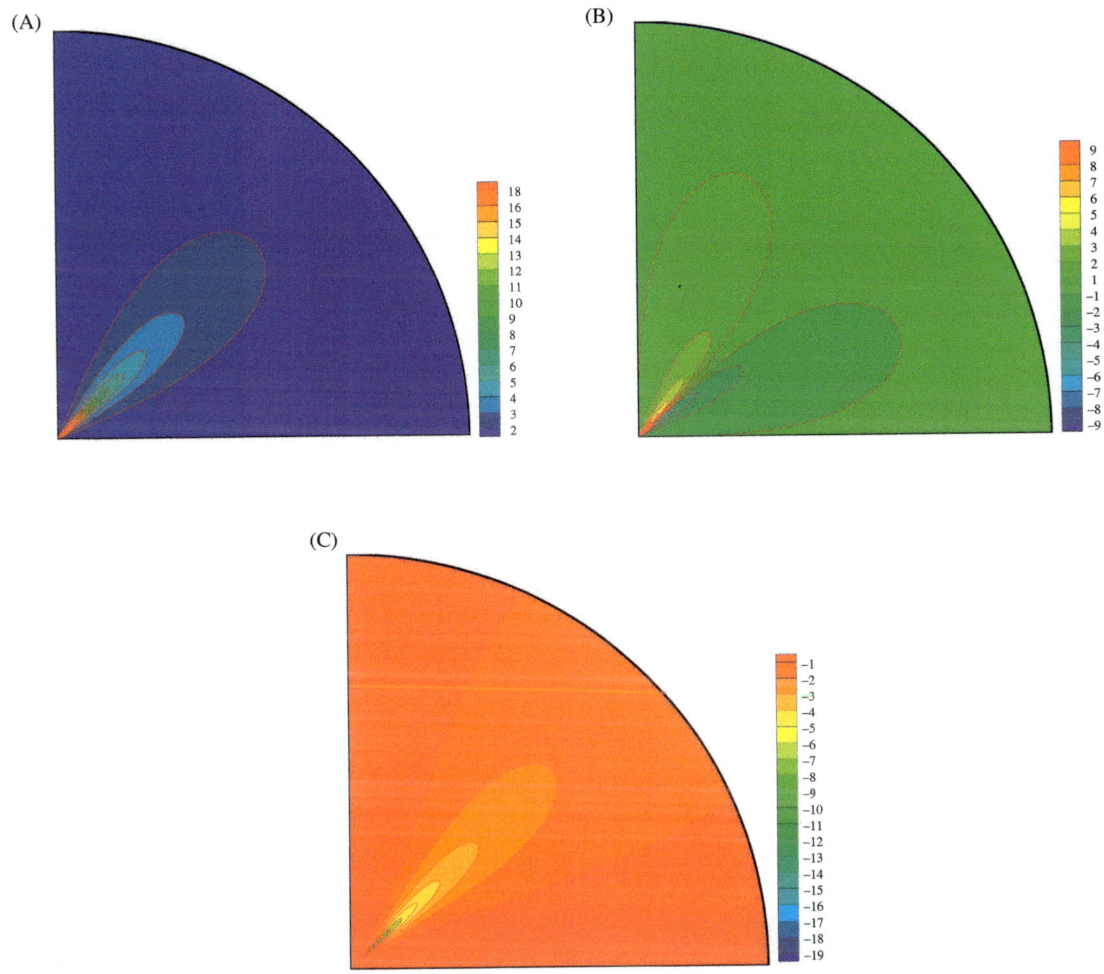

FIGURE 8.46 Contours of the (A) magnetic field strength H; (B) magnetic field intensity component in x direction Hx; (C) magnetic field intensity component in y direction Hy. (A) $H(x, y)$, (B) $Hx(x, y)$, (C) $Hy(x, y)$.

$$\alpha_{nf} = \frac{k_{nf}}{(\rho C_p)_{nf}}, \tag{8.155}$$

$$\beta_{nf} = \beta_f(1-\phi) + \beta_s\phi. \tag{8.156}$$

$$\mu_{nf} = \frac{\mu_f}{(1-\phi)^{2.5}}. \tag{8.157}$$

$$k_{nf} = k_f\left(\frac{k_s - 2\phi(k_f - k_s) + 2k_f}{k_s + \phi(k_f - k_s) + 2k_f}\right), \tag{8.158}$$

$$\sigma_{nf} = \sigma_f\left[\frac{3(\sigma 1 - 1)\phi}{(\sigma 1 + 2) - (\sigma 1 - 1)\phi} + 1\right], \qquad \sigma 1 = \sigma_s/\sigma_f \tag{8.159}$$

Dimensionless parameters are defined as:

$$(b, a) = \frac{(\bar{b}, \bar{a})}{L}, (H_y, H_x, H) = \frac{(\overline{H_y}, \overline{H_x}, \overline{H})}{\overline{H_0}}, P = \frac{p}{\rho_f(\alpha_f/L)^2}$$
$$U = \frac{uL}{\alpha_f}, V = \frac{vL}{\alpha_f}, \Theta = \frac{T - T_c}{(T_h - T_c)}, (X, Y) = \frac{(x, y)}{L} \tag{8.160}$$

So equations change to:

$$\frac{\partial V}{\partial Y}+\frac{\partial U}{\partial X}=0, \tag{8.161}$$

$$U\frac{\partial U}{\partial X}+\frac{\partial U}{\partial Y}V=\Pr\left[\frac{\mu_{nf}/\mu_f}{\rho_{nf}/\rho_f}\right]\left(\frac{\partial^2 U}{\partial Y^2}+\frac{\partial^2 U}{\partial X^2}\right)-Ha^2\Pr\left[\frac{\sigma_{nf}/\sigma_f}{\rho_{nf}/\rho_f}\right]\left(H_y^2U-H_xH_yV\right)-\frac{\partial P}{\partial X}, \tag{8.162}$$

$$\begin{aligned}V\frac{\partial V}{\partial Y}+U\frac{\partial V}{\partial X}&=\Pr\left(\frac{\partial^2 V}{\partial Y^2}+\frac{\partial^2 V}{\partial X^2}\right)\left[\frac{\mu_{nf}/\mu_f}{\rho_{nf}/\rho_f}\right]\\&-Ha^2\Pr\left[\frac{\sigma_{nf}/\sigma_f}{\rho_{nf}/\rho_f}\right](H_x^2V-H_xH_yU)-\frac{\partial P}{\partial Y}+Ra\Pr\left[\frac{\beta_{nf}}{\beta_f}\right]\Theta,\end{aligned} \tag{8.163}$$

$$\begin{aligned}V\frac{\partial \Theta}{\partial Y}+U\frac{\partial \Theta}{\partial X}&=\left[\frac{k_{nf}\,(\rho C_P)_f}{k_f\,(\rho C_P)_{nf}}\right]\left(\frac{\partial^2 \Theta}{\partial Y^2}+\frac{\partial^2 \Theta}{\partial X^2}\right)+Ha^2\,Ec\left(\frac{(\rho C_P)_f}{(\rho C_P)_{nf}}\frac{\sigma_{nf}}{\sigma_f}\right)\{V\,H_x-U\,H_y\}^2\\&+\left[\frac{\frac{\mu_{nf}}{\mu_f}}{\frac{(\rho C_P)_{nf}}{(\rho C_P)_f}}\right]Ec\left\{2\left(\frac{\partial U}{\partial X}\right)^2+2\left(\frac{\partial V}{\partial Y}\right)^2+\left(\frac{\partial U}{\partial Y}+\frac{\partial V}{\partial X}\right)^2\right\}+\left(\frac{4}{3}Rd\left(\frac{k_{nf}}{k_f}\right)^{-1}\right)\frac{\partial^2\Theta}{\partial Y^2}\end{aligned} \tag{8.164}$$

and dimensionless parameters are

$$\begin{aligned}&Ra_f=g\beta_f L^3\Delta T/(\alpha_f\upsilon_f),\Pr_f=\upsilon_f/\alpha_f,Ha=L\mu_0H_0\sqrt{\sigma_f/\mu_f},\\&Ec=\left(\mu_f\alpha_f\right)/\left[(\rho C_P)_f\Delta T\,L^2\right],\ \ Rd=4\sigma_e T_c^3/(\beta_R k_f)\end{aligned} \tag{8.165}$$

The thermophysical properties of Fe_3O_4 and water are presented in Table 8.1. The pressure gradient source terms are discarded by vorticity stream function.

$$\Omega=\frac{\omega L^2}{\alpha_f},\Psi=\frac{\psi}{\alpha_f},\omega=-\frac{\partial u}{\partial y}+\frac{\partial v}{\partial x},(u,v)=\left(\frac{\partial\psi}{\partial y},-\frac{\partial\psi}{\partial x}\right) \tag{8.166}$$

According to Fig. 8.45, boundary conditions are

$$\begin{array}{ll}\text{on left wall} & \theta=1.0\\ \text{on all walls} & \Psi=0.0\\ \text{on other walls} & \theta=0.0\end{array} \tag{8.167}$$

Nu_{local}, Nu_{ave} along curve wall are:

$$Nu_{loc}=\left(\frac{k_{nf}}{k_f}\right)\left(1+\frac{4}{3}Rd\left(\frac{k_{nf}}{k_f}\right)^{-1}\right)\frac{\partial\Theta}{\partial r} \tag{8.168}$$

$$Nu_{ave}=\frac{1}{0.5\pi}\int_0^{0.5\pi}Nu_{loc}(\zeta)\,d\zeta \tag{8.169}$$

8.8.3 Effects of Active Parameters

The influence of thermal radiation on convection heat transfer in a curved cavity is reported. The roles of Radiation parameter (*Rd*), Hartmann number (*Ha*), Rayleigh number (*Ra*), and volume fraction of Fe_3O_4 (ϕ) are presented. *Ec* and Pr are 10^{-5} and 6.8, respectively.

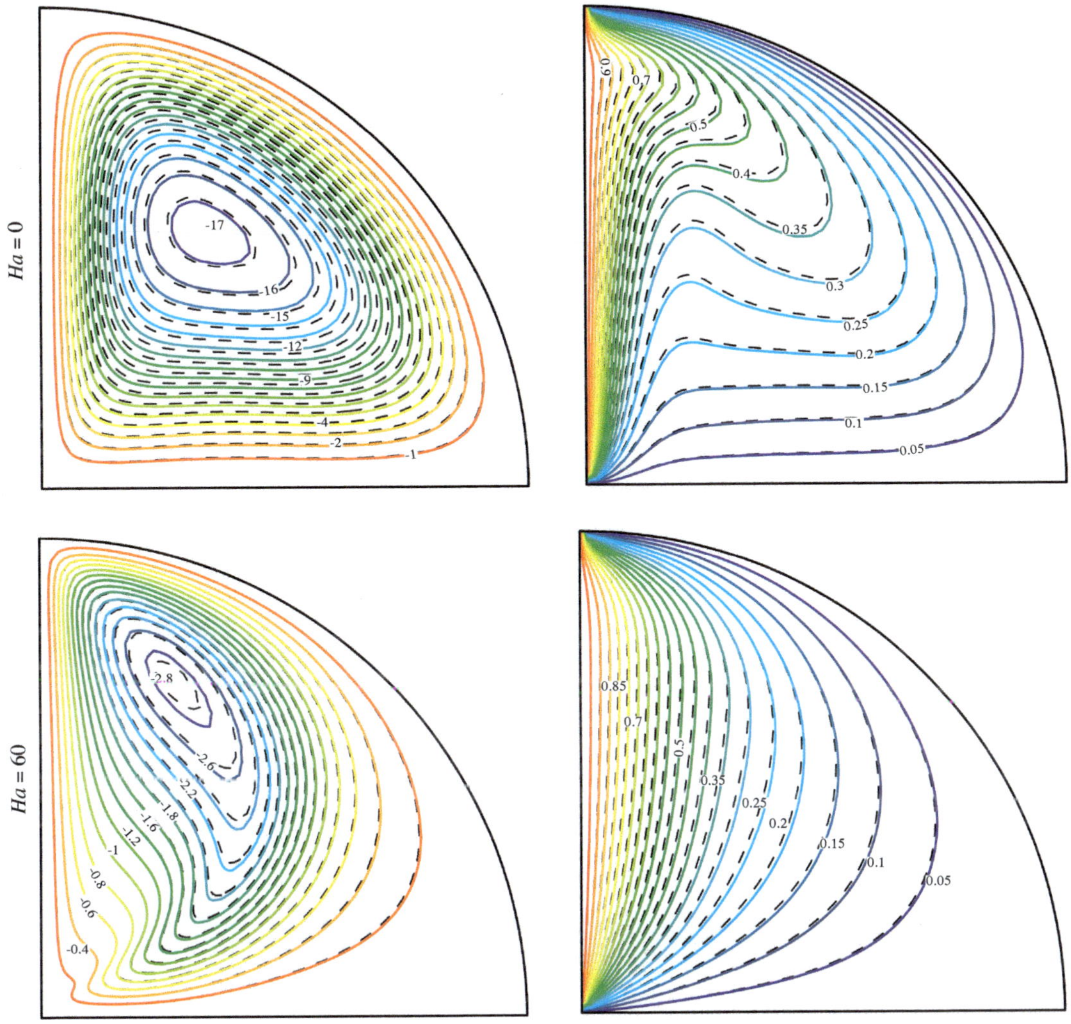

FIGURE 8.47 Influence of nanofluid volume fraction on streamline (left) and isotherm (right) contours (nanofluid ($\phi = 0.04$) (—) and pure fluid ($\phi = 0$) (- - -)) when $Ra = 10^5, Rd = 0$.

Fig. 8.47 illustrates the influence of Fe_3O_4 nanoparticles on hydrothermal behavior. The nanofluid velocity augments due to enhancing the solid movements. The temperature gradient increase with the rise of the volume fraction Fe_3O_4.

The influence of *Rd, Ha,* and *Ra* on streamlines and isotherms are demonstrated in Figs. 8.48–8.50. In all cases, one main clockwise eddy appears in the enclosure. Isotherms are parallel together in low *Ra*. As *Ha* augments the primary eddy stretches vertically and moves upward. Enhancing the radiation effect makes isotherms go far away from the curve boundary. Therefore temperature boundary layer thickness near the curve wall augments and this way the temperature gradient reduces. Also $|\Psi_{\max}|$ increases with the rise of *Rd*. The convection mode becomes stronger with the increase of *Ra* and a thermal plume is generated on the hot wall. As the Lorentz force augments, $|\Psi_{\max}|$ reduces and this force diminishes the thermal plume. The impact of *Rd* on hydrothermal behavior is more sensible at high *Ra*.

Figs. 8.51 and 8.52 demonstrate the impact of ϕ, *Ha*, *Rd*, and *Ra* on Nu_{ave}, Nu_{loc}. The correlation for Nu_{ave} corresponding active parameters is:

$$\begin{aligned} Nu_{ave} &= 10.01 - 5.23\log(Ra) - 0.95Rd + 0.06Ha - 5.4\phi \\ &+ 0.63\log(Ra)Rd - 0.025\log(Ra)\,Ha + 2.05\log(Ra)\phi \\ &- 0.013Rd\,Ha + 4.6\,Rd\,\phi - 0.04\,Ha\,\phi \\ &+ 0.84(\log(Ra))^2 - 0.37Rd^2 + 3.8\times10^{-4}Ha^2 + 47.7\phi^2 \end{aligned} \tag{8.170}$$

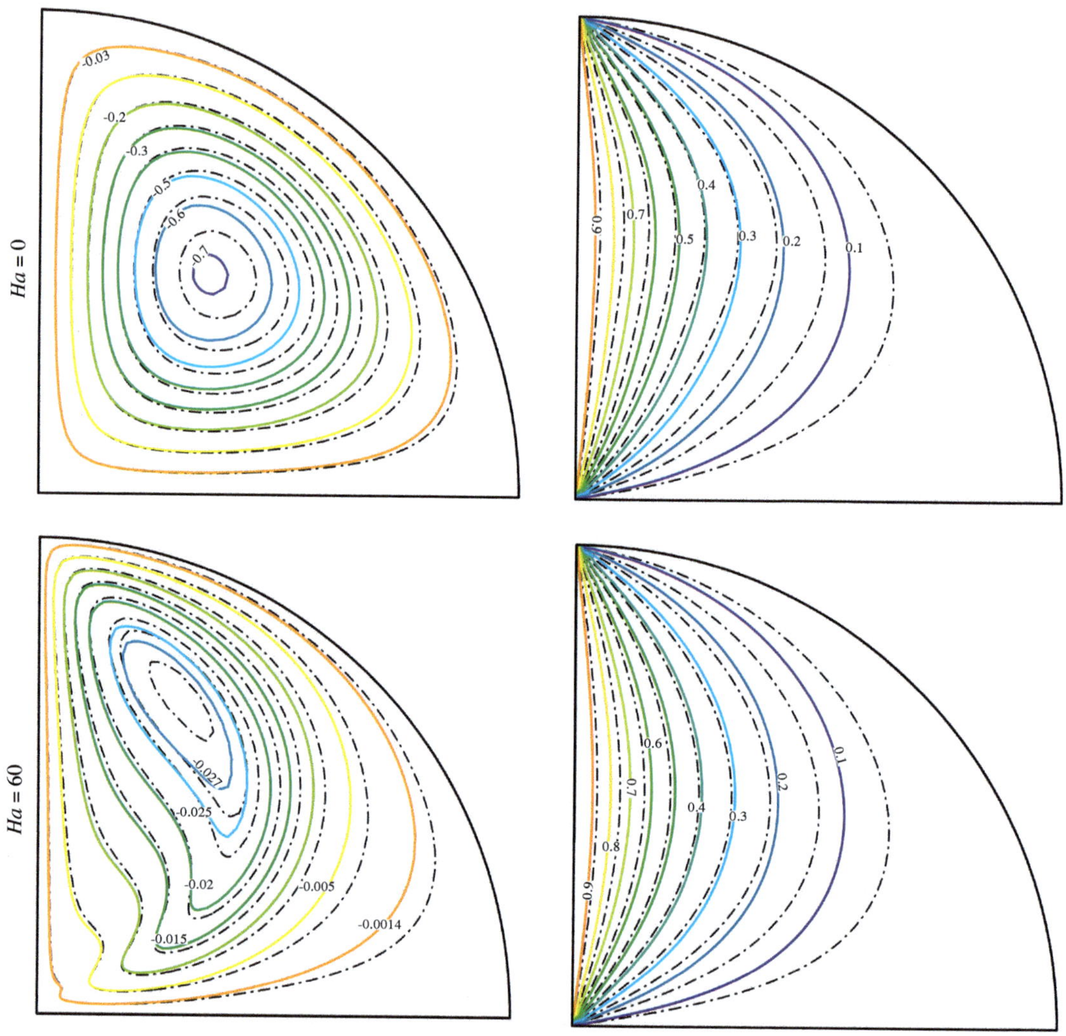

FIGURE 8.48 Influence of *Rd*, *Ha* on streamline (left) and isotherm (right) contours ($Rd = 0.8$ (—) and $Rd = 0$ (– · –)) when $\phi = 0.04$, $Ra = 10^3$.

Nu_{ave} augments with augment of *Rd*. This observation can be explain by Eq. (8.186), the temperature gradient drop is much smaller than the first terms of the definition. As buoyancy forces rises, the temperature gradient near the curve wall augments and in turn Nu_{ave} enhances with the increase of the Rayleigh number. Enhancing Lorentz forces leads to reducing Nu_{ave} due to the domination of the conduction mode. Adding Fe_3O_4 nanoparticles into water makes Nu_{ave} enhance. Table 8.8 illustrates the impacts of *Ha*, *Rd*, and Ra on heat transfer improvement, which is defined as: $E = 100^* \left(Nu_{ave}\big|_{\phi=0.04} - Nu_{ave}\big|_{\phi=0}\right) / Nu_{ave}\big|_{\phi=0}$. *E* is relevant to variation of thermal conductivity and *k* has more change when the temperature boundary layer thickness is larger. Therefore, *E* enhances with the rise of *Ha* while it reduces with the enhancement of *Ra*, *Rd*.

8.9 FERROFLUID CONVECTIVE HEAT TRANSFER UNDER THE INFLUENCE OF EXTERNAL MAGNETIC SOURCE

8.9.1 Problem Definition

An enclosure with hot inner walls is considered. Fig. 8.53A depicts the boundary conditions. To obtain the shape of the left sinusoidal wall profile, the following equation should be used:

$$\frac{x}{H} = 1 - \left\{ a\left(1 + \sin\left(\pi \frac{y}{H} - \pi/2\right)\right) \right\} \tag{8.171}$$

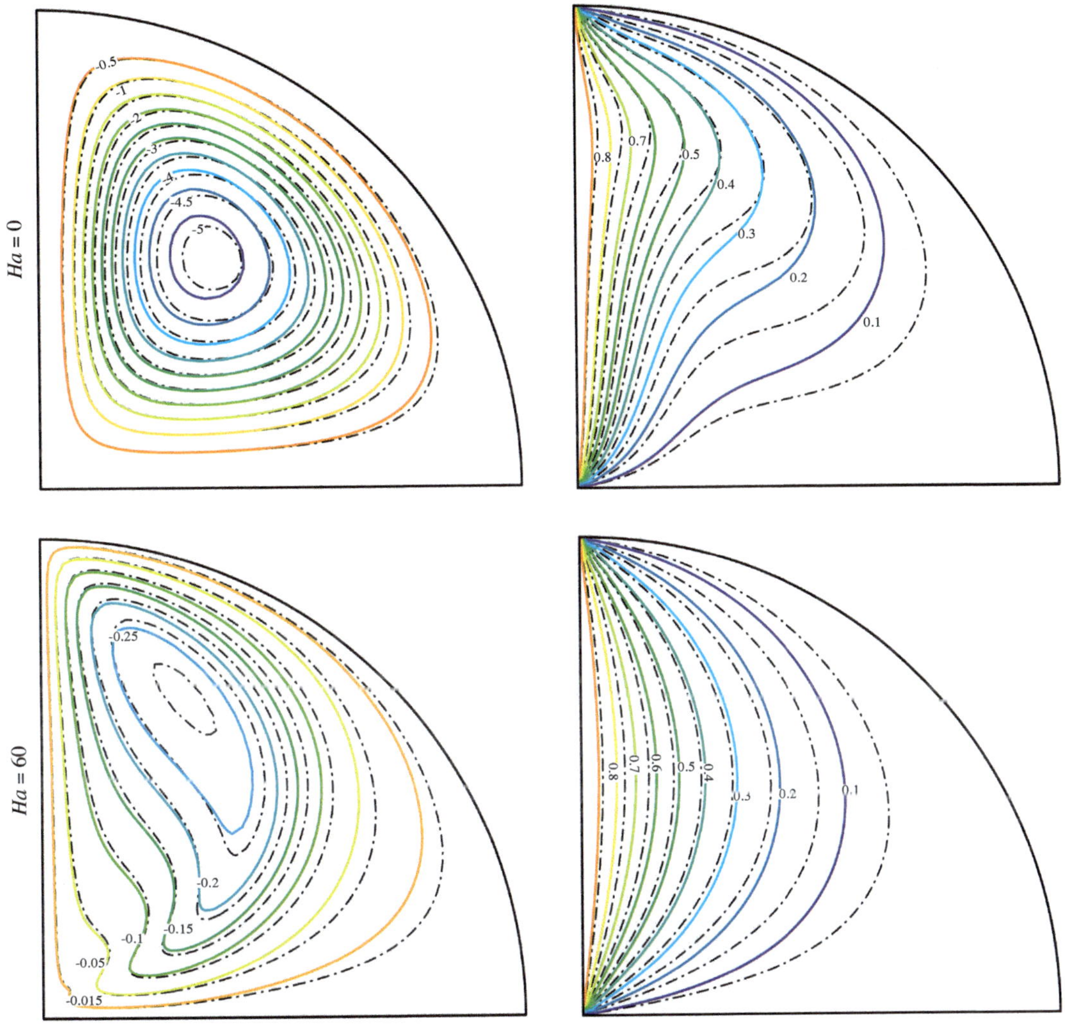

FIGURE 8.49 Influence of *Rd, Ha* on streamline (left) and isotherm (right) contours ($Rd=0.8$ (—) and $Rd=0$ (– · –)) when $\phi=0.04,\ Ra=10^4$.

Magnetic source has been considered as shown in Fig. 8.54. $\overline{H_x}, \overline{H_y}, \overline{H}$ can be calculated as follow:

$$\overline{H_x} = \left(y-\overline{b}\right)\left[(\overline{a}-x)^2+\left(\overline{b}-y\right)^2\right]^{-1}\frac{\gamma}{2\pi},$$
$$\overline{H_y} = (\overline{a}-x)\left[(\overline{a}-x)^2+\left(\overline{b}-y\right)^2\right]^{-1}\frac{\gamma}{2\pi}, \tag{8.172}$$

$$\overline{H} = \sqrt{\overline{H}_x^2+\overline{H}_y^2} \tag{8.173}$$

8.9.2 Governing Equation

2D laminar nanofluid flow and forced convective heat transfer is taken into account. The governing PDEs can be considered as:

$$\frac{\partial u}{\partial x}+\frac{\partial v}{\partial y}=0, \tag{8.174}$$

$$\left(\frac{\partial u}{\partial x}u+\frac{\partial u}{\partial y}v\right)=\left(\rho_{nf}\right)^{-1}\left[\left(\frac{\partial^2 u}{\partial y^2}+\frac{\partial^2 u}{\partial x^2}\right)\mu_{nf}-\frac{\partial P}{\partial x}-\sigma_{nf}B_y^2u+\sigma_{nf}B_xB_yv\right] \tag{8.175}$$

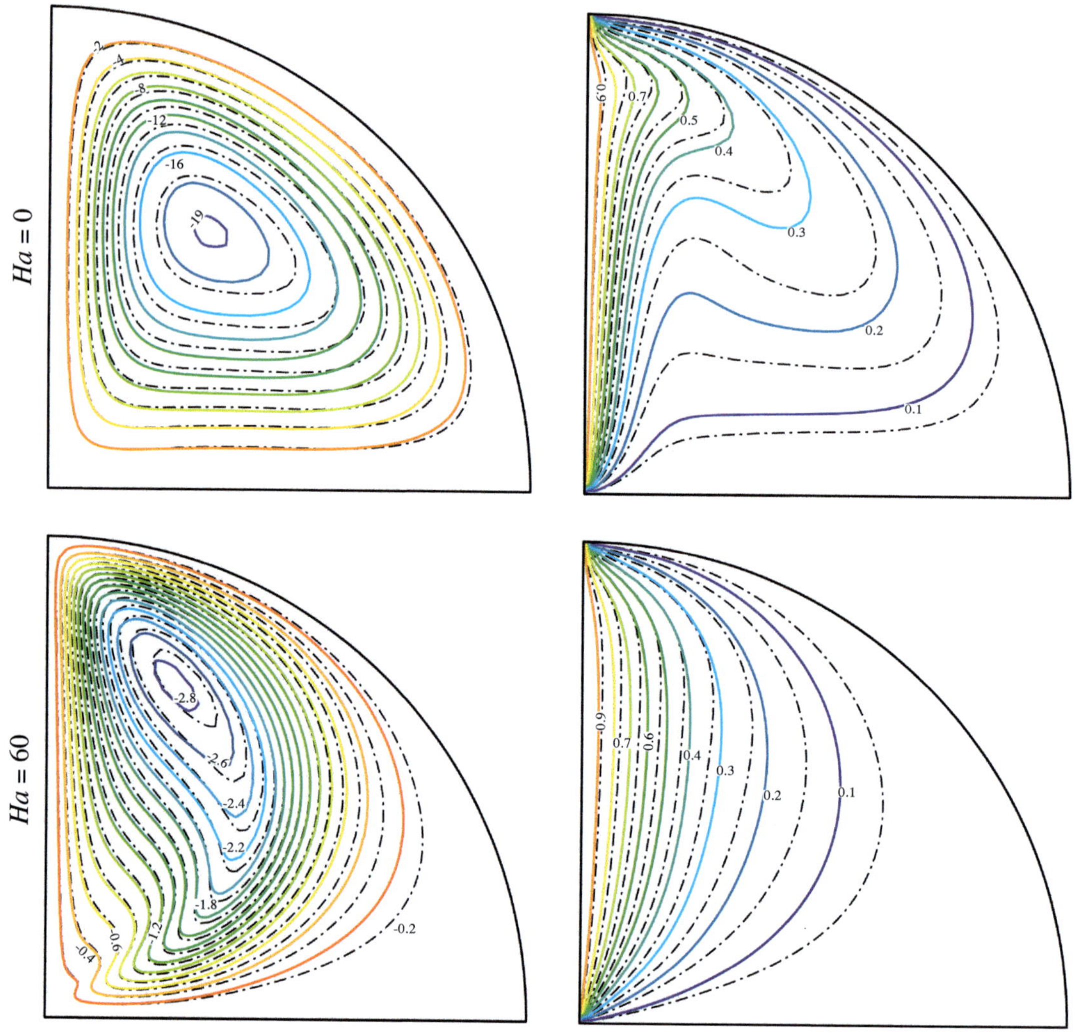

FIGURE 8.50 Influence of *Rd*, *Ha* on streamline (left) and isotherm (right) contours ($Rd = 0.8$ (—) and $Rd = 0$ (– · –)) when $\phi = 0.04$, $Ra = 10^5$.

$$\rho_{nf}\left(\frac{\partial v}{\partial x}u + \frac{\partial v}{\partial y}v\right) = +\mu_{nf}\left(\frac{\partial^2 v}{\partial x^2} + \frac{\partial^2 v}{\partial y^2}\right) - \frac{\partial P}{\partial y} + B_y\sigma_{nf}B_x u - B_x\sigma_{nf}B_x v + (T - T_c)\beta_{nf}g\rho_{nf} \tag{8.176}$$

$$(\rho C_p)_{nf}\left(v\frac{\partial T}{\partial y} + u\frac{\partial T}{\partial x}\right) = \sigma_{nf}(B_x v - B_y u)^2 + k_{nf}\left(\frac{\partial^2 T}{\partial x^2} + \frac{\partial^2 T}{\partial y^2}\right) + \mu_{nf}\left\{2\left(\frac{\partial u}{\partial x}\right)^2 + 2\left(\frac{\partial v}{\partial x}\right)^2 + \left(\frac{\partial u}{\partial x} + \frac{\partial v}{\partial y}\right)^2\right\}. \tag{8.177}$$

ρ_{nf}, $(\rho C_p)_{nf}$, α_{nf}, β_{nf}, μ_{nf}, k_{nf}, and σ_{nf} are calculated as

$$\rho_{nf} = \rho_f(1 - \phi) + \rho_s\phi, \tag{8.178}$$

$$(\rho C_p)_{nf} = (\rho C_p)_f(1 - \phi) + (\rho C_p)_s\phi, \tag{8.179}$$

$$\alpha_{nf} = \frac{k_{nf}}{(\rho C_p)_{nf}}, \tag{8.180}$$

$$\beta_{nf} = \beta_f(1 - \phi) + \beta_s\phi. \tag{8.181}$$

$$\mu_{nf} = \frac{\mu_f}{(1-\phi)^{2.5}}. \tag{8.182}$$

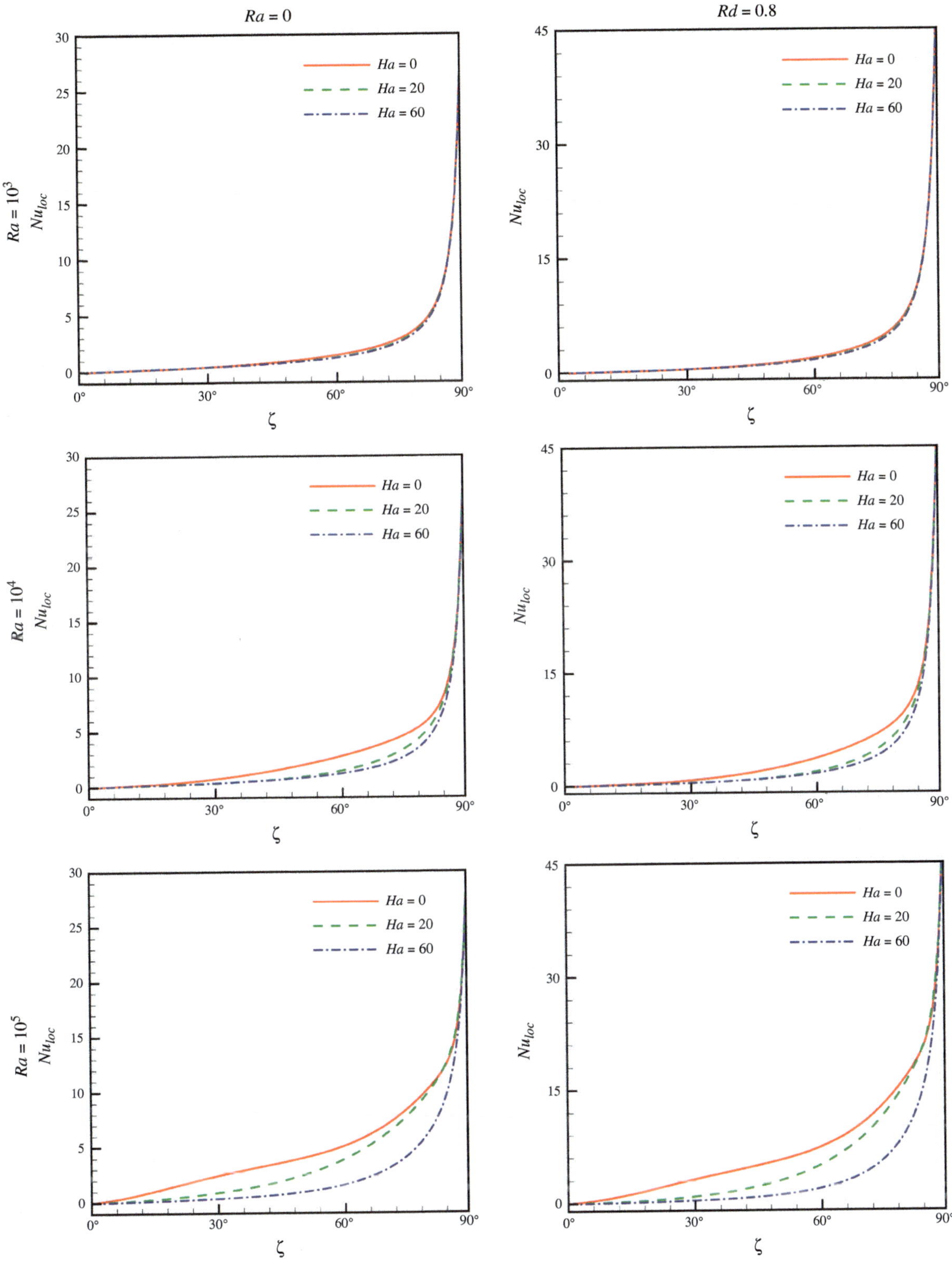

FIGURE 8.51 Effects of *Ha*, *Ra*, and *Rd* on local Nusselt number.

$$k_{nf} = \left(\frac{k_s + 2k_f - 2\phi(k_f - k_s)}{k_s + 2k_f + \phi(k_f - k_s)}\right)k_f, \tag{8.183}$$

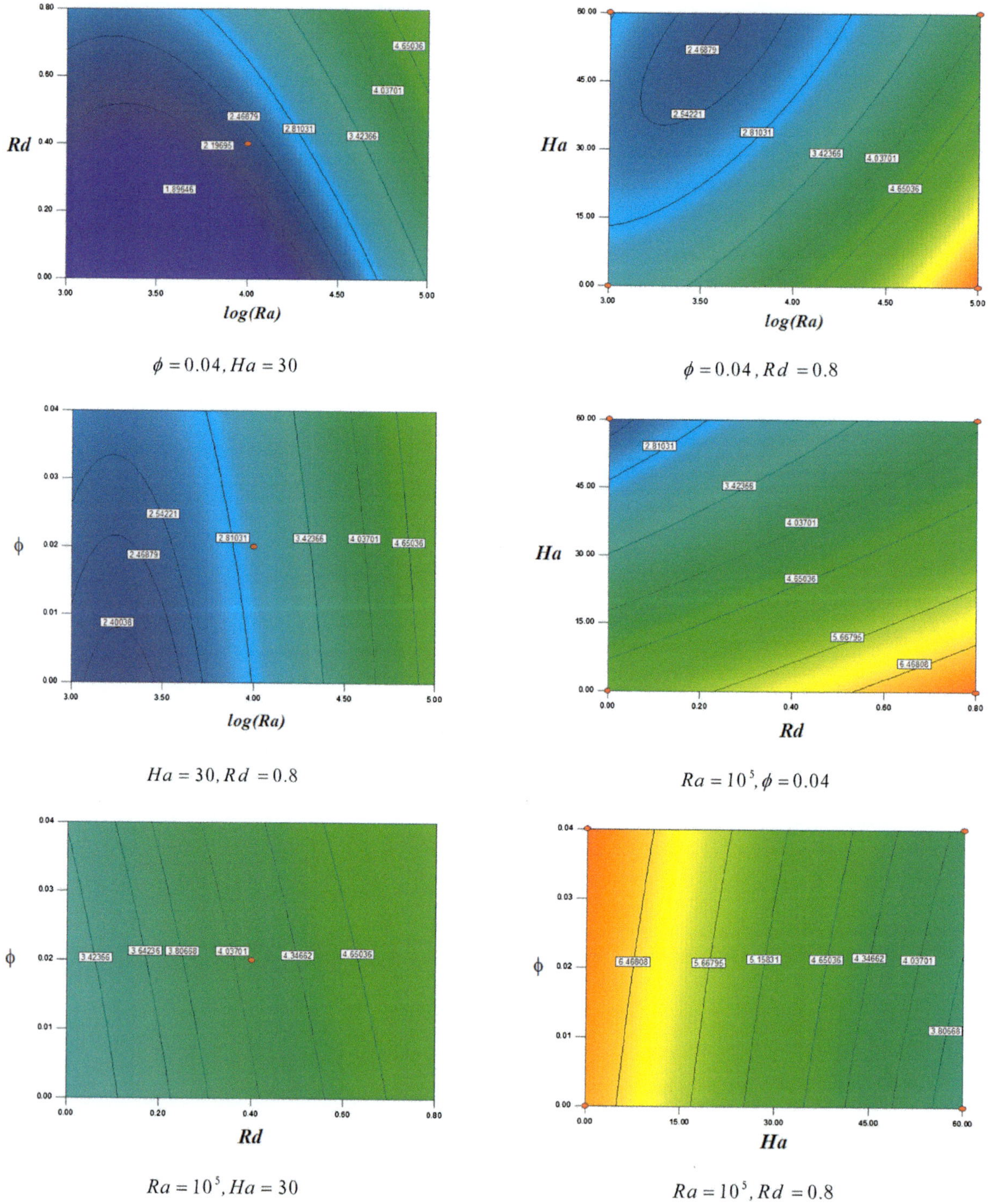

FIGURE 8.52 Effects of ϕ, *Ha*, *Rd*, and *Ra* on average Nusselt number.

$$\frac{\sigma_{nf}}{\sigma_f} = 1 + \frac{3\left(\frac{\sigma_s}{\sigma_f} - 1\right)\phi}{\left(\frac{\sigma_s}{\sigma_f} + 2\right) - \left(\frac{\sigma_s}{\sigma_f} - 1\right)\phi}. \tag{8.184}$$

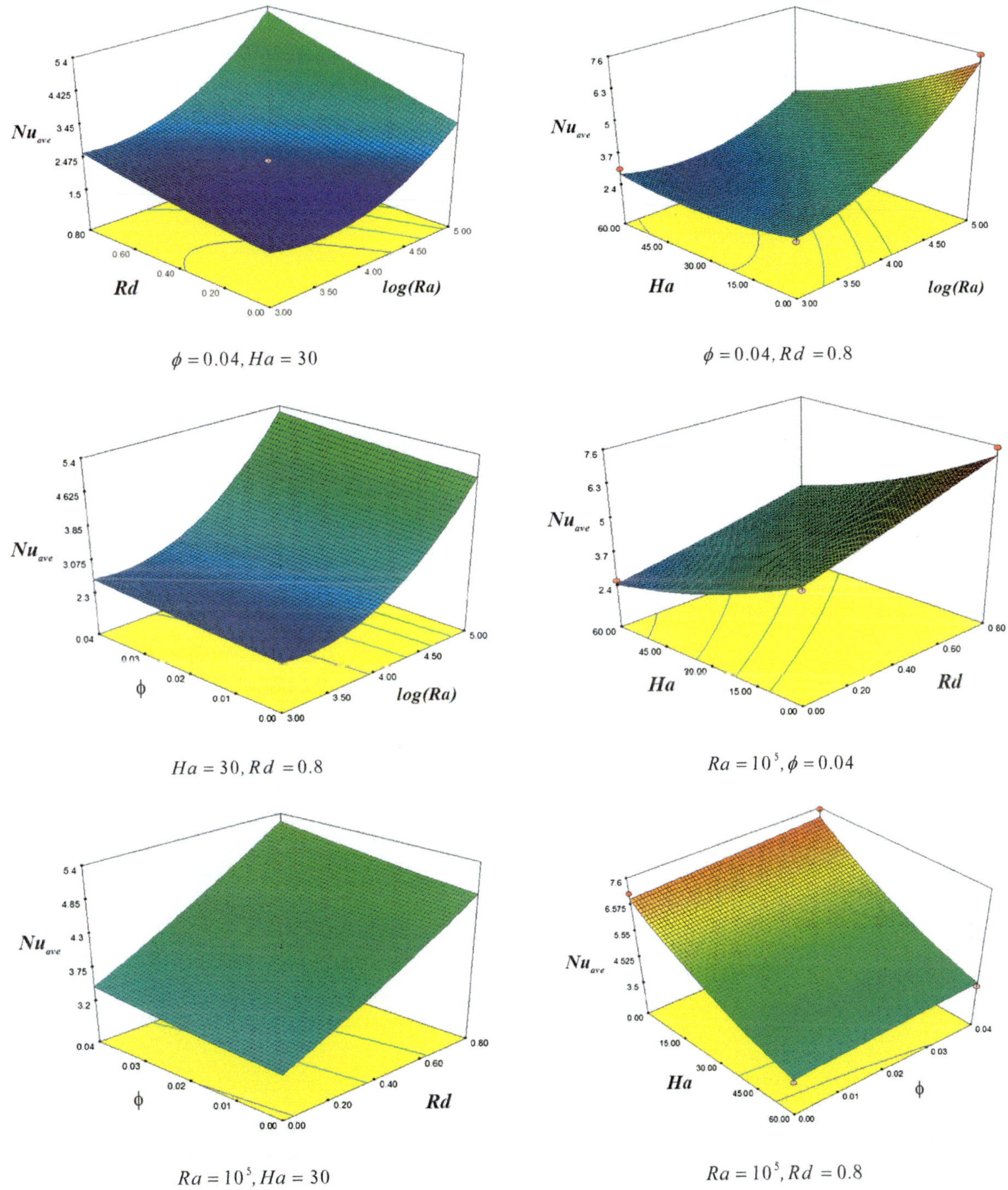

FIGURE 8.52 (Continued).

Dimensionless parameters are defined as:

$$(a,b) = \frac{(\bar{a},\bar{b})}{L},\ (H,H_x,H_y) = \frac{(\overline{H},\overline{H_x},\overline{H_y})}{\overline{H_0}},\ (X,Y) = \frac{(x,y)}{L}$$
$$U = \frac{uL}{\alpha_f}, V = \frac{vL}{\alpha_f},\ \Theta = \frac{T - T_c}{(q''L/k_f)}, P = \frac{p}{\rho_f(\alpha_f/L)^2} \tag{8.185}$$

TABLE 8.8 Effects of *Ha*, *Rd*, and *Ra* on Heat Transfer Enhancement

Ra	*Rd*	*Ha*	E
10^3	0	0	8.69526
10^3	0	20	9.106752
10^3	0	60	9.202467
10^3	0.2	0	7.440591
10^3	0.2	20	7.787079
10^3	0.2	60	7.869225
10^4	0	0	7.61383
10^4	0	20	8.146552
10^4	0	60	9.036499
10^4	0.2	0	6.431219
10^4	0.2	20	12.66447
10^4	0.2	60	7.719316

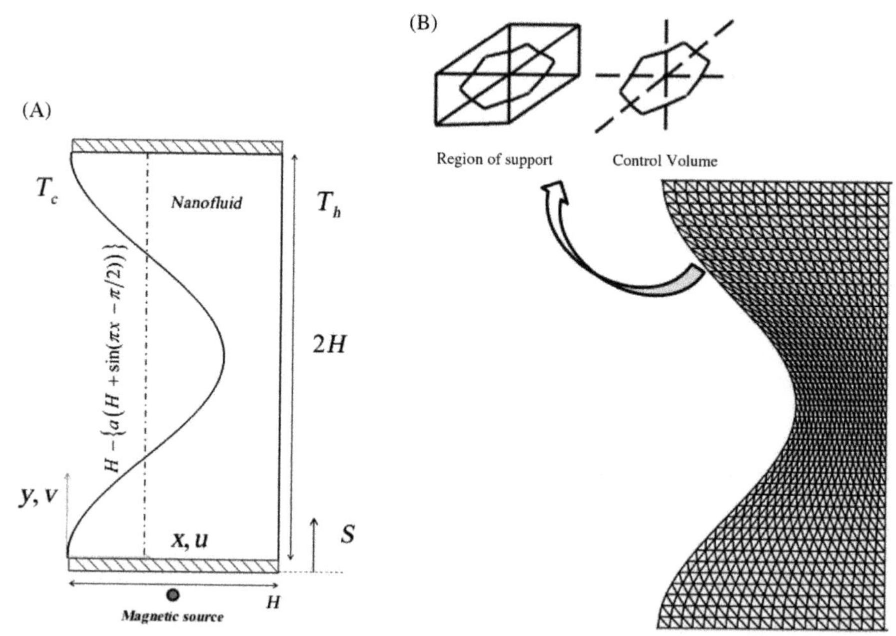

FIGURE 8.53 (A) Geometry and the boundary conditions; (B) the mesh of enclosure considered in this work.

So final equations are:

$$\frac{\partial U}{\partial X} + \frac{\partial V}{\partial Y} = 0, \tag{8.186}$$

$$\frac{\partial U}{\partial X}U + \frac{\partial U}{\partial Y}V = \Pr\left[\frac{\mu_{nf}/\mu_f}{\rho_{nf}/\rho_f}\right]\left(\frac{\partial^2 U}{\partial Y^2} + \frac{\partial^2 U}{\partial X^2}\right) - Ha^2\Pr\left[\frac{\sigma_{nf}/\sigma_f}{\rho_{nf}/\rho_f}\right]\left(H_y^2 U - H_x H_y V\right) - \frac{\partial P}{\partial X}, \tag{8.187}$$

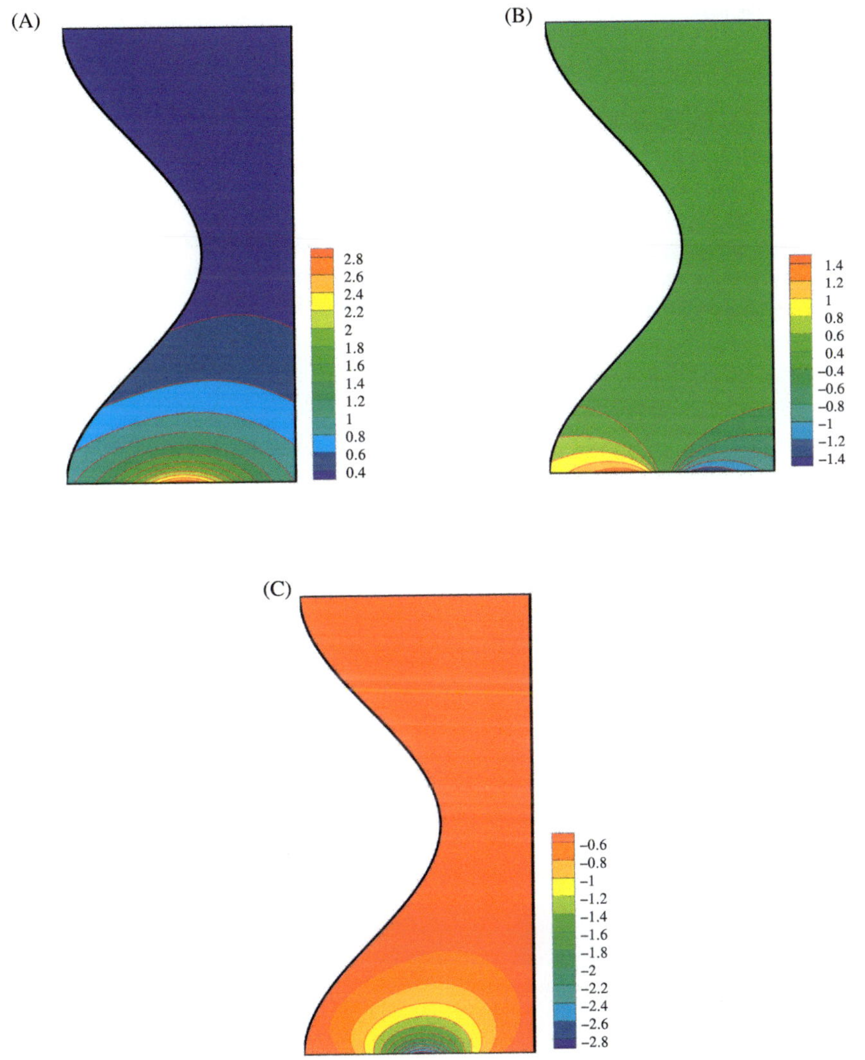

FIGURE 8.54 Contours of the (A) magnetic field strength H; (B) magnetic field intensity component in x direction Hx; (C) magnetic field intensity component in y direction Hy. (A) $H(x,y)$, (B) $Hx(x,y)$, (C) $Hy(x,y)$.

$$V\frac{\partial V}{\partial Y} + U\frac{\partial V}{\partial X} = \Pr\left(\frac{\partial^2 V}{\partial Y^2} + \frac{\partial^2 V}{\partial X^2}\right)\left[\frac{\mu_{nf}/\mu_f}{\rho_{nf}/\rho_f}\right] - Ha^2\Pr\left[\frac{\sigma_{nf}/\sigma_f}{\rho_{nf}/\rho_f}\right](H_x^2 V - H_x H_y U) - \frac{\partial P}{\partial Y} + Ra\Pr\left[\frac{\beta_{nf}}{\beta_f}\right]\Theta, \tag{8.188}$$

$$V\frac{\partial \Theta}{\partial Y} + U\frac{\partial \Theta}{\partial X} = \left(\frac{\partial^2 \Theta}{\partial Y^2} + \frac{\partial^2 \Theta}{\partial X^2}\right)\left[\frac{\frac{(\rho C_P)_{nf}}{(\rho C_P)_f}}{\frac{k_{nf}}{k_f}}\right]^{-1} + Ha^2\, Ec\left[\frac{\frac{\sigma_{nf}}{\sigma_f}}{\frac{(\rho C_P)_{nf}}{(\rho C_P)_f}}\right]\{V\,H_x - U\,H_y\}^2 + \left[\frac{\frac{\mu_{nf}}{\mu_f}}{\frac{(\rho C_P)_{nf}}{(\rho C_P)_f}}\right] Ec\left\{2\left(\frac{\partial U}{\partial X}\right)^2 + 2\left(\frac{\partial V}{\partial Y}\right)^2 + \left(\frac{\partial U}{\partial Y} + \frac{\partial V}{\partial X}\right)^2\right\}, \tag{8.189}$$

and dimensionless parameters are

$$Ra_f = g\beta_f L^3 (q''L/k_f)/(\alpha_f \upsilon_f), \mathrm{Pr}_f = \upsilon_f/\alpha_f, Ha = L\mu_0 H_0 \sqrt{\sigma_f/\mu_f}, Ec = \left(\mu_f \alpha_f\right)/\left[(\rho C_P)_f \Delta T\, L^2\right] \tag{8.190}$$

The thermophysical properties of Fe_3O_4 and water are presented in Table 8.1. Pressure gradient source terms are discarded by vorticity stream function.

$$\Omega = \frac{\omega L^2}{\alpha_f}, \Psi = \frac{\psi}{\alpha_f}, \omega = -\frac{\partial u}{\partial y} + \frac{\partial v}{\partial x}, (u, v) = \left(\frac{\partial \psi}{\partial y}, -\frac{\partial \psi}{\partial x}\right) \tag{8.191}$$

According to Fig. 8.53, boundary conditions are

$$\begin{array}{ll} \text{on left wall} & \Theta = 0.0 \\ \text{on all walls} & \Psi = 0.0 \\ \text{on right wall} & \Theta = 1.0 \\ \text{on other walls} & \dfrac{\partial \Theta}{\partial n} = 0.0 \end{array} \tag{8.192}$$

Nu_{local}, Nu_{ave} along left wall are :

$$Nu_{loc} = \left(\frac{k_{nf}}{k_f}\right)\frac{\partial \Theta}{\partial n} \tag{8.193}$$

$$Nu_{ave} = \frac{1}{2}\int_0^2 Nu_{loc}\, dS \tag{8.194}$$

8.9.3 Effects of Active Parameters

A cavity filled with nanofluid in the presence of a magnetic source is studied. The cold wall has a sinusoidal shape. Various levels of Hartmann number ($Ha = 0$ to 20), Rayleigh number ($Ra = 10^3, 10^4$ and 10^5), amplitude of the sinusoidal wall ($a = 0.1$ to 0.3), and volume fraction of Fe_3O_4 ($\phi = 0$ and 0.04) have been considered. Ec and Pr are 10^{-5} and 6.8, respectively.

Figs. 8.55–8.57 depict the impacts of amplitude of the sinusoidal wall, Rayleigh and Hartmann numbers on streamlines and isotherms. At a = 0.1, one main eddy appears. As the amplitude of the sinusoidal wall augments, the main eddy turns into two eddies. By applying a magnetic field, Lorentz forces are generated. These forces cause the nanofluid flow to retard. Increasing Lorentz forces causes eddies to stretch vertically. In weak buoyancy forces, the domination mode is conduction. So reducing the distance between cold and hot walls, the rate of heat transfer augments but the opposite behavior can be seen for stronger buoyancy forces. As the buoyancy forces enhances, eddies become stronger and isotherms become nonparallel together. So temperature gradient near the right wall enhances. As Lorentz force augments, the velocity reduces and this force make isotherms to become parallel together.

Figs. 8.58 and 8.59 depict the influence of the Fe_3O_4 volume fraction, amplitude of the sinusoidal wall, Lorentz and buoyancy forces on Nu_{loc}, Nu_{ave}. The correlation for Nu_{ave} corresponding to active parameters is:

$$\begin{aligned} Nu_{ave} &= -2.85 + 9.12a + 1.33(\log(Ra)) - 4.34\,\phi + 1.79\times 10^{-4} Ha \\ &- 2.6\,a\,(\log(Ra)) - 0.13\,a\,\phi - 1.6\times 10^{-16} a\,Ha + 1.9(\log(Ra))\phi - 1.15\times 10^{-17}(\log(Ra))Ha \\ &- 1.18\times 10^{-15}\phi Ha + 3.5a^2 - 7.5\times 10^{-4}(\log(Ra))^2 - 0.45\phi^2 - 8.9\times 10^{-6} Ha^2 \end{aligned} \tag{8.195}$$

As buoyancy forces augment, the temperature gradient near the right wall enhances and in turn Nu_{ave} enhances with the increase of Rayleigh number. Enhancing Lorentz forces leads to reducing Nu_{ave} due to the domination of the conduction mode. Adding Fe_3O_4 nanoparticles into water makes the Nusselt number enhance. Nu_{ave} enhances with the rise of a in low Ra but opposite changes occur for high Ra.

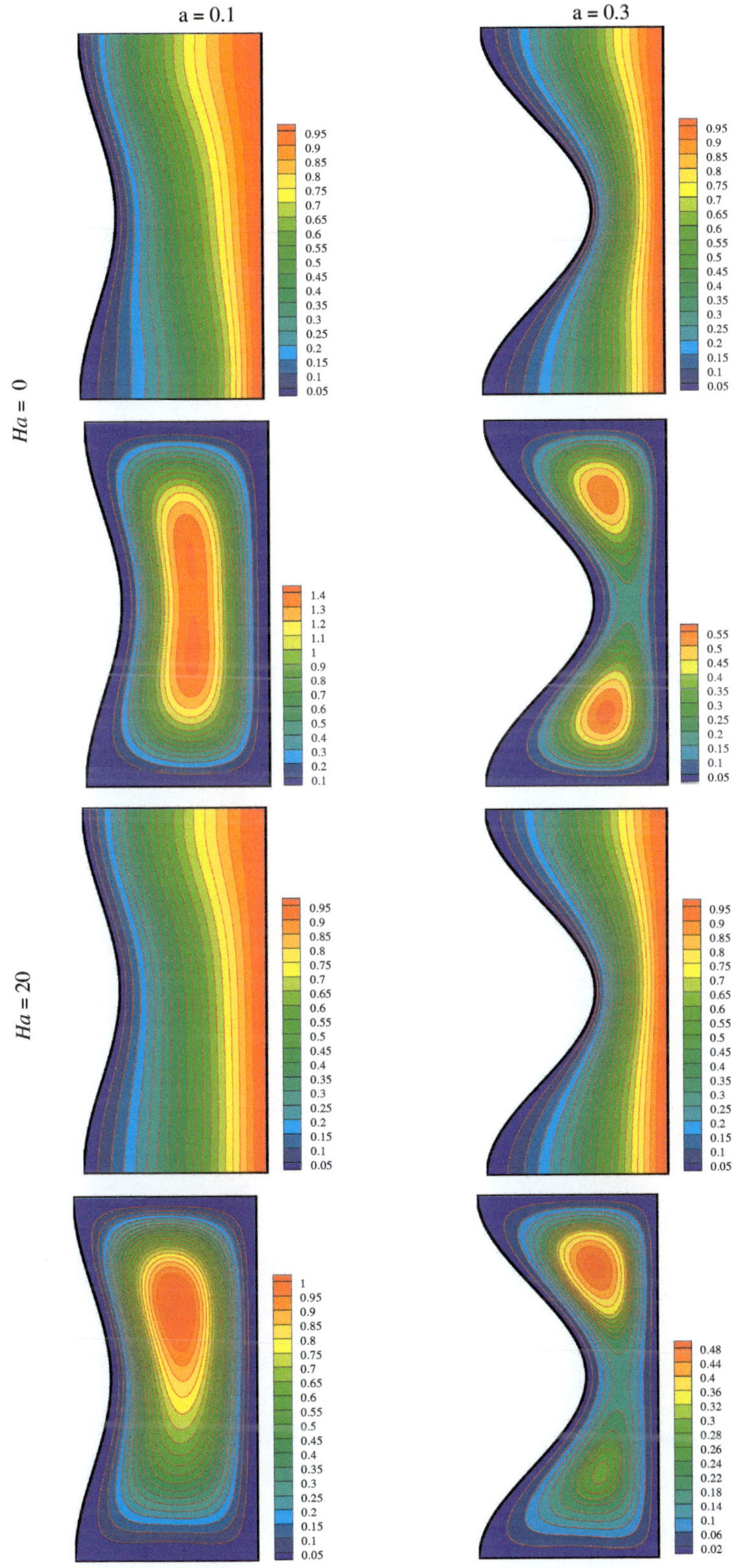

FIGURE 8.55 Isotherm (up) and streamline (down) contours for different values of Hartmann number and amplitude of the sinusoidal wall when $Ra = 10^3$, $\phi = 0.04$.

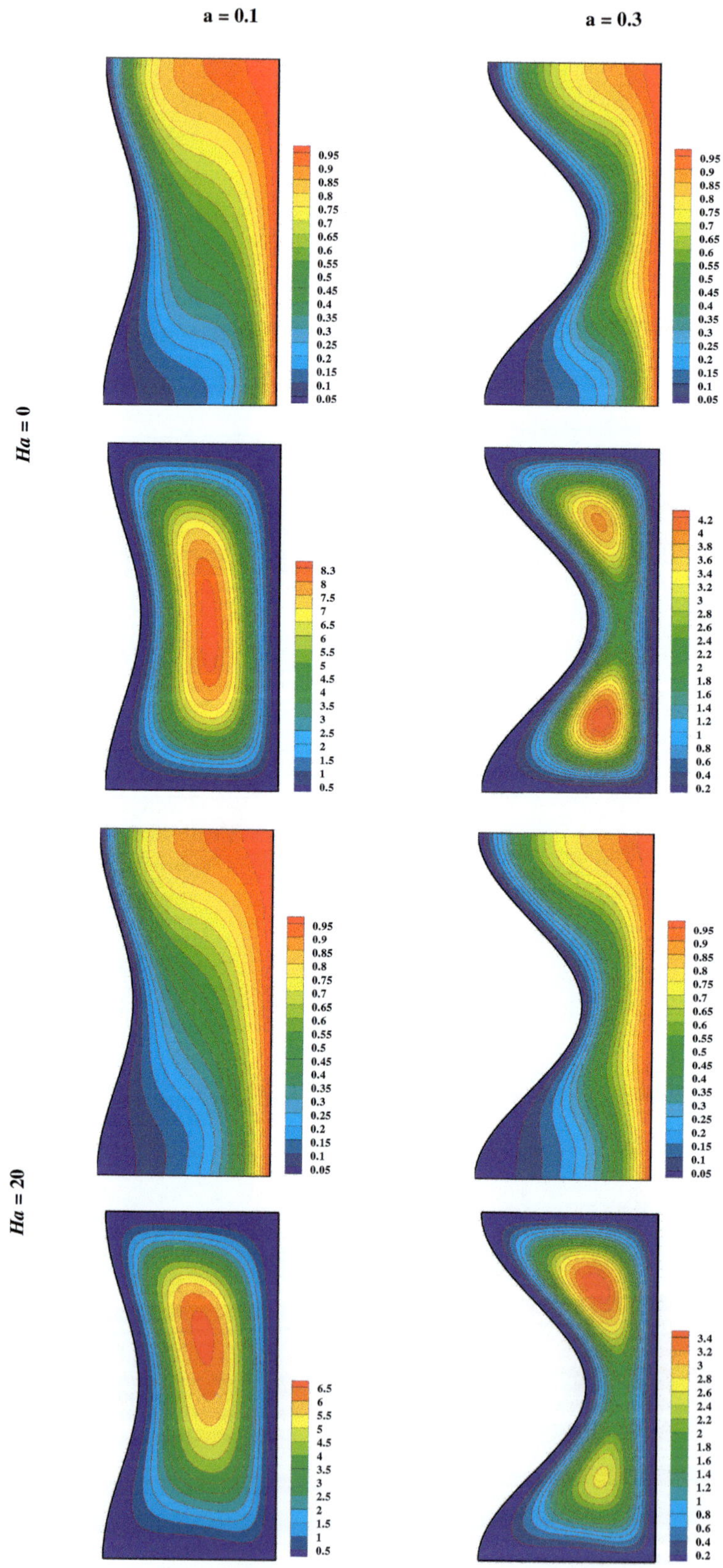

FIGURE 8.56 Isotherm (up) and streamline (down) contours for different values of Hartmann number and amplitude of the sinusoidal wall when $Ra = 10^4, \phi = 0.04$.

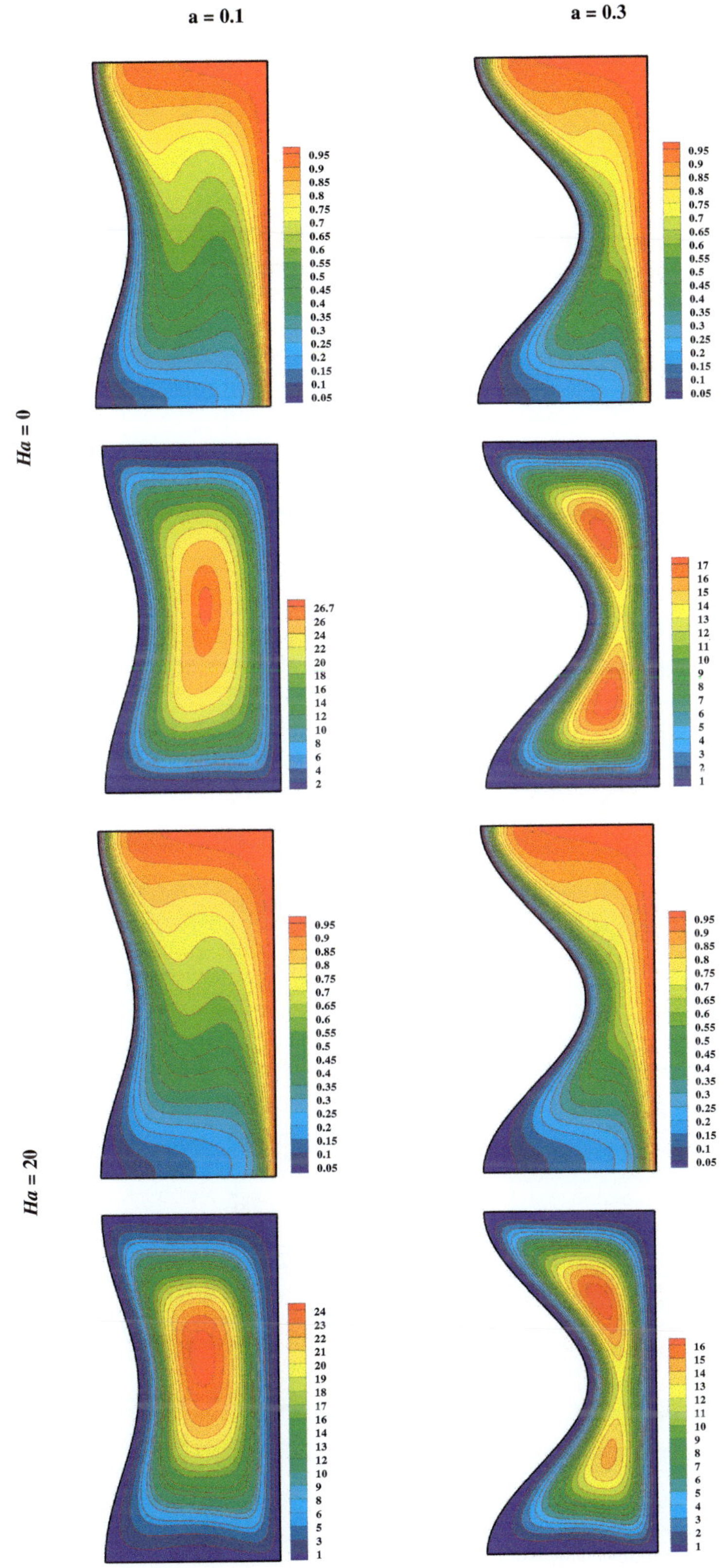

FIGURE 8.57 Isotherm (up) and streamline (down) contours for different values of Hartmann number and amplitude of the sinusoidal wall when $Ra = 10^5$, $\phi = 0.04$.

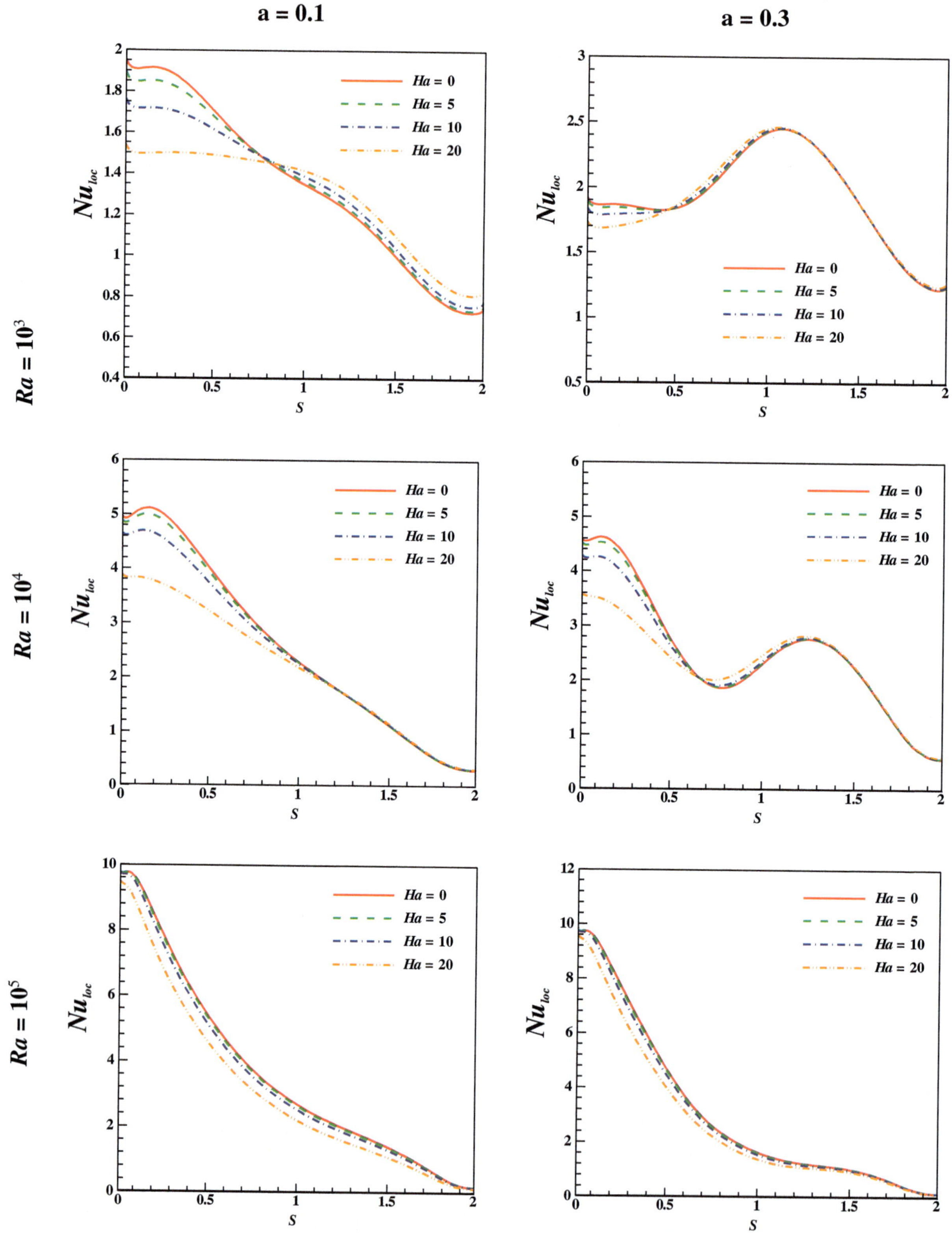

FIGURE 8.58 Effects of Hartmann number, amplitude of the sinusoidal wall, and Rayleigh number on local Nusselt number Nu_{loc} along right wall.

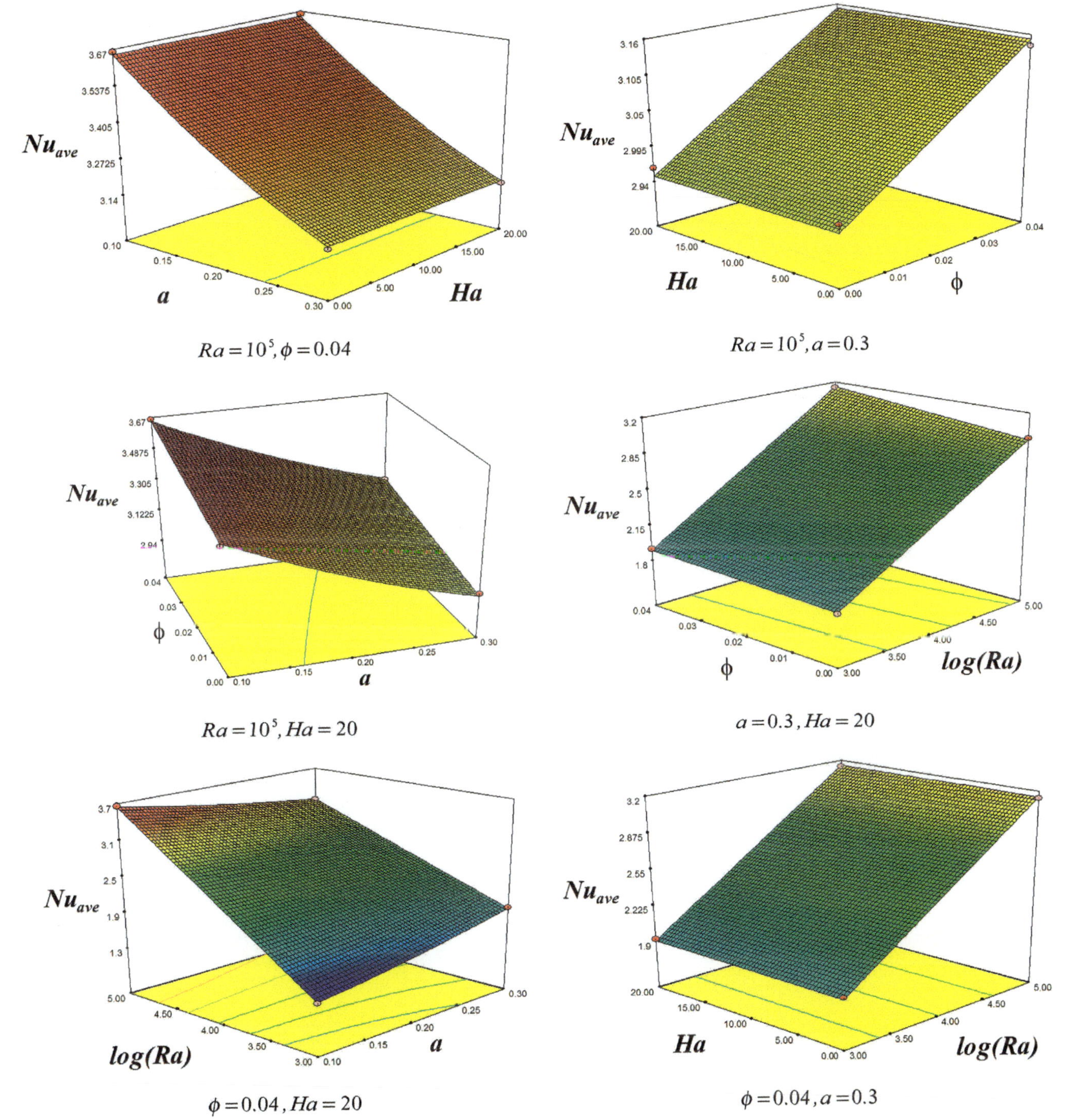

FIGURE 8.59 Effects of Hartmann number, amplitude of the sinusoidal wall, and Rayleigh number on local Nusselt number Nu_{ave} along right wall.

8.10 NANOFLUID HYDROTHERMAL TREATMENT IN A CAVITY WITH VARIABLE MAGNETIC FIELD

8.10.1 Problem Definition

The geometry for our numerical model is a two-dimensional square cavity with length of side walls equal to L, which represents the characteristic length for the problem (see Fig. 8.60A). The heat source is centrally located

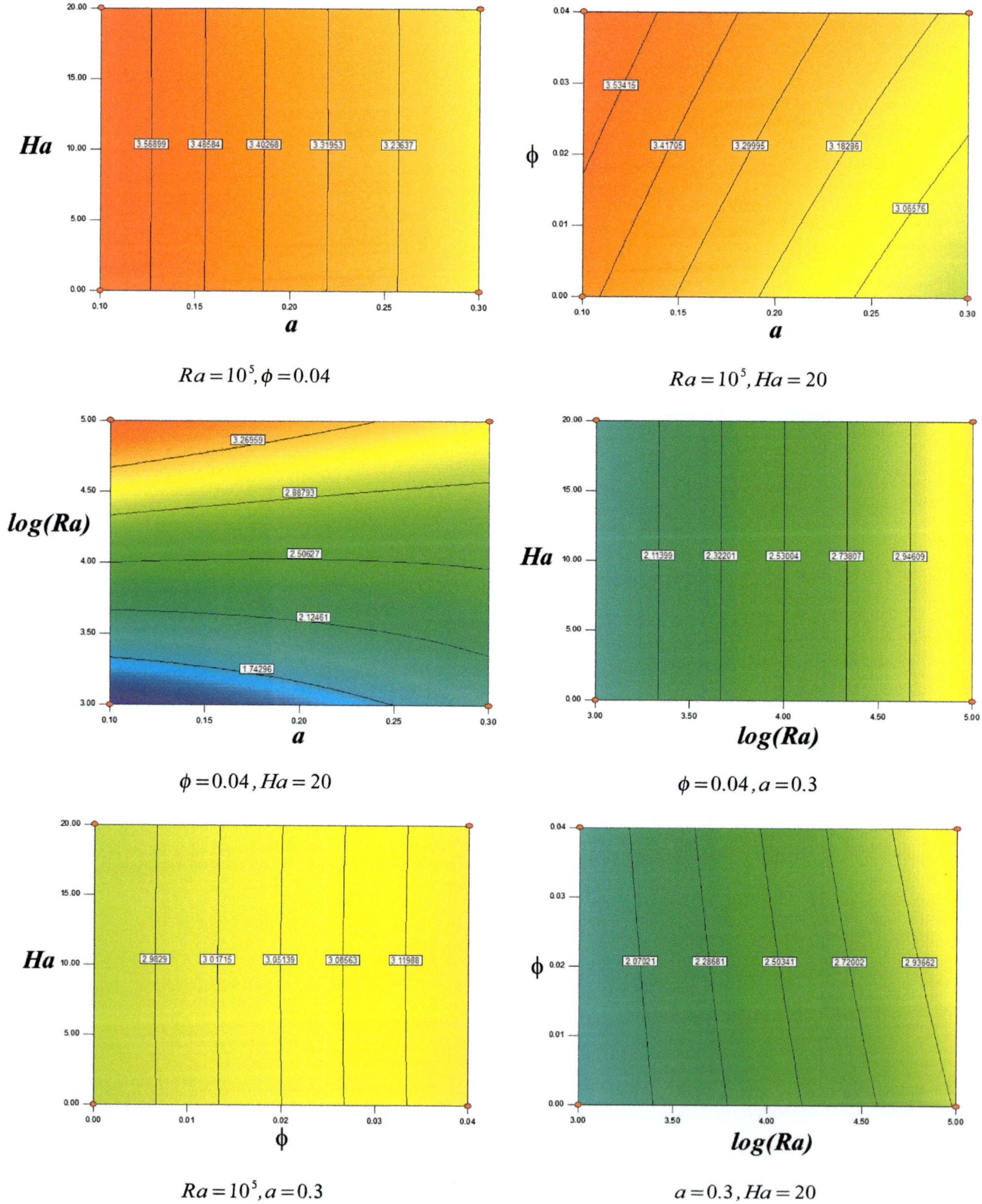

FIGURE 8.59 (Continued).

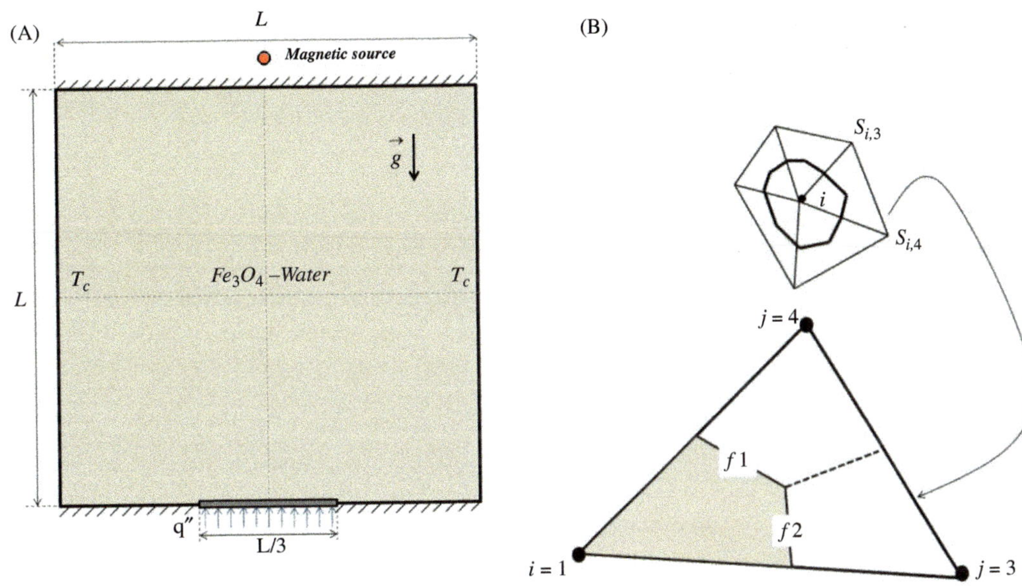

FIGURE 8.60 (A) Geometry and the boundary conditions; (B) a sample triangular element and its corresponding control volume.

on the bottom surface with length L/3. The cooling is achieved by the two vertical walls. The heat source has constant heat flux q'' while the cooling walls have a constant temperature T_c; all the other surfaces are adiabatic. A magnetic source is considered which is represented by a magnetic wire placed vertically to the x-y plane at the point $(\bar{a}, \bar{b})$. The components of the magnetic field intensity $(\overline{H}_x, \overline{H}_y)$, and the magnetic field strength $(\overline{H})$ are considered in the following form:

$$\overline{H}_x = \frac{\gamma}{2\pi} \frac{1}{(x-\bar{a})^2 + (y-\bar{b})^2}(y - \bar{b}), \tag{8.196}$$

$$\overline{H}_y = \frac{\gamma}{2\pi} \frac{1}{(x-\bar{a})^2 + (y-\bar{b})^2}(x - \bar{a}), \tag{8.197}$$

$$\overline{H} = \sqrt{\overline{H}_x^2 + \overline{H}_y^2} = \frac{\gamma}{2\pi} \frac{1}{\sqrt{(x-\bar{a})^2 + (y-\bar{b})^2}}, \tag{8.198}$$

where γ is the magnetic field strength at the source (the wire) and $(\bar{a}, \bar{b})$ is the position where the source is located. The contours of the magnetic field strength are shown in Fig. 8.61. In this study the magnetic source is located at (0.5 *cols*, 1.05 *rows*).

8.10.2 Governing Equation

Let us consider a steady incompressible two-dimensional laminar flow. The magnetic Reynolds number is assumed to be small, compared to the applied magnetic field so that the induced magnetic field can be neglected. Using the Boussinesq approximation, the governing equations of fluid flow and heat transfer for a nanofluid can be written as

$$\frac{\partial v}{\partial y} + \frac{\partial u}{\partial x} = 0, \tag{8.199}$$

$$\rho_{nf}\left(v\frac{\partial u}{\partial y} + u\frac{\partial u}{\partial x}\right) = -\frac{\partial P}{\partial x} + \mu_{nf}\left(\frac{\partial^2 u}{\partial x^2} + \frac{\partial^2 u}{\partial y^2}\right) - \sigma_{nf}B_y^2 u + \sigma_{nf}B_x B_y v, \tag{8.200}$$

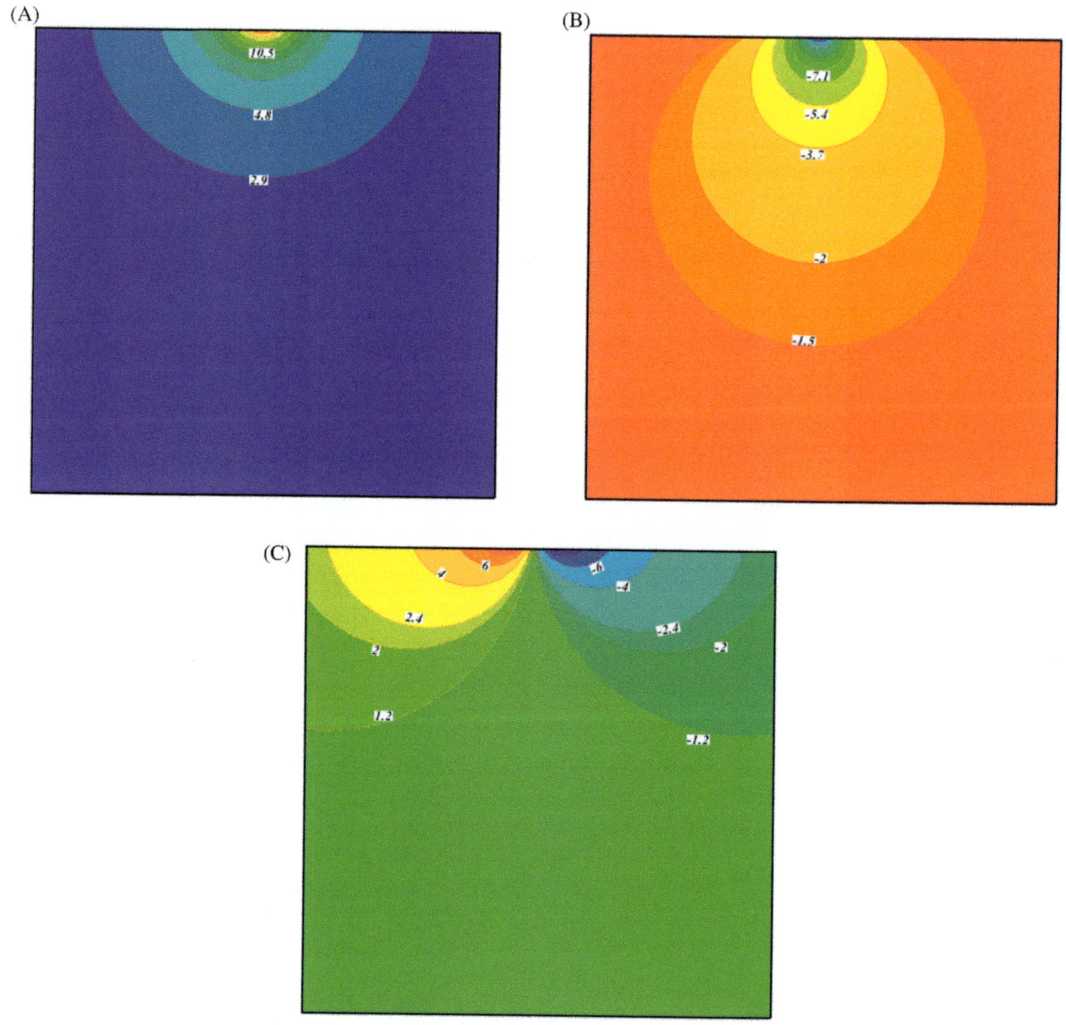

FIGURE 8.61 Contours of the (A) magnetic field strength H; (B) magnetic field intensity component in x direction Hx; (C) magnetic field intensity component in y direction Hy. (A) $H(x,y)$, (B) $Hx(x,y)$, (C) $Hy(x,y)$.

$$\rho_{nf}\left(v\frac{\partial v}{\partial y}+u\frac{\partial v}{\partial x}\right)=-\frac{\partial P}{\partial y}+\rho_{nf}\beta_{nf}g(T-T_c)+\mu_{nf}\left(\frac{\partial^2 v}{\partial x^2}+\frac{\partial^2 v}{\partial y^2}\right)+\sigma_{nf}B_xB_yu-\sigma_{nf}B_x^2v, \tag{8.201}$$

$$\begin{aligned}(\rho C_p)_{nf}\left(v\frac{\partial T}{\partial y}+u\frac{\partial T}{\partial x}\right)&=k_{nf}\left(\frac{\partial^2 T}{\partial x^2}+\frac{\partial^2 T}{\partial y^2}\right)+\sigma_{nf}(uB_y-vB_x)^2\\&+\mu_{nf}\left\{2\left(\frac{\partial u}{\partial x}\right)^2+2\left(\frac{\partial v}{\partial x}\right)^2+\left(\frac{\partial u}{\partial x}+\frac{\partial v}{\partial y}\right)^2\right\},\end{aligned} \tag{8.202}$$

$$B_y=\mu_0H_y,$$
$$B_x=\mu_0H_x,$$

where

$\rho_{nf}, (\rho C_p)_{nf}, \alpha_{nf}, \beta_{nf}, \mu_{nf}, k_{nf}$ and σ_{nf} are defined as

$$\rho_{nf}=\rho_s\phi+\rho_f(1-\phi), \tag{8.203}$$

$$(\rho C_p)_{nf}=(\rho C_p)_s\phi+(\rho C_p)_f(1-\phi), \tag{8.204}$$

$$\alpha_{nf} = \frac{k_{nf}}{(\rho C_p)_{nf}}, \tag{8.205}$$

$$\beta_{nf} = \beta_f(1-\phi) + \beta_s\phi. \tag{8.206}$$

$$\mu_{nf} = \frac{\mu_f}{(1-\phi)^{2.5}}. \tag{8.207}$$

$$\frac{k_{nf}}{k_f} = \frac{-2\phi(k_f - k_s) + 2k_f + k_s}{\phi(k_f - k_s) + 2k_f + k_s}, \tag{8.208}$$

$$\frac{\sigma_{nf}}{\sigma_f} = 1 + \frac{3\left(\frac{\sigma_s}{\sigma_f} - 1\right)\phi}{\left(\frac{\sigma_s}{\sigma_f} + 2\right) - \left(\frac{\sigma_s}{\sigma_f} - 1\right)\phi}. \tag{8.209}$$

The following dimensionless variables can be introduced:

$$\begin{gathered}\Theta = \frac{T - T_c}{(q''L/k_f)}, U = \frac{uL}{\alpha_f}, V = \frac{vL}{\alpha_f}, a = \frac{\overline{a}}{L}, b = \frac{\overline{b}}{L}, Y = \frac{y}{L}, X = \frac{x}{L}, \\ P = \frac{pL^2}{\rho_f(\alpha_f)^2}, (H, H_x, H_y) = \frac{(\overline{H}, \overline{H_x}, \overline{H_y})}{\overline{H_0}}, \overline{H_0} = \overline{H}(\overline{a}, 0) = \frac{\gamma}{2\pi|b|}.\end{gathered} \tag{8.210}$$

Using (8.210) Eqs. (8.199)–(8.202) can be written in dimensionless form as

$$\frac{\partial V}{\partial Y} + \frac{\partial U}{\partial X} = 0, \tag{8.211}$$

$$\begin{aligned}V\frac{\partial U}{\partial Y} + U\frac{\partial U}{\partial X} &= +\Pr\left[\frac{\mu_{nf}/\mu_f}{\rho_{nf}/\rho_f}\right]\left(\frac{\partial^2 U}{\partial X^2} + \frac{\partial^2 U}{\partial Y^2}\right) \\ &- Ha^2\Pr\left[\frac{\sigma_{nf}/\sigma_f}{\rho_{nf}/\rho_f}\right]\left(H_y^2 U - H_x H_y V\right) - \frac{\partial P}{\partial X},\end{aligned} \tag{8.212}$$

$$\begin{aligned}V\frac{\partial V}{\partial Y} + U\frac{\partial V}{\partial X} &= \Pr\left[\frac{\mu_{nf}/\mu_f}{\rho_{nf}/\rho_f}\right]\left(\frac{\partial^2 V}{\partial X^2} + \frac{\partial^2 V}{\partial Y^2}\right) \\ &- \frac{\partial P}{\partial Y} - Ha^2\Pr\left[\frac{\sigma_{nf}/\sigma_f}{\rho_{nf}/\rho_f}\right](H_x^2 V - H_x H_y U) + Ra\Pr\left[\frac{\beta_{nf}}{\beta_f}\right]\Theta,\end{aligned} \tag{8.213}$$

$$\begin{aligned}U\frac{\partial \Theta}{\partial X} + V\frac{\partial \Theta}{\partial Y} &= \left[\frac{\frac{k_{nf}}{k_f}}{\frac{(\rho C_P)_{nf}}{(\rho C_P)_f}}\right]\left(\frac{\partial^2 \Theta}{\partial X^2} + \frac{\partial^2 \Theta}{\partial Y^2}\right) + Ha^2\,Ec\left[\frac{\frac{\sigma_{nf}}{\sigma_f}}{\frac{(\rho C_P)_{nf}}{(\rho C_P)_f}}\right]\{U\,H_y - V\,H_x\}^2 \\ &+ Ec\left[\frac{\frac{\mu_{nf}}{\mu_f}}{\frac{(\rho C_P)_{nf}}{(\rho C_P)_f}}\right]\left\{2\left(\frac{\partial U}{\partial X}\right)^2 + 2\left(\frac{\partial V}{\partial Y}\right)^2 + \left(\frac{\partial U}{\partial Y} + \frac{\partial V}{\partial X}\right)^2\right\},\end{aligned} \tag{8.214}$$

with dimensionless parameters

$$Ra_f = \frac{g\beta_f L^3((q''L/k_f))}{(\alpha_f \upsilon_f)}, \mathrm{Pr} = \frac{\upsilon_f}{\alpha_f}, Ha = L\mu_0 H_0\sqrt{\frac{\sigma_f}{\mu_f}}, Ec = \frac{(\alpha_f \mu_f)}{\left[(\rho C_P)_f((q''L/k_f))\, L^2\right]}. \quad (8.215)$$

The thermophysical properties of Fe_3O_4 and water are presented in Table 8.1. The stream function and vorticity are defined as

$$\Omega = \frac{\omega L^2}{\alpha_f}, \quad \Psi = \frac{\psi}{\alpha_f}, v = -\frac{\partial \psi}{\partial x}, u = \frac{\partial \psi}{\partial y}, \quad \omega = -\frac{\partial u}{\partial y} + \frac{\partial v}{\partial x}. \quad (8.216)$$

The appropriate boundary conditions for the problem are

$$\begin{aligned} &\text{on the left and right} && \Theta = 0.0, \\ &\text{on all walls} && \Psi = 0.0, \\ &\text{on all the other walls} && \partial\Theta/\partial n = 0.0, \\ &\text{on the heat source} && \partial\Theta/\partial n = -1.0. \end{aligned} \quad (8.217)$$

The values of vorticity on the boundary of the enclosure can be obtained using the stream function formulation and the known velocity conditions during the iterative solution procedure. The local Nusselt number of the nanofluid along the heat source can be expressed as

$$Nu_{local} = \left(\frac{k_{nf}}{k_f}\right)\frac{1}{\theta}\Bigg|_{Y=0,\ L/3<X<2L/3}. \quad (8.218)$$

The average Nusselt number on a hot circular wall can be evaluated by

$$Nu_{ave} = \frac{3}{L}\int_{L/3}^{2L/3} Nu_{loc}(X)\, dX. \quad (8.219)$$

To estimate the enhancement of heat transfer between the nanofluid case and the pure fluid (base fluid) case, the enhancement is defined as

$$En = \frac{Nu(\phi = 0.04) - Nu(\phi = 0)}{Nu(\phi = 0)} \times 100. \quad (8.220)$$

8.10.3 Effects of Active Parameters

Effects of a variable magnetic field on Fe_3O_4-water nanofluid flow in a cavity are presented. The cavity is heated by a uniform heat flux. Different values for volume fraction of Fe_3O_4 ($\phi = 0$ and 0.04), Hartmann number ($Ha = 0$ to 10), and Rayleigh number ($Ra = 10^3$, 10^4 and 10^5) are considered. Also, Ec and Pr are equal to 10^{-5} and 6.8, respectively.

Influences of Rayleigh and Hartmann numbers on streamlines and isotherms are depicted in Fig. 8.62 when $\phi = 0.04$. According to this figure, the mode of heat transfer changes from conduction to convection as the buoyancy force increases. So, temperature gradient near the heater reduces with the augment of Rayleigh number. Furthermore, the strength of the main cell augments with the rise of the convection heat transfer. As a magnetic field is applied to the domain, Lorentz forces affect the nanofluid flow. As the Lorentz forces augment, the axial velocity reduces and, in turn, the temperature boundary layer thickness increases. Also the core of the main eddy shifts downward with an increase in the Lorentz forces. Besides, the distortion of isotherms reduces in the presence of magnetic field.

Figs. 8.63 and 8.64 demonstrate the effects of the Lorentz and buoyancy forces on local and average Nusselt numbers. As the Rayleigh number enhances, the distortion of isotherms increases. Thus the temperature gradient increases with an increase in the buoyancy force. Lorentz forces retard the nanofluid flow and eliminate the thermal plume. So, enhancement of the Lorentz force leads to a reduction in the Nusselt number. Fig. 8.65 depicts

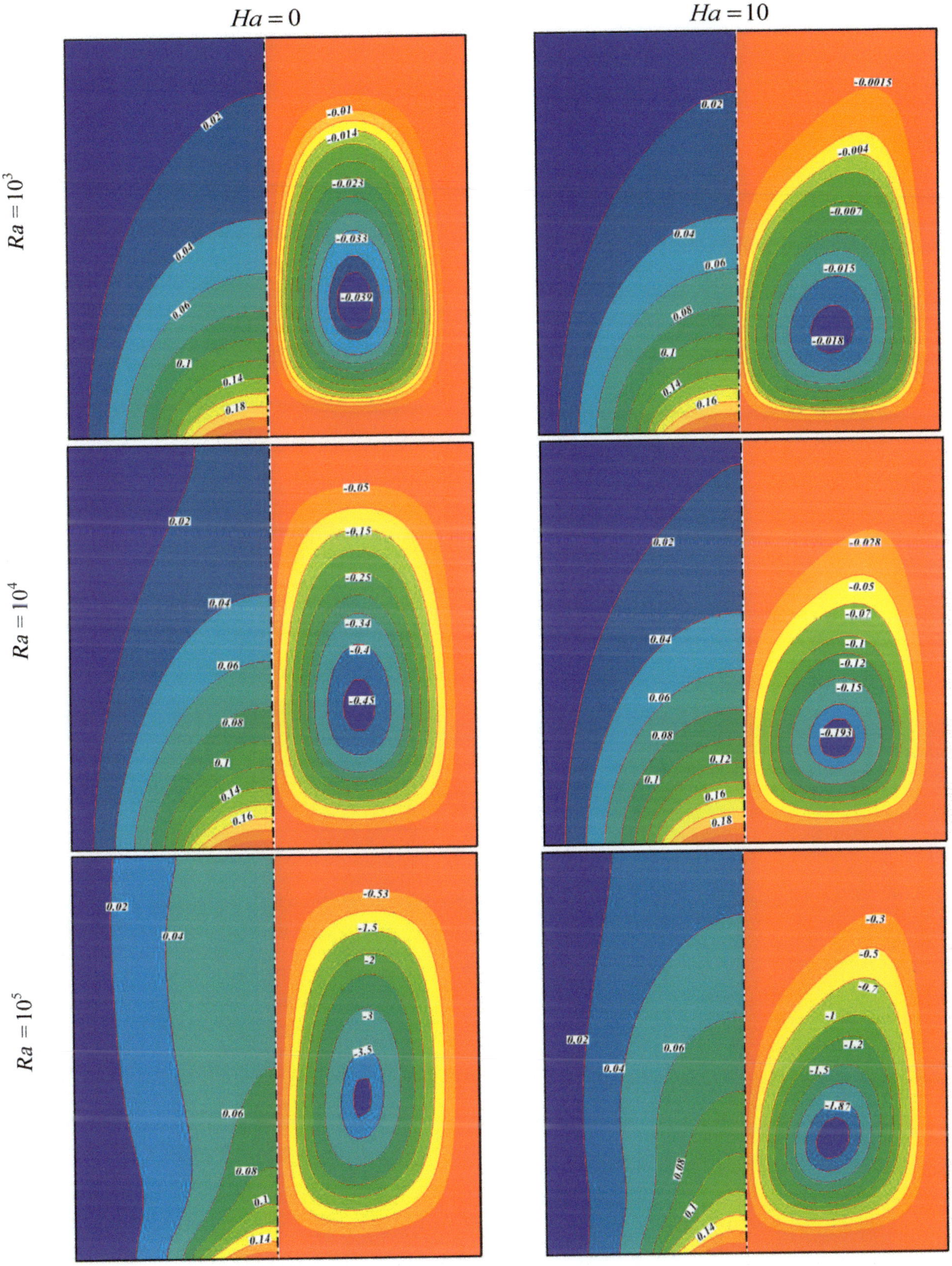

FIGURE 8.62 Isotherm (left) and streamline (right) contours for different values of Hartmann number and Raleigh number when $\phi = 0.04$, $Pr = 6.8$.

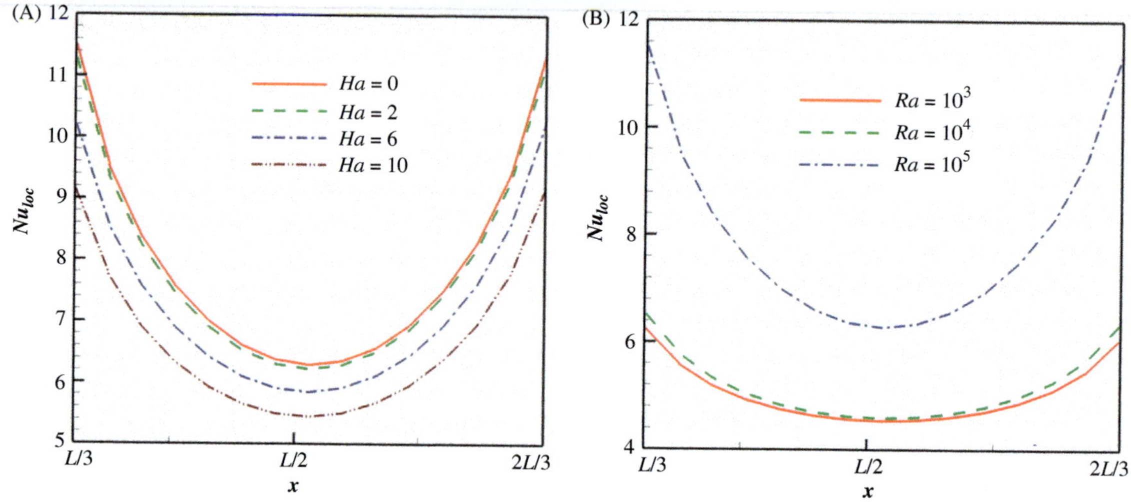

FIGURE 8.63 Effects of the Hartmann number and the Rayleigh number on the local Nusselt number when $\phi = 0.04$, $\Pr = 6.2$. (A) $Ra = 10^5$, (B) $Ha = 10$.

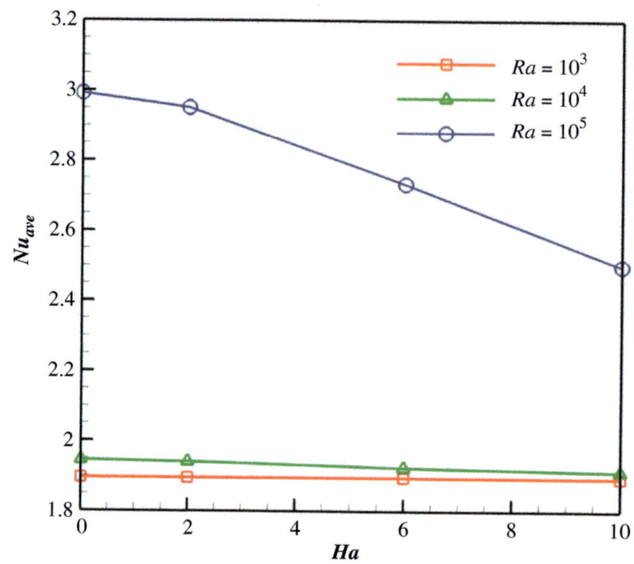

FIGURE 8.64 Effects of the Hartmann number and the Rayleigh number on the verage Nusselt number when $\phi = 0.04$, $\Pr = 6.2$.

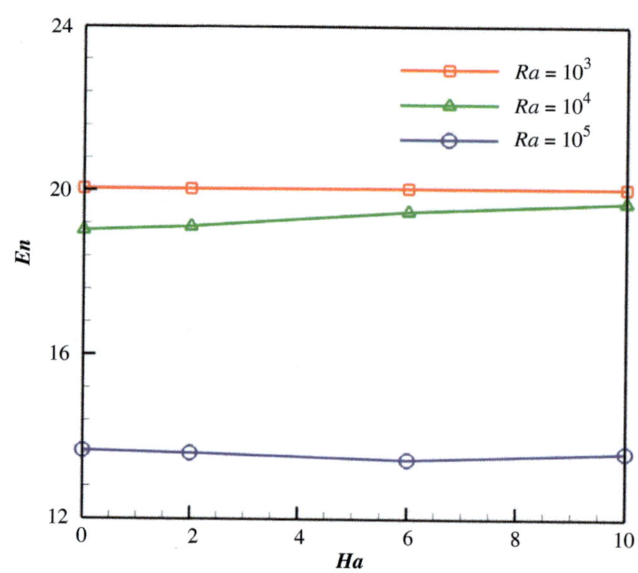

FIGURE 8.65 Effects of the Hartmann number and the Rayleigh number on the ratio of heat transfer enhancement when $\Pr = 6.2$.

the effects of the Lorentz and buoyancy forces on the heat transfer improvement. By adding Fe_3O_4 nanoparticles into the water, the temperature gradient enhances. This increment is more sensible in the presence of a magnetic field. In the conduction mode, addition of Fe_3O_4 affects significantly the thermal conductivity. Hence, the heat transfer (En) increases with an increase in the Lorentz force; but quite the opposite is true with the buoyancy force. It means that as convective heat transfer becomes stronger, the impact of adding nanoparticles into the base fluid becomes weaker.

8.11 MAGNETIC SOURCE IMPACT ON MAGNETIC NANOFLUID CONVECTIVE HEAT TRANSFER

8.11.1 Problem Definition

An enclosure with hot inner walls is considered. Fig. 8.66A depicts the boundary conditions. To obtain the shape of outer square and inner circular boundary, the following equation should be used:

$$\left(\frac{X}{a}\right)^{2\hat{n}} + \left(\frac{Y}{b}\right)^{2\hat{n}} = 1 \tag{8.221}$$

As $\hat{n}$ augments from 1 the geometry would approach a square for $a = b$. Magnetic source has been considered as shown in Fig. 8.67. $\overline{H_x}, \overline{H_y}, \overline{H}$ can be calculated as follows:

$$\begin{aligned} \overline{H_x} &= \left(y - \overline{b}\right)\left[\left(\overline{a} - x\right)^2 + \left(\overline{b} - y\right)^2\right]^{-1} \frac{\gamma}{2\pi}, \\ \overline{H_y} &= \left(\overline{a} - x\right)\left[\left(\overline{a} - x\right)^2 + \left(\overline{b} - y\right)^2\right]^{-1} \frac{\gamma}{2\pi}, \end{aligned} \tag{8.222}$$

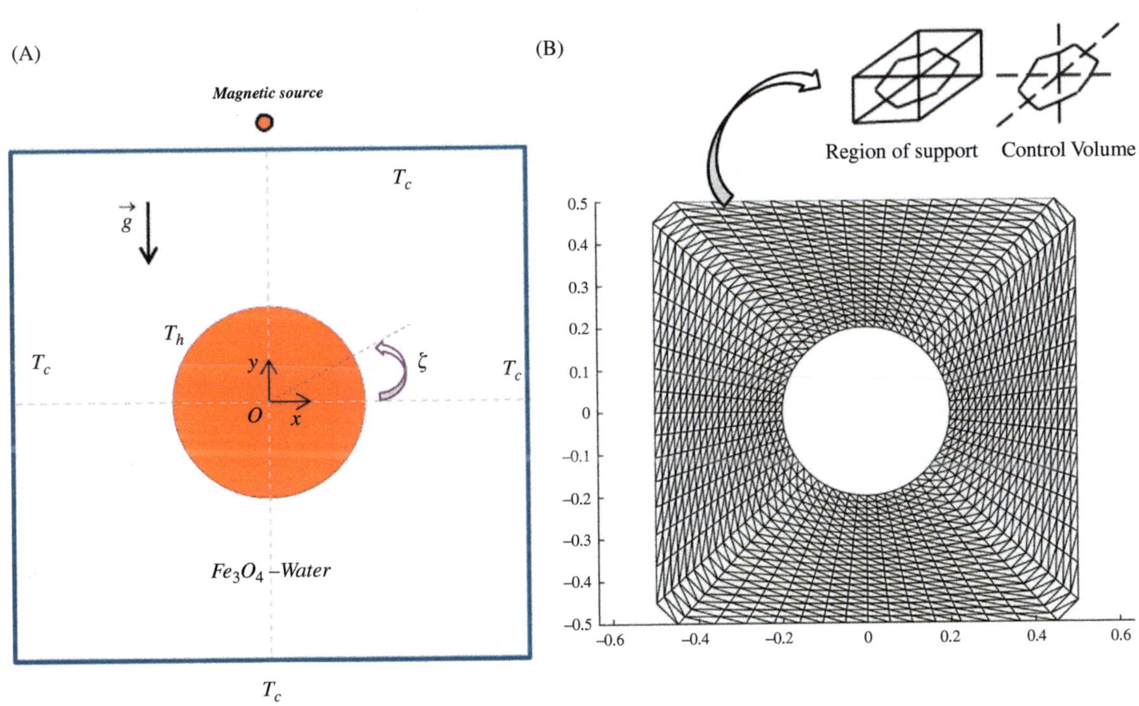

FIGURE 8.66 (A) Geometry and the boundary conditions with (B) the mesh of geometry considered in this work.

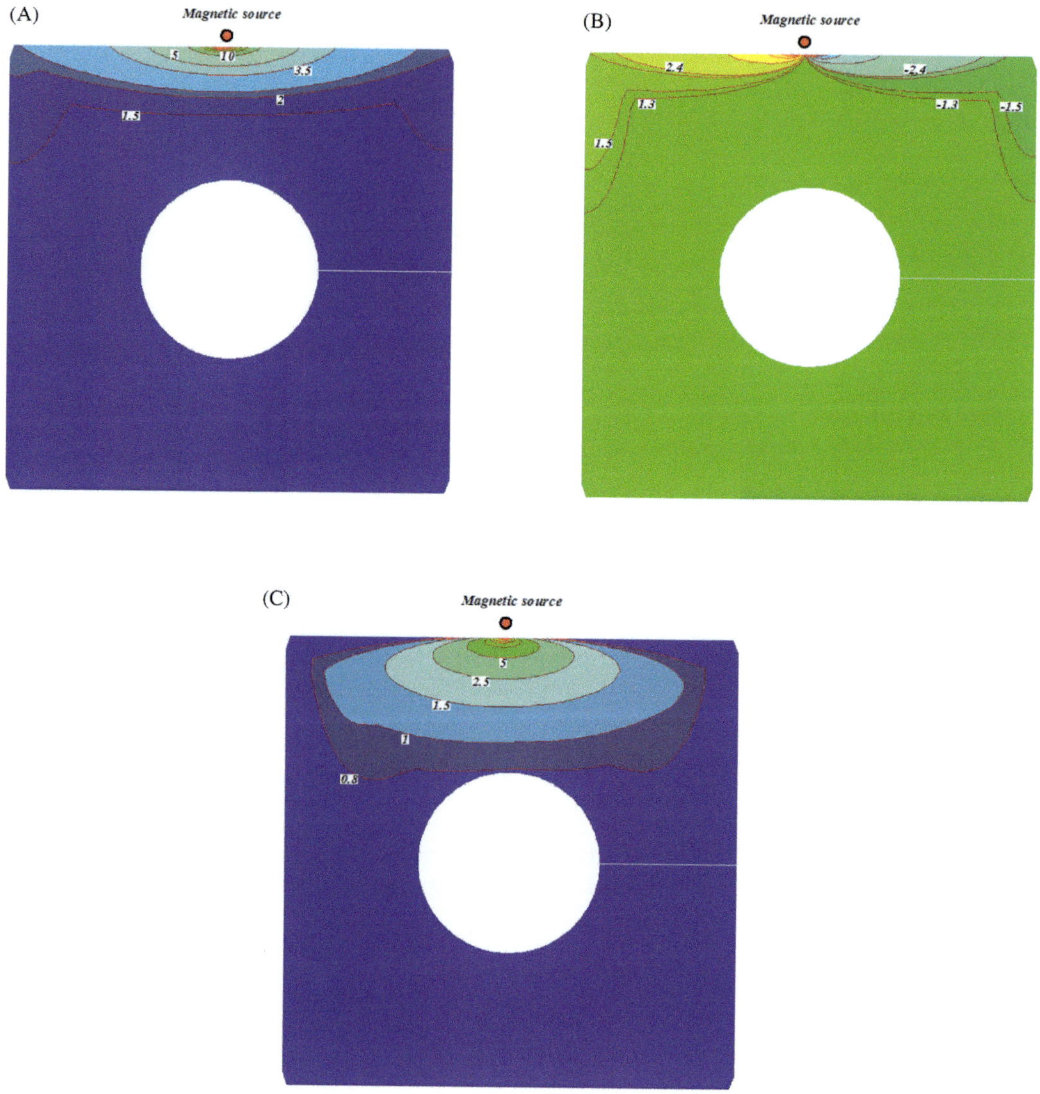

FIGURE 8.67 Contours of the (A) magnetic field strength H; (B) magnetic field intensity component in x direction Hx; (C) magnetic field intensity component in y direction Hy. (A) $H(x,y)$, (B) $Hx(x,y)$, (C) $Hy(x,y)$.

$$\overline{H} = \sqrt{\overline{H}_x^2 + \overline{H}_y^2} \tag{8.223}$$

8.11.2 Governing Equation

2D laminar nanofluid flow and forced convective heat transfer is taken into account. The governing PDEs can be considered as:

$$\frac{\partial u}{\partial x} + \frac{\partial v}{\partial y} = 0, \tag{8.224}$$

$$\left(\frac{\partial u}{\partial x}u + \frac{\partial u}{\partial y}v\right) = \left(\rho_{nf}\right)^{-1}\left[\left(\frac{\partial^2 u}{\partial y^2} + \frac{\partial^2 u}{\partial x^2}\right)\mu_{nf} - \frac{\partial P}{\partial x} - \sigma_{nf}B_y^2 u + \sigma_{nf}B_x B_y v\right] \tag{8.225}$$

$$\rho_{nf}\left(\frac{\partial v}{\partial x}u+\frac{\partial v}{\partial y}v\right)=+\mu_{nf}\left(\frac{\partial^2 v}{\partial x^2}+\frac{\partial^2 v}{\partial y^2}\right)-\frac{\partial P}{\partial y}+B_y\sigma_{nf}B_x u-B_x\sigma_{nf}B_x v+(T-T_c)\beta_{nf}g\rho_{nf}, \tag{8.226}$$

$$B=\mu_0 H$$

$$\begin{aligned}(\rho C_p)_{nf}\left(v\frac{\partial T}{\partial y}+u\frac{\partial T}{\partial x}\right)&=\sigma_{nf}\left(B_x v-B_y u\right)^2+k_{nf}\left(\frac{\partial^2 T}{\partial x^2}+\frac{\partial^2 T}{\partial y^2}\right)\\&+\mu_{nf}\left\{2\left(\frac{\partial u}{\partial x}\right)^2+2\left(\frac{\partial v}{\partial y}\right)^2+\left(\frac{\partial u}{\partial y}+\frac{\partial v}{\partial x}\right)^2\right\}-\frac{1}{(\rho C_p)_{nf}}\frac{\partial q_r}{\partial y},\left[q_r=-\frac{4\sigma_e}{3\beta_R}\frac{\partial T^4}{\partial y},\quad T^4\cong 4T_c^3T-3T_c^4\right]\end{aligned} \tag{8.227}$$

ρ_{nf}, $(\rho C_p)_{nf}$, α_{nf}, β_{nf}, μ_{nf}, k_{nf}, and σ_{nf} are calculated as

$$\rho_{nf}=\rho_f(1-\phi)+\rho_s\phi, \tag{8.228}$$

$$(\rho C_p)_{nf}=(\rho C_p)_f(1-\phi)+(\rho C_p)_s\phi, \tag{8.229}$$

$$\alpha_{nf}=\frac{k_{nf}}{(\rho C_p)_{nf}}, \tag{8.230}$$

$$\beta_{nf}=\beta_f(1-\phi)+\beta_s\phi. \tag{8.231}$$

$$\mu_{nf}=\frac{\mu_f}{(1-\phi)^{2.5}}. \tag{8.232}$$

$$k_{nf}=\left(\frac{k_s+2k_f-2\phi(k_f-k_s)}{k_s+2k_f+\phi(k_f-k_s)}\right)k_f, \tag{8.233}$$

$$\frac{\sigma_{nf}}{\sigma_f}=1+\frac{3\left(\frac{\sigma_s}{\sigma_f}-1\right)\phi}{\left(\frac{\sigma_s}{\sigma_f}+2\right)-\left(\frac{\sigma_s}{\sigma_f}-1\right)\phi}. \tag{8.234}$$

Dimensionless parameters are defined as:

$$\begin{aligned}&(a,b)=\frac{(\bar{a},\bar{b})}{L},\ (H,H_x,H_y)=\frac{(\overline{H},\overline{H_x},\overline{H_y})}{\overline{H_0}},(X,Y)=\frac{(x,y)}{L}\\&U=\frac{uL}{\alpha_f},V=\frac{vL}{\alpha_f},\ \Theta=\frac{T-T_c}{T_h-T_c},P=\frac{p}{\rho_f(\alpha_f/L)^2}\end{aligned} \tag{8.235}$$

So final equations are:

$$\frac{\partial U}{\partial X}+\frac{\partial V}{\partial Y}=0, \tag{8.236}$$

$$\frac{\partial U}{\partial X}U+\frac{\partial U}{\partial Y}V=\Pr\left[\frac{\mu_{nf}/\mu_f}{\rho_{nf}/\rho_f}\right]\left(\frac{\partial^2 U}{\partial Y^2}+\frac{\partial^2 U}{\partial X^2}\right)-Ha^2\Pr\left[\frac{\sigma_{nf}/\sigma_f}{\rho_{nf}/\rho_f}\right]\left(H_y^2U-H_xH_yV\right)-\frac{\partial P}{\partial X}, \tag{8.237}$$

$$V\frac{\partial V}{\partial Y}+U\frac{\partial V}{\partial X}=\Pr\left(\frac{\partial^2 V}{\partial Y^2}+\frac{\partial^2 V}{\partial X^2}\right)\left[\frac{\mu_{nf}/\mu_f}{\rho_{nf}/\rho_f}\right]-Ha^2\Pr\left[\frac{\sigma_{nf}/\sigma_f}{\rho_{nf}/\rho_f}\right](H_x^2V-H_xH_yU)-\frac{\partial P}{\partial Y}+Ra\Pr\left[\frac{\beta_{nf}}{\beta_f}\right]\Theta, \tag{8.238}$$

$$V\frac{\partial \Theta}{\partial Y}+U\frac{\partial \Theta}{\partial X}=\left[\frac{k_{nf}}{k_f}\frac{(\rho C_P)_f}{(\rho C_P)_{nf}}\right]\left(\frac{\partial^2 \Theta}{\partial Y^2}+\frac{\partial^2 \Theta}{\partial X^2}\right)+Ha^2\,Ec\left(\frac{(\rho C_P)_f}{(\rho C_P)_{nf}}\frac{\sigma_{nf}}{\sigma_f}\right)\{V\,H_x-U\,H_y\}^2$$
$$+\left[\frac{\dfrac{\mu_{nf}}{\mu_f}}{\dfrac{(\rho C_P)_{nf}}{(\rho C_P)_f}}\right]Ec\left\{2\left(\frac{\partial U}{\partial X}\right)^2+2\left(\frac{\partial V}{\partial Y}\right)^2+\left(\frac{\partial U}{\partial Y}+\frac{\partial V}{\partial X}\right)^2\right\}+\left(\frac{4}{3}Rd\left(\frac{k_{nf}}{k_f}\right)^{-1}\right)\frac{\partial^2\Theta}{\partial Y^2} \tag{8.239}$$

and dimensionless parameters are

$$Ra_f=g\beta_fL^3(T_h-T_c)/(\alpha_f\upsilon_f),\ \Pr_f=\upsilon_f/\alpha_f,\ Ha=L\mu_0H_0\sqrt{\sigma_f/\mu_f},$$
$$Ec=\left(\mu_f\alpha_f\right)/\left[(\rho C_P)_f\Delta T\,L^2\right],\quad Rd=4\sigma_eT_c^3/(\beta_Rk_f) \tag{8.240}$$

The thermophysical properties of Fe_3O_4 and water are presented in Table 8.1. Pressure gradient source terms are discarded by the vorticity stream function.

$$\Omega=\frac{\omega L^2}{\alpha_f},\Psi=\frac{\psi}{\alpha_f},\omega=-\frac{\partial u}{\partial y}+\frac{\partial v}{\partial x},(u,v)=\left(\frac{\partial\psi}{\partial y},-\frac{\partial\psi}{\partial x}\right) \tag{8.241}$$

According to Fig. 8.66, boundary conditions are

$$\begin{aligned}&\text{on outer wall } \Theta=0.0\\&\text{on all walls } \Psi=0.0\\&\text{on inner wall } \Theta=1.0\end{aligned} \tag{8.242}$$

Nu_{loc}, Nu_{ave} along hot wall are :

$$Nu_{loc}=\left(\frac{k_{nf}}{k_f}\right)\left(1+\frac{4}{3}Rd\left(\frac{k_{nf}}{k_f}\right)^{-1}\right)\frac{\partial\Theta}{\partial r} \tag{8.243}$$

$$Nu_{ave}=\frac{1}{2\pi}\int_0^{2\pi}Nu_{loc}\,d\zeta \tag{8.244}$$

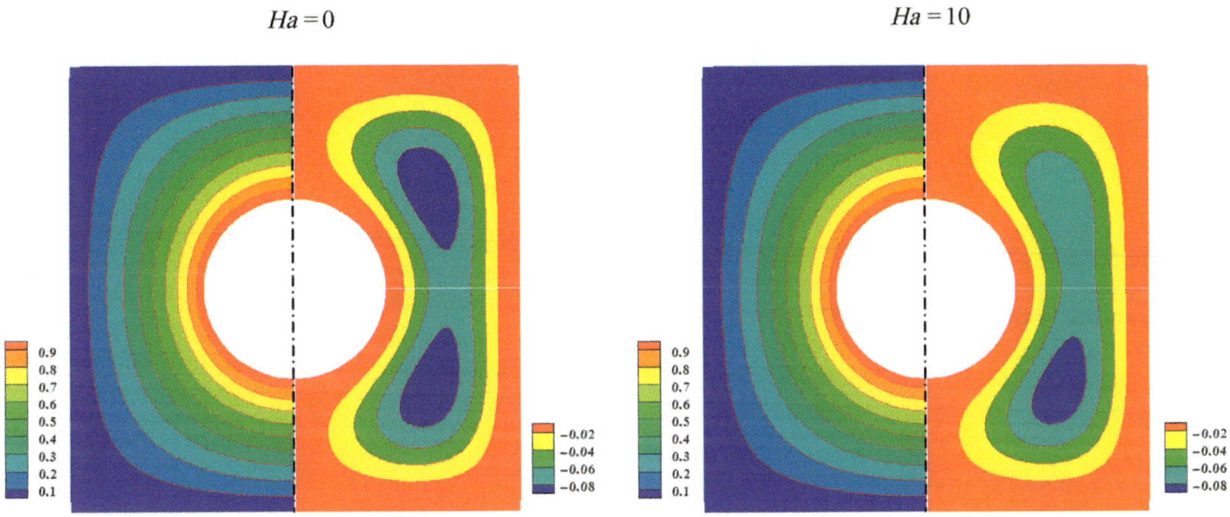

FIGURE 8.68 Isotherm (left) and streamline (right) contours for different values of Hartmann number when $Ra = 10^3, \phi = 0.04, Rd = 0.8$.

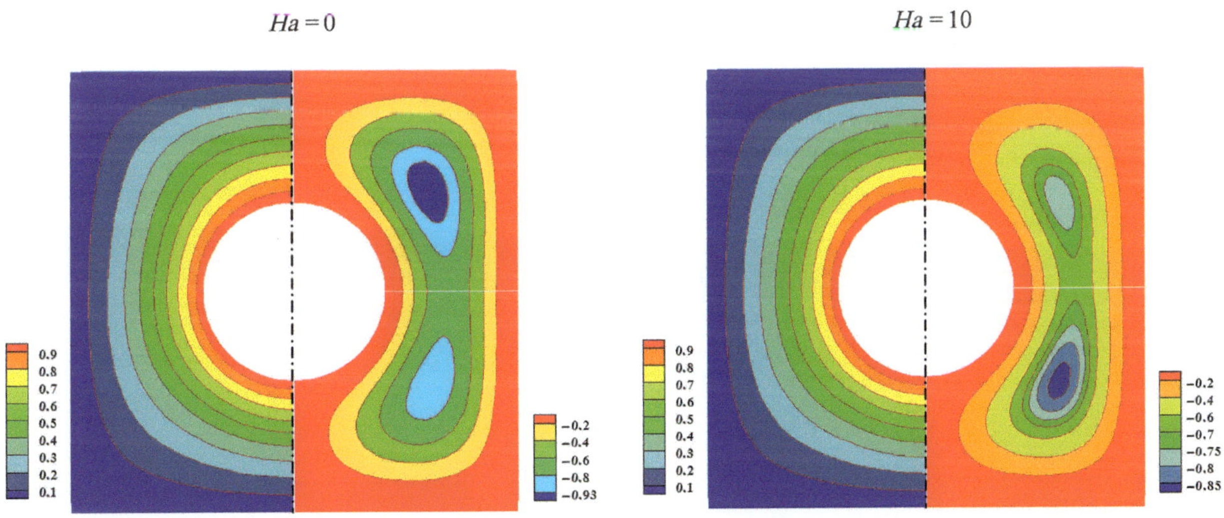

FIGURE 8.69 Isotherm (left) and streamline (right) contours for different values of Hartmann number when $Ra = 10^4, \phi = 0.04, Rd = 0.8$.

8.11.3 Effects of Active Parameters

Ferrofluid natural convection and radiation heat transfer in the presence of a magnetic source impact in a cavity with a circular hot cylinder is simulated. Various values of Rayleigh number ($Ra = 10^3, 10^4$ and 10^5), Hartmann number ($Ha = 0$ to 10), and radiation parameter ($Rd = 0$ to 0.8) are reported. ϕ, Ec, and Pr are 0.04, 10^{-5}, and 6.8, respectively.

Figs. 8.68–8.70 depict the impacts of radiation parameter, Rayleigh, and Hartmann numbers on isotherms and streamlines. In a dominant conduction mechanism, two main vortexes are generated in the absence of Lorentz forces. As the Hartmann number increases, the upper vortex vanishes. As the Rayleigh number augments, a thermal plume appears near the centerline and vortexes become stronger. Increasing the Hartmann number causes the thermal plume to vanish and changes the heat transfer mechanism from convection to conduction. So, the Nusselt number decreases with the increase of Lorentz forces.

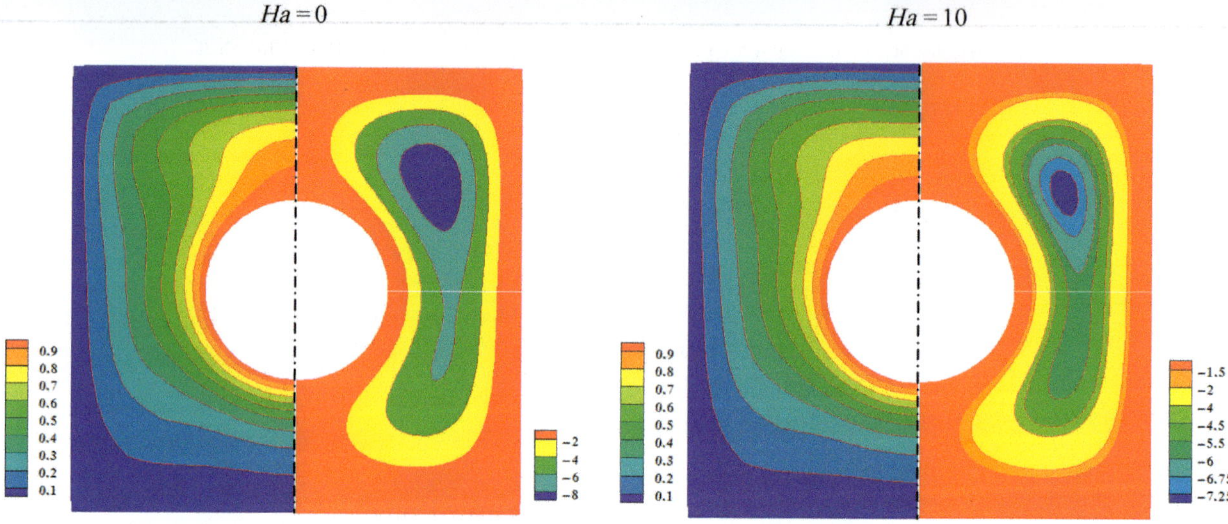

FIGURE 8.70 Isotherm (left) and streamline (right) contours for different values of Hartmann number when $Ra = 10^5$, $\phi = 0.04$, $Rd = 0.8$.

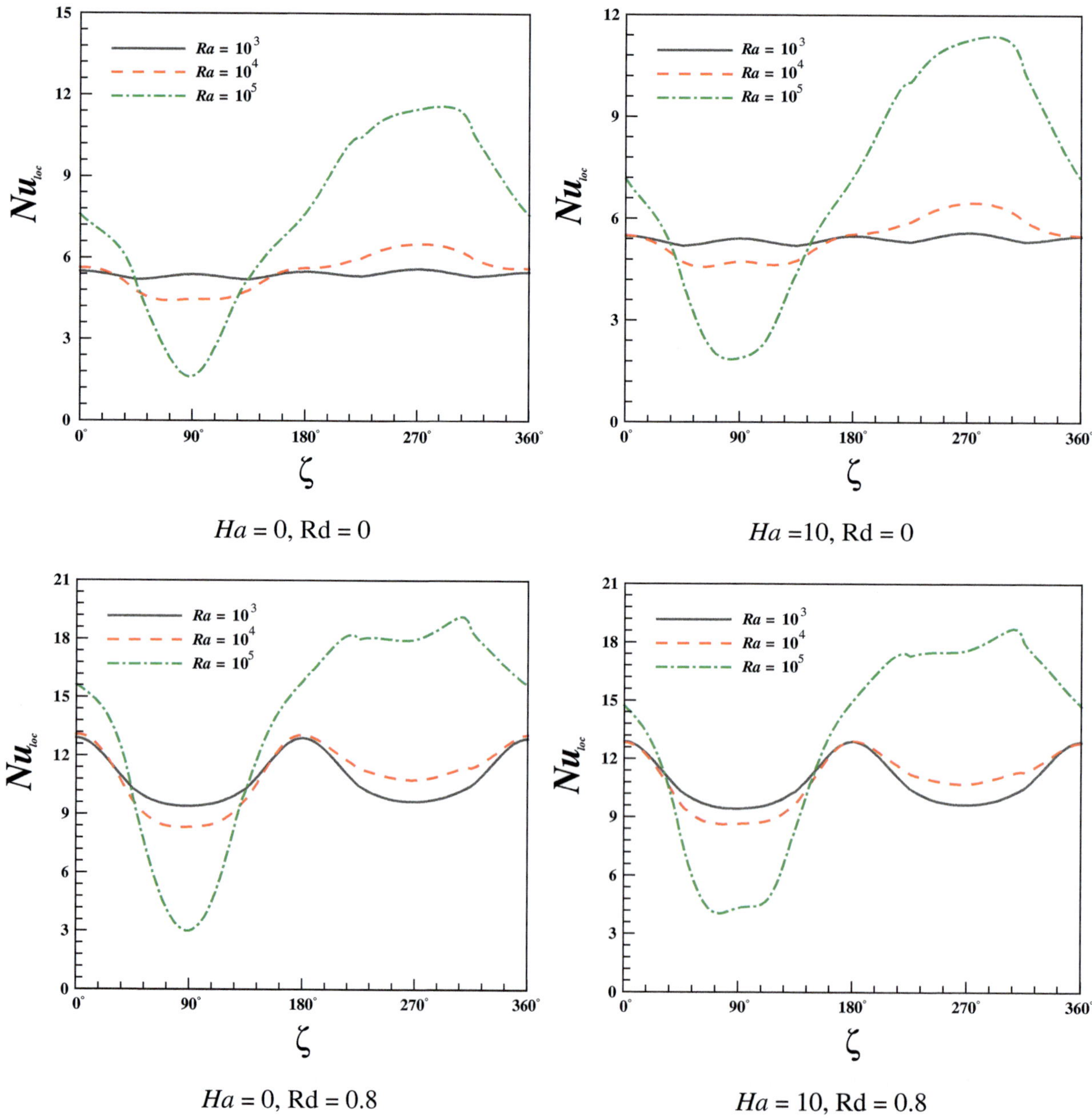

FIGURE 8.71 Effects of radiation parameter, Hartmann number, and Rayleigh number on local Nusselt number Nu_{loc} along hot wall when $\phi = 0.04$.

The influences of the radiation parameter, Lorentz, and buoyancy forces on Nu_{loc}, Nu_{ave} are shown in Figs. 8.71 and 8.72. The formula for Nu_{ave} corresponding active parameters is:

$$\begin{aligned} Nu_{ave} &= 16.32 - 1.07Rd - 6.39(\log(Ra)) + 0.04Ha + 0.48Rd(\log(Ra)) \\ &- 0.03Rd\,Ha - 0.01(\log(Ra))\,Ha + 8.02\,Rd^2 + 0.9(\log(Ra))^2 + 1.2\times10^{-3}Ha^2 \end{aligned} \tag{8.245}$$

The extreme points which are shown in Nu_{loc} are due to the thermal plume which is generated at high Rayleigh number. As buoyancy forces increase, the temperature gradient near the circular wall augments, and in turn Nu_{ave} is enhanced with the increase of the Rayleigh number. Enhancing Lorentz forces leads to reducing Nu_{ave} due to the domination of the conduction mode. The Nusselt number increases with the rise of the radiation parameter.

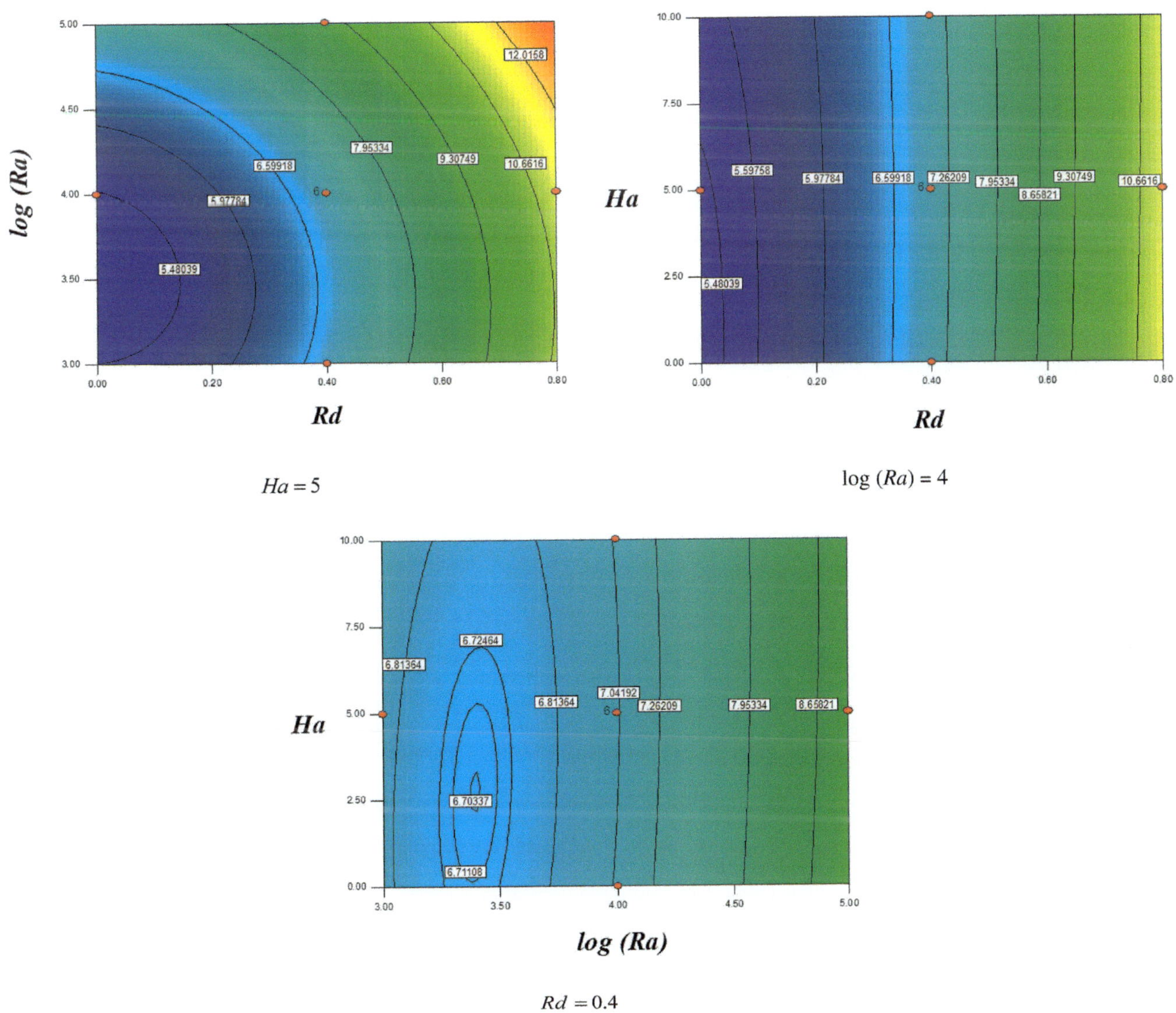

FIGURE 8.72 Effects of radiation parameter, Hartmann number, and Rayleigh number on average Nusselt number Nu_{ave} along hot wall when $\phi = 0.04$.

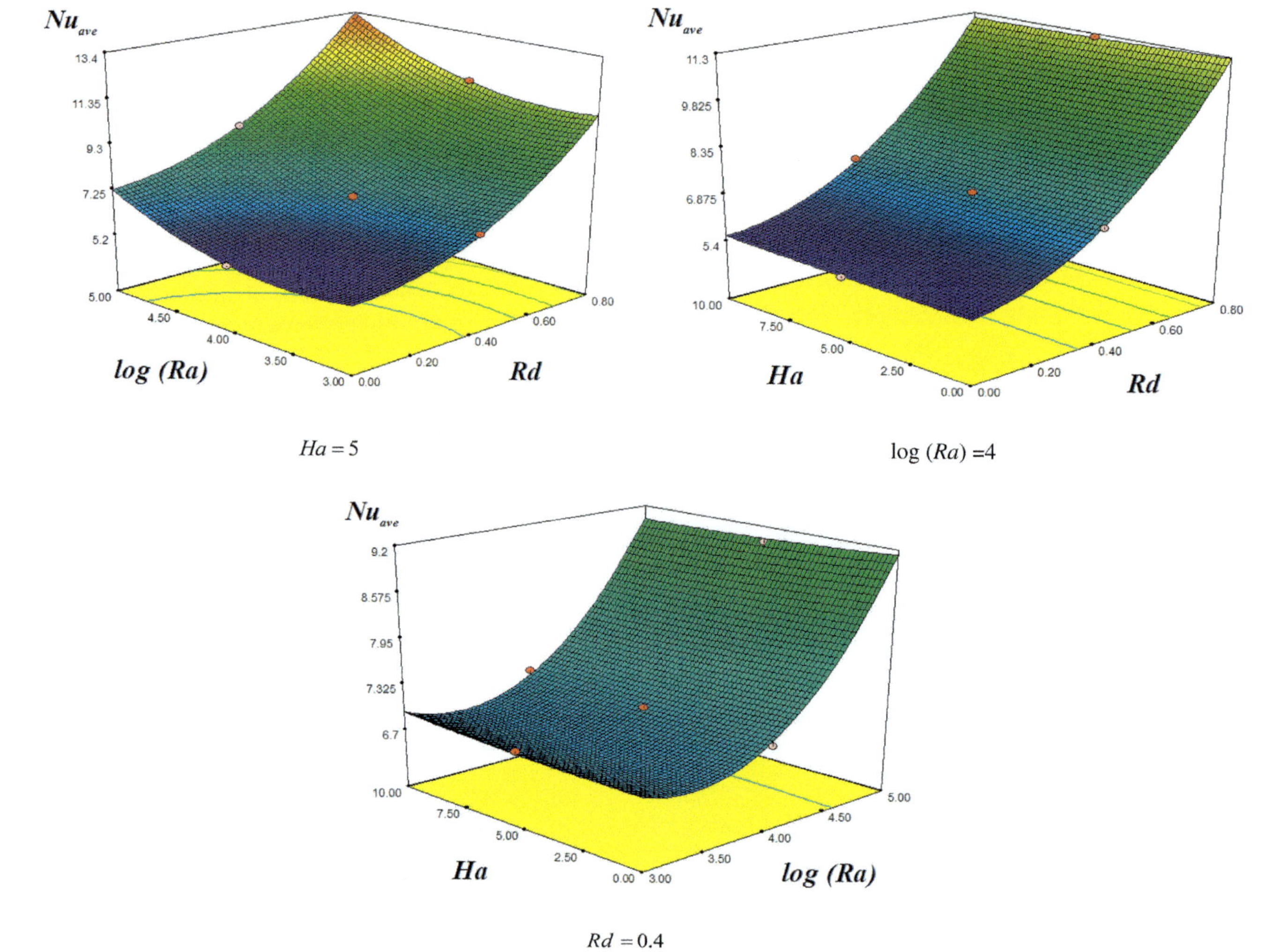

FIGURE 8.72 (Continued).

References

[1] M. Sheikholeslami, D.D. Ganji, Numerical investigation for two phase modeling of nanofluid in a rotating system with permeable sheet, J. Mol. Liq. 194 (2014) 13–19.

[2] T.C.H. Jue, Analysis of combined thermal and magnetic convection ferrofluid flow in a cavity, Int. Commun. Heat Mass Transfer 33 (2006) 846–852.

[3] F. Selimefendigil, H.F. Oztop, Effect of a rotating cylinder in forced convection of ferrofluid over a backward facing step, Int. J. Heat Mass Transfer 71 (2014) 142–148.

[4] C.E. Nanjundappa, I.S. Shivakumara, M.M. Ravisha, The onset of ferroconvection in a horizontal ferrofluid saturated porous layer heated from below and cooled from above with constant heat flux subject to MFD viscosity, Int. Commun. Heat Mass Transfer 37 (2010) 1246–1250.

[5] X. Zhang, H. Huang, Effect of magnetic obstacle on fluid flow and heat transfer in a rectangular duct, Int. Commun. Heat Mass Transfer 51 (2014) 31–38.

[6] M. Sheikholeslami, D.D. Ganji, Magnetohydrodynamic flow in a permeable channel filled with nanofluid, Scientia Iranica B 21 (1) (2014) 203–212.

[7] M. Sheikholeslami, D.D. Ganji, Three dimensional heat and mass transfer in a rotating system using nanofluid, Powder Technol. 253 (2014) 789–796.

[8] M. Sheikholeslami, D.D. Ganji, Heat transfer of Cu-water nanofluid flow between parallel plates, Powder Technol. 235 (2013) 873–879.

[9] M. Sheikholeslami, Influence of Lorentz forces on nanofluid flow in a porous cylinder considering Darcy model, J. Mol. Liq. 225 (2017) 903–912.

[10] M. Sheikholeslami, CVFEM for magnetic nanofluid convective heat transfer in a porous curved enclosure, Eur. Phys. J. Plus 131 (2016) 413. Available from: https://doi.org/10.1140/epjp/i2016-16413-y.

[11] M. Sheikholeslami, Influence of Coulomb forces on Fe3O4-H2O nanofluid thermal improvement, Int. J. Hydrogen Energy 42 (2017) 821–829.
[12] M. Sheikholeslami, Magnetic field influence on nanofluid thermal radiation in a cavity with tilted elliptic inner cylinder, J. Mol. Liq. 229 (2017) 137–147.
[13] M. Sheikholeslami, Numerical simulation of magnetic nanofluid natural convection in porous media, Phys. Lett. A 381 (2017) 494–503.
[14] M. Sheikholeslami, Magnetic source impact on nanofluid heat transfer using CVFEM, Neural Comput. Applicat. (2016). Available from: https://doi.org/10.1007/s00521-016-2740-7.
[15] M. Sheikholeslami, CuO-water nanofluid free convection in a porous cavity considering Darcy law, Europ. Phys. J. Plus 132 (2017) 55. Available from: https://doi.org/10.1140/epjp/i2017-11330-3.
[16] M. Sheikholeslami, Numerical investigation of MHD nanofluid free convective heat transfer in a porous tilted enclosure, Eng. Comput. 34 (6) (2017) 1939–1955.
[17] M.S. Kandelousi, Effect of spatially variable magnetic field on ferrofluid flow and heat transfer considering constant heat flux boundary condition, Europ. Phys. J. Plus (2014) 129–248. Available from: https://doi.org/10.1140/epjp/i2014-14248-2.
[18] M. Sheikholeslami, Numerical simulation for external magnetic field influence on Fe_3O_4-water nanofluid forced convection, Engineering Computations, Inpress.
[19] M. Sheikholeslami, Influence of Lorentz forces on nanofluid flow in a porous cavity by means of Non- Darcy model, Eng. Comput. 34 (8) (2017) 2651–2667. Available from: https://doi.org/10.1108/EC-01-2017-0008.
[20] M. Sheikholeslami, H.B. Rokni, Nanofluid convective heat transfer intensification in a porous circular cylinder, Chem. Eng. Process. Process Intensif. 120 (2017) 93–104.
[21] M. Sheikholeslami, M.K. Sadoughi, Numerical modeling for Fe3O4 -water nanofluid flow in porous medium considering MFD viscosity, J. Mol. Liq. 242 (2017) 255–264.
[22] M. Sheikholeslami, M. Shamlooei, Magnetic source influence on nanofluid flow in porous medium considering shape factor effect, Phys. Lett. A 381 (2017) 3071–3078.
[23] M. Sheikholeslami, M. Shamlooei, Convective flow of nanofluid inside a lid driven porous cavity using CVFEM, Phys. B: Condensed Matter 521 (2017) 239–250.
[24] M. Sheikholeslami, K. Vajravelu, Forced convection heat transfer in Fe3O4- ethylene glycol nanofluid under the influence of Coulomb force, J. Mol. Liq. 233 (2017) 203–210.
[25] M. Sheikholeslami, H.B. Rokni, Influence of EFD viscosity on nanofluid forced convection in a cavity with sinusoidal wall, J. Mol. Liq. 232 (2017) 390–395.
[26] M. Sheikholeslami, H.B. Rokni, Magnetic nanofluid natural convection in the presence of thermal radiation considering variable viscosity, Europ. Phys. J. Plus 132 (2017) 238. Available from: https://doi.org/10.1140/epjp/i2017-11498-4.
[27] M. Sheikholeslami, H.B. Rokni, Numerical simulation for impact of Coulomb force on nanofluid heat transfer in a porous enclosure in presence of thermal radiation, Int. J. Heat Mass Transfer 118 (2018) 823–831.
[28] M. Sheikholeslami, M. Shamlooei, R. Moradi, Fe3O4- Ethylene glycol nanofluid forced convection inside a porous enclosure in existence of Coulomb force, J. Mol. Liq. 249 (2018) 429–437.
[29] M. Sheikholeslami, S.A. Shehzad, Numerical analysis of Fe_3O_4 $-H_2O$ nanofluid flow in permeable media under the effect of external magnetic source, Int. J. Heat Mass Transfer 118 (2018) 182–192.
[30] M. Sheikholeslami, S.A. Shehzad, Magnetohydrodynamic nanofluid convection in a porous enclosure considering heat flux boundary condition, Int. J. Heat Mass Transfer 106 (2017) 1261–1269.
[31] M. Sheikholeslami, S.A. Shehzad, Thermal radiation of ferrofluid in existence of Lorentz forces considering variable viscosity, Int. J. Heat Mass Transfer 109 (2017) 82–92.
[32] M. Sheikholeslami, H.B. Rokni, Numerical modeling of nanofluid natural convection in a semi annulus in existence of Lorentz force, Comput. Methods Appl. Mech. Eng. 317 (2017) 419–430.
[33] M. Sheikholeslami, K. Vajravelu, Nanofluid flow and heat transfer in a cavity with variable magnetic field, Appl. Math. Comput. 298 (2017) 272–282.
[34] M. Sheikholeslami, H.B. Rokni, Magnetohydrodynamic CuO-water nanofluid in a porous complex shaped enclosure, ASME, J. Thermal Sci. Eng. Applicat. 9 (4) (2017) 041007. Available from: https://doi.org/10.1115/1.4035973.
[35] M. Sheikholeslami, H.B. Rokni, Nanofluid two phase model analysis in existence of induced magnetic field, Int. J. Heat Mass Transfer 107 (2017) 288–299.
[36] M. Sheikholeslami, T. Hayat, A. Alsaedi, S. Abelman, Numerical analysis of EHD nanofluid force convective heat transfer considering electric field dependent viscosity, Int. J. Heat Mass Transfer 108 (2017) 2558–2565.
[37] M. Sheikholeslami, Magnetic field influence on CuO -H2O nanofluid convective flow in a permeable cavity considering various shapes for nanoparticles, Int. J. Hydrogen Energy 42 (2017) 19611–19621.
[38] M. Sheikholeslami, M.M. Bhatti, Forced convection of nanofluid in presence of constant magnetic field considering shape effects of nanoparticles, Int. J. Heat Mass Transfer 111 (2017) 1039–1049.
[39] M. Sheikholeslami, A. Zeeshan, Analysis of flow and heat transfer in water based nanofluid due to magnetic field in a porous enclosure with constant heat flux using CVFEM, Comput. Methods Appl. Mech. Eng. 320 (2017) 68–81.
[40] M. Sheikholeslami, M. Shamlooei, Fe3O4- H2O nanofluid natural convection in presence of thermal radiation, Int. J. Hydrogen Energy 42 (9) (2017) 5708–5718.
[41] M. Sheikholeslami, A.J. Chamkha, Flow and convective heat transfer of a ferro-nanofluid in a double-sided lid-driven cavity with a wavy wall in the presence of a variable magnetic field, Numer. Heat Transfer, Part A VOL. 69 (10) (2016) 1186–1200. Available from: https://doi.org/10.1080/10407782.2015.1125709.
[42] M. Sheikholeslami, A.J. Chamkha, Electrohydrodynamic free convection heat transfer of a nanofluid in a semi-annulus enclosure with a sinusoidal wall, Numer. Heat Transfer, Part A VOL. 69 (7) (2016) 781–793. Available from: https://doi.org/10.1080/10407782.2015.1090819.

[43] M. Sheikholeslami, M.K. Sadoughi, Simulation of CuO- water nanofluid heat transfer enhancement in presence of melting surface, Int. J. Heat Mass Transfer 116 (2018) 909–919.
[44] M. Sheikholeslami, H.B. Rokni, Simulation of nanofluid heat transfer in presence of magnetic field: a review, Int. J. Heat Mass Transfer 115 (2017) 1203–1233.
[45] M. Sheikholeslami, H.B. Rokni, Melting heat transfer influence on nanofluid flow inside a cavity in existence of magnetic field, Int. J. Heat Mass Transfer 114 (2017) 517–526.
[46] M. Sheikholeslami, S.A. Shehzad, Magnetohydrodynamic nanofluid convective flow in a porous enclosure by means of LBM, Int. J. Heat Mass Transfer 113 (2017) 796–805.
[47] M. Sheikholeslami, Magnetohydrodynamic nanofluid forced convection in a porous lid driven cubic cavity using Lattice Boltzmann Method, J. Mol. Liq. 231 (2017) 555–565.
[48] M.S. Kandelousi, KKL correlation for simulation of nanofluid flow and heat transfer in a permeable channel, Phys. Lett. A 378 (45) (2014) 3331–3339.
[49] M. Sheikholeslami, Effect of uniform suction on nanofluid flow and heat transfer over a cylinder, J. Braz. Soc. Mech. Sci. Eng. 37 (2015) 1623–1633.
[50] M. Sheikholeslami, Lattice Boltzmann Method simulation of MHD non-Darcy nanofluid free convection, Phys. B 516 (2017) 55–71.
[51] M. Sheikholeslami, Influence of magnetic field on nanofluid free convection in an open porous cavity by means of Lattice Boltzmann Method, J. Mol. Liq. 234 (2017) 364–374.
[52] M. Sheikholeslami, Ali J. Chamkha, Influence of Lorentz forces on nanofluid forced convection considering Marangoni convection, J. Mol. Liq. 225 (2017) 750–757.
[53] M. Sheikholeslami, T. Hayat, A. Alsaedi, Numerical simulation of nanofluid forced convection heat transfer improvement in existence of magnetic field using Lattice Boltzmann Method, Int. J. Heat Mass Transfer 108 (2017) 1870–1883.
[54] M. Sheikholeslami, R. Ellahi, Three dimensional mesoscopic simulation of magnetic field effect on natural convection of nanofluid, Int. J. Heat Mass Transfer 89 (2015) 799–808.
[55] M. Sheikholeslami, S. Abelman, Two phase simulation of nanofluid flow and heat transfer in an annulus in the presence of an axial magnetic field, IEEE Trans. Nanotechnol. 14 (3) (2015) 561–569.
[56] M. Sheikholeslami, R. Ellahi, Simulation of ferrofluid flow for magnetic drug targeting using Lattice Boltzmann method, J. Zeitschrift Fur Naturforschung A 70 (2) (2015) 115–124.
[57] M. Sheikholeslami, M. Seyednezhad, Lattice Boltzmann Method simulation for CuO-water nanofluid flow in a porous enclosure with hot obstacle, J. Mol. Liq. 243 (2017) 249–256.
[58] M. Sheikholeslami, M. Sadoughi, Mesoscopic method for MHD nanofluid flow inside a porous cavity considering various shapes of nanoparticles, Int. J. Heat Mass Transfer 113 (2017) 106–114.
[59] M. Sheikholeslami, A. Zeeshan, Mesoscopic simulation of CuO-H2O nanofluid in a porous enclosure with elliptic heat source, Int. J. Hydrogen Energy 42 (22) (2017) 15393–15402.
[60] M. Sheikholeslami, S.A. Shehzad, CVFEM for influence of external magnetic source on Fe3O4 – H2O nanofluid behavior in a permeable cavity considering shape effect, Int. J. Heat Mass Transfer 115 (2017) 180–191.
[61] M. Sheikholeslami, M. Seyednezhad, Nanofluid heat transfer in a permeable enclosure in presence of variable magnetic field by means of CVFEM, Int. J. Heat Mass Transfer 114 (2017) 1169–1180.
[62] M. Sheikholeslami, H.B. Rokni, Free convection of CuO-H2O nanofluid in a curved porous enclosure using mesoscopic approach, Int. J. Hydrogen Energy 42 (22) (2017) 14942–14949.
[63] M. Sheikholeslami, Numerical investigation for CuO-H_2O nanofluid flow in a porous channel with magnetic field using mesoscopic method, J. Mol. Liq. 249 (2018) 739–746.
[64] M. Sheikholeslami, H.B. Rokni, Influence of melting surface on MHD nanofluid flow by means of two phase model, Chin. J. Phys. 55 (2017) 1352–1360.
[65] M. Sheikholeslami, H.B. Rokni, Effect of melting heat transfer on nanofluid flow in existence of magnetic field considering Buongiorno Model, Chin. J. Phys. 55 (2017) 1115–1126.
[66] M. Sheikholeslami, T. Hayat, A. Alsaedi, MHD free convection of Al2O3–water nanofluid considering thermal radiation: a numerical study, Int. J. Heat Mass Transfer 96 (2016) 513–524.
[67] M. Sheikholeslami, K. Vajravelu, M.M. Rashidi, Forced convection heat transfer in a semi annulus under the influence of a variable magnetic field, Int. J. Heat Mass Transfer 92 (2016) 339–348.
[68] M. Sheikholeslami, D.D. Ganji, Ferrohydrodynamic and Magnetohydrodynamic effects on ferrofluid flow and convective heat transfer, Energy 75 (2014) 400–410.
[69] M. Sheikholeslami, D.D. Ganji, M.M. Rashidi, Ferrofluid flow and heat transfer in a semi annulus enclosure in the presence of magnetic source considering thermal radiation, J. Taiwan Inst. Chem. Eng. 47 (2015) 6–17.
[70] M. Sheikholeslami, M.M. Rashidi, Effect of space dependent magnetic field on free convection of Fe3O4-water nanofluid, J. Taiwan Inst. Chem. Eng. 56 (2015) 6–15.
[71] M. Sheikholeslami, M.M. Rashidi, Non-uniform magnetic field effect on nanofluid hydrothermal treatment considering Brownian motion and thermophoresis effects, J Braz. Soc. Mech. Sci. Eng. 38 (2016) 1171–1184.
[72] M. Sheikholeslami, D.D. Ganji, Numerical investigation of nanofluid transportation in a curved cavity in existence of magnetic source, Chem. Phys. Lett. 667 (2017) 307–316.
[73] M. Sheikholeslami, D.D. Ganji, Ferrofluid convective heat transfer under the influence of external magnetic source, Alexandria Eng. J. (2018). Available from: http://dx.doi.org/10.1016/j.aej.2016.11.007.
[74] M. Sheikholeslami, D.D. Ganji, Free convection of Fe3O4-water nanofluid under the influence of an external magnetic source, J. Mol. Liq. 229 (2017) 530–540.
[75] M. Sheikholeslami, R. Ellahi, K. Vafai, Study of Fe3O4-water nanofluid with convective heat transfer in the presence of magnetic source, Alexandria Eng. J. (2017). Available from: http://dx.doi.org/10.1016/j.aej.2017.01.027.

[76] M. Sheikholeslami, H. Oztop, MHD free convection of nanofluid in a cavity with sinusoidal walls by using CVFEM, Chin. J. Phys. 55 (6) (2017) 2291–2304.
[77] M. Sheikholeslami, Z. Li, M. Shamlooei, Nanofluid MHD natural convection through a porous complex shaped cavity considering thermal radiation, Phys. Lett. A 382 (2018) 1615–1632.
[78] M. Sheikholeslami, H.B. Rokni, Magnetic nanofluid flow and convective heat transfer in a porous cavity considering Brownian motion effects, Phys. Fluids 30 (1) (2018). Available from: https://doi.org/10.1063/1.5012517.
[79] M. Sheikholeslami, S.A. Shehzad, Simulation of water based nanofluid çonvective flow inside a porous enclosure via non-equilibrium model, Int. J. Heat Mass Transfer 120 (2018) 1200–1212.
[80] M. Sheikholeslami, S.A. Shehzad, Non-Darcy free convection of Fe_3O_4-water nanoliquid in a complex shaped enclosure under impact of uniform Lorentz force, Chin. J. Phys 56 (2018) 270–281.
[81] M. Sheikholeslami, M. Seyednezhad, Simulation of nanofluid flow and natural convection in a porous media under the influence of electric field using CVFEM, Int. J. Heat Mass Transfer 120 (2018) 772–781.
[82] M. Sheikholeslami, M. Shamlooei, R. Moradi, Numerical simulation for heat transfer intensification of nanofluid in a permeable curved enclosure considering shape effect of Fe_3O_4 nanoparticles, Chem. Eng. Process. Process Intensif. 124 (2018) 71–82.
[83] M. Sheikholeslami, T. Hayat, T. Muhammad, A. Alsaedi, MHD forced convection flow of nanofluid in a porous cavity with hot elliptic obstacle by means of Lattice Boltzmann method, Int. J. Mech. Sci. 135 (2018) 532–540.
[84] M. Sheikholeslami, Numerical investigation of nanofluid free convection under the influence of electric field in a porous enclosure, J. Mol. Liq. 249 (2018) 1212–1221.
[85] M. Sheikholeslami, CuO-water nanofluid flow due to magnetic field inside a porous media considering Brownian motion, J. Mol. Liq. 249 (2018) 921–929.
[86] M. Sheikholeslami, Numerical investigation for CuO-H_2O nanofluid flow in a porous channel with magnetic field using mesoscopic method, J. Mol. Liq. 249 (2018) 739–746.
[87] M. Sheikholeslami, M. Shamlooei, R. Moradi, Fe_3O_4- Ethylene glycol nanofluid forced convection inside a porous enclosure in existence of Coulomb force, J. Mol. Liq. 249 (2018) 429–437.
[88] M. Sheikholeslami, T. Hayat, A. Alsaedi, Numerical simulation for forced convection flow of MHD CuO-H_2O nanofluid inside a cavity by means of LBM, J. Mol. Liq. 249 (2018) 941–948.
[89] M. Sheikholeslami, H.R. Kataria, A.S. Mittal, Effect of thermal diffusion and heat-generation on MHD nanofluid flow past an oscillating vertical plate through porous medium, J. Mol. Liq. 257 (2018) 12–25.
[90] M. Sheikholeslami, S.A. Shehzad, Z. Li, Nanofluid heat transfer intensification in a permeable channel due to magnetic field using Lattice Boltzmann method, Phys. B (2018). Available from: https://doi.org/10.1016/j.physb.2018.03.036.
[91] M. Sheikholeslami, A. Zeeshan, Numerical simulation of Fe_3O_4 -water nanofluid flow in a non-Darcy porous media, Int. J. Numer. Methods Heat Fluid Flow 28 (3) (2018) 641–660. Available from: https://doi.org/10.1108/HFF-04-2017-0160.
[92] M. Sheikholeslami, D.D. Ganji, Influence of electric field on Fe_3O_4- water nanofluid radiative and convective heat transfer in a permeable enclosure, J. Mol. Liq. 250 (2018) 404–412.
[93] M. Sheikholeslami, M. Barzegar Gerdroodbary, S. Valiallah Mousavi, D.D. Ganji, R. Moradi, Heat transfer enhancement of ferrofluid inside an 90o elbow channel by non-uniform magnetic field, J. Magnet. Magnet. Mater. 460 (2018) 302–311.
[94] M. Sheikholeslami, M.M. Bhatti, Active method for nanofluid heat transfer enhancement by means of EHD, Int. J. Heat Mass Transfer 109 (2017) 115–122.
[95] M. Sheikholeslami, T. Hayat, A. Alsaedi, Numerical study for external magnetic source influence on water based nanofluid convective heat transfer, Int. J. Heat Mass Transfer 106 (2017) 745–755.
[96] M. Sheikholeslami, Numerical simulation for solidification in a LHTESS by means of nano-enhanced PCM, J. Taiwan Inst. Chem. Eng. 86 (2018) 25–41.
[97] M. Sheikholeslami, Numerical modeling of nano enhanced PCM solidification in an enclosure with metallic fin, J. Mol. Liq. 259 (2018) 424–438.
[98] M. Sheikholeslami, A. Ghasemi, Solidification heat transfer of nanofluid in existence of thermal radiation by means of FEM, Int. J. Heat Mass Transfer 123 (2018) 418–431.
[99] M. Sheikholeslami, S.A. Shehzad, CVFEM simulation for nanofluid migration in a porous medium using Darcy model, Int. J. Heat Mass Transfer 122 (2018) 1264–1271.
[100] M. Sheikholeslami, M. Darzi, M.K. Sadoughi, Heat transfer improvement and pressure drop during condensation of refrigerant-based nanofluid; an experimental procedure, Int. J. Heat Mass Transfer 122 (2018) 643–650.
[101] M. Sheikholeslami, H.B. Rokni, CVFEM for effect of Lorentz forces on nanofluid flow in a porous complex shaped enclosure by means of non-equilibrium model, J. Mol. Liq. 254 (2018) 446–462.
[102] M. Sheikholeslami, S.A. Shehzad, Z. Li, Water based nanofluid free convection heat transfer in a three dimensional porous cavity with hot sphere obstacle in existence of Lorenz forces, Int. J. Heat Mass Transfer 125 (2018) 375–386.
[103] M. Sheikholeslami, M. Jafaryar, D.D. Ganji, Z. Li, Exergy loss analysis for nanofluid forced convection heat transfer in a pipe with modified turbulators, J. Mol. Liq. 262 (2018) 104–110.
[104] M. Sheikholeslami, M. Jafaryar, Z. Li, Nanofluid turbulent convective flow in a circular duct with helical turbulators considering CuO nanoparticles, Int. J. Heat Mass Transfer 124 (2018) 980–989.

CHAPTER

9

Nanofluid Forced Convective Heat Transfer in Presence of Variable Magnetic Field Using CVFEM

9.1 INTRODUCTION

A nanofluid is a new class of heat transfer fluids engineered by dispersing metallic or nonmetallic nanoparticles with a typical size of less than 100 nm in the base fluid (e.g., water, ethylene glycol and oil). The poor thermophysical properties of liquids may have led to the use of suspended solid particles as an additive to enhance the thermal properties and improve the heat transfer characteristics of liquids. The key idea is to improve the thermal conductivity. Because the solid particles have a larger thermal conductivity than the liquids, solid particles suspended in the liquid will improve the thermal conductivity of liquid. For many years, suspensions of millimeter- or micrometer-sized solid particles have been tested to enhance the thermal conductivity of conventional fluids, but problems of sedimentation led to increased pressure drop in the flow channel, as reported by Lee et al. [1].

The effect of magnetic field on fluids is worth investigating due to the numerous applications in a wide range of fields. The study of interaction of the magnetic field or the electromagnetic field with fluids have been documented, e.g., among chemical engineering, medicine, high speed noiseless printing, and transformer cooling. Sheikholeslami et al. [2] presented the influence of variable magnetic field on forced convection heat transfer of Fe_3O_4-water nanofluid. Sheikholeslami et al. [3] utilized two-phase model for simulation of nanofluid flow in an enclosure with lid wall in presence of magnetic field. Forced convection heat transfer in a semiannulus under the influence of a variable magnetic field has been presented by Sheikholeslami et al. [4]. Sheikholeslami and Chamkha [5] studied the nanofluid flow and convective heat transfer in a double-sided lid-driven cavity with a wavy wall in the presence of a variable magnetic field. Sheikholeslami et al. [6] investigated Numerical investigation of forced convective heat transfer of Fe_3O_4-water nanofluid in the presence of an external magnetic source. Sheikholeslami and Rokni [7] studied the influence of EFD viscosity on nanofluid forced convection in a cavity with sinusoidal wall. Effect of external forces on nanofluid behavior has been investigated in various publications [8–93].

9.2 EFFECT OF NONUNIFORM MAGNETIC FIELD ON FORCED CONVECTION HEAT TRANSFER OF FE_3O_4-WATER NANOFLUID

9.2.1 Problem Definition

The schematic diagram and the mesh of the semiannulus enclosure used in the present CVFEM program are shown in Fig. 9.1. The inner wall is maintained at constant temperatures T_h and the other walls are maintained at constant temperature $T_c (T_h > T_c)$. For the expression of the magnetic field strength it can be considered that the magnetic source represents a magnetic wire placed vertically to the x-y plane at the point $(\bar{a}, \bar{b})$. The components of the magnetic field intensity $(\overline{H_x}, \overline{H_y})$ and the magnetic field strength $(\overline{H})$ can be considered as:

$$\overline{H_x} = \frac{\gamma'}{2\pi} \frac{1}{(x-\bar{a})^2 + (y-\bar{b})^2} (y - \bar{b}) \tag{9.1}$$

Application of Control Volume based Finite Element Method (CVFEM) for Nanofluid Flow and Heat Transfer.
DOI: https://doi.org/10.1016/B978-0-12-814152-6.00009-6

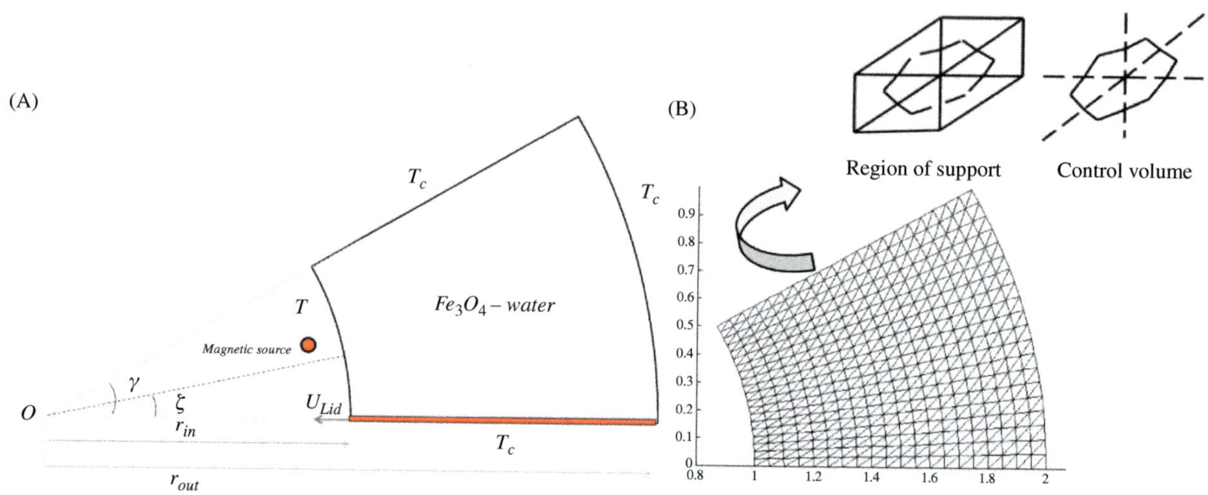

FIGURE 9.1 (A) Geometry and the boundary conditions; (B) the mesh of enclosure considered in this work.

$$\overline{H_y} = -\frac{\gamma'}{2\pi}\frac{1}{(x-\overline{a})^2+(y-\overline{b})^2}(x-\overline{a}) \tag{9.2}$$

$$\overline{H} = \sqrt{\overline{H}_x^2+\overline{H}_y^2} = \frac{\gamma'}{2\pi}\frac{1}{\sqrt{(x-\overline{a})^2+(y-\overline{b})^2}} \tag{9.3}$$

where γ' the magnetic field strength at the source (of the wire) and $(\overline{a},\overline{b})$ is the position where the source is located. The contours of the magnetic field strength are shown in Fig. 9.2. In this section magnetic source is located at $(-0.01cols, 0.5rows)$. The lower wall is lid-driven with velocity of U_{Lid}. In this study γ is equal to 45°.

9.2.2 Governing Equation

The flow is two-dimensional, laminar, and incompressible. The magnetic Reynolds number is assumed to be small so that the induced magnetic field can be neglected compared to the applied magnetic field. The flow is considered to be steady, two-dimensional, and laminar. Using the Boussinesq approximation, the governing equations of heat transfer and fluid flow for nanofluid can be obtained as follows:

$$\frac{\partial u}{\partial x}+\frac{\partial v}{\partial y}=0 \tag{9.4}$$

$$\rho_{nf}\left(u\frac{\partial u}{\partial x}+v\frac{\partial u}{\partial y}\right) = -\frac{\partial P}{\partial x}+\mu_{nf}\left(\frac{\partial^2 u}{\partial x^2}+\frac{\partial^2 u}{\partial y^2}\right)-\sigma_{nf}B_y^2u+\sigma_{nf}B_xB_yv \tag{9.5}$$

$$\rho_{nf}\left(u\frac{\partial v}{\partial x}+v\frac{\partial v}{\partial y}\right) = -\frac{\partial P}{\partial y}+\mu_{nf}\left(\frac{\partial^2 v}{\partial x^2}+\frac{\partial^2 v}{\partial y^2}\right)-\sigma_{nf}B_x^2v+\sigma_{nf}B_xB_yu \tag{9.6}$$

$$\begin{aligned}(\rho C_p)_{nf}\left(u\frac{\partial T}{\partial x}+v\frac{\partial T}{\partial y}\right) &= k_{nf}\left(\frac{\partial^2 T}{\partial x^2}+\frac{\partial^2 T}{\partial y^2}\right)+\sigma_{nf}(uB_y-vB_x)^2\\ &-\mu_0T\frac{\partial M}{\partial T}\left(u\frac{\partial \overline{H}}{\partial x}+v\frac{\partial \overline{H}}{\partial y}\right)+\mu_{nf}\left\{2\left(\frac{\partial u}{\partial x}\right)^2+2\left(\frac{\partial v}{\partial x}\right)^2+\left(\frac{\partial u}{\partial x}+\frac{\partial v}{\partial y}\right)^2\right\}\end{aligned} \tag{9.7}$$

The terms $-\sigma_{nf}B_y^2u+\sigma_{nf}B_xB_yv$ and $-\sigma_{nf}B_x^2v+\sigma_{nf}B_xB_yu$ appearing in (9.5) and (9.6), respectively, represent the Lorentz force per unit volume towards the x and y directions and arise due to the electrical conductivity of the fluid. These two terms are known in MHD. The term $\mu_0T\frac{\partial M}{\partial T}\left(u\frac{\partial \overline{H}}{\partial x}+v\frac{\partial \overline{H}}{\partial y}\right)$ in Eq. (9.7) represents the thermal

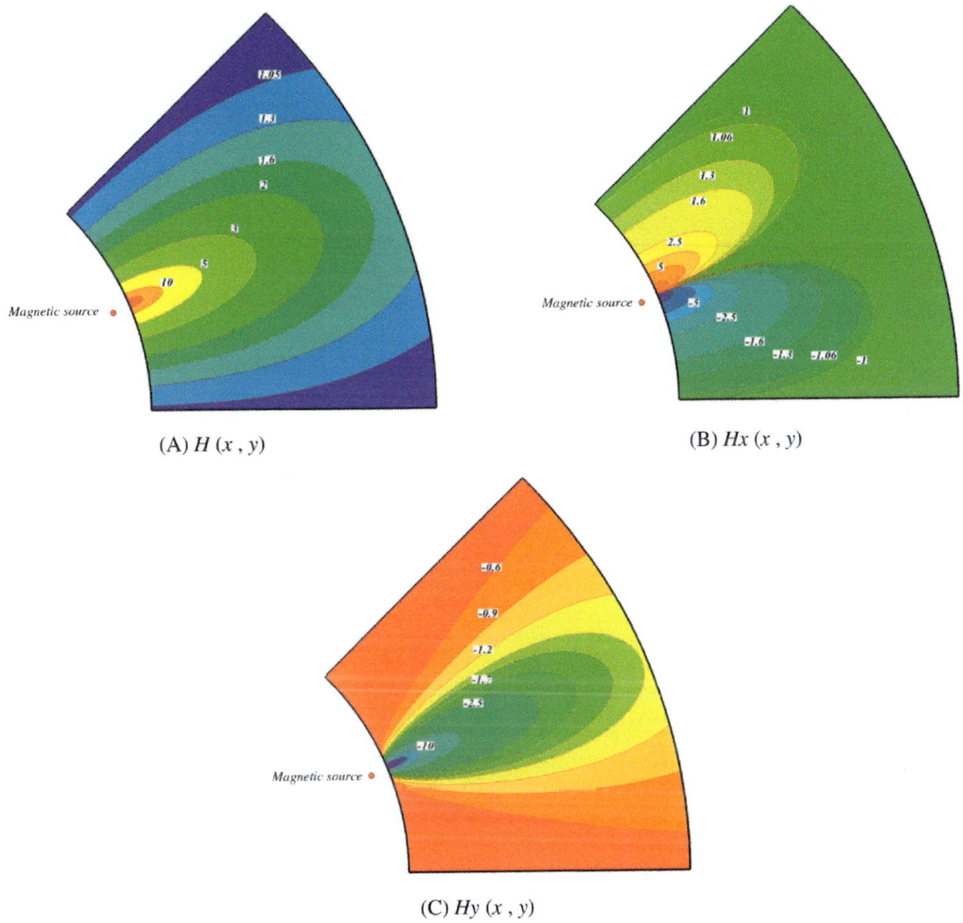

FIGURE 9.2 Contours of the (A) magnetic field strength H; (B) magnetic field intensity component in x direction Hx; (C) magnetic field intensity component in y direction Hy. (A) $H(x,y)$, (B) $Hx(x,y)$, (C) $Hy(x,y)$.

power per unit volume due to the magneto caloric effect. Also the term $\sigma_{nf}(uB_y - vB_x)^2$ in (9.7) represents the Joule heating. For the variation of the magnetization M, with the magnetic field intensity $\overline{H}$ and temperature T, the following relation derived experimentally is considered:

$$M = K'\overline{H}(T_c' - T) \tag{9.8}$$

where K' is a constant and T_c' is the Curie temperature.

In the above equations, μ_0 is the magnetic permeability of vacuum $(4\pi \times 10^{-7} Tm/A)$, $\overline{H}$ is the magnetic field strength, $\overline{B}$ is the magnetic induction $(\overline{B} = \mu_0 \overline{H})$, and the bar above the quantities denotes that they are dimensional. The effective density (ρ_{nf}) and heat capacitance $(\rho C_p)_{nf}$ of the nanofluid are defined as:

$$\rho_{nf} = \rho_f(1 - \phi) + \rho_s \phi \tag{9.9}$$

$$(\rho C_p)_{nf} = (\rho C_p)_f(1 - \phi) + (\rho C_p)_s \phi \tag{9.10}$$

where ϕ is the solid volume fraction of nanoparticles. Thermal diffusivity of the nanofluid is

$$\alpha_{nf} = \frac{k_{nf}}{(\rho C_p)_{nf}} \tag{9.11}$$

The dynamic viscosity of the nanofluid given by Brinkman is

$$\mu_{nf} = \frac{\mu_f}{(1-\phi)^{2.5}} \tag{9.12}$$

The effective thermal conductivity of the nanofluid can be approximated by the Maxwell–Garnetts (MG) model as:

$$\frac{k_{nf}}{k_f}=\frac{k_s+2k_f-2\phi(k_f-k_s)}{k_s+2k_f+\phi(k_f-k_s)} \tag{9.13}$$

and the effective electrical conductivity of nanofluid was presented by Maxwell as below:

$$\frac{\sigma_{nf}}{\sigma_f}=1+\frac{3\left(\frac{\sigma_s}{\sigma_f}-1\right)\phi}{\left(\frac{\sigma_s}{\sigma_f}+2\right)-\left(\frac{\sigma_s}{\sigma_f}-1\right)\phi} \tag{9.14}$$

The stream function and vorticity are defined as:

$$u=\frac{\partial\psi}{\partial y},\ v=-\frac{\partial\psi}{\partial x},\ \omega=\frac{\partial v}{\partial x}-\frac{\partial u}{\partial y} \tag{9.15}$$

The stream function satisfies the continuity Eq. (9.4). The vorticity equation is obtained by eliminating the pressure between the two momentum equations, i.e., by taking the y-derivative of Eq. (9.6) and subtracting from it the x-derivative of Eq. (9.5). By introducing the following nondimensional variables:

$$X=\frac{x}{L},\quad Y=\frac{y}{L},\quad \Omega=\frac{\omega L}{U_{Lid}},\ \Psi=\frac{\psi}{LU_{Lid}},\ \Theta=\frac{T-T_c}{T_h-T_c},\ U=\frac{u}{U_{Lid}},\ V=\frac{v}{U_{Lid}},$$
$$H=\frac{\overline{H}}{\overline{H_0}},\ H_x=\frac{\overline{H_x}}{\overline{H_0}},\ H_y=\frac{\overline{H_y}}{\overline{H_0}} \tag{9.16}$$

where in Eq. (9.16) $\overline{H_0}=\overline{H}(\overline{a},0)=\frac{\gamma}{2\pi|b|}$ and $L=r_{out}-r_{in}=r_{in}$. Using the dimensionless parameters, the equations now become:

$$\begin{aligned}\frac{\partial\Psi}{\partial Y}\frac{\partial\Omega}{\partial X}-\frac{\partial\Psi}{\partial X}\frac{\partial\Omega}{\partial Y}=&\frac{1}{\mathrm{Re}}\left[\frac{\mu_{nf}/\mu_f}{\rho_{nf}/\rho_f}\right]\left(\frac{\partial^2\Omega}{\partial X^2}+\frac{\partial^2\Omega}{\partial Y^2}\right)-\frac{Ha^2}{\mathrm{Re}}\left[\frac{\sigma_{nf}/\sigma_f}{\rho_{nf}/\rho_f}\right]\\ &\times\left\{\frac{\partial V}{\partial X}H_x^2+V\left(2H_x\frac{\partial H_x}{\partial X}\right)-\frac{\partial U}{\partial X}H_xH_y-U\frac{\partial H_x}{\partial X}H_y-U\frac{\partial H_y}{\partial X}H_x\right.\\ &\left.-\frac{\partial U}{\partial Y}H_y^2-U\left(2H_y\frac{\partial H_y}{\partial Y}\right)+\frac{\partial V}{\partial Y}H_xH_y+V\frac{\partial H_x}{\partial Y}H_y+V\frac{\partial H_y}{\partial Y}H_x\right\}\end{aligned} \tag{9.17}$$

$$\begin{aligned}\frac{\partial\Psi}{\partial Y}\frac{\partial\Theta}{\partial X}-\frac{\partial\Psi}{\partial X}\frac{\partial\Theta}{\partial Y}=&\frac{1}{\mathrm{Pr\,Re}}\left(\frac{k_{nf}}{k_f}\right)\Big/\left(\frac{(\rho C_P)_{nf}}{(\rho C_P)_f}\right)\left(\frac{\partial^2\Theta}{\partial X^2}+\frac{\partial^2\Theta}{\partial Y^2}\right)\\ &+Ha^2\frac{Ec}{\mathrm{Re}}\left(\frac{\sigma_{nf}}{\sigma_f}\right)\Big/\left(\frac{(\rho C_P)_{nf}}{(\rho C_P)_f}\right)\{U\,H_y-V\,H_x\}^2+Mn_F Ec\frac{(\rho C_P)_f}{(\rho C_P)_{nf}}\left\{U\frac{\partial H}{\partial X}+V\frac{\partial H}{\partial Y}\right\}H(\varepsilon_1+\Theta)\\ &+\frac{Ec}{\mathrm{Re}}\left(\frac{\mu_{nf}}{\mu_f}\right)\Big/\left(\frac{(\rho C_P)_{nf}}{(\rho C_P)_f}\right)\left\{2\left(\frac{\partial U}{\partial X}\right)^2+2\left(\frac{\partial V}{\partial X}\right)^2+\left(\frac{\partial U}{\partial X}+\frac{\partial V}{\partial Y}\right)^2\right\}\end{aligned} \tag{9.18}$$

$$\frac{\partial^2\Psi}{\partial X^2}+\frac{\partial^2\Psi}{\partial Y^2}=-\Omega \tag{9.19}$$

TABLE 9.1 Thermophysical Properties of Water and Nanoparticles

	$\rho(kg/m^3)$	$C_p(j/kgk)$	$k(W/m.k)$	$\beta\times10^5(K^{-1})$	$d_p(nm)$	$\sigma(\Omega\cdot m)^{-1}$
Pure water	997.1	4179	0.613	21	-	0.05
Fe_3O_4	5200	670	6	1.3	47	25000

where $\mathrm{Re}=\frac{\rho_f L\, U_{Lid}}{\mu_f}, Ha=LH_0\mu_0\sqrt{\sigma_f/\mu_f}, \varepsilon_1=T_1/\Delta T$ and $Ec=\left(\rho_f U_{Lid}{}^2\right)/\left[(\rho C_P)_f\Delta T\right]$ are the Reynolds number, Hartmann number, temperature number, and Eckert number for the base fluid, respectively. The thermophysical properties of the nanofluid are given in Table 9.1. The boundary conditions as shown in Fig. 9.1 are:

$$\begin{aligned}&\Theta=1.0 && \text{on the inner circular boundary}\\ &\Theta=0.0 && \text{on the other walls}\end{aligned} \tag{9.20}$$

The values of vorticity on the boundary of the enclosure can be obtained using the stream function formulation and the known velocity conditions during the iterative solution procedure.

The local Nusselt number of the nanofluid along the hot wall can be expressed as:

$$Nu_{loc}=\left(\frac{k_{nf}}{k_f}\right)\frac{\partial\Theta}{\partial r} \tag{9.21}$$

where r is the radial direction. The average Nusselt number on the hot circular wall is evaluated as:

$$Nu_{ave}=\frac{1}{\gamma}\int_0^{\gamma} Nu_{loc}(\zeta)\,d\zeta \tag{9.22}$$

To estimate the enhancement of heat transfer between the case of $\phi=0.04$ and the pure fluid (base fluid) case, the heat transfer enhancement is defined as:

$$En=\frac{Nu(\phi=0.04)-Nu(basefluid)}{Nu(basefluid)}\times100 \tag{9.23}$$

9.2.3 Effects of Active Parameters

In this study, forced convection heat transfer of ferrofluid in the presence of variable magnetic field is investigated using CVFEM. The working fluid is Fe_3O_4-water nanofluid. Calculations are made for various values of volume fraction of nanoparticles ($\phi=0\%$ and 4%), Reynolds number ($Re=10, 100$, and 1000), and Hartmann number ($Ha=0$, 5, 10, and 20). In all calculations, the Prandtl number (Pr), temperature number (ε_1) and Eckert number (Ec) are set to 6.8, 0.0, and 10^{-5}, respectively.

Isotherm and streamline contours for different values of Reynolds number and Hartmann are depicted in Figs. 9.3–9.5. At low Reynolds number, two clockwise eddies exist in streamline which are symmetric with respect to $\zeta=22.5^\circ$. By applying magnetic field, Lorentz forces retard the flow and another eddy generates which is rotated counterclockwise. As the Reynolds number increases, two main eddies turn into three eddies in which the middle one rotates in the reverse direction. So a thermal plume appears near the $\zeta=18^\circ$ and in turn the minimum rate of heat transfer occurs at this point. Increasing the Reynolds number causes the thermal boundary layer thickness to decrease. As the Hartmann number increases the strength of the middle eddy decreases and in turn the thermal plume becomes weaker. So Lorentz forces decrease the rate of heat transfer.

The effects of Hartmann number and Reynolds number on local and average Nusselt number are shown in Figs. 9.6 and 9.7. Thermal boundary layer thickness increases with the increase of the Hartmann number while it decreases with the increase of the Reynolds number. So, the Nusselt number increases with the increase of the Reynolds number while it decreases with the increase of the Hartmann number. The existence of thermal plumes leads to appearance of extremes for the local Nusselt number profile.

The effects of Reynolds number and Hartmann number on heat transfer enhancement are shown in Fig. 9.8. The effect of nanoparticles is more pronounced at high Hartmann number than at low Hartmann number

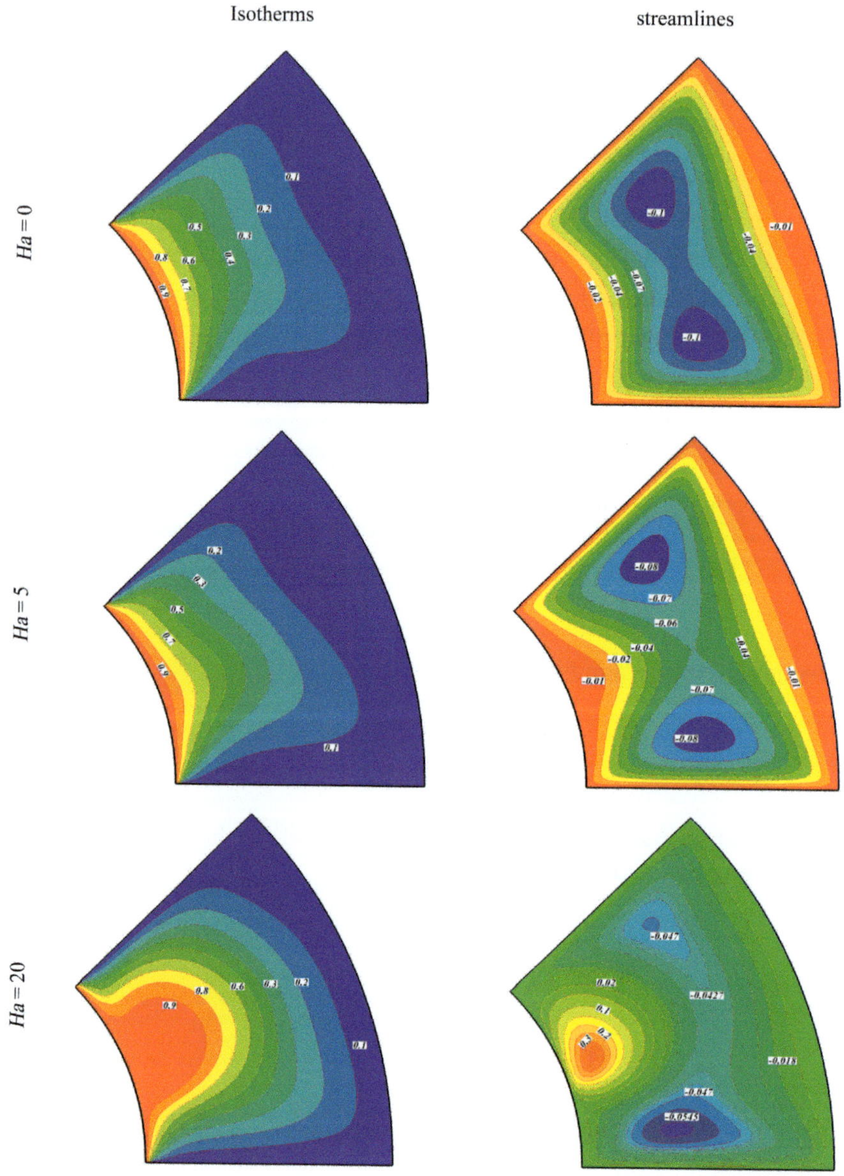

FIGURE 9.3 Isotherm (left) and streamline (right) contours for different values of Hartmann number when Re = 10.

because of the greater enhancement rate. This observation can be explained by noting that at high Hartmann number the heat transfer is dominant by conduction. Therefore, the addition of high thermal conductivity nanoparticles will increase the conduction and make the enhancement more effective. The heat transfer enhancement decreases with increases of Reynolds number.

9.3 MAGNETIC NANOFLUID FORCED CONVECTIVE HEAT TRANSFER IN THE PRESENCE OF VARIABLE MAGNETIC FIELD USING TWO-PHASE MODEL

9.3.1 Problem Definition

The schematic diagram and the mesh of the semiannulus enclosure used in the present CVFEM program are shown in Fig. 9.9. The inner wall is maintained at constant temperatures T_h and the other walls are maintained at constant temperature $T_c (T_h > T_c)$. The boundary conditions for concentration are similar to those for temperature. For the expression of the magnetic field strength it can be considered that the magnetic source represents a

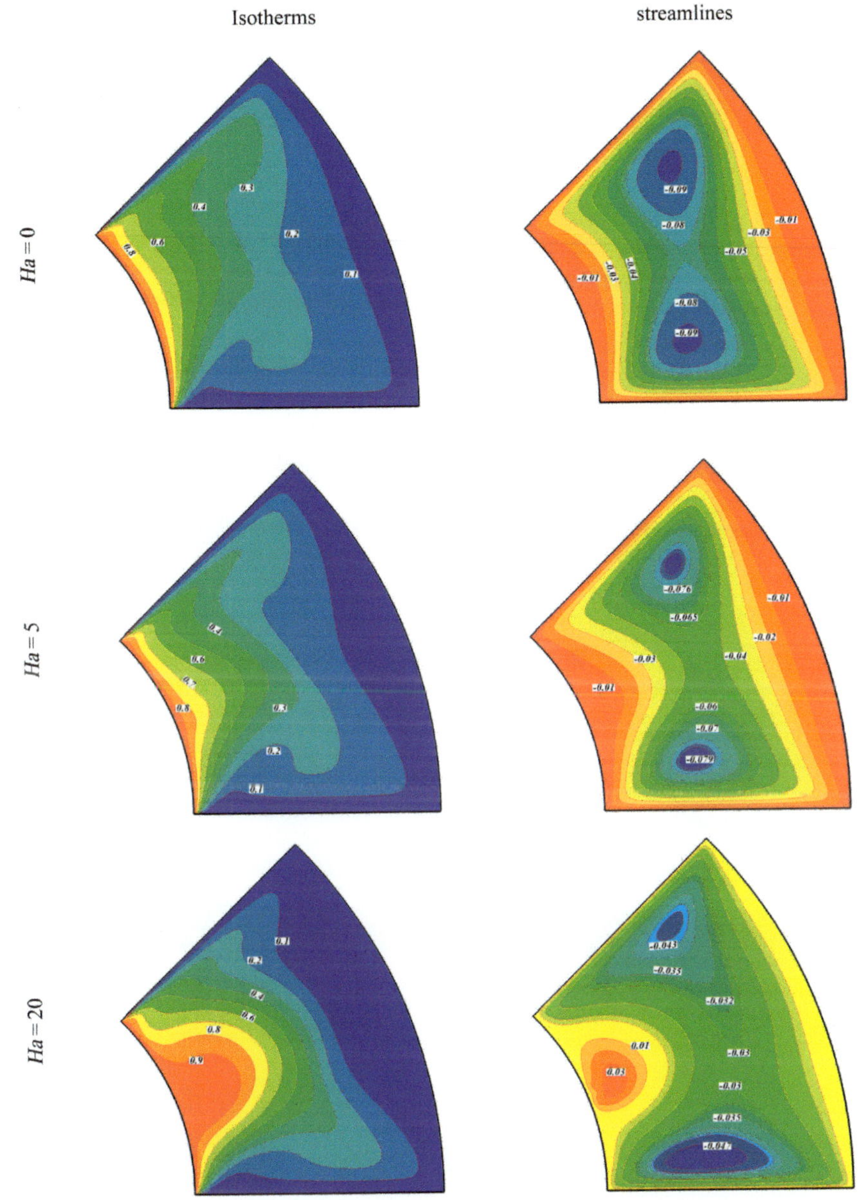

FIGURE 9.4 Isotherm (left) and streamline (right) contours for different values of Hartmann number when Re = 100.

magnetic wire placed vertically to the x-y plane at the point $(\bar{a}, \bar{b})$. The components of the magnetic field intensity $(\overline{H_x}, \overline{H_y})$ and the magnetic field strength $(\overline{H})$ can be considered as:

$$\overline{H_x} = \frac{\gamma'}{2\pi} \frac{1}{(x-\bar{a})^2 + (y-\bar{b})^2} (y - \bar{b}) \tag{9.24}$$

$$\overline{H_y} = -\frac{\gamma'}{2\pi} \frac{1}{(x-\bar{a})^2 + (y-\bar{b})^2} (x - \bar{a}) \tag{9.25}$$

$$\overline{H} = \sqrt{\overline{H}_x^2 + \overline{H}_y^2} = \frac{\gamma'}{2\pi} \frac{1}{\sqrt{(x-\bar{a})^2 + (y-\bar{b})^2}} \tag{9.26}$$

where γ' the magnetic field strength at the source (of the wire) and $(\bar{a}, \bar{b})$ is the position where the source is located. The contours of the magnetic field strength are shown in Fig. 9.10. In this section magnetic source is located at $(-0.01cols, 0.5rows)$. The upper wall is lid-driven with velocity of U_{Lid}. In this study γ is equal to 45°.

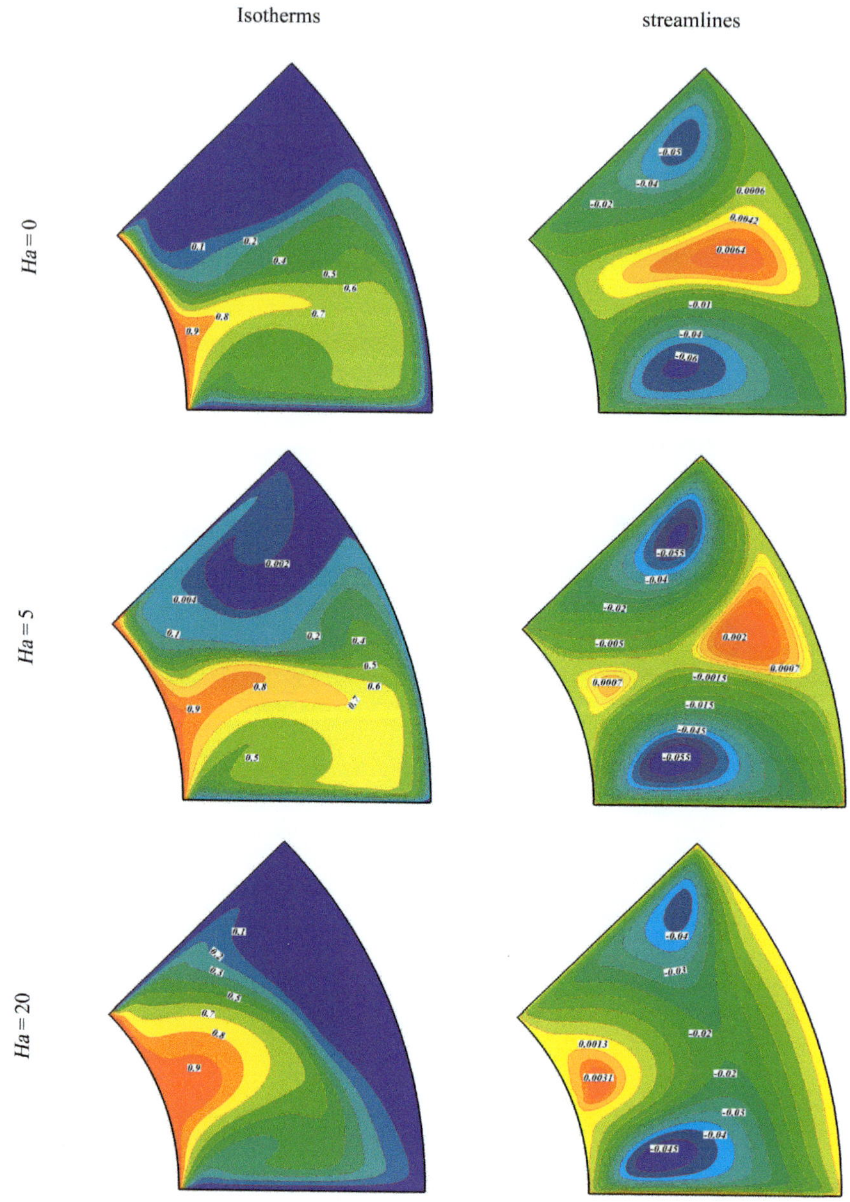

FIGURE 9.5 Isotherm (left) and streamline (right) contours for different values of Hartmann number when Re = 1000.

9.3.2 Governing Equation

The continuity, momentum under Boussinesq approximation, and energy equations for the laminar and steady state natural convection in a two-dimensional enclosure can be written in dimensional form as follows:

$$\frac{\partial u}{\partial x} + \frac{\partial v}{\partial y} = 0 \tag{9.27}$$

$$\rho_f\left\{u\frac{\partial u}{\partial x} + v\frac{\partial u}{\partial y}\right\} = -\frac{\partial P}{\partial x} + \mu\left(\frac{\partial^2 u}{\partial x^2} + \frac{\partial^2 u}{\partial y^2}\right) - \sigma B_y^2 u + \sigma B_x B_y v \tag{9.28}$$

$$\rho_f\left\{u\frac{\partial v}{\partial x} + v\frac{\partial v}{\partial y}\right\} = -\frac{\partial P}{\partial y} + \mu\left(\frac{\partial^2 v}{\partial x^2} + \frac{\partial^2 v}{\partial y^2}\right) - \sigma B_x^2 v + \sigma B_x B_y u \tag{9.29}$$

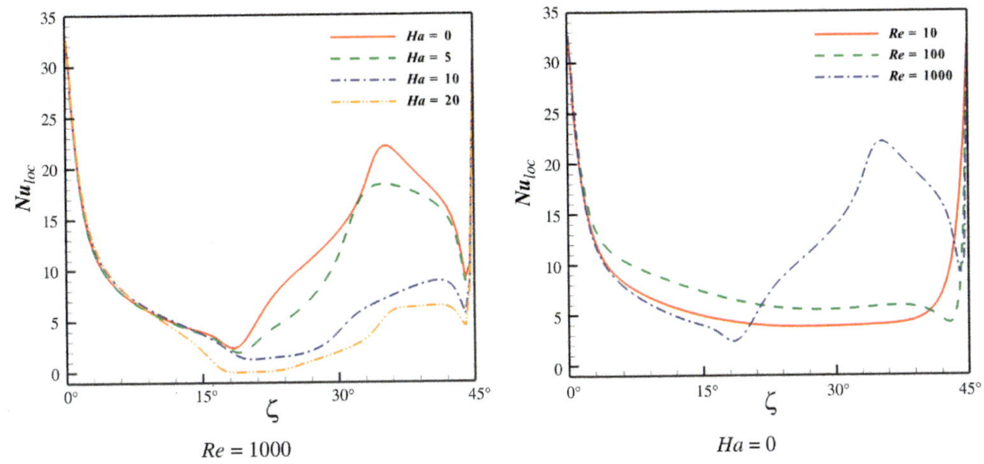

FIGURE 9.6 Effects of Hartmann number and Reynolds number on local Nusselt number Nu_{loc} along hot wall when $\phi = 0.04$.

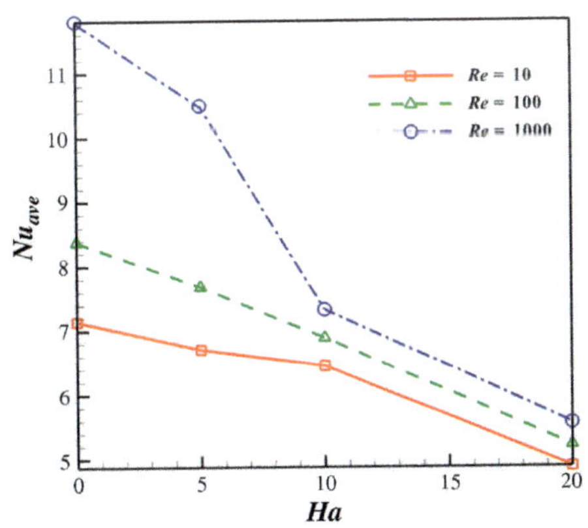

FIGURE 9.7 Effects of Hartmann number and Reynolds number on local Nusselt number Nu_{ave} along hot wall when $\phi = 0.04$.

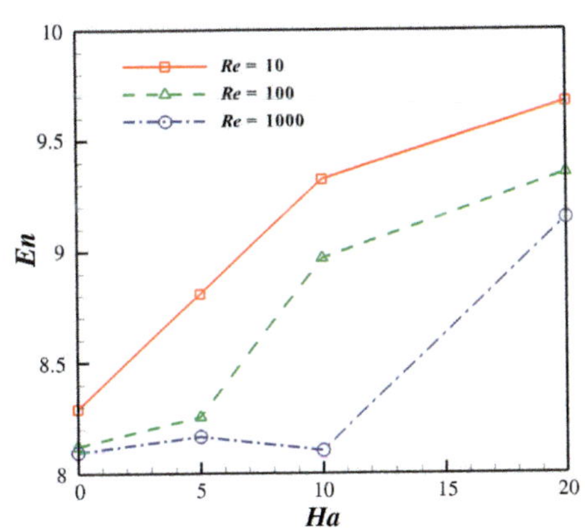

FIGURE 9.8 Effects of Reynolds number and Hartmann number on heat transfer enhancement.

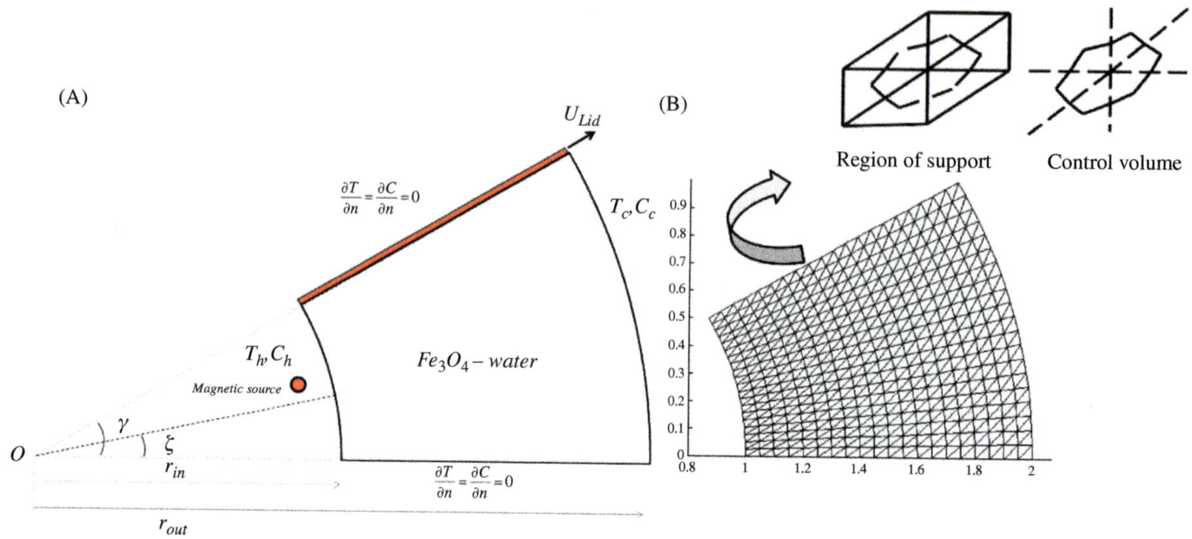

FIGURE 9.9 (A) Geometry and the boundary conditions; (B) the mesh of enclosure considered in this work.

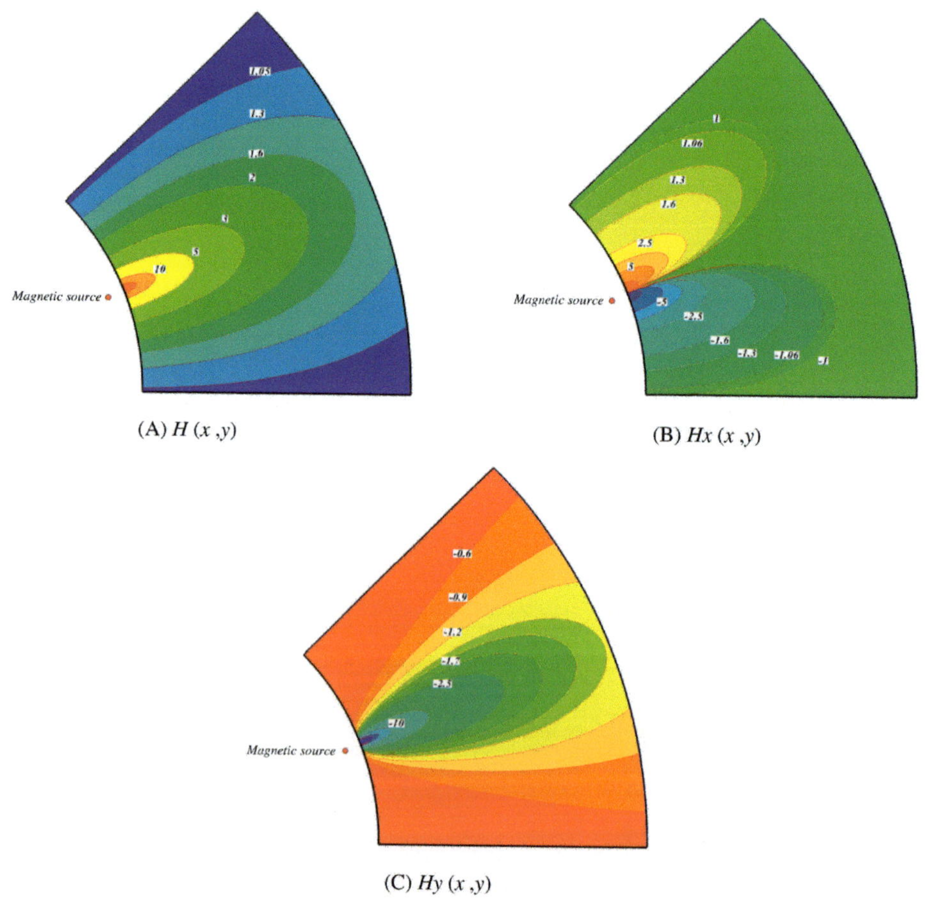

FIGURE 9.10 Contours of the (A) magnetic field strength H; (B) magnetic field intensity component in x direction Hx; (C) magnetic field intensity component in y direction Hy. (A) $H(x,y)$, (B) $Hx(x,y)$, (C) $Hy(x,y)$.

$$u\frac{\partial T}{\partial x}+v\frac{\partial T}{\partial y}=\alpha\left(\frac{\partial^2 T}{\partial x^2}+\frac{\partial^2 T}{\partial y^2}\right)+\frac{(\rho c)_p}{(\rho c)_f}\left[\begin{array}{l}D_B\left\{\frac{\partial \phi}{\partial x}\cdot\frac{\partial T}{\partial x}+\frac{\partial \phi}{\partial y}\cdot\frac{\partial T}{\partial y}\right\}\\ +(D_T/T_c)\left\{\left(\frac{\partial T}{\partial x}\right)^2+\left(\frac{\partial T}{\partial y}\right)^2\right\}\end{array}\right] + \sigma\left(uB_y-vB_x\right)^2+\mu\left\{2\left(\frac{\partial u}{\partial x}\right)^2+2\left(\frac{\partial v}{\partial x}\right)^2+\left(\frac{\partial u}{\partial x}+\frac{\partial v}{\partial y}\right)^2\right\} \tag{9.30}$$

$$u\frac{\partial \phi}{\partial x}+v\frac{\partial \phi}{\partial y}=D_B\left\{\frac{\partial^2 \phi}{\partial x^2}+\frac{\partial^2 \phi}{\partial y^2}\right\}+\left(\frac{D_T}{T_c}\right)\left\{\frac{\partial^2 T}{\partial x^2}+\frac{\partial^2 T}{\partial y^2}\right\} \tag{9.31}$$

The stream function and vorticity are defined as follows:

$$u=\frac{\partial \psi}{\partial y}, v=-\frac{\partial \psi}{\partial x}, \omega=\frac{\partial v}{\partial x}-\frac{\partial u}{\partial y} \tag{9.32}$$

The stream function satisfies the continuity Eq. (9.27). The vorticity equation is obtained by eliminating the pressure between the two momentum equations, i.e., by taking the y-derivative of Eq. (9.28) and subtracting from it the x-derivative of Eq. (9.29). Also the following nondimensional variables should be introduced:

$$X=\frac{x}{L},\quad Y=\frac{y}{L},\quad \Omega=\frac{\omega L}{U_{Lid}},\Psi=\frac{\psi}{U_{Lid}L},\Theta=\frac{T-T_c}{T_h-T_c},\quad \Phi=\frac{\phi-\phi_c}{\phi_h-\phi_c},U=\frac{u}{U_{Lid}},V=\frac{v}{U_{Lid}}$$
$$H=\frac{\overline{H}}{\overline{H_0}},\quad H_x=\frac{\overline{H_x}}{\overline{H_0}},\quad H_y=\frac{\overline{H_y}}{\overline{H_0}} \tag{9.33}$$

By using these dimensionless parameters the equations become:

$$\left[\frac{\partial \Psi}{\partial Y}\frac{\partial \Omega}{\partial X}-\frac{\partial \Psi}{\partial X}\frac{\partial \Omega}{\partial Y}\right]=\frac{1}{\mathrm{Re}}\left(\frac{\partial^2 \Omega}{\partial X^2}+\frac{\partial^2 \Omega}{\partial Y^2}\right) -\frac{Ha^2}{\mathrm{Re}}\times\left\{\frac{\partial V}{\partial X}H_x^2+V\left(2H_x\frac{\partial H_x}{\partial X}\right)-\frac{\partial U}{\partial X}H_xH_y-U\frac{\partial H_x}{\partial X}H_y-U\frac{\partial H_y}{\partial X}H_x -\frac{\partial U}{\partial Y}H_y^2-U\left(2H_y\frac{\partial H_y}{\partial Y}\right)+\frac{\partial V}{\partial Y}H_xH_y+V\frac{\partial H_x}{\partial Y}H_y+V\frac{\partial H_y}{\partial Y}H_x\right\} \tag{9.34}$$

$$\frac{\partial \Psi}{\partial Y}\frac{\partial \Theta}{\partial X}-\frac{\partial \Psi}{\partial X}\frac{\partial \Theta}{\partial Y}=\frac{1}{\mathrm{Re}\,\mathrm{Pr}}\left(\frac{\partial^2 \Theta}{\partial X^2}+\frac{\partial^2 \Theta}{\partial Y^2}\right)+\frac{Nb}{\mathrm{Re}}\left(\frac{\partial \Phi}{\partial X}\frac{\partial \Theta}{\partial X}+\frac{\partial \Phi}{\partial Y}\frac{\partial \Theta}{\partial Y}\right)+\frac{Nt}{\mathrm{Re}}\left(\left(\frac{\partial \Theta}{\partial X}\right)^2+\left(\frac{\partial \Theta}{\partial Y}\right)^2\right) + Ha^2\,Ec\left\{U\,H_y-V\,H_x\right\}^2+Ec\left\{2\left(\frac{\partial U}{\partial X}\right)^2+2\left(\frac{\partial V}{\partial X}\right)^2+\left(\frac{\partial U}{\partial X}+\frac{\partial V}{\partial Y}\right)^2\right\} \tag{9.35}$$

$$\frac{\partial \Psi}{\partial Y}\frac{\partial \Phi}{\partial X}-\frac{\partial \Psi}{\partial X}\frac{\partial \Phi}{\partial Y}=\frac{1}{Le\,\mathrm{Re}}\left(\frac{\partial^2 \Phi}{\partial X^2}+\frac{\partial^2 \Phi}{\partial Y^2}\right)+\frac{Nt}{Nb\;Le\,\mathrm{Re}}\left(\frac{\partial^2 \Theta}{\partial X^2}+\frac{\partial^2 \Theta}{\partial Y^2}\right) \tag{9.36}$$

$$\frac{\partial^2 \Psi}{\partial X^2}+\frac{\partial^2 \Psi}{\partial Y^2}=-\Omega \tag{9.37}$$

where Prandtl number, the Brownian motion parameter, the thermophoretic parameter, Lewis number, Hartmann number, Eckert number, and Reynolds number are defined as:

$\Pr = \mu/\rho_f\alpha$, $Nb = (\rho c)_p D_B(\phi_h - \phi_c)/\left((\rho c)_f\alpha\right)$, $Nt = (\rho c)_p D_T(T_h - T_c)/[(\rho c)_f \alpha T_c]$, $Le = \alpha/D_B$, $Ha = LH_0\mu_0\sqrt{\sigma/\mu}$, $Ec = (U_{Lid}^2)/[(C_P)\Delta T]$, and $\mathrm{Re} = \frac{\rho_f L\, U_{Lid}}{\mu}$, respectively.

The local Nusselt number of the nanofluid along the hot wall can be expressed as:

$$Nu_{loc} = \frac{\partial \Theta}{\partial r} \tag{9.38}$$

where r is the radial direction. The average Nusselt number on the hot circular wall is evaluated as:

$$Nu_{ave} = \frac{1}{\gamma}\int_0^{\gamma} Nu_{loc}(\zeta)\, d\zeta \tag{9.39}$$

The heatlines are adequate tools for visualization and analysis of 2D convection heat transfer, through an extension of the heat flux line concept to include the advection terms. Heat function (H) is defined in terms of the energy equation as

$$\frac{\partial H}{\partial Y} = U\Theta - \frac{\partial \Theta}{\partial X}, \quad -\frac{\partial H}{\partial X} = V\Theta - \frac{\partial \Theta}{\partial Y} \tag{9.40}$$

9.3.3 Effects of Active Parameters

In this study, forced convection heat transfer of ferrofluid in presence of variable magnetic field is investigated using CVFEM. Two-phase model is used to simulate nanofluid. Calculations are made for various values of Reynolds numbers ($\mathrm{Re} = 10, 100$, and 500), Lewis number ($Le = 2, 4$, and 8), and Hartmann number ($Ha = 0$, 5, 10, and 20). In all calculations, the Prandtl number (Pr), temperature number (ε_1), and Eckert number (Ec) are set to 6.8, 0.0, and 10^{-5}, respectively.

The effects of Reynolds number and Hartmann number on isotherm, streamline, isoconcentration, and heatline contours are shown in Figs. 9.11–9.13. At low Reynolds number, one main eddy exists in streamline. By increasing Reynolds number, another small eddy generates near the bottom wall. As the Reynolds number increases up to 500, the second eddy become stronger. So the enclosure divides into two regions with respect to $\zeta = 22.5^\circ$. Due to the existing two eddies which are rotating in reverse direction, the thermal plume is generated near the hot wall. As the Hartmann number increases, Lorentz forces suppress the flow and diminish the thermal plume. Isoconcentration becomes more distributed at high Reynolds number and this effect is reduced in the presence of magnetic field. Heatline contours have two regions at low Reynolds number while there are three regions at high Reynolds number. Applying magnetic field leads to the generation of a small passive region at bottom right corner.

Figs. 9.14 and 9.15 show the effects of Hartmann number, Reynolds number, and Lewis number on local and average Nusselt number. As Reynolds number increases the thermal boundary layer thickness near the hot wall decreases and in turn the rate of heat transfer increases with the increase of Reynolds number. As the Hartmann number increases Lorentz forces become stronger and suppress the flow. So the thermal boundary layer thickness increases with the increase of Hartmann number and in turn the Nusselt number decreases with the increase of Hartmann number. Increasing the Lewis number leads to a decrease in the rate of heat transfer. Due to the existing thermal plume, a maximum or minimum point appears in local Nusselt number profile.

9.4 FORCED CONVECTION HEAT TRANSFER IN A SEMIANNULUS UNDER THE INFLUENCE OF A VARIABLE MAGNETIC FIELD

9.4.1 Problem Definition

The physical model along with the needed geometrical parameters and the mesh of the enclosure used in the present CVFEM program are shown in Fig. 9.16. The outer wall is maintained at constant temperature T_h and the other walls are maintained at a constant temperature T_c. The shape of inner cylinder profile is assumed to mimic the following form:

$$r = r_{in} + A\cos(N(\zeta)), \tag{9.41}$$

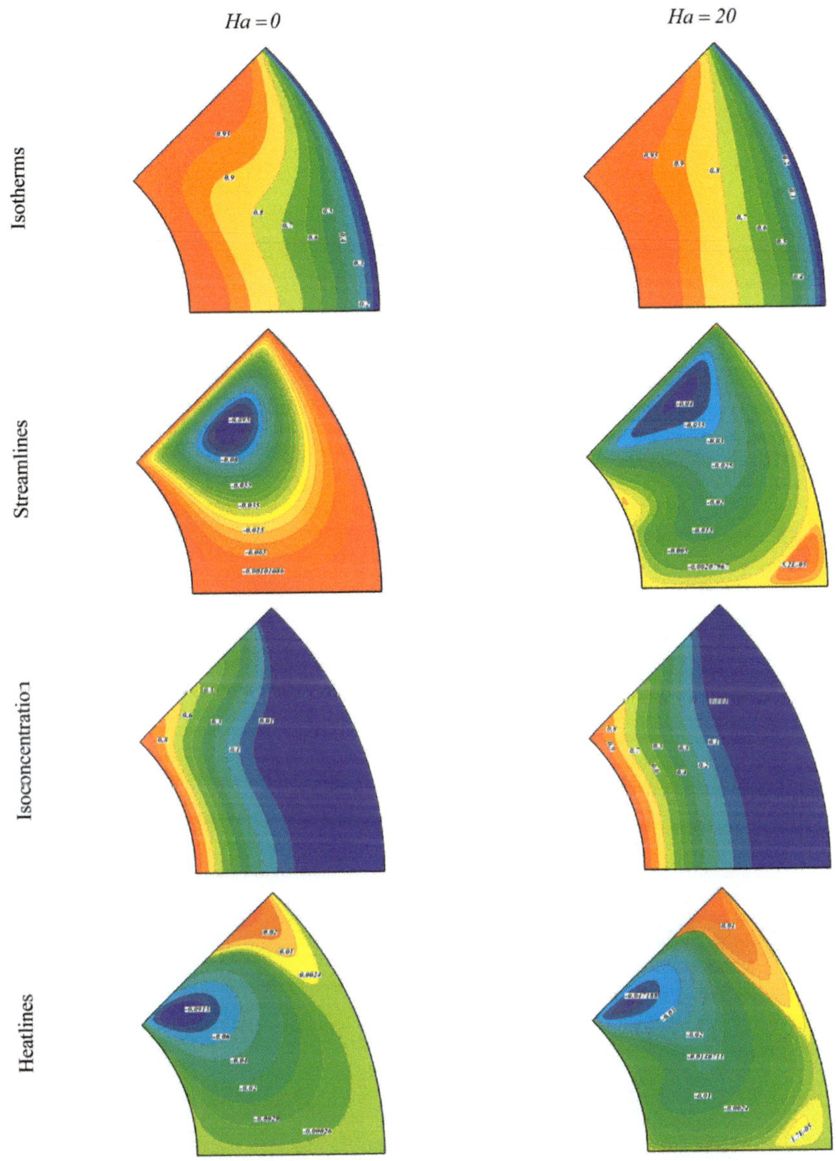

FIGURE 9.11 Comparison of the isotherm, streamline, isoconcentration, and heatline contours for different values of Hartmann number when $Le = 4, Nt = Nb = 0.5, \mathrm{Re} = 10,$ and $\mathrm{Pr} = 6.85$.

in which r_{in} is the base circle radius, r_{out} is the radius of the outer cylinder, A and N are amplitude and number of undulations, respectively, and ζ is the rotation angle. In this study A and N equal to 0.2 and 4, respectively. For the expression of the magnetic field strength it can be considered that the magnetic source represents a magnetic wire placed vertically to the xy plane at the point $(\overline{a}, \overline{b})$. The components of the magnetic field intensity $(\overline{H_x}, \overline{H_y})$ and the magnetic field strength $(\overline{H})$ can be considered as:

$$\overline{H_x} = \frac{\gamma'}{2\pi} \frac{1}{(x-\overline{a})^2 + (y-\overline{b})^2} (y - \overline{b}), \tag{9.42}$$

$$\overline{H_y} = -\frac{\gamma'}{2\pi} \frac{1}{(x-\overline{a})^2 + (y-\overline{b})^2} (x - \overline{a}), \tag{9.43}$$

$$\overline{H} = \sqrt{\overline{H}_x^2 + \overline{H}_y^2} = \frac{\gamma'}{2\pi} \frac{1}{\sqrt{(x-\overline{a})^2 + (y-\overline{b})^2}}, \tag{9.44}$$

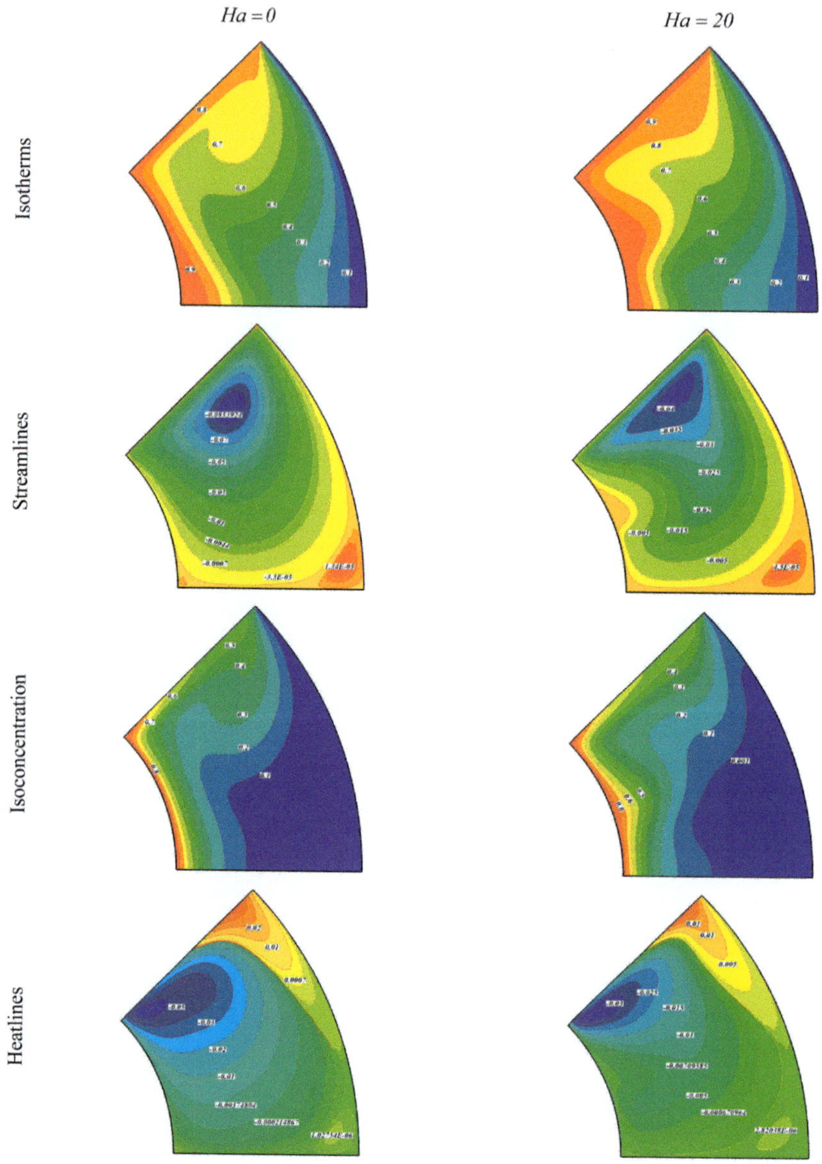

FIGURE 9.12 Comparison of the isotherm, streamline, isoconcentration, and heatline contours for different values of Hartmann number when $Le = 4, Nt = Nb = 0.5, \mathrm{Re} = 100$, and $\mathrm{Pr} = 6.85$.

where γ' the magnetic field strength at the source (of the wire) and $(\bar{a}, \bar{b})$ is the position where the source is located. The contours of the magnetic field strength are shown in Fig. 9.17. In this study magnetic source is located at $(-0.05cols, 0.5rows)$.

9.4.2 Governing Equation

The flow is steady, two-dimensional, laminar, and incompressible. The magnetic Reynolds number is assumed to be small so that the induced magnetic field can be neglected compared to the applied magnetic field. Using the Boussinesq approximation, the governing equations of heat transfer, and fluid flow for nanofluid can be obtained as follows:

$$\frac{\partial u}{\partial x} + \frac{\partial v}{\partial y} = 0, \qquad (9.45)$$

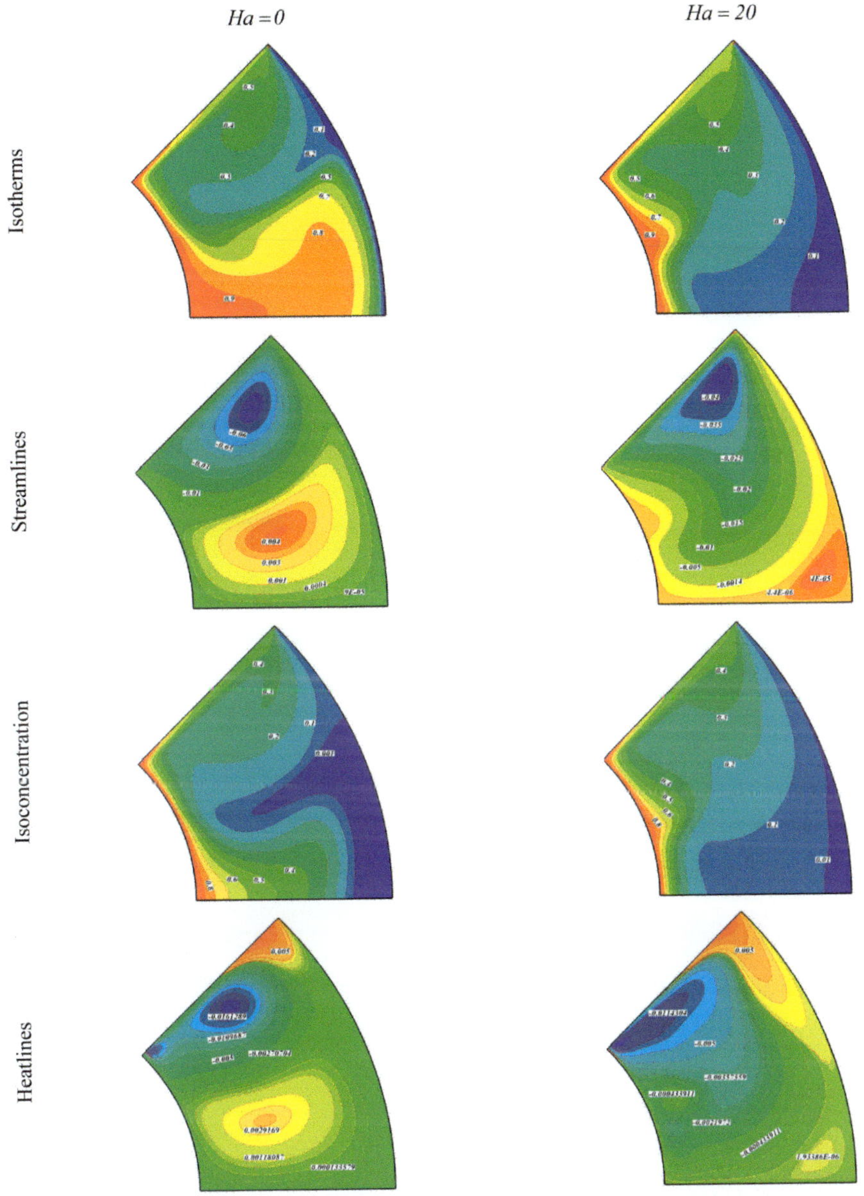

FIGURE 9.13 Comparison of the isotherm, streamline, isoconcentration, and heatline contours for different values of Hartmann number when $Le = 4, Nt = Nb = 0.5, \mathrm{Re} = 500,$ and $\mathrm{Pr} = 6.85$.

$$\rho_{nf}\left(u\frac{\partial u}{\partial x}+v\frac{\partial u}{\partial y}\right)=-\frac{\partial p}{\partial x}+\mu_{nf}\left(\frac{\partial^2 u}{\partial x^2}+\frac{\partial^2 u}{\partial y^2}\right)+\mu_0 M\frac{\partial \overline{H}}{\partial x}-\sigma_{nf}B_y^2 u+\sigma_{nf}B_x B_y v, \tag{9.46}$$

$$\rho_{nf}\left(u\frac{\partial v}{\partial x}+v\frac{\partial v}{\partial y}\right)=-\frac{\partial p}{\partial y}+\mu_{nf}\left(\frac{\partial^2 v}{\partial x^2}+\frac{\partial^2 v}{\partial y^2}\right)+\mu_0 M\frac{\partial \overline{H}}{\partial y}-\sigma_{nf}B_x^2 v+\sigma_{nf}B_x B_y u, \tag{9.47}$$

$$\begin{aligned}(\rho C_p)_{nf}\left(u\frac{\partial T}{\partial x}+v\frac{\partial T}{\partial y}\right)&=k_{nf}\left(\frac{\partial^2 T}{\partial x^2}+\frac{\partial^2 T}{\partial y^2}\right)+\sigma_{nf}(uB_y-vB_x)^2\\&-\mu_0 T\frac{\partial M}{\partial T}\left(u\frac{\partial \overline{H}}{\partial x}+v\frac{\partial \overline{H}}{\partial y}\right)+\mu_{nf}\left\{2\left(\frac{\partial u}{\partial x}\right)^2+2\left(\frac{\partial v}{\partial y}\right)^2+\left(\frac{\partial u}{\partial y}+\frac{\partial v}{\partial x}\right)^2\right\}.\end{aligned} \tag{9.48}$$

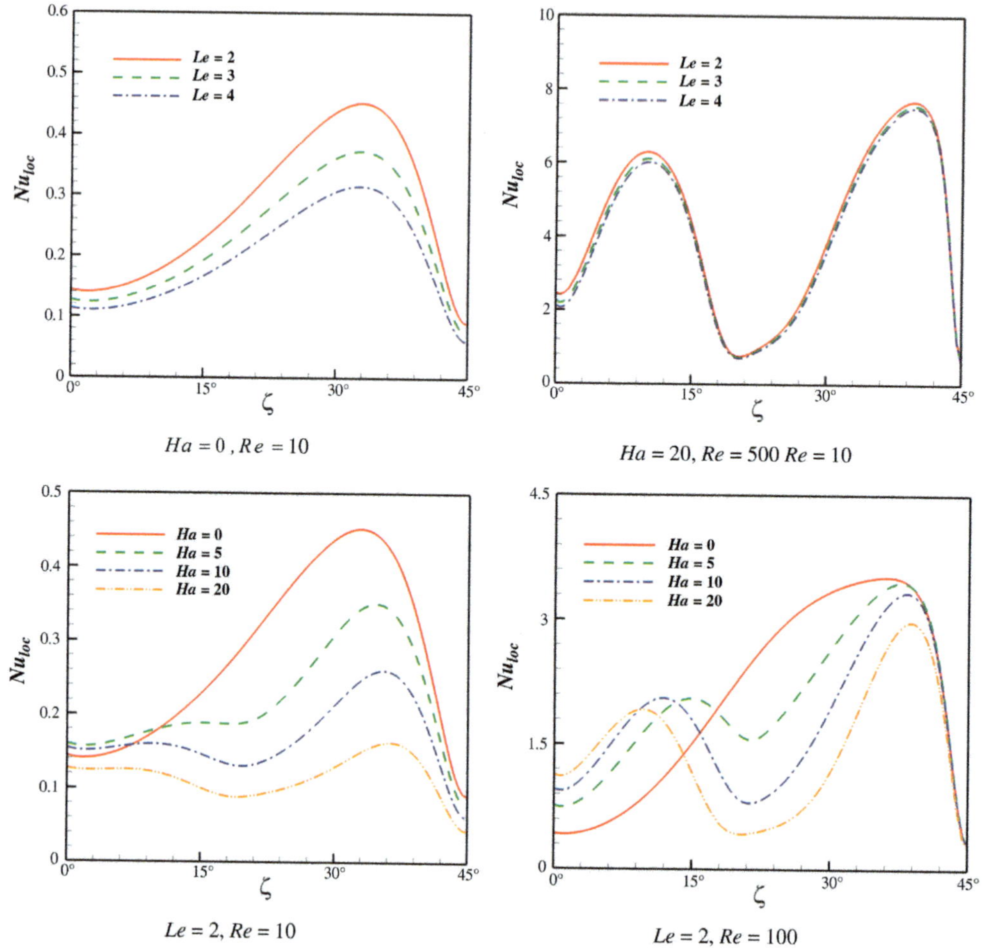

FIGURE 9.14 Effects of Hartmann number, Reynolds number, and Lewis number on local Nusselt number Nu_{loc} along hot wall.

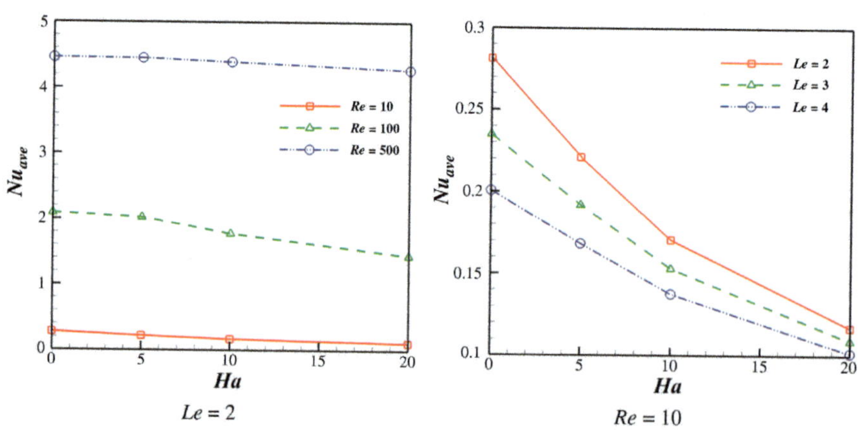

FIGURE 9.15 Effects of Hartmann number, Reynolds number, and Lewis number on average Nusselt number Nu_{ave} along hot wall.

The terms $\mu_0 M \frac{\partial \overline{H}}{\partial x}$ and $\mu_0 M \frac{\partial \overline{H}}{\partial y}$ in (9.46) and (9.47), respectively, represent the components of magnetic force per unit volume, and depend on the existence of the magnetic gradient on the corresponding x and y directions. These two terms are well known from FHD which is the so-called the Kelvin force. The terms $-\sigma_{nf} B_y^2 u + \sigma_{nf} B_x B_y v$ and $-\sigma_{nf} B_x^2 v + \sigma_{nf} B_x B_y u$ appearing in (9.46) and (9.47), respectively, represent the Lorentz force per unit volume towards the x and y directions, and arise due to the electrical conductivity of the fluid. These two terms are known in MHD. The principles of MHD and FHD are combined in the mathematical model

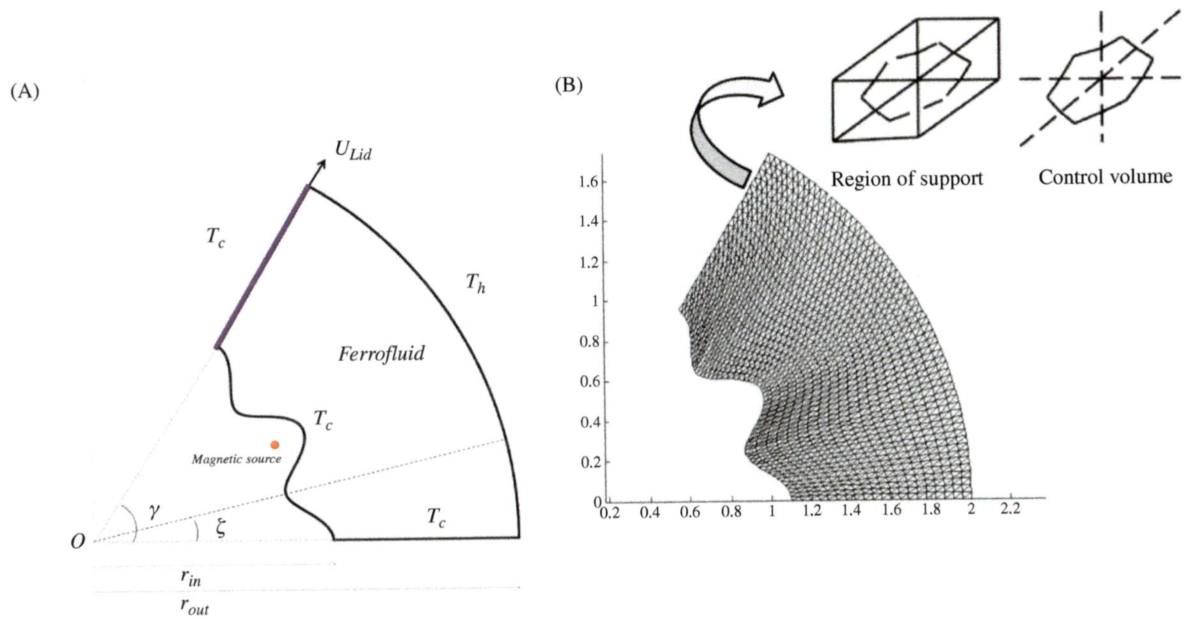

FIGURE 9.16 (A) Geometry and the boundary conditions; (B) the mesh of enclosure considered in this work.

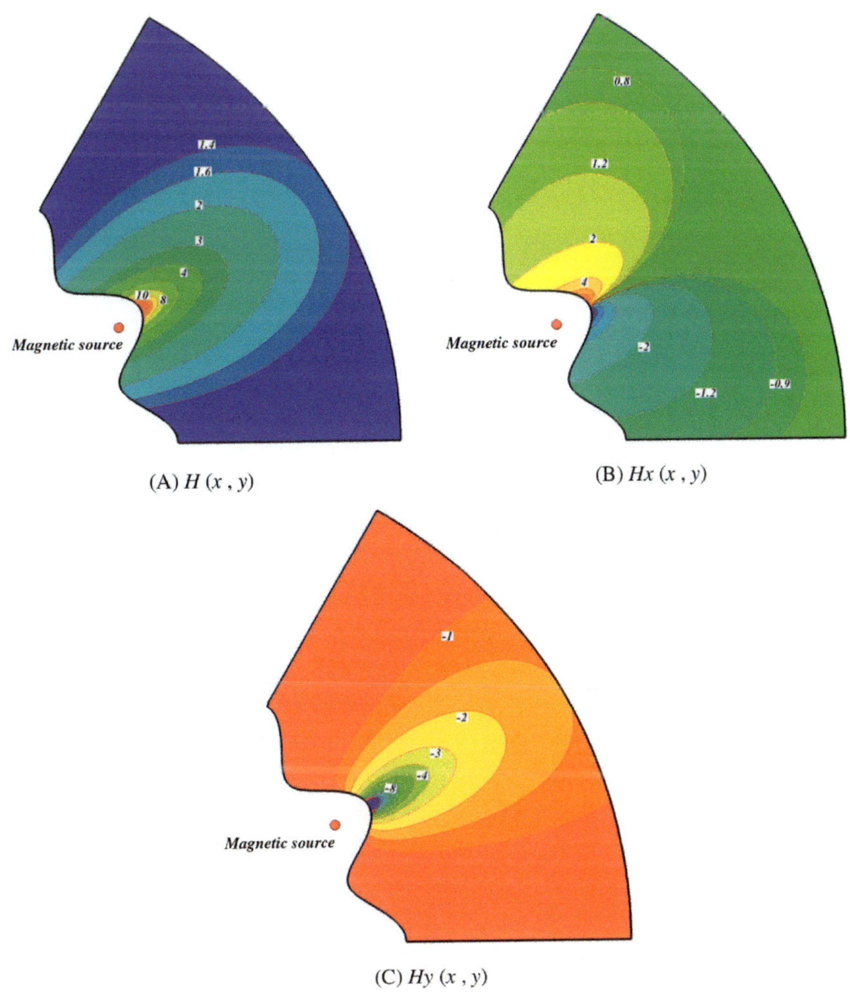

FIGURE 9.17 Contours of the (A) magnetic field strength H; (B) magnetic field intensity component in x direction Hx; (C) magnetic field intensity component in y direction Hy. (A) $H(x,y)$, (B) $Hx(x,y)$, (C) $Hy(x,y)$

presented and the above mentioned terms arise together in the governing equations (9.46) and (9.47). The term $\mu_0 T\frac{\partial M}{\partial T}\left(u\frac{\partial \overline{H}}{\partial x}+v\frac{\partial \overline{H}}{\partial y}\right)$ in Eq. (9.48) represents the thermal power per unit volume due to the magneto caloric effect. Also the term $\sigma_{nf}(uB_y - vB_x)^2$ in (9.48) represents the Joule heating. For the variation of the magnetization M, with the magnetic field intensity $\overline{H}$ and temperature T, the following relation derived is considered:

$$M = K'\overline{H}(T_c' - T) \tag{9.49}$$

where K' is a constant and T_c' is the Curie temperature.

In the above equations, μ_0 is the magnetic permeability of vacuum $(4\pi\times10^{-7}Tm/A)$, $\overline{H}$ is the magnetic field strength, $\overline{B}$ is the magnetic induction $(\overline{B}=\mu_0\overline{H})$, and the bar above the quantities denotes that they are dimensional. The effective density (ρ_{nf}) and heat capacitance $(\rho C_p)_{nf}$ of the nanofluid are defined as

$$\rho_{nf} = \rho_f(1-\phi) + \rho_s\phi, \tag{9.50}$$

$$(\rho C_p)_{nf} = (\rho C_p)_f(1-\phi) + (\rho C_p)_s\phi, \tag{9.51}$$

where ϕ is the solid volume fraction of nanoparticles. Thermal diffusivity of the nanofluids is

$$\alpha_{nf} = \frac{k_{nf}}{(\rho C_p)_{nf}}, \tag{9.52}$$

The dynamic viscosity of the nanofluids given by Brinkman is

$$\mu_{nf} = \frac{\mu_f}{(1-\phi)^{2.5}}. \tag{9.53}$$

The effective thermal conductivity of the nanofluid can be approximated by the Maxwell–Garnetts (MG) model as [4]:

$$\frac{k_{nf}}{k_f} = \frac{k_s + 2k_f - 2\phi(k_f - k_s)}{k_s + 2k_f + \phi(k_f - k_s)}, \tag{9.54}$$

and the effective electrical conductivity of nanofluid was presented by Maxwell as below:

$$\frac{\sigma_{nf}}{\sigma_f} = 1 + \frac{3\left(\frac{\sigma_s}{\sigma_f}-1\right)\phi}{\left(\frac{\sigma_s}{\sigma_f}+2\right)-\left(\frac{\sigma_s}{\sigma_f}-1\right)\phi}. \tag{9.55}$$

By introducing the following nondimensional variables:

$$a=\frac{\overline{a}}{L}, b=\frac{\overline{b}}{L}, X=\frac{x}{L}, \quad Y=\frac{y}{L}, \quad \Theta=\frac{T-T_c}{T_h-T_c}, U=\frac{u}{u_r}, V=\frac{v}{u_r},$$
$$H=\frac{\overline{H}}{\overline{H_0}}, H_x=\frac{\overline{H_x}}{\overline{H_0}}, H_y=\frac{\overline{H_y}}{\overline{H_0}}, P=\frac{p}{\rho_f(u_r)^2}, \tag{9.56}$$

where in Eq. (9.56) $\overline{H_0}=\overline{H}(\overline{a},0)=\frac{\gamma}{2\pi|b|}$ and $L=r_{out}-r_{in}$; and using them in basic equations, we get

$$\frac{\partial U}{\partial X}+\frac{\partial V}{\partial Y}=0, \tag{9.57}$$

$$U\frac{\partial U}{\partial X}+V\frac{\partial U}{\partial Y} = -\frac{\partial P}{\partial X}+\frac{1}{\text{Re}}\left[\frac{\mu_{nf}/\mu_f}{\rho_{nf}/\rho_f}\right]\left(\frac{\partial^2 U}{\partial X^2}+\frac{\partial^2 U}{\partial Y^2}\right)$$
$$+Mn_F\left(\frac{\rho_f}{\rho_{nf}}\right)(\varepsilon_2-\varepsilon_1-\Theta)H\frac{\partial H}{\partial X}-\frac{Ha^2}{\text{Re}}\left[\frac{\sigma_{nf}/\sigma_f}{\rho_{nf}/\rho_f}\right]\left(H_y^2U-H_xH_yV\right), \tag{9.58}$$

$$U\frac{\partial V}{\partial X}+V\frac{\partial V}{\partial Y}=-\frac{\partial P}{\partial X}+\frac{1}{\mathrm{Re}}\left[\frac{\mu_{nf}/\mu_f}{\rho_{nf}/\rho_f}\right]\left(\frac{\partial^2 V}{\partial X^2}+\frac{\partial^2 V}{\partial Y^2}\right)$$
$$+Mn_F\left(\frac{\rho_f}{\rho_{nf}}\right)(\varepsilon_2-\varepsilon_1-\Theta)H\frac{\partial H}{\partial Y}-\frac{Ha^2}{\mathrm{Re}}\left[\frac{\sigma_{nf}/\sigma_f}{\rho_{nf}/\rho_f}\right](H_x^2V-H_xH_yU), \tag{9.59}$$

$$U\frac{\partial\Theta}{\partial X}+V\frac{\partial\Theta}{\partial Y}=\left[\frac{\frac{k_{nf}}{k_f}}{\frac{(\rho C_P)_{nf}}{(\rho C_P)_f}}\right]\left(\frac{\partial^2\Theta}{\partial X^2}+\frac{\partial^2\Theta}{\partial Y^2}\right)+Ha^2\,Ec\left[\frac{\frac{\sigma_{nf}}{\sigma_f}}{\frac{(\rho C_P)_{nf}}{(\rho C_P)_f}}\right]\{U\,H_y-V\,H_x\}^2$$
$$+Mn_F Ec\frac{(\rho C_P)_f}{(\rho C_P)_{nf}}\left\{U\frac{\partial H}{\partial X}+V\frac{\partial H}{\partial Y}\right\}H(\varepsilon_1+\Theta)+Ec\left[\frac{\frac{\mu_{nf}}{\mu_f}}{\frac{(\rho C_P)_{nf}}{(\rho C_P)_f}}\right]\left\{2\left(\frac{\partial U}{\partial X}\right)^2+2\left(\frac{\partial V}{\partial Y}\right)^2+\left(\frac{\partial U}{\partial Y}+\frac{\partial V}{\partial X}\right)^2\right\}, \tag{9.60}$$

where $\mathrm{Re}=\frac{\rho_f L\,u_r}{\mu_f}$, $Ha=LH_0\mu_0\sqrt{\sigma_f/\mu_f}$, $\varepsilon_1=T_1/\Delta T$, $\varepsilon_2=T_c'/\Delta T$, $Ec=\left(\rho_f\,u_r^2\right)/\left[(\rho C_P)_f\Delta T\right]$, and $Mn_F=\mu_0H_0^2K'(T_h-T_c)/\left(\mu_f\,u_r^2\right)$ are the Reynolds number, the Hartman number, the temperature number, the Curie temperature number, Eckert number, and the magnetic number arising from FHD the for the base fluid, respectively. The thermophysical properties of the nanofluid are given in Table 9.1. The stream function and vorticity are defined as

$$u=\frac{\partial\psi}{\partial y},\ v=-\frac{\partial\psi}{\partial x},\ \omega=\frac{\partial v}{\partial x}-\frac{\partial u}{\partial y},$$
$$\Omega=\frac{\omega L}{u_r},\ \Psi=\frac{\psi}{Lu_r}. \tag{9.61}$$

The stream function satisfies the continuity Eq. (9.57). The vorticity equation is obtained by eliminating the pressure between the two momentum equations. The appropriate boundary conditions can be written as

$$\begin{aligned}&\Theta=1.0 &&\text{on the outer circular boundary,}\\&\Theta=0.0 &&\text{other walls boundaries, and}\\&\Psi=0.0 &&\text{on all solid boundaries.}\end{aligned} \tag{9.62}$$

The values of vorticity on the boundary of the enclosure can be obtained using the stream function formulation and the known velocity conditions during the iterative solution procedure.

The local Nusselt number of the nanofluid along the hot wall can be expressed as

$$Nu_{loc}=\left(\frac{k_{nf}}{k_f}\right)\frac{\partial\Theta}{\partial r}, \tag{9.63}$$

where r is the radial direction. The average Nusselt number on the cold circular wall is evaluated as:

$$Nu_{ave}=\frac{1}{\pi/3}\int_0^{\pi/3}Nu_{loc}(\zeta)\,d\zeta. \tag{9.64}$$

To estimate the enhancement of heat transfer between the case of $\phi=0.04$ and the pure fluid (base fluid) case, the heat transfer enhancement is defined as

$$En=\frac{Nu(\phi=0.04)-Nu(basefluid)}{Nu(basefluid)}\times 100. \tag{9.65}$$

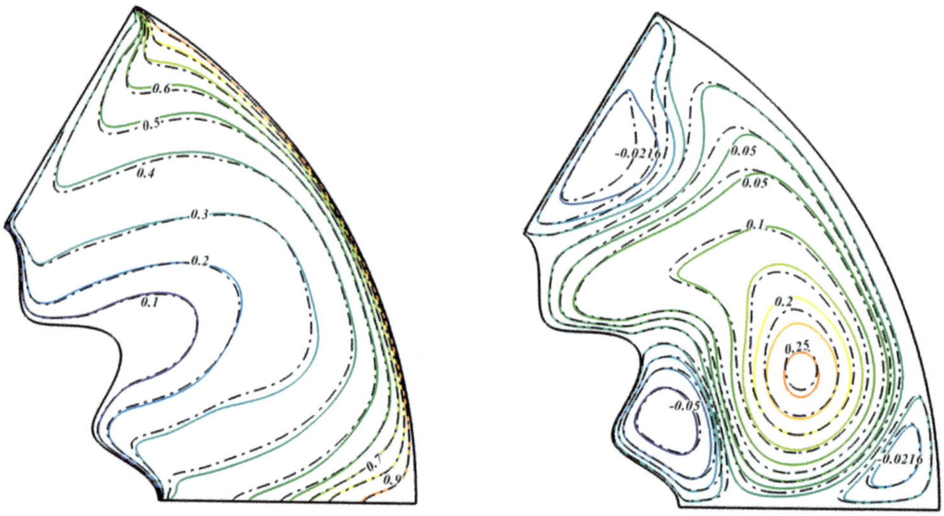

FIGURE 9.18 Comparison of the isotherm and streamlines between nanofluid ($\phi = 0.04$) (– · – · –) and pure fluid ($\phi = 0$) (—) when $Mn_F = 10, Ha = 10, \mathrm{Re} = 600$ and $\mathrm{Pr} = 6.8$.

9.4.3 Effects of Active Parameters

Forced convection heat transfer of ferrofluid in a lid-driven semiannulus enclosure in the presence of external magnetic field is studied. Ferrohydrodynamics (FHD) and Magnetohydrodynamics (MHD) are combined to obtain a suitable mathematical model. Calculations are made for various values of the volume fraction of the nanoparticles ($\phi = 0\%$ and 4%), the Reynolds number ($Ra = 10$ to 600), the magnetic number arising from FHD ($Mn_F = 0, 2, 6$, and 10), and the Hartmann number arising from MHD ($Ha = 0, 2, 6$, and 10). In all calculations, the Prandtl number (Pr), the temperature number (ε_1), and the Eckert number (Ec) are set to 6.8, 0.0, and 10^{-5}, respectively. The effects of nanoparticles on the isotherms and the streamlines are shown in Fig. 9.18. Thermal boundary layer thickness increases with an increase in the nanofluid volume fraction. Also, velocity increase with the addition of nanoparticles.

Effects of the Reynolds number, the Hartmann number, and the magnetic number on isotherms and streamlines are shown in Figs. 9.19 and 9.20. An increase in the Lorentz force leads to an increase in the thermal boundary layer thickness and a reduction in the velocity field. Thermal boundary layer thickness decreases with an increase in the Kelvin forces, and a thermal plume appears over the outer layer. As the Reynolds number increases the absolute value of the stream function increases, and a thermal plume is generated due to two strong eddies. The thermal plume disappears as the Hartmann number increases. It is an interesting observation that as the magnetic number increases and when the Reynolds number is large the main eddies turn into four smaller eddies. Also, it can be seen that the effect of magnetic number on isotherms is more pronounced for higher values of Reynolds number.

Effects of the magnetic number, the Hartmann number, and the Reynolds number on the local and the average Nusselt numbers are shown in Figs. 9.21 and 9.22, respectively. A local maximum and a minimum in local Nusselt number profile exist due to the existence of a thermal plume. Nusselt number increases with increase in the Reynolds number and the magnetic number; but it decreases with the augmentation of the Hartmann number. Effects of the Reynolds number, the Hartmann number, and the magnetic number on heat transfer are shown in Fig. 9.23. Heat transfer enhances with an increase in the Reynolds number and the magnetic number; while the opposite trend is observed with the Hartmann number.

9.5 FLOW AND CONVECTIVE HEAT TRANSFER OF A FERRONANOFLUID IN A DOUBLE-SIDED LID-DRIVEN CAVITY WITH A WAVY WALL IN THE PRESENCE OF A VARIABLE MAGNETIC FIELD

9.5.1 Problem Definition

The schematic of the problem and the related boundary conditions as well as the mesh of the enclosure which is used in the present CVFEM program are shown in Fig. 9.24. The enclosure has a width/height aspect ratio of

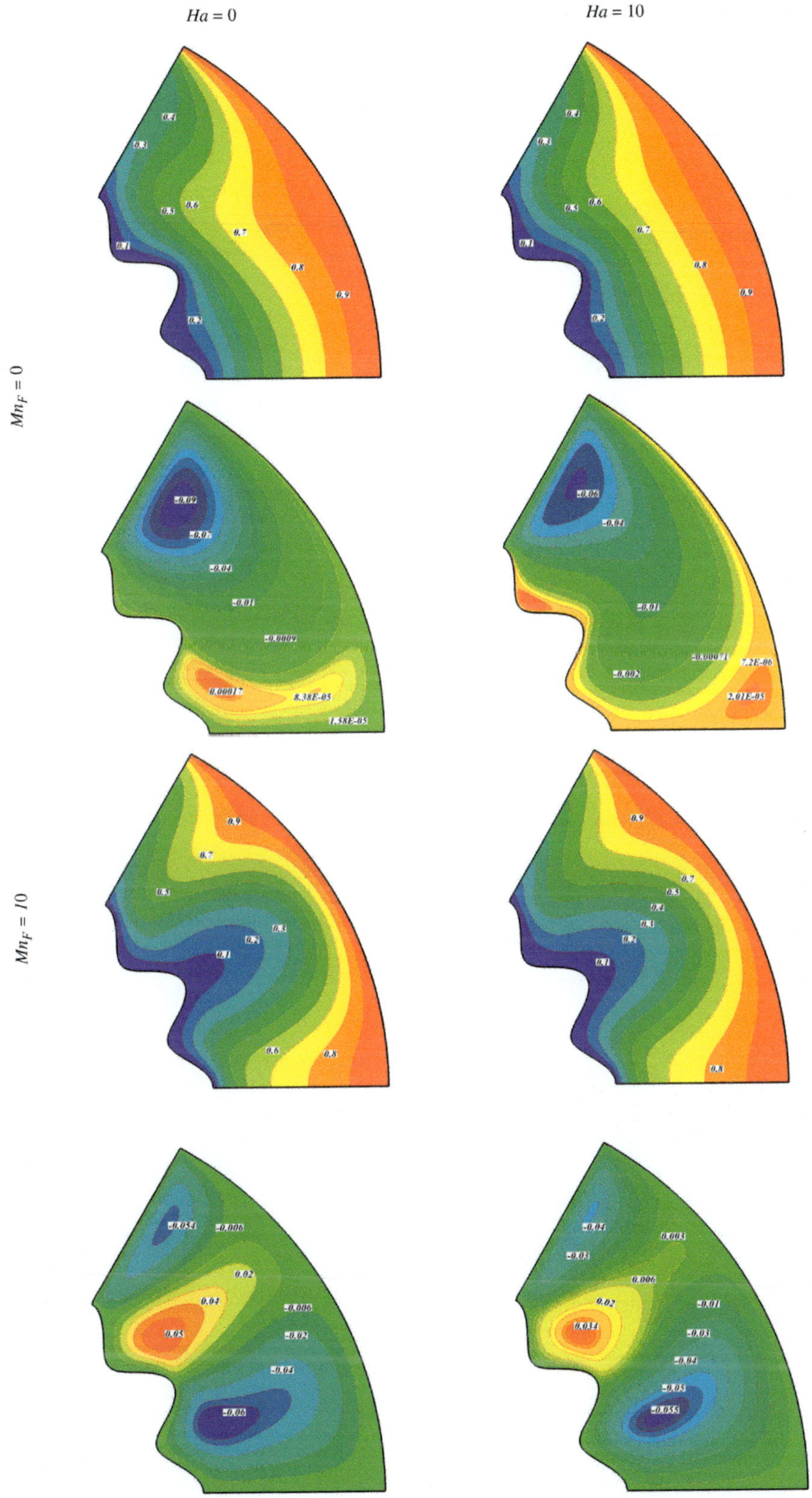

FIGURE 9.19 Isotherm (up) and streamline (down) contours for different values of Hartmann number and Magnetic number when Re = 10, $\phi = 0.04$, Pr = 6.8.

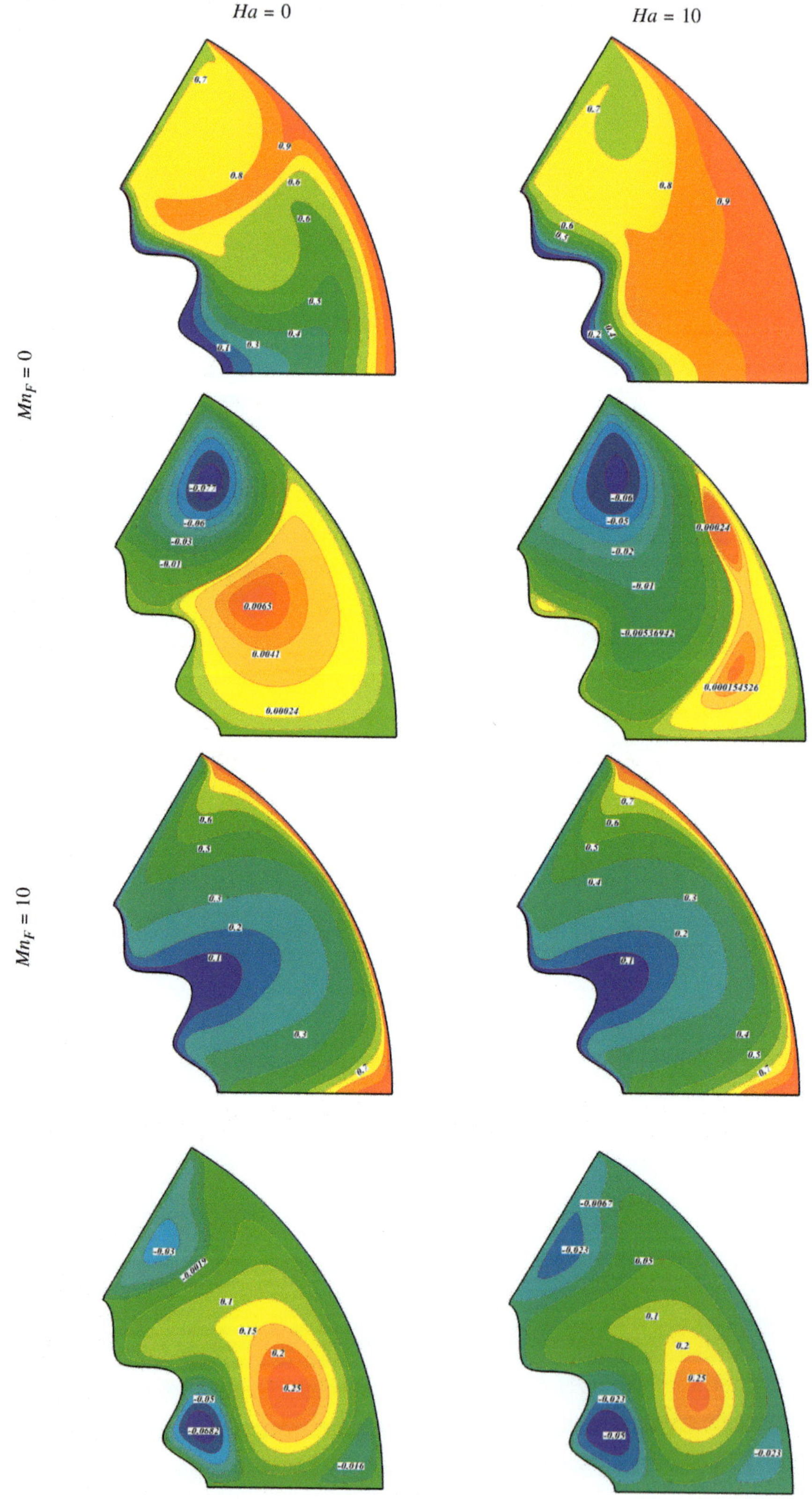

FIGURE 9.20 Isotherm (up) and streamline (down) contours for different values of Hartmann number and magnetic number when $Re = 600$, $\phi = 0.04$, $Pr = 6.8$.

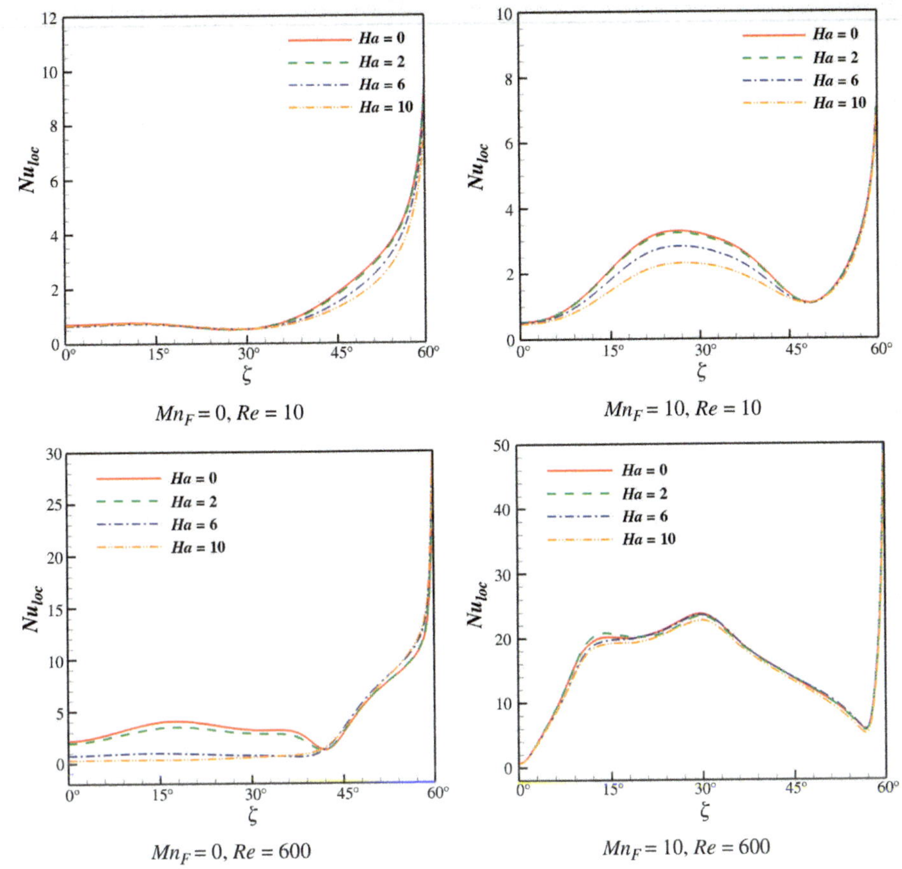

FIGURE 9.21 Effects of magnetic number, Hartmann number, and Reynolds number on local Nusselt number Nu_{loc} along hot wall when $\phi = 0.04$.

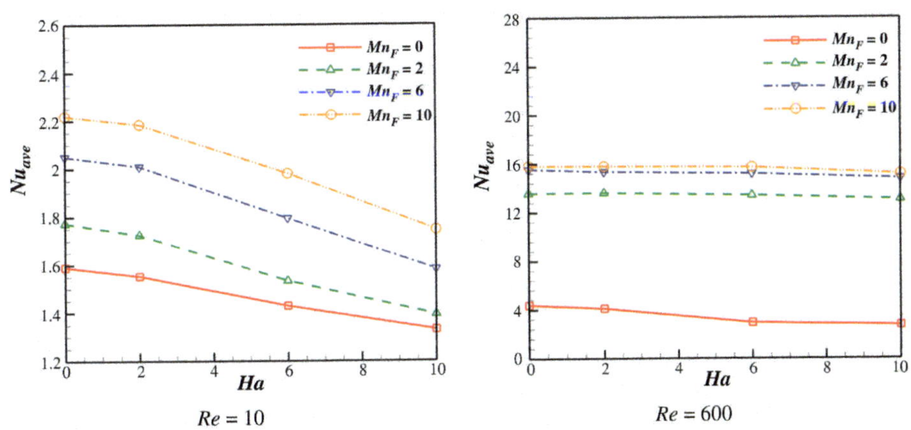

FIGURE 9.22 Effects of Reynolds number, Hartmann number, and magnetic number on average Nusselt number when $\phi = 0.04$

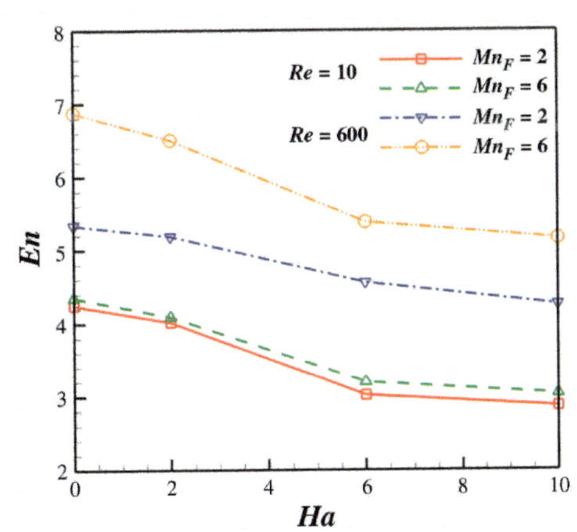

FIGURE 9.23 Effects of Reynolds number, Hartmann number, and magnetic number on heat transfer enhancement.

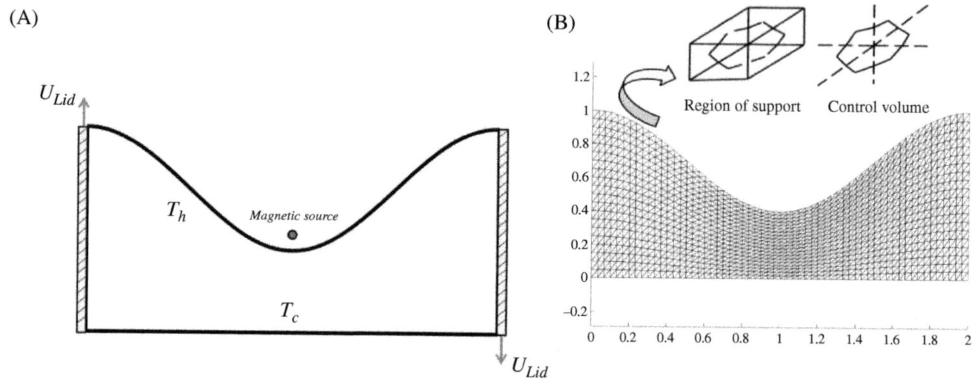

FIGURE 9.24 (A) Geometry and the boundary conditions; (B) the mesh of enclosure considered in this work.

two. The two sidewalls with length H are thermally insulated whereas the lower flat and upper sinusoidal walls are maintained at constant temperatures T_c and T_h, respectively. Under all circumstances, the condition $T_h > T_c$ is maintained. The shape of the upper sinusoidal wall profile is assumed to mimic the following pattern:

$$Y = H - \{a(H + \sin(\pi x - \pi/2))\} \tag{9.66}$$

where a is the dimensionless amplitude of the sinusoidal wall. For the expression of the magnetic field strength, it can be considered that the magnetic source represents a magnetic wire placed vertically in the x-y plane at the point $(\bar{a}, \bar{b})$. The components of the magnetic field intensity $(\overline{H_x}, \overline{H_y})$ and the magnetic field strength $(\overline{H})$ can be considered as:

$$\overline{H_x} = \frac{\gamma'}{2\pi} \frac{1}{(x-\bar{a})^2 + (y-\bar{b})^2}(y - \bar{b}) \tag{9.67}$$

$$\overline{H_y} = -\frac{\gamma'}{2\pi} \frac{1}{(x-\bar{a})^2 + (y-\bar{b})^2}(x - \bar{a}) \tag{9.68}$$

$$\overline{H} = \sqrt{\overline{H}_x^2 + \overline{H}_y^2} = \frac{\gamma'}{2\pi} \frac{1}{\sqrt{(x-\bar{a})^2 + (y-\bar{b})^2}} \tag{9.69}$$

where γ' is the magnetic field strength at the source (of the wire) and $(\bar{a}, \bar{b})$ is the position where the source is located. The contours of the magnetic field strength are shown in Fig. 9.25. In this section, the magnetic source is located at $(1.01cols, 0.5rows)$. The upper wall is the driven lid with a velocity of U_{Lid}.

9.5.2 Governing Equation

The flow is two-dimensional, laminar, and incompressible. The magnetic Reynolds number is assumed to be small so that the induced magnetic field can be neglected compared to the applied magnetic field. The flow is considered to be steady, two-dimensional, and laminar. Using the Boussinesq approximation, the governing equations for the nanofluid flow and heat transfer can be written as follows:

$$\frac{\partial u}{\partial x} + \frac{\partial v}{\partial y} = 0 \tag{9.70}$$

$$\rho_{nf}\left(u\frac{\partial u}{\partial x} + v\frac{\partial u}{\partial y}\right) = -\frac{\partial P}{\partial x} + \mu_{nf}\left(\frac{\partial^2 u}{\partial x^2} + \frac{\partial^2 u}{\partial y^2}\right) + \mu_0 M\frac{\partial \overline{H}}{\partial x} - \sigma_{nf}B_y^2 u + \sigma_{nf}B_x B_y v \tag{9.71}$$

$$\rho_{nf}\left(u\frac{\partial v}{\partial x} + v\frac{\partial v}{\partial y}\right) = -\frac{\partial P}{\partial y} + \mu_{nf}\left(\frac{\partial^2 v}{\partial x^2} + \frac{\partial^2 v}{\partial y^2}\right) + \mu_0 M\frac{\partial \overline{H}}{\partial y} - \sigma_{nf}B_x^2 v + \sigma_{nf}B_x B_y u \tag{9.72}$$

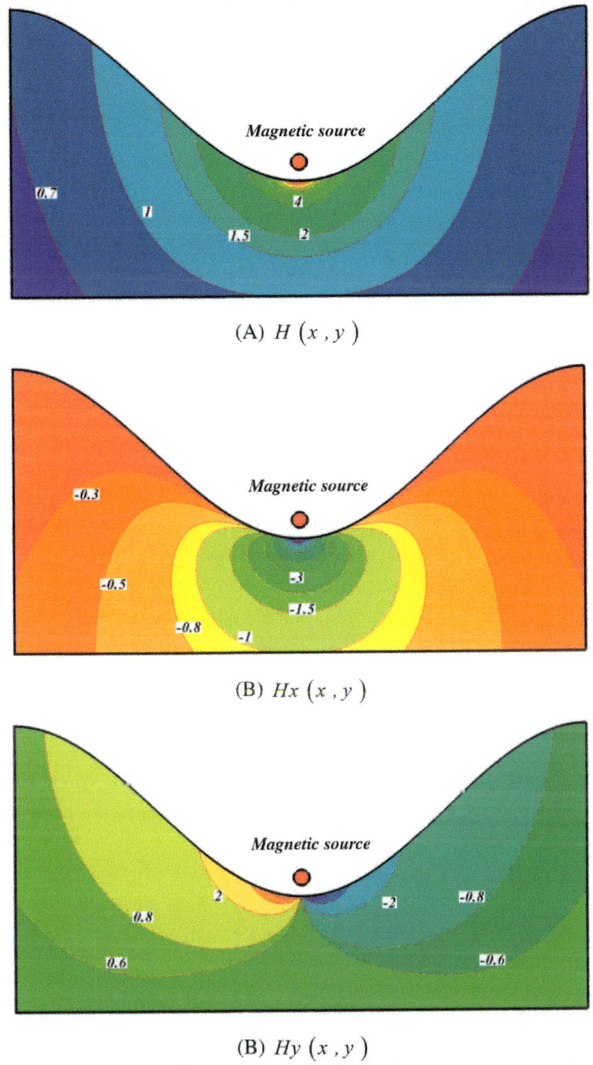

FIGURE 9.25 Contours of the (A) magnetic field strength H; (B) magnetic field intensity component in x direction Hx; (C) magnetic field intensity component in y direction Hy. (A) $H(x,y)$, (B) $Hx(x,y)$, (C) $Hy(x,y)$.

$$(\rho C_p)_{nf}\left(u\frac{\partial T}{\partial x}+v\frac{\partial T}{\partial y}\right)=k_{nf}\left(\frac{\partial^2 T}{\partial x^2}+\frac{\partial^2 T}{\partial y^2}\right)+\sigma_{nf}(uB_y-vB_x)^2$$
$$-\mu_0 T\frac{\partial M}{\partial T}\left(u\frac{\partial \overline{H}}{\partial x}+v\frac{\partial \overline{H}}{\partial y}\right)+\mu_{nf}\left\{2\left(\frac{\partial u}{\partial x}\right)^2+2\left(\frac{\partial v}{\partial x}\right)^2+\left(\frac{\partial u}{\partial x}+\frac{\partial v}{\partial y}\right)^2\right\} \tag{9.73}$$

$$M=K'\overline{H}(T_c'-T) \tag{9.74}$$

where K' is a constant and T_c' is the Curie temperature.

In the above equations, μ_0 is the magnetic permeability of vacuum $(4\pi\times 10^{-7}Tm/A)$, $\overline{H}$ is the magnetic field strength, $\overline{B}$ is the magnetic induction $(\overline{B}=\mu_0\overline{H})$, and the bar above the quantities denotes that they are dimensional. The effective density (ρ_{nf}) and heat capacitance $(\rho C_p)_{nf}$ of the nanofluid are defined as:

$$\rho_{nf}=\rho_f(1-\phi)+\rho_s\phi \tag{9.75}$$

$$(\rho C_p)_{nf}=(\rho C_p)_f(1-\phi)+(\rho C_p)_s\phi \tag{9.76}$$

where ϕ is the solid volume fraction of nanoparticles. The thermal diffusivity of the nanofluids is

$$\alpha_{nf} = \frac{k_{nf}}{(\rho C_p)_{nf}} \tag{9.77}$$

and the thermal expansion coefficient of the nanofluid can be determined as

$$\beta_{nf} = \beta_f(1-\phi) + \beta_s\phi \tag{9.78}$$

The dynamic viscosity of the nanofluids given by Brinkman is

$$\mu_{nf} = \frac{\mu_f}{(1-\phi)^{2.5}} \tag{9.79}$$

The effective thermal conductivity of the nanofluid can be approximated by the Maxwell–Garnetts (MG) model as:

$$\frac{k_{nf}}{k_f} = \frac{k_s + 2k_f - 2\phi(k_f - k_s)}{k_s + 2k_f + \phi(k_f - k_s)} \tag{9.80}$$

and the effective electrical conductivity of the nanofluid was presented by Maxwell as below:

$$\frac{\sigma_{nf}}{\sigma_f} = 1 + \frac{3\left(\frac{\sigma_s}{\sigma_f} - 1\right)\phi}{\left(\frac{\sigma_s}{\sigma_f} + 2\right) - \left(\frac{\sigma_s}{\sigma_f} - 1\right)\phi} \tag{9.81}$$

The stream function and vorticity are defined as:

$$u = \frac{\partial\psi}{\partial y},\quad v = -\frac{\partial\psi}{\partial x},\quad \omega = \frac{\partial v}{\partial x} - \frac{\partial u}{\partial y} \tag{9.82}$$

The stream function satisfies the continuity equation (9.70). The vorticity equation is obtained by eliminating the pressure between the two momentum equations, i.e., by taking the y-derivative of Eq. (9.71) and subtracting from it the x-derivative of Eq. (9.72). By introducing the following nondimensional variables:

$$\begin{gathered} X = \frac{x}{L},\quad Y = \frac{y}{L},\quad \Omega = \frac{\omega L}{U_{Lid}},\ \Psi = \frac{\psi}{LU_{Lid}},\ \Theta = \frac{T - T_c}{T_h - T_c},\ U = \frac{u}{U_{Lid}},\ V = \frac{v}{U_{Lid}}, \\ H = \frac{\overline{H}}{\overline{H_0}},\quad H_x = \frac{\overline{H_x}}{\overline{H_0}},\quad H_y = \frac{\overline{H_y}}{\overline{H_0}} \end{gathered} \tag{9.83}$$

where in Eq. (9.83) $\overline{H_0} = \overline{H}(\overline{a}, 0) = \frac{\gamma}{2\pi|b|}$ and $L = r_{out} - r_{in} = r_{in}$. Using the dimensionless parameters, the governing equations now become:

$$\begin{aligned} \frac{\partial\Psi}{\partial Y}\frac{\partial\Omega}{\partial X} - \frac{\partial\Psi}{\partial X}\frac{\partial\Omega}{\partial Y} &= \frac{1}{\mathrm{Re}}\left[\frac{\mu_{nf}/\mu_f}{\rho_{nf}/\rho_f}\right]\left(\frac{\partial^2\Omega}{\partial X^2} + \frac{\partial^2\Omega}{\partial Y^2}\right) \\ &+ Mn_F\left(\frac{\rho_f}{\rho_{nf}}\right)H\left\{\frac{\partial H}{\partial X}\frac{\partial\Theta}{\partial Y} - \frac{\partial H}{\partial Y}\frac{\partial\Theta}{\partial X}\right\} - \frac{Ha^2}{\mathrm{Re}}\left[\frac{\sigma_{nf}/\sigma_f}{\rho_{nf}/\rho_f}\right] \\ &\times\left\{\frac{\partial V}{\partial X}H_x^2 + V\left(2H_x\frac{\partial H_x}{\partial X}\right) - \frac{\partial U}{\partial X}H_xH_y - U\frac{\partial H_x}{\partial X}H_y - U\frac{\partial H_y}{\partial X}H_x\right. \\ &\left. - \frac{\partial U}{\partial Y}H_y^2 - U\left(2H_y\frac{\partial H_y}{\partial Y}\right) + \frac{\partial V}{\partial Y}H_xH_y + V\frac{\partial H_x}{\partial Y}H_y + V\frac{\partial H_y}{\partial Y}H_x\right\} \end{aligned} \tag{9.84}$$

$$\begin{aligned}\frac{\partial \Psi}{\partial Y}\frac{\partial \Theta}{\partial X}-\frac{\partial \Psi}{\partial X}\frac{\partial \Theta}{\partial Y}&=\frac{1}{\Pr \text{Re}}\left(\frac{k_{nf}}{k_f}\right)\Big/\left(\frac{(\rho C_P)_{nf}}{(\rho C_P)_f}\right)\left(\frac{\partial^2\Theta}{\partial X^2}+\frac{\partial^2\Theta}{\partial Y^2}\right)\\&+Ha^2\frac{Ec}{\text{Re}}\left(\frac{\sigma_{nf}}{\sigma_f}\right)\Big/\left(\frac{(\rho C_P)_{nf}}{(\rho C_P)_f}\right)\{U H_y - V H_x\}^2\\&+Mn_F Ec\frac{(\rho C_P)_f}{(\rho C_P)_{nf}}\left\{U\frac{\partial H}{\partial X}+V\frac{\partial H}{\partial Y}\right\}H(\varepsilon_1+\Theta)\\&+\frac{Ec}{\text{Re}}\left(\frac{\mu_{nf}}{\mu_f}\right)\Big/\left(\frac{(\rho C_P)_{nf}}{(\rho C_P)_f}\right)\left\{2\left(\frac{\partial U}{\partial X}\right)^2+2\left(\frac{\partial V}{\partial X}\right)^2+\left(\frac{\partial U}{\partial X}+\frac{\partial V}{\partial Y}\right)^2\right\}\end{aligned} \tag{9.85}$$

$$\frac{\partial^2\Psi}{\partial X^2}+\frac{\partial^2\Psi}{\partial Y^2}=-\Omega \tag{9.86}$$

where $\text{Re}=\frac{\rho_f L\, U_{\text{Lid}}}{\mu_f}, Ha=LH_0\mu_0\sqrt{\sigma_f/\mu_f}, \varepsilon_1=T_1/\Delta T, Ec=\left(\rho_f U_{Lid}{}^2\right)/\left[(\rho C_P)_f \Delta T\right],$ and $Mn_F=\mu_0 H_0^2 K'(T_h-T_c)/\left(\rho_f\, U_{Lid}{}^2\right)$ are the Reynolds number, Hartmann number arising from MHD, temperature number, Eckert number, and magnetic number arising from FHD for the base fluid, respectively. The thermophysical properties of the nanofluid are given in Table 9.1. The boundary conditions as shown in Fig. 9.24 are:

$$\begin{aligned}&\Theta=1.0 \quad \text{on the inner circular boundary}\\&\Theta=0.0 \quad \text{on the other walls}\end{aligned} \tag{9.87}$$

The values of vorticity on the boundary of the enclosure can be obtained using the stream function formulation and the known velocity conditions during the iterative solution procedure.

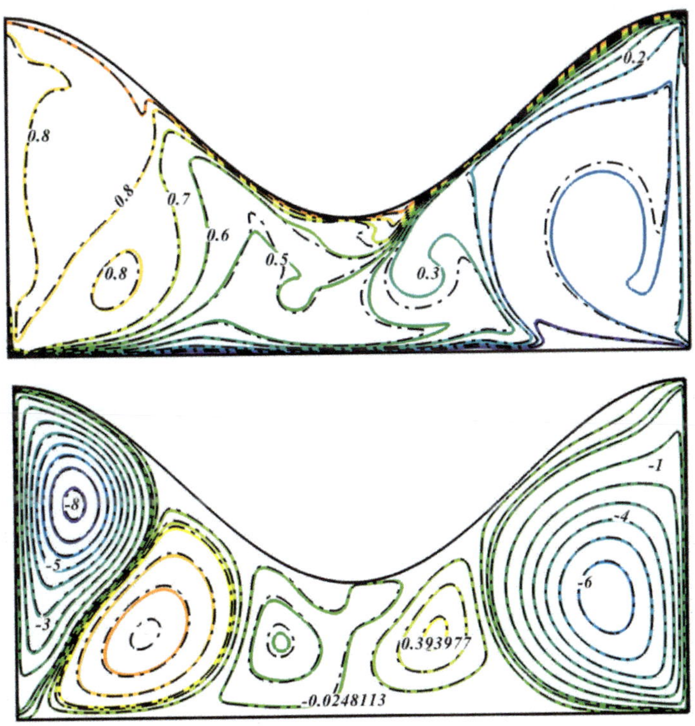

FIGURE 9.26 Comparison of the streamlines between nanofluid ($\phi=0.04$) (—) and pure fluid ($\phi=0$) (– · – · –) when $Re=100$, $Mn_F=10$, $Ha=10$, $Ec=10^{-5}$, $\varepsilon_1=0$, $\Pr=6.8$.

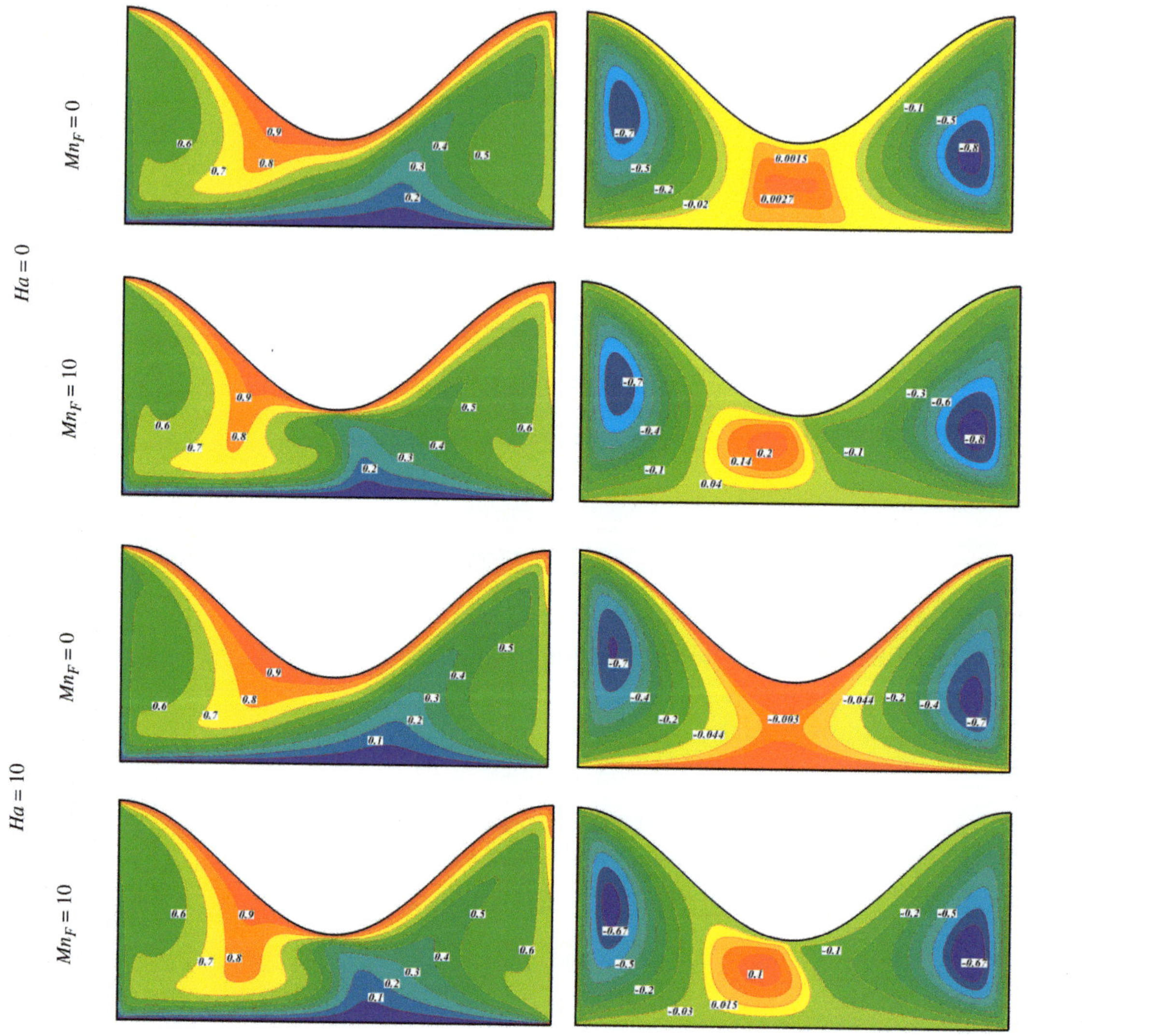

FIGURE 9.27 Isotherm (left) and streamline (right) contours for different values of Hartmann number and Magnetic number when Re = 10.

The local Nusselt number of the nanofluid along the cold wall can be expressed as:

$$Nu_{loc} = -\left(\frac{k_{nf}}{k_f}\right)\frac{\partial\Theta}{\partial Y} \tag{9.88}$$

where Y is the nondimensional vertical direction. The average Nusselt number on the cold wall is evaluated as:

$$Nu_{ave} = \frac{1}{2}\int_0^2 Nu_{loc}(X)\,dX \tag{9.89}$$

To estimate the enhancement of heat transfer between the case of $\phi = 0.04$ and the pure fluid (base fluid) case, the heat transfer enhancement is defined as:

$$En = \frac{Nu(\phi = 0.04) - Nu(basefluid)}{Nu(basefluid)} \times 100 \tag{9.90}$$

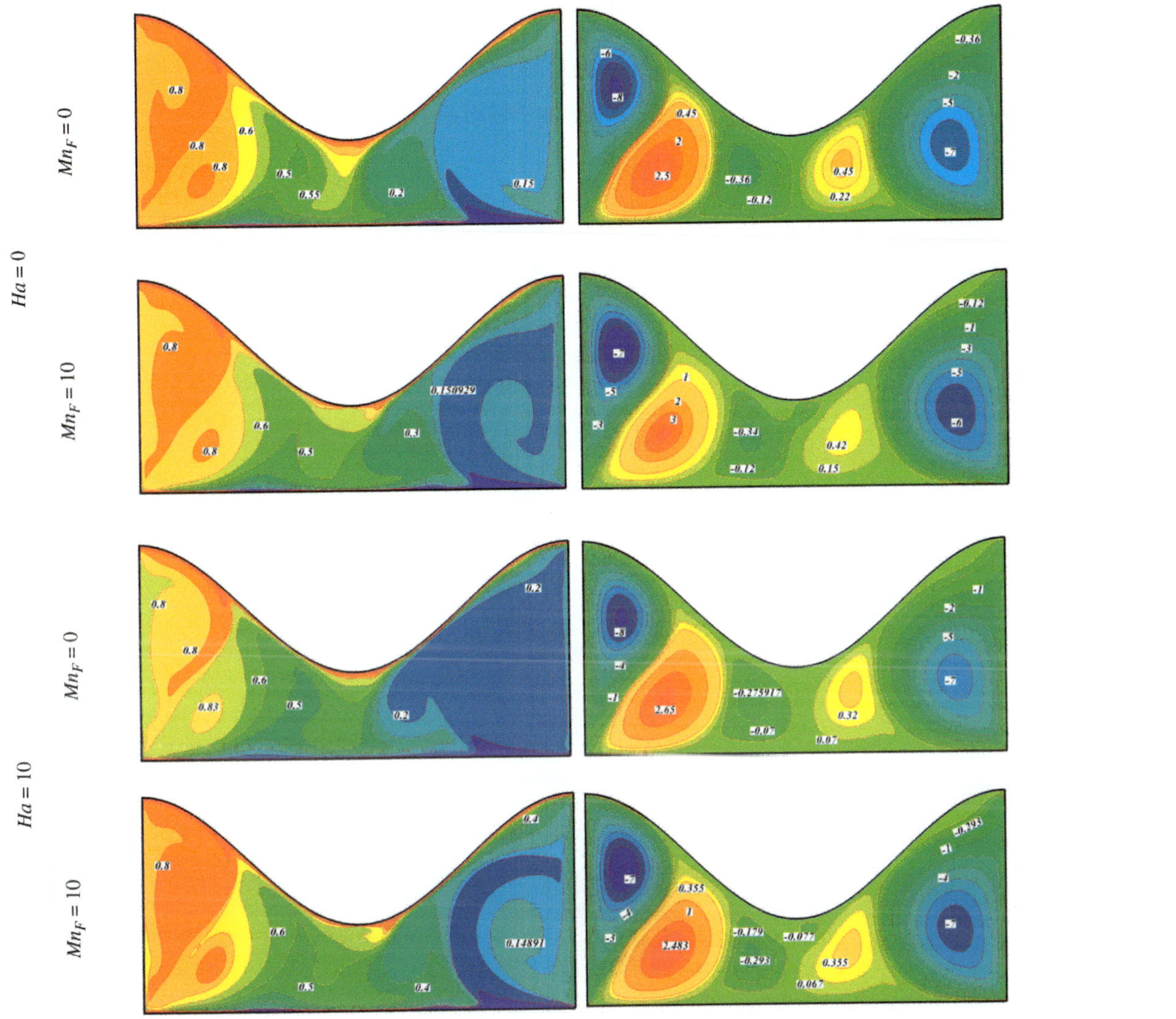

FIGURE 9.28 Isotherm (left) and streamline (right) contours for different values of Hartmann number and Magnetic number when Re = 100.

9.5.3 Effects of Active Parameters

Forced convection heat transfer of a ferronanofluid in the presence of a variable magnetic field is studied using CVFEM. A combination of both the ferrohydrodynamic and the magnetohydrodynamic laws are used to obtain the mathematical model. The enclosure is filled with a Fe_3O_4-water nanofluid. Calculations are made for various values of the Reynolds numbers (Re = 10 and 100), volume fraction of nanoparticles ($\phi = 0\%$ and 4%), Hartmann number ($Ha = 0,\ 5$, and 10), and magnetic number ($Mn_F = 0, 2, 4, 6$, and 10). In all calculations, the Prandtl number (Pr), temperature number (ε_1), and the Eckert number (Ec) are set to be 6.8, 0.0, and 10^{-5}, respectively.

A comparison of the streamlines between the nanofluid case and the pure fluid case is shown in Fig. 9.26. Adding nanoparticles in the base fluid causes the energy transport to increase. Also, the thermal boundary layer thickness increases with the increase of the volume fraction of the nanofluid. The isotherm and streamline contours for different values of the Reynolds, Hartmann, and the magnetic numbers are shown in Figs. 9.27 and 9.28. In the absence of the magnetic field at Re = 10, three rotating eddies exist in which the middle one rotates in the clockwise direction while the other two rotate in the counterclockwise direction. Two thermal plumes are predicted to generate over the hot and cold walls. As the Kelvin force increases, the middle eddy stretches to the left side. In turn, the thermal plumes slant to the left side. However, increasing the Lorentz force in the absence of the Kelvin force causes the middle eddy to disappear. As the Reynolds number increases, the thermal boundary layer thickness decreases. The middle eddy splits into three small vortices such that the isotherms become more complex at high Reynolds numbers.

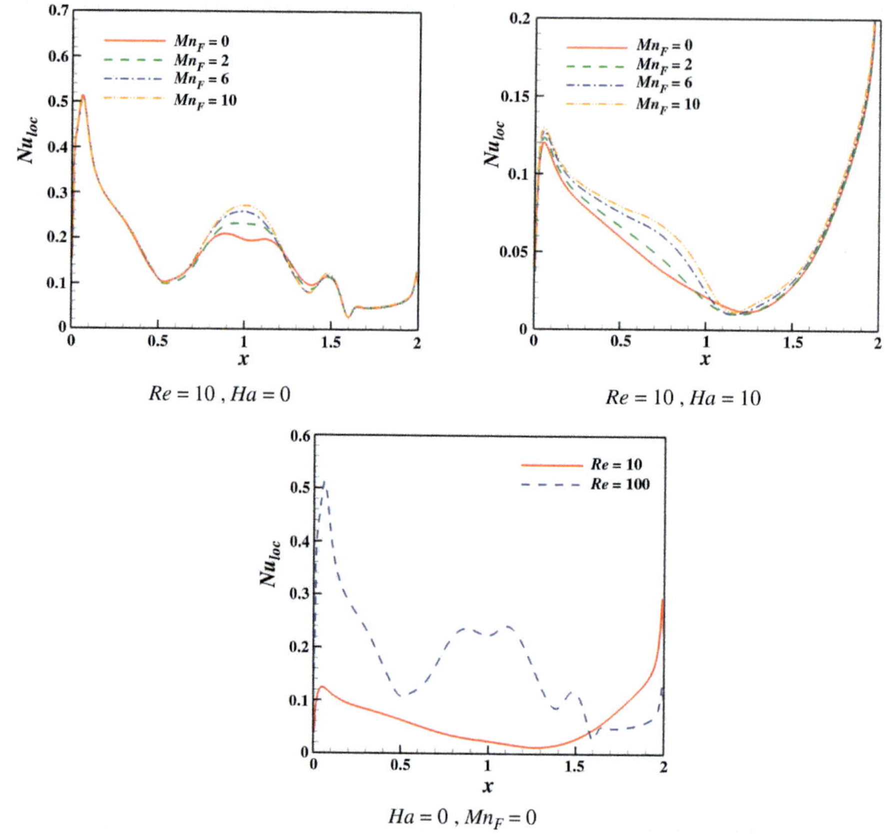

FIGURE 9.29 Effects of Hartmann number, Magnetic number, and Reynolds number on local Nusselt number Nu_{loc} along cold wall.

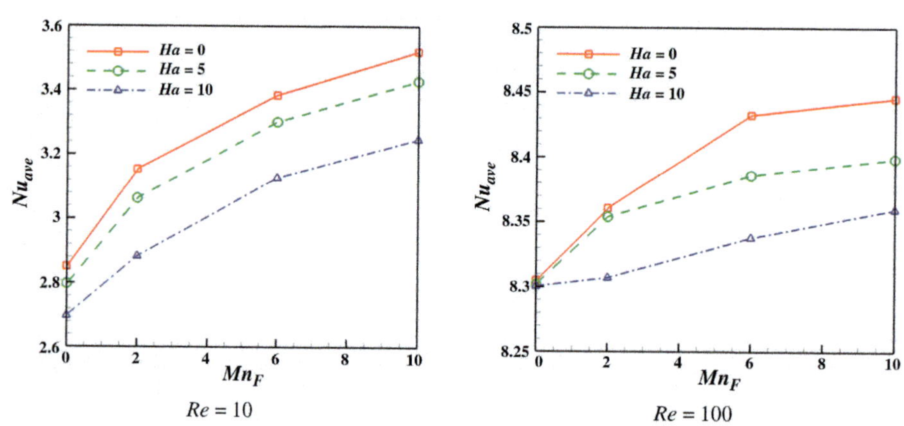

FIGURE 9.30 Effects of Hartmann number, Magnetic number, and Reynolds number on local Nusselt number Nu_{ave} along cold wall.

Figs. 9.29 and 9.30 depict that the effects of the magnetic number, Hartmann number, and the Reynolds number on the local and average Nusselt numbers. The thermal boundary layer thickness increases with the increase of the Hartmann number while it decreases with the augmentation of the magnetic number and the Reynolds number. As a result, the average Nusselt number increases with the increase of the Reynolds number and the magnetic number while it decreases with the increase of the Hartmann number and the magnetic number. For higher values of the Reynolds number, the effect of the Lorentz force on the Nusselt number becomes more sensible at higher values of the magnetic number. The effects of the magnetic number, Reynolds number, and the Hartmann number on the heat transfer enhancement are shown in Fig. 9.31. The heat transfer enhancement decreases with the increase in the magnetic number. The effect of nanoparticles is more pronounced at high

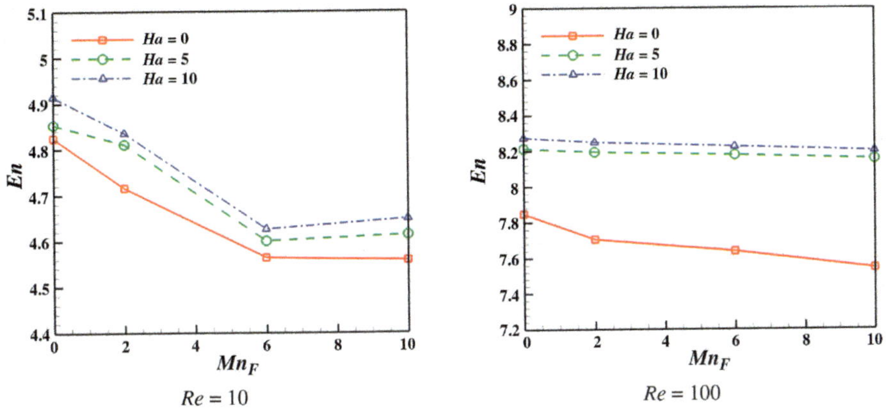

FIGURE 9.31 Effects of Hartmann number, Magnetic number, and Reynolds number on heat transfer enhancement.

Hartmann and Reynolds numbers. This observation can be explained by noting that at high values of these parameters, the heat transfer is dominant by conduction. Therefore, the addition of high thermal conductivity nanoparticles increases the conduction heat transfer and makes the heat transfer enhancement more effective.

References

[1] S. Lee, S.U.S. Choi, S. Li, J. Eastman, Measuring thermal conductivity of fluids containing oxide nanoparticles, ASME J. Heat Transfer 121 (1999) 280–289.

[2] M. Sheikholeslami, M.M. Rashidi, D.D. Ganji, Effect of non-uniform magnetic field on forced convection heat transfer of Fe_3O_4- water nanofluid, Comput. Methods Appl. Mech. Engrg. 294 (2015) 299–312.

[3] M. Sheikholeslami, M.M. Rashidi, D.D. Ganji, Numerical investigation of magnetic nanofluid forced convective heat transfer in existence of variable magnetic field using two phase model, J. Mol. Liq. 212 (2015) 117–126.

[4] M. Sheikholeslami, K. Vajravelu, M.M. Rashidi, Forced convection heat transfer in a semi annulus under the influence of a variable magnetic field, Int. J. Heat Mass Transfer 92 (2016) 339–348.

[5] M. Sheikholeslami, A.J. Chamkha, Flow and convective heat transfer of a ferro-nanofluid in a double-sided lid-driven cavity with a wavy wall in the presence of a variable magnetic field, Numer. Heat Transfer, Part A 69 (10) (2016) 1186–1200. Available from: https://doi.org/10.1080/10407782.2015.1125709.

[6] M. Sheikholeslami, M. Barzegar Gerdroodbary, D.D. Ganji, Numerical investigation of forced convective heat transfer of Fe_3O_4-water nanofluid in presence of external magnetic source, Comput. Methods Appl. Mech. Eng. 315 (2017) 831–845.

[7] M. Sheikholeslami, H.B. Rokni, Influence of EFD viscosity on nanofluid forced convection in a cavity with sinusoidal wall, J. Mol. Liq. 232 (2017) 390–395.

[8] M. Sheikholeslami, Influence of Lorentz forces on nanofluid flow in a porous cylinder considering Darcy model, J. Mol. Liq. 225 (2017) 903–912.

[9] M. Sheikholeslami, CVFEM for magnetic nanofluid convective heat transfer in a porous curved enclosure, Eur. Phys. J. Plus 131 (2016) 413. Available from: https://doi.org/10.1140/epjp/i2016-16413-y.

[10] M. Sheikholeslami, Influence of Coulomb forces on Fe_3O_4-H_2O nanofluid thermal improvement, Int. J. Hydrogen Energy 42 (2017) 821–829.

[11] M. Sheikholeslami, Magnetic field influence on nanofluid thermal radiation in a cavity with tilted elliptic inner cylinder, J. Mol. Liq. 229 (2017) 137–147.

[12] M. Sheikholeslami, Numerical simulation of magnetic nanofluid natural convection in porous media, Phys. Lett. A 381 (2017) 494–503.

[13] M. Sheikholeslami, Magnetic source impact on nanofluid heat transfer using CVFEM, Neural Comput. Applicat. (2016). Available from: https://doi.org/10.1007/s00521-016-2740-7.

[14] M. Sheikholeslami, Numerical investigation of MHD nanofluid free convective heat transfer in a porous tilted enclosure, Eng. Comput. 34 (6) (2017) 1939–1955.

[15] M.S. Kandelousi, Effect of spatially variable magnetic field on ferrofluid flow and heat transfer considering constant heat flux boundary condition, Europ. Phys. J. Plus (2014) 129–248. Available from: https://doi.org/10.1140/epjp/i2014-14248-2.

[16] M. Sheikholeslami, Numerical simulation for external magnetic field influence on Fe_3O_4-water nanofluid forced convection, Engineering Computations, Inpress.

[17] M. Sheikholeslami, Influence of Lorentz forces on nanofluid flow in a porous cavity by means of Non- Darcy model, Eng. Comput. 34 (8) (2017) 2651–2667. https://doi.org/10.1108/EC-01-2017-0008.

[18] M. Sheikholeslami, H.B. Rokni, Nanofluid convective heat transfer intensification in a porous circular cylinder, Chem. Eng. Proces. Process Intensi. 120 (2017) 93–104.

[19] M. Sheikholeslami, M.K. Sadoughi, Numerical modeling for Fe_3O_4 -water nanofluid flow in porous medium considering MFD viscosity, J. Mol. Liq. 242 (2017) 255–264.

[20] M. Sheikholeslami, M. Shamlooei, Magnetic source influence on nanofluid flow in porous medium considering shape factor effect, Phys. Lett. A 381 (2017) 3071–3078.

[21] M. Sheikholeslami, M. Shamlooei, Convective flow of nanofluid inside a lid driven porous cavity using CVFEM, Phys. B: Condensed Matter 521 (2017) 239–250.

[22] M. Sheikholeslami, K. Vajravelu, Forced convection heat transfer in Fe_3O_4- ethylene glycol nanofluid under the influence of Coulomb force, J. Mol. Liq. 233 (2017) 203–210.

[23] M. Sheikholeslami, H.B. Rokni, Magnetic nanofluid natural convection in the presence of thermal radiation considering variable viscosity, Europ. Phys. J. Plus 132 (2017) 238. Available from: https://doi.org/10.1140/epjp/i2017-11498-4.

[24] M. Sheikholeslami, Houman B. Rokni, Numerical simulation for impact of Coulomb force on nanofluid heat transfer in a porous enclosure in presence of thermal radiation, Int. J. Heat Mass Transfer 118 (2018) 823–831.

[25] M. Sheikholeslami, M. Shamlooei, R. Moradi, Fe_3O_4- Ethylene glycol nanofluid forced convection inside a porous enclosure in existence of Coulomb force, J. Mol. Liq. 249 (2018) 429–437.

[26] M. Sheikholeslami, S.A. Shehzad, Numerical analysis of Fe_3O_4 $-H_2O$ nanofluid flow in permeable media under the effect of external magnetic source, Int. J. Heat Mass Transfer 118 (2018) 182–192.

[27] M. Sheikholeslami, S.A. Shehzad, Magnetohydrodynamic nanofluid convection in a porous enclosure considering heat flux boundary condition, Int. J. Heat Mass Transfer 106 (2017) 1261–1269.

[28] M. Sheikholeslami, S.A. Shehzad, Thermal radiation of ferrofluid in existence of Lorentz forces considering variable viscosity, Int. J. Heat Mass Transfer 109 (2017) 82–92.

[29] M. Sheikholeslami, H.B. Rokni, Numerical modeling of nanofluid natural convection in a semi annulus in existence of Lorentz force, Comput. Methods Appl. Mech. Eng. 317 (2017) 419–430.

[30] M. Sheikholeslami, K. Vajravelu, Nanofluid flow and heat transfer in a cavity with variable magnetic field, Appl. Math. Comput. 298 (2017) 272–282.

[31] M. Sheikholeslami, H.B. Rokni, Magnetohydrodynamic CuO-water nanofluid in a porous complex shaped enclosure, ASME, J. Ther. Sci. Eng. Applicat. 9 (4) (2017) 041007. Available from: https://doi.org/10.1115/1.4035973.

[32] M. Sheikholeslami, H.B. Rokni, Nanofluid two phase model analysis in existence of induced magnetic field, Int. J. Heat Mass Transfer 107 (2017) 288–299.

[33] M. Sheikholeslami, T. Hayat, A. Alsaedi, S. Abelman, Numerical analysis of EHD nanofluid force convective heat transfer considering electric field dependent viscosity, Int. J. Heat Mass Transfer 108 (2017) 2558–2565.

[34] M. Sheikholeslami, Magnetic field influence on CuO $-H_2O$ nanofluid convective flow in a permeable cavity considering various shapes for nanoparticles, Int. J. Hydrogen Energy 42 (2017) 19611–19621.

[35] M. Sheikholeslami, M.M. Bhatti, Forced convection of nanofluid in presence of constant magnetic field considering shape effects of nanoparticles, Int. J. Heat Mass Transfer 111 (2017) 1039–1049.

[36] M. Sheikholeslami, A. Zeeshan, Analysis of flow and heat transfer in water based nanofluid due to magnetic field in a porous enclosure with constant heat flux using CVFEM, Comput. Methods Appl. Mech. Eng. 320 (2017) 68–81.

[37] M. Sheikholeslami, M. Shamlooei, Fe_3O_4- H_2O nanofluid natural convection in presence of thermal radiation, Int. J. Hydrogen Energy 42 (9) (2017) 5708–5718.

[38] M. Sheikholeslami, M.M. Bhatti, Active method for nanofluid heat transfer enhancement by means of EHD, Int. J. Heat Mass Transfer 109 (2017) 115–122.

[39] M. Sheikholeslami, T. Hayat, A. Alsaedi, Numerical study for external magnetic source influence on water based nanofluid convective heat transfer, Int. J. Heat Mass Transfer 106 (2017) 745–755.

[40] M. Sheikholeslami, A.J. Chamkha, Electrohydrodynamic free convection heat transfer of a nanofluid in a semi-annulus enclosure with a sinusoidal wall, Numer. Heat Transfer, Part A 69 (7) (2016) 781–793. https://doi.org/10.1080/10407782.2015.1090819.

[41] M. Sheikholeslami, M.K. Sadoughi, Simulation of CuO- water nanofluid heat transfer enhancement in presence of melting surface, Int. J. Heat Mass Transfer 116 (2018) 909–919.

[42] M. Sheikholeslami, H.B. Rokni, Simulation of nanofluid heat transfer in presence of magnetic field: a review, Int. J. Heat Mass Transfer 115 (2017) 1203–1233.

[43] M. Sheikholeslami, H.B. Rokni, Melting heat transfer influence on nanofluid flow inside a cavity in existence of magnetic field, Int. J. Heat Mass Transfer 114 (2017) 517–526.

[44] M. Sheikholeslami, S.A. Shehzad, Magnetohydrodynamic nanofluid convective flow in a porous enclosure by means of LBM, Int. J. Heat Mass Transfer 113 (2017) 796–805.

[45] M. Sheikholeslami, Magnetohydrodynamic nanofluid forced convection in a porous lid driven cubic cavity using Lattice Boltzmann Method, J. Mol. Liq. 231 (2017) 555–565.

[46] M.S. Kandelousi, KKL correlation for simulation of nanofluid flow and heat transfer in a permeable channel, Phys. Lett. A 378 (45) (2014) 3331–3339.

[47] M. Sheikholeslami, Effect of uniform suction on nanofluid flow and heat transfer over a cylinder, J Braz. Soc. Mech. Sci. Eng. 37 (2015) 1623–1633.

[48] M. Sheikholeslami, Lattice Boltzmann Method simulation of MHD non-Darcy nanofluid free convection, Physica B 516 (2017) 55–71.

[49] M. Sheikholeslami, Influence of magnetic field on nanofluid free convection in an open porous cavity by means of Lattice Boltzmann Method, J. Mol. Liq. 234 (2017) 364–374.

[50] M. Sheikholeslami, A.J. Chamkha, Influence of Lorentz forces on nanofluid forced convection considering Marangoni convection, J. Mol. Liq. 225 (2017) 750–757.

[51] M. Sheikholeslami, T. Hayat, A. Alsaedi, Numerical simulation of nanofluid forced convection heat transfer improvement in existence of magnetic field using Lattice Boltzmann Method, Int. J. Heat Mass Transfer 108 (2017) 1870–1883.

[52] M. Sheikholeslami, R. Ellahi, Three dimensional mesoscopic simulation of magnetic field effect on natural convection of nanofluid, Int. J. Heat Mass Transfer 89 (2015) 799–808.

[53] M. Sheikholeslami, S. Abelman, Two phase simulation of nanofluid flow and heat transfer in an annulus in the presence of an axial magnetic field, IEEE Trans. Nanotechnol. 14 (3) (2015) 561–569.
[54] M. Sheikholeslami, R. Ellahi, Simulation of ferrofluid flow for magnetic drug targeting using Lattice Boltzmann method, J. Zeitschrift Fur Naturforschung A 70 (2) (2015) 115–124.
[55] M. Sheikholeslami, M. Seyednezhad, Lattice Boltzmann Method simulation for CuO-water nanofluid flow in a porous enclosure with hot obstacle, J. Mol. Liq. 243 (2017) 249–256.
[56] M. Sheikholeslami, M. Sadoughi, Mesoscopic method for MHD nanofluid flow inside a porous cavity considering various shapes of nanoparticles, Int. J. Heat Mass Transfer 113 (2017) 106–114.
[57] M. Sheikholeslami, A. Zeeshan, Mesoscopic simulation of CuO-H_2O nanofluid in a porous enclosure with elliptic heat source, Int. J. Hydrogen Energy 42 (22) (2017) 15393–15402.
[58] M. Sheikholeslami, S.A. Shehzad, CVFEM for influence of external magnetic source on $Fe_3O_4 - H_2O$ nanofluid behavior in a permeable cavity considering shape effect, Int. J. Heat Mass Transfer 115 (2017) 180–191.
[59] M. Sheikholeslami, M. Seyednezhad, Nanofluid heat transfer in a permeable enclosure in presence of variable magnetic field by means of CVFEM, Int. J. Heat Mass Transfer 114 (2017) 1169–1180.
[60] M. Sheikholeslami, H.B. Rokni, Free convection of CuO-H_2O nanofluid in a curved porous enclosure using mesoscopic approach, Int. J. Hydrogen Energy 42 (22) (2017) 14942–14949.
[61] M. Sheikholeslami, Numerical investigation for CuO-H_2O nanofluid flow in a porous channel with magnetic field using mesoscopic method, J. Mol. Liq. 249 (2018) 739–746.
[62] M. Sheikholeslami, H.B. Rokni, Influence of melting surface on MHD nanofluid flow by means of two phase model, Chin. J. Phys. 55 (2017) 1352–1360.
[63] M. Sheikholeslami, H.B. Rokni, Effect of melting heat transfer on nanofluid flow in existence of magnetic field considering Buongiorno Model, Chin. J. Phys. 55 (2017) 1115–1126.
[64] M. Sheikholeslami, T. Hayat, A. Alsaedi, MHD free convection of Al_2O_3–water nanofluid considering thermal radiation: a numerical study, Int. J. Heat Mass Transfer 96 (2016) 513–524.
[65] Milad Darzi, M.K. Sadoughi, M. Sheikholeslami, Condensation of nano-refrigerant inside a horizontal tube, Phys. B: Condens. Matter 537 (2018) 33–39.
[66] Rakesh Kumar, Ravinder Kumar, Sabir Ali Shehzad, M. Sheikholeslami, Rotating frame analysis of radiating and reacting ferro-nanofluid considering Joule heating and viscous dissipation, Int. J. Heat Mass Transfer 120 (2018) 540–551.
[67] M. Sheikholeslami, D.D. Ganji, Numerical modeling of magnetohydrodynamic CuO-water transportation inside a porous cavity considering shape factor effect, Colloids Surf. A 529 (2017) 705–714.
[68] M. Sheikholeslami, Numerical simulation for solidification in a LHTESS by means of nano-enhanced PCM, J. Taiwan Inst. Chem. Eng. 86 (2018) 25–41.
[69] M. Sheikholeslami, Numerical modeling of nano enhanced PCM solidification in an enclosure with metallic fin, J. Mol. Liq. 259 (2018) 424–438.
[70] M. Sheikholeslami, A. Ghasemi, Solidification heat transfer of nanofluid in existence of thermal radiation by means of FEM, Int. J. Heat Mass Transfer 123 (2018) 418–431.
[71] M. Sheikholeslami, S.A. Shehzad, CVFEM simulation for nanofluid migration in a porous medium using Darcy model, Int. J. Heat Mass Transfer 122 (2018) 1264–1271.
[72] M. Sheikholeslami, M. Darzi, M.K. Sadoughi, Heat transfer improvement and pressure drop during condensation of refrigerant-based nanofluid; an experimental procedure, Int. J. Heat Mass Transfer 122 (2018) 643–650.
[73] M. Sheikholeslami, H.B. Rokni, CVFEM for effect of Lorentz forces on nanofluid flow in a porous complex shaped enclosure by means of non-equilibrium model, J. Mol. Liq. 254 (2018) 446–462.
[74] M. Sheikholeslami, S.A. Shehzad, Z. Li, Water based nanofluid free convection heat transfer in a three dimensional porous cavity with hot sphere obstacle in existence of Lorenz forces, Int. J. Heat Mass Transfer 125 (2018) 375–386.
[75] M. Sheikholeslami, M. Jafaryar, D.D. Ganji, Z. Li, Exergy loss analysis for nanofluid forced convection heat transfer in a pipe with modified turbulators, J. Mol. Liq. 262 (2018) 104–110.
[76] M. Sheikholeslami, M. Jafaryar, Z. Li, Nanofluid turbulent convective flow in a circular duct with helical turbulators considering CuO nanoparticles, Int. J. Heat Mass Transfer 124 (2018) 980–989.
[77] M. Sheikholeslami, Z. Li, M. Shamlooei, Nanofluid MHD natural convection through a porous complex shaped cavity considering thermal radiation, Phys. Lett. A 382 (2018) 1615–1632.
[78] M. Sheikholeslami, H.B. Rokni, Magnetic nanofluid flow and convective heat transfer in a porous cavity considering Brownian motion effects, Phys. Fluids 30 (1) (2018). Available from: https://doi.org/10.1063/1.5012517.
[79] M. Sheikholeslami, S.A. Shehzad, Simulation of water based nanofluid convective flow inside a porous enclosure via non-equilibrium model, Int. J. Heat Mass Transfer 120 (2018) 1200–1212.
[80] M. Sheikholeslami, S.A. Shehzad, Non-Darcy free convection of Fe_3O_4-water nanoliquid in a complex shaped enclosure under impact of uniform Lorentz force, Chin. J. Phys. 56 (2018) 270–281.
[81] M. Sheikholeslami, M. Seyednezhad, Simulation of nanofluid flow and natural convection in a porous media under the influence of electric field using CVFEM, Int. J. Heat Mass Transfer 120 (2018) 772–781.
[82] M. Sheikholeslami, M. Shamlooei, R. Moradi, Numerical simulation for heat transfer intensification of nanofluid in a permeable curved enclosure considering shape effect of Fe_3O_4 nanoparticles, Chem. Eng. Process. Process Intensif. 124 (2018) 71–82.
[83] M. Sheikholeslami, T. Hayat, T. Muhammad, A. Alsaedi, MHD forced convection flow of nanofluid in a porous cavity with hot elliptic obstacle by means of Lattice Boltzmann method, Int. J. Mech. Sci. 135 (2018) 532–540.
[84] M. Sheikholeslami, Numerical investigation of nanofluid free convection under the influence of electric field in a porous enclosure, J. Mol. Liq. 249 (2018) 1212–1221.

[85] M. Sheikholeslami, CuO-water nanofluid flow due to magnetic field inside a porous media considering Brownian motion, J. Mol. Liq. 249 (2018) 921–929.
[86] M. Sheikholeslami, M. Barzegar Gerdroodbary, S. Valiallah Mousavi, D.D. Ganji, R. Moradi, Heat transfer enhancement of ferrofluid inside an 90o elbow channel by non-uniform magnetic field, J. Magnet. Magnet. Mater. 460 (2018) 302–311.
[87] M. Sheikholeslami, CuO-water nanofluid free convection in a porous cavity considering Darcy law, Eur. Phys. J. Plus 132 (2017) 55. Available from: https://doi.org/10.1140/epjp/i2017-11330-3.
[88] M. Sheikholeslami, T. Hayat, A. Alsaedi, Numerical simulation for forced convection flow of MHD CuO-H_2O nanofluid inside a cavity by means of LBM, J. Mol. Liq. 249 (2018) 941–948.
[89] M. Sheikholeslami, H.R. Kataria, A.S. Mittal, Effect of thermal diffusion and heat-generation on MHD nanofluid flow past an oscillating vertical plate through porous medium, J. Mol. Liq. 257 (2018) 12–25.
[90] M. Sheikholeslami, S.A. Shehzad, Z. Li, Nanofluid heat transfer intensification in a permeable channel due to magnetic field using Lattice Boltzmann method, Phys. B (2018). Available from: https://doi.org/10.1016/j.physb.2018.03.036.
[91] M. Sheikholeslami, A. Zeeshan, Numerical simulation of Fe_3O_4 -water nanofluid flow in a non-Darcy porous media, Int. J. Numer. Methods Heat Fluid Flow 28 (3) (2018) 641–660. Available from: https://doi.org/10.1108/HFF-04-2017-0160.
[92] M. Sheikholeslami, D.D. Ganji, Influence of electric field on Fe_3O_4- water nanofluid radiative and convective heat transfer in a permeable enclosure, J. Mol. Liq. 250 (2018) 404–412.
[93] M. Sheikholeslami, D.D. Ganji, Numerical approach for magnetic nanofluid flow in a porous cavity using CuO nanoparticles, Mater. Des. 120 (2017) 382–393.

CHAPTER

10

Influence of Shape Factor on Nanofluid Heat Transfer Improvement Using CVFEM

10.1 INTRODUCTION

Most conventional heat transfer fluids, such as water, ethylene glycol and engine oil, have limited capabilities in terms of thermal properties, which in turn, may impose severe restrictions in many thermal applications. And in spite of considerable research and efforts deployed, a clear and urgent need does exist to date to develop new strategies in order to improve the effective thermal behaviors of these fluids. On the other hand, most solids, in particular metals, have thermal conductivities much higher, say by 1–3 orders of magnitude, compared to that of liquids. Hence, one can then expect that fluid containing solid particles may significantly increase its conductivity.

Sheikholeslami [1] investigated the effect of nanoparticle shape on convective flow in a permeable cavity. Sheikholeslami and Bhatti [2] studied the forced convection of nanofluid in the presence of a constant magnetic field considering shape effects of nanoparticles. Sheikholeslami and Shamlooei [3] reported the magnetic source influence on nanofluid flow in a porous medium considering shape factor effect. Sheikholeslami and Ganji [4] simulated the shape factor effect on magnetohydrodynamic CuO-water transportation inside a porous cavity. In recent decade, various methods have been utilized to show nanofluid behavior [5–86].

10.2 FORCED CONVECTION OF NANOFLUID IN THE PRESENCE OF CONSTANT MAGNETIC FIELD CONSIDERING SHAPE EFFECTS OF NANOPARTICLES

10.2.1 Problem Definition

Fig. 10.1 depicts the geometry, boundary condition, and sample element. The lower wall has the velocity of U_{Lid} and others are stationary. The lower wall has constant temperature T_h and the temperature of other walls is T_c. Horizontal magnetic field has been applied. Nanofluid forced convection heat transfer in a porous semiannulus is investigated.

10.2.2 Governing Equation

2D steady convective flow of nanofluid in a porous media is considered in the presence of a constant magnetic field. The PDEs equations are:

$$\frac{\partial u}{\partial x} + \frac{\partial v}{\partial y} = 0 \tag{10.1}$$

$$\rho_{nf}\left(v\frac{\partial u}{\partial y} + \frac{\partial u}{\partial x}u\right) = \left[-\sigma_{nf}B_y^2 u + \sigma_{nf}B_x B_y v + \left(\frac{\partial^2 u}{\partial y^2} + \frac{\partial^2 u}{\partial x^2}\right)\mu_{nf} - \frac{\partial P}{\partial x}\right] - \frac{\mu_{nf}}{K}u \tag{10.2}$$

Application of Control Volume based Finite Element Method (CVFEM) for Nanofluid Flow and Heat Transfer.
DOI: https://doi.org/10.1016/B978-0-12-814152-6.00010-2

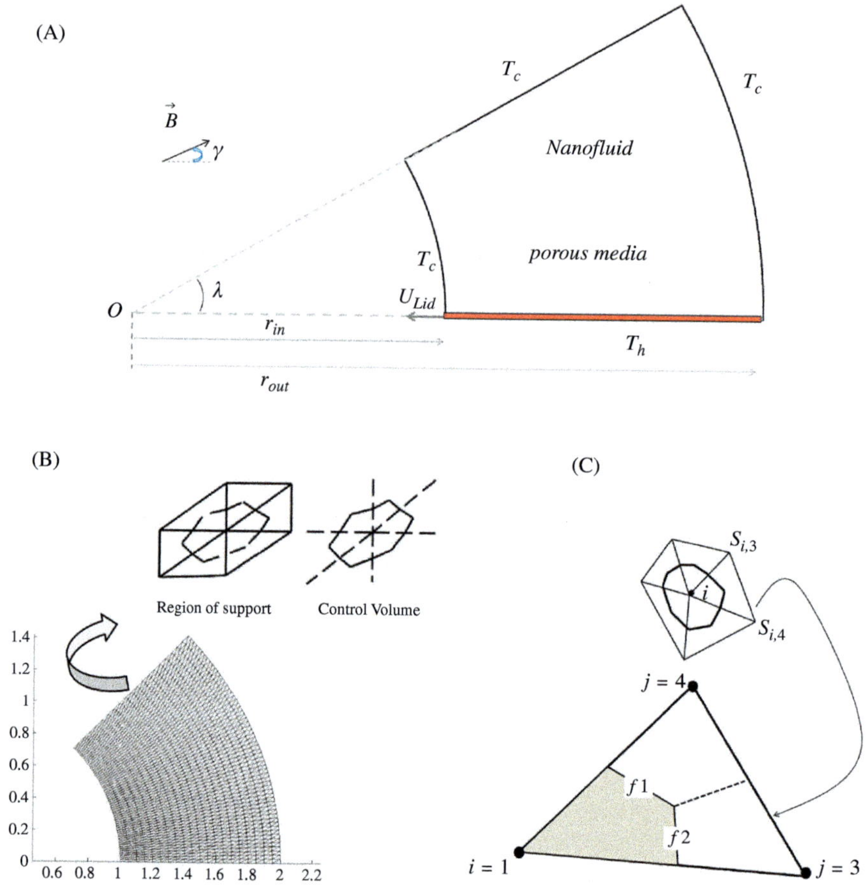

FIGURE 10.1 (A) Geometry and the boundary conditions with (B) the mesh of Geometry considered in this work; (C) a sample triangular element and its corresponding control volume.

$$\rho_{nf}\left(\frac{\partial v}{\partial x}u+\frac{\partial v}{\partial y}v\right)=+\mu_{nf}\left(\frac{\partial^2 v}{\partial x^2}+\frac{\partial^2 v}{\partial y^2}\right)-\frac{\partial P}{\partial y}+B_y\sigma_{nf}B_x u-B_x\sigma_{nf}B_x v-\frac{\mu_{nf}}{K}v, \tag{10.3}$$

$$B_x=B_o\cos\lambda,\;\; B_y=B_o\sin\lambda$$

$$(\rho C_p)_{nf}\left(v\frac{\partial T}{\partial y}+u\frac{\partial T}{\partial x}\right)=k_{nf}\left(\frac{\partial^2 T}{\partial x^2}+\frac{\partial^2 T}{\partial y^2}\right), \tag{10.4}$$

$(\rho C_p)_{nf}$, ρ_{nf}, and σ_{nf} are defined as:

$$(\rho C_p)_{nf}=(\rho C_p)_f(1-\phi)+(\rho C_p)_s\phi \tag{10.5}$$

$$\rho_{nf}=\rho_f(1-\phi)+\rho_s\phi \tag{10.6}$$

$$\frac{\sigma_{nf}}{\sigma_f}=1+\frac{3\left(\frac{\sigma_s}{\sigma_f}-1\right)\phi}{\left(\frac{\sigma_s}{\sigma_f}+2\right)-\left(\frac{\sigma_s}{\sigma_f}-1\right)\phi} \tag{10.7}$$

The KKL (Koo–Kleinstreuer–Li) correlation has been utilized for viscosity of nanofluid

$$\mu_{eff} = \mu_{static} + \mu_{Brownian} = \mu_{static} + \frac{k_{Brownian}}{k_f} \times \frac{\mu_f}{\Pr_f}$$

$$k_{Brownian} = 5 \times 10^4 \phi \rho_f c_{p,f} \sqrt{\frac{\kappa_b T}{\rho_p d_p}} g'(T, \phi, d_p) \tag{10.8}$$

$$g'(T, \phi, d_p) = \left(a_1 + a_2 \ln(d_p) + a_3 \ln(\phi) + a_4 \ln(\phi) \ln(d_p) + a_5 \ln(d_p)^2\right) \ln(T) + \left(a_6 + a_7 \ln(d_p) + a_8 \ln(\phi) + a_9 \ln(\phi) \ln(d_p) + a_{10} \ln(d_p)^2\right)$$

The related coefficient and properties of CuO-water nanofluid are presented in Tables 10.1 and 10.2. The Maxwell model and Hamilton–Crosser model for irregular particle geometries by introducing a shape factor can be expressed as

$$\frac{k_{nf}}{k_f} = \frac{k_p + (m+1)k_f - (m+1)\phi(k_f - k_p)}{k_p + (m+1)k_f + \phi(k_f - k_p)} \tag{10.9}$$

in which k_p and k_f are the conductivities of the particle material and the base fluid. In this equation "m" is shaper factor. Table 10.3 shows the different values of shape factors for various shapes of nanoparticles. Vorticity and stream function should be used to eliminate pressure source terms:

$$\omega + \frac{\partial u}{\partial y} - \frac{\partial v}{\partial x} = 0, \quad \frac{\partial \psi}{\partial x} = -v, \quad \frac{\partial \psi}{\partial y} = u \tag{10.10}$$

Introducing dimensionless quantities:

$$U = \frac{u}{U_{Lid}}, V = \frac{v}{U_{Lid}}, \ \theta = \frac{T - T_c}{\Delta T}, \Delta T = T_h - T_c, (X, Y) = \frac{(x, y)}{L}, \Psi = \frac{\psi L}{U_{Lid}}, \Omega - \frac{\omega}{LU_{Lid}} \tag{10.11}$$

TABLE 10.1 The Coefficient Values of CuO – Water Nanofluid

Coefficient Values	***CuO* – Water**
a_1	−26.5933108
a_2	−0.403818333
a_3	−33.3516805
a_4	−1.915825591
a_5	6.421858E-02
a_6	48.40336955
a_7	−9.787756683
a_8	190.245610009
a_9	10.9285386565
a_{10}	−0.72009983664

TABLE 10.2 Thermophysical Properties of Water and Nanoparticles

	$\rho(kg/m^3)$	$C_p(j/kgk)$	$k(W/m.k)$	$\beta \times 10^5(K^{-1})$	$d_p(nm)$	$\sigma(\Omega \cdot m)^{-1}$
Water	997.1	4179	0.613	21	-	0.05
CuO	6500	540	18	29	45	10^{-10}

TABLE 10.3 The Values of Shape Factor of Different Shapes of Nanoparticles

	Spherical	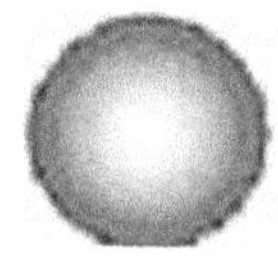	3
m	Platelet		5.7
	Cylinder		4.8
	Brick	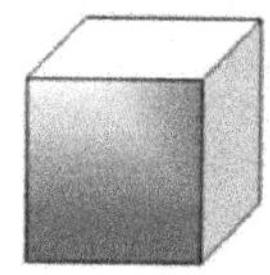	3.7

The final formulae are:

$$\frac{\partial^2 \Psi}{\partial Y^2} + \frac{\partial^2 \Psi}{\partial X^2} = -\Omega, \tag{10.12}$$

$$U\frac{\partial \Omega}{\partial X} + \frac{\partial \Omega}{\partial Y}V = \frac{1}{\mathrm{Re}}\frac{A_5}{A_1}\left(\frac{\partial^2 \Omega}{\partial Y^2} + \frac{\partial^2 \Omega}{\partial X^2}\right) + \frac{Ha^2}{\mathrm{Re}}\frac{A_6}{A_1}\left(\frac{\partial U}{\partial X}B_x B_y - \frac{\partial V}{\partial X}B_x{}^2 + \frac{\partial U}{\partial Y}B_y^2 - \frac{\partial V}{\partial Y}B_x B_y\right) - \frac{1}{\mathrm{Re}\,Da}\frac{A_5}{A_1}\Omega, \tag{10.13}$$

$$\frac{\partial \theta}{\partial X}U + \frac{\partial \theta}{\partial Y}V = \frac{1}{\mathrm{Re}\,\mathrm{Pr}}\frac{A_4}{A_2}\left(\frac{\partial^2 \theta}{\partial X^2} + \frac{\partial^2 \theta}{\partial Y^2}\right) \tag{10.14}$$

where dimensionless and constants parameters are illustrated as:

$$\mathrm{Re} = \frac{\rho_f U_{Lid} L}{\mu_f},\ Ha = LB_0\sqrt{\sigma_f/\mu_f},\ A_1 = \frac{\rho_{nf}}{\rho_f},\ A_2 = \frac{(\rho C_p)_{nf}}{(\rho C_p)_f},\ A_4 = \frac{k_{nf}}{k_f},\ A_5 = \frac{\mu_{nf}}{\mu_f},\ A_6 = \frac{\sigma_{nf}}{\sigma_f},\ Da = \frac{K}{L^2} \tag{10.15}$$

and boundary conditions are:

$$\begin{aligned} &\theta = 1.0 && \text{on bottom wall} \\ &\theta = 0.0 && \text{on other walls} \\ &\Psi = 0.0 && \text{on all walls} \end{aligned} \tag{10.16}$$

Local and average Nusselt over the hot wall can be calculated as:

$$Nu_{loc} = \frac{\partial \theta}{\partial y}\left(\frac{k_{nf}}{k_f}\right) \tag{10.17}$$

$$Nu_{ave} = \frac{1}{L}\int_{r_{in}}^{r_{out}} Nu_{loc}\,dx \tag{10.18}$$

10.2.3 Effects of Active Parameters

In this chapter, the effect of uniform magnetic field on nanofluid forced convection in a porous semiannulus is investigated using the Control volume-based finite element method. Numerical outputs are illustrated for various values of Hartmann number ($Ha = 0$ to 40), Darcy number ($Da = 0.01$ to 100), Reynolds number ($\mathrm{Re} = 100$ to 600), and volume fraction of CuO ($\phi = 0\%$ to 4%). Table 10.4 shows the shape of nanoparticle effect on Nusselt number. Maximum values of Nusselt number are obtained if Platelet shape is selected for nanoparticles. So, Platelet shape is selected for further investigation.

TABLE 10.4 Nusselt Number for Various Shapes of Nanoparticles When Re = 600, $\phi = 0.04$

Da	Ha	Spherical	Brick	Cylinder	Platelet
0.01	0	14.45798	14.77039	15.25743	15.65244
0.01	40	12.44094	12.70772	13.12349	13.46058
100	0	15.1156	15.44261	15.95241	16.3659
100	40	12.71177	12.98315	13.40598	13.74869

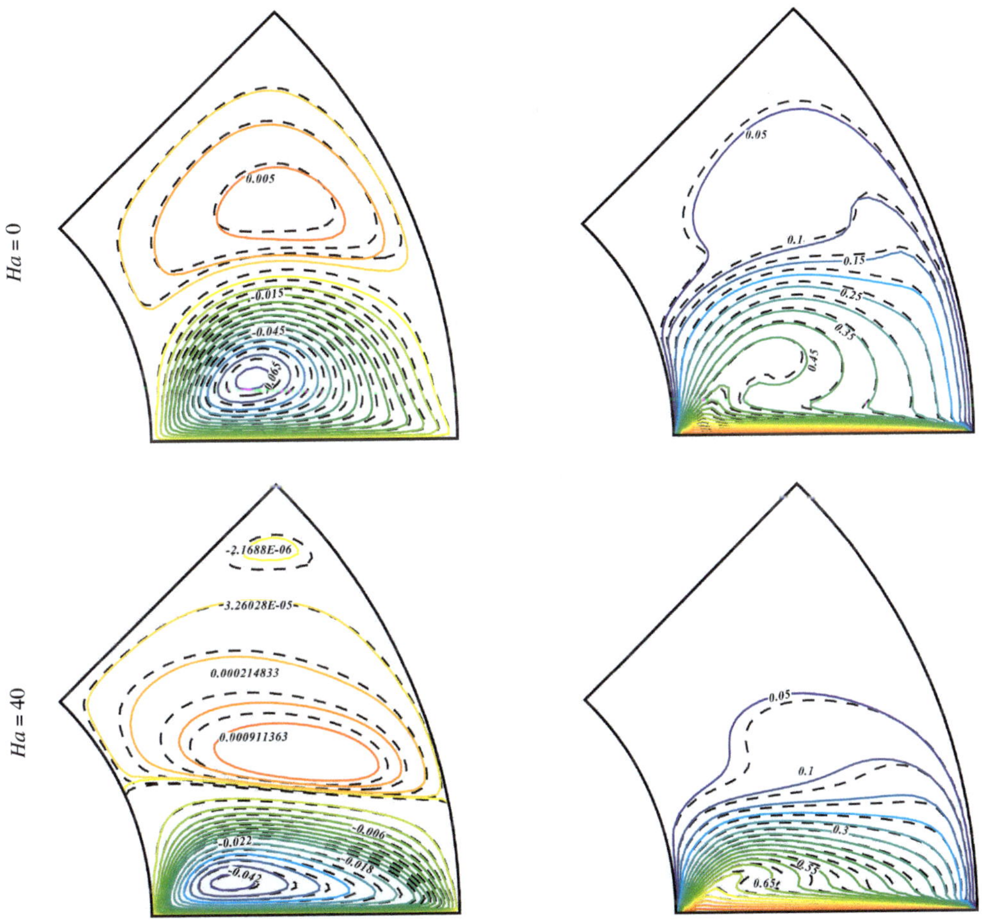

FIGURE 10.2 Influence of nanofluid volume fraction on streamline (left) and isotherm (right) contours (nanofluid ($\phi = 0.04$) (—) and pure fluid($\phi = 0$) (- - -)) when Re = 600, $Da = 100$.

Fig. 10.2 demonstrates the effect of nanofluid volume fraction on isotherms and streamlines. Augmenting nanofluid volume fraction leads to an increase in thermal boundary layer thickness. The nanofluid velocity increases due to the increase in the solid movements. Figs. 10.3–10.5 demonstrate the influence of Darcy, Hartmann, and Reynolds numbers on hydrothermal treatment. In the absence of magnetic field, when Darcy and Reynolds number are low, only one eddy exists in streamline and isotherms are parallel together. As the Hartmann number increases, the main eddy converts to two smaller ones. As the Darcy number increases, the convective heat transfer enhances and three layers are generated for streamlines in the presence of a magnetic field. As the Reynolds number enhances, isotherms become denser to the lid wall due to increment of convective mode. Also the absolute values of stream function augment with the rise of Reynolds number. At Re = 600, two eddies are generated in the absence of a magnetic field. The upper one stretches to the right side with an increase of Darcy number. These eddies convert to three eddies in the presence of a magnetic field. Increasing Lorentz forces make the isotherms becomes less dense. Also velocity decreases with rise of Hartmann number.

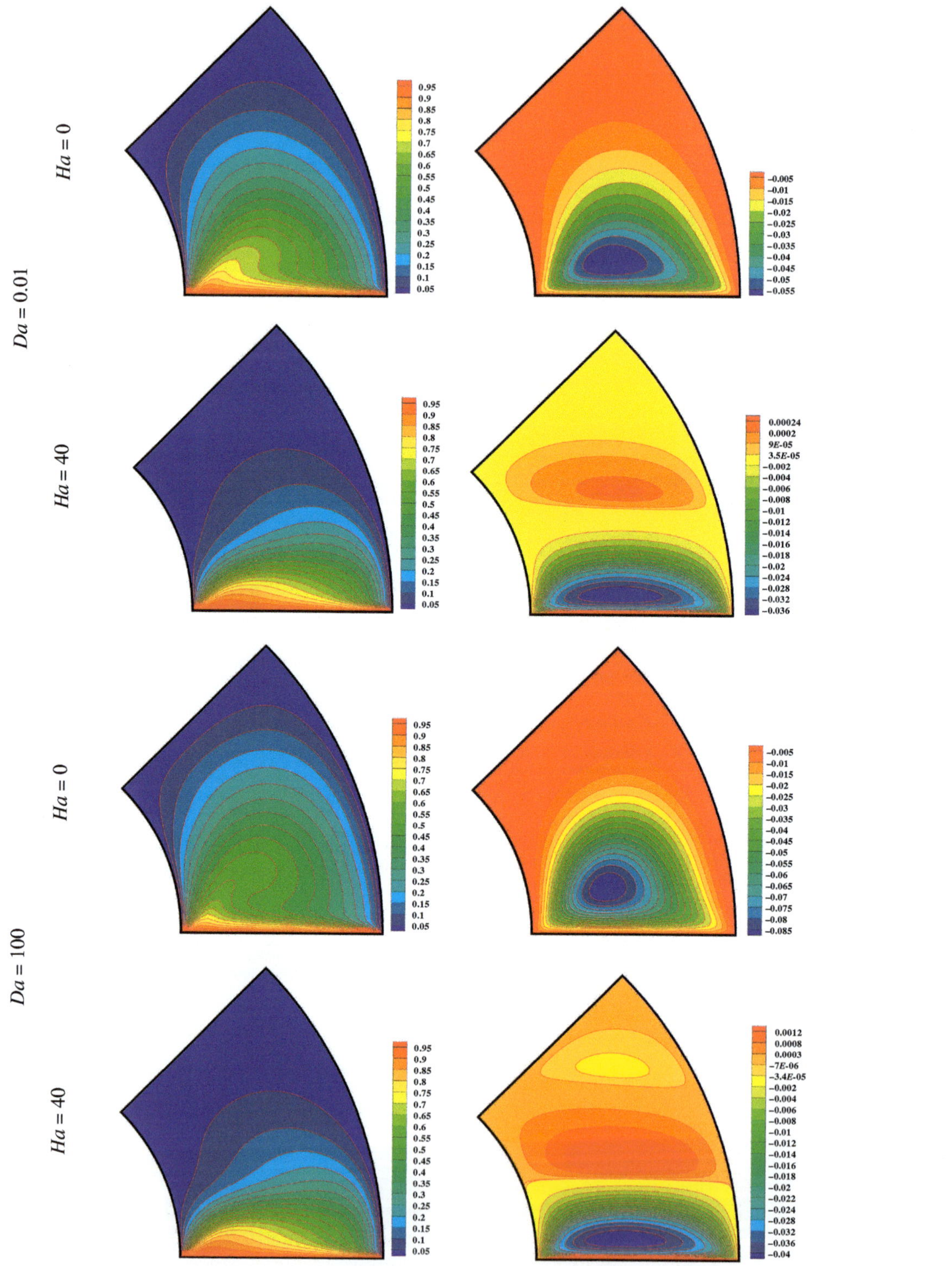

FIGURE 10.3 Effects of Darcy and Hartmann numbers on isotherm (left) and streamlines (right) contours when $\phi = 0.04$, Re = 100.

The influence of important parameters on average Nusselt number is shown in Fig. 10.6. The correlation for average Nusselt number is as follows:

$$Nu_{ave} = 10.74 + 0.7Da^* + 1.9\mathrm{Re}^* - 0.34Ha^* - 0.045Da^*\,\mathrm{Re}^* - 0.028Da^*\,Ha^* + 0.05\mathrm{Re}^*\,Ha^* + 0.28Da^{*2} - 0.18\mathrm{Re}^{*2} - 0.09Ha^{*2} \tag{10.19}$$

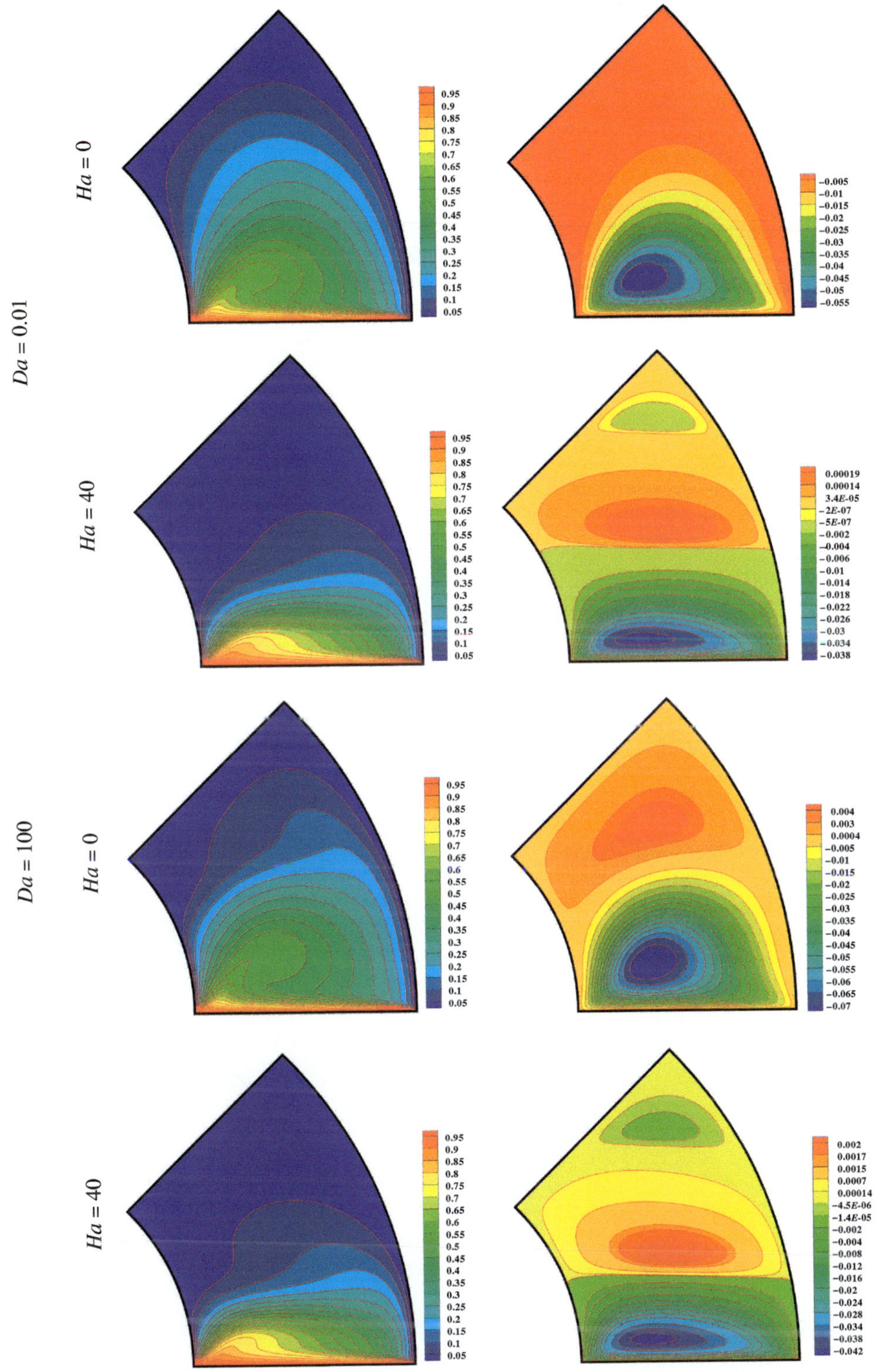

FIGURE 10.4 Effects of Darcy and Hartmann numbers on isotherm (left) and streamline (right) contours when $\phi = 0.04, \mathrm{Re} = 300$

where $\mathrm{Re}^* = 0.01\mathrm{Re}, Da^* = 0.01Da,\ Ha^* = 0.1Ha$. Rate of heat transfer enhances with the rise of Reynolds and Darcy numbers because of the increase in convective heat transfer. Adding nanoparticles causes the Nusselt number to increase due to the increase in thermal conductivity of nanofluid. As the Hartmann number increases, the temperature gradient decreases and in turn the Nusselt number reduces with the rise of Lorentz forces.

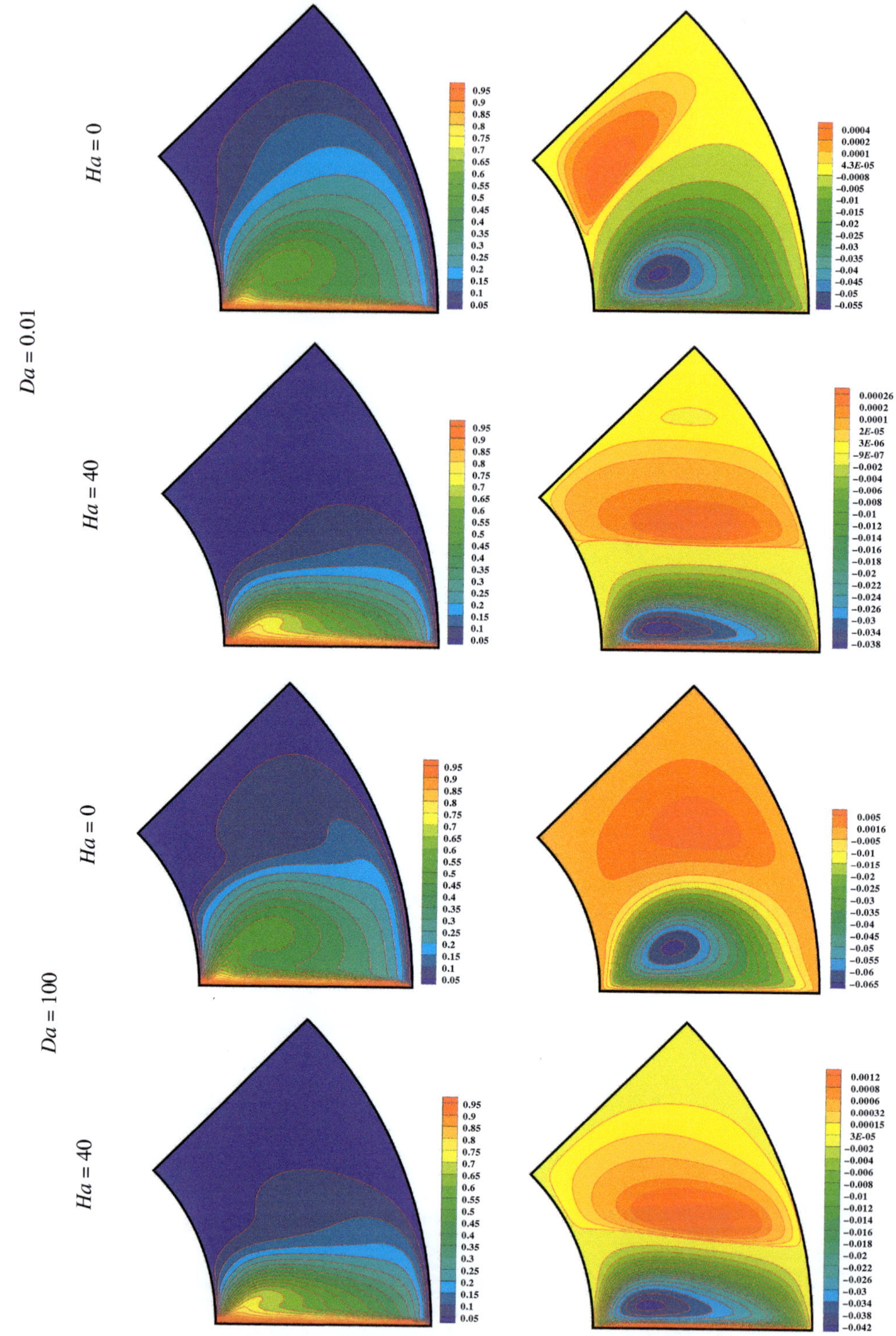

FIGURE 10.5 Effects of Darcy and Hartmann numbers on isotherm (left) and streamline (right) contours when $\phi = 0.04$, Re = 600

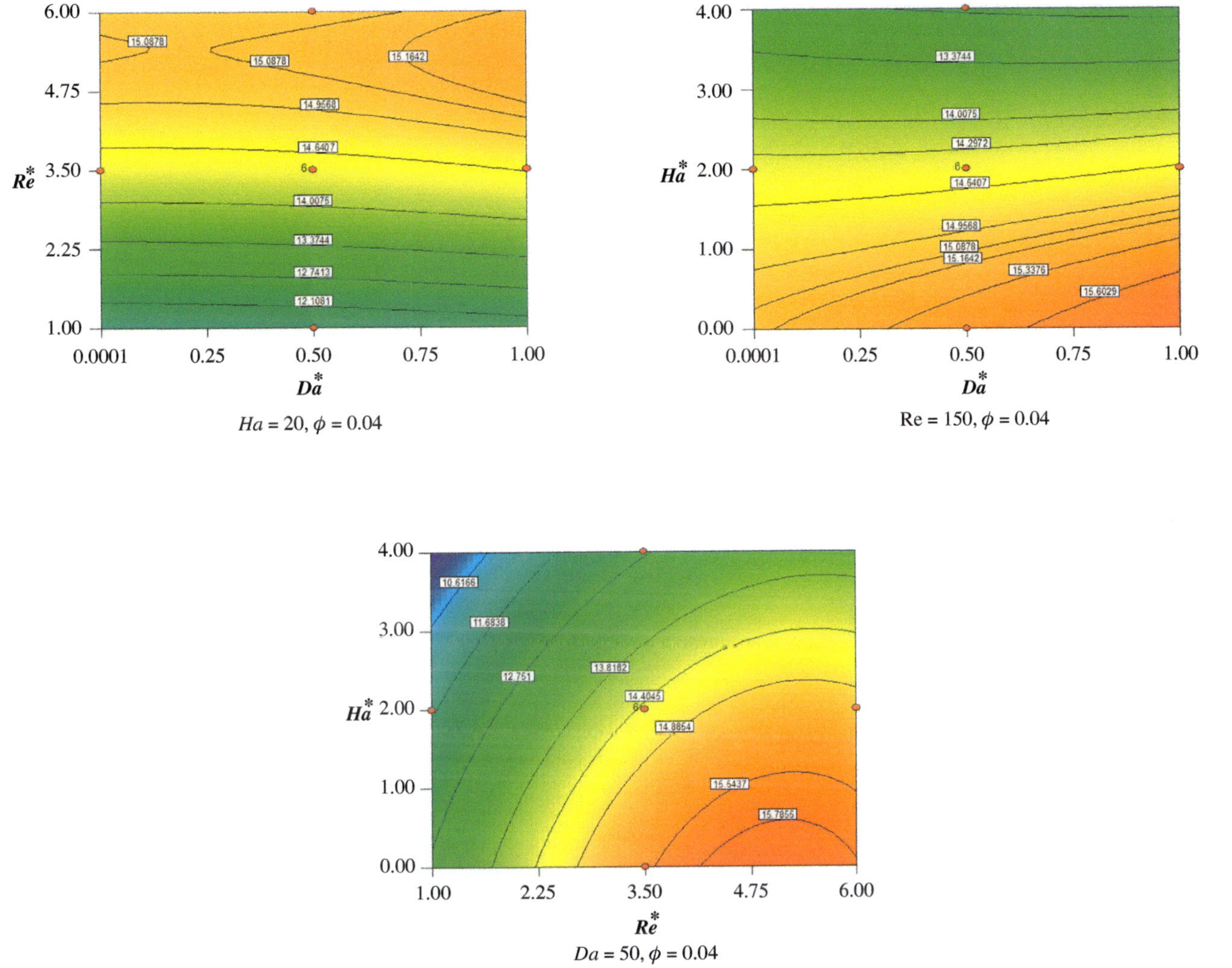

FIGURE 10.6 Influences of the Darcy, Reynolds number, and Hartmann numbers on average Nusselt number.

10.3 EFFECT OF SHAPE FACTOR ON FE_3O_4-WATER NANOFLUID FORCED CONVECTION IN THE PRESENCE OF EXTERNAL MAGNETIC FIELD

10.3.1 Problem Definition

Sample element, boundary condition, and geometry are demonstrated in Fig. 10.7. The bottom wall is hot and the other walls are cold. Also the bottom wall can move horizontally. The porous cavity is filled with nanofluid and affected by a horizontal magnetic field.

10.3.2 Governing Equation

2D steady convective non-Darcy flow of nanofluid is considered in the presence of a constant magnetic field. The PDEs are:

$$\frac{\partial v}{\partial y} + \frac{\partial u}{\partial x} = 0 \tag{10.20}$$

$$\left[-\sigma_{nf}B_y^2 u + \sigma_{nf}B_xB_y v + \left(\frac{\partial^2 u}{\partial y^2} + \frac{\partial^2 u}{\partial x^2}\right)\mu_{nf} - \frac{\partial P}{\partial x}\right] - \frac{\mu_{nf}}{K}u = \rho_{nf}\left(v\frac{\partial u}{\partial y} + \frac{\partial u}{\partial x}u\right) \tag{10.21}$$

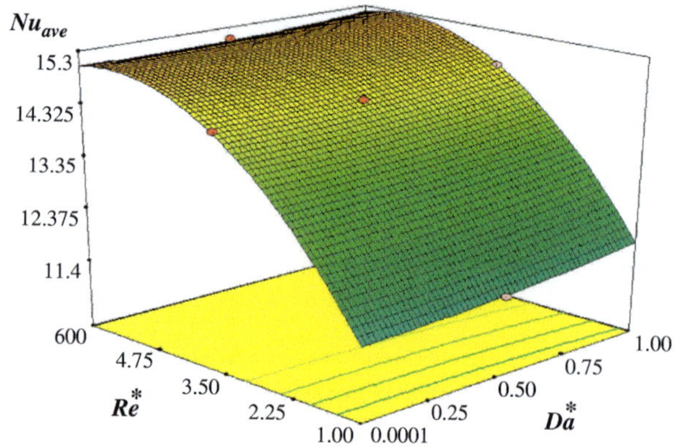

$Ha = 20, \phi = 0.04$

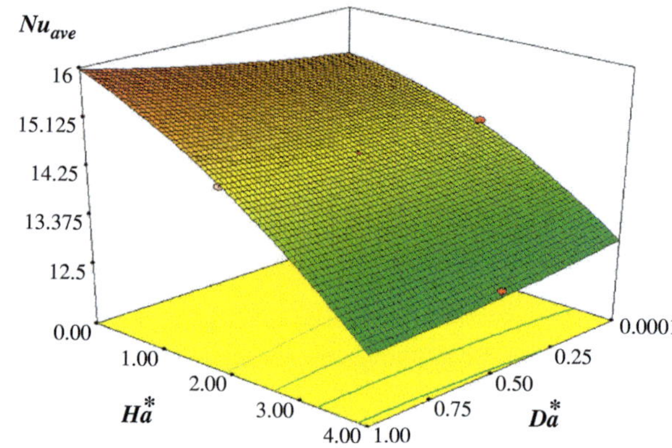

$\mathrm{Re} = 150, \phi = 0.04$

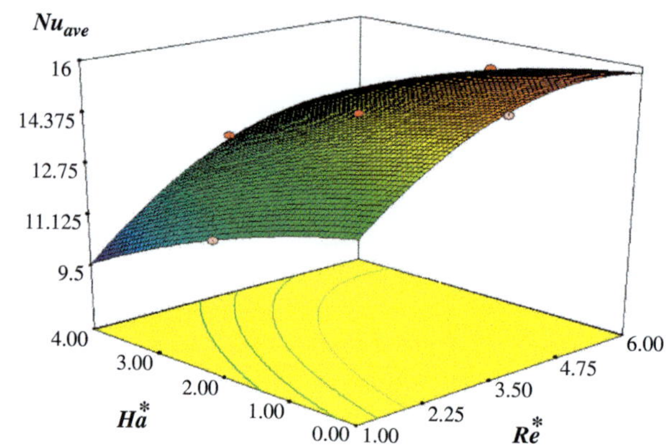

$Da = 50, \phi = 0.04$

FIGURE 10.6 (Continued).

$$\mu_{nf}\left(\frac{\partial^2 v}{\partial y^2}+\frac{\partial^2 v}{\partial x^2}\right)-\frac{\partial P}{\partial y}-B_x\sigma_{nf}B_x v+B_y\sigma_{nf}B_x u-\frac{\mu_{nf}}{K}v=\rho_{nf}\left(\frac{\partial v}{\partial x}u+\frac{\partial v}{\partial y}v\right), \quad (10.22)$$
$$B_x = B_o\cos\lambda, \quad B_y = B_o\sin\lambda$$

$$(\rho C_p)_{nf}\left(v\frac{\partial T}{\partial y}+u\frac{\partial T}{\partial x}\right)=k_{nf}\left(\frac{\partial^2 T}{\partial x^2}+\frac{\partial^2 T}{\partial y^2}\right), \quad (10.23)$$

$(\rho C_p)_{nf}$, ρ_{nf}, k_{nf}, and σ_{nf} are defined as:

$$(\rho C_p)_{nf}=(\rho C_p)_f(1-\phi)+(\rho C_p)_s\phi \quad (10.24)$$

$$\rho_{nf}=\rho_f(1-\phi)+\rho_s\phi \quad (10.25)$$

$$\frac{\sigma_{nf}}{\sigma_f}=1+\frac{3\left(\frac{\sigma_s}{\sigma_f}-1\right)\phi}{\left(\frac{\sigma_s}{\sigma_f}+2\right)-\left(\frac{\sigma_s}{\sigma_f}-1\right)\phi} \quad (10.26)$$

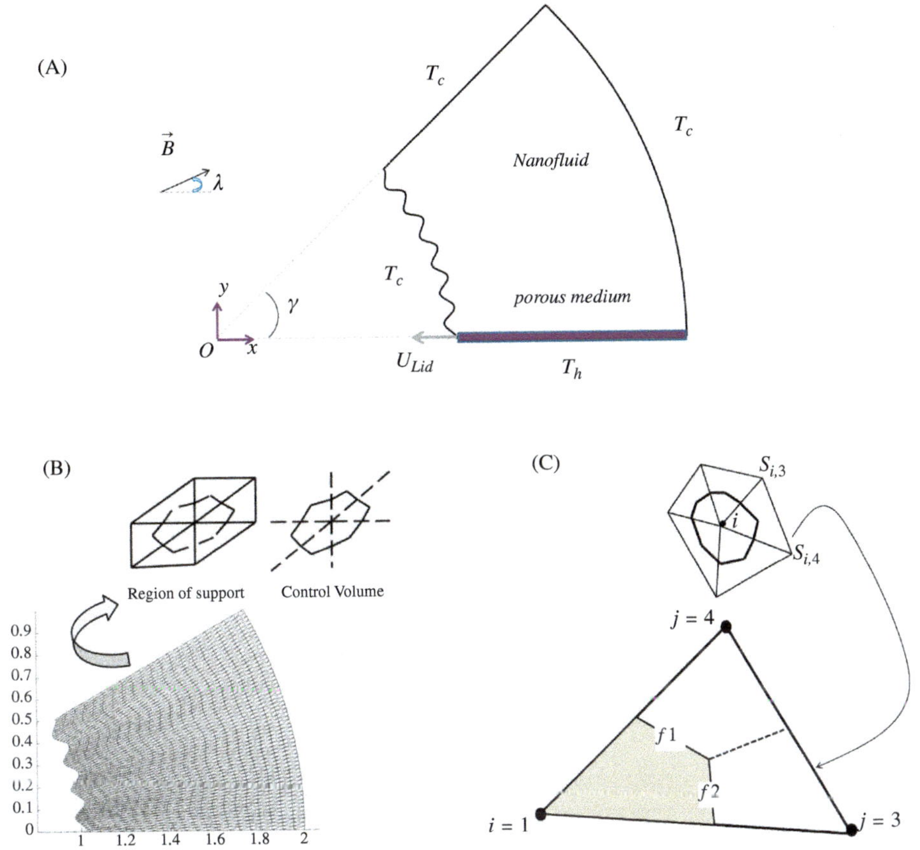

FIGURE 10.7 (A) Geometry and the boundary conditions with (B) the mesh of Geometry considered in this work; (C) a sample triangular element and its corresponding control volume.

TABLE 10.5 Thermophysical Properties of Water and Nanoparticles

	$\rho(kg/m^3)$	$C_p(j/kgk)$	$k(W/m.k)$	$d_p(nm)$	$\sigma(\Omega\cdot m)^{-1}$
Pure water	997.1	4179	0.613	-	0.05
Fe_3O_4	5200	670	6	7.5	25,000

According to experimental outputs, μ_{nf} is obtained as follows:

$$\mu_{nf} = \left(0.035B^2 + 3.1B - 27886.4807\phi^2 + 4263.02\phi + 316.0629\right)e^{-0.01T} \tag{10.27}$$

Properties of nanoparticles and base fluid are provided in Table 10.5. k_{nf} can be expressed as:

$$\frac{k_{nf}}{k_f} = \frac{-m(k_f - k_p)\phi + (k_p - k_f)\phi + mk_f + k_p + k_f}{mk_f + (k_f - k_p)\phi + k_f + k_p} \tag{10.28}$$

Different values of shape factors for various shapes of nanoparticles are illustrated in Table 10.3. Vorticity and stream function should be employed to eliminate pressure source terms:

$$\omega + \frac{\partial u}{\partial y} - \frac{\partial v}{\partial x} = 0, \quad \frac{\partial \psi}{\partial x} = -v, \quad \frac{\partial \psi}{\partial y} = u \tag{10.29}$$

Introducing dimensionless quantities:

$$U=\frac{u}{U_{Lid}},V=\frac{v}{U_{Lid}},\ \theta=\frac{T-T_c}{\Delta T},\Delta T=T_h-T_c,(X,Y)=\frac{(x,y)}{L},\Psi=\frac{\psi L}{U_{Lid}},\Omega=\frac{\omega}{LU_{Lid}} \tag{10.30}$$

The final formulae are:

$$\frac{\partial^2\Psi}{\partial X^2}+\Omega+\frac{\partial^2\Psi}{\partial Y^2}=0, \tag{10.31}$$

$$\begin{aligned}&\frac{\partial\Omega}{\partial Y}V+\frac{\partial\Omega}{\partial X}U=\frac{1}{\mathrm{Re}}\frac{A_5}{A_1}\left(\frac{\partial^2\Omega}{\partial Y^2}+\frac{\partial^2\Omega}{\partial X^2}\right)\\&+\frac{Ha^2}{\mathrm{Re}}\frac{A_6}{A_1}\left(\frac{\partial U}{\partial X}B_xB_y-\frac{\partial V}{\partial X}{B_x}^2+\frac{\partial U}{\partial Y}B_y^2-\frac{\partial V}{\partial Y}B_xB_y\right)-\frac{1}{\mathrm{Re}\,Da}\frac{A_5}{A_1}\Omega,\end{aligned} \tag{10.32}$$

$$\frac{A_4}{A_2}\left(\frac{\partial^2\theta}{\partial Y^2}+\frac{\partial^2\theta}{\partial X^2}\right)=\mathrm{PrRe}\left(\frac{\partial\theta}{\partial X}U+\frac{\partial\theta}{\partial Y}V\right) \tag{10.33}$$

where dimensionless and constants parameters are illustrated as:

$$\begin{aligned}&\mathrm{Re}=\frac{\rho_f U_{Lid}L}{\mu_f},\ Ha=LB_0\sqrt{\sigma_f/\mu_f},\\&A_1=\frac{\rho_{nf}}{\rho_f},A_5=\frac{\mu_{nf}}{\mu_f},\\&A_2=\frac{(\rho C_p)_{nf}}{(\rho C_p)_f},A_4=\frac{k_{nf}}{k_f},A_6=\frac{\sigma_{nf}}{\sigma_f}\end{aligned} \tag{10.34}$$

and boundary conditions are:

$$\begin{aligned}&\theta=1.0 \quad \text{on bottom wall}\\&\theta=0.0 \quad \text{on other walls}\\&\Psi=0.0 \quad \text{on all walls}\end{aligned} \tag{10.35}$$

Nu_{loc} and Nu_{ave} over the hot wall can be calculated as:

$$Nu_{loc}=\frac{\partial\theta}{\partial y}\left(\frac{k_{nf}}{k_f}\right) \tag{10.36}$$

$$Nu_{ave}=\frac{1}{L}\int_{r_{in}}^{r_{out}}Nu_{loc}dx \tag{10.37}$$

10.3.3 Effects of Active Parameters

In this article, the impact of Lorentz forces on forced convection of nanofluid in a lid-driven porous cavity is presented. The working fluid is Fe_3O_4-water and its viscosity is a function of nanofluid volume fraction and magnetic field. CVFEM is utilized to find the effects of Darcy number ($Da=0.01$ to 100), Reynolds number ($\mathrm{Re}=100$ to 500), volume fraction of Fe_3O_4 ($\phi=0\%$ to 4%), shape of nanoparticle, and Hartmann number ($Ha=0$ to 40).

Impacts of shape of the nanoparticles on Nu_{ave} are discussed in Table 10.6. The maximum Nu is caused by Platelet, followed by Cylinder, Brick, and Spherical. Therefore, Platelet nanoparticle has been selected for further investigation.

The impact of adding nanoparticles in the water on velocity and temperature contours is depicted in Fig. 10.8. Temperature gradient reduces with the increase of ϕ. $|\Psi_{\max}|$ increases with the addition of nanoparticles because of the increase in the solid movements.

TABLE 10.6 Effect of Shape of Nanoparticles on Nusselt Number When Re = 600, $\phi = 0.04$

Da	Ha	Spherical	Brick	Cylinder	Platelet
0.01	0	12.90812	13.19201	13.635	13.99464
0.01	40	11.22557	11.47157	11.85536	12.16689
100	0	13.42214	13.71721	14.1776	14.55133
100	40	11.43061	11.67944	12.06752	12.38241

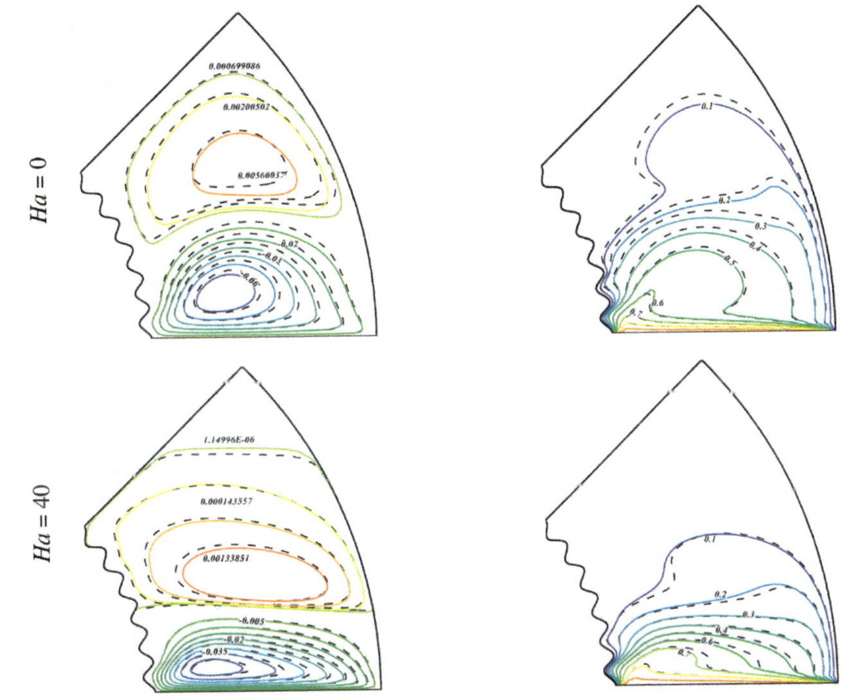

FIGURE 10.8 Influence of nanofluid volume fraction on streamline (left) and isotherm (right) contours (nanofluid ($\phi = 0.04$)(—) and pure fluid($\phi = 0$) (- - -)) when Re = 500, $Da = 100$

Figs. 10.9–10.11 demonstrate the impact of Da, Re, Ha on isotherms and streamlines. At $Da = 0.01, \mathrm{Re} = 100, Ha = 0$, only one eddy exists in streamline. By applying a magnetic field, two new eddies with opposite direction are generated above the main eddy. The distortion of isotherms becomes less than before. As the Darcy number increases, convective mode becomes stronger and the primary eddy converts to two eddies. Increasing the Hartmann number leads to generation of the third eddy. As Re augments, the strength of the new eddy enhances. In the absence of a magnetic field at Re = 500, a counterclockwise eddy appears above the main eddy. Impacts of Darcy and Hartmann number on isotherms and streamlines at Re = 500 is similar to those of Re = 100 except for the stream function values. It means that the absolute stream function values increase with the increase of the Reynolds number.

The influence of significant parameters on Nu_{ave} is illustrated in Fig. 10.12. The correlation for Nu_{ave} is as follows:

$$\begin{aligned} Nu_{ave} &= 9.94 + 1.69\mathrm{Re}^* + 0.66Da^* - 0.18Ha^* \\ &\quad - 0.057Da^*\,\mathrm{Re}^* + 0.051\mathrm{Re}^*\,Ha^* - 0.13Da^*\,Ha^* \\ &\quad - 0.18\mathrm{Re}^{*2} + 0.49Da^{*2} - 0.13Ha^{*2} \end{aligned} \tag{10.38}$$

where $\mathrm{Re}^* = 0.01\mathrm{Re}, Ha* = 0.1Ha, Da* = 0.01Da$. Increasing the Reynolds number can enhance the heat transfer rate because of a reduction in the thermal boundary layer thickness. A similar trend is seen for Darcy number. Temperature gradient decreases with augment of Lorentz forces. So, Nu_{ave} decreases with the rise of Hartmann number.

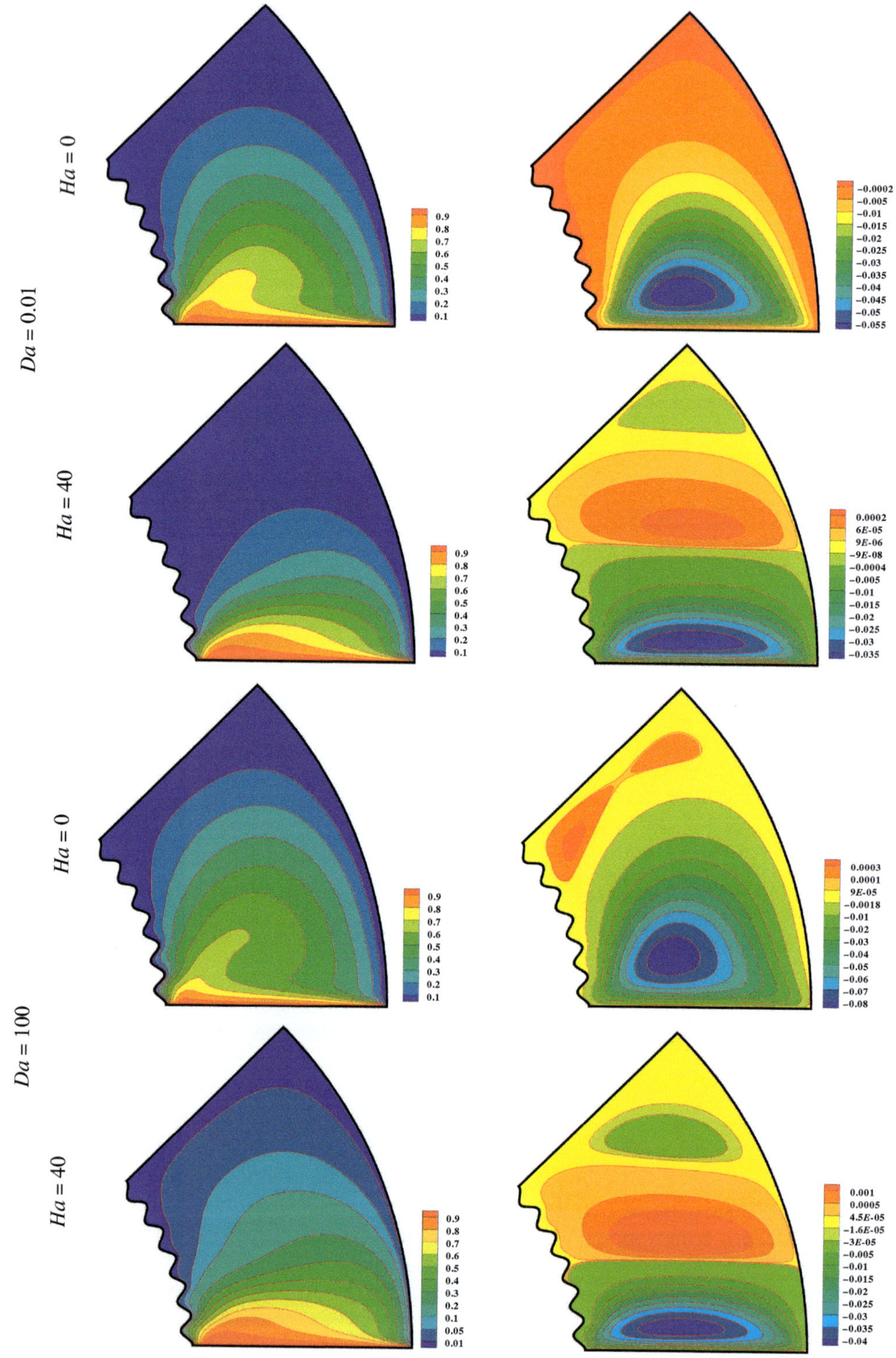

FIGURE 10.9 Effect of Darcy and Hartmann numbers on isotherm (left) and streamline (right) contours when $\phi = 0.04$, Re = 100.

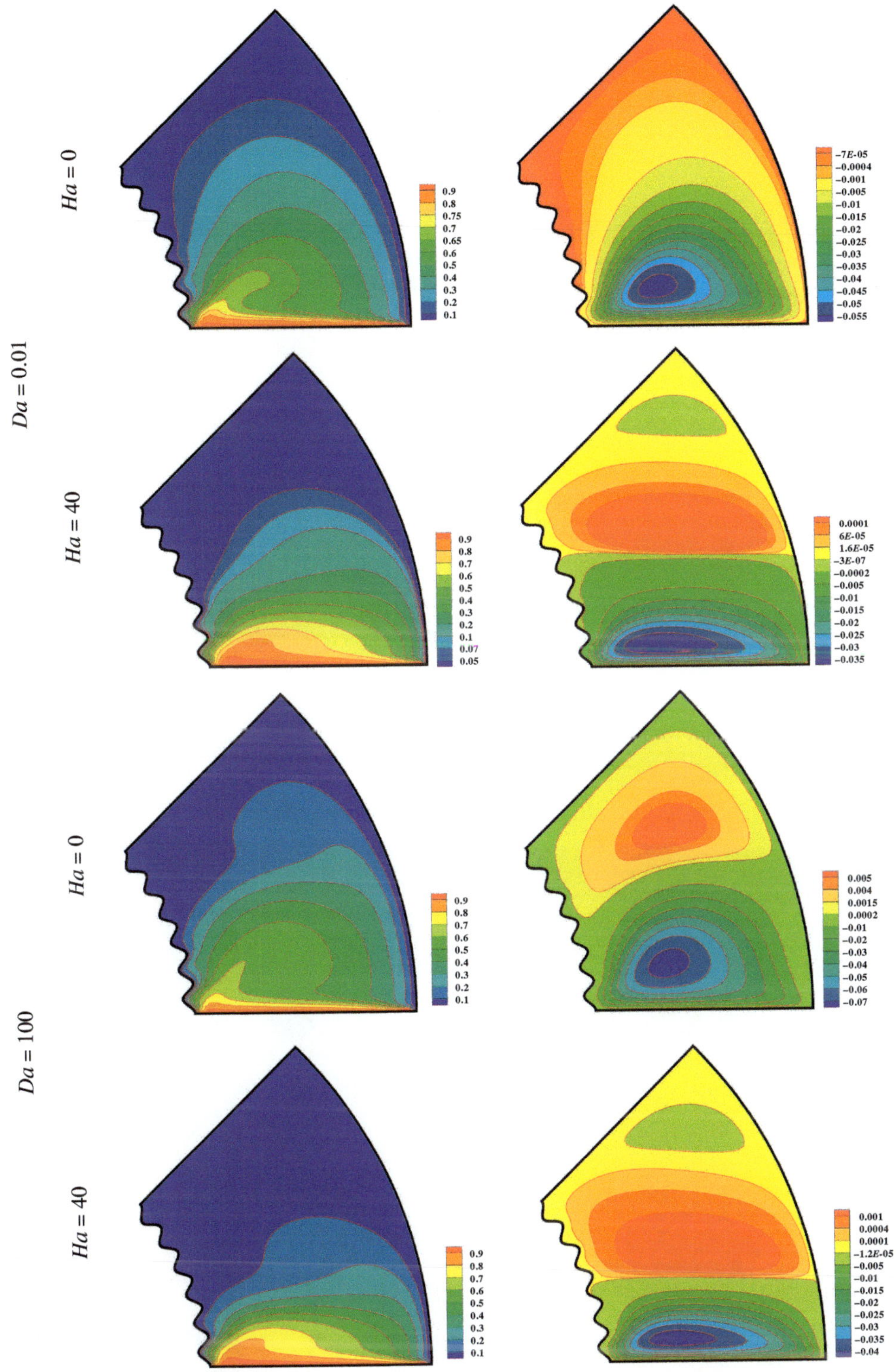

FIGURE 10.10 Effect of Darcy and Hartmann numbers on isotherm (left) and streamline (right) contours when $\phi = 0.04$, $Re = 300$.

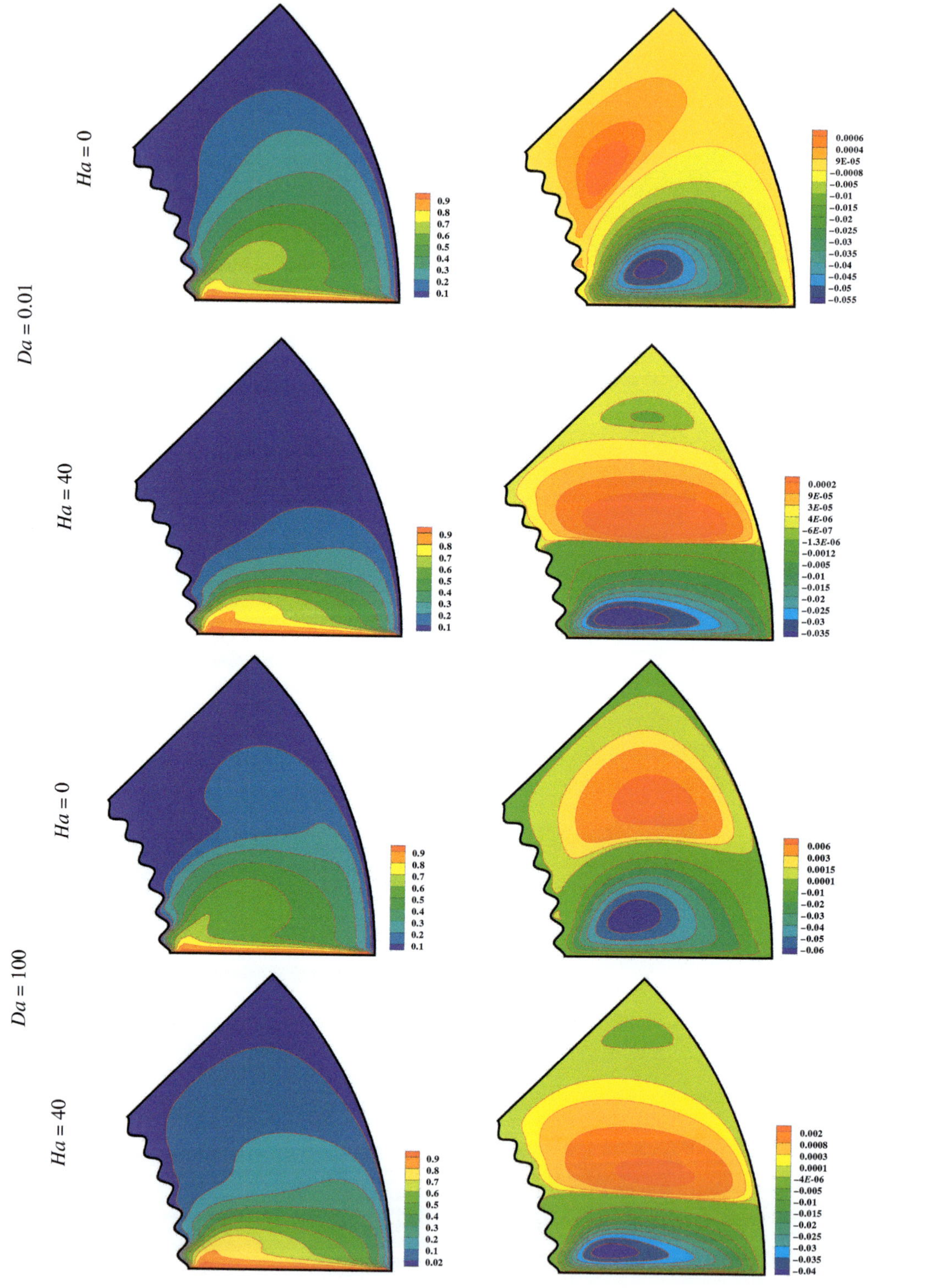

FIGURE 10.11 Effect of Darcy and Hartmann numbers on isotherm (left) and streamline (right) contours when $\phi = 0.04$, Re = 500.

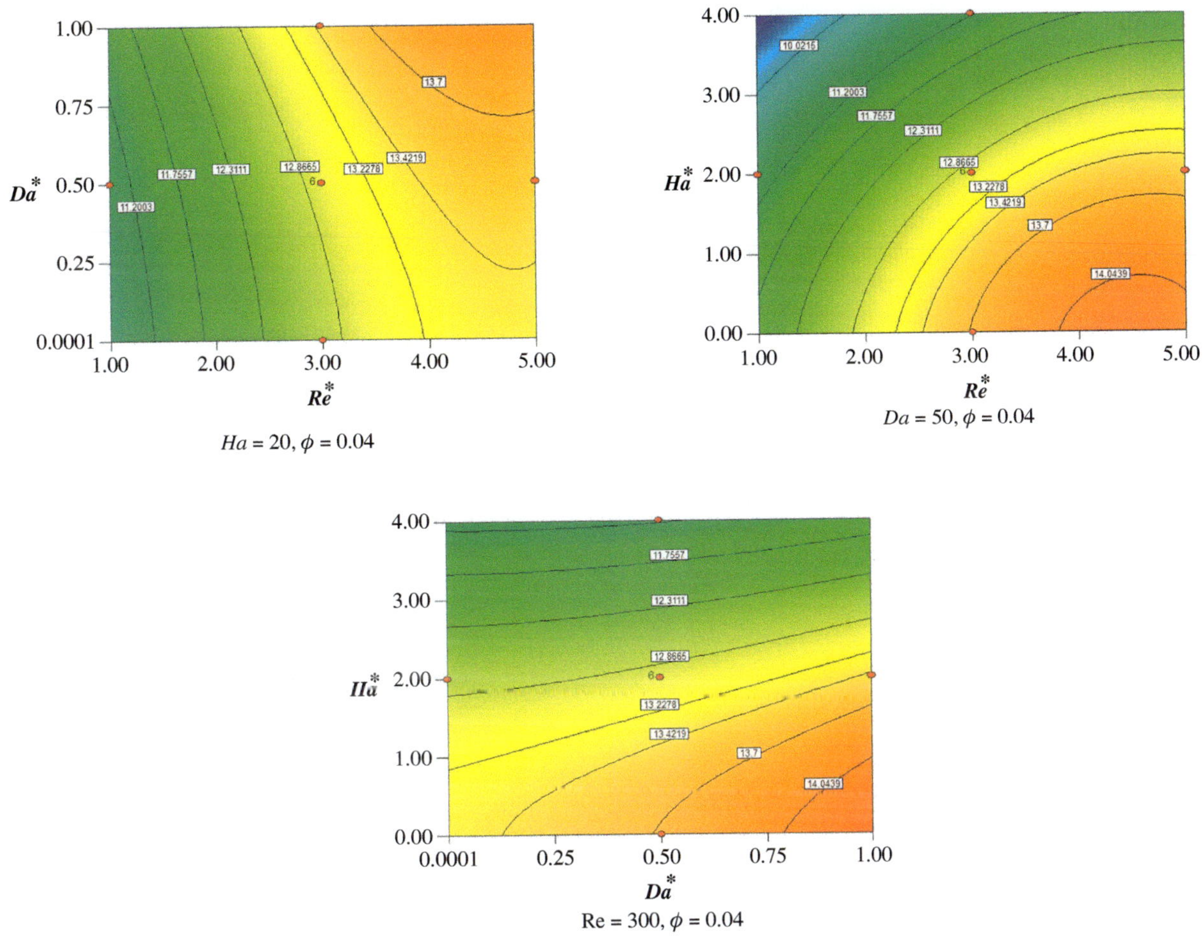

FIGURE 10.12 Influences of the Darcy, Reynolds, and Hartmann numbers on average Nusselt number.

10.4 MAGNETIC SOURCE EFFECT ON NANOFLUID FLOW IN POROUS MEDIUM CONSIDERING VARIOUS SHAPE OF NANOPARTICLES

10.4.1 Problem Definition

Geometry, boundary condition, and sample element are illustrated in Fig. 10.13. An external magnetic source is applied (see Fig. 10.14). $\overline{H}, \overline{H_x}, \overline{H_y}$ are:

$$\overline{H_y} = (\bar{a} - x)\frac{\gamma}{2\pi}\left[(\bar{b}-y)^2 + (\bar{a}-x)^2\right]^{-1}, \tag{10.39}$$

$$\overline{H_x} = (y - \bar{b})\frac{\gamma}{2\pi}\left[(\bar{b}-y)^2 + (\bar{a}-x)^2\right]^{-1}, \tag{10.40}$$

$$\overline{H} = \left(\overline{H}_y^2 + \overline{H}_x^2\right)^{0.5}. \tag{10.41}$$

10.4.2 Governing Equation

2D steady convective non-Darcy flow of nanofluid is considered in the presence of a variable magnetic field. The PDEs are:

$$\frac{\partial u}{\partial x} + \frac{\partial v}{\partial y} = 0, \tag{10.42}$$

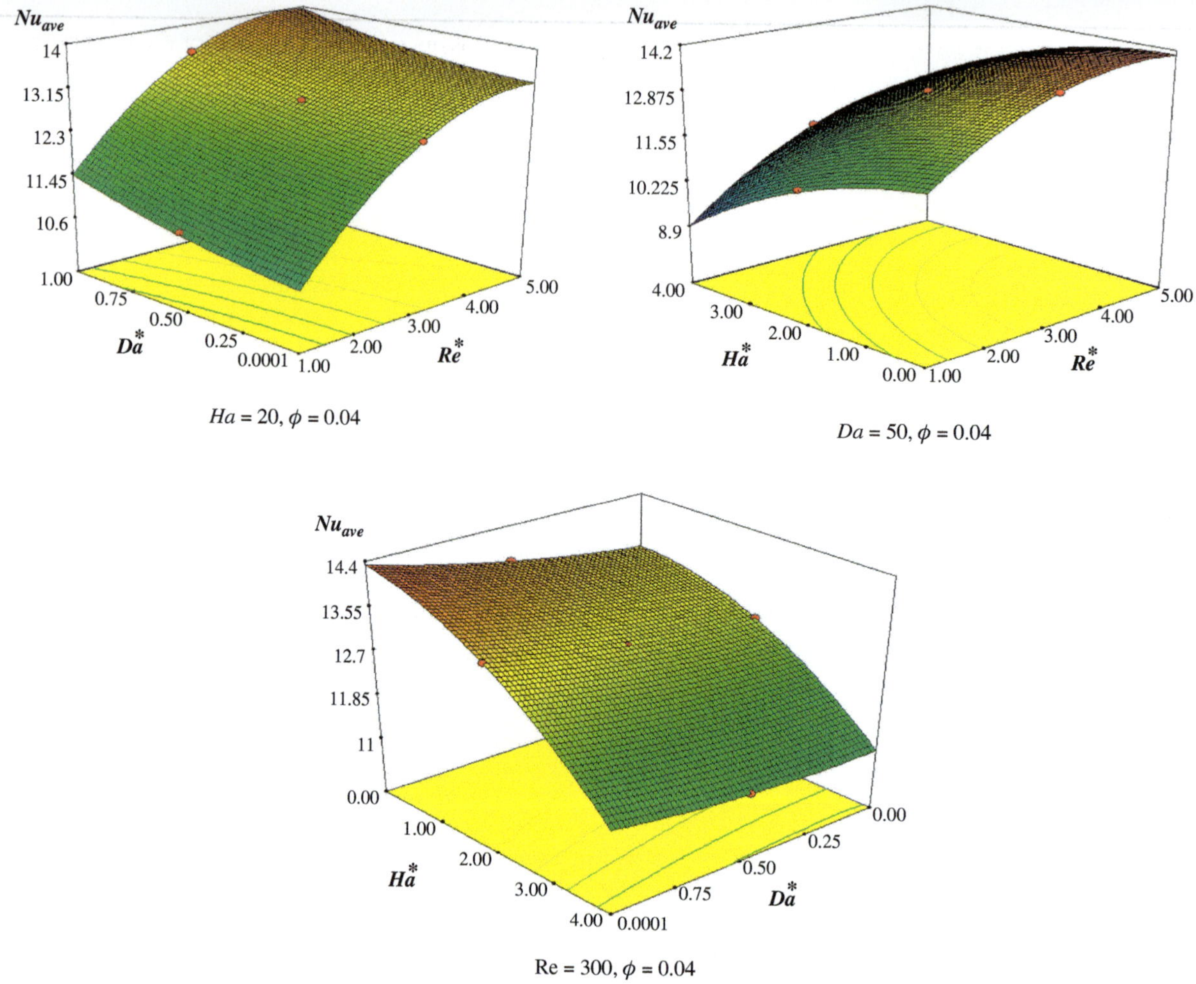

FIGURE 10.12 (Continued).

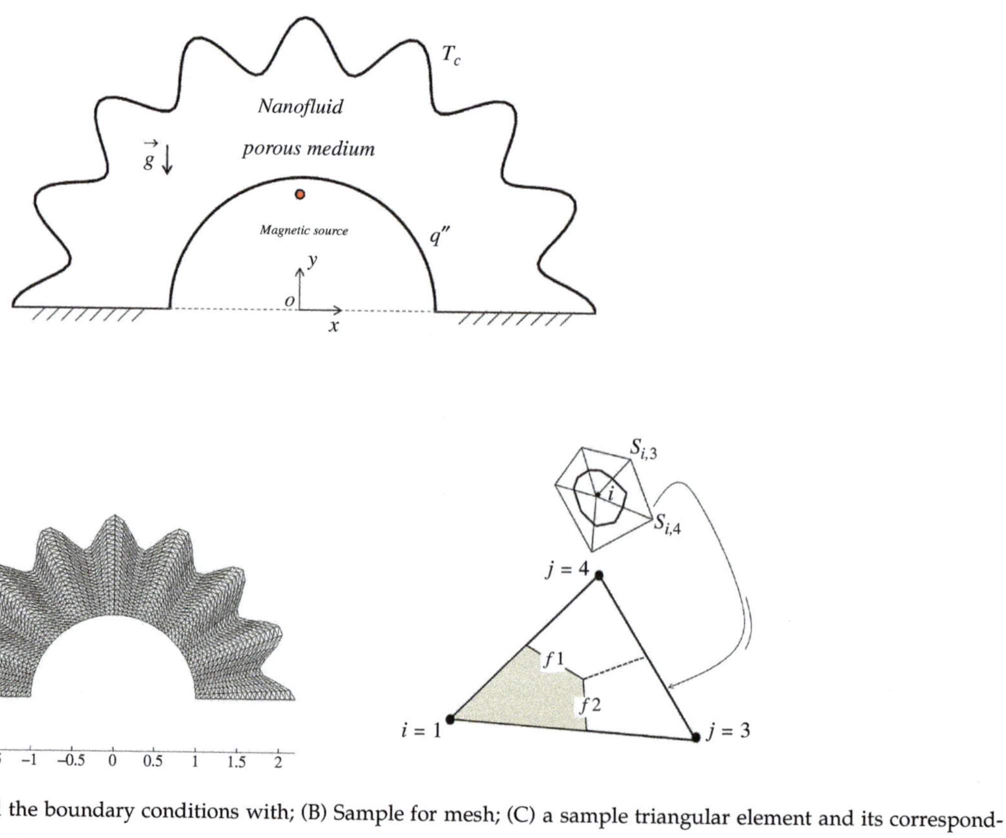

FIGURE 10.13 (A) Geometry and the boundary conditions with; (B) Sample for mesh; (C) a sample triangular element and its corresponding control volume.

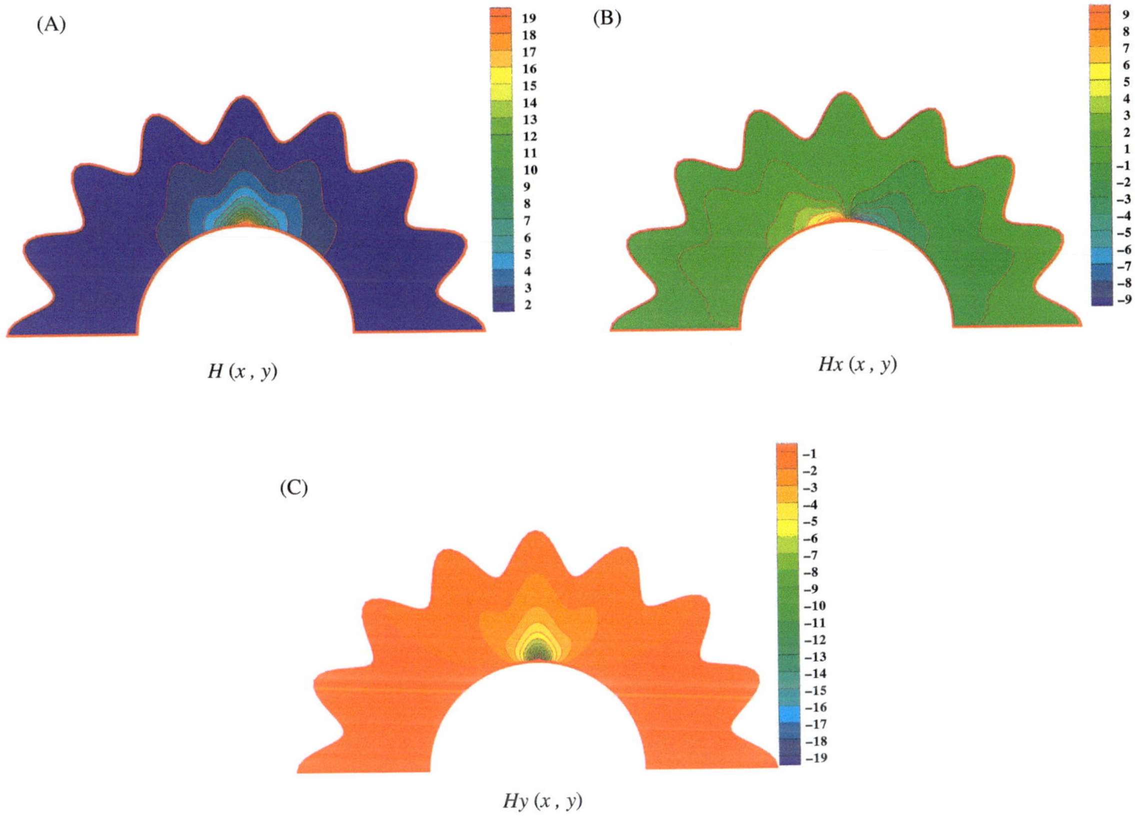

FIGURE 10.14 Contours of the (A) magnetic field strength H; (B) magnetic field intensity component in x direction Hx; (C) magnetic field intensity component in y direction Hy. (A) $H(x,y)$, (B) $Hx(x,y)$, (C) $Hy(x,y)$.

$$\left(\frac{\partial^2 u}{\partial y^2}+\frac{\partial^2 u}{\partial x^2}\right)\mu_{nf}-\frac{\partial P}{\partial x}-\mu_0^2\sigma_{nf}\overline{H}_y^2 u+\sigma_{nf}\mu_0^2\overline{H}_x\overline{H}_y v-\frac{\mu_{nf}}{K}u=\left(\rho_{nf}\right)\left(\frac{\partial u}{\partial x}u+\frac{\partial u}{\partial y}v\right), \tag{10.43}$$

$$\mu_{nf}\left(\frac{\partial^2 v}{\partial x^2}+\frac{\partial^2 v}{\partial y^2}\right)-\frac{\partial P}{\partial y}+\mu_0^2\overline{H}_y\sigma_{nf}\overline{H}_x u-\mu_0^2\overline{H}_x\sigma_{nf}\overline{H}_x v-\frac{\mu_{nf}}{K}v+(T-T_c)\beta_{nf}g\rho_{nf}=\rho_{nf}\left(\frac{\partial v}{\partial x}u+\frac{\partial v}{\partial y}v\right), \tag{10.44}$$

$$k_{nf}\left(\frac{\partial^2 T}{\partial y^2}+\frac{\partial^2 T}{\partial x^2}\right)-\frac{\partial q_r}{\partial y}=\left(v\frac{\partial T}{\partial y}+u\frac{\partial T}{\partial x}\right)(\rho C_p)_{nf}, \quad \left[q_r=-\frac{4\sigma_e}{3\beta_R}\frac{\partial T^4}{\partial y},\quad T^4\cong 4T_c^3T-3T_c^4\right]. \tag{10.45}$$

$(\rho C_p)_{nf}, (\rho\beta)_{nf}, \rho_{nf}, k_{nf}$, and σ_{nf} are calculated as:

$$(\rho C_p)_{nf}=(\rho C_p)_f(1-\phi)+(\rho C_p)_s\phi \tag{10.46}$$

$$\rho_{nf}=\rho_f(1-\phi)+\rho_s\phi \tag{10.47}$$

$$(\rho\beta)_{nf}=(\rho\beta)_f(1-\phi)+(\rho\beta)_s\phi, \tag{10.48}$$

$$\sigma_{nf}=\sigma_f\left[\frac{3\phi(\sigma 1-1)}{(1-\sigma 1)\phi+(2+\sigma 1)}+1\right], \qquad \sigma 1=\sigma_s/\sigma_f. \tag{10.49}$$

μ_{nf} is obtained as follows:

$$\mu_{nf}=\left(0.035\mu_0^2H^2+3.1\mu_0 H-27886.4807\phi^2+4263.02\phi+316.0629\right)e^{-0.01T} \tag{10.50}$$

k_{nf} can be expressed as:

$$\frac{k_{nf}}{k_f} = \frac{-m(k_f - k_p)\phi + (k_p - k_f)\phi + mk_f + k_p + k_f}{mk_f + (k_f - k_p)\phi + k_f + k_p} \tag{10.51}$$

Vorticity and stream function should be used to eliminate pressure source terms:

$$\omega + \frac{\partial u}{\partial y} - \frac{\partial v}{\partial x} = 0, \quad \frac{\partial \psi}{\partial x} = -v, \quad \frac{\partial \psi}{\partial y} = u. \tag{10.52}$$

Dimensionless parameters are defined as:

$$\begin{aligned} &(H_y, H_x, H) = \frac{(\overline{H_y}, \overline{H_x}, \overline{H})}{\overline{H_0}}, \ (b, a) = \frac{(\overline{b}, \overline{a})}{L}, \\ &U = \frac{uL}{\alpha_{nf}}, V = \frac{vL}{\alpha_{nf}}, \ (X, Y) = \frac{(x, y)}{L}, \theta = \frac{T - T_c}{\Delta T}, \Delta T = q'' L / k_f, \\ &\Psi = \frac{\psi}{\alpha_{nf}}, \Omega = \frac{\omega L^2}{\alpha_{nf}}. \end{aligned} \tag{10.53}$$

So equations change to:

$$\frac{\partial^2 \Psi}{\partial Y^2} + \Omega + \frac{\partial^2 \Psi}{\partial X^2} = 0, \tag{10.54}$$

$$\begin{aligned} &\frac{\partial \Omega}{\partial X} U + V \frac{\partial \Omega}{\partial Y} = \mathrm{Pr} \frac{A_5}{A_1} \frac{A_2}{A_4} \left(\frac{\partial^2 \Omega}{\partial Y^2} + \frac{\partial^2 \Omega}{\partial X^2} \right) \\ &+ \mathrm{Pr} Ha^2 \frac{A_6}{A_1} \frac{A_2}{A_4} \left(\frac{\partial U}{\partial X} H_y H_x - \frac{\partial V}{\partial X} {H_x}^2 + \frac{\partial U}{\partial Y} H_y^2 - \frac{\partial V}{\partial Y} H_y H_x \right) + \mathrm{Pr}\ Ra \frac{A_3 A_2^2}{A_1 A_4^2} \frac{\partial \theta}{\partial X} - \frac{\mathrm{Pr}}{Da} \frac{A_5}{A_1} \frac{A_2}{A_4}, \end{aligned} \tag{10.55}$$

$$\frac{\partial \theta}{\partial X} U + V \frac{\partial \theta}{\partial Y} = \left(\frac{\partial^2 \theta}{\partial X^2} + \frac{\partial^2 \theta}{\partial Y^2} \right) + \frac{4}{3} \frac{1}{A_4} Rd \frac{\partial^2 \theta}{\partial Y^2}. \tag{10.56}$$

and dimensionless parameters are:

$$\begin{aligned} &Ra_f = g\beta_f L^3 \Delta T / (\alpha_f \upsilon_f), \mathrm{Pr}_f = \upsilon_f / \alpha_f, Ha = L\mu_0 H_0 \sqrt{\sigma_f / \mu_f}, \\ &Da = K / L^2, Ec = \left(\mu_f \alpha_f\right) / \left[(\rho C_P)_f \Delta T\, L^2\right], \\ &A_1 = \frac{\rho_{nf}}{\rho_f}, A_2 = \frac{(\rho C_p)_{nf}}{(\rho C_p)_f}, A_3 = \frac{(\rho\beta)_{nf}}{(\rho\beta)_f}, \\ &A_4 = \frac{k_{nf}}{k_f}, A_5 = \frac{\mu_{nf}}{\mu_f}, A_6 = \frac{\sigma_{nf}}{\sigma_f}, Rd = 4\sigma_e T_c^3 / (\beta_R k_f) \end{aligned} \tag{10.57}$$

and boundary conditions are:

$$\begin{aligned} &\text{on inner wall} && \frac{\partial \theta}{\partial n} = 1.0 \\ &\text{on outer wall} && \theta = 0.0 \\ &\text{on other walls} && \frac{\partial \theta}{\partial n} = 0 \\ &\text{on all walls} && \Psi = 0.0 \end{aligned} \tag{10.58}$$

Nu_{loc}, Nu_{ave} over the hot wall can be calculated as:

$$Nu_{loc} = \left(\frac{k_{nf}}{k_f}\right)\left(1 + \left(\frac{k_{nf}}{k_f}\right)^{-1}\frac{4Rd}{3}\right)\frac{1}{\theta}, \tag{10.59}$$

$$Nu_{ave} = \frac{1}{\pi}\int_0^{\pi} Nu_{loc}\, d\xi. \tag{10.60}$$

10.4.3 Effects of Active Parameters

Fe_3O_4-water nanofluid flow in a permeable enclosure in the presence of variable Lorentz force is investigated numerically. Shape factor and magnetic field influences on nanofluid properties have been considered. CVFEM is employed to find the influences of radiation parameter ($Rd = 0$ *to* 0.8), Darcy number ($Da = 0.01$ to 100), Rayleigh number ($Ra = 10^3, 10^4, 10^5$), volume fraction of Fe_3O_4 ($\phi = 0\%$ to 4%), shape of nanoparticle, and Hartmann number ($Ha = 0$ to 10).

Table 10.7 demonstrates the impact of shape factor on rate of heat transfer. The maximum and minimum values of Nusselt number are obtained for Platelet and Spherical shapes, respectively. So, Platelet shape is selected for further investigation. The influence of adding nanoparticles to water on flow and heat transfer is depicted in Fig. 10.15. $|\Psi_{max}|$ increases with the addition of nanoparticles because of an increase in the solid movements. Isotherms become less dense with the increase of ϕ. Fig. 10.16 demonstrates the influence of Rd on streamline and isotherm contours. Thermal boundary layer thickness increases with increase of radiation parameter. By increasing Lorentz forces, the impact of radiation parameter on streamlines becomes not significant. Figs. 10.17–10.19 demonstrate the impact of Da, Ra, Ha on isotherms and streamlines. Only one eddy appears in streamline. By applying a magnetic field, the main eddy moves downward. The distortion of isotherms becomes less than before. As Ra enhances, a thermal plume is generated near the vertical symmetric line. Increasing Lorentz forces makes $|\Psi_{max}|$ decrease and shifts the thermal plume to the left side. As Darcy number increases, the convective mode becomes stronger and the primary eddy converts to two eddies.

Effects of important parameters on Nu_{ave} are demonstrated in Fig. 10.20. The correlation for Nu_{ave} is as follows:

$$\begin{aligned} Nu_{ave} = {} & 27.97 - 3.84Rd - 13.2\log(Ra) - 0.6Da^* + 1.5Ha^* \\ & + 2.32Rd\log(Ra) + 0.5Rd\,Da^* - 0.79Rd\,Ha^* \\ & + 0.25Da^*\log(Ra) - 0.4Ha^*\log(Ra) - 0.64Da^*Ha^* \\ & - 3.11Rd^2 + 1.72(\log(Ra))^2 + 0.33(Da^*)^2 - 0.22(Ha^*)^2 \end{aligned} \tag{10.61}$$

where $Da* = 0.01Da$, $Ha* = 0.1Ha$. As convective heat transfer increases, the Nusselt number increases. So, the Nusselt number has a direct relationship with the permeability of porous media and buoyancy forces. Lorentz forces make the conduction mechanism improve. So the Nusselt number has an inverse relationship with the Hartmann number.

TABLE 10.7 Effect of Shape of Nanoparticles on Nusselt Number When $Da = 100, Ra = 10^5, Rd = 0.8, \phi = 0.04$

	Ha	
	0	**10**
Spherical	10.80306	8.304584
Brick	10.87294	8.344276
Cylinder	10.98493	8.407705
Platelet	11.07831	8.460315

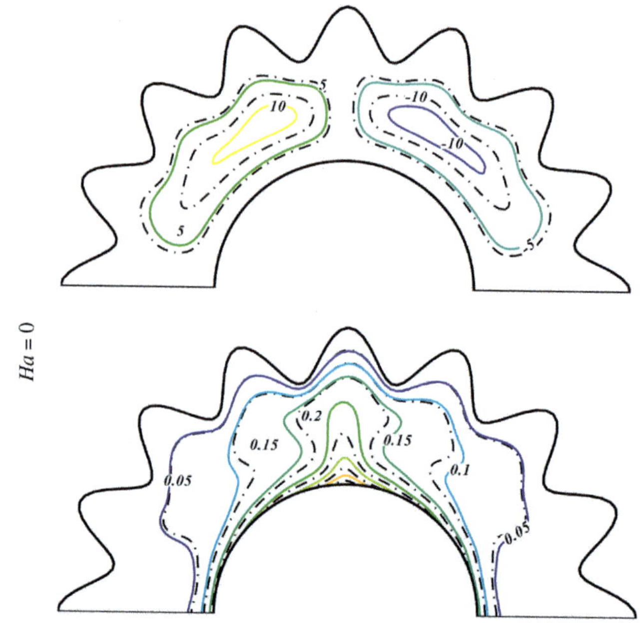

FIGURE 10.15 Impact of nanofluid volume fraction on streamline (top) and isotherm (bottom) contours (nanofluid ($\phi = 0.04$)(—) and pure fluid($\phi = 0$) (– · –)) when $Ra = 10^5, Da = 100, Rd = 0.8$.

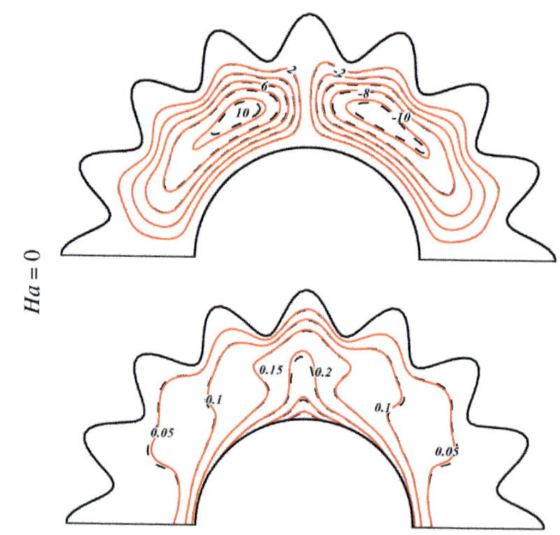

FIGURE 10.16 Impact of radiation parameter on streamline (top) and isotherm (bottom) contour ($Rd = 0.8$ (—), $Rd = 0$ (- - -)) when $Ra = 10^5, Da = 100, \phi = 0.04$.

10.5 MAGNETOHYDRODYNAMIC CUO-WATER TRANSPORTATION INSIDE A POROUS CAVITY CONSIDERING SHAPE FACTOR EFFECT

10.5.1 Problem Definition

Geometry, boundary condition, and sample element are depicted (see Fig. 10.21). The bottom wall is hot and other walls are cold. Also the bottom wall can move horizontally. A porous cavity is filled with nanofluid and affected by horizontal magnetic field.

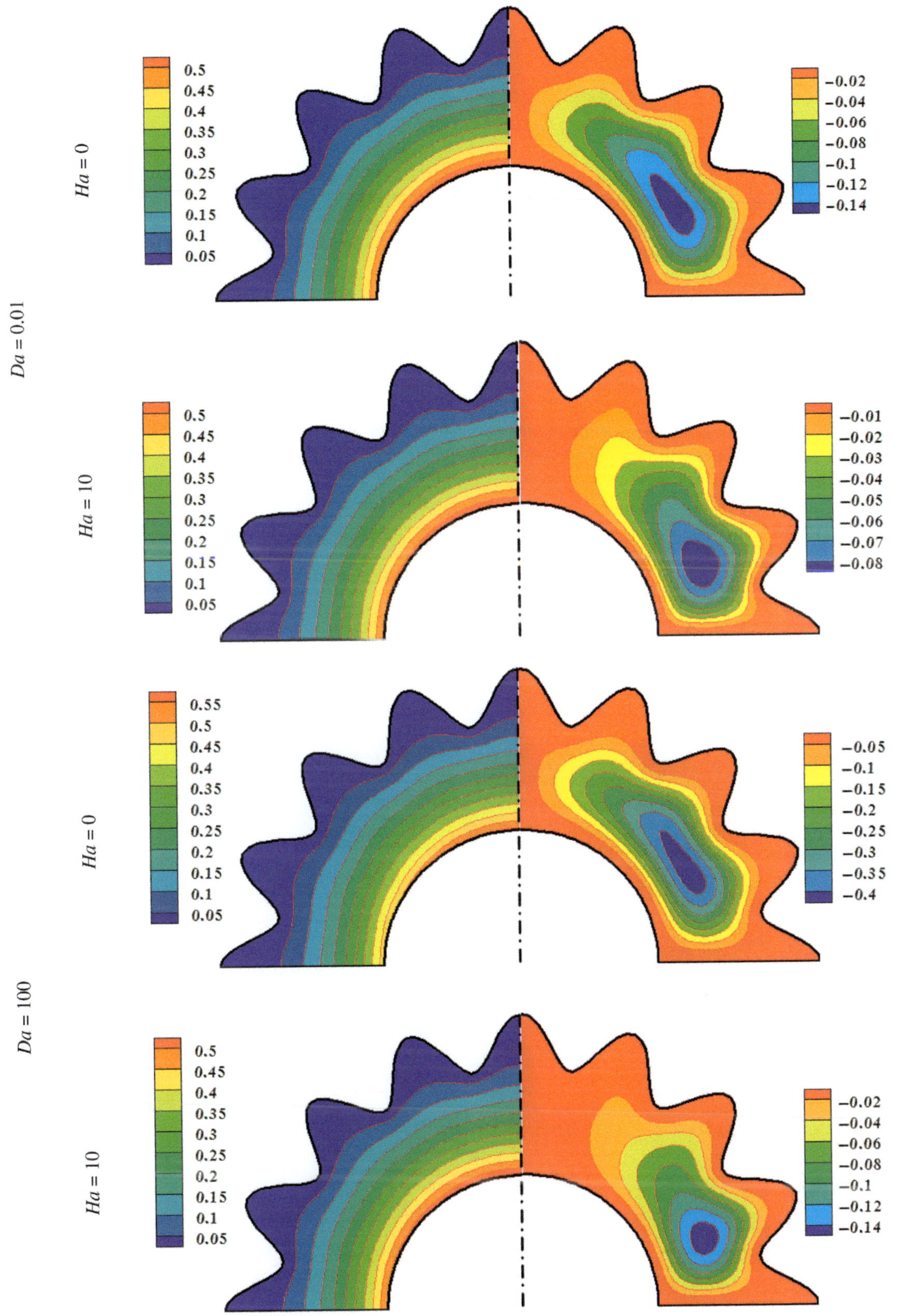

FIGURE 10.17 Influence of *Da*, *Ha* on streamline (left) and isotherm (right) contours when $\phi = 0.04$, $Ra = 10^3$.

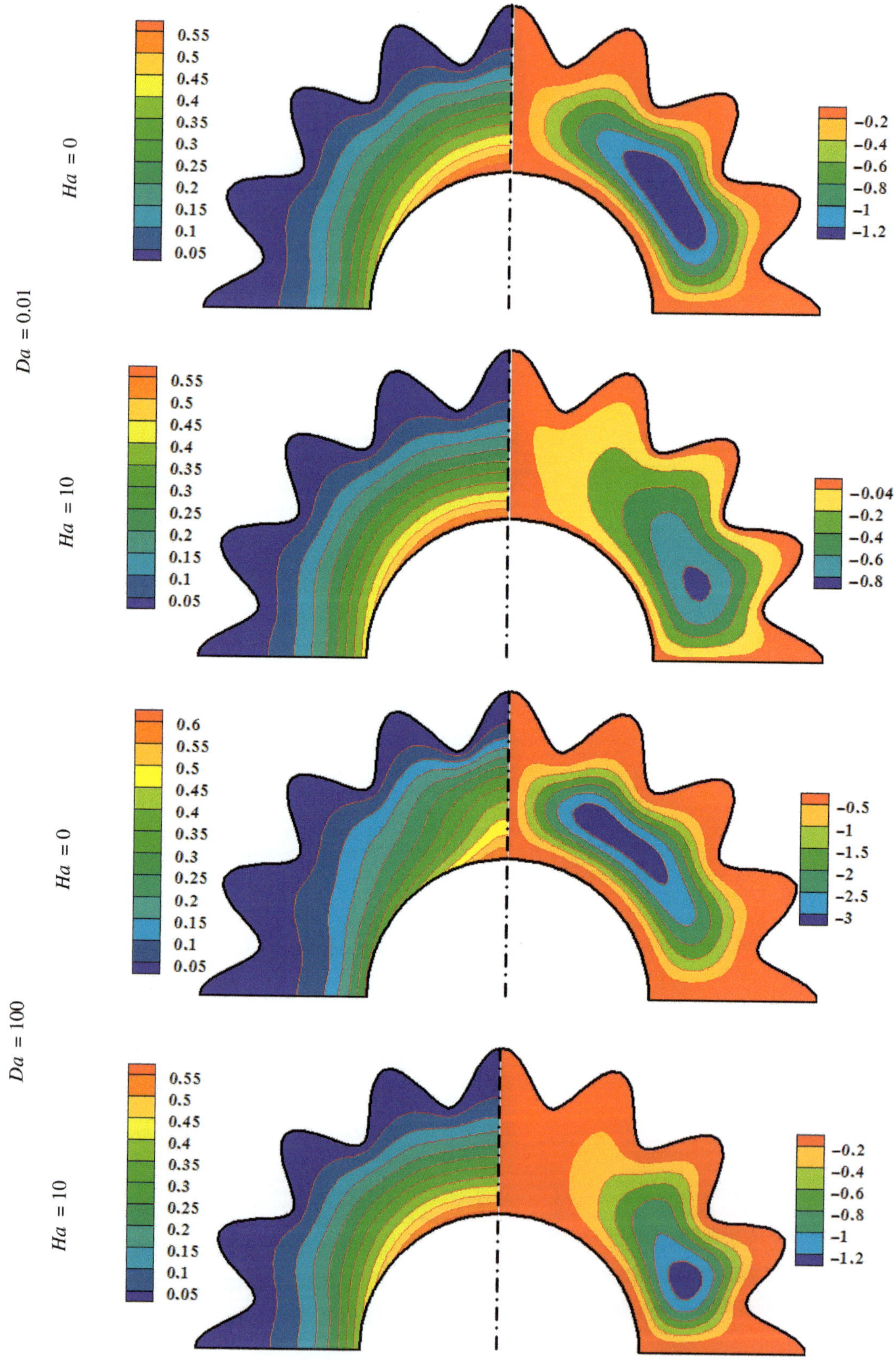

FIGURE 10.18 Influence of *Da*, *Ha* on streamline (left) and isotherm (right) contours when $\phi = 0.04$, $Ra = 10^4$.

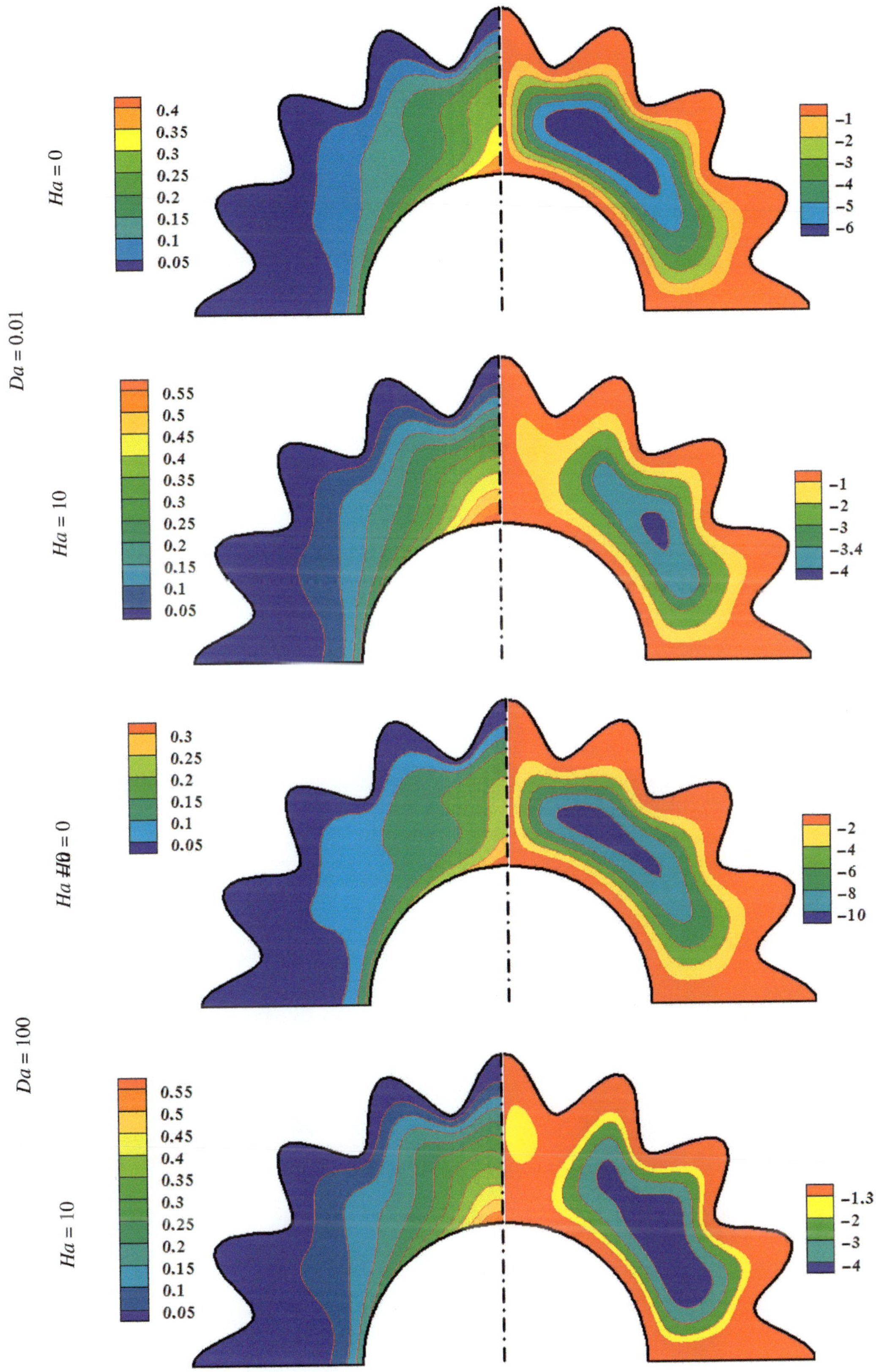

FIGURE 10.19 Influence of *Da*, *Ha* on streamline (left) and isotherm (right) contours when $\phi = 0.04$, $Ra = 10^5$.

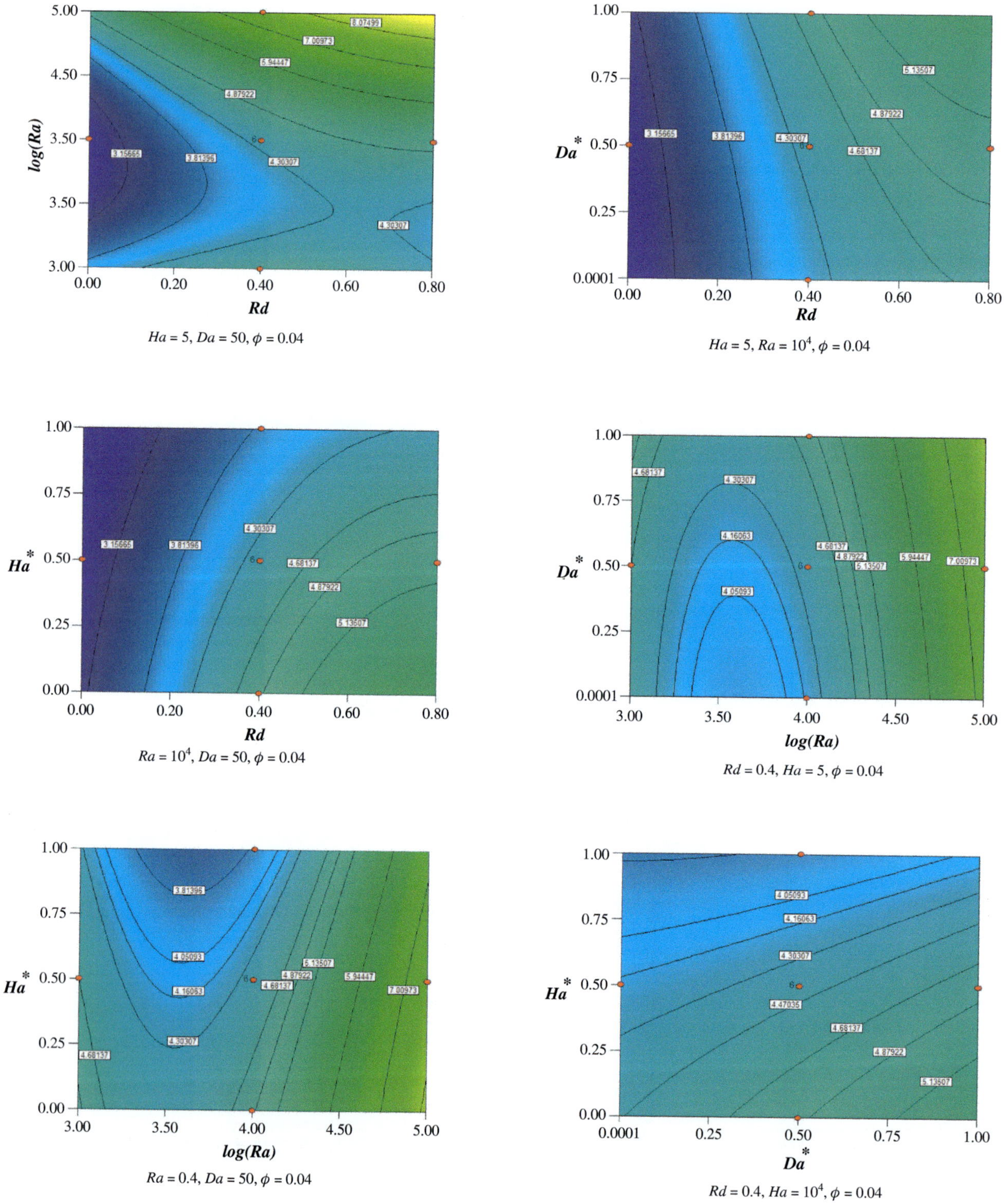

FIGURE 10.20 Effects of *Da*, *Ha*, and *Ra* on average Nusselt number.

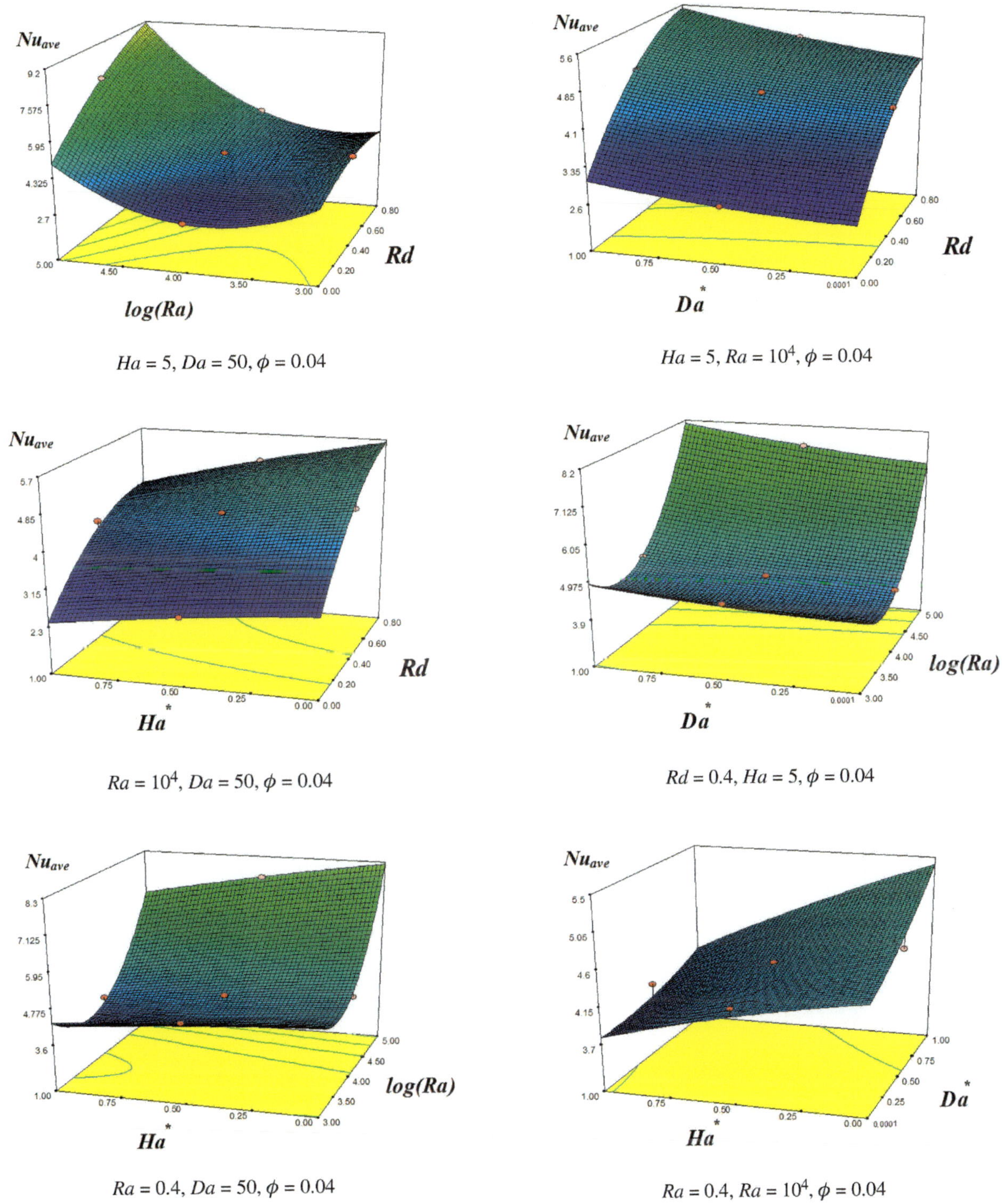

FIGURE 10.20 (Continued).

10.5.2 Governing Equation

2D steady convective non-Darcy flow of nanofluid is considered in the presence of a constant magnetic field. The PDEs are:

$$\frac{\partial v}{\partial y} + \frac{\partial u}{\partial x} = 0 \tag{10.62}$$

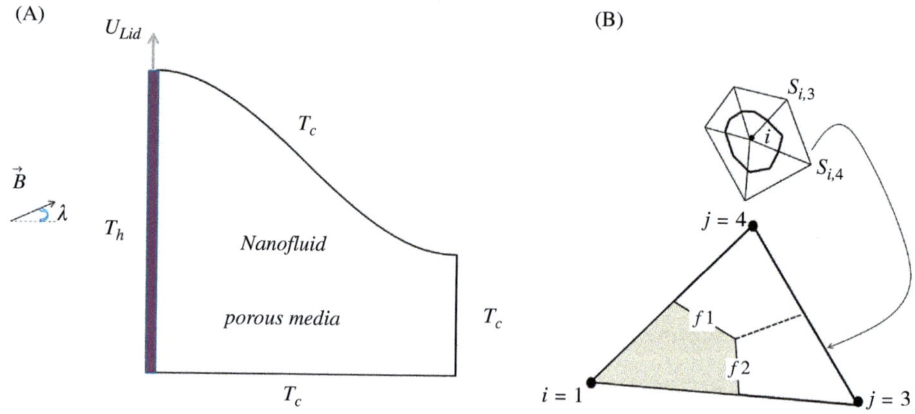

FIGURE 10.21 (A) Geometry and the boundary conditions with (B) a sample triangular element and its corresponding control volume.

$$\left[-\sigma_{nf}B_y^2 u + \sigma_{nf}B_x B_y v + \left(\frac{\partial^2 u}{\partial y^2} + \frac{\partial^2 u}{\partial x^2}\right)\mu_{nf} - \frac{\partial P}{\partial x}\right] - \frac{\mu_{nf}}{K}u = \rho_{nf}\left(v\frac{\partial u}{\partial y} + \frac{\partial u}{\partial x}u\right) \tag{10.63}$$

$$\mu_{nf}\left(\frac{\partial^2 v}{\partial y^2} + \frac{\partial^2 v}{\partial x^2}\right) - \frac{\partial P}{\partial y} - B_x\sigma_{nf}B_x v + B_y\sigma_{nf}B_x u - \frac{\mu_{nf}}{K}v = \rho_{nf}\left(\frac{\partial v}{\partial x}u + \frac{\partial v}{\partial y}v\right), B_x = B_o\cos\lambda,\ \ B_y = B_o\sin\lambda \tag{10.64}$$

$$(\rho C_p)_{nf}\left(v\frac{\partial T}{\partial y} + u\frac{\partial T}{\partial x}\right) = k_{nf}\left(\frac{\partial^2 T}{\partial x^2} + \frac{\partial^2 T}{\partial y^2}\right), \tag{10.65}$$

$(\rho C_p)_{nf}$, ρ_{nf}, and σ_{nf} are defined as:

$$(\rho C_p)_{nf} = (\rho C_p)_f(1-\phi) + (\rho C_p)_s\phi \tag{10.66}$$

$$\rho_{nf} = \rho_f(1-\phi) + \rho_s\phi \tag{10.67}$$

$$\frac{\sigma_{nf}}{\sigma_f} = 1 + \frac{3\left(\frac{\sigma_s}{\sigma_f} - 1\right)\phi}{\left(\frac{\sigma_s}{\sigma_f} + 2\right) - \left(\frac{\sigma_s}{\sigma_f} - 1\right)\phi} \tag{10.68}$$

μ_{nf} is calculated via KKL model:

$$\begin{aligned}
&\mu_{eff} = \mu_{static} + \mu_{Brownian} = \mu_{static} + \frac{k_{Brownian}}{k_f} \times \frac{\mu_f}{\Pr_f} \\
&k_{Brownian} = 5\times 10^4 g'(\phi, T, d_p)\phi\rho_f c_{p,f}\sqrt{\frac{\kappa_b T}{d_p\rho_p}} \\
&g'(\phi, d_p, T) = Ln(T)\left(a_1 + a_3 Ln(\phi) + a_2 Ln(d_p) + a_5 Ln(d_p)^2 + a_4 \ln(d_p)Ln(\phi)\right) \\
&\quad + \left(a_6 + a_8 Ln(\phi) + a_7 Ln(d_p) + a_{10} Ln(d_p)^2 + a_9 Ln(\phi)\ln(d_p)\right)
\end{aligned} \tag{10.69}$$

k_{nf} can be expressed as:

$$\frac{k_{nf}}{k_f} = \frac{-m(k_f - k_p)\phi + (k_p - k_f)\phi + mk_f + k_p + k_f}{mk_f + (k_f - k_p)\phi + k_f + k_p} \tag{10.70}$$

Vorticity and stream function should be used to eliminate pressure source terms:

$$\omega + \frac{\partial u}{\partial y} - \frac{\partial v}{\partial x} = 0,\quad \frac{\partial \psi}{\partial x} = -v,\quad \frac{\partial \psi}{\partial y} = u \tag{10.71}$$

Introducing dimensionless quantities:

$$U = \frac{u}{U_{Lid}}, V = \frac{v}{U_{Lid}}, \ \theta = \frac{T - T_c}{\Delta T}, \Delta T = T_h - T_c, (X, Y) = \frac{(x, y)}{L}, \Psi = \frac{\psi L}{U_{Lid}}, \Omega = \frac{\omega}{L U_{Lid}} \tag{10.72}$$

The final formulae are:

$$\frac{\partial^2 \Psi}{\partial X^2} + \Omega + \frac{\partial^2 \Psi}{\partial Y^2} = 0, \tag{10.73}$$

$$\begin{aligned} \frac{\partial \Omega}{\partial Y} V + \frac{\partial \Omega}{\partial X} U &= \frac{1}{\mathrm{Re}} \frac{A_5}{A_1} \left(\frac{\partial^2 \Omega}{\partial Y^2} + \frac{\partial^2 \Omega}{\partial X^2} \right) \\ &+ \frac{Ha^2}{\mathrm{Re}} \frac{A_6}{A_1} \left(\frac{\partial U}{\partial X} B_x B_y - \frac{\partial V}{\partial X} {B_x}^2 + \frac{\partial U}{\partial Y} B_y^2 - \frac{\partial V}{\partial Y} B_x B_y \right) - \frac{1}{\mathrm{Re}\, Da} \frac{A_5}{A_1} \Omega, \end{aligned} \tag{10.74}$$

$$\frac{A_4}{A_2} \left(\frac{\partial^2 \theta}{\partial Y^2} + \frac{\partial^2 \theta}{\partial X^2} \right) = \mathrm{PrRe} \left(\frac{\partial \theta}{\partial X} U + \frac{\partial \theta}{\partial Y} V \right) \tag{10.75}$$

where dimensionless and constants parameters are illustrated as:

$$\begin{aligned} &\mathrm{Re} = \frac{\rho_f U_{Lid} L}{\mu_f}, \ Ha = L B_0 \sqrt{\sigma_f / \mu_f}, \\ &A_1 = \frac{\rho_{nf}}{\rho_f}, A_5 = \frac{\mu_{nf}}{\mu_f}, \\ &A_2 = \frac{(\rho C_p)_{nf}}{(\rho C_p)_f}, A_4 - \frac{k_{nf}}{k_f}, A_6 = \frac{\sigma_{nf}}{\sigma_f} \end{aligned} \tag{10.76}$$

and boundary conditions are:

$$\begin{aligned} &\theta = 1.0 \quad \text{on bottom wall} \\ &\theta = 0.0 \quad \text{on other walls} \\ &\Psi = 0.0 \quad \text{on all walls} \end{aligned} \tag{10.77}$$

Nu_{loc} and Nu_{ave} over the hot wall can be calculated as:

$$Nu_{loc} = \frac{\partial \theta}{\partial y} \left(\frac{k_{nf}}{k_f} \right) \tag{10.78}$$

$$Nu_{ave} = \frac{1}{L} \int_{r_{in}}^{r_{out}} Nu_{loc} dx \tag{10.79}$$

10.5.3 Effects of Active Parameters

In this research, the impact of Lorentz forces on forced convection of nanofluid in a lid-driven porous cavity is illustrated. μ_{nf} of CuO-water nanofluid are estimated by means of KKL model. CVFEM is utilized to find the effects of Darcy number ($Da = 0.01$ to 100), Reynolds number ($\mathrm{Re} = 100$ to 500), volume fraction of CuO ($\phi = 0\%$ to 4%), shape of nanoparticle, and Hartmann number ($Ha = 0$ to 40). Influences of shape of the nanoparticles on Nu_{ave} are shown in Table 10.8. The maximum Nu is caused by Platelet, followed by Cylinder, Brick, and Spherical. So, Platelet nanoparticle has been selected for further investigation. The impact of adding nanoparticles in the base fluid on velocity and temperature contours is depicted in Fig. 10.22. The temperature gradient decreases with the increase of ϕ. $|\Psi_{\max}|$ increases with the addition of nanoparticles because of the increase in the solid movements. Figs. 10.23–10.25 demonstrate the impact of Da, Re, Ha on isotherms and streamlines. At $Da = 0.01, \mathrm{Re} = 100, Ha = 0$, only one eddy exists in streamline. By applying a magnetic field, two new eddies with opposite direction are generated above the main eddy. The distortion of isotherms becomes less than before. As the Darcy number increases, the convective mode becomes stronger and the primary eddy converts to two eddies. Increasing the Hartmann number leads to the generation of the third eddy. As Re increases, the strength

TABLE 10.8 Effect of Shape of Nanoparticles on Nusselt Number When $Da = 100, \mathrm{Re} = 600, \phi = 0.04$

	Ha	
	0	**40**
Spherical	2.587961	0.572717
Brick	2.61641	0.579895
Cylinder	2.659818	0.59086
Platelet	2.694209	0.599562

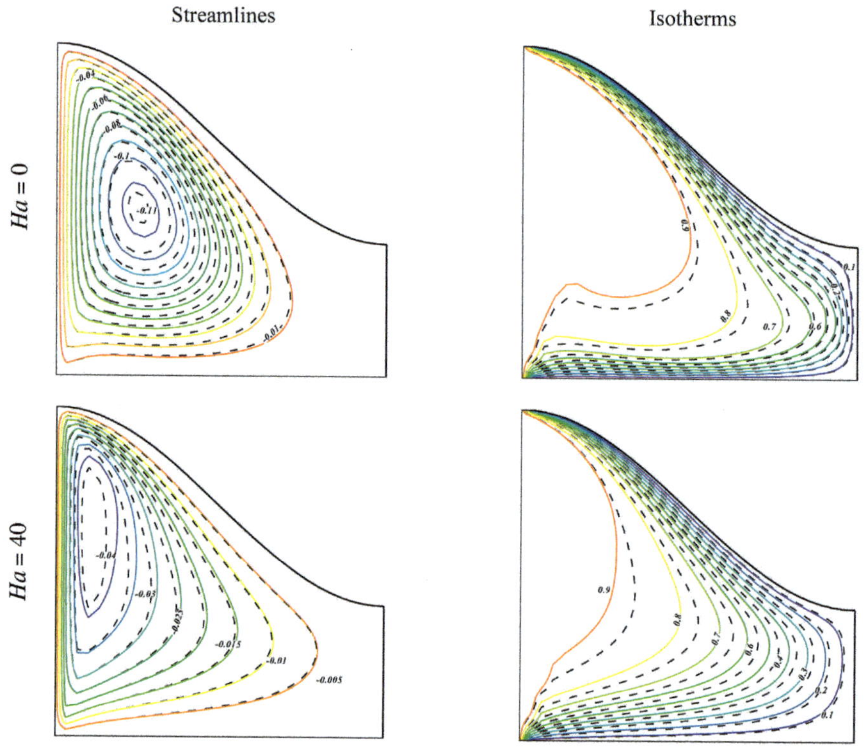

FIGURE 10.22 Influence of nanofluid volume fraction on streamline (left) and isotherm (right) contours (nanofluid ($\phi = 0.04$)(——) and pure fluid ($\phi = 0$) (- - -)) when $\mathrm{Re} = 500, Da = 100$.

of the secondary eddy is enhanced. In the absence of a magnetic field at $Re = 500$, a counterclockwise eddy appears above the main eddy. The impacts of Darcy and Hartmann number on isotherms and streamlines at $Re = 500$ are similar to those of $Re = 100$ except for the stream function values. It means that the absolute stream function values increase with the increase of the Reynolds number.

The influence of significant parameters on Nu_{ave} is depicted in Fig. 10.26. The correlation for Nu_{ave} is as follows:

$$\begin{aligned} Nu_{ave} = & -0.25 + 0.29\mathrm{Re}^* - 0.1Da^* + 0.2Ha^* \\ & + 0.12Da^*\,\mathrm{Re}^* - 0.07\mathrm{Re}^*\,Ha^* - 0.12Da^*\,Ha^* \\ & + 0.006\mathrm{Re}^{*2} + 0.26Da^{*2} - 0.023Ha^{*2} \end{aligned} \tag{10.80}$$

where $\mathrm{Re}^* = 0.01\mathrm{Re}, Ha* = 0.1Ha, Da* = 0.01Da$. Increasing the Reynolds number can enhance the heat transfer rate because of the reduction in the thermal boundary layer thickness. A similar trend is seen for Darcy number. The temperature gradient decreases with the increase of Lorentz forces. So, Nu_{ave} decreases with increase of Hartmann number.

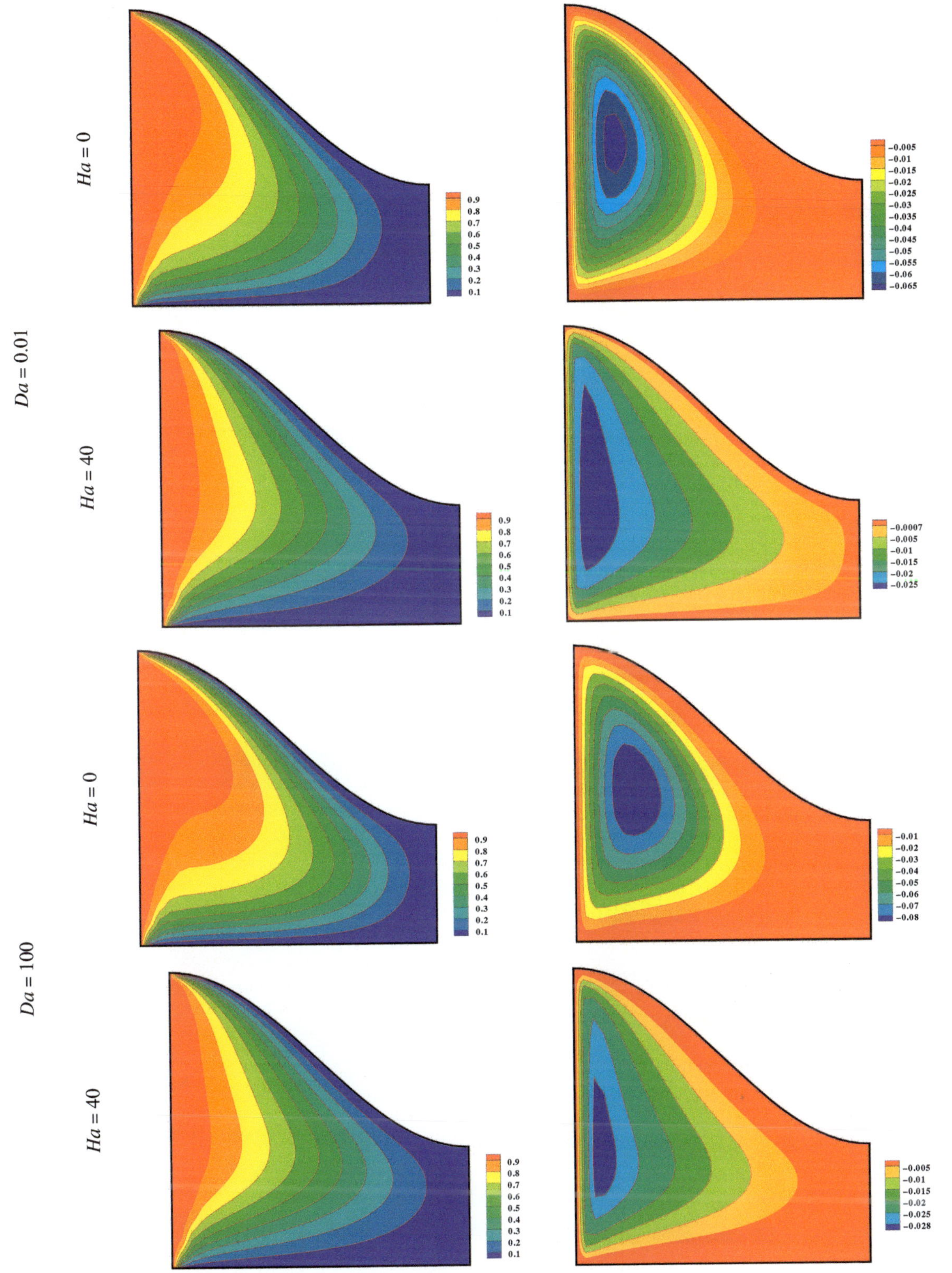

FIGURE 10.23 Effect of Darcy and Hartmann numbers on isotherm (left) and streamline (right) contours when $\phi = 0.04$, Re = 100.

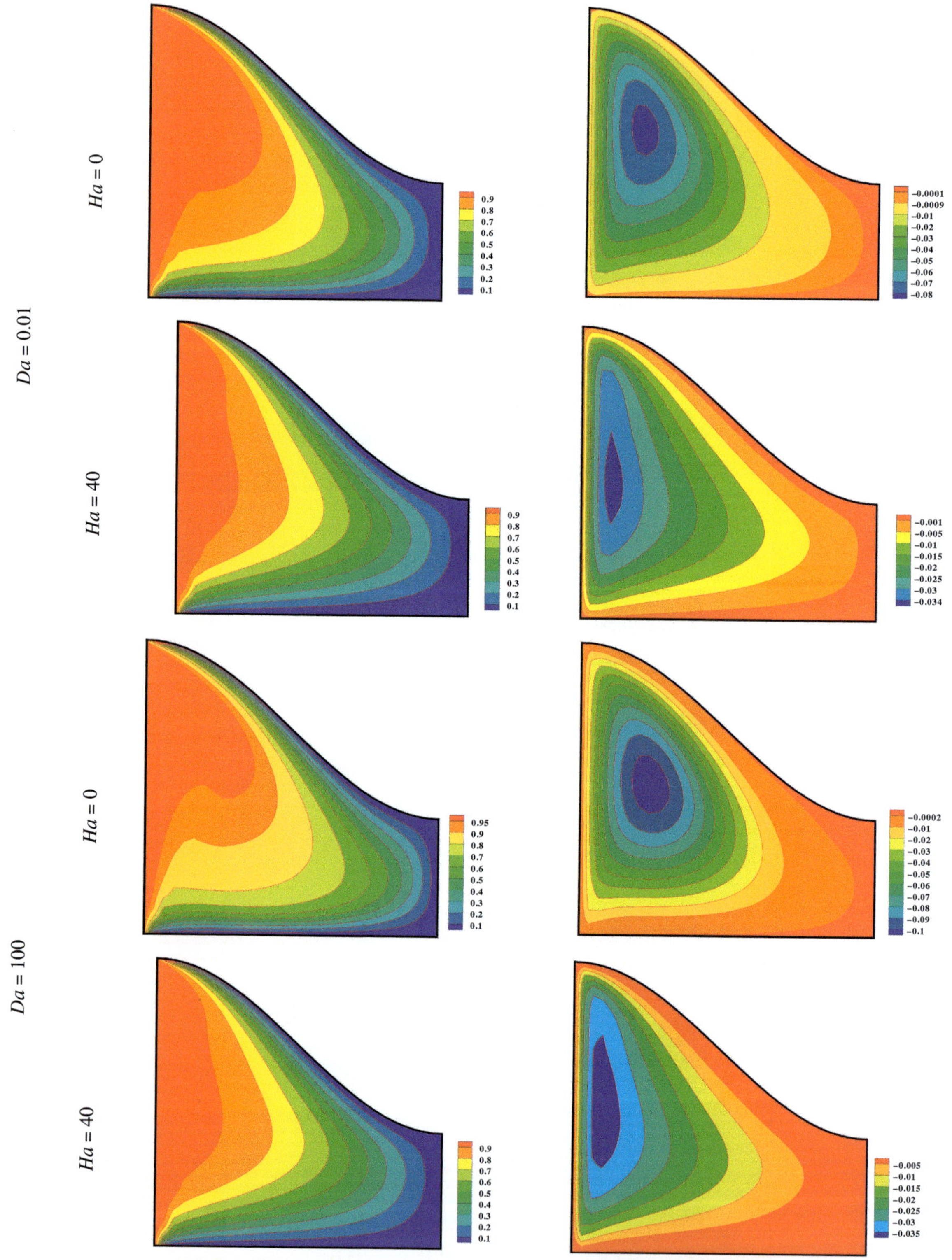

FIGURE 10.24 Effect of Darcy and Hartmann numbers on isotherm (left) and streamline (right) contours when $\phi = 0.04$, $\mathrm{Re} = 300$.

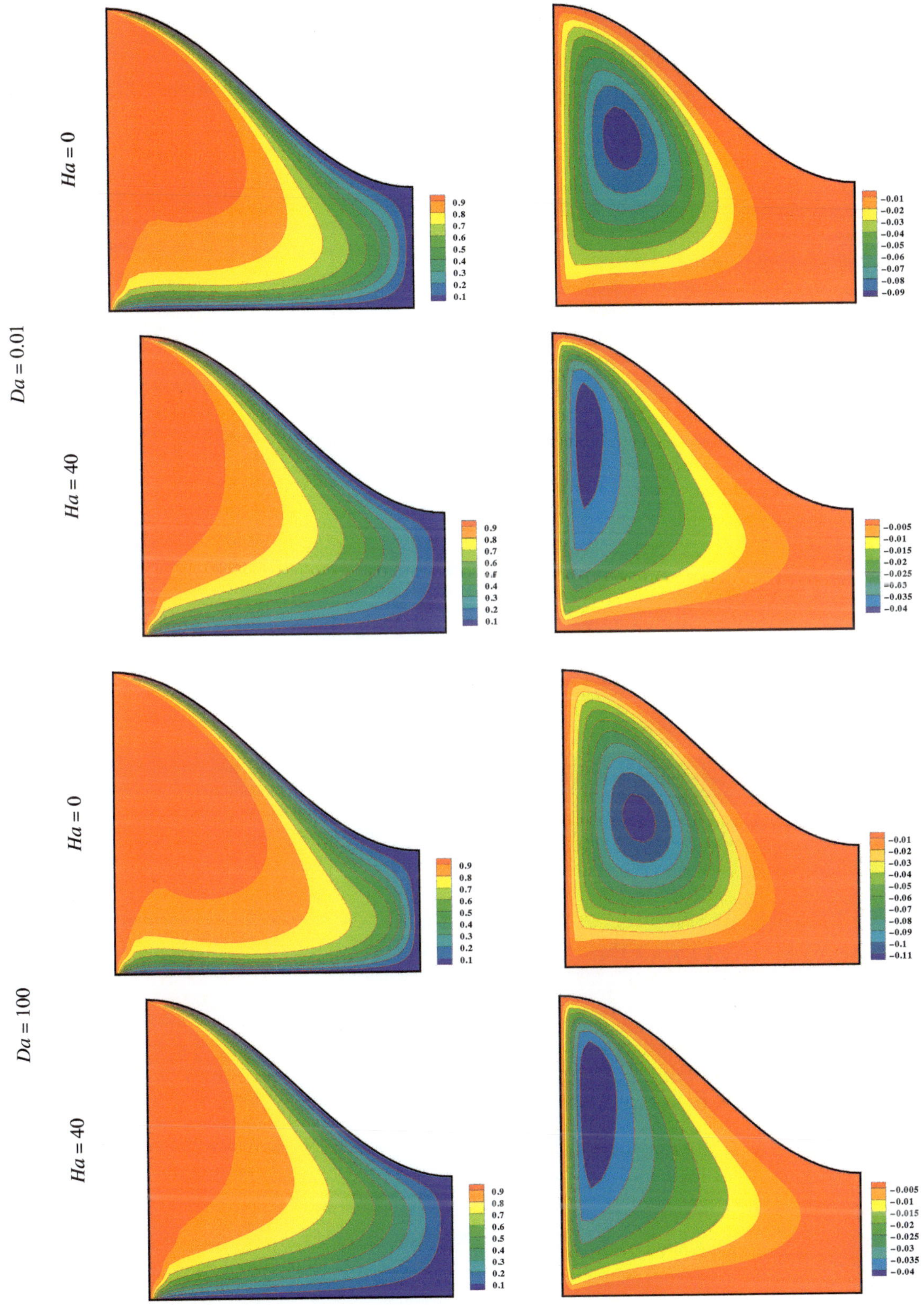

FIGURE 10.25 Effect of Darcy and Hartmann numbers on isotherm (left) and streamline (right) contours when $\phi = 0.04, \mathrm{Re} = 500$.

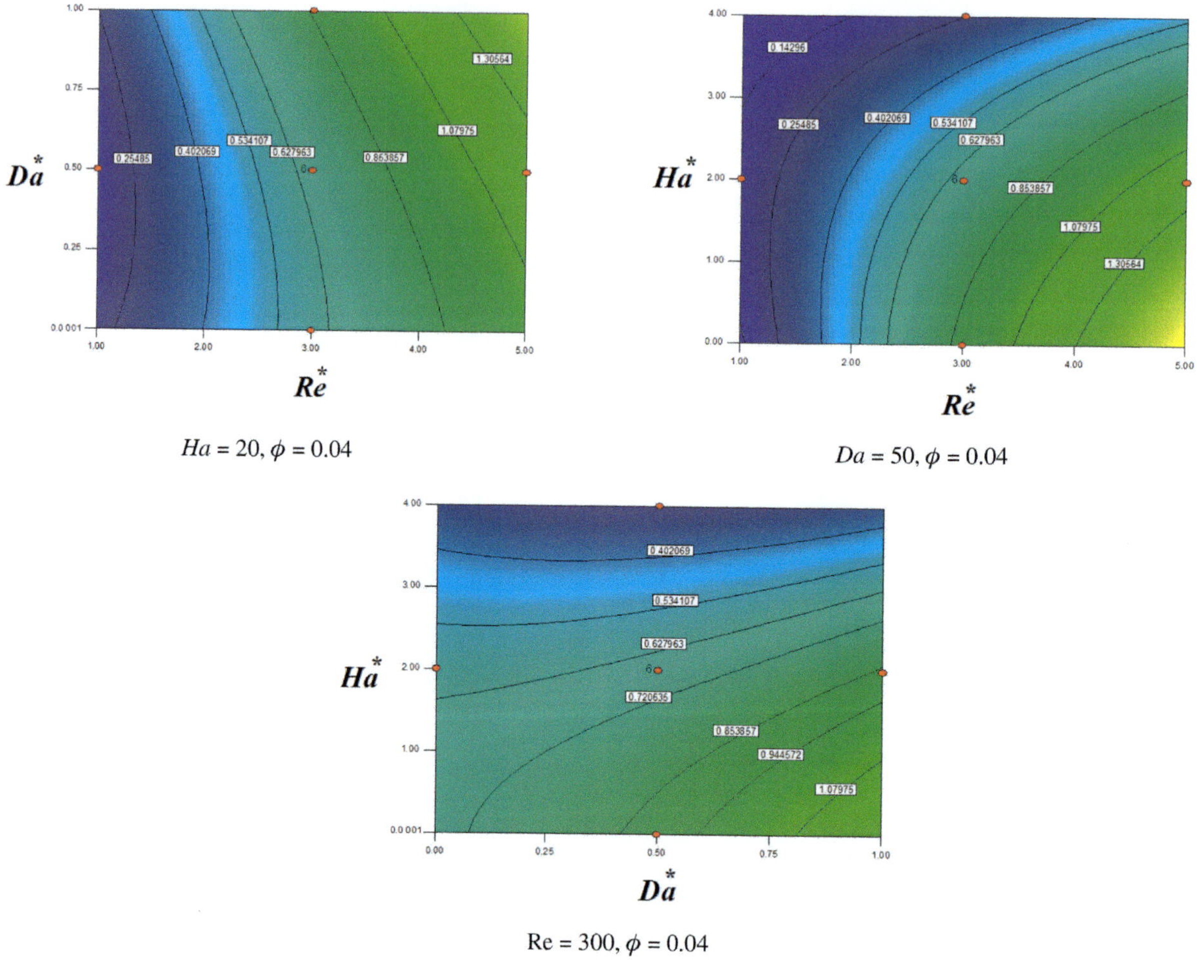

FIGURE 10.26 Influences of the Darcy, Reynolds, and Hartmann numbers on average Nusselt number.

10.6 MAGNETIC FIELD INFLUENCE ON CUO-H_2O NANOFLUID CONVECTIVE FLOW IN A PERMEABLE CAVITY CONSIDERING VARIOUS SHAPES FOR NANOPARTICLES

10.6.1 Problem Definition

Geometry, boundary conditions, and sample element are depicted in Fig. 10.27. The bottom wall is hot and the other walls are cold. Also the bottom wall can move horizontally. A porous cavity is filled with nanofluid and affected by horizontal magnetic field.

10.6.2 Governing Equation

2D steady convective non-Darcy flow of nanofluid is considered in the presence of a constant magnetic field. The PDEs equations are:

$$\frac{\partial v}{\partial y} + \frac{\partial u}{\partial x} = 0, \tag{10.81}$$

$$\left[-\sigma_{nf} B_y^2 u + \sigma_{nf} B_x B_y v + \left(\frac{\partial^2 u}{\partial y^2} + \frac{\partial^2 u}{\partial x^2}\right)\mu_{nf} - \frac{\partial P}{\partial x}\right] - \frac{\mu_{nf}}{K} u = \rho_{nf}\left(v\frac{\partial u}{\partial y} + \frac{\partial u}{\partial x} u\right), \tag{10.82}$$

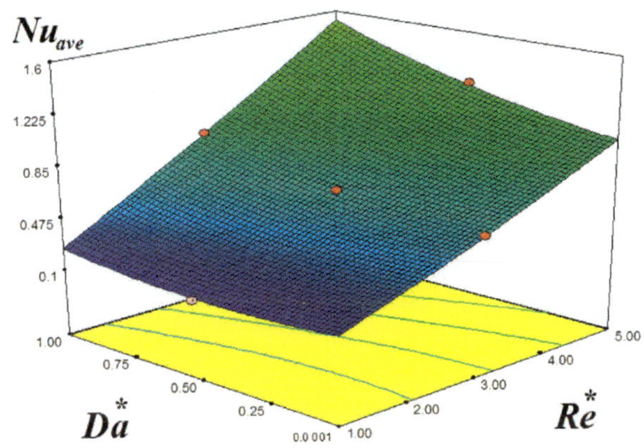

$Ha = 20$, $\phi = 0.04$

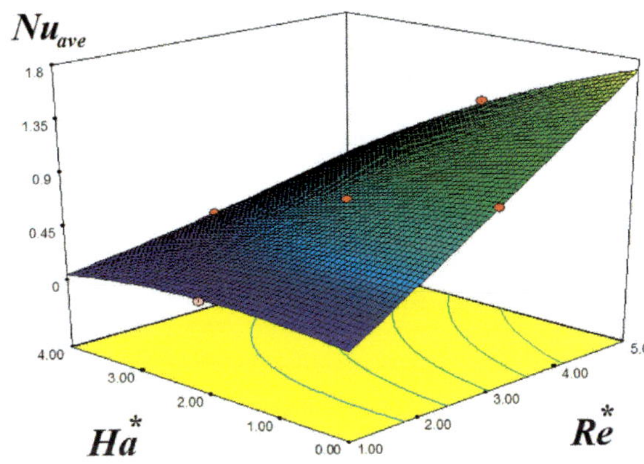

$Da = 50$, $\phi = 0.04$

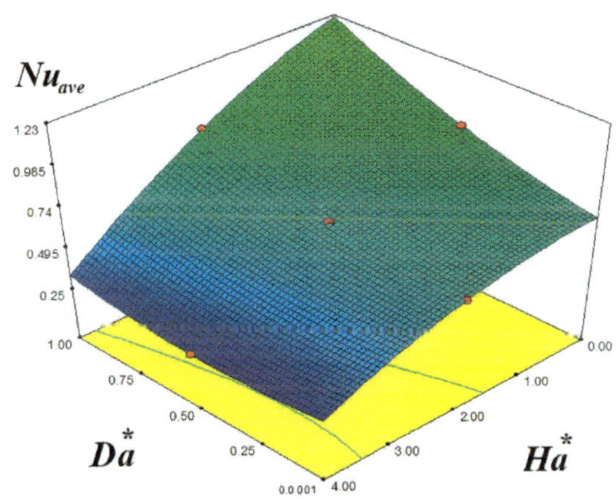

$\mathrm{Re} = 300$, $\phi = 0.04$

FIGURE 10.26 (Continued).

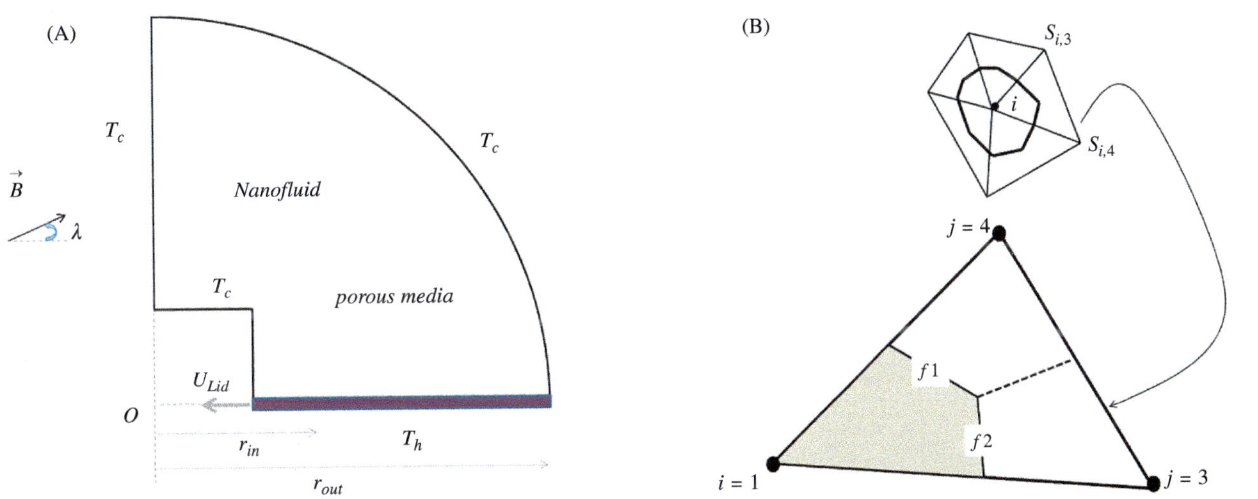

FIGURE 10.27 (A) Geometry and the boundary conditions with (B) a sample triangular element and its corresponding control volume.

$$\mu_{nf}\left(\frac{\partial^2 v}{\partial y^2}+\frac{\partial^2 v}{\partial x^2}\right)-\frac{\partial P}{\partial y}-B_x\sigma_{nf}B_x v+B_y\sigma_{nf}B_x u-\frac{\mu_{nf}}{K}v=\rho_{nf}\left(\frac{\partial v}{\partial x}u+\frac{\partial v}{\partial y}v\right),$$
$$B_x=B_o\cos\lambda,\ \ B_y=B_o\sin\lambda, \tag{10.83}$$

$$(\rho C_p)_{nf}\left(v\frac{\partial T}{\partial y}+u\frac{\partial T}{\partial x}\right)=k_{nf}\left(\frac{\partial^2 T}{\partial x^2}+\frac{\partial^2 T}{\partial y^2}\right). \tag{10.84}$$

$(\rho C_p)_{nf}$, ρ_{nf}, and σ_{nf} are defined as:

$$(\rho C_p)_{nf}=(\rho C_p)_f(1-\phi)+(\rho C_p)_s\phi, \tag{10.85}$$

$$\rho_{nf}=\rho_f(1-\phi)+\rho_s\phi, \tag{10.86}$$

$$\frac{\sigma_{nf}}{\sigma_f}=1+\frac{3\left(\frac{\sigma_s}{\sigma_f}-1\right)\phi}{\left(\frac{\sigma_s}{\sigma_f}+2\right)-\left(\frac{\sigma_s}{\sigma_f}-1\right)\phi}, \tag{10.87}$$

μ_{nf} is calculated via KKL model:

$$\mu_{eff}=\mu_{static}+\mu_{Brownian}=\mu_{static}+\frac{k_{Brownian}}{k_f}\times\frac{\mu_f}{\Pr_f},$$
$$k_{Brownian}=5\times 10^4 g'(\phi,T,d_p)\phi\rho_f c_{p,f}\sqrt{\frac{\kappa_b T}{d_p\rho_p}},$$
$$g'(\phi,d_p,T)=Ln(T)\left(a_1+a_3Ln(\phi)+a_2Ln(d_p)+a_5Ln(d_p)^2+a_4\ln(d_p)Ln(\phi)\right)$$
$$+\left(a_6+a_8Ln(\phi)+a_7Ln(d_p)+a_{10}Ln(d_p)^2+a_9Ln(\phi)\ln(d_p)\right). \tag{10.88}$$

k_{nf} can be expressed as:

$$\frac{k_{nf}}{k_f}=\frac{-m(k_f-k_p)\phi+(k_p-k_f)\phi+mk_f+k_p+k_f}{mk_f+(k_f-k_p)\phi+k_f+k_p} \tag{10.89}$$

Vorticity and stream function should be used to eliminate pressure source terms:

$$\omega+\frac{\partial u}{\partial y}-\frac{\partial v}{\partial x}=0,\ \ \frac{\partial\psi}{\partial x}=-v,\ \ \frac{\partial\psi}{\partial y}=u \tag{10.90}$$

Introducing dimensionless quantities:

$$U=\frac{u}{U_{Lid}},V=\frac{v}{U_{Lid}},\ \theta=\frac{T-T_c}{\Delta T},\Delta T=T_h-T_c,(X,Y)=\frac{(x,y)}{L},\Psi=\frac{\psi L}{U_{Lid}},\Omega=\frac{\omega}{LU_{Lid}}. \tag{10.91}$$

The final formulae are:

$$\frac{\partial^2\Psi}{\partial X^2}+\Omega+\frac{\partial^2\Psi}{\partial Y^2}=0, \tag{10.92}$$

$$\frac{\partial\Omega}{\partial Y}V+\frac{\partial\Omega}{\partial X}U=\frac{1}{\mathrm{Re}}\frac{A_5}{A_1}\left(\frac{\partial^2\Omega}{\partial Y^2}+\frac{\partial^2\Omega}{\partial X^2}\right)$$
$$+\frac{Ha^2}{\mathrm{Re}}\frac{A_6}{A_1}\left(\frac{\partial U}{\partial X}B_xB_y-\frac{\partial V}{\partial X}B_x{}^2+\frac{\partial U}{\partial Y}B_y^2-\frac{\partial V}{\partial Y}B_xB_y\right)-\frac{1}{\mathrm{Re}\,Da}\frac{A_5}{A_1}\Omega, \tag{10.93}$$

$$\frac{A_4}{A_2}\left(\frac{\partial^2\theta}{\partial Y^2}+\frac{\partial^2\theta}{\partial X^2}\right)=\mathrm{PrRe}\left(\frac{\partial\theta}{\partial X}U+\frac{\partial\theta}{\partial Y}V\right). \tag{10.94}$$

where dimensionless and constants parameters are illustrated as:

$$\mathrm{Re}=\frac{\rho_f U_{Lid} L}{\mu_f},\ Ha=LB_0\sqrt{\sigma_f/\mu_f},$$
$$A_1=\frac{\rho_{nf}}{\rho_f},A_5=\frac{\mu_{nf}}{\mu_f}, \tag{10.95}$$
$$A_2=\frac{(\rho C_p)_{nf}}{(\rho C_p)_f},A_4=\frac{k_{nf}}{k_f},A_6=\frac{\sigma_{nf}}{\sigma_f}.$$

and boundary conditions are:

$$\begin{array}{ll}\theta=1.0 & \text{on bottom wall}\\ \theta=0.0 & \text{on other walls}\\ \Psi=0.0 & \text{on all walls}\end{array} \tag{10.96}$$

Nu_{loc} and Nu_{ave} over the hot wall can be calculated as:

$$Nu_{loc}=\frac{\partial\theta}{\partial y}\left(\frac{k_{nf}}{k_f}\right) \tag{10.97}$$

$$Nu_{ave}=\frac{1}{L}\int_{r_{in}}^{r_{out}}Nu_{loc}dx \tag{10.98}$$

10.6.3 Effects of Active Parameters

In this paper, the effect of magnetic field on forced convection of CuO-H_2O nanofluid in a lid-driven porous cavity is demonstrated. μ_{nf} of CuO-H_2O nanofluid are estimated by means of KKL model. CVFEM is utilized to find the effects of Darcy number ($Da=0.01$ to 100), Reynolds number ($\mathrm{Re}=100$ to 500), volume fraction of CuO ($\phi=0\%$ to 4%), shape of nanoparticle, and Hartmann number ($Ha=0$ to 40). Influences of shape of the nanoparticles on Nu_{ave} are shown in Table 10.9. The maximum Nu is caused by Platelet, followed by Cylinder, Brick,

TABLE 10.9 Effect of Shape of Nanoparticles on Nusselt Number When $Da=100$, $\mathrm{Re}=600$, $\phi=0.04$

	Ha	
	0	**40**
Spherical	8.654815	7.765007
Brick	8.85174	7.947995
Cylinder	9.159554	8.234346
Platelet	9.409924	8.467536

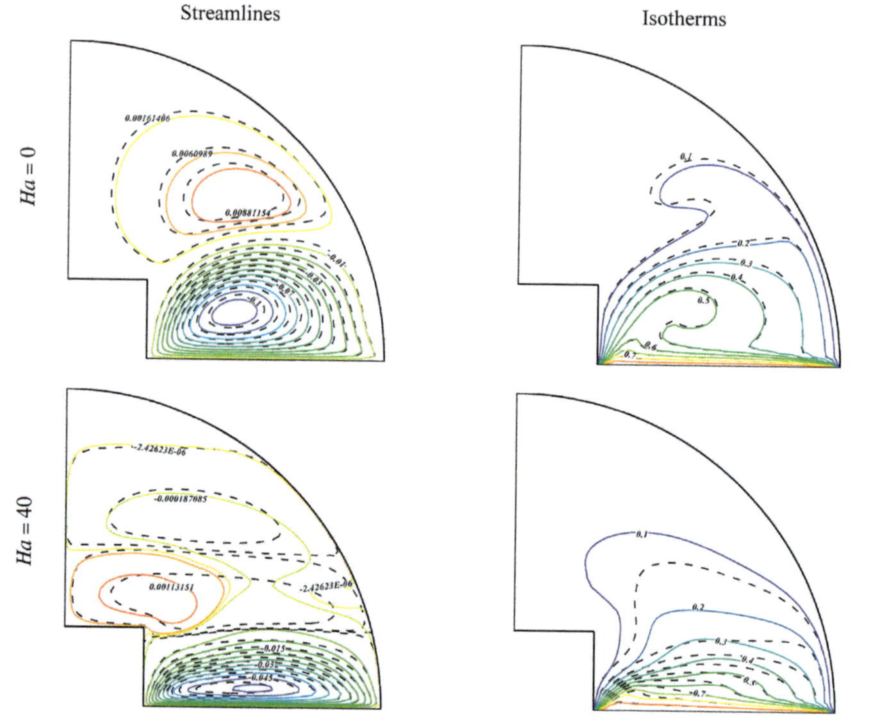

FIGURE 10.28 Influence of nanofluid volume fraction on streamline (left) and isotherm (right) contours (nanofluid ($\phi = 0.04$)(—) and pure fluid($\phi = 0$) (- - -)) when $\mathrm{Re} = 500, Da = 100$.

and Spherical. Therefore, Platelet nanoparticle has been selected for further investigation. The impact of adding nanoparticles in the base fluid on velocity and temperature contours is depicted in Fig. 10.28. The temperature gradient decreases with the increase of ϕ. $|\Psi_{\max}|$ increases with the addition of nanoparticles because of the increase in the solid movements. Figs. 10.29–10.31 demonstrate the impact of Da, Re, Ha on isotherms and streamlines. At $Da = 0.01, \mathrm{Re} = 100, Ha = 0$, only one eddy exists in streamline. By applying a magnetic field, two new eddies with opposite direction are generated above the main eddy. The distortion of isotherms becomes less than before. As the Darcy number increases, the convective mode becomes stronger, and the primary eddy converts to two eddies. Increasing the Hartmann number leads to the generation of the third eddy. As Re augments, the strength of secondary eddy, which is generated due to the increase of the Darcy number, enhances. In the absence of a magnetic field at $Re = 500$, a counterclockwise eddy appears above the main eddy. The impacts of Darcy and Hartmann numbers on isotherms and streamlines at $Re = 500$ is similar to those of $Re = 100$ except for the stream function values. It means that the absolute stream function values increase with the rise of the Reynolds number. The influence of significant parameters on Nu_{ave} is illustrated in Fig. 10.32. The correlation for Nu_{ave} is as follows:

$$\begin{aligned} Nu_{ave} = {} & 6.54 + 0.52\mathrm{Re}^* + 1.56Da^* + 0.05Ha^* \\ & + 0.12Da^*\,\mathrm{Re}^* + 0.03\mathrm{Re}^*\,Ha^* - 0.15Da^*\,Ha^* \\ & - 0.04\mathrm{Re}^{*2} + 0.58Da^{*2} - 0.14Ha^{*2} \end{aligned} \tag{10.99}$$

where $\mathrm{Re}^* = 0.01\mathrm{Re}, Ha* = 0.1Ha, Da* = 0.01Da$. Increasing the Reynolds number can enhance the heat transfer rate because of a reduction in the thermal boundary layer thickness. A similar trend is seen for the Darcy number. The temperature gradient decreases with the increase of Lorentz forces. So, Nu_{ave} decreases with the rise of the Hartmann number.

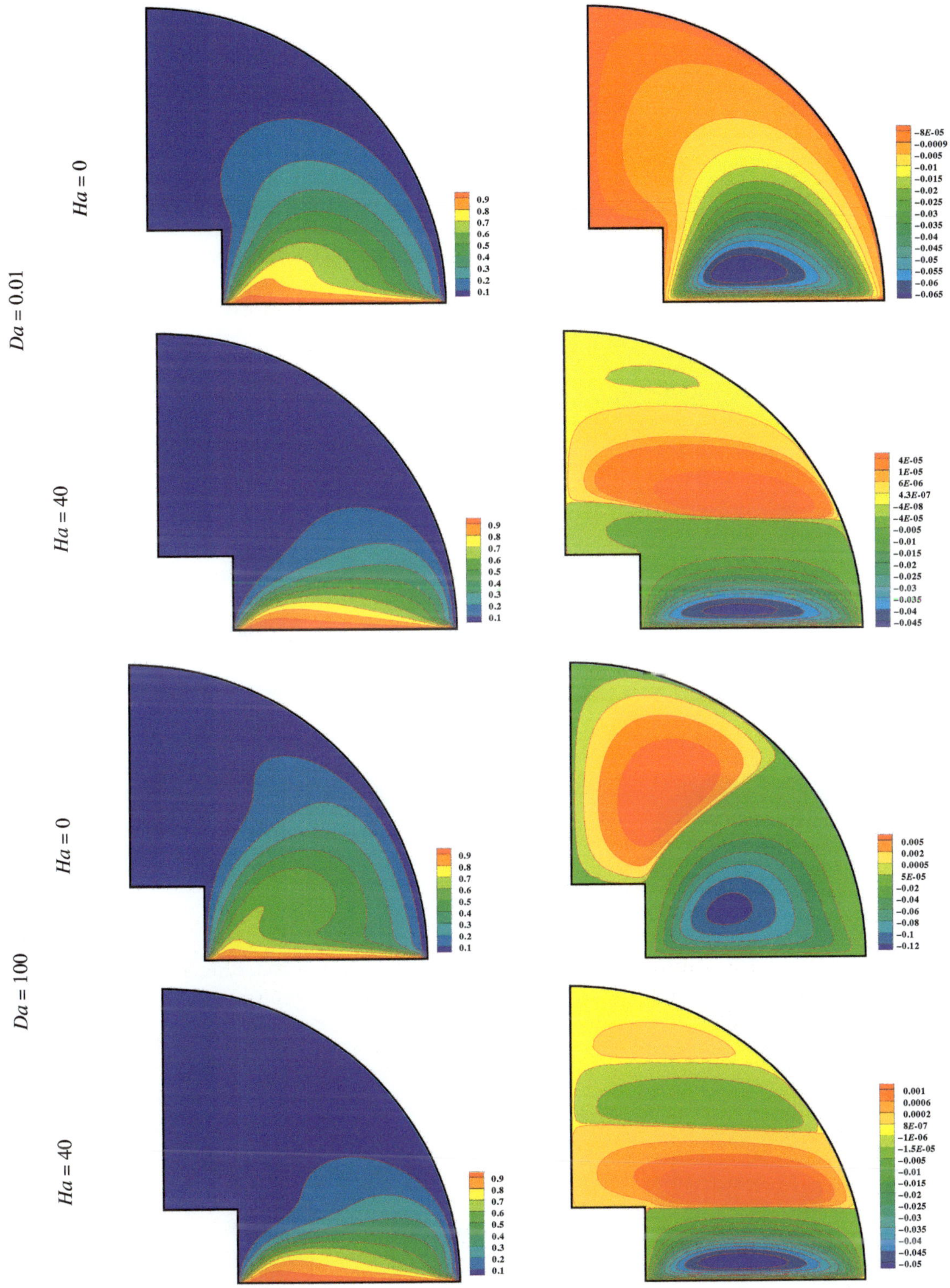

FIGURE 10.29 Effect of Darcy and Hartmann numbers on isotherm (left) and streamline (right) contours when $\phi = 0.04, \mathrm{Re} = 100$.

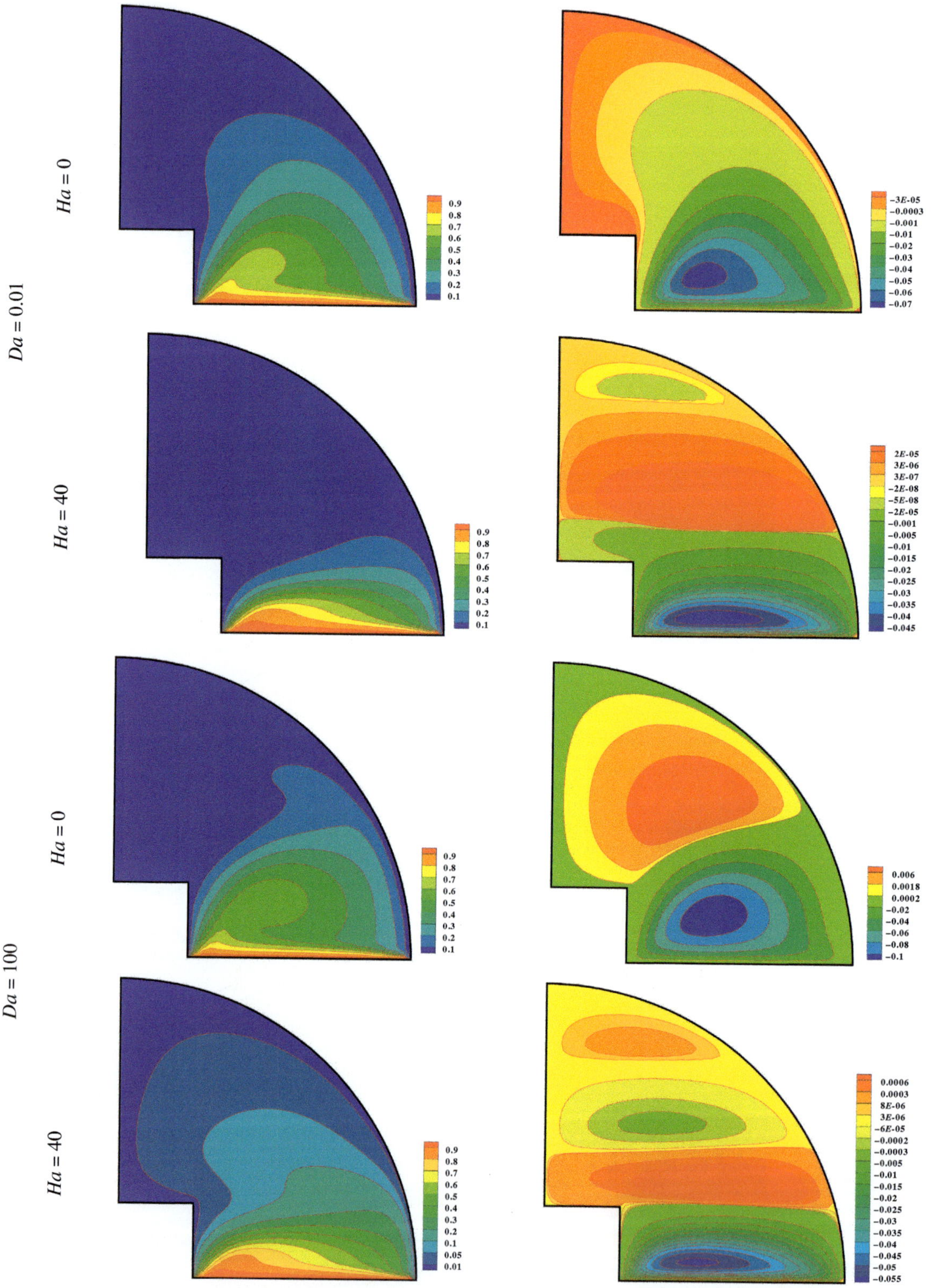

FIGURE 10.30 Effect of Darcy and Hartmann numbers on isotherm (left) and streamline (right) contours when $\phi = 0.04$, $\mathrm{Re} = 300$.

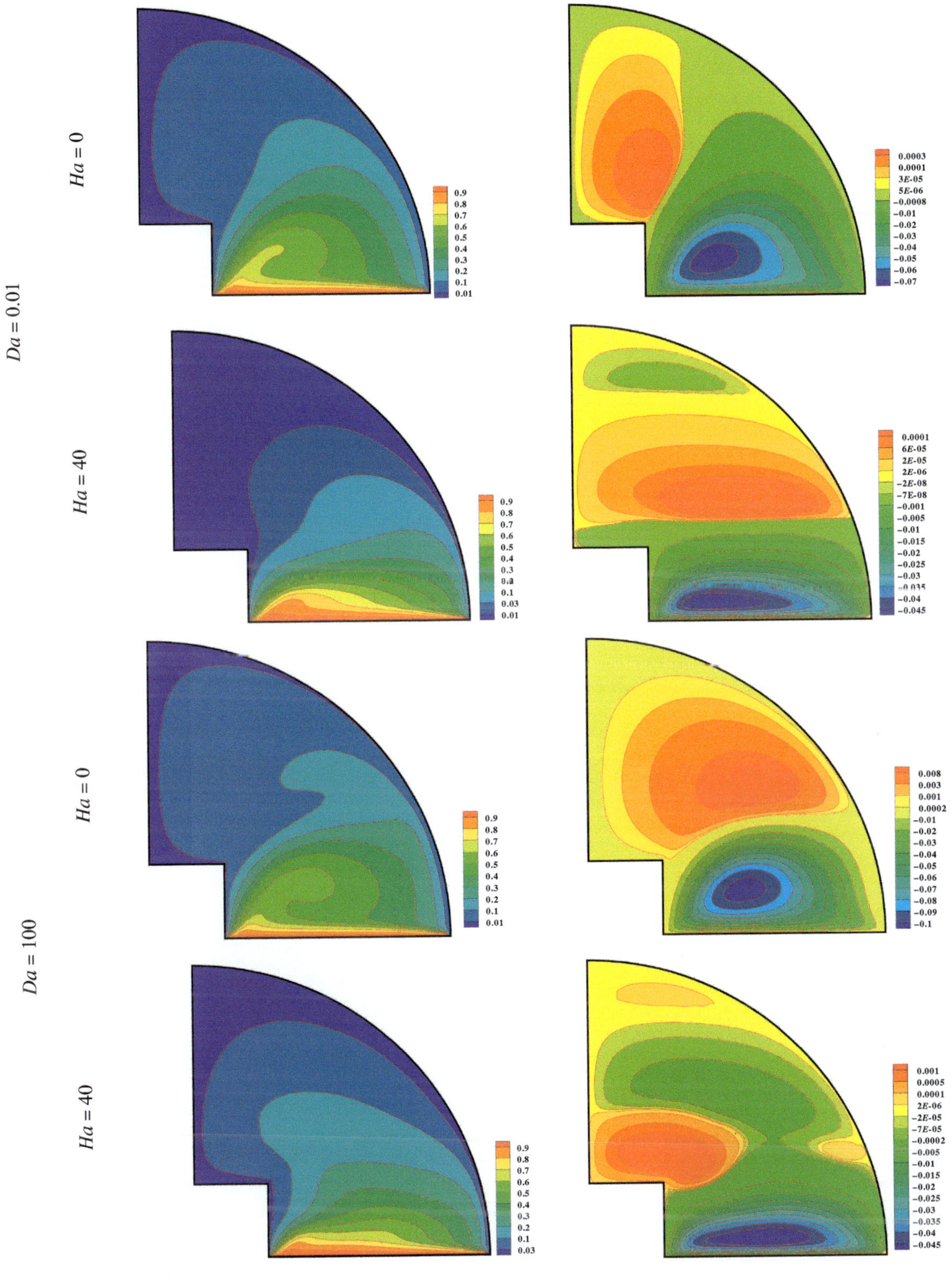

FIGURE 10.31 Effect of Darcy and Hartmann numbers on isotherm (left) and streamline (right) contours when $\phi = 0.04$, Re = 500.

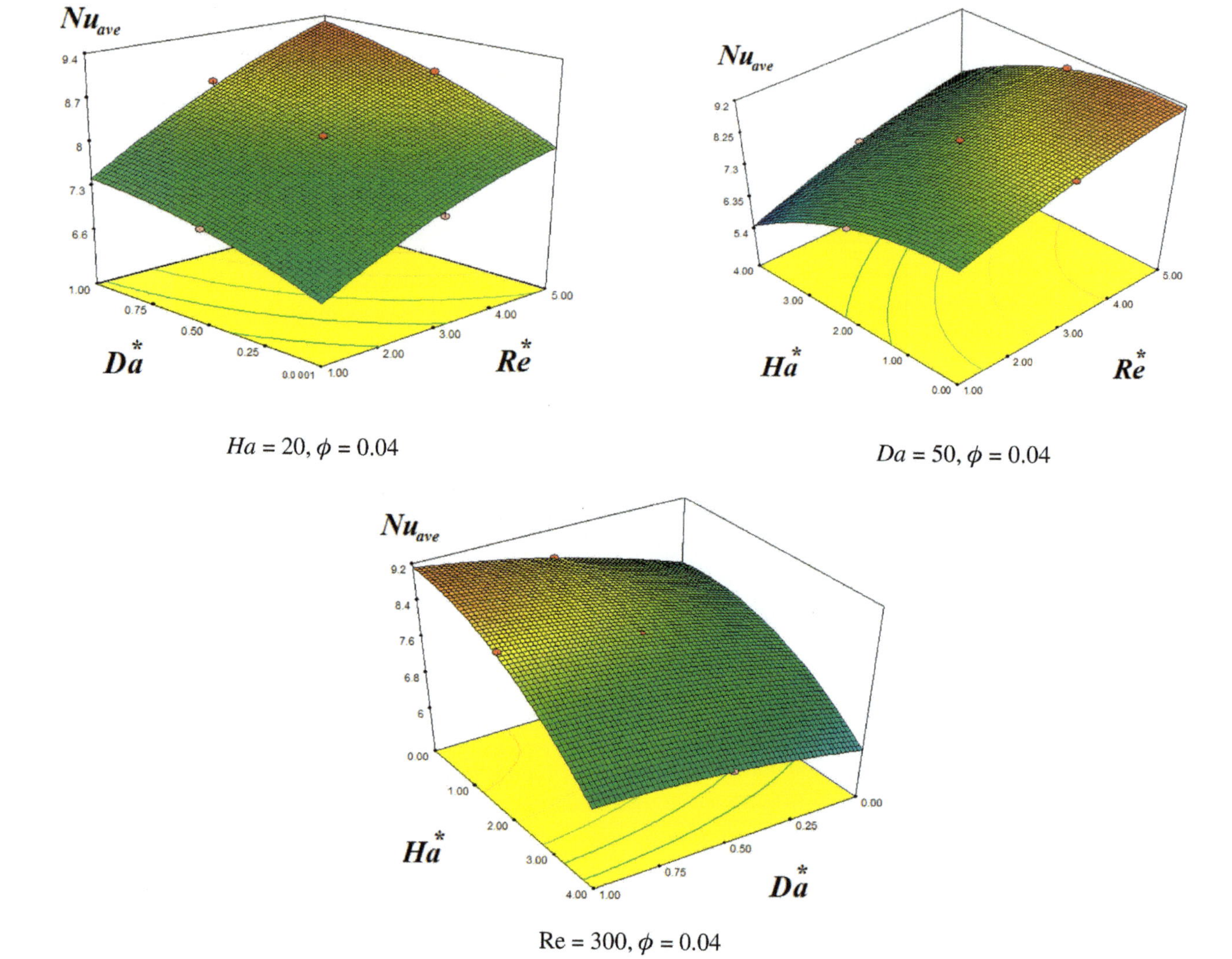

FIGURE 10.32 Influences of the Darcy, Reynolds, and Hartmann numbers on average Nusselt number.

References

[1] M. Sheikholeslami, Magnetic field influence on CuO -H_2O nanofluid convective flow in a permeable cavity considering various shapes for nanoparticles, Int. J. Hyd. Energy 42 (2017) 19611–19621.

[2] M. Sheikholeslami, M.M. Bhatti, Forced convection of nanofluid in presence of constant magnetic field considering shape effects of nanoparticles, Int. J. Heat Mass Transfer 111 (2017) 1039–1049.

[3] M. Sheikholeslami, M. Shamlooei, Magnetic source influence on nanofluid flow in porous medium considering shape factor effect, Phys. Lett. A 381 (2017) 3071–3078.

[4] M. Sheikholeslami, D.D. Ganji, Numerical modeling of magnetohydrodynamic CuO -water transportation inside a porous cavity considering shape factor effect, Colloids Surfaces A: Physicochem. Eng. Aspects 529 (2017) 705–714.

[5] M. Sheikholeslami, Influence of Lorentz forces on nanofluid flow in a porous cylinder considering Darcy model, J. Mol. Liq. 225 (2017) 903–912.

[6] M. Sheikholeslami, CVFEM for magnetic nanofluid convective heat transfer in a porous curved enclosure, Eur. Phys. J. Plus 131 (2016) 413. Available from: https://doi.org/10.1140/epjp/i2016-16413-y.

[7] M. Sheikholeslami, Influence of Coulomb forces on Fe_3O_4-H_2O nanofluid thermal improvement, Int. J. Hydrogen Energy 42 (2017) 821–829.

[8] M. Sheikholeslami, Magnetic field influence on nanofluid thermal radiation in a cavity with tilted elliptic inner cylinder, J. Mol. Liq. 229 (2017) 137–147.

[9] M. Sheikholeslami, Numerical simulation of magnetic nanofluid natural convection in porous media, Phys. Lett. A 381 (2017) 494–503.

[10] M. Sheikholeslami, Magnetic source impact on nanofluid heat transfer using CVFEM, Neural Comput. Applicat. (2016). Available from: https://doi.org/10.1007/s00521-016-2740-7.

[11] M. Sheikholeslami, CuO-water nanofluid free convection in a porous cavity considering Darcy law, Europ. Phys. J. Plus 132 (2017) 55. Available from: https://doi.org/10.1140/epjp/i2017-11330-3.
[12] M. Sheikholeslami, Numerical investigation of MHD nanofluid free convective heat transfer in a porous tilted enclosure, Eng. Comput. 34 (6) (2017) 1939–1955.
[13] M.S. Kandelousi, Effect of spatially variable magnetic field on ferrofluid flow and heat transfer considering constant heat flux boundary condition, Europ. Phys. J. Plus (2014) 129–248. Available from: https://doi.org/10.1140/epjp/i2014-14248-2.
[14] M. Sheikholeslami, Numerical simulation for external magnetic field influence on Fe3O4-water nanofluid forced convection, Engineering Computations, Inpress.
[15] M. Sheikholeslami, Influence of Lorentz forces on nanofluid flow in a porous cavity by means of Non- Darcy model, Eng. Comput. 34 (8) (2017) 2651–2667. Available from: https://doi.org/10.1108/EC-01-2017-0008.
[16] M. Sheikholeslami, H.B. Rokni, Nanofluid convective heat transfer intensification in a porous circular cylinder, Chem. Eng. Process. Process Intensif. 120 (2017) 93–104.
[17] M. Sheikholeslami, M.K. Sadoughi, Numerical modeling for Fe3O4 -water nanofluid flow in porous medium considering MFD viscosity, J. Mol. Liq. 242 (2017) 255–264.
[18] M. Sheikholeslami, M. Shamlooei, Convective flow of nanofluid inside a lid driven porous cavity using CVFEM, Phys. B: Condensed Matter 521 (2017) 239–250.
[19] M. Sheikholeslami, K. Vajravelu, Forced convection heat transfer in Fe3O4- ethylene glycol nanofluid under the influence of Coulomb force, J. Mol. Liq. 233 (2017) 203–210.
[20] M. Sheikholeslami, H.B. Rokni, Influence of EFD viscosity on nanofluid forced convection in a cavity with sinusoidal wall, J. Mol. Liq. 232 (2017) 390–395.
[21] M. Sheikholeslami, H.B. Rokni, Magnetic nanofluid natural convection in the presence of thermal radiation considering variable viscosity, Europ. Phys. J. Plus 132 (2017) 238. Available from: https://doi.org/10.1140/epjp/i2017-11498-4.
[22] M. Sheikholeslami, H.B. Rokni, Numerical simulation for impact of Coulomb force on nanofluid heat transfer in a porous enclosure in presence of thermal radiation, Int. J. Heat Mass Transfer 118 (2018) 823–831.
[23] M. Sheikholeslami, M. Shamlooei, R. Moradi, Fe3O4- Ethylene glycol nanofluid forced convection inside a porous enclosure in existence of Coulomb force, J. Mol. Liq. 249 (2018) 429–437.
[24] M. Sheikholeslami, S.A. Shehzad, Numerical analysis of Fe_3O_4 $-H_2O$ nanofluid flow in permeable media under the effect of external magnetic source, Int. J. Heat Mass Transfer 118 (2018) 182–192.
[25] M. Sheikholeslami, S.A. Shehzad, Magnetohydrodynamic nanofluid convection in a porous enclosure considering heat flux boundary condition, Int. J. Heat Mass Transfer 106 (2017) 1261–1269.
[26] M. Sheikholeslami, S.A. Shehzad, Thermal radiation of ferrofluid in existence of Lorentz forces considering variable viscosity, Int. J. Heat Mass Transfer 109 (2017) 82–92.
[27] M. Sheikholeslami, H.B. Rokni, Numerical modeling of nanofluid natural convection in a semiannulus in existence of Lorentz force, Comput. Methods Appl. Mech. Eng. 317 (2017) 419–430.
[28] M. Sheikholeslami, K. Vajravelu, Nanofluid flow and heat transfer in a cavity with variable magnetic field, Appl. Math. Comput. 298 (2017) 272–282.
[29] M. Sheikholeslami, H.B. Rokni, Magnetohydrodynamic CuO-water nanofluid in a porous complex shaped enclosure, ASME, J. Ther. Sci. Eng. Applicat. 9 (4) (2017) 041007. Available from: https://doi.org/10.1115/1.4035973.
[30] M. Sheikholeslami, H.B. Rokni, Nanofluid two phase model analysis in existence of induced magnetic field, Int. J. Heat Mass Transfer 107 (2017) 288–299.
[31] M. Sheikholeslami, T. Hayat, A. Alsaedi, S. Abelman, Numerical analysis of EHD nanofluid force convective heat transfer considering electric field dependent viscosity, Int. J. Heat Mass Transfer 108 (2017) 2558–2565.
[32] M. Sheikholeslami, A. Zeeshan, Analysis of flow and heat transfer in water based nanofluid due to magnetic field in a porous enclosure with constant heat flux using CVFEM, Comput. Methods Appl. Mech. Eng. 320 (2017) 68–81.
[33] M. Sheikholeslami, M. Shamlooei, Fe_3O_4- H_2O nanofluid natural convection in presence of thermal radiation, Int. J. Hydrogen Energy 42 (9) (2017) 5708–5718.
[34] M. Sheikholeslami, M.M. Bhatti, Active method for nanofluid heat transfer enhancement by means of EHD, Int. J. Heat Mass Transfer 109 (2017) 115–122.
[35] M. Sheikholeslami, T. Hayat, A. Alsaedi, Numerical study for external magnetic source influence on water based nanofluid convective heat transfer, Int. J. Heat Mass Transfer 106 (2017) 745–755.
[36] M. Sheikholeslami, A.J. Chamkha, Flow and convective heat transfer of a ferro-nanofluid in a double-sided lid-driven cavity with a wavy wall in the presence of a variable magnetic field, Numer. Heat Transfer, Part A 69 (10) (2016) 1186–1200. Available from: https://doi.org/10.1080/10407782.2015.1125709.
[37] M. Sheikholeslami, A.J. Chamkha, Electrohydrodynamic free convection heat transfer of a nanofluid in a semi-annulus enclosure with a sinusoidal wall, Numer. Heat Transfer, Part A 69 (7) (2016) 781–793. Available from: http://dx.doi.org/10.1080/10407782.2015.1090819.
[38] M. Sheikholeslami, M.K. Sadoughi, Simulation of CuO- water nanofluid heat transfer enhancement in presence of melting surface, Int. J. Heat Mass Transfer 116 (2018) 909–919.
[39] M. Sheikholeslami, H.B. Rokni, Simulation of nanofluid heat transfer in presence of magnetic field: a review, Int. J. Heat Mass Transfer 115 (2017) 1203–1233.
[40] M. Sheikholeslami, H.B. Rokni, Melting heat transfer influence on nanofluid flow inside a cavity in existence of magnetic field, Int. J. Heat Mass Transfer 114 (2017) 517–526.
[41] M. Sheikholeslami, S.A. Shehzad, Magnetohydrodynamic nanofluid convective flow in a porous enclosure by means of LBM, Int. J. Heat Mass Transfer 113 (2017) 796–805.
[42] M. Sheikholeslami, Magnetohydrodynamic nanofluid forced convection in a porous lid driven cubic cavity using Lattice Boltzmann Method, J. Mol. Liq. 231 (2017) 555–565.

[43] M.S. Kandelousi, KKL correlation for simulation of nanofluid flow and heat transfer in a permeable channel, Phys. Lett. A 378 (45) (2014) 3331–3339.

[44] M. Sheikholeslami, Effect of uniform suction on nanofluid flow and heat transfer over a cylinder, J. Braz. Soc. Mech. Sci. Eng. 37 (2015) 1623–1633.

[45] M. Sheikholeslami, Lattice Boltzmann Method simulation of MHD non-Darcy nanofluid free convection, Physica B 516 (2017) 55–71.

[46] M. Sheikholeslami, Influence of magnetic field on nanofluid free convection in an open porous cavity by means of Lattice Boltzmann Method, J. Mol. Liq. 234 (2017) 364–374.

[47] M. Sheikholeslami, A.J. Chamkha, Influence of Lorentz forces on nanofluid forced convection considering Marangoni convection, J. Mol. Liq. 225 (2017) 750–757.

[48] M. Sheikholeslami, T. Hayat, A. Alsaedi, Numerical simulation of nanofluid forced convection heat transfer improvement in existence of magnetic field using Lattice Boltzmann Method, Int. J. Heat Mass Transfer 108 (2017) 1870–1883.

[49] M. Sheikholeslami, R. Ellahi, Three dimensional mesoscopic simulation of magnetic field effect on natural convection of nanofluid, Int. J. Heat Mass Transfer 89 (2015) 799–808.

[50] M. Sheikholeslami, S. Abelman, Two phase simulation of nanofluid flow and heat transfer in an annulus in the presence of an axial magnetic field, IEEE Trans. Nanotechnol. 14 (3) (2015) 561–569.

[51] M. Sheikholeslami, R. Ellahi, Simulation of ferrofluid flow for magnetic drug targeting using Lattice Boltzmann method, J. Zeitschrift Fur Naturforschung A 70 (2) (2015) 115–124.

[52] M. Sheikholeslami, M. Seyednezhad, Lattice Boltzmann Method simulation for CuO-water nanofluid flow in a porous enclosure with hot obstacle, J. Mol. Liq. 243 (2017) 249–256.

[53] M. Sheikholeslami, M. Sadoughi, Mesoscopic method for MHD nanofluid flow inside a porous cavity considering various shapes of nanoparticles, Int. J. Heat Mass Transfer 113 (2017) 106–114.

[54] M. Sheikholeslami, A. Zeeshan, Mesoscopic simulation of CuO-H_2O nanofluid in a porous enclosure with elliptic heat source, Int. J. Hydrogen Energy 42 (22) (2017) 15393–15402.

[55] M. Sheikholeslami, S.A. Shehzad, CVFEM for influence of external magnetic source on $Fe_3O_4 - H_2O$ nanofluid behavior in a permeable cavity considering shape effect, Int. J. Heat Mass Transfer 115 (2017) 180–191.

[56] M. Sheikholeslami, M. Seyednezhad, Nanofluid heat transfer in a permeable enclosure in presence of variable magnetic field by means of CVFEM, Int. J. Heat Mass Transfer 114 (2017) 1169–1180.

[57] M. Sheikholeslami, H.B. Rokni, Free convection of CuO-H_2O nanofluid in a curved porous enclosure using mesoscopic approach, Int. J. Hydrogen Energy 42 (22) (2017) 14942–14949.

[58] M. Sheikholeslami, Numerical investigation for CuO-H_2O nanofluid flow in a porous channel with magnetic field using mesoscopic method, J. Mol. Liq. 249 (2018) 739–746.

[59] M. Sheikholeslami, H.B. Rokni, Influence of melting surface on MHD nanofluid flow by means of two phase model, Chin. J. Phys. 55 (2017) 1352–1360.

[60] M. Sheikholeslami, H.B. Rokni, Effect of melting heat transfer on nanofluid flow in existence of magnetic field considering Buongiorno Model, Chin. J. Phys. 55 (2017) 1115–1126.

[61] M. Sheikholeslami, T. Hayat, A. Alsaedi, MHD free convection of Al_2O_3–water nanofluid considering thermal radiation: a numerical study, Int. J. Heat Mass Transfer 96 (2016) 513–524.

[62] M. Sheikholeslami, K. Vajravelu, M.M. Rashidi, Forced convection heat transfer in a semiannulus under the influence of a variable magnetic field, Int. J. Heat Mass Transfer 92 (2016) 339–348.

[63] M. Sheikholeslami, T. Hayat, A. Alsaedi, Numerical simulation for forced convection flow of MHD CuO-H_2O nanofluid inside a cavity by means of LBM, J. Mol. Liq. 249 (2018) 941–948.

[64] M. Sheikholeslami, H.R. Kataria, Akhil, S. Mittal, Effect of thermal diffusion and heat-generation on MHD nanofluid flow past an oscillating vertical plate through porous medium, J. Mol. Liq. 257 (2018) 12–25.

[65] M. Sheikholeslami, S.A. Shehzad, Z. Li, Nanofluid heat transfer intensification in a permeable channel due to magnetic field using Lattice Boltzmann method, Phys. B (2018). Available from: https://doi.org/10.1016/j.physb.2018.03.036.

[66] M. Sheikholeslami, A. Zeeshan, Numerical simulation of Fe_3O_4 -water nanofluid flow in a non-Darcy porous media, Int. J. Numer. Methods Heat Fluid Flow Vol. 28 (Issue: 3) (2018) 641–660. Available from: https://doi.org/10.1108/HFF-04-2017-0160.

[67] M. Sheikholeslami, M. Barzegar Gerdroodbary, S. Valiallah Mousavi, D.D. Ganji, R. Moradi, Heat transfer enhancement of ferrofluid inside an 90o elbow channel by non-uniform magnetic field, J. Magnet. Magnet. Mater. 460 (2018) 302–311.

[68] M. Sheikholeslami, Numerical simulation for solidification in a LHTESS by means of nano-enhanced PCM, J. Taiwan Inst. Chem. Eng. 86 (2018) 25–41.

[69] M. Sheikholeslami, Numerical modeling of Nano enhanced PCM solidification in an enclosure with metallic fin, J. Mol. Liq. 259 (2018) 424–438.

[70] M. Sheikholeslami, A. Ghasemi, Solidification heat transfer of nanofluid in existence of thermal radiation by means of FEM, Int. J. Heat Mass Transfer 123 (2018) 418–431.

[71] M. Sheikholeslami, S.A. Shehzad, CVFEM simulation for nanofluid migration in a porous medium using Darcy model, Int. J. Heat Mass Transfer 122 (2018) 1264–1271.

[72] M. Sheikholeslami, M. Darzi, M.K. Sadoughi, Heat transfer improvement and pressure drop during condensation of refrigerant-based nanofluid; an experimental procedure, Int. J. Heat Mass Transfer 122 (2018) 643–650.

[73] M. Sheikholeslami, H.B. Rokni, CVFEM for effect of Lorentz forces on nanofluid flow in a porous complex shaped enclosure by means of non-equilibrium model, J. Mol. Liq. 254 (2018) 446–462.

[74] M. Sheikholeslami, S.A. Shehzad, Z. Li, Water based nanofluid free convection heat transfer in a three dimensional porous cavity with hot sphere obstacle in existence of Lorenz forces, Int. J. Heat Mass Transfer 125 (2018) 375–386.

[75] M. Sheikholeslami, M. Jafaryar, D.D. Ganji, Z. Li, Exergy loss analysis for nanofluid forced convection heat transfer in a pipe with modified turbulators, J. Mol. Liq. 262 (2018) 104–110.

[76] M. Sheikholeslami, M. Jafaryar, Z. Li, Nanofluid turbulent convective flow in a circular duct with helical turbulators considering CuO nanoparticles, Int. J. Heat Mass Transfer 124 (2018) 980–989.
[77] M. Sheikholeslami, Z. Li, M. Shamlooei, Nanofluid MHD natural convection through a porous complex shaped cavity considering thermal radiation, Phys. Lett. A 382 (2018) 1615–1632.
[78] M. Sheikholeslami, H.B. Rokni, Magnetic nanofluid flow and convective heat transfer in a porous cavity considering Brownian motion effects, Phys. Fluids, 30, 1, https://doi.org/10.1063/1.5012517
[79] M. Sheikholeslami, S.A. Shehzad, Simulation of water based nanofluid convective flow inside a porous enclosure via non-equilibrium model, Int. J. Heat Mass Transfer 120 (2018) 1200–1212.
[80] M. Sheikholeslami, S.A. Shehzad, Non-Darcy free convection of Fe_3O_4-water nanoliquid in a complex shaped enclosure under impact of uniform Lorentz force, Chin. J. Phys. 56 (2018) 270–281.
[81] M. Sheikholeslami, M. Seyednezhad, Simulation of nanofluid flow and natural convection in a porous media under the influence of electric field using CVFEM, Int. J. Heat Mass Transfer 120 (2018) 772–781.
[82] M. Sheikholeslami, M. Shamlooei, R. Moradi, Numerical simulation for heat transfer intensification of nanofluid in a permeable curved enclosure considering shape effect of Fe_3O_4 nanoparticles, Chem. Eng. Process. Process Intensif. 124 (2018) 71–82.
[83] M. Sheikholeslami, T. Hayat, T. Muhammad, A. Alsaedi, MHD forced convection flow of nanofluid in a porous cavity with hot elliptic obstacle by means of Lattice Boltzmann method, Int. J. Mech. Sci. 135 (2018) 532–540.
[84] M. Sheikholeslami, Numerical investigation of nanofluid free convection under the influence of electric field in a porous enclosure, J. Mol. Liq. 249 (2018) 1212–1221.
[85] M. Sheikholeslami, CuO-water nanofluid flow due to magnetic field inside a porous media considering Brownian motion, J. Mol. Liq. 249 (2018) 921–929.
[86] M. Sheikholeslami, D.D. Ganji, Influence of electric field on Fe_3O_4- water nanofluid radiative and convective heat transfer in a permeable enclosure, J. Mol. Liq. 250 (2018) 404–412.

CHAPTER

11

Electrohydrodynamic Nanofluid Natural Convection Using CVFEM

11.1 INTRODUCTION

It is well known that an externally applied electric field across a plane layer of dielectric liquid can induce secondary flows. This phenomenon is called the Electrohydrodynamic (EHD) effect and can result in heat transfer enhancement. Such EHD enhancement is particularly attractive for enhancing convective heat transfer of a weakly conducting fluid through a narrow space at low Reynolds number, where the application of any conventional passive enhancement methods is neither easy nor effective [1]. In the last 30 years, many experimental studies on EHD enhancement of heat transfer have been carried out, not only for forced convection but also for two-phase heat transfer [2]. Among these studies, EHD enhancement of nucleate boiling has drawn researcher's particular attention in order to pursue the potential applications of such EHD techniques in renewable energy sources, in which heat exchangers with small temperature difference are normally required [3]. Sheikholeslami et al. [4] investigated EHD free convection heat transfer of a nanofluid in a semiannulus enclosure with a sinusoidal wall. Recently, electric field and magnetic field effect have been investigated by several scientists [5–90].

11.2 ELECTROHYDRODYNAMIC FREE CONVECTION HEAT TRANSFER OF A NANOFLUID IN A SEMIANNULUS ENCLOSURE WITH A SINUSOIDAL WALL

11.2.1 Problem Definition

Fig. 11.1 illustrates the physical geometry along with the important parameters and mesh of the semiannulus enclosure. The lower wall has a constant temperature T_1 and the temperatures of the other walls are T_0. Also, the remaining boundary conditions are depicted in Fig. 11.1A. The shape of the inner cylindrical profile is assumed to mimic the following pattern:

$$r = r_{in} + A\cos(N(\zeta)) \tag{11.1}$$

where r_{in} is the base circle radius, r_{out} is the radius of the outer cylinder, A and N are the amplitude and the number of undulations, respectively. ζ is the rotation angle. In this study, A and N are equal to 0.025 and 48, respectively.

11.2.2 Governing Equation

In order to simulate the nanofluid hydrothermal treatment in the existence of an electric field, we should combine equations of electric fields with those of hydrothermal treatment. The formulas of the electric field are:

$$\nabla.\left(\vec{E}\,\varepsilon\right) = q \tag{11.2}$$

$$(-\nabla\varphi) = \vec{E} \tag{11.3}$$

Application of Control Volume based Finite Element Method (CVFEM) for Nanofluid Flow and Heat Transfer.
DOI: https://doi.org/10.1016/B978-0-12-814152-6.00011-4

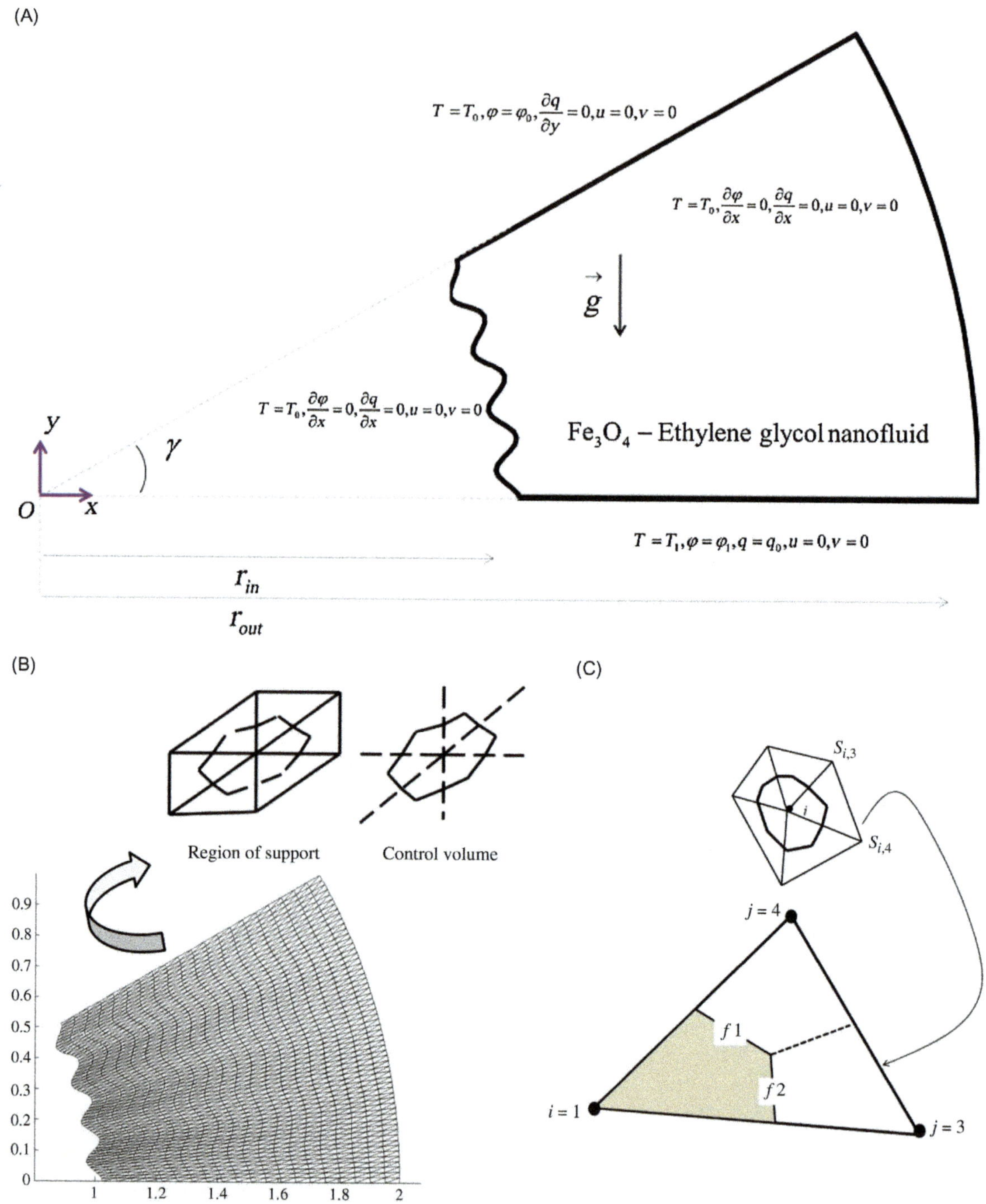

FIGURE 11.1 (A) Geometry of the problem and boundary conditions; (B) the mesh of enclosure considered in this work; (C) a sample triangular element and its corresponding control volume.

$$\nabla.\vec{J}+\frac{\partial q}{\partial t}=0 \tag{11.4}$$

There exist two models for charge distribution: (1) conductivity model and (2) mobility model. In the first model, the electroconvection relies on the temperature gradient. But in the second model, the electroconvection is independent of the temperature gradient in the liquid. In the case of free charge origination, the second model is more acceptable according to experimental results. The electric current density can be defined as:

$$\vec{J}=\sigma\vec{E}-D\nabla q+q\vec{V} \tag{11.5}$$

where $\sigma\vec{E}$ represents the ionic mobility, $D\nabla q$ represents the diffusion, $q\vec{V}$ represents the convection.

TABLE 11.1 Thermophysical Properties of Water and Nanoparticles

	$\rho\,(kg/m^3)$	$C_p\,(j/kgk)$	$\beta \times 10^{-5}(1/k)$	$k\,(W/m.k)$
Ethylene glycol	1110	2400	65	0.26
Fe_3O_4	5200	670	1.3	6

According to Eqs. (11.4) and (11.5), the equation for the electric charge density can be obtained as follows:

$$\frac{\partial \overline{q}}{\partial \overline{t}} + \overline{v}\frac{\partial \overline{q}}{\partial \overline{y}} + \overline{u}\frac{\partial \overline{q}}{\partial \overline{x}} + \frac{\Pr}{\Pr_E}\left[\overline{q}\left(\frac{\partial \overline{E}_y}{\partial \overline{y}} + \frac{\partial \overline{E}_x}{\partial \overline{x}}\right) + \overline{E}_y\frac{\partial \overline{q}}{\partial \overline{y}} + \overline{E}_x\frac{\partial \overline{q}}{\partial \overline{x}}\right] = \frac{\Pr}{D_e}\left(\frac{\partial^2 \overline{q}}{\partial \overline{y}^2} + \frac{\partial^2 \overline{q}}{\partial \overline{x}^2}\right) \tag{11.6}$$

According to [4] the diffusion term can be taken as negligible. Also $D\nabla q$ in Eq. (11.5) can be taken as negligible and $\sigma = bq$ [4]. Therefore, Eq. (11.5) can be considered as:

$$\vec{J} = q\vec{V} + qb\vec{E} \tag{11.7}$$

In the presence of an electric field, the Coulomb forces should be added to the momentum equation and Joule heating effect should be added in the energy equation. So, we have:

$$\begin{cases} \nabla.\vec{V} = 0 \\ \rho_{nf}\left(\dfrac{\partial \vec{V}}{\partial t} + \left(\vec{V}.\nabla\right)\vec{V}\right) = -\nabla p + \mu_{nf}\nabla^2\vec{V} + q\vec{E} - \beta\vec{g}(T - T_0) \\ \left(\rho C_p\right)_{nf}\left(\dfrac{\partial T}{\partial t} + \left(\vec{V}.\nabla\right)T\right) = k_{nf}\nabla^2 T + \vec{J}.\vec{E} \\ \nabla.\vec{J} + \dfrac{\partial q}{\partial t} = 0 \\ \nabla.\varepsilon\vec{E} = q \\ \vec{E} = -\nabla\varphi \end{cases} \tag{11.8}$$

ρ_{nf}, $\left(\rho C_p\right)_{nf}$, α_{nf}, β_{nf}, μ_{nf}, and k_{nf} are defined as [4]:

$$\rho_{nf} = \rho_s\phi + \rho_f(1-\phi) \tag{11.9}$$

$$\left(\rho C_p\right)_{nf} = \left(\rho C_p\right)_s\phi + \left(\rho C_p\right)_f(1-\phi) \tag{11.10}$$

$$\alpha_{nf} = k_{nf}/\left(\rho C_p\right)_{nf} \tag{11.11}$$

$$\beta_{nf} = \beta_s\phi + \beta_f(1-\phi) \tag{11.12}$$

$$k_{nf} = k_f\frac{-2\phi(k_f - k_s) + 2k_f + k_s}{+\phi(k_f - k_s)_s + 2k_f + k} \tag{11.13}$$

The thermophysical properties of the working fluid are given in Table 11.1. The electric field-dependent viscosity of Fe_3O_4-ethylene glycol nanofluid can be obtained as follows [4]:

$$\mu = A_1 + A_2(\Delta\varphi) + A_3(\Delta\varphi)^2 + A_4(\Delta\varphi)^3 \tag{11.14}$$

Table 11.2 shows the coefficient values of Eq. (11.14). Nondimensional parameters are introduced as follows:

$$\begin{gathered} \overline{t} = \frac{t\alpha_f}{L^2}, \overline{p} = \frac{P}{\rho\left(\alpha_f/L\right)^2}, \overline{y} = \frac{y}{L}, \overline{x} = \frac{x}{L}, \overline{v} = \frac{v\,L}{\alpha_f}, \overline{u} = \frac{u\,L}{\alpha_f}, \\ \theta = \frac{T - T_0}{\nabla T}, \nabla T = T_1 - T_0, \overline{\varphi} = \frac{\varphi - \varphi_0}{\nabla\varphi}, \nabla\varphi = \varphi_1 - \varphi_0, \overline{q} = \frac{q}{q_0}, \overline{E} = \frac{E}{E_0} \end{gathered} \tag{11.15}$$

TABLE 11.2 The Coefficient Values of Eq. (11.13)

Coefficient Values	$\phi=0$	$\phi=0.05$
A_1	1.0603E + 001	9.5331
A_2	−2.698E-003	−3.4119E-003
A_3	2.9082E-006	5.5228E-006
A_4	−1.1876E-008	−4.1344E-008

where ∇T and $\nabla\varphi$ are $(T_1 - T_0)$ and $(\varphi_1 - \varphi_0)$, respectively. In order to reach a clear formulation, the over bar will be deleted in the next equations. So, the governing equations can be considered as follows:

$$\begin{cases} \nabla.\left(\vec{V}\right)=0 \\ \left(\dfrac{\partial \vec{V}}{\partial t}+\left(\vec{V}.\nabla\right)\vec{V}\right)=-\nabla p+\dfrac{\mu_{nf}/\mu_f}{\rho_{nf}/\rho_f}\mathrm{Pr}\nabla^2\vec{V}+\dfrac{S_E}{\rho_{nf}/\rho_f}q\vec{E}-Ra\mathrm{Pr}\left[\dfrac{\beta_{nf}}{\beta_f}\right]\theta \\ \left(\dfrac{\partial \theta}{\partial t}+\left(\vec{V}.\nabla\right)\theta\right)=\dfrac{k_{nf}/k_f}{(\rho C_p)_{nf}/(\rho C_p)_f}\nabla^2\theta+\dfrac{S_E Ec}{(\rho C_p)_{nf}/(\rho C_p)_f}\left(\vec{J}.\vec{E}\right) \\ \nabla.\vec{J}+\dfrac{\partial q}{\partial t}=0 \\ \nabla.\varepsilon\vec{E}=q \\ \vec{E}=-\nabla\varphi \end{cases} \tag{11.16}$$

The formulas of the vorticity and the stream function are:

$$\begin{aligned} \omega &= \frac{\partial v}{\partial x}-\frac{\partial u}{\partial y}, \\ v &= -\frac{\partial \psi}{\partial x},\ u=\frac{\partial \psi}{\partial y}, \\ \Omega &= \frac{\omega L^2}{\alpha},\ \Psi=\frac{\psi}{\alpha} \end{aligned} \tag{11.17}$$

It should be mentioned that the continuity equation has been satisfied by the stream function. By eliminating pressure between the x-momentum and the y-momentum equations, the vorticity equation can be obtained.

The local Nusselt number Nu_{loc} and the average Nusselt number Nu_{ave} along the hot wall can be obtained as:

$$Nu_{loc} = \left(\frac{k_{nf}}{k_f}\right)\frac{\partial \Theta}{\partial Y} \tag{11.18}$$

$$Nu_{ave} = \frac{1}{L}\int_{r_{in}}^{r_{out}} Nu_{loc}\, dX \tag{11.19}$$

11.2.3 Effects of Active Parameters

The effect of nonuniform electric field on a nanofluid free convection heat transfer in an enclosure with a sinusoidal wall is presented. The working nanofluid is a mixture of ethylene glycol and Fe_3O_4. The viscosity of the nanofluid relies on the strength of the electric field. Calculations are prepared for various values of the supplied voltage ($\Delta\varphi = 0, 2, 4,$ and $6kV$), volume fraction of nanoparticles ($\phi = 0\%$ and 5%), and the Rayleigh number (Ra = 50, 100, and 500). In all calculations, the Prandtl number (Pr) and the Eckert number (Ec) are set to 149.54 and 10^{-6}, respectively. Fig. 11.2 depicts the distribution of the electric density distribution injected by the bottom

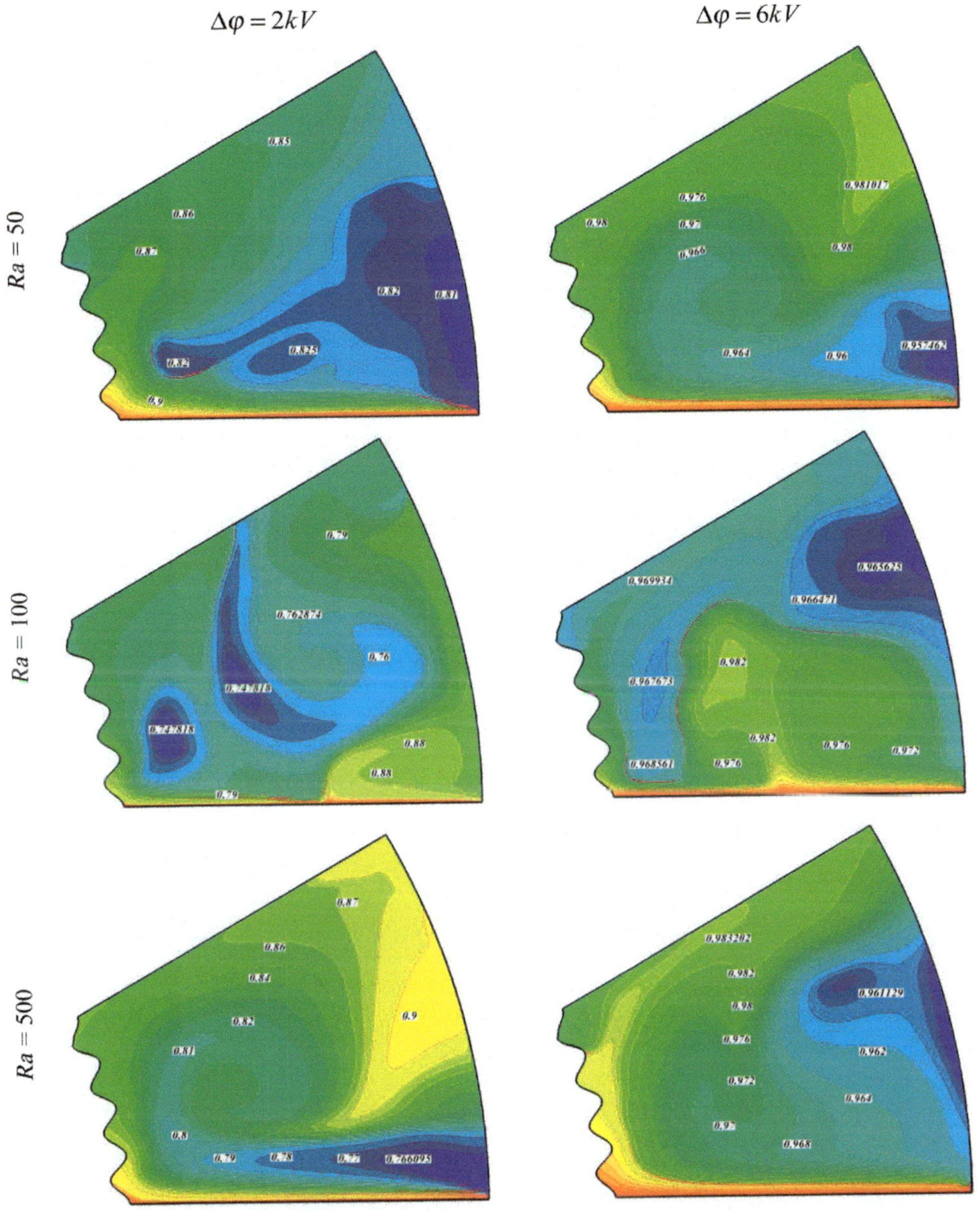

FIGURE 11.2 Electric density distribution injected by the bottom electrode.

electrode for different Rayleigh numbers and supplied voltages. It is observed that the electric density contours become more disturbed for higher values of the supplied voltage. The influence of the Rayleigh number and the supplied voltage on the streamlines and the isotherms are shown in Figs. 11.3–11.5. At $Ra = 50$, two main eddies exist in the streamlines which rotate in the reverse direction. As the electric field is applied, the two main eddies combine and turn into one eddy. Also, the isotherm become denser near the hot wall due to the existence of these eddies. As the Rayleigh number increases, the buoyancy forces increase and in turn, the rate of heat transfer increases. In order to show the effect of the magnetic field, we consider the low Rayleigh number, so that the effect of the Rayleigh number on the flow and heat transfer is not sensible. Thermal plumes generate because of the existence of different vortices near the hot wall. Figs. 11.6 and 11.7 depict the influence of $\Delta\varphi$ and Ra on the values of Nu_{loc} and Nu_{ave} along the hot wall. As the Rayleigh number increases, the Nusselt number increases due to a decrease in the thermal boundary layer thickness. Increasing the supplied voltage makes the isotherms more distorted. The local Nusselt number profiles show extreme values at higher values of the supplied voltage because of the existence of the thermal plumes. The Nusselt number is an increasing function of the supplied voltage. In the absence of the electric field, the Nusselt number for $Ra = 500$ is higher than that corresponding to

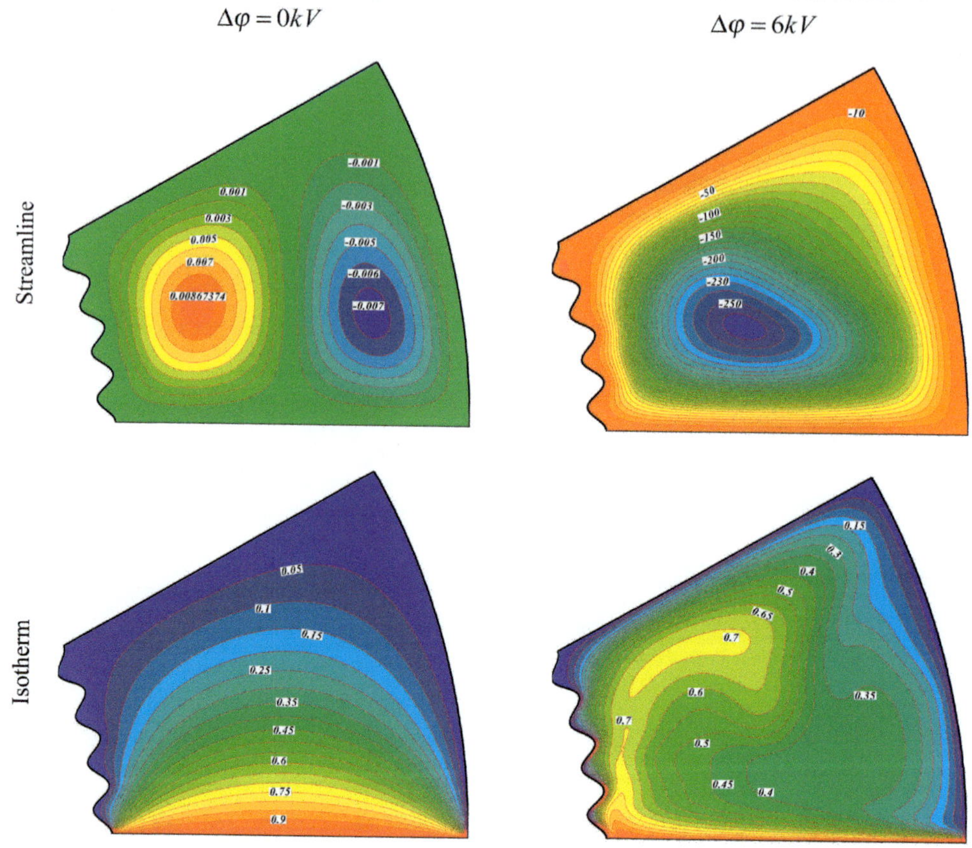

FIGURE 11.3 Effect of supplied voltage on streamlines and isotherms when Ra = 50, $\phi = 0.05$.

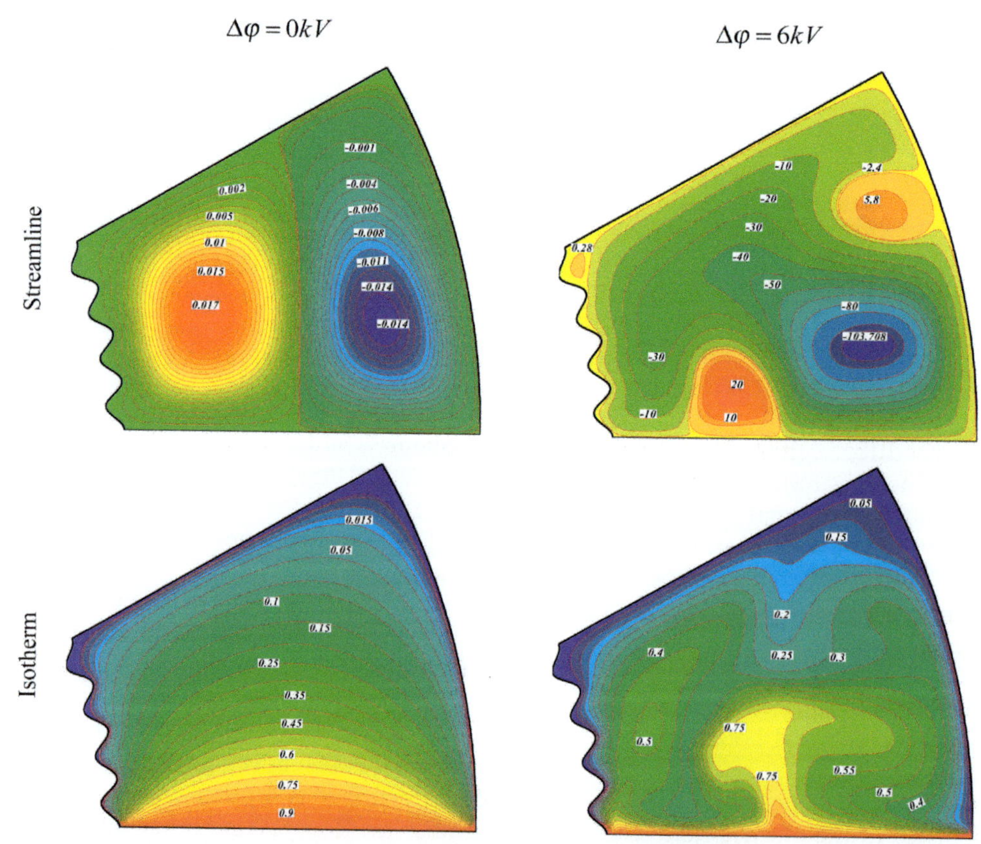

FIGURE 11.4 Effect of supplied voltage on streamlines and isotherms when Ra = 100, $\phi = 0.05$.

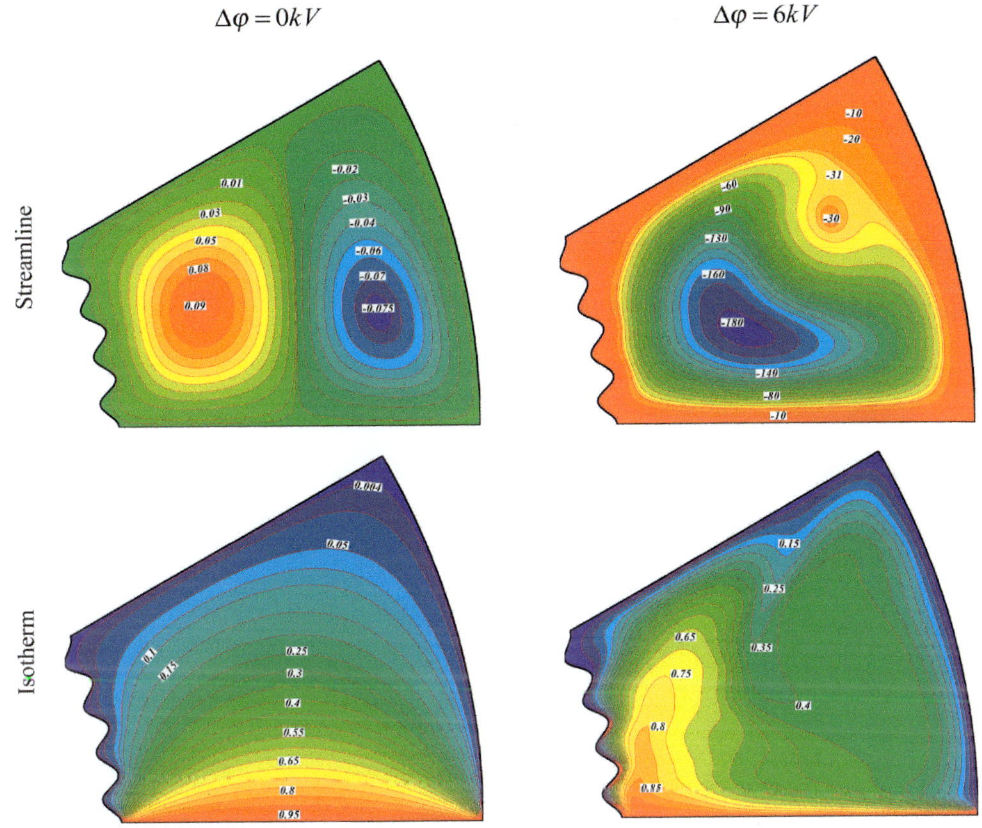

FIGURE 11.5 Effect of supplied voltage on streamlines and isotherms when Ra = 500, $\phi = 0.05$.

$Ra = 50$ while in the presence of an electric field ($\Delta\varphi = 6$), the opposite trend is observed. Also, it can be concluded that the Nusselt number at $\Delta\varphi = 6$ for $Ra = 50$, 100, and 500 are 3.13665, 2.972727, and 2.598383 times higher than those obtained at $\Delta\varphi = 0$. This observation confirms that the impact of the electric field is more marked for lower Reynolds numbers.

11.3 FREE CONVECTION OF NANOFLUID UNDER THE EFFECT OF ELECTRIC FIELD IN A POROUS ENCLOSURE

11.3.1 Problem Definition

Fig. 11.8 depicts the porous enclosure and its boundary conditions. Ethylene glycol–Fe_3O_4 nanofluid is utilized. All walls are stationary except for the bottom wall. The influence of Darcy and Reynolds numbers on the contour of q is demonstrated in Fig. 11.9. Effects of Re on q are less sensible than Da. As the Darcy number augments the distortion of isoelectric density lines become greater.

11.3.2 Governing Equation

The definition of electric field is:

$$\vec{E} = -\nabla\varphi \tag{11.20}$$

$$q = \nabla.\varepsilon\vec{E} \tag{11.21}$$

$$\vec{J} = q\vec{V} - D\nabla q + \sigma\vec{E} \tag{11.22}$$

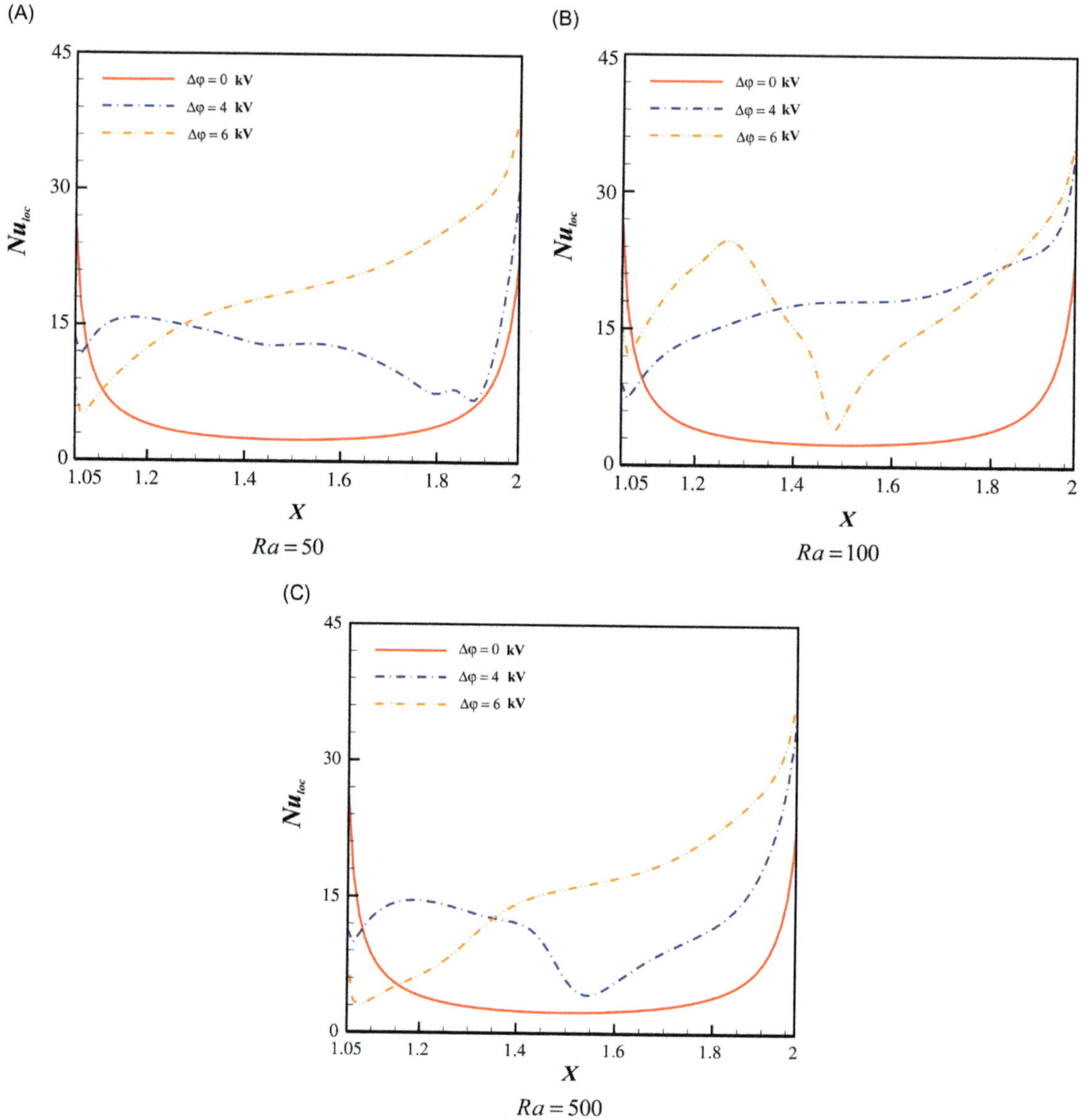

FIGURE 11.6 Effects of Rayleigh number and supplied voltage on local Nusselt number when $\phi = 0.05$. (A) $Ra = 50$, (B) $Ra = 100$, (C) $Ra = 500$.

$$\nabla . \vec{J} + \frac{\partial q}{\partial t} = 0 \tag{11.23}$$

The governing formulae are:

$$\begin{cases} \nabla . \vec{V} = 0, \\ \rho_{nf}\left(\dfrac{\partial \vec{V}}{\partial t} + \left(\vec{V}.\nabla\right)\vec{V}\right) = -\nabla p + \mu_{nf}\nabla^2\vec{V} + q\vec{E} - (\rho\beta)_{nf}\vec{g}(T - T_0) - \dfrac{\mu_{nf}}{K}\vec{V}, \\ \left(\left(\vec{V}.\nabla\right)T + \dfrac{\partial T}{\partial t}\right) = \dfrac{k_{nf}}{(\rho C_p)_{nf}}\nabla^2 T + \dfrac{\vec{J}.\vec{E}}{(\rho C_p)_{nf}} - \dfrac{1}{(\rho C_p)_{nf}}\dfrac{\partial q_r}{\partial y}, \quad \left[q_r = -\dfrac{4\sigma_e}{3\beta_R}\dfrac{\partial T^4}{\partial y}, \quad T^4 \cong 4T_c^3 T - 3T_c^4\right], \\ \nabla\varphi = -\vec{E}, \\ \dfrac{\partial q}{\partial t} = -\nabla . \vec{J}, \\ q = \nabla . \varepsilon \vec{E} \end{cases} \tag{11.24}$$

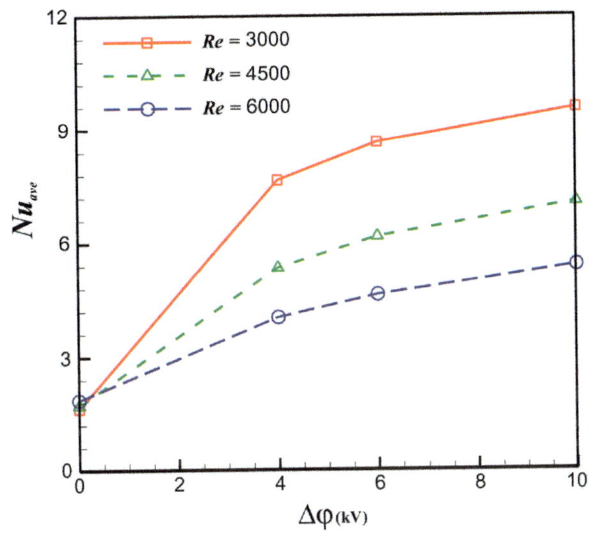

FIGURE 11.7 Effects of Rayleigh number and supplied voltage on average Nusselt number.

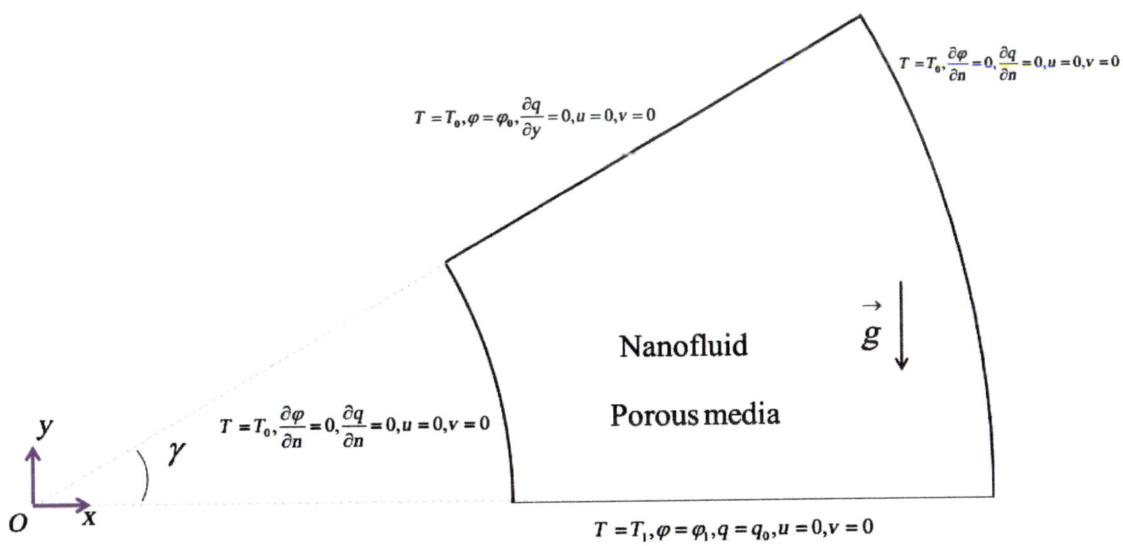

FIGURE 11.8 Geometry and the boundary conditions.

$(\rho C_p)_{nf}$, $(\beta)_{nf}$, μ_{nf}, and ρ_{nf} can be obtained as:

$$\begin{aligned} &(\rho C_p)_{nf} = (\rho C_p)_f(1-\phi) + (\rho C_p)_s\phi, \quad (\beta)_{nf} = (\beta)_f(1-\phi) + (\beta)_s\phi, \\ &\mu_{nf} = A_1 + A_2(\Delta\varphi) + A_3(\Delta\varphi)^2 + A_4(\Delta\varphi)^3, \\ &\rho_{nf} = \rho_f(1-\phi) + \rho_s\phi \end{aligned} \tag{11.25}$$

Properties of Fe_3O_4 and ethylene glycol are illustrated in Table 11.1. Table 11.2 illustrates the coefficient values of this formula.

k_{nf} can be expressed as:

$$\frac{k_{nf}}{k_f} = \frac{-m(k_f - k_p)\phi + (k_p - k_f)\phi + mk_f + k_p + k_f}{mk_f + (k_f - k_p)\phi + k_f + k_p} \tag{11.26}$$

Different values of shape factors for various shapes of nanoparticles are illustrated in Table 11.3.

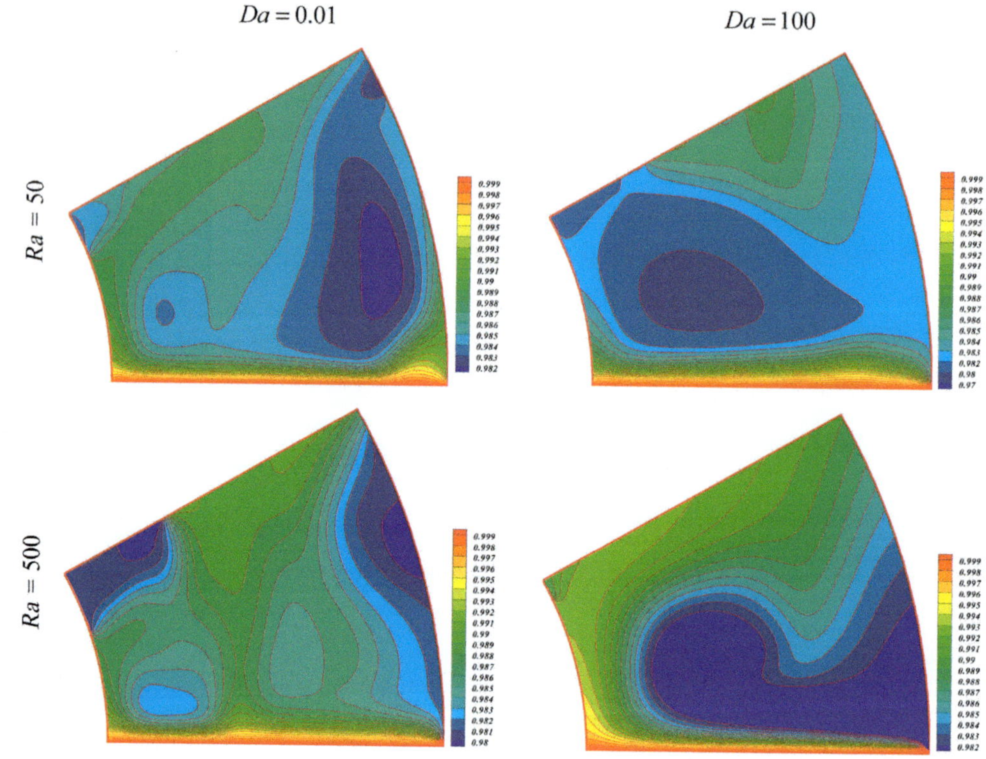

FIGURE 11.9 Electric density distribution injected by the bottom electrode when $\Delta\varphi = 6kV, \phi = 0.05, Rd = 0.8$.

TABLE 11.3 The Values of Shape Factor of Different Shapes of Nanoparticles

	Spherical	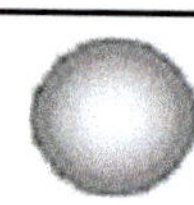	3
m	Platelet		5.7
	Cylinder		4.8
	Brick	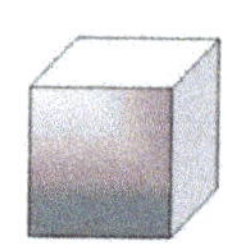	3.7

So, the final PDE in the presence of thermal radiation and electric field in porous media are:

$$\begin{cases} \nabla.\vec{V} = 0, \\ \left(\dfrac{\partial \vec{V}}{\partial t} + \left(\vec{V}.\nabla\right)\vec{V}\right) = -\nabla p + \dfrac{\mu_{nf}/\mu_f}{\rho_{nf}/\rho_f}\Pr\nabla^2\vec{V} + \dfrac{S_E}{\rho_{nf}/\rho_f}q\,\vec{E} - Ra\Pr\left[\dfrac{\beta_{nf}}{\beta_f}\right]\theta - \dfrac{\Pr}{Da}\dfrac{\mu_{nf}/\mu_f}{\rho_{nf}/\rho_f}\vec{V} \\ \left(\left(\vec{V}.\nabla\right)\theta + \dfrac{\partial\theta}{\partial t}\right) = \dfrac{k_{nf}/k_f}{(\rho C_p)_{nf}/(\rho C_p)_f}\nabla^2\theta + \dfrac{S_E Ec}{(\rho C_p)_{nf}/(\rho C_p)_f}\left(\vec{J}\,.\vec{E}\right) + \dfrac{4}{3}\dfrac{k_{nf}/k_f}{(\rho C_p)_{nf}/(\rho C_p)_f}\left(\dfrac{k_{nf}}{k_f}\right)^{-1} Rd\dfrac{\partial^2\theta}{\partial Y^2} \\ \vec{E} = -\nabla\varphi \\ q = \nabla.\varepsilon\vec{E} \\ \nabla.\vec{J} = -\dfrac{\partial q}{\partial t} \end{cases} \tag{11.27}$$

where

$$\bar{t} = \frac{t\alpha_f}{L^2}, \bar{p} = \frac{P}{\rho(\alpha_f/L)^2}, \bar{y} = \frac{y}{L}, \bar{x} = \frac{x}{L}, \bar{v} = \frac{v\,L}{\alpha_f}, \bar{u} = \frac{u\,L}{\alpha_f},$$
$$\theta = \frac{T - T_0}{\nabla T}, \nabla T = T_1 - T_0, \bar{\varphi} = \frac{\varphi - \varphi_0}{\nabla\varphi}, \nabla\varphi = \varphi_1 - \varphi_0, \bar{q} = \frac{q}{q_0}, \bar{E} = \frac{E}{E_0} \quad (11.28)$$

Vorticity and stream function should be employed in order to diminish the pressure gradient:

$$v = -\frac{\partial\psi}{\partial x}, \frac{\partial\psi}{\partial y} = u, \Psi = \frac{\psi}{\alpha_f}, \Omega = \frac{\omega L^2}{\alpha_f}, \omega = \frac{\partial v}{\partial x} - \frac{\partial u}{\partial y} \quad (11.29)$$

Nu_{loc} and Nu_{ave} along the bottom wall are calculated as:

$$Nu_{loc} = \left(\frac{k_{nf}}{k_f}\right)\left(1 + \frac{4}{3}Rd\left(\frac{k_{nf}}{k_f}\right)^{-1}\right)\frac{\partial\theta}{\partial X} \quad (11.30)$$

$$Nu_{ave} = \frac{1}{L}\int_0^L Nu_{loc}\,dY \quad (11.31)$$

11.3.3 Effects of Active Parameters

Effect of electric field on nanofluid free convection heat transfer is investigated in the presence of thermal radiation. Electric field dependent viscosity is considered for nanofluid. The porous enclosure is filled with Fe_3O_4–ethylene glycol and has one lid wall. Roles of the Darcy number ($Da = 10^{-2}$ to 10^2), Radiation parameter ($Rd = 0$ to 0.8), supplied voltage ($\Delta\varphi = 0$ to $6kV$), volume fraction of Fe_3O_4 ($\phi = 0\%$ to 5%), and Rayleigh number ($Ra = 50$ to 500) are illustrated graphically.

At first, the influence of shape factor on rate of heat transfer is reported in Table 11.4. In this table, various shapes of nanoparticles are utilized. The maximum Nu is caused by Platelet, followed by Cylinder, Brick, and Spherical. So, Platelet nanoparticle has been utilized for further investigation.

Impacts of $Da, \Delta\varphi$, and Ra on isotherm and streamlines are demonstrated in Figs. 11.10–11.13. At low Rayleigh number, there are two eddies which rotate in opposite directions. In the presence of an electric field, eddies are stretched and thermal plume generates. Isotherms become more disturbed in the presence of an electric field. As buoyancy forces increase, another eddy is generated. At high values of the Rayleigh number, increasing the Darcy number leads to the conversion of all eddies to one clockwise eddy and a thermal plume shifts to the left side.

Nu_{ave} versus Ra, Da, Rd, and $\Delta\varphi$ is depicted in Fig. 11.14. The related formula is:

$$\begin{aligned} Nu_{ave} &= 4.07 - 0.53\Delta\varphi + 2.39Ra^* + 0.35Da^* + 0.92Rd \\ &+ 0.06\Delta\varphi Ra^* + 0.49\Delta\varphi Da^* \\ &- 0.98\Delta\varphi Rd - 0.61Ra^* Da^* \\ &+ 0.41Ra^* Rd + 3.2Da^* Rd \\ &+ 0.17\Delta\varphi^2 - 0.39(Ra^*)^2 + 0.82(Da^*)^2 + 2.5Rd^2 \end{aligned} \quad (11.32)$$

TABLE 11.4 Effect of Shape of Nanoparticles on Nusselt Number When $Rd = 0.8, Ra = 500, \Delta\varphi = 6, \phi = 0.05$

	Da	
	10^{-2}	**10^{2}**
Spherical	18.15263	21.44078
Brick	18.94481	21.67239
Cylinder	13.05472	22.01274
Platelet	19.70303	22.27031

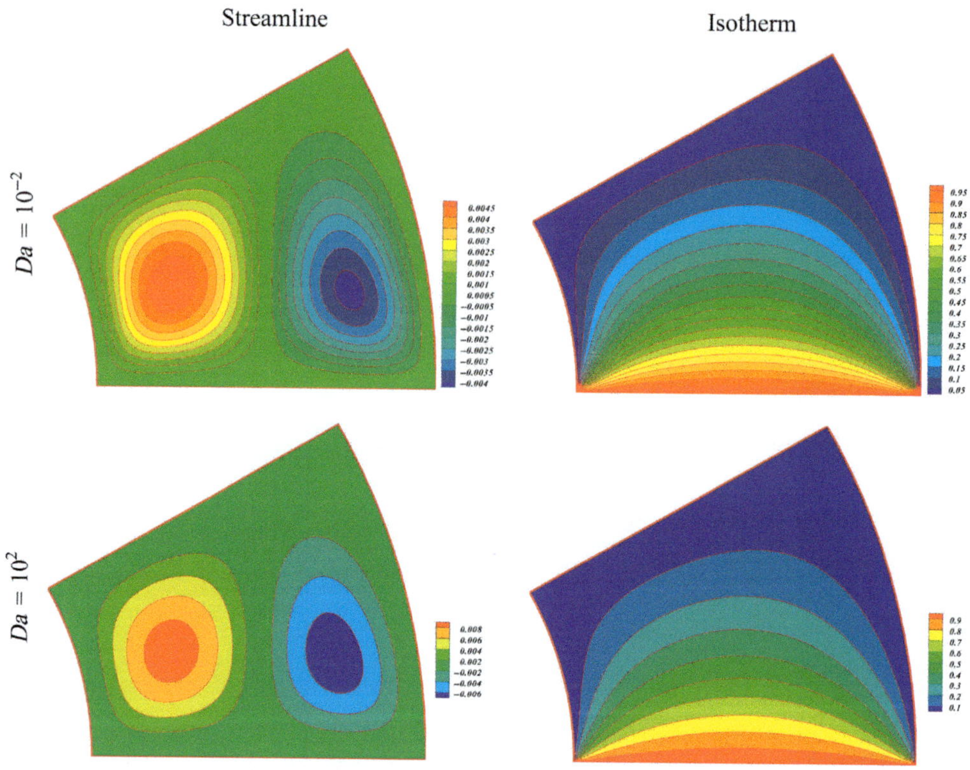

FIGURE 11.10 Effect of Darcy number on streamlines and isotherms when $Ra = 50, \Delta\varphi = 0kV, \phi = 0.05, Rd = 0.8$.

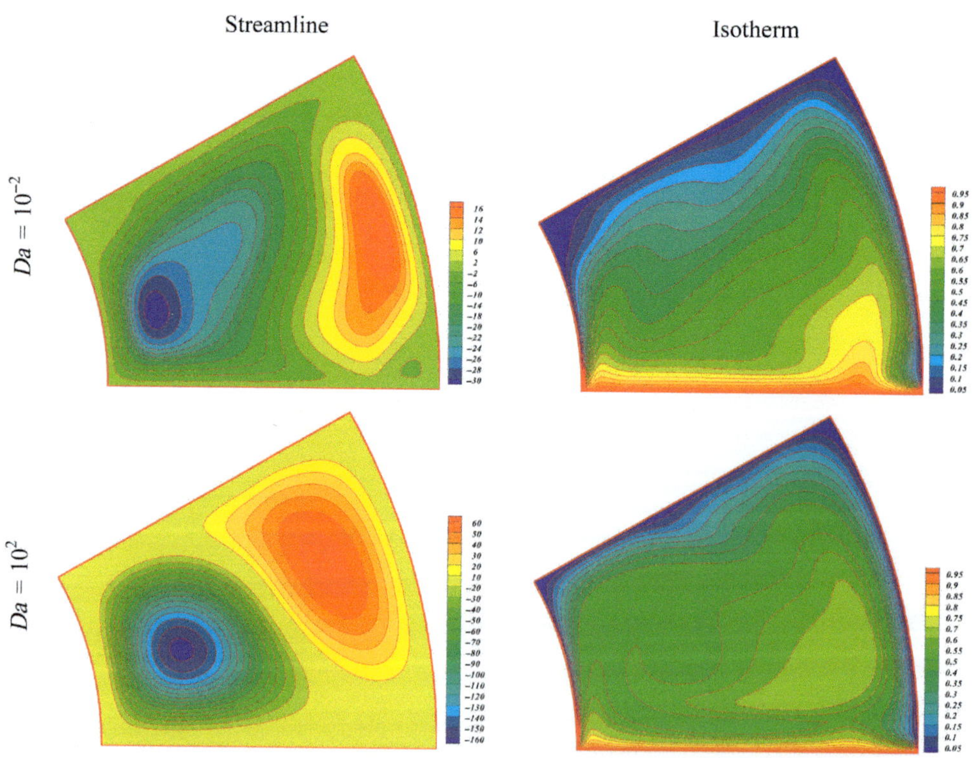

FIGURE 11.11 Effect of Darcy number on streamlines and isotherms when $Ra = 50, \Delta\varphi = 6kV, \phi = 0.05, Rd = 0.8$.

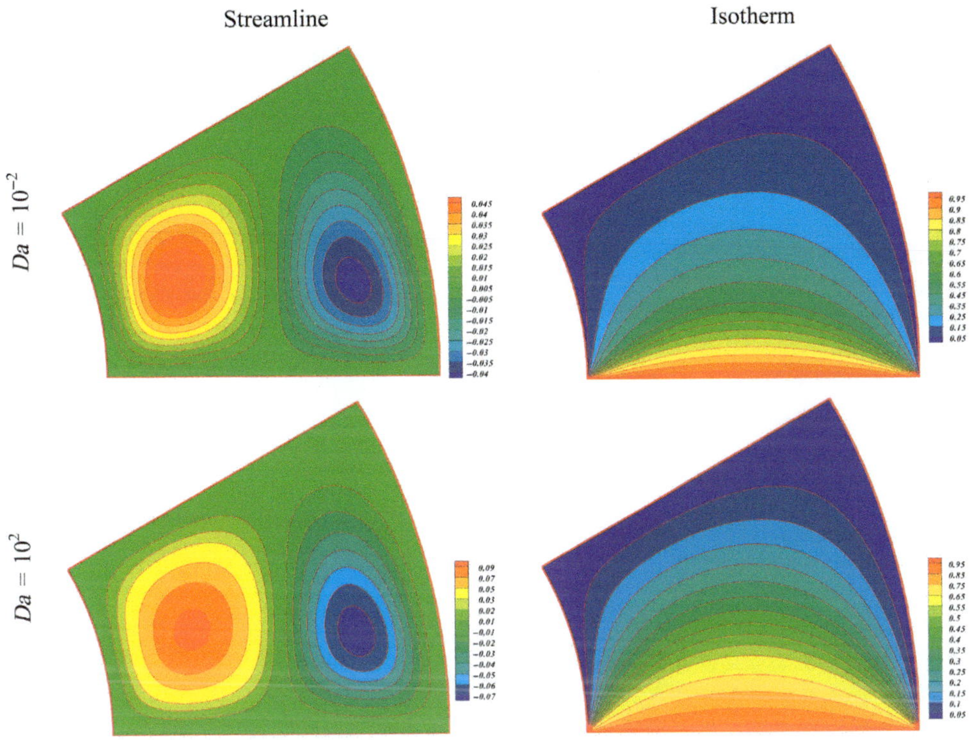

FIGURE 11.12 Effect of Darcy number on streamlines and isotherms when $Ra = 500, \Delta\varphi = 0kV, \phi = 0.05, Rd = 0.8$.

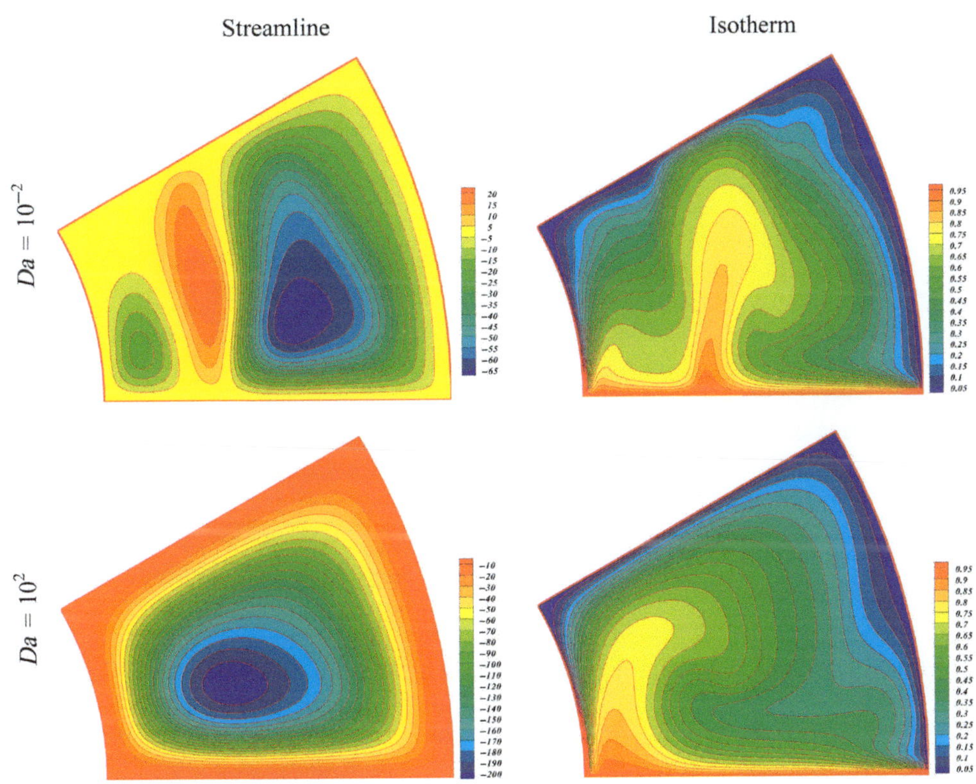

FIGURE 11.13 Effect of Darcy number on streamlines and isotherms when $Ra = 500, \Delta\varphi = 6kV, \phi = 0.05, Rd = 0.8$.

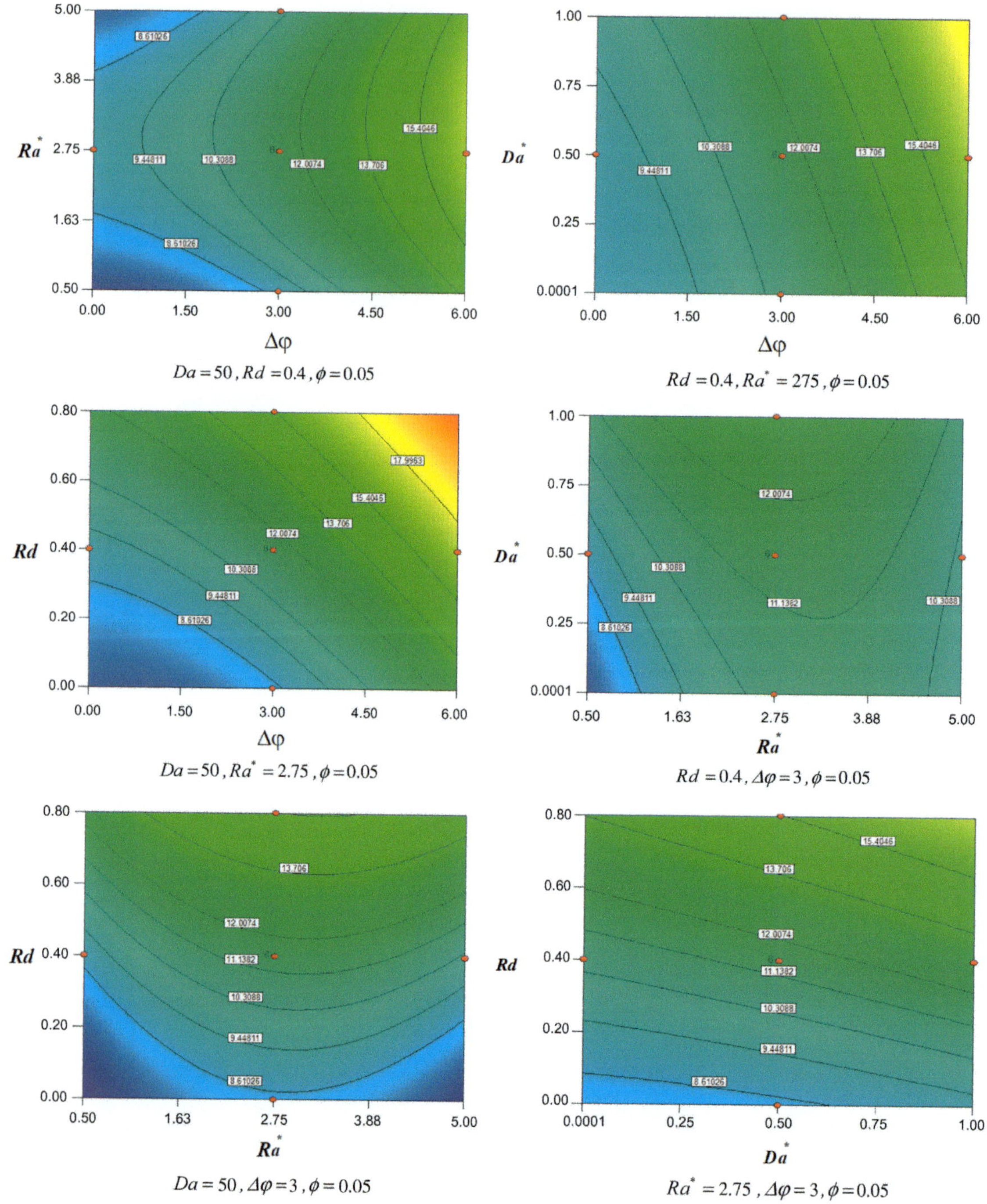

FIGURE 11.14 Effects of *Da*, $\Delta\varphi$, *Rd*, and *Ra* on average Nusselt number.

where $Ra^* = 0.01Ra$, $Da^* = 0.01Da$, and $\Delta\varphi$ is voltage supply in kilovolt. Rate of heat transfer enhances with the rise of the Rayleigh number. The electric field helps the convective mode to enhance. So, Nu_{ave} augments with the increase of $\Delta\varphi$. Thermal radiation enhances the temperature gradient near the lid wall. Then influence of the Darcy number is the same as the radiation parameter. Therefore, Nu_{ave} is an increasing function of *Rd*, *Da*.

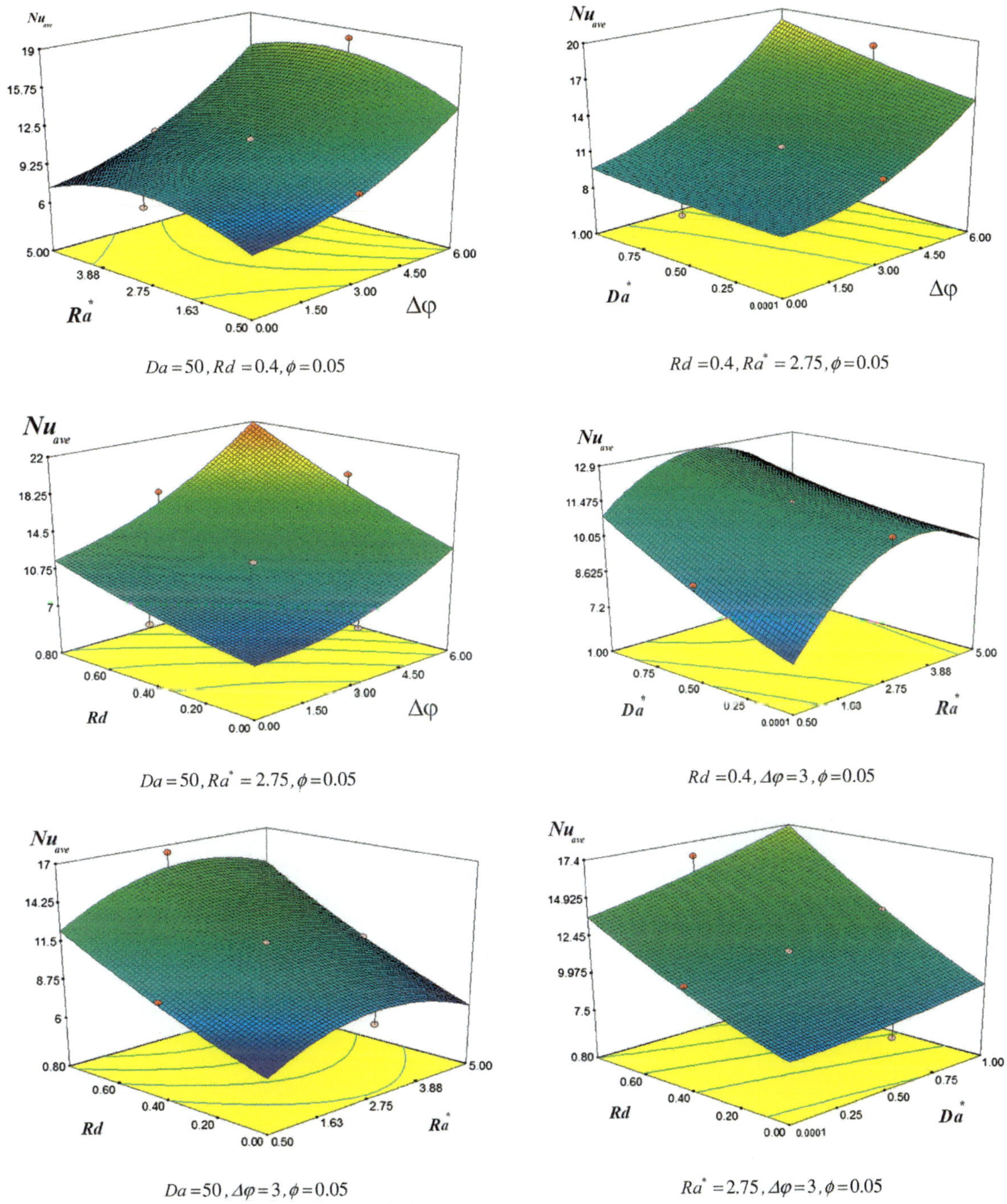

FIGURE 11.14 (Continued).

11.4 NANOFLUID NATURAL CONVECTION UNDER THE INFLUENCE OF COULOMB FORCE IN A POROUS ENCLOSURE

11.4.1 Problem Definition

Fig. 11.15 depicts the porous enclosure and its boundary conditions. Ethylene glycol–Fe_3O_4 nanofluid is utilized. All walls are stationary except for the bottom wall. The influence of the Darcy and Reynolds numbers on contours of q is demonstrated in Fig. 11.16. The effect of Re on q is less sensible than Da. As the Darcy number augments the distortion of isoelectric density lines become greater.

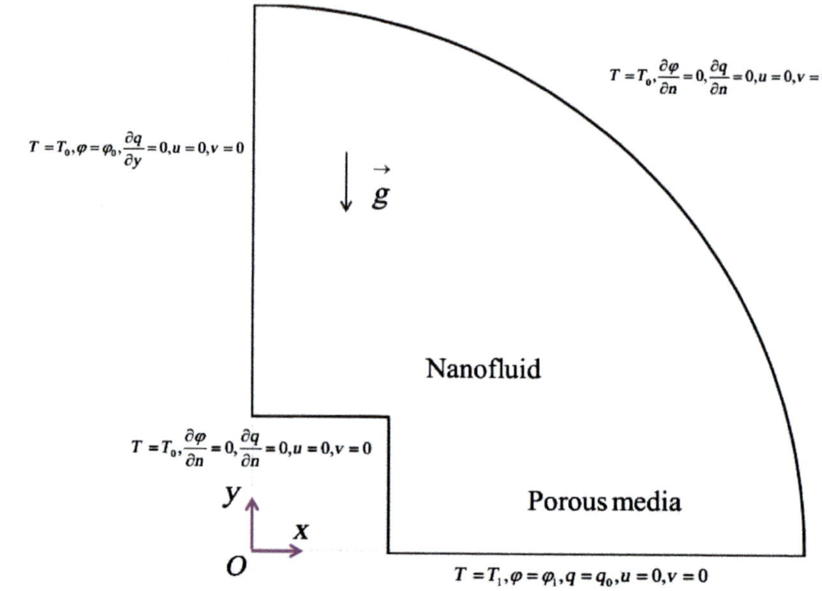

FIGURE 11.15 Geometry and the boundary conditions.

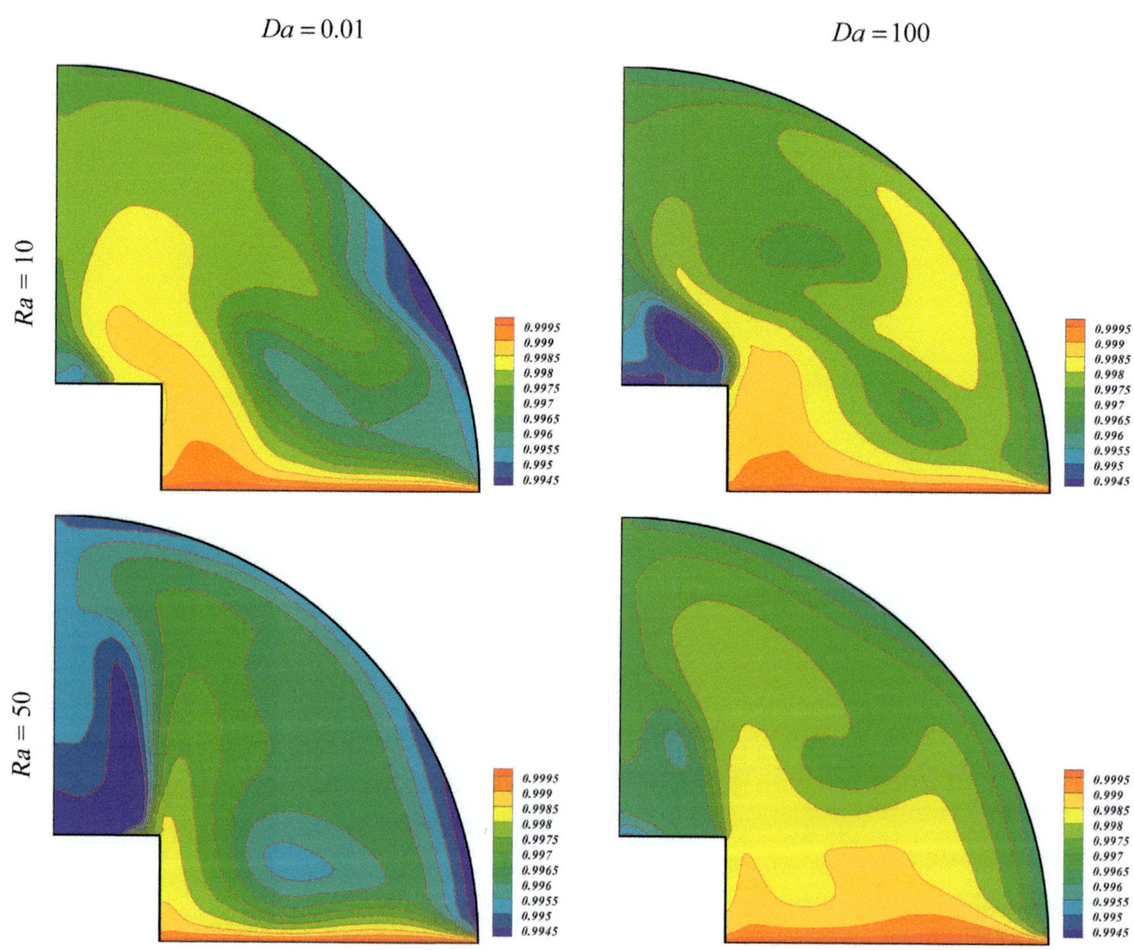

FIGURE 11.16 Electric density distribution injected by the bottom electrode when $\Delta\varphi=6kV$, $\phi=0.05$, $Rd=0.8$.

11.4.2 Governing Equation

The definition of electric field is:

$$\vec{E} = -\nabla\varphi \tag{11.33}$$

$$q = \nabla.\varepsilon\,\vec{E} \tag{11.34}$$

$$\vec{J} = q\vec{V} - D\nabla q + \sigma\vec{E} \tag{11.35}$$

$$\nabla.\vec{J} + \frac{\partial q}{\partial t} = 0 \tag{11.36}$$

TABLE 11.5 Effect of Shape of Nanoparticles on Nusselt Number When $Rd = 0.8, Ra = 500, \Delta\varphi = 6, \phi = 0.05$

	Da	
	10^{-2}	10^{2}
Spherical	6.541472	8.142749
Brick	6.926098	9.461129
Cylinder	7.568405	9.903384
Platelet	8.607858	10.30412

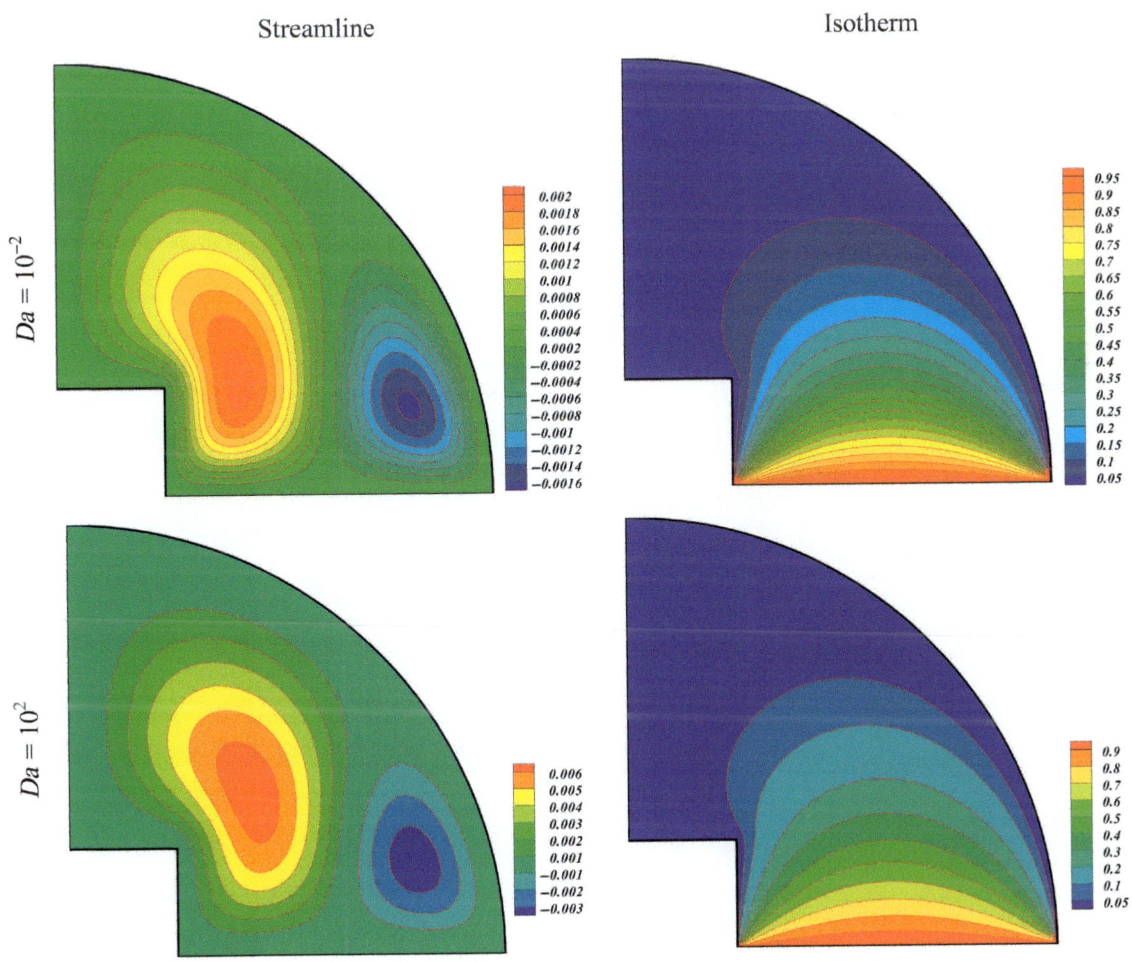

FIGURE 11.17 Effect of Darcy number on streamlines and isotherms when $Ra = 10, \Delta\varphi = 0kV, \phi = 0.05, Rd = 0.8$.

The governing formulae are:

$$\begin{cases} \nabla.\vec{V}=0, \\ \rho_{nf}\left(\dfrac{\partial \vec{V}}{\partial t}+\left(\vec{V}.\nabla\right)\vec{V}\right)=-\nabla p+\mu_{nf}\nabla^2\vec{V}+q\vec{E}-(\rho\beta)_{nf}\vec{g}(T-T_0)-\dfrac{\mu_{nf}}{K}\vec{V}, \\ \left(\left(\vec{V}.\nabla\right)T+\dfrac{\partial T}{\partial t}\right)=\dfrac{k_{nf}}{\left(\rho C_p\right)_{nf}}\nabla^2 T+\dfrac{\vec{J}.\vec{E}}{\left(\rho C_p\right)_{nf}}-\dfrac{1}{\left(\rho C_p\right)_{nf}}\dfrac{\partial q_r}{\partial y}, \quad \left[q_r=-\dfrac{4\sigma_e}{3\beta_R}\dfrac{\partial T^4}{\partial y}, \quad T^4\cong 4T_c^3T-3T_c^4\right], \\ \nabla\varphi=-\vec{E}, \\ \dfrac{\partial q}{\partial t}=-\nabla.\vec{J}, \\ q=\nabla.\varepsilon\vec{E} \end{cases} \tag{11.37}$$

$\left(\rho C_p\right)_{nf}$, $(\beta)_{nf}$, μ_{nf}, and ρ_{nf} can be obtained as:

$$\begin{aligned} &\left(\rho C_p\right)_{nf}=\left(\rho C_p\right)_f(1-\phi)+\left(\rho C_p\right)_s\phi, \quad (\beta)_{nf}=(\beta)_f(1-\phi)+(\beta)_s\phi, \\ &\mu_{nf}=A_1+A_2(\Delta\varphi)+A_3(\Delta\varphi)^2+A_4(\Delta\varphi)^3, \\ &\rho_{nf}=\rho_f(1-\phi)+\rho_s\phi \end{aligned} \tag{11.38}$$

Properties of Fe_3O_4 and ethylene glycol are illustrated in Table 11.1.

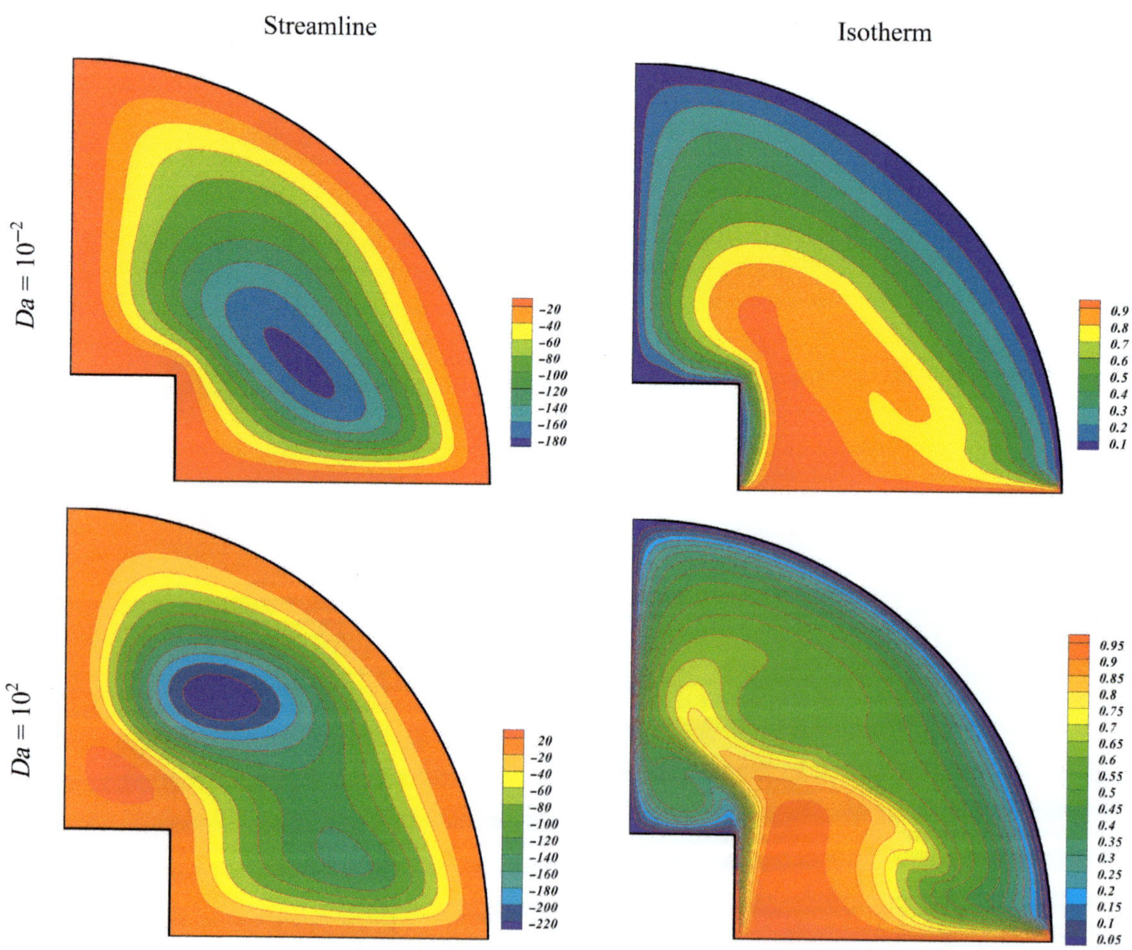

FIGURE 11.18 Effect of Darcy number on streamlines and isotherms when $Ra=10, \Delta\varphi=6kV, \phi=0.05, Rd=0.8$.

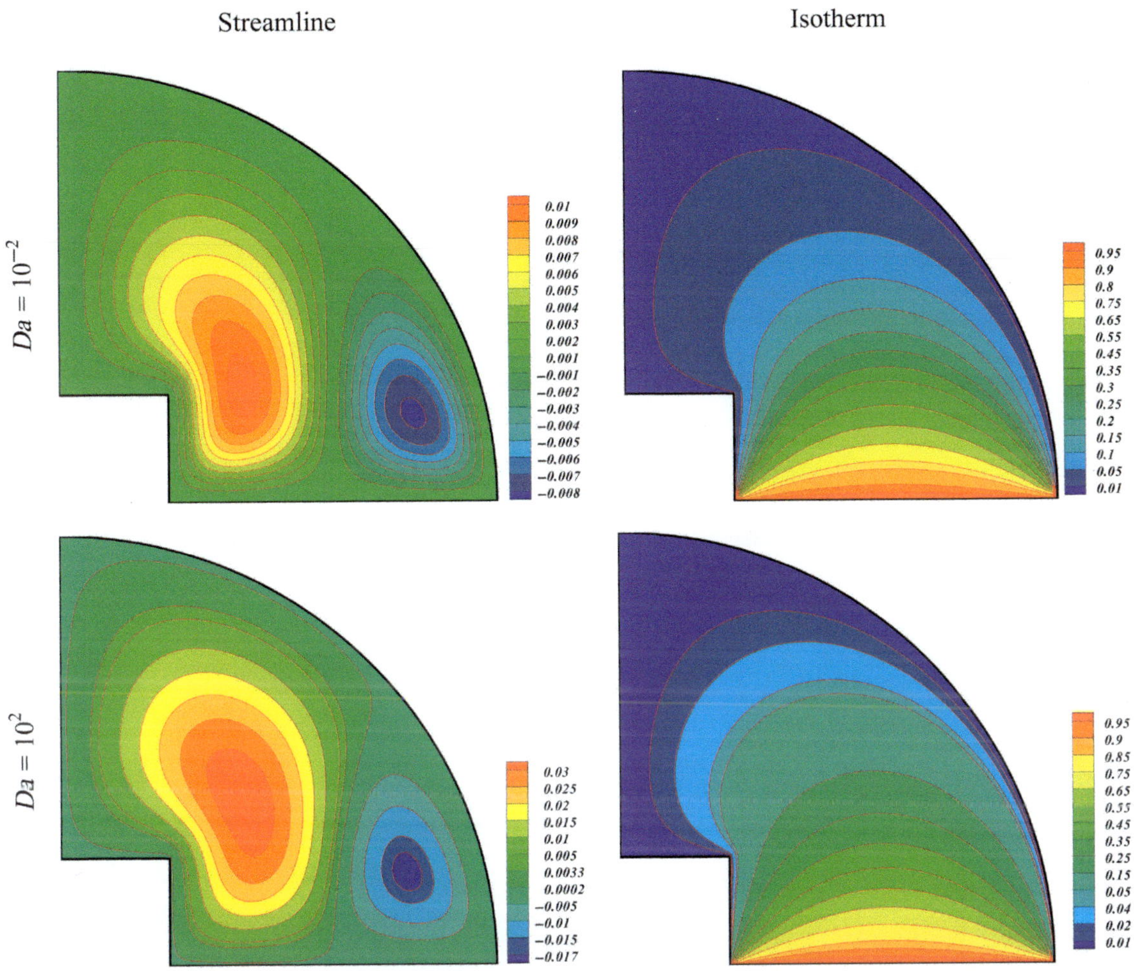

FIGURE 11.19 Effect of Darcy number on streamlines and isotherms when $Ra = 50, \Delta\varphi = 0kV, \phi = 0.05, Rd = 0.8$.

k_{nf} can be expressed as:

$$\frac{k_{nf}}{k_f} = \frac{-m(k_f - k_p)\phi + (k_p - k_f)\phi + mk_f + k_p + k_f}{mk_f + (k_f - k_p)\phi + k_f + k_p} \tag{11.39}$$

Different values of shape factors for various shapes of nanoparticles are illustrated in Table 11.3. So, the final PDE in the presence of thermal radiation and electric field in porous media are:

$$\begin{cases} \nabla.\vec{V} = 0, \\ \left(\dfrac{\partial \vec{V}}{\partial t} + \left(\vec{V}.\nabla\right)\vec{V}\right) = -\nabla p + \dfrac{\mu_{nf}/\mu_f}{\rho_{nf}/\rho_f}\mathrm{Pr}\nabla^2\vec{V} + \dfrac{S_E}{\rho_{nf}/\rho_f}q\,\vec{E} - Ra\mathrm{Pr}\left[\dfrac{\beta_{nf}}{\beta_f}\right]\theta - \dfrac{\mathrm{Pr}}{Da}\dfrac{\mu_{nf}/\mu_f}{\rho_{nf}/\rho_f}\vec{V} \\ \left(\left(\vec{V}.\nabla\right)\theta + \dfrac{\partial\theta}{\partial t}\right) = \dfrac{k_{nf}/k_f}{(\rho C_p)_{nf}/(\rho C_p)_f}\nabla^2\theta + \dfrac{S_E Ec}{(\rho C_p)_{nf}/(\rho C_p)_f}\left(\vec{J}\,.\,\vec{E}\right) + \dfrac{4}{3}\dfrac{k_{nf}/k_f}{(\rho C_p)_{nf}/(\rho C_p)_f}\left(\frac{k_{nf}}{k_f}\right)^{-1}Rd\dfrac{\partial^2\theta}{\partial Y^2} \\ \vec{E} = -\nabla\varphi \\ q = \nabla.\varepsilon\vec{E} \\ \nabla.\vec{J} = -\dfrac{\partial q}{\partial t} \end{cases} \tag{11.40}$$

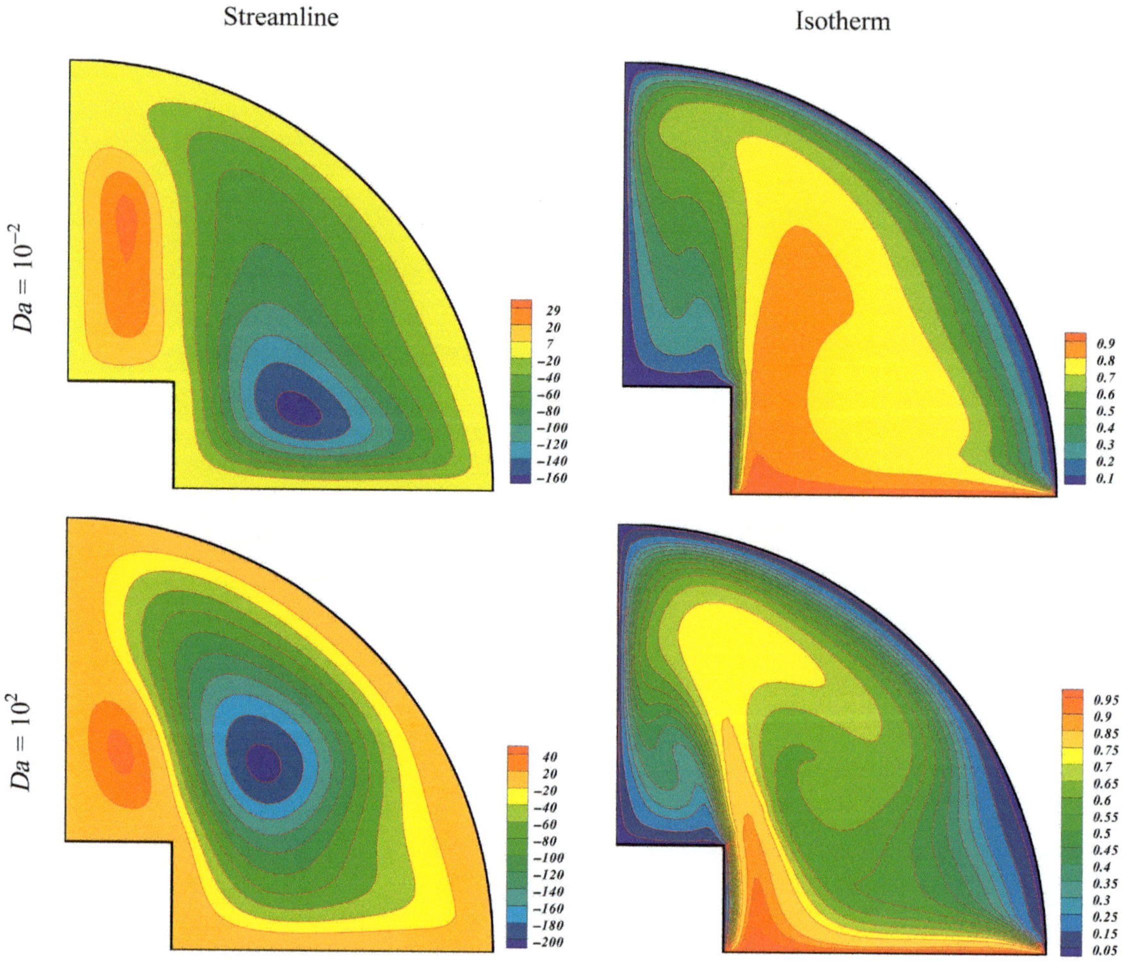

FIGURE 11.20 Effect of Darcy number on streamlines and isotherms when $Ra = 50, \Delta\varphi = 6kV, \phi = 0.05, Rd = 0.8$.

where

$$\bar{t} = \frac{t\alpha_f}{L^2}, \bar{p} = \frac{P}{\rho\left(\alpha_f/L\right)^2}, \bar{y} = \frac{y}{L}, \bar{x} = \frac{x}{L}, \bar{v} = \frac{v\,L}{\alpha_f}, \bar{u} = \frac{u\,L}{\alpha_f},$$
$$\theta = \frac{T - T_0}{\nabla T}, \nabla T = T_1 - T_0, \bar{\varphi} = \frac{\varphi - \varphi_0}{\nabla\varphi}, \nabla\varphi = \varphi_1 - \varphi_0, \bar{q} = \frac{q}{q_0}, \bar{E} = \frac{E}{E_0} \tag{11.41}$$

Vorticity and stream function should be employed in order to diminish the pressure gradient:

$$v = -\frac{\partial\psi}{\partial x}, \frac{\partial\psi}{\partial y} = u, \Psi = \frac{\psi}{\alpha_f}, \Omega = \frac{\omega L^2}{\alpha_f}, \omega = \frac{\partial v}{\partial x} - \frac{\partial u}{\partial y} \tag{11.42}$$

Nu_{loc} and Nu_{ave} along the bottom wall are calculated as:

$$Nu_{loc} = \left(\frac{k_{nf}}{k_f}\right)\left(1 + \frac{4}{3}Rd\left(\frac{k_{nf}}{k_f}\right)^{-1}\right)\frac{\partial\theta}{\partial X} \tag{11.43}$$

$$Nu_{ave} = \frac{1}{L}\int_0^L Nu_{loc}\, dY \tag{11.44}$$

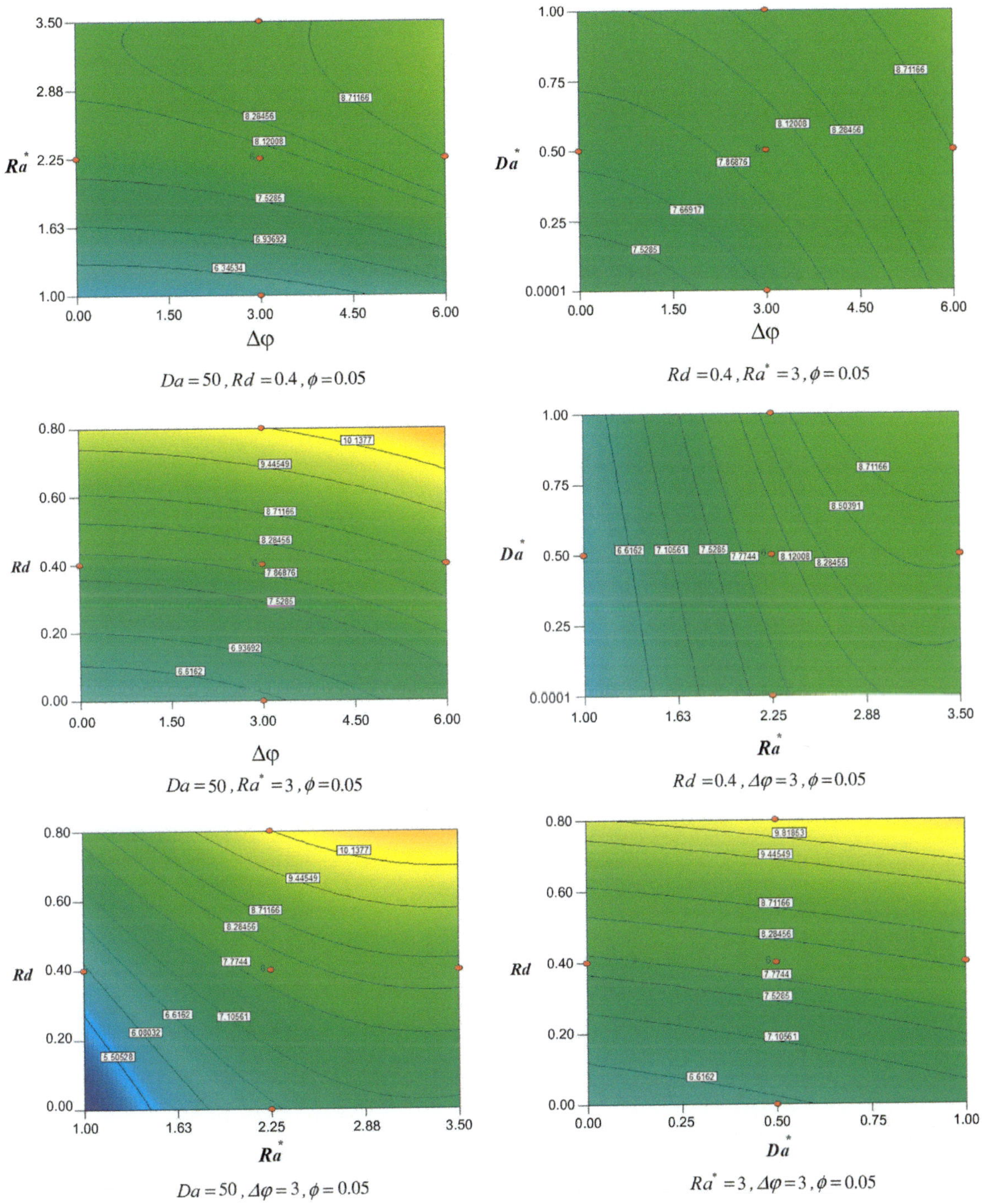

FIGURE 11.21 Effects of *Da*, $\Delta\varphi$, *Rd*, and *Ra* on average Nusselt number.

11.4.3 Effects of Active Parameters

EHD nanofluid free convection heat transfer through the porous enclosure is investigated in the presence of thermal radiation. Electric field dependent viscosity is considered for nanofluid. The porous enclosure is filled with Fe_3O_4–ethylene glycol and has one lid wall. Roles of Darcy number ($Da = 10^{-2}$ to 10^2), Radiation parameter ($Rd = 0$ to 0.8), supplied voltage ($\Delta\varphi = 0$ to $6kV$), volume fraction of Fe_3O_4 ($\phi = 0\%$ to 5%), and Rayleigh number ($Ra = 50$ to 500) are illustrated graphically.

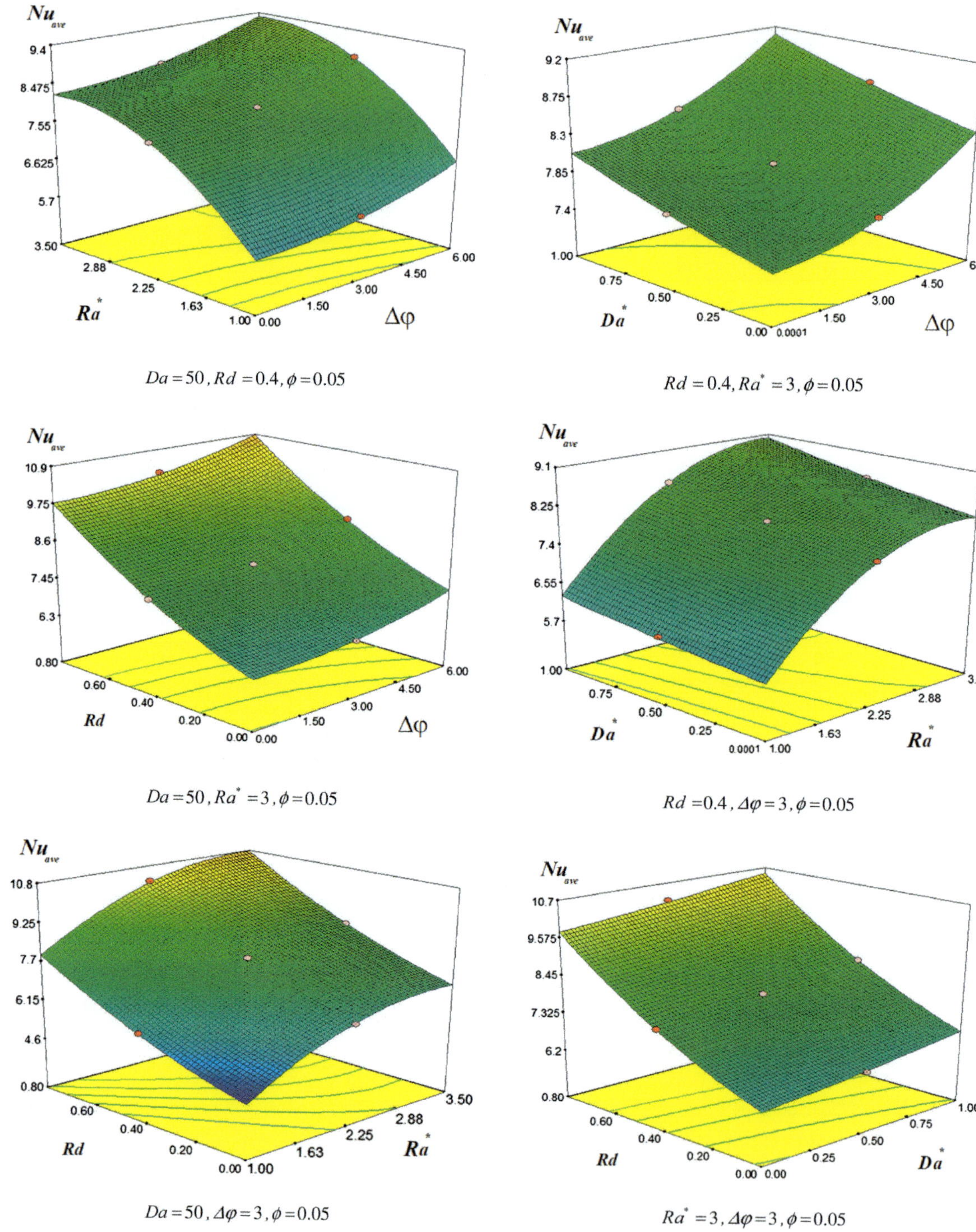

FIGURE 11.21 (Continued).

At first, the influence of shape factor on rate of heat transfer is reported in Table 11.5. In this table, various shapes of nanoparticles are utilized. The maximum *Nu* is caused by Platelet, followed by Cylinder, Brick, and Spherical. So, Platelet nanoparticle has been utilized for further investigation.

Impacts of $Da, \Delta\varphi$, and Ra on isotherms and streamlines are demonstrated in Figs. 11.17–11.20. At low Rayleigh number, there are two eddies which rotate in opposite directions. In the presence of an electric field, two eddies convert to one clockwise eddy. Isotherms become more disturbed in the presence of an electric field and a thermal plume is generated. Increasing the Darcy number converts the clockwise eddy to two rotating eddies. As the buoyancy forces increase, the shapes of the streamlines become more complex. At high values of the Rayleigh number, by increasing the Darcy number the secondary eddy becomes stronger. Nu_{ave} versus $Ra, Da, Rd,$ and $\Delta\varphi$ is depicted in Fig. 11.21. The related formula is:

$$\begin{aligned} Nu_{ave} &= 1.84 - 0.017\Delta\varphi + 2.9Ra^* + 0.11Da^* + 2.07Rd \\ &+ 2.07\Delta\varphi Ra^* + 8.03\times10^{-3}\Delta\varphi Da^* \\ &+ 0.034\Delta\varphi Rd + 0.158Ra^*Da^* \\ &+ 0.21Ra^*Rd + 0.14Da^*Rd \\ &+ 0.024\Delta\varphi^2 - 0.46(Ra^*)^2 + 0.14(Da^*)^2 + 2.15Rd^2 \end{aligned} \tag{11.13}$$

where $Ra^* = 0.01Ra, Da^* = 0.01Da,$ and $\Delta\varphi$ is voltage supply in Kilovolt. Rate of heat transfer enhances with rise of Rayleigh number. Electric field helps the convective mode to enhance. So, Nu_{ave} augments with augment of $\Delta\varphi$. Thermal radiation enhances the temperature gradient near the lid wall. The influence of Darcy number is the same as the radiation parameter. Therefore, Nu_{ave} is an increasing function of Rd, Da.

References

[1] A.E. Bergles, Enhancement of heat transfer, Proc. 6th Int. Heat Transfer Conf. vol. 1 (6) (1978) 89–108. Hemisphere, Washington, DC.

[2] P.H.G. Allen, T.G. Karayiannis, Electrohydrodynamic enhancement of heat transfer and fluid flow, Heat Recov. Syst. CHP 15 (15) (1995) 389–423.

[3] Y.Y. Yan, T.G. Karayiannis, P.H.G. Allen, M.W. Collins, R.S. Neve, and R.K. Al-Dadah, EHD Enhanced Heat Exchangers New Challenging Equipment for Renewable Energy Applications, Proc. 1st Int. Congress of Energy and Environment, Shanghai, China, (1995) 398–405.

[4] M. Sheikholeslami, A.J. Chamkha, Electrohydrodynamic free convection heat transfer of a nanofluid in a semi-annulus enclosure with a sinusoidal wall, Numer. Heat Transfer Part A 69 (7) (2016) 781–793. Available from: https://doi.org/10.1080/10407782.2015.1090819.

[5] M. Sheikholeslami, Influence of Lorentz forces on nanofluid flow in a porous cylinder considering Darcy model, J. Mol. Liq. 225 (2017) 903–912.

[6] M. Sheikholeslami, CVFEM for magnetic nanofluid convective heat transfer in a porous curved enclosure, Eur. Phys. J. Plus 131 (2016) 413. Available from: https://doi.org/10.1140/epjp/i2016-16413-y.

[7] M. Sheikholeslami, Influence of Coulomb forces on Fe_3O_4-H_2O nanofluid thermal improvement, Int. J. Hydrogen Energy 42 (2017) 821–829.

[8] M. Sheikholeslami, Magnetic field influence on nanofluid thermal radiation in a cavity with tilted elliptic inner cylinder, J. Mol. Liq. 229 (2017) 137–147.

[9] M. Sheikholeslami, Numerical simulation of magnetic nanofluid natural convection in porous media, Phys. Lett. A 381 (2017) 494–503.

[10] M. Sheikholeslami, Magnetic source impact on nanofluid heat transfer using CVFEM, Neural Comput. Applicat. (2016). Available from: https://doi.org/10.1007/s00521-016-2740-7.

[11] M. Sheikholeslami, CuO-water nanofluid free convection in a porous cavity considering Darcy law, Europ. Phys. J. Plus 132 (2017) 55. Available from: https://doi.org/10.1140/epjp/i2017-11330-3.

[12] M. Sheikholeslami, Numerical investigation of MHD nanofluid free convective heat transfer in a porous tilted enclosure, Eng. Comput. 34 (6) (2017) 1939–1955.

[13] M.S. Kandelousi, Effect of spatially variable magnetic field on ferrofluid flow and heat transfer considering constant heat flux boundary condition, Europ. Phys. J. Plus (2014) 129–248. Available from: https://doi.org/10.1140/epjp/i2014-14248-2.

[14] M. Sheikholeslami, Numerical simulation for external magnetic field influence on Fe_3O_4-water nanofluid forced convection, Engineering Computations, Inpress.

[15] M. Sheikholeslami, Influence of Lorentz forces on nanofluid flow in a porous cavity by means of Non- Darcy model, Eng. Comput. 34 (8) (2017) 2651–2667. Available from: https://doi.org/10.1108/EC-01-2017-0008.

[16] M. Sheikholeslami, H.B. Rokni, Nanofluid convective heat transfer intensification in a porous circular cylinder, Chem. Eng. Process. Process Intensif. 120 (2017) 93–104.

[17] M. Sheikholeslami, M.K. Sadoughi, Numerical modeling for Fe_3O_4 -water nanofluid flow in porous medium considering MFD viscosity, J. Mol. Liq. 242 (2017) 255–264.

[18] M. Sheikholeslami, M. Shamlooei, Magnetic source influence on nanofluid flow in porous medium considering shape factor effect, Phys. Lett. A 381 (2017) 3071–3078.

[19] M. Sheikholeslami, M. Shamlooei, Convective flow of nanofluid inside a lid driven porous cavity using CVFEM, Phys. B: Condensed Matter 521 (2017) 239–250.

[20] M. Sheikholeslami, K. Vajravelu, Forced convection heat transfer in Fe_3O_4- ethylene glycol nanofluid under the influence of Coulomb force, J. Mol. Liq. 233 (2017) 203–210.

[21] M. Sheikholeslami, H.B. Rokni, Influence of EFD viscosity on nanofluid forced convection in a cavity with sinusoidal wall, J. Mol. Liq. 232 (2017) 390–395.

[22] M. Sheikholeslami, H.B. Rokni, Magnetic nanofluid natural convection in the presence of thermal radiation considering variable viscosity, Europ. Phys. J. Plus 132 (2017) 238. Available from: https://doi.org/10.1140/epjp/i2017-11498-4.

[23] M. Sheikholeslami, H.B. Rokni, Numerical simulation for impact of Coulomb force on nanofluid heat transfer in a porous enclosure in presence of thermal radiation, Int. J. Heat Mass Transfer 118 (2018) 823–831.

[24] M. Sheikholeslami, M. Shamlooei, R. Moradi, Fe_3O_4- Ethylene glycol nanofluid forced convection inside a porous enclosure in existence of Coulomb force, J. Mol. Liq. 249 (2018) 429–437.

[25] M. Sheikholeslami, S.A. Shehzad, Numerical analysis of Fe_3O_4–H_2O nanofluid flow in permeable media under the effect of external magnetic source, Int. J. Heat Mass Transfer 118 (2018) 182–192.

[26] M. Sheikholeslami, S.A. Shehzad, Magnetohydrodynamic nanofluid convection in a porous enclosure considering heat flux boundary condition, Int. J. Heat Mass Transfer 106 (2017) 1261–1269.

[27] M. Sheikholeslami, S.A. Shehzad, Thermal radiation of ferrofluid in existence of Lorentz forces considering variable viscosity, Int. J. Heat Mass Transfer 109 (2017) 82–92.

[28] M. Sheikholeslami, H.B. Rokni, Numerical modeling of nanofluid natural convection in a semi annulus in existence of Lorentz force, Comput. Methods Appl. Mech. Eng. 317 (2017) 419–430.

[29] M. Sheikholeslami, K. Vajravelu, Nanofluid flow and heat transfer in a cavity with variable magnetic field, Appl. Math. Comput. 298 (2017) 272–282.

[30] M. Sheikholeslami, H.B. Rokni, Magnetohydrodynamic CuO-water nanofluid in a porous complex shaped enclosure, ASME, J. Therm. Sci. Eng. Applicat. 9 (4) (2017) 041007. Available from: https://doi.org/10.1115/1.4035973.

[31] M. Sheikholeslami, H.B. Rokni, Nanofluid two phase model analysis in existence of induced magnetic field, Int. J. Heat Mass Transfer 107 (2017) 288–299.

[32] M. Sheikholeslami, T. Hayat, A. Alsaedi, S. Abelman, Numerical analysis of EHD nanofluid force convective heat transfer considering electric field dependent viscosity, Int. J. Heat Mass Transfer 108 (2017) 2558–2565.

[33] M. Sheikholeslami, Magnetic field influence on CuO -H_2O nanofluid convective flow in a permeable cavity considering various shapes for nanoparticles, Int. J. Hydrogen Energy 42 (2017) 19611–19621.

[34] M. Sheikholeslami, M.M. Bhatti, Forced convection of nanofluid in presence of constant magnetic field considering shape effects of nanoparticles, Int. J. Heat Mass Transfer 111 (2017) 1039–1049.

[35] M. Sheikholeslami, A. Zeeshan, Analysis of flow and heat transfer in water based nanofluid due to magnetic field in a porous enclosure with constant heat flux using CVFEM, Comput. Methods Appl. Mech. Eng. 320 (2017) 68–81.

[36] M. Sheikholeslami, M. Shamlooei, Fe_3O_4- H_2O nanofluid natural convection in presence of thermal radiation, Int. J. Hydrogen Energy 42 (9) (2017) 5708–5718.

[37] M. Sheikholeslami, M.M. Bhatti, Active method for nanofluid heat transfer enhancement by means of EHD, Int. J. Heat Mass Transfer 109 (2017) 115–122.

[38] M. Sheikholeslami, T. Hayat, A. Alsaedi, Numerical study for external magnetic source influence on water based nanofluid convective heat transfer, Int. J. Heat Mass Transfer 106 (2017) 745–755.

[39] M. Sheikholeslami, A.J. Chamkha, Flow and convective heat transfer of a ferro-nanofluid in a double-sided lid-driven cavity with a wavy wall in the presence of a variable magnetic field, Numer. Heat Transfer Part A 69 (10) (2016) 1186–1200. Available from: https://doi.org/10.1080/10407782.2015.1125709.

[40] M. Sheikholeslami, M.K. Sadoughi, Simulation of CuO- water nanofluid heat transfer enhancement in presence of melting surface, Int. J. Heat Mass Transfer 116 (2018) 909–919.

[41] M. Sheikholeslami, H.B. Rokni, Simulation of nanofluid heat transfer in presence of magnetic field: a review, Int. J. Heat Mass Transfer 115 (2017) 1203–1233.

[42] M. Sheikholeslami, R. Ellahi, Simulation of ferrofluid flow for magnetic drug targeting using Lattice Boltzmann method, J. Zeitschrift Fur Naturforschung A 70 (2) (2015) 115–124.

[43] M. Sheikholeslami, M. Seyednezhad, Lattice Boltzmann Method simulation for CuO-water nanofluid flow in a porous enclosure with hot obstacle, J. Mol. Liq. 243 (2017) 249–256.

[44] M. Sheikholeslami, M. Sadoughi, Mesoscopic method for MHD nanofluid flow inside a porous cavity considering various shapes of nanoparticles, Int. J. Heat Mass Transfer 113 (2017) 106–114.

[45] M. Sheikholeslami, A. Zeeshan, Mesoscopic simulation of CuO-H_2O nanofluid in a porous enclosure with elliptic heat source, Int. J. Hydrogen Energy 42 (22) (2017) 15393–15402.

[46] M. Sheikholeslami, S.A. Shehzad, CVFEM for influence of external magnetic source on Fe_3O_4 – H_2O nanofluid behavior in a permeable cavity considering shape effect, Int. J. Heat Mass Transfer 115 (2017) 180–191.

[47] M. Sheikholeslami, M. Seyednezhad, Nanofluid heat transfer in a permeable enclosure in presence of variable magnetic field by means of CVFEM, Int. J. Heat Mass Transfer 114 (2017) 1169–1180.

[48] M. Sheikholeslami, H.B. Rokni, Melting heat transfer influence on nanofluid flow inside a cavity in existence of magnetic field, Int. J. Heat Mass Transfer 114 (2017) 517–526.

[49] M. Sheikholeslami, S.A. Shehzad, Magnetohydrodynamic nanofluid convective flow in a porous enclosure by means of LBM, Int. J. Heat Mass Transfer 113 (2017) 796–805.

[50] M. Sheikholeslami, Magnetohydrodynamic nanofluid forced convection in a porous lid driven cubic cavity using Lattice Boltzmann Method, J. Mol. Liq. 231 (2017) 555–565.

[51] M. Sheikholeslami Kandelousi, KKL correlation for simulation of nanofluid flow and heat transfer in a permeable channel, Phys. Lett. A 378 (45) (2014) 3331–3339.

[52] M. Sheikholeslami, Effect of uniform suction on nanofluid flow and heat transfer over a cylinder, J. Braz. Soc. Mech. Sci. Eng. 37 (2015) 1623–1633.

[53] M. Sheikholeslami, Lattice Boltzmann Method simulation of MHD non-Darcy nanofluid free convection, Phys. B 516 (2017) 55–71.

[54] M. Sheikholeslami, Influence of magnetic field on nanofluid free convection in an open porous cavity by means of Lattice Boltzmann Method, J. Mol. Liq. 234 (2017) 364–374.
[55] M. Sheikholeslami, A.J. Chamkha, Influence of Lorentz forces on nanofluid forced convection considering Marangoni convection, J. Mol. Liq. 225 (2017) 750–757.
[56] M. Sheikholeslami, T. Hayat, A. Alsaedi, Numerical simulation of nanofluid forced convection heat transfer improvement in existence of magnetic field using Lattice Boltzmann Method, Int. J. Heat Mass Transfer 108 (2017) 1870–1883.
[57] M. Sheikholeslami, R. Ellahi, Three dimensional mesoscopic simulation of magnetic field effect on natural convection of nanofluid, Int. J. Heat Mass Transfer 89 (2015) 799–808.
[58] M. Sheikholeslami, S. Abelman, Two phase simulation of nanofluid flow and heat transfer in an annulus in the presence of an axial magnetic field, IEEE Trans. Nanotechnol. 14 (3) (2015) 561–569.
[59] M. Sheikholeslami, H.B. Rokni, Free convection of CuO-H_2O nanofluid in a curved porous enclosure using mesoscopic approach, Int. J. Hydrogen Energy 42 (22) (2017) 14942–14949.
[60] M. Sheikholeslami, Numerical investigation for CuO–H_2O nanofluid flow in a porous channel with magnetic field using mesoscopic method, J. Mol. Liq. 249 (2018) 739–746.
[61] M. Sheikholeslami, H.B. Rokni, Influence of melting surface on MHD nanofluid flow by means of two phase model, Chin. J. Phys. 55 (2017) 1352–1360.
[62] M. Sheikholeslami, H.B. Rokni, Effect of melting heat transfer on nanofluid flow in existence of magnetic field considering Buongiorno Model, Chin. J. Phys. 55 (2017) 1115–1126.
[63] M. Sheikholeslami, T. Hayat, A. Alsaedi, MHD free convection of Al_2O_3–water nanofluid considering thermal radiation: a numerical study, Int. J. Heat Mass Transfer 96 (2016) 513–524.
[64] M. Sheikholeslami, K. Vajravelu, M.M. Rashidi, Forced convection heat transfer in a semi annulus under the influence of a variable magnetic field, Int. J. Heat Mass Transfer 92 (2016) 339–348.
[65] M. Sheikholeslami, T. Hayat, A. Alsaedi, Numerical simulation for forced convection flow of MHD CuO-H_2O nanofluid inside a cavity by means of LBM, J. Mol. Liq. 249 (2018) 941–948.
[66] M. Sheikholeslami, H.R. Kataria, A.S. Mittal, Effect of thermal diffusion and heat-generation on MHD nanofluid flow past an oscillating vertical plate through porous medium, J. Mol. Liq. 257 (2018) 12–25.
[67] M. Sheikholeslami, M. Barzegar Gerdroodbary, S. Valiallah Mousavi, D.D. Ganji, R. Moradi, Heat transfer enhancement of ferrofluid inside an 90° elbow channel by non-uniform magnetic field, J. Magn. Magn. Mater. 460 (2018) 302–311.
[68] M. Sheikholeslami, Numerical simulation for solidification in a LHTESS by means of nano-enhanced PCM, J. Taiwan Inst. Chem. Eng. 86 (2018) 25–41.
[69] M. Sheikholeslami, Numerical modeling of nano enhanced PCM solidification in an enclosure with metallic fin, J. Mol. Liq. 259 (2018) 424–438.
[70] M. Sheikholeslami, A. Ghasemi, Solidification heat transfer of nanofluid in existence of thermal radiation by means of FEM, Int. J. Heat Mass Transfer 123 (2018) 418–431.
[71] M. Sheikholeslami, S.A. Shehzad, CVFEM simulation for nanofluid migration in a porous medium using Darcy model, Int. J. Heat Mass Transfer 122 (2018) 1264–1271.
[72] M. Sheikholeslami, M. Darzi, M.K. Sadoughi, Heat transfer improvement and Pressure Drop during condensation of refrigerant-based nanofluid: an experimental procedure, Int. J. Heat Mass Transfer 122 (2018) 643–650.
[73] M. Sheikholeslami, H.B. Rokni, CVFEM for effect of Lorentz forces on nanofluid flow in a porous complex shaped enclosure by means of non-equilibrium model, J. Mol. Liq. 254 (2018) 446–462.
[74] M. Sheikholeslami, S.A. Shehzad, Z. Li, Water based nanofluid free convection heat transfer in a three dimensional porous cavity with hot sphere obstacle in existence of Lorenz forces, Int. J. Heat Mass Transfer 125 (2018) 375–386.
[75] M. Sheikholeslami, M. Jafaryar, D.D. Ganji, Z. Li, Exergy loss analysis for nanofluid forced convection heat transfer in a pipe with modified turbulators, J. Mol. Liq. 262 (2018) 104–110.
[76] M. Sheikholeslami, M. Jafaryar, Z. Li, Nanofluid turbulent convective flow in a circular duct with helical turbulators considering CuO nanoparticles, Int. J. Heat Mass Transfer 124 (2018) 980–989.
[77] M. Sheikholeslami, Z. Li, M. Shamlooei, Nanofluid, MHD natural convection through a porous complex shaped cavity considering thermal radiation, Phys. Lett. A 382 (2018) 1615–1632.
[78] M. Sheikholeslami, H. B. Rokni, Magnetic nanofluid flow and convective heat transfer in a porous cavity considering Brownian motion effects, Phys. Fluids, 30, 1, https://doi.org/10.1063/1.5012517.
[79] M. Sheikholeslami, S.A. Shehzad, Simulation of water based nanofluid convective flow inside a porous enclosure via non-equilibrium model, Int. J. Heat Mass Transfer 120 (2018) 1200–1212.
[80] M. Sheikholeslami, S.A. Shehzad, Non-Darcy free convection of Fe_3O_4-water nanoliquid in a complex shaped enclosure under impact of uniform Lorentz force, Chin. J. Phys. 56 (2018) 270–281.
[81] M. Sheikholeslami, M. Seyednezhad, Simulation of nanofluid flow and natural convection in a porous media under the influence of electric field using CVFEM, Int. J. Heat Mass Transfer 120 (2018) 772–781.
[82] M. Sheikholeslami, M. Shamlooei, R. Moradi, Numerical simulation for heat transfer intensification of nanofluid in a permeable curved enclosure considering shape effect of Fe_3O_4 nanoparticles, Chem. Eng. Process.: Process Intensif. 124 (2018) 71–82.
[83] M. Sheikholeslami, T. Hayat, T. Muhammad, A. Alsaedi, MHD forced convection flow of nanofluid in a porous cavity with hot elliptic obstacle by means of Lattice Boltzmann method, Int. J. Mech. Sci. 135 (2018) 532–540.
[84] M. Sheikholeslami, Numerical investigation of nanofluid free convection under the influence of electric field in a porous enclosure, J. Mol. Liq. 249 (2018) 1212–1221.
[85] M. Sheikholeslami, CuO-water nanofluid flow due to magnetic field inside a porous media considering Brownian motion, J. Mol. Liq. 249 (2018) 921–929.

[86] M. Sheikholeslami, D.D. Ganji, Influence of electric field on Fe_3O_4-water nanofluid radiative and convective heat transfer in a permeable enclosure, J. Mol. Liq. 250 (2018) 404–412.
[87] M. Sheikholeslami, S.A. Shehzad, Z. Li, Nanofluid heat transfer intensification in a permeable channel due to magnetic field using Lattice Boltzmann method, Physica B (2018). Available from: https://doi.org/10.1016/j.physb.2018.03.036.
[88] M. Sheikholeslami, A. Zeeshan, Numerical simulation of Fe_3O_4 -water nanofluid flow in a non-Darcy porous media, Int. J. Num. Methods Heat Fluid Flow 28 (3) (2018) 641–660. Available from: https://doi.org/10.1108/HFF-04-2017-0160.
[89] M. Sheikholeslami, D.D. Ganji, Numerical approach for magnetic nanofluid flow in a porous cavity using CuO nanoparticles, Mater. Des. 120 (2017) 382–393.
[90] M. Sheikholeslami, D.D. Ganji, Numerical modeling of magnetohydrodynamic CuO-water transportation inside a porous cavity considering shape factor effect, Colloids Surf. A 529 (2017) 705–714.

CHAPTER

12

Forced Convection of Nanofluid in Existence of Electric Field Using CVFEM

12.1 INTRODUCTION

Thermal conductivity of fluid can be improved by adding nanosize metal particles into the base fluid. This method can enhance heat transfer. Electric field is one of the effective active methods of enhancing heat transfer. Shu and Lai [1] applied electric field on induced flow. Kasayapanand et al. [2] investigated the effect of electrode arrangements on Nusselt number. Sheikholeslami et al. [3] presented HD nanofluid force convective heat transfer. They considered electric field dependent viscosity. Sheikholeslami and Ellahi [4] investigated electrohydrodynamic nanofluid hydrothermal treatment in an enclosure with sinusoidal upper wall. Sheikholeslami et al. [5] studied the effect of electric field on nanofluid flow in a complex geometry. Sheikholeslami [6] investigated Fe_3O_4-H_2O nanofluid thermal improvement inexistence of Coulomb forces. Sheikholeslami and Ganji [7] considered variable viscosity for simulation of nanofluid forced convection in the presence of an electric field. Sheikholeslami and Bhatti [8] presented an active method for nanofluid heat transfer enhancement by means of EHD. Sheikholeslami and Rokni [9] presented the impact of EFD viscosity on nanofluid forced convection in a cavity with sinusoidal wall. Sheikholeslami and Vajravelu [10] reported forced convection heat transfer in Fe_3O_4-ethylene glycol nanofluid under the influence of Coulomb force. In the recent decade, applications of nanofluid for various engineering problems were presented [11–92].

12.2 EHD NANOFLUID FORCE CONVECTIVE HEAT TRANSFER CONSIDERING ELECTRIC FIELD DEPENDENT VISCOSITY

12.2.1 Problem Definition

Geometry and boundary conditions are shown in Fig. 12.1. All walls are stationary except the lower one. The lid wall is hot wall and the others are cold. Distribution of electric density is depicted in Fig. 12.2.

12.2.2 Governing Equation

Electric field equations should be combined with hydrothermal equations. Electric field equations are:

$$q = \nabla.\left(\vec{E}\,\varepsilon\right) \tag{12.1}$$

$$\vec{E} = (-\nabla\varphi) \tag{12.2}$$

$$\frac{\partial q}{\partial t} = -\nabla.\vec{J} \tag{12.3}$$

Application of Control Volume based Finite Element Method (CVFEM) for Nanofluid Flow and Heat Transfer.
DOI: https://doi.org/10.1016/B978-0-12-814152-6.00012-6

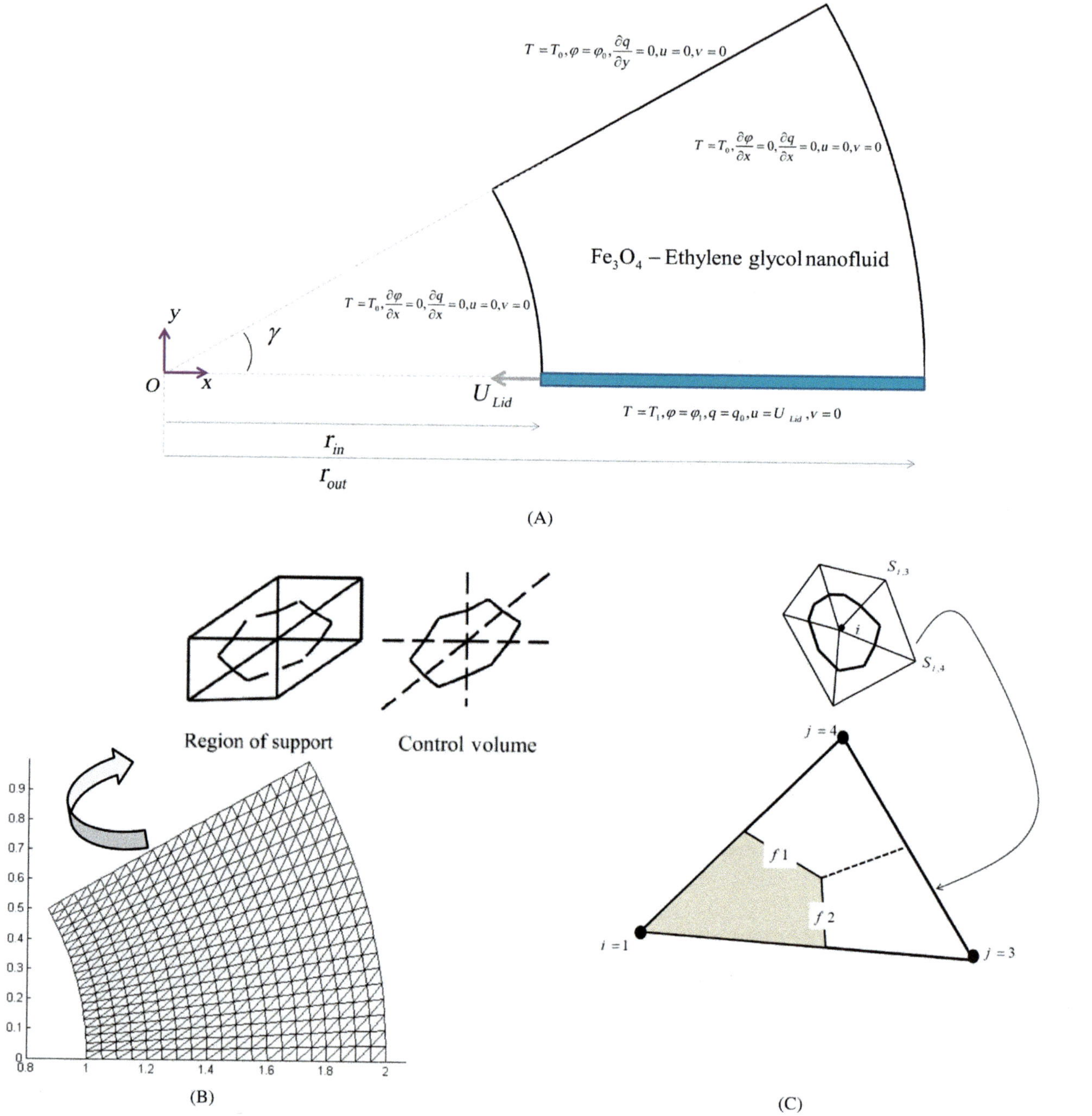

FIGURE 12.1 (A) Geometry of the problem and boundary conditions; (B) the mesh of enclosure considered in this work; (C) a sample triangular element and its corresponding control volume.

Charge distribution can be modeled in two ways: mobility and conductivity models. The equation of electric current density is:

$$\vec{J} = q\vec{V} + \sigma\vec{E} - D\nabla q \tag{12.4}$$

Using the above equations leads to the following equation:

$$\frac{\partial \overline{q}}{\partial t} + \overline{u}\frac{\partial \overline{q}}{\partial \overline{x}} + \overline{v}\frac{\partial \overline{q}}{\partial \overline{y}} + \frac{1}{\text{Re}\,\text{Pr}_E}\left[\overline{q}\left(\frac{\partial \overline{E}_y}{\partial \overline{y}} + \frac{\partial \overline{E}_x}{\partial \overline{x}}\right) + \overline{E}_y\frac{\partial \overline{q}}{\partial \overline{y}} + \overline{E}_x\frac{\partial \overline{q}}{\partial \overline{x}}\right] = \frac{\rho_{nf}/\rho_f}{\mu_{nf}/\mu_f}\frac{1}{\text{Re}\,D_e}\left(\frac{\partial^2 \overline{q}}{\partial \overline{y}^2} + \frac{\partial^2 \overline{q}}{\partial \overline{x}^2}\right) \tag{12.5}$$

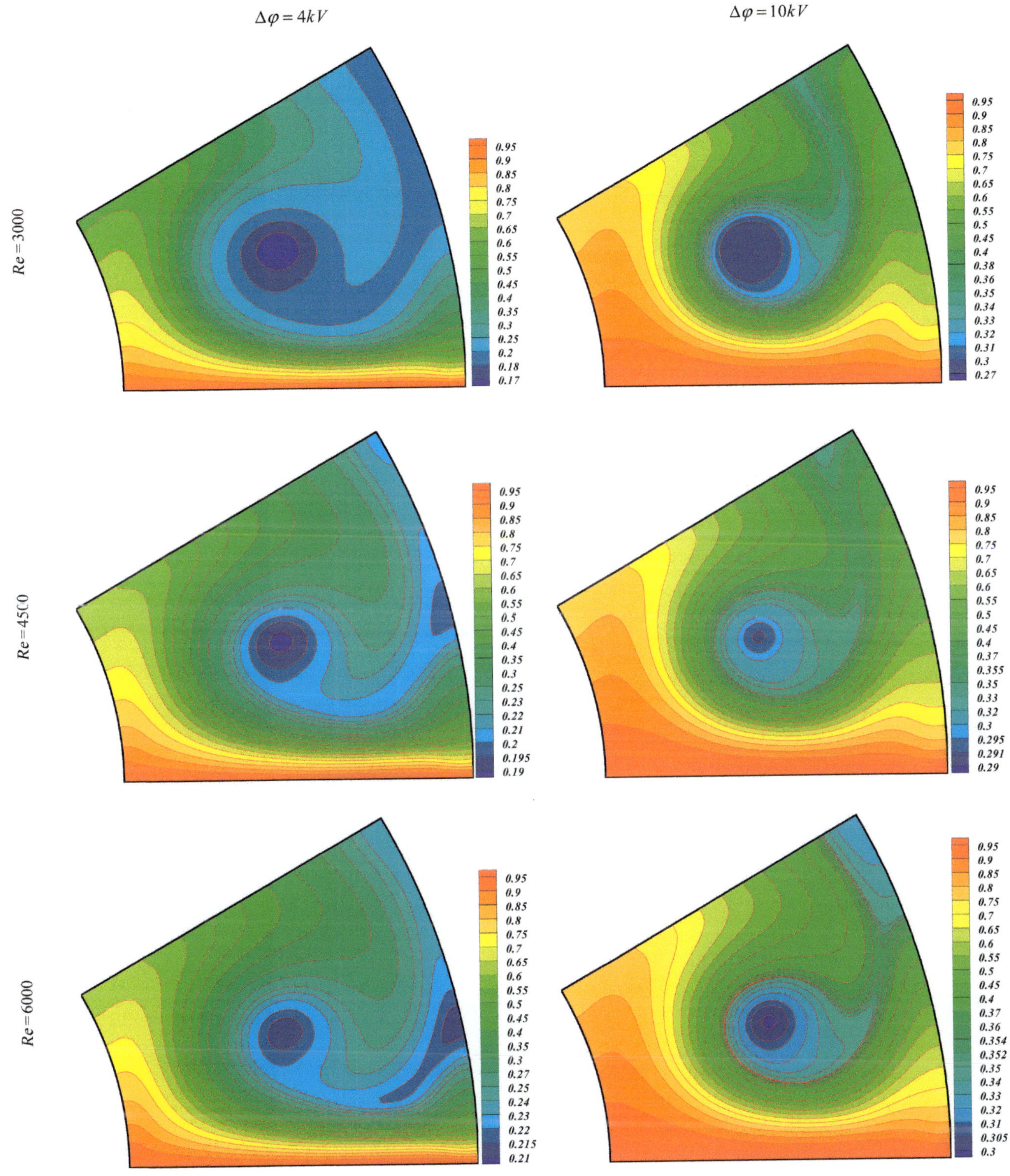

FIGURE 12.2 Electric density distribution injected by the bottom electrode.

Diffusion term is small, Eq. (12.4) can be changed to:

$$\vec{J} = qb\vec{E} + q\vec{V} \tag{12.6}$$

In the presence of an electric field, Coulomb forces should be added to momentum equations:

$$\begin{cases} \nabla.\vec{V} = 0 \\ \rho_{nf}\left(\dfrac{\partial \vec{V}}{\partial t} + \left(\vec{V}.\nabla\right)\vec{V}\right) = -\nabla p + \mu_{nf}\nabla^2 V^{\rightarrow} + q\vec{E} \\ (\rho C_p)_{nf}\left(\dfrac{\partial T}{\partial t} + \left(\vec{V}.\nabla\right)T\right) = k_{nf}\nabla^2 T + \vec{J}.\vec{E} \\ \nabla.\vec{J} + \dfrac{\partial q}{\partial t} = 0 \\ \nabla.\varepsilon\vec{E} = q \\ \vec{E} = -\nabla\varphi \end{cases} \tag{12.7}$$

ρ_{nf}, $(\rho C_p)_{nf}$, α_{nf}, β_{nf}, μ_{nf}, and k_{nf} are defined as:

$$\rho_{nf} = \rho_s\phi + \rho_f(1-\phi) \tag{12.8}$$

$$(\rho C_p)_{nf} = (\rho C_p)_s\phi + (\rho C_p)_f(1-\phi) \tag{12.9}$$

$$\alpha_{nf} = k_{nf}/(\rho C_p)_{nf} \tag{12.10}$$

$$\beta_{nf} = \beta_s\phi + \beta_f(1-\phi) \tag{12.11}$$

$$k_{nf} = k_f\frac{-2\phi(k_f - k_s) + 2k_f + k_s}{+\phi(k_f - k_s)_s + 2k_f + k} \tag{12.12}$$

Table 12.1 illustrate the properties of the base fluid and nanoparticles. The effect of electric field on viscosity of nanofluid has been taken into account:

$$\mu = A_1 + A_2(\Delta\varphi) + A_3(\Delta\varphi)^2 + A_4(\Delta\varphi)^3 \tag{12.13}$$

Table 12.2 shows the coefficient values of this equation. Nondimensional parameters are presented as follows:

$$\begin{gathered} \bar{t} = \frac{tU_{Lid}}{L}, \bar{p} = \frac{P}{\rho U_{Lid}{}^2}, \bar{y} = \frac{y}{L}, \bar{x} = \frac{x}{L}, \bar{v} = \frac{v}{U_{Lid}}, \bar{u} = \frac{u}{U_{Lid}}, \\ \theta = \frac{T - T_0}{\nabla T}, \nabla T = T_1 - T_0, \bar{\varphi} = \frac{\varphi - \varphi_0}{\nabla\varphi}, \nabla\varphi = \varphi_1 - \varphi_0, \bar{q} = \frac{q}{q_0}, \bar{E} = \frac{E}{E_0} \end{gathered} \tag{12.14}$$

TABLE 12.1 Thermophysical Properties of Water and Nanoparticles

	$\rho\ (kg/m^3)$	$C_p\ (j/kgk)$	$k\ (W/m.k)$	$\mu\ (Pa.s)$
Ethylene glycol	1110	2400	0.26	0.0162
Fe_3O_4	5200	670	6	–

TABLE 12.2 The Coefficient Values of Eq. (12.13)

Coefficient values	$\phi=0$	$\phi=0.05$
A_1	1.0603E + 001	9.5331
A_2	−2.698E-003	−3.4119E-003
A_3	2.9082E-006	5.5228E-006
A_4	−1.1876E-008	−4.1344E-008

where ∇T and $\nabla\varphi$ are (T_1-T_0) and $(\varphi_1-\varphi_0)$, respectively. By eliminating the over bar, the equations are:

$$\begin{cases} \nabla.\left(\vec{V}\right)=0 \\ \left(\dfrac{\partial\vec{V}}{\partial t}+\left(\vec{V}.\nabla\right)\vec{V}\right)=-\nabla p+\dfrac{\rho_{nf}/\rho_f}{\mu_{nf}/\mu_f}\dfrac{1}{\mathrm{Re}}\nabla^2\vec{V}+\dfrac{S_E}{\rho_{nf}/\rho_f}q\vec{E} \\ \left(\dfrac{\partial\theta}{\partial t}+\left(\vec{V}.\nabla\right)\theta\right)=\dfrac{k_{nf}/k_f}{\left(\rho C_p\right)_{nf}/\left(\rho C_p\right)_f}\dfrac{1}{\mathrm{RePr}}\nabla^2\theta+\dfrac{S_E Ec}{\left(\rho C_p\right)_{nf}/\left(\rho C_p\right)_f}\left(\vec{J}.\vec{E}\right) \\ \nabla.\vec{J}+\dfrac{\partial q}{\partial t}=0 \\ \nabla.\varepsilon\vec{E}=q,\quad \vec{E}=-\nabla\varphi \end{cases} \tag{12.15}$$

Stream function and vorticity can be defined as:

$$v=-\frac{\partial\psi}{\partial x},u=\frac{\partial\psi}{\partial y},\omega=\frac{\partial v}{\partial x}-\frac{\partial u}{\partial y},\quad \Omega=\frac{\omega L}{U_{Lid}},\Psi=\frac{\psi}{LU_{Lid}} \tag{12.16}$$

Stream function can satisfy the continuity equation. Vorticity equation can be derived by eliminating pressure sources.

Nu_{loc} and Nu_{ave} along the lid wall can be obtained as:

$$Nu_{loc}=\left(\frac{k_{nf}}{k_f}\right)\frac{\partial\Theta}{\partial Y} \tag{12.17}$$

$$Nu_{ave}=\frac{1}{L}\int_{r_{in}}^{r_{out}}Nu_{loc}\,dX \tag{12.18}$$

12.2.3 Effects of Active Parameters

Nanofluid flow in a semi lid-driven enclosure in existence of electric field has been investigated. The working fluid is ethylene glycol and Fe_3O_4. Viscosity of nanofluid relies on the strength of the electric field. The effects of volume fraction of nanoparticles ($\phi=0\%$ and 5%), Reynolds number ($\mathrm{Re}=3000, 4500$, and 6000), and supplied voltage ($\Delta\varphi=0,4,6,$ and $10kV$) are examined. In all cases Pr and $Ec$are equal to 149.54 and 10^{-6}, respectively.

Figs. 12.3–12.5 depict the impacts of supplied voltage and Reynolds number on isotherm and streamlines.

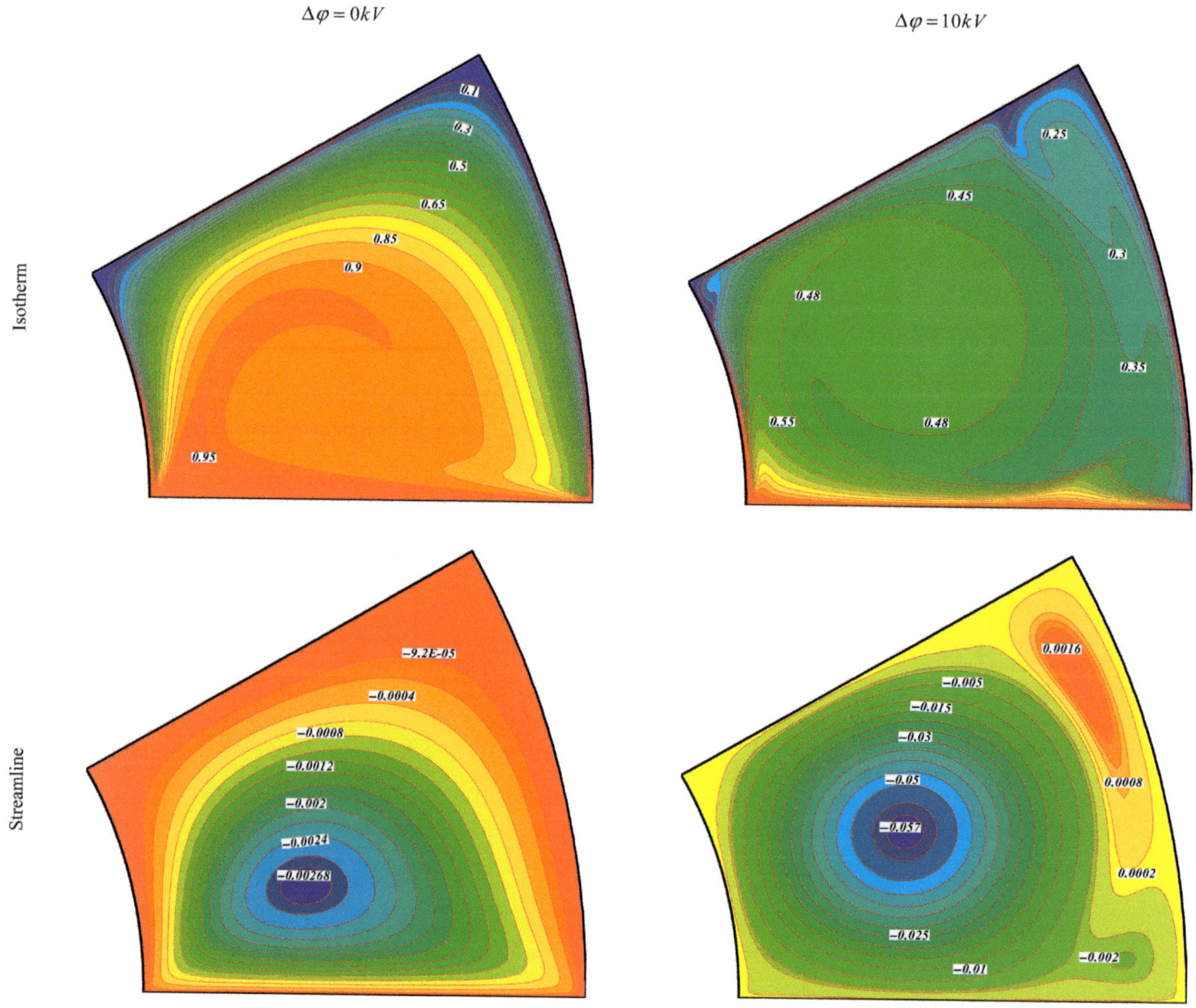

FIGURE 12.3 Effect of supplied voltage on streamlines and isotherm when Re = 3000, $\phi = 0.05$.

At low Reynolds number, streamline has one main eddy and two very small vortexes at upper corners of the cavity. By increasing the electric field the main eddy converts to two smaller eddies. So the thermal boundary layer thickness decreases near the moving wall and in turn the temperature gradient reduces. As the Reynolds number increases, the thermal boundary layer thickness decreases. So the rate of heat transfer enhances with the rise of the Reynolds number.

The effect of $\Delta\varphi$ and Re on local and average Nusselt number are shown in Figs. 12.6 and 12.7. Enhancing supplied voltage and Reynolds number leads to more distortion of isotherms. So Nusselt number augments with rise of these parameters. By applying the electric field stronger mixing leads to the destruction of the thermal boundary layer and creates vortex flow, thinning the boundary layer, and therefore causes the rise in heat transfer rate. Existence of extremums in the local Nusselt number is due to the generation of thermal plumes. The effect of coulomb forces on rate of heat transfer is more sensible for lower Reynolds number. This is due to this fact that the boundary layer thickness at low Reynolds number is thicker than that at high Reynolds number. So the electric field can be more effective t low Reynolds number.

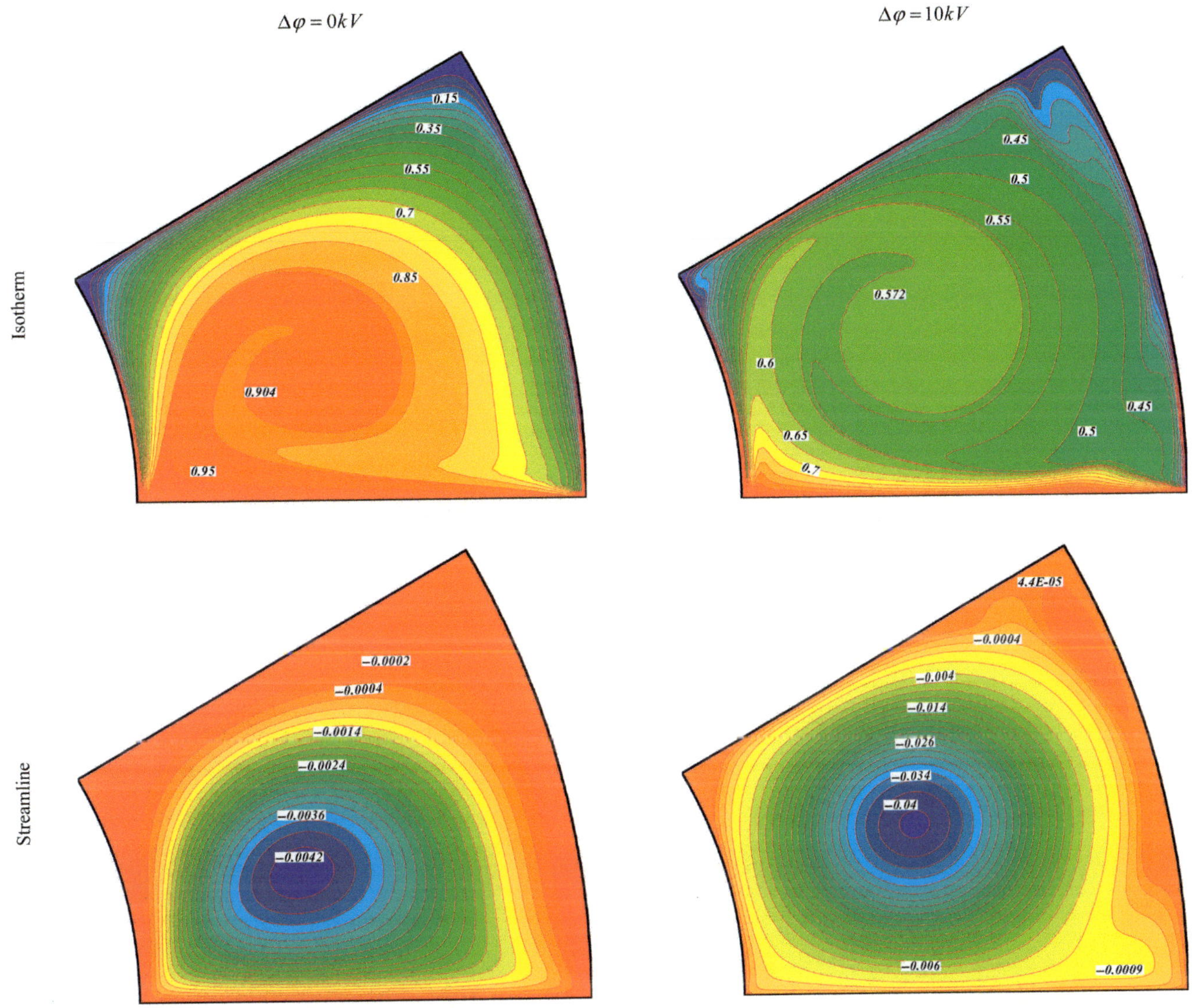

FIGURE 12.4 Effect of supplied voltage on streamlines and isotherm when Re = 4500, $\phi = 0.05$.

12.3 ELECTROHYDRODYNAMIC NANOFLUID HYDROTHERMAL TREATMENT IN AN ENCLOSURE WITH SINUSOIDAL UPPER WALL

12.3.1 Problem Definition

Fig. 12.8 illustrates the physical geometry along with the important parameters and mesh of the enclosure. The lower wall has the velocity of U_{Lid} and others are stationary. The lower wall has constant temperature T_1 and the temperature of other walls is T_0. Also the retained boundary conditions are depicted in Fig. 12.8A. The formula of the upper sinusoidal wall is:

$$Y = 1 - \{0.03(1 + \sin(2\pi x - \pi/2))\} \tag{12.19}$$

Fig. 12.9 depicts the distribution of electric density for different Reynolds number and supplied voltage.

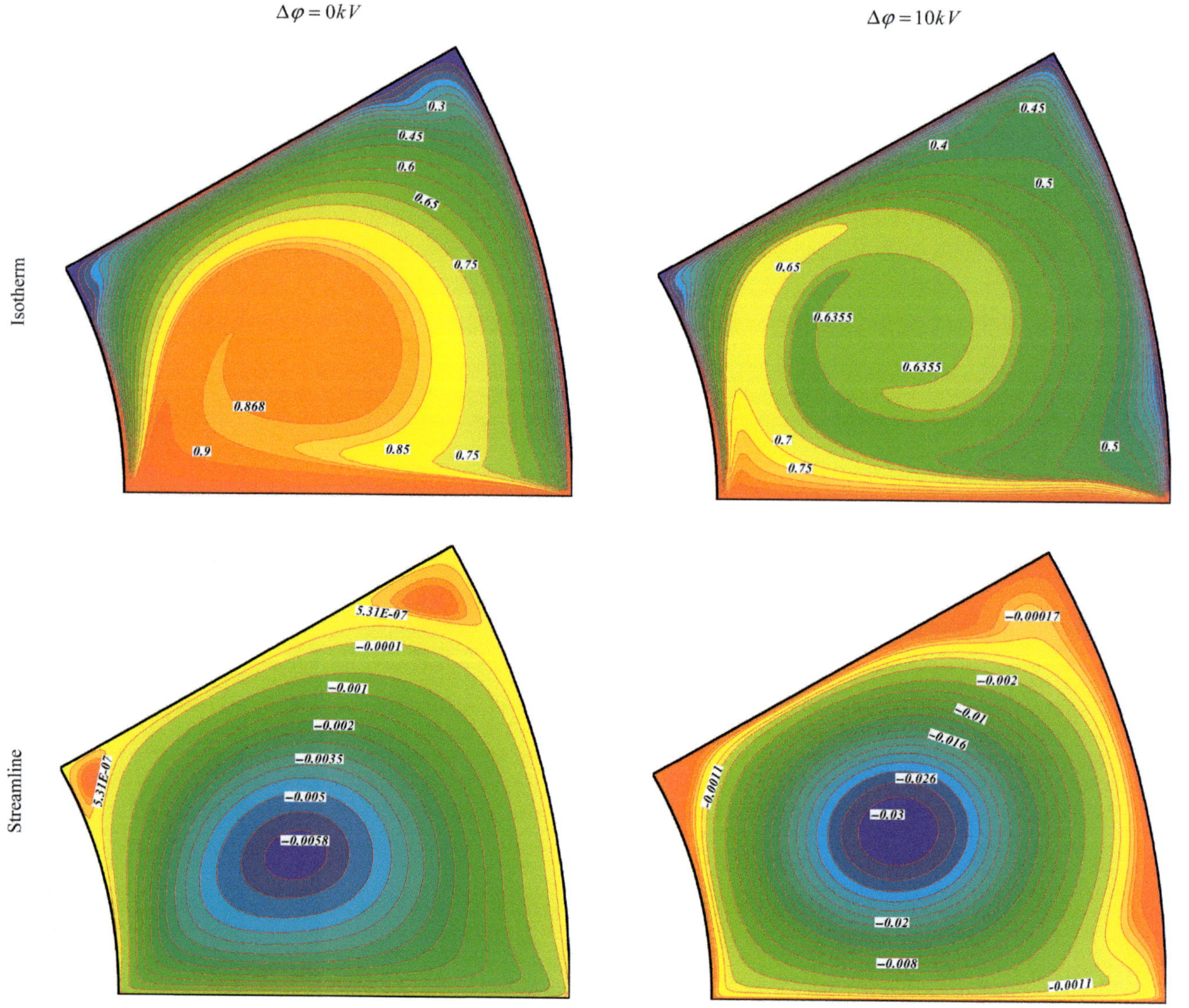

FIGURE 12.5 Effect of supplied voltage on streamlines and isotherm when Re = 6000, $\phi = 0.05$.

12.3.2 Governing Equation

In order to simulate nanofluid hydrothermal treatment in the presence of an electric field, we should combine the equations of electric fields with those of hydrothermal. The formulas of electric field are:

$$\nabla.\left(\vec{E}\,\varepsilon\right) = q \tag{12.20}$$

$$(-\nabla\varphi) = \vec{E} \tag{12.21}$$

$$\nabla.\vec{J} + \frac{\partial q}{\partial t} = 0 \tag{12.22}$$

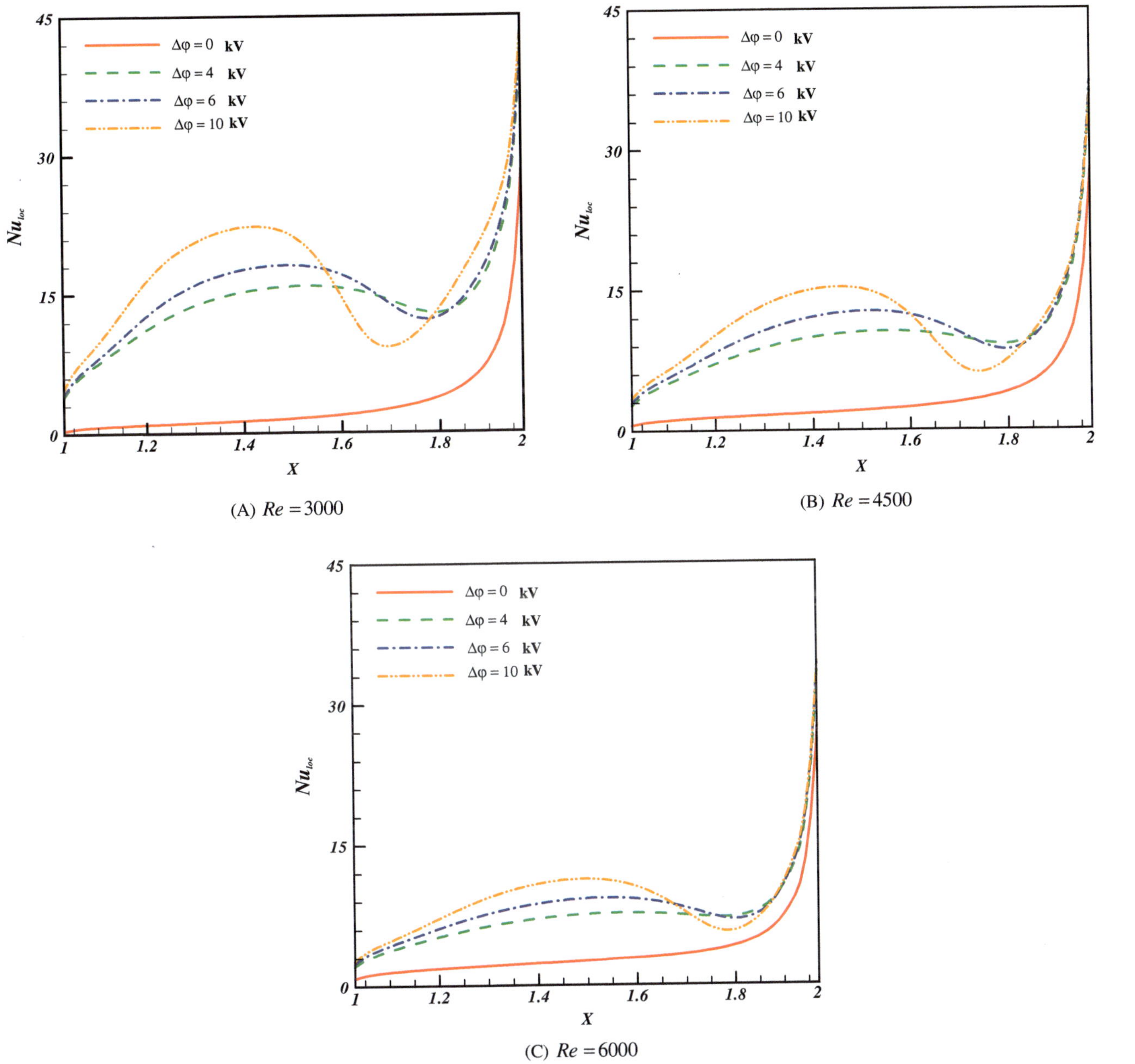

FIGURE 12.6 Effects of Reynolds number and supplied voltage on local Nusselt number when $\phi = 0.05$. (A) Re = 3000, (B) Re = 4500, (C) Re = 6000.

There exist two models for charge distribution: (1) conductivity model and (2) mobility model. In the first model, electroconvection relies on a temperature gradient. But in the second model, electroconvection is independent of the temperature gradient in the liquid. In the case of free charge origination, the second model is more acceptable according to experimental results. Electric current density can be defined as:

$$\vec{J} = \sigma \vec{E} - D\nabla q + q\vec{V} \tag{12.23}$$

where $\sigma\vec{E}$ is ionic mobility, $D\nabla q$ is diffusion, $q\vec{V}$ is convection.

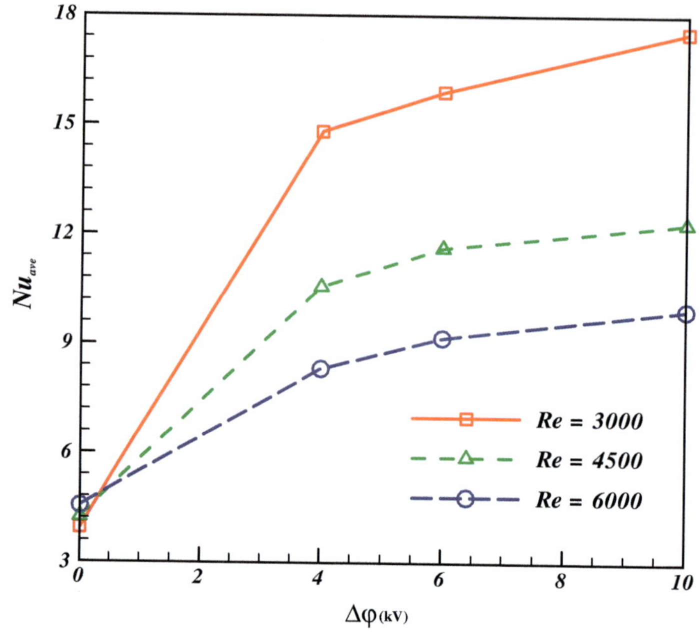

FIGURE 12.7 Effects of Reynolds number and supplied voltage on average Nusselt number.

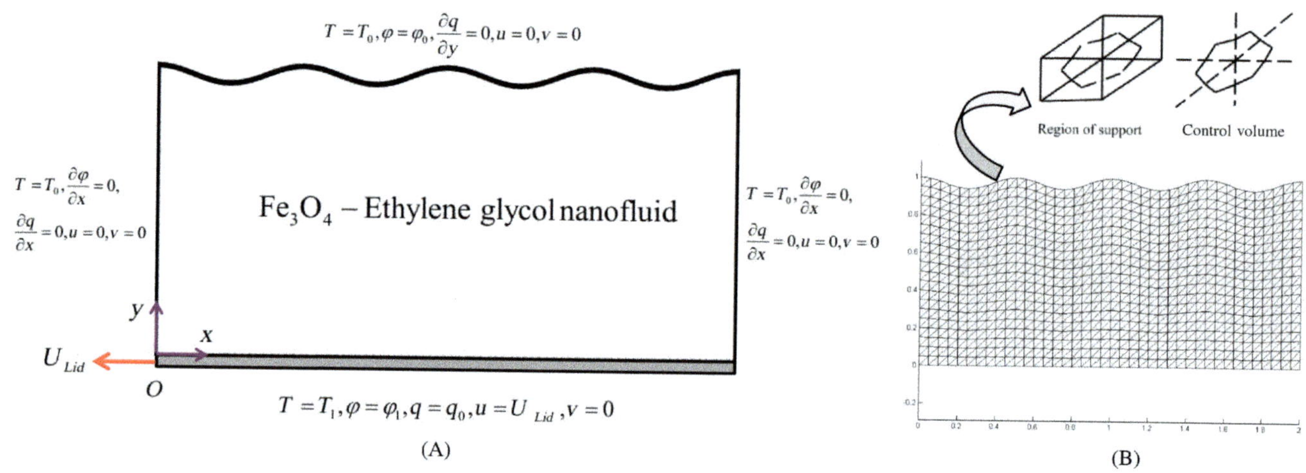

FIGURE 12.8 (A) Geometry of the problem and boundary conditions; (B) the mesh of enclosure considered in this work.

According to Eqs. (12.22) and (12.23), the equation for electric charge density can be obtained as follows:

$$\begin{aligned}
&\frac{\partial \overline{q}}{\partial \overline{t}} + \overline{v}\frac{\partial \overline{q}}{\partial \overline{y}} + \overline{u}\frac{\partial \overline{q}}{\partial \overline{x}} \\
&+ \frac{1}{\mathrm{Re}\,\mathrm{Pr}_E}\left[\overline{q}\left(\frac{\partial \overline{E_y}}{\partial \overline{y}} + \frac{\partial \overline{E_x}}{\partial \overline{x}}\right) + \overline{E_y}\frac{\partial \overline{q}}{\partial \overline{y}} + \overline{E_x}\frac{\partial \overline{q}}{\partial \overline{x}}\right] \\
&= \frac{\rho_{nf}/\rho_f}{\mu_{nf}/\mu_f}\frac{1}{\mathrm{Re}\,D_e}\left(\frac{\partial^2 \overline{q}}{\partial \overline{y}^2} + \frac{\partial^2 \overline{q}}{\partial \overline{x}^2}\right)
\end{aligned} \tag{12.24}$$

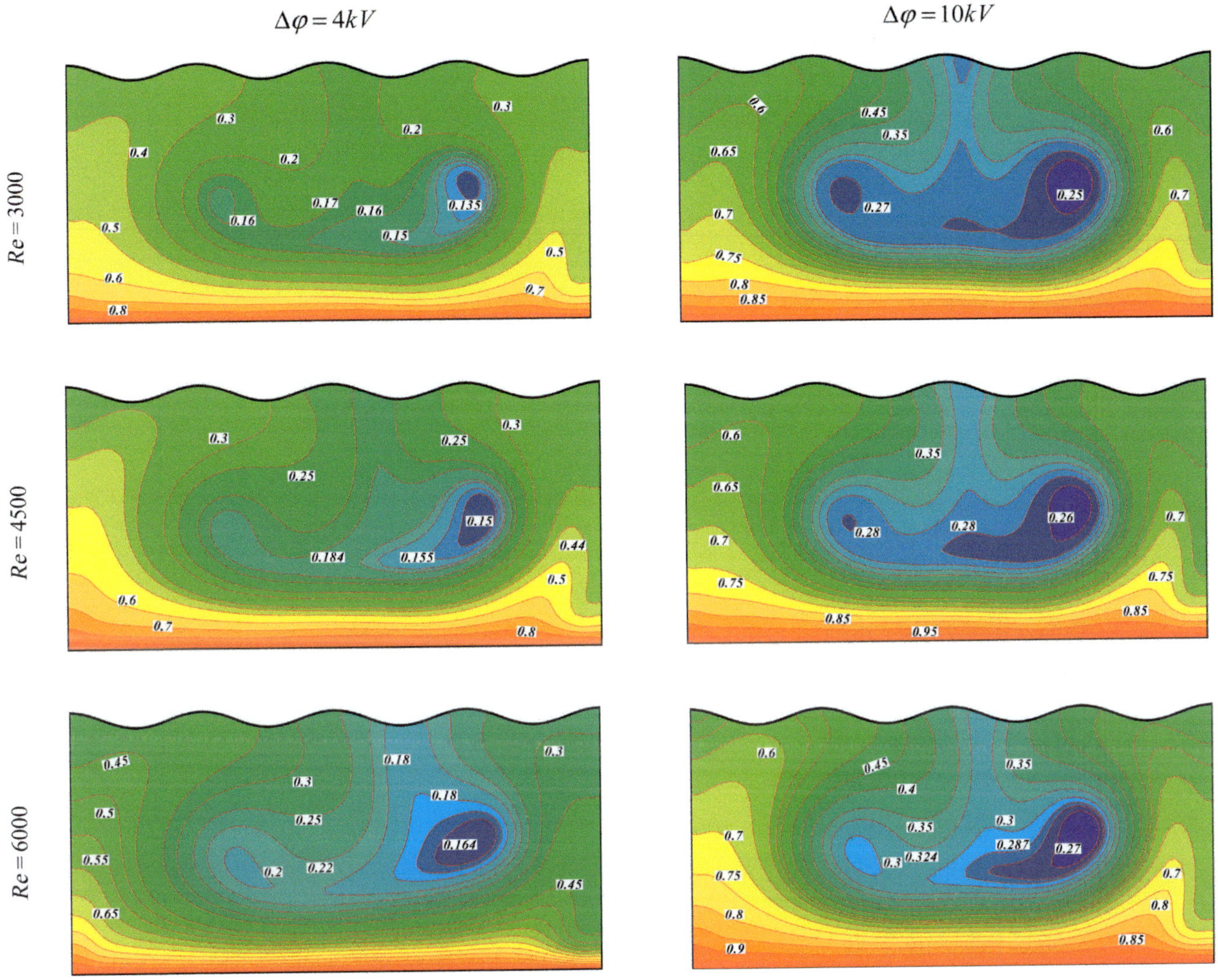

FIGURE 12.9 Electric density distribution injected by the bottom electrode.

The diffusion term can be taken negligible. Also $D\nabla q$ in Eq. (12.23) can be taken as negligible and $\sigma = bq$. So Eq. (12.23) can be considered as:

$$\vec{J} = q\vec{V} + qb\vec{E} \tag{12.25}$$

In presence of electric field Coulomb forces should be added to the momentum equation and the Joule heating effect should be added in the energy equation. So we have:

$$\begin{cases} \nabla.\vec{V} = 0 \\ \rho_{nf}\left(\dfrac{\partial \vec{V}}{\partial t} + \left(\vec{V}.\nabla\right)\vec{V}\right) = -\nabla p + \mu_{nf}\nabla^2\vec{V} + q\vec{E} \\ \left(\rho C_p\right)_{nf}\left(\dfrac{\partial T}{\partial t} + \left(\vec{V}.\nabla\right)T\right) = k_{nf}\nabla^2 T + \vec{J}.\vec{E} \\ \nabla.\vec{J} + \dfrac{\partial q}{\partial t} = 0 \\ \nabla.\varepsilon\vec{E} = q \\ \vec{E} = -\nabla\varphi \end{cases} \tag{12.26}$$

ρ_{nf}, $(\rho C_p)_{nf}$, α_{nf}, μ_{nf}, and k_{nf} are defined as:

$$\rho_{nf} = \rho_s\phi + \rho_f(1-\phi) \tag{12.27}$$

$$(\rho C_p)_{nf} = (\rho C_p)_s\phi + (\rho C_p)_f(1-\phi) \tag{12.28}$$

$$\alpha_{nf} = k_{nf}/(\rho C_p)_{nf} \tag{12.29}$$

$$\mu_{nf} = \frac{\mu_f}{(1-\phi)^{2.5}} \tag{12.30}$$

$$k_{nf} = k_f \frac{-2\phi(k_f - k_s) + 2k_f + k_s}{+\phi(k_f - k_s)_s + 2k_f + k} \tag{12.31}$$

The thermophysical properties of the working fluid are given in Table 12.1. Nondimensional parameters are introduced as follow:

$$\begin{gathered} \bar{t} = \frac{tU_{Lid}}{L}, \bar{p} = \frac{P}{\rho U_{Lid}{}^2}, \bar{y} = \frac{y}{L}, \bar{x} = \frac{x}{L}, \\ \bar{v} = \frac{v}{U_{Lid}}, \bar{u} = \frac{u}{U_{Lid}}, \theta = \frac{T - T_0}{\nabla T}, \nabla T = T_1 - T_0, \\ \bar{\varphi} = \frac{\varphi - \varphi_0}{\nabla\varphi}, \nabla\varphi = \varphi_1 - \varphi_0, \bar{q} = \frac{q}{q_0}, \bar{E} = \frac{E}{E_0} \end{gathered} \tag{12.32}$$

In order to reach a clear formulation, the over bar will be deleted in the next equations. So, the governing equations can be considered as follows:

$$\begin{cases} \nabla.\left(\vec{V}\right) = 0 \\ \left(\frac{\partial \vec{V}}{\partial t} + \left(\vec{V}.\nabla\right)\vec{V}\right) = -\nabla p + \frac{\rho_{nf}/\rho_f}{\mu_{nf}/\mu_f}\frac{1}{\text{Re}}\nabla^2\vec{V} + \frac{S_E}{\rho_{nf}/\rho_f}q\,\vec{E} \\ \left(\frac{\partial \theta}{\partial t} + \left(\vec{V}.\nabla\right)\theta\right) = \frac{k_{nf}/k_f}{(\rho C_p)_{nf}/(\rho C_p)_f}\frac{1}{\text{RePr}}\nabla^2\theta + \frac{S_E Ec}{(\rho C_p)_{nf}/(\rho C_p)_f}\left(\vec{J}.\vec{E}\right) \\ \nabla.\vec{J} + \frac{\partial q}{\partial t} = 0 \\ \nabla.\varepsilon\vec{E} = q \\ \vec{E} = -\nabla\varphi \end{cases} \tag{12.33}$$

The formulas of vorticity and stream function are:

$$\begin{gathered} \omega = \frac{\partial v}{\partial x} - \frac{\partial u}{\partial y}, \\ v = -\frac{\partial \psi}{\partial x}, u = \frac{\partial \psi}{\partial y}, \\ \Omega = \frac{\omega L}{U_{Lid}}, \Psi = \frac{\psi}{LU_{Lid}} \end{gathered} \tag{12.34}$$

Continuity equation has been satisfied by the stream function. By eliminating pressure between x-momentum and y- momentum, the vorticity equation can be obtained.

Nu_{loc} and Nu_{ave} along the lid wall can be obtained as:

$$Nu_{loc} = \left(\frac{k_{nf}}{k_f}\right)\frac{\partial \Theta}{\partial Y} \tag{12.35}$$

$$Nu_{ave} = \frac{1}{L}\int_0^L Nu_{loc}\, dX \tag{12.36}$$

12.3.3 Effects of Active Parameters

The electric field effect on hydrothermal behavior of nanofluid in an enclosure with moving lower and sinusoidal upper walls is presented. Table 12.1 illustrates the properties of ethylene glycol and Fe_3O_4. Calculations are prepared for various values of supplied voltage ($\Delta\varphi = 0, 4, 6,$ and $10kV$), volume fraction of nanoparticles ($\phi = 0\%$ and 4%) and Reynolds number numbers (Re = 3000, 4500, and 6000). In all calculations, the Prandtl number (Pr) and Eckert number (Ec) are set to 149.54, 0.0, and 10^{-6}.

The influence of Reynolds number and supplied voltage on streamlines and isotherms are shown in Figs. 12.10–12.12. At $Re = 3000$, one main eddy and two very small eddies at the upper corners of the enclosure exist in streamline. As the electric field is applied the main eddy turns into two smaller ones. Also the isotherms become denser near the hot wall due to the existence of these eddies. As the Reynolds number increases up to 6000, the upper right eddy becomes stronger and isotherms become denser near the bottom wall. Since convective heat transfer is helped more successfully at a greater flow rate, the Nusselt number is improved at greater Reynolds numbers. So the isotherms become more distorted at greater values of Reynolds number. As the electric field increases, the main cell converts to two eddies which rotate in opposite directions.

Figs. 12.13 and 12.14 depict the influences of $\Delta\varphi$ and Re on Nu_{loc} and Nu_{ave} along the lid wall. As the Reynolds number increases, the Nusselt number increases due to the decrease in thermal boundary layer thickness. Increasing supplied voltage makes the isotherms more distorted. Local Nusselt number profiles have extremums at higher values of supplied voltage because of the presence of thermal plumes. The Nusselt number is an

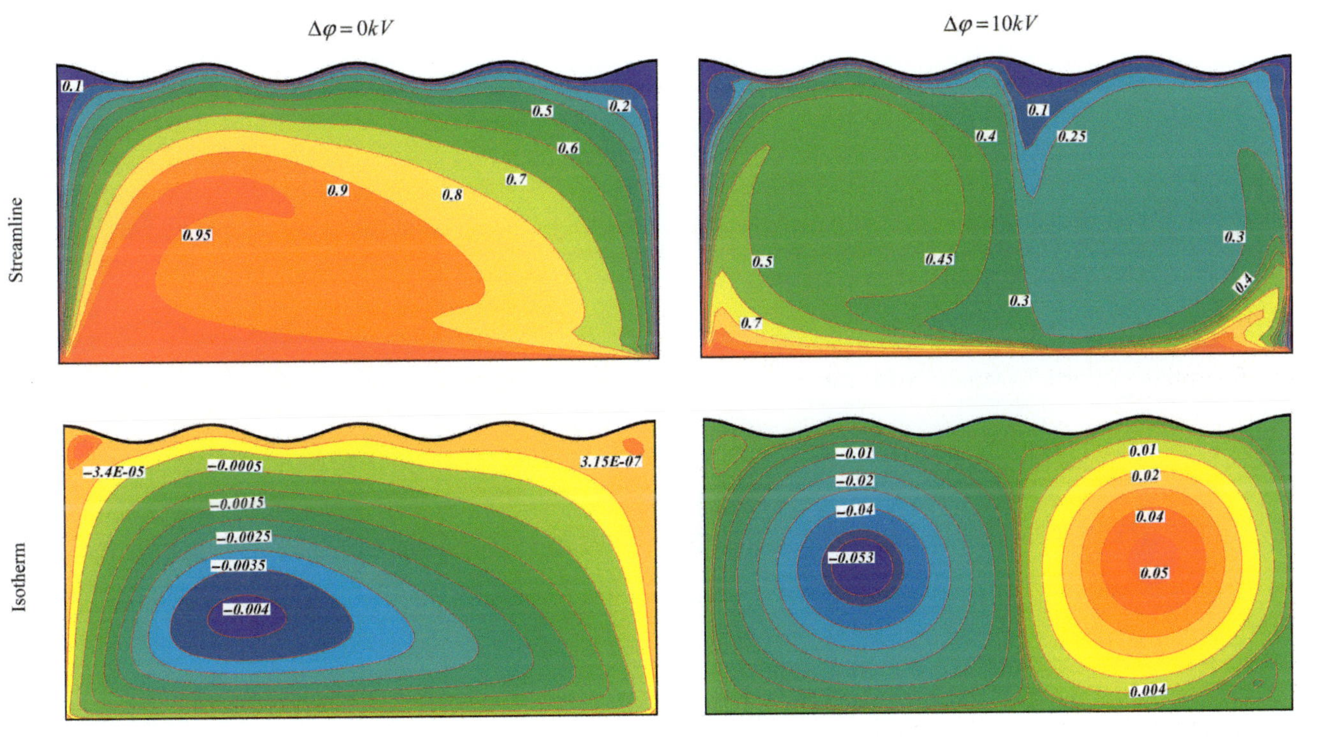

FIGURE 12.10 Effect of supplied voltage on streamlines and isotherms when Re = 3000, $\phi = 0.04$.

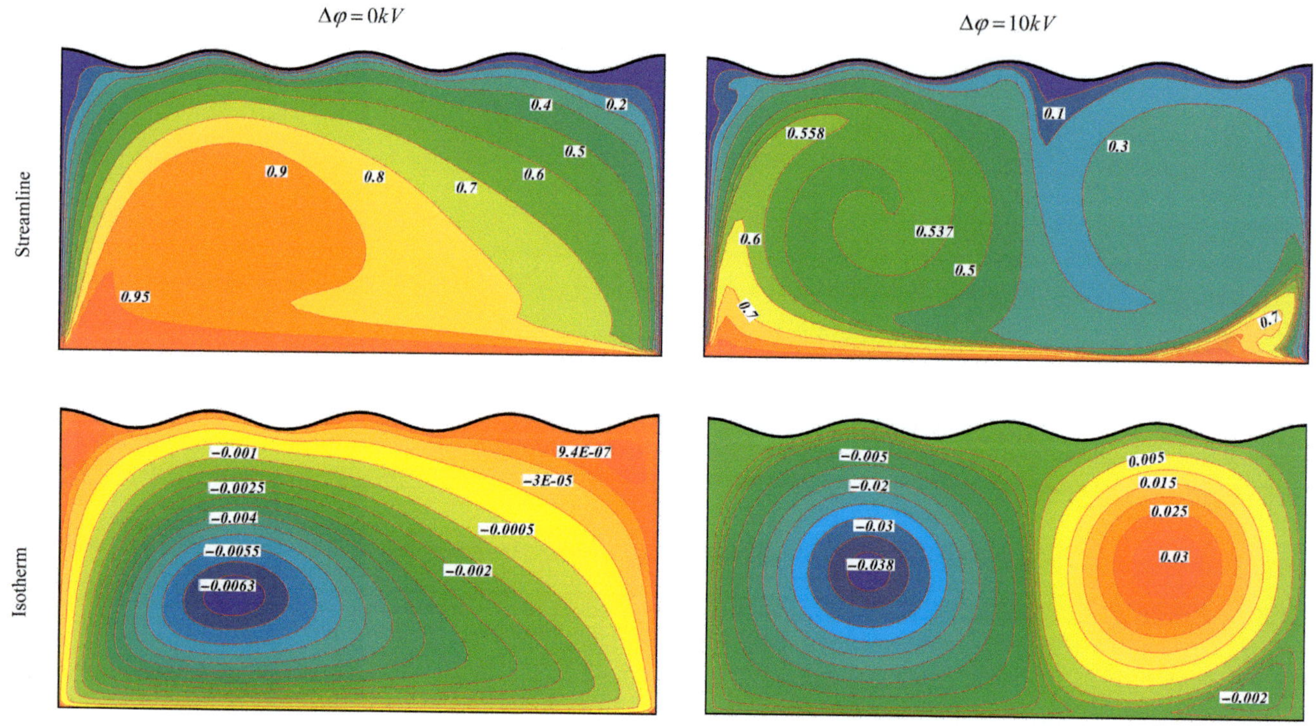

FIGURE 12.11 Effect of supplied voltage on streamlines and isotherms when Re = 4500, $\phi = 0.04$.

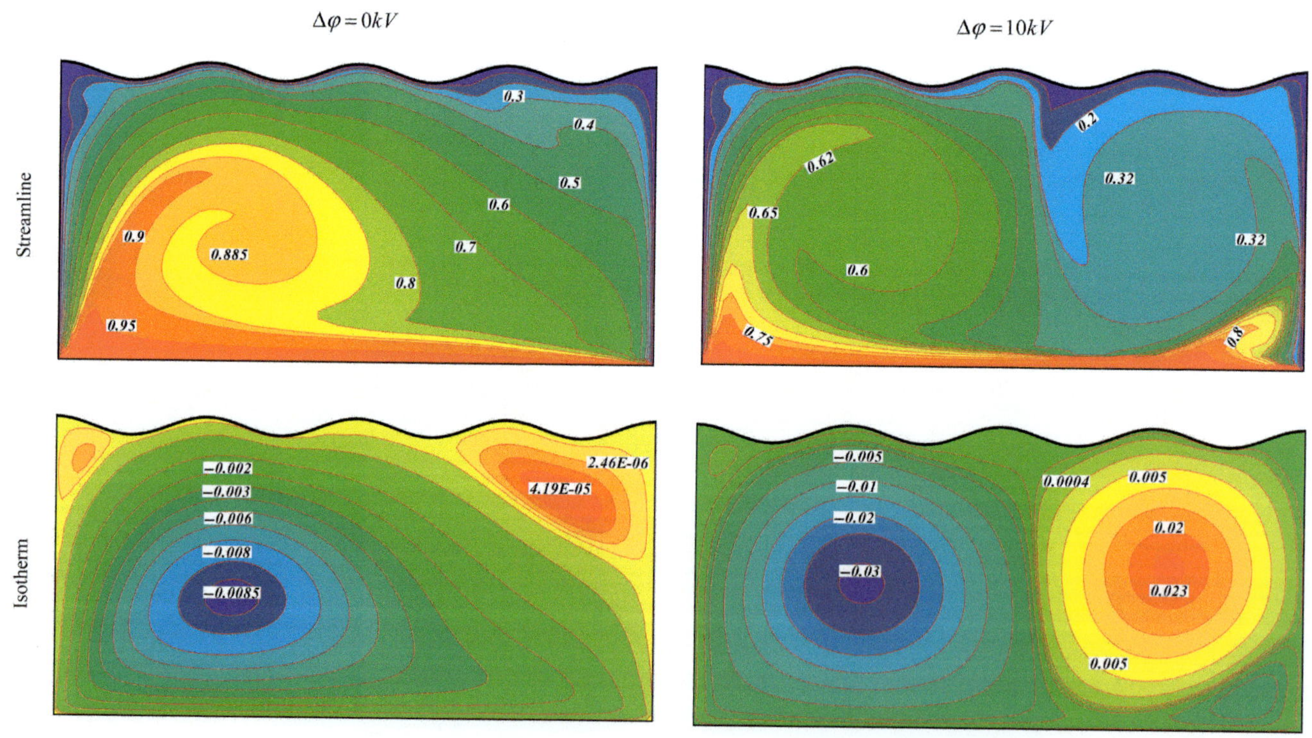

FIGURE 12.12 Effect of supplied voltage on streamlines and isotherms when Re = 6000, $\phi = 0.04$.

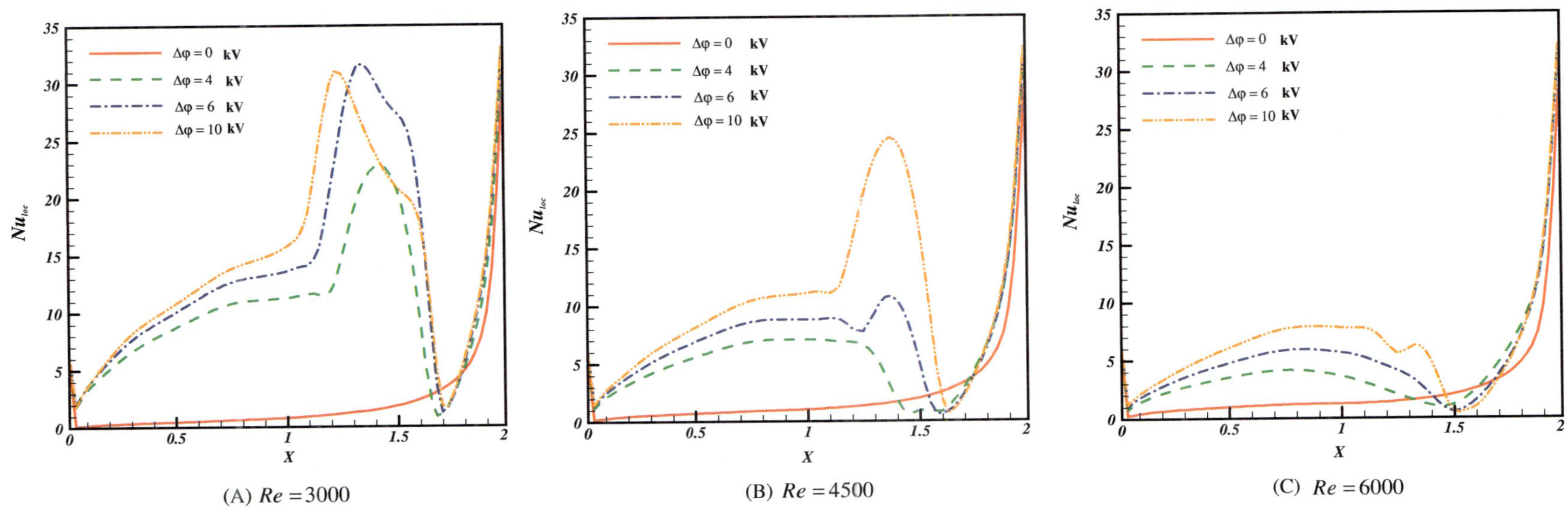

FIGURE 12.13 Effects of Reynolds number and supplied voltage on local Nusselt number when $\phi = 0.04$. (A) Re = 3000, (B) Re = 4500, (C) Re = 6000.

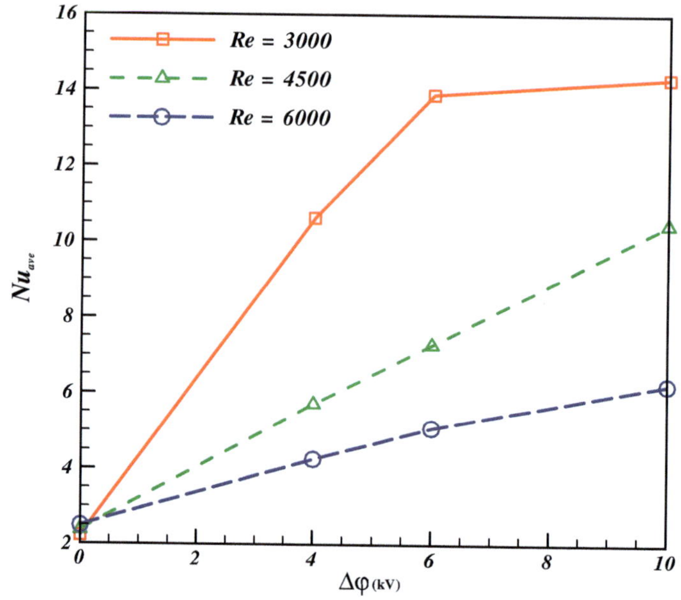

FIGURE 12.14 Effects of Reynolds number and supplied voltage on average Nusselt number.

increasing function of supplied voltage. In absence of electric field, the Nusselt number for $Re = 6000$ is 1.117 times higher than that of $Re = 3000$ while in the presence of electric field ($\Delta\varphi = 10$), the Nusselt number for $Re = 6000$ is 2.298385 times lower than that of $Re = 3000$. Also it can be concluded that the Nusselt number at $\Delta\varphi = 10$ for $Re = 3000$, 4500, and 6000 are higher than those of obtained at $\Delta\varphi = 0$. This observation confirms that the impact of electric field is more marked for a lower Reynolds number.

12.4 EFFECT OF ELECTRIC FIELD ON HYDROTHERMAL BEHAVIOR OF NANOFLUID IN A COMPLEX GEOMETRY

12.4.1 Problem Definition

Fig. 12.15 illustrates the physical geometry along with the important parameters and mesh of the enclosure. The lower wall has the velocity of U_{Lid} and others are stationary. The lower wall has constant temperature T_1 and the temperature of other walls is T_0. Also the retained boundary conditions are depicted in Fig. 12.15A. The shape of the inner cylinder profile is assumed to mimic the following pattern

$$r = r_{in} + A\cos(N(\zeta)) \tag{12.37}$$

in which r_{in} is the base circle radius, r_{out} is the radius of outer cylinder, A and N are amplitude and number of undulations, respectively. ζ is the rotation angle. In this study A and N are equal to 0.025 and 48, respectively. Fig. 12.16 depicts the distribution of electric density for different Reynolds number and supplied voltage.

12.4.2 Governing Equation

In order to simulate nanofluid hydrothermal treatment in the presence of an electric field, we should combine the equations of electric fields with those of hydrothermal. The formulas of the electric field are:

$$\nabla.\left(\vec{E}\,\varepsilon\right) = q \tag{12.38}$$

$$(-\nabla\varphi) = \vec{E} \tag{12.39}$$

$$\nabla.\vec{J} + \frac{\partial q}{\partial t} = 0 \tag{12.40}$$

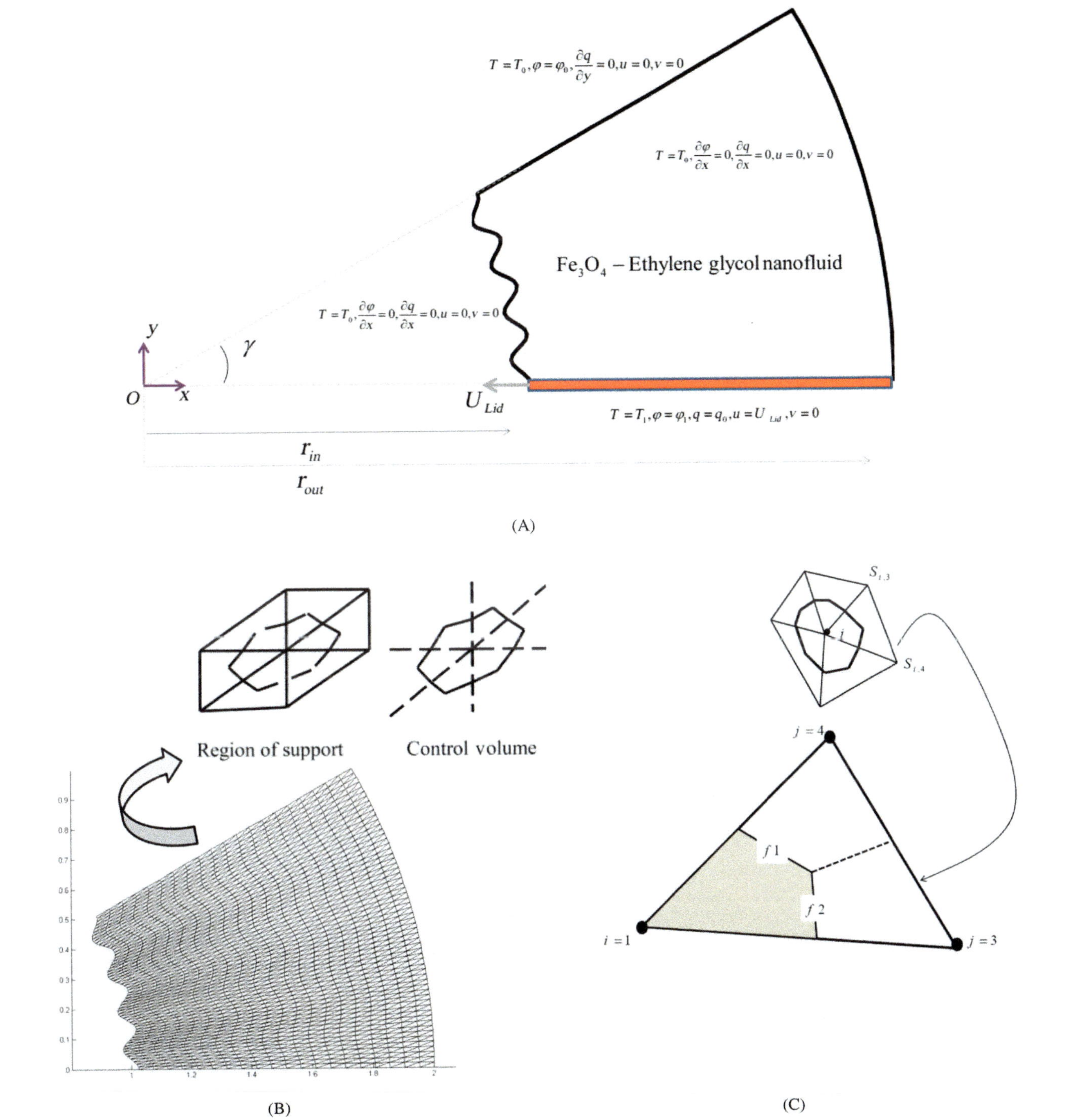

FIGURE 12.15 (A) Geometry of the problem and boundary conditions; (B) the mesh of enclosure considered in this work; (C) a sample triangular element and its corresponding control volume.

$$\vec{J} = \sigma \vec{E} - D\nabla q + q\vec{V} \tag{12.41}$$

$$\frac{\partial \overline{q}}{\partial \overline{t}} + \overline{v}\frac{\partial \overline{q}}{\partial \overline{y}} + \overline{u}\frac{\partial \overline{q}}{\partial \overline{x}} + \frac{1}{\mathrm{Re}\,\mathrm{Pr}_E}\left[\overline{q}\left(\frac{\partial \overline{E_y}}{\partial \overline{y}} + \frac{\partial \overline{E_x}}{\partial \overline{x}}\right) + \overline{E_y}\frac{\partial \overline{q}}{\partial \overline{y}} + \overline{E_x}\frac{\partial \overline{q}}{\partial \overline{x}}\right] = \frac{\rho_{nf}/\rho_f}{\mu_{nf}/\mu_f}\frac{1}{\mathrm{Re}\,D_e}\left(\frac{\partial^2 \overline{q}}{\partial \overline{y}^2} + \frac{\partial^2 \overline{q}}{\partial \overline{x}^2}\right) \tag{12.42}$$

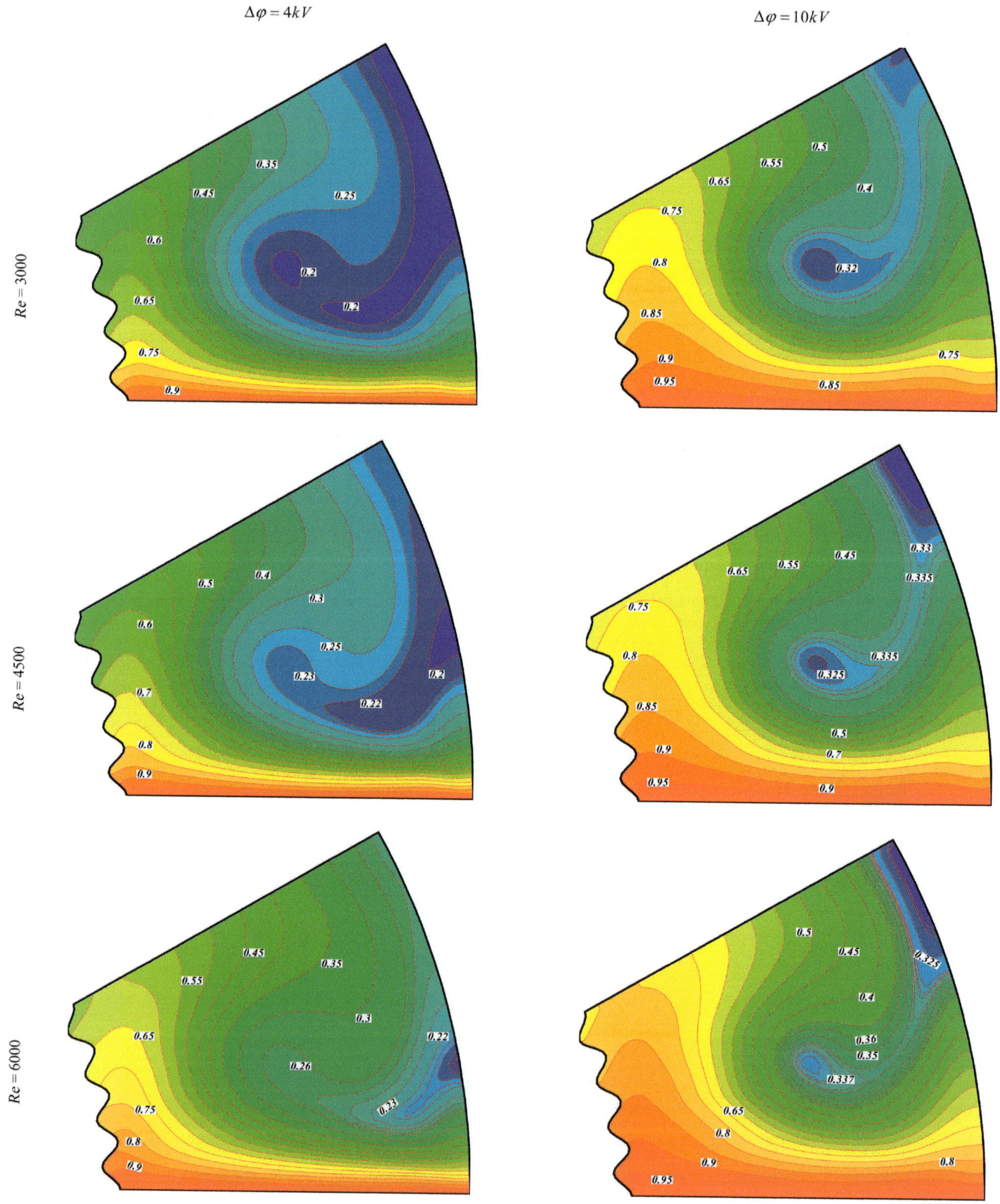

FIGURE 12.16 Electric density distribution injected by the bottom electrode.

Diffusion term can be taken as negligible. Also $D\nabla q$ in Eq. (12.41) can be taken as negligible and $\sigma = bq$ [51]. So Eq. (12.5) can be considered as:

$$\vec{J} = q\vec{V} + qb\vec{E} \tag{12.43}$$

In the presence of an electric field Coulomb forces should be added to the momentum equation and the Joule heating effect should be added in the energy equation. So we have:

$$\begin{cases} \nabla.\vec{V} = 0 \\ \rho_{nf}\left(\dfrac{\partial \vec{V}}{\partial t} + \left(\vec{V}.\nabla\right)\vec{V}\right) = -\nabla p + \mu_{nf}\nabla^2\vec{V} + q\vec{E} \\ \left(\rho C_p\right)_{nf}\left(\dfrac{\partial T}{\partial t} + \left(\vec{V}.\nabla\right)T\right) = k_{nf}\nabla^2 T + \vec{J}.\vec{E} \\ \nabla.\vec{J} + \dfrac{\partial q}{\partial t} = 0 \\ \nabla.\varepsilon\vec{E} = q \\ \vec{E} = -\nabla\varphi \end{cases} \tag{12.44}$$

$\rho_{nf}, \left(\rho C_p\right)_{nf}, \alpha_{nf}, k_{nf}$, and μ_{nf} are defined as:

$$\rho_{nf} = \rho_s\phi + \rho_f(1-\phi) \tag{12.45}$$

$$\left(\rho C_p\right)_{nf} = \left(\rho C_p\right)_s\phi + \left(\rho C_p\right)_f(1-\phi) \tag{12.46}$$

$$\alpha_{nf} = k_{nf}/\left(\rho C_p\right)_{nf} \tag{12.47}$$

$$k_{nf} = k_f\frac{-2\phi(k_f - k_s) + 2k_f + k_s}{+\phi(k_f - k_s)_s + 2k_f + k} \tag{12.48}$$

$$\mu_{nf} = \frac{\mu_f}{(1-\phi)^{2.5}} \tag{12.49}$$

Nondimensional parameters are introduced as follows:

$$\begin{gathered} \bar{t} = \frac{tU_{Lid}}{L}, \bar{p} = \frac{P}{\rho U_{Lid}{}^2}, \bar{y} = \frac{y}{L}, \bar{x} = \frac{x}{L}, \bar{v} = \frac{v}{U_{Lid}}, \bar{u} = \frac{u}{U_{Lid}}, \\ \theta = \frac{T - T_0}{\nabla T}, \nabla T = T_1 - T_0, \bar{\varphi} = \frac{\varphi - \varphi_0}{\nabla\varphi}, \nabla\varphi = \varphi_1 - \varphi_0, \bar{q} = \frac{q}{q_0}, \bar{E} = \frac{E}{E_0} \end{gathered} \tag{12.50}$$

where ∇T and $\nabla\varphi$ are $(T_1 - T_0)$ and $(\varphi_1 - \varphi_0)$, respectively. In order to reach a clear formulation, the over bar will be deleted in the next equations. So, the governing equations can be considered as follows:

$$\begin{cases} \nabla.\left(\vec{V}\right) = 0 \\ \left(\dfrac{\partial \vec{V}}{\partial t} + \left(\vec{V}.\nabla\right)\vec{V}\right) = -\nabla p + \dfrac{\rho_{nf}/\rho_f}{\mu_{nf}/\mu_f}\dfrac{1}{\mathrm{Re}}\nabla^2\vec{V} + \dfrac{S_E}{\rho_{nf}/\rho_f}q\vec{E} \\ \left(\dfrac{\partial \theta}{\partial t} + \left(\vec{V}.\nabla\right)\theta\right) = \dfrac{k_{nf}/k_f}{\left(\rho C_p\right)_{nf}/\left(\rho C_p\right)_f}\dfrac{1}{\mathrm{RePr}}\nabla^2\theta + \dfrac{S_E Ec}{\left(\rho C_p\right)_{nf}/\left(\rho C_p\right)_f}\left(\vec{J}.\vec{E}\right) \\ \nabla.\vec{J} + \dfrac{\partial q}{\partial t} = 0 \\ \nabla.\varepsilon\vec{E} = q \\ \vec{E} = -\nabla\varphi \end{cases} \tag{12.51}$$

The formulas of vorticity and stream function are:

$$\omega = \frac{\partial v}{\partial x} - \frac{\partial u}{\partial y},$$
$$v = -\frac{\partial \psi}{\partial x}, u = \frac{\partial \psi}{\partial y}, \tag{12.52}$$
$$\Omega = \frac{\omega L}{U_{Lid}}, \Psi = \frac{\psi}{LU_{Lid}}$$

Continuity equation has been satisfied by the stream function. By eliminating pressure between x-momentum and y-momentum, the vorticity equation can be obtained.

Nu_{loc} and Nu_{ave} along the lid wall can be obtained as:

$$Nu_{loc} = \left(\frac{k_{nf}}{k_f}\right)\frac{\partial \Theta}{\partial Y} \tag{12.53}$$

$$Nu_{ave} = \frac{1}{L}\int_{r_{in}}^{r_{out}} Nu_{loc}\, dX \tag{12.54}$$

12.4.3 Effects of Active Parameters

Effect of nonuniform electric field on nanofluid hydrothermal behavior in an enclosure with sinusoidal wall is presented. The working fluid is ethylene glycol and Fe_3O_4. Calculations are prepared for various values of supplied voltage ($\Delta\varphi = 0, 4, 6,$ and $10kV$), volume fraction of nanoparticles ($\phi = 0\%$ and 5%) and Reynolds number numbers ($\text{Re} = 3000, 4500$, and 6000). In all calculations, the Prandtl number (Pr) and Eckert number (Ec) are set to 149.54 and 10^{-6}.

Influence of Reynolds number and supplied voltage on streamlines and isotherm are shown in Figs. 12.17–12.19. At $\text{Re} = 3000$, one main eddy and two very small eddies at the upper corners of the enclosure exist in the streamlines. As the electric field is applied the main eddy turns into two smaller ones. Also the isotherms become denser near the hot wall due to the existence of these eddies. As the Reynolds number increases up to 6000, the upper right eddy becomes stronger and isotherms become denser near the bottom wall. Since convective heat transfer is helped more successfully at a greater flow rate, the Nusselt number is improved at greater Reynolds numbers. So the isotherms become more distorted at greater values of Reynolds number. As the electric field increases, the main cell converts to two eddies which rotate in opposite directions.

Figs. 12.20 and 12.21 depict the influences of $\Delta\varphi$ and Re on Nu_{loc} and Nu_{ave} along the lid wall. As the Reynolds number increases, the Nusselt number increases due to a decrease in thermal boundary layer thickness. Increasing the supplied voltage makes the isotherms more distorted. The local Nusselt number profiles have extremums at higher values of supplied voltage because of the existence of thermal plumes. The Nusselt number is an increasing function of supplied voltage. In the absence of an electric field, the Nusselt number for $Re = 6000$ is 1.135921 times higher than that of $Re = 3000$ while in the presence of electric field ($\Delta\varphi = 10$), the Nusselt number for $Re = 6000$ is 1.768835 times lower than that of $Re = 3000$. Also it can be concluded that Nusselt number at $\Delta\varphi = 10$ for $Re = 3000$, 4500, and 6000 are 5.816335, 4.112412, and 2.894771 times higher than those of obtained at $\Delta\varphi = 0$. This observation confirms that the impact of an electric field is more marked for a lower Reynolds number.

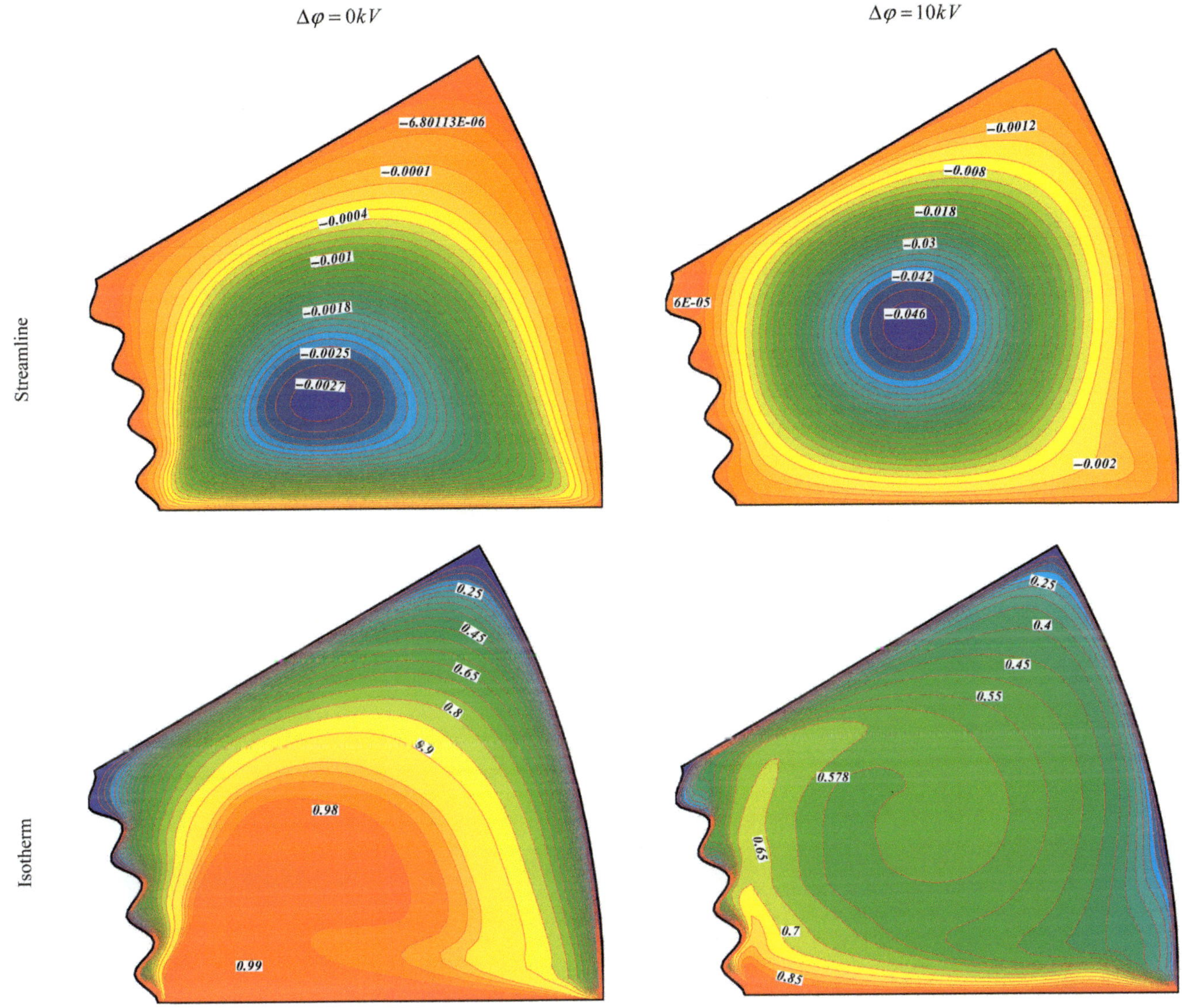

FIGURE 12.17 Effect of supplied voltage on streamlines and isotherms when Re = 3000, $\phi = 0.05$.

12.5 EFFECT OF COULOMB FORCES ON FE_3O_4-H2O NANOFLUID THERMAL IMPROVEMENT

12.5.1 Problem Definition

Fig. 12.22 demonstrates the schematic of this problem and its boundary conditions. The influence of an electric field on Fe_3O_4-H_2O nanofluid is considered. Only the top wall can move. The contours of electric density for various values of active parameters is depicted in Fig. 12.23.

12.5.2 Governing Equation

According to Gauss's law and Maxwell's relation, the electric field can be defined as:

$$q = \nabla .\varepsilon \vec{E} \tag{12.55}$$

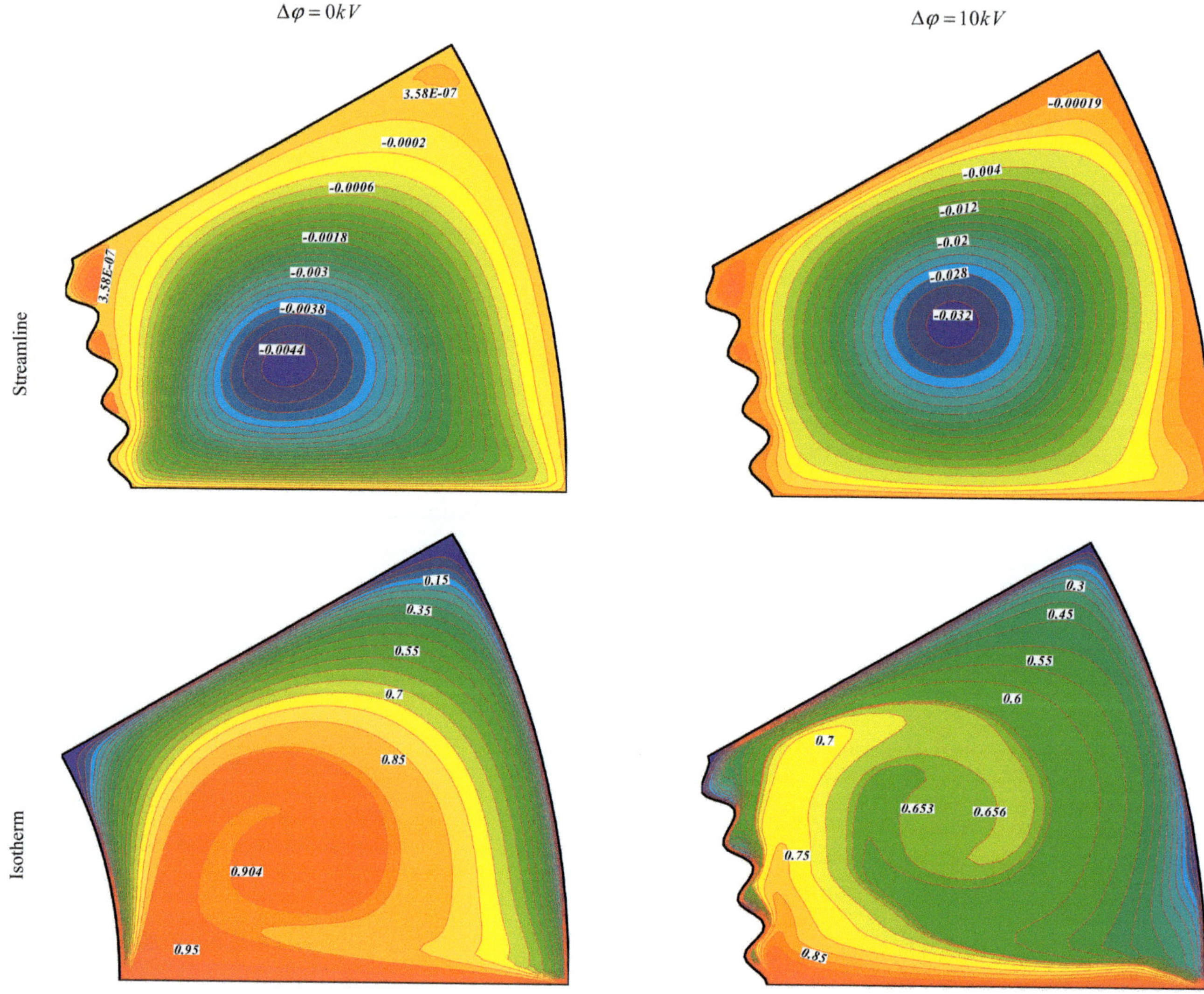

FIGURE 12.18 Effect of supplied voltage on streamlines and isotherms when $Re = 4500, \phi = 0.05$.

$$\vec{E} = -\nabla\varphi \tag{12.56}$$

$$\nabla.\vec{J} + \frac{\partial q}{\partial t} = 0 \tag{12.57}$$

$$\vec{J} = q\vec{V} - D\nabla q + \sigma\vec{E} \tag{12.58}$$

The governing equations are as follows:

$$\begin{cases} \nabla.\vec{V} = 0 \\ \rho_{nf}\left(\left(\vec{V}.\nabla\right)\vec{V} + \frac{\partial \vec{V}}{\partial t}\right) = q\vec{E} + \mu_{nf}\nabla^2\vec{V} - \nabla p \\ \left(\left(\vec{V}.\nabla\right)T + \frac{\partial T}{\partial t}\right) = \frac{k_{nf}}{(\rho C_p)_{nf}}\nabla^2 T + \frac{\vec{J}.\vec{E}}{(\rho C_p)_{nf}} \\ \nabla.\vec{J} + \frac{\partial q}{\partial t} = 0 \\ \nabla.\varepsilon\vec{E} = q \\ \vec{E} = -\nabla\varphi \end{cases} \tag{12.59}$$

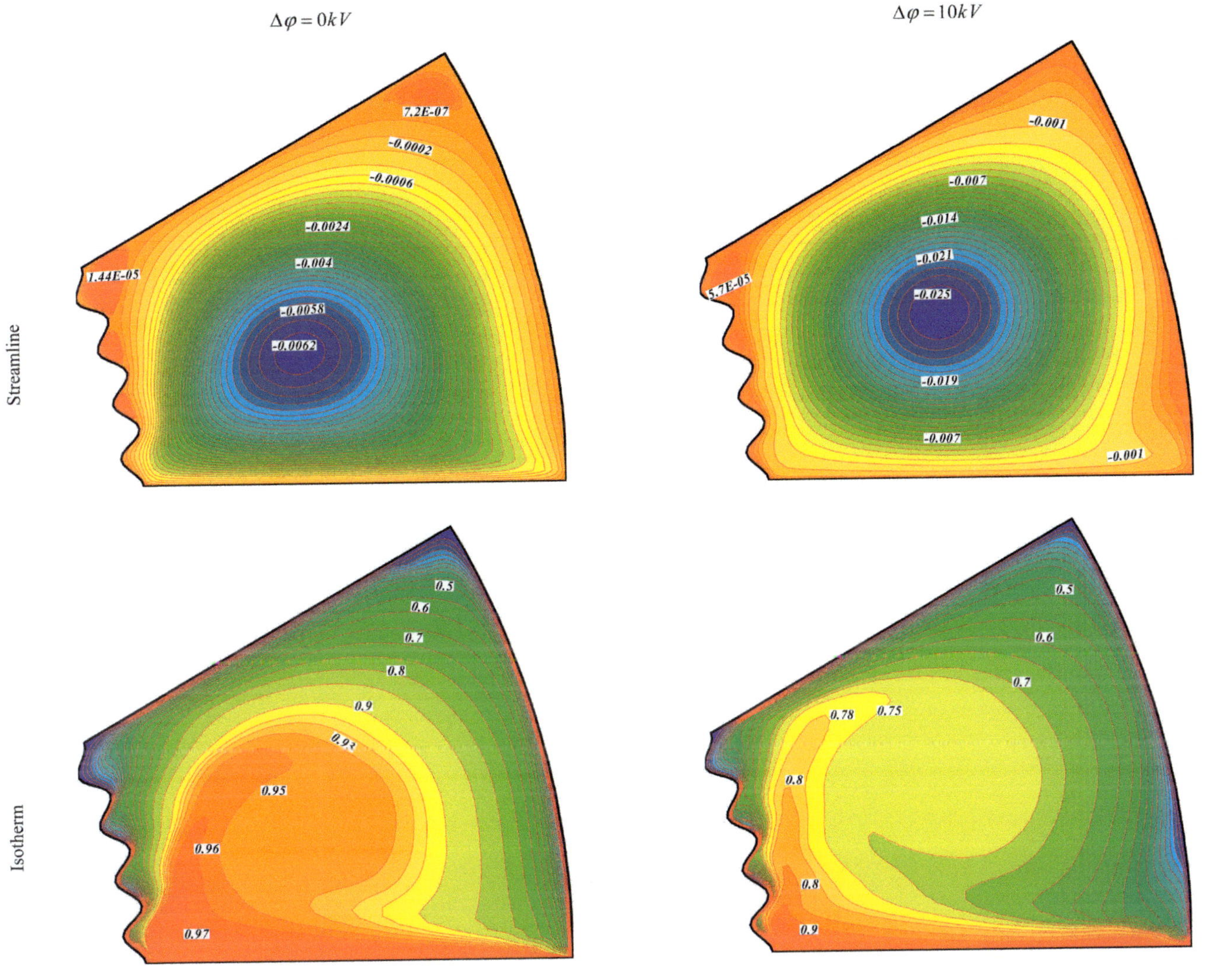

FIGURE 12.19 Effect of supplied voltage on streamlines and isotherms when Re = 6000, $\phi = 0.05$.

ρ_{nf}, $(\rho C_p)_{nf}$, μ_{nf}, and k_{nf} can be obtained as:

$$\begin{aligned} &\rho_{nf} = \rho_f(1-\phi) + \rho_s\phi, \quad &(\rho C_p)_{nf} = (\rho C_p)_f(1-\phi) + (\rho C_p)_s\phi, \\ &\mu_{nf} = \frac{\mu_f}{(1-\phi)^{2.5}}, \quad &\frac{k_{nf}}{k_f} = \frac{k_s + 2k_f - 2\phi(k_f - k_s)}{k_s + 2k_f + \phi(k_f - k_s)} \end{aligned} \tag{12.60}$$

Properties of Fe_3O_4 and H_2O are illustrated in Table 12.3.

So, the final dimensionless equations in the presence of an electric field are:

$$\begin{cases} \nabla.\vec{V} = 0 \\ \left(\left(\vec{V}.\nabla\right)\vec{V} + \dfrac{\partial \vec{V}}{\partial t}\right) = \dfrac{S_E}{\rho_{nf}/\rho_f} q\vec{E} + \dfrac{\rho_{nf}/\rho_f}{\mu_{nf}/\mu_f}\dfrac{1}{\mathrm{Re}}\nabla^2\vec{V} - \nabla p \\ \left(\left(\vec{V}.\nabla\right)\theta + \dfrac{\partial \theta}{\partial t}\right) = \dfrac{k_{nf}/k_f}{(\rho C_p)_{nf}/(\rho C_p)_f}\dfrac{1}{\mathrm{RePr}}\nabla^2\theta + \dfrac{S_E Ec}{(\rho C_p)_{nf}/(\rho C_p)_f}\left(\vec{J}\,.\,\vec{E}\right) \\ \vec{E} = -\nabla\varphi \\ q = \nabla.\varepsilon\vec{E} \\ \nabla.\vec{J} = -\dfrac{\partial q}{\partial t} \end{cases} \tag{12.61}$$

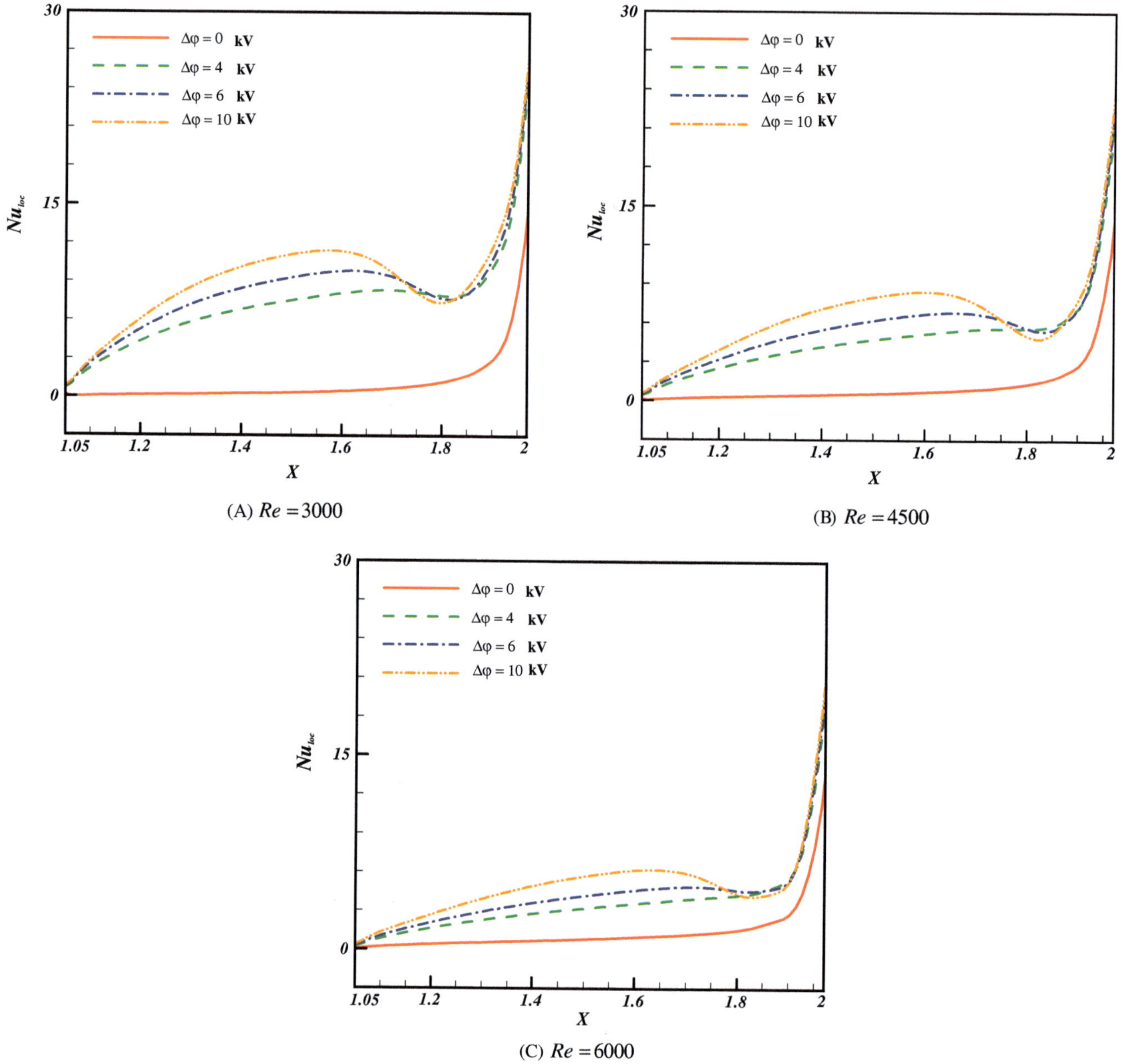

FIGURE 12.20 Effects of Reynolds number and supplied voltage on the local Nusselt number when $\phi = 0.05$. (A) Re = 3000, (B) Re = 4500, (C) Re = 6000.

where

$$\begin{aligned}
&\overline{u} = \frac{u}{U_{Lid}}, \overline{v} = \frac{v}{U_{Lid}}, \overline{y} = \frac{y}{L}, \overline{x} = \frac{x}{L}, \\
&\overline{t} = \frac{tU_{Lid}}{L}, \overline{p} = \frac{P}{\rho {U_{Lid}}^2} \overline{q} = \frac{q}{q_0}, \overline{E} = \frac{E}{E_0}, \\
&\nabla T = T_1 - T_0, \theta = \frac{T - T_0}{\nabla T}, \nabla \varphi = \varphi_1 - \varphi_0, \overline{\varphi} = \frac{\varphi - \varphi_0}{\nabla \varphi}
\end{aligned} \tag{12.62}$$

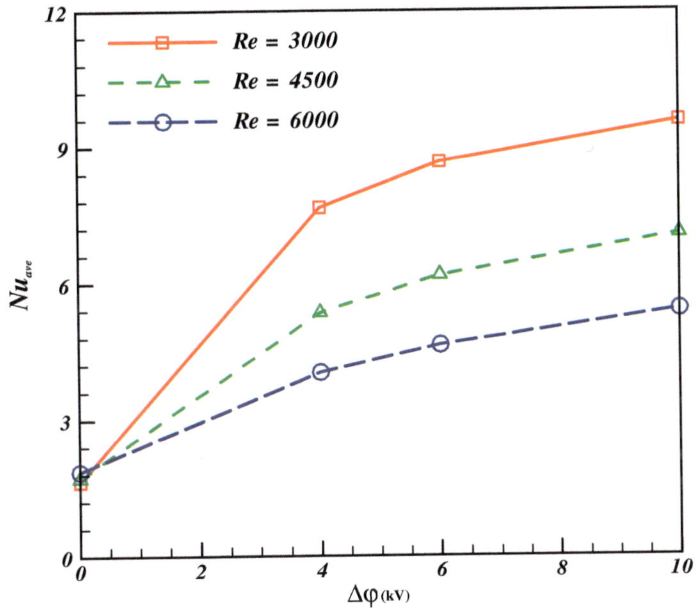

FIGURE 12.21 Effects of Reynolds number and supplied voltage on the average Nusselt number.

By eliminating the pressure gradient, the vorticity and stream function can be introduced as:

$$\omega = \frac{\partial v}{\partial x} - \frac{\partial u}{\partial y}, -\frac{\partial \psi}{\partial x} = v, \frac{\partial \psi}{\partial y} = u, \Psi = \frac{\psi}{\alpha_f}, \Omega = \frac{\omega L^2}{\alpha_f} \tag{12.63}$$

Nu_{loc} and Nu_{ave} along the left wall are defined as:

$$Nu_{loc} = \frac{\partial \Theta}{\partial X}\left(\frac{k_{nf}}{k_f}\right) \tag{12.64}$$

$$Nu_{ave} = \frac{1}{L}\int_0^L Nu_{loc}\, dY \tag{12.65}$$

12.5.3 Effects of Active Parameters

Coulomb forces impact on Fe_3O_4–H_2O hydrothermal behavior in a cavity with moving wall is studied. Various values of active parameters are examined, such as Reynolds number ($Re = 3000, 4000, 5000$, and 6000), volume fraction of solid particle ($\phi = 0\%$ and 4%), and supplied voltage ($\Delta\varphi = 0, 5, 8$, and $10kV$). Impacts of $\Delta\varphi$ and Re on isotherms and streamlines are demonstrated in Figs. 12.24–12.27. At low Re, one main vortex exists in the streamlines. As the Coulomb forces augment the main vortex turns into three smaller vortexes and the temperature gradient near the hot walls is enhanced.

As lid velocity increases, another eddy appears at the bottom right of the enclosure. As the voltage enhances, the previous cell alters to two vortexes which revolve in opposite directions. A further increase in the Reynolds number leads to the generation of another small eddy at the bottom left of the enclosure. Temperature gradient augments at high Reynolds numbers due to stronger convective heat transfer.

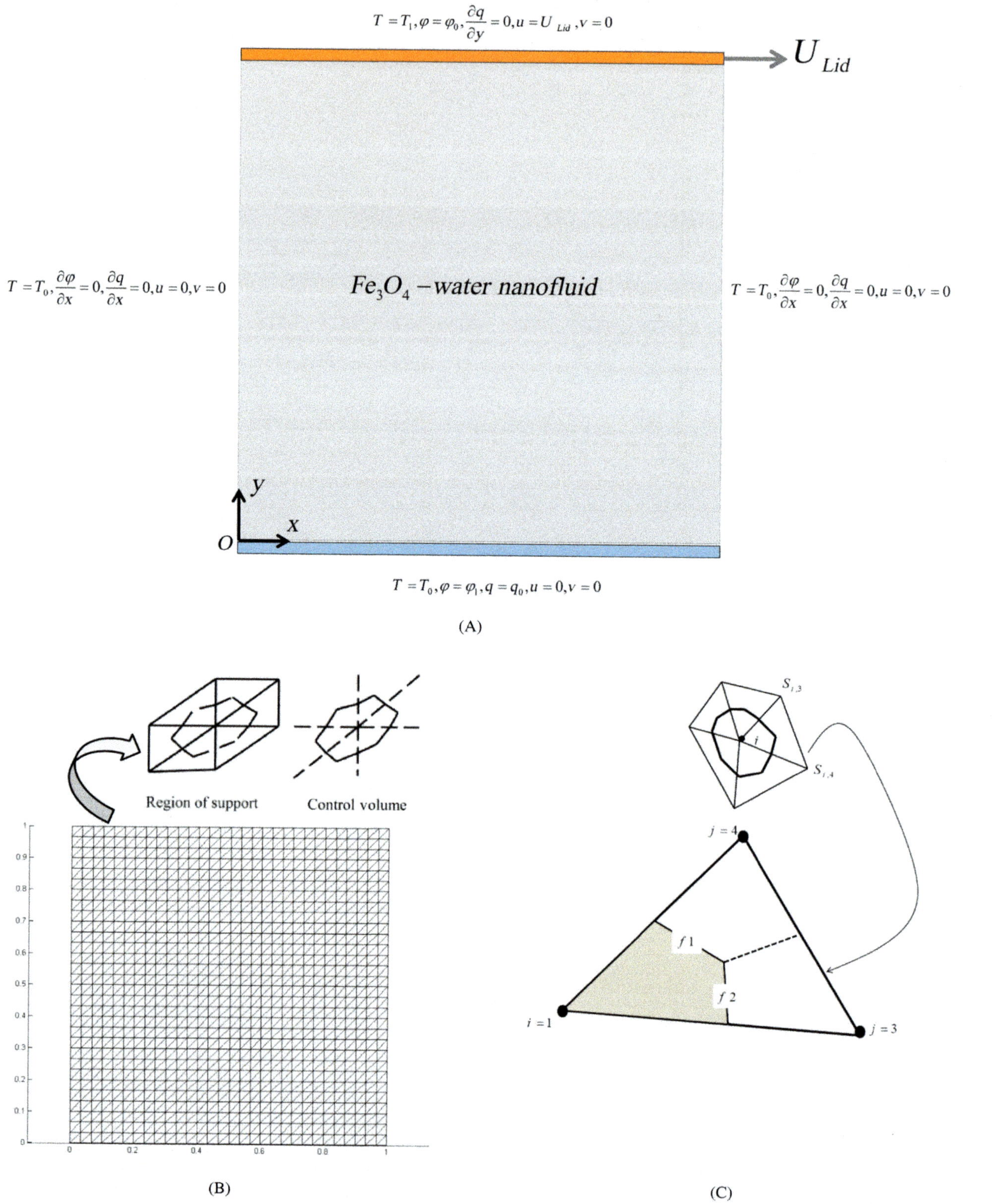

FIGURE 12.22 (A) Geometry of the problem and boundary conditions; (B) the mesh of enclosure considered in this work; (C) a sample triangular element and its corresponding control volume.

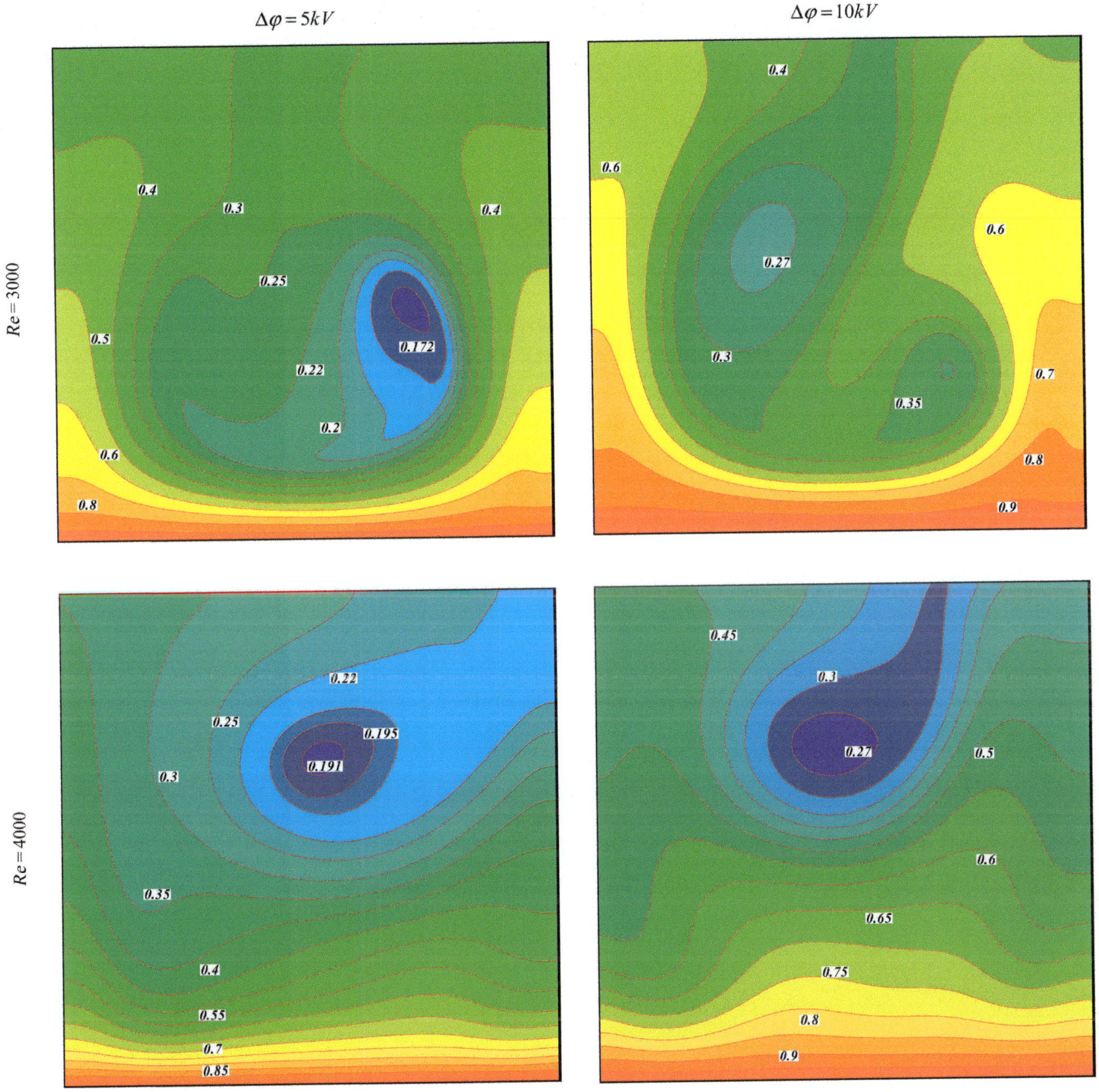

FIGURE 12.23 Electric density distribution injected by the bottom electrode.

Impacts of $\Delta\varphi$ and Re on Nu_{loc} and Nu_{ave} are demonstrated in Figs. 12.28 and 12.29. As the lid wall velocity augments, the Nusselt number enhances due to a rise in the temperature gradient. Augmenting the Coulomb forces causes the isotherms near the left wall to be denser. The presence of thermal plumes for high values of supplied voltage leads to the existence of extremums for Nu_{loc} profiles. The Nusselt numbers in the presence of electric field with supplied voltage $\Delta\varphi = 10kV$ of $Re = 3000$, 4000, 5000, and 6000 are, respectively, 1.09, 2.15, 8.96, and 1.76 times higher than those in the absence of an electric field.

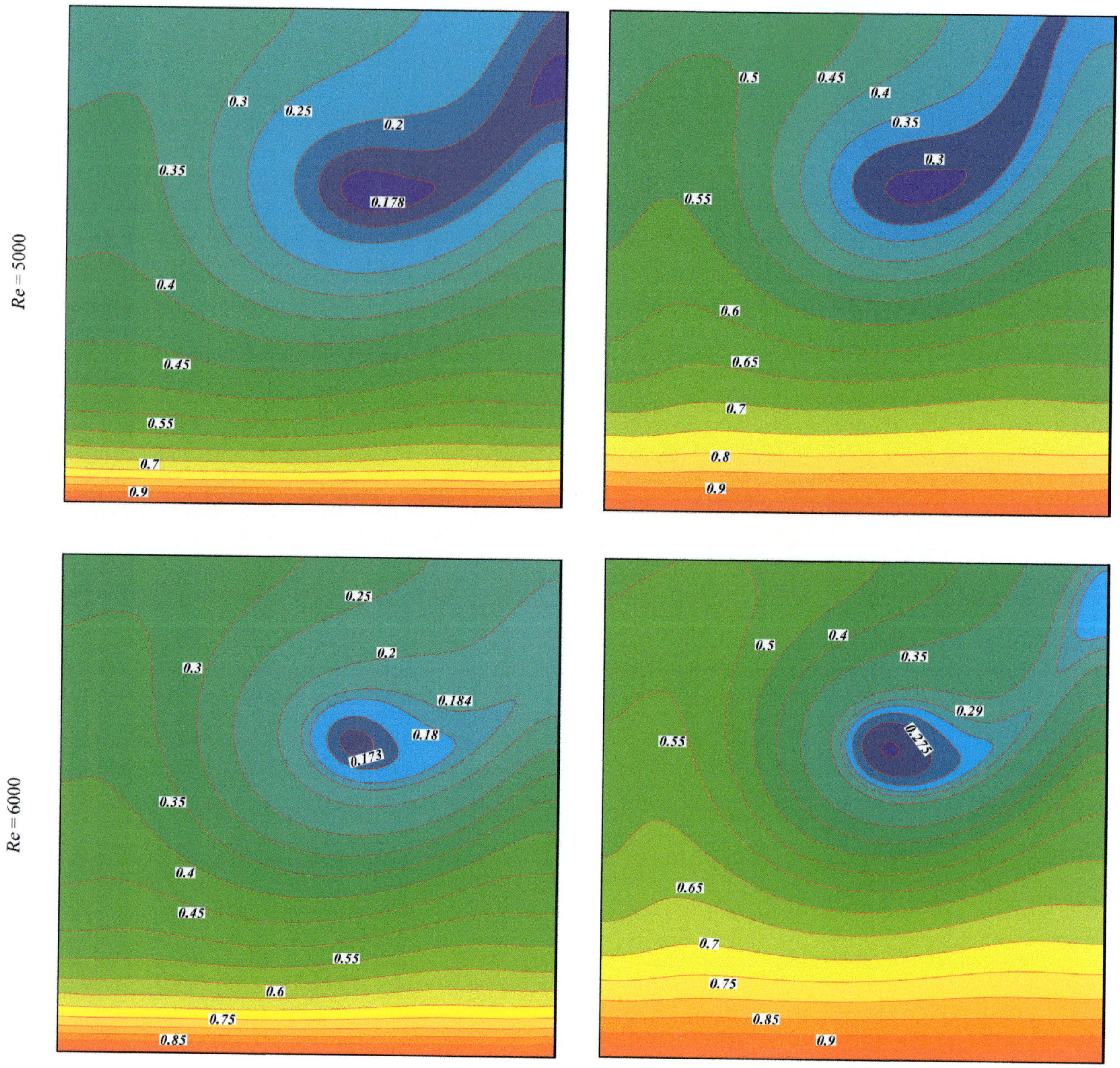

FIGURE 12.23 (Continued).

TABLE 12.3 Thermophysical Properties of Water and Nanoparticles

	$\rho\ (kg/m^3)$	$C_p\ (j/kgk)$	$k\ (W/m.k)$
H_2O	997.1	4179	0.613
Fe_3O_4	5200	670	6

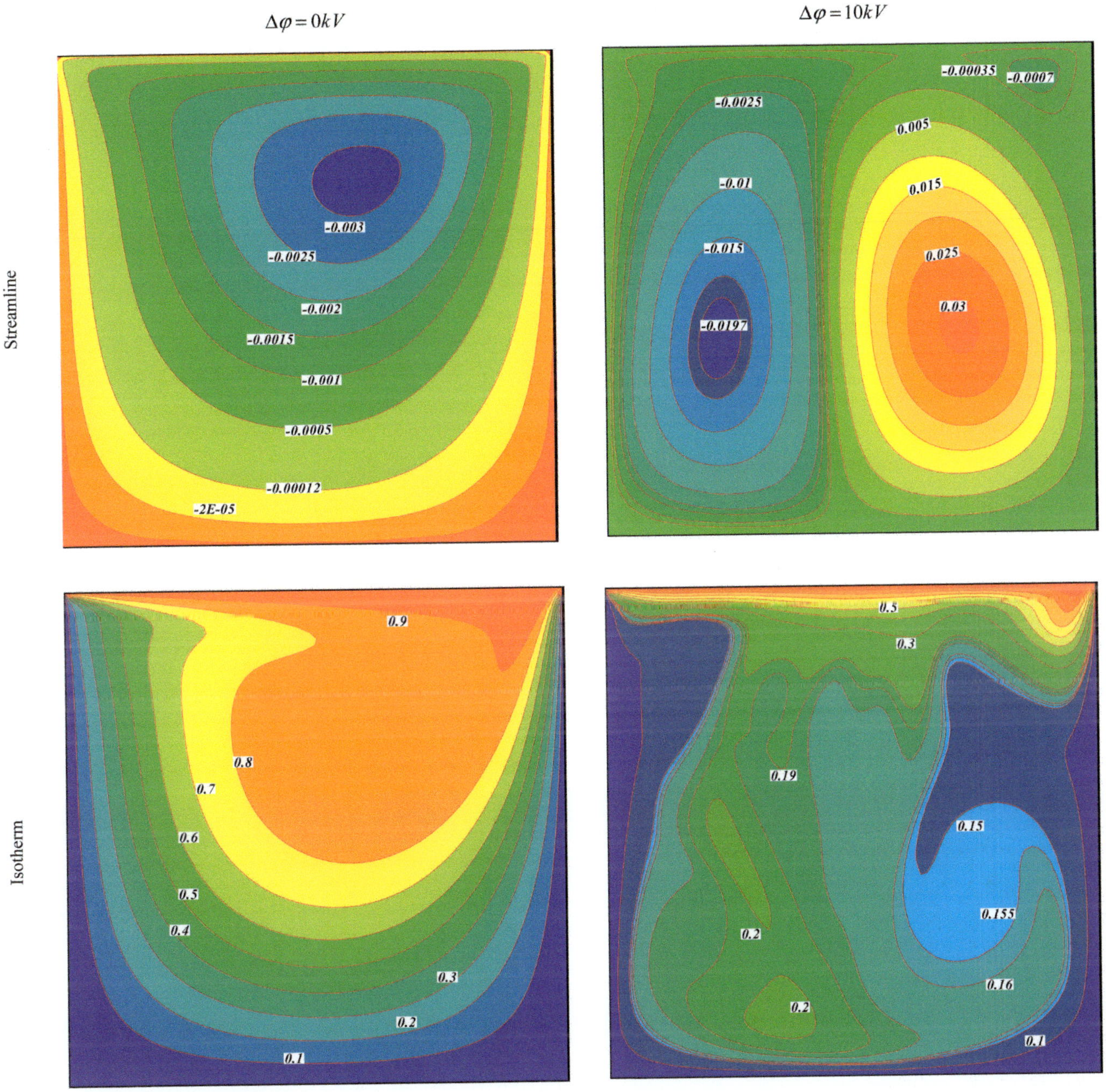

FIGURE 12.24 Effect of supplied voltage on streamlines and isotherms when Re = 3000, $\phi = 0.04$.

12.6 ACTIVE METHOD FOR NANOFLUID HEAT TRANSFER ENHANCEMENT BY MEANS OF EHD

12.6.1 Problem Definition

Fig. 12.30 illustrates the schematic of this problem and its boundary conditions. The influence of an electric field on Fe_3O_4-ethylene glycol nanofluid is considered. Only the bottom wall can move. Fig. 12.31 illustrates the contour of q for various values of $\nabla\varphi$ and Re. As $\nabla\varphi$ increases the distortion of isoelectric density lines become greater and one cell appears on the right side. The influence of $\nabla\varphi$ on q is more sensible than Re.

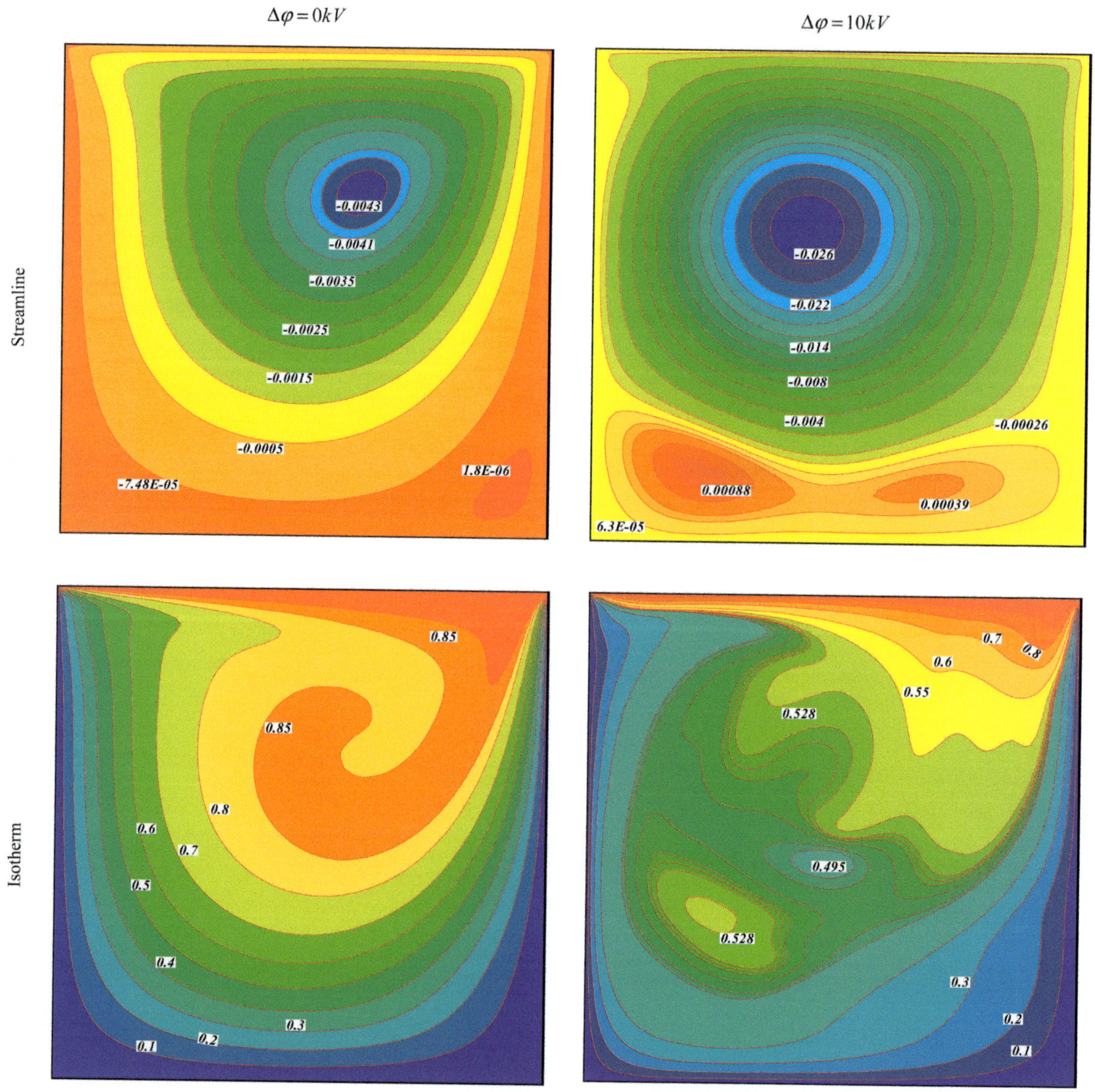

FIGURE 12.25 Effect of supplied voltage on streamlines and isotherms when Re = 4000, $\phi = 0.04$.

12.6.2 Governing Equation

According to Gauss's law and Maxwell's relation, the electric field can be defined as:

$$q = \nabla.\varepsilon\,\vec{E} \tag{12.66}$$

$$\vec{E} = -\nabla\varphi \tag{12.67}$$

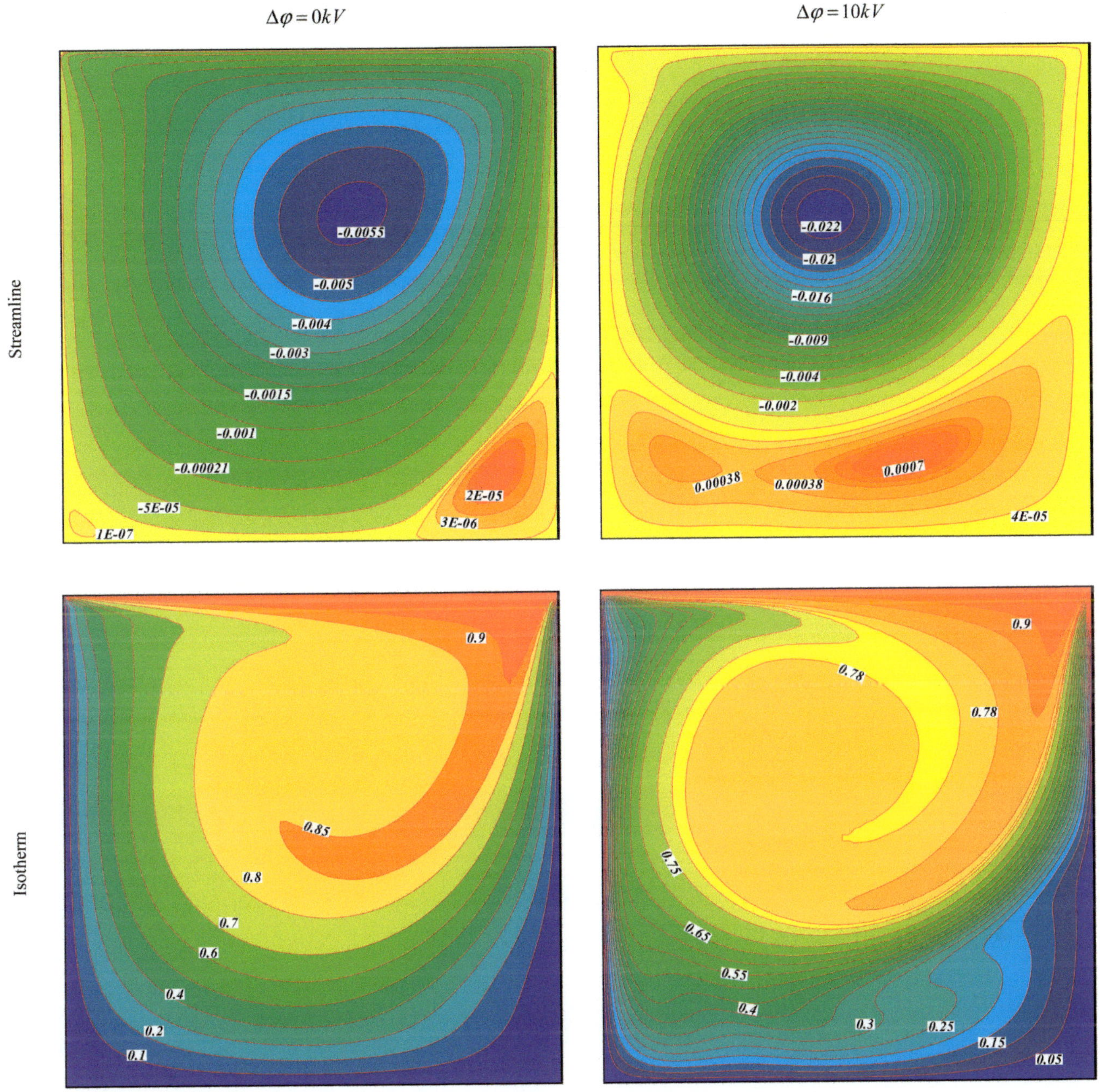

FIGURE 12.26 Effect of supplied voltage on streamlines and isotherms when Re = 5000, $\phi = 0.04$.

$$\nabla . \vec{J} + \frac{\partial q}{\partial t} = 0 \tag{12.68}$$

$$\vec{J} = q\vec{V} - D\nabla q + \sigma \vec{E} \tag{12.69}$$

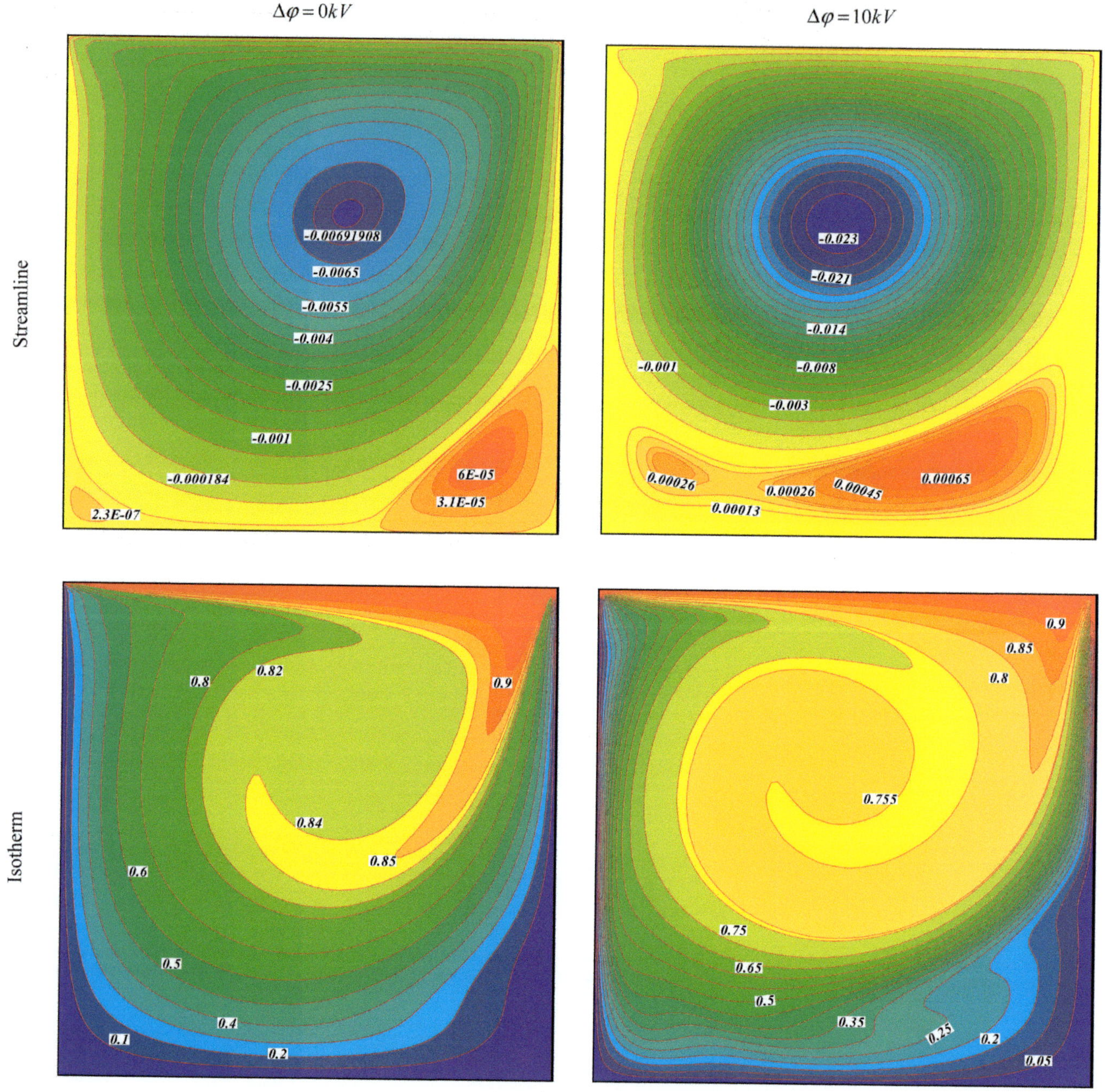

FIGURE 12.27 Effect of supplied voltage on streamlines and isotherms when Re = 6000, $\phi = 0.04$.

The governing equations are:

$$\begin{cases} \nabla.\vec{V} = 0 \\ \left(\left(\vec{V}.\nabla \right) \vec{V} + \dfrac{\partial \vec{V}}{\partial t} \right) = \dfrac{q\vec{E}}{\rho_{nf}} + \dfrac{\mu_{nf}}{\rho_{nf}} \nabla^2 \vec{V} - \dfrac{\nabla p}{\rho_{nf}} \\ \left(\left(\vec{V}.\nabla \right) T + \dfrac{\partial T}{\partial t} \right) = \dfrac{k_{nf}}{\left(\rho C_p\right)_{nf}} \nabla^2 T + \dfrac{\vec{J}.\vec{E}}{\left(\rho C_p\right)_{nf}} \\ \nabla\varphi = -\vec{E} \\ \dfrac{\partial q}{\partial t} = -\nabla.\vec{J} \\ q = \nabla.\varepsilon\vec{E} \end{cases} \tag{12.70}$$

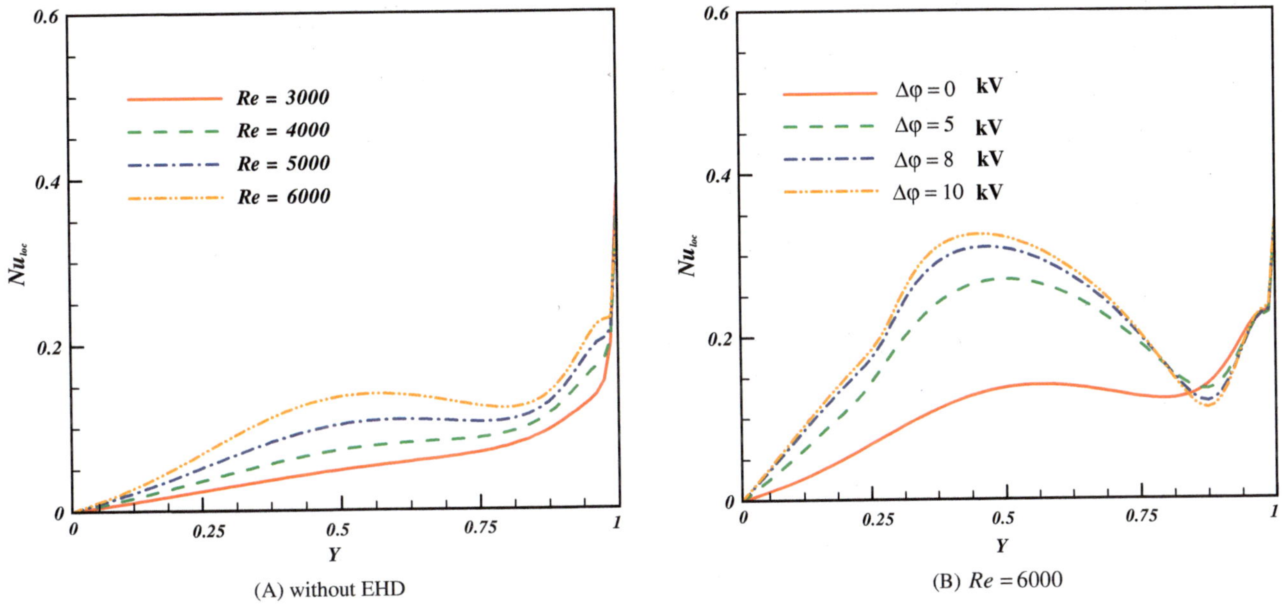

FIGURE 12.28 Effects of Reynolds number and supplied voltage on the local Nusselt number when $\phi = 0.04$. (A) without EHD, (B) Re = 6000.

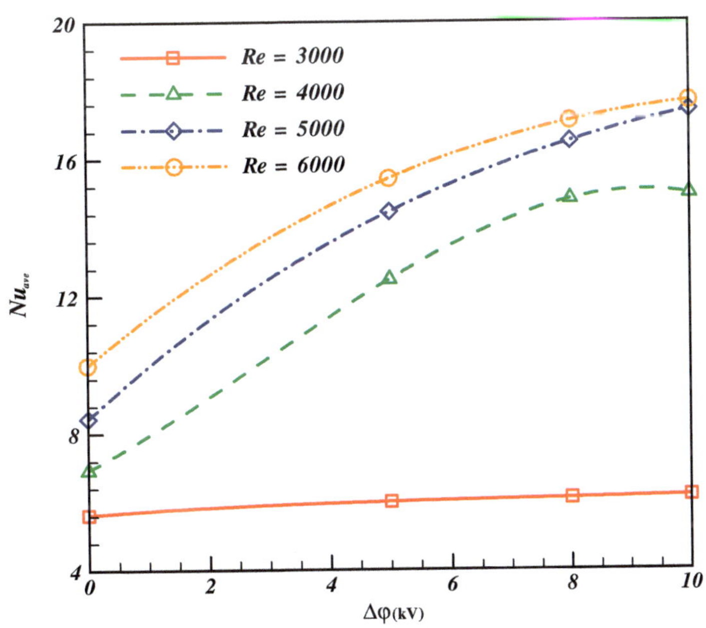

FIGURE 12.29 Effects of Reynolds number and supplied voltage on the average Nusselt number.

k_{nf}, $(\rho C_p)_{nf}$, μ_{nf}, and ρ_{nf} can be obtained as:

$$\begin{aligned}
&\frac{k_{nf}}{k_f} = \frac{-2\phi(k_f - k_s) + k_s + 2k_f}{\phi(k_f - k_s) + k_s + 2k_f},\\
&(\rho C_p)_{nf} = (\rho C_p)_f(1-\phi) + (\rho C_p)_s\phi,\\
&\mu = A_1 + A_2(\Delta\varphi) + A_3(\Delta\varphi)^2 + A_4(\Delta\varphi)^3,\\
&\rho_{nf} = \rho_f(1-\phi) + \rho_s\phi
\end{aligned} \tag{12.71}$$

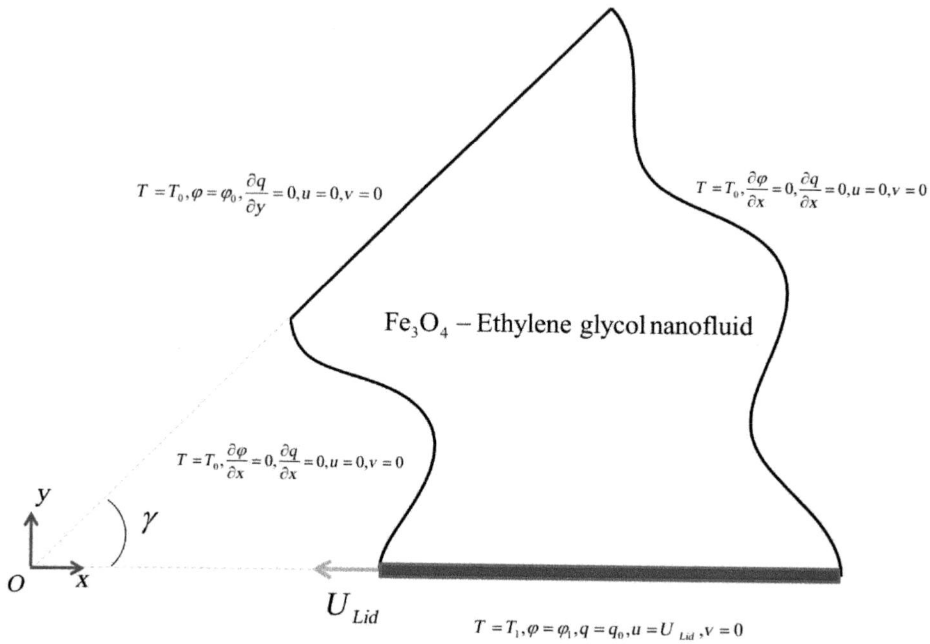

FIGURE 12.30 Geometry of the problem and boundary conditions.

Properties of Fe_3O_4 and ethylene glycol are illustrated in Table 12.1. So, the final dimensionless equations in the presence of an electric field are:

$$\begin{cases} \nabla.\vec{V}=0 \\ \left(\left(\vec{V}.\nabla\right)\vec{V}+\frac{\partial \vec{V}}{\partial t}\right)=\frac{1}{\mathrm{Re}}\frac{\rho_{nf}/\rho_f}{\mu_{nf}/\mu_f}\nabla^2\vec{V}-\nabla p+\frac{S_E}{\rho_{nf}/\rho_f}q\vec{E} \\ \left(\left(\vec{V}.\nabla\right)\theta+\frac{\partial \theta}{\partial t}\right)=\frac{1}{Pr\mathrm{Re}}\frac{k_{nf}/k_f}{(\rho C_p)_{nf}/(\rho C_p)_f}\nabla^2\theta+\frac{1}{(\rho C_p)_{nf}/(\rho C_p)_f}S_E Ec\left(\vec{J}.\vec{E}\right) \\ \vec{E}=-\nabla\varphi \\ q=\nabla.\varepsilon\vec{E} \\ \nabla.\vec{J}=-\frac{\partial q}{\partial t} \end{cases} \tag{12.72}$$

where

$$(\overline{u},\overline{v})=\frac{(u,v)}{U_{Lid}},\overline{\varphi}=\frac{\varphi-\varphi_0}{\nabla\varphi},(\overline{y},\overline{x})=\frac{(y,x)}{L},\theta=\frac{T-T_0}{\nabla T},$$

$$\overline{t}=\frac{tU_{Lid}}{L},\overline{p}=\frac{P}{\rho U_{Lid}{}^2}\overline{q}=\frac{q}{q_0},\overline{E}=\frac{E}{E_0}, \tag{12.73}$$

$$\nabla T=T_1-T_0,\nabla\varphi=\varphi_1-\varphi_0.$$

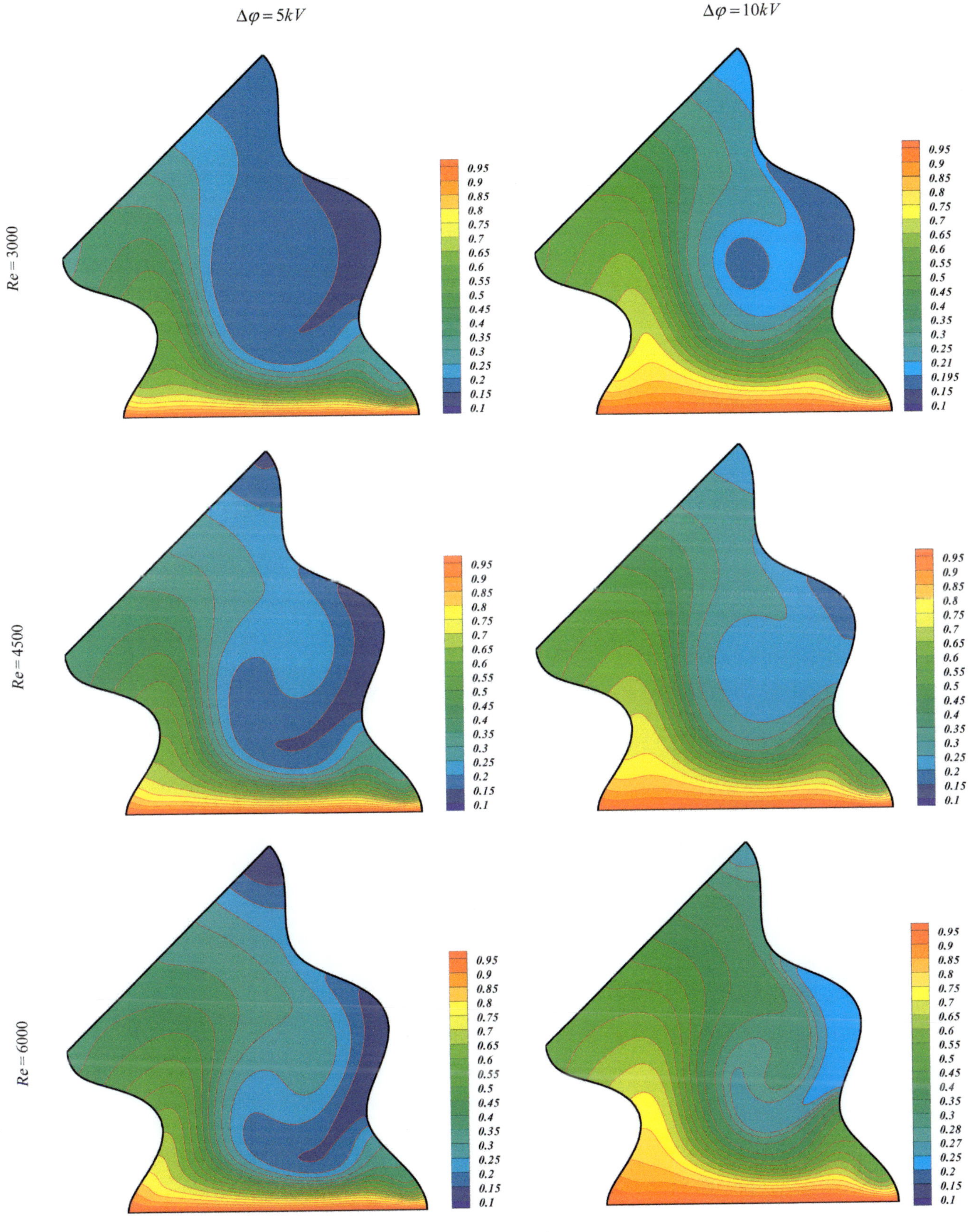

FIGURE 12.31 Electric density distribution injected by the bottom electrode.

By eliminating pressure gradient, vorticity and stream function can be introduced as:

$$v = -\frac{\partial\psi}{\partial x}, \frac{\partial\psi}{\partial y} = u, \Psi = \frac{\psi}{\alpha_f}, \Omega = \frac{\omega L^2}{\alpha_f}, \omega = \frac{\partial v}{\partial x} - \frac{\partial u}{\partial y} \tag{12.74}$$

Nu_{loc} and Nu_{ave} along the hot wall are calculated as:

$$Nu_{loc} = \left(\frac{k_{nf}}{k_f}\right)\frac{\partial\Theta}{\partial X} \tag{12.75}$$

$$Nu_{ave} = \frac{1}{L}\int_0^L Nu_{loc}\, dY \tag{12.76}$$

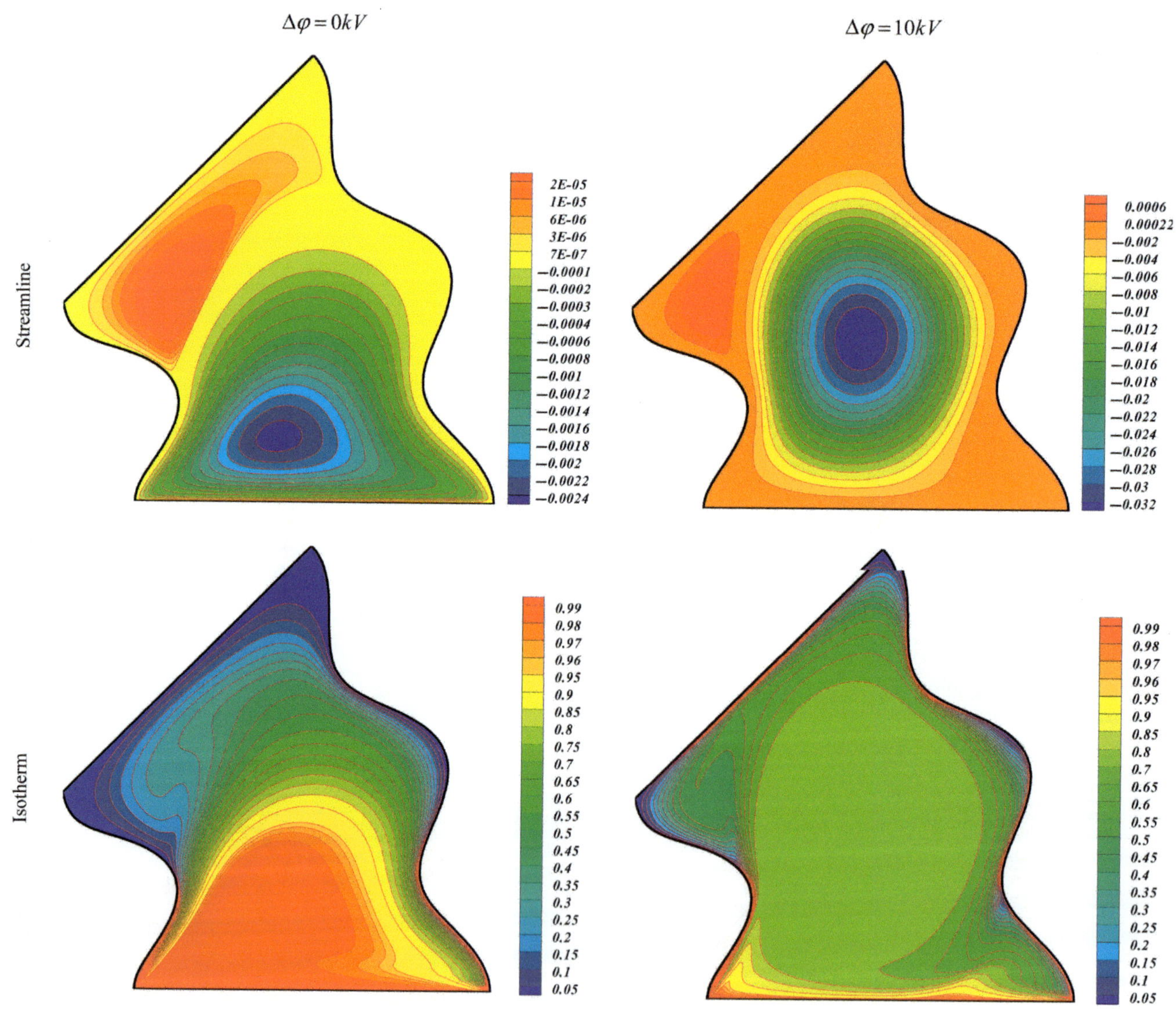

FIGURE 12.32 Effect of supplied voltage on streamlines and isotherms when Re = 3000, $\phi = 0.05$.

12.6.3 Effects of Active Parameters

Fe_3O_4-ethylene glycol nanofluid forced convection in an enclosure is examined in the presence of an electric field. The bottom wall is the moving lid. The roles of supplied voltage ($\Delta\varphi = 0$ to $10kV$), Reynolds number ($Re = 3000$ to 6000), and volume fraction of Fe_3O_4 ($\phi = 0\%$ to 5%) are presented graphically.

The effects of $\Delta\varphi$ and Re on streamlines and isotherms are illustrated in Figs. 12.32–12.34. Two eddies appear in streamlines which rotate in opposite directions. The counterclockwise eddy is stronger and located at the left side upper of the square obstacle. The presence of such eddies leads to the generation of two thermal plumes at the left and upper side of the square obstacle. As an electric field is applied, the clockwise eddy becomes stronger and the other one converts to two smaller counterclockwise eddies which are located at the left side of enclosure. The isotherms become more disturbed by augmenting $\Delta\varphi$. By increasing Re, the strength of the rotating eddies is enhanced and the distortion of the isotherms becomes greater than before. The temperature gradient enhances at high Reynolds numbers due to stronger convective heat transfer. At high Reynolds

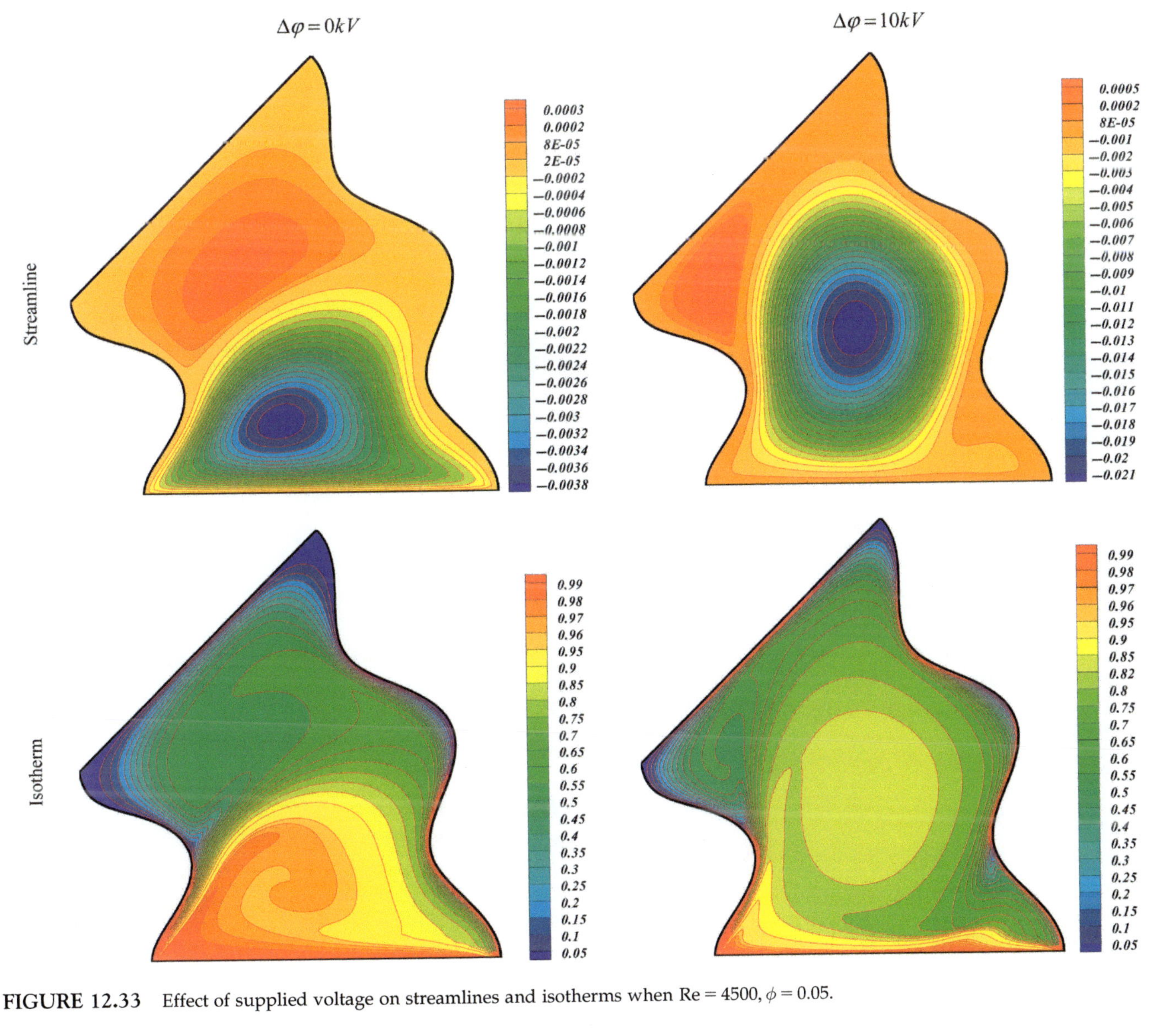

FIGURE 12.33 Effect of supplied voltage on streamlines and isotherms when $Re = 4500$, $\phi = 0.05$.

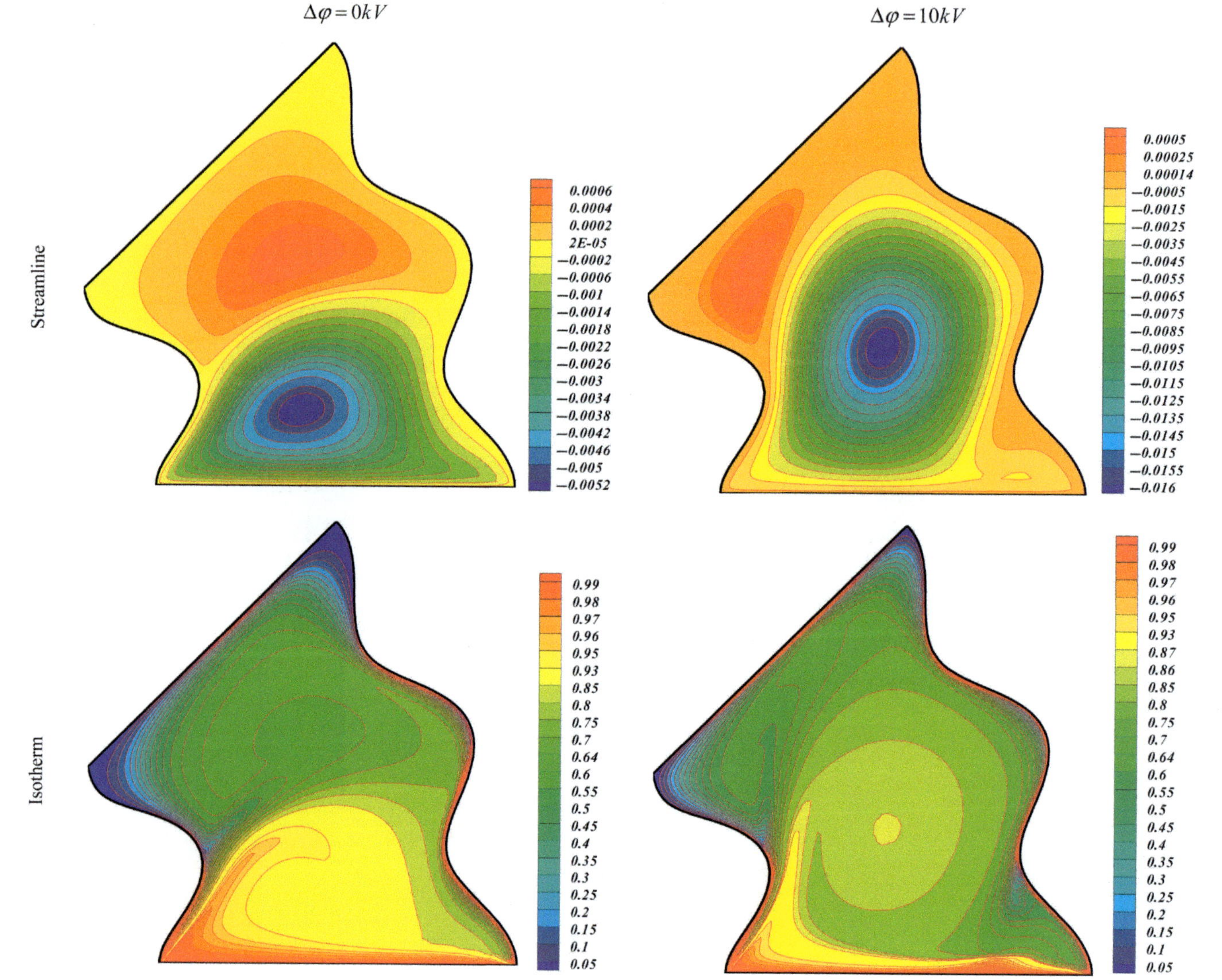

FIGURE 12.34 Effect of supplied voltage on streamlines and isotherms when Re = 6000, $\phi = 0.05$.

number, the Coulomb force is weaker than the viscous force, so the impact of adding an electric field becomes weaker with the increase of Re.

The influences of Re and $\Delta\varphi$ on Nu_{ave} are illustrated in Fig. 12.35. A correlation for average Nusselt number is presented as:

$$Nu_{ave} = 4.86 - 1.52\mathrm{Re}^* + 1.01\Delta\varphi - 0.12\mathrm{Re}^*\Delta\varphi + 0.17(\mathrm{Re}^*)^2 - 0.02(\Delta\varphi)^2 \tag{12.77}$$

where $\mathrm{Re}^* = \mathrm{Re} \times 10^{-3}$ and $\Delta\varphi$ is voltage supply in Kilovolt. In the absence of an electric field, as the velocity of the bottom wall augments, convective heat transfer becomes stronger and the temperature gradient enhances with rise of Re. Coulomb force helps the convention mode to enhance. So Nu_{ave} augments with the rise of Re and $\Delta\varphi$. In the presence of a high electric field enhancing *Re* shows the reverse behavior and the Nusselt number decreases with the increase of Re. These outputs proved that using an electric field is more useful at low Reynolds number.

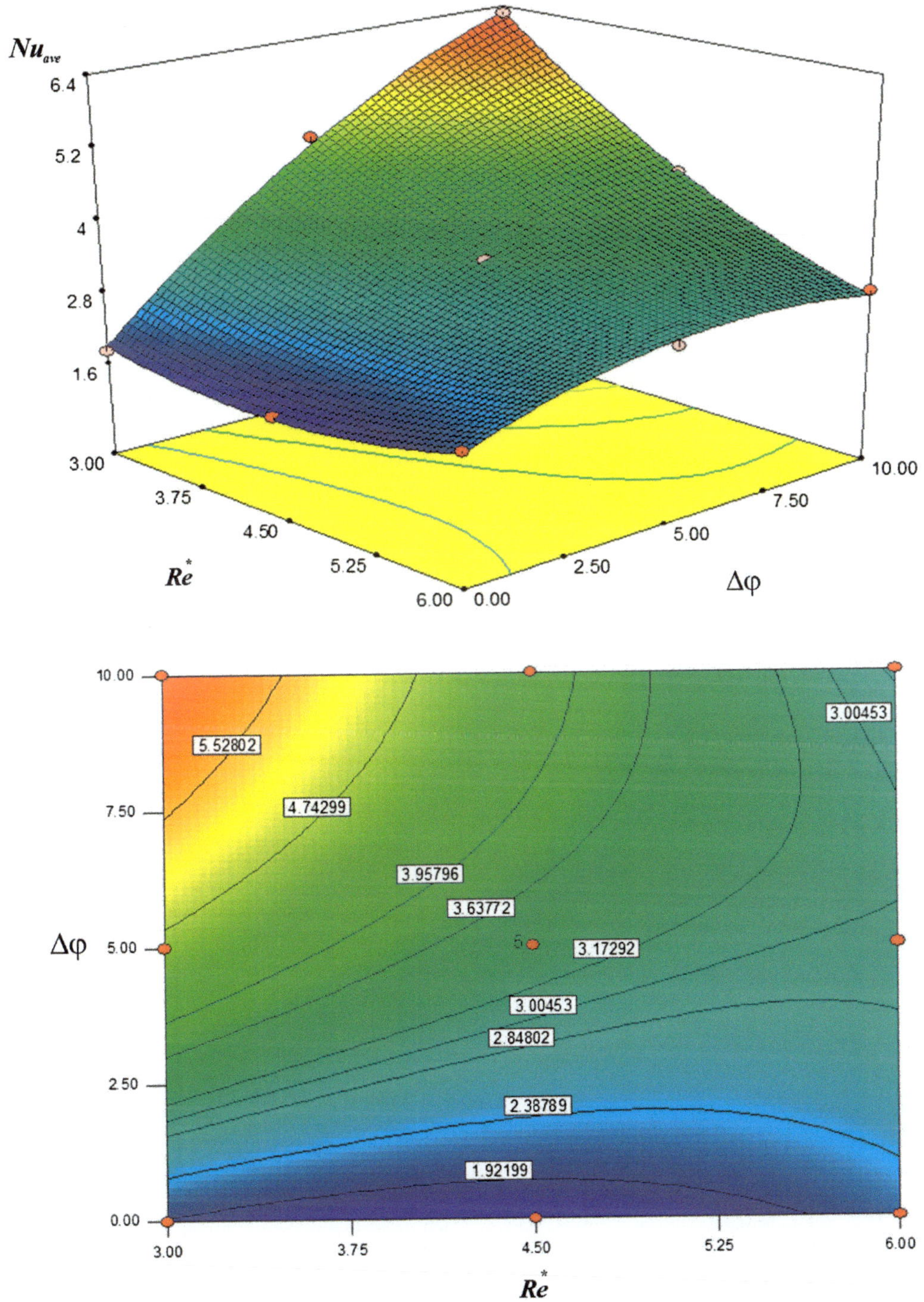

FIGURE 12.35 Effects of Reynolds number and supplied voltage on average Nusselt number when $\phi = 0.05$.

References:

[1] H.S. Shu, F.C. Lai, Effect of Electrical Field on Buoyancy-Induced Flows in an Enclosure, Conference Record—IAS Ann. Meet. (IEEE Ind. Applicat. Soc.) vol. 2 (1995) 1465–1471.

[2] N. Kasayapanand, J. Tiansuwan, W. Asvapoositkul, N. Vorayos, T. Kiatsiriroat, Effect of the electrode arrangements in tube bank on the characteristic of electrohydrodynamic heat transfer enhancement: low Reynolds number, J. Enhanced Heat Transf. 9 (2002) 229–242.

[3] M. Sheikholeslami, T. Hayat, A. Alsaedi, S. Abelman, EHD nanofluid force convective heat transfer considering electric field dependent viscosity, Int. J. Heat Mass Transfer 108 (2017).

[4] M. Sheikholeslami, R. Ellahi, Electrohydrodynamic nanofluid hydrothermal treatment in an enclosure with sinusoidal upper wall, Appl. Sci. 5 (2015) 294–306. Available from: https://doi.org/10.3390/app5030294.

[5] M. Sheikholeslami, S. Soleimani, D.D. Ganji, Effect of electric field on hydrothermal behavior of nanofluid in a complex geometry, J. Mol. Liq. 213 (2016) 153–161.
[6] M. Sheikholeslami, Influence of Coulomb forces on Fe3O4-H2O nanofluid thermal improvement, Int. J. Hydrogen Energy 42 (2017) (2016) 821–829.
[7] M. Sheikholeslami, D.D. Ganji, Impact of electric field on nanofluid forced convection heat transfer with considering variable properties, J. Mol. Liq. 229 (2017) 566–573. Available from: https://doi.org/10.1016/j.molliq.2016.12.107.
[8] M. Sheikholeslami, M.M. Bhatti, Active method for nanofluid heat transfer enhancement by means of EHD, Int. J. Heat Mass Transfer 109 (2017) 115–122.
[9] M. Sheikholeslami, H.B. Rokni, Influence of EFD viscosity on nanofluid forced convection in a cavity with sinusoidal wall, J. Mol. Liq. 232 (2017) 390–395.
[10] M. Sheikholeslami, K. Vajravelu, Forced convection heat transfer in Fe3O4- ethylene glycol nanofluid under the influence of Coulomb force, J. Mol. Liq. 233 (2017) 203–210.
[11] M. Sheikholeslami, Influence of Lorentz forces on nanofluid flow in a porous cylinder considering Darcy model, J. Mol. Liq. 225 (2017) 903–912.
[12] M. Sheikholeslami, CVFEM for magnetic nanofluid convective heat transfer in a porous curved enclosure, Eur. Phys. J. Plus 131 (2016) 413. Available from: https://doi.org/10.1140/epjp/i2016-16413-y.
[13] M. Sheikholeslami, Magnetic field influence on nanofluid thermal radiation in a cavity with tilted elliptic inner cylinder, J. Mol. Liq. 229 (2017) 137–147.
[14] M. Sheikholeslami, Numerical simulation for external magnetic field influence on Fe3O4-water nanofluid forced convection, Eng. Comput., Inpress.
[15] M. Sheikholeslami, Influence of Lorentz forces on nanofluid flow in a porous cavity by means of Non- Darcy model, Eng. Comput. 34 (8) (2017) 2651–2667. Available from: https://doi.org/10.1108/EC-01-2017-0008.
[16] M. Sheikholeslami, H.B. Rokni, Nanofluid convective heat transfer intensification in a porous circular cylinder, Chem. Eng. Process. Process Intensif. 120 (2017) 93–104.
[17] M. Sheikholeslami, M.K. Sadoughi, Numerical modeling for Fe3O4 -water nanofluid flow in porous medium considering MFD viscosity, J. Mol. Liq. 242 (2017) 255–264.
[18] M. Sheikholeslami, M. Shamlooei, Magnetic source influence on nanofluid flow in porous medium considering shape factor effect, Phys. Lett. A 381 (2017) 3071–3078.
[19] M. Sheikholeslami, M. Shamlooei, Convective flow of nanofluid inside a lid driven porous cavity using CVFEM, Phys. B: Condensed Matter 521 (2017) 239–250.
[20] M. Sheikholeslami, H.B. Rokni, Magnetic nanofluid natural convection in the presence of thermal radiation considering variable viscosity, Europ. Phys. J. Plus 132 (2017) 238. Available from: https://doi.org/10.1140/epjp/i2017-11498-4.
[21] M. Sheikholeslami, H.B. Rokni, Numerical simulation for impact of Coulomb force on nanofluid heat transfer in a porous enclosure in presence of thermal radiation, Int. J. Heat Mass Transfer 118 (2018) 823–831.
[22] M. Sheikholeslami, Numerical simulation of magnetic nanofluid natural convection in porous media, Phys. Lett. A 381 (2017) 494–503.
[23] M. Sheikholeslami, Magnetic source impact on nanofluid heat transfer using CVFEM, Neural Comput. Applicat. (2016). Available from: https://doi.org/10.1007/s00521-016-2740-7.
[24] M. Sheikholeslami, CuO-water nanofluid free convection in a porous cavity considering Darcy law, Europ. Phys. J. Plus 132 (2017) 55. Available from: https://doi.org/10.1140/epjp/i2017-11330-3.
[25] M. Sheikholeslami, Numerical investigation of MHD nanofluid free convective heat transfer in a porous tilted enclosure, Eng. Comput. 34 (6) (2017) 1939–1955.
[26] M.S. Kandelousi, Effect of spatially variable magnetic field on ferrofluid flow and heat transfer considering constant heat flux boundary condition, Europ. Phys. J. Plus (2014) 129–248. Available from: https://doi.org/10.1140/epjp/i2014-14248-2.
[27] M. Sheikholeslami, S.A. Shehzad, Thermal radiation of ferrofluid in existence of Lorentz forces considering variable viscosity, Int. J. Heat Mass Transfer 109 (2017) 82–92.
[28] M. Sheikholeslami, H.B. Rokni, Numerical modeling of nanofluid natural convection in a semi annulus in existence of Lorentz force, Comput. Methods Appl. Mech. Eng. 317 (2017) 419–430.
[29] M. Sheikholeslami, K. Vajravelu, Nanofluid flow and heat transfer in a cavity with variable magnetic field, Appl. Math. Comput. 298 (2017) 272–282.
[30] M. Sheikholeslami, H.B. Rokni, Magnetohydrodynamic CuO-water nanofluid in a porous complex shaped enclosure, ASME, J. Ther. Sci. Eng. Applicat. 9 (4) (2017) 041007. Available from: https://doi.org/10.1115/1.4035973.
[31] M. Sheikholeslami, H.B. Rokni, Nanofluid two phase model analysis in existence of induced magnetic field, Int. J. Heat Mass Transfer 107 (2017) 288–299.
[32] M. Sheikholeslami, Magnetic field influence on CuO -H2O nanofluid convective flow in a permeable cavity considering various shapes for nanoparticles, Int. J. Hydrogen Energy 42 (2017) 19611–19621.
[33] M. Sheikholeslami, M. Shamlooei, R. Moradi, Fe3O4- Ethylene glycol nanofluid forced convection inside a porous enclosure in existence of Coulomb force, J. Mol. Liq. 249 (2018) 429–437.
[34] M. Sheikholeslami, S.A. Shehzad, Numerical analysis of Fe_3O_4–H_2O nanofluid flow in permeable media under the effect of external magnetic source, Int. J. Heat Mass Transfer 118 (2018) 182–192.
[35] M. Sheikholeslami, S.A. Shehzad, Magnetohydrodynamic nanofluid convection in a porous enclosure considering heat flux boundary condition, Int. J. Heat Mass Transfer 106 (2017) 1261–1269.
[36] M. Sheikholeslami, H.B. Rokni, Effect of melting heat transfer on nanofluid flow in existence of magnetic field considering Buongiorno Model, Chin. J. Phys. 55 (2017) 1115–1126.
[37] M. Sheikholeslami, T. Hayat, A. Alsaedi, MHD free convection of Al2O3–water nanofluid considering thermal radiation: a numerical study, Int. J. Heat Mass Transfer 96 (2016) 513–524.

[38] M. Sheikholeslami, K. Vajravelu, M.M. Rashidi, Forced convection heat transfer in a semi annulus under the influence of a variable magnetic field, Int. J. Heat Mass Transfer 92 (2016) 339–348.
[39] M. Sheikholeslami, M.M. Bhatti, Forced convection of nanofluid in presence of constant magnetic field considering shape effects of nanoparticles, Int. J. Heat Mass Transfer 111 (2017) 1039–1049.
[40] M. Sheikholeslami, A. Zeeshan, Analysis of flow and heat transfer in water based nanofluid due to magnetic field in a porous enclosure with constant heat flux using CVFEM, Comput. Methods Appl. Mech. Eng. 320 (2017) 68–81.
[41] M. Sheikholeslami, M. Shamlooei, Fe3O4- H2O nanofluid natural convection in presence of thermal radiation, Int. J. Hydrogen Energy 42 (9) (2017) 5708–5718.
[42] M. Sheikholeslami, M.K. Sadoughi, Simulation of CuO- water nanofluid heat transfer enhancement in presence of melting surface, Int. J. Heat Mass Transfer 116 (2018) 909–919.
[43] M. Sheikholeslami, H.B. Rokni, Simulation of nanofluid heat transfer in presence of magnetic field: a review, Int. J. Heat Mass Transfer 115 (2017) 1203–1233.
[44] M. Sheikholeslami, H.B. Rokni, Melting heat transfer influence on nanofluid flow inside a cavity in existence of magnetic field, Int. J. Heat Mass Transfer 114 (2017) 517–526.
[45] M. Sheikholeslami, S.A. Shehzad, Magnetohydrodynamic nanofluid convective flow in a porous enclosure by means of LBM, Int. J. Heat Mass Transfer 113 (2017) 796–805.
[46] M. Sheikholeslami, Magnetohydrodynamic nanofluid forced convection in a porous lid driven cubic cavity using Lattice Boltzmann Method, J. Mol. Liq. 231 (2017) 555–565.
[47] M.S. Kandelousi, KKL correlation for simulation of nanofluid flow and heat transfer in a permeable channel, Phys. Lett. A 378 (45) (2014) 3331–3339.
[48] M. Sheikholeslami, Effect of uniform suction on nanofluid flow and heat transfer over a cylinder, J. Braz. Soc. Mech. Sci. Eng. 37 (2015) 1623–1633.
[49] M. Sheikholeslami, T. Hayat, A. Alsaedi, Numerical study for external magnetic source influence on water based nanofluid convective heat transfer, Int. J. Heat Mass Transfer 106 (2017) 745–755.
[50] M. Sheikholeslami, A.J. Chamkha, Flow and convective heat transfer of a ferro-nanofluid in a double-sided lid-driven cavity with a wavy wall in the presence of a variable magnetic field, Numer. Heat Transfer, Part A 69 (10) (2016) 1186–1200. Available from: https://doi.org/10.1080/10407782.2015.1125709.
[51] M. Sheikholeslami, A.J. Chamkha, Electrohydrodynamic free convection heat transfer of a nanofluid in a semi-annulus enclosure with a sinusoidal wall, Numer. Heat Transfer, Part A 69 (7) (2016) 781–793. Available from: http://dx.doi.org/10.1080/10407782.2015.1090819.
[52] M. Sheikholeslami, Lattice Boltzmann Method simulation of MHD non-Darcy nanofluid free convection, Physica B 516 (2017) 55–71.
[53] M. Sheikholeslami, Influence of magnetic field on nanofluid free convection in an open porous cavity by means of Lattice Boltzmann Method, J. Mol. Liq. 234 (2017) 364–374.
[54] M. Sheikholeslami, A.J. Chamkha, Influence of Lorentz forces on nanofluid forced convection considering Marangoni convection, J. Mol. Liq. 225 (2017) 750–757.
[55] M. Sheikholeslami, T. Hayat, A. Alsaedi, Numerical simulation of nanofluid forced convection heat transfer improvement in existence of magnetic field using Lattice Boltzmann Method, Int. J. Heat Mass Transfer 108 (2017) 1870–1883.
[56] M. Sheikholeslami, M. Sadoughi, Mesoscopic method for MHD nanofluid flow inside a porous cavity considering various shapes of nanoparticles, Int. J. Heat Mass Transfer 113 (2017) 106–114.
[57] M. Sheikholeslami, A. Zeeshan, Mesoscopic simulation of CuO-H2O nanofluid in a porous enclosure with elliptic heat source, Int. J. Hydrogen Energy 42 (22) (2017) 15393–15402.
[58] M. Sheikholeslami, S.A. Shehzad, CVFEM for influence of external magnetic source on Fe3O4 – H2O nanofluid behavior in a permeable cavity considering shape effect, Int. J. Heat Mass Transfer 115 (2017) 180–191.
[59] M. Sheikholeslami, M. Seyednezhad, Nanofluid heat transfer in a permeable enclosure in presence of variable magnetic field by means of CVFEM, Int. J. Heat Mass Transfer 114 (2017) 1169–1180.
[60] M. Sheikholeslami, H.B. Rokni, Free convection of CuO-H2O nanofluid in a curved porous enclosure using mesoscopic approach, Int. J. Hydrogen Energy 42 (22) (2017) 14942–14949.
[61] M. Sheikholeslami, Numerical investigation for CuO–H_2O nanofluid flow in a porous channel with magnetic field using mesoscopic method, J. Mol. Liq. 249 (2018) 739–746.
[62] M. Sheikholeslami, R. Ellahi, Three dimensional mesoscopic simulation of magnetic field effect on natural convection of nanofluid, Int. J. Heat Mass Transfer 89 (2015) 799–808.
[63] M. Sheikholeslami, S. Abelman, Two phase simulation of nanofluid flow and heat transfer in an annulus in the presence of an axial magnetic field, IEEE Trans. Nanotechnol. 14 (3) (2015) 561–569.
[64] M. Sheikholeslami, R. Ellahi, Simulation of ferrofluid flow for magnetic drug targeting using Lattice Boltzmann method, J. Zeitschrift Fur Naturforschung A 70 (2) (2015) 115–124.
[65] M. Sheikholeslami, M. Seyednezhad, Lattice Boltzmann Method simulation for CuO-water nanofluid flow in a porous enclosure with hot obstacle, J. Mol. Liq. 243 (2017) 249–256.
[66] M. Sheikholeslami, H.B. Rokni, Influence of melting surface on MHD nanofluid flow by means of two phase model, Chin. J. Phys. 55 (2017) 1352–1360.
[67] M. Sheikholeslami, M. Barzegar Gerdroodbary, S. Valiallah Mousavi, D.D. Ganji, R. Moradi, Heat transfer enhancement of ferrofluid inside an 90o elbow channel by non-uniform magnetic field, J. Magn. Magn. Mater. 460 (2018) 302–311.
[68] M. Sheikholeslami, Numerical simulation for solidification in a LHTESS by means of nano-enhanced PCM, J. Taiwan Inst. Chem. Eng. 86 (2018) 25–41.
[69] M. Sheikholeslami, Numerical modeling of nano enhanced PCM solidification in an enclosure with metallic fin, J. Mol. Liq. 259 (2018) 424–438.

[70] M. Sheikholeslami, A. Ghasemi, Solidification heat transfer of nanofluid in existence of thermal radiation by means of FEM, Int. J. Heat Mass Transfer 123 (2018) 418–431.

[71] M. Sheikholeslami, S.A. Shehzad, CVFEM simulation for nanofluid migration in a porous medium using Darcy model, Int. J. Heat Mass Transfer 122 (2018) 1264–1271.

[72] M. Sheikholeslami, M. Darzi, M.K. Sadoughi, Heat transfer improvement and pressure drop during condensation of refrigerant-based nanofluid: an experimental procedure, Int. J. Heat Mass Transfer 122 (2018) 643–650.

[73] M. Sheikholeslami, H.B. Rokni, CVFEM for effect of Lorentz forces on nanofluid flow in a porous complex shaped enclosure by means of non-equilibrium model, J. Mol. Liq. 254 (2018) 446–462.

[74] M. Sheikholeslami, S.A. Shehzad, Z. Li, Water based nanofluid free convection heat transfer in a three dimensional porous cavity with hot sphere obstacle in existence of Lorenz forces, Int. J. Heat Mass Transfer 125 (2018) 375–386.

[75] M. Sheikholeslami, M. Jafaryar, D.D. Ganji, Z. Li, Exergy loss analysis for nanofluid forced convection heat transfer in a pipe with modified turbulators, J. Mol. Liq. 262 (2018) 104–110.

[76] M. Sheikholeslami, M. Jafaryar, Z. Li, Nanofluid turbulent convective flow in a circular duct with helical turbulators considering CuO nanoparticles, Int. J. Heat Mass Transfer 124 (2018) 980–989.

[77] M. Sheikholeslami, Z. Li, M. Shamlooei, Nanofluid MHD natural convection through a porous complex shaped cavity considering thermal radiation, Phys. Lett. A 382 (2018) 1615–1632.

[78] M. Sheikholeslami, H.B. Rokni, Magnetic nanofluid flow and convective heat transfer in a porous cavity considering Brownian motion effects, Phys. Fluids, 30, 1, https://doi.org/10.1063/1.5012517.

[79] M. Sheikholeslami, S.A. Shehzad, Simulation of water based nanofluid convective flow inside a porous enclosure via non-equilibrium model, Int. J. Heat Mass Transfer 120 (2018) 1200–1212.

[80] M. Sheikholeslami, S.A. Shehzad, Non-Darcy free convection of Fe_3O_4-water nanoliquid in a complex shaped enclosure under impact of uniform Lorentz force, Chin. J. Phys. 56 (2018) 270–281.

[81] M. Sheikholeslami, M. Seyednezhad, Simulation of nanofluid flow and natural convection in a porous media under the influence of electric field using CVFEM, Int. J. Heat Mass Transfer 120 (2018) 772–781.

[82] M. Sheikholeslami, M. Shamlooei, R. Moradi, Numerical simulation for heat transfer intensification of nanofluid in a permeable curved enclosure considering shape effect of Fe_3O_4 nanoparticles, Chem. Eng. Process: Process Intensif. 124 (2018) 71–82.

[83] M. Sheikholeslami, T. Hayat, T. Muhammad, A. Alsaedi, MHD forced convection flow of nanofluid in a porous cavity with hot elliptic obstacle by means of Lattice Boltzmann method, Int. J. Mech. Sci. 135 (2018) 532–540.

[84] M. Sheikholeslami, Numerical investigation of nanofluid free convection under the influence of electric field in a porous enclosure, J. Mol. Liq. 249 (2018) 1212–1221.

[85] M. Sheikholeslami, CuO-water nanofluid flow due to magnetic field inside a porous media considering Brownian motion, J. Mol. Liq. 249 (2018) 921–929.

[86] M. Sheikholeslami, D.D. Ganji, Influence of electric field on Fe_3O_4-water nanofluid radiative and convective heat transfer in a permeable enclosure, J. Mol. Liq. 250 (2018) 404–412.

[87] M. Sheikholeslami, S.A. Shehzad, Z. Li, Nanofluid heat transfer intensification in a permeable channel due to magnetic field using Lattice Boltzmann method, Physica B (2018). Available from: https://doi.org/10.1016/j.physb.2018.03.036.

[88] M. Sheikholeslami, A. Zeeshan, Numerical simulation of Fe_3O_4-water nanofluid flow in a non-Darcy porous media, Int. J. Num. Methods Heat Fluid Flow 28 (3) (2018) 641–660. Available from: https://doi.org/10.1108/HFF-04-2017-0160.

[89] M. Sheikholeslami, D.D. Ganji, Numerical approach for magnetic nanofluid flow in a porous cavity using CuO nanoparticles, Mater. Des. 120 (2017) 382–393.

[90] M. Sheikholeslami, D.D. Ganji, Numerical modeling of magnetohydrodynamic CuO-water transportation inside a porous cavity considering shape factor effect, Colloids Surf. A 529 (2017) 705–714.

[91] M. Sheikholeslami, T. Hayat, A. Alsaedi, Numerical simulation for forced convection flow of MHD CuO-H_2O nanofluid inside a cavity by means of LBM, J. Mol. Liq. 249 (2018) 941–948.

[92] M. Sheikholeslami, H.R. Kataria, A.S. Mittal, Effect of thermal diffusion and heat-generation on MHD nanofluid flow past an oscillating vertical plate through porous medium, J. Mol. Liq. 257 (2018) 12–25.

CHAPTER

13

Darcy Model for Nanofluid Flow in a Porous Media by Means of CVFEM

13.1 INTRODUCTION

Convective flows in porous media have occupied the central stage in many fundamental heat transfer analyses and have received considerable attention over the last few decades. This interest is due to their wide range of applications in high performance insulation for buildings, chemical catalytic reactors, packed sphere beds, grain storage, and such geophysical problems as frost heave. Porous media are also of interest in relation to the underground spread of pollutants, solar power collectors, and to geothermal energy systems. The literature concerning convective flows in porous media is abundant and representative studies in this may be found in the books by Oosthuizen and Naylor [1].

Darcy's law can be used for porous media. Because of the complexity of pore geometries in a porous medium, Darcy's law has to be used to obtain any meaningful insights into the physics of flow in porous media. Darcy's law has been verified by the results of many experiments. Theoretical backing for it has been obtained in various ways, with the aid of either deterministic or statistical models [2]. Vafai [3] have described that Darcy's approximation is reasonable for the geothermal flows, except perhaps near boreholes [3]. Darcy's law is valid only when the pore Reynolds number is of the order of 1, and for many practical applications Darcy's law is not valid, and boundary and inertial effects need to be accounted for [3]. For example, modeling fluid flows in the fuel cells. This is because the fuel cell consists of a number of distinct layers of different porosities. Darcy's law assumes no effect of boundaries and the fluid velocity in Darcy's equation is determined by the permeability of the matrix. Therefore at the interface between the regions of different porosity in the fuel cell, particularly between the free fluid flow region, such as a gas flow channel, and a permeable medium, then a discontinuity in the fluid velocity and/or the shear stress could emerge [4]. Sheikholeslami and Rokni [5] studied the magnetic field effect on nanofluid flow in a porous complex shaped enclosure. Sheikholeslami and Zeeshan [6] analyzed flow and heat transfer of water-based nanofluid due to magnetic field in a porous enclosure with constant heat flux. Sheikholeslami and Shehzad [7] studied magnetohydrodynamic nanofluid convection in a porous enclosure considering the heat flux boundary condition. Sheikholeslami [8] employed Darcy's law for porous enclosure filled with nanofluid. Sheikholeslami and Ganji [9] investigated the transportation of MHD nanofluid free convection in a porous semiannulus using numerical approach. Sheikholeslami [10] presented the CuO-water nanofluid free convection in a porous cavity considering Darcy's law. Sheikholeslami et al. [11] simulated nanofluid flow in a porous media in the presence of Lorentz forces. Recently, several simulation works have been published in various journals [12–93].

13.2 MAGNETOHYDRODYNAMIC CUO-WATER NANOFLUID IN A POROUS COMPLEX SHAPED ENCLOSURE

13.2.1 Problem Definition

Fig. 13.1 shows the geometry, boundary condition, and sample element. The formula of the inner cylinder is:

$$r = r_{in} + A\cos\left(N(\zeta - \zeta_0)\right), \quad A = 0.5 \tag{13.1}$$

in which r_{in}, r_{out} are radius of the base circle and outer cylinder.

Application of Control Volume based Finite Element Method (CVFEM) for Nanofluid Flow and Heat Transfer.
DOI: https://doi.org/10.1016/B978-0-12-814152-6.00013-8

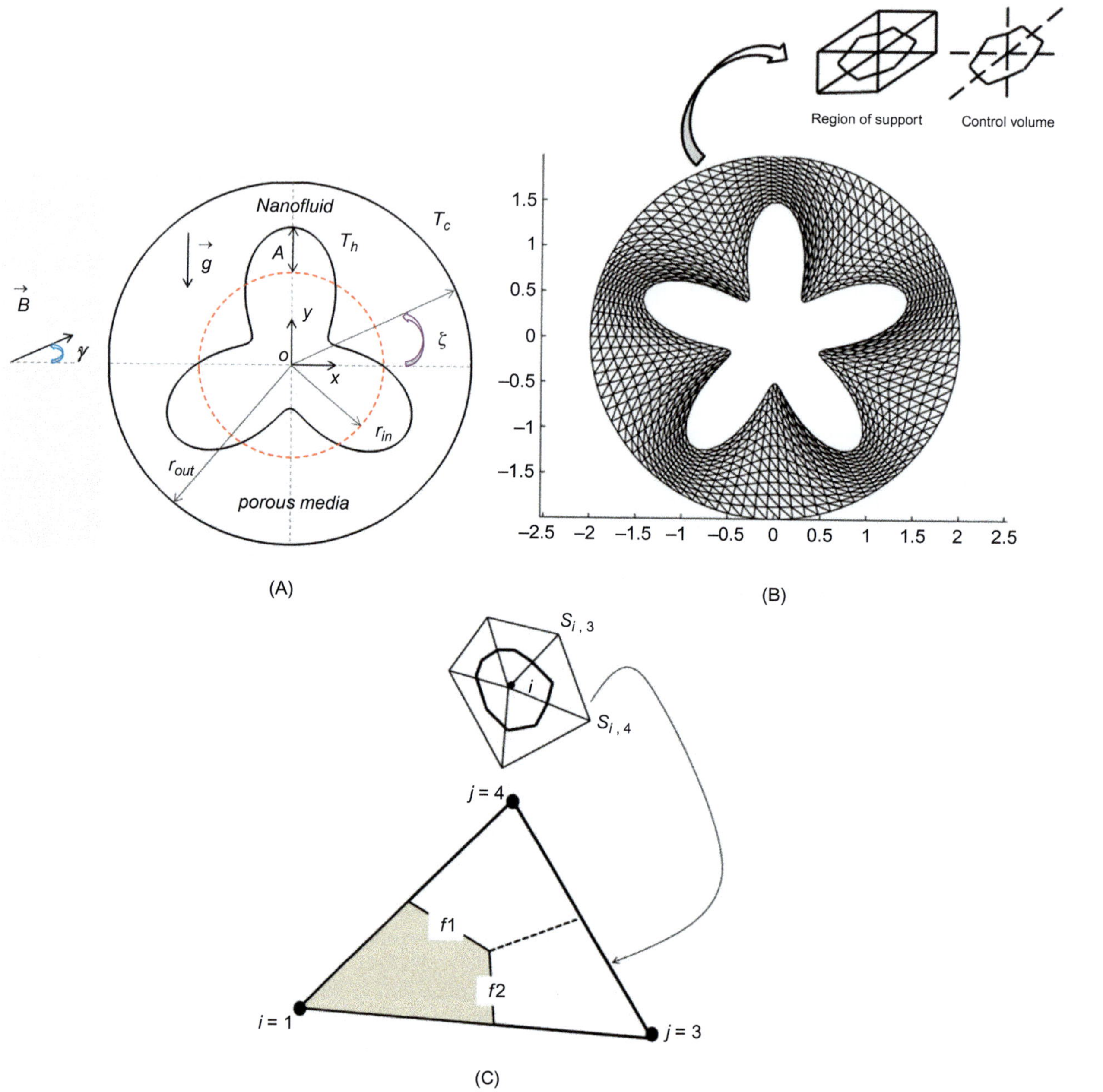

FIGURE 13.1 (A) Geometry and the boundary conditions with (B) the mesh of geometry considered in this work; (C) a sample triangular element and its corresponding control volume.

13.2.2 Governing Equation

2D steady convective flow of nanofluid in a porous media is considered in the presence of a constant horizontal magnetic field. The PDEs equations are:

$$\nabla.\vec{V} = 0 \tag{13.2}$$

$$\frac{\mu_{nf}}{K}\vec{V} = \left(\nabla p + \vec{I} \times \vec{B} + \rho_{nf}\,\vec{g}\right) \tag{13.3}$$

$$(\rho C_p)_{nf}\left(\vec{V}.\nabla\right)T = k_{nf}\,\nabla^2 T \tag{13.4}$$

$$\nabla.\vec{I} = 0 \tag{13.5}$$

$$\sigma_{nf}\left(\vec{V} \times \vec{B} - \nabla\varphi\right) = \vec{I} \tag{13.6}$$

In Eq. (13.2), Darcy's model is used for a porous medium. Eqs. (13.5) and (13.6) reduce to $\nabla^2\varphi = 0$. So the electric field can be neglected. So the above equations turn into:

$$\frac{\partial v}{\partial y} + \frac{\partial u}{\partial x} = 0 \tag{13.7}$$

$$-\frac{K}{\mu_{nf}}\frac{\partial p}{\partial x} - \frac{\sigma_{nf}KB_0^2}{\mu_{nf}}\left(-u\sin^2\gamma + v\sin\gamma\cos\gamma\right) = u \tag{13.8}$$

$$-\frac{K}{\mu_{nf}}\frac{\partial p}{\partial y} - \frac{\sigma_{nf}KB_0^2}{\mu_{nf}}\left(-v\cos^2\gamma + u\sin\gamma\cos\gamma\right) + g\frac{(\rho\beta)_{nf}K}{\mu_{nf}}(T - T_c) = v \tag{13.9}$$

$$\left(\rho C_p\right)_{nf}\left(u\frac{\partial T}{\partial x} + v\frac{\partial T}{\partial y}\right) = k_{nf}\left(\frac{\partial^2 T}{\partial x^2} + \frac{\partial^2 T}{\partial y^2}\right) \tag{13.10}$$

$\left(\rho C_p\right)_{nf}$, $(\rho\beta)_{nf}$, ρ_{nf}, and σ_{nf} are defined as:

$$\left(\rho C_p\right)_{nf} = \left(\rho C_p\right)_f(1-\phi) + \left(\rho C_p\right)_s\phi \tag{13.11}$$

$$(\rho\beta)_{nf} = (\rho\beta)_f(1-\phi) + (\rho\beta)_s\phi \tag{13.12}$$

$$\rho_{nf} = \rho_f(1-\phi) + \rho_s\phi \tag{13.13}$$

$$\frac{\sigma_{nf}}{\sigma_f} = 1 + \frac{3\left(\frac{\sigma_s}{\sigma_f} - 1\right)\phi}{\left(\frac{\sigma_s}{\sigma_f} + 2\right) - \left(\frac{\sigma_s}{\sigma_f} - 1\right)\phi} \tag{13.14}$$

k_{nf}, μ_{nf} are obtained according to the Koo–Kleinstreuer–Li (KKL) model:

$$k_{nf} = 1 + \frac{3\left(\frac{k_p}{k_f} - 1\right)\phi}{\left(\frac{k_p}{k_f} + 2\right) - \left(\frac{k_p}{k_f} - 1\right)\phi} + 5\times10^4 g'(\phi, T, d_p)\phi\rho_f c_{p,f}\sqrt{\frac{\kappa_b T}{\rho_p d_p}}$$

$$g'(\phi, T, d_p) = \left(a_1 + a_2Ln(d_p) + a_3Ln(\phi) + a_4Ln(\phi)\ln(d_p) + a_5Ln(d_p)^2\right)Ln(T)$$
$$+ \left(a_6 + a_7Ln(d_p) + a_8Ln(\phi) + a_9\ln(d_p)Ln(\phi) + a_{10}Ln(d_p)^2\right) \tag{13.15}$$
$$R_f = d_p/k_{p,eff} - d_p/k_p,\ R_f = 4\times10^{-8}km^2/W$$

$$\mu_{nf} = \frac{\mu_f}{(1-\phi)^{2.5}} + \frac{k_{Brownian}}{k_f} \times \frac{\mu_f}{\Pr} \tag{13.16}$$

All required coefficients and properties are illustrated in Tables 13.1 and 13.2.
Introducing dimensionless quantities:

$$\Psi = \psi/\alpha_{nf},\ (X, Y) = (x, y)/L,\ \theta = \frac{T - T_c}{T_h - T_c} \tag{13.17}$$

By discarding the pressure, the final equations are:

$$\frac{\partial^2\Psi}{\partial X^2} + \frac{\partial^2\Psi}{\partial Y^2} = -\frac{A_6}{A_5}Ha\left[\frac{\partial^2\Psi}{\partial Y^2}(\sin^2\gamma) + \frac{\partial^2\Psi}{\partial X^2}(\cos^2\gamma) + 2\frac{\partial^2\Psi}{\partial X\,\partial Y}(\sin\gamma)(\cos\gamma)\right]$$
$$-\frac{A_3}{A_4}\frac{A_2}{A_5}\frac{\partial\theta}{\partial X}Ra \tag{13.18}$$

TABLE 13.1 The Coefficient values Of CuO – Water Nanofluid

Coefficient values	CuO – Water
a_1	– 26.593310846
a_2	– 0.403818333
a_3	– 33.3516805
a_4	– 1.915825591
a_5	6.42185846658E-02
a_6	48.40336955
a_7	– 9.787756683
a_8	190.245610009
a_9	10.9285386565
a_{10}	– 0.72009983664

TABLE 13.2 Thermophysical Properties of Water and Nanoparticles

	$\rho(kg/m^3)$	$C_p(j/kgk)$	$k(W/m.k)$	$\beta\times10^5(K^{-1})$	$d_p(nm)$	$\sigma(\Omega\cdot m)^{-1}$
Water	997.1	4179	0.613	21	-	0.05
CuO	6500	540	18	29	10^{-10}	6500

$$\left(\frac{\partial^2\theta}{\partial Y^2}+\frac{\partial^2\theta}{\partial X^2}\right)=\frac{\partial\theta}{\partial X}\frac{\partial\Psi}{\partial Y}-\frac{\partial\theta}{\partial Y}\frac{\partial\Psi}{\partial X} \tag{13.19}$$

where $Ra=\frac{g\,K\,(\rho\beta)_f L\,\Delta T}{\mu_f\,\alpha_f}$ and $Ha=\frac{\sigma_f K\,B_0^2}{\mu_f}$ are the Rayleigh and Hartmann numbers for the porous media. Also $A_i(i=1..6)$ are constants parameters which are obtained as:

$$\begin{aligned}A_1&=\frac{\rho_{nf}}{\rho_f},\quad A_3=\frac{(\rho\beta)_{nf}}{(\rho\beta)_f},\quad A_5=\frac{\mu_{nf}}{\mu_f},\\ A_2&=\frac{(\rho C_P)_{nf}}{(\rho C_P)_f},\quad A_4=\frac{k_{nf}}{k_f},\quad A_6=\frac{\sigma_{nf}}{\sigma_f}\end{aligned} \tag{13.20}$$

and boundary conditions are:

$$\begin{aligned}\Theta&=1.0 \quad \text{on inner wall}\\ \Theta&=0.0 \quad \text{on outer wall}\\ \Psi&=0.0 \quad \text{on all walls}\end{aligned} \tag{13.21}$$

Local and average Nusselt numbers over the cold cylinder can be calculated as:

$$Nu_{loc}=A_4\frac{\partial\Theta}{\partial r} \tag{13.22}$$

$$Nu_{ave}=\frac{1}{\pi}\int_{\pi/2}^{3\pi/2}Nu_{loc}(\zeta)\,d\zeta \tag{13.23}$$

13.2.3 Effects of Active Parameters

Lorentz forces' influence on nanofluid flow in a porous enclosure with complex hot cylinder is investigated. Darcy's model is taken into account to present the governing equations. Numerical simulations are examined for various Rayleigh numbers for porous medium ($Ra = 10^2, 250,$ and 10^3), number of undulations ($N = 3, 4, 5,$ and 6), volume fraction of CuO-water ($\phi = 0$ and 0.04), and Hartmann number for porous medium ($Ha = 0$ to 20).

Figs. 13.2–13.4 demonstrate the impacts of number of undulations, Rayleigh and Hartmann numbers for porous medium on streamlines and isotherms. As nanofluid temperature augments, the nanofluid begins moving from the warm surface to the outer one and dropping along the circular cylinder, afterwards rising again at the inner cylinder, generating a clockwise revolving vortex inside the cavity. Under the dominance of the conduction mode, isotherms follow the shape of cylinders. $|\Psi_{max}|$ augments as buoyancy force augments and it reduces as Lorentz force enhances. For odd values of N, a very small eddy is generated around $\zeta = 90^\circ$. This phenomenon is observed because of the opposition of the crest in contradiction of the flow movement between the hot and cold walls. But there is no secondary eddy near $\zeta = 90^\circ$ for even values of N because the crest is parallel to the gravity force. As buoyancy forces enhances, the eddies become stronger and the thermal plume is generated near the centerline. Increasing the Hartmann number causes the thermal plume to diminish and change the heat transfer mechanism from convection to conduction. So, Nu reduces with the rise of Lorentz forces. In the absence of Lorentz force when N = 3, Ra = 1000, two vortexes rotating in opposite directions appear near the vertical centerline. So, a strong thermal plume appears in this region. As Lorentz force augments, these two vortexes merge together and the thermal plume vanishes.

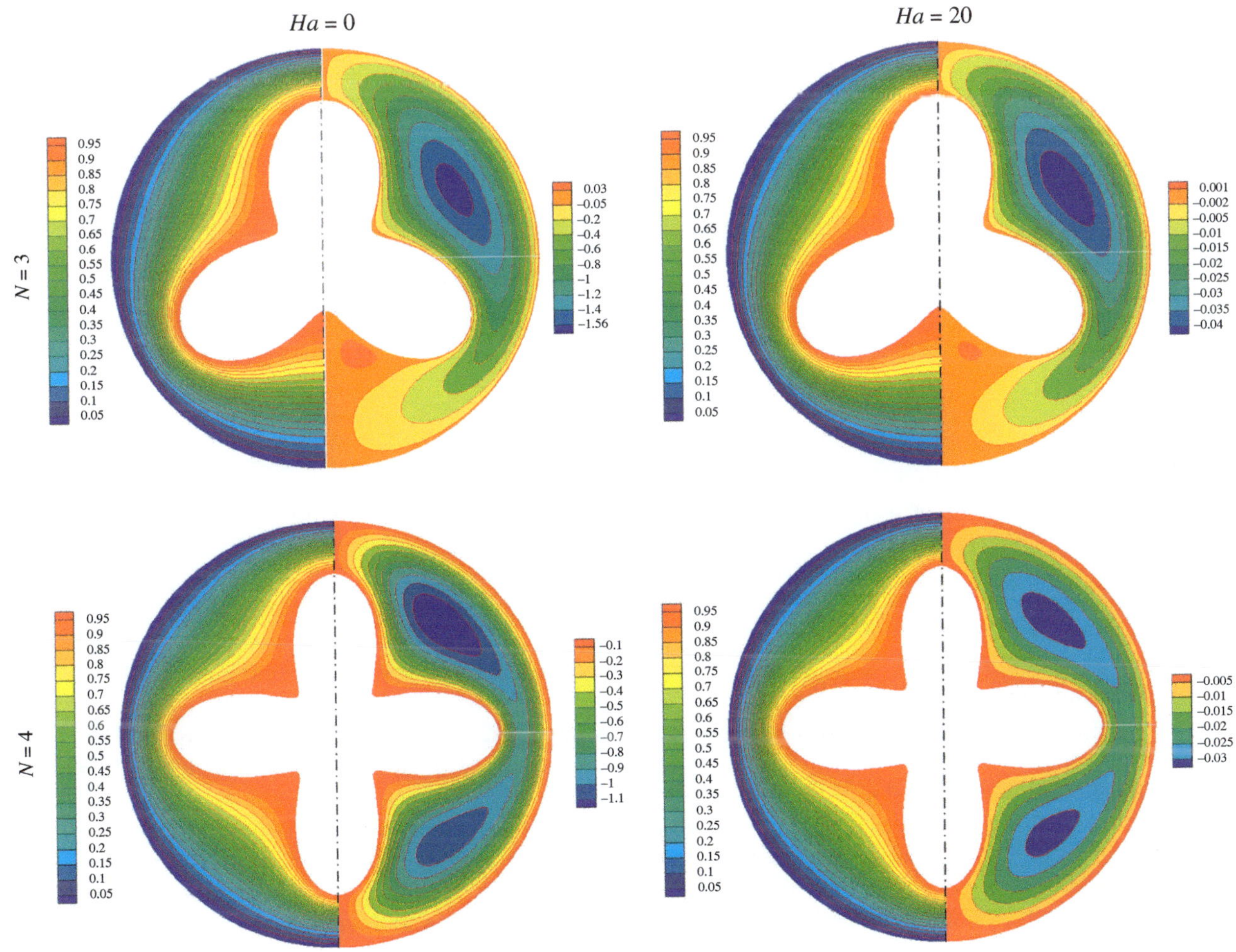

FIGURE 13.2 Isotherm (left) and streamline (right) contours for different values of number of undulations and Hartmann number for porous medium when $\phi = 0.04$, $Ra = 100$.

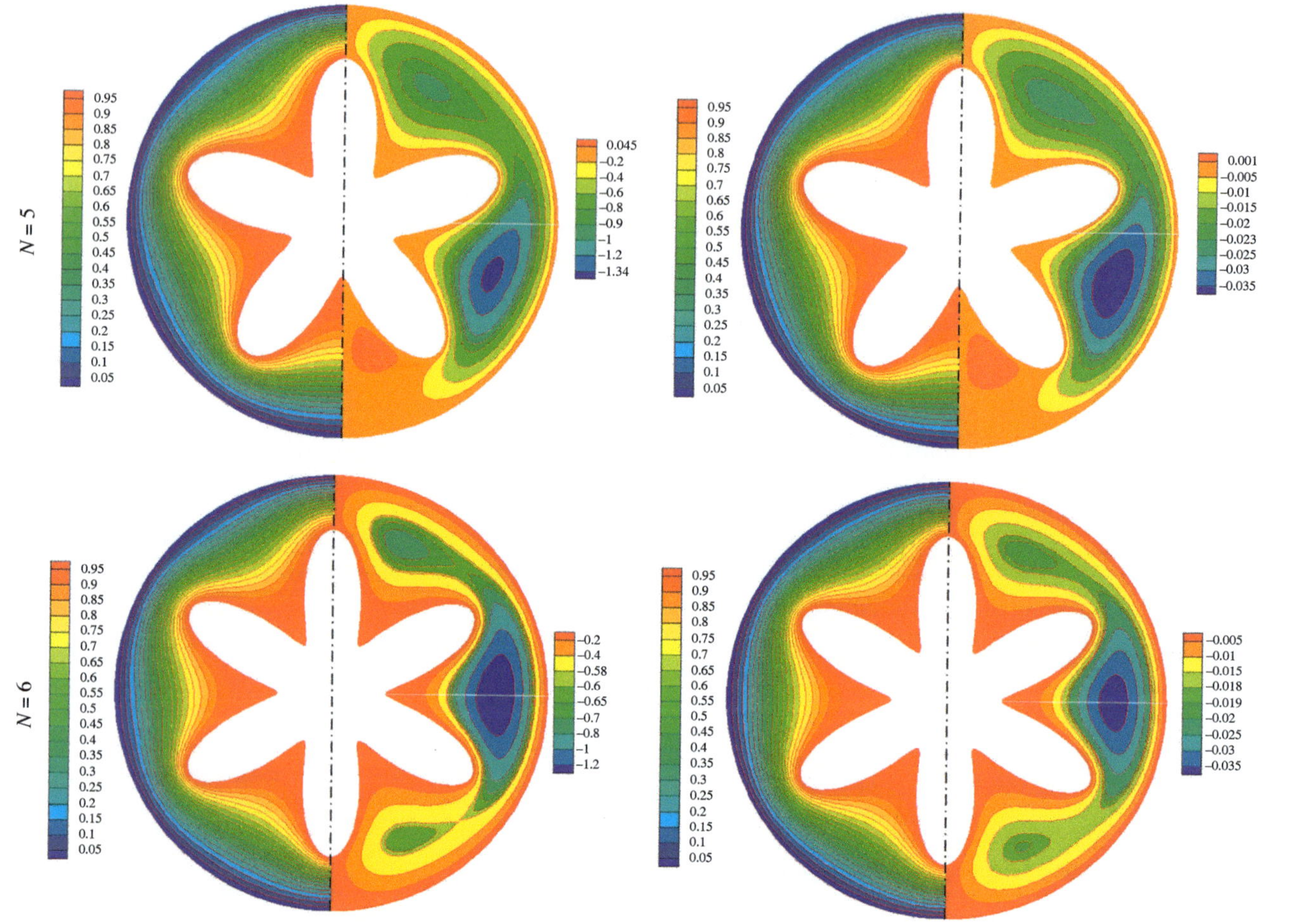

FIGURE 13.2 (Continued).

Figs. 13.5 and 13.6 depict the influence of N, ϕ, Ra, and Ha for porous media on Nu_{loc}, Nu_{ave}. The correlation for Nu_{ave} corresponding active parameters is:

$$\begin{aligned} Nu_{ave} &= 0.98 + 5.5\times10^{-4}Ra - 0.016N + 0.51\phi - 0.021Ha - 3.2\times10^{-6}Ra\,N \\ &- 1.71\times10^{-5}Ra\,\phi - 3.6Ra\,Ha + 0.15\,N\phi + 3.15\times10^{-4}N\,Ha + 0.016\phi\,Ha \\ &+ 1.4\times10^{-7}Ra^2 + 7.9\times10^{-3}N^2 + 46.5\phi^2 + 1.16\times10^{-3}Ha^2 \end{aligned} \tag{13.24}$$

The number of extremum in in Nu_{loc} profile is matching to existence of thermal plume and number of undulation. As the buoyancy force increases, the temperature gradient increases and in turn Nu_{ave} enhances with the rise of buoyancy forces. As the nanofluid volume fraction augments, the temperature of the fluid is enhanced and Nu_{ave} increases with the enhancement of the thermal conductivity of the fluid. Enhancing the Lorentz force causes the nanofluid flow to retard and the Nusselt number reduces.

The impacts of number of undulations, buoyancy, and Lorentz forces on heat transfer improvement are demonstrated in Table 13.3 and Fig. 13.7. Adding CuO nanoparticle into water can enhance the thermal conductivity of fluid. So, this passive method is more effective when a conduction mechanism is dominant. According to this fact, heat transfer improvement increases with the rise of Lorentz forces and it reduces with the enhancement of buoyancy forces. As the number of undulations increases, E is enhanced. The maximum value of E is 12.43, which is obtained at N = 6, Ha = 20, Ra = 100.

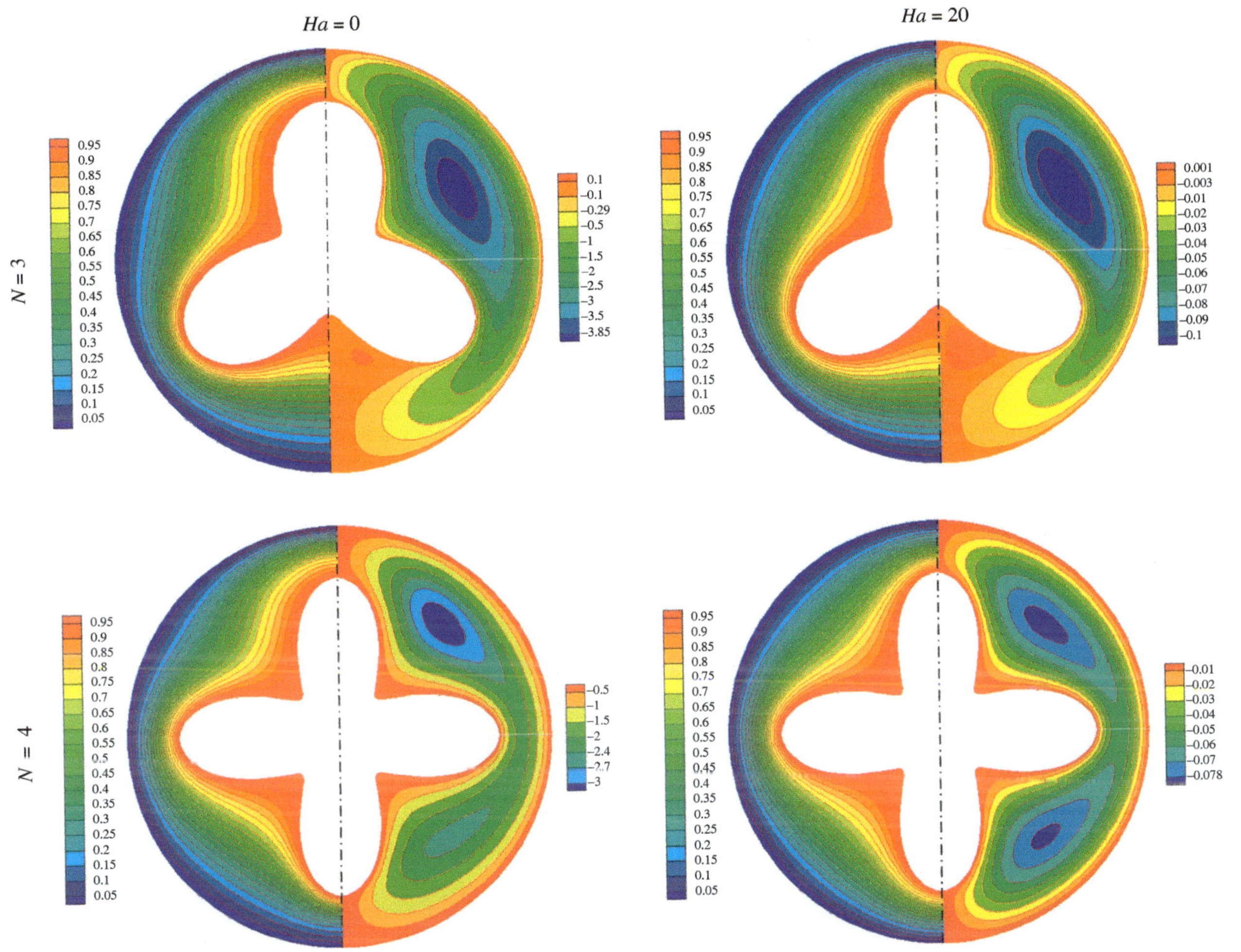

FIGURE 13.3 Isotherm (left) and streamline (right) contours for different values of number of undulations and Hartmann number for porous medium when $\phi = 0.04$, $Ra = 250$.

13.3 ANALYSIS OF WATER-BASED NANOFLUID FLOW AND HEAT TRANSFER DUE TO MAGNETIC FIELD IN A POROUS ENCLOSURE

13.3.1 Problem Definition

Fig. 13.8 depicts the geometry, boundary condition, and sample element. The formula of the outer cylinder is:

$$b = \sqrt{1 - \varepsilon^2} \, . \, a \tag{13.25}$$

where a, b, ε are the major, minor axis of elliptic cylinder, and eccentricity for the inner cylinder. The inner cylinder has constant heat flux condition.

13.3.2 Governing Equation

2D steady convective flow of nanofluid in a porous medium is considered in the presence of a constant magnetic field. The PDEs equations are:

$$\nabla . \vec{V} = 0 \tag{13.26}$$

$$\frac{\mu_{nf}}{K} \vec{V} = \left(-\nabla p + \vec{I} \times \vec{B} + \rho_{nf} \, \vec{g} \right) \tag{13.27}$$

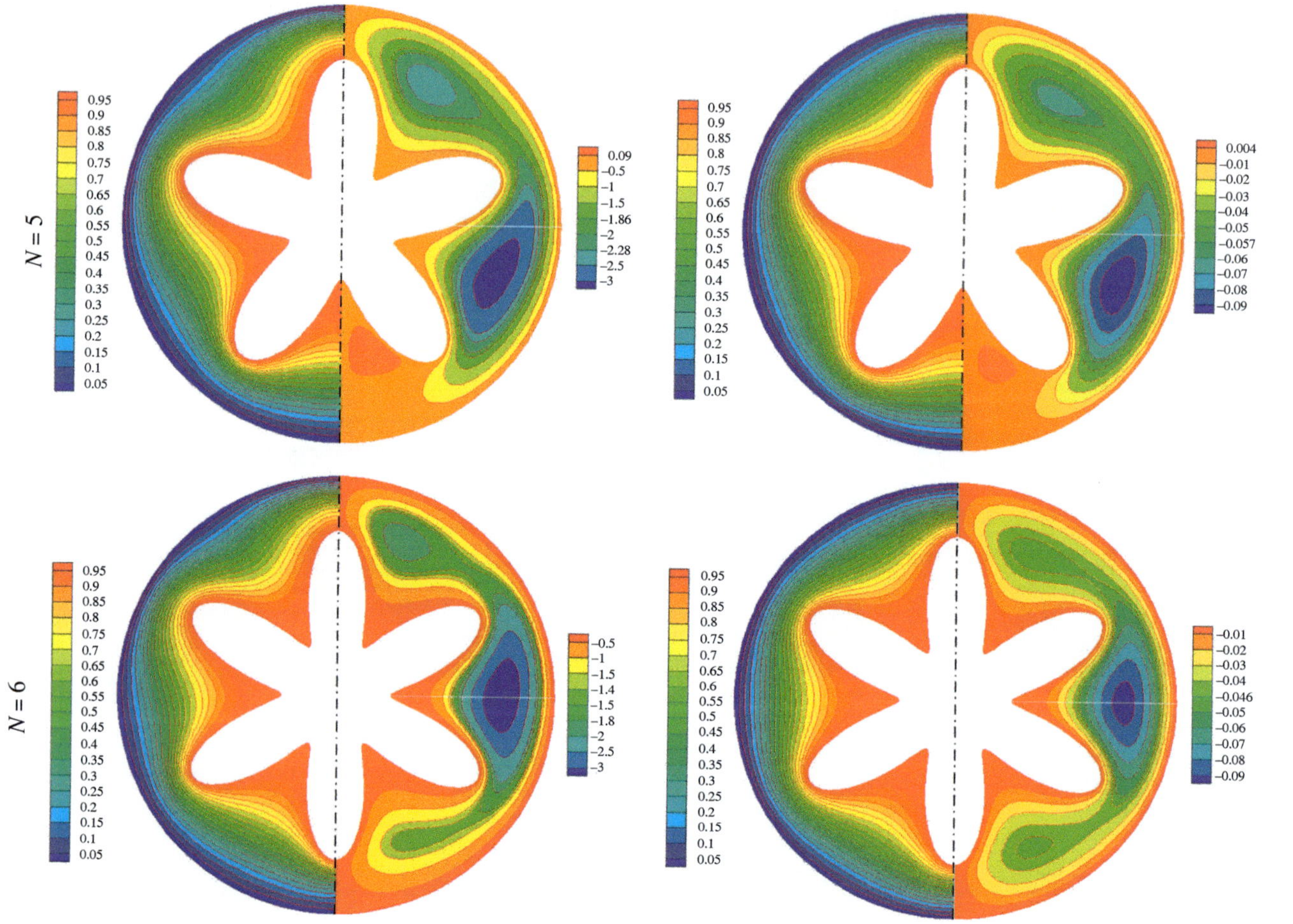

FIGURE 13.3 (Continued).

$$(\rho C_p)_{nf}\left(\vec{V}.\nabla\right)T = k_{nf}\,\nabla^2 T \tag{13.28}$$

$$\nabla.\vec{I} = 0 \tag{13.29}$$

$$\sigma_{nf}\left(\vec{V}\times\vec{B} - \nabla\varphi\right) = \vec{I} \tag{13.30}$$

In Eq. (13.26), Darcy's model is used for a porous medium. Eqs. (13.29) and (13.30) reduce to $\nabla^2\varphi = 0$. So the electric field can be neglected. So the above equations turn into:

$$\frac{\partial v}{\partial y} + \frac{\partial u}{\partial x} = 0 \tag{13.31}$$

$$-\frac{K}{\mu_{nf}}\frac{\partial p}{\partial x} - \frac{\sigma_{nf}KB_0^2}{\mu_{nf}}\left(-u\sin^2\gamma + v\sin\gamma\cos\gamma\right) = u \tag{13.32}$$

$$-\frac{K}{\mu_{nf}}\frac{\partial p}{\partial y} - \frac{\sigma_{nf}KB_0^2}{\mu_{nf}}\left(-v\cos^2\gamma + u\sin\gamma\cos\gamma\right) + g\frac{(\rho\beta)_{nf}K}{\mu_{nf}}(T - T_c) = v \tag{13.33}$$

$$(\rho C_p)_{nf}\left(u\frac{\partial T}{\partial x} + v\frac{\partial T}{\partial y}\right) = k_{nf}\left(\frac{\partial^2 T}{\partial x^2} + \frac{\partial^2 T}{\partial y^2}\right) \tag{13.34}$$

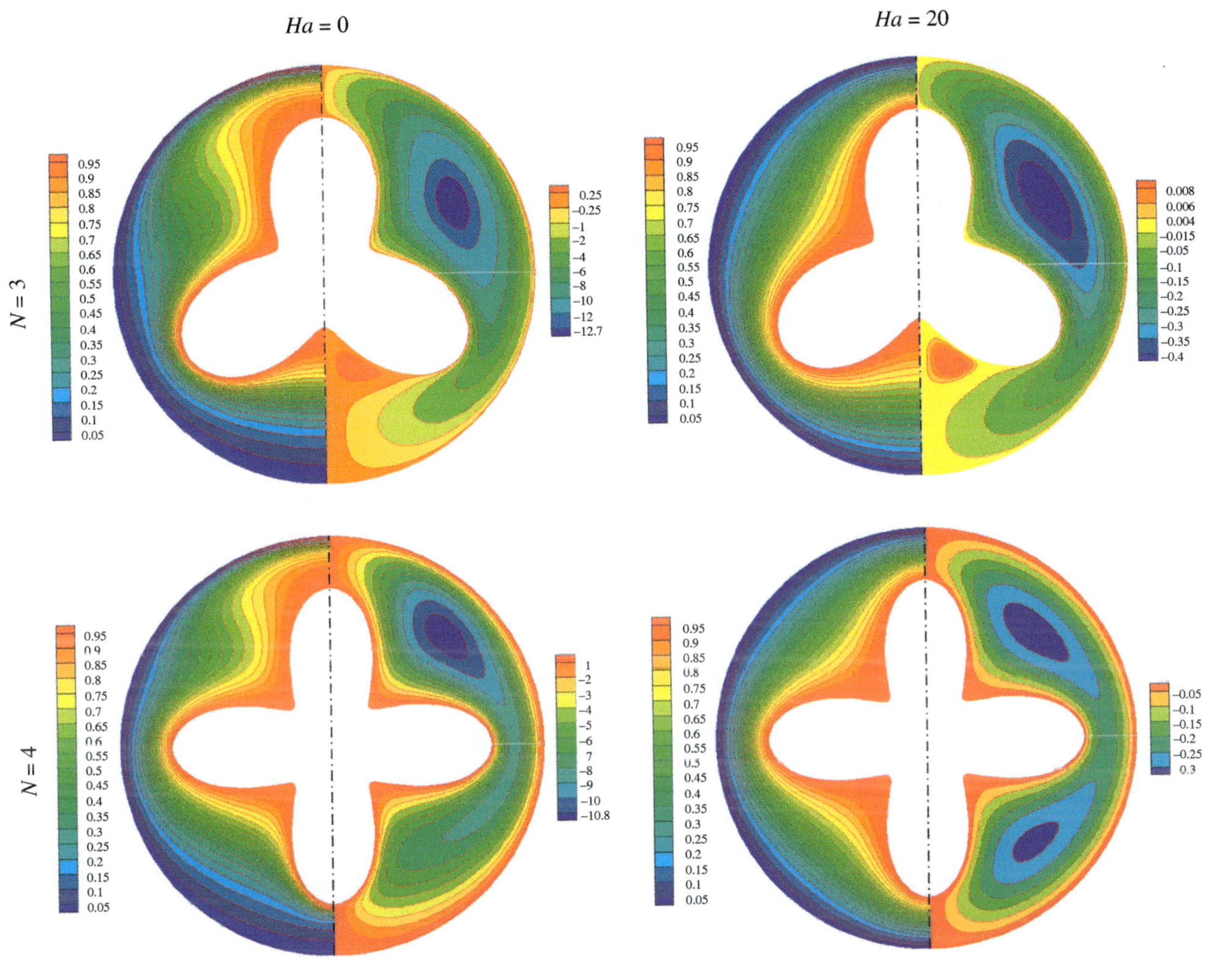

FIGURE 13.4 Isotherm (left) and streamline (right) contours for different values of number of undulations and Hartmann number for porous medium when $\phi = 0.04, Ra = 1000$.

$(\rho C_p)_{nf}, (\rho\beta)_{nf}, \rho_{nf}$, and σ_{nf} are defined as:

$$(\rho C_p)_{nf} = (\rho C_p)_f(1-\phi) + (\rho C_p)_s\phi \tag{13.35}$$

$$(\rho\beta)_{nf} = (\rho\beta)_f(1-\phi) + (\rho\beta)_s\phi \tag{13.36}$$

$$\rho_{nf} = \rho_f(1-\phi) + \rho_s\phi \tag{13.37}$$

$$\frac{\sigma_{nf}}{\sigma_f} = 1 + \frac{3\left(\frac{\sigma_s}{\sigma_f} - 1\right)\phi}{\left(\frac{\sigma_s}{\sigma_f} + 2\right) - \left(\frac{\sigma_s}{\sigma_f} - 1\right)\phi} \tag{13.38}$$

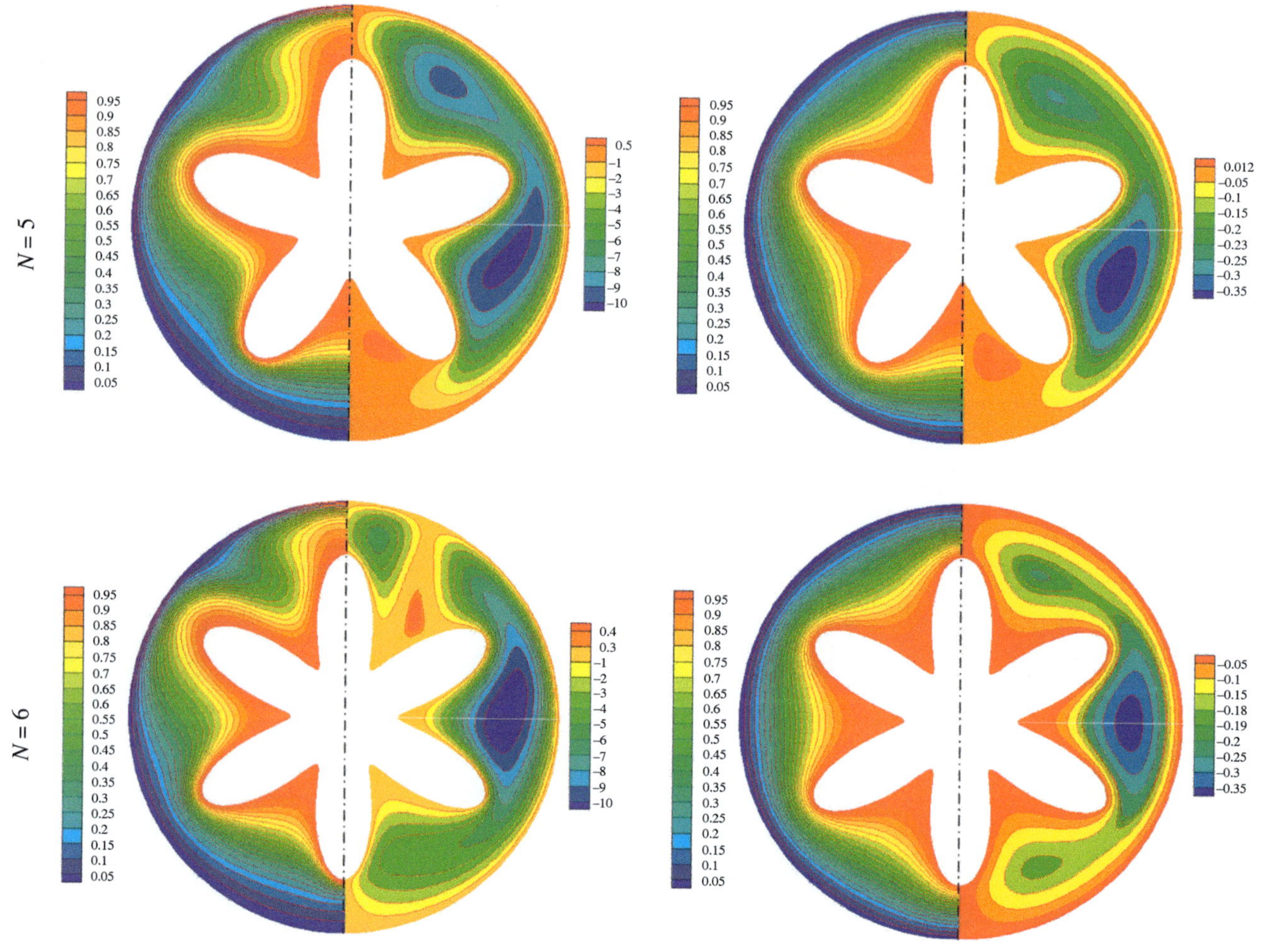

FIGURE 13.4 (Continued).

k_{nf}, μ_{nf} are obtained according to the Koo–Kleinstreuer–Li (KKL) model:

$$k_{nf} = 1 + \frac{3\left(\frac{k_p}{k_f} - 1\right)\phi}{\left(\frac{k_p}{k_f} + 2\right) - \left(\frac{k_p}{k_f} - 1\right)\phi} + 5\times 10^4 g'(\phi, T, d_p)\phi\rho_f c_{p,f}\sqrt{\frac{\kappa_b T}{\rho_p d_p}}$$
$$g'(\phi, T, d_p) = \left(a_1 + a_2 Ln(d_p) + a_3 Ln(\phi) + a_4 Ln(\phi)\ln(d_p) + a_5 Ln(d_p)^2\right)Ln(T)$$
$$+ \left(a_6 + a_7 Ln(d_p) + a_8 Ln(\phi) + a_9 \ln(d_p) Ln(\phi) + a_{10} Ln(d_p)^2\right)$$
$$R_f = d_p/k_{p,eff} - d_p/k_p,\ R_f = 4\times 10^{-8} km^2/W \tag{13.39}$$

$$\mu_{nf} = \frac{\mu_f}{(1-\phi)^{2.5}} + \frac{k_{Brownian}}{k_f} \times \frac{\mu_f}{\Pr} \tag{13.40}$$

All required coefficients and properties are illustrated in Tables 13.1 and 13.2. Introducing dimensionless quantities:

$$\Psi = \psi/\alpha_{nf},\ \ (X, Y) = (x, y)/L,\ \ \theta = \frac{T - T_c}{\Delta T}, \Delta T = \frac{q''L}{k_f} \tag{13.41}$$

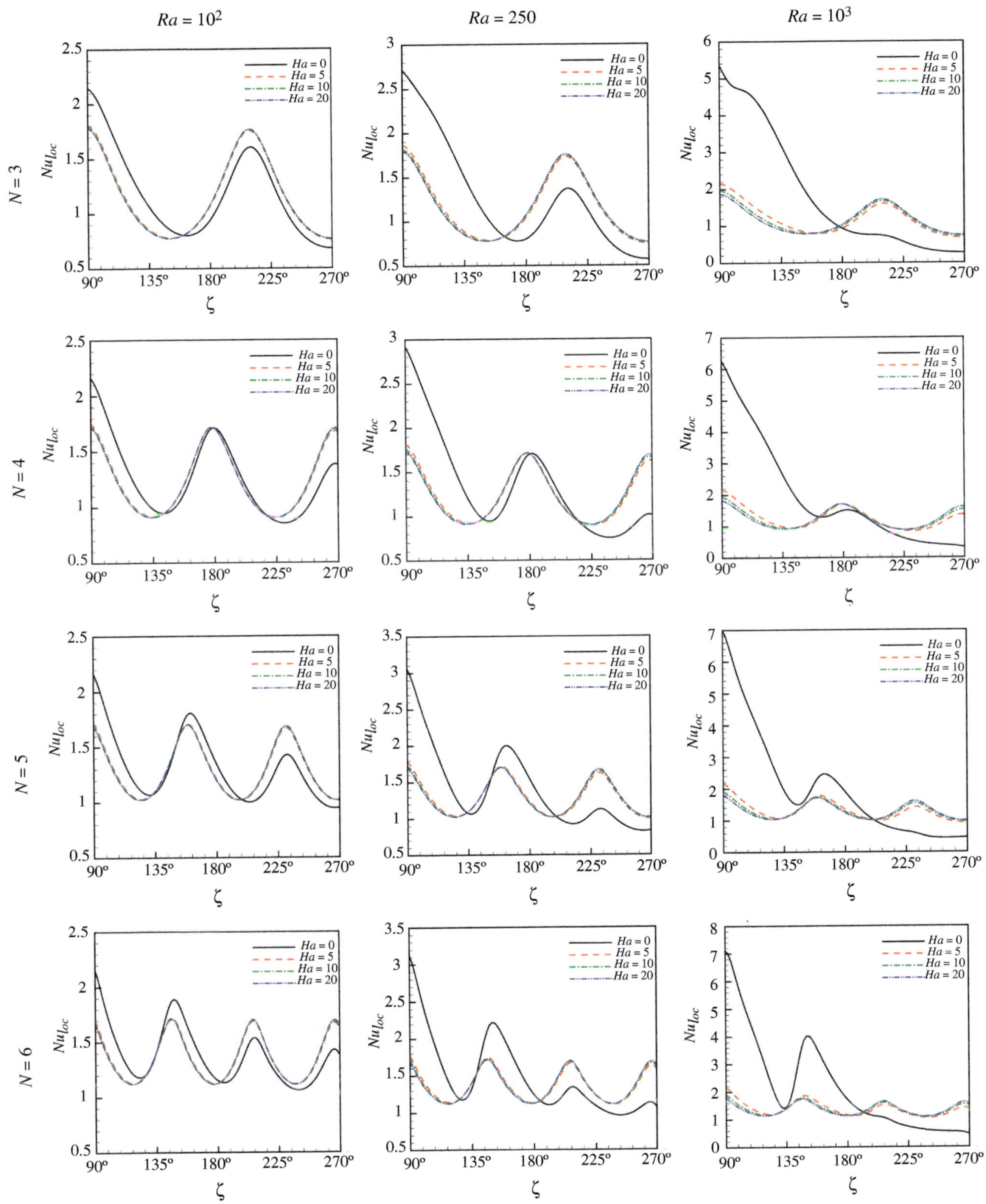

FIGURE 13.5 Effects of the number of undulations, Rayleigh number, and Hartmann number for porous medium on local Nusselt number at $\phi = 0.04$.

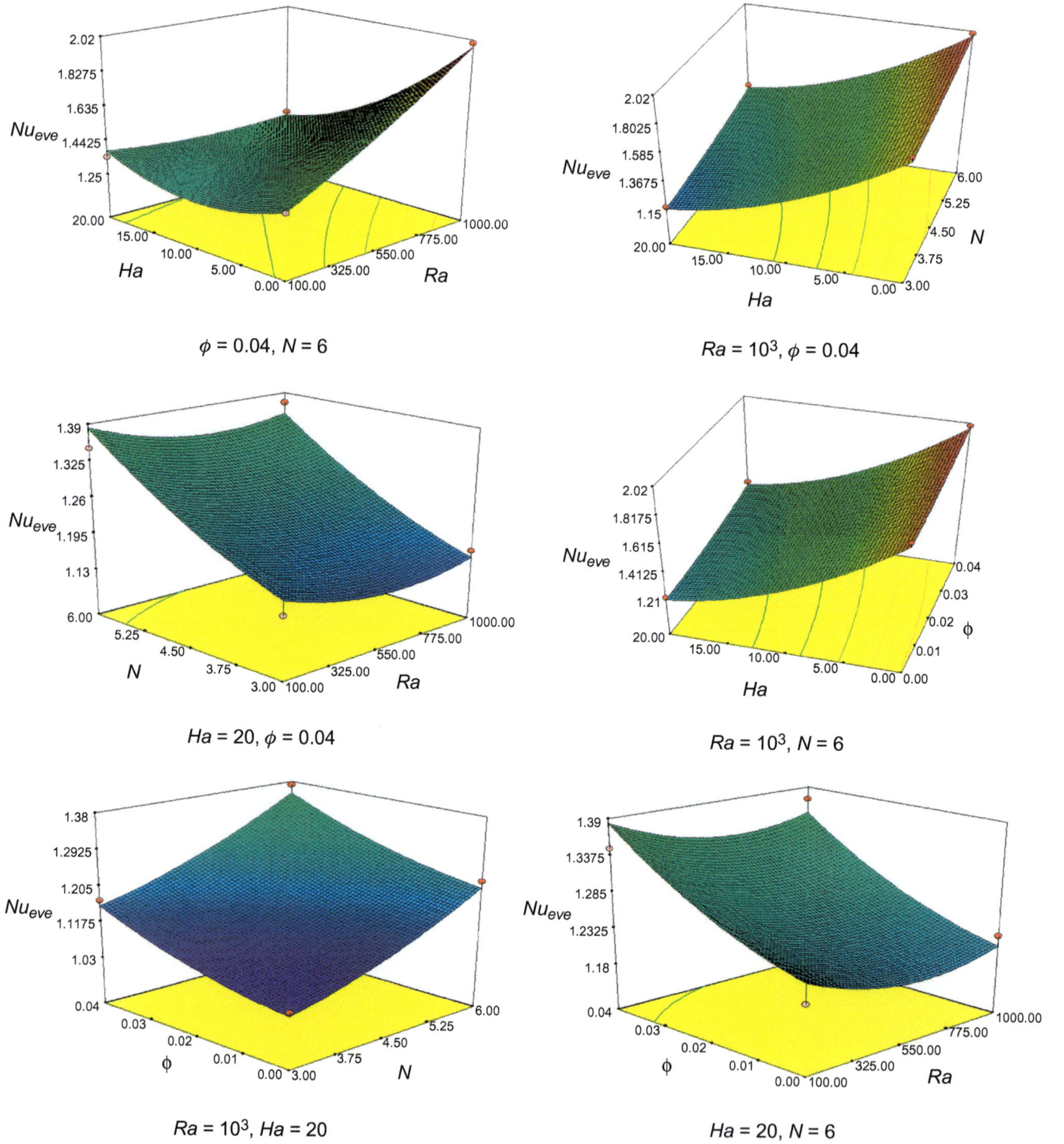

FIGURE 13.6 Effects of the number of undulations, nanoparticle volume fraction, Rayleigh number, and Hartmann number for porous medium on average Nusselt number.

By discarding the pressure, the final equations are:

$$\frac{\partial^2 \Psi}{\partial X^2} + \frac{\partial^2 \Psi}{\partial Y^2} = -\frac{A_6}{A_5} Ha \left[\frac{\partial^2 \Psi}{\partial Y^2} (\sin^2 \gamma) + \frac{\partial^2 \Psi}{\partial X^2} (\cos^2 \gamma) + 2 \frac{\partial^2 \Psi}{\partial X \, \partial Y} (\sin \gamma)(\cos \gamma) \right] - \frac{A_3}{A_4} \frac{A_2}{A_5} \frac{\partial \theta}{\partial X} Ra \tag{13.42}$$

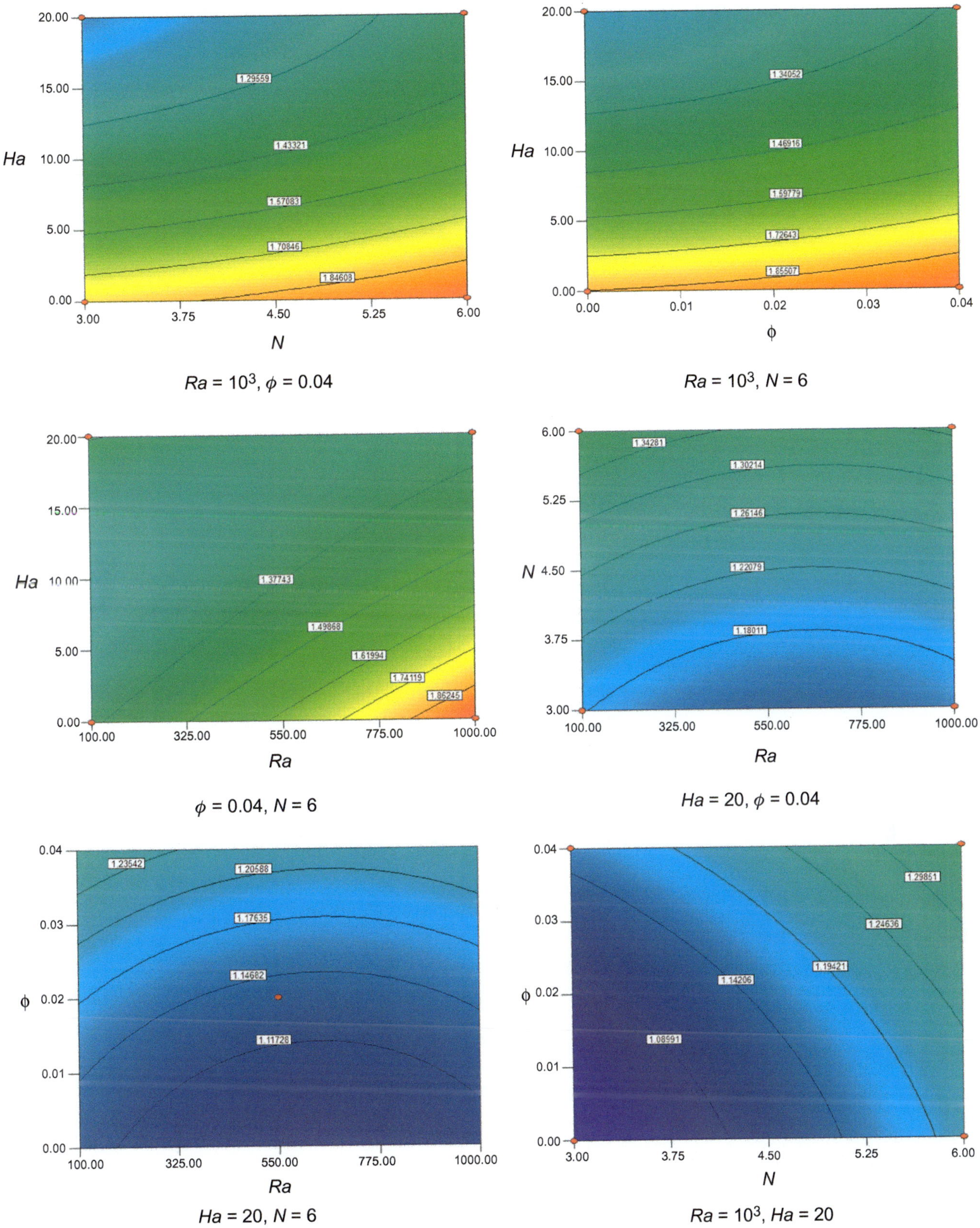

FIGURE 13.6 (Continued).

TABLE 13.3 Effects of the Number of Undulations and Hartmann Number on the Ratio of Heat Transfer Enhancement When $Ra = 10^2$

	N		
Ha	3	4	6
0	11.49639	11.62185	11.70252
5	12.43029	12.43048	12.42186
10	12.43759	12.4375	12.43239
20	12.43968	12.43959	12.43681

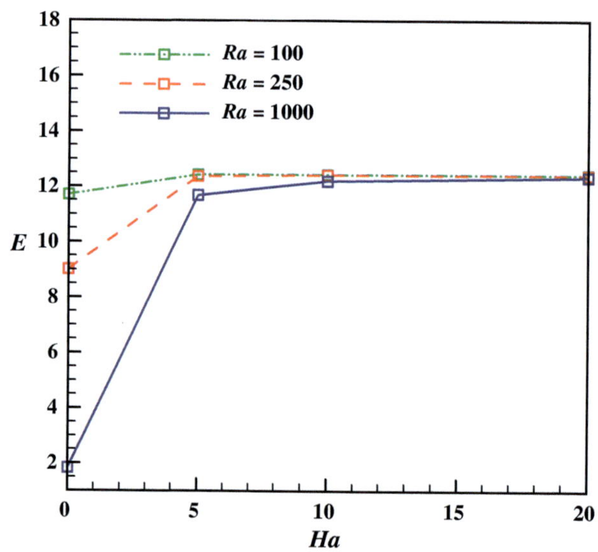

FIGURE 13.7 Effects of the Hartmann number and Rayleigh number for porous medium on the ratio of heat transfer enhancement due to addition of nanoparticles when $N = 6$.

$$\left(\frac{\partial^2\theta}{\partial Y^2} + \frac{\partial^2\theta}{\partial X^2}\right) = -\frac{\partial\theta}{\partial Y}\frac{\partial\Psi}{\partial X} + \frac{\partial\Psi}{\partial Y}\frac{\partial\theta}{\partial X} \tag{13.43}$$

where $Ra = \frac{g\,K\,(\rho\beta)_f L\,\Delta T}{\mu_f\,\alpha_f}$ and $Ha = \frac{\sigma_f K\,B_0^2}{\mu_f}$ are the Rayleigh and Hartmann numbers for the porous media. Also $A_i(i = 1..6)$ are constants parameters which are obtained as:

$$A_1 = \frac{\rho_{nf}}{\rho_f},\quad A_3 = \frac{(\rho\beta)_{nf}}{(\rho\beta)_f},\quad A_5 = \frac{\mu_{nf}}{\mu_f},$$
$$A_2 = \frac{(\rho C_P)_{nf}}{(\rho C_P)_f},\quad A_4 = \frac{k_{nf}}{k_f},\quad A_6 = \frac{\sigma_{nf}}{\sigma_f} \tag{13.44}$$

and boundary conditions are:

$$\frac{\partial\theta}{\partial n} = 1.0 \quad \text{on inner wall}$$
$$\theta = 0.0 \quad \text{on outer wall} \tag{13.45}$$
$$\Psi = 0.0 \quad \text{on all walls}$$

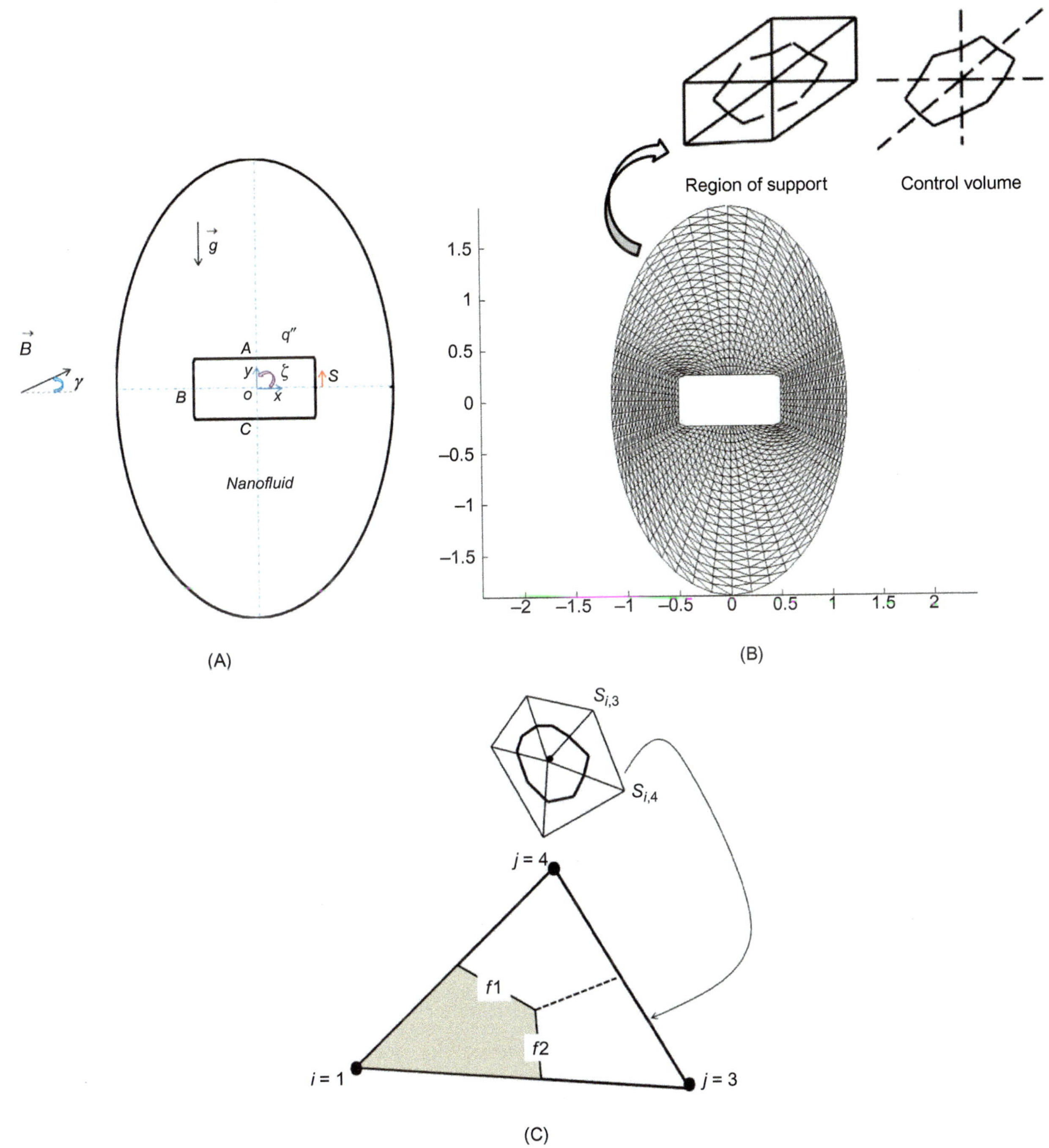

FIGURE 13.8 (A) Geometry and the boundary conditions with (B) the mesh of geometry considered in this work; (C) a sample triangular element and its corresponding control volume.

Local and average Nusselt numbers over the inner cylinder can be obtained as:

$$Nu_{loc} = \frac{1}{\theta}\left(\frac{k_{nf}}{k_f}\right) \tag{13.46}$$

$$Nu_{ave} = \frac{1}{S}\int_0^S Nu_{loc}\, ds \tag{13.47}$$

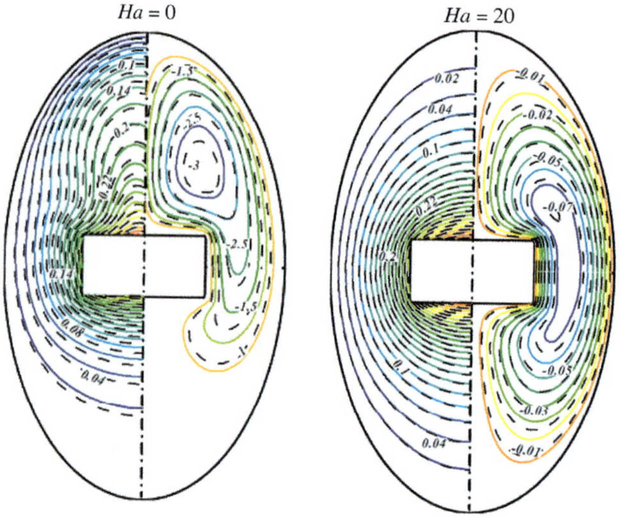

FIGURE 13.9 Impact of nanofluid volume fraction on streamline (right) and isotherm (left) contours (nanofluid ($\phi = 0.04$)(–) and pure fluid ($\phi = 0$) (- - -)) when $Ra = 10^5$.

13.3.3 Effects of Active Parameters

CuO-water nanofluid motion in permeable media in the presence of Lorentz forces is demonstrated. KKL and Darcy models are utilized for nanofluid and permeable media, respectively. CVFEM is utilized to present the effects of active parameters such as Rayleigh number for porous medium ($Ra = 10^2, 250$ and 10^3), inclination angle ($\xi = 0^\circ$ to 90°), Hartmann number for porous medium ($Ha = 0$ to 20), and volume fraction of CuO-water ($\phi = 0$ and 0.04).

Fig. 13.9 illustrates the impact of ϕ on hydrothermal behavior. As ϕ increases, nanoparticles interaction increases and in turn nanofluid motion increases. Also nanofluid velocity enhances with the rise of ϕ. The impact of ϕ on isotherms is more sensible in the absence of Lorentz forces. Figs. 13.10 and 13.11 show the influences of inclination angle, Hartmann and Rayleigh numbers for porous medium on streamlines and isotherms. A single clockwise eddy exists in streamlines. At low Rayleigh number, isotherms follow the shape of cylinders. $|\Psi_{\max}|$ enhances with the increase of Ra. As Ra increases, the strength of the eddy enhances and the center of this eddy moves upward. Also a thermal plume is generated near the vertical centerline. As Ha increases the thermal plume vanishes and $|\Psi_{\max}|$ reduces. As the outer cylinder rotates, the main eddy is stretched vertically and by increasing Ra, the center of the eddy moves to the top middle. In the presence of Lorentz forces, the distribution of isotherms reduces and the eddy moves downward.

Figs. 13.12 and 13.13 depict the influence of ξ, Ra and Ha on Nu_{loc}, Nu_{ave}. The correlation for Nu_{ave} corresponding to active parameters is:

$$\begin{aligned} Nu_{ave} &= 2.19 - 0.03\xi + 0.81 \times 10^{-4} Ra - 0.05 Ha \\ &- 7.55 \times 10^{-5} \xi\, Ra + 3.9 \times 10^{-3} \xi\, Ha - 5.58 \times 10^{-5} Ra\, Ha \\ &- 7.6 \times 10^{-3} \xi^2 + 2.96 \times 10^{-7} Ra^2 + 7.6 Ha^2 \end{aligned} \tag{13.48}$$

The root mean squared error of this formula is equal to 0.98. The presence of the thermal plume makes the Nu_{loc} have extremum points.Nu_{ave} increases with the increase of the Rayleigh number and nanofluid volume fraction. Increasing ξ, Ha leads to changing the convection mechanism to conduction mode, so Nu_{ave} detracts with the enhancement of ξ, Ha . Table 13.4 illustrates the influences of inclination angle, Lorentz and buoyancy forces on heat transfer augmentation $\left(E = 100 \times (Nu_{\phi=0.04} - Nu_{\phi=0})/Nu_{\phi=0}\right)$. Adding CuO nanoparticle into H_2O can augment the thermal conductivity of fluid. So, this technique is more efficient when the conduction mode is dominant. Therefore E increases with the rise of the inclination angle and Lorentz forces but it decreases with the rise of buoyancy forces.

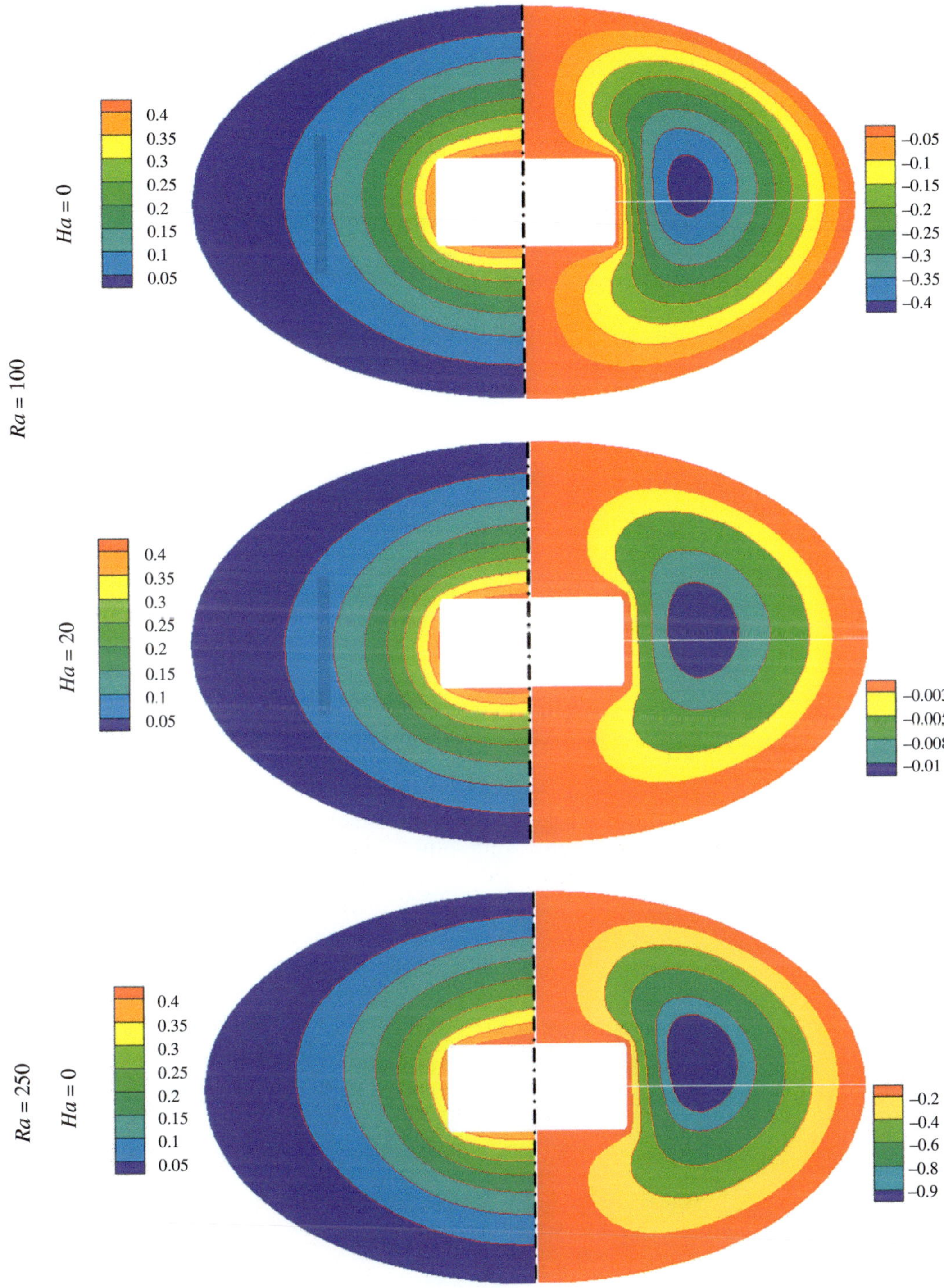

FIGURE 13.10 Isotherm (left) and streamline (right) contours for different values Rayleigh and Hartmann numbers for porous medium when $\phi = 0.04$, $\xi = 0^{\circ}$.

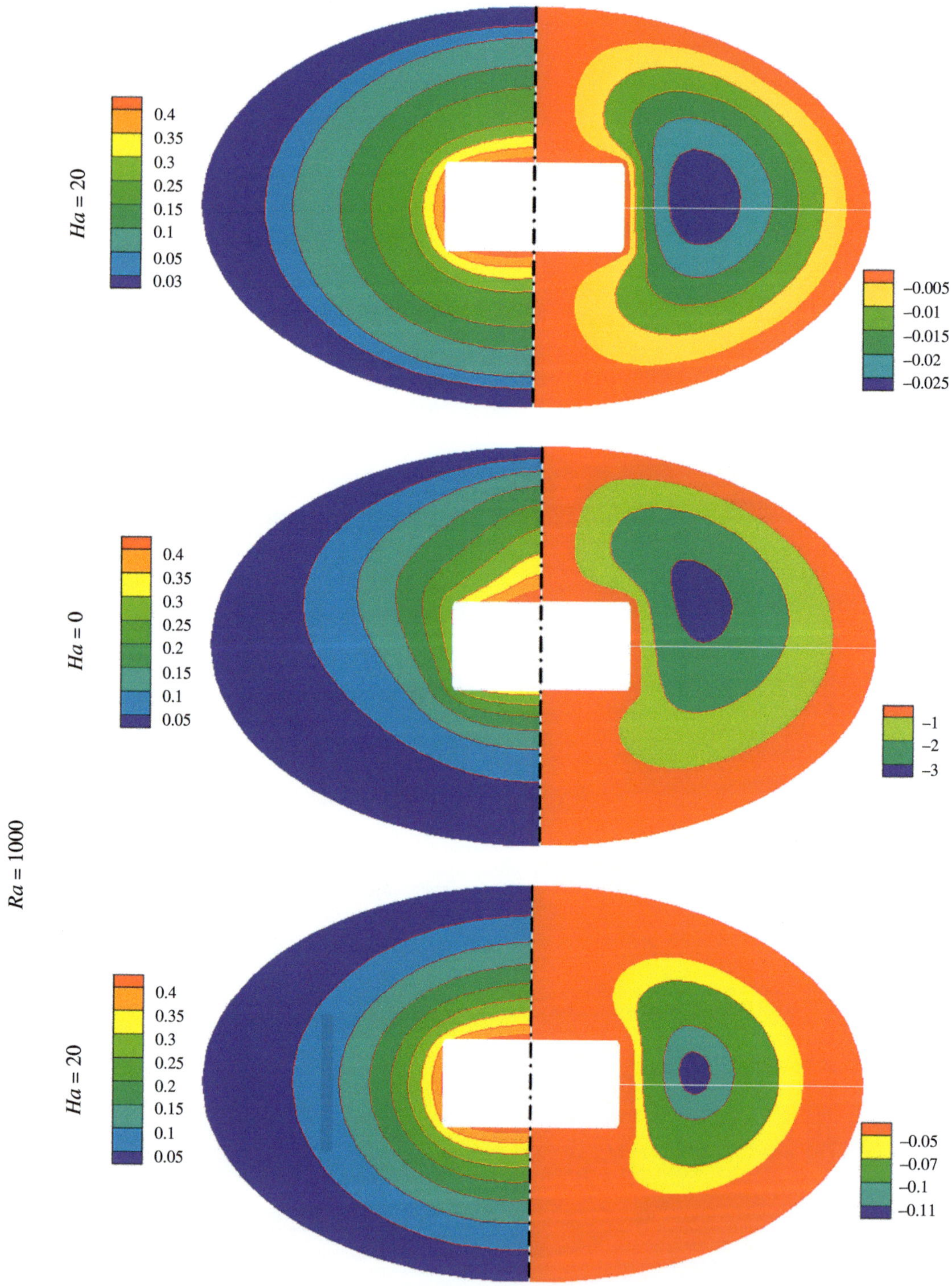

FIGURE 13.10 (Continued).

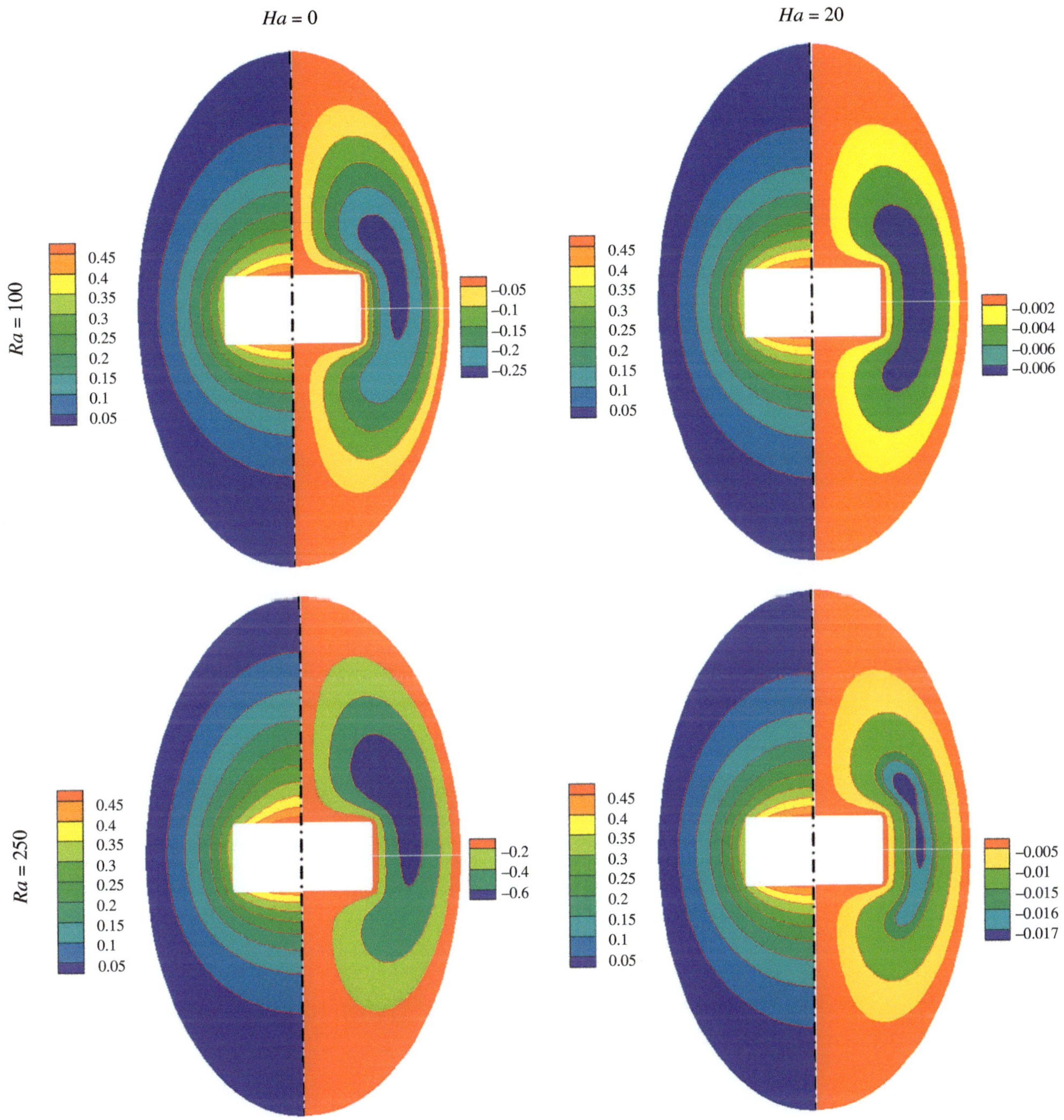

FIGURE 13.11 Isotherm (left) and streamline (right) contours for different values of Rayleigh and Hartmann numbers for porous medium when $\phi = 0.04$, $\xi = 90^{\circ}$.

13.4 MAGNETOHYDRODYNAMIC NANOFLUID CONVECTION IN A POROUS ENCLOSURE CONSIDERING HEAT FLUX BOUNDARY CONDITION

13.4.1 Problem Definition

Fig. 13.14 shows the geometry, boundary condition, and sample element. The formula of the inner cylinder is:

$$\left(\frac{X}{a}\right)^{2\hat{n}} + \left(\frac{Y}{b}\right)^{2\hat{n}} = 1 \tag{13.49}$$

where $\hat{n} > 1$ geometry is square when $a = b$.

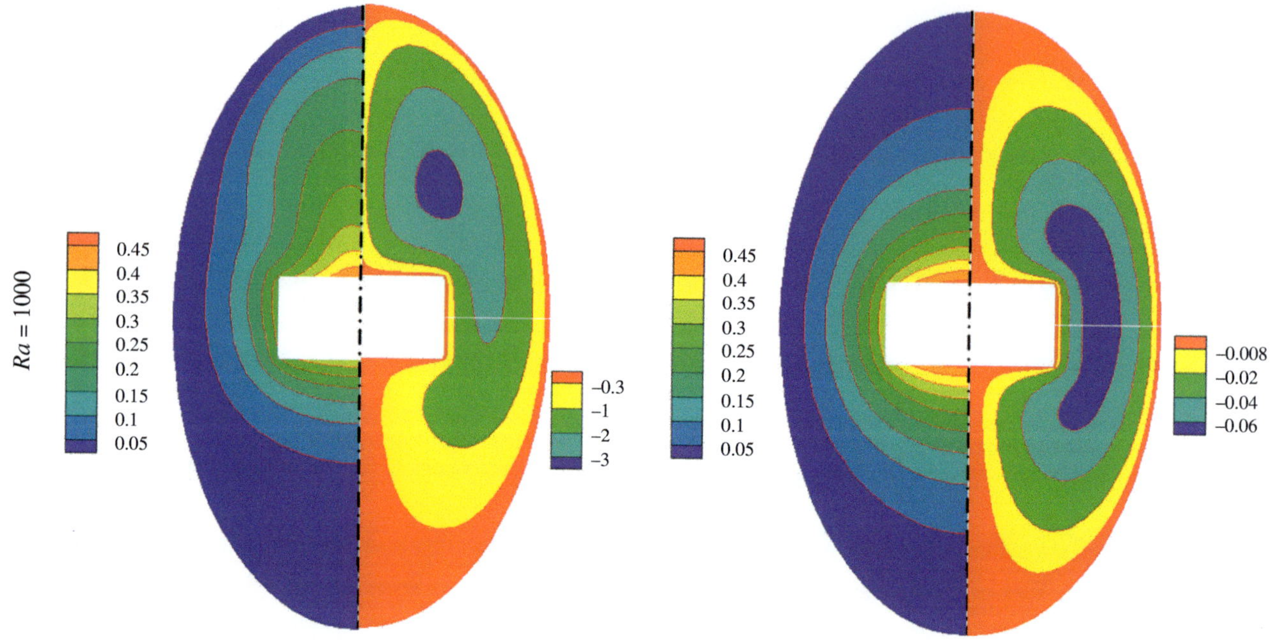

FIGURE 13.11 (Continued).

13.4.2 Governing Equation

Two-dimensional steady convective flow of nanofluid in a porous media is considered in the presence of a constant horizontal magnetic field. The PDEs equations are:

$$\nabla.\vec{V} = 0 \tag{13.50}$$

$$\frac{\mu_{nf}}{K}\vec{V} = \left(\nabla p + \vec{I} \times \vec{B} + \rho_{nf}\,\vec{g}\right) \tag{13.51}$$

$$(\rho C_p)_{nf}\left(\vec{V}.\nabla\right)T = k_{nf}\,\nabla^2 T \tag{13.52}$$

$$\nabla.\vec{I} = 0 \tag{13.53}$$

$$\sigma_{nf}\left(\vec{V} \times \vec{B} - \nabla\varphi\right) = \vec{I} \tag{13.54}$$

In Eq. (13.51), Darcy's model is utilized for porous medium. Eqs. (13.53) and (13.54) reduce to $\nabla^2\varphi = 0$. So the electric field can be neglected. Therefore the governing equations are:

$$\frac{\partial v}{\partial y} + \frac{\partial u}{\partial x} = 0 \tag{13.55}$$

$$-\frac{K}{\mu_{nf}}\frac{\partial p}{\partial x} - \frac{\sigma_{nf} K B_0^2}{\mu_{nf}}\left(-u\sin^2\gamma + v\sin\gamma\cos\gamma\right) = u \tag{13.56}$$

$$-\frac{K}{\mu_{nf}}\frac{\partial p}{\partial y} - \frac{\sigma_{nf} K B_0^2}{\mu_{nf}}\left(-v\cos^2\gamma + u\sin\gamma\cos\gamma\right) + g\frac{(\rho\beta)_{nf}K}{\mu_{nf}}(T - T_c) = v \tag{13.57}$$

$$(\rho C_p)_{nf}\left(u\frac{\partial T}{\partial x} + v\frac{\partial T}{\partial y}\right) = k_{nf}\left(\frac{\partial^2 T}{\partial x^2} + \frac{\partial^2 T}{\partial y^2}\right) \tag{13.58}$$

$(\rho C_p)_{nf}$, $(\rho\beta)_{nf}$, ρ_{nf}, and σ_{nf} are defined as:

$$(\rho C_p)_{nf} = (\rho C_p)_f(1-\phi) + (\rho C_p)_s\phi \tag{13.59}$$

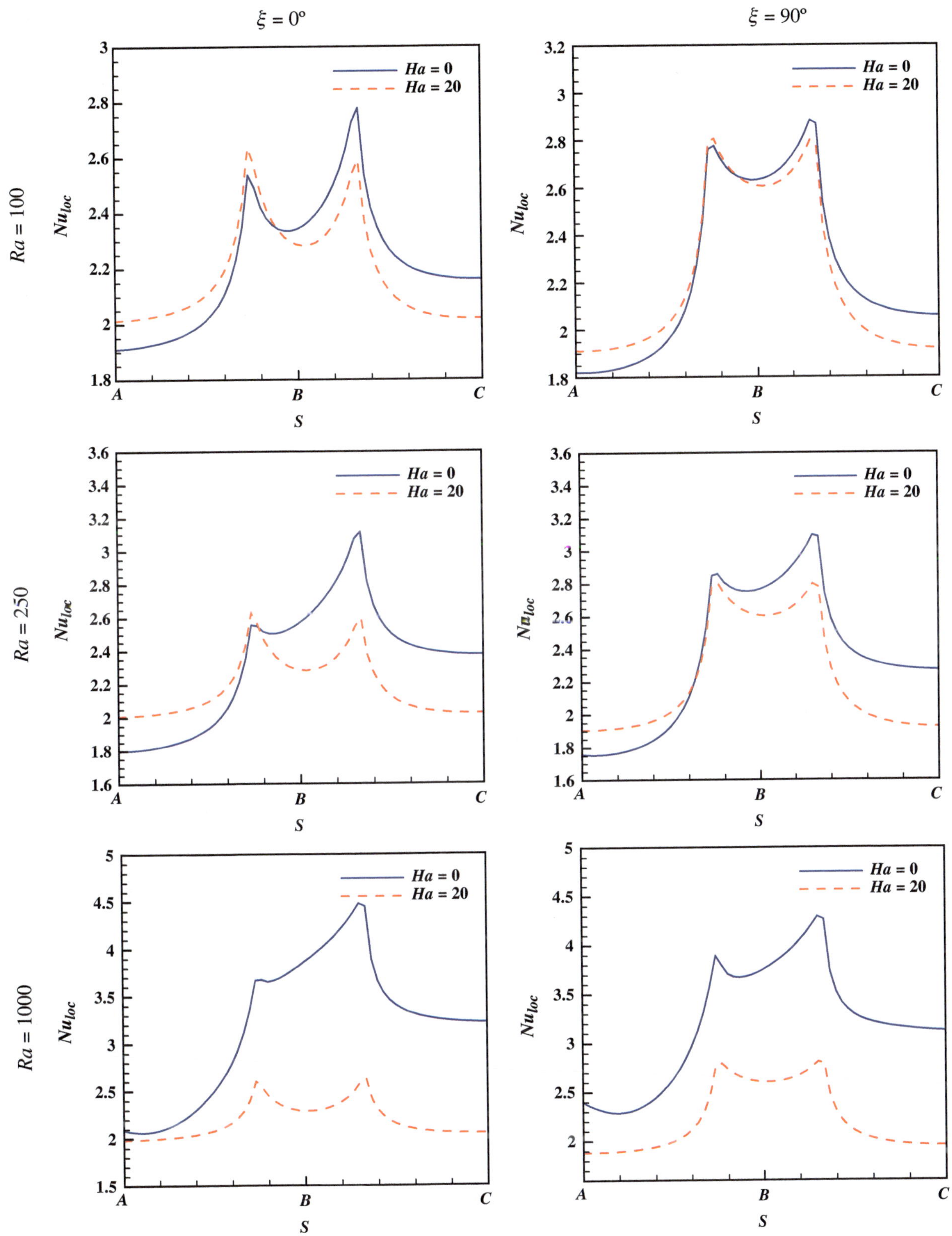

FIGURE 13.12 Effects of the inclination angle, Rayleigh number, and Hartmann number for porous medium on local Nusselt number at $\phi = 0.04$.

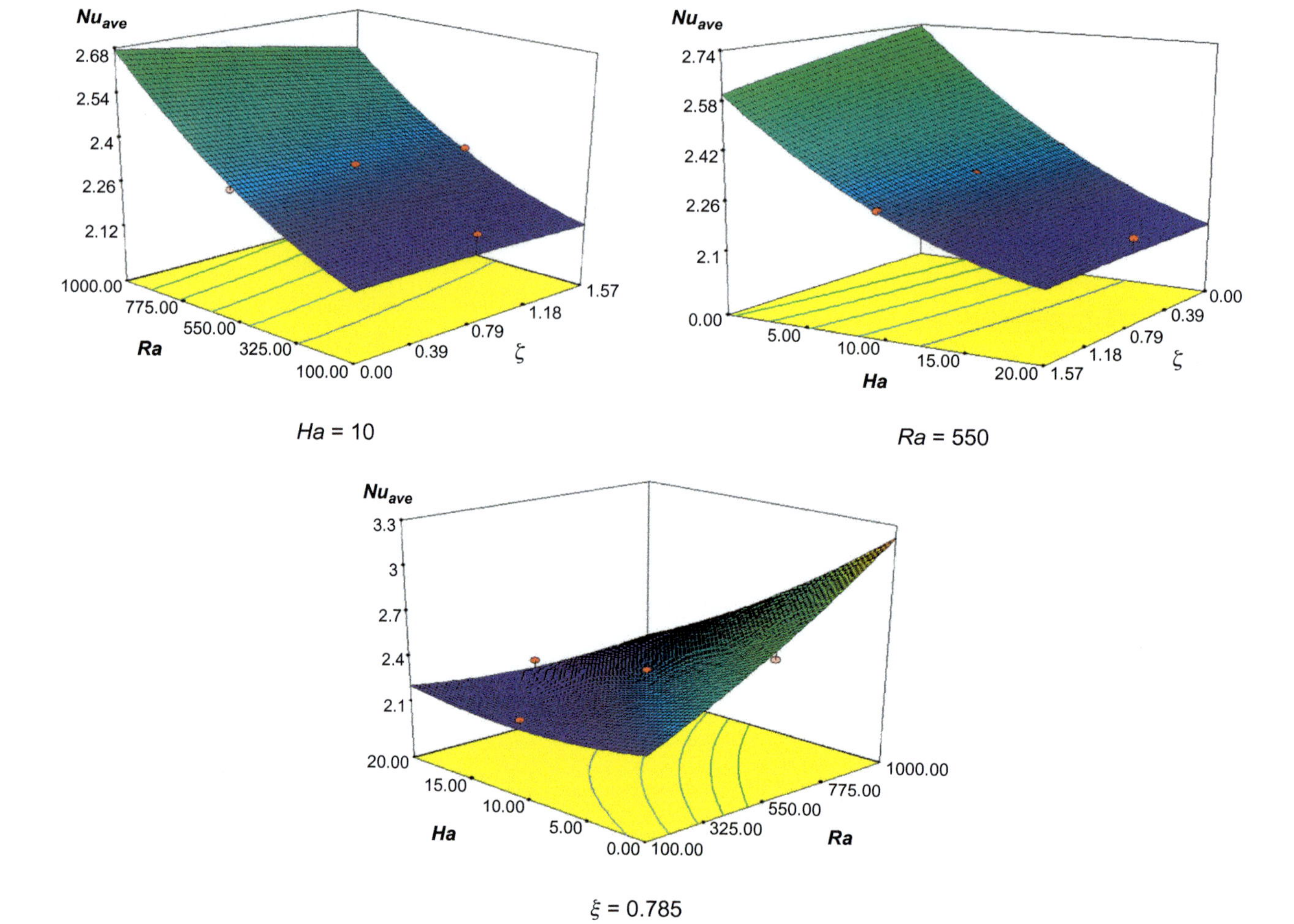

FIGURE 13.13 Effects of the inclination angle, Rayleigh number, and Hartmann number for porous medium on average Nusselt number.

$$(\rho\beta)_{nf} = (\rho\beta)_f(1-\phi) + (\rho\beta)_s\phi \tag{13.60}$$

$$\rho_{nf} = \rho_f(1-\phi) + \rho_s\phi \tag{13.61}$$

$$\frac{\sigma_{nf}}{\sigma_f} = 1 + \frac{3\left(\frac{\sigma_s}{\sigma_f} - 1\right)\phi}{\left(\frac{\sigma_s}{\sigma_f} + 2\right) - \left(\frac{\sigma_s}{\sigma_f} - 1\right)\phi} \tag{13.62}$$

k_{nf}, μ_{nf} are obtained according to the Koo–Kleinstreuer–Li (KKL) model:

$$\begin{aligned}
k_{nf} &= 1 + \frac{3\left(\frac{k_p}{k_f} - 1\right)\phi}{\left(\frac{k_p}{k_f} + 2\right) - \left(\frac{k_p}{k_f} - 1\right)\phi} + 5\times 10^4 g'(\phi, T, d_p)\phi\rho_f c_{p,f}\sqrt{\frac{\kappa_b T}{\rho_p d_p}} \\
g'(\phi, T, d_p) &= \left(a_1 + a_2 Ln(d_p) + a_3 Ln(\phi) + a_4 Ln(\phi)\ln(d_p) + a_5 Ln(d_p)^2\right)Ln(T) \\
&+ \left(a_6 + a_7 Ln(d_p) + a_8 Ln(\phi) + a_9 \ln(d_p)Ln(\phi) + a_{10} Ln(d_p)^2\right) \\
R_f &= d_p/k_{p,eff} - d_p/k_p,\ R_f = 4\times 10^{-8} km^2/W
\end{aligned} \tag{13.63}$$

$$\mu_{nf} = \frac{\mu_f}{(1-\phi)^{2.5}} + \frac{k_{Brownian}}{k_f} \times \frac{\mu_f}{\Pr} \tag{13.64}$$

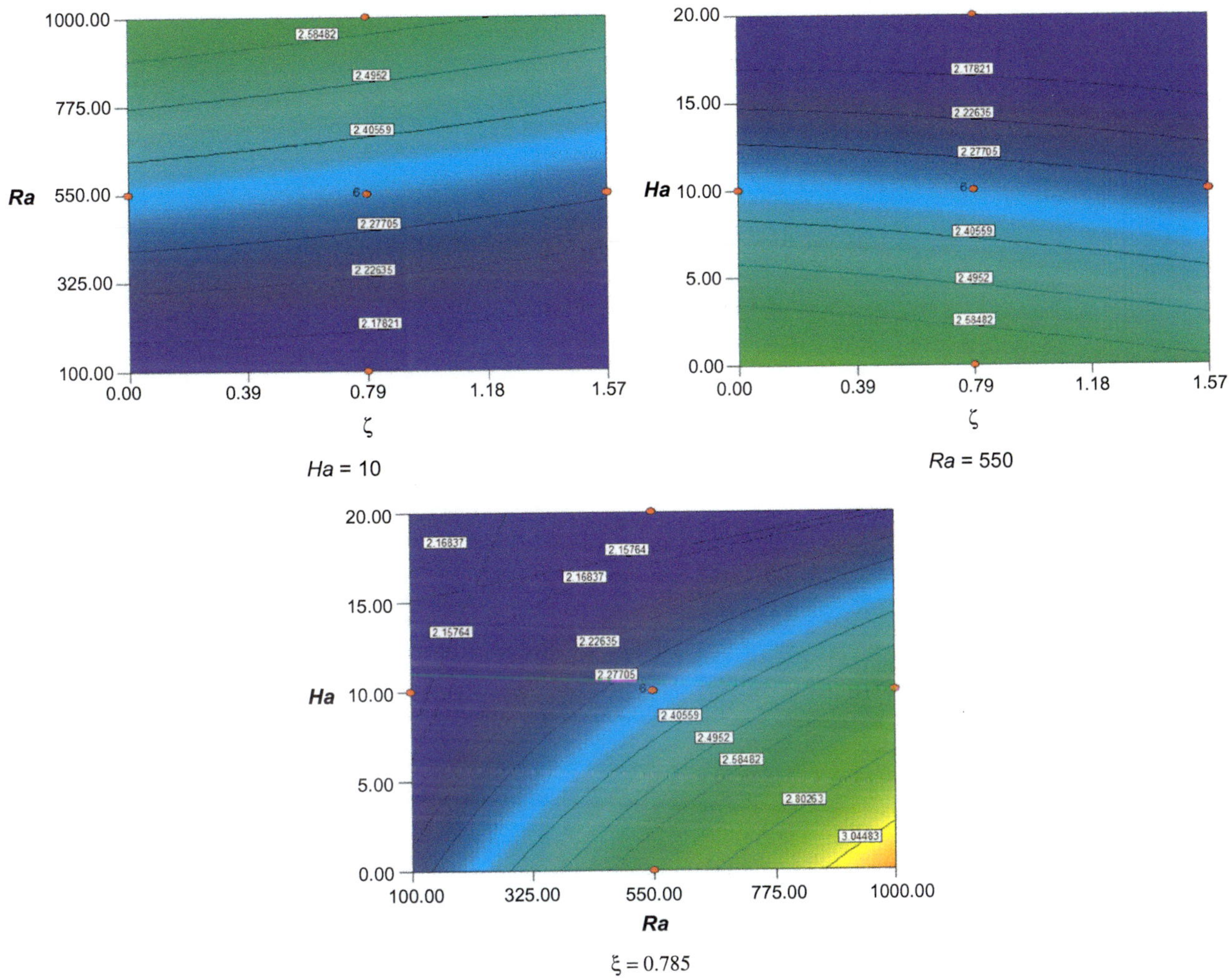

FIGURE 13.13 (Continued).

TABLE 13.4 Effects of ξ, Ha and Ra on Heat Transfer Enhancement

Ra	ξ	*Ha*	*E*
100	0	0	11.44969
100	0	20	12.43305
250	0	0	8.840375
250	0	20	12.41954
1000	0	0	3.717265
1000	0	20	12.31083
100	1.57	0	11.73324
100	1.57	20	12.43413
250	1.57	0	9.087008
250	1.57	20	12.42324
1000	1.57	0	4.142252
1000	1.57	20	12.34446

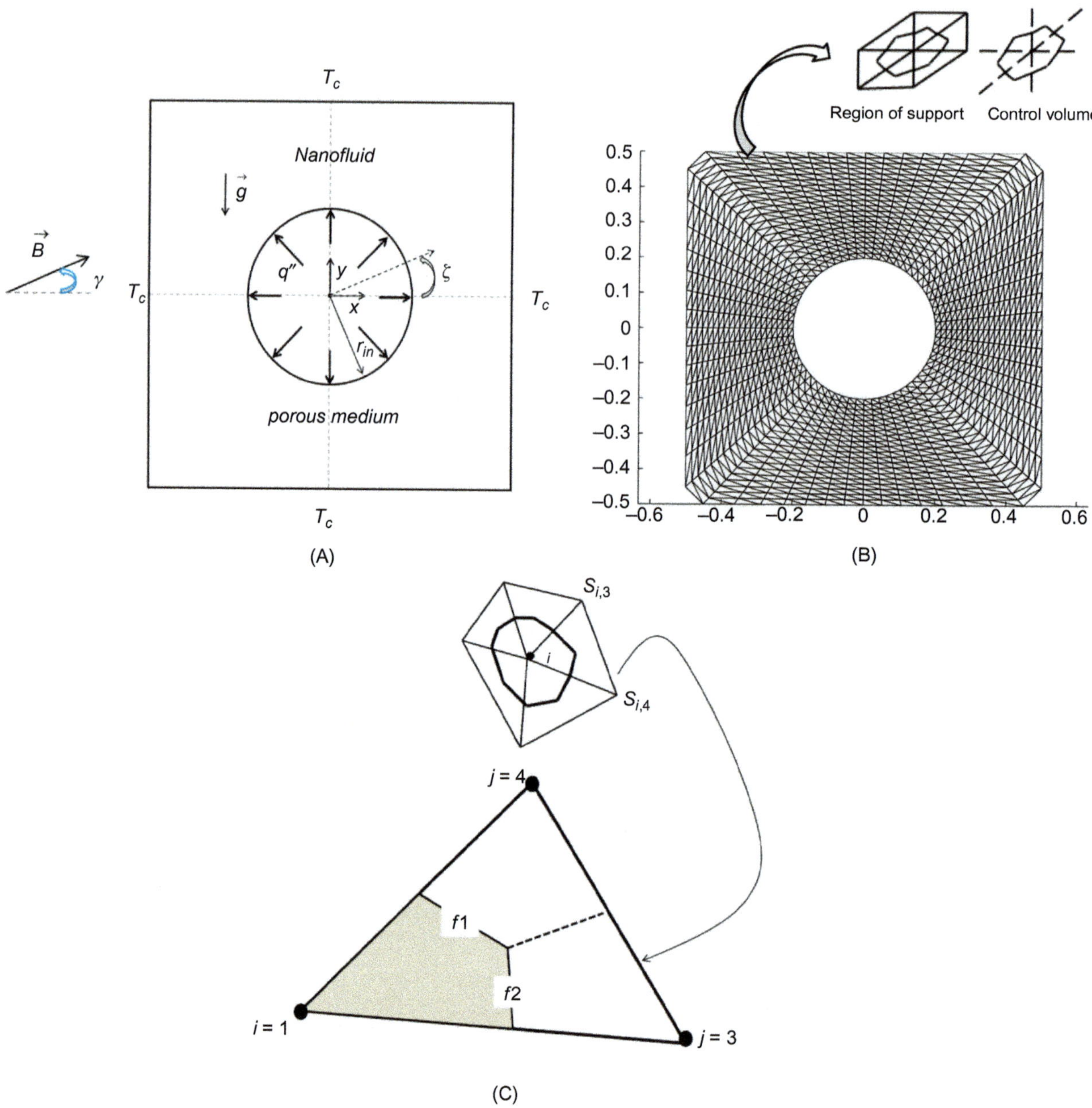

FIGURE 13.14 (A) Geometry and the boundary conditions with (B) the mesh of geometry considered in this work; (C) a sample triangular element and its corresponding control volume.

All required coefficients and properties are illustrated in Tables 13.1 and 13.2. Introducing dimensionless quantities:

$$\Psi = \psi/\alpha_{nf}, \quad (X, Y) = (x, y)/L, \quad \theta = \frac{T - T_c}{T_h - T_c} \tag{13.65}$$

By discarding the pressure, the final equations are:

$$\frac{\partial^2 \Psi}{\partial X^2} + \frac{\partial^2 \Psi}{\partial Y^2} = -\frac{A_6}{A_5} Ha \left[\frac{\partial^2 \Psi}{\partial Y^2} (\sin^2\gamma) + \frac{\partial^2 \Psi}{\partial X^2} (\cos^2\gamma) + 2 \frac{\partial^2 \Psi}{\partial X\, \partial Y} (\sin\gamma)\,(\cos\gamma) \right] - \frac{A_3}{A_4} \frac{A_2}{A_5} \frac{\partial \theta}{\partial X} Ra \tag{13.66}$$

$$\left(\frac{\partial^2\theta}{\partial Y^2}+\frac{\partial^2\theta}{\partial X^2}\right)=\frac{\partial\theta}{\partial X}\frac{\partial\Psi}{\partial Y}-\frac{\partial\theta}{\partial Y}\frac{\partial\Psi}{\partial X} \tag{13.67}$$

where $Ra=\frac{g\,K\,(\rho\beta)_f L\,\Delta T}{\mu_f\,\alpha_f}$ and $Ha=\frac{\sigma_f K\,B_0^2}{\mu_f}$ are the Rayleigh and Hartmann numbers for the porous media. Also $A_i(i=1..6)$ are constants parameters which are obtained as:

$$A_1=\frac{\rho_{nf}}{\rho_f},\quad A_3=\frac{(\rho\beta)_{nf}}{(\rho\beta)_f},\quad A_5=\frac{\mu_{nf}}{\mu_f},$$
$$A_2=\frac{(\rho C_P)_{nf}}{(\rho C_P)_f},A_4=\frac{k_{nf}}{k_f},\quad A_6=\frac{\sigma_{nf}}{\sigma_f} \tag{13.68}$$

and boundary conditions are:

$$\frac{\partial\theta}{\partial n}=1.0 \quad \text{on inner wall}$$
$$\theta=0.0 \quad \text{on other wall} \tag{13.69}$$
$$\Psi=0.0 \quad \text{on all walls}$$

Nu_{local}, Nu_{ave} over the inner cylinder and E can be calculated as:

$$Nu_{loc}=A_4\frac{1}{\theta}\bigg|_{inner\ wall}$$
$$Nu_{ave}=\frac{1}{2\pi}\int_0^{2\pi}Nu_{loc}\,d\zeta \tag{13.70}$$

$$E=\frac{Nu(\phi=0.04)-Nu(basefluid)}{Nu(basefluid)}\times 100 \tag{13.71}$$

13.4.3 Effects of Active Parameters

Effect of Lorentz forces on CuO-water nanofluid flow in a porous semiannulus with constant heat flux is examined using Control Volume-based Finite Element Method. KKL and Darcy's models are utilized to obtain the governing equations. Numerical simulations are presented for various Rayleigh numbers for porous medium ($Ra=10^2, 250$ and 10^3), radius of inner cylinder ($r_{in}=0.2$ to 0.4), CuO-water volume fraction ($\phi=0$ to 0.04), and Hartmann number for porous medium ($Ha=0$ to 20).

Figs. 13.15 and 13.16 illustrate the influence of $r_{in}, Ra,$ and Ha for porous medium on isotherms and streamlines. At low Rayleigh number, isotherms follow the shape of cylinders and one main vortex can be observed in the streamlines. As the Rayleigh number increases, $|\Psi_{\max}|$ increases and the center of the vortex moves upward. Also a thermal plume is generated near $\zeta=90^\circ$. By applying magnetic forces, the center of the vortex returns to the centerline and the thermal plume vanishes. At a high value of r_{in}, there is not enough space for the nanofluid to rotate, so the conduction mode becomes dominant.

Figs. 13.17 and 13.18 illustrate the influence of $\phi, r_{in}, Ra,$ and Ha on Nu_{loc}, Nu_{ave}. The correlation for Nu_{ave} corresponding effective parameters is:

$$\begin{aligned}Nu_{ave}&=2.17+3.7\times10^{-4}Ra+12.85r_{in}+13.42\phi-0.018Ha-4.5\times10^{-4}Ra\,r_{in}\\&-3.82\times10^{-4}Ra\,\phi-1.45\times10^{-5}Ra\,Ha+3.45\,r_{in}\phi+0.04r_{in}\,Ha+0.036\phi\,Ha\\&+5.6\times10^{-8}Ra^2+9.44{r_{in}}^2+27.9\phi^2+3.4710^{-4}Ha^2\end{aligned} \tag{13.72}$$

The number of extremum in the Nu_{loc} profile matches the presence of thermal plumes. The influence of adding a magnetic field is more pronounced at lower values of r_{in}. As buoyancy force increases, the temperature

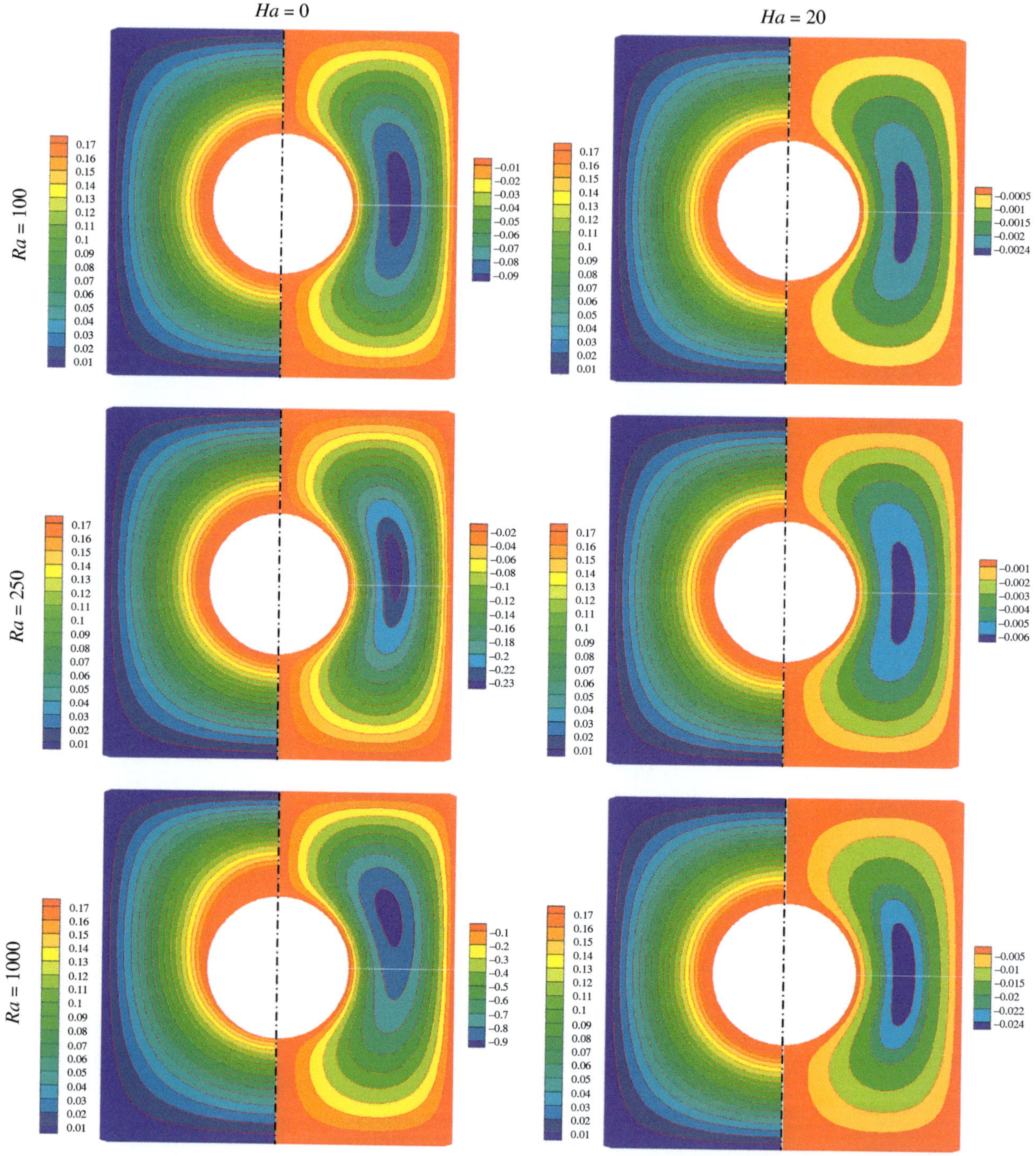

FIGURE 13.15 Isotherm (left) and streamline (right) contours for different values of Rayleigh number and Hartmann number for porous medium when $\phi = 0.04$, $r_{in} = 0.2$.

gradient augments and in turn Nu_{ave} is enhanced with the increase of buoyancy forces. A similar effect is observed for ϕ, r_{in}. Enhancing the Lorentz force causes the nanofluid flow to retard and the Nusselt number reduces.

The impacts of Ra and Ha on heat transfer improvement are depicted in Fig. 13.19. Thermal conductivity has an important role in conduction mode and, in turn, the addition of nanoparticles is more sensible in this condition. According to this fact, heat transfer improvement augments with rise of Lorentz forces and it reduces with the enhancement of buoyancy forces.

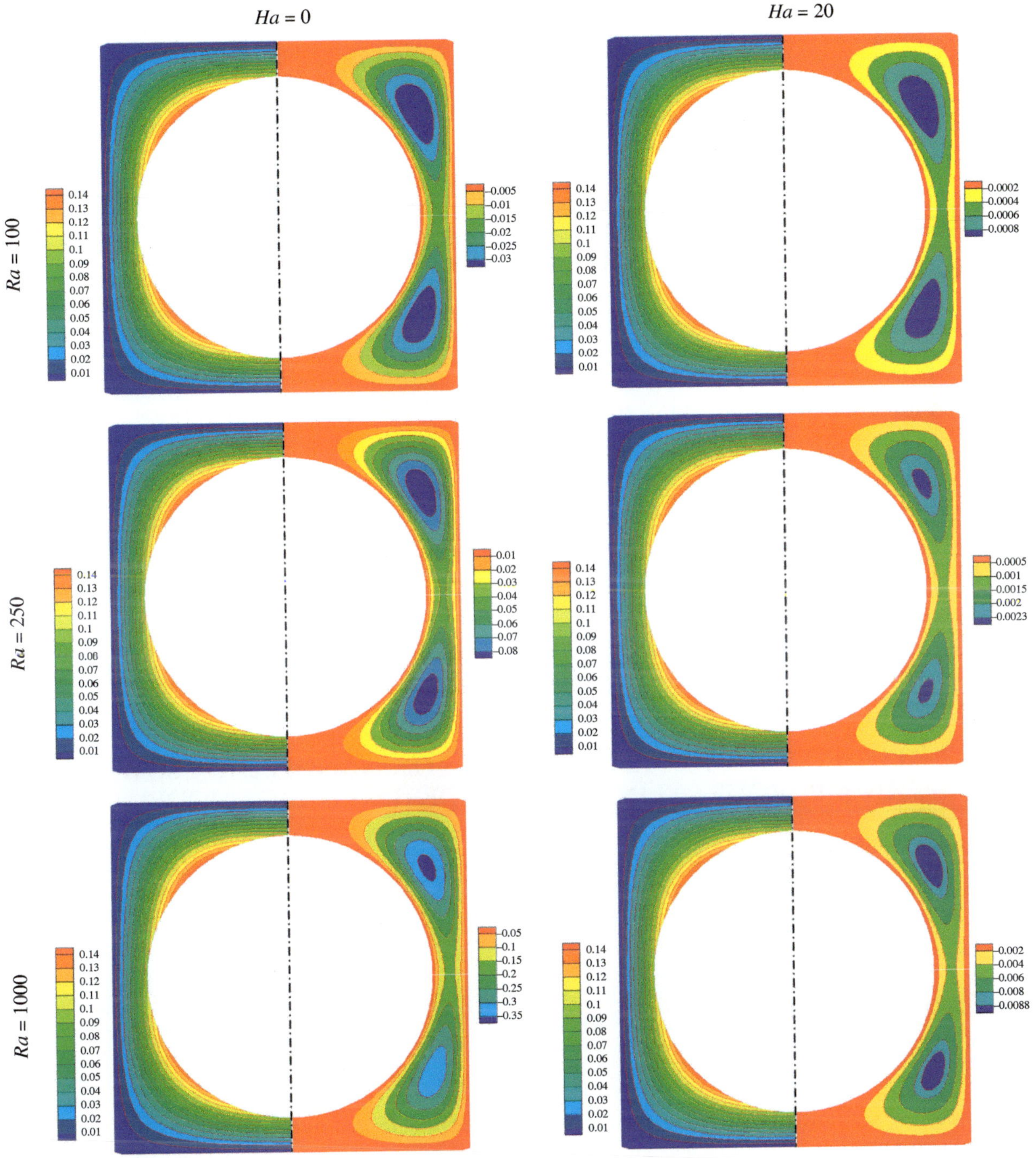

FIGURE 13.16 Isotherm (left) and streamline (right) contours for different values of Rayleigh number and Hartmann number for porous medium when $\phi = 0.04$, $r_{in} = 0.4$.

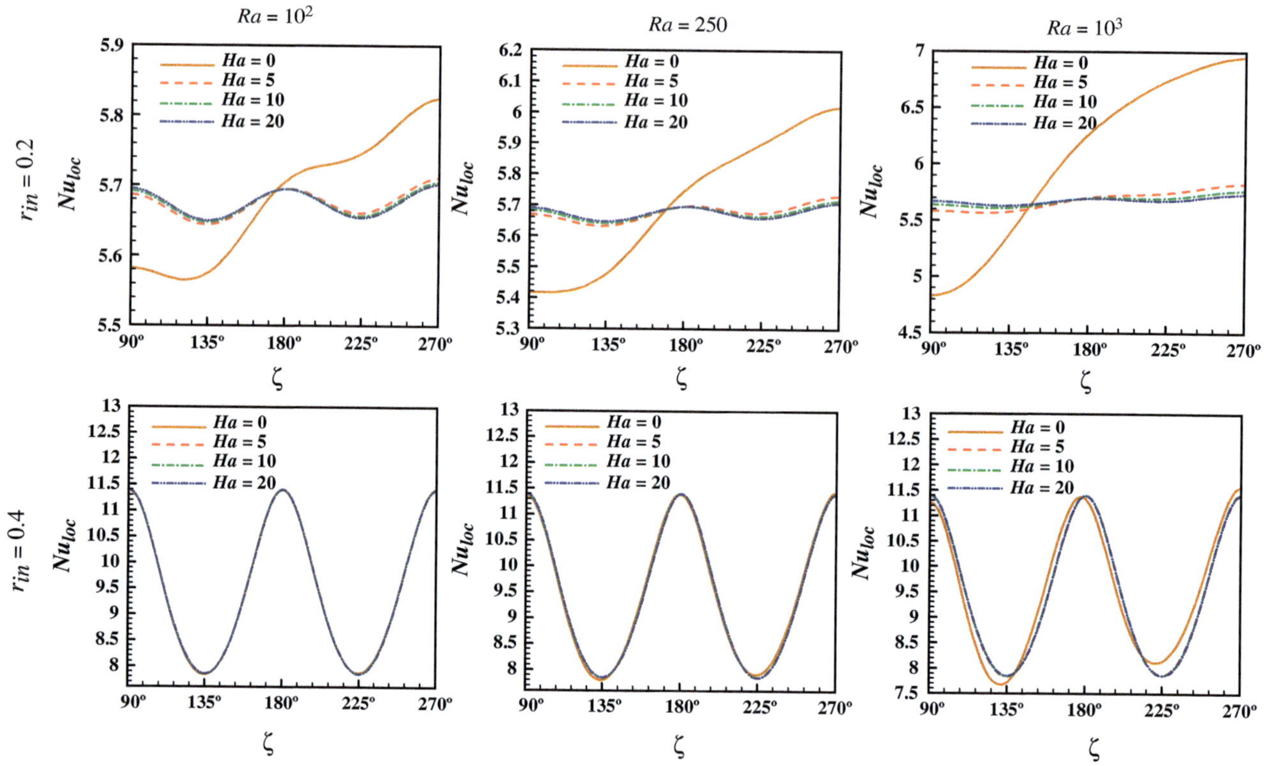

FIGURE 13.17 Effects of r_{in}, Ra and Ha for porous medium on local Nusselt number at $\phi = 0.04$.

13.5 EFFECT OF LORENTZ FORCES ON NANOFLUID FLOW IN A POROUS CYLINDER CONSIDERING DARCY MODEL

13.5.1 Problem Definition

Fig. 13.20 shows the geometry, boundary condition, and sample element. The formula of the inner cylinder is:

$$\left(\frac{X}{a}\right)^{2\hat{n}} + \left(\frac{Y}{b}\right)^{2\hat{n}} = 1 \tag{13.73}$$

where $\hat{n} > 1$ geometry is square when $a = b$.

13.5.2 Governing Equation

2D steady convective flow of nanofluid in a porous media is considered in the presence of a constant horizontal magnetic field. The PDEs equations are:

$$\nabla . \vec{V} = 0 \tag{13.74}$$

$$\frac{\mu_{nf}}{K}\vec{V} = \left(\nabla p + \vec{I} \times \vec{B} + \rho_{nf}\,\vec{g}\,\right) \tag{13.75}$$

$$(\rho C_p)_{nf}\left(\vec{V}.\nabla\right)T = k_{nf}\,\nabla^2 T \tag{13.76}$$

$$\nabla . \vec{I} = 0 \tag{13.77}$$

$$\sigma_{nf}\left(\vec{V} \times \vec{B} - \nabla\varphi\right) = \vec{I} \tag{13.78}$$

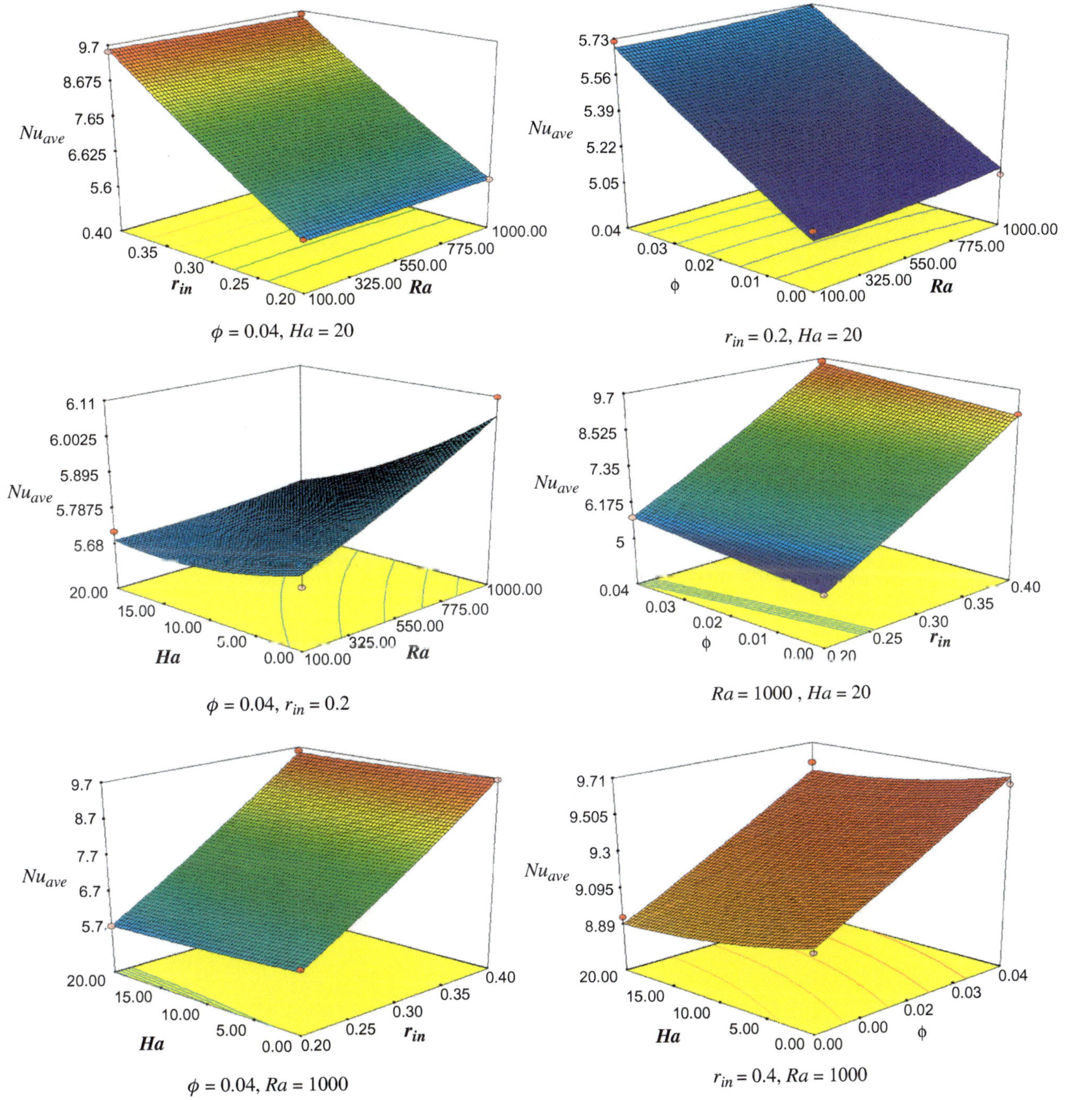

FIGURE 13.18 Effects of ϕ, r_{in}, Ra, and Ha for porous medium on average Nusselt number.

Electric field can be neglected. So the above equations turn into:

$$\frac{\partial v}{\partial y} + \frac{\partial u}{\partial x} = 0 \tag{13.79}$$

$$-\frac{K}{\mu_{nf}}\frac{\partial p}{\partial x} - \frac{\sigma_{nf} K B_0^2}{\mu_{nf}}\left(-u \sin^2\gamma + v \sin\gamma \cos\gamma\right) = u \tag{13.80}$$

$$-\frac{K}{\mu_{nf}}\frac{\partial p}{\partial y} - \frac{\sigma_{nf} K B_0^2}{\mu_{nf}}\left(-v \cos^2\gamma + u \sin\gamma \cos\gamma\right) + g\frac{(\rho\beta)_{nf} K}{\mu_{nf}}(T - T_c) = v \tag{13.81}$$

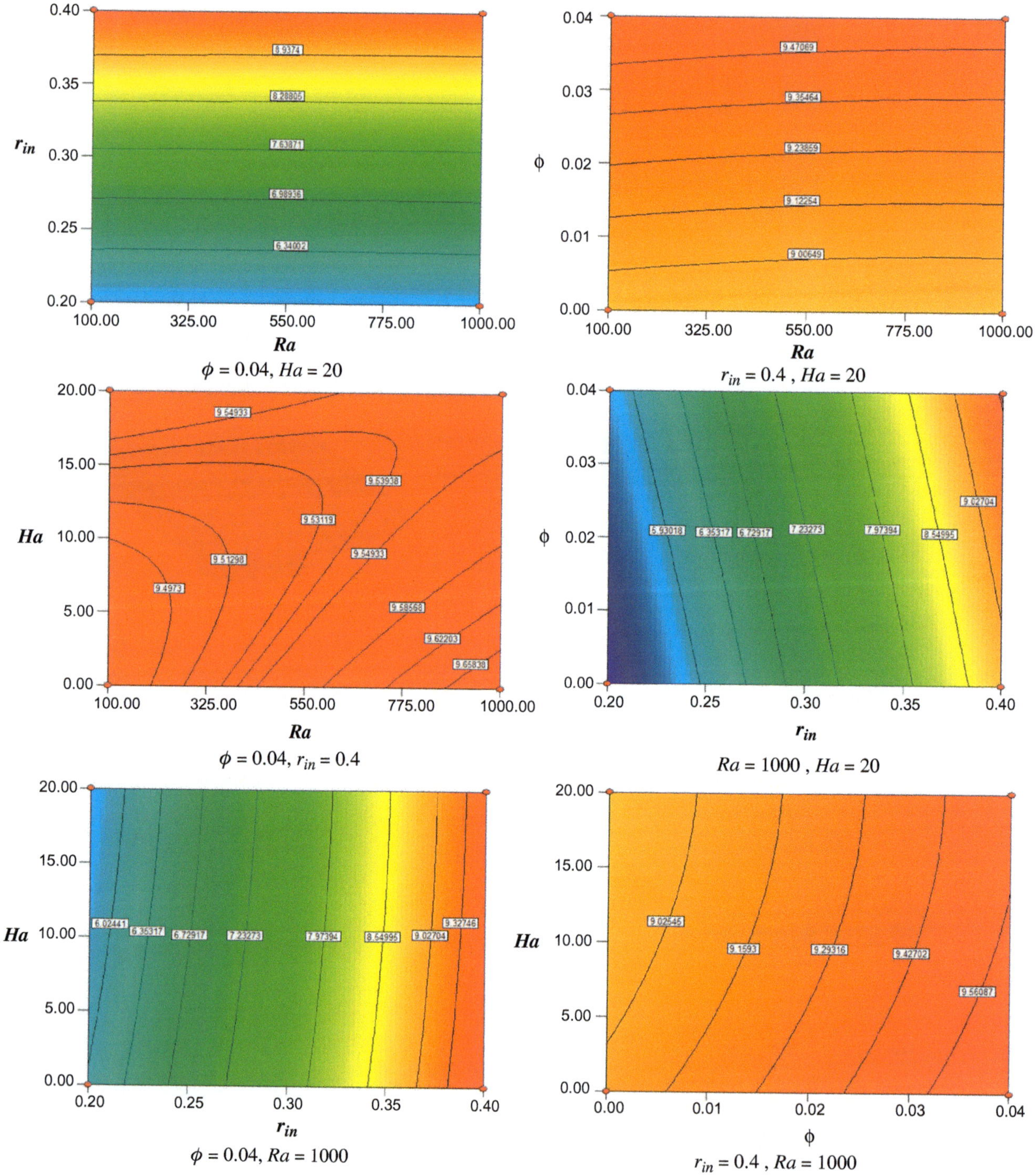

FIGURE 13.18 (Continued).

$$\left(\rho C_p\right)_{nf}\left(u\frac{\partial T}{\partial x}+v\frac{\partial T}{\partial y}\right)=k_{nf}\left(\frac{\partial^2 T}{\partial x^2}+\frac{\partial^2 T}{\partial y^2}\right) \tag{13.82}$$

$\left(\rho C_p\right)_{nf}$, $(\rho\beta)_{nf}$, ρ_{nf}, and σ_{nf} are defined as:

$$\left(\rho C_p\right)_{nf}=\left(\rho C_p\right)_f(1-\phi)+\left(\rho C_p\right)_s\phi \tag{13.83}$$

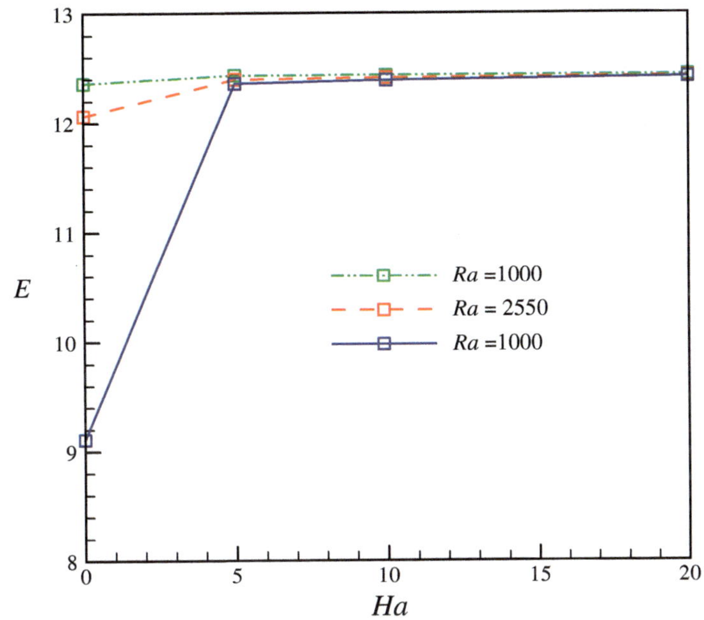

FIGURE 13.19 Effects of the *Ra* and *Ha* for porous medium on the ratio of heat transfer enhancement due to addition of nanoparticles when $r_{in} = 0.2$.

$$(\rho\beta)_{nf} = (\rho\beta)_f(1-\phi) + (\rho\beta)_s\phi \tag{13.84}$$

$$\rho_{nf} = \rho_f(1-\phi) + \rho_s\phi \tag{13.85}$$

$$\frac{\sigma_{nf}}{\sigma_f} = 1 + \frac{3\left(\frac{\sigma_s}{\sigma_f} - 1\right)\phi}{\left(\frac{\sigma_s}{\sigma_f} + 2\right) - \left(\frac{\sigma_s}{\sigma_f} - 1\right)\phi} \tag{13.86}$$

k_{nf}, μ_{nf} are obtained according to the Koo–Kleinstreuer–Li (KKL) model:

$$\begin{gathered} k_{nf} = 1 + \frac{3\left(\frac{k_p}{k_f} - 1\right)\phi}{\left(\frac{k_p}{k_f} + 2\right) - \left(\frac{k_p}{k_f} - 1\right)\phi} + 5\times 10^4 g'(\phi, T, d_p)\phi\rho_f c_{p,f}\sqrt{\frac{\kappa_b T}{\rho_p d_p}} \\ g'(\phi, T, d_p) = \left(a_1 + a_2 Ln(d_p) + a_3 Ln(\phi) + a_4 Ln(\phi)\ln(d_p) + a_5 Ln(d_p)^2\right)Ln(T) \\ + \left(a_6 + a_7 Ln(d_p) + a_8 Ln(\phi) + a_9 \ln(d_p)Ln(\phi) + a_{10} Ln(d_p)^2\right) \\ R_f = d_p/k_{p,eff} - d_p/k_p, R_f = 4\times 10^{-8} km^2/W \end{gathered} \tag{13.87}$$

$$\mu_{nf} = \frac{\mu_f}{(1-\phi)^{2.5}} + \frac{k_{Brownian}}{k_f} \times \frac{\mu_f}{\Pr} \tag{13.88}$$

All required coefficients and properties are illustrated in Tables 13.1 and 13.2.
Introducing dimensionless quantities:

$$\Psi = \psi/\alpha_{nf}, \quad (X, Y) = (x, y)/L, \quad \theta = \frac{T - T_c}{T_h - T_c} \tag{13.89}$$

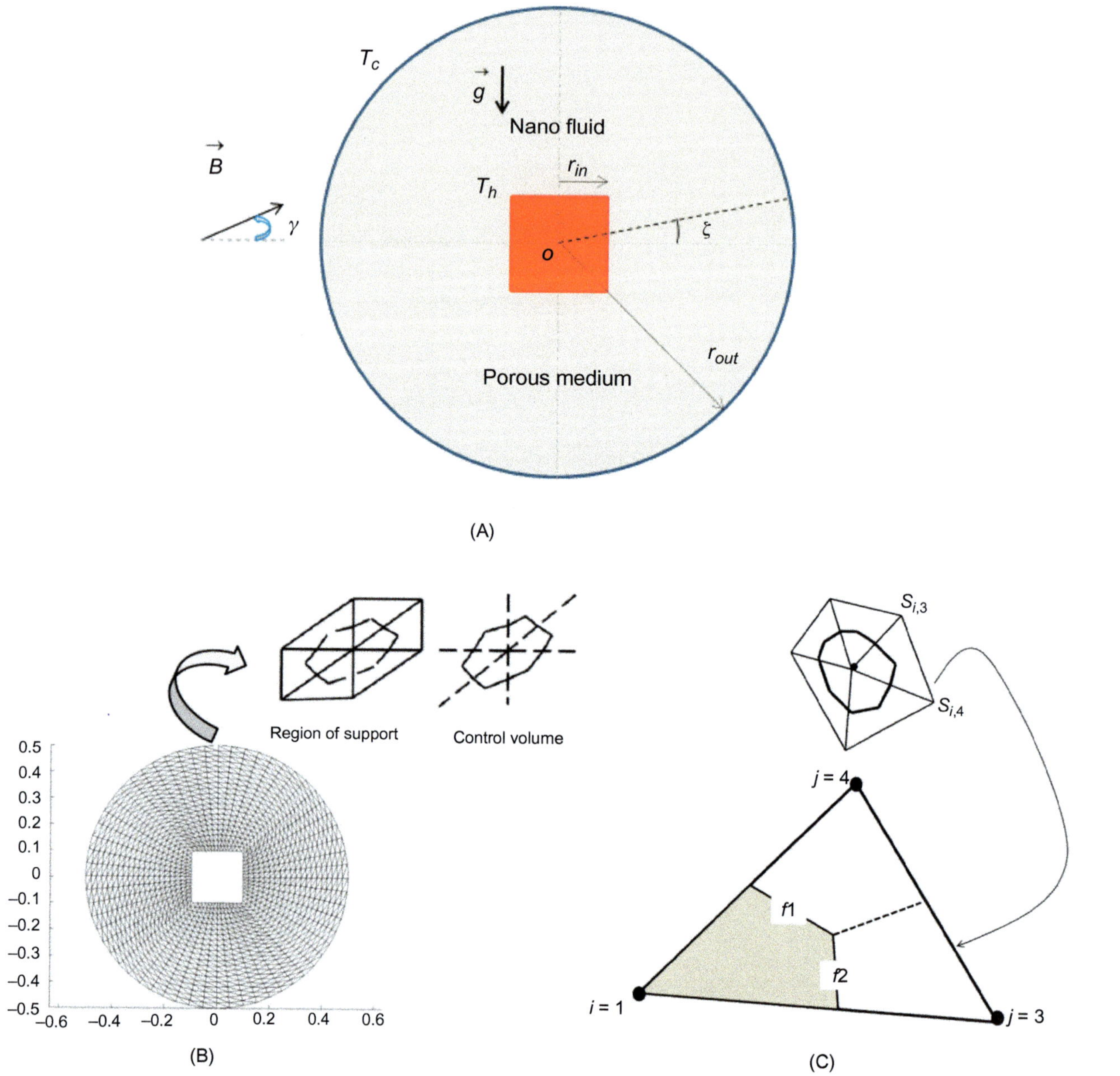

FIGURE 13.20 (A) Geometry and the boundary conditions with (B) the mesh of geometry considered in this work; (C) a sample triangular element and its corresponding control volume.

By discarding the pressure, the final equations are:

$$\frac{\partial^2\Psi}{\partial X^2}+\frac{\partial^2\Psi}{\partial Y^2}=-\frac{A_6}{A_5}Ha\left[\frac{\partial^2\Psi}{\partial Y^2}(\sin^2\gamma)+\frac{\partial^2\Psi}{\partial X^2}(\cos^2\gamma)+2\frac{\partial^2\Psi}{\partial X\,\partial Y}(\sin\gamma)(\cos\gamma)\right]-\frac{A_3}{A_4}\frac{A_2}{A_5}\frac{\partial\theta}{\partial X}Ra \tag{13.90}$$

$$\left(\frac{\partial^2\theta}{\partial Y^2}+\frac{\partial^2\theta}{\partial X^2}\right)=\frac{\partial\theta}{\partial X}\frac{\partial\Psi}{\partial Y}-\frac{\partial\theta}{\partial Y}\frac{\partial\Psi}{\partial X} \tag{13.91}$$

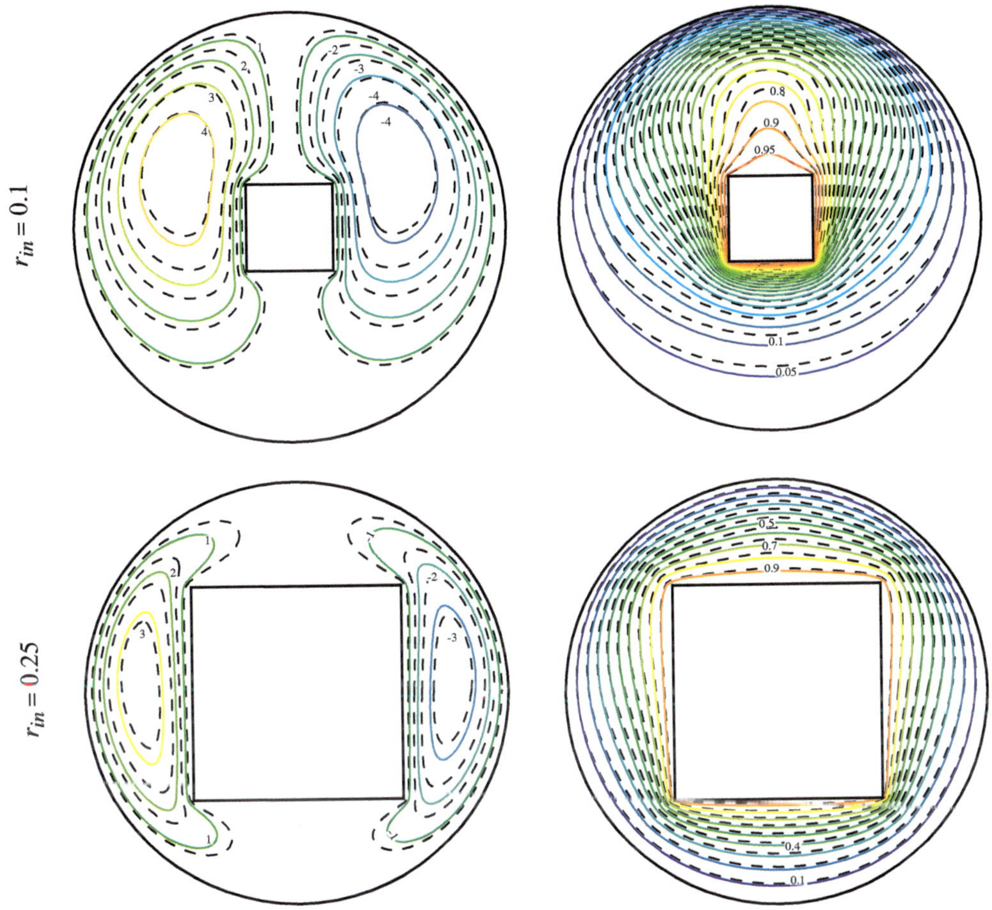

FIGURE 13.21 Comparison of the streamline (left) and isotherm (right) contours between nanofluid ($\phi = 0.04$)(–) and pure fluid ($\phi = 0$) (- - -) when $Ra = 10^3, Ha = 0$

where $Ra = \frac{g\ K\ (\rho\beta)_f L\ \Delta T}{\mu_f\ \alpha_f}$ and $Ha = \frac{\sigma_f K\ B_0^2}{\mu_f}$ are the Rayleigh and Hartmann numbers for the porous media. Also $A_i (i = 1..6)$ are constants parameters which are obtained as:

$$A_1 = \frac{\rho_{nf}}{\rho_f}, \quad A_3 = \frac{(\rho\beta)_{nf}}{(\rho\beta)_f}, \quad A_5 = \frac{\mu_{nf}}{\mu_f},$$
$$A_2 = \frac{(\rho C_P)_{nf}}{(\rho C_P)_f}, A_4 = \frac{k_{nf}}{k_f}, \quad A_6 = \frac{\sigma_{nf}}{\sigma_f} \tag{13.92}$$

and boundary conditions are:

$$\begin{aligned} \theta &= 1.0 \quad \text{on inner wall} \\ \theta &= 0.0 \quad \text{on outer wall} \\ \Psi &= 0.0 \quad \text{on both walls} \end{aligned} \tag{13.93}$$

Nu_{local}, Nu_{ave} over the inner cylinder and E can be calculated as:

$$Nu_{loc} = A_4 \frac{\partial \theta}{\partial r}$$
$$Nu_{ave} = \frac{1}{2\pi} \int_0^{2\pi} Nu_{loc}\ d\zeta \tag{13.94}$$

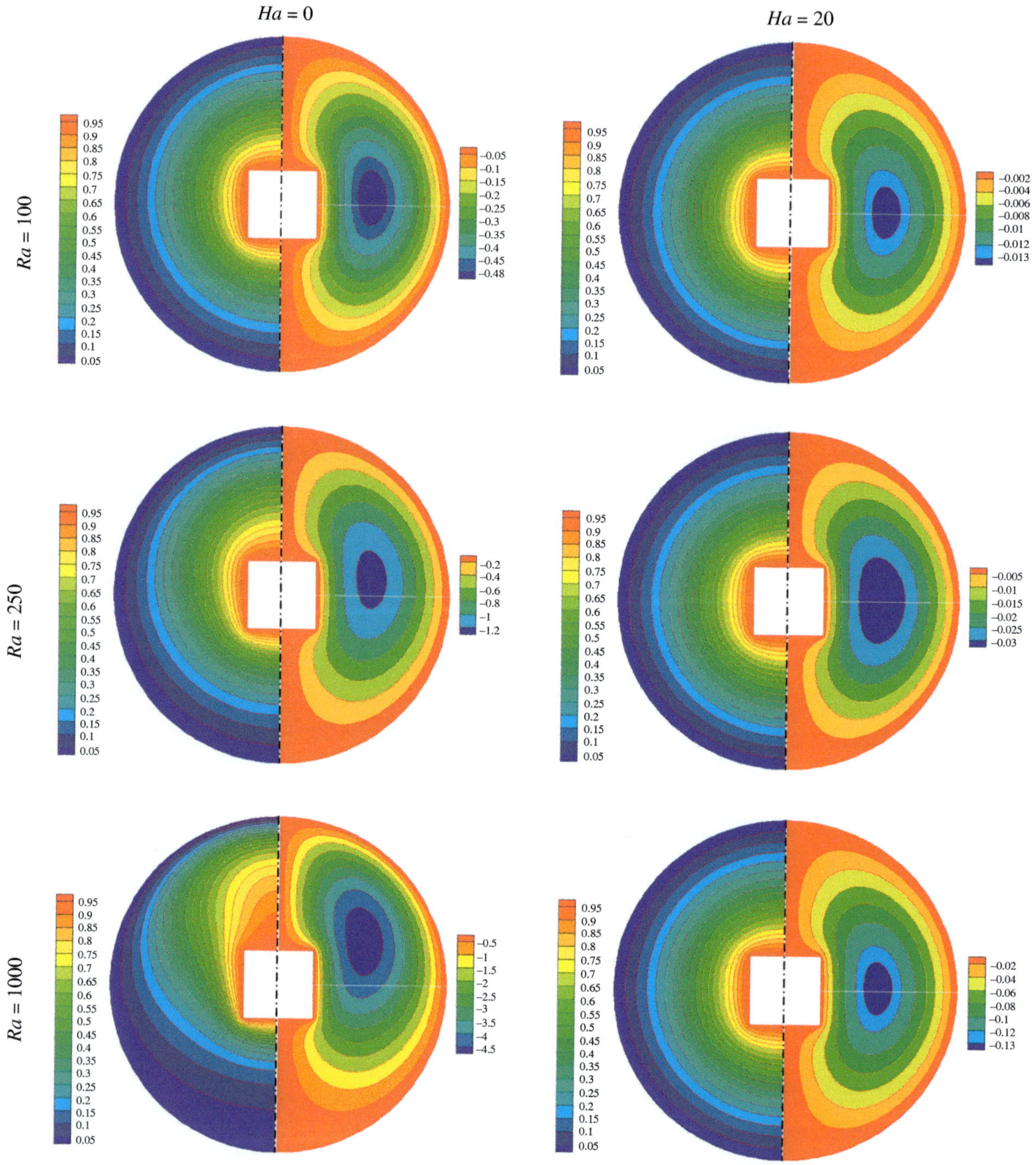

FIGURE 13.22 Isotherms (left) and streamline (right) contours for different values of Rayleigh and Hartmann numbers for porous medium when $\phi = 0.04, r_{in} = 0.1$.

$$E = \frac{Nu(\phi = 0.04) - Nu(basefluid)}{Nu(basefluid)} \times 100 \tag{13.95}$$

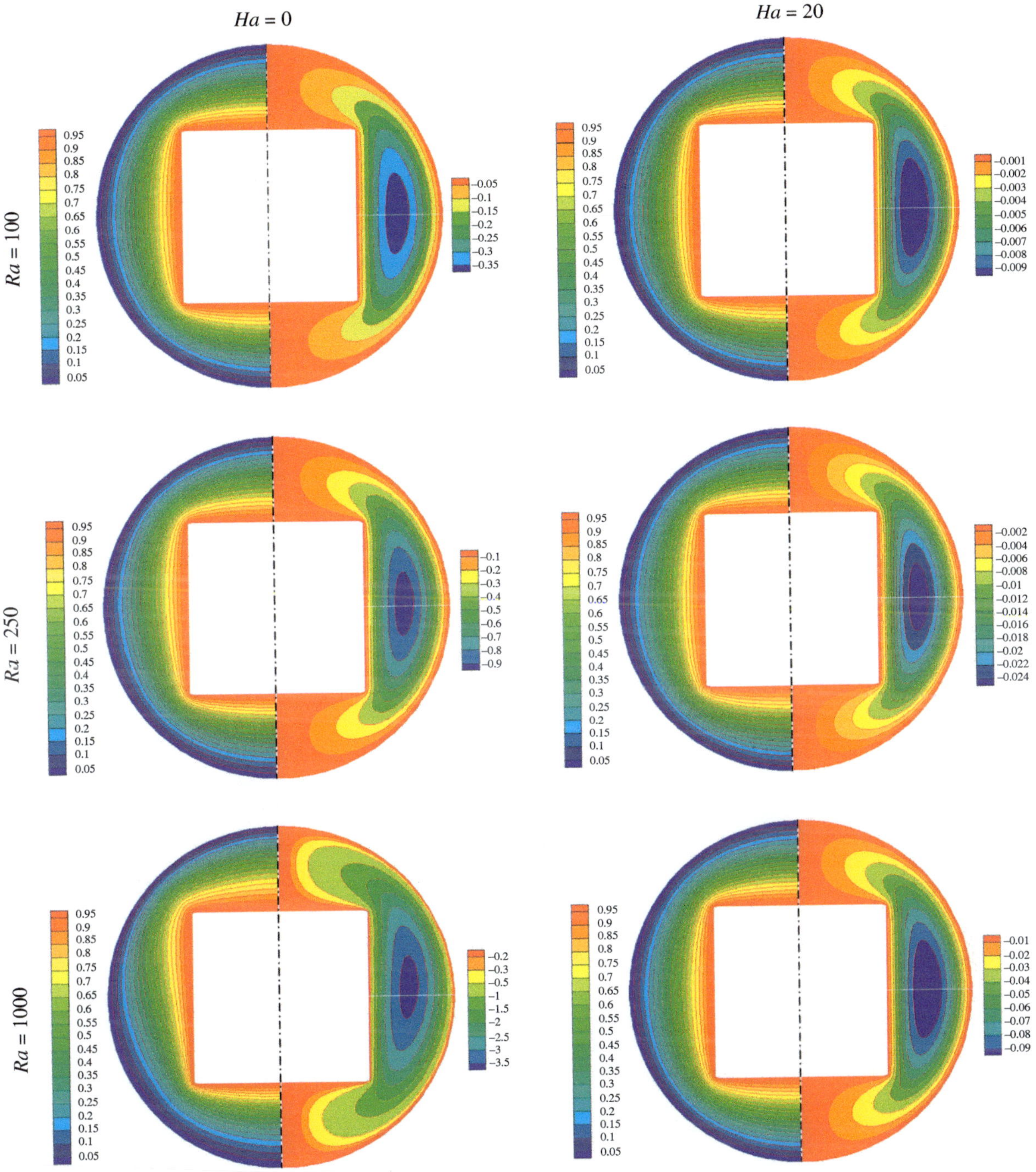

FIGURE 13.23 Isotherm (left) and streamline (right) contours for different values of Rayleigh and Hartmann numbers for porous medium when $\phi = 0.04, r_{in} = 0.25$.

13.5.3 Effects of Active Parameters

The effect of magnetic field on nanofluid natural convection in a circular permeable cylinder with hot square cylinder is studied. The names of the models which are applied for porous media and nanofluid are Darcy and KKL, respectively. The roles of Rayleigh number for porous medium ($Ra = 10^2, 250,$ and 10^3), length of inner cylinder ($r_{in} = 0.1$ to 0.25), CuO-water volume fraction ($\phi = 0$ to 0.04) and Hartmann number for porous medium ($Ha = 0$ to 20) are illustrated as figures.

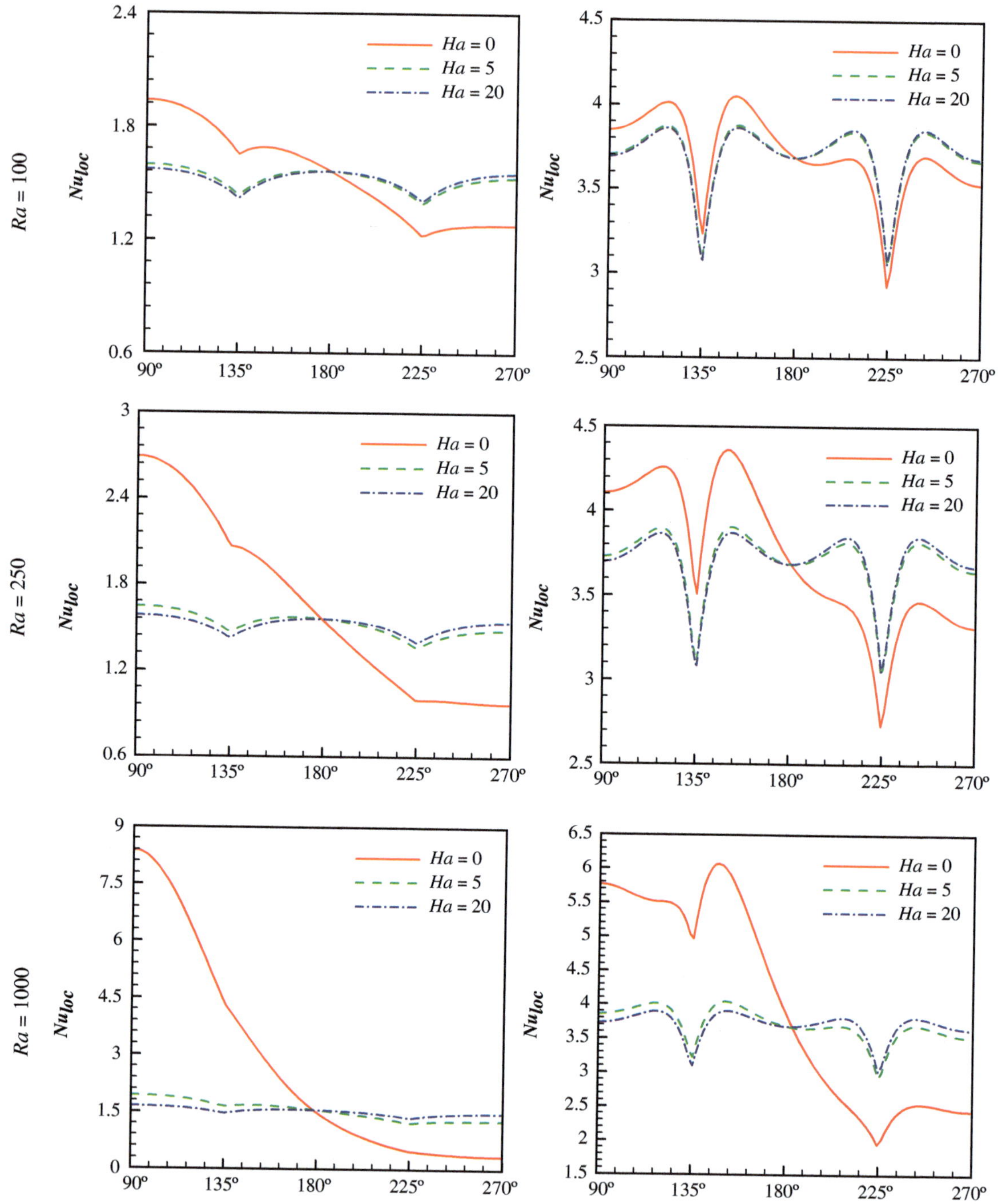

FIGURE 13.24 Effects of Ha, r_{in}, and Ra on local Nusselt number along the cold wall.

Fig. 13.21 depicts the effect of adding nanoparticles on hydrothermal treatment. Isotherms move away from the hot cylinder on the addition of nanoparticles. $|\Psi_{\max}|$ increases with the increase of ϕ. The impacts of r_{in}, Ra, and Ha for porous medium on streamlines and isotherms are shown in Figs. 13.22 and 13.23. At low Ra, one main vortex appears in the streamlines and isotherms follow the shape of cylinders. As Ra augments, a thermal plume generates at $\zeta = 90^\circ$ and the main vortex moves upward. As the Hartmann number augments, the thermal plume diminishes and the strength of the vortex reduces. As r_{in} augments, the main vortex stretches vertically and isotherms become denser near the outer wall. So the temperature gradient is enhanced with increasing r_{in}.

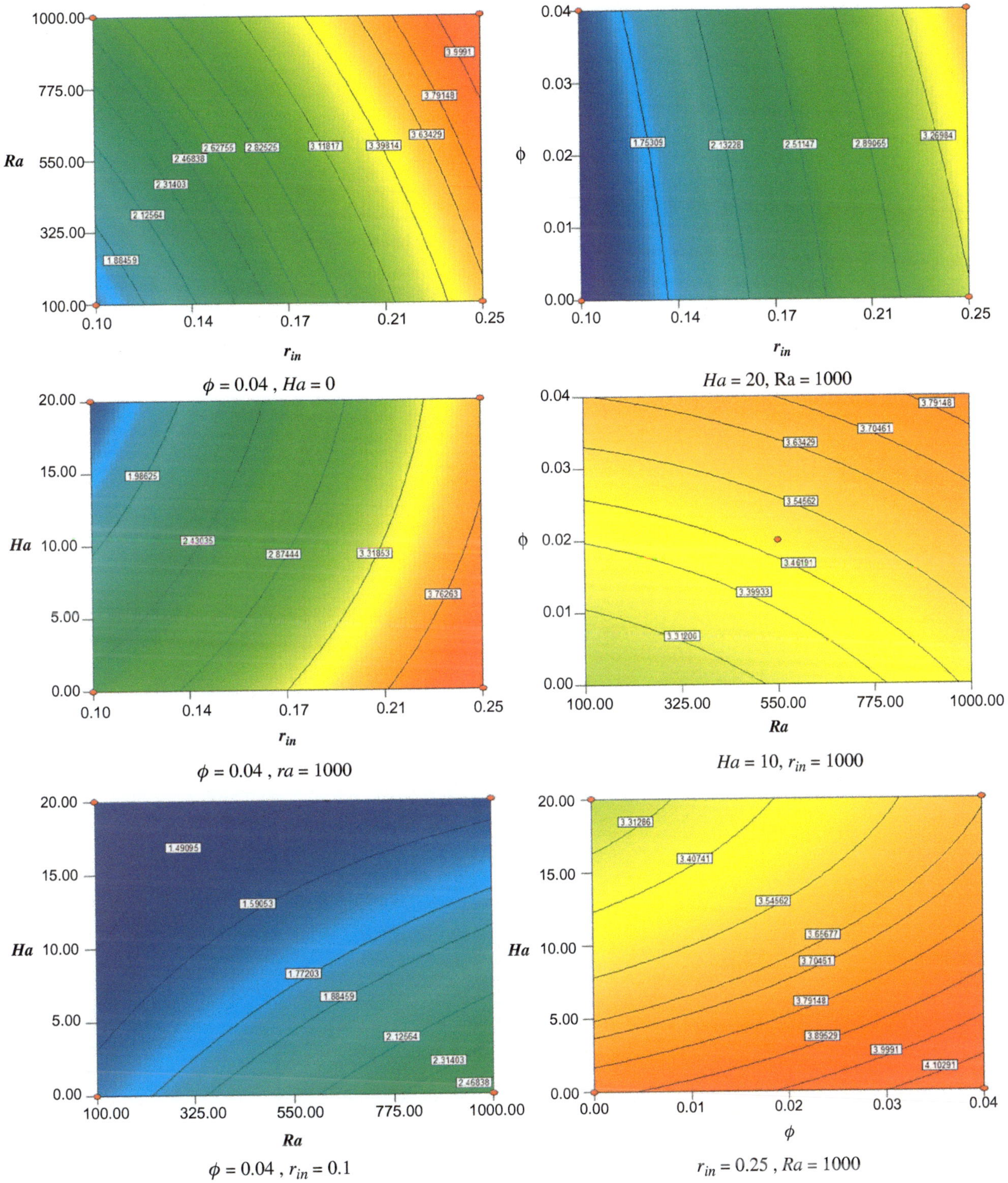

FIGURE 13.25 Effects of ϕ, Ha, r_{in}, and Ra on average Nusselt number along the cold wall.

Figs. 13.24 and 13.25 depict the impact of $\phi, r_{in}, Ra,$ and Ha on Nu_{loc}, Nu_{ave}. Table 13.5 shows the heat transfer improvement for various cases. The correlation for Nu_{ave} corresponding to effective parameters is:

$$\begin{aligned} Nu_{ave} &= +0.52 + 8.18 r_{in} + 1.4 \times 10^{-3} Ra - 3.02\,\phi - 0.04 Ha \\ &- 1.9 \times 10^{-3} r_{in}\ Ra + 38.4\ r_{in}\ \phi + 0.14 r_{in}\ Ha - 1.01 \times 10^{-3} Ra\ \phi - 0.45 \times 10^{-3} Ra\ Ha \\ &+ 0.078\phi\ Ha + 9.4 {r_{in}}^2 + 1.45 \times 10^{-7} Ra^2 + 70.85\phi^2 - 8.6 \times 10^{-4} Ha^2 \end{aligned} \quad (13.96)$$

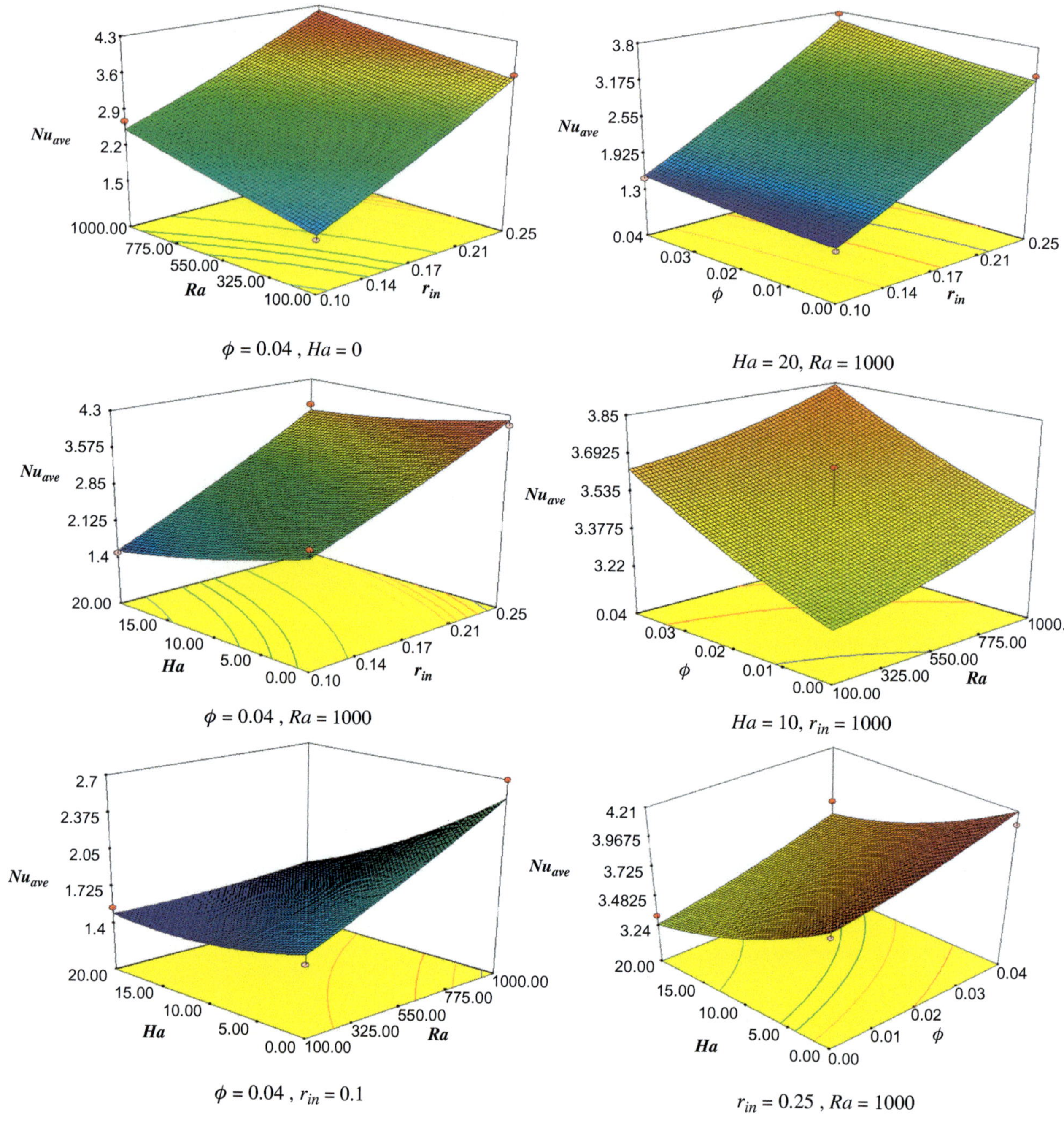

FIGURE 13.25 (Continued).

The number of extremum in the Nu_{loc} profile matches the presence of thermal plumes. As the buoyancy force increases, the temperature gradient augments and in turn Nu_{ave} increases with the increase of buoyancy forces. A similar influence is seen in ϕ, r_{in}. Enhancing the Lorentz force causes the nanofluid flow to retard and the Nusselt number decreases. Heat transfer improvement increase with the rise of r_{in} and Lorentz forces, and it reduces with enhancement of buoyancy forces due to more variations in thermal conductivity.

TABLE 13.5 Effects of Ha, r_{in} and Ra on Heat Transfer Enhancement

r_{in}	Ra	Ha	E
0.1	100	0	11.5852
0.1	100	5	12.42115
0.1	100	20	12.43677
0.1	250	0	8.189697
0.1	250	5	12.36346
0.1	250	20	12.4291
0.1	1000	0	4.420415
0.1	1000	5	11.58128
0.1	1000	20	12.35301
0.25	100	0	12.3712
0.25	100	5	12.43895
0.25	100	20	12.44018
0.25	250	0	12.05567
0.25	250	5	12.43441
0.25	250	20	12.43958
0.25	1000	0	8.280751
0.25	1000	5	12.37197
0.25	1000	20	12.4336

References

[1] P.H. Oosthuizen, D. Naylor, An Introduction to Convection Heat Transfer Analysis, McGraw–Hill, Singapure, 1999.
[2] D.A. Nield, A. Bejan, Convection in Porous Media, third ed, Springer, New York, 2006.
[3] K. Vafai, Handbook of Porous Media, second ed, Taylor and Francis, New York, 2005.
[4] D. Ingham, I. Pop, Transport Phenomena in Porous Media III, Elsevier Science, Oxford, 2005.
[5] M. Sheikholeslami, H.B. Rokni, Magnetohydrodynamic CuO-water nanofluid in a porous complex shaped enclosure, ASME, J. Therm. Sci. Eng. Applicat. 9 (4) (2017) 041007. Available from: https://doi.org/10.1115/1.4035973.
[6] M. Sheikholeslami, A. Zeeshan, Analysis of flow and heat transfer in water based nanofluid due to magnetic field in a porous enclosure with constant heat flux using CVFEM, Comput. Methods Appl. Mech. Eng. 320 (2017) 68–81.
[7] M. Sheikholeslami, S.A. Shehzad, Magnetohydrodynamic nanofluid convection in a porous enclosure considering heat flux boundary condition, Int. J. Heat Mass Transfer 106 (2017) 1261–1269.
[8] M. Sheikholeslami, Influence of Lorentz forces on nanofluid flow in a porous cylinder considering Darcy model, J. Mol. Liq. 225 (2017) 903–912.
[9] M. Sheikholeslami, D.D. Ganji, Transportation of MHD nanofluid free convection in a porous semi annulus using numerical approach, Chem. Phys. Lett. 669 (2017) 202–210.
[10] M. Sheikholeslami, CuO-water nanofluid free convection in a porous cavity considering Darcy law, Eur. Phys. J. Plus 132 (2017) 55. Available from: https://doi.org/10.1140/epjp/i2017-11330-3.
[11] M. Sheikholeslami, Z. Ziabakhsh, D.D. Ganji, Transport of Magnetohydrodynamic nanofluid in a porous media, Colloids Surfaces A: Physicochem. Eng. Asp. 520 (2017) 201–212.
[12] M. Sheikholeslami, CVFEM for magnetic nanofluid convective heat transfer in a porous curved enclosure, Eur. Phys. J. Plus 131 (2016) 413. Available from: https://doi.org/10.1140/epjp/i2016-16413-y.
[13] M. Sheikholeslami, Influence of Coulomb forces on Fe_3O_4-H_2O nanofluid thermal improvement, Int. J. Hyd. Energy 42 (2017) 821–829.
[14] M. Sheikholeslami, Magnetic field influence on nanofluid thermal radiation in a cavity with tilted elliptic inner cylinder, J. Mol. Liquids 229 (2017) 137–147.
[15] M. Sheikholeslami, Numerical simulation of magnetic nanofluid natural convection in porous media, Phys. Lett. A 381 (2017) 494–503.
[16] M. Sheikholeslami, Magnetic source impact on nanofluid heat transfer using CVFEM, Neural Comput. Applicat. (2016). Available from: https://doi.org/10.1007/s00521-016-2740-7.
[17] M. Sheikholeslami, Numerical investigation of MHD nanofluid free convective heat transfer in a porous tilted enclosure, Eng. Computat. 34 (6) (2017) 1939–1955.
[18] M.S. Kandelousi, Effect of spatially variable magnetic field on ferrofluid flow and heat transfer considering constant heat flux boundary condition, Europ. Phys. J. Plus (2014) 129–248. Available from: https://doi.org/10.1140/epjp/i2014-14248-2.

[19] M. Sheikholeslami, Numerical simulation for external magnetic field influence on Fe_3O_4-water nanofluid forced convection, Engineering Computations, Inpress.
[20] M. Sheikholeslami, Influence of Lorentz forces on nanofluid flow in a porous cavity by means of Non- Darcy model, Eng. Comput. 34 (8) (2017) 2651–2667. Available from: https://doi.org/10.1108/EC-01-2017-0008.
[21] M. Sheikholeslami, H.B. Rokni, Nanofluid convective heat transfer intensification in a porous circular cylinder, Chem. Eng. Proc. Process Intensif. 120 (2017) 93–104.
[22] M. Sheikholeslami, M.K. Sadoughi, Numerical modeling for Fe_3O_4 -water nanofluid flow in porous medium considering MFD viscosity, J. Mol. Liq. 242 (2017) 255–264.
[23] M. Sheikholeslami, M. Shamlooei, Magnetic source influence on nanofluid flow in porous medium considering shape factor effect, Phys. Lett. A 381 (2017) 3071–3078.
[24] M. Sheikholeslami, M. Shamlooei, Convective flow of nanofluid inside a lid driven porous cavity using CVFEM, Physica B: Condensed Matter 521 (2017) 239–250.
[25] M. Sheikholeslami, K. Vajravelu, Forced convection heat transfer in Fe_3O_4- ethylene glycol nanofluid under the influence of Coulomb force, J. Mol. Liq. 233 (2017) 203–210.
[26] M. Sheikholeslami, H.B. Rokni, Influence of EFD viscosity on nanofluid forced convection in a cavity with sinusoidal wall, J. Mol. Liq. 232 (2017) 390–395.
[27] M. Sheikholeslami, H.B. Rokni, Magnetic nanofluid natural convection in the presence of thermal radiation considering variable viscosity, Europ. Phys. J. Plus 132 (2017) 238. Available from: https://doi.org/10.1140/epjp/i2017-11498-4.
[28] M. Sheikholeslami, H.B. Rokni, Numerical simulation for impact of Coulomb force on nanofluid heat transfer in a porous enclosure in presence of thermal radiation, Int. J. Heat Mass Transfer 118 (2018) 823–831.
[29] M. Sheikholeslami, M. Shamlooei, R. Moradi, Fe_3O_4- Ethylene glycol nanofluid forced convection inside a porous enclosure in existence of Coulomb force, J. Mol. Liq. 249 (2018) 429–437.
[30] M. Sheikholeslami, S.A. Shehzad, Numerical analysis of Fe_3O_4–H_2O nanofluid flow in permeable media under the effect of external magnetic source, Int. J. Heat Mass Transfer 118 (2018) 182–192.
[31] M. Sheikholeslami, S.A. Shehzad, Thermal radiation of ferrofluid in existence of Lorentz forces considering variable viscosity, Int. J. Heat Mass Transfer 109 (2017) 82–92.
[32] M. Sheikholeslami, H.B. Rokni, Numerical modeling of nanofluid natural convection in a semi annulus in existence of Lorentz force, Comput. Methods Appl. Mech. Eng. 317 (2017) 419–430.
[33] M. Sheikholeslami, K. Vajravelu, Nanofluid flow and heat transfer in a cavity with variable magnetic field, Appl. Math. Comput. 298 (2017) 272–282.
[34] M. Sheikholeslami, H.B. Rokni, Nanofluid two phase model analysis in existence of induced magnetic field, Int. J. Heat Mass Transfer 107 (2017) 288–299.
[35] M. Sheikholeslami, T. Hayat, A. Alsaedi, S. Abelman, Numerical analysis of EHD nanofluid force convective heat transfer considering electric field dependent viscosity, Int. J. Heat Mass Transfer 108 (2017) 2558–2565.
[36] M. Sheikholeslami, Magnetic field influence on CuO-H_2O nanofluid convective flow in a permeable cavity considering various shapes for nanoparticles, Int. J. Hydrogen Energy 42 (2017) 19611–19621.
[37] M. Sheikholeslami, M.M. Bhatti, Forced convection of nanofluid in presence of constant magnetic field considering shape effects of nanoparticles, Int. J. Heat Mass Transfer 111 (2017) 1039–1049.
[38] M. Sheikholeslami, M. Shamlooei, Fe_3O_4-H_2O nanofluid natural convection in presence of thermal radiation, Int. J. Hydrogen Energy 42 (9) (2017) 5708–5718.
[39] M. Sheikholeslami, M.M. Bhatti, Active method for nanofluid heat transfer enhancement by means of EHD, Int. J. Heat Mass Transfer 109 (2017) 115–122.
[40] M. Sheikholeslami, T. Hayat, A. Alsaedi, Numerical study for external magnetic source influence on water based nanofluid convective heat transfer, Int. J. Heat Mass Transfer 106 (2017) 745–755.
[41] M. Sheikholeslami, A.J. Chamkha, Flow and convective heat transfer of a ferro-nanofluid in a double-sided lid-driven cavity with a wavy wall in the presence of a variable magnetic field, Numer. Heat Transfer, Part A 69 (10) (2016) 1186–1200. Available from: https://doi.org/10.1080/10407782.2015.1125709.
[42] M. Sheikholeslami, A.J. Chamkha, Electrohydrodynamic free convection heat transfer of a nanofluid in a semi-annulus enclosure with a sinusoidal wall, Numer. Heat Transfer, Part A 69 (7) (2016) 781–793. https://doi.org/10.1080/10407782.2015.1090819.
[43] M. Sheikholeslami, M.K. Sadoughi, Simulation of CuO- water nanofluid heat transfer enhancement in presence of melting surface, Int. J. Heat Mass Transfer 116 (2018) 909–919.
[44] M. Sheikholeslami, H.B. Rokni, Simulation of nanofluid heat transfer in presence of magnetic field: a review, Int. J. Heat Mass Transfer 115 (2017) 1203–1233.
[45] M. Sheikholeslami, H.B. Rokni, Melting heat transfer influence on nanofluid flow inside a cavity in existence of magnetic field, Int. J. Heat Mass Transfer 114 (2017) 517–526.
[46] M. Sheikholeslami, S.A. Shehzad, Magnetohydrodynamic nanofluid convective flow in a porous enclosure by means of LBM, Int. J. Heat Mass Transfer 113 (2017) 796–805.
[47] M. Sheikholeslami, Magnetohydrodynamic nanofluid forced convection in a porous lid driven cubic cavity using Lattice Boltzmann Method, J. Mol. Liq. 231 (2017) 555–565.
[48] M. Sheikholeslami Kandelousi, KKL correlation for simulation of nanofluid flow and heat transfer in a permeable channel, Phys. Lett. A 378 (45) (2014) 3331–3339.
[49] M. Sheikholeslami, Effect of uniform suction on nanofluid flow and heat transfer over a cylinder, J. Braz. Soc. Mech. Sci. Eng. 37 (2015) 1623–1633.
[50] M. Sheikholeslami, Lattice Boltzmann Method simulation of MHD non-Darcy nanofluid free convection, Phys. B 516 (2017) 55–71.

[51] M. Sheikholeslami, Influence of magnetic field on nanofluid free convection in an open porous cavity by means of Lattice Boltzmann Method, J. Mol. Liq. 234 (2017) 364–374.
[52] M. Sheikholeslami, A.J. Chamkha, Influence of Lorentz forces on nanofluid forced convection considering Marangoni convection, J. Mol. Liq. 225 (2017) 750–757.
[53] M. Sheikholeslami, T. Hayat, A. Alsaedi, Numerical simulation of nanofluid forced convection heat transfer improvement in existence of magnetic field using Lattice Boltzmann Method, Int. J. Heat Mass Transfer 108 (2017) 1870–1883.
[54] M. Sheikholeslami, R. Ellahi, Three dimensional mesoscopic simulation of magnetic field effect on natural convection of nanofluid, Int. J. Heat Mass Transfer 89 (2015) 799–808.
[55] M. Sheikholeslami, S. Abelman, Two phase simulation of nanofluid flow and heat transfer in an annulus in the presence of an axial magnetic field, IEEE Trans. Nanotechnol. 14 (3) (2015) 561–569.
[56] M. Sheikholeslami, R. Ellahi, Simulation of ferrofluid flow for magnetic drug targeting using Lattice Boltzmann method, J. Zeitschrift Fur Naturforschung A 70 (2) (2015) 115–124.
[57] M. Sheikholeslami, M. Seyednezhad, Lattice Boltzmann Method simulation for CuO-water nanofluid flow in a porous enclosure with hot obstacle, J. Mol. Liq. 243 (2017) 249–256.
[58] M. Sheikholeslami, M. Sadoughi, Mesoscopic method for MHD nanofluid flow inside a porous cavity considering various shapes of nanoparticles, Int. J. Heat Mass Transfer 113 (2017) 106–114.
[59] M. Sheikholeslami, A. Zeeshan, Mesoscopic simulation of CuO-H_2O nanofluid in a porous enclosure with elliptic heat source, Int. J. Hydrogen Energy 42 (22) (2017) 15393–15402.
[60] M. Sheikholeslami, S.A. Shehzad, CVFEM for influence of external magnetic source on Fe_3O_4–H_2O nanofluid behavior in a permeable cavity considering shape effect, Int. J. Heat Mass Transfer 115 (2017) 180–191.
[61] M. Sheikholeslami, M. Seyednezhad, Nanofluid heat transfer in a permeable enclosure in presence of variable magnetic field by means of CVFEM, Int. J. Heat Mass Transfer 114 (2017) 1169–1180.
[62] M. Sheikholeslami, H.B. Rokni, Free convection of CuO-H_2O nanofluid in a curved porous enclosure using mesoscopic approach, Int. J. Hydrogen Energy 42 (22) (2017) 14942–14949.
[63] M. Sheikholeslami, Numerical investigation for CuO-H_2O nanofluid flow in a porous channel with magnetic field using mesoscopic method, J. Mol. Liq. 249 (2018) 739–746.
[64] M. Sheikholeslami, H.B. Rokni, Influence of melting surface on MHD nanofluid flow by means of two phase model, Chin. J. Phys. 55 (2017) 1352–1360.
[65] M. Sheikholeslami, H.B. Rokni, Effect of melting heat transfer on nanofluid flow in existence of magnetic field considering Buongiorno Model, Chin. J. Phys. 55 (2017) 1115–1126.
[66] M. Sheikholeslami, T. Hayat, A. Alsaedi, MHD free convection of Al_2O_3–water nanofluid considering thermal radiation: a numerical study, Int. J. Heat Mass Transfer 96 (2016) 513–524.
[67] M. Sheikholeslami, K. Vajravelu, M.M. Rashidi, Forced convection heat transfer in a semi annulus under the influence of a variable magnetic field, Int. J. Heat Mass Transfer 92 (2016) 339–348.
[68] M. Sheikholeslami, Numerical simulation for solidification in a LHTESS by means of nano-enhanced PCM, J. Taiwan Inst. Chem. Eng. 86 (2018) 25–41.
[69] M. Sheikholeslami, Numerical modeling of nano enhanced PCM solidification in an enclosure with metallic fin, J. Mol. Liq. 259 (2018) 424–438.
[70] M. Sheikholeslami, A. Ghasemi, Solidification heat transfer of nanofluid in existence of thermal radiation by means of FEM, Int. J. Heat Mass Transfer 123 (2018) 418–431.
[71] M. Sheikholeslami, S.A. Shehzad, CVFEM simulation for nanofluid migration in a porous medium using Darcy model, Int. J. Heat Mass Transfer 122 (2018) 1264–1271.
[72] M. Sheikholeslami, M. Darzi, M.K. Sadoughi, Heat transfer improvement and pressure drop during condensation of refrigerant-based nanofluid: an experimental procedure, Int. J. Heat Mass Transfer 122 (2018) 643–650.
[73] M. Sheikholeslami, H.B. Rokni, CVFEM for effect of Lorentz forces on nanofluid flow in a porous complex shaped enclosure by means of non-equilibrium model, J. Mol. Liq. 254 (2018) 446–462.
[74] M. Sheikholeslami, S.A. Shehzad, Z. Li, Water based nanofluid free convection heat transfer in a three dimensional porous cavity with hot sphere obstacle in existence of Lorenz forces, Int. J. Heat Mass Transfer 125 (2018) 375–386.
[75] M. Sheikholeslami, M. Jafaryar, D.D. Ganji, Z. Li, Exergy loss analysis for nanofluid forced convection heat transfer in a pipe with modified turbulators, J. Mol. Liq. 262 (2018) 104–110.
[76] M. Sheikholeslami, M. Jafaryar, Z. Li, Nanofluid turbulent convective flow in a circular duct with helical turbulators considering CuO nanoparticles, Int. J. Heat Mass Transfer 124 (2018) 980–989.
[77] M. Sheikholeslami, Z. Li, M. Shamlooei, Nanofluid MHD natural convection through a porous complex shaped cavity considering thermal radiation, Phys. Lett. A 382 (2018) 1615–1632.
[78] M. Sheikholeslami, H. B. Rokni, Magnetic nanofluid flow and convective heat transfer in a porous cavity considering Brownian motion effects, Phys. Fluids, 30, 1, https://doi.org/10.1063/1.5012517.
[79] M. Sheikholeslami, S.A. Shehzad, Simulation of water based nanofluid convective flow inside a porous enclosure via non-equilibrium model, Int. J. Heat Mass Transfer 120 (2018) 1200–1212.
[80] M. Sheikholeslami, S.A. Shehzad, Non-Darcy free convection of Fe_3O_4-water nanoliquid in a complex shaped enclosure under impact of uniform Lorentz force, Chin. J. Phys. 56 (2018) 270–281.
[81] M. Sheikholeslami, M. Seyednezhad, Simulation of nanofluid flow and natural convection in a porous media under the influence of electric field using CVFEM, Int. J. Heat Mass Transfer 120 (2018) 772–781.
[82] M. Sheikholeslami, M. Shamlooei, R. Moradi, Numerical simulation for heat transfer intensification of nanofluid in a permeable curved enclosure considering shape effect of Fe_3O_4 nanoparticles, Chem. Eng. Process. Process Intensif. 124 (2018) 71–82.

[83] M. Sheikholeslami, T. Hayat, T. Muhammad, A. Alsaedi, MHD forced convection flow of nanofluid in a porous cavity with hot elliptic obstacle by means of Lattice Boltzmann method, Int. J. Mech. Sci. 135 (2018) 532–540.

[84] M. Sheikholeslami, Numerical investigation of nanofluid free convection under the influence of electric field in a porous enclosure, J. Mol. Liq. 249 (2018) 1212–1221.

[85] M. Sheikholeslami, CuO-water nanofluid flow due to magnetic field inside a porous media considering Brownian motion, J. Mol. Liq. 249 (2018) 921–929.

[86] M. Sheikholeslami, D.D. Ganji, Influence of electric field on Fe_3O_4-water nanofluid radiative and convective heat transfer in a permeable enclosure, J. Mol. Liq. 250 (2018) 404–412.

[87] M. Sheikholeslami, S.A. Shehzad, Z. Li, Nanofluid heat transfer intensification in a permeable channel due to magnetic field using Lattice Boltzmann method, Physica B (2018). Available from: https://doi.org/10.1016/j.physb.2018.03.036.

[88] M. Sheikholeslami, A. Zeeshan, Numerical simulation of Fe_3O_4-water nanofluid flow in a non-Darcy porous media, Int. J. Num. Methods Heat Fluid Flow 28 (3) (2018) 641–660. Available from: https://doi.org/10.1108/HFF-04-2017-0160.

[89] M. Sheikholeslami, D.D. Ganji, Numerical approach for magnetic nanofluid flow in a porous cavity using CuO nanoparticles, Mater. Des. 120 (2017) 382–393.

[90] M. Sheikholeslami, D.D. Ganji, Numerical modeling of magnetohydrodynamic CuO-water transportation inside a porous cavity considering shape factor effect, Colloids Surf. A 529 (2017) 705–714.

[91] M. Sheikholeslami, T. Hayat, A. Alsaedi, Numerical simulation for forced convection flow of MHD CuO-H_2O nanofluid inside a cavity by means of LBM, J. Mol. Liq. 249 (2018) 941–948.

[92] M. Sheikholeslami, H.R. Kataria, A.S. Mittal, Effect of thermal diffusion and heat-generation on MHD nanofluid flow past an oscillating vertical plate through porous medium, J. Mol. Liq. 257 (2018) 12–25.

[93] M. Sheikholeslami, M. Barzegar Gerdroodbary, S. Valiallah Mousavi, D.D. Ganji, R. Moradi, Heat transfer enhancement of ferrofluid inside an 90° elbow channel by non-uniform magnetic field, J. Magn. Magn. Mater. 460 (2018) 302–311.

CHAPTER

14

Non-Darcy Model for Nanofluid Hydrothermal Treatment in a Porous Medium Using CVFEM

14.1 INTRODUCTION

The knowledge of free or forced convection heat transfer inside geometries of irregular shape (e.g., wavy channel and pipe bend) for porous media has many significant engineering applications; e.g., geothermal engineering, solar-collectors, performance of cold storage, and thermal insulation of buildings. A considerable number of published articles are available that deal with flow characteristics, heat transfer, flow and heat transfer instability, transition to turbulence, design aspects, etc. For a non-Darcy porous medium, Kumar and Gupta [1] reported the flow and thermal fields' characteristics in wavy cavities. Sheikholeslami [2] investigated MHD nanofluid free convective heat transfer in a porous tilted enclosure by means of a non-Darcy porous medium. Sheikholeslami and Ganji [3] studied the magnetic nanofluid flow in a porous cavity using CuO nanoparticles. Sheikholeslami and Ganji [4] investigated the nanofluid transportation in porous media under the influence of an external magnetic source. Sheikholeslami and Rokni [5] reported nanofluid convective heat transfer intensification in a porous circular cylinder. Sheikholeslami and Shamlooei [6] utilized CVFEM for convective flow of nanofluid inside a lid-driven porous cavity. Sheikholeslami and Seyednezhad [7] simulated the nanofluid heat transfer in a permeable enclosure in the presence of a variable magnetic field. Sheikholeslami [8] demonstrated the influence of Lorentz forces on nanofluid flow in a porous cavity by means of a non-Darcy model. Sheikholeslami and Zeeshan [9] presented the numerical simulation of Fe3O4-water nanofluid flow in a non-Darcy porous media. Nanofluid flows in various mediums were studied in recent years [10–92].

14.2 MHD NANOFLUID FREE CONVECTIVE HEAT TRANSFER IN A POROUS TILTED ENCLOSURE

14.2.1 Problem Definition

Fig. 14.1 illustrates the important geometric parameters of current geometry. Also a sample mesh is presented. The inner and outer cylinders are considered as hot and cold walls, respectively. A horizontal magnetic field has been considered.

14.2.2 Governing Equation

2D steady convective flow of nanofluid in a porous media is considered in the presence of a constant magnetic field. The PDEs equations are:

$$\frac{\partial v}{\partial y} + \frac{\partial u}{\partial x} = 0 \tag{14.1}$$

Application of Control Volume based Finite Element Method (CVFEM) for Nanofluid Flow and Heat Transfer.
DOI: https://doi.org/10.1016/B978-0-12-814152-6.00014-X

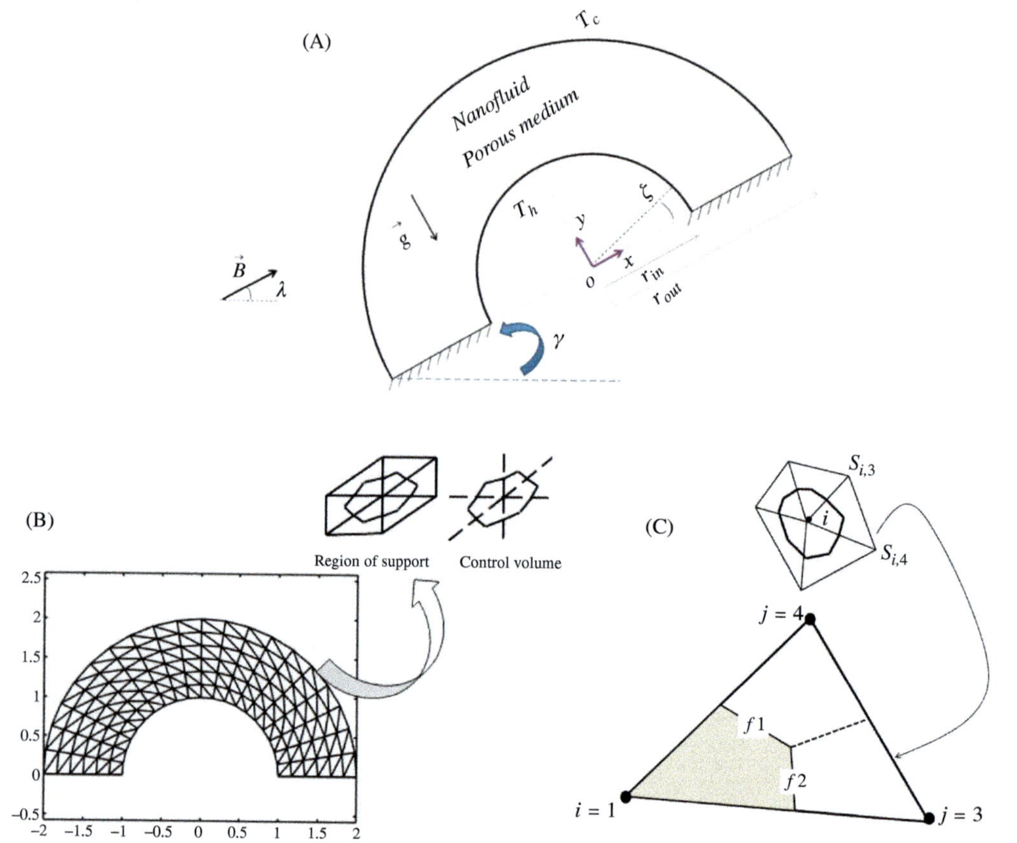

FIGURE 14.1 (A) Geometry and the boundary conditions with (B) the mesh of half-annulus enclosure considered in this work; (C) a sample triangular element and its corresponding control volume.

$$\frac{\mu_{nf}}{\rho_{nf}}\left(\frac{\partial^2 u}{\partial x^2}+\frac{\partial^2 u}{\partial y^2}\right)-\frac{1}{\rho_{nf}}\frac{\partial P}{\partial x}-\frac{1}{\rho_{nf}}\frac{\mu_{nf}}{K}u-(T_c-T)\beta_{nf}g\sin\gamma + \sigma_{nf}B_0^2\left[-u(\sin\lambda)^2+v(\sin\lambda)(\cos\lambda)\right]=v\frac{\partial u}{\partial y}+u\frac{\partial u}{\partial x} \tag{14.2}$$

$$\frac{\mu_{nf}}{\rho_{nf}}\left(\frac{\partial^2 v}{\partial x^2}+\frac{\partial^2 v}{\partial y^2}\right)-(T_c-T)\beta_{nf}g\cos\gamma-\frac{\partial P}{\partial y}\frac{1}{\rho_{nf}}-\frac{1}{\rho_{nf}}\frac{\mu_{nf}}{K}v + \sigma_{nf}B_0^2\left[-v(\cos\lambda)^2+u(\sin\lambda)(\cos\lambda)\right]=v\frac{\partial v}{\partial y}+u\frac{\partial v}{\partial x} \tag{14.3}$$

$$(\rho C_p)_{nf}\left(v\frac{\partial T}{\partial y}+u\frac{\partial T}{\partial x}\right)=k_{nf}\left(\frac{\partial^2 T}{\partial x^2}+\frac{\partial^2 T}{\partial y^2}\right) \tag{14.4}$$

$(\rho C_p)_{nf}$, $(\rho\beta)_{nf}$, ρ_{nf}, and σ_{nf} are defined as:

$$(\rho C_p)_{nf}=(\rho C_p)_f(1-\phi)+(\rho C_p)_s\phi \tag{14.5}$$

$$(\rho\beta)_{nf}=(\rho\beta)_f(1-\phi)+(\rho\beta)_s\phi, \tag{14.6}$$

$$\rho_{nf}=\rho_f(1-\phi)+\rho_s\phi \tag{14.7}$$

$$\frac{\sigma_{nf}}{\sigma_f}=1+\frac{3\left(\frac{\sigma_s}{\sigma_f}-1\right)\phi}{\left(\frac{\sigma_s}{\sigma_f}+2\right)-\left(\frac{\sigma_s}{\sigma_f}-1\right)\phi} \tag{14.8}$$

k_{nf}, μ_{nf} are obtained according to the Koo–Kleinstreuer–Li (KKL) model:

$$k_{nf} = k_{static} + k_{Brownian}$$
$$\frac{k_{static}}{k_f} = 1 + \frac{3\left(\frac{k_p}{k_f} - 1\right)\phi}{\left(\frac{k_p}{k_f} + 2\right) - \left(\frac{k_p}{k_f} - 1\right)\phi}$$
$$\frac{k_{Brownian}}{k_f} = 5 \times 10^4 g'(\phi, T, d_p)\phi\rho_f c_{p,f}\sqrt{\frac{\kappa_b T}{\rho_p d_p}}$$
$$g'(\phi, T, d_p) = \left(a_1 + a_2 Ln(d_p) + a_3 Ln(\phi) + a_4 Ln(\phi)\ln(d_p) + a_5 Ln(d_p)^2\right)Ln(T)$$
$$+ \left(a_6 + a_7 Ln(d_p) + a_8 Ln(\phi) + a_9 \ln(d_p)Ln(\phi) + a_{10} Ln(d_p)^2\right)$$
$$R_f = d_p/k_{p,eff} - d_p/k_p, R_f = 4 \times 10^{-8} km^2/W \tag{14.9}$$

$$\mu_{nf} = \frac{\mu_f}{(1-\phi)^{2.5}} + \frac{k_{Brownian}}{k_f} \times \frac{\mu_f}{\Pr} \tag{14.10}$$

All required coefficients and properties are illustrated in Tables 14.1 and 14.2.
Vorticity and stream function should be used to eliminate pressure source terms:

$$\omega + \frac{\partial u}{\partial y} - \frac{\partial v}{\partial x} = 0, \quad \frac{\partial \psi}{\partial x} = -v, \quad \frac{\partial \psi}{\partial y} = u \tag{14.11}$$

TABLE 14.1 The Coefficient Values of CuO – Water Nanofluid

Coefficient values	CuO – Water
a_1	– 26.5933108
a_2	– 0.403818333
a_3	– 33.3516805
a_4	– 1.915825591
a_5	6.421858E-02
a_6	48.40336955
a_7	– 9.787756683
a_8	190.245610009
a_9	10.9285386565
a_{10}	– 0.72009983664

TABLE 14.2 Thermophysical Properties of Water and Nanoparticles

	$\rho(kg/m^3)$	$C_p(j/kgk)$	$k(W/m.k)$	$\beta \times 10^5 (K^{-1})$	$d_p(nm)$	$\sigma(\Omega \cdot m)^{-1}$
Water	997.1	4179	0.613	21	-	0.05
CuO	6500	540	18	29	45	10^{-10}

Introducing dimensionless quantities:

$$U = \frac{uL}{\alpha_{nf}}, V = \frac{vL}{\alpha_{nf}}, \ \theta = \frac{T - T_c}{\Delta T}, \Delta T = T_h - T_c, (X, Y) = \frac{(x, y)}{L},$$
$$\Psi = \frac{\psi}{\alpha_{nf}}, \Omega = \frac{\omega L^2}{\alpha_{nf}} \tag{14.12}$$

The final formulae are:

$$\frac{\partial^2 \Psi}{\partial Y^2} + \frac{\partial^2 \Psi}{\partial X^2} = -\Omega, \tag{14.13}$$

$$U\frac{\partial \Omega}{\partial X} + \frac{\partial \Omega}{\partial Y}V = \Pr\frac{A_5\,A_2}{A_1\,A_4}\left(\frac{\partial^2\Omega}{\partial Y^2} + \frac{\partial^2\Omega}{\partial X^2}\right)$$
$$+ \Pr Ha^2\frac{A_6\,A_2}{A_1\,A_4}\left(\frac{\partial U}{\partial X}\cos\lambda\sin\lambda - \frac{\partial V}{\partial X}(\cos\lambda)^2 + \frac{\partial U}{\partial Y}(\sin\lambda)^2 - \frac{\partial V}{\partial Y}\cos\lambda\sin\lambda\right) \tag{14.14}$$
$$+ \Pr\,Ra\frac{A_3 A_2^2}{A_1 A_4^2}\left(\frac{\partial\theta}{\partial X}\cos\gamma - \frac{\partial\theta}{\partial Y}\sin\gamma\right) - \frac{\Pr\;A_5\,A_2}{Da\,A_1\,A_4}\Omega,$$

$$\frac{\partial\theta}{\partial X}U + \frac{\partial\theta}{\partial Y}V = \left(\frac{\partial^2\theta}{\partial X^2} + \frac{\partial^2\theta}{\partial Y^2}\right) \tag{14.15}$$

where dimensionless and constants parameters are defined as:

$$\Pr = \upsilon_f/\alpha_f, Ra = g(\rho\beta)_f\Delta T L^3/\left(\mu_f\alpha_f\right), \ Ha = LB_0\sqrt{\sigma_f/\mu_f}$$
$$A_1 = \frac{\rho_{nf}}{\rho_f}, A_2 = \frac{(\rho C_p)_{nf}}{(\rho C_p)_f}, A_3 = \frac{(\rho\beta)_{nf}}{(\rho\beta)_f}, \tag{14.16}$$
$$A_4 = \frac{k_{nf}}{k_f}, A_5 = \frac{\mu_{nf}}{\mu_f}, A_6 = \frac{\sigma_{nf}}{\sigma_f}$$

and boundary conditions are:

$$\begin{array}{ll} \theta = 1.0 & \text{on inner wall} \\ \theta = 0.0 & \text{on outer wall} \\ \dfrac{\partial\theta}{\partial n} = 0.0 & \text{on other walls} \\ \Psi = 0.0 & \text{on all walls} \end{array} \tag{14.17}$$

Local and average Nusselt numbers over the hot wall can be calculated as:

$$Nu_{loc} = A_4\frac{\partial\theta}{\partial r} \tag{14.18}$$

$$Nu_{ave} = \frac{1}{0.5\pi}\int_0^{0.5\pi} Nu_{loc}(\zeta)\,d\zeta \tag{14.19}$$

14.2.3 Effects of Active Parameters

In this chapter, magnetohydrodynamic nanofluid flow and convective heat transfer in a porous tilted annulus is investigated. CVFEM is utilized to obtain the outputs for various values of Hartmann number ($Ha = 0$ to 40), Rayleigh number ($Ra = 10^3, 10^4$, and 10^5), tilted angle ($\gamma = 0^\circ$ to 90°), Darcy number ($Da = 0.01$ to 100), and volume fraction of CuO ($\phi = 0\%$ and 4%).

Figs. 14.2 and 14.3 demonstrate the influences of Hartmann, Rayleigh, Darcy numbers, and tilted angle on hydrothermal behavior. At $\gamma = 0^{\circ}$, in conduction mode, there are two vortexes in streamlines which rotate in opposite directions. As buoyancy forces increase, the vortexes become stronger and their centers move upward. Then the thermal plume appears at $\xi = 90^{\circ}$. Appling a magnetic field reduces the strength of the vortexes and the thermal plume. As the tilted angle increases, the convective mode becomes less than before. At $\gamma = 45^{\circ}$, the main vortexes convert to new two vortexes in which the upper one rotates clockwise. As the Rayleigh number increases, the primary vortexes become stronger and the thermal plume appears in the region between the two vortexes. At $\gamma = 90^{\circ}$, only one vortex exists in the absence of a magnetic field. As the buoyancy forces increase, the main vortex converts to two vortexes and the thermal plume is generated. Increasing the Darcy number makes the convective heat transfer increase. It is an interesting observation that in the presence of a magnetic field at high values of *Ra* and *Da*, two thermal plumes are generated at the upper region due to the existence of three rotating vortexes.

Impacts of significant parameters on Nu_{loc} and Nu_{ave} are illustrated in Figs. 14.4 and 14.5. The correlation for Nu_{ave} is as follows:

$$\begin{aligned} Nu_{ave} &= 5.54 + 0.29\gamma - 0.47Da^* - 3.18\log(Ra) + 0.02Ha^* \\ &- 0.02\gamma Da^* - 0.086\gamma Ha^* + 0.014\gamma\log(Ra) + 0.2Da^*Ha^* - 0.13\log(Ra)Ha^* \\ &- 0.039\gamma^2 - 0.015(Da^*)^2 + 0.53(\log(Ra))^2 + 0.097(Ha^*)^2 \end{aligned} \tag{14.20}$$

where $Ha^* = 0.1Ha$, $Da^* = 0.01Da$. Due to symmetric geometry and boundary conditions, Nu_{loc} profiles are symmetric respect to $\zeta = 90^{\circ}$ when $\gamma = 0^{\circ}$. The Nusselt number increases with the increase of Darcy and Rayleigh numbers. The rate of heat transfer reduces with the increase of the tilted angle. Lorenz forces have a reverse effect on the Nusselt number due to the increase in thermal boundary layer thickness with the increase of the Hartmann number.

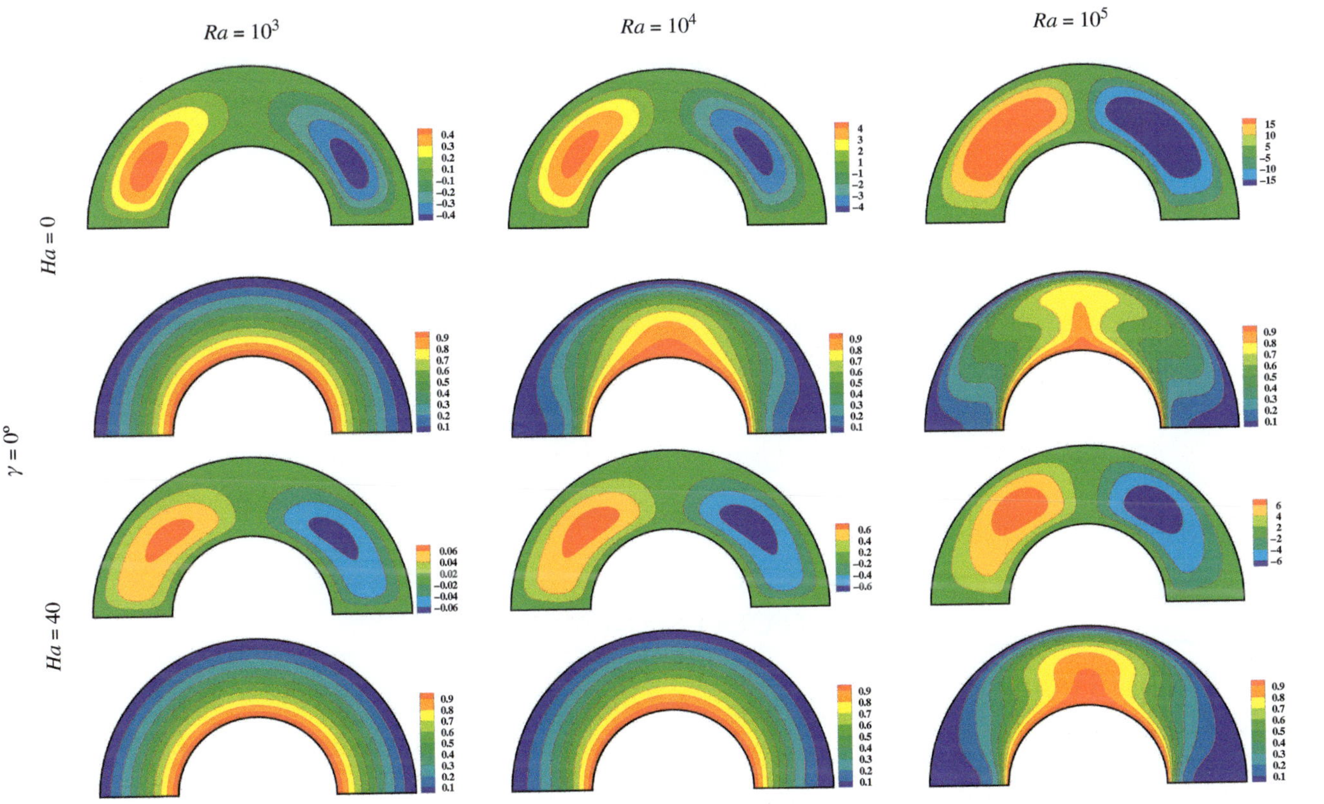

FIGURE 14.2 Isotherm (bottom) and streamline (top) contours for different values of Rayleigh number and Hartmann number when $Da = 0.01$.

FIGURE 14.2 (Continued)

14.3 MAGNETIC NANOFLUID FLOW IN A POROUS CAVITY USING CUO NANOPARTICLES

14.3.1 Problem Definition

Fig. 14.6 demonstrates the important geometric parameters of current geometry. Also a sample mesh is presented. Constant heat flux is introduced from the inner wall. The outer wall is cold and the other walls are adiabatic. A horizontal magnetic field is taken into account. The radiation effect is considered in a porous medium.

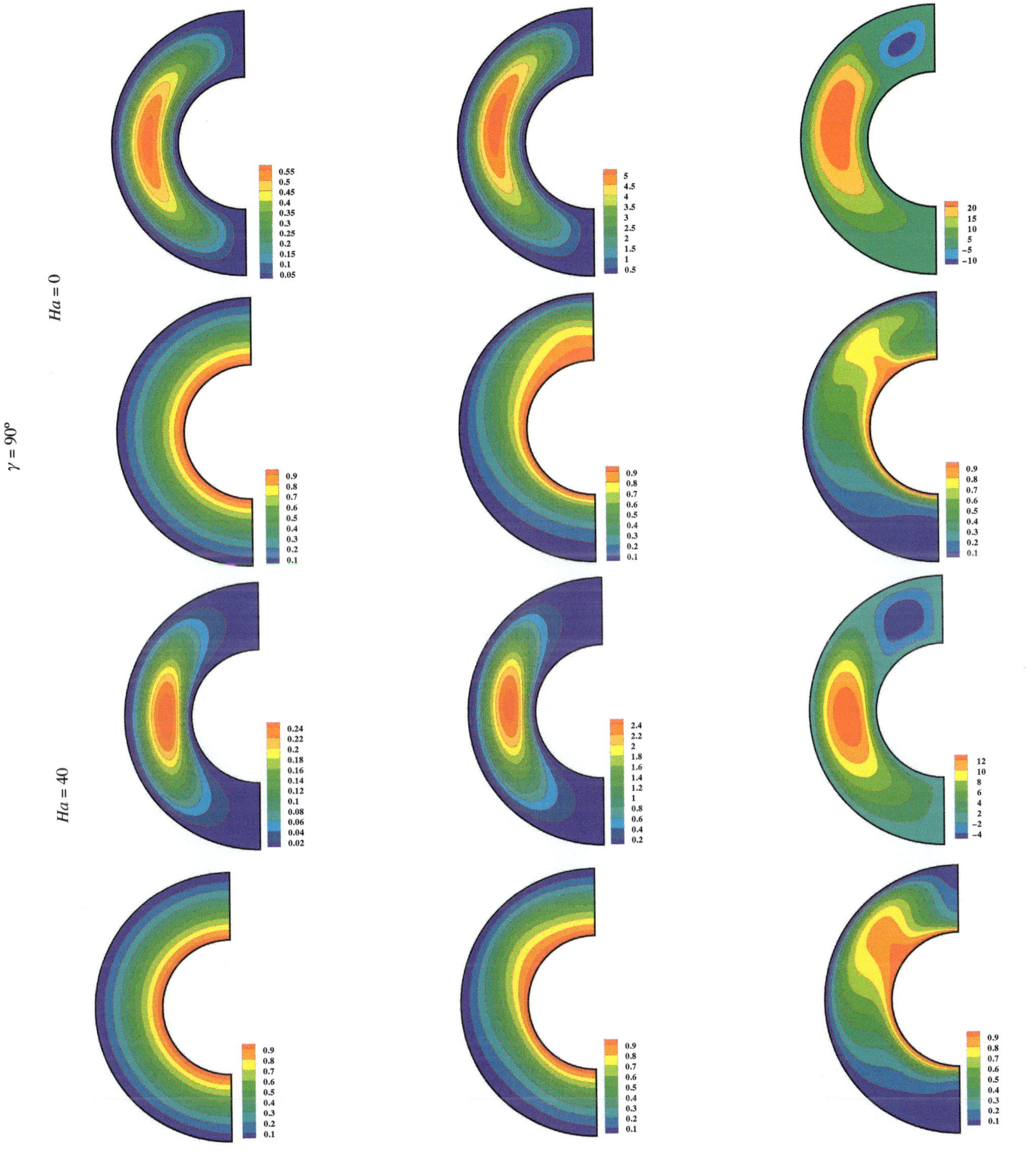

FIGURE 14.2 (Continued)

14.3.2 Governing Equation

Free convective MHD nanofluid flow in a porous media is considered. A non-Darcy model is used for porous media. The PDEs equations are:

$$\frac{\partial v}{\partial y} + \frac{\partial u}{\partial x} = 0 \tag{14.21}$$

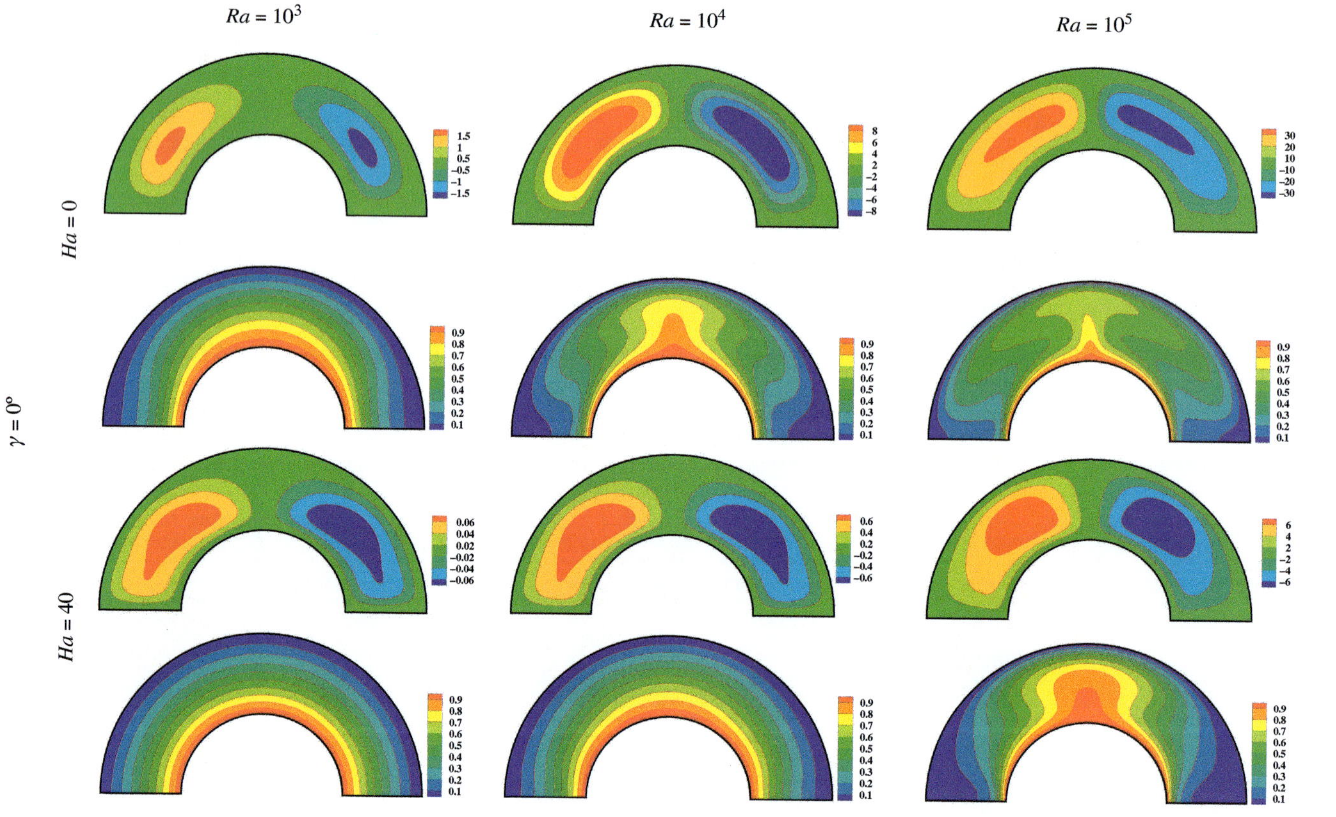

FIGURE 14.3 Isotherm (bottom) and streamline (top) contours for different values of Rayleigh number and Hartmann number when $Da = 100$.

$$\mu_{nf}\left(\frac{\partial^2 u}{\partial x^2}+\frac{\partial^2 u}{\partial y^2}\right)-\frac{\partial P}{\partial x}-\frac{\mu_{nf}}{K}u + \sigma_{nf}B_0^2\left[-u(\sin\lambda)^2+v(\sin\lambda)(\cos\lambda)\right]=\rho_{nf}\left(v\frac{\partial u}{\partial y}+u\frac{\partial u}{\partial x}\right) \tag{14.22}$$

$$\mu_{nf}\left(\frac{\partial^2 v}{\partial x^2}+\frac{\partial^2 v}{\partial y^2}\right)-(T_c-T)(\rho\beta)_{nf}g-\frac{\partial P}{\partial y}-\frac{\mu_{nf}}{K}v + \sigma_{nf}B_0^2\left[-v(\cos\lambda)^2+u(\sin\lambda)(\cos\lambda)\right]=\rho_{nf}\left(v\frac{\partial v}{\partial y}+u\frac{\partial v}{\partial x}\right) \tag{14.23}$$

$$(\rho C_p)_{nf}\left(v\frac{\partial T}{\partial y}+u\frac{\partial T}{\partial x}\right)=k_{nf}\left(\frac{\partial^2 T}{\partial x^2}+\frac{\partial^2 T}{\partial y^2}\right)-\frac{\partial q_r}{\partial y}, \quad \left[q_r=-\frac{4\sigma_e}{3\beta_R}\frac{\partial T^4}{\partial y}, \quad T^4\cong 4T_c^3T-3T_c^4\right] \tag{14.24}$$

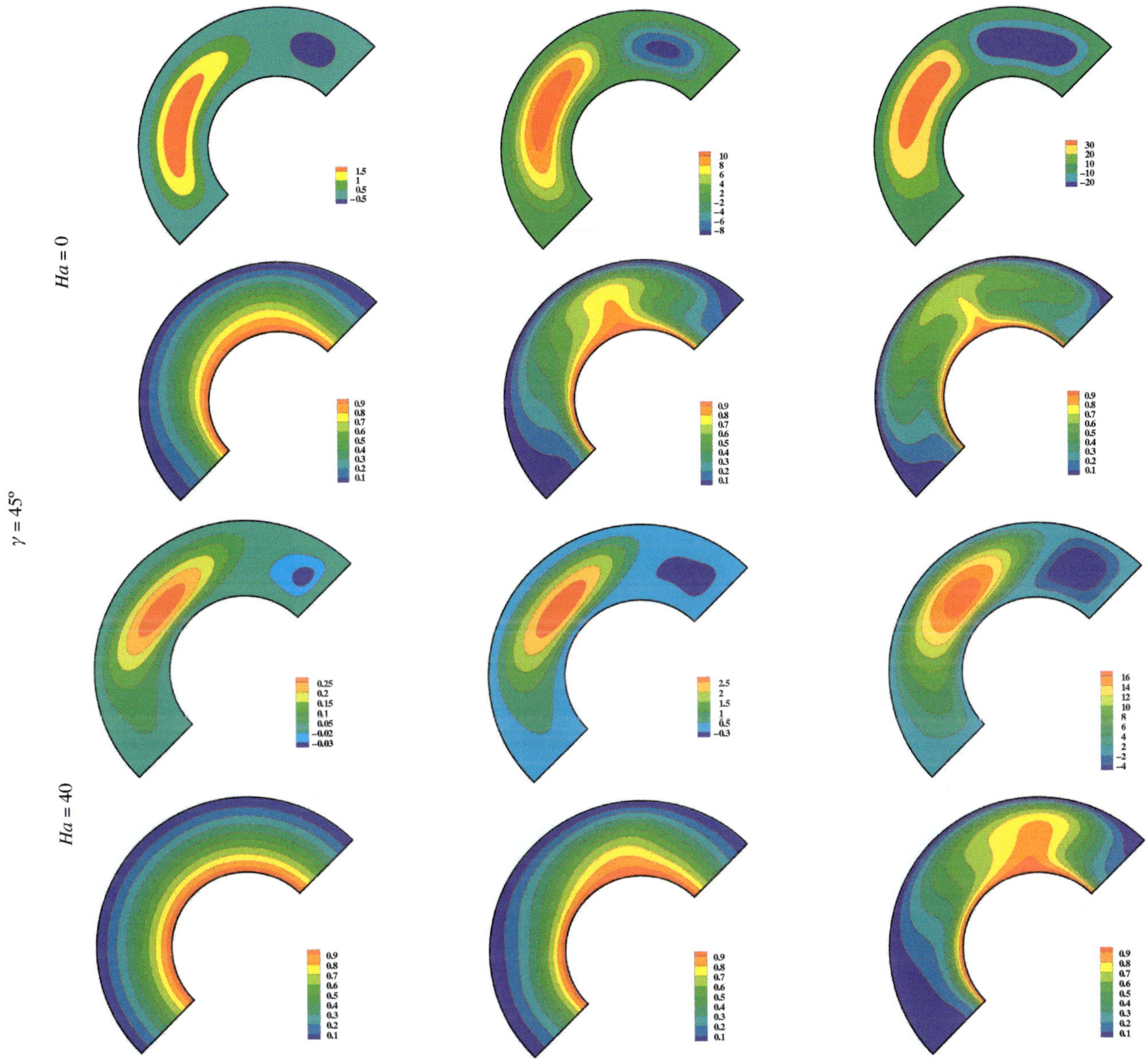

FIGURE 14.3 (Continued)

$(\rho C_p)_{nf}$, $(\rho\beta)_{nf}$, ρ_{nf}, and σ_{nf} are defined as:

$$(\rho C_p)_{nf} = (\rho C_p)_f(1-\phi) + (\rho C_p)_s\phi \tag{14.25}$$

$$(\rho\beta)_{nf} = (\rho\beta)_f(1-\phi) + (\rho\beta)_s\phi, \tag{14.26}$$

$$\rho_{nf} = \rho_f(1-\phi) + \rho_s\phi \tag{14.27}$$

$$\frac{\sigma_{nf}}{\sigma_f} = 1 + \frac{3\left(\frac{\sigma_s}{\sigma_f}-1\right)\phi}{\left(\frac{\sigma_s}{\sigma_f}+2\right) - \left(\frac{\sigma_s}{\sigma_f}-1\right)\phi} \tag{14.28}$$

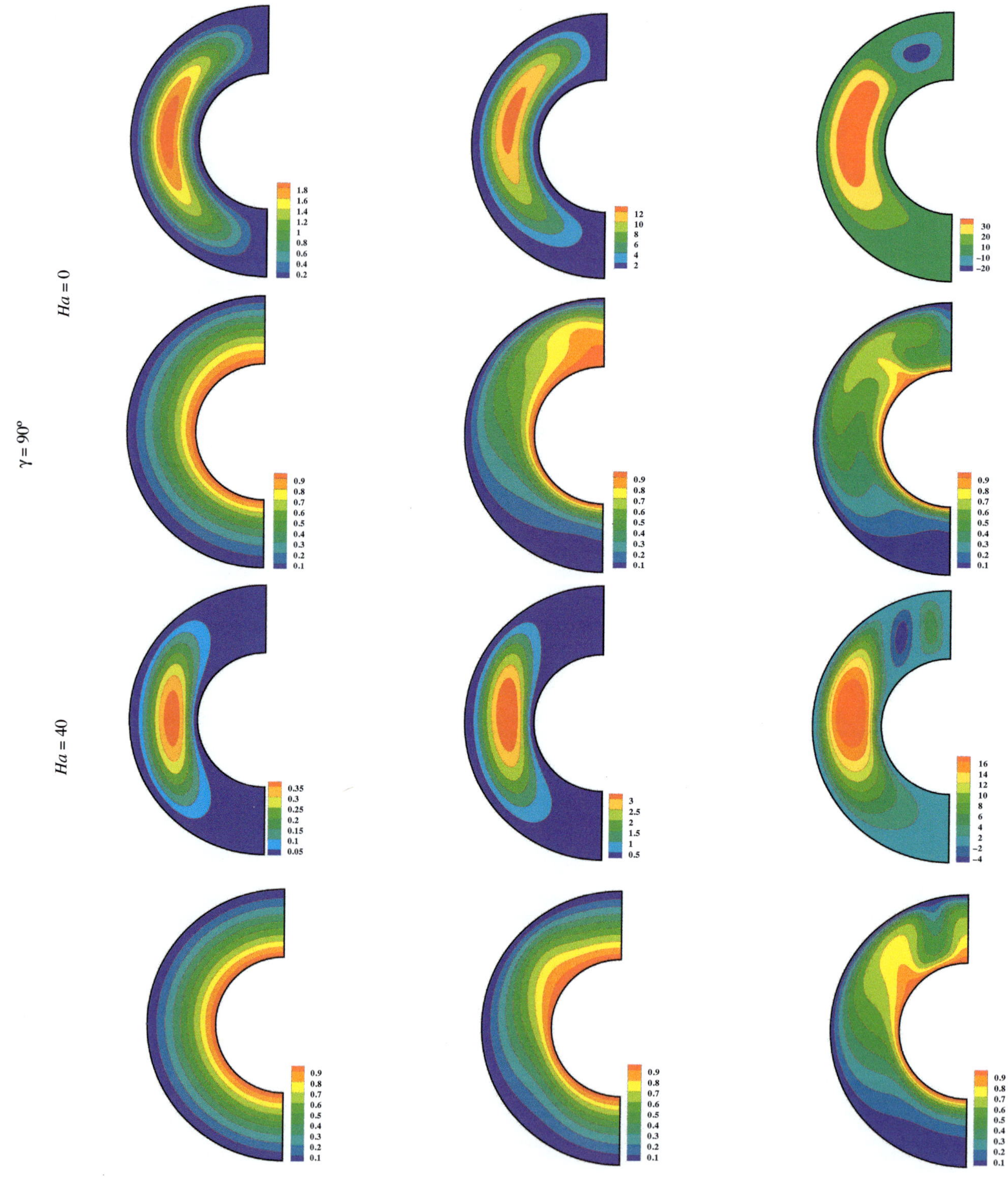

FIGURE 14.3 (Continued)

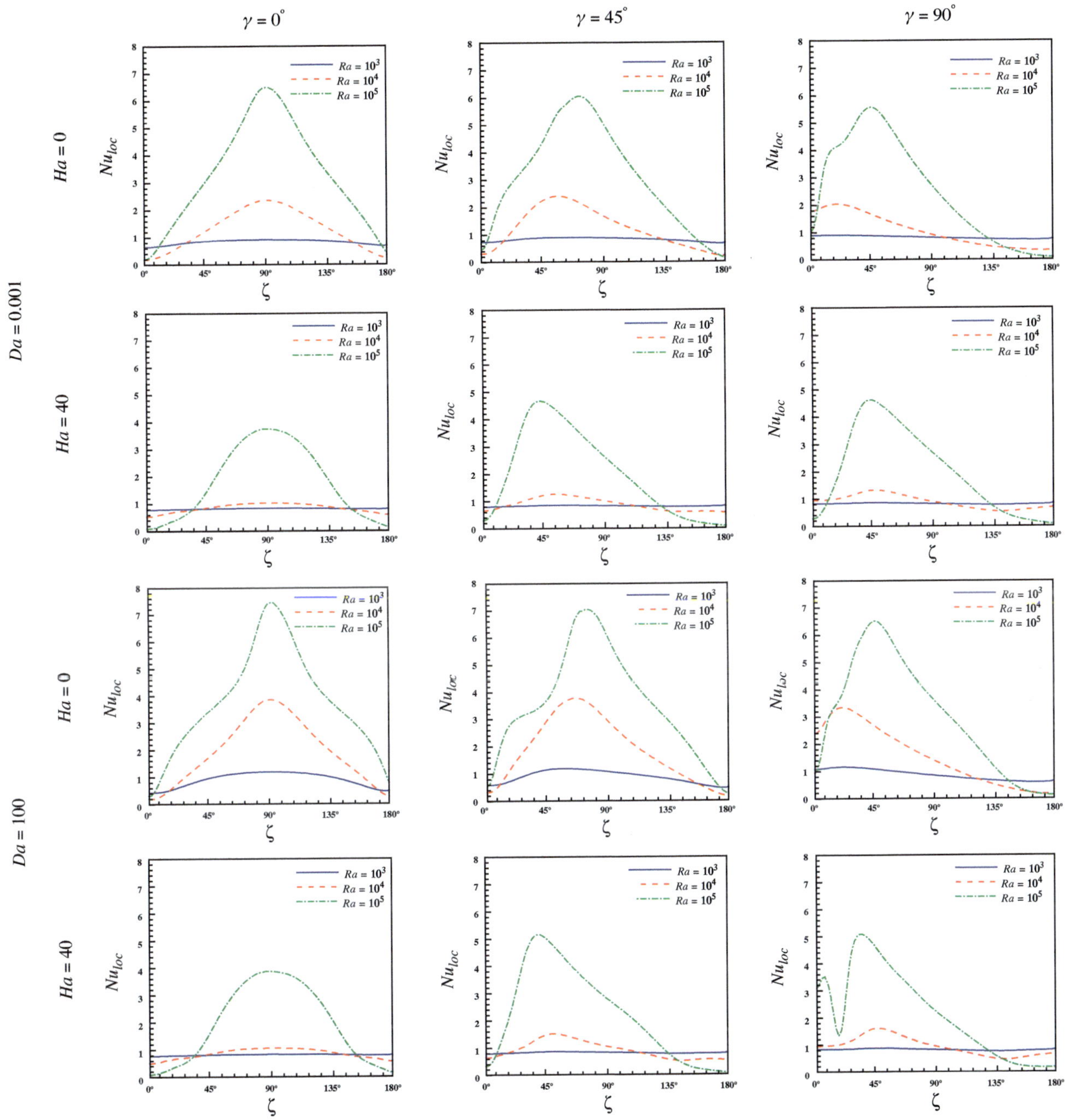

FIGURE 14.4 Effects of the Hartmann number, Rayleigh number, and tilted angle on Local Nusselt number.

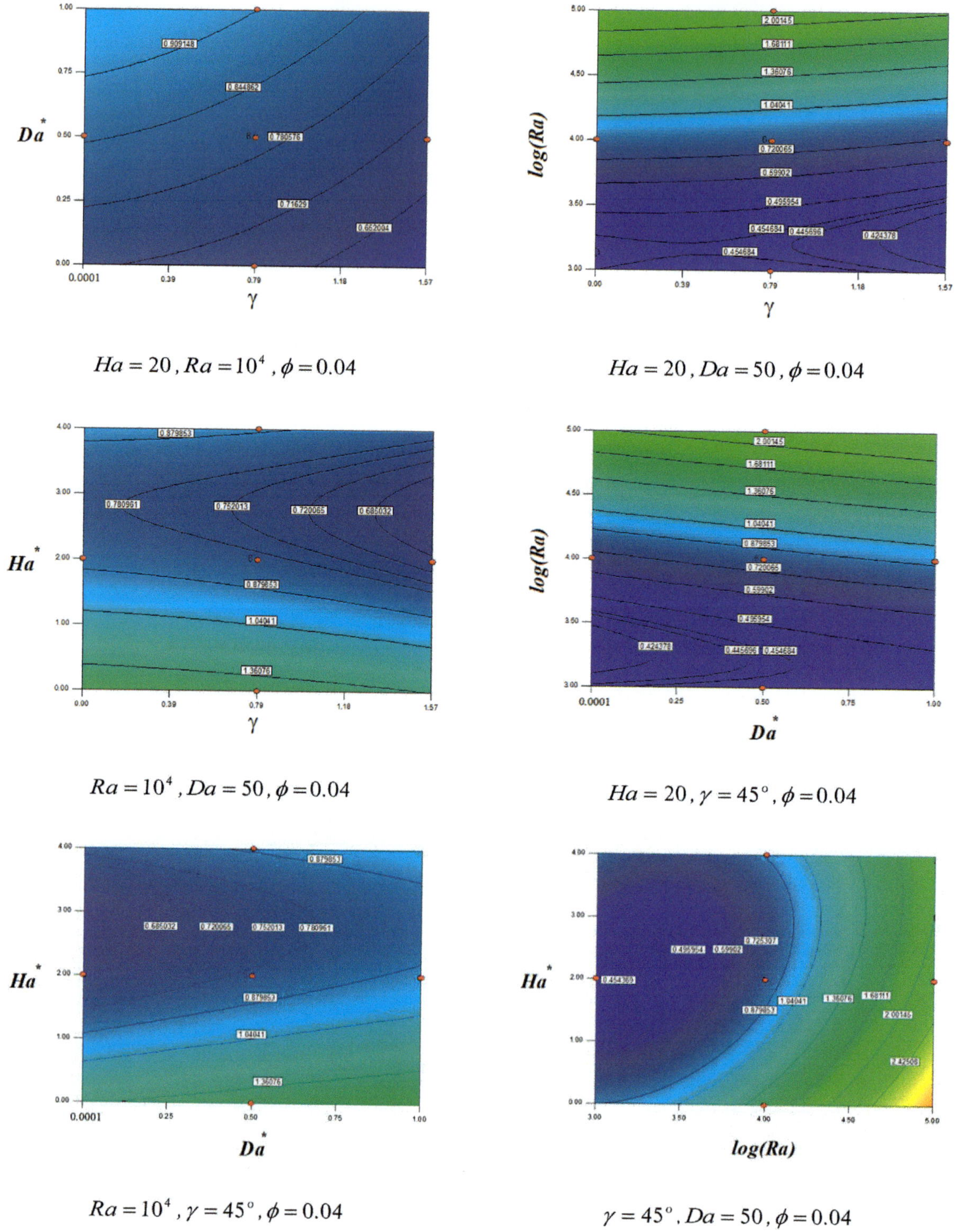

FIGURE 14.5 Effects of the Hartmann number, Rayleigh number, and tilted angle on average Nusselt number.

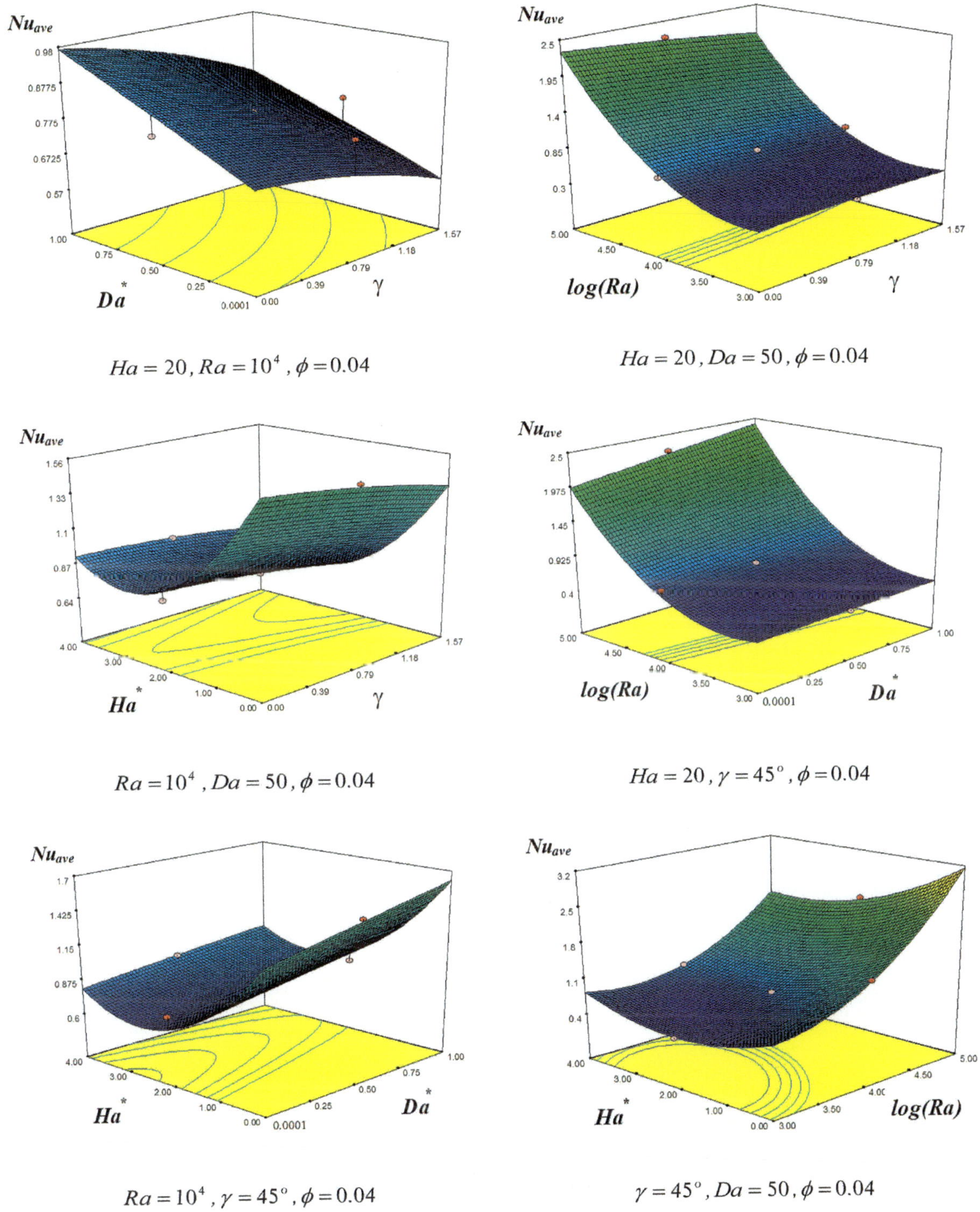

FIGURE 14.5 (Continued)

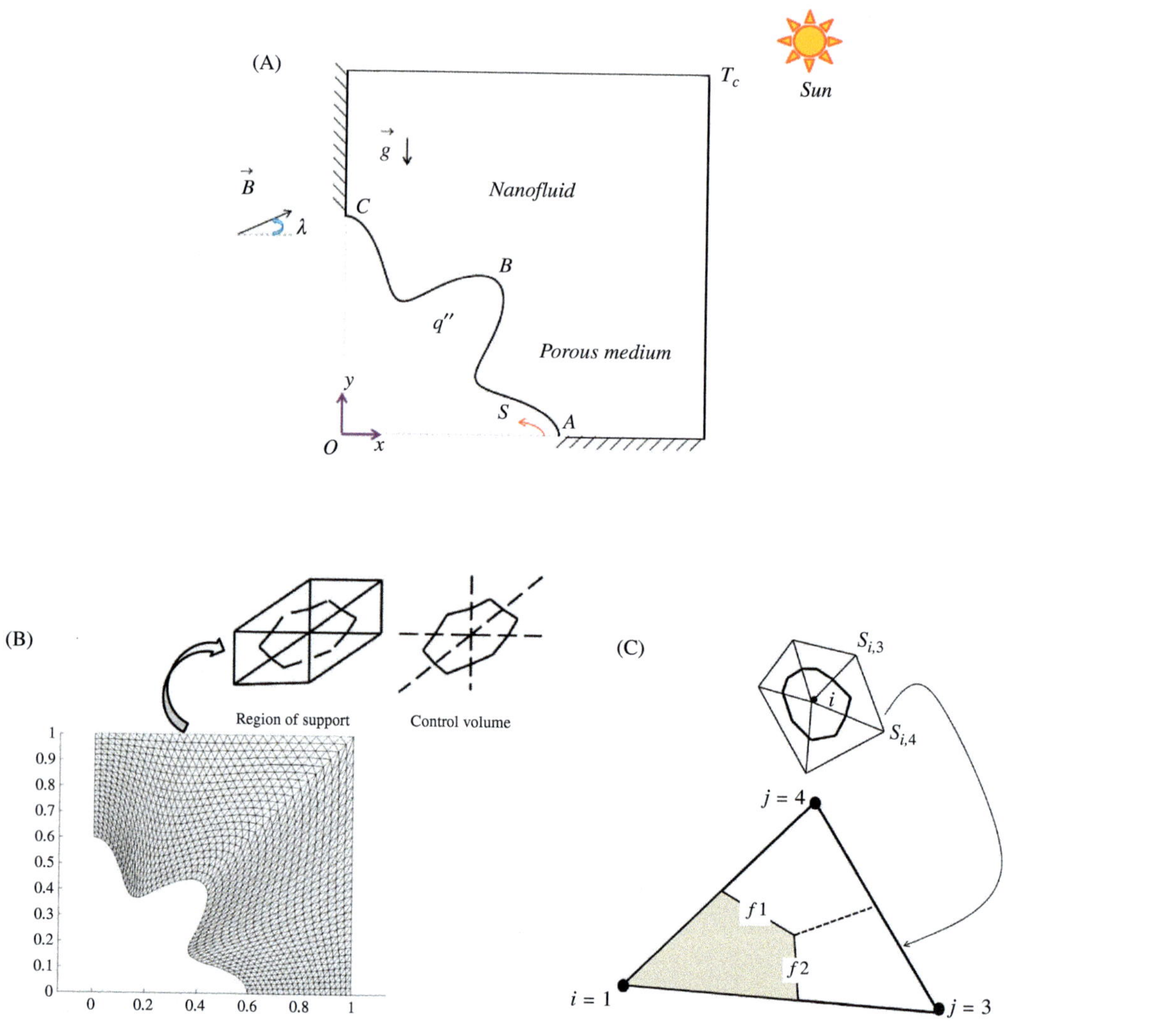

FIGURE 14.6 (A) Geometry and the boundary conditions with (B) the mesh of geometry considered in this work; (C) a sample triangular element and its corresponding control volume.

k_{nf}, μ_{nf} are obtained according to the Koo–Kleinstreuer–Li (KKL) model:

$$\begin{aligned}
&k_{nf} = k_{static} + k_{Brownian} \\
&\frac{k_{static}}{k_f} = 1 + \frac{3\left(\frac{k_p}{k_f} - 1\right)\phi}{\left(\frac{k_p}{k_f} + 2\right) - \left(\frac{k_p}{k_f} - 1\right)\phi} \\
&\frac{k_{Brownian}}{k_f} = 5 \times 10^4 g'(\phi, T, d_p)\phi\rho_f c_{p,f}\sqrt{\frac{\kappa_b T}{\rho_p d_p}} \\
&g'(\phi, T, d_p) = \left(a_1 + a_2 Ln(d_p) + a_3 Ln(\phi) + a_4 Ln(\phi)\ln(d_p) + a_5 Ln(d_p)^2\right)Ln(T) \\
&\quad + \left(a_6 + a_7 Ln(d_p) + a_8 Ln(\phi) + a_9 \ln(d_p)Ln(\phi) + a_{10} Ln(d_p)^2\right) \\
&R_f = d_p/k_{p,eff} - d_p/k_p,\ R_f = 4 \times 10^{-8} km^2/W
\end{aligned} \tag{14.29}$$

$$\mu_{nf} = \frac{\mu_f}{(1-\phi)^{2.5}} + \frac{k_{Brownian}}{k_f} \times \frac{\mu_f}{\Pr} \tag{14.30}$$

All required coefficients and properties are illustrated in Tables 14.1 and 14.2. Vorticity and stream function should be used to eliminate pressure source terms:

$$\omega + \frac{\partial u}{\partial y} - \frac{\partial v}{\partial x} = 0, \quad \frac{\partial \psi}{\partial x} = -v, \quad \frac{\partial \psi}{\partial y} = u \tag{14.31}$$

Introducing dimensionless quantities:

$$\begin{aligned} &U = \frac{uL}{\alpha_{nf}}, V = \frac{vL}{\alpha_{nf}}, \ \theta = \frac{T - T_c}{\Delta T}, \Delta T = q''L/k_f, (X, Y) = \frac{(x, y)}{L}, \\ &\Psi = \frac{\psi}{\alpha_{nf}}, \Omega = \frac{\omega L^2}{\alpha_{nf}} \end{aligned} \tag{14.32}$$

The final formulae are:

$$\frac{\partial^2 \Psi}{\partial Y^2} + \frac{\partial^2 \Psi}{\partial X^2} = -\Omega, \tag{14.33}$$

$$\begin{aligned} &U\frac{\partial \Omega}{\partial X} + \frac{\partial \Omega}{\partial Y}V = \mathrm{Pr}\frac{A_5\,A_2}{A_1\,A_4}\left(\frac{\partial^2 \Omega}{\partial Y^2} + \frac{\partial^2 \Omega}{\partial X^2}\right) \\ &+ \mathrm{P_1}IIa^2\frac{A_6\,A_2}{A_1\,A_4}\left(\frac{\partial U}{\partial X}\cos\lambda\sin\lambda - \frac{\partial V}{\partial X}(\cos\lambda)^2 + \frac{\partial U}{\partial Y}(\sin\lambda)^2 - \frac{\partial V}{\partial Y}\cos\lambda\sin\lambda\right) \\ &+ \mathrm{Pr}\,Ra\frac{A_3 A_2^2}{A_1 A_4^2}\left(\frac{\partial \theta}{\partial X}\right) - \frac{\mathrm{Pr}\,A_5\,A_2}{Da\,A_1\,A_4}\Omega, \end{aligned} \tag{14.34}$$

$$\frac{\partial \theta}{\partial X}U + \frac{\partial \theta}{\partial Y}V = \left(\frac{\partial^2 \theta}{\partial X^2} + \frac{\partial^2 \theta}{\partial Y^2}\right) + \frac{4}{3}\frac{1}{A_4}Rd\frac{\partial^2 \theta}{\partial Y^2} \tag{14.35}$$

where dimensionless and constants parameters are defined as:

$$\begin{aligned} &\mathrm{Pr} = \upsilon_f/\alpha_f, Ra = g(\rho\beta)_f \Delta T L^3/\left(\mu_f \alpha_f\right), \ Ha = LB_0\sqrt{\sigma_f/\mu_f}, Rd = 4\sigma_e T_c^3/\left(\beta_R k_f\right) \\ &A_1 = \frac{\rho_{nf}}{\rho_f}, A_2 = \frac{(\rho C_p)_{nf}}{(\rho C_p)_f}, A_3 = \frac{(\rho\beta)_{nf}}{(\rho\beta)_f}, \\ &A_4 = \frac{k_{nf}}{k_f}, A_5 = \frac{\mu_{nf}}{\mu_f}, A_6 = \frac{\sigma_{nf}}{\sigma_f} \end{aligned} \tag{14.36}$$

and boundary conditions are:

$$\begin{aligned} &\frac{\partial \theta}{\partial n} = 1.0 && \text{on inner wall} \\ &\theta = 0.0 && \text{on outer wall} \\ &\frac{\partial \theta}{\partial n} = 0.0 && \text{on other walls} \\ &\Psi = 0.0 && \text{on all walls} \end{aligned} \tag{14.37}$$

Local and average Nusselt numbers over the hot wall can be calculated as:

$$Nu_{loc} = \frac{1}{\theta}\left(\frac{k_{nf}}{k_f}\right)\left(1 + \frac{4}{3}Rd\left(\frac{k_{nf}}{k_f}\right)^{-1}\right) \tag{14.38}$$

$$Nu_{ave} = \frac{1}{S}\int_0^S Nu_{loc}\,ds \tag{14.39}$$

14.3.3 Effects of Active Parameters

A non-Darcy model is applied for the natural convection of nanofluid in a porous enclosure. The influence of thermal radiation and magnetic field are taken into account. The KKL model is utilized for estimating viscosity and thermal conductivity of CuO-water nanofluid. The numerical procedure is conducted by means of CVFEM. Effects of Darcy number ($Da = 0.01$ to 100), radiation parameter ($Rd = 0$ to 0.8), Rayleigh number ($Ra = 10^3$ to 10^5), Hartmann number ($Ha = 0$ to 40), and volume fraction of nanofluid ($\phi = 0$ to 0.04) are examined.

Fig. 14.7 demonstrates the effect of adding nanoparticle into the base fluid on hydrothermal behavior. The thermal boundary layer thickness reduces with the addition of nanoparticles. So, the rate of heat transfer enhances with the increase of the volume fraction of nanofluid. Also the nanofluid velocity is greater than the base fluid velocity due to the increase in nanoparticles' motion. Besides, the influence of adding nanoparticles is more sensible in the presence of a magnetic field. The influence of the radiation parameter on nanofluid flow and heat transfer is shown in Fig. 14.8. As the radiation parameter increases, the thermal boundary layer thickness is enhanced. $|\Psi_{max}|$ augments with the increase of the radiation parameter.

The impacts of Darcy, Hartmann, and Rayleigh numbers on hydrothermal behavior of nanofluid are reported in Figs. 14.9–14.11. In low Darcy and Rayleigh numbers, the conduction mechanism can be seen, so the isotherms follow the shape of the cylinders. As the Rayleigh number increases, the buoyancy forces enhance the convection heat transfer. So the isotherms become more disturbed with the increase of Ra. Also $|\Psi_{max}|$ increases with the rise of Ra. As the Darcy number increases the permeability of the medium increases and the convective mechanism is enhanced. So, the rate of heat transfer and absolute values of stream function are enhanced with the rise of Da. As the magnetic field increases, the Lorentz forces are generated and these forces reduce the velocity of the nanofluid. Also the rate of heat transfer reduces with the rise of the Hartmann number.

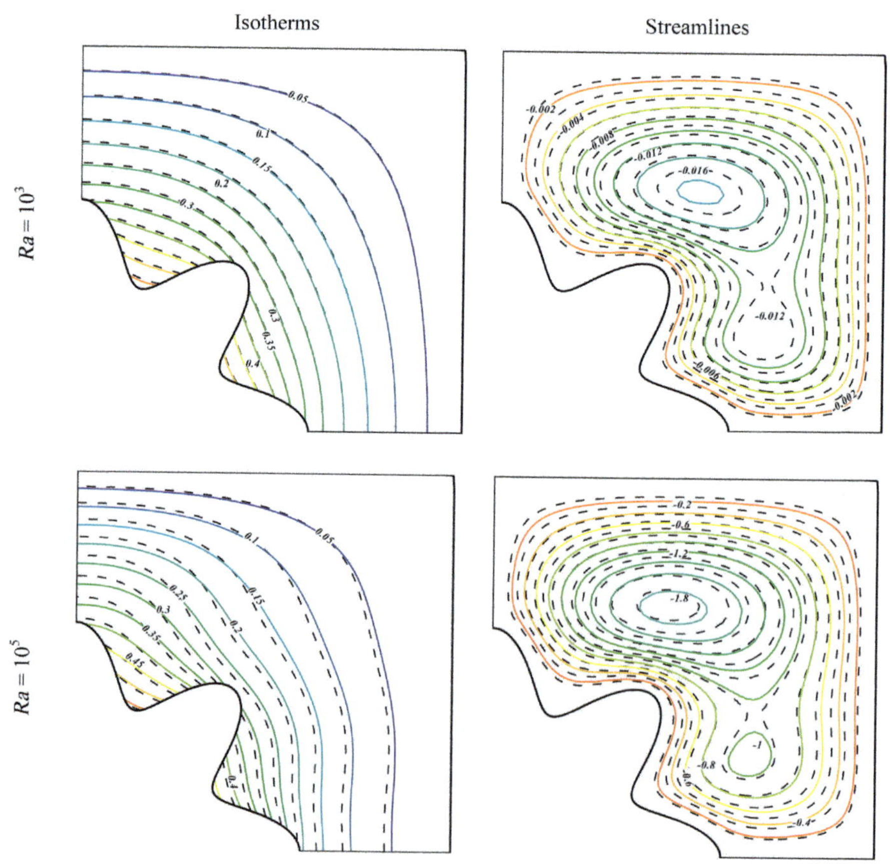

FIGURE 14.7 Influence of nanofluid volume fraction on streamlines (left) and isotherms (right) contours (nanofluid ($\phi = 0.04$)(—) and pure fluid ($\phi = 0$) (- - -)) when $Ha = 40, Da = 100, Rd = 0.8$.

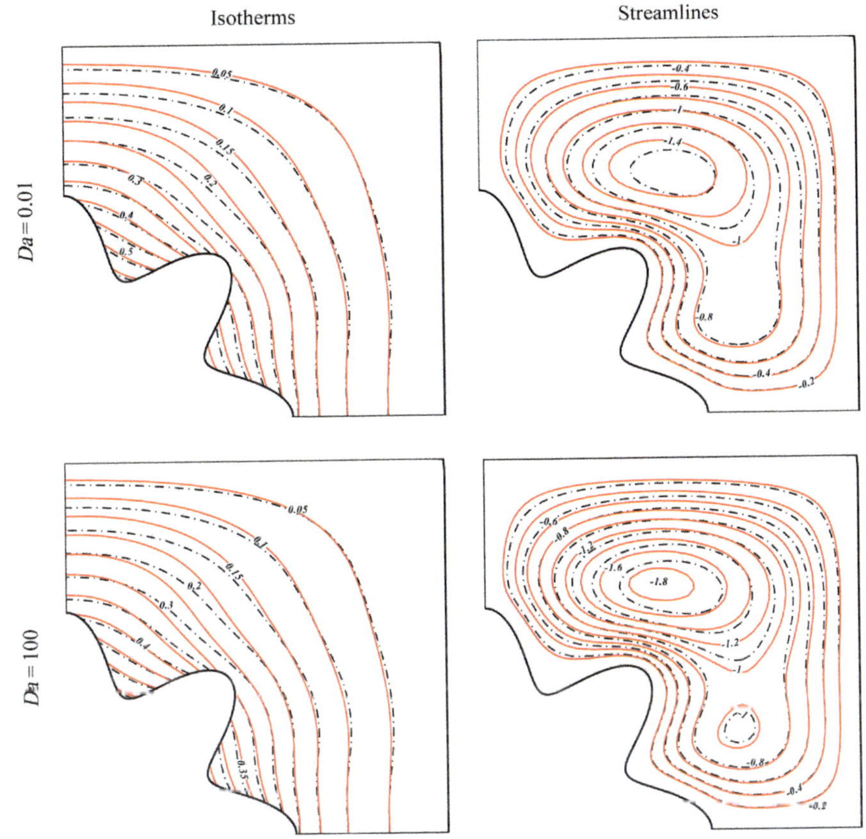

FIGURE 14.8 Influence of radiation parameter on streamline (left) and isotherm (right) contours (nanofluid ($Rd = 0.8$)(——) and pure fluid ($Rd = 0$) (— · —)) when $Ra = 10^5$, $Ha = 0$.

Figs. 14.12 and 14.13 demonstrate the influence of Rd, Da, Ra , and Ha on Nu_{loc}, Nu_{ave}. The formula for Nu_{ave} is:

$$\begin{aligned} Nu_{ave} &= 4.15 - 0.816Rd - 1.45\log(Ra) - 0.67Da^* + 0.79Ha^* \\ &+ 0.67Rd\log(Ra) + 0.14Rd\,Da^* - 0.16Rd\,Ha^* + 0.039\log(Ra)Da^* \\ &- 0.28\log(Ra)Ha^* - 0.009Da^*Ha^* + 0.29Rd^2 + 0.29(\log(Ra))^2 \\ &+ 0.66(Da^*)^2 + 0.046(Ha^*)^2 \end{aligned} \tag{14.40}$$

where $Ha^* = 0.1Ha$, $Da^* = 0.01Da$. The root mean squared error of this formula is equal to 0.98. The existence of extremum points on the local Nusselt number is relevant to the presence of the undulation of the inner wall and the thermal plume. The Nusselt number increases with the increase of Rayligh and Darcy numbers due to the increase of the convective heat transfer mechanism. Also Fig. 14.13 indicates that rate of heat transfer increases with the increase of the radiation parameter. Futhermore, Lorentz forces reduce the convective heat transfer mode. So, the Nusselt number decreases with the enhancement of the Hartmann number.

14.4 NANOFLUID TRANSPORTATION IN POROUS MEDIA UNDER THE INFLUENCE OF EXTERNAL MAGNETIC SOURCE

14.4.1 Problem Definition

Boundary conditions are depicted in Fig. 14.14. The inner elliptic wall has constant temperature and is considered as a hot wall. The outer circular wall is a cold wall, the others are adiabatic. A magnetic source has been considered as shown in Fig. 15.15. $\overline{H_x}, \overline{H_y}, \overline{H}$ can be calculated as follows:

$$\overline{H_y} = \left[(\bar{b}-y)^2+(\bar{a}-x)^2\right]^{-1}\frac{\gamma}{2\pi}(\bar{a}-x), \tag{14.41}$$

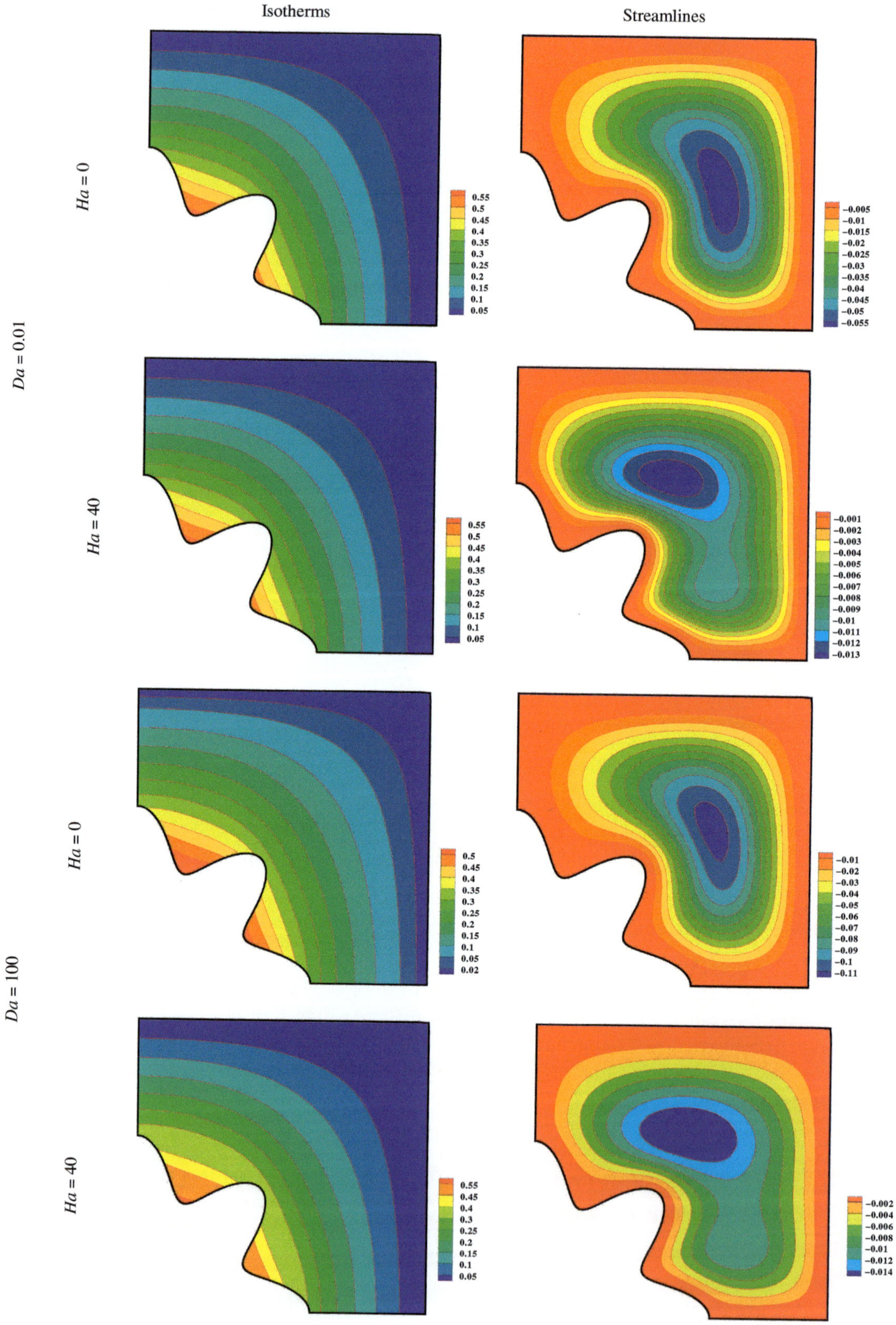

FIGURE 14.9 Isotherm (left) and streamline (right) contours for different values of Darcy and Hartmann numbers when $Ra = 10^3, \phi = 0.04, Rd = 0.8$.

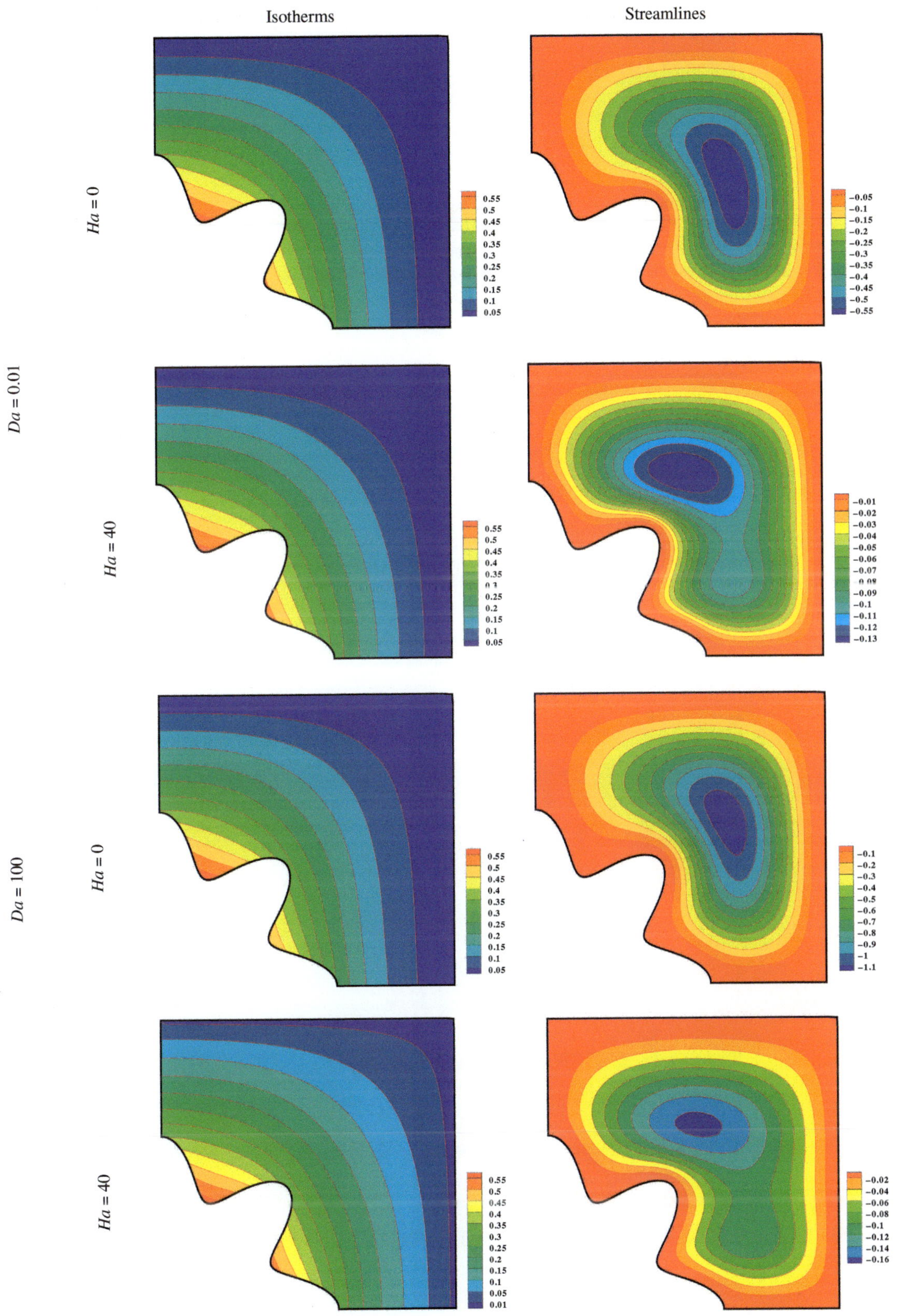

FIGURE 14.10 Isotherm (left) and streamline (right) contours for different values of Darcy and Hartmann numbers when $Ra = 10^4, \phi = 0.04, Rd = 0.8$.

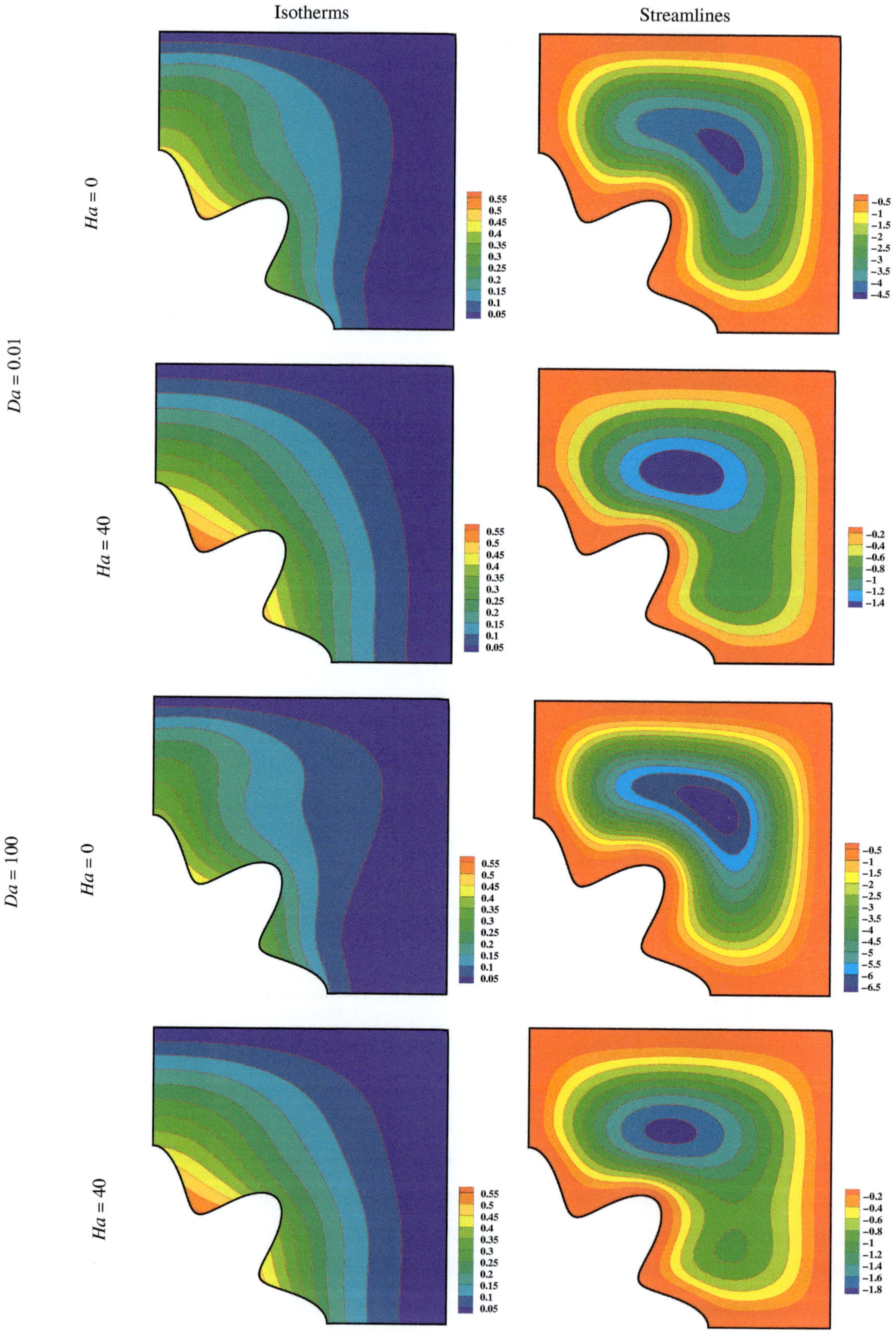

FIGURE 14.11 Isotherm (left) and streamline (right) contours for different values of Darcy and Hartmann numbers when $Ra = 10^5$, $\phi = 0.04$, $Rd = 0.8$.

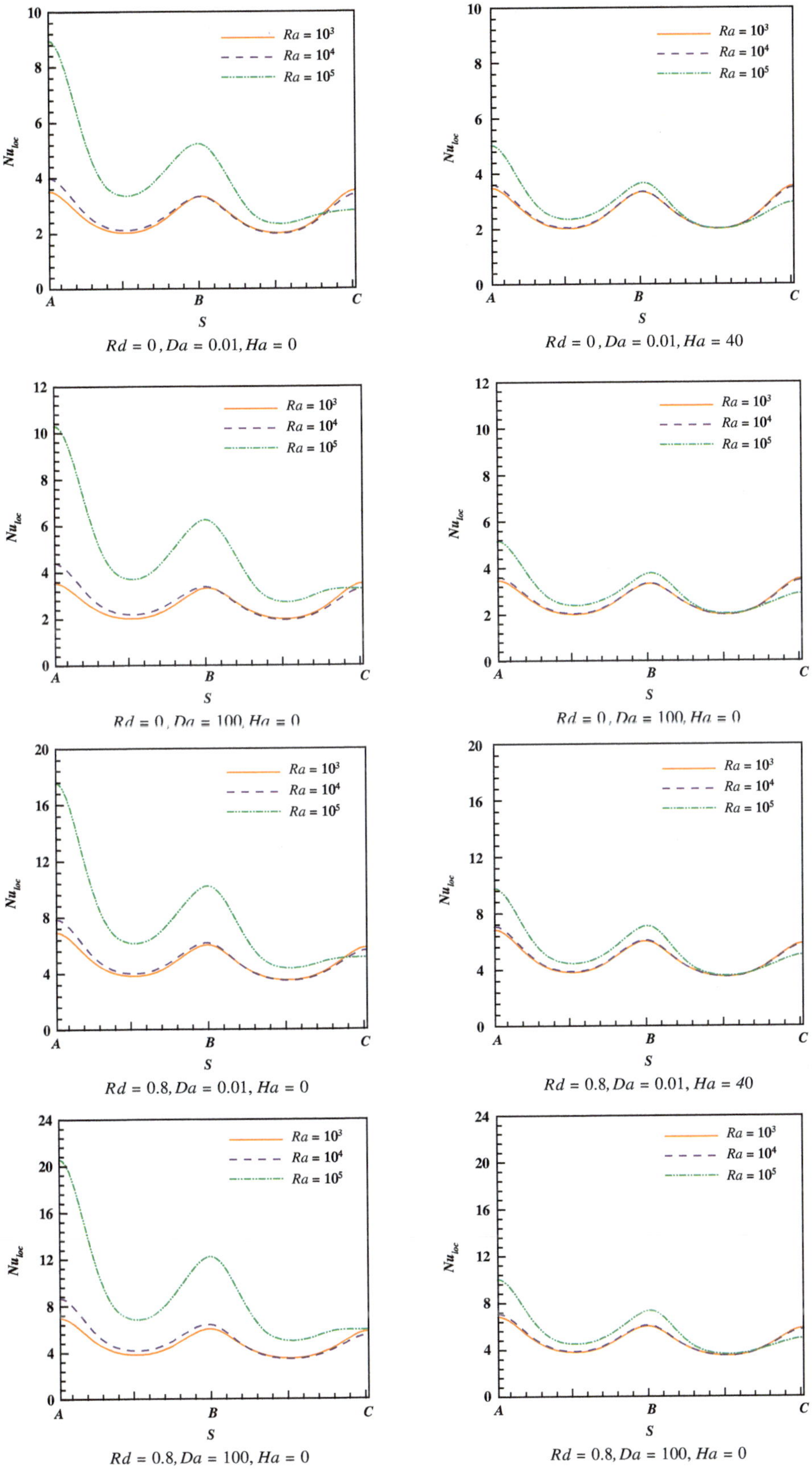

FIGURE 14.12 Effects of radiation parameter, Darcy, Rayleigh, and Hartmann numbers on local Nusselt number at $\phi = 0.04$.

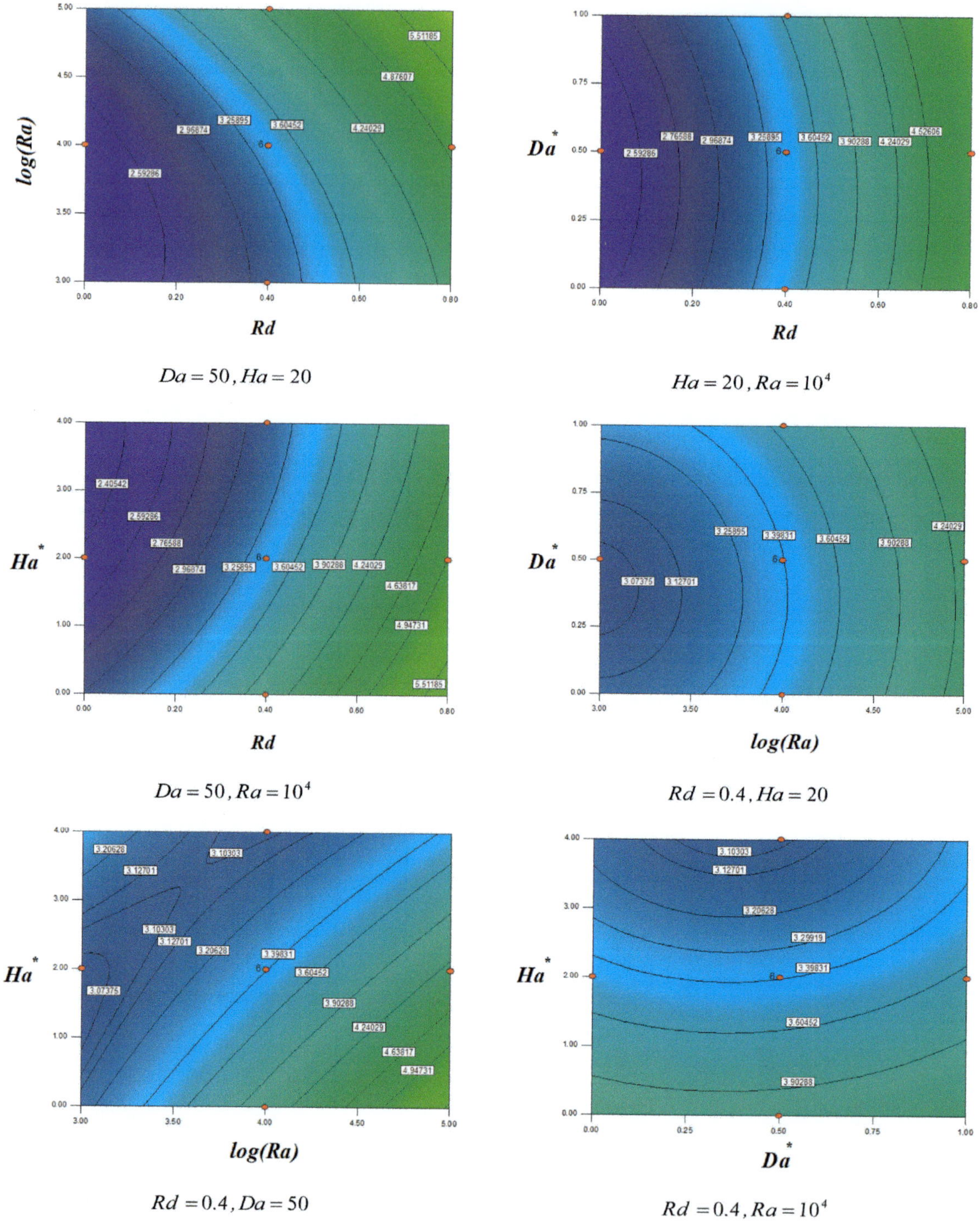

FIGURE 14.13 Effects of radiation parameter, Darcy, Rayleigh, and Hartmann numbers on local Nusselt number at $\phi = 0.04$.

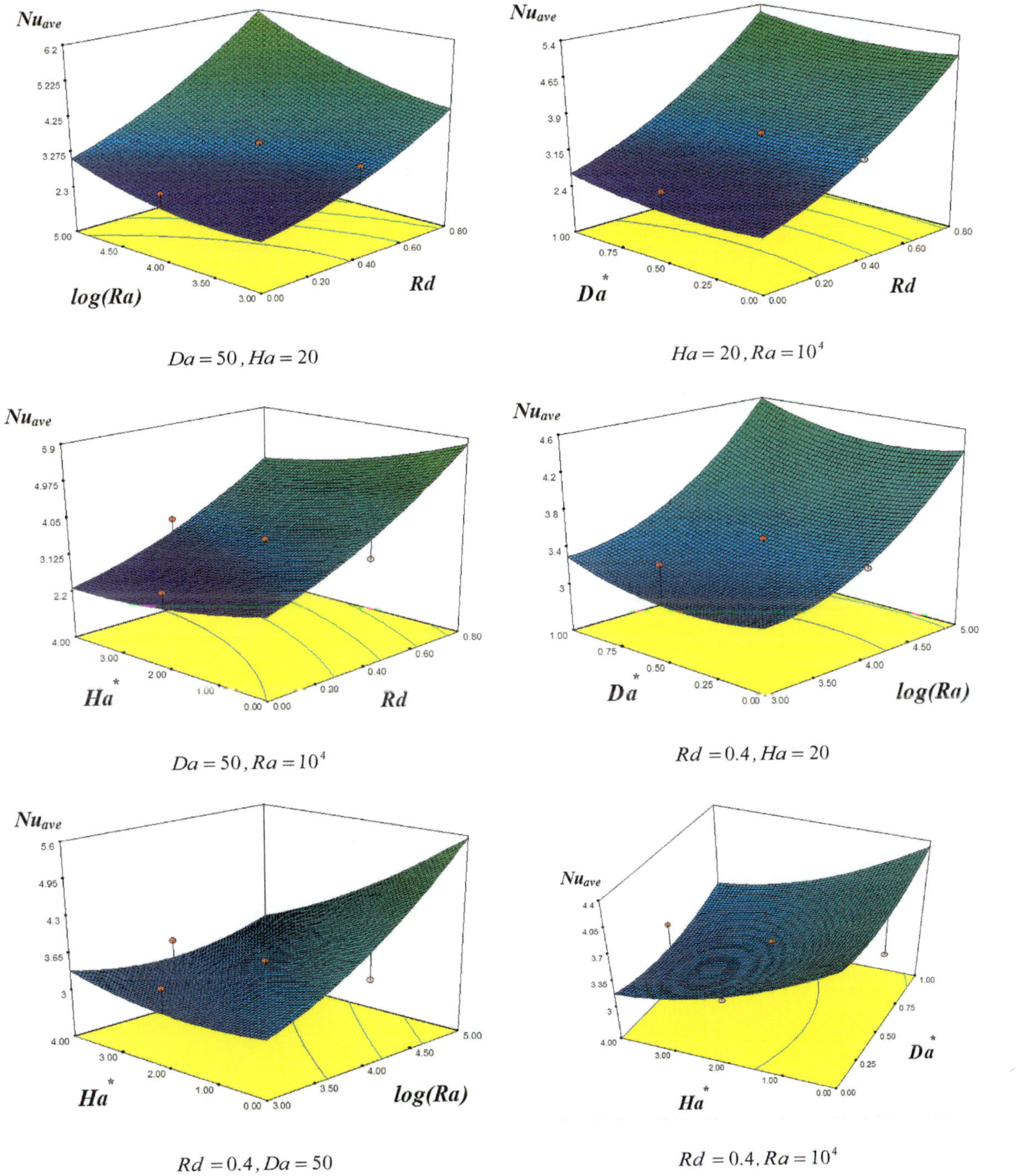

FIGURE 14.13 (Continued)

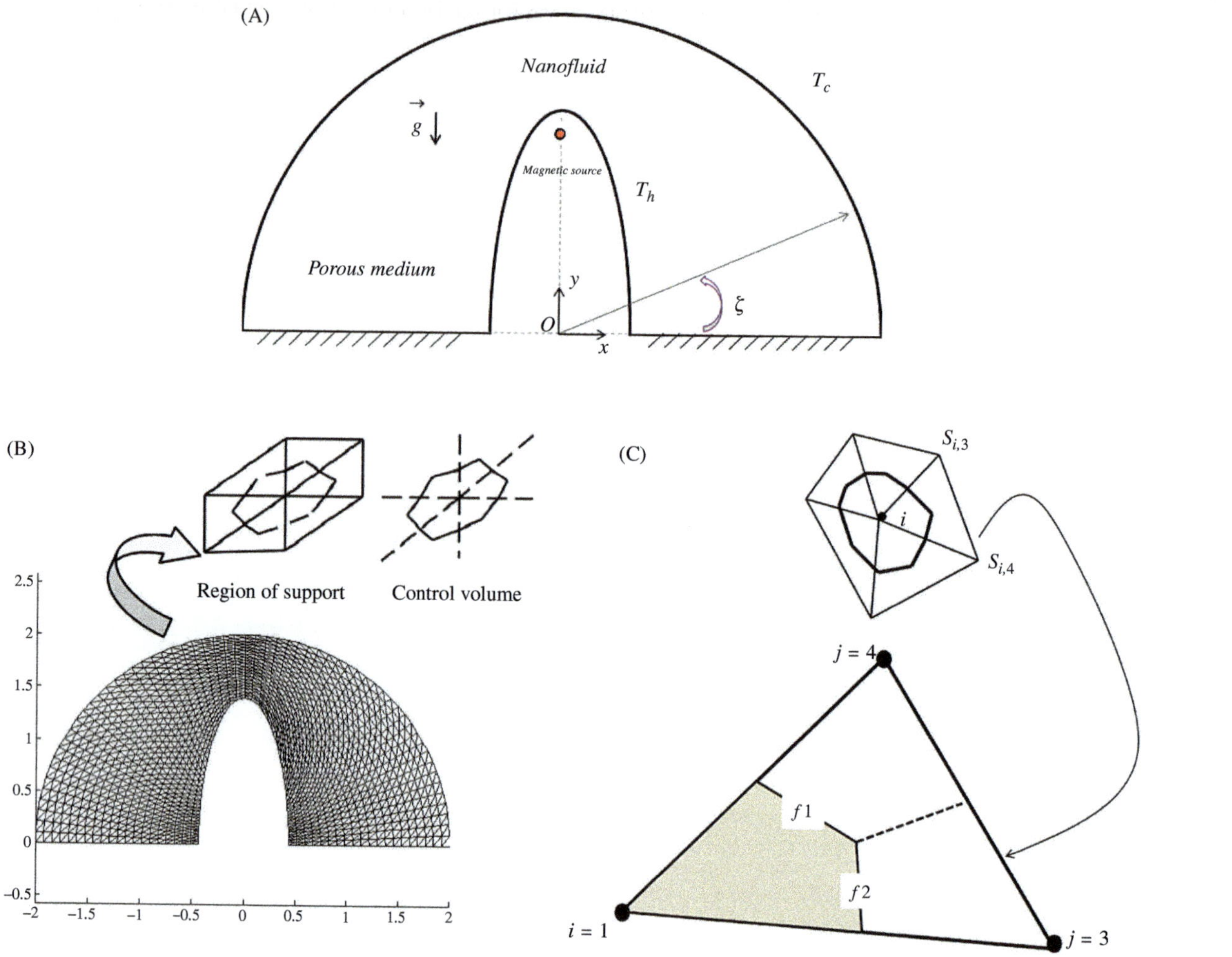

FIGURE 14.14 (A) Geometry and the boundary conditions with (B) the mesh of geometry considered in this work; (C) a sample triangular element and its corresponding control volume.

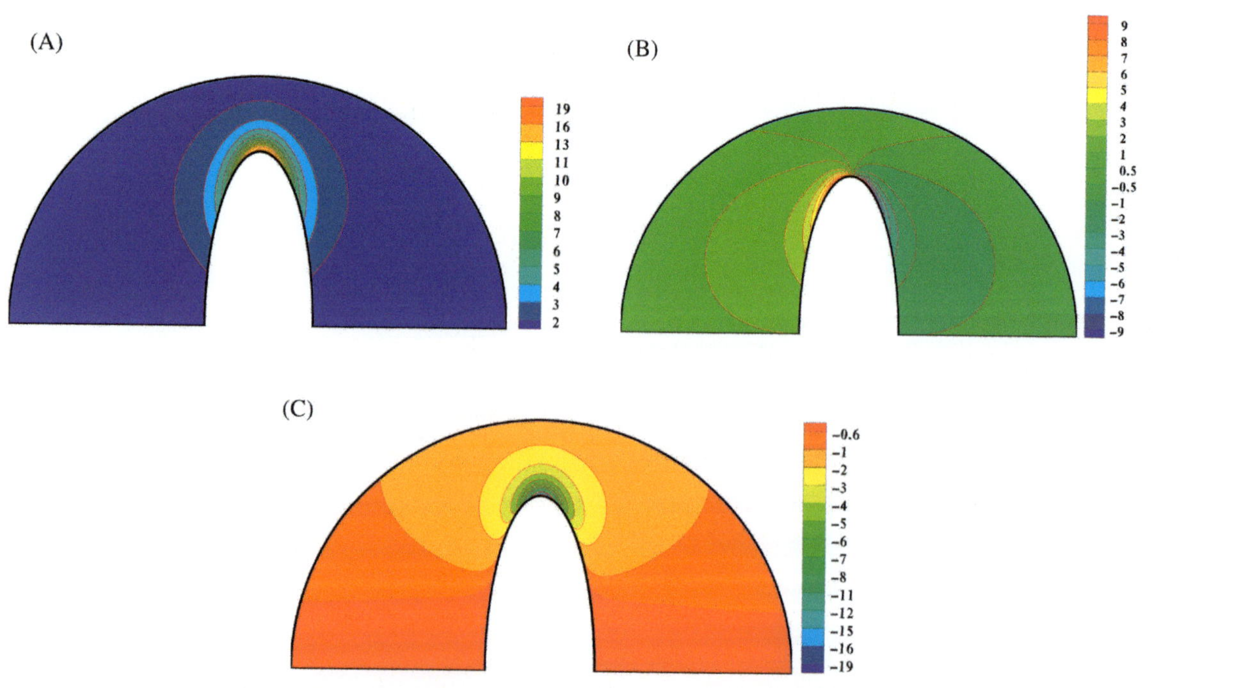

FIGURE 14.15 Contours of the (A) magnetic field strength H; (B) magnetic field intensity component in x direction Hx; (C) magnetic field intensity component in y direction Hy. (A) $H(x,y)$, (B) $Hx(x,y)$, (C) $Hy(x,y)$.

$$\overline{H_x} = \left[(\overline{b}-y)^2+(\overline{a}-x)^2\right]^{-1}\frac{\gamma}{2\pi}(y-\overline{b}), \tag{14.42}$$

$$\overline{H} = \sqrt{\overline{H}_x^2+\overline{H}_y^2}. \tag{14.43}$$

14.4.2 Governing Equation

2D laminar nanofluid flow and free convective heat transfer is taken into account. The governing PDEs are:

$$\frac{\partial v}{\partial y}+\frac{\partial u}{\partial x}=0, \tag{14.44}$$

$$\begin{aligned}\left(\rho_{nf}\right)\left(u\frac{\partial u}{\partial x}+\frac{\partial u}{\partial y}v\right) &= \left(\frac{\partial^2 u}{\partial y^2}+\frac{\partial^2 u}{\partial x^2}\right)\mu_{nf}-\frac{\partial P}{\partial x}\\ &-\mu_0^2\sigma_{nf}H_y^2u+\sigma_{nf}\mu_0^2H_xH_yv-\frac{\mu_{nf}}{K}u,\end{aligned} \tag{14.45}$$

$$\begin{aligned}\rho_{nf}\left(\frac{\partial v}{\partial x}u+\frac{\partial v}{\partial y}v\right) &= +\mu_{nf}\left(\frac{\partial^2 v}{\partial x^2}+\frac{\partial^2 v}{\partial y^2}\right)-\frac{\partial P}{\partial y}\\ &+\mu_0^2H_y\sigma_{nf}H_xu-\mu_0^2H_x\sigma_{nf}H_xv-\frac{\mu_{nf}}{K}v\\ &+(T-T_c)\beta_{nf}g\rho_{nf},\end{aligned} \tag{14.46}$$

$$\begin{aligned}(\rho C_p)_{nf}\left(v\frac{\partial T}{\partial y}+u\frac{\partial T}{\partial x}\right) &= \sigma_{nf}\mu_0^2(H_xv-H_yu)^2+k_{nf}\left(\frac{\partial^2 T}{\partial x^2}+\frac{\partial^2 T}{\partial y^2}\right)\\ &+\mu_{nf}\left\{2\left(\frac{\partial u}{\partial x}\right)^2+2\left(\frac{\partial v}{\partial y}\right)^2+\left(\frac{\partial u}{\partial y}+\frac{\partial v}{\partial x}\right)^2\right\},\end{aligned} \tag{14.47}$$

ρ_{nf}, $(\rho C_p)_{nf}$, β_{nf}, k_{nf}, and σ_{nf} are calculated as

$$\rho_{nf}=\rho_f(1-\phi)+\rho_s\phi, \tag{14.48}$$

$$(\rho C_p)_{nf}=(\rho C_p)_f(1-\phi)+(\rho C_p)_s\phi, \tag{14.49}$$

$$\beta_{nf}=\beta_f(1-\phi)+\beta_s\phi, \tag{14.50}$$

$$k_{nf}=k_f\left(\frac{k_s-2\phi(k_f-k_s)+2k_f}{k_s+\phi(k_f-k_s)+2k_f}\right), \tag{14.51}$$

$$\sigma_{nf}=\sigma_f\left[\frac{3(\sigma 1-1)\phi}{(\sigma 1+2)-(\sigma 1-1)\phi}+1\right], \qquad \sigma 1=\sigma_s/\sigma_f. \tag{14.52}$$

μ_{nf} is obtained as follows:

$$\mu_{nf}=(0.035\mu_0^2H^2+3.1\mu_0H-27886.4807\phi^2+4263.02\phi+316.0629)e^{-0.01T} \tag{14.53}$$

Dimensionless parameters are defined as:

$$\begin{aligned}&(b,a)=\frac{(\overline{b},\overline{a})}{L},\ (H_y,H_x,H)=\frac{(\overline{H_y},\overline{H_x},\overline{H})}{\overline{H_0}},\ P=\frac{p}{\rho_f(\alpha_f/L)^2}\\ &U=\frac{uL}{\alpha_f},\ V=\frac{vL}{\alpha_f},\ \Theta=\frac{T-T_c}{(T_h-T_c)},\ (X,Y)=\frac{(x,y)}{L}.\end{aligned} \tag{14.54}$$

So equations change to:

$$\frac{\partial V}{\partial Y} + \frac{\partial U}{\partial X} = 0, \tag{14.55}$$

$$\begin{aligned} U\frac{\partial U}{\partial X} + \frac{\partial U}{\partial Y}V &= \Pr\left[\frac{\mu_{nf}/\mu_f}{\rho_{nf}/\rho_f}\right]\left(\frac{\partial^2 U}{\partial Y^2} + \frac{\partial^2 U}{\partial X^2}\right) \\ &- Ha^2 \Pr\left[\frac{\sigma_{nf}/\sigma_f}{\rho_{nf}/\rho_f}\right]\left(H_y^2 U - H_x H_y V\right) - \frac{\partial P}{\partial X} - \frac{\Pr}{Da}\left[\frac{\mu_{nf}/\mu_f}{\rho_{nf}/\rho_f}\right]U, \end{aligned} \tag{14.56}$$

$$\begin{aligned} V\frac{\partial V}{\partial Y} + U\frac{\partial V}{\partial X} &= \Pr\left(\frac{\partial^2 V}{\partial Y^2} + \frac{\partial^2 V}{\partial X^2}\right)\left[\frac{\mu_{nf}/\mu_f}{\rho_{nf}/\rho_f}\right] \\ &- Ha^2 \Pr\left[\frac{\sigma_{nf}/\sigma_f}{\rho_{nf}/\rho_f}\right](H_x^2 V - H_x H_y U) \\ &- \frac{\partial P}{\partial Y} + Ra\Pr\left[\frac{\beta_{nf}}{\beta_f}\right]\Theta - \frac{\Pr}{Da}\left[\frac{\mu_{nf}/\mu_f}{\rho_{nf}/\rho_f}\right]V, \end{aligned} \tag{14.57}$$

$$\begin{aligned} V\frac{\partial \Theta}{\partial Y} + U\frac{\partial \Theta}{\partial X} &= \left[\frac{k_{nf}}{k_f}\frac{(\rho C_P)_f}{(\rho C_P)_{nf}}\right]\left(\frac{\partial^2 \Theta}{\partial Y^2} + \frac{\partial^2 \Theta}{\partial X^2}\right) \\ &+ Ha^2\, Ec\left(\frac{(\rho C_P)_f}{(\rho C_P)_{nf}}\frac{\sigma_{nf}}{\sigma_f}\right)\{V\, H_x - U\, H_y\}^2 \\ &+ \left[\frac{\dfrac{\mu_{nf}}{\mu_f}}{\dfrac{(\rho C_P)_{nf}}{(\rho C_P)_f}}\right] Ec\left\{2\left(\frac{\partial U}{\partial X}\right)^2 + 2\left(\frac{\partial V}{\partial Y}\right)^2 + \left(\frac{\partial U}{\partial Y} + \frac{\partial V}{\partial X}\right)^2\right\} \end{aligned} \tag{14.58}$$

and dimensionless parameters are

$$\begin{aligned} &Ra_f = g\beta_f L^3 \Delta T/(\alpha_f \upsilon_f), \Pr_f = \upsilon_f/\alpha_f, Ha = L\mu_0 H_0\sqrt{\sigma_f/\mu_f}, \\ &Ec = \left(\mu_f \alpha_f\right)/\left[(\rho C_P)_f \Delta T\, L^2\right],\ Da = K/L^2. \end{aligned} \tag{14.59}$$

The thermophysical properties of Fe_3O_4 and water are presented in Table 14.3. The pressure gradient source terms are discarded by the vorticity stream function.

$$\Omega = \frac{\omega L^2}{\alpha_f}, \Psi = \frac{\psi}{\alpha_f}, \omega = -\frac{\partial u}{\partial y} + \frac{\partial v}{\partial x}, (u, v) = \left(\frac{\partial \psi}{\partial y}, -\frac{\partial \psi}{\partial x}\right). \tag{14.60}$$

TABLE 14.3 Thermophysical Properties of Water and Nanoparticles

	$\rho(kg/m^3)$	$C_p(j/kgk)$	$k(W/m.k)$	$d_p(nm)$	$\sigma(\Omega\cdot m)^{-1}$
Pure water	997.1	4179	0.613	-	0.05
Fe_3O_4	5200	670	6	47	25000

According to Fig. 14.14, boundary conditions are:

$$\begin{aligned} &\text{on inner wall} && \theta = 1.0 \\ &\text{on outer wall} && \theta = 0.0 \\ &\text{on other walls} && \frac{\partial \theta}{\partial y} = 0 \\ &\text{on all walls} && \Psi = 0.0 \end{aligned} \tag{14.61}$$

Nu_{loc}, Nu_{ave} along the cold wall are :

$$Nu_{loc} = \left| \left(\frac{k_{nf}}{k_f} \right) \frac{\partial \Theta}{\partial r} \right|, \tag{14.62}$$

$$Nu_{ave} = \frac{1}{0.5\pi} \int_0^{0.5\pi} Nu_{loc}(\zeta)\, d\zeta, \tag{14.63}$$

14.4.3 Effects of Active Parameters

In this section, the influence of magnetic field on Fe_3O_4-water nanofluid in a porous enclosure is reported. The governing equations have been solved via CVFEM and the outputs are depicted in several plots for the influence of various parameters on the flow and heat transfer. These parameters are Darcy number (Da), Rayleigh number (Ra), Hartmann number (Ha), and volume fraction of Fe_3O_4 (ϕ). Pr and Ec are 6.8 and 10^{-5}, respectively.

Fig. 14.16 demonstrates the influence of adding Fe_3O_4 into water on the hydrothermal characteristics. This figure depicts that an increase in nanoparticle volume fraction results in an increase in nanofluid velocity. It is also found that the thermal boundary layer thickness of water-based nanofluid is higher than pure fluid.

Figs. 14.17–14.19 illustrate the impact of Darcy, Hartmann, and Rayleigh numbers on isotherms and streamlines. In dominant conduction modes, one main clockwise cell appears in half of the enclosure. An increase in magnetic field results in the generation of a secondary cell near the vertical centerline. As the permeability of porous media enhances, the convective heat transfer improves and a thermal plume appears. An increase in buoyancy forces result in an increase in strength of the main eddy and a thermal plume is generated near the vertical centerline. As Lorentz forces increase, the position of the thermal plume becomes far from the

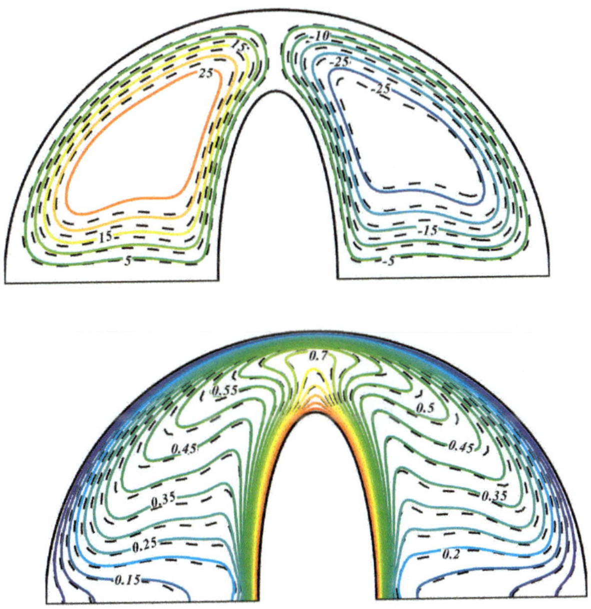

FIGURE 14.16 Impact of nanofluid volume fraction on streamline (top) and isotherm (bottom) contours (nanofluid ($\phi = 0.04$)(–) and pure fluid ($\phi = 0$) (- - -)) when $Ra = 10^5, Da = 100,\ Ha = 0$.

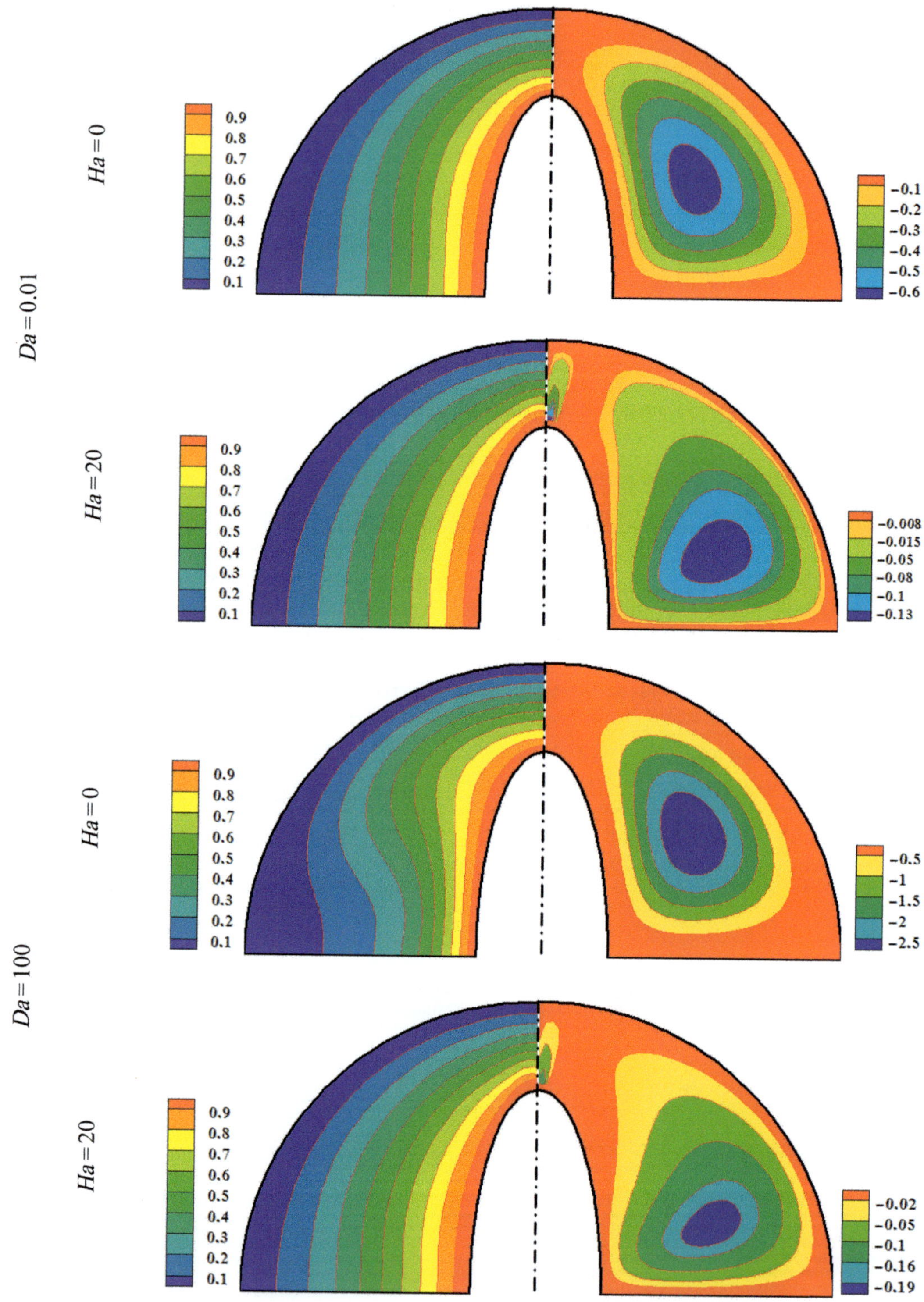

FIGURE 14.17 Influence of *Da*, *Ha* on streamline (left) and isotherm (right) contours when $\phi = 0.04$, $Ra = 10^3$.

vertical centerline. It is a fantastic observation that in the case of Da = 100, Ra = 10^5, $|\Psi_{max}|$ reaches to its maximum value and the main eddy stretches horizontally. Also one powerful thermal plume generates near the $\zeta = 90^{\circ}$. Applying a magnetic field for such a case converts the main eddy to three smaller ones. The middle one rotates counterclockwise. The existence of such eddies generates two thermal plumes over the hot elliptic wall.

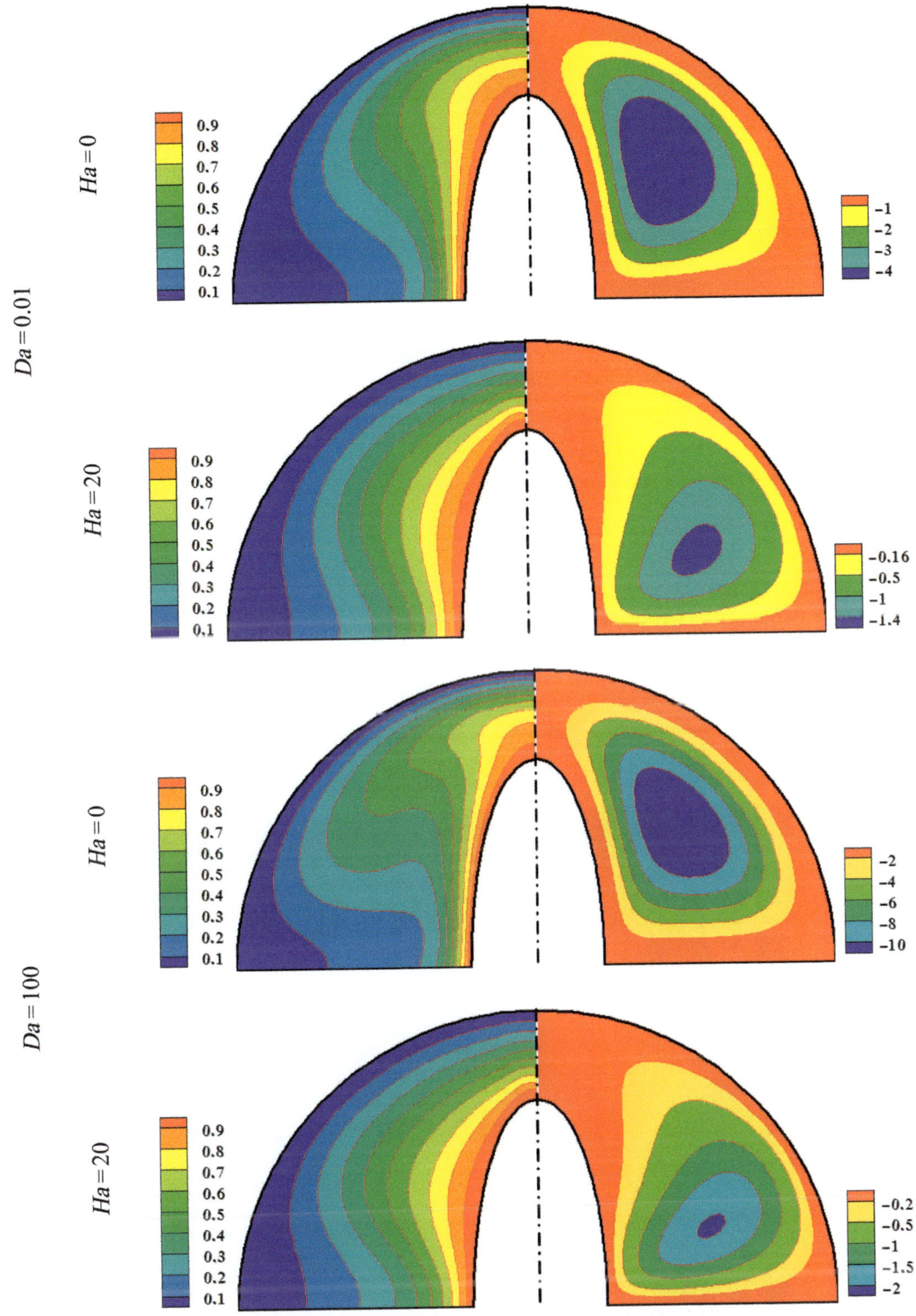

FIGURE 14.18 Influence of *Da*, *Ha* on streamline (left) and isotherm (right) contours when $\phi = 0.04$, $Ra = 10^4$.

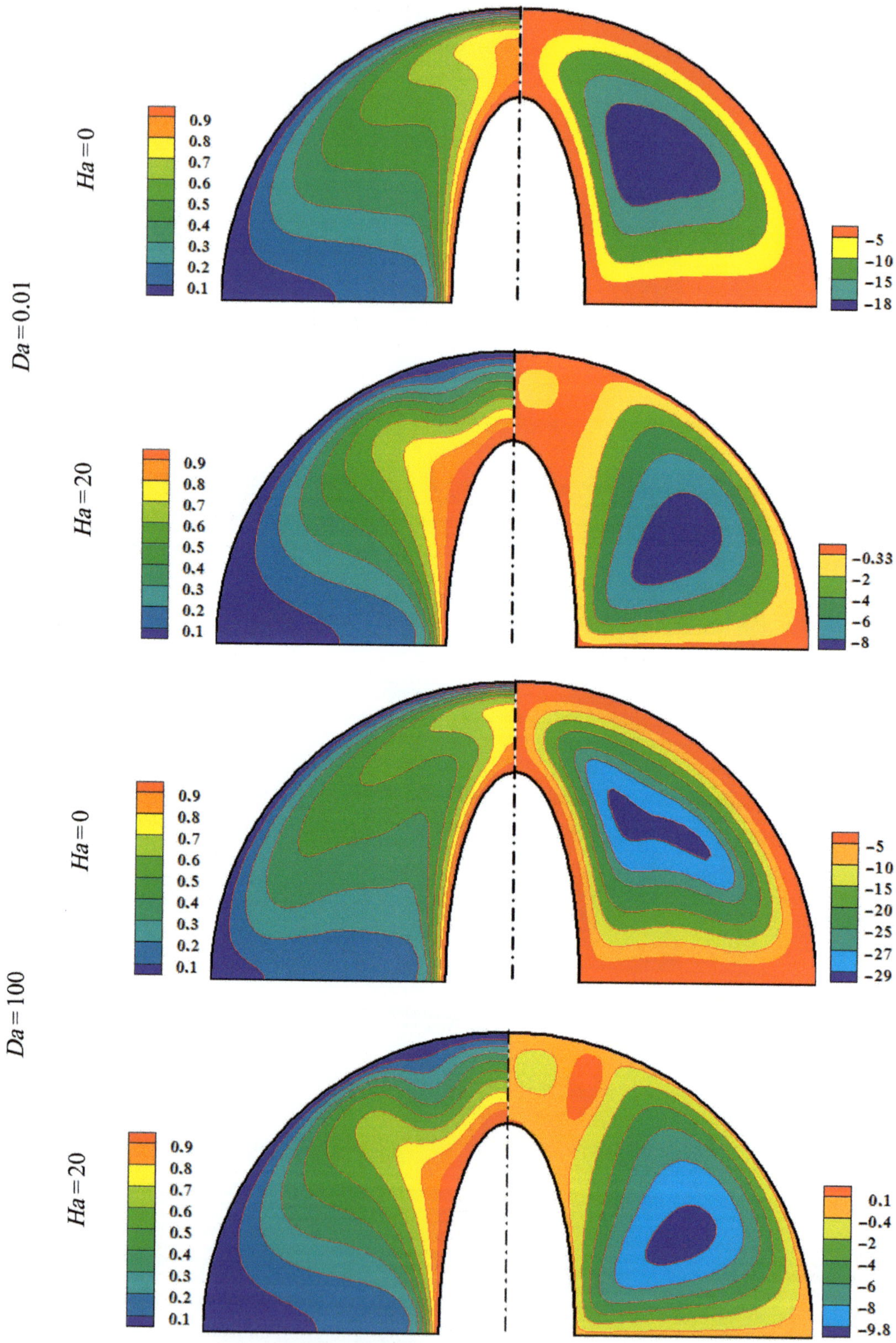

FIGURE 14.19 Influence of *Da*, *Ha* on streamline (left) and isotherm (right) contours when $\phi = 0.04$, $Ra = 10^5$.

The rate of heat transfer is depicted in Fig. 14.20. The formula of Nu_{ave} corresponding to important parameters is:

$$\begin{aligned} Nu_{ave} &= 4.37 + 0.15Da^* - 2.77\log(Ra) + 0.03Ha^* \\ &+ 0.03Da^*\log(Ra) - 0.21Da^*Ha^* - 0.39\log(Ra)Ha^* \\ &+ 0.07Da^{*2} + 0.51\left(\log(Ra)\right)^2 + 0.61Ha^{*2} \end{aligned} \tag{14.64}$$

where $Da^* = 0.01Da$, $Ha^* = 0.1Ha$. Increasing the permeability of the porous media results in an increase in the rate of heat transfer. Enhancing the Rayleigh number makes the convective heat transfer increase. So this nondimension parameter has a similar effect on the Nusselt number to that of obtained for Darcy number. As the Lorentz force increases, Nu_{ave} reduces due to the domination of the conduction mode. Adding Fe_3O_4 nanoparticles into base fluid enhances the Nusselt number. Table 14.4 demonstrates the influence of Da, Ha, and Ra on heat transfer improvement. This output is defined as $E = 100 * \left(Nu_{ave}|_{\phi=0.04} - Nu_{ave}|_{\phi=0}\right)/Nu_{ave}|_{\phi=0}$. In conduction mode, the influence of adding nanoparticles has more benefit because of more changes in thermal conductivity. Therefore, heat transfer improvement enhances with enhance of Hartmann number but it decreases with the rise of Darcy and Rayleigh numbers.

14.5 NANOFLUID CONVECTIVE HEAT TRANSFER INTENSIFICATION IN A POROUS CIRCULAR CYLINDER

14.5.1 Problem Definition

Fig. 14.21 depicts the geometry, boundary condition, and sample element. The inner cylinder has a constant heat flux condition and the outer cylinder is cold. A horizontal magnetic field has been applied in this porous media. A non-Darcy model is utilized for the porous media.

14.5.2 Governing Equation

Steady convective nanofluid flow in a porous enclosure is considered in the presence of a uniform magnetic field. The PDEs equations are:

$$\frac{\partial v}{\partial y} + \frac{\partial u}{\partial x} = 0 \tag{14.65}$$

$$\left(v\frac{\partial u}{\partial y} + u\frac{\partial u}{\partial x}\right) = \left[\sigma_{nf}B_xB_yv - \sigma_{nf}B_y^2 + \left(\frac{\partial^2 u}{\partial y^2} + \frac{\partial^2 u}{\partial x^2}\right)\mu_{nf} - \frac{\partial P}{\partial x} - \frac{\mu_{nf}}{K}u\right]\left(\rho_{nf}\right)^{-1} \tag{14.66}$$

$$\begin{aligned} &\rho_{nf}\left(\frac{\partial v}{\partial x}u + \frac{\partial v}{\partial y}v\right) = \mu_{nf}\left(\frac{\partial^2 v}{\partial x^2} + \frac{\partial^2 v}{\partial y^2}\right) - \frac{\partial P}{\partial y} - B_x\sigma_{nf}B_xv + B_y\sigma_{nf}B_xu - \frac{\mu_{nf}}{K}v \\ &+ (T - T_c)\beta_{nf}g\rho_{nf}, \\ &B_x = B_o\cos\lambda, \quad B_y = B_o\sin\lambda \end{aligned} \tag{14.67}$$

$$\left(\rho C_p\right)_{nf}\left(v\frac{\partial T}{\partial y} + u\frac{\partial T}{\partial x}\right) = k_{nf}\left(\frac{\partial^2 T}{\partial x^2} + \frac{\partial^2 T}{\partial y^2}\right) \tag{14.68}$$

$\left(\rho C_p\right)_{nf}$, $(\rho\beta)_{nf}$, ρ_{nf}, k_{nf}, and σ_{nf} are defined as:

$$\left(\rho C_p\right)_{nf} = \phi\left(\rho C_p\right)_s + (1 - \phi)\left(\rho C_p\right)_f \tag{14.69}$$

$$\frac{(\rho\beta)_{nf}}{(\rho\beta)_f} = \phi\frac{(\rho\beta)_s}{(\rho\beta)_f} + (1 - \phi) \tag{14.70}$$

$$\rho_{nf} = \rho_f(1 - \phi) + \rho_s\phi \tag{14.71}$$

$$k_{nf} = k_f\left(\frac{k_s + 2k_f + 2\phi(k_s - k_f)}{k_s - \phi(k_s - k_f) + 2k_f}\right) \tag{14.72}$$

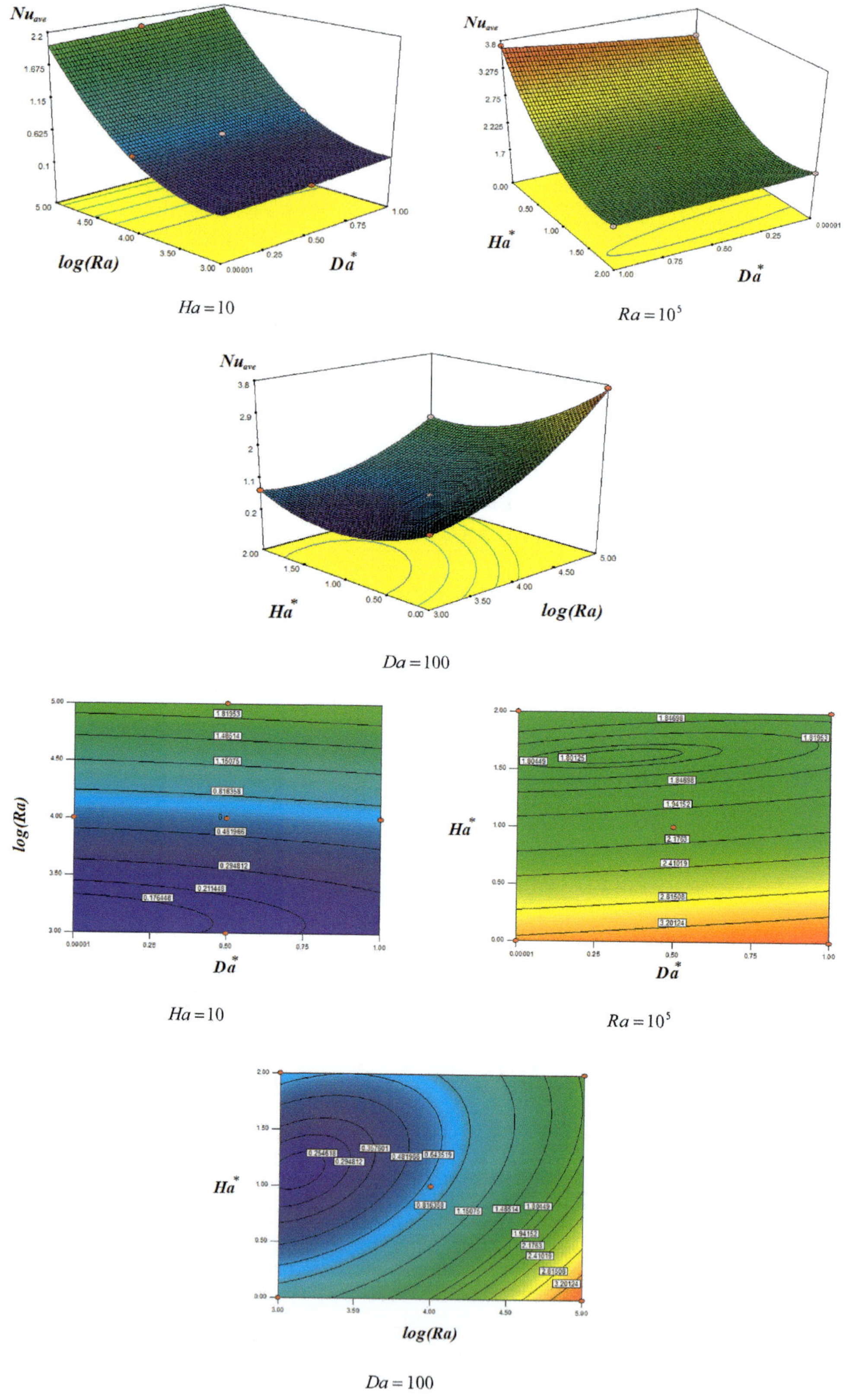

FIGURE 14.20 Effects of *Da*, *Ha*, and *Ra* on average Nusselt number.

TABLE 14.4 Effects of *Da*, *Ha*, and *Ra* on Heat Transfer Enhancement

Ra	*Da*	*Ha*	E
10^3	0.01	0	11.09238
10^3	0.01	20	11.42401
10^4	0.01	0	4.20269
10^4	0.01	20	7.652818
10^5	0.01	0	2.613391
10^5	0.01	20	3.245394
10^3	100	0	5.874633
10^3	100	20	7.615688
10^4	100	0	3.787017
10^4	100	20	6.464888
10^5	100	0	3.757416
10^5	100	20	1.560289

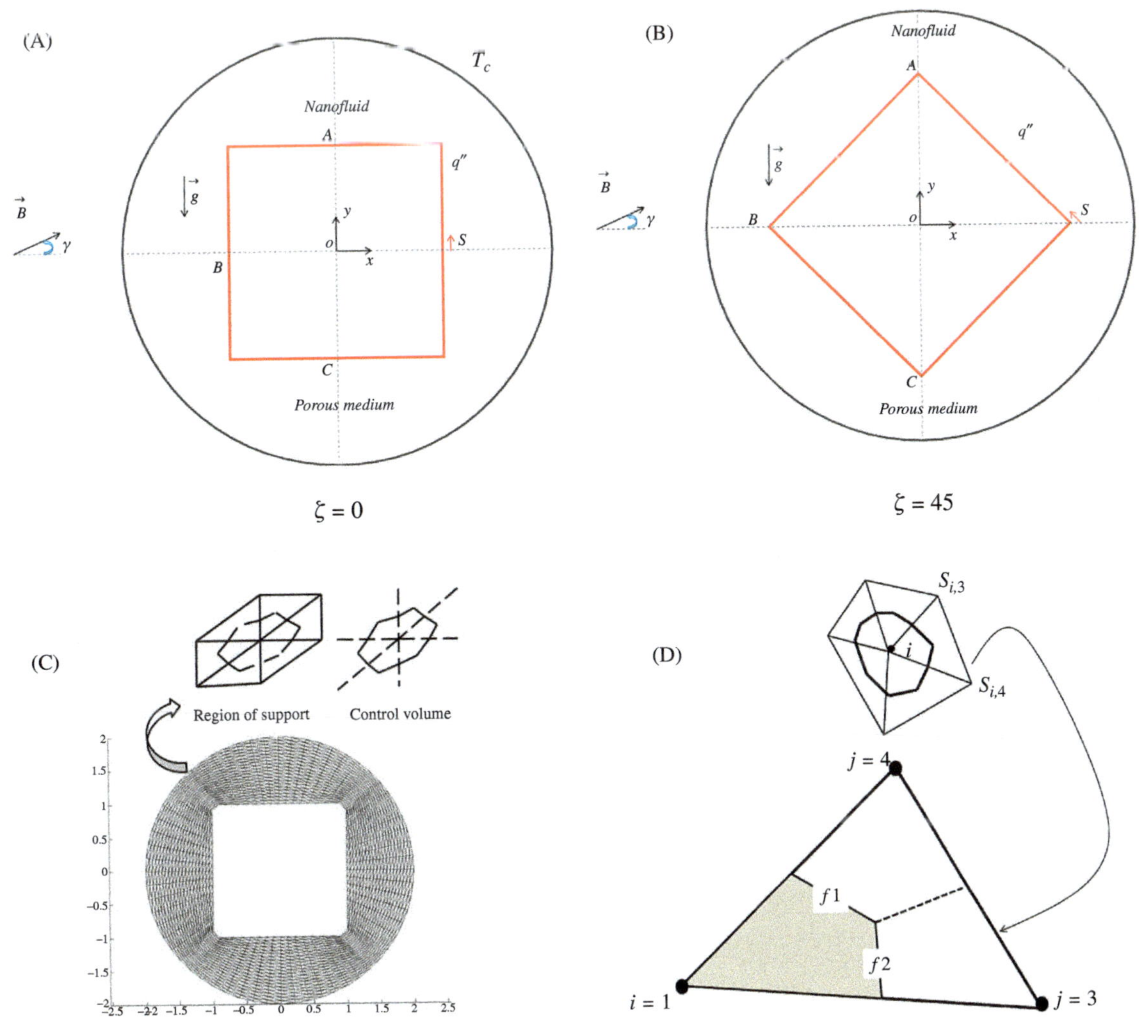

FIGURE 14.21 (A, B) Geometry and the boundary conditions with (C) the mesh of geometry considered in this work; (D) a sample triangular element and its corresponding control volume.

$$\sigma_{nf} = \left[\left(\frac{(2+\sigma_s/\sigma_f)-(\sigma_s/\sigma_f-1)\phi}{3\phi(-1+\sigma_s/\sigma_f)}\right)^{-1}+1\right]\sigma_f \tag{14.73}$$

μ_{nf} is obtained as follows:

$$\mu_{nf} = \left(0.035B^2+3.1B-27886.4807\phi^2+4263.02\phi+316.0629\right)e^{-0.01T} \tag{14.74}$$

The properties of nanofluid are provided in Table 14.1. Vorticity and stream function should be used to eliminate the pressure source terms:

$$\omega+\frac{\partial u}{\partial y}-\frac{\partial v}{\partial x}=0,\quad \frac{\partial\psi}{\partial x}=-v,\quad \frac{\partial\psi}{\partial y}=u \tag{14.75}$$

Introducing dimensionless quantities:

$$(Y,X)=(y,x)/L, P=\frac{p}{\rho_{nf}\left(\alpha_{nf}/L\right)^2}, U=\frac{uL}{\alpha_{nf}}, V=\frac{vL}{\alpha_{nf}},\ \theta=\frac{T-T_c}{\Delta T}, \Delta T=q''L/k_f \tag{14.76}$$

The final formulae are:

$$\frac{\partial^2\Psi}{\partial X^2}+\Omega+\frac{\partial^2\Psi}{\partial Y^2}=0, \tag{14.77}$$

$$\begin{aligned}\frac{\partial\Omega}{\partial X}U+\frac{\partial\Omega}{\partial Y}V &= \Pr\frac{A_5}{A_1}\frac{A_2}{A_4}\left(\frac{\partial^2\Omega}{\partial Y^2}+\frac{\partial^2\Omega}{\partial X^2}\right)\\ &+\Pr Ha^2\frac{A_6}{A_1}\frac{A_2}{A_4}\left(\frac{\partial U}{\partial X}B_xB_y-\frac{\partial V}{\partial X}B_x{}^2+\frac{\partial U}{\partial Y}B_y^2-\frac{\partial V}{\partial Y}B_xB_y\right)\\ &+\Pr Ra\frac{A_3A_2^2}{A_1A_4^2}\frac{\partial\theta}{\partial X}-\frac{\Pr}{Da}\left(\frac{A_5A_2}{A_1A_4}\right)\Omega,\end{aligned} \tag{14.78}$$

$$V\frac{\partial\theta}{\partial Y}+\frac{\partial\theta}{\partial X}U=\left(\frac{\partial^2\theta}{\partial Y^2}+\frac{\partial^2\theta}{\partial X^2}\right). \tag{14.79}$$

where dimensionless and constants parameters are illustrated as:

$$\begin{gathered}\Pr=\upsilon_f/\alpha_f, Ra=g\beta_f q''L^4/\left(k_f\upsilon_f\alpha_f\right),\ Ha=LB_0\sqrt{\sigma_f/\mu_f},\ Da=\frac{K}{L^2},\\ A_1=\frac{\rho_{nf}}{\rho_f},\quad A_3=\frac{(\rho\beta)_{nf}}{(\rho\beta)_f},\quad A_5=\frac{\mu_{nf}}{\mu_f},\quad A_2=\frac{(\rho C_P)_{nf}}{(\rho C_P)_f}, A_4=\frac{k_{nf}}{k_f},\quad A_6=\frac{\sigma_{nf}}{\sigma_f}\end{gathered} \tag{14.80}$$

and boundary conditions are:

$$\begin{aligned}&\frac{\partial\theta}{\partial n}=1.0 && \text{on inner wall}\\ &\theta=0.0 && \text{on outer wall}\\ &\Psi=0.0 && \text{on all walls}\end{aligned} \tag{14.81}$$

Local and average Nusselt numbers over the inner cylinder can be calculated as:

$$Nu_{loc}=\frac{1}{\theta}\left(\frac{k_{nf}}{k_f}\right) \tag{14.82}$$

$$Nu_{ave}=\frac{1}{S}\int_0^S Nu_{loc}\,ds \tag{14.83}$$

14.5.3 Effects of Active Parameters

The influence of magnetic field on nanofluid transportation in a porous cylinder with inner inclined square obstacle is presented. The working fluid is considered as Fe_3O_4-water and its viscosity is a function of ϕ and Ha. Results are demonstrated for several values of volume fraction of Fe_3O_4-water ($\phi = 0$ to 0.04), Darcy number ($Da = 0.001$ to 100), Hartmann number ($Ha = 0$ to 40), Rayleigh number ($Ra = 10^3$ to 10^5), and inclination angle ($\xi = 0°$ and $45°$).

Fig. 14.22 demonstrates the impact of ϕ on isotherms and streamlines. Augmenting the nanofluid volume fraction leads to an increase in the temperature boundary layer thickness. The nanofluid velocity increase because of the increase of ϕ. The impacts of Ha, ξ, Da, and Ra on hydrothermal behavior are demonstrated in Figs. 14.23–14.25. As nanofluid temperature increases, the nanofluid initiates, moving from the inner cylinder to the cold one and dropping along the outer cylinder. Conduction mode is dominant at low Rayleigh and Darcy numbers. So isotherms follow the shape of the enclosure. At $\xi = 0°$, one main eddy exists and when the inner cylinder is inclined ($\xi = 45°$) the main eddy converts to two similar ones. The strength of this main eddy is enhanced with the rise of convective heat transfer. So $|\Psi_{max}|$ rises with the increase of Da, Ra. Also a thermal plume appears near the vertical centerline when convection mode is dominating. As ξ increases, the distortion of the isotherms is enhanced. As the magnetic field increases, $|\Psi_{max}|$ reduces and the center of the main eddy moves upward. Also the Lorentz force makes the thermal plume diminish.

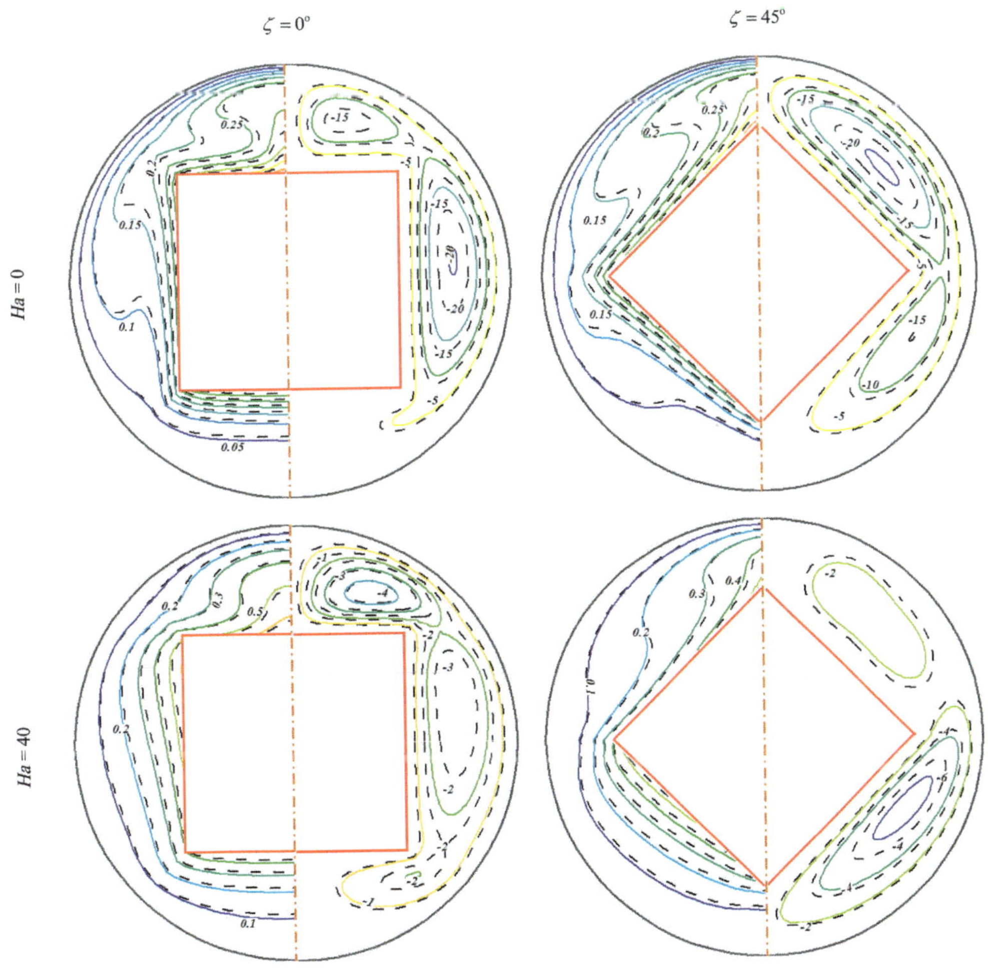

FIGURE 14.22 Effect of volume fraction of nanofluid on isotherm (left) and streamline (right) contours ($\phi = 0$ (−−), $\phi = 0.04$ (−)) when $Da = 100, Ra = 10^5$.

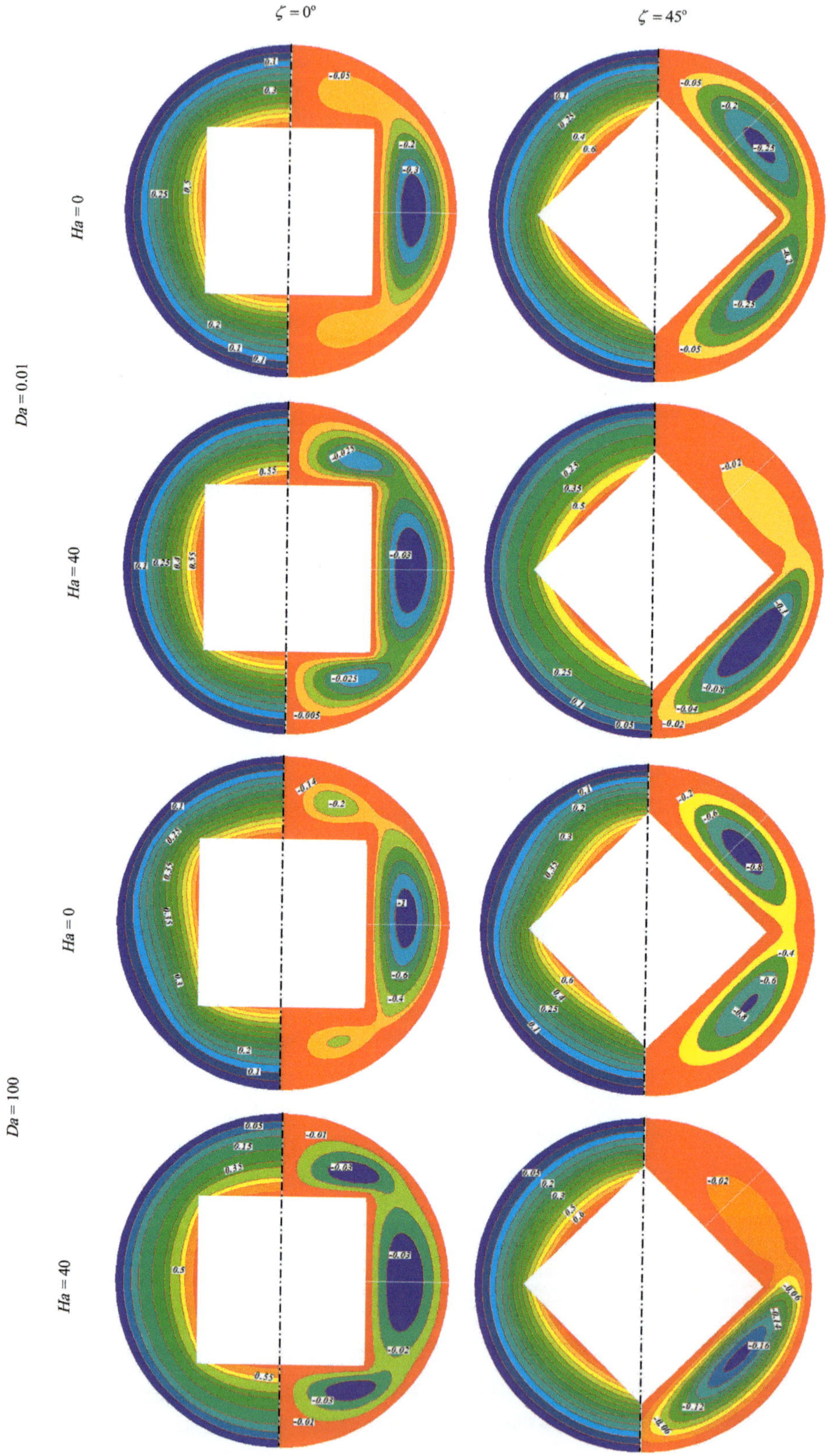

FIGURE 14.23 Effect of Darcy and Hartmann numbers on isotherm (left) and streamline (right) contours when $\phi = 0.04, Ra = 10^3$.

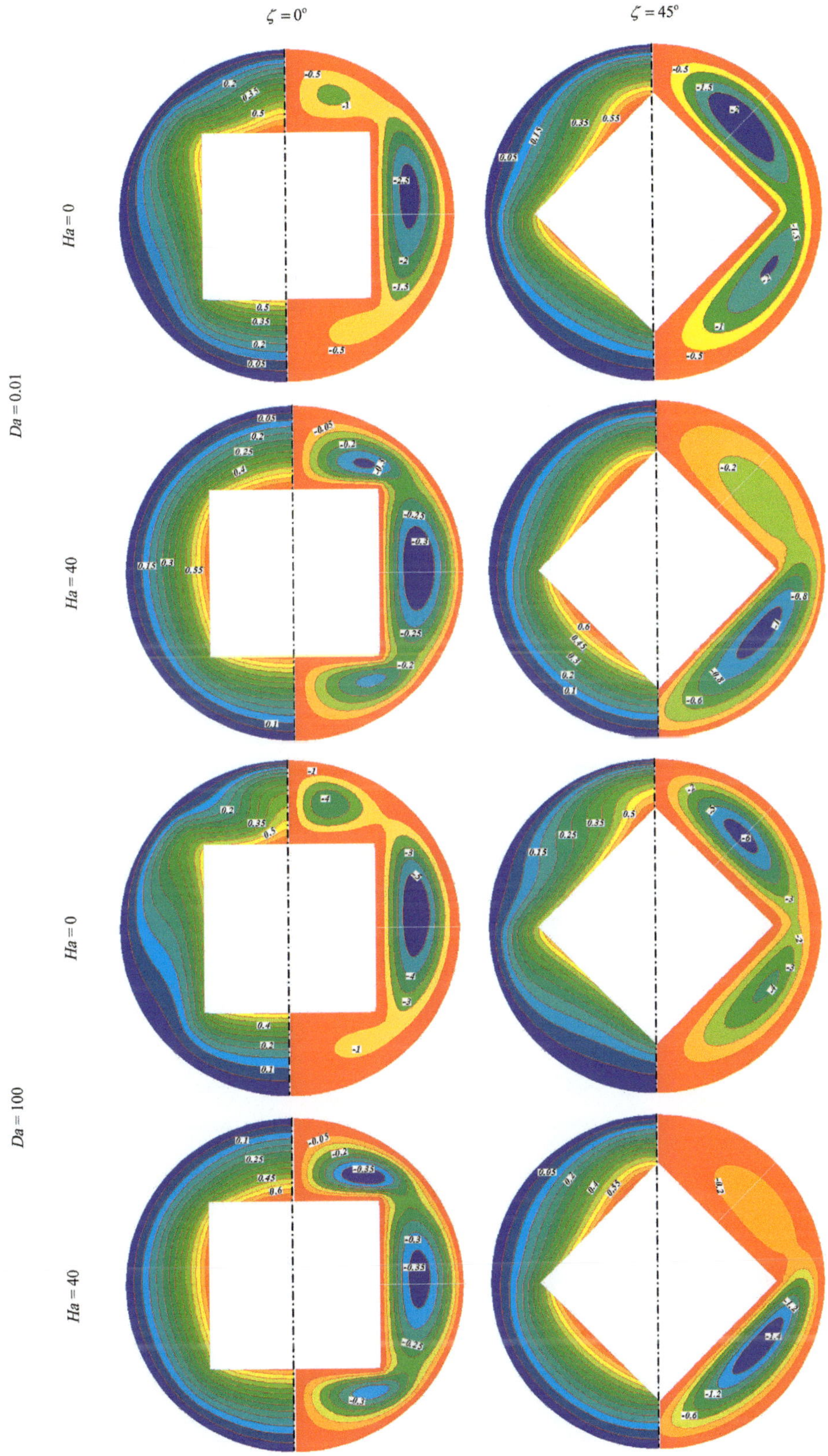

FIGURE 14.24 Effect of Darcy and Hartmann numbers on isotherm (left) and streamline (right) contours when $\phi = 0.04$, $Ra = 10^4$.

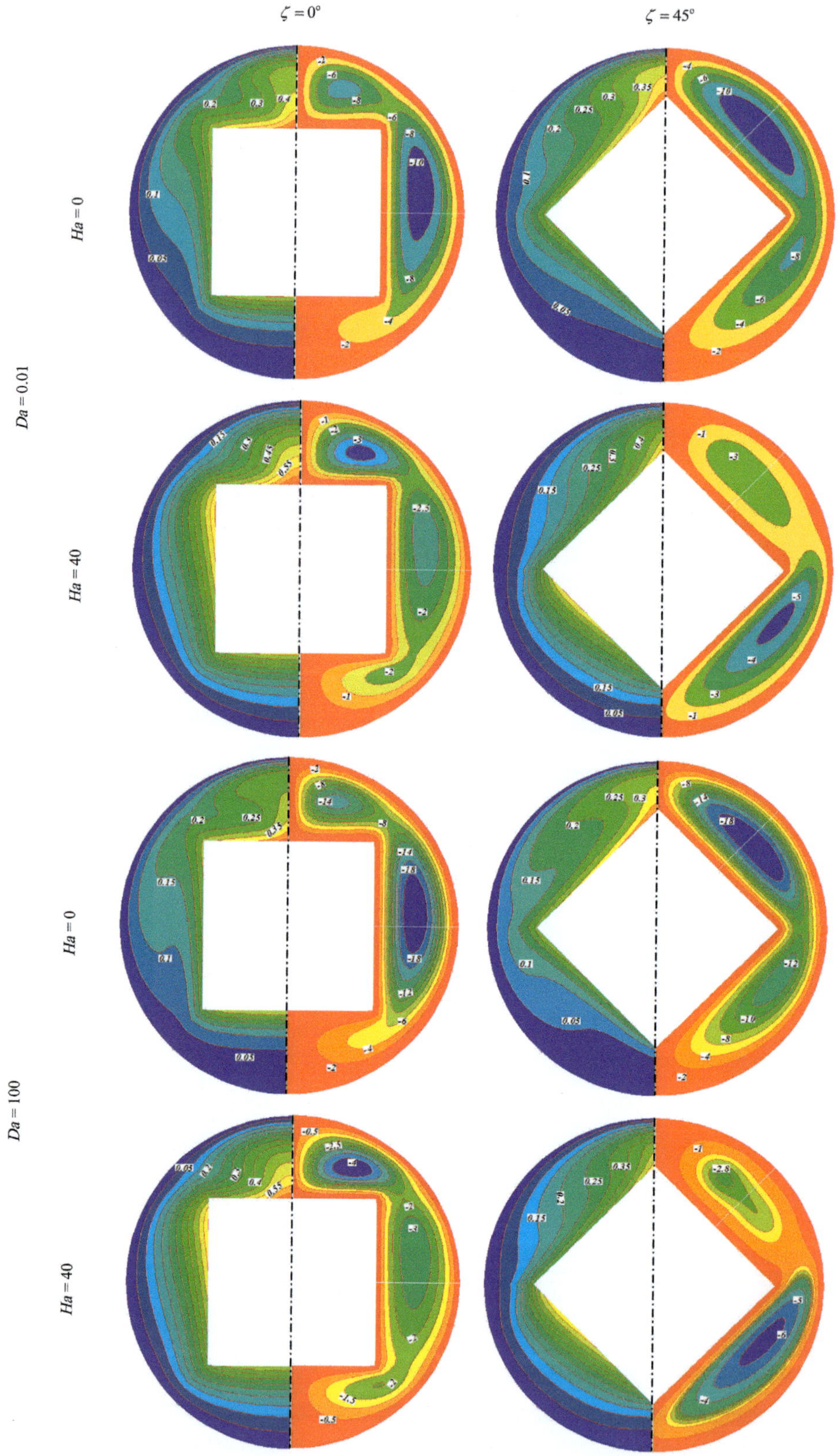

FIGURE 14.25 Effect of Darcy and Hartmann numbers on isotherm (left) and streamline (right) contours when $\phi = 0.04$, $Ra = 10^5$.

Figs. 14.26 and 14.27 illustrate the impact of ξ, Da, Ra, and Ha on Nu_{ave}, Nu_{loc}. With respect to the active parameters, the following equation can be obtained:

$$\begin{aligned} Nu_{ave} &= 4.58 - 0.66\xi - 2\log(Ra) - 0.58Da^* + 0.18Ha^* \\ &+ 0.19\xi\,\log(Ra) + 0.09\xi Da^* - 0.04\xi Ha^* + 0.18\log(Ra)Da^* \\ &- 0.15\log(Ra)\,Ha^* - 0.41Da^*\,Ha^* + 0.17\xi^2 + 0.33(\log(Ra))^2 + 0.1Rd^2 \\ &+ 0.08Ha^{*2} \end{aligned} \tag{14.84}$$

where $Ha^* = 0.1Ha$, $Da^* = 0.01Da$. The number of extremum in Nu_{loc} matches the presence of the corners of the square cylinder and thermal plume. Nu_{ave} increases with the increase of Da, Ra, ξ but it decreases with the rise of Ha.

14.6 CONVECTIVE FLOW OF NANOFLUID INSIDE A LID-DRIVEN POROUS CAVITY

14.6.1 Problem Definition

Sample element, boundary condition, and geometry are depicted in Fig. 14.28. The south wall is hot and the others are cold. Also the south wall can move horizontally. A porous cavity is filled with nanofluid and affected by a horizontal magnetic field.

14.6.2 Governing Equation

Nanofluid forced convective non-Darcy flow is taken into account in the presence of a uniform magnetic field. The equations are:

$$\frac{\partial v}{\partial y} + \frac{\partial u}{\partial x} = 0 \tag{14.85}$$

$$\left[B_y\sigma_{nf}B_x v - B_y^2\sigma_{nf}u + \left(\frac{\partial^2 u}{\partial y^2} + \frac{\partial^2 u}{\partial x^2}\right)\mu_{nf} - \frac{\partial P}{\partial x}\right] - \frac{\mu_{nf}}{K}u = \rho_{nf}\left(v\frac{\partial u}{\partial y} + \frac{\partial u}{\partial x}u\right) \tag{14.86}$$

$$\mu_{nf}\left(\frac{\partial^2 v}{\partial y^2} + \frac{\partial^2 v}{\partial x^2}\right) - \frac{\partial P}{\partial y} + B_y\sigma_{nf}uB_x - B_x\sigma_{nf}B_x v - \frac{\mu_{nf}}{K}v = \rho_{nf}\left(\frac{\partial v}{\partial x}u + \frac{\partial v}{\partial y}v\right), \tag{14.87}$$

$$B_x = B_o\cos\lambda,\quad B_y = B_o\sin\lambda$$

$$k_{nf}\left(\frac{\partial^2 T}{\partial x^2} + \frac{\partial^2 T}{\partial y^2}\right) = (\rho C_p)_{nf}\left(\frac{\partial T}{\partial x}u + \frac{\partial T}{\partial y}v\right), \tag{14.88}$$

σ_{nf}, $(\rho C_p)_{nf}$ and ρ_{nf} are :

$$\frac{\sigma_{nf}}{\sigma_f} = 1 + \frac{3\left(\frac{\sigma_s}{\sigma_f} - 1\right)\phi}{\left(\frac{\sigma_s}{\sigma_f} + 2\right) - \left(\frac{\sigma_s}{\sigma_f} - 1\right)\phi} \tag{14.89}$$

$$(\rho C_p)_{nf} = \phi(\rho C_p)_s + (1-\phi)(\rho C_p)_f \tag{14.90}$$

$$\rho_{nf} = \rho_f(1-\phi) + \rho_s\phi, \tag{14.91}$$

μ_{nf}, k_{nf} can be presented as:

$$\mu_{nf} = \frac{k_{Brownian}}{k_f} \times \frac{\mu_f}{\Pr} + \mu_f(1-\phi)^{-2.5} \tag{14.92}$$

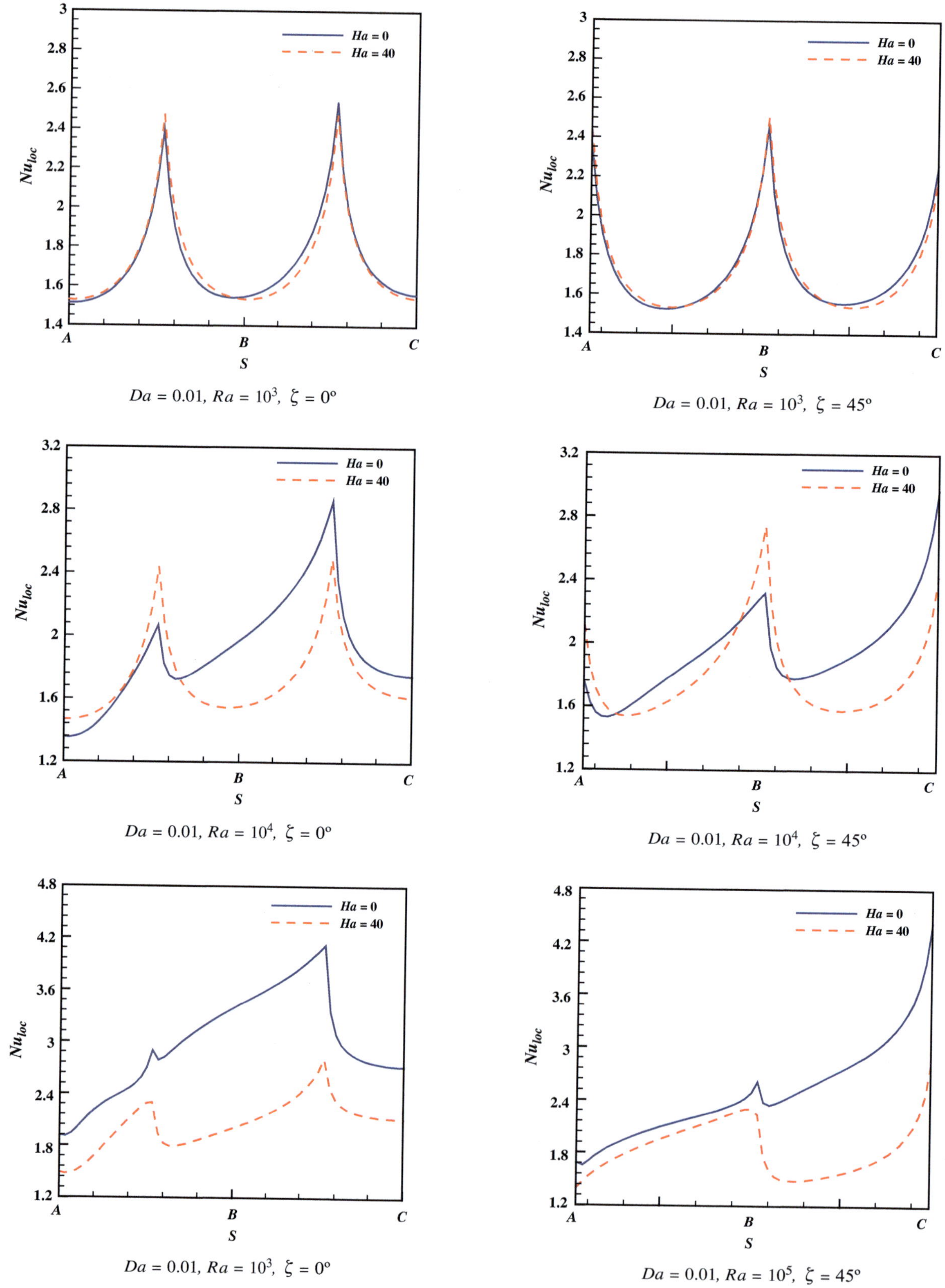

FIGURE 14.26 Effects of inclination angle, Darcy, Rayleigh, and Hartmann numbers on local Nusselt number when $\phi = 0.04$.

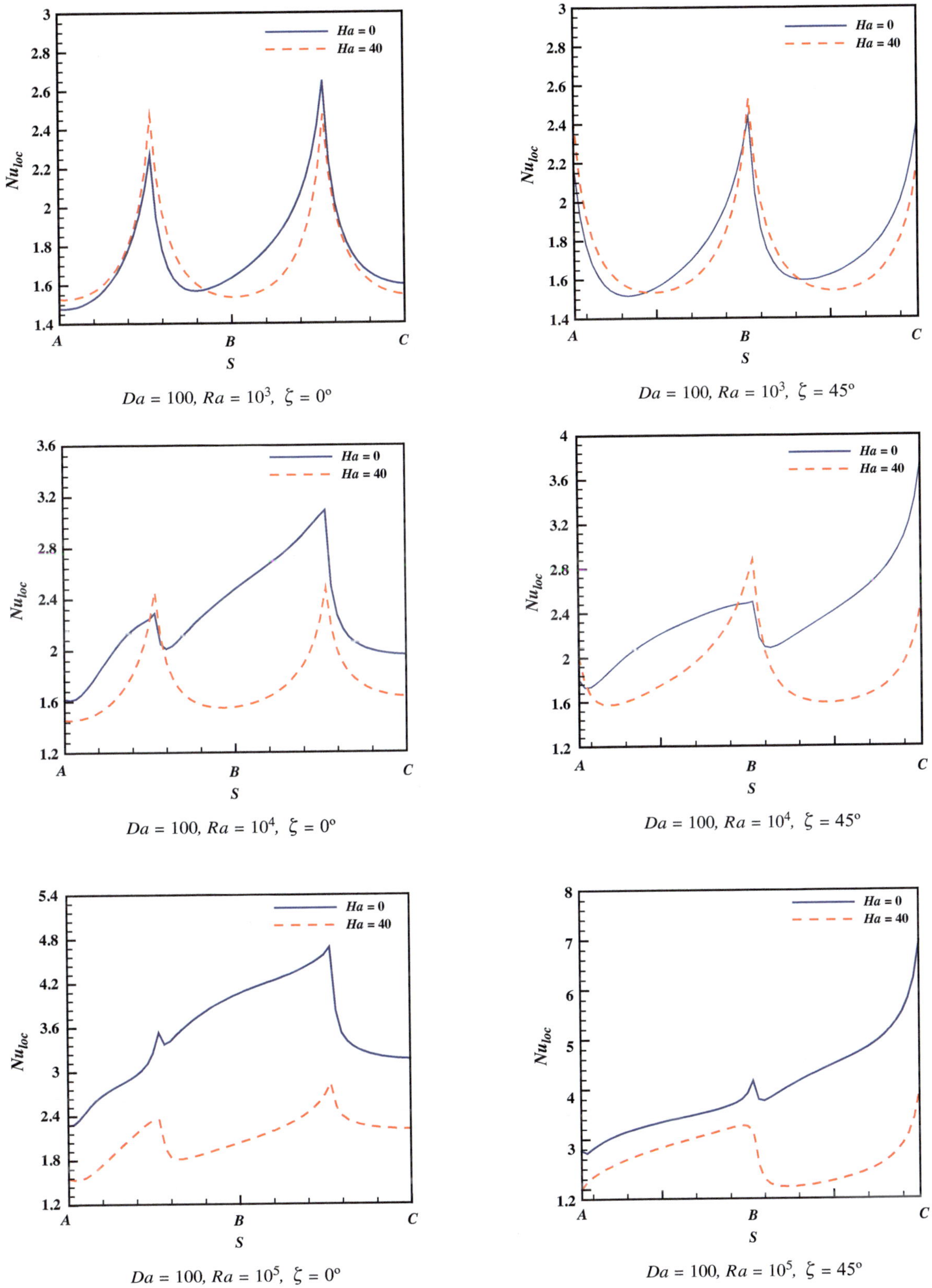

FIGURE 14.26 (Continued)

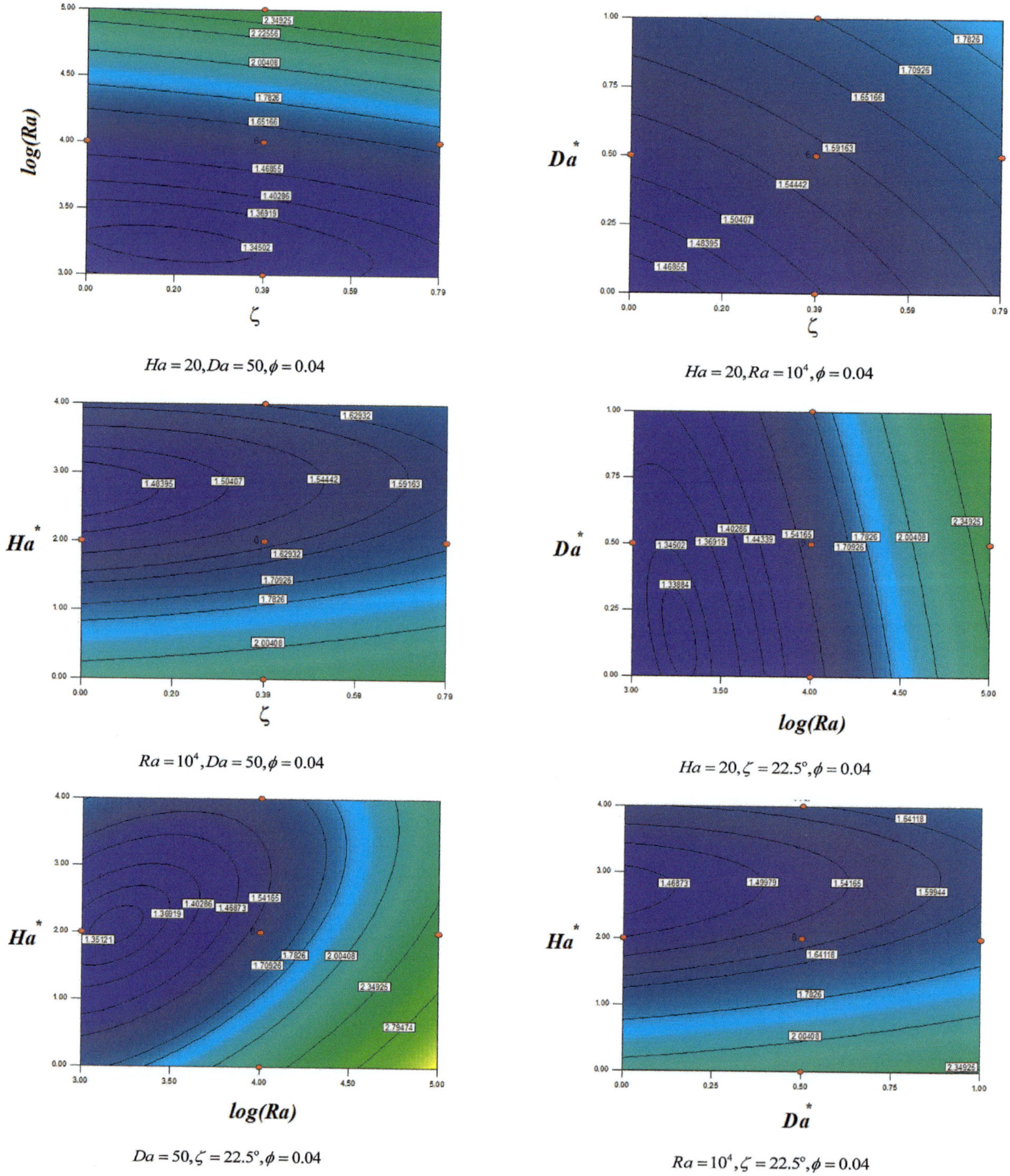

FIGURE 14.27 Effects of inclination angle, Darcy, Rayleigh, and Hartmann numbers on average Nusselt number.

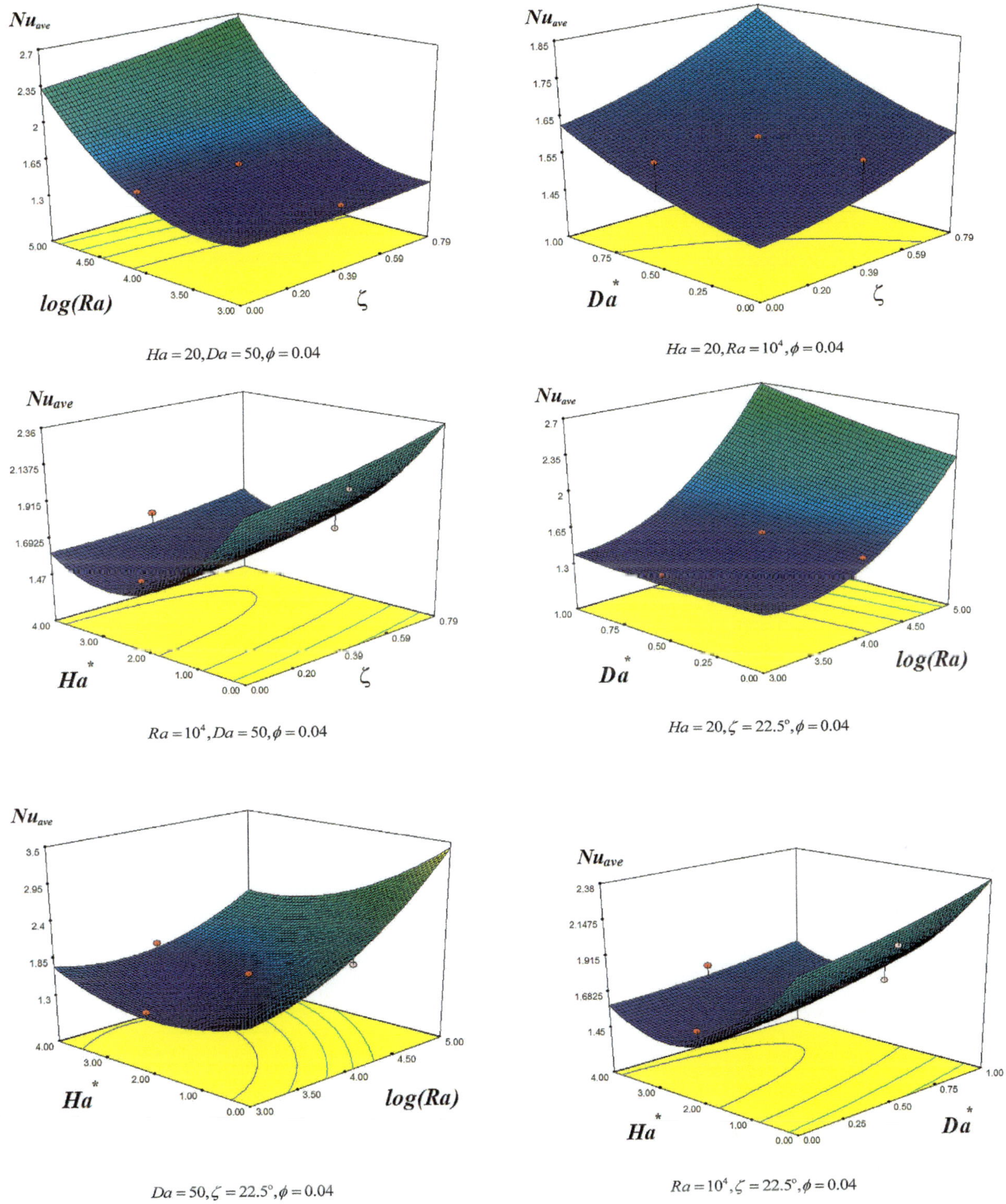

FIGURE 14.27 Effects of inclination angle, Darcy, Rayleigh, and Hartmann numbers on average Nusselt number.

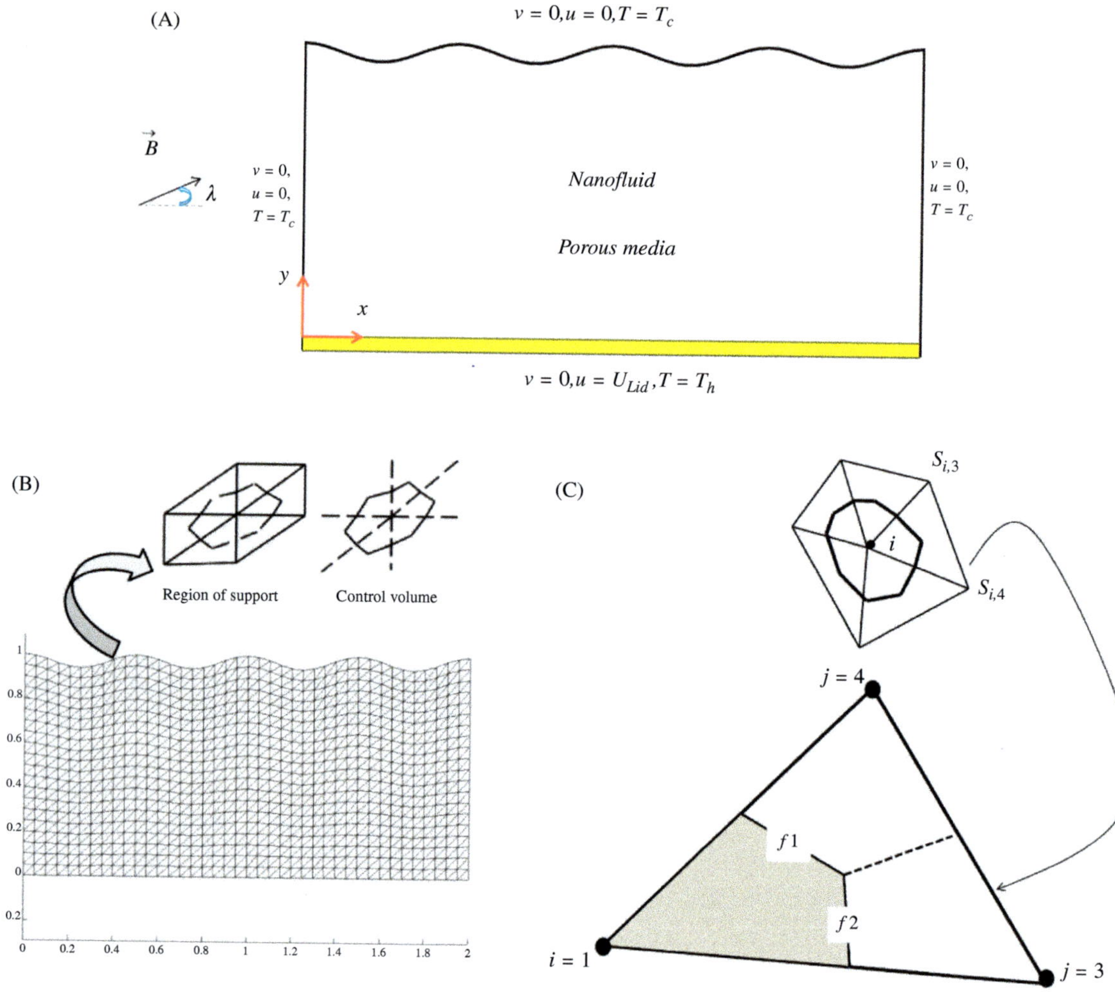

FIGURE 14.28 (A) Geometry and the boundary conditions with (B) the mesh of geometry considered in this work; (C) a sample triangular element and its corresponding control volume.

$$
\begin{aligned}
&k_{nf} = k_{Brownian} + k_{static} \\
&\frac{k_{Brownian}}{k_f} = 5\times 10^4 g'(d_p,T,\phi)\phi\sqrt{\frac{\kappa_b T}{\rho_p d_p}}\rho_f c_{p,f},\ \frac{k_{static}}{k_f} = 1 + \frac{3(-1+k_p/k_f)\phi}{\left(\frac{k_p}{k_f}+2\right) - \left(\frac{k_p}{k_f}-1\right)\phi}, \\
&g'(d_p,T,\phi) = Ln(T)\left(Ln(d_p)a_2 + a_5 Ln(d_p)^2 + a_1 + a_3 Ln(\phi) + Ln(\phi)Ln(d_p)a_4\right) \\
&+\left(a_7 Ln(d_p) + a_6 + Ln(d_p)a_9 Ln(\phi) + a_8 Ln(\phi) + a_{10} Ln(d_p)^2\right),\ R_f = 4\times 10^{-8} km^2/W, \\
&R_f = \left[-k_p{}^{-1} + k_{p,eff}{}^{-1}\right]d_p,
\end{aligned}
\tag{14.93}
$$

Properties and needed parameters are provided in Tables 14.1 and 14.2.
ψ, ω can be defined as:

$$
\omega + \frac{\partial u}{\partial y} - \frac{\partial v}{\partial x} = 0,\ \frac{\partial \psi}{\partial x} = -v,\ \frac{\partial \psi}{\partial y} = u \tag{14.94}
$$

Introducing dimensionless quantities:

$$\theta = \frac{T - T_c}{\Delta T}, \Delta T = T_h - T_c, (X, Y) = \frac{(x, y)}{L}, \Psi = \frac{\psi L}{U_{Lid}},$$
$$V = \frac{v}{U_{Lid}}, \Omega = \frac{\omega}{L U_{Lid}}, U = \frac{u}{U_{Lid}} \tag{14.95}$$

The final formulae are:

$$\frac{\partial^2 \Psi}{\partial X^2} + \Omega + \frac{\partial^2 \Psi}{\partial Y^2} = 0, \tag{14.96}$$

$$\frac{\partial \Omega}{\partial Y} V + \frac{\partial \Omega}{\partial X} U = \frac{1}{\text{Re}} \frac{A_5}{A_1} \left(\frac{\partial^2 \Omega}{\partial Y^2} + \frac{\partial^2 \Omega}{\partial X^2} \right)$$
$$+ \frac{Ha^2}{\text{Re}} \frac{A_6}{A_1} \left(\frac{\partial U}{\partial Y} B_y^2 - \frac{\partial V}{\partial X} B_x{}^2 + B_y B_x \frac{\partial U}{\partial X} - B_x \frac{\partial V}{\partial Y} B_y \right) \tag{14.97}$$
$$- \frac{1}{\text{Re}\, Da} \frac{A_5}{A_1} \Omega,$$

$$\frac{A_4}{A_2} \left(\frac{\partial^2 \theta}{\partial X^2} + \frac{\partial^2 \theta}{\partial Y^2} \right) = \text{PrRe} \left(V \frac{\partial \theta}{\partial Y} + \frac{\partial \theta}{\partial X} U \right) \tag{14.98}$$

where dimensionless and constants parameters are:

$$\text{Re} = \frac{\rho_f U_{Lid} L}{\mu_f}, \; Ha = L B_0 \sqrt{\sigma_f / \mu_f},$$
$$A_1 = \frac{\rho_{nf}}{\rho_f}, A_5 = \frac{\mu_{nf}}{\mu_f}, \tag{14.99}$$
$$A_2 = \frac{(\rho C_p)_{nf}}{(\rho C_p)_f}, A_4 = \frac{k_{nf}}{k_f}, A_6 = \frac{\sigma_{nf}}{\sigma_f}$$

and boundary conditions are:

$$\begin{aligned} \Psi &= 0.0 \quad \text{on all walls} \\ \theta &= 1.0 \quad \text{on south wall} \\ \theta &= 0.0 \quad \text{on other walls} \end{aligned} \tag{14.100}$$

Nu_{loc} and Nu_{ave} over the south wall are:

$$Nu_{loc} = \left(\frac{k_{nf}}{k_f} \right) \frac{\partial \theta}{\partial y} \tag{14.101}$$

$$Nu_{ave} = \frac{1}{L} \int_{r_{in}}^{r_{out}} Nu_{loc} dx \tag{14.102}$$

14.6.3 Effects of Active Parameters

Nanofluid MHD forced convection in a porous sinusoidal enclosure is examined numerically in this section. Numerical outputs are presented for various Darcy numbers ($Da = 0.01$ to 100), Hartmann numbers ($Ha = 0$ to 20), Reynolds numbers ($\text{Re} = 100$ to 600), and volume fraction of CuO ($\phi = 0\%$ to 4%).

Fig. 14.29 illustrates the influence of ϕ on isotherms and streamlines. Increasing ϕ leads to an increase in thermal boundary layer thickness. The nanofluid velocity is enhanced by adding nanoparticles.

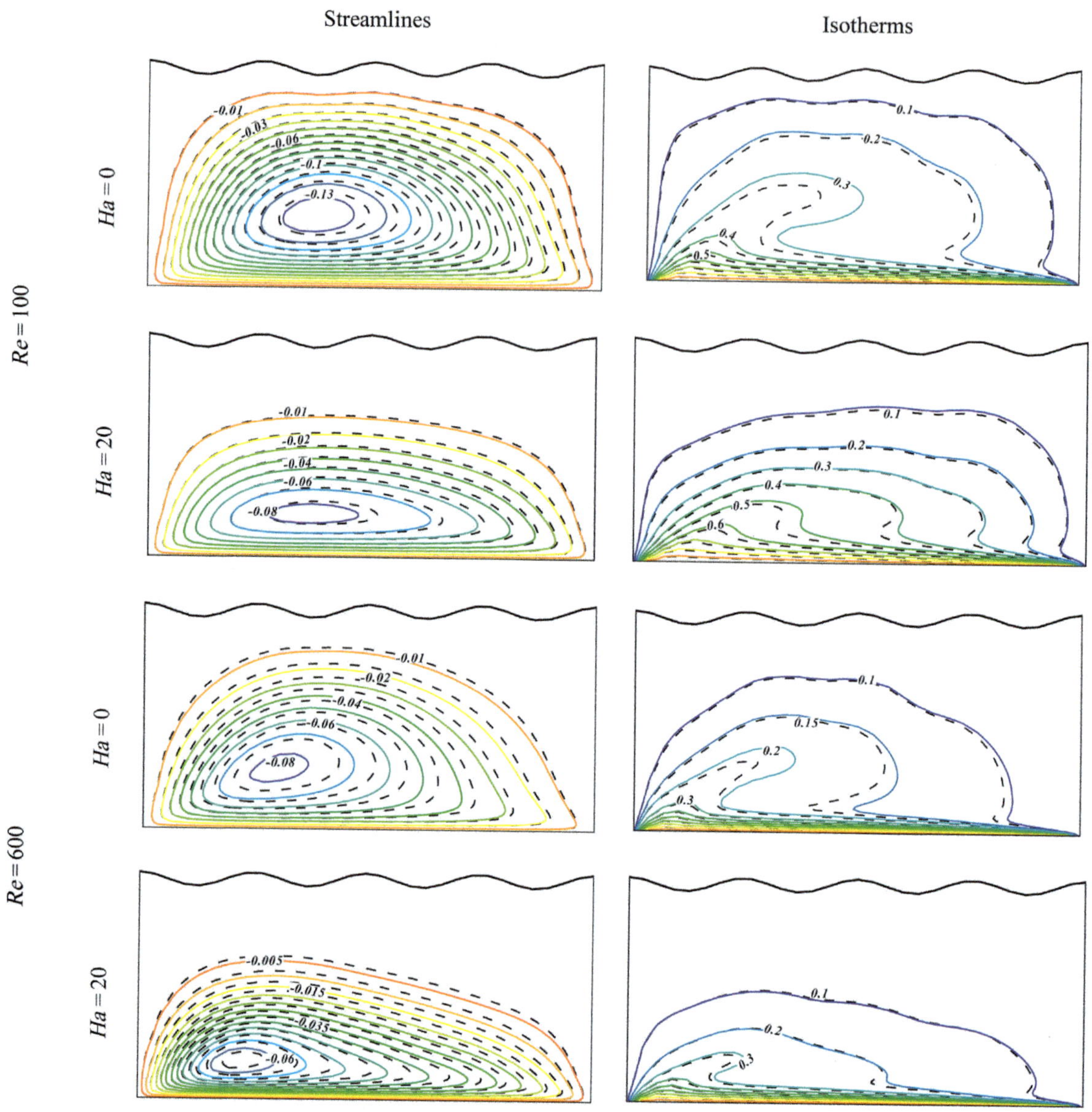

FIGURE 14.29 Influence of nanofluid volume fraction on streamlines (left) and isotherm (right) contours (nanofluid ($\phi = 0.04$)(–) and pure fluid ($\phi = 0$) (- - -)) when $Da = 100$.

Figs. 14.30 and 14.31 illustrate the effect of the Darcy, Hartmann, and Reynolds numbers on hydrothermal behavior. In the absence of a magnetic field, when Darcy and Reynolds number are low, only one eddy exists in streamlines and isotherms are parallel to each other. As the Reynolds number increases, the isotherms become denser at the lid wall due to the increase of convective mode. Also $|\Psi_{\max}|$ augments with the rise of the Reynolds number. As the Darcy number increases, the convective mode becomes stronger due to an increase in the permeability of the medium. So the temperature gradient over the hot wall increases with the increase of the Darcy number. Increasing the Lorentz forces makes the isotherms becomes less dense. Also velocity reduces with the rise of the Hartmann number.

The influence of important parameters on Nu_{ave} is depicted in Fig. 14.32. The correlation for average Nusselt number is as follows:

$$\begin{aligned} Nu_{ave} &= 7.56 + 4.19\mathrm{Re}^* + 0.45Da^* - 0.1Ha^* + 25.6\phi \\ &+ 0.017\mathrm{Re}^*Da^* + 0.048\mathrm{Re}^*Ha^* + 0.45\phi\mathrm{Re}^* - 0.0018Da^*Ha^* \\ &+ 0.6Da^*\phi - 0.048Ha^*Da^* - 0.38\mathrm{Re}^{*2} + 0.52Da^{*2} - 0.22Ha^{*2} - 1.35\phi^2 \end{aligned} \tag{14.103}$$

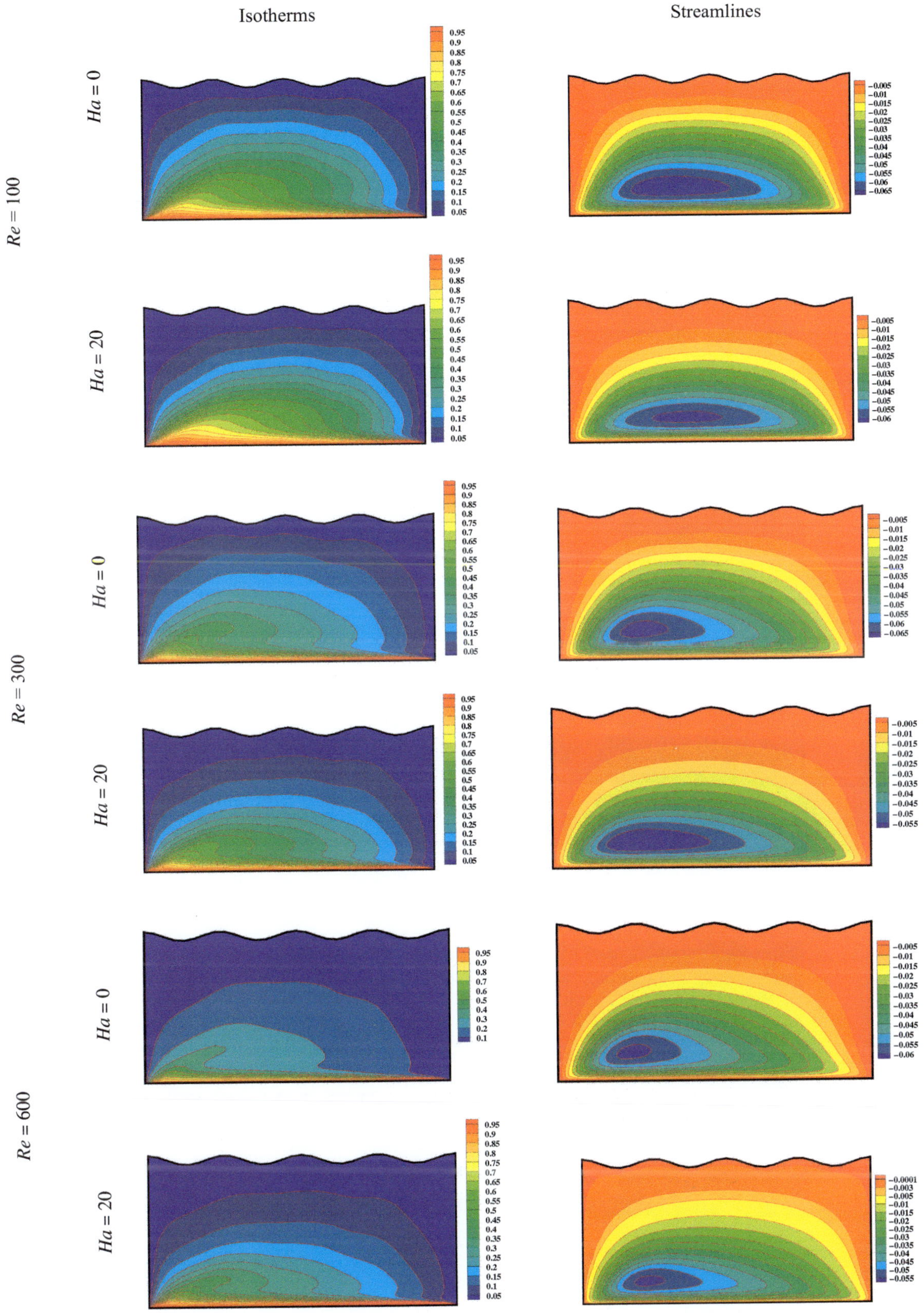

FIGURE 14.30 Isotherm (left) and streamline (right) contours for different values of Reynolds and Hartmann numbers when $Da = 0.01, \phi = 0.04$.

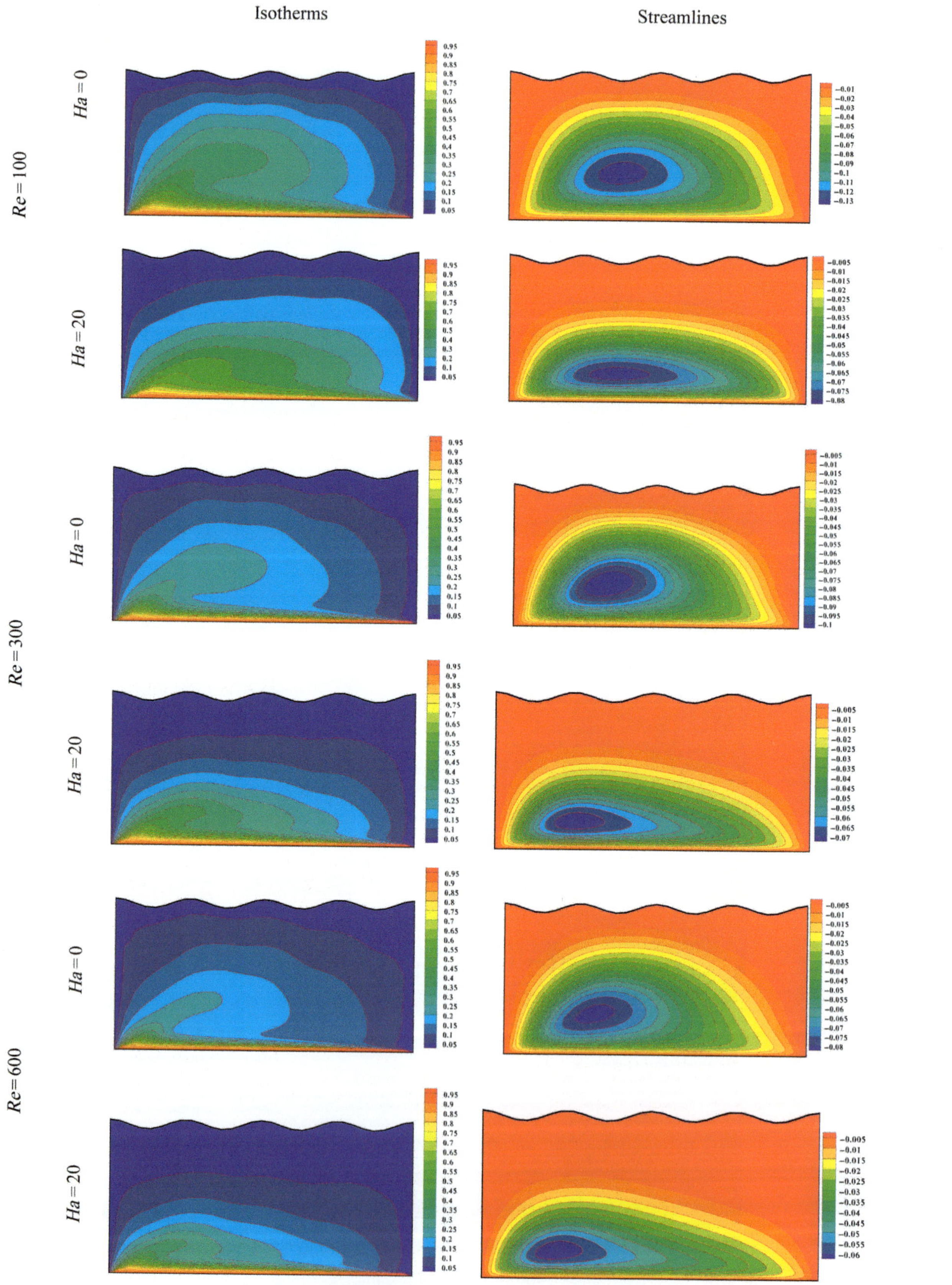

FIGURE 14.31 Isotherm (left) and streamline (right) contours for different values of Reynolds and Hartmann numbers when $Da = 100, \phi = 0.04$.

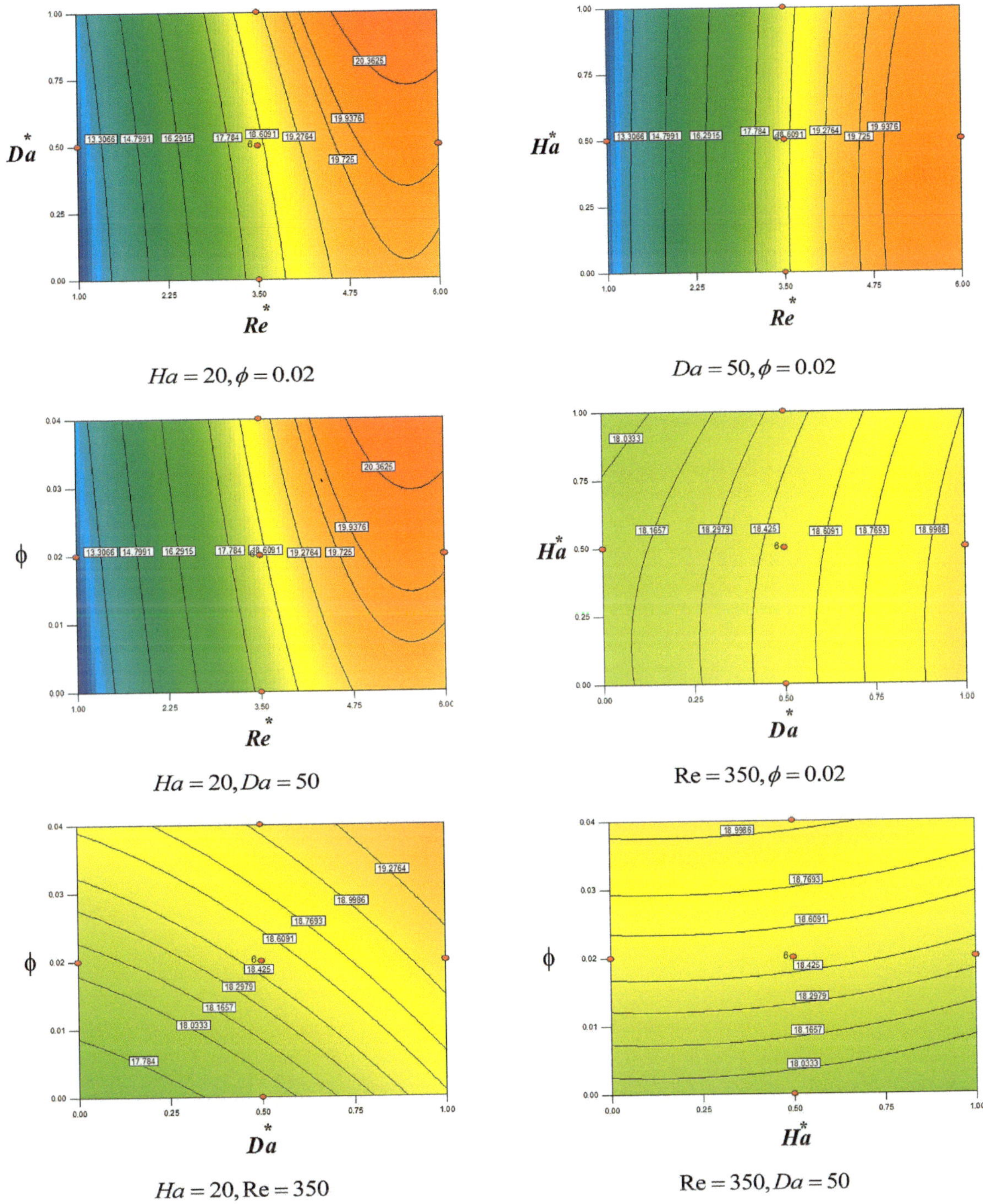

FIGURE 14.32 Influences of the volume fraction of nanofluid and Darcy, Reynolds, and Hartmann numbers on average Nusselt number.

where $\mathrm{Re}^* = 0.01\mathrm{Re}$, $Ha^* = 0.1Ha$. Adding nanoparticles makes the Nusselt number increase due to the increase in k_{nf}. Nu_{ave} improves with the increase of Darcy and Reynolds numbers because of the increase in convective heat transfer. As Ha improves, the temperature gradient decreases and in turn the Reynolds number is reduced with the increase of Lorentz forces.

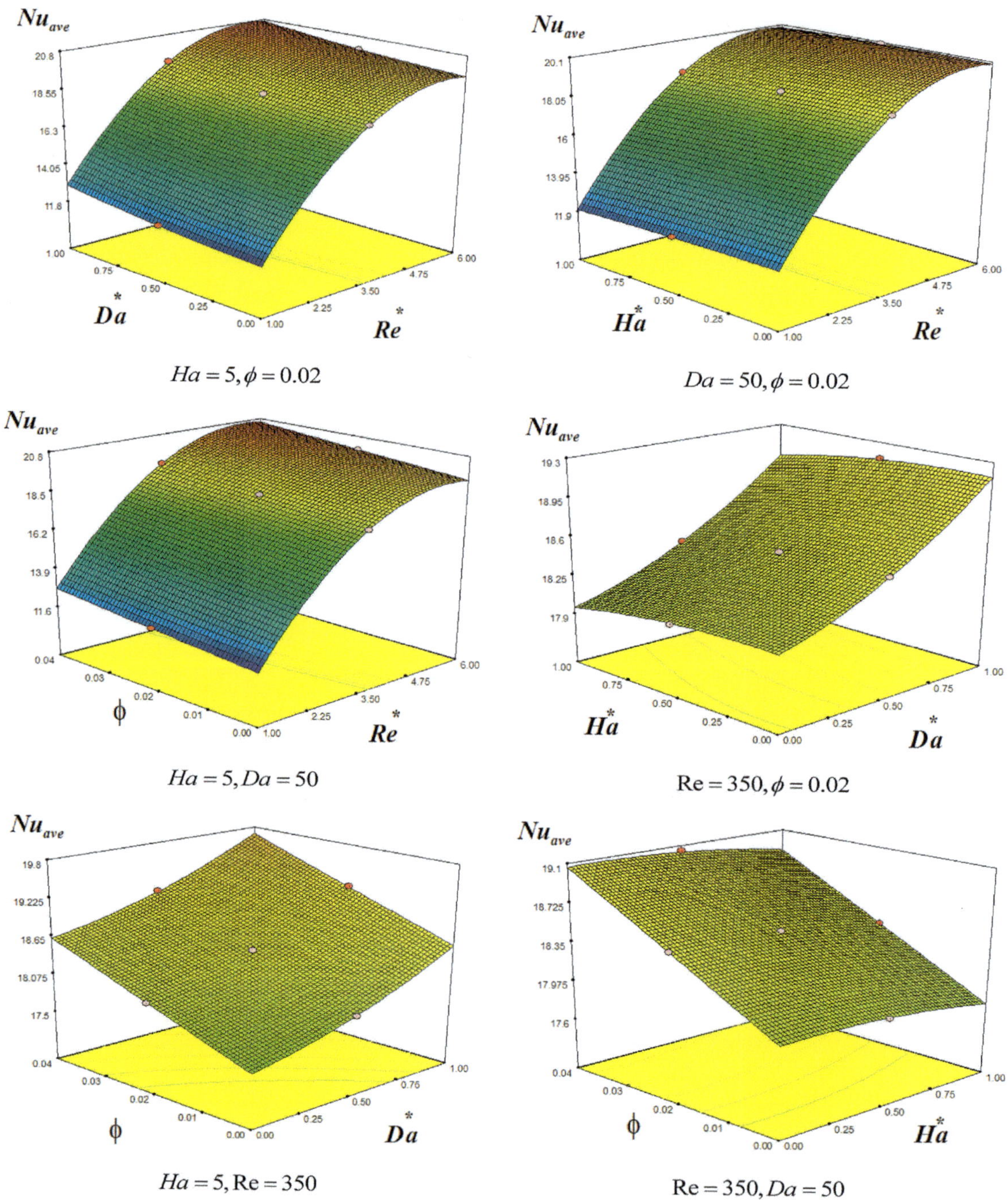

FIGURE 14.32 (Continued)

14.7 NANOFLUID HEAT TRANSFER IN A PERMEABLE ENCLOSURE IN PRESENCE OF VARIABLE MAGNETIC FIELD

14.7.1 Problem Definition

Geometry, boundary condition, and sample element are demonstrated in Fig. 14.33. External magnetic source is applied (see Fig. 14.34). $\overline{H}, \overline{H_x}, \overline{H_y}$ are:

$$\overline{H_y} = (\bar{a} - x)\frac{\gamma}{2\pi}\left[(\bar{b}-y)^2+(\bar{a}-x)^2\right]^{-1}, \tag{14.104}$$

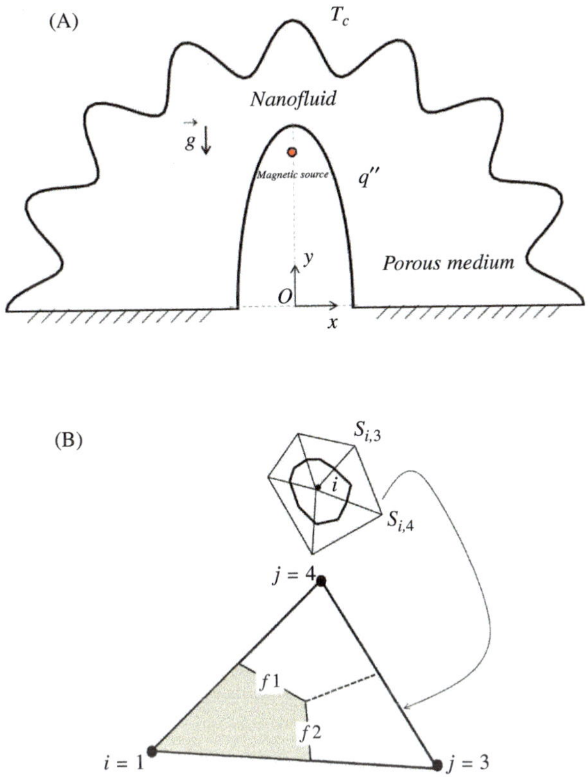

FIGURE 14.33 (A) Geometry and the boundary conditions with; (B) a sample triangular element and its corresponding control volume.

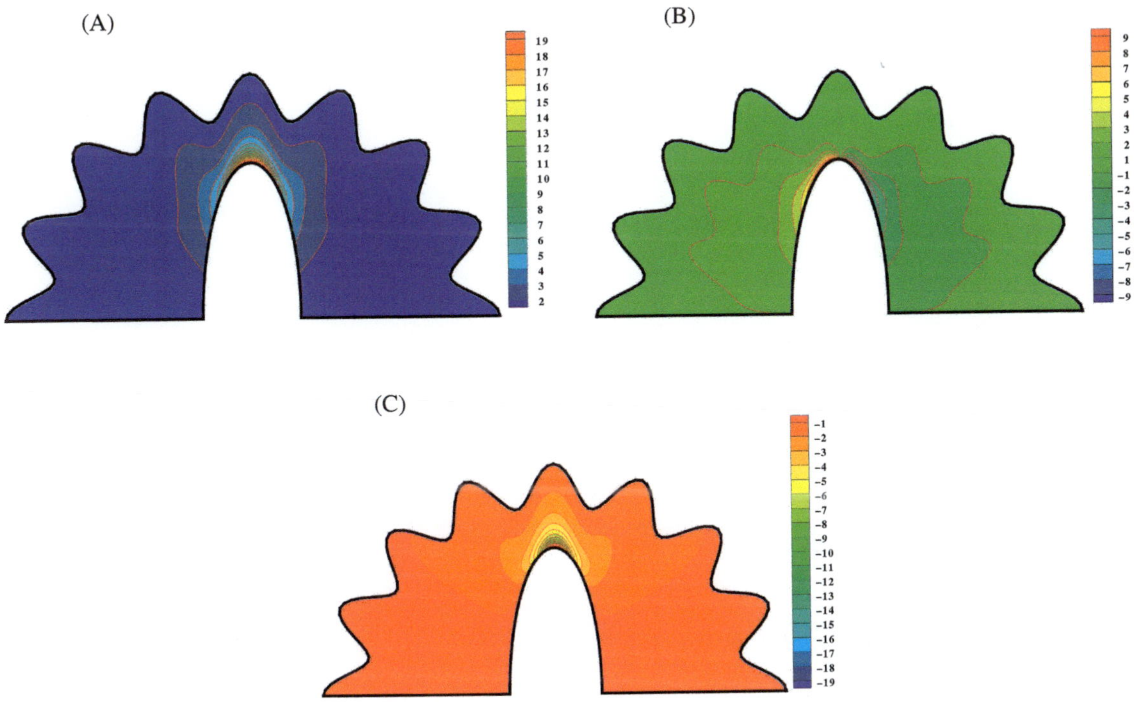

FIGURE 14.34 Contours of the (A) magnetic field strength H; (B) magnetic field intensity component in x direction Hx; (C) magnetic field intensity component in y direction Hy. (A) $H(x, y)$, (B) $Hx(x, y)$, (C) $Hy(x, y)$.

$$\overline{H_x} = \left(y - \overline{b}\right)\frac{\gamma}{2\pi}\left[\left(\overline{b} - y\right)^2 + \left(\overline{a} - x\right)^2\right]^{-1}, \tag{14.105}$$

$$\overline{H} = \left(\overline{H}_y^2 + \overline{H}_x^2\right)^{0.5}. \tag{14.106}$$

14.7.2 Governing Equation

Two-dimensional convective non-Darcy flow of nanofluid is considered in the presence of an external magnetic source. The governing equations are:

$$\frac{\partial u}{\partial x} = -\frac{\partial v}{\partial y}, \tag{14.107}$$

$$\begin{aligned}&\left(\frac{\partial^2 u}{\partial y^2} + \frac{\partial^2 u}{\partial x^2}\right)\mu_{nf} - \frac{\partial P}{\partial x} - \mu_0^2\sigma_{nf}\overline{H}_y^2 u + \sigma_{nf}\mu_0^2\overline{H}_x\overline{H}_y v \\ &- \frac{\mu_{nf}}{K}u = \left(\rho_{nf}\right)\left(\frac{\partial u}{\partial x}u + \frac{\partial u}{\partial y}v\right),\end{aligned} \tag{14.108}$$

$$\begin{aligned}&\mu_{nf}\left(\frac{\partial^2 v}{\partial x^2} + \frac{\partial^2 v}{\partial y^2}\right) - \frac{\partial P}{\partial y} + \mu_0^2\overline{H}_y\sigma_{nf}\overline{H}_x u - \mu_0^2\overline{H}_x\sigma_{nf}\overline{H}_x v - \frac{\mu_{nf}}{K}v \\ &+ (T - T_c)\beta_{nf}g\rho_{nf} = \rho_{nf}\left(\frac{\partial v}{\partial x}u + \frac{\partial v}{\partial y}v\right),\end{aligned} \tag{14.109}$$

$$\begin{aligned}&k_{nf}\left(\frac{\partial^2 T}{\partial y^2} + \frac{\partial^2 T}{\partial x^2}\right) - \frac{\partial q_r}{\partial y} = \left(v\frac{\partial T}{\partial y} + u\frac{\partial T}{\partial x}\right)\left(\rho C_p\right)_{nf}, \\ &\left[q_r = -\frac{4\sigma_e}{3\beta_R}\frac{\partial T^4}{\partial y}, \quad T^4 \cong 4T_c^3 T - 3T_c^4\right].\end{aligned} \tag{14.110}$$

$\left(\rho C_p\right)_{nf}$, ρ_{nf}, $(\rho\beta)_{nf}$, and σ_{nf} are defined as:

$$\left(\rho C_p\right)_{nf} = \left(\rho C_p\right)_f(1 - \phi) + \left(\rho C_p\right)_s\phi \tag{14.111}$$

$$\rho_{nf} = \rho_f(1 - \phi) + \rho_s\phi \tag{14.112}$$

$$(\rho\beta)_{nf} = (\rho\beta)_f(1 - \phi) + (\rho\beta)_s\phi, \tag{14.113}$$

$$\sigma_{nf} = \sigma_f\left[\frac{3\phi(\sigma 1 - 1)}{(1 - \sigma 1)\phi + (2 + \sigma 1)} + 1\right], \quad \sigma 1 = \sigma_s/\sigma_f. \tag{14.114}$$

μ_{nf} is calculated as follows:

$$\mu_{nf} = \left(0.035\mu_0^2\overline{H}^2 + 3.1\mu_0\overline{H} - 27886.4807\phi^2 + 4263.02\phi + 316.0629\right)e^{-0.01T} \tag{14.115}$$

k_{nf} can be calculated as:

$$\frac{k_{nf}}{k_f} = \frac{-m\left(k_f - k_p\right)\phi + \left(k_p - k_f\right)\phi + mk_f + k_p + k_f}{mk_f + \left(k_f - k_p\right)\phi + k_f + k_p} \tag{14.116}$$

TABLE 14.5 The Values of Shape Factor of Different Shapes of Nanoparticles

m	Spherical	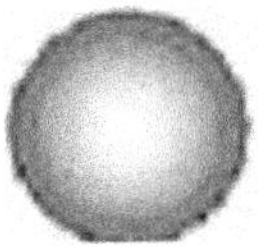	3
	Platelet		5.7
	Cylinder		4.8
	Brick	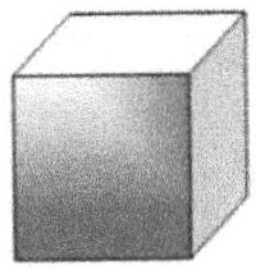	3.7

The properties of the nanofluid are depicted in Table 14.3.
Different values of shape factors for various shapes of nanoparticles are illustrated in Table 14.5.
Vorticity and stream function should be used to eliminate pressure source terms:

$$\omega + \frac{\partial u}{\partial y} - \frac{\partial v}{\partial x} = 0, \quad \frac{\partial \psi}{\partial x} = -v, \quad \frac{\partial \psi}{\partial y} = u. \tag{14.117}$$

Dimensionless parameters are defined as:

$$\begin{aligned} &(H_y, H_x, H) = \frac{(\overline{H_y}, \overline{H_x}, \overline{H})}{\overline{H_0}}, \quad (b, a) = \frac{(\bar{b}, \bar{a})}{L}, \\ &U = \frac{uL}{\alpha_{nf}}, V = \frac{vL}{\alpha_{nf}}, \quad (X, Y) = \frac{(x, y)}{L}, \theta = \frac{T - T_c}{\Delta T}, \Delta T = q''L/k_f, \\ &\Psi = \frac{\psi}{\alpha_{nf}}, \Omega = \frac{\omega L^2}{\alpha_{nf}}. \end{aligned} \tag{14.118}$$

So equations change to:

$$\frac{\partial^2 \Psi}{\partial Y^2} + \Omega + \frac{\partial^2 \Psi}{\partial X^2} = 0, \tag{14.119}$$

$$\begin{aligned} &\frac{\partial \Omega}{\partial X} U + V \frac{\partial \Omega}{\partial Y} = \Pr \frac{A_5}{A_1} \frac{A_2}{A_4} \left(\frac{\partial^2 \Omega}{\partial Y^2} + \frac{\partial^2 \Omega}{\partial X^2} \right) \\ &+ \Pr Ha^2 \frac{A_6}{A_1} \frac{A_2}{A_4} \left(\frac{\partial U}{\partial X} H_y H_x - \frac{\partial V}{\partial X} H_x{}^2 + \frac{\partial U}{\partial Y} H_y^2 - \frac{\partial V}{\partial Y} H_y H_x \right) \\ &+ \Pr Ra \frac{A_3 A_2^2}{A_1 A_4^2} \frac{\partial \theta}{\partial X} - \frac{\Pr}{Da} \frac{A_5}{A_1} \frac{A_2}{A_4}, \end{aligned} \tag{14.120}$$

$$\frac{\partial \theta}{\partial X} U + V \frac{\partial \theta}{\partial Y} = \left(\frac{\partial^2 \theta}{\partial X^2} + \frac{\partial^2 \theta}{\partial Y^2} \right) + \frac{4}{3} \frac{1}{A_4} Rd \frac{\partial^2 \theta}{\partial Y^2}. \tag{14.121}$$

and dimensionless parameters are:

$$
\begin{aligned}
&Ra_f = g\beta_f L^3 \Delta T / (\alpha_f \upsilon_f), \Pr_f = \upsilon_f/\alpha_f, Ha = L\mu_0 H_0 \sqrt{\sigma_f/\mu_f},\\
&Da = K/L^2, Ec = (\mu_f \alpha_f) / \left[(\rho C_P)_f \Delta T\, L^2\right],\\
&A_1 = \frac{\rho_{nf}}{\rho_f}, A_2 = \frac{(\rho C_p)_{nf}}{(\rho C_p)_f}, A_3 = \frac{(\rho\beta)_{nf}}{(\rho\beta)_f},\\
&A_4 = \frac{k_{nf}}{k_f}, A_5 = \frac{\mu_{nf}}{\mu_f}, A_6 = \frac{\sigma_{nf}}{\sigma_f},\\
&Rd = 4\sigma_e T_c^3 / (\beta_R k_f)
\end{aligned}
\tag{14.122}
$$

and boundary conditions are:

$$
\begin{aligned}
&\text{on inner wall} && \frac{\partial\theta}{\partial n} = 1.0\\
&\text{on outer wall} && \theta = 0.0\\
&\text{on other walls} && \frac{\partial\theta}{\partial n} = 0\\
&\text{on all walls} && \Psi = 0.0
\end{aligned}
\tag{14.123}
$$

Nu_{loc}, Nu_{ave} over the hot wall can be calculated as:

$$
Nu_{loc} = \left(\frac{k_{nf}}{k_f}\right)\left(1 + \left(\frac{k_{nf}}{k_f}\right)^{-1}\frac{4Rd}{3}\right)\frac{1}{\theta}, \tag{14.124}
$$

$$
Nu_{ave} = \frac{1}{S}\int_0^S Nu_{loc}\, ds. \tag{14.125}
$$

14.7.3 Effects of Active Parameters

The impact of nonuniform magnetic field on Fe_3O_4-water flow in a permeable enclosure is simulated. Nanofluid viscosity is estimated according to previous experimental data. The shape effect of nanoparticles on k_{nf} is taken into consideration. CVFEM is utilized to find the effects of radiation parameter ($Rd = 0$ *to* 0.8), Darcy number ($Da = 0.01$ to 100), Rayleigh number ($Ra = 10^3, 10^4, 10^5$), volume fraction of Fe_3O_4-water ($\phi = 0\%$ to 4%), shape of nanoparticle, and Hartmann number ($Ha = 0$ to 10).

The impacts of shape of the nanoparticles on Nu_{ave} are presented in Table 14.6. The maximum Nu_{ave} is obtained for Platelet, followed by Cylinder, Brick, and Spherical. So, Platelet nanoparticle has been selected to complete this section.

Fig. 14.35 shows the impact of adding nanoparticles to water on the hydrothermal treatment. The temperature gradient reduces with the increase of ϕ. Velocity increases with the addition of nanoparticles because of an increase in the solid movements. Fig. 14.36 demonstrates the effect of the radiation parameter on streamline and

TABLE 14.6 Effect of Shape of Nanoparticles on Nusselt Number When $Da = 100, Ra = 10^5, Rd = 0.8, \phi = 0.04$

	Ha	
	0	**10**
Spherical	12.32892	10.1131
Brick	12.40613	10.15818
Cylinder	12.52868	10.23001
Platelet	12.62989	10.28956

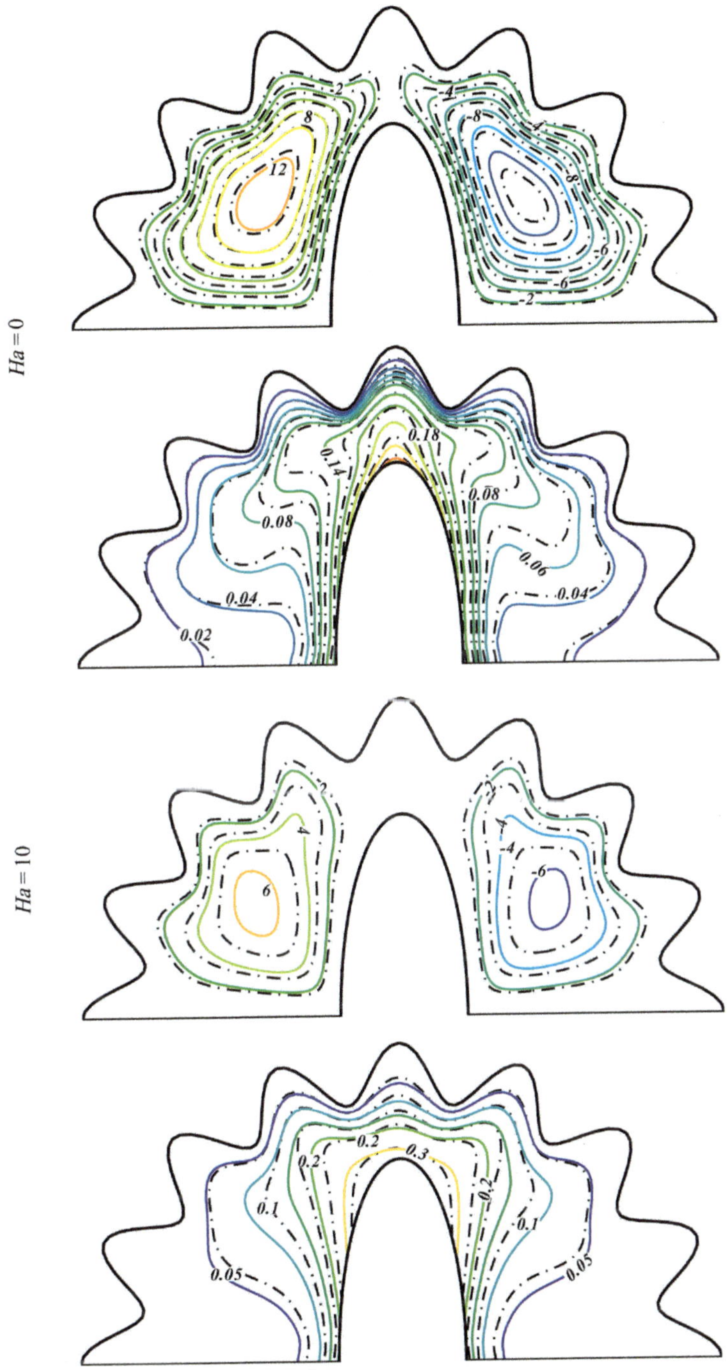

FIGURE 14.35 Impact of nanofluid volume fraction on streamline (top) and isotherm (bottom) contours (nanofluid ($\phi = 0.04$)(–) and pure fluid ($\phi = 0$) (– · –)) when $Ra = 10^5, Da = 100, Rd = 0.8$.

isotherm contours. The thermal boundary layer thickness increases with the increase of *Rd*. By adding a magnetic field, the impact of the radiation parameter on streamlines becomes not significant. Figs. 14.37–14.39 illustrate the effects of *Da, Ra, Ha* on isotherms and streamlines. Only one eddy appears in streamlines. By increasing the Hartmann number, the main eddy moves downward and the distortion of isotherms becomes less than before. As the buoyancy forces increase, a thermal plume is generated near the vertical symmetric line. Increasing the Lorentz forces shifts the thermal plume to left and reduces $|\Psi_{max}|$. As the permeability of the media increases, the convective mode becomes stronger and the shape of isotherms becomes more complicated.

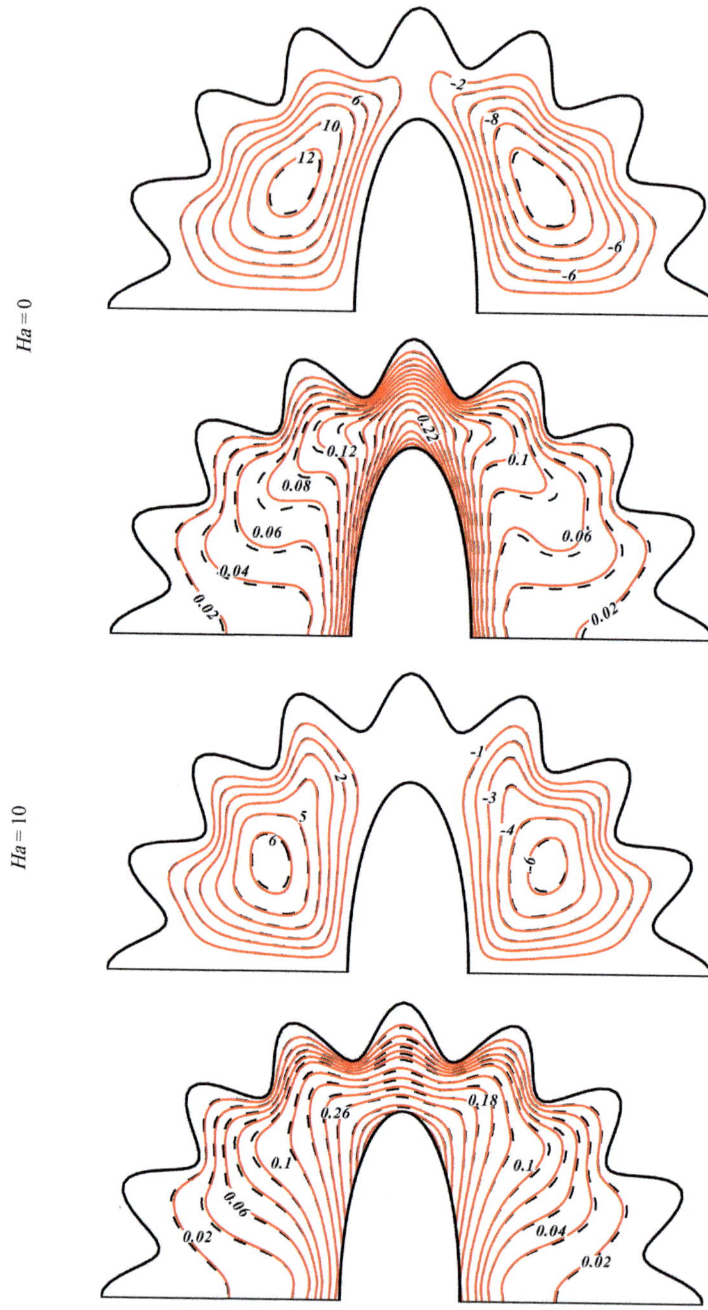

FIGURE 14.36 Impact of radiation parameter on streamline (top) and isotherm (bottom) contours ($Rd = 0.8$ (–), $Rd = 0$ (- - -)) when $Ra = 10^5, Da = 100, \phi = 0.04$.

Effects of significant parameters on Nu_{ave} are depicted in Fig. 14.40. The correlation for Nu_{ave} is:

$$\begin{aligned} Nu_{ave} &= 31.9 - 5.2Rd - 15.3\log(Ra) - 0.6Da^* + 1.25Ha^* \\ &+ 2.93Rd\log(Ra) + 0.44Rd\,Da^* - 0.57Rd\,Ha^* \\ &+ 0.28Da^*\log(Ra) - 0.37Ha^*\log(Ra) - 0.69Da^*Ha^* \\ &- 4.13Rd^2 + 2.02(\log(Ra))^2 + 0.47(Da^*)^2 - 0.27(Ha^*)^2 \end{aligned} \tag{14.126}$$

where $Da^* = 0.01Da$, $Ha^* = 0.1Ha$. Heat transfer rate increases with the increase in permeability of the porous media. A similar treatment is reported for the Rayleigh number. The temperature gradient is reduced with the increase of the Hartmann number.

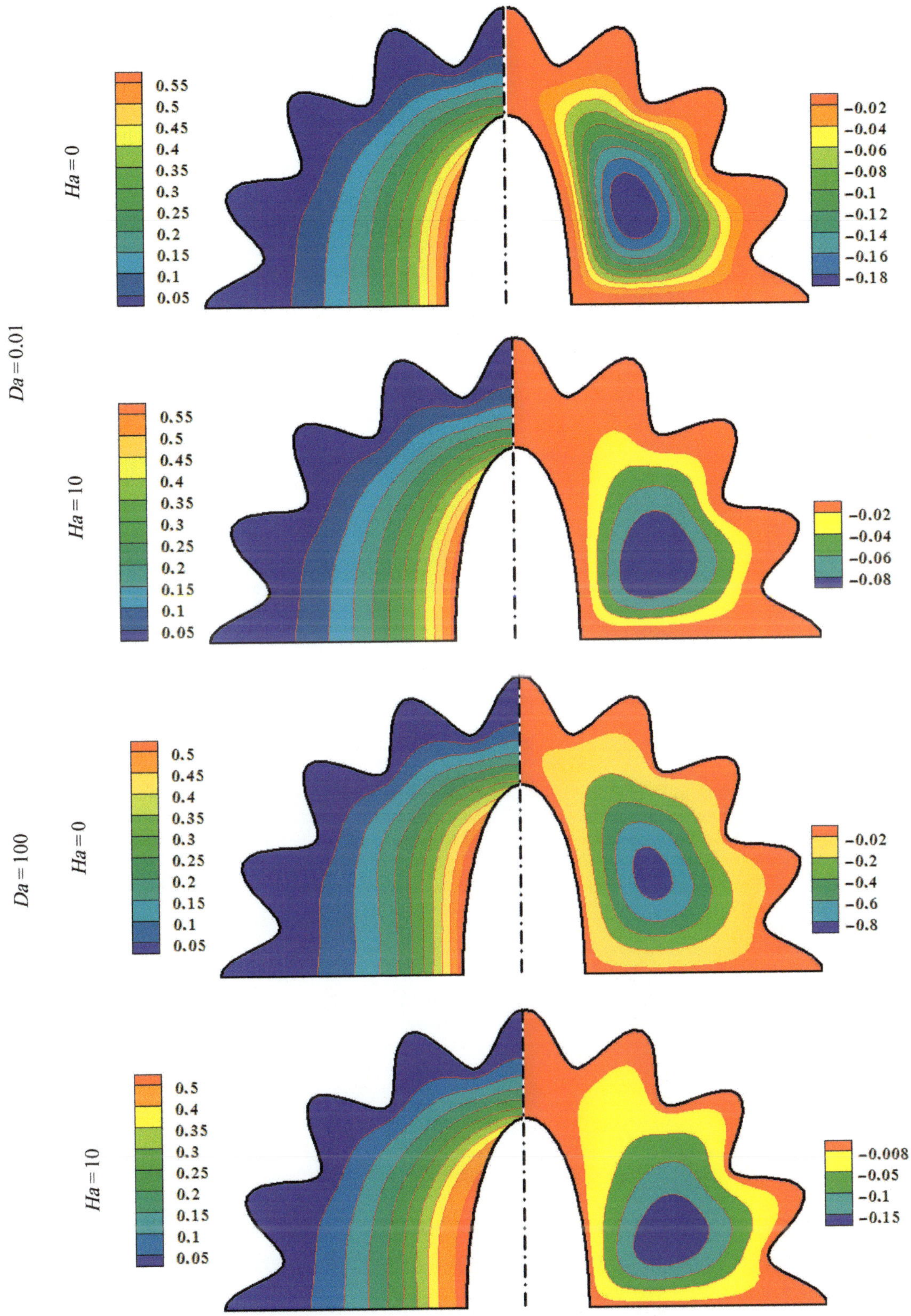

FIGURE 14.37 Influence of *Da*, *Ha* on streamline (right) and isotherm (left) contours when $\phi = 0.04$, $Ra = 10^3$.

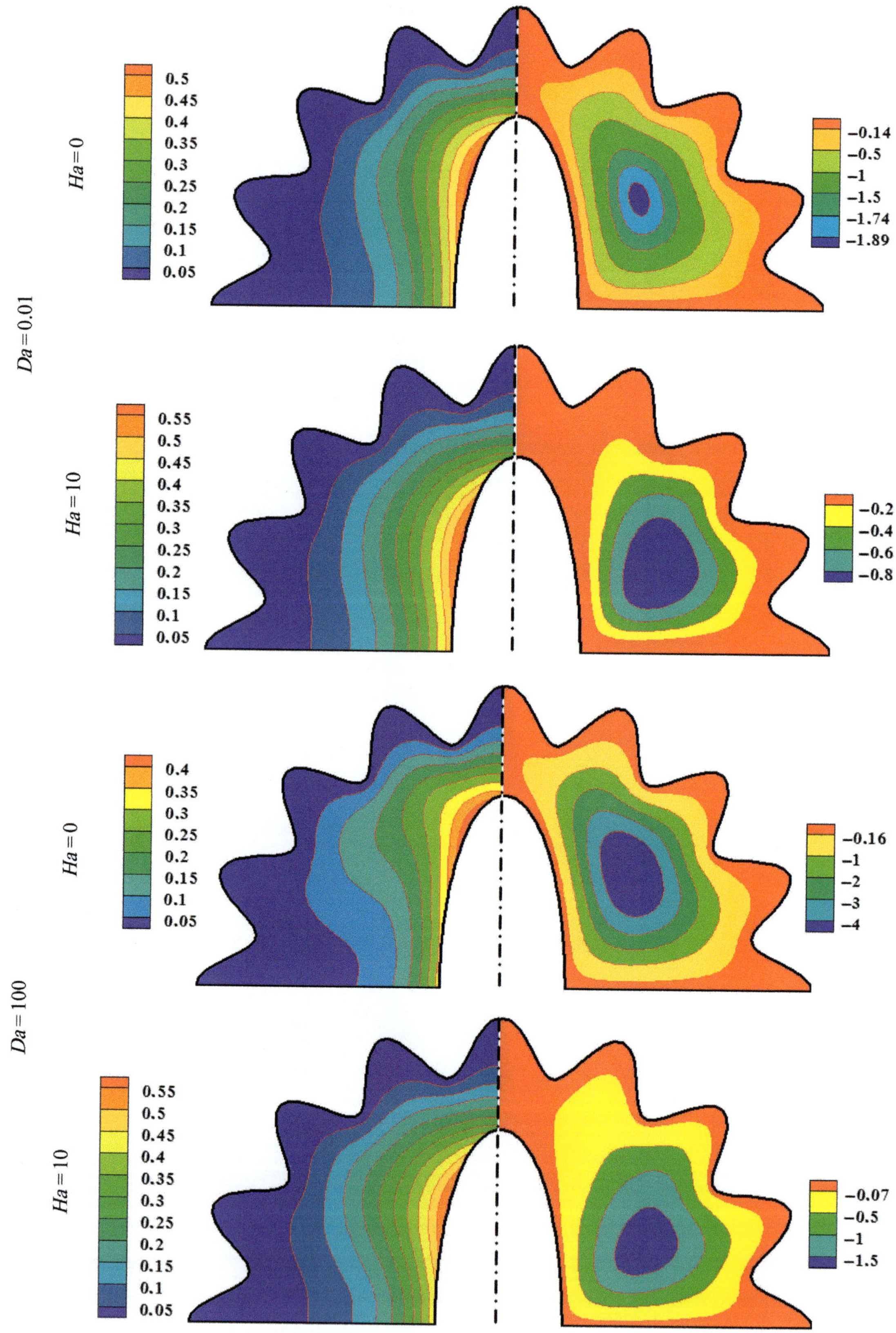

FIGURE 14.38 Influence of *Da*, *Ha* on streamline (right) and isotherm (left) contours when $\phi = 0.04$, $Ra = 10^4$.

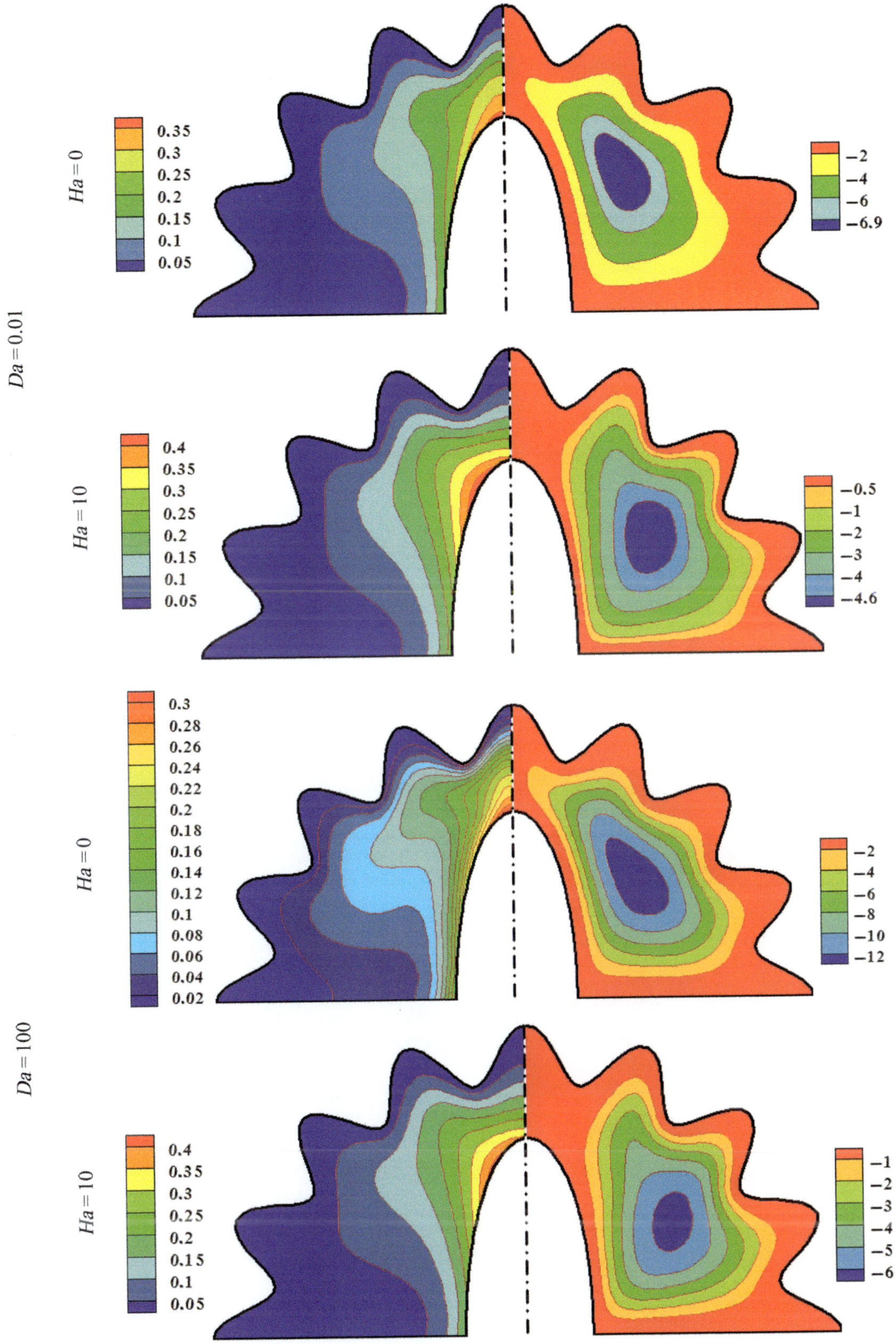

FIGURE 14.39 Influence of *Da*, *Ha* on streamline (right) and isotherm (left) contours when $\phi = 0.04$, $Ra = 10^5$.

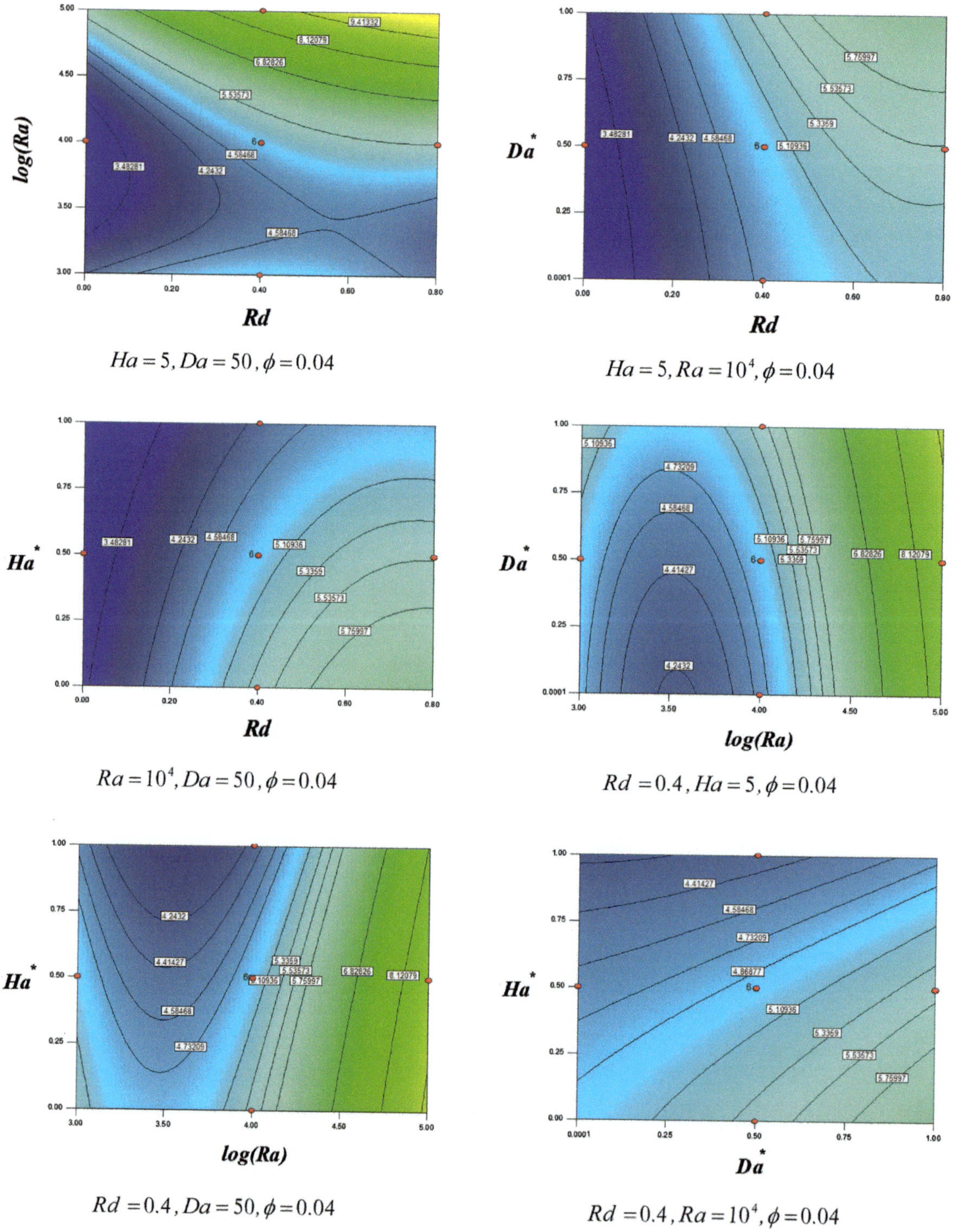

FIGURE 14.40 Effects of *Da*, *Ha*, *Rd*, and *Ra* on average Nusselt number.

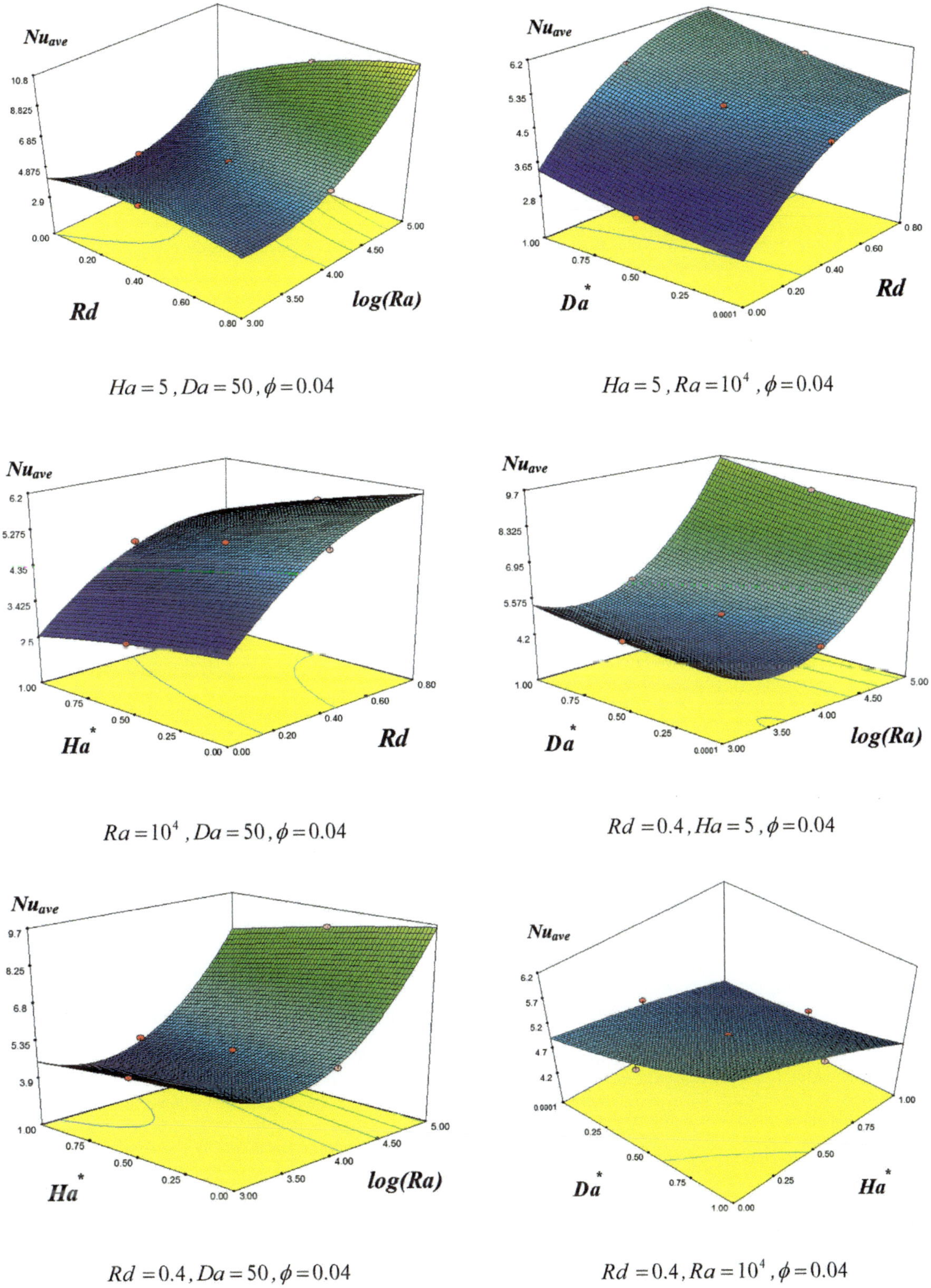

FIGURE 14.40 (Continued)

References

[1] B.V.R. Kumar, S. Gupta, Free convection in a non-Darcian wavy porous enclosure, Int. J. Eng. Sci. 41 (2003) 1827–1848.

[2] M. Sheikholeslami, Numerical investigation of MHD nanofluid free convective heat transfer in a porous tilted enclosure, Eng. Computat. 34 (2017) 1939–1955.

[3] M. Sheikholeslami, D.D. Ganji, Numerical approach for magnetic nanofluid flow in a porous cavity using CuO nanoparticles, Mater. Design 120 (2017) 382–393.

[4] M. Sheikholeslami, D.D. Ganji, Numerical analysis of nanofluid transportation in porous media under the influence of external magnetic source, J. Mol. Liq. 233 (2017) 499–507.

[5] M. Sheikholeslami, H.B. Rokni, Nanofluid convective heat transfer intensification in a porous circular cylinder, Chem. Eng. Proc. Process Intensif. 120 (2017) 93–104.

[6] M. Sheikholeslami, M. Shamlooei, Convective flow of nanofluid inside a lid driven porous cavity using CVFEM, Phys. B: Condensed Matter 521 (2017) 239–250.

[7] M. Sheikholeslami, M. Seyednezhad, Nanofluid heat transfer in a permeable enclosure in presence of variable magnetic field by means of CVFEM, Int. J. Heat Mass Transfer 114 (2017) 1169–1180.

[8] M. Sheikholeslami, Influence of Lorentz forces on nanofluid flow in a porous cavity by means of Non- Darcy model, Eng. Comput. 34 (2017) 2651–2667.

[9] M. Sheikholeslami, A. Zeeshan, Numerical simulation of Fe_3O_4 -water nanofluid flow in a non-Darcy porous media, Int. J. Numer. Methods Heat Fluid Flow, In press.

[10] M. Sheikholeslami, Influence of Lorentz forces on nanofluid flow in a porous cylinder considering Darcy model, J. Mol. Liq. 225 (2017) 903–912.

[11] M. Sheikholeslami, CVFEM for magnetic nanofluid convective heat transfer in a porous curved enclosure, Eur. Phys. J. Plus 131 (2016) 413. Available from: https://doi.org/10.1140/epjp/i2016-16413-y.

[12] M. Sheikholeslami, Influence of Coulomb forces on Fe_3O_4-H_2O nanofluid thermal improvement, Int. J. Hydrogen Energy 42 (2017) 821–829.

[13] M. Sheikholeslami, Magnetic field influence on nanofluid thermal radiation in a cavity with tilted elliptic inner cylinder, J. Mol. Liq. 229 (2017) 137–147.

[14] M. Sheikholeslami, Numerical simulation of magnetic nanofluid natural convection in porous media, Phys. Lett. A 381 (2017) 494–503.

[15] M. Sheikholeslami, Magnetic source impact on nanofluid heat transfer using CVFEM, Neur. Comput. Applicat. (2016). Available from: https://doi.org/10.1007/s00521-016-2740-7.

[16] M. Sheikholeslami, CuO-water nanofluid free convection in a porous cavity considering Darcy law, Europ. Phys. J. Plus 132 (2017) 55. Available from: https://doi.org/10.1140/epjp/i2017-11330-3.

[17] M.S. Kandelousi, Effect of spatially variable magnetic field on ferrofluid flow and heat transfer considering constant heat flux boundary condition, Europ. Phys. J. Plus (2014) 129–248. Available from: https://doi.org/10.1140/epjp/i2014-14248-2.

[18] M. Sheikholeslami, Numerical simulation for external magnetic field influence on Fe_3O_4-water nanofluid forced convection, Engineering Computations, Inpress.

[19] M. Sheikholeslami, M.K. Sadoughi, Numerical modeling for Fe_3O_4 -water nanofluid flow in porous medium considering MFD viscosity, J. Mol. Liq. 242 (2017) 255–264.

[20] M. Sheikholeslami, M. Shamlooei, Magnetic source influence on nanofluid flow in porous medium considering shape factor effect, Phys. Lett. A 381 (2017) 3071–3078.

[21] M. Sheikholeslami, K. Vajravelu, Forced convection heat transfer in Fe_3O_4- ethylene glycol nanofluid under the influence of Coulomb force, J. Mol. Liq. 233 (2017) 203–210.

[22] M. Sheikholeslami, H.B. Rokni, Influence of EFD viscosity on nanofluid forced convection in a cavity with sinusoidal wall, J. Mol. Liq. 232 (2017) 390–395.

[23] M. Sheikholeslami, H.B. Rokni, Magnetic nanofluid natural convection in the presence of thermal radiation considering variable viscosity, Europ. Phys. J. Plus 132 (2017) 238. Available from: https://doi.org/10.1140/epjp/i2017-11498-4.

[24] M. Sheikholeslami, H.B. Rokni, Numerical simulation for impact of Coulomb force on nanofluid heat transfer in a porous enclosure in presence of thermal radiation, Int. J. Heat Mass Transfer 118 (2018) 823–831.

[25] M. Sheikholeslami, M. Shamlooei, R. Moradi, Fe_3O_4- Ethylene glycol nanofluid forced convection inside a porous enclosure in existence of Coulomb force, J. Mol. Liq. 249 (2018) 429–437.

[26] M. Sheikholeslami, S.A. Shehzad, Numerical analysis of Fe_3O_4–H_2O nanofluid flow in permeable media under the effect of external magnetic source, Int. J. Heat Mass Transfer 118 (2018) 182–192.

[27] M. Sheikholeslami, S.A. Shehzad, Magnetohydrodynamic nanofluid convection in a porous enclosure considering heat flux boundary condition, Int. J. Heat Mass Transfer 106 (2017) 1261–1269.

[28] M. Sheikholeslami, S.A. Shehzad, Thermal radiation of ferrofluid in existence of Lorentz forces considering variable viscosity, Int. J. Heat Mass Transfer 109 (2017) 82–92.

[29] M. Sheikholeslami, H.B. Rokni, Numerical modeling of nanofluid natural convection in a semi annulus in existence of Lorentz force, Comput. Methods Appl. Mech. Eng. 317 (2017) 419–430.

[30] M. Sheikholeslami, T. Hayat, A. Alsaedi, Numerical study for external magnetic source influence on water based nanofluid convective heat transfer, Int. J. Heat Mass Transfer 106 (2017) 745–755.

[31] M. Sheikholeslami, A.J. Chamkha, Flow and convective heat transfer of a ferro-nanofluid in a double-sided lid-driven cavity with a wavy wall in the presence of a variable magnetic field, Numer. Heat Transfer, Part A 69 (2016) 1186–1200. Available from: https://doi.org/10.1080/10407782.2015.1125709.

[32] M. Sheikholeslami, A.J. Chamkha, Electrohydrodynamic free convection heat transfer of a nanofluid in a semi-annulus enclosure with a sinusoidal wall, Numer. Heat Transfer, Part A 69 (2016) 781–793. Available from: http://dx.doi.org/10.1080/10407782.2015.1090819.

[33] M. Sheikholeslami, M.K. Sadoughi, Simulation of CuO- water nanofluid heat transfer enhancement in presence of melting surface, Int. J. Heat Mass Transfer 116 (2018) 909–919.
[34] M. Sheikholeslami, H.B. Rokni, Simulation of nanofluid heat transfer in presence of magnetic field: a review, Int. J. Heat Mass Transfer 115 (2017) 1203–1233.
[35] M. Sheikholeslami, H.B. Rokni, Melting heat transfer influence on nanofluid flow inside a cavity in existence of magnetic field, Int. J. Heat Mass Transfer 114 (2017) 517–526.
[36] M. Sheikholeslami, S.A. Shehzad, Magnetohydrodynamic nanofluid convective flow in a porous enclosure by means of LBM, Int. J. Heat Mass Transfer 113 (2017) 796–805.
[37] M. Sheikholeslami, Magnetohydrodynamic nanofluid forced convection in a porous lid driven cubic cavity using Lattice Boltzmann Method, J. Mol. Liq. 231 (2017) 555–565.
[38] M.S. Kandelousi, KKL correlation for simulation of nanofluid flow and heat transfer in a permeable channel, Phys. Lett. A 378 (2014) 3331–3339.
[39] M. Sheikholeslami, Effect of uniform suction on nanofluid flow and heat transfer over a cylinder, J. Braz. Soc. Mech. Sci. Eng. 37 (2015) 1623–1633.
[40] M. Sheikholeslami, Lattice Boltzmann Method simulation of MHD non-Darcy nanofluid free convection, Physica B 516 (2017) 55–71.
[41] M. Sheikholeslami, Influence of magnetic field on nanofluid free convection in an open porous cavity by means of Lattice Boltzmann Method, J. Mol. Liq. 234 (2017) 364–374.
[42] M. Sheikholeslami, A.J. Chamkha, Influence of Lorentz forces on nanofluid forced convection considering Marangoni convection, J. Mol. Liq. 225 (2017) 750–757.
[43] M. Sheikholeslami, T. Hayat, A. Alsaedi, Numerical simulation of nanofluid forced convection heat transfer improvement in existence of magnetic field using Lattice Boltzmann Method, Int. J. Heat Mass Transfer 108 (2017) 1870–1883.
[44] M. Sheikholeslami, R. Ellahi, Three dimensional mesoscopic simulation of magnetic field effect on natural convection of nanofluid, Int. J. Heat Mass Transfer 89 (2015) 799–808.
[45] M. Sheikholeslami, S. Abelman, Two phase simulation of nanofluid flow and heat transfer in an annulus in the presence of an axial magnetic field, IEEE Trans. Nanotechnol. 14 (2015) 561–569.
[46] M. Sheikholeslami, R. Ellahi, Simulation of ferrofluid flow for magnetic drug targeting using Lattice Boltzmann method, J. Zeitschrift Fur Naturforschung A 70 (2015) 115–124.
[47] M. Sheikholeslami, M. Seyednezhad, Lattice Boltzmann Method simulation for CuO-water nanofluid flow in a porous enclosure with hot obstacle, J. Mol. Liq. 243 (2017) 249–256.
[48] M. Sheikholeslami, M. Sadoughi, Mesoscopic method for MHD nanofluid flow inside a porous cavity considering various shapes of nanoparticles, Int. J. Heat Mass Transfer 113 (2017) 106–114.
[49] M. Sheikholeslami, A. Zeeshan, Mesoscopic simulation of CuO-H_2O nanofluid in a porous enclosure with elliptic heat source, Int. J. Hydrogen Energy 42 (2017) 15393–15402.
[50] M. Sheikholeslami, S.A. Shehzad, CVFEM for influence of external magnetic source on $Fe_3O_4 - H_2O$ nanofluid behavior in a permeable cavity considering shape effect, Int. J. Heat Mass Transfer 115 (2017) 180–191.
[51] M. Sheikholeslami, H.B. Rokni, Free convection of CuO-H_2O nanofluid in a curved porous enclosure using mesoscopic approach, Int. J. Hydrogen Energy 42 (2017) 14942–14949.
[52] M. Sheikholeslami, Numerical investigation for CuO-H_2O nanofluid flow in a porous channel with magnetic field using mesoscopic method, J. Mol. Liq. 249 (2018) 739–746.
[53] M. Sheikholeslami, H.B. Rokni, Influence of melting surface on MHD nanofluid flow by means of two phase model, Chin. J. Phys. 55 (2017) 1352–1360.
[54] M. Sheikholeslami, H.B. Rokni, Effect of melting heat transfer on nanofluid flow in existence of magnetic field considering Buongiorno Model, Chin. J. Phys. 55 (2017) 1115–1126.
[55] M. Sheikholeslami, T. Hayat, A. Alsaedi, MHD free convection of Al_2O_3–water nanofluid considering thermal radiation: a numerical study, Int. J. Heat Mass Transfer 96 (2016) 513–524.
[56] M. Sheikholeslami, K. Vajravelu, M.M. Rashidi, Forced convection heat transfer in a semi annulus under the influence of a variable magnetic field, Int. J. Heat Mass Transfer 92 (2016) 339–348.
[57] M. Sheikholeslami, K. Vajravelu, Nanofluid flow and heat transfer in a cavity with variable magnetic field, Appl. Math. Comput. 298 (2017) 272–282.
[58] M. Sheikholeslami, H.B. Rokni, Magnetohydrodynamic CuO-water nanofluid in a porous complex shaped enclosure, ASME, J. Thermal Sci. Eng. Applicat. 9 (2017) 041007. Available from: https://doi.org/10.1115/1.4035973.
[59] M. Sheikholeslami, H.B. Rokni, Nanofluid two phase model analysis in existence of induced magnetic field, Int. J. Heat Mass Transfer 107 (2017) 288–299.
[60] M. Sheikholeslami, T. Hayat, A. Alsaedi, S. Abelman, Numerical analysis of EHD nanofluid force convective heat transfer considering electric field dependent viscosity, Int. J. Heat Mass Transfer 108 (2017) 2558–2565.
[61] M. Sheikholeslami, Magnetic field influence on CuO -H_2O nanofluid convective flow in a permeable cavity considering various shapes for nanoparticles, Int. J. Hydrogen Energy 42 (2017) 19611–19621.
[62] M. Sheikholeslami, M.M. Bhatti, Forced convection of nanofluid in presence of constant magnetic field considering shape effects of nanoparticles, Int. J. Heat Mass Transfer 111 (2017) 1039–1049.
[63] M. Sheikholeslami, A. Zeeshan, Analysis of flow and heat transfer in water based nanofluid due to magnetic field in a porous enclosure with constant heat flux using CVFEM, Comput. Methods Appl. Mech. Eng. 320 (2017) 68–81.
[64] M. Sheikholeslami, M. Shamlooei, Fe_3O_4- H_2O nanofluid natural convection in presence of thermal radiation, Int. J. Hydrogen Energy 42 (2017) 5708–5718.
[65] M. Sheikholeslami, M.M. Bhatti, Active method for nanofluid heat transfer enhancement by means of EHD, Int. J. Heat Mass Transfer 109 (2017) 115–122.

[66] M. Sheikholeslami, H.R. Kataria, A.S. Mittal, Effect of thermal diffusion and heat-generation on MHD nanofluid flow past an oscillating vertical plate through porous medium, J. Mol. Liq. 257 (2018) 12–25.
[67] M. Sheikholeslami, M. Barzegar Gerdroodbary, S. Valiallah Mousavi, D.D. Ganji, R. Moradi, Heat transfer enhancement of ferrofluid inside an 90° elbow channel by non-uniform magnetic field, J. Magn. Magn. Mater. 460 (2018) 302–311.
[68] M. Sheikholeslami, Numerical simulation for solidification in a LHTESS by means of nano-enhanced PCM, J. Taiwan Inst. Chem. Eng. 86 (2018) 25–41.
[69] M. Sheikholeslami, Numerical modeling of nano enhanced PCM solidification in an enclosure with metallic fin, J. Mol. Liq. 259 (2018) 424–438.
[70] M. Sheikholeslami, A. Ghasemi, Solidification heat transfer of nanofluid in existence of thermal radiation by means of FEM, Int. J. Heat Mass Transfer 123 (2018) 418–431.
[71] M. Sheikholeslami, S.A. Shehzad, CVFEM simulation for nanofluid migration in a porous medium using Darcy model, Int. J. Heat Mass Transfer 122 (2018) 1264–1271.
[72] M. Sheikholeslami, M. Darzi, M.K. Sadoughi, Heat transfer improvement and pressure drop during condensation of refrigerant-based nanofluid: an experimental procedure, Int. J. Heat Mass Transfer 122 (2018) 643–650.
[73] M. Sheikholeslami, H.B. Rokni, CVFEM for effect of Lorentz forces on nanofluid flow in a porous complex shaped enclosure by means of non-equilibrium model, J. Mol. Liq. 254 (2018) 446–462.
[74] M. Sheikholeslami, S.A. Shehzad, Z. Li, Water based nanofluid free convection heat transfer in a three dimensional porous cavity with hot sphere obstacle in existence of Lorenz forces, Int. J. Heat Mass Transfer 125 (2018) 375–386.
[75] M. Sheikholeslami, M. Jafaryar, D.D. Ganji, Z. Li, Exergy loss analysis for nanofluid forced convection heat transfer in a pipe with modified turbulators, J. Mol. Liq. 262 (2018) 104–110.
[76] M. Sheikholeslami, M. Jafaryar, Z. Li, Nanofluid turbulent convective flow in a circular duct with helical turbulators considering CuO nanoparticles, Int. J. Heat Mass Transfer 124 (2018) 980–989.
[77] M. Sheikholeslami, Z. Li, M. Shamlooei, Nanofluid, MHD natural convection through a porous complex shaped cavity considering thermal radiation, Phys. Lett. A 382 (2018) 1615–1632.
[78] M. Sheikholeslami, H.B. Rokni, Magnetic nanofluid flow and convective heat transfer in a porous cavity considering Brownian motion effects, Phys. Fluids, Volume 30, Issue 1, https://doi.org/10.1063/1.5012517.
[79] M. Sheikholeslami, S.A. Shehzad, Simulation of water based nanofluid convective flow inside a porous enclosure via non-equilibrium model, Int. J. Heat Mass Transfer 120 (2018) 1200–1212.
[80] M. Sheikholeslami, S.A. Shehzad, Non-Darcy free convection of Fe_3O_4-water nanoliquid in a complex shaped enclosure under impact of uniform Lorentz force, Chin. J. Phys. 56 (2018) 270–281.
[81] M. Sheikholeslami, M. Seyednezhad, Simulation of nanofluid flow and natural convection in a porous media under the influence of electric field using CVFEM, Int. J. Heat Mass Transfer 120 (2018) 772–781.
[82] M. Sheikholeslami, M. Shamlooei, R. Moradi, Numerical simulation for heat transfer intensification of nanofluid in a permeable curved enclosure considering shape effect of Fe_3O_4 nanoparticles, Chem. Eng. Process. : Process Intensif. 124 (2018) 71–82.
[83] M. Sheikholeslami, T. Hayat, T. Muhammad, A. Alsaedi, MHD forced convection flow of nanofluid in a porous cavity with hot elliptic obstacle by means of Lattice Boltzmann method, Int. J. Mech. Sci. 135 (2018) 532–540.
[84] M. Sheikholeslami, Numerical investigation of nanofluid free convection under the influence of electric field in a porous enclosure, J. Mol. Liq. 249 (2018) 1212–1221.
[85] M. Sheikholeslami, CuO-water nanofluid flow due to magnetic field inside a porous media considering Brownian motion, J. Mol. Liq. 249 (2018) 921–929.
[86] M. Sheikholeslami, D.D. Ganji, Influence of electric field on Fe_3O_4-water nanofluid radiative and convective heat transfer in a permeable enclosure, J. Mol. Liq. 250 (2018) 404–412.
[87] M. Sheikholeslami, S.A. Shehzad, Z. Li, Nanofluid heat transfer intensification in a permeable channel due to magnetic field using Lattice Boltzmann method, Physica B (2018). Available from: https://doi.org/10.1016/j.physb.2018.03.036.
[88] M. Sheikholeslami, A. Zeeshan, Numerical simulation of Fe_3O_4-water nanofluid flow in a non-Darcy porous media, Int. J, Num. Methods Heat Fluid Flow Vol. 28 (Issue: 3) (2018) 641–660. Available from: https://doi.org/10.1108/HFF-04-2017-0160.
[89] M. Sheikholeslami, D.D. Ganji, Numerical approach for magnetic nanofluid flow in a porous cavity using CuO nanoparticles, Mater. Des. 120 (2017) 382–393.
[90] M. Darzi, M.K. Sadoughi, M. Sheikholeslami, Condensation of nano-refrigerant inside a horizontal tube, Phys. B: Condens. Matter 537 (2018) 33–39.
[91] M. Sheikholeslami, T. Hayat, A. Alsaedi, Numerical simulation for forced convection flow of MHD CuO-H_2O nanofluid inside a cavity by means of LBM, J. Mol. Liq. 249 (2018) 941–948.
[92] M. Sheikholeslami, S. Soleimani, D.D. Ganji, Effect of electric field on hydrothermal behavior of nanofluid in a complex geometry, J. Mol. Liq. 213 (2016) 153–161.

CHAPTER

15

Thermal Nonequilibrium Model for Nanofluid Flow in a Porous Enclosure by Means of CVFEM

15.1 INTRODUCTION

The study of fluid flow and heat transfer in a porous medium has been a subject of continuous interest for the past several decades because of the wide range of applications, such as geothermal systems, drying technologies, production of thermal isolators, the control of pollutant spread in groundwater, the insulation of buildings, solar power collectors, design of nuclear reactors, and compact heat exchangers, etc. A comprehensive literature survey on this subject can be found in the books by Nield and Bejan [1]. In porous media applications, such as the environmental impact of buried nuclear heat generating waste, chemical reactors, thermal energy transport/storage systems, the cooling of electronic devices, etc., a temperature discrepancy between the solid matrix and the saturating fluid has been observed and recognized. Therefore an analysis of separate energy equations for the two phases, i.e., a local thermal nonequilibrium (LTNE) model, has led to the increasing importance of such approach. Problems on boundary layer flows and heat transfer past a vertical flat plate and a horizontal circular cylinder embedded in a porous medium using LTNE have been studied by several authors and references on this topic may be found in the review papers by Kuznetsov [2]. Convective flow in porous enclosures using the LTE model is a topic studied very intensively during the last several decades and reviews can be found in the books mentioned above and in the review papers by Bejan [3], and Nield [4]. However, it appears that Baytas and Pop [5] were the first to investigate LTNE for the problem of steady free convection in a square porous cavity in order to modify the flow behavior and heat transfer rates. Unsteady laminar natural convection flow in a rectangular enclosure formed by nonisothermal walls, filled with a fluid saturated porous medium in the case of LTNE condition was considered by Hossain and Wilson [6]. In recent years, nanofluid flow in a porous medium has been developed [7–101].

15.2 SIMULATION OF NANOFLUID FLOW INSIDE A POROUS ENCLOSURE VIA NONEQUILIBRIUM MODEL

15.2.1 Problem definition

Fig. 15.1 demonstrates the boundary condition and geometry. The hot wall formulation is:

$$b = \sqrt{1 - \varepsilon_1^2}.a \tag{15.1}$$

where a, b, ε_1 are the major, minor axis of elliptic cylinder, and eccentricity for the inner cylinder. A horizontal magnetic field is employed. CuO-water nanofluid is a working fluid. The enclosure has porous media.

Application of Control Volume based Finite Element Method (CVFEM) for Nanofluid Flow and Heat Transfer.
DOI: https://doi.org/10.1016/B978-0-12-814152-6.00015-1

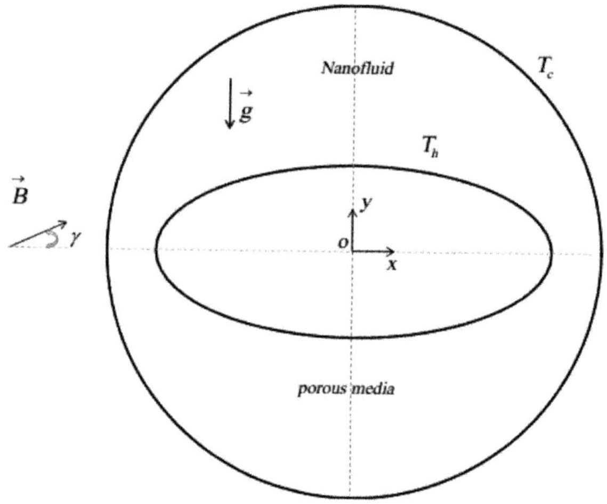

FIGURE 15.1 Geometry and the boundary conditions.

15.2.2 Governing equation

The thermal nonequilibrium model and Boussinesq–Darcy law for flow are employed. So, the two-temperature model is utilized. Considering these conditions, the governing PDEs are:

$$\nabla.\vec{V} = 0 \tag{15.2}$$

$$-\frac{\mu_{nf}}{K} - (\rho\beta)_{nf}(T_{nf} - T_c)\,\vec{g} - \sigma_{nf}\left(\vec{V} \times \vec{B}\right) - \nabla p = 0 \tag{15.3}$$

$$\frac{k_s}{(\rho C_p)_s}\nabla^2 T_s + \frac{h_{nfs}}{(1-\varepsilon)(\rho C_p)_s}(T_{nf} - T_s) = 0 \tag{15.4}$$

$(\rho C_p)_{nf}, (\rho\beta)_{nf}, \rho_{nf}$, and σ_{nf} can be defined as:

$$(\rho C_p)_{nf} = \phi(\rho C_p)_p + (1-\phi)(\rho C_p)_f \tag{15.5}$$

$$(\rho\beta)_{nf} = (1-\phi)(\rho\beta)_f + \phi(\rho\beta)_p \tag{15.6}$$

$$\rho_{nf} = \rho_f(1-\phi) + \rho_p\phi \tag{15.7}$$

$$\frac{\sigma_{nf}}{\sigma_f} = 3\frac{(MM-1)\phi}{\phi(1-MM)+(MM+2)} + 1,\ MM = \frac{\sigma_p}{\sigma_f} \tag{15.8}$$

k_{nf}, μ_{nf} can be estimated via the Koo–Kleinstreuer–Li (KKL) model:

$$\begin{gathered}\frac{k_{nf}}{k_f} = 1 - 3\frac{\left(1-\frac{k_p}{k_f}\right)\phi}{\left(1-\frac{k_p}{k_f}\right)\phi + \left(\frac{k_p}{k_f}+2\right)} + 5\times 10^4 g'(d_p, T, \phi)\rho_f\phi\sqrt{\frac{\kappa_b T}{\rho_p d_p}}c_{p,f} \\ g'(d_p, T, \phi) = Ln(T)\left(a_2 Ln(d_p) + a_5 Ln(d_p)^2 + a_1 + a_3 Ln(\phi) + a_4 Ln(d_p)Ln(\phi)\right) \\ + \left(a_7 Ln(d_p) + a_6 + a_8 Ln(\phi) + a_9 \ln(d_p)Ln(\phi) + a_{10} Ln(d_p)^2\right) \\ R_f = d_p/k_{p,eff} - d_p/k_p,\ R_f = 4\times 10^{-8} km^2/W\end{gathered} \tag{15.9}$$

$$\mu_{nf} = \frac{\mu_f}{(1-\phi)^{2.5}} + \frac{\mu_f}{k_f}\left(k_{Brownian}/\Pr\right) \tag{15.10}$$

Properties of CuO-water and related coefficients are depicted in Tables 15.1 and 15.2. Nondimensional quantities are:

$$\begin{gathered} v = -\frac{\partial \psi}{\partial x}, u = \frac{\partial \psi}{\partial y}, \\ \theta_s = (T_s - T_c)/(T_h - T_c), \theta_{nf} = \left(T_{nf} - T_c\right)/(T_h - T_c), \\ (X, Y) = (x, y)/L, \quad \Psi = \psi/\alpha_{nf} \end{gathered} \tag{15.11}$$

According to the above definitions, we have:

$$\begin{aligned} \frac{\partial^2 \Psi}{\partial X^2} + \frac{\partial^2 \Psi}{\partial Y^2} = & -\frac{A_6}{A_5} Ha \left[\frac{\partial^2 \Psi}{\partial Y^2}\left(\sin^2\gamma\right) + \frac{\partial^2 \Psi}{\partial X^2}\left(\cos^2\gamma\right) + 2\frac{\partial^2 \Psi}{\partial X\, \partial Y}(\sin\gamma)\,(\cos\gamma)\right] \\ & - \frac{A_3}{A_4}\frac{A_2}{A_5}\frac{\partial \theta_{nf}}{\partial X} Ra \end{aligned} \tag{15.12}$$

$$\varepsilon\left(\frac{\partial^2 \theta_{nf}}{\partial Y^2} + \frac{\partial^2 \theta_{nf}}{\partial X^2}\right) + Nhs\left(\theta_s - \theta_{nf}\right) = -\frac{\partial \theta_{nf}}{\partial Y}\frac{\partial \Psi}{\partial X} + \frac{\partial \Psi}{\partial Y}\frac{\partial \theta_{nf}}{\partial X} \tag{15.13}$$

$$\varepsilon\left(\frac{\partial^2 \theta_s}{\partial Y^2} + \frac{\partial^2 \theta_s}{\partial X^2}\right) + Nhs\ \delta_s\left(\theta_{nf} - \theta_s\right) = 0 \tag{15.14}$$

TABLE 15.1 The Coefficient Values of CuO – Water Nanofluid

Coefficient values	**CuO – Water**
a_1	– 26.593310846
a_2	– 0.403818333
a_3	– 33.3516805
a_4	– 1.915825591
a_5	6.42185846658E-02
a_6	48.40336955
a_7	– 9.787756683
a_8	190.245610009
a_9	10.9285386565
a_{10}	-0.72009983664

TABLE 15.2 Thermophysical Properties of Water and Nanoparticles

	$\rho(kg/m^3)$	$C_p(j/kgk)$	$k(W/m.k)$	$\beta \times 10^5(K^{-1})$	$d_p(nm)$	$\sigma(\Omega \cdot m)^{-1}$
Water	997.1	4179	0.613	21	-	0.05
CuO	6500	540	18	29	29	10^{-10}

where the constant and dimensionless parameters are:

$$\begin{aligned}
&A_1 = \frac{\rho_{nf}}{\rho_f}, \quad A_3 = \frac{(\rho\beta)_{nf}}{(\rho\beta)_f}, \quad A_5 = \frac{\mu_{nf}}{\mu_f}, \\
&A_2 = \frac{(\rho C_P)_{nf}}{(\rho C_P)_f}, A_6 = \frac{\sigma_{nf}}{\sigma_f}, A_4 = \frac{k_{nf}}{k_f}, \\
&Ra = \frac{g\,K\,(\rho\beta)_f L\,\Delta T}{\mu_f\,\alpha_f}, Ha = \frac{\sigma_f K\,B_0^2}{\mu_f}, \\
&Nhs = h_{nfs}L^2/k_{nf}, \quad \delta_s = k_{nf}/[k_s(1-\varepsilon)]
\end{aligned} \tag{15.15}$$

The inner wall is hot.

$$\begin{aligned}
&\Psi = 0.0 && \text{on all walls} \\
&\theta_{nf} = \theta_s = 0.0 && \text{on outer wall} \\
&\theta_{nf} = \theta_s = 1.0 && \text{on inner wall}
\end{aligned} \tag{15.16}$$

Nu_{loc}, Nu_{ave} along the cold wall are:

$$Nu_{loc} = \left(\frac{k_{nf}}{k_f}\right)\frac{\partial\theta_{nf}}{\partial r} \tag{15.17}$$

$$Nu_{ave} = \frac{1}{2\pi}\int_0^{2\pi} Nu_{loc}\,dr \tag{15.18}$$

15.2.3 Effects of active parameters

As in the application of CVFEM, nanofluid natural convection in porous media is analyzed by means of the Nonequilibrium model. Simulations are depicted for the impacts of Rayleigh number ($Ra = 100, 500$, and 10^3), the porosity ($\varepsilon = 0.3$ to 0.9), the solid–nanofluid interface heat transfer parameter ($Nhs = 10$ to 1000), and Hartmann number ($Ha = 0$ to 20) on nanofluid hydrothermal behavior.

Figs. 15.2–15.7 depict the impacts of Ra, Nhs, ε, and Ha on streamlines, isotherms for solid and nanofluid. In conduction mode, the influences of other parameters are not significant. So, isotherms for the nanofluid (θ_{nf}) and the solid (θ_s) have the same shapes. As buoyancy forces are enhanced, the thermal plume can be observed near the vertical centerline and isotherms for the nanofluid (θ_{nf}) have a complex shape while isotherms for the solid (θ_s) are parallel to each other. Augmenting the Lorentz forces reduces the nanofluid velocity and the temperature fields are stratified. As Nhs enhances, convective flow become stronger and (Ψ_{max}) augments. But the temperature gradient of the nanofluid reduces and in turn Nu_{ave} decreases. As porosity of porous medium augments, the pore volume inside the cavity increases. Therefore, improvement in the convective flow can be observed. Also, the effect of ε on isotherms is similar to that of Nhs. Fig. 15.8 depicts the influence of Ra, Ha, ε, Nhs on average Nusselt number. According to obtained data Nu_{ave} can be expressed as:

$$\begin{aligned}
Nu_{ave} &= 2.63 + 1.42Ra^* - 0.16Ha^* - 0.12\varepsilon - 0.11Nhs^* \\
&- 0.35Ra^*Ha^* - 0.56Ra^*\varepsilon - 0.63Ra^*Nhs^* + 0.12Ha^*\varepsilon + 0.14Ha^*Nhs^* + 0.29\varepsilon Nhs^* \\
&- 0.05(Ra^*)^2 + 0.018(Ha^*)^2 - 0.29(\varepsilon)^2 - 0.15(Nhs^*)^2
\end{aligned} \tag{15.19}$$

where $Ra^* = 0.001Ra, Ha^* = 0.1Ha, Nhs^* = 0.001Nhs$. Increasing the porous medium porosity leads to a decrease in Nusselt number and a similar impact is shown for the Hartmann number. The Nhs has a direct relationship with the rate of heat transfer because the thermal boundary layer thickness reduces with the rise of Nhs. Nu_{ave} is enhanced with the increase of buoyancy forces.

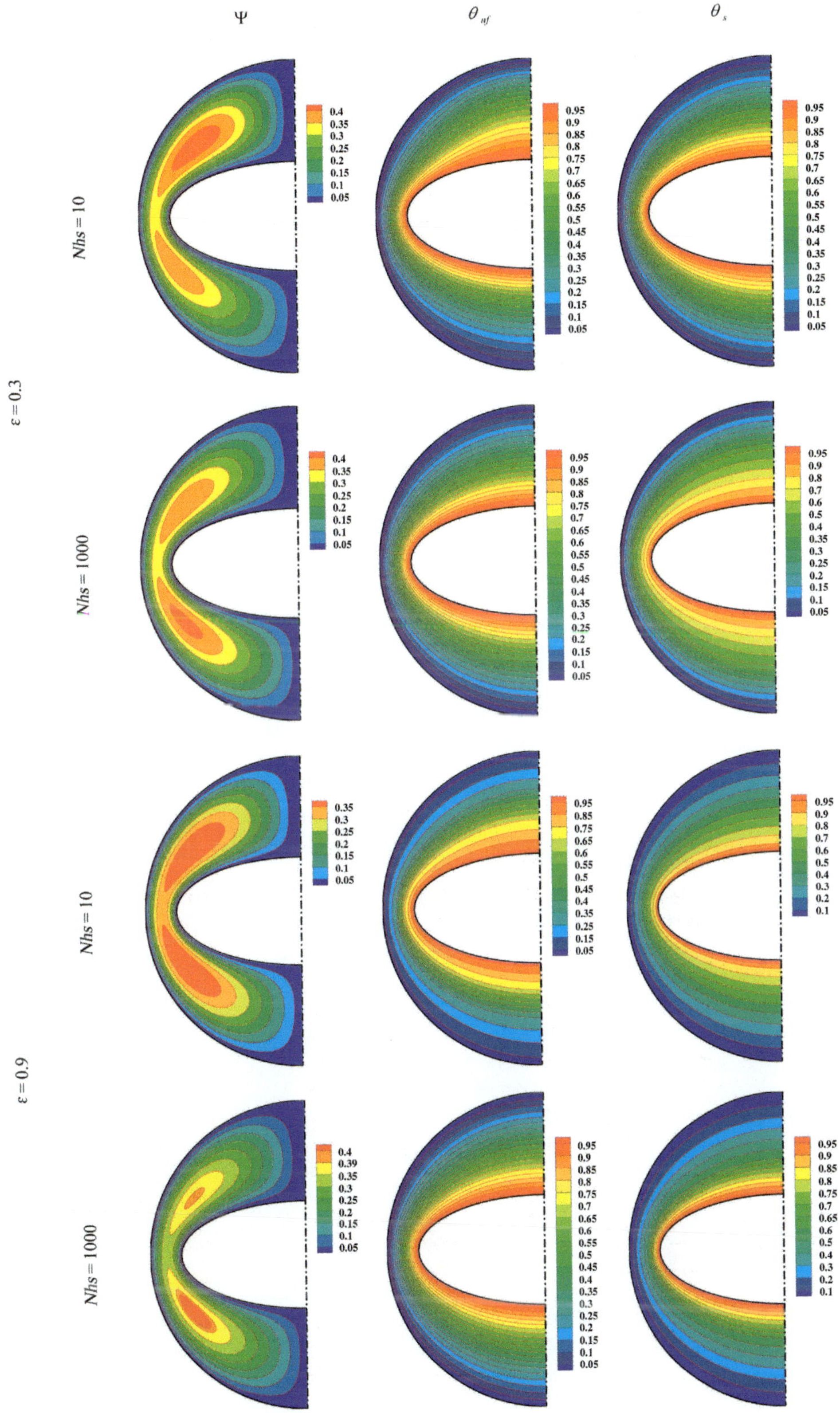

FIGURE 15.2 Streamlines (Ψ), isotherms for the nanofluid (θ_{nf}) and the solid (θ_s) at $Ra = 100, Ha = 0, \phi = 0.04$.

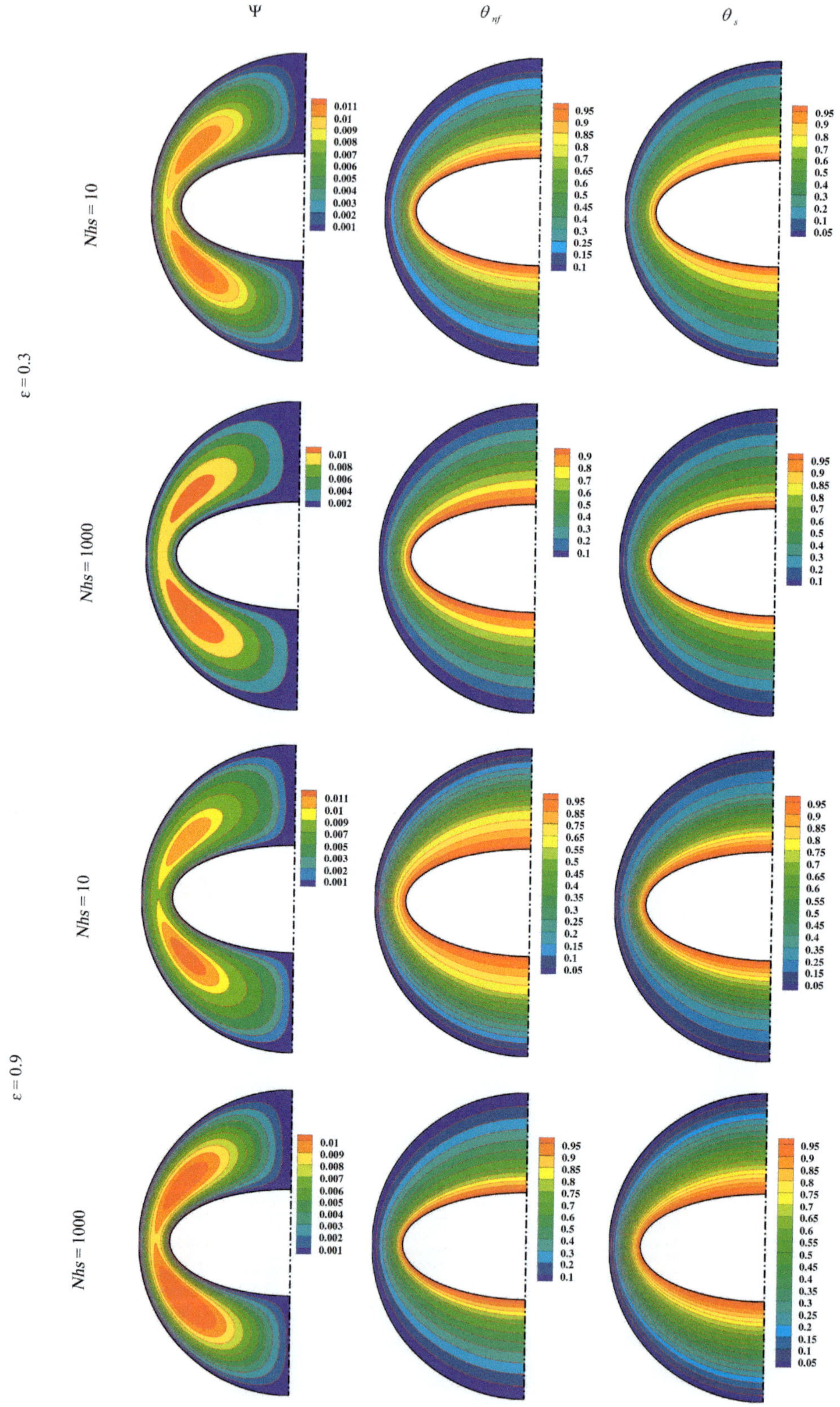

FIGURE 15.3 Streamlines (Ψ), isotherms for the nanofluid (θ_{nf}) and the solid (θ_s) at $Ra = 100, Ha = 20, \phi = 0.04$.

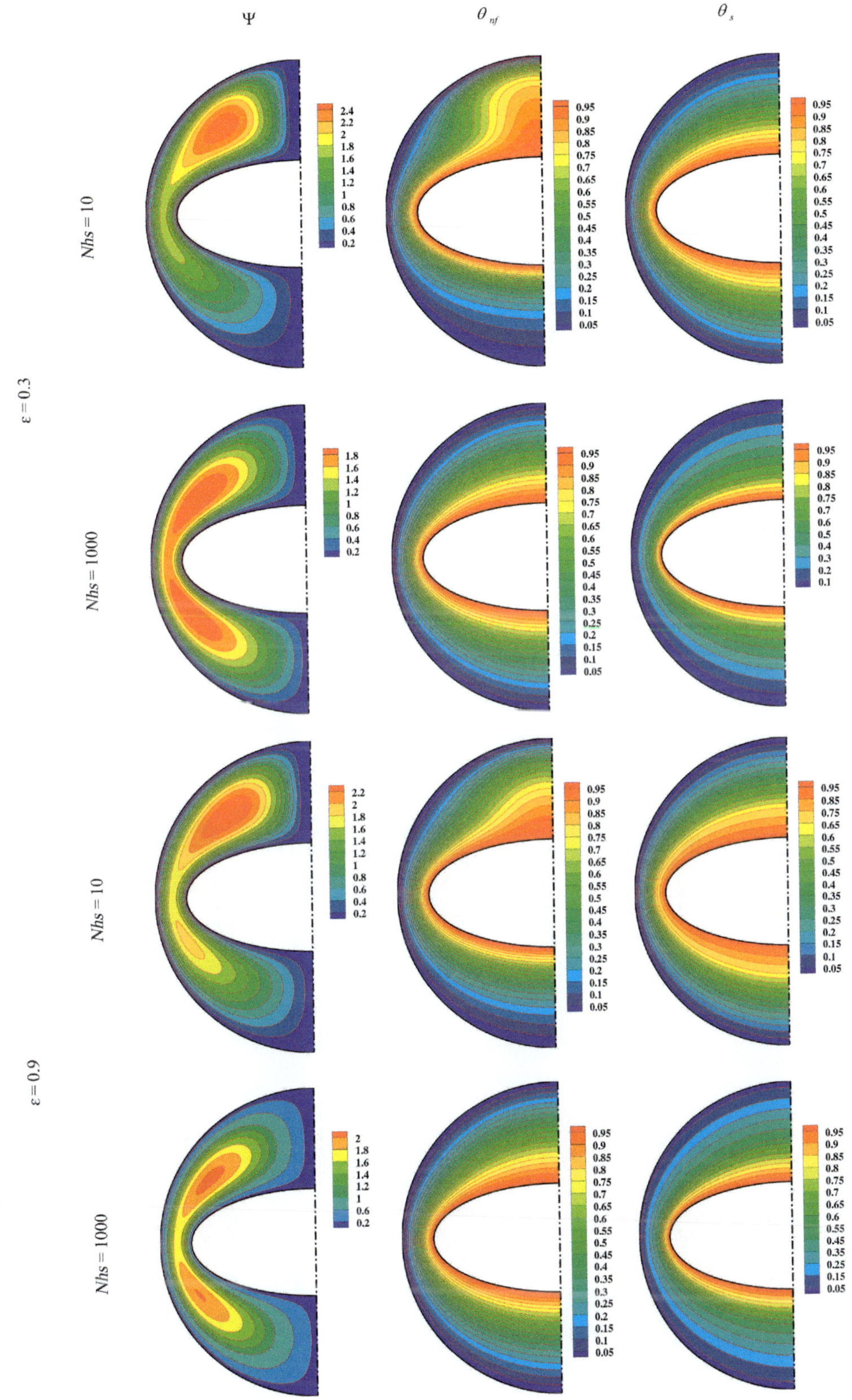

FIGURE 15.4 Streamlines (Ψ), isotherms for the nanofluid (θ_{nf}) and the solid (θ_s) at $Ra = 500, Ha = 0, \phi = 0.04$.

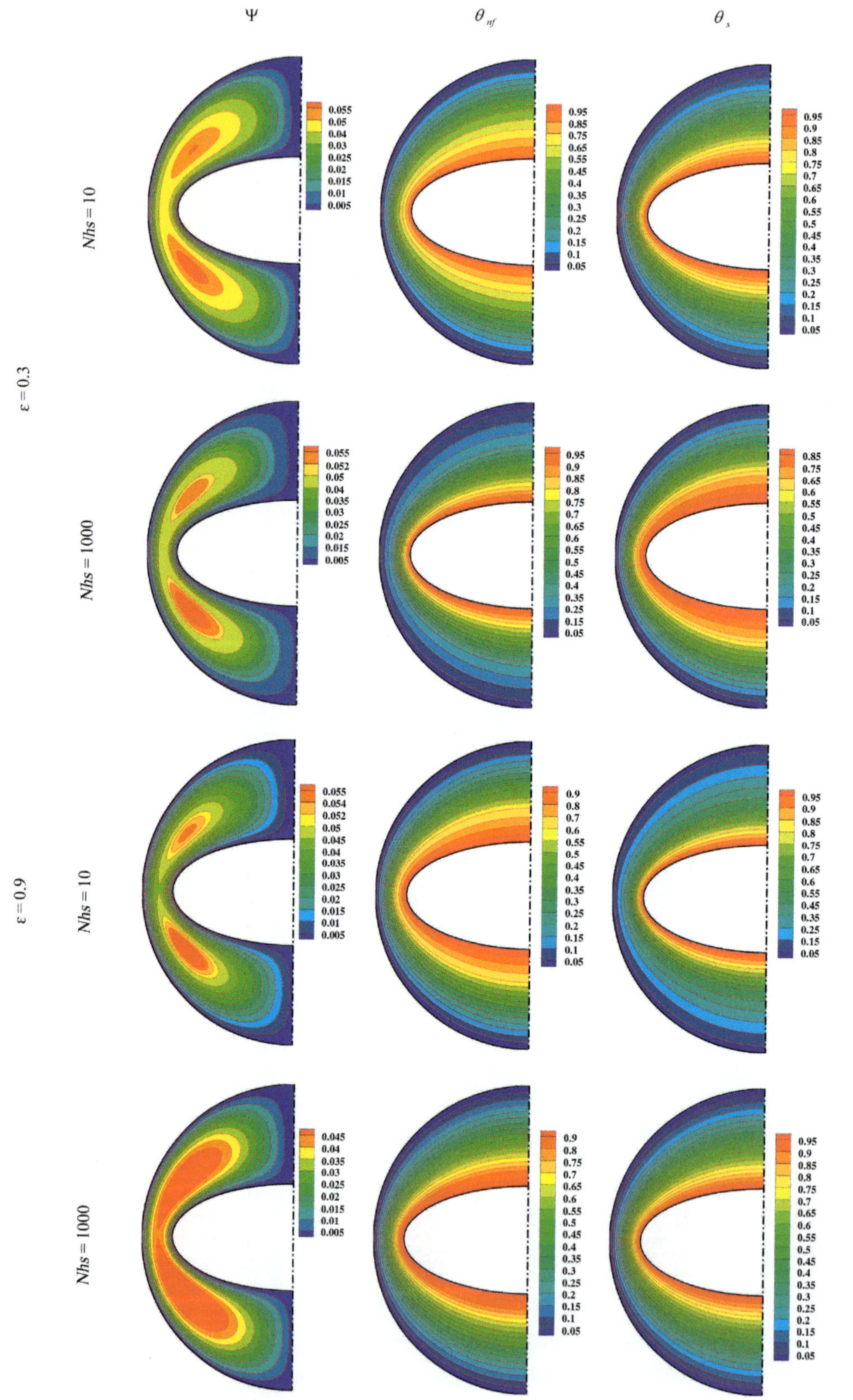

FIGURE 15.5 Streamlines (Ψ), isotherms for the nanofluid (θ_{nf}) and the solid (θ_s) at $Ra = 500, Ha = 20, \phi = 0.04$.

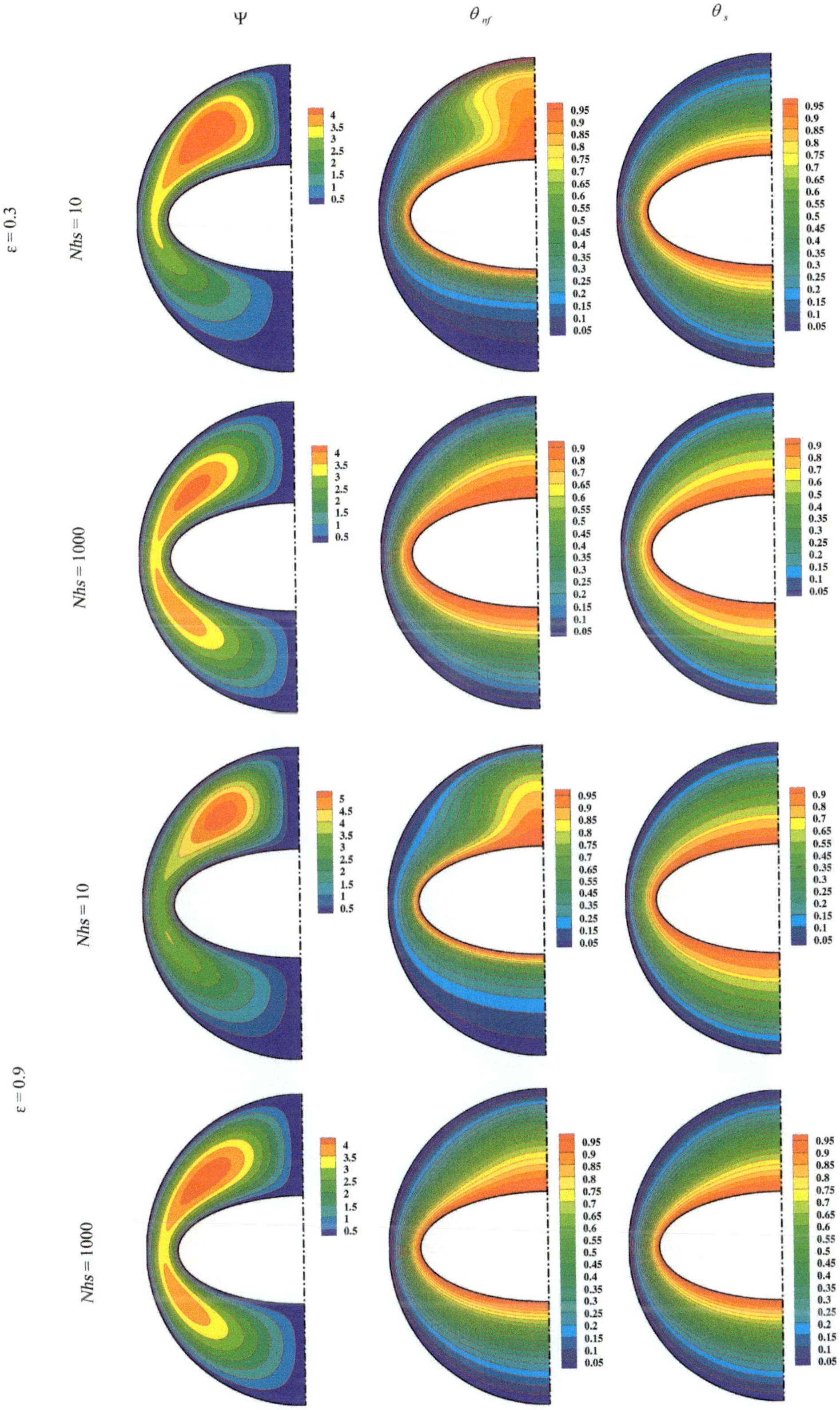

FIGURE 15.6 Streamlines (Ψ), isotherms for the nanofluid (θ_{nf}) and the solid (θ_s) at $Ra = 1000, Ha = 0, \phi = 0.04$.

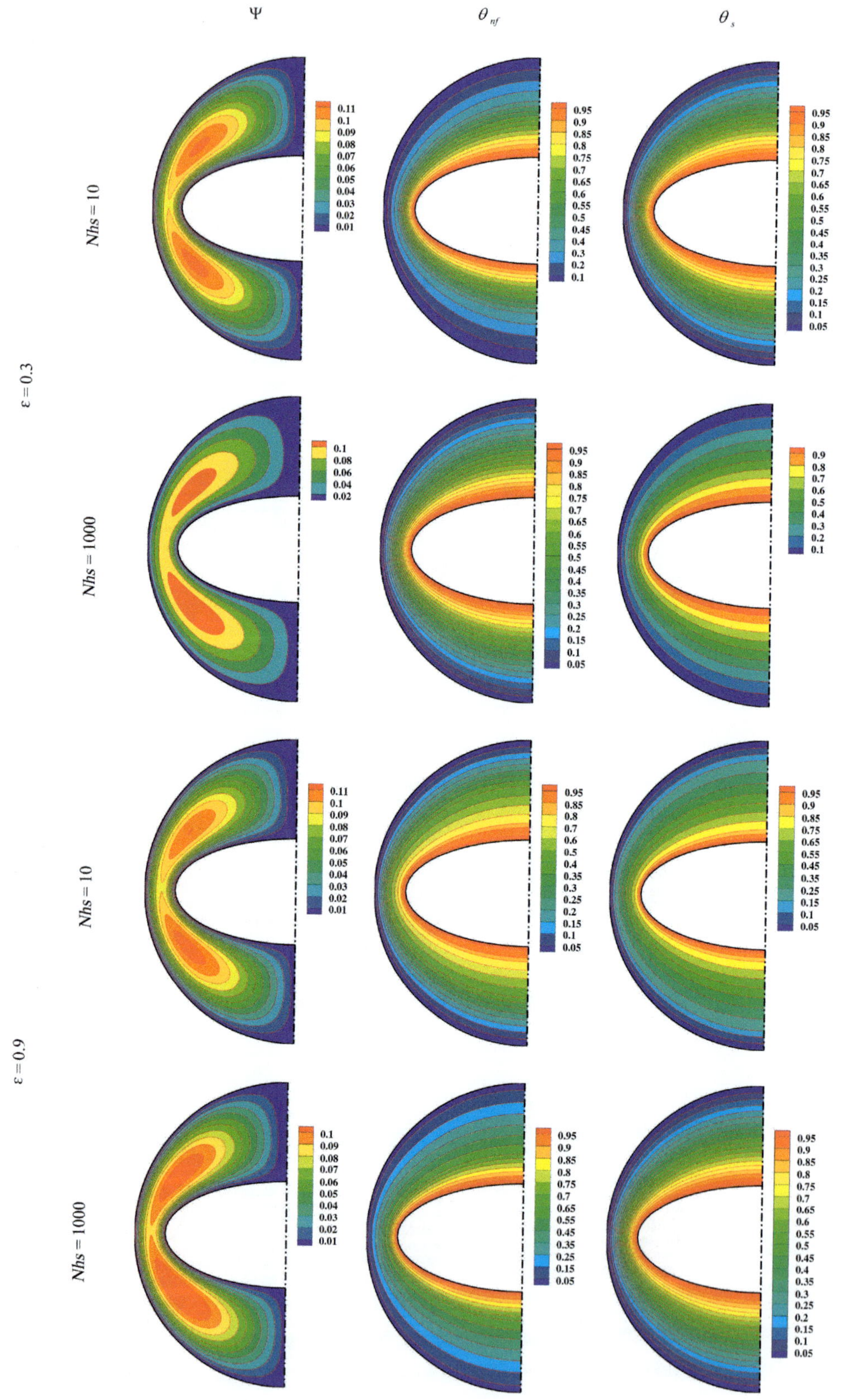

FIGURE 15.7 Streamlines (Ψ), isotherms for the nanofluid (θ_{nf}) and the solid (θ_s) at $Ra = 1000, Ha = 20, \phi = 0.04$.

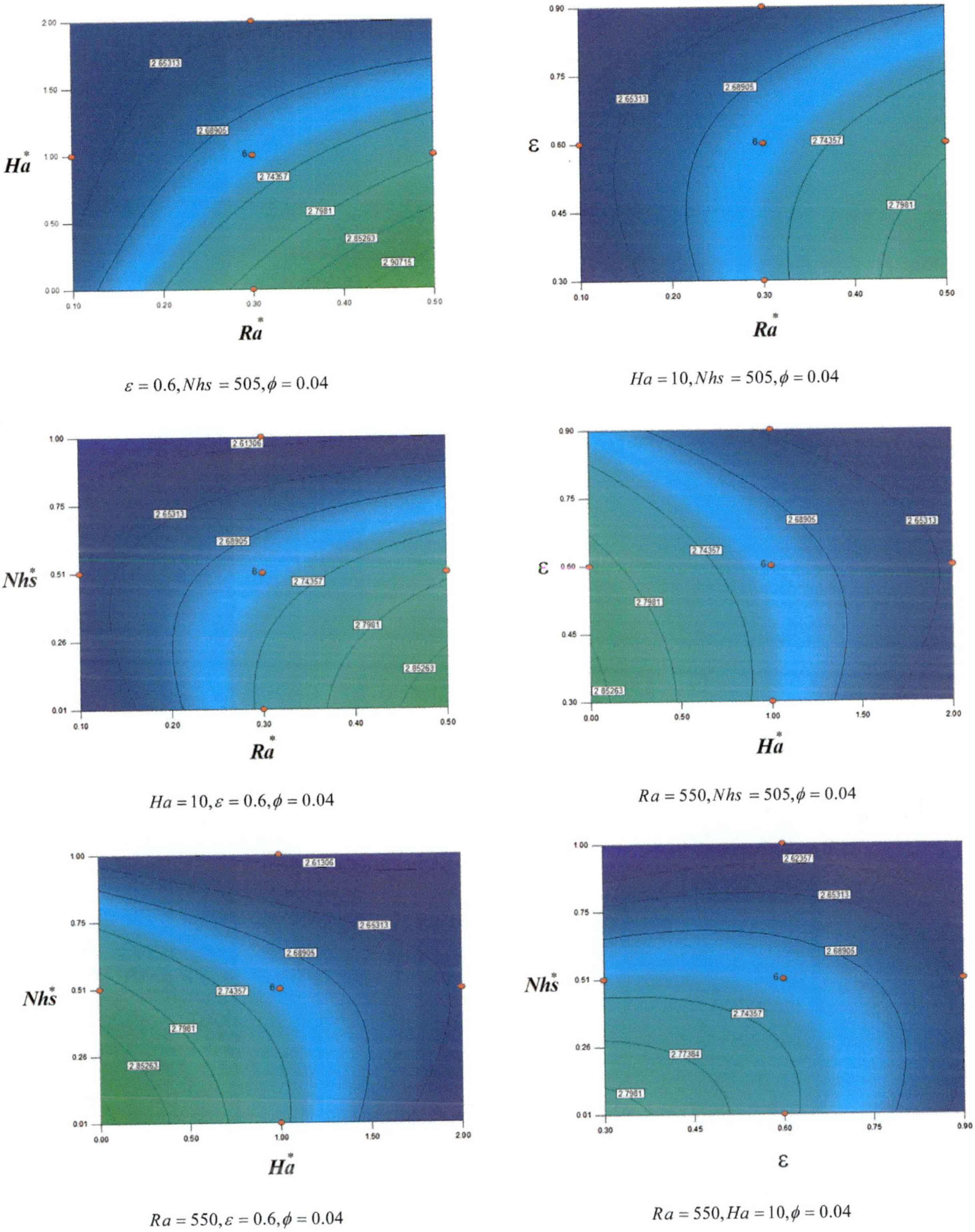

FIGURE 15.8 Effects of Ra, Ha, ε, Nhs on average Nusselt number.

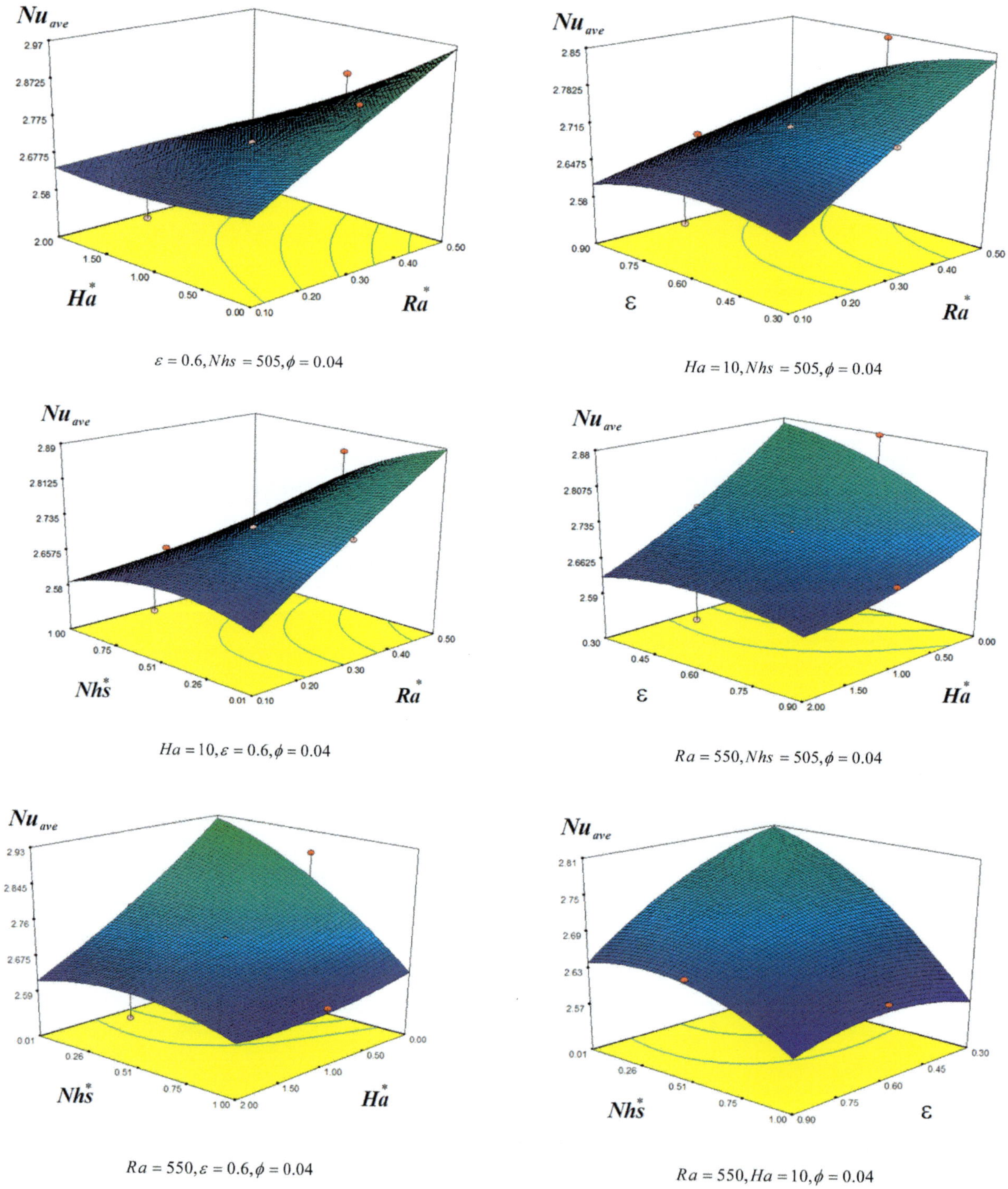

FIGURE 15.8 (Continued).

15.3 NANOFLUID FREE CONVECTION IN A POROUS CAVITY CONSIDERING THE TWO-TEMPERATURE MODEL

15.3.1 Problem definition

Fig. 15.9 demonstrates the boundary conditions and geometry. A horizontal magnetic field is employed. CuO-water nanofluid is a working fluid. The enclosure has porous media.

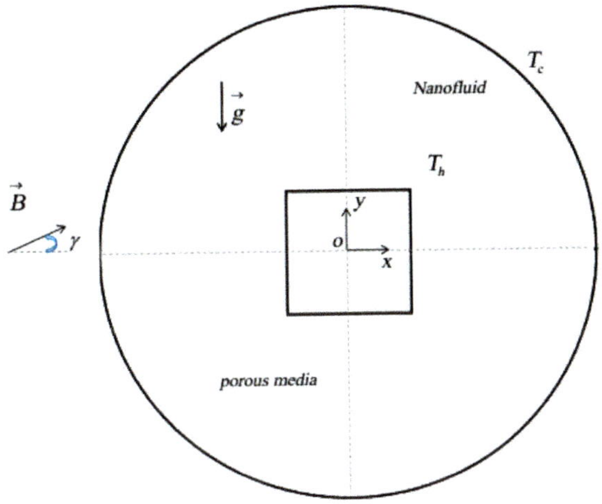

FIGURE 15.9 Geometry and the boundary conditions.

15.3.2 Governing equation

The thermal nonequilibrium model and Boussinesq–Darcy law for flow are employed. So, the two-temperature model is utilized. Considering these conditions, the governing PDEs are:

$$\nabla.\vec{V} = 0 \tag{15.20}$$

$$-\frac{\mu_{nf}}{K} - (\rho\beta)_{nf}(T_{nf} - T_c)\,\vec{g} - \sigma_{nf}\left(\vec{V} \times \vec{B}\right) - \nabla p = 0 \tag{15.21}$$

$$\frac{k_s}{(\rho C_p)_s}\nabla^2 T_s + \frac{h_{nfs}}{(1-\varepsilon)(\rho C_p)_s}(T_{nf} - T_s) = 0 \tag{15.22}$$

$(\rho C_p)_{nf}$, $(\rho\beta)_{nf}$, ρ_{nf}, and σ_{nf} can defined as:

$$(\rho C_p)_{nf} = \phi(\rho C_p)_p + (1-\phi)(\rho C_p)_f \tag{15.23}$$

$$(\rho\beta)_{nf} = (1-\phi)(\rho\beta)_f + \phi(\rho\beta)_p \tag{15.24}$$

$$\rho_{nf} = \rho_f(1-\phi) + \rho_p\phi \tag{15.25}$$

$$\frac{\sigma_{nf}}{\sigma_f} = 3\frac{(MM-1)\phi}{\phi(1-MM)+(MM+2)} + 1,\ MM = \frac{\sigma_p}{\sigma_f} \tag{15.26}$$

k_{nf}, μ_{nf} can be estimated via the KKL model:

$$\frac{k_{nf}}{k_f} = 1 - 3\frac{\left(1-\frac{k_p}{k_f}\right)\phi}{\left(1-\frac{k_p}{k_f}\right)\phi + \left(\frac{k_p}{k_f}+2\right)} + 5\times10^4 g'(d_p,T,\phi)\rho_f\phi\sqrt{\frac{\kappa_b T}{\rho_p d_p}}c_{p,f}$$

$$g'(d_p,T,\phi) = Ln(T)\left(a_2Ln(d_p) + a_5Ln(d_p)^2 + a_1 + a_3Ln(\phi) + a_4Ln(d_p)Ln(\phi)\right)$$
$$+\left(a_7Ln(d_p) + a_6 + a_8Ln(\phi) + a_9\ln(d_p)Ln(\phi) + a_{10}Ln(d_p)^2\right)$$
$$R_f = d_p/k_{p,eff} - d_p/k_p,\ R_f = 4\times10^{-8}km^2/W \tag{15.27}$$

$$\mu_{nf} = \frac{\mu_f}{(1-\phi)^{2.5}} + \frac{\mu_f}{k_f}(k_{Brownian}/\Pr) \tag{15.28}$$

Properties of CuO-water and related coefficients are depicted in Tables 15.1 and 15.2.

Nondimensional quantities are:

$$\begin{aligned} &v = -\frac{\partial \psi}{\partial x}, u = \frac{\partial \psi}{\partial y}, \\ &\theta_s = (T_s - T_c)/(T_h - T_c), \theta_{nf} = (T_{nf} - T_c)/(T_h - T_c), \\ &(X, Y) = (x, y)/L, \quad \Psi = \psi/\alpha_{nf} \end{aligned} \tag{15.29}$$

According to the above definitions, we have:

$$\begin{aligned} &\frac{\partial^2 \Psi}{\partial X^2} + \frac{\partial^2 \Psi}{\partial Y^2} = -\frac{A_6}{A_5} Ha \left[\frac{\partial^2 \Psi}{\partial Y^2} (\sin^2 \gamma) + \frac{\partial^2 \Psi}{\partial X^2} (\cos^2 \gamma) + 2 \frac{\partial^2 \Psi}{\partial X \, \partial Y} (\sin \gamma) (\cos \gamma) \right] \\ &- \frac{A_3}{A_4} \frac{A_2}{A_5} \frac{\partial \theta_{nf}}{\partial X} Ra \end{aligned} \tag{15.30}$$

$$\varepsilon \left(\frac{\partial^2 \theta_{nf}}{\partial Y^2} + \frac{\partial^2 \theta_{nf}}{\partial X^2} \right) + Nhs(\theta_s - \theta_{nf}) = -\frac{\partial \theta_{nf}}{\partial Y} \frac{\partial \Psi}{\partial X} + \frac{\partial \Psi}{\partial Y} \frac{\partial \theta_{nf}}{\partial X} \tag{15.31}$$

$$\varepsilon \left(\frac{\partial^2 \theta_s}{\partial Y^2} + \frac{\partial^2 \theta_s}{\partial X^2} \right) + Nhs \, \delta_s (\theta_{nf} - \theta_s) = 0 \tag{15.32}$$

where the constant and dimensionless parameters are:

$$\begin{aligned} &A_1 = \frac{\rho_{nf}}{\rho_f}, \quad A_3 = \frac{(\rho\beta)_{nf}}{(\rho\beta)_f}, \quad A_5 = \frac{\mu_{nf}}{\mu_f}, \\ &A_2 = \frac{(\rho C_P)_{nf}}{(\rho C_P)_f}, A_6 = \frac{\sigma_{nf}}{\sigma_f}, A_4 = \frac{k_{nf}}{k_f}, \\ &Ra = \frac{g \, K \, (\rho\beta)_f L \, \Delta T}{\mu_f \, \alpha_f}, Ha = \frac{\sigma_f K \, B_0^2}{\mu_f}, \\ &Nhs = h_{nfs} L^2 / k_{nf}, \quad \delta_s = k_{nf} / [k_s (1 - \varepsilon)] \end{aligned} \tag{15.33}$$

The inner wall is hot.

$$\begin{aligned} &\Psi = 0.0 && \text{on all walls} \\ &\theta_{nf} = \theta_s = 0.0 && \text{on outer wall} \\ &\theta_{nf} = \theta_s = 1.0 && \text{on inner wall} \end{aligned} \tag{15.34}$$

Nu_{loc}, Nu_{ave} along the cold wall are:

$$Nu_{loc} = \left(\frac{k_{nf}}{k_f} \right) \frac{\partial \theta_{nf}}{\partial r} \tag{15.35}$$

$$Nu_{ave} = \frac{1}{2\pi} \int_0^{2\pi} Nu_{loc} \, dr \tag{15.36}$$

15.3.3 Effects of active parameters

The nonequilibrium model is employed for the porous media to analyze nanofluid flow and natural convection. This section is an application of CVFEM. The KKL model is selected for nanofluid. Simulations are shown for the influences of Rayleigh number ($Ra = 100, 500$ and 10^3), the porosity ($\varepsilon = 0.3$ to 0.9), the solid–nanofluid interface heat transfer parameter ($Nhs = 10$ to 1000), and Hartmann number ($Ha = 0$ to 20) on nanofluid hydrothermal behavior. Effects of Ra, Nhs, ε, and Ha on streamlines (Ψ), isotherms for solid (θ_s) and nanofluid (θ_{nf}) are shown in Figs. 15.10–15.15. In low Ra, effects of other parameters are not visible because of conduction mode dominance. In this case, isotherms for the nanofluid (θ_{nf}) and the solid (θ_s) have the same shapes. As the buoyancy forces increase, the thermal plume can be observed near the vertical centerline and isotherms for the

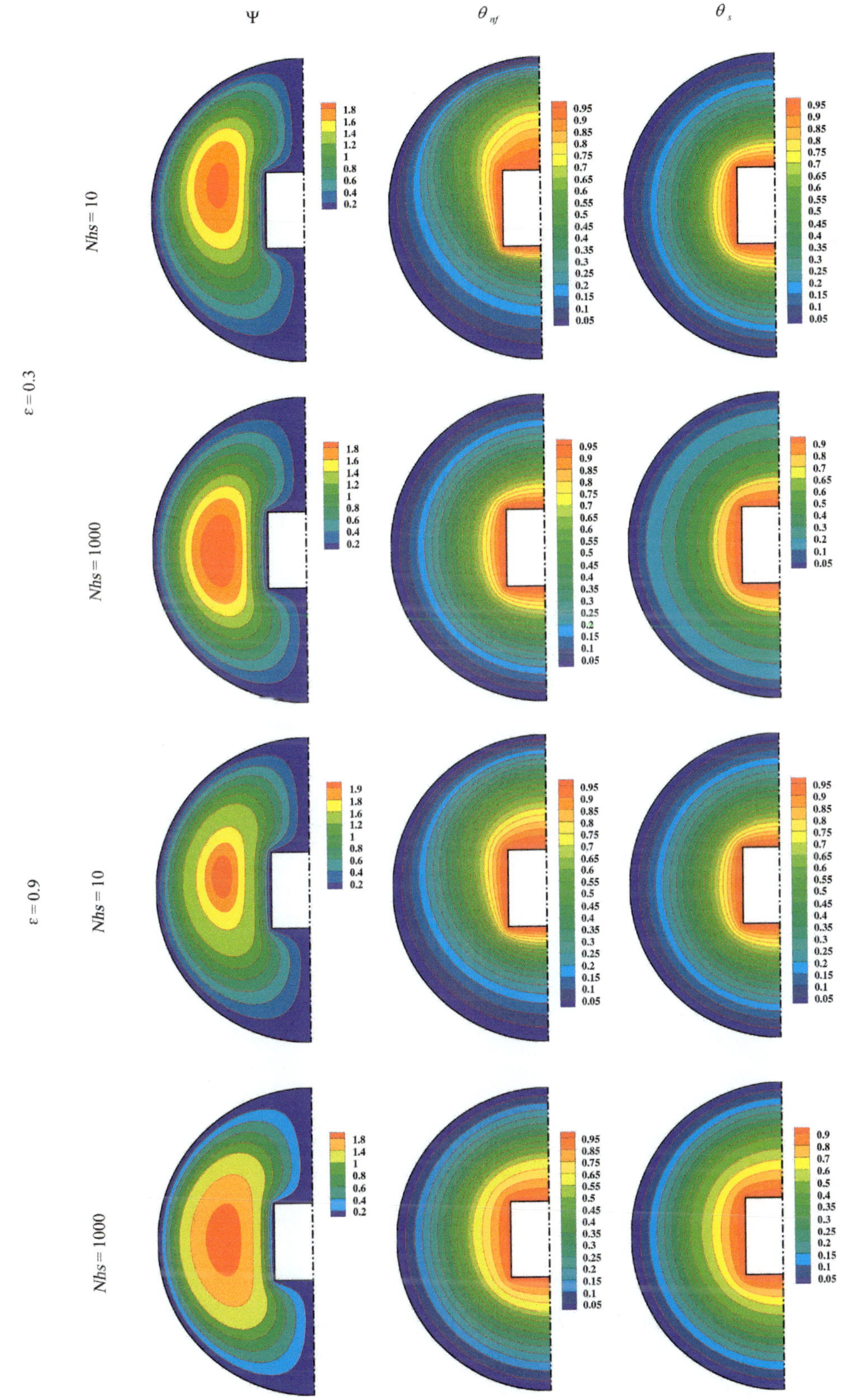

FIGURE 15.10 Streamlines (Ψ), isotherms for the nanofluid (θ_{nf}) and the solid (θ_s) at $Ra = 100, Ha = 0, \phi = 0.04$.

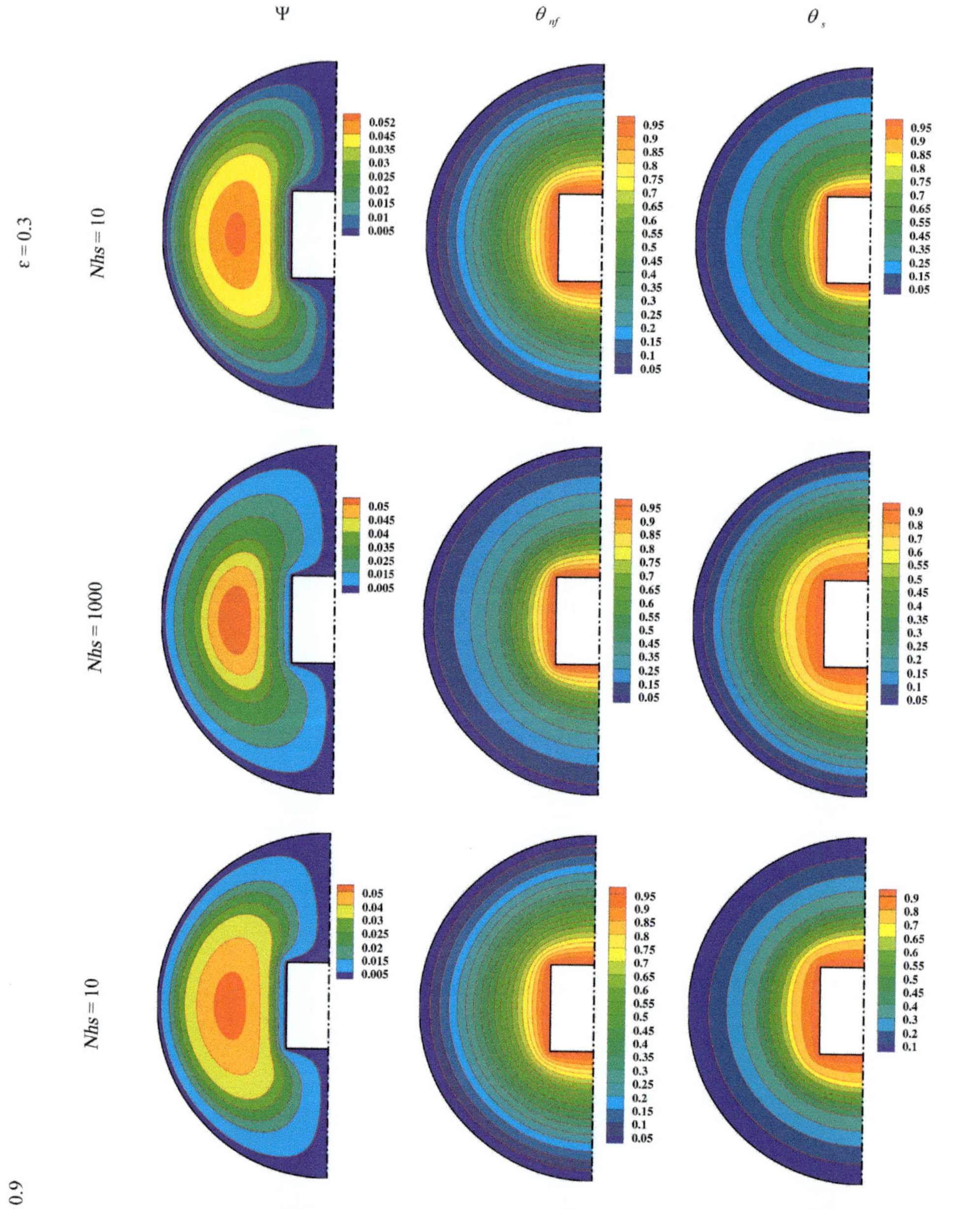

FIGURE 15.11 Streamlines (Ψ), isotherms for the nanofluid (θ_{nf}) and the solid (θ_s) at $Ra = 100, Ha = 20, \phi = 0.04$.

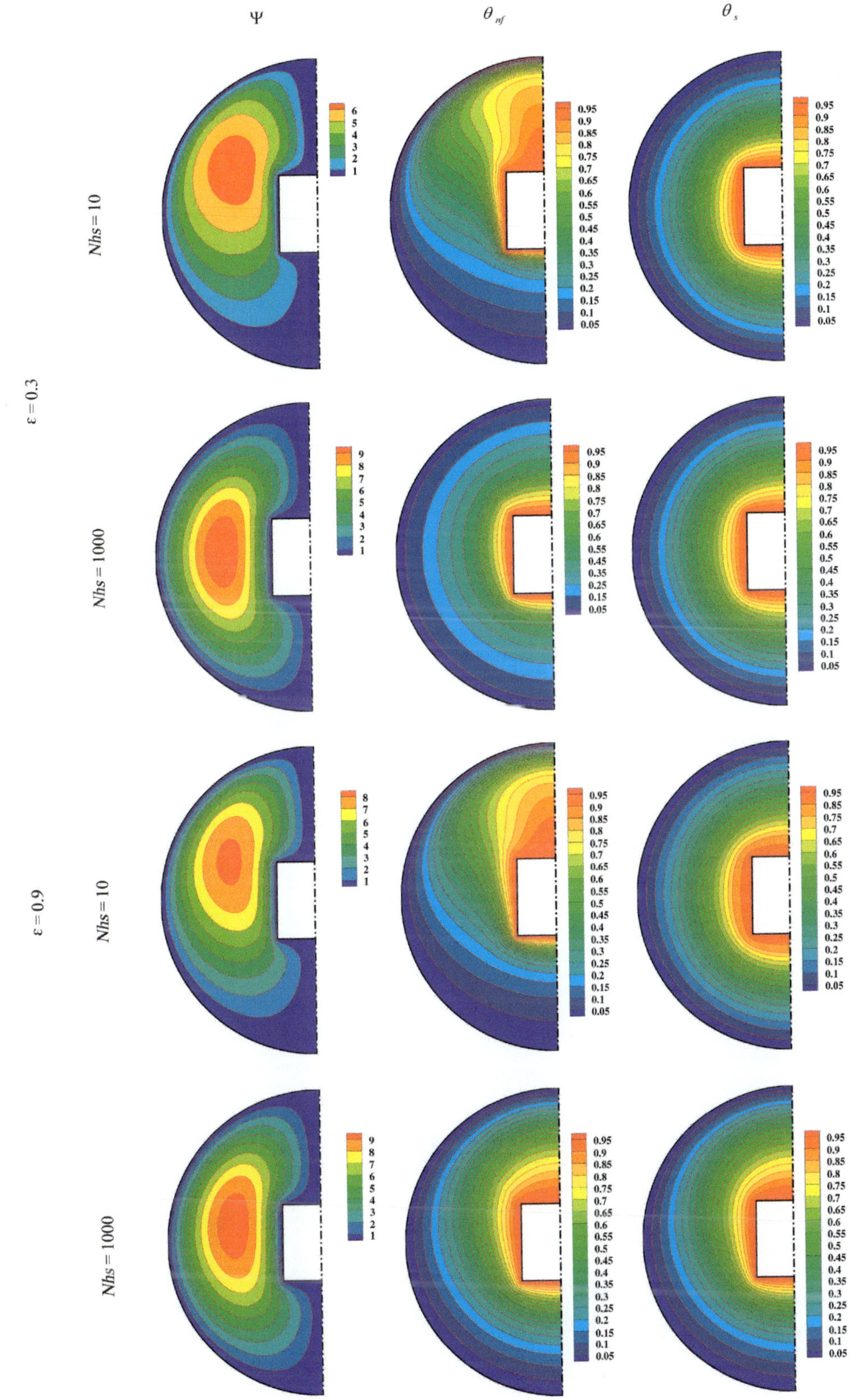

FIGURE 15.12 Streamlines (Ψ), isotherms for the nanofluid (θ_{nf}) and the solid (θ_s) at $Ra = 500, Ha = 0, \phi = 0.04$.

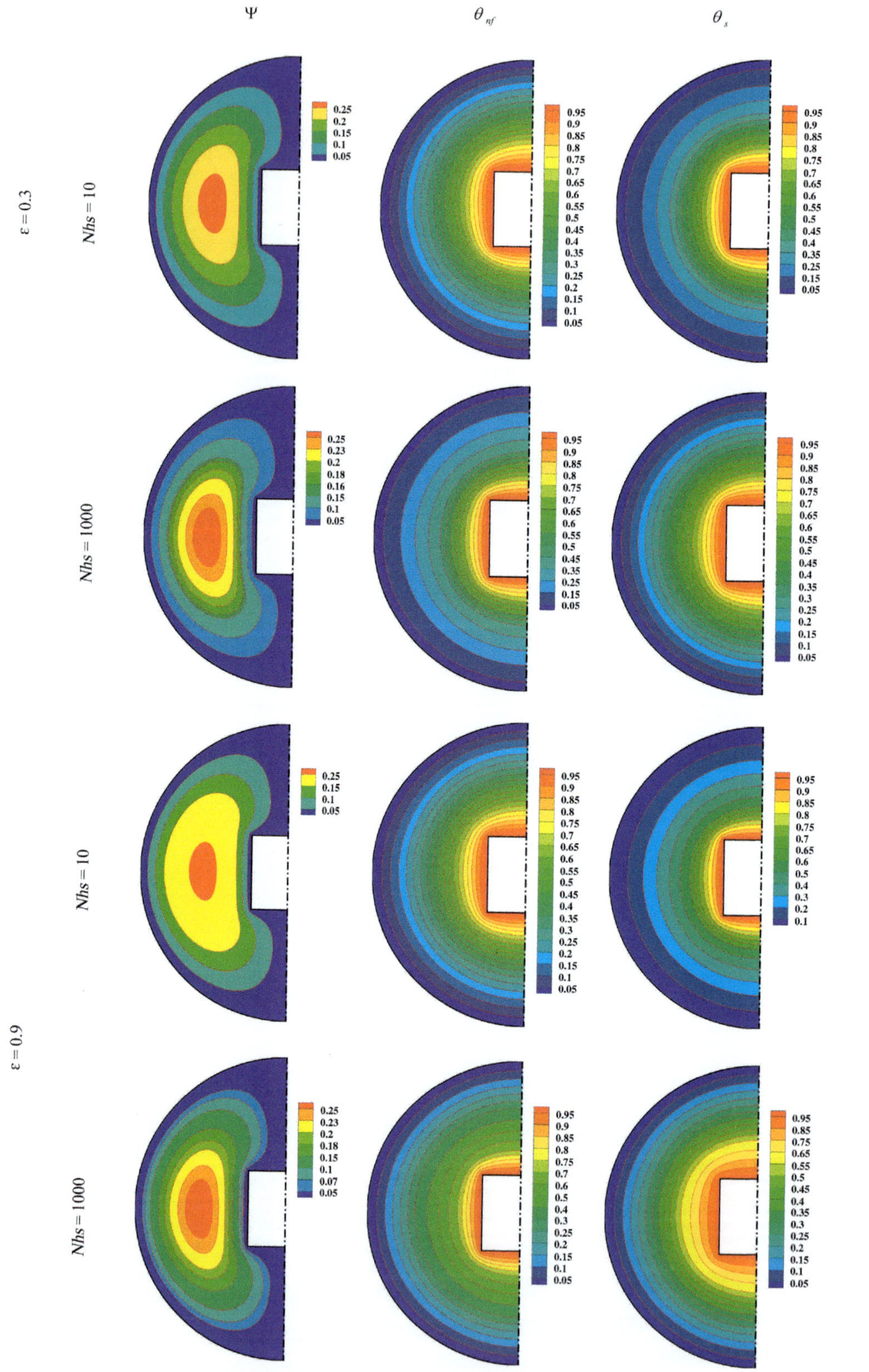

FIGURE 15.13 Streamlines (Ψ), isotherms for the nanofluid $\left(\theta_{nf}\right)$ and the solid (θ_s) at $Ra = 500, Ha = 20, \phi = 0.04$.

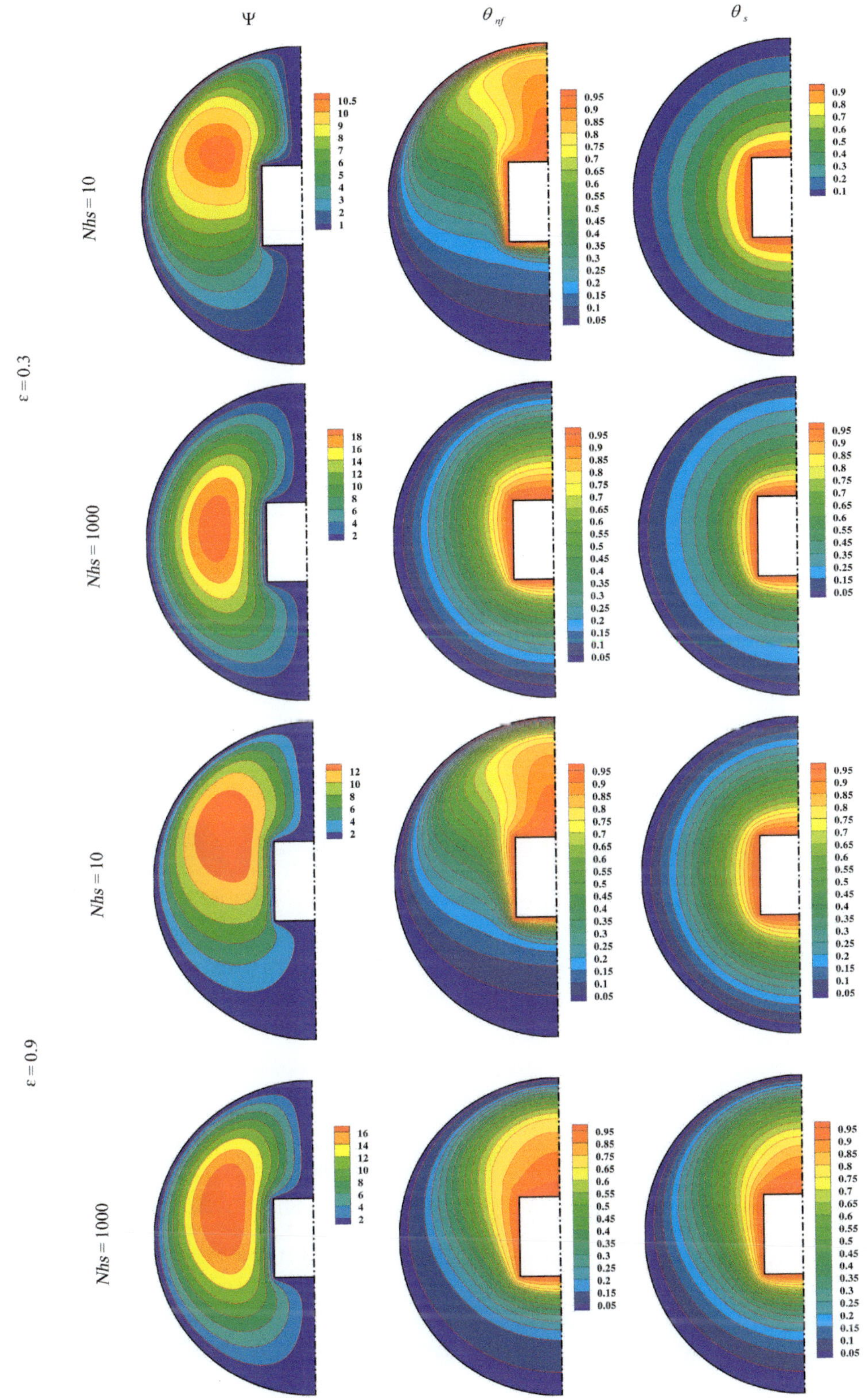

FIGURE 15.14 Streamlines (Ψ), isotherms for the nanofluid (θ_{nf}) and the solid (θ_s) at $Ra = 1000, Ha = 0, \phi = 0.04$.

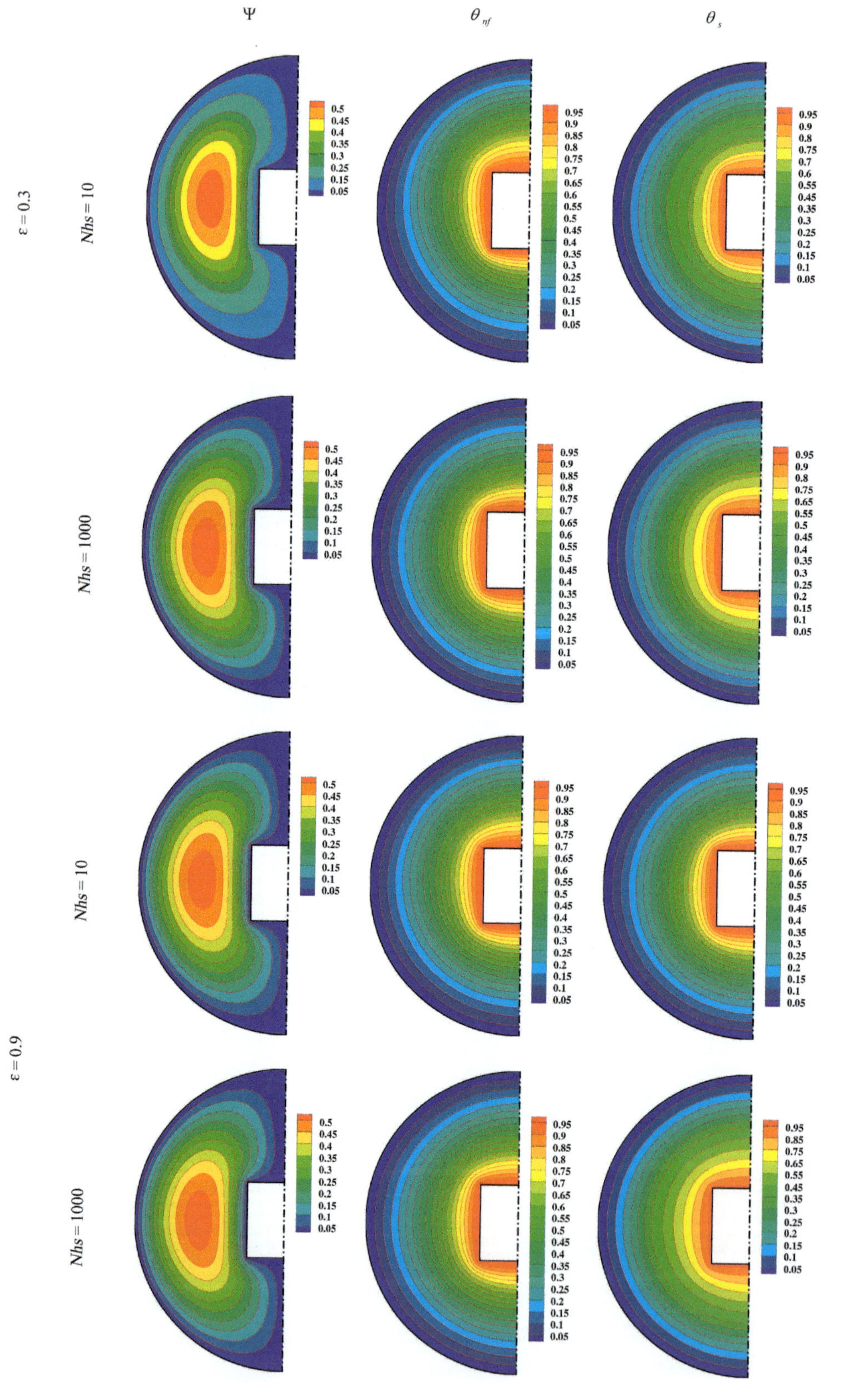

FIGURE 15.15 Streamlines (Ψ), isotherms for the nanofluid (θ_{nf}) and the solid (θ_s) at $Ra = 1000, Ha = 20, \phi = 0.04$.

nanofluid (θ_{nf}) have complex shapes while isotherms for the solid (θ_s) are parallel to each other. Increasing the Lorentz forces reduces the nanofluid velocity and the temperature fields are stratified. As *Nhs* enhances, convective flow becomes stronger and (Ψ_{max}) augments. But the temperature gradient of nanofluid reduces and in turn Nu_{ave} decreases. As the porous medium porosity enhances, the pore volume inside the cavity increases. Therefore, an improvement in the convective flow can be observed. Also, the effect of ε on isotherms is similar to that of *Nhs*. Fig. 15.16 depicts the impact of Ra, Ha, ε, Nhs on average Nusselt number. According to obtained data, Nu_{ave} can be expressed as:

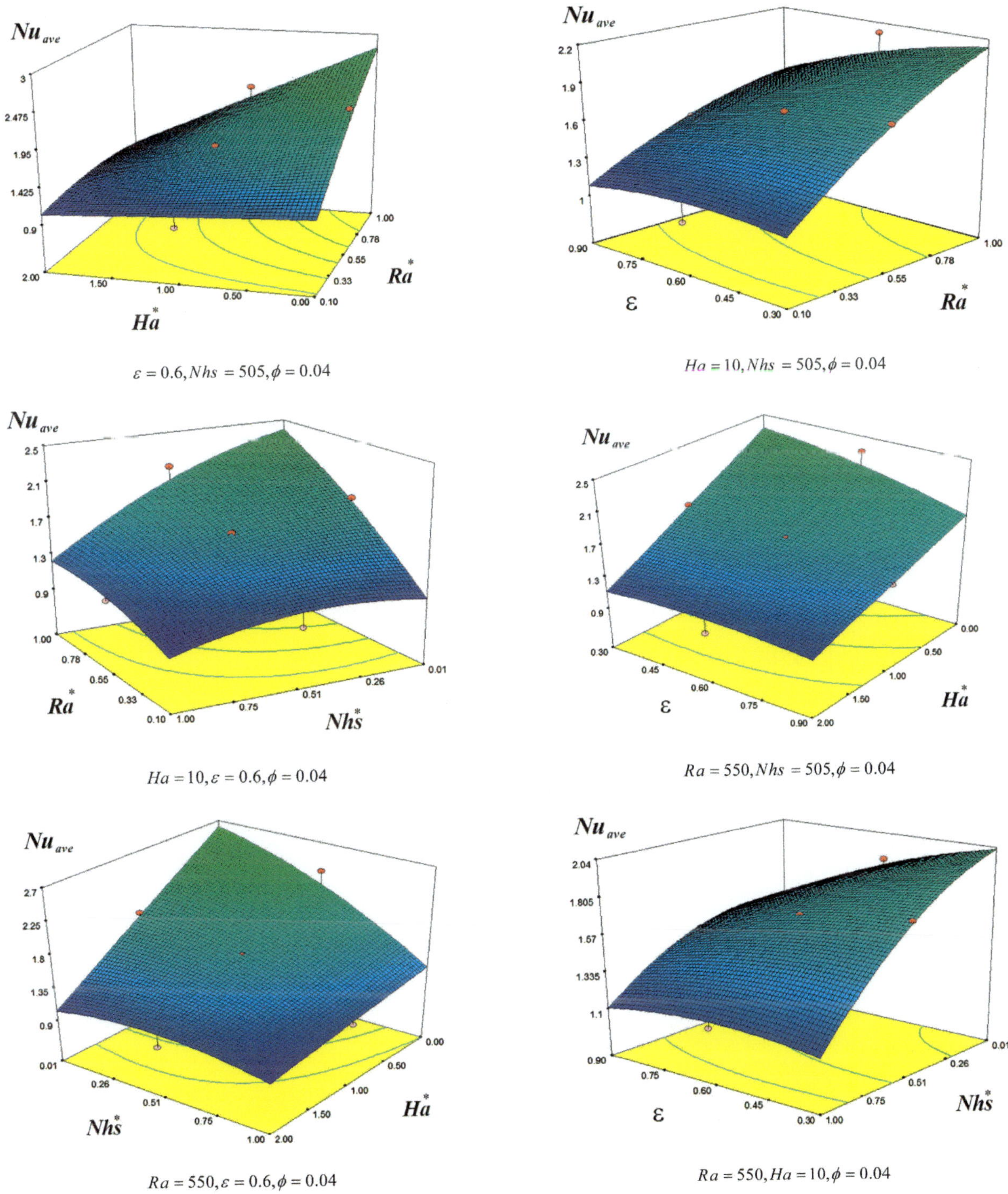

FIGURE 15.16 Effects of Ra, Ha, ε, Nhs on average Nusselt number.

$$\begin{aligned}Nu_{ave} &= 1.48 + 3.24Ra^* - 0.57Ha^* - 0.04\varepsilon - 0.49Nhs^* \\ &- 0.81Ra^*Ha^* - 0.76Ra^*\varepsilon - 1.15Ra^*Nhs^* + 0.4Ha^*\varepsilon + 0.62Ha^*Nhs^* + 0.8\varepsilon Nhs^* \\ &- 0.46(Ra^*)^2 + 0.007(Ha^*)^2 - 0.66(\varepsilon)^2 - 0.63(Nhs^*)^2\end{aligned} \tag{19}$$

where $Ra^* = 0.001Ra, Ha^* = 0.1Ha, Nhs^* = 0.001Nhs$.

The Nhs has a direct relationship with the rate of heat transfer because the thermal boundary layer thickness reduces with the rise of Nhs. Increasing the porous medium porosity leads to a decrease in Nusselt number and a similar impact is shown for the Hartmann number. Nu_{ave} is enhanced with the increase of buoyancy forces.

15.4 NANOFLUID FLOW IN A POROUS SINUSOIDAL CAVITY CONSIDERING THERMAL NONEQUILIBRIUM MODEL

15.4.1 Problem definition

Fig. 15.17 demonstrates the boundary conditions and geometry. The sinusoidal wall formulation is:

$$r_{out} = r_{in} + A\cos(N(\zeta - \zeta_0)) \tag{15.37}$$

A horizontal magnetic field is employed. CuO-water nanofluid is a working fluid. The enclosure has porous media.

15.4.2 Governing equation

The thermal nonequilibrium model and Boussinesq–Darcy law for flow are employed. So, the two-temperature model is utilized. Considering these conditions, the governing PDEs are:

$$\nabla.\vec{V} = 0 \tag{15.38}$$

$$-\frac{\mu_{nf}}{K} - (\rho\beta)_{nf}(T_{nf} - T_c)\,\vec{g} - \sigma_{nf}\left(\vec{V} \times \vec{B}\right) - \nabla p = 0 \tag{15.39}$$

$$\frac{k_s}{(\rho C_p)_s}\nabla^2 T_s + \frac{h_{nfs}}{(1-\varepsilon)(\rho C_p)_s}(T_{nf} - T_s) = 0 \tag{15.40}$$

$(\rho C_p)_{nf}, (\rho\beta)_{nf}, \rho_{nf}$, and σ_{nf} can defined as:

$$(\rho C_p)_{nf} = \phi(\rho C_p)_p + (1-\phi)(\rho C_p)_f \tag{15.41}$$

$$(\rho\beta)_{nf} = (1-\phi)(\rho\beta)_f + \phi(\rho\beta)_p \tag{15.42}$$

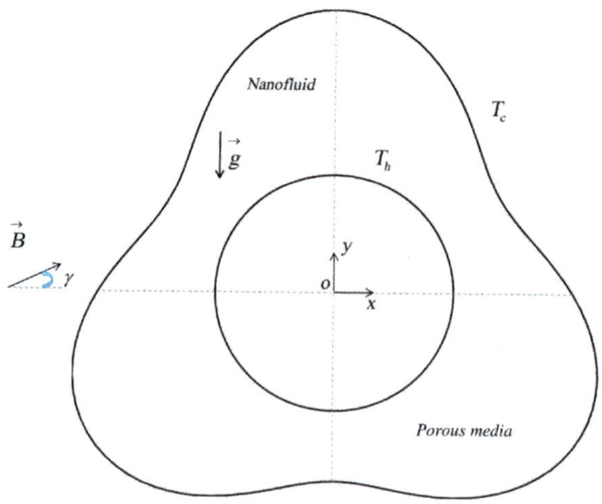

FIGURE 15.17 Geometry and the boundary.

$$\rho_{nf} = \rho_f(1-\phi) + \rho_p\phi \tag{15.43}$$

$$\frac{\sigma_{nf}}{\sigma_f} = 3\frac{(MM-1)\phi}{\phi(1-MM)+(MM+2)} + 1,\ MM = \frac{\sigma_p}{\sigma_f} \tag{15.44}$$

k_{nf}, μ_{nf} can be estimated via the KKL model:

$$\frac{k_{nf}}{k_f} = 1 - 3\frac{\left(1-\frac{k_p}{k_f}\right)\phi}{\left(1-\frac{k_p}{k_f}\right)\phi + \left(\frac{k_p}{k_f}+2\right)} + 5\times10^4 g'(d_p,T,\phi)\rho_f\phi\sqrt{\frac{\kappa_b T}{\rho_p d_p}}c_{p,f}$$
$$g'(d_p,T,\phi) = Ln(T)\left(a_2Ln(d_p) + a_5Ln(d_p)^2 + a_1 + a_3Ln(\phi) + a_4Ln(d_p)Ln(\phi)\right)$$
$$+\left(a_7Ln(d_p) + a_6 + a_8Ln(\phi) + a_9\ln(d_p)Ln(\phi) + a_{10}Ln(d_p)^2\right)$$
$$R_f = d_p/k_{p,eff} - d_p/k_p,\ R_f = 4\times10^{-8}km^2/W \tag{15.45}$$

$$\mu_{nf} = \frac{\mu_f}{(1-\phi)^{2.5}} + \frac{\mu_f}{k_f}(k_{Brownian}/\Pr) \tag{15.46}$$

Properties of CuO-water and related coefficients are depicted in Tables 15.1 and 15.2. Nondimensional quantities are:

$$v = -\frac{\partial\psi}{\partial x},\ u = \frac{\partial\psi}{\partial y},$$
$$\theta_s = (T_s - T_c)/(T_h - T_c),\ \theta_{nf} - (T_{nf} - T_c)/(T_h - T_c),$$
$$(X,Y) = (x,y)/L,\quad \Psi = \psi/\alpha_{nf} \tag{15.47}$$

According to the above definitions, we have:

$$\frac{\partial^2\Psi}{\partial X^2} + \frac{\partial^2\Psi}{\partial Y^2} = -\frac{A_6}{A_5}Ha\left[\frac{\partial^2\Psi}{\partial Y^2}(\sin^2\gamma) + \frac{\partial^2\Psi}{\partial X^2}(\cos^2\gamma) + 2\frac{\partial^2\Psi}{\partial X\,\partial Y}(\sin\gamma)(\cos\gamma)\right]$$
$$-\frac{A_3}{A_4}\frac{A_2}{A_5}\frac{\partial\theta_{nf}}{\partial X}Ra \tag{15.48}$$

$$\varepsilon\left(\frac{\partial^2\theta_{nf}}{\partial Y^2} + \frac{\partial^2\theta_{nf}}{\partial X^2}\right) + Nhs(\theta_s - \theta_{nf}) = -\frac{\partial\theta_{nf}}{\partial Y}\frac{\partial\Psi}{\partial X} + \frac{\partial\Psi}{\partial Y}\frac{\partial\theta_{nf}}{\partial X} \tag{15.49}$$

$$\varepsilon\left(\frac{\partial^2\theta_s}{\partial Y^2} + \frac{\partial^2\theta_s}{\partial X^2}\right) + Nhs\ \delta_s(\theta_{nf} - \theta_s) = 0 \tag{15.50}$$

where the constant and dimensionless parameters are:

$$A_1 = \frac{\rho_{nf}}{\rho_f},\quad A_3 = \frac{(\rho\beta)_{nf}}{(\rho\beta)_f},\quad A_5 = \frac{\mu_{nf}}{\mu_f},$$
$$A_2 = \frac{(\rho C_P)_{nf}}{(\rho C_P)_f},\ A_6 = \frac{\sigma_{nf}}{\sigma_f},\ A_4 = \frac{k_{nf}}{k_f},$$
$$Ra = \frac{g\,K\,(\rho\beta)_f L\,\Delta T}{\mu_f\,\alpha_f},\ Ha = \frac{\sigma_f K\,B_0^2}{\mu_f},$$
$$Nhs = h_{nfs}L^2/k_{nf},\quad \delta_s = k_{nf}/[k_s(1-\varepsilon)] \tag{15.51}$$

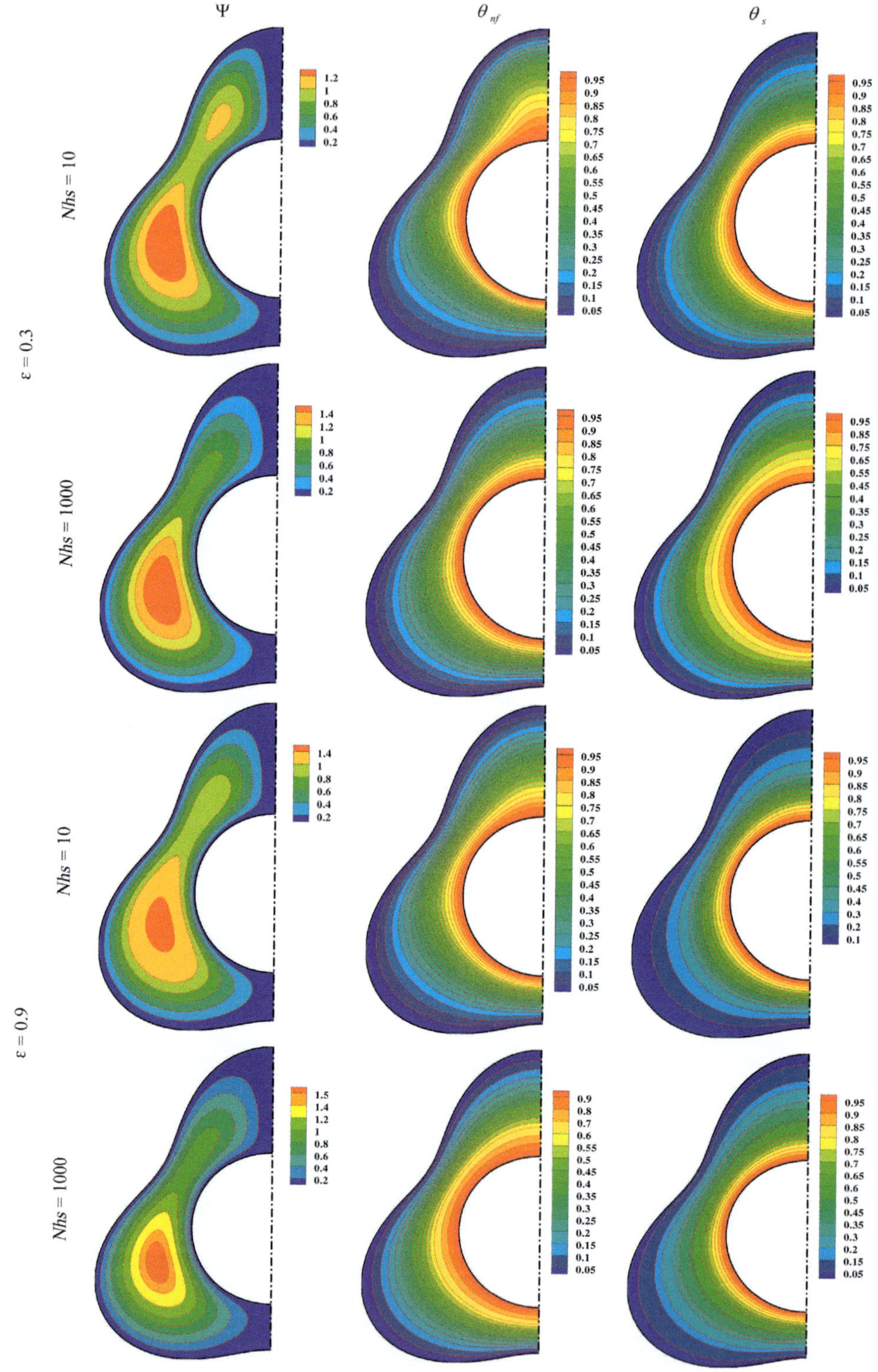

FIGURE 15.18 Streamlines (Ψ), isotherms for the nanofluid (θ_{nf}) and the solid (θ_s) at $Ra = 100, Ha = 0, \phi = 0.04$.

The inner wall is hot.

$$\begin{array}{ll} \Psi = 0.0 & \text{on all walls} \\ \theta_{nf} = \theta_s = 0.0 & \text{on outer wall} \\ \theta_{nf} = \theta_s = 1.0 & \text{on inner wall} \end{array} \tag{15.52}$$

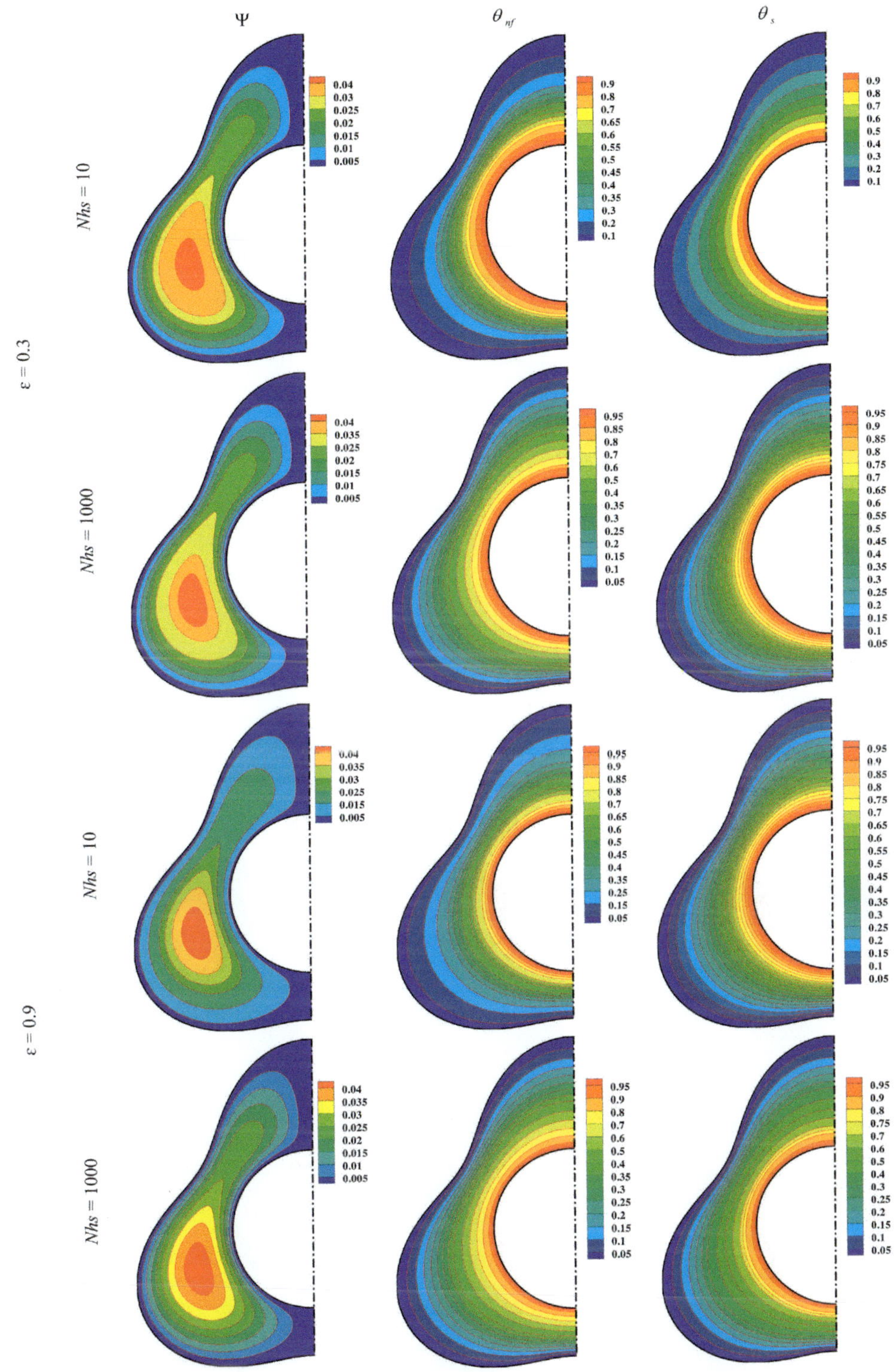

FIGURE 15.19 Streamlines (Ψ), isotherms for the nanofluid (θ_{nf}) and the solid (θ_s) at $Ra = 100, Ha = 20, \phi = 0.04$.

Nu_{loc}, Nu_{ave} along the cold wall are:

$$Nu_{loc} = \left(\frac{k_{nf}}{k_f}\right)\frac{\partial \theta_{nf}}{\partial r} \tag{15.53}$$

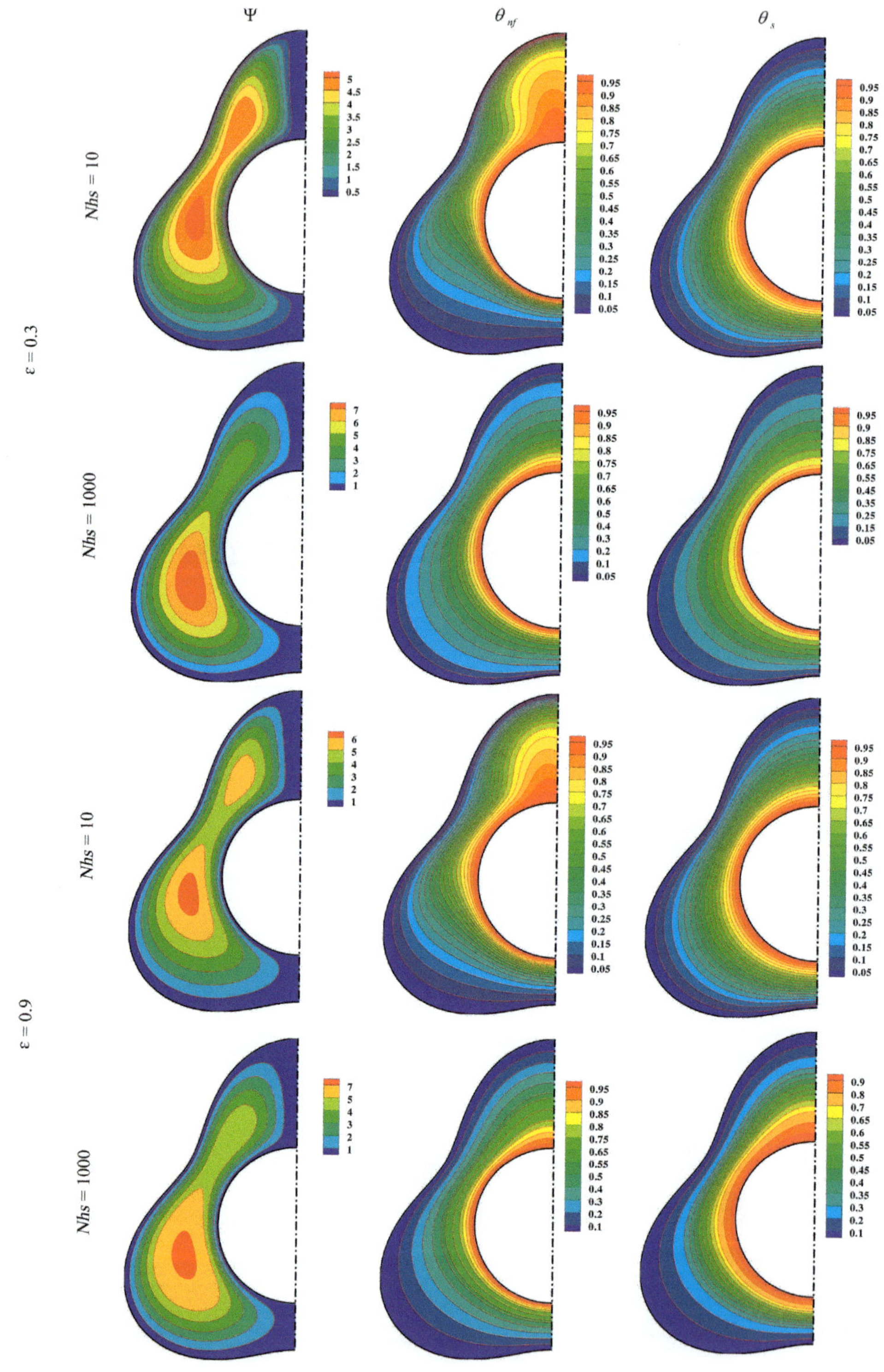

FIGURE 15.20 Streamlines (Ψ), isotherms for the nanofluid (θ_{nf}) and the solid (θ_s) at $Ra = 500, Ha = 0, \phi = 0.04$.

$$Nu_{ave} = \frac{1}{2\pi}\int_0^{2\pi} Nu_{loc}\, dr \tag{15.54}$$

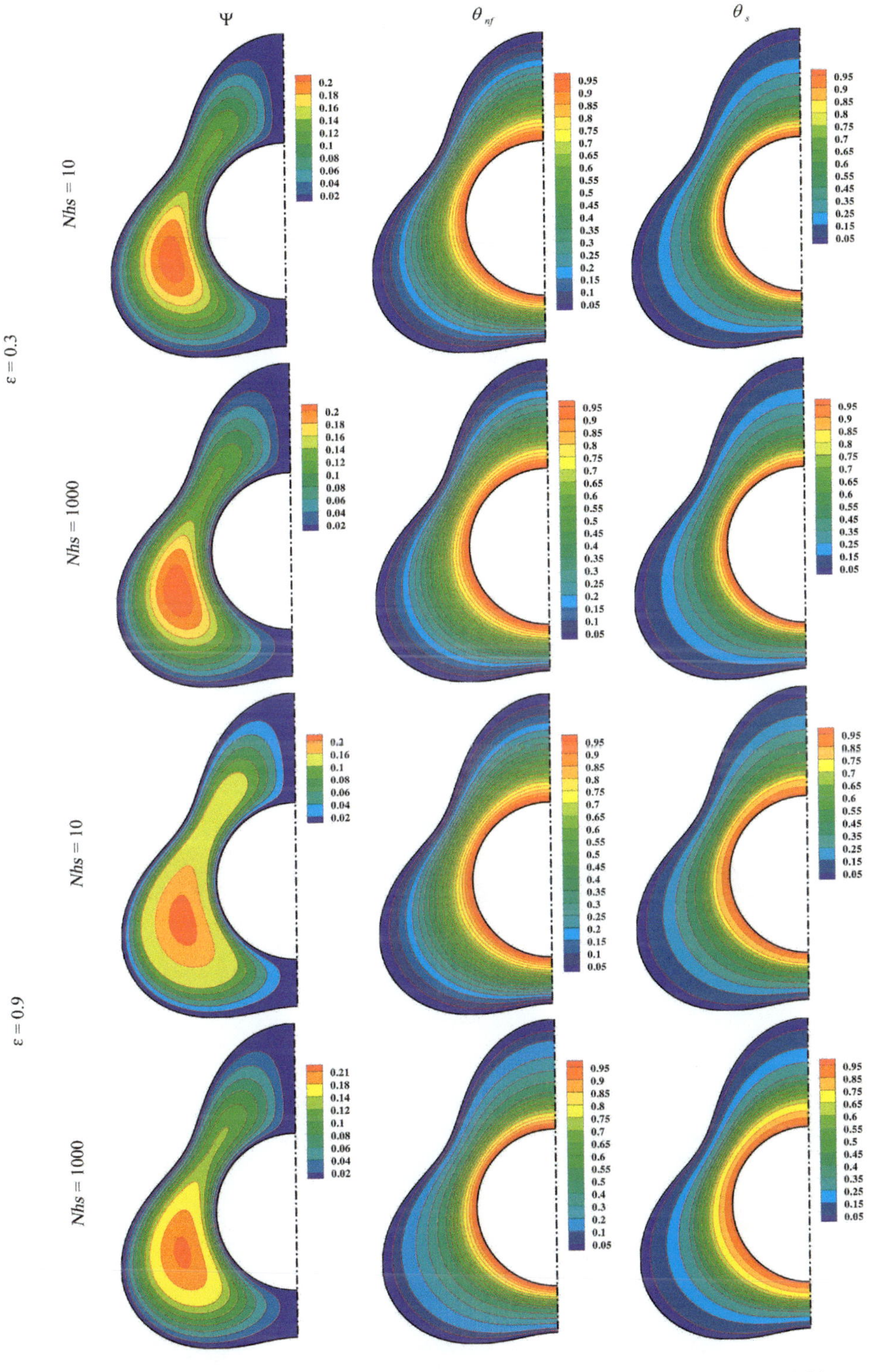

FIGURE 15.21 Streamlines (Ψ), isotherms for the nanofluid (θ_{nf}) and the solid (θ_s) at $Ra = 500, Ha = 20, \phi = 0.04$.

15.4.3 Effects of active parameters

The nonequilibrium model is employed for porous media to analyze nanofluid flow and natural convection. This section is an application of CVFEM. The KKL model is selected for nanofluid. Simulations are shown for the influences of Rayleigh number ($Ra = 100, 500$ and 10^3), the porosity ($\varepsilon = 0.3$ to 0.9), the solid–nanofluid interface heat transfer parameter ($Nhs = 10$ to 1000), and Hartmann number ($Ha = 0$ to 20) on nanofluid hydrothermal

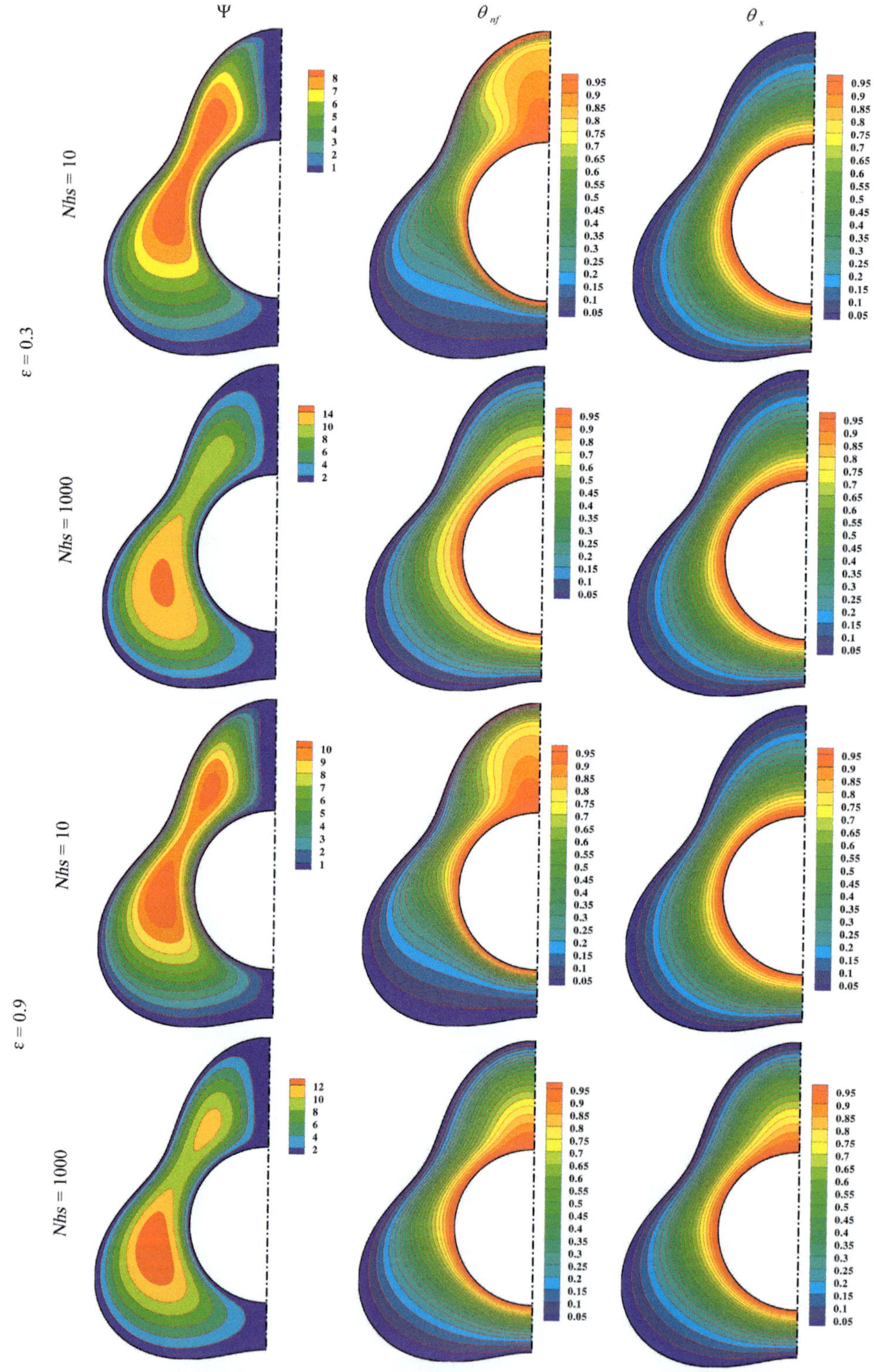

FIGURE 15.22 Streamlines (Ψ), isotherms for the nanofluid (θ_{nf}) and the solid (θ_s) at $Ra = 1000, Ha = 0, \phi = 0.04$.

behavior. Effects of $Ra, Nhs, \varepsilon,$ and Ha on streamlines (Ψ), isotherms for solid (θ_s) and nanofluid (θ_{nf}) are shown in Figs. 15.18–15.23. In low Ra, the effects of other parameters are not visible because of conduction mode dominance. In this case, isotherms for the nanofluid (θ_{nf}) and the solid (θ_s) have the same shapes. As buoyancy forces increase, a thermal plume can be observed near the vertical centerline and isotherms for the nanofluid (θ_{nf}) have

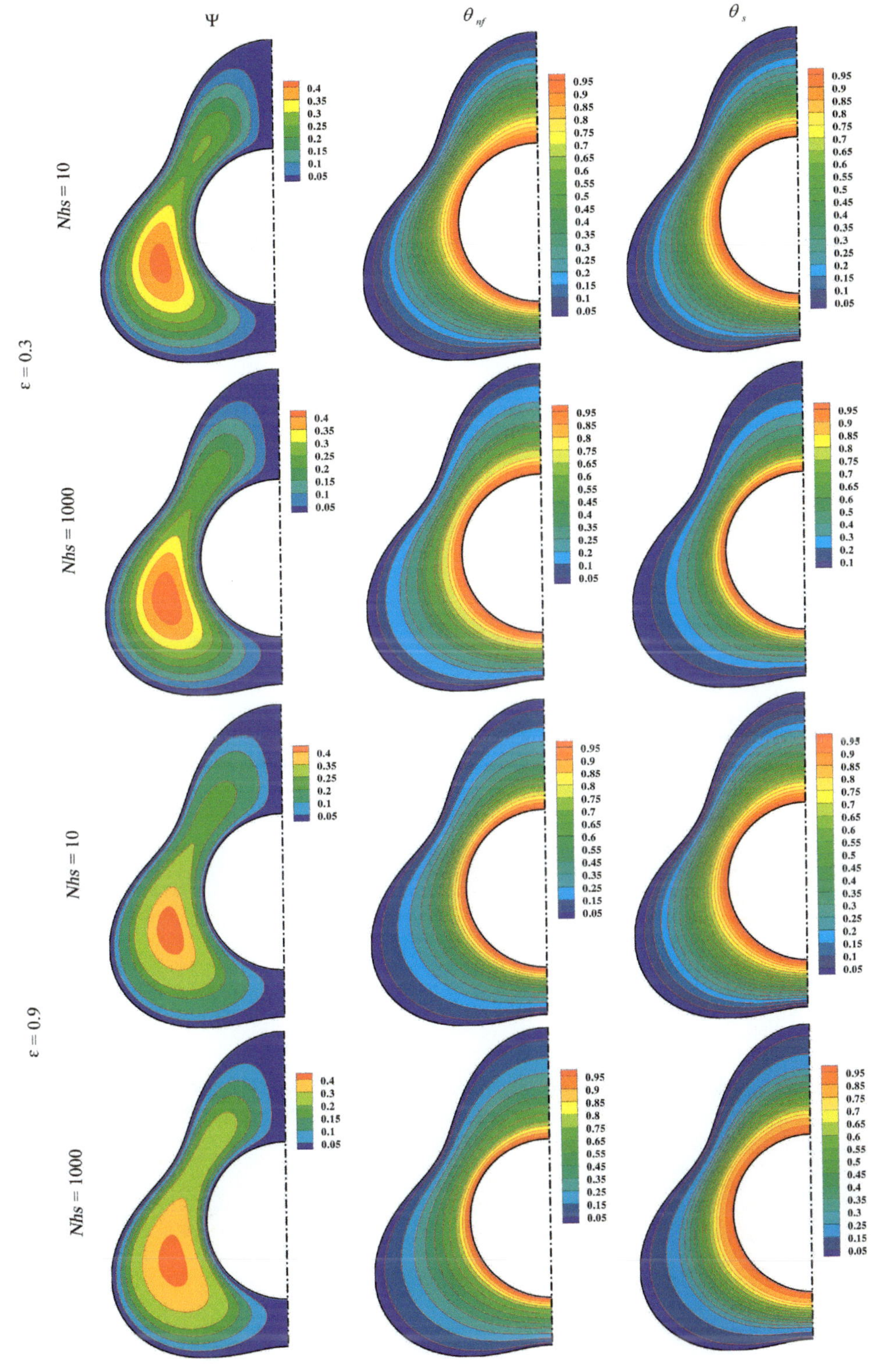

FIGURE 15.23 Streamlines (Ψ), isotherms for the nanofluid (θ_{nf}) and the solid (θ_s) at $Ra = 1000, Ha = 20, \phi = 0.04$.

complex shapes while isotherms for the solid (θ_s) are parallel to each other. Increasing the Lorentz forces reduces the nanofluid velocity and the temperature fields are stratified. As *Nhs* enhances, convective flow become stronger and (Ψ_{max}) augments. But the temperature gradient of the nanofluid is reduced and in turn Nu_{ave} decreases. As the porous medium porosity enhances, the pore volume inside the cavity increases. Therefore, improvement

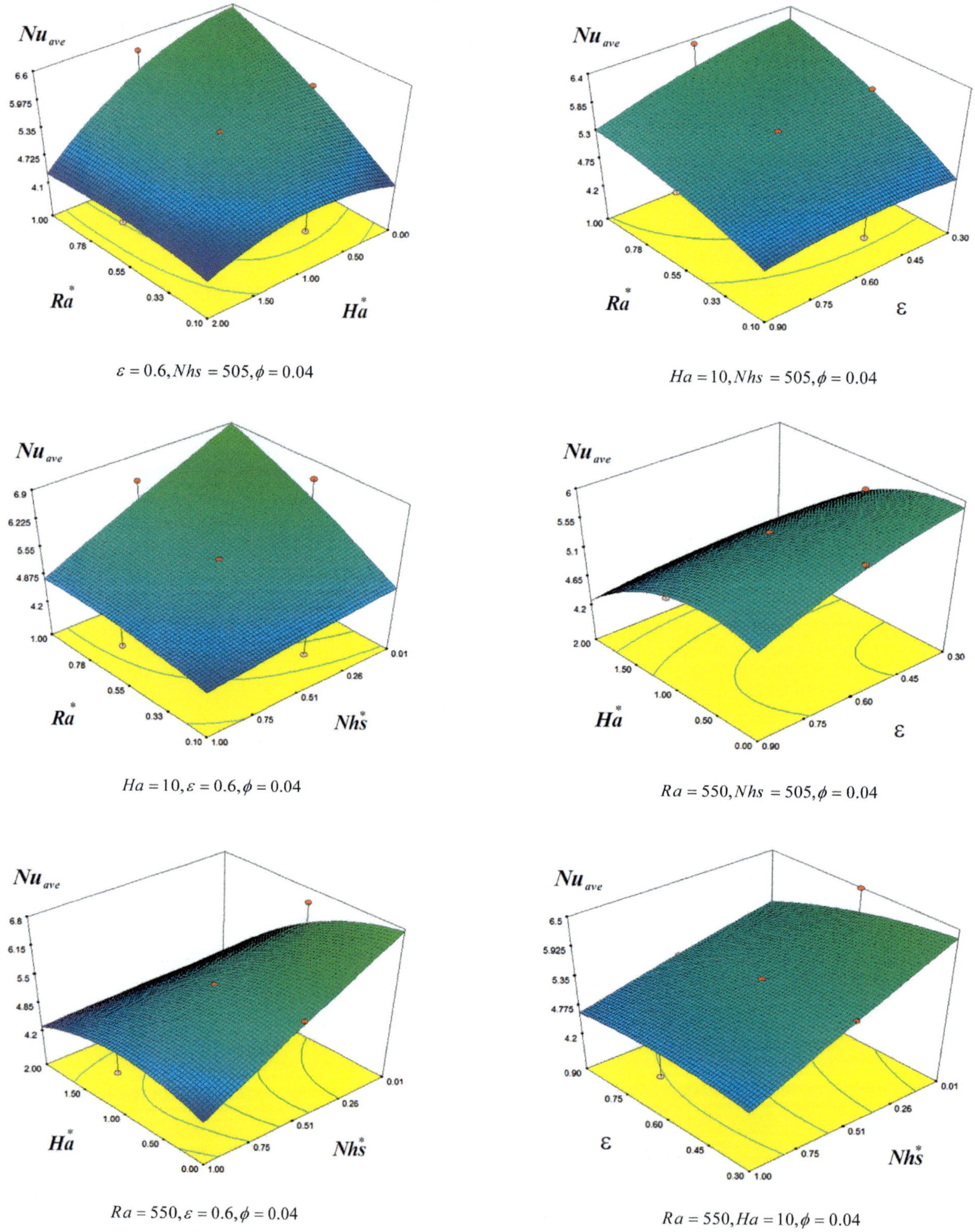

FIGURE 15.24 Effects of Ra, Ha, ε, Nhs on average Nusselt number.

in the convective flow can be observed. Also, the effect of ε on isotherms is similar to that of *Nhs*. Fig. 15.24 depicts the impact of Ra, Ha, ε, Nhs on average Nusselt number. According to obtained data, Nu_{ave} can be expressed as:

$$\begin{aligned} Nu_{ave} = {} & 5.04 + 4.95Ra^* - 0.07Ha^* + 0.24\varepsilon - 2.04Nhs^* \\ & - 1.13Ra^*Ha^* - 1.38Ra^*\varepsilon - 1.88Ra^*Nhs^* + 0.69Ha^*\varepsilon + 1.11Ha^*Nhs^* + 1.36\varepsilon Nhs^* \\ & - 0.69(Ra^*)^2 - 0.44(Ha^*)^2 - 1.33(\varepsilon)^2 - 0.12(Nhs^*)^2 \end{aligned} \tag{15.55}$$

where $Ra^* = 0.001Ra, Ha^* = 0.1Ha, Nhs^* = 0.001Nhs$.

The *Nhs* has direct relationship with rate of heat transfer because the thermal boundary layer thickness reduces with the increase of *Nhs*. Increasing the porous medium porosity leads to a decrease in Nusselt number and a similar impact is shown for the Hartmann number. Nu_{ave} is enhanced with the increase of buoyancy forces.

References

[1] D.A. Nield, A. Bejan, Convection in Porous Media, fourth ed., Springer, New York, USA, 2013.
[2] A.V. Kuznetsov, Thermal nonequilibrium forced convection in porous media, in: D.B. Ingham, I. Pop (Eds.), Transport Phenomena in Porous Media, Vol. 1, Pergamon Press, Oxford, UK, 2005, pp. 103–129.
[3] A. Bejan, Convective heat transfer in porous mediaChapter 16 in: S. Kakaç, R.K. Shah, W. Aung (Eds.), Handbook of Single-Phase Convective Heat Transfer, Wiley, New York, 1987.
[4] D.A. Nield, Modelling fluid flow in saturated porous media and at interfaces, in: D.B. Ingham, I. Pop (Eds.), Transport Phenomena in Porous Media., Vol. 2, Pergamon Press, Oxford, UK, 2002, pp. 1–19.
[5] A.C. Baytas, I. Pop, Free convection in a square porous cavity using a thermal non-equilibrium model, Int. J. Thermal Sci. 41 (2002) 861–870.
[6] M.A. Hossain, M. Wilson, Natural convection flow in a fluid-saturated porous medium enclosed by non-isothermal walls with heat generation, Int. J. Thermal Sci. 41 (2002) 447–454.
[7] M. Sheikholeslami, Influence of Lorentz forces on nanofluid flow in a porous cylinder considering Darcy model, J. Mol. Liq. 225 (2017) 903–912.
[8] M. Sheikholeslami, CVFEM for magnetic nanofluid convective heat transfer in a porous curved enclosure, Eur. Phys. J. Plus 131 (2016) 413. Available from: https://doi.org/10.1140/epjp/i2016-16413-y.
[9] M. Sheikholeslami, Influence of Coulomb forces on Fe3O4-H2O nanofluid thermal improvement, Int. J. Hydrogen Energy 42 (2017) 821–829.
[10] M. Sheikholeslami, Magnetic field influence on nanofluid thermal radiation in a cavity with tilted elliptic inner cylinder, J. Mol. Liq. 229 (2017) 137–147.
[11] M. Sheikholeslami, Numerical simulation of magnetic nanofluid natural convection in porous media, Phys. Lett. A 381 (2017) 494–503.
[12] M. Sheikholeslami, Magnetic source impact on nanofluid heat transfer using CVFEM, Neur. Comput. Applicat. (2016). Available from: https://doi.org/10.1007/s00521-016-2740-7.
[13] M. Sheikholeslami, CuO-water nanofluid free convection in a porous cavity considering Darcy law, Europ. Phys. J. Plus 132 (2017) 55. Available from: https://doi.org/10.1140/epjp/i2017-11330-3.
[14] M. Sheikholeslami, Numerical investigation of MHD nanofluid free convective heat transfer in a porous tilted enclosure, Eng. Comput. 34 (6) (2017) 1939–1955.
[15] M.S. Kandelousi, Effect of spatially variable magnetic field on ferrofluid flow and heat transfer considering constant heat flux boundary condition, Europ. Phys. J. Plus (2014) 129–248. Available from: https://doi.org/10.1140/epjp/i2014-14248-2.
[16] M. Sheikholeslami, Numerical simulation for external magnetic field influence on Fe_3O_4-water nanofluid forced convection, Eng. Comput., Inpress.
[17] M. Sheikholeslami, Influence of Lorentz forces on nanofluid flow in a porous cavity by means of Non- Darcy model, Eng. Comput. 34 (8) (2017) 2651–2667. Available from: https://doi.org/10.1108/EC-01-2017-0008.
[18] M. Sheikholeslami, H.B. Rokni, Nanofluid convective heat transfer intensification in a porous circular cylinder, Chem. Eng. Proc. Process Intensif. 120 (2017) 93–104.
[19] M. Sheikholeslami, M.K. Sadoughi, Numerical modeling for Fe3O4 -water nanofluid flow in porous medium considering MFD viscosity, J. Mol. Liq. 242 (2017) 255–264.
[20] M. Sheikholeslami, M. Shamlooei, Magnetic source influence on nanofluid flow in porous medium considering shape factor effect, Phys. Lett. A 381 (2017) 3071–3078.
[21] M. Sheikholeslami, M. Shamlooei, Convective flow of nanofluid inside a lid driven porous cavity using CVFEM, Phys. B: Condensed Matter 521 (2017) 239–250.
[22] M. Sheikholeslami, K. Vajravelu, Forced convection heat transfer in Fe3O4- ethylene glycol nanofluid under the influence of Coulomb force, J. Mol. Liq. 233 (2017) 203–210.
[23] M. Sheikholeslami, H.B. Rokni, Influence of EFD viscosity on nanofluid forced convection in a cavity with sinusoidal wall, J. Mol. Liq. 232 (2017) 390–395.
[24] M. Sheikholeslami, H.B. Rokni, Magnetic nanofluid natural convection in the presence of thermal radiation considering variable viscosity, Europ. Phys. J. Plus 132 (2017) 238. Available from: https://doi.org/10.1140/epjp/i2017-11498-4.
[25] M. Sheikholeslami, H.B. Rokni, Numerical simulation for impact of Coulomb force on nanofluid heat transfer in a porous enclosure in presence of thermal radiation, Int. J. Heat Mass Transfer (2018). Available from: https://doi.org/10.1016/j.ijheatmasstransfer.2017.11.041.

[26] M. Sheikholeslami, M. Shamlooei, R. Moradi, Fe3O4- Ethylene glycol nanofluid forced convection inside a porous enclosure in existence of Coulomb force, J. Mol. Liq. 249 (2018) 429–437.
[27] M. Sheikholeslami, S.A. Shehzad, Numerical analysis of Fe3O4 –H2O nanofluid flow in permeable media under the effect of external magnetic source, Int. J. Heat Mass Transfer, In press.
[28] M. Sheikholeslami, S.A. Shehzad, Magnetohydrodynamic nanofluid convection in a porous enclosure considering heat flux boundary condition, Int. J. Heat Mass Transfer 106 (2017) 1261–1269.
[29] M. Sheikholeslami, S.A. Shehzad, Thermal radiation of ferrofluid in existence of Lorentz forces considering variable viscosity, Int. J. Heat Mass Transfer 109 (2017) 82–92.
[30] M. Sheikholeslami, H.B. Rokni, Numerical modeling of nanofluid natural convection in a semi annulus in existence of Lorentz force, Comput. Methods Appl. Mech. Eng. 317 (2017) 419–430.
[31] M. Sheikholeslami, K. Vajravelu, Nanofluid flow and heat transfer in a cavity with variable magnetic field, Appl. Math. Comput. 298 (2017) 272–282.
[32] M. Sheikholeslami, H.B. Rokni, Magnetohydrodynamic CuO-water nanofluid in a porous complex shaped enclosure, ASME, J. Thermal Sci. Eng. Applicat. 9 (4) (2017) 041007. Available from: https://doi.org/10.1115/1.4035973.
[33] M. Sheikholeslami, H.B. Rokni, Nanofluid two phase model analysis in existence of induced magnetic field, Int. J. Heat Mass Transfer 107 (2017) 288–299.
[34] M. Sheikholeslami, T. Hayat, A. Alsaedi, S. Abelman, Numerical analysis of EHD nanofluid force convective heat transfer considering electric field dependent viscosity, Int. J. Heat Mass Transfer 108 (2017) 2558–2565.
[35] M. Sheikholeslami, Magnetic field influence on CuO -H2O nanofluid convective flow in a permeable cavity considering various shapes for nanoparticles, Int. J. Hydrogen Energy 42 (2017) 19611–19621.
[36] M. Sheikholeslami, M.M. Bhatti, Forced convection of nanofluid in presence of constant magnetic field considering shape effects of nanoparticles, Int. J. Heat Mass Transfer 111 (2017) 1039–1049.
[37] M. Sheikholeslami, A. Zeeshan, Analysis of flow and heat transfer in water based nanofluid due to magnetic field in a porous enclosure with constant heat flux using CVFEM, Comput. Methods Appl. Mech. Eng. 320 (2017) 68–81.
[38] M. Sheikholeslami, M. Shamlooei, Fe3O4- H2O nanofluid natural convection in presence of thermal radiation, Int. J. Hydrogen Energy 42 (9) (2017) 5708–5718.
[39] M. Sheikholeslami, M.M. Bhatti, Active method for nanofluid heat transfer enhancement by means of EHD, Int. J. Heat Mass Transfer 109 (2017) 115–122.
[40] M. Sheikholeslami, A.J. Chamkha, Flow and convective heat transfer of a ferro-nanofluid in a double-sided lid-driven cavity with a wavy wall in the presence of a variable magnetic field, Numer. Heat Transfer, Part A 69 (10) (2016) 1186–1200. Available from: https://doi.org/10.1080/10407782.2015.1125709.
[41] M. Sheikholeslami, A.J. Chamkha, Electrohydrodynamic free convection heat transfer of a nanofluid in a semi-annulus enclosure with a sinusoidal wall, Numer. Heat Transfer, Part A 69 (7) (2016) 781–793. Available from: http://dx.doi.org/10.1080/10407782.2015.1090819.
[42] M. Sheikholeslami, M.K. Sadoughi, Simulation of CuO- water nanofluid heat transfer enhancement in presence of melting surface, Int. J. Heat Mass Transfer 116 (2018) 909–919.
[43] M. Sheikholeslami, H.B. Rokni, Simulation of nanofluid heat transfer in presence of magnetic field: a review, Int. J. Heat Mass Transfer 115 (2017) 1203–1233.
[44] M. Sheikholeslami, H.B. Rokni, Melting heat transfer influence on nanofluid flow inside a cavity in existence of magnetic field, Int. J. Heat Mass Transfer 114 (2017) 517–526.
[45] M. Sheikholeslami, S.A. Shehzad, Magnetohydrodynamic nanofluid convective flow in a porous enclosure by means of LBM, Int. J. Heat Mass Transfer 113 (2017) 796–805.
[46] M. Sheikholeslami, Magnetohydrodynamic nanofluid forced convection in a porous lid driven cubic cavity using Lattice Boltzmann Method, J. Mol. Liq. 231 (2017) 555–565.
[47] M.S. Kandelousi, KKL correlation for simulation of nanofluid flow and heat transfer in a permeable channel, Phys. Lett. A 378 (45) (2014) 3331–3339.
[48] M. Sheikholeslami, Effect of uniform suction on nanofluid flow and heat transfer over a cylinder, J. Braz. Soc. Mech. Sci. Eng. 37 (2015) 1623–1633.
[49] M. Sheikholeslami, Lattice Boltzmann Method simulation of MHD non-Darcy nanofluid free convection, Phys. B 516 (2017) 55–71.
[50] M. Sheikholeslami, Influence of magnetic field on nanofluid free convection in an open porous cavity by means of Lattice Boltzmann Method, J. Mol. Liq. 234 (2017) 364–374.
[51] M. Sheikholeslami, Ali J. Chamkha, Influence of Lorentz forces on nanofluid forced convection considering Marangoni convection, J. Mol. Liq. 225 (2017) 750–757.
[52] M. Sheikholeslami, T. Hayat, A. Alsaedi, Numerical simulation of nanofluid forced convection heat transfer improvement in existence of magnetic field using Lattice Boltzmann Method, Int. J. Heat Mass Transfer 108 (2017) 1870–1883.
[53] M. Sheikholeslami, R. Ellahi, Three dimensional mesoscopic simulation of magnetic field effect on natural convection of nanofluid, Int. J. Heat Mass Transfer 89 (2015) 799–808.
[54] M. Sheikholeslami, S. Abelman, Two phase simulation of nanofluid flow and heat transfer in an annulus in the presence of an axial magnetic field, IEEE Trans. Nanotechnol. 14 (3) (2015) 561–569.
[55] M. Sheikholeslami, R. Ellahi, Simulation of ferrofluid flow for magnetic drug targeting using Lattice Boltzmann method, J. Zeitschrift Fur Naturforschung A 70 (2) (2015) 115–124.
[56] M. Sheikholeslami, M. Seyednezhad, Lattice Boltzmann Method simulation for CuO-water nanofluid flow in a porous enclosure with hot obstacle, J. Mol. Liq. 243 (2017) 249–256.
[57] M. Sheikholeslami, M. Sadoughi, Mesoscopic method for MHD nanofluid flow inside a porous cavity considering various shapes of nanoparticles, Int. J. Heat Mass Transfer 113 (2017) 106–114.

[58] M. Sheikholeslami, A. Zeeshan, Mesoscopic simulation of CuO-H2O nanofluid in a porous enclosure with elliptic heat source, Int. J. Hydrogen Energy 42 (22) (2017) 15393–15402.
[59] M. Sheikholeslami, S.A. Shehzad, CVFEM for influence of external magnetic source on Fe3O4 – H2O nanofluid behavior in a permeable cavity considering shape effect, Int. J. Heat Mass Transfer 115 (2017) 180–191.
[60] M. Sheikholeslami, M. Seyednezhad, Nanofluid heat transfer in a permeable enclosure in presence of variable magnetic field by means of CVFEM, Int. J. Heat Mass Transfer 114 (2017) 1169–1180.
[61] M. Sheikholeslami, H.B. Rokni, Free convection of CuO-H2O nanofluid in a curved porous enclosure using mesoscopic approach, Int. J. Hydrogen Energy 42 (22) (2017) 14942–14949.
[62] M. Sheikholeslami, Numerical investigation for CuO-H2O nanofluid flow in a porous channel with magnetic field using mesoscopic method, J. Mol. Liq., In press.
[63] M. Sheikholeslami, H.B. Rokni, Influence of melting surface on MHD nanofluid flow by means of two phase model, Chin. J. Phys. 55 (2017) 1352–1360.
[64] M. Sheikholeslami, H.B. Rokni, Effect of melting heat transfer on nanofluid flow in existence of magnetic field considering Buongiorno Model, Chin. J. Phys. 55 (2017) 1115–1126.
[65] M. Sheikholeslami, T. Hayat, A. Alsaedi, MHD free convection of Al2O3–water nanofluid considering thermal radiation: a numerical study, Int. J. Heat Mass Transfer 96 (2016) 513–524.
[66] M. Sheikholeslami, K. Vajravelu, M.M. Rashidi, Forced convection heat transfer in a semi annulus under the influence of a variable magnetic field, Int. J. Heat Mass Transfer 92 (2016) 339–348.
[67] M. Sheikholeslami, S. Soleimani, D.D. Ganji, Effect of electric field on hydrothermal behavior of nanofluid in a complex geometry, J. Mol. Liq. 213 (2016) 153–161.
[68] M. Sheikholeslami, Numerical simulation for solidification in a LHTESS by means of nano-enhanced PCM, J. Taiwan Inst. Chem. Eng. 86 (2018) 25–41.
[69] M. Sheikholeslami, Numerical modeling of nano enhanced PCM solidification in an enclosure with metallic fin, J. Mol. Liq. 259 (2018) 424–438.
[70] M. Sheikholeslami, A. Ghasemi, Solidification heat transfer of nanofluid in existence of thermal radiation by means of FEM, Int. J. Heat Mass Transfer 123 (2018) 418–431.
[71] M. Sheikholeslami, S.A. Shehzad, CVFEM simulation for nanofluid migration in a porous medium using Darcy model, Int. J.Heat Mass Transfer 122 (2018) 1264–1271.
[72] M. Sheikholeslami, M. Darzi, M.K. Sadoughi, Heat transfer improvement and pressure drop during condensation of refrigerant-based nanofluid: an experimental procedure, Int. J. Heat Mass Transfer 122 (2018) 643–650.
[73] M. Sheikholeslami, H.B. Rokni, CVFEM for effect of Lorentz forces on nanofluid flow in a porous complex shaped enclosure by means of Non-equilibrium model, J. Mol. Liq. 254 (2018) 446–462.
[74] M. Sheikholeslami, S.A. Shehzad, Z. Li, Water based nanofluid free convection heat transfer in a three dimensional porous cavity with hot sphere obstacle in existence of Lorenz forces, Int. J. Heat Mass Transfer 125 (2018) 375–386.
[75] M. Sheikholeslami, M. Jafaryar, D.D. Ganji, Z. Li, Exergy loss analysis for nanofluid forced convection heat transfer in a pipe with modified turbulators, J. Mol. Liq. 262 (2018) 104–110.
[76] M. Sheikholeslami, M. Jafaryar, Z. Li, Nanofluid turbulent convective flow in a circular duct with helical turbulators considering CuO nanoparticles, Int. J. Heat Mass Transfer 124 (2018) 980–989.
[77] M. Sheikholeslami, Z. Li, M. Shamlooei, Nanofluid MHD natural convection through a porous complex shaped cavity considering thermal radiation, Phys. Lett. A 382 (2018) 1615–1632.
[78] M. Sheikholeslami, H.B. Rokni, Magnetic nanofluid flow and convective heat transfer in a porous cavity considering Brownian motion effects, Phys. Fluids, Volume 30, Issue 1. Available from: https://doi.org/10.1063/1.5012517.
[79] M. Sheikholeslami, S.A. Shehzad, Simulation of water based nanofluid convective flow inside a porous enclosure via non-equilibrium model, Int. J. Heat Mass Transfer 120 (2018) 1200–1212.
[80] M. Sheikholeslami, S.A. Shehzad, Non-Darcy free convection of Fe_3O_4-water nanoliquid in a complex shaped enclosure under impact of uniform Lorentz force, Chin. J. Phys. 56 (2018) 270–281.
[81] M. Sheikholeslami, M. Seyednezhad, Simulation of nanofluid flow and natural convection in a porous media under the influence of electric field using CVFEM, Int. J. Heat Mass Transfer 120 (2018) 772–781.
[82] M. Sheikholeslami, M. Shamlooei, R. Moradi, Numerical simulation for heat transfer intensification of nanofluid in a permeable curved enclosure considering shape effect of Fe_3O_4 nanoparticles, Chem. Eng, Process. : Process Intensif. 124 (2018) 71–82.
[83] M. Sheikholeslami, T. Hayat, T. Muhammad, A. Alsaedi, MHD forced convection flow of nanofluid in a porous cavity with hot elliptic obstacle by means of Lattice Boltzmann method, Int. J. Mech. Sci. 135 (2018) 532–540.
[84] M. Sheikholeslami, Numerical investigation of nanofluid free convection under the influence of electric field in a porous enclosure, J. Mol. Liq. 249 (2018) 1212–1221.
[85] M. Sheikholeslami, CuO-water nanofluid flow due to magnetic field inside a porous media considering Brownian motion, J. Mol. Liq. 249 (2018) 921–929.
[86] M. Sheikholeslami, D.D. Ganji, Influence of electric field on Fe_3O_4-water nanofluid radiative and convective heat transfer in a permeable enclosure, J. Mol. Liq. 250 (2018) 404–412.
[87] M. Sheikholeslami, S.A. Shehzad, Z. Li, Nanofluid heat transfer intensification in a permeable channel due to magnetic field using Lattice Boltzmann method, Physica B (2018). Available from: https://doi.org/10.1016/j.physb.2018.03.036.
[88] M. Sheikholeslami, A. Zeeshan, Numerical simulation of Fe_3O_4-water nanofluid flow in a non-Darcy porous media, Int. J. Num. Methods Heat Fluid Flow Vol. 28 (Issue: 3) (2018) 641–660. Available from: https://doi.org/10.1108/HFF-04-2017-0160.
[89] M. Sheikholeslami, D.D. Ganji, Numerical approach for magnetic nanofluid flow in a porous cavity using CuO nanoparticles, Mater. Des. 120 (2017) 382–393.

[90] M. Sheikholeslami, D.D. Ganji, Numerical modeling of magnetohydrodynamic CuO-water transportation inside a porous cavity considering shape factor effect, Colloids Surf. A 529 (2017) 705–714.
[91] M. Sheikholeslami, T. Hayat, A. Alsaedi, Numerical simulation for forced convection flow of MHD CuO-H_2O nanofluid inside a cavity by means of LBM, J. Mol. Liq. 249 (2018) 941–948.
[92] M. Sheikholeslami, H.R. Kataria, Akhil, S. Mittal, Effect of thermal diffusion and heat-generation on MHD nanofluid flow past an oscillating vertical plate through porous medium, J. Mol. Liq. 257 (2018) 12–25.
[93] M. Sheikholeslami, M. Barzegar Gerdroodbary, S. Valiallah Mousavi, D.D. Ganji, R. Moradi, Heat transfer enhancement of ferrofluid inside an 90° elbow channel by non-uniform magnetic field, J. Magn. Magn. Mater. 460 (2018) 302–311.
[94] M. Sheikholeslami, T. Hayat, A. Alsaedi, Numerical study for external magnetic source influence on water based nanofluid convective heat transfer, Int. J. Heat Mass Transfer 106 (2017) 745–755.
[95] M. Sheikholeslami, M. Jafaryar, Ahmad Shafee, Zhixiong Li, Investigation of second law and hydrothermal behavior of nanofluid through a tube using passive methods, Journal of Molecular Liquids 269 (2018) 407–416.
[96] Mohsen Sheikholeslami, Ahmad Zeeshan, Aaqib Majeed, Control volume based finite element simulation of magnetic nanofluid flow and heat transport in non-Darcy medium, Journal of Molecular Liquids 268 (2018) 354–364.
[97] M. Sheikholeslami, S.A. Shehzad, Zhixiong Li, Ahmad Shafee, Numerical modeling for Alumina nanofluid magnetohydrodynamic convective heat transfer in a permeable medium using Darcy law, International Journal of Heat and Mass Transfer 127 (2018) 614–622.
[98] M. Sheikholeslami, Application of Darcy law for nanofluid flow in a porous cavity under the impact of Lorentz forces, Journal of Molecular Liquids 266 (2018) 495–503.
[99] M. Sheikholeslami, Zhixiong Li, Ahmad Shafee, Lorentz forces effect on NEPCM heat transfer during solidification in a porous energy storage system, International Journal of Heat and Mass Transfer 127 (2018) 665–674.
[100] M. Sheikholeslami, Finite element method for PCM solidification in existence of CuO nanoparticles, Journal of Molecular Liquids 265 (2018) 347–355.
[101] M. Sheikholeslami, M. Jafaryar, S. Saleem, Zhixiong Li, Ahmad Shafee, Yu Jiang, Nanofluid heat transfer augmentation and exergy loss inside a pipe equipped with innovative turbulators, International Journal of Heat and Mass Transfer 126 (2018) 156–163.

CHAPTER

16

Nonuniform Magnetic Field Effect on Nanofluid Convective Flow in a Porous Cavity

16.1 INTRODUCTION

Free convection in porous media occurs in many systems and in nature, including in geophysical, environmental, and technological problems. Such problems are of great interest, e.g., in the geothermal, high-performance building insulation, postaccident heat removal from pebble-bed nuclear reactors, multishield structures used in the insulation of nuclear reactors, pollutant dispersion in aquifers, solar power collector, etc. The state-of-the-art has been very well summarized in the recent books by Pop and Ingham [1]. The problem of free convection flow in differentially heated cavities, with top and bottom walls insulated, and filled with Darcian or non-Darcian fluid-saturated porous media, is of fundamental interest to many technological applications in the modern industry. Sheikholeslami [2] utilized CVFEM for magnetic nanofluid convective heat transfer in a porous curved enclosure. Magnetic nanofluid natural convection in porous media has been investigated by Sheikholeslami [3]. Sheikholeslami et al. [4] presented the heat transfer of Fe3O4-water nanofluid in a permeable medium with thermal radiation in the presence of constant heat flux. Sheikholeslami and Shehzad [5] investigated the influence of external magnetic source on Fe3O4-H2O nanofluid behavior in a permeable cavity considering shape effect. Natural convection of nanofluid was analyzed in several articles [6–90].

16.2 EFFECT OF VARIABLE MAGNETIC FIELD ON NANOFLUID CONVECTIVE HEAT TRANSFER IN A POROUS CURVED ENCLOSURE

16.2.1 Problem Definition

A curved cavity with hot left wall is considered. Boundary conditions are depicted in Fig. 16.1. A magnetic source exists near the left bottom corner. Magnetic source has been considered as shown in Fig. 16.2. $\overline{H_x}, \overline{H_y}, \overline{H}$ can be calculated as follow:

$$\overline{H_y} = \left[(\overline{b}-y)^2 + (\overline{a}-x)^2\right]^{-1} \frac{\gamma}{2\pi}(\overline{a}-x), \tag{16.1}$$

$$\overline{H_x} = \left[(\overline{b}-y)^2 + (\overline{a}-x)^2\right]^{-1} \frac{\gamma}{2\pi}(y-\overline{b}), \tag{16.2}$$

$$\overline{H} = \sqrt{\overline{H}_x^2 + \overline{H}_y^2} \tag{16.3}$$

16.2.2 Governing Equation

2D laminar nanofluid flow and forced convective heat transfer is taken into account. The governing PDEs are:

$$\frac{\partial v}{\partial y} + \frac{\partial u}{\partial x} = 0, \tag{16.4}$$

Application of Control Volume based Finite Element Method (CVFEM) for Nanofluid Flow and Heat Transfer.
DOI: https://doi.org/10.1016/B978-0-12-814152-6.00016-3

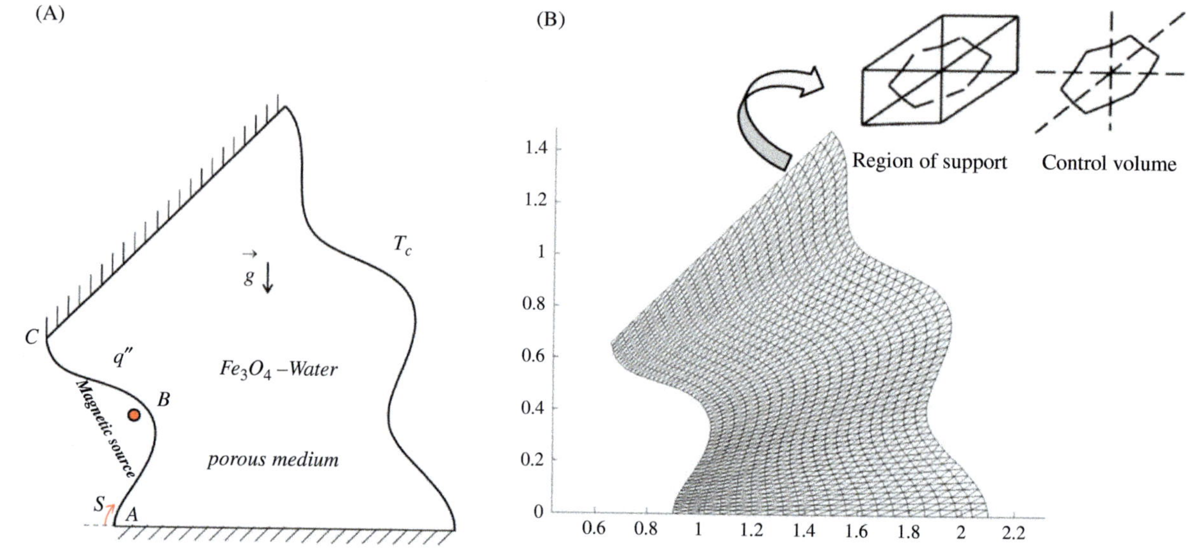

FIGURE 16.1 (A) Geometry and the boundary conditions (B) sample mesh.

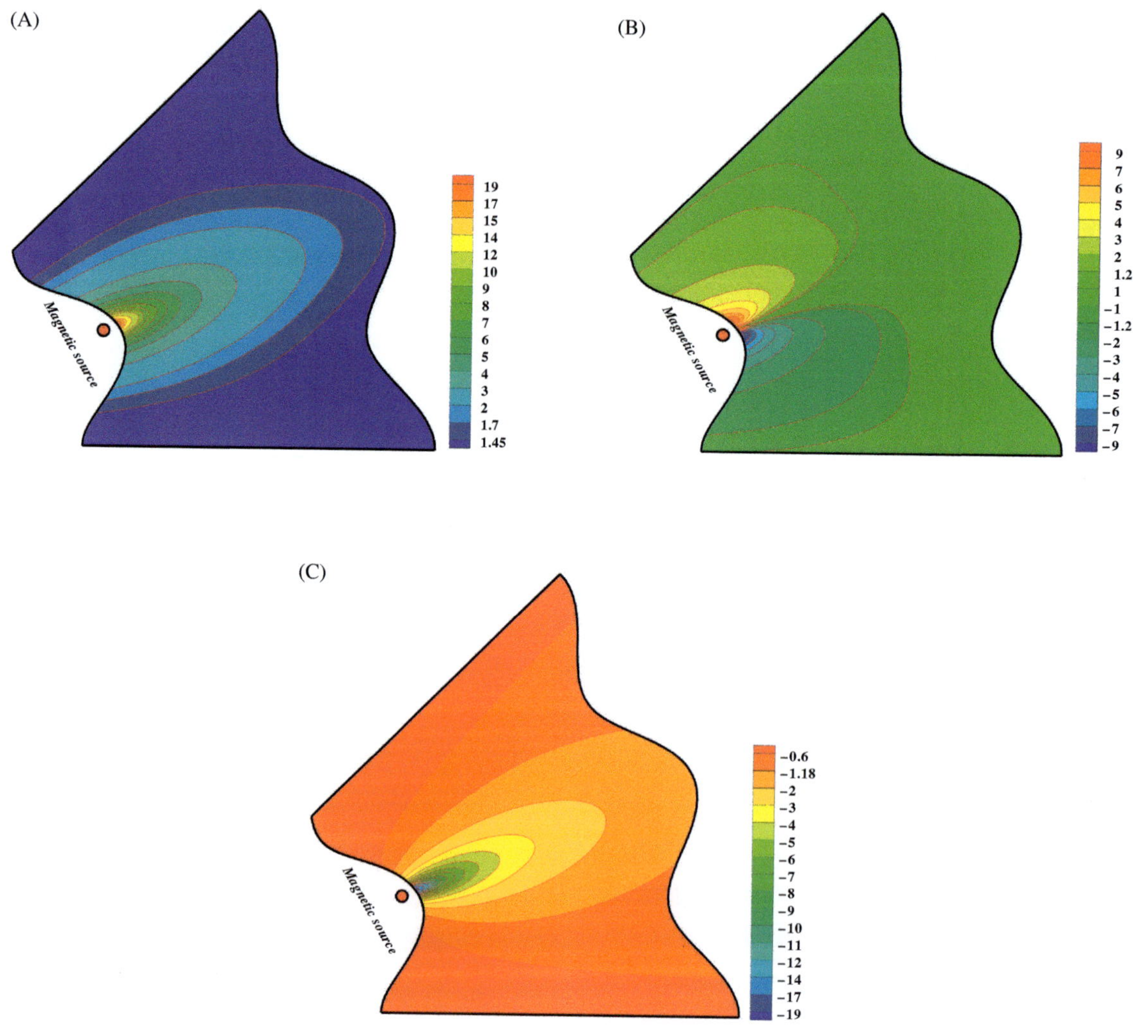

FIGURE 16.2 Contours of the (A) magnetic field strength H; (B) magnetic field intensity component in x direction Hx; (C) magnetic field intensity component in y direction Hy. (A) $H(x,y)$, (B) $Hx(x,y)$, (C) $Hy(x,y)$.

$$\left(u\frac{\partial u}{\partial x}+\frac{\partial u}{\partial y}v\right)=\left(\rho_{nf}\right)^{-1}\left[\left(\frac{\partial^2 u}{\partial y^2}+\frac{\partial^2 u}{\partial x^2}\right)\mu_{nf}-\frac{\partial P}{\partial x}-\mu_0^2\sigma_{nf}H_y^2u+\sigma_{nf}\mu_0^2H_xH_yv-\frac{\mu_{nf}}{K}u\right] \tag{16.5}$$

$$\begin{aligned}\rho_{nf}\left(\frac{\partial v}{\partial x}u+\frac{\partial v}{\partial y}v\right)=&+\mu_{nf}\left(\frac{\partial^2 v}{\partial x^2}+\frac{\partial^2 v}{\partial y^2}\right)-\frac{\partial P}{\partial y}+\mu_0^2H_y\sigma_{nf}H_xu-\mu_0^2H_x\sigma_{nf}H_xv-\frac{\mu_{nf}}{K}v\\&+(T-T_c)\beta_{nf}g\rho_{nf}\end{aligned} \tag{16.6}$$

$$\begin{aligned}(\rho C_p)_{nf}\left(v\frac{\partial T}{\partial y}+u\frac{\partial T}{\partial x}\right)=&\ \sigma_{nf}\mu_0^2(H_xv-H_yu)^2+k_{nf}\left(\frac{\partial^2 T}{\partial x^2}+\frac{\partial^2 T}{\partial y^2}\right)\\&+\mu_{nf}\left\{2\left(\frac{\partial u}{\partial x}\right)^2+2\left(\frac{\partial v}{\partial y}\right)^2+\left(\frac{\partial u}{\partial y}+\frac{\partial v}{\partial x}\right)^2\right\},\end{aligned} \tag{16.7}$$

ρ_{nf}, $(\rho C_p)_{nf}$, α_{nf}, β_{nf}, μ_{nf}, k_{nf}, and σ_{nf} are calculated as

$$\rho_{nf}=\rho_f(1-\phi)+\rho_s\phi, \tag{16.8}$$

$$(\rho C_p)_{nf}=(\rho C_p)_f(1-\phi)+(\rho C_p)_s\phi, \tag{16.9}$$

$$\alpha_{nf}=\frac{k_{nf}}{(\rho C_p)_{nf}}, \tag{16.10}$$

$$\beta_{nf}=\beta_f(1-\phi)+\beta_s\phi. \tag{16.11}$$

$$\mu_{nf}=\frac{\mu_f}{(1-\phi)^{2.5}}. \tag{16.12}$$

$$k_{nf}=k_f\left(\frac{k_s-2\phi(k_f-k_s)+2k_f}{k_s+\phi(k_f-k_s)+2k_f}\right), \tag{16.13}$$

$$\sigma_{nf}=\sigma_f\left[\frac{3(\sigma 1-1)\phi}{(\sigma 1+2)-(\sigma 1-1)\phi}+1\right],\qquad \sigma 1=\sigma_s/\sigma_f \tag{16.14}$$

Dimensionless parameters are defined as:

$$\begin{aligned}&(b,a)=\frac{(\overline{b},\overline{a})}{L},\ (H_y,H_x,H)=\frac{(\overline{H_y},\overline{H_x},\overline{H})}{\overline{H_0}},\ P=\frac{p}{\rho_f(\alpha_f/L)^2}\\&U=\frac{uL}{\alpha_f},\ V=\frac{vL}{\alpha_f},\ \Theta=\frac{T-T_c}{\Delta T},\ \Delta T=q''L/k_f,\ (X,Y)=\frac{(x,y)}{L}\end{aligned} \tag{16.15}$$

So equations change to:

$$\frac{\partial V}{\partial Y}+\frac{\partial U}{\partial X}=0, \tag{16.16}$$

$$\begin{aligned}U\frac{\partial U}{\partial X}+\frac{\partial U}{\partial Y}V=&\ \Pr\left[\frac{\mu_{nf}/\mu_f}{\rho_{nf}/\rho_f}\right]\left(\frac{\partial^2 U}{\partial Y^2}+\frac{\partial^2 U}{\partial X^2}\right)\\&-Ha^2\Pr\left[\frac{\sigma_{nf}/\sigma_f}{\rho_{nf}/\rho_f}\right]\left(H_y^2U-H_xH_yV\right)-\frac{\partial P}{\partial X}-\frac{\Pr}{Da}\left[\frac{\mu_{nf}/\mu_f}{\rho_{nf}/\rho_f}\right]U,\end{aligned} \tag{16.17}$$

$$V\frac{\partial V}{\partial Y}+U\frac{\partial V}{\partial X}=\Pr\left(\frac{\partial^2 V}{\partial Y^2}+\frac{\partial^2 V}{\partial X^2}\right)\left[\frac{\mu_{nf}/\mu_f}{\rho_{nf}/\rho_f}\right]$$
$$-Ha^2\Pr\left[\frac{\sigma_{nf}/\sigma_f}{\rho_{nf}/\rho_f}\right]\left(H_x^2 V-H_xH_yU\right) \quad (16.18)$$
$$-\frac{\partial P}{\partial Y}+Ra\Pr\left[\frac{\beta_{nf}}{\beta_f}\right]\Theta-\frac{\Pr}{Da}\left[\frac{\mu_{nf}/\mu_f}{\rho_{nf}/\rho_f}\right]V,$$

$$V\frac{\partial \Theta}{\partial Y}+U\frac{\partial \Theta}{\partial X}=\left[\frac{k_{nf}}{k_f}\frac{(\rho C_P)_f}{(\rho C_P)_{nf}}\right]\left(\frac{\partial^2 \Theta}{\partial Y^2}+\frac{\partial^2 \Theta}{\partial X^2}\right)$$
$$+Ha^2\,Ec\left(\frac{(\rho C_P)_f}{(\rho C_P)_{nf}}\frac{\sigma_{nf}}{\sigma_f}\right)\{V\,H_x-U\,H_y\}^2 \quad (16.19)$$
$$+\left[\frac{\dfrac{\mu_{nf}}{\mu_f}}{\dfrac{(\rho C_P)_{nf}}{(\rho C_P)_f}}\right]Ec\left\{2\left(\frac{\partial U}{\partial X}\right)^2+2\left(\frac{\partial V}{\partial Y}\right)^2+\left(\frac{\partial U}{\partial Y}+\frac{\partial V}{\partial X}\right)^2\right\}$$

and dimensionless parameters are

$$Ra_f=g\beta_f L^3\Delta T/(\alpha_f \upsilon_f),\Pr_f=\upsilon_f/\alpha_f, Ha=L\mu_0 H_0\sqrt{\sigma_f/\mu_f},$$
$$Ec=\left(\mu_f\alpha_f\right)/\left[(\rho C_P)_f\Delta T\,L^2\right],\quad Da=K/L^2 \quad (16.20)$$

The thermophysical properties of Fe_3O_4 and water are presented in Table 16.1. Pressure gradient source terms are discarded by the vorticity stream function.

$$\Omega=\frac{\omega L^2}{\alpha_f},\Psi=\frac{\psi}{\alpha_f},\omega=-\frac{\partial u}{\partial y}+\frac{\partial v}{\partial x},(u,v)=\left(\frac{\partial \psi}{\partial y},-\frac{\partial \psi}{\partial x}\right) \quad (16.21)$$

According to Fig. 16.1, boundary conditions are

$$\begin{aligned}&\text{on left wall} && \frac{\partial \Theta}{\partial n}=1.0\\ &\text{on all walls} && \Psi=0.0\\ &\text{on right wall} && \Theta=0.0\\ &\text{on other walls} && \frac{\partial \Theta}{\partial n}=0.0\end{aligned} \quad (16.22)$$

TABLE 16.1 Thermophysical Properties of Water and Nanoparticles

	ρ(kg/m³)	Cp(j/kgk)	k(W/m.k)	dp(nm)	σ(Ω·m)⁻¹
Pure water	997.1	4179	0.613	-	0.05
Fe_3O_4	5200	670	6	47	25,000

Nu_{local}, Nu_{ave} along curve wall are:

$$Nu_{loc} = \left(\frac{k_{nf}}{k_f}\right)\frac{1}{\theta} \tag{16.23}$$

$$Nu_{ave} = \frac{1}{S}\int_0^s Nu_{loc}\, ds \tag{16.24}$$

16.2.3 Effects of Active Parameters

Influences of magnetic field on nanofluid free convection are simulated. Influences of Darcy number (Da), Hartmann number (Ha), Rayleigh number (Ra), and volume fraction of Fe_3O_4 (ϕ) are presented. Pr and Ec are 6.8 and 10^{-5}, respectively.

The impact of Fe_3O_4 nanoparticles on hydrothermal treatment is depicted in Fig. 16.3. Temperature gradient enhances with the rise of volume fraction Fe_3O_4. The nanofluid motion augments due to enhancement of the solid movements.

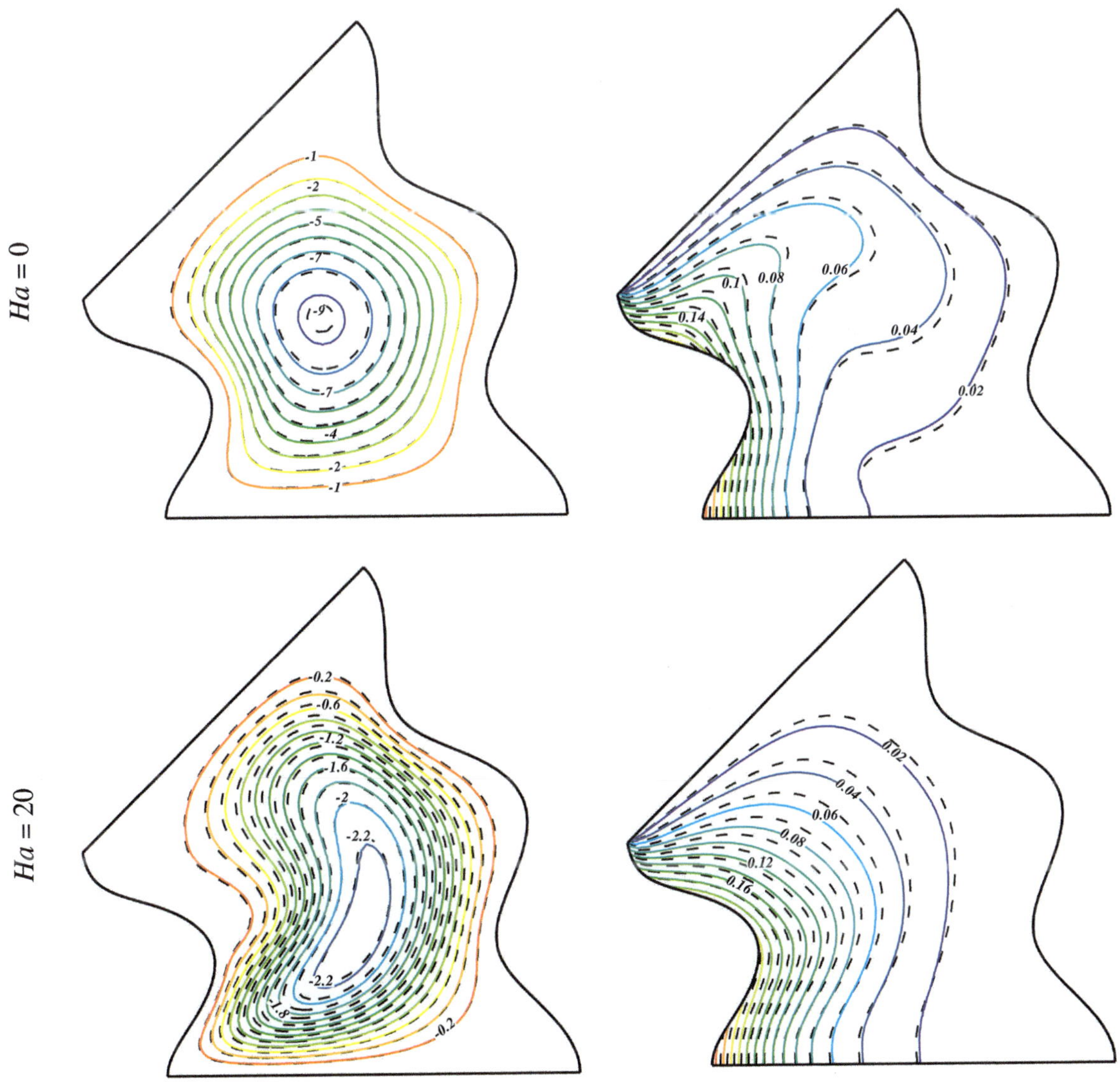

FIGURE 16.3 Influence of nanofluid volume fraction on streamline (left) and isotherm (right) contours (nanofluid ($\phi = 0.04$)(—) and pure fluid ($\phi = 0$) (- - -)) when $Ra = 10^5, Da = 100$.

Figs. 16.4–16.6 illustrate the impacts of $Da, Ha,$ and Ra on isotherms and streamlines. One clockwise vortex can be seen in streamline. At low Rayleigh number, isotherms are parallel to each other. As Ha increases the primary eddy stretches vertically and moves downward. Increasing the Darcy number makes the isotherms become

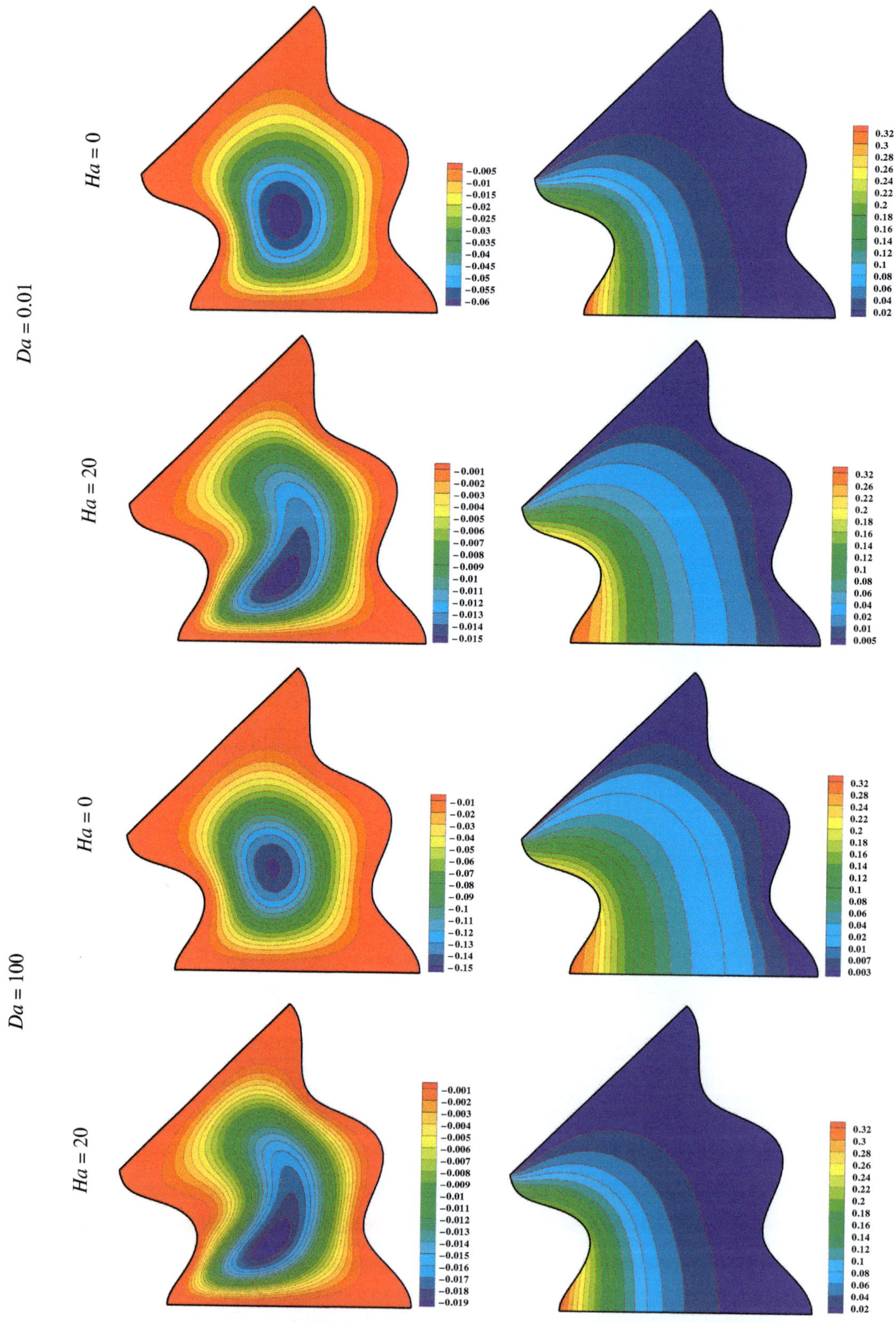

FIGURE 16.4 Influence of Da, Ha on streamline (left) and isotherm (right) contours when $\phi = 0.04$, $Ra = 10^3$.

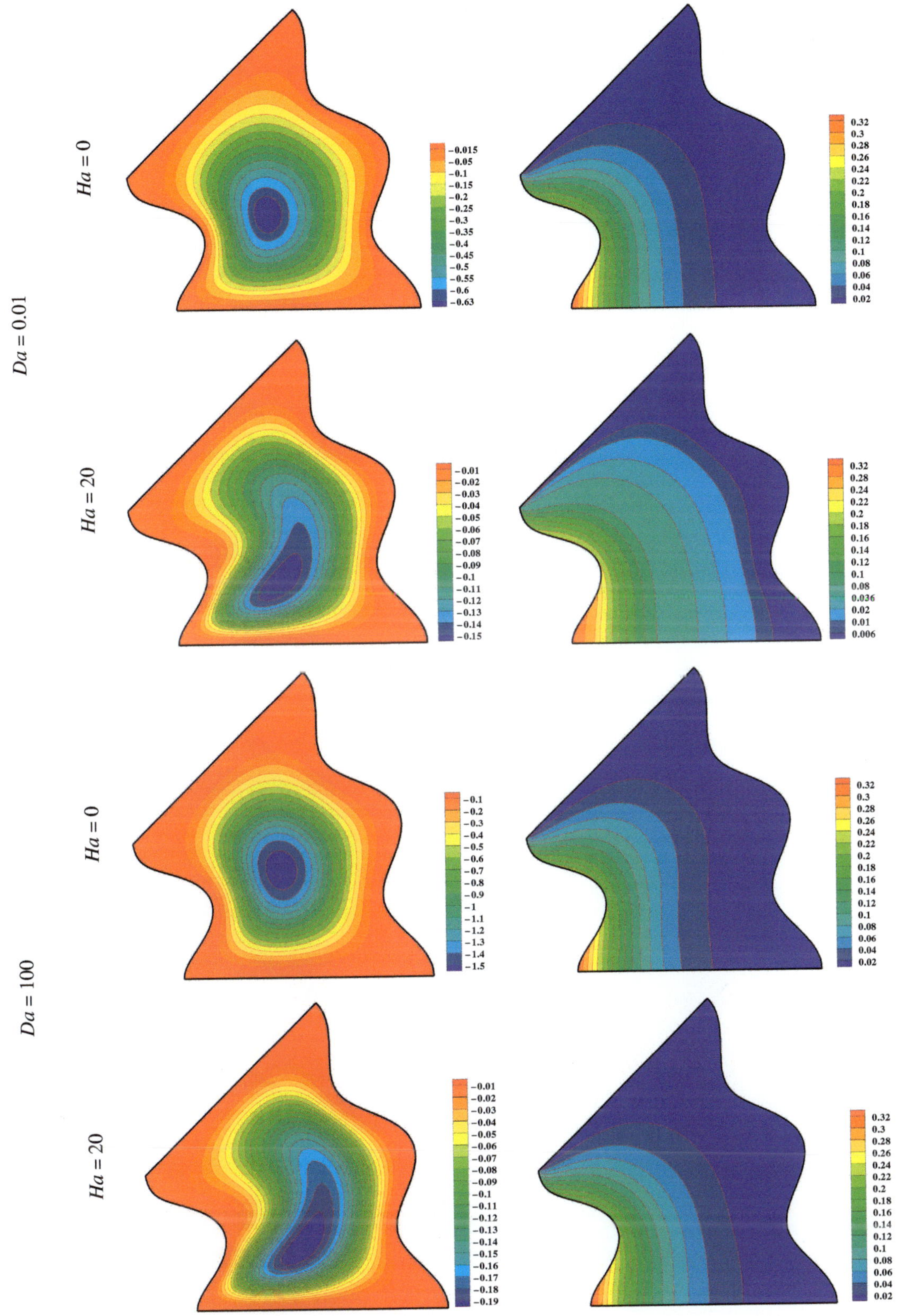

FIGURE 16.5 Influence of *Da*, *Ha* on streamline (left) and isotherm (right) contours when $\phi = 0.04$, $Ra = 10^4$.

denser and nanofluid flow is enhanced. Also $|\Psi_{max}|$ increases with the increase of *Da*. Convection mode becomes stronger with the rise of *Ra*, *Da* and isotherms become more distorted. As the Lorentz force is enhanced, the distortion of isotherms reduces and $|\Psi_{max}|$ reduces.

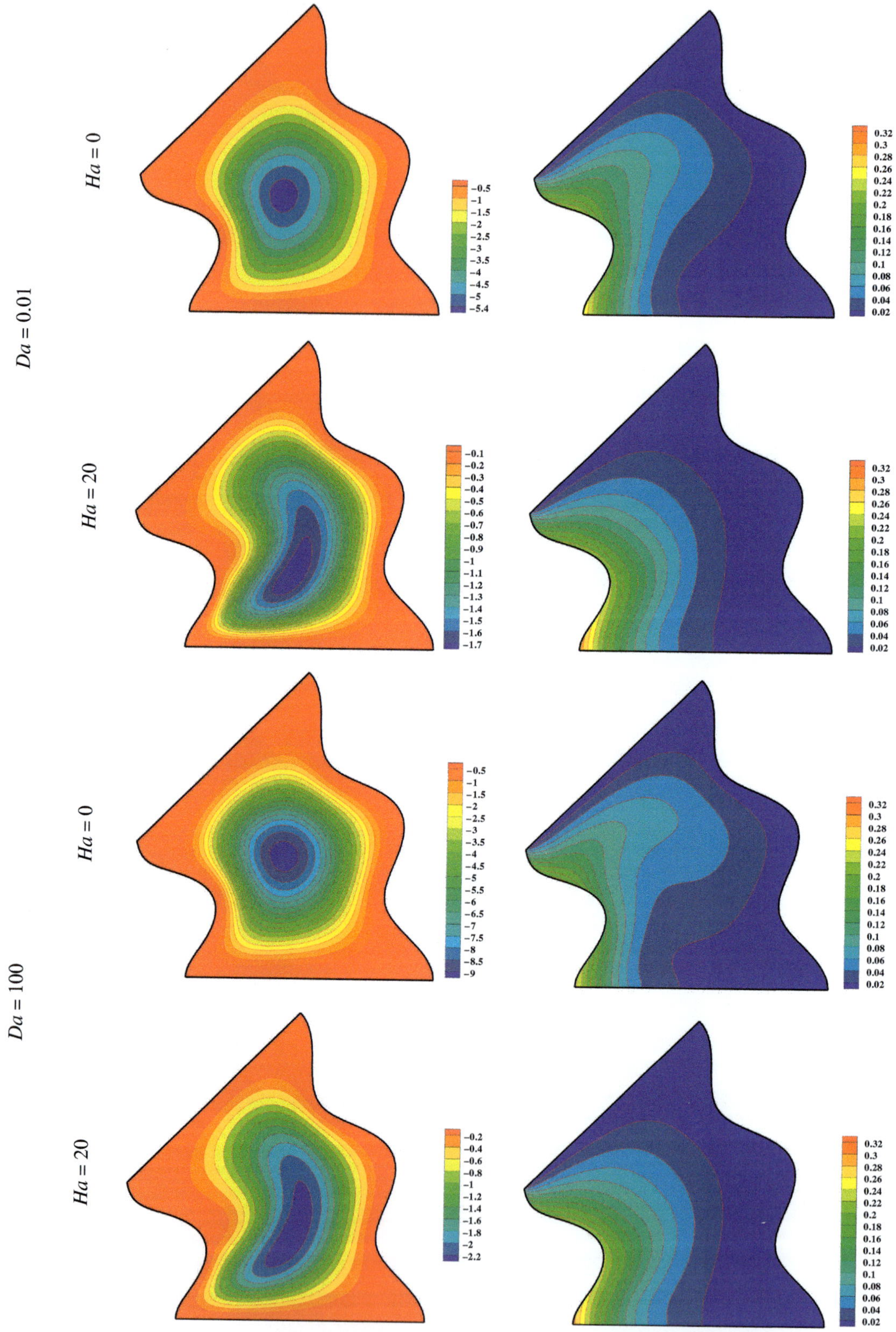

FIGURE 16.6 Influence of *Da*, *Ha* on streamline (left) and isotherm (right) contours when $\phi = 0.04$, $Ra = 10^5$.

Impacts of ϕ, Ha, Rd, and Ra on Nu_{ave}, Nu_{loc} are depicted in Figs. 16.7 and 16.8. The formula for Nu_{ave} is:

$$\begin{aligned} Nu_{ave} &= 7.68 - 0.22Da^* - 1.6\log(Ra) + 0.96Ha^* \\ &+ 0.17Da^*\log(Ra) - 0.1Da^*Ha^* - 0.23\log(Ra)Ha^* \\ &- 0.28Da^{*2} + 0.25(\log(Ra))^2 - 0.093Ha^{*2} \end{aligned} \tag{16.25}$$

where $Da^* = 0.01Da$, $Ha^* = 0.1Ha$. Nusselt number rises with the increase of Darcy number due to the domination of convection. As the buoyancy force rises, the temperature gradient near the hot wall increases and in turn Nu_{ave} is enhanced with the increase of Rayleigh number. Augmenting the Lorentz forces leads to a reduction of Nu_{ave} due to domination of conduction mode. Adding Fe_3O_4 nanoparticles into water makes Nu_{ave} increase. Table 16.2 depicts the impacts of Ha, Rd, and Ra on heat transfer improvement which is defined as: $E = 100 * \left(Nu_{ave}|_{\phi=0.04} - Nu_{ave}|_{\phi=0}\right)/Nu_{ave}|_{\phi=0}$. As thermal conductivity changes, this parameter changes. Therefore, the impact of adding nanoparticles is greater in the conduction mechanism. E is enhanced with the rise of Ha while it is reduced with the increase of Da, Ra.

16.3 NANOFLUID NATURAL CONVECTION IN POROUS MEDIA IN THE PRESENCE OF A MAGNETIC SOURCE

16.3.1 Problem Definition

A curved cavity with a hot left wall is considered. Boundary conditions are depicted in Fig. 16.9. A magnetic source exists near the left bottom corner. Magnetic source has been considered as shown in Fig. 16.10. $\overline{H_x}, \overline{H_y}, \overline{H}$ can be calculated as follows:

$$\overline{H_y} = \left[(\overline{b}-y)^2 + (\overline{a}-x)^2\right]^{-1} \frac{\gamma}{2\pi}(\overline{a} - x), \tag{16.26}$$

$$\overline{H_x} = \left[(\overline{b}-y)^2 + (\overline{a}-x)^2\right]^{-1} \frac{\gamma}{2\pi}(y - \overline{b}), \tag{16.27}$$

$$\overline{H} = \sqrt{\overline{H}_x^2 + \overline{H}_y^2}. \tag{16.28}$$

16.3.2 Governing Equation

2D laminar nanofluid flow and forced convective heat transfer are taken into account. The governing PDEs are:

$$\frac{\partial v}{\partial y} + \frac{\partial u}{\partial x} = 0, \tag{16.29}$$

$$\left(u\frac{\partial u}{\partial x} + \frac{\partial u}{\partial y}v\right) = \left(\rho_{nf}\right)^{-1}\left[\left(\frac{\partial^2 u}{\partial y^2} + \frac{\partial^2 u}{\partial x^2}\right)\mu_{nf} - \frac{\partial P}{\partial x} - \mu_0^2\sigma_{nf}H_y^2 u + \sigma_{nf}\mu_0^2 H_x H_y v - \frac{\mu_{nf}}{K}u\right], \tag{16.30}$$

$$\begin{aligned} \rho_{nf}\left(\frac{\partial v}{\partial x}u + \frac{\partial v}{\partial y}v\right) &= +\mu_{nf}\left(\frac{\partial^2 v}{\partial x^2} + \frac{\partial^2 v}{\partial y^2}\right) - \frac{\partial P}{\partial y} + \mu_0^2 H_y \sigma_{nf} H_x u - \mu_0^2 H_x \sigma_{nf} H_x v - \frac{\mu_{nf}}{K}v \\ &+ (T - T_c)\beta_{nf} g \rho_{nf}, \end{aligned} \tag{16.31}$$

$$\begin{aligned} (\rho C_p)_{nf}\left(v\frac{\partial T}{\partial y} + u\frac{\partial T}{\partial x}\right) &= \sigma_{nf}\mu_0^2(H_x v - H_y u)^2 + k_{nf}\left(\frac{\partial^2 T}{\partial x^2} + \frac{\partial^2 T}{\partial y^2}\right) \\ &+ \mu_{nf}\left\{2\left(\frac{\partial u}{\partial x}\right)^2 + 2\left(\frac{\partial v}{\partial y}\right)^2 + \left(\frac{\partial u}{\partial y} + \frac{\partial v}{\partial x}\right)^2\right\}, \end{aligned} \tag{16.32}$$

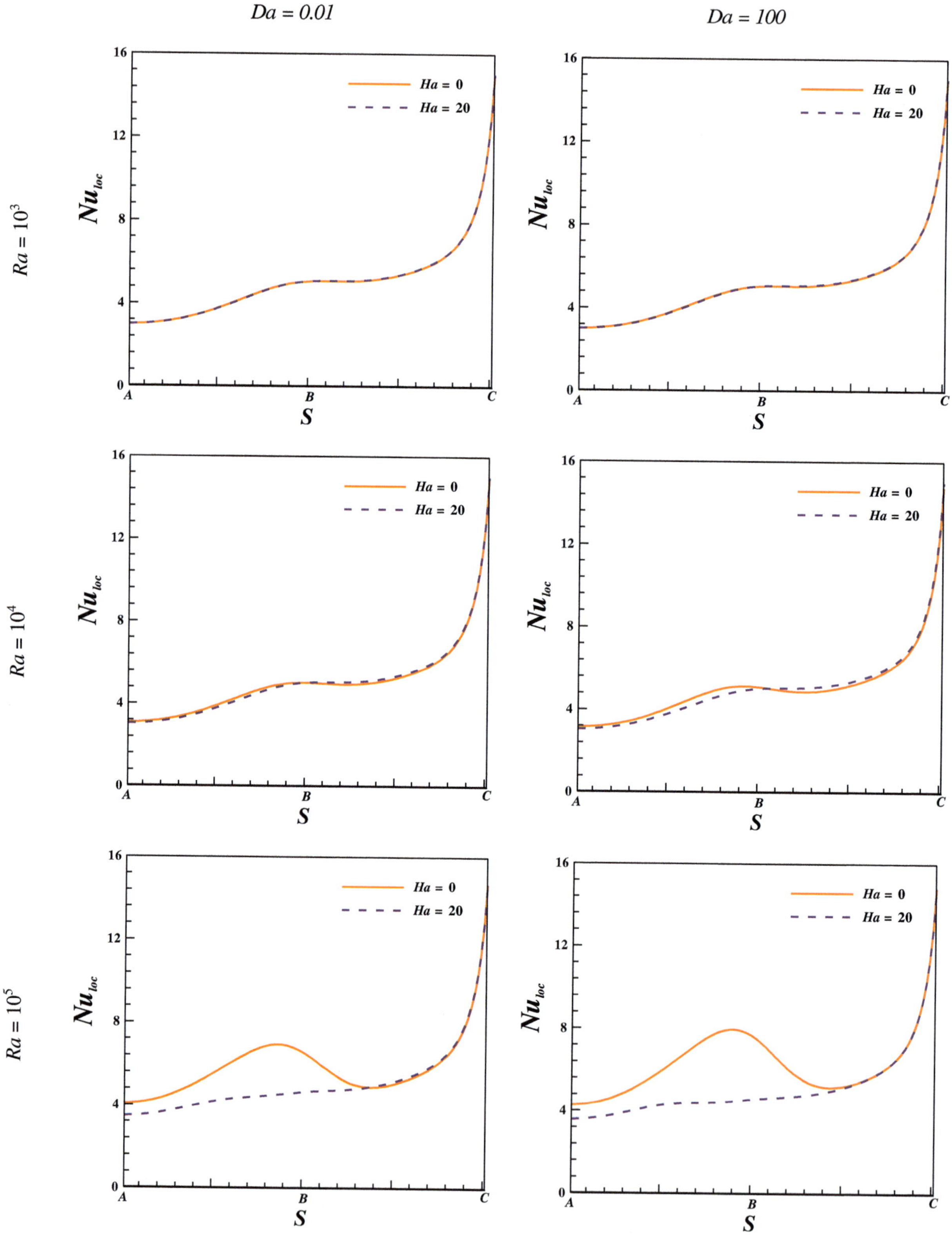

FIGURE 16.7 Effects of *Ha*, *Ra*, and *Da* on local Nusselt number.

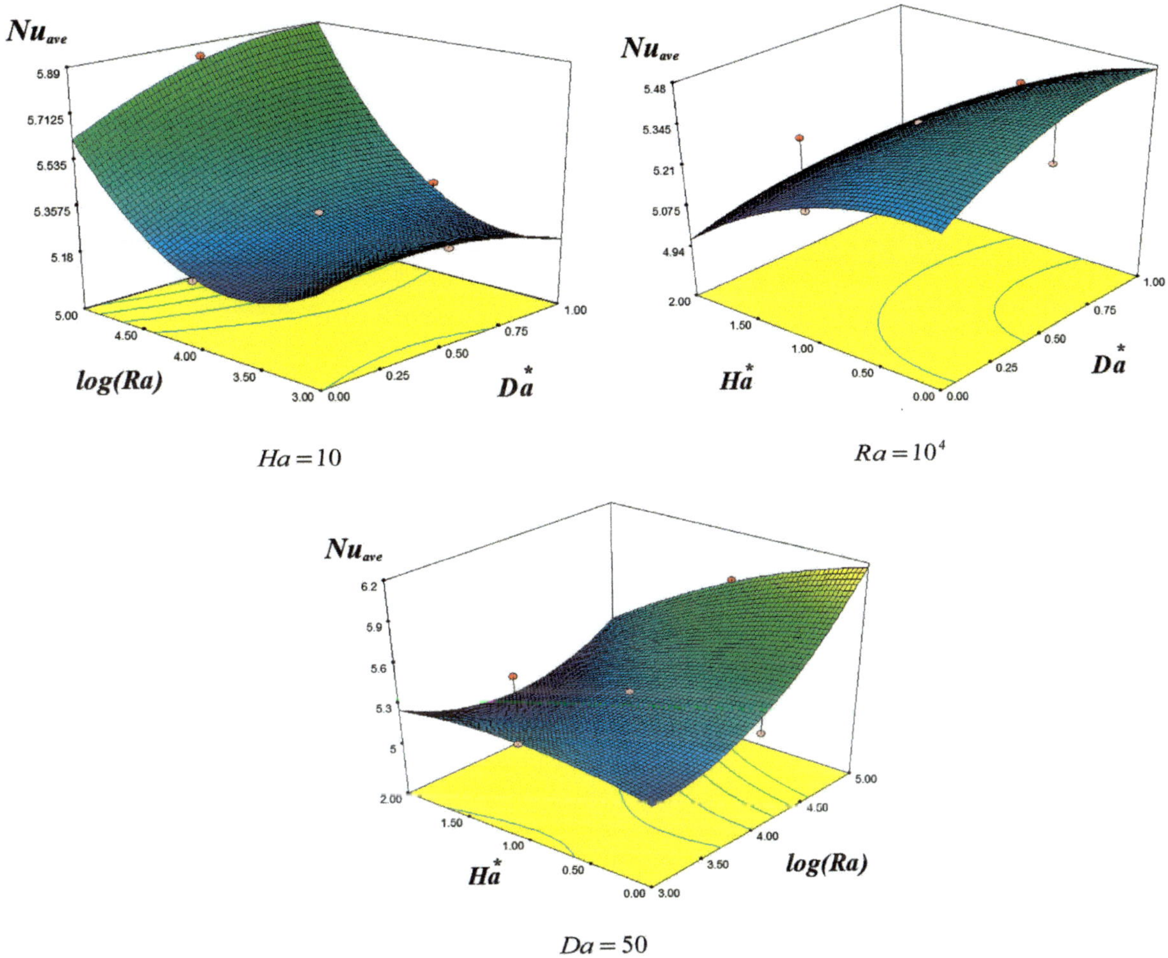

FIGURE 16.8 Effects of *Da*, *Ha*, and *Ra* on average Nusselt number when $\phi = 0.04$.

TABLE 16.2 Effects of Da, Ha, and Ra on Heat Transfer Enhancement

Ra	Da	Ha	E
10^3	0.01	0	19.5528
10^3	0.01	20	19.55272
10^5	0.01	0	16.3129
10^5	0.01	20	18.64209
10^3	100	0	19.55295
10^3	100	20	19.5528
10^5	100	0	15.36061
10^5	100	20	18.13567

$\rho_{nf}, (\rho C_p)_{nf}, \alpha_{nf}, \beta_{nf}, \mu_{nf}, k_{nf},$ and σ_{nf} are calculated as

$$\rho_{nf} = \rho_f(1-\phi) + \rho_s\phi, \tag{16.33}$$

$$(\rho C_p)_{nf} = (\rho C_p)_f(1-\phi) + (\rho C_p)_s\phi, \tag{16.34}$$

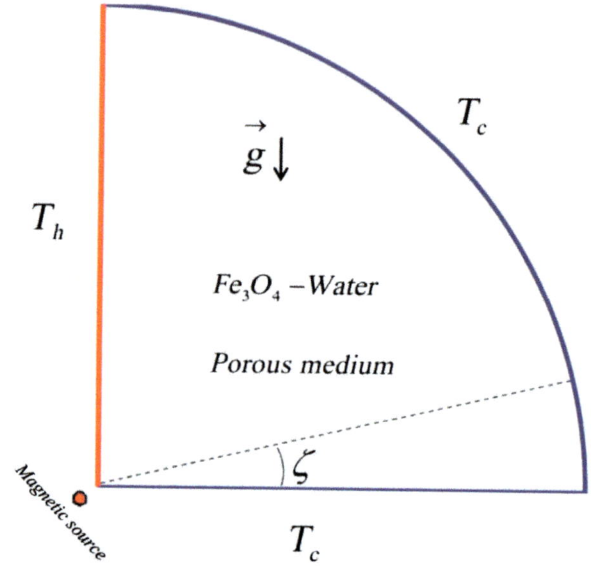

FIGURE 16.9 Geometry and the boundary conditions.

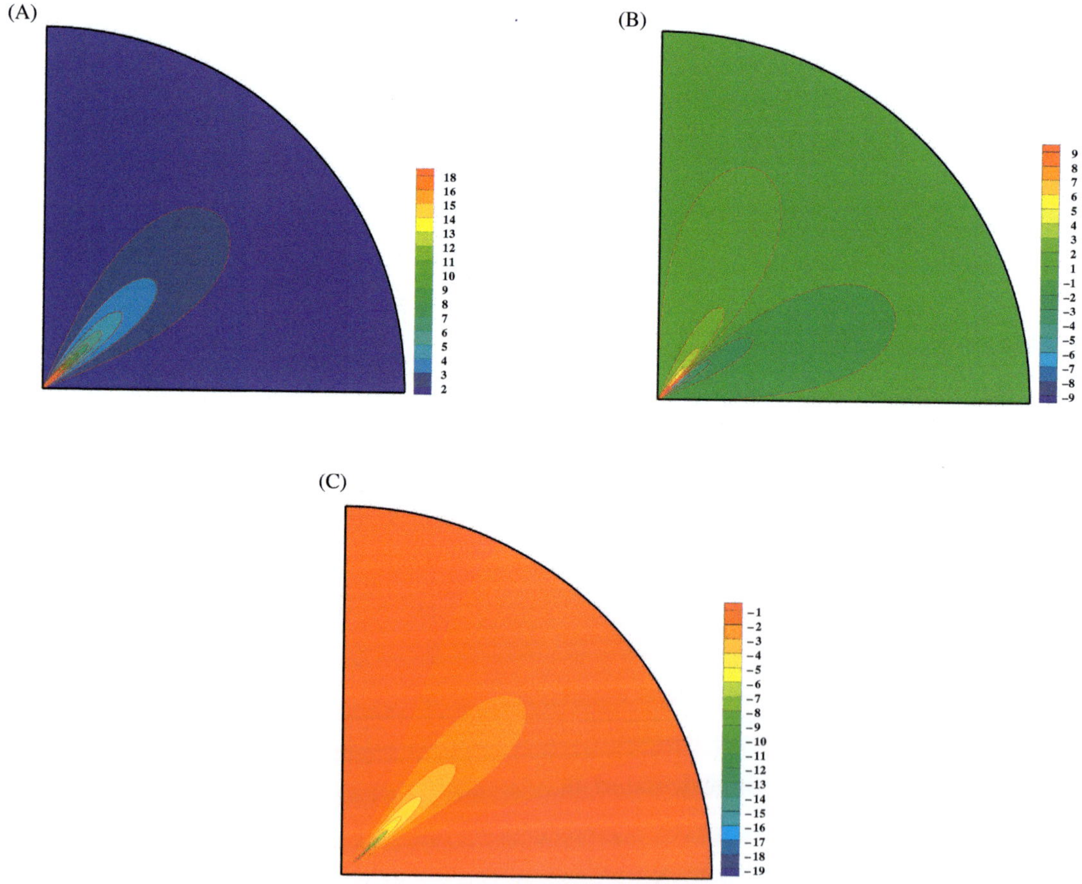

FIGURE 16.10 Contours of the (A) magnetic field strength H; (B) magnetic field intensity component in x direction Hx; (C) magnetic field intensity component in y direction Hy. (A) $H(x, y)$, (B) $Hx(x, y)$, (C) $Hy(x, y)$.

$$\alpha_{nf} = \frac{k_{nf}}{(\rho C_p)_{nf}}, \tag{16.35}$$

$$\beta_{nf} = \beta_f(1-\phi) + \beta_s\phi, \tag{16.36}$$

$$\mu_{nf} = \frac{\mu_f}{(1-\phi)^{2.5}}, \tag{16.37}$$

$$k_{nf} = k_f\left(\frac{k_s - 2\phi(k_f - k_s) + 2k_f}{k_s + \phi(k_f - k_s) + 2k_f}\right), \tag{16.38}$$

$$\sigma_{nf} = \sigma_f\left[\frac{3(\sigma 1 - 1)\phi}{(\sigma 1 + 2) - (\sigma 1 - 1)\phi} + 1\right], \qquad \sigma 1 = \sigma_s/\sigma_f. \tag{16.39}$$

Dimensionless parameters are defined as:

$$(b,a) = \frac{(\bar{b},\bar{a})}{L}, (H_y, H_x, H) = \frac{(\overline{H_y}, \overline{H_x}, \overline{H})}{\overline{H_0}}, P = \frac{p}{\rho_f(\alpha_f/L)^2}$$
$$U = \frac{uL}{\alpha_f}, V = \frac{vL}{\alpha_f}, \Theta = \frac{T - T_c}{(T_h - T_c)}, (X,Y) = \frac{(x,y)}{L}. \tag{16.40}$$

So equations change to:

$$\frac{\partial V}{\partial Y} + \frac{\partial U}{\partial X} = 0, \tag{16.41}$$

$$U\frac{\partial U}{\partial X} + \frac{\partial U}{\partial Y}V = \Pr\left[\frac{\mu_{nf}/\mu_f}{\rho_{nf}/\rho_f}\right]\left(\frac{\partial^2 U}{\partial Y^2} + \frac{\partial^2 U}{\partial X^2}\right)$$
$$- Ha^2\Pr\left[\frac{\sigma_{nf}/\sigma_f}{\rho_{nf}/\rho_f}\right]\left(H_y^2 U - H_x H_y V\right) - \frac{\partial P}{\partial X} - \frac{\Pr}{Da}\left[\frac{\mu_{nf}/\mu_f}{\rho_{nf}/\rho_f}\right]U, \tag{16.42}$$

$$V\frac{\partial V}{\partial Y} + U\frac{\partial V}{\partial X} = \Pr\left(\frac{\partial^2 V}{\partial Y^2} + \frac{\partial^2 V}{\partial X^2}\right)\left[\frac{\mu_{nf}/\mu_f}{\rho_{nf}/\rho_f}\right]$$
$$- Ha^2\Pr\left[\frac{\sigma_{nf}/\sigma_f}{\rho_{nf}/\rho_f}\right](H_x^2 V - H_x H_y U) \tag{16.43}$$
$$- \frac{\partial P}{\partial Y} + Ra\Pr\left[\frac{\beta_{nf}}{\beta_f}\right]\Theta - \frac{\Pr}{Da}\left[\frac{\mu_{nf}/\mu_f}{\rho_{nf}/\rho_f}\right]V,$$

$$V\frac{\partial \Theta}{\partial Y} + U\frac{\partial \Theta}{\partial X} = \left[\frac{k_{nf}}{k_f}\frac{(\rho C_P)_f}{(\rho C_P)_{nf}}\right]\left(\frac{\partial^2 \Theta}{\partial Y^2} + \frac{\partial^2 \Theta}{\partial X^2}\right)$$
$$+ Ha^2\, Ec\left(\frac{(\rho C_P)_f}{(\rho C_P)_{nf}}\frac{\sigma_{nf}}{\sigma_f}\right)\{V H_x - U H_y\}^2 \tag{16.44}$$
$$+ \left[\frac{\dfrac{\mu_{nf}}{\mu_f}}{\dfrac{(\rho C_P)_{nf}}{(\rho C_P)_f}}\right] Ec\left\{2\left(\frac{\partial U}{\partial X}\right)^2 + 2\left(\frac{\partial V}{\partial Y}\right)^2 + \left(\frac{\partial U}{\partial Y} + \frac{\partial V}{\partial X}\right)^2\right\}$$

and dimensionless parameters are

$$Ra_f = g\beta_f L^3 \Delta T/(\alpha_f \upsilon_f), \Pr_f = \upsilon_f/\alpha_f, Ha = L\mu_0 H_0 \sqrt{\sigma_f/\mu_f},$$
$$Ec = (\mu_f \alpha_f)/[(\rho C_P)_f \Delta T\, L^2], \quad Da = K/L^2. \tag{16.45}$$

The thermophysical properties of Fe_3O_4 and water are presented in Table 6.11. Pressure gradient source terms are discarded by the vorticity stream function.

$$\Omega = \frac{\omega L^2}{\alpha_f}, \Psi = \frac{\psi}{\alpha_f}, \omega = -\frac{\partial u}{\partial y} + \frac{\partial v}{\partial x}, (u, v) = \left(\frac{\partial \psi}{\partial y}, -\frac{\partial \psi}{\partial x}\right). \tag{16.46}$$

According to Fig. 16.9, boundary conditions are:

$$\begin{aligned} &\text{on left wall} && \theta = 1.0 \\ &\text{on all walls} && \Psi = 0.0 \\ &\text{on other walls} && \theta = 0.0 \end{aligned} \tag{16.47}$$

Nu_{local}, Nu_{ave} along the curved wall are:

$$Nu_{loc} = \left(\frac{k_{nf}}{k_f}\right)\frac{\partial \Theta}{\partial r}, \tag{16.48}$$

$$Nu_{ave} = \frac{1}{0.5\pi}\int_0^{0.5\pi} Nu_{loc}(\zeta)\, d\zeta, \tag{16.49}$$

16.3.3 Effects of Active Parameters

Fe_3O_4-water ferrofluid convective heat transfer in permeable medium is investigated. Impacts of Darcy number (Da), Rayleigh number (Ra), Hartmann number (Ha), and volume fraction of Fe_3O_4 (ϕ) are presented. Ec and Pr are 10^{-5} and 6.8, respectively. Fig. 16.11 depicts the impact of Fe_3O_4 nanoparticles on hydrothermal characteristic. The nanofluid motion is enhanced due to the increase in solid particles interaction. The thermal boundary layer thickness increases with the rise of the volume fraction of Fe_3O_4. Impacts of $Da, Ha,$ and Ra on isotherms and streamlines are illustrated in Figs. 16.12–16.14. In all conditions, one main clockwise vortex is generated in the cavity. At low Darcy and Rayleigh number the main heat transfer mechanism is conduction. Isotherms are parallel together in this mode. As Ha enhances the main vortex moves upward. As the Rayleigh number increases, the convective heat transfer becomes stronger. So isotherms become more disturbed and $|\Psi_{\max}|$ increases. As Darcy number is enhanced the nanofluid flow becomes easier. The impact of Da on hydrothermal behavior is similar to that of Ra. Therefore the temperature boundary layer thickness near the curved wall is reduced and in this way the temperature gradient is enhanced. As Lorentz forces increase, $|\Psi_{\max}|$ is reduced and the distribution of isotherms decreases. Figs. 16.15 and 16.16 demonstrate the impact of Da, Ha, and Ra on Nu_{ave}, Nu_{loc}. The correlation for Nu_{ave} corresponding to important parameters is:

$$\begin{aligned} Nu_{ave} &= 5.96 - 0.35Da^* - 2.9\log(Ra) + 0.77Ha^* \\ &+ 0.23Da^*\log(Ra) - 0.05Da^*Ha^* - 0.17\log(Ra)Ha^* \\ &- 0.22Da^{*2} + 0.5(\log(Ra))^2 - 0.06Ha^{*2}. \end{aligned} \tag{16.50}$$

where $Da^* = 0.01Da, \; Ha^* = 0.1Ha$.

Nu_{ave} is enhanced with the increase of Da due to an increase in the temperature gradient near the curved wall. A similar trend is seen for buoyancy forces, so Nu_{ave} increases with the rise of Rayleigh number. As Lorentz forces increase, Nu_{ave} is reduced due to the domination of the conduction mode. Adding Fe_3O_4 nanoparticles into water makes Nu_{ave} be enhanced. Table 16.3 depicts the influence of Da, Ha, and Ra on heat transfer augmentation, which is defined as: $E = 100 * (Nu_{ave}|_{\phi=0.04} - Nu_{ave}|_{\phi=0})/Nu_{ave}|_{\phi=0}$. E is relevant to the variation of thermal conductivity and k has more changes when the temperature boundary layer thickness is larger. So, heat transfer augmentation is enhanced with the rise of the Hartmann number while it decreases with the enhancement of Ra, Da.

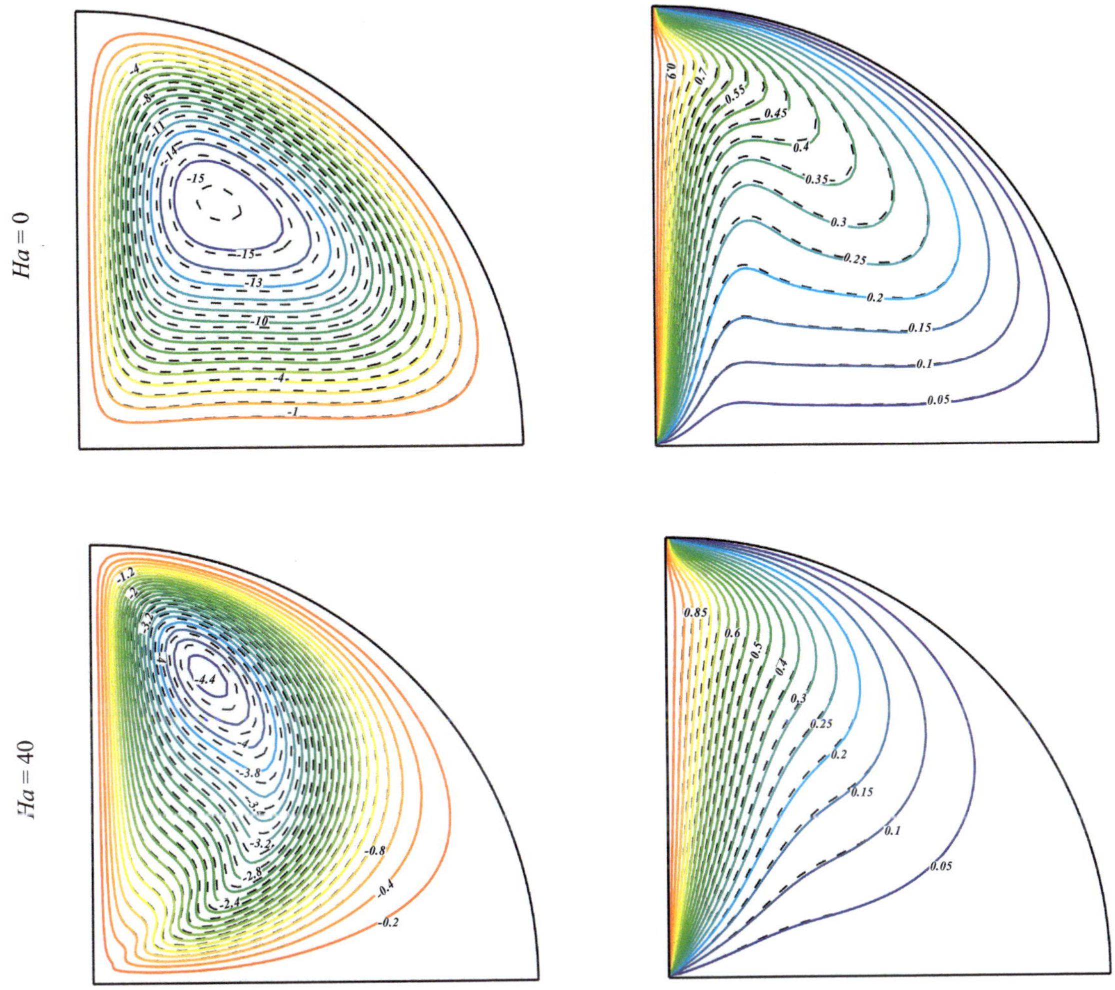

FIGURE 16.11 Impact of nanofluid volume fraction on streamline (left) and isotherm (right) contours (nanofluid ($\phi = 0.04$) (—) and pure fluid ($\phi = 0$) (- - -)) when $Ra = 10^5, Da = 100$.

16.4 HEAT TRANSFER OF FE$_3$O$_4$-WATER NANOFLUID IN A PERMEABLE MEDIUM WITH THERMAL RADIATION

16.4.1 Problem Definition

Sample element, boundary condition, and geometry are depicted in Fig. 16.17. Fig. 16.18 shows the external magnetic source. $\overline{H}, \overline{H_x}, \overline{H_y}$ are:

$$\overline{H_y} = (\overline{a} - x)\frac{\gamma}{2\pi}\left[(\overline{b}-y)^2 + (\overline{a}-x)^2\right]^{-1}, \tag{16.51}$$

$$\overline{H_x} = (y - \overline{b})\frac{\gamma}{2\pi}\left[(\overline{b}-y)^2 + (\overline{a}-x)^2\right]^{-1}, \tag{16.52}$$

$$\overline{H} = \left(\overline{H}_y^2 + \overline{H}_x^2\right)^{0.5}. \tag{16.53}$$

16.4.2 Governing Equation

Two-dimensional steady convective non-Darcy flow of nanofluid is considered in the presence of a constant magnetic field. The PDEs are:

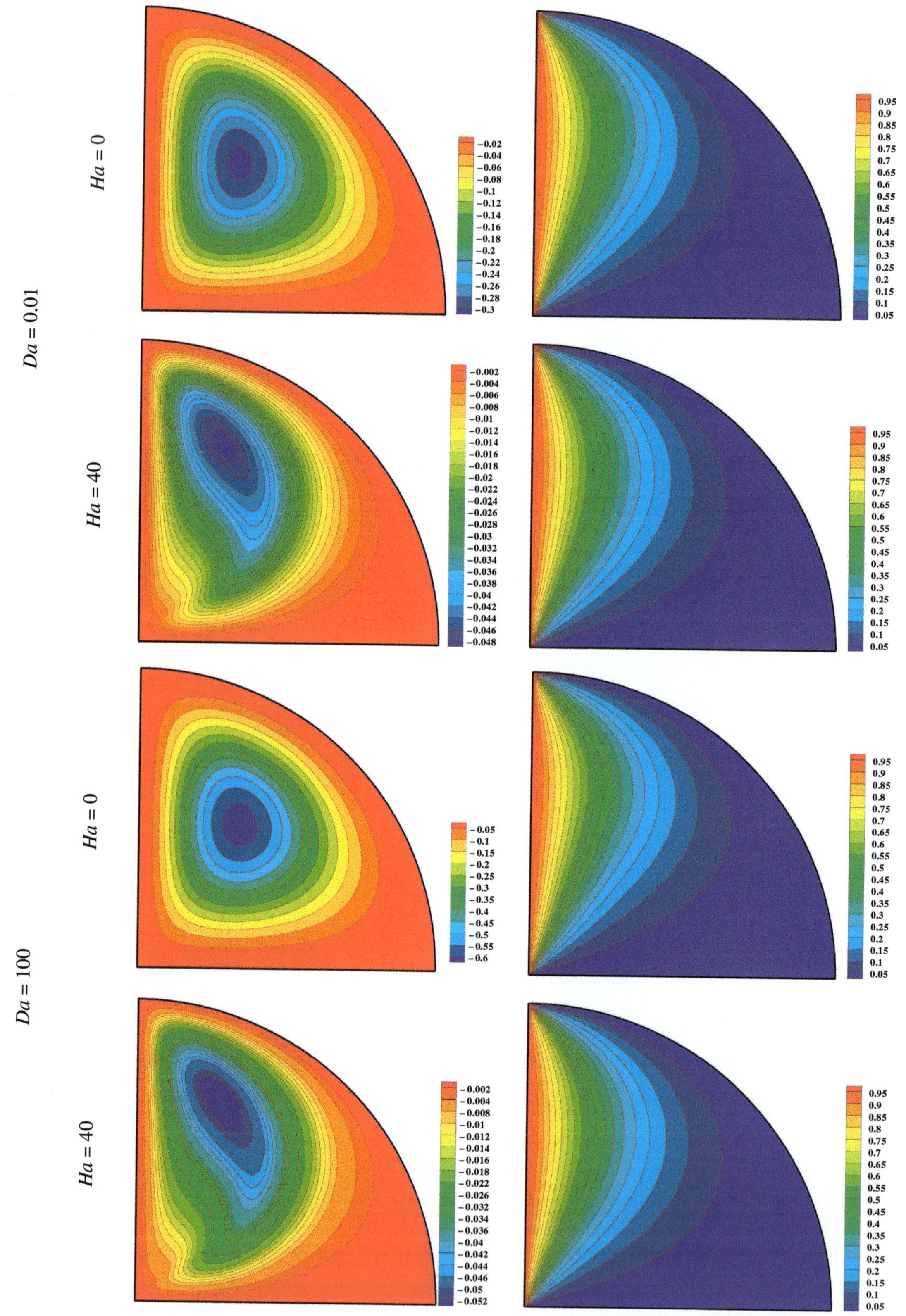

FIGURE 16.12 Influence of *Da*, *Ha* on streamline (left) and isotherm (right) contours when $\phi = 0.04$, $Ra = 10^3$.

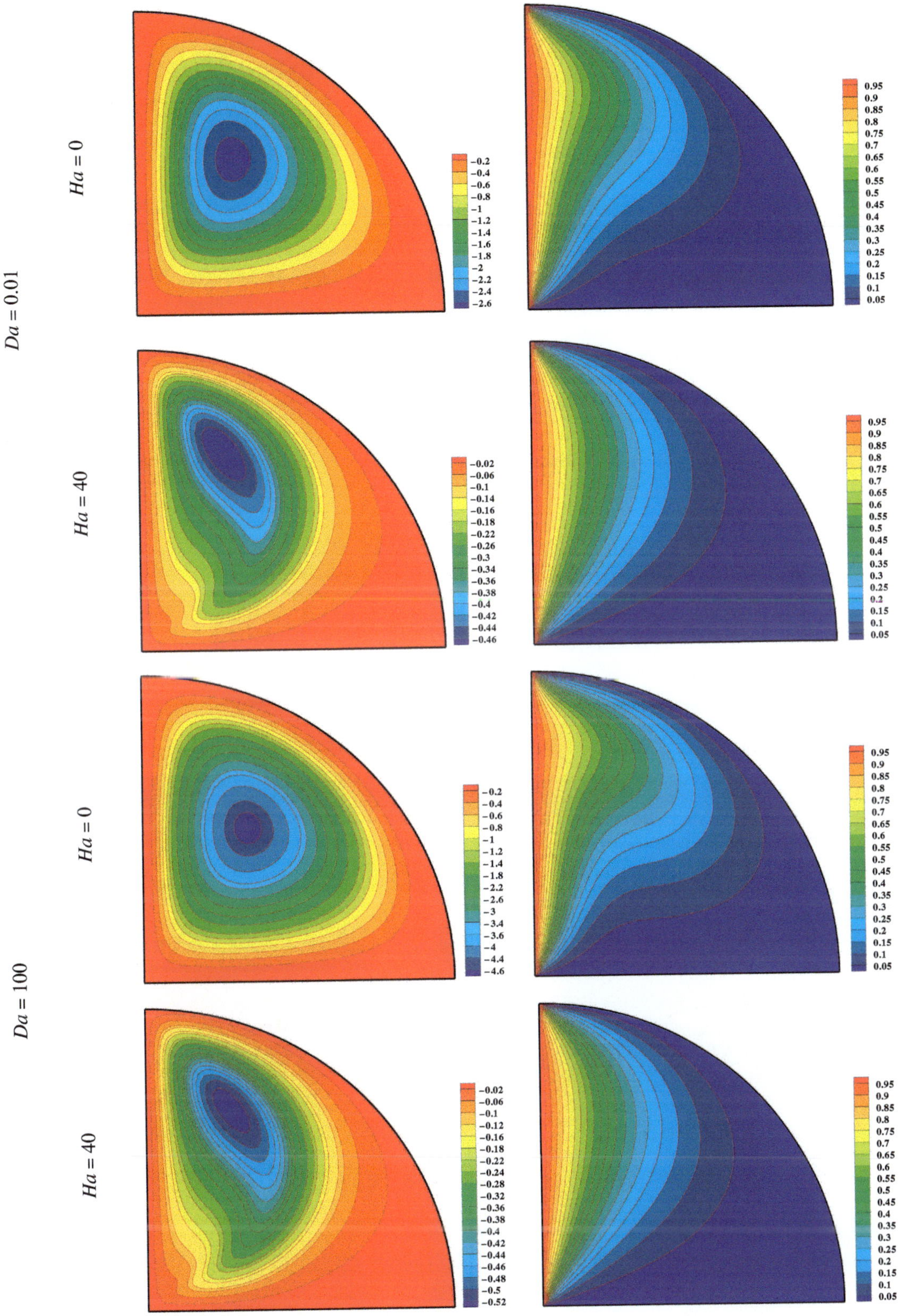

FIGURE 16.13 Influence of *Da*, *Ha* on streamline (left) and isotherm (right) contours when $\phi = 0.04$, $Ra = 10^4$.

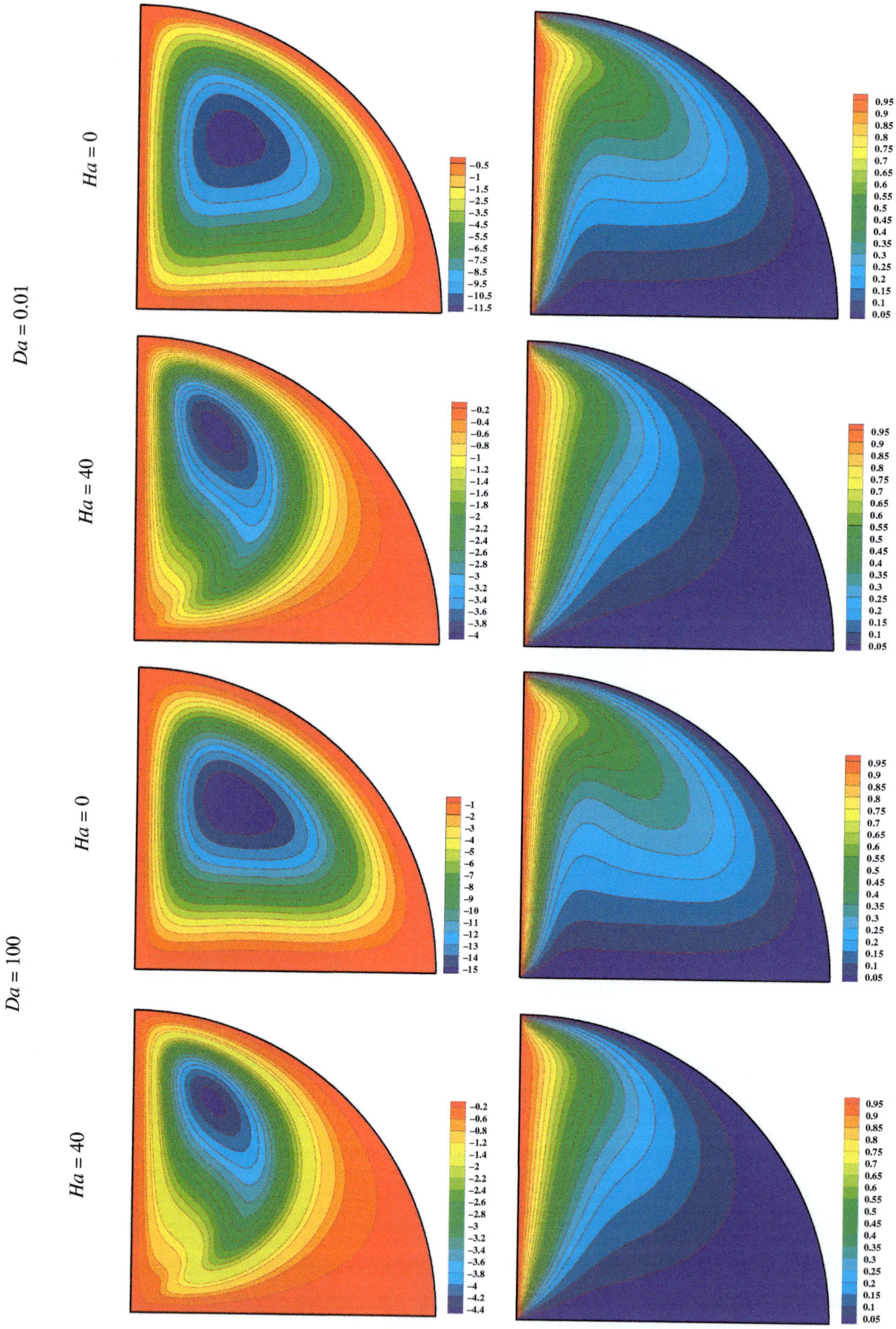

FIGURE 16.14 Influence of *Da*, *Ha* on streamline (left) and isotherm (right) contours when $\phi = 0.04$, $Ra = 10^5$.

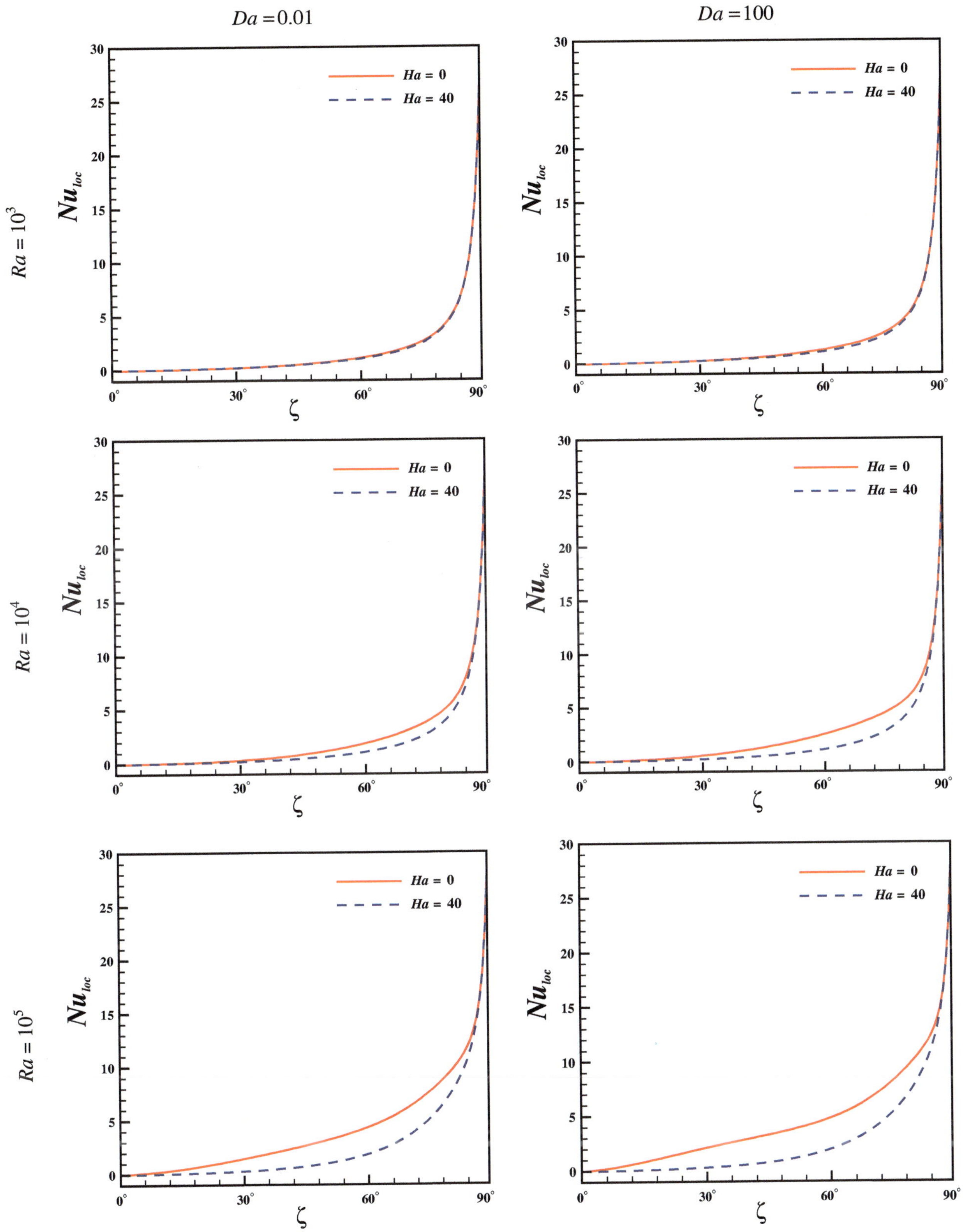

FIGURE 16.15 Influence of *Ha*, *Ra*, and *Da* on local Nusselt number.

$$\frac{\partial u}{\partial x} = -\frac{\partial v}{\partial y}, \tag{16.54}$$

$$\left(\frac{\partial^2 u}{\partial y^2} + \frac{\partial^2 u}{\partial x^2}\right)\mu_{nf} - \frac{\partial P}{\partial x} - \mu_0^2 \sigma_{nf} \overline{H}_y^2 u + \sigma_{nf} \mu_0^2 \overline{H}_x \overline{H}_y v - \frac{\mu_{nf}}{K} u = \left(\rho_{nf}\right)\left(\frac{\partial u}{\partial x} u + \frac{\partial u}{\partial y} v\right), \tag{16.55}$$

$$\mu_{nf}\left(\frac{\partial^2 v}{\partial x^2} + \frac{\partial^2 v}{\partial y^2}\right) - \frac{\partial P}{\partial y} + \mu_0^2 \overline{H}_y \sigma_{nf} \overline{H}_x u - \mu_0^2 \overline{H}_x \sigma_{nf} \overline{H}_x v - \frac{\mu_{nf}}{K} v + (T - T_c)\beta_{nf} g \rho_{nf} = \rho_{nf}\left(\frac{\partial v}{\partial x} u + \frac{\partial v}{\partial y} v\right), \tag{16.56}$$

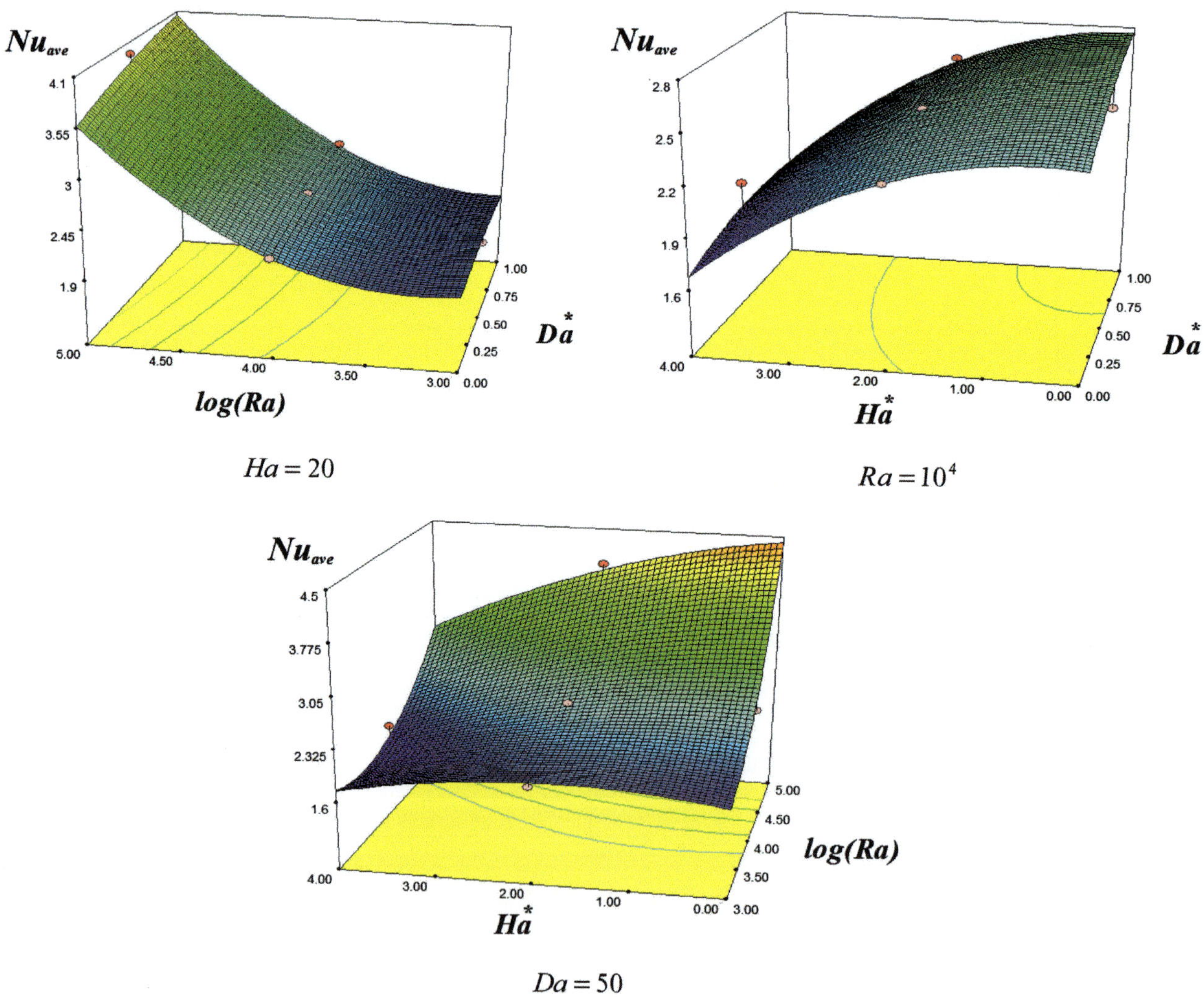

FIGURE 16.16 Effects of *Da*, *Ha*, and *Ra* on average Nusselt number.

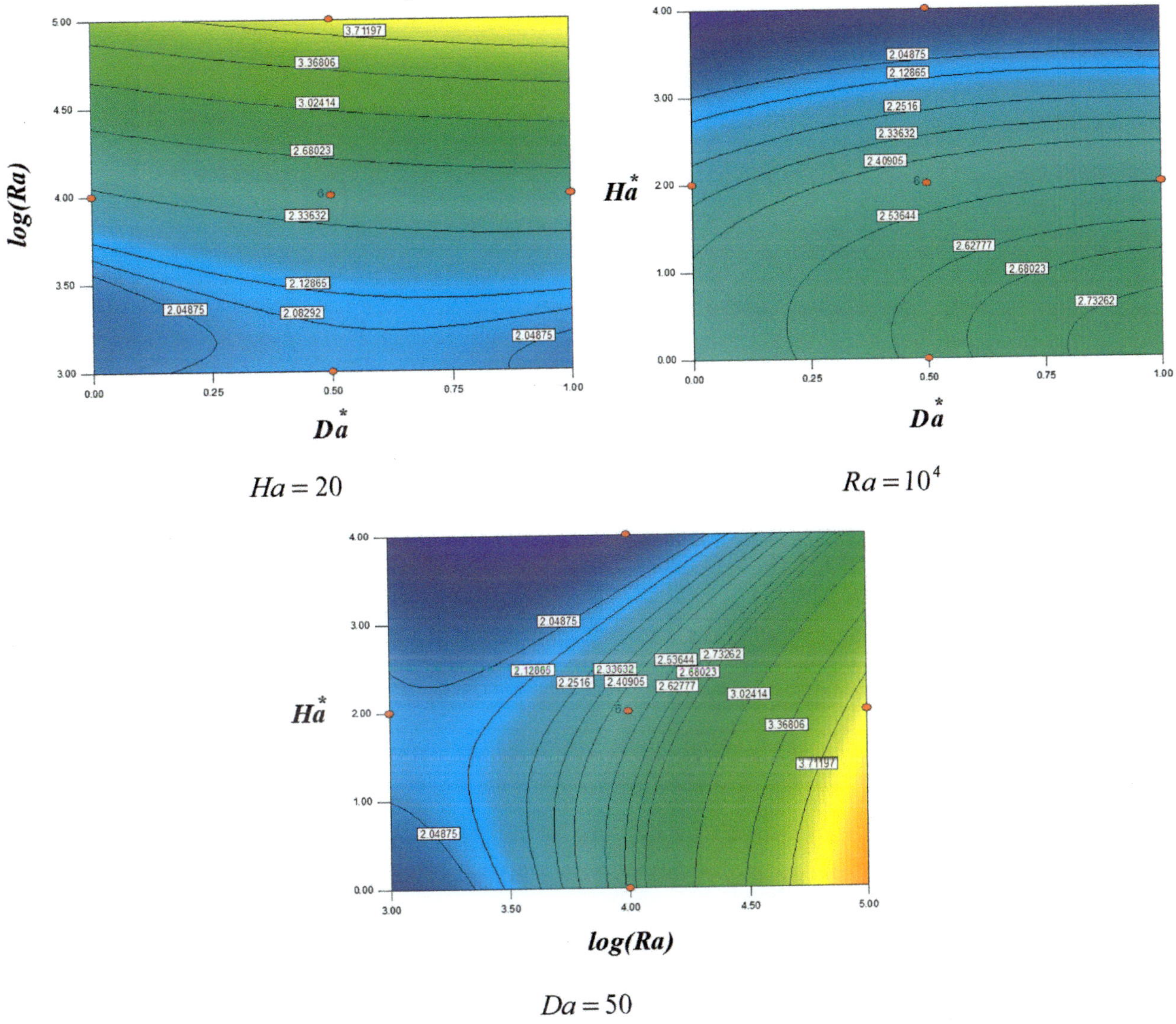

FIGURE 16.16 *(Continued).*

TABLE 16.3 Effects of *Da*, *Ha*, and *Ra* on Heat Transfer Enhancement

Ra	*Da*	*Ha*	E
10^3	0.01	0	9.113388
10^3	0.01	40	9.180014
10^5	0.01	0	9.396332
10^5	0.01	40	8.235231
10^3	100	0	8.706959
10^3	100	40	9.170092
10^5	100	0	9.150151
10^5	100	40	8.044214

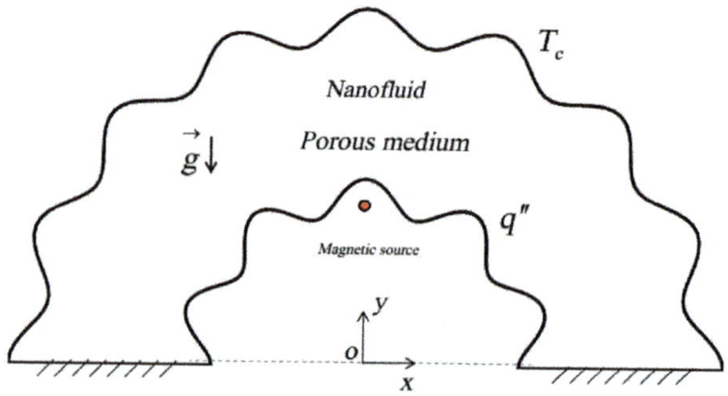

FIGURE 16.17 Geometry and the boundary conditions.

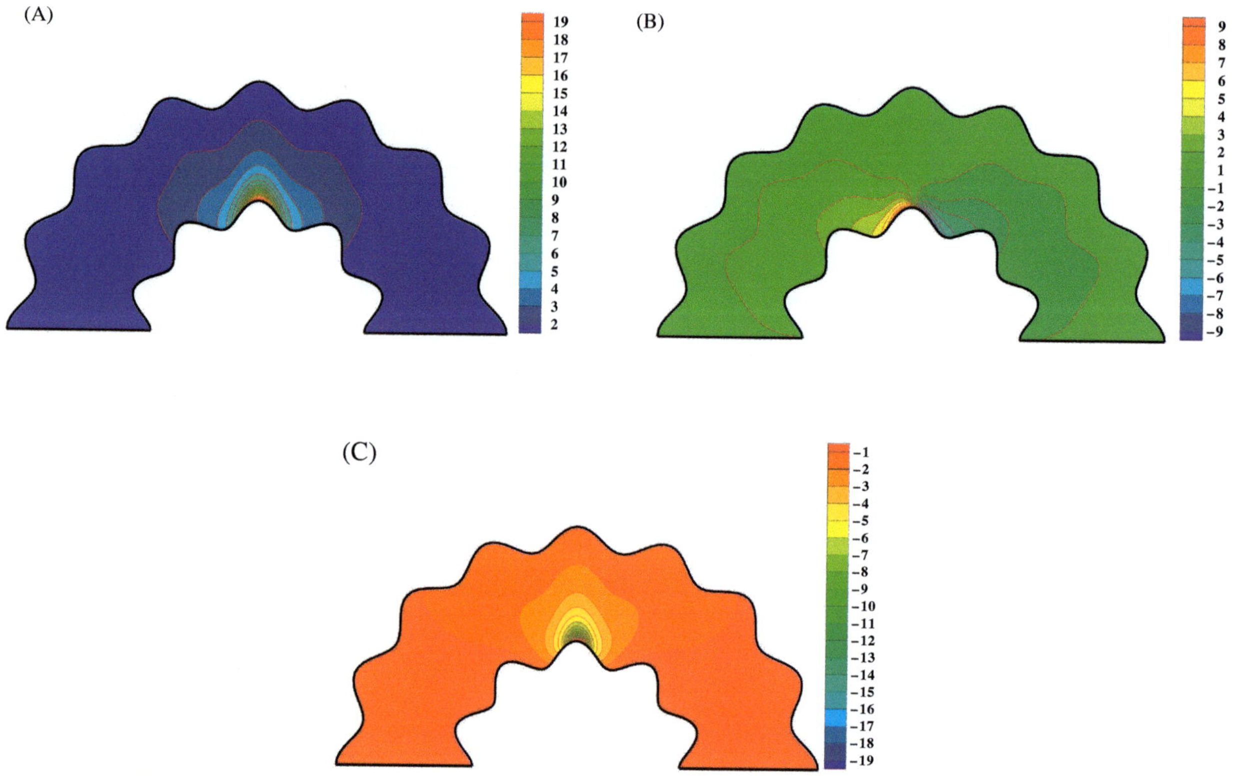

FIGURE 16.18 Contours of the (A) magnetic field strength H; (B) magnetic field intensity component in x direction Hx; (C) magnetic field intensity component in y direction Hy. (A) $H(x,y)$, (B) $Hx(x,y)$, (C) $Hy(x,y)$.

$$k_{nf}\left(\frac{\partial^2 T}{\partial y^2}+\frac{\partial^2 T}{\partial x^2}\right)-\frac{\partial q_r}{\partial y}=\left(v\frac{\partial T}{\partial y}+u\frac{\partial T}{\partial x}\right)(\rho C_p)_{nf},$$
$$\left[q_r=-\frac{4\sigma_e}{3\beta_R}\frac{\partial T^4}{\partial y},\quad T^4\cong 4T_c^3T-3T_c^4\right]. \tag{16.57}$$

$(\rho C_p)_{nf}, \rho_{nf}, (\rho\beta)_{nf}, \sigma_{nf}$ are:

$$(\rho C_p)_{nf}=(\rho C_p)_f(1-\phi)+(\rho C_p)_s\phi \tag{16.58}$$

$$\rho_{nf} = \rho_f(1-\phi) + \rho_s\phi \tag{16.59}$$

$$(\rho\beta)_{nf} = (\rho\beta)_f(1-\phi) + (\rho\beta)_s\phi, \tag{16.60}$$

$$\sigma_{nf} = \sigma_f\left[\frac{3\phi(\sigma 1 - 1)}{(1-\sigma 1)\phi + (2+\sigma 1)} + 1\right], \qquad \sigma 1 = \sigma_s/\sigma_f. \tag{16.61}$$

μ_{nf} is obtained as follows:

$$\mu_{nf} = \left(0.035\mu_0^2\overline{H}^2 + 3.1\mu_0\overline{H} - 27886.4807\phi^2 + 4263.02\phi + 316.0629\right)e^{-0.01T} \tag{16.62}$$

k_{nf} can be obtained as:

$$\frac{k_{nf}}{k_f} = \frac{-m(k_f-k_p)\phi + (k_p-k_f)\phi + mk_f + k_p + k_f}{mk_f + (k_f-k_p)\phi + k_f + k_p} \tag{16.63}$$

The properties of nanofluid are depicted in Table 16.1.
Different values of shape factors for various shapes of nanoparticles are illustrated in Table 16.4.
Vorticity and stream function should be used to eliminate pressure source terms:

$$\omega + \frac{\partial u}{\partial y} - \frac{\partial v}{\partial x} = 0, \quad \frac{\partial \psi}{\partial x} = -v, \quad \frac{\partial \psi}{\partial y} = u. \tag{16.64}$$

Dimensionless parameters are defined as:

$$\begin{aligned} &(H_y, H_x, H) = \frac{(\overline{H_y}, \overline{H_x}, \overline{H})}{\overline{H_0}}, \ (b,a) = \frac{(\bar{b}, \bar{a})}{L}, \\ &U = \frac{uL}{\alpha_{nf}}, V = \frac{vL}{\alpha_{nf}}, \ (X,Y) = \frac{(x,y)}{L}, \theta = \frac{T-T_c}{\Delta T}, \Delta T = q''L/k_f, \\ &\Psi = \frac{\psi}{\alpha_{nf}}, \Omega = \frac{\omega L^2}{\alpha_{nf}}. \end{aligned} \tag{16.65}$$

So equations change to:

$$\frac{\partial^2\Psi}{\partial Y^2} + \Omega + \frac{\partial^2\Psi}{\partial X^2} = 0, \tag{16.66}$$

TABLE 16.4 The Values of the Shape Factor of Different Shapes of Nanoparticles

	Spherical	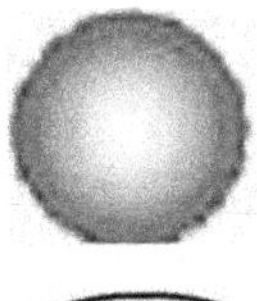	3
m	Platelet		5.7
	Cylinder		4.8
	Brick	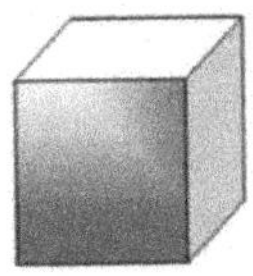	3.7

$$\frac{\partial \Omega}{\partial X}U + V\frac{\partial \Omega}{\partial Y} = \Pr\frac{A_5}{A_1}\frac{A_2}{A_4}\left(\frac{\partial^2 \Omega}{\partial Y^2} + \frac{\partial^2 \Omega}{\partial X^2}\right)$$
$$+ \Pr Ha^2\frac{A_6}{A_1}\frac{A_2}{A_4}\left(\frac{\partial U}{\partial X}H_yH_x - \frac{\partial V}{\partial X}H_x{}^2 + \frac{\partial U}{\partial Y}H_y^2 - \frac{\partial V}{\partial Y}H_yH_x\right) \tag{16.67}$$
$$+ \Pr Ra\frac{A_3A_2^2}{A_1A_4^2}\frac{\partial \theta}{\partial X} - \frac{\Pr}{Da}\frac{A_5}{A_1}\frac{A_2}{A_4}\Omega,$$

$$\frac{\partial \theta}{\partial X}U + V\frac{\partial \theta}{\partial Y} = \left(\frac{\partial^2 \theta}{\partial X^2} + \frac{\partial^2 \theta}{\partial Y^2}\right) + \frac{4}{3}\frac{1}{A_4}Rd\frac{\partial^2 \theta}{\partial Y^2}. \tag{16.68}$$

and dimensionless parameters are

$$Ra_f = g\beta_f L^3 \Delta T/(\alpha_f \upsilon_f), \Pr_f = \upsilon_f/\alpha_f, Ha = L\mu_0 H_0\sqrt{\sigma_f/\mu_f},$$
$$Da = K/L^2, Ec = (\mu_f\alpha_f)/\left[(\rho C_P)_f \Delta T\, L^2\right],$$
$$A_1 = \frac{\rho_{nf}}{\rho_f}, A_2 = \frac{(\rho C_p)_{nf}}{(\rho C_p)_f}, A_3 = \frac{(\rho\beta)_{nf}}{(\rho\beta)_f}, \tag{16.69}$$
$$A_4 = \frac{k_{nf}}{k_f}, A_5 = \frac{\mu_{nf}}{\mu_f}, A_6 = \frac{\sigma_{nf}}{\sigma_f},$$
$$Rd = 4\sigma_e T_c^3/(\beta_R k_f).$$

and boundary conditions are:

$$\begin{aligned} &\text{on inner wall} && \frac{\partial \theta}{\partial n} = 1.0 \\ &\text{on outer wall} && \theta = 0.0 \\ &\text{on other walls} && \frac{\partial \theta}{\partial n} = 0 \\ &\text{on all walls} && \Psi = 0.0 \end{aligned} \tag{16.70}$$

Nu_{loc}, Nu_{ave} over the hot wall can be calculated as:

$$Nu_{loc} = \left(\frac{k_{nf}}{k_f}\right)\left(1 + \left(\frac{k_{nf}}{k_f}\right)^{-1}\frac{4Rd}{3}\right)\frac{1}{\theta}, \tag{16.71}$$

$$Nu_{ave} = \frac{1}{S}\int_0^S Nu_{loc}\, ds. \tag{16.72}$$

16.4.3 Effects of Active Parameters

The impact of nonuniform Lorentz forces on Fe_3O_4-water nanofluid treatment in a permeable medium is modeled. The shape effect of nanoparticles on k_{nf} is taken into consideration. μ_{nf} of Fe_3O_4-water are estimated according to previous experimental data. CVFEM is utilized to find the influences of radiation parameter ($Rd = 0$ to 0.8), Darcy number ($Da = 0.01$ to 100), Rayleigh number ($Ra = 10^3, 10^4, 10^5$), volume fraction of Fe_3O_4 ($\phi = 0\%$ to 4%), shape of nanoparticle, and Hartmann number ($Ha = 0$ to 10). Impacts of shape of the nanoparticles on Nu_{ave} are reported in Table 16.5. The maximum Nu is caused by Platelet, followed by Cylinder, Brick, and Spherical. Therefore, Platelet nanoparticle has been selected for further investigation. Fig. 16.19 demonstrates the effect of adding nanoparticles in the water on velocity and temperature contours. The temperature gradient is reduced

TABLE 16.5 Effect of Shape of Nanoparticles on Nusselt Number When $Da = 100, Ra = 10^5$, $Rd = 0.8, \phi = 0.04$

	Ha	
	0	**10**
Spherical	8.754093	6.893915
Brick	8.802073	6.930307
Cylinder	8.87772	6.987786
Platelet	8.940132	7.034996

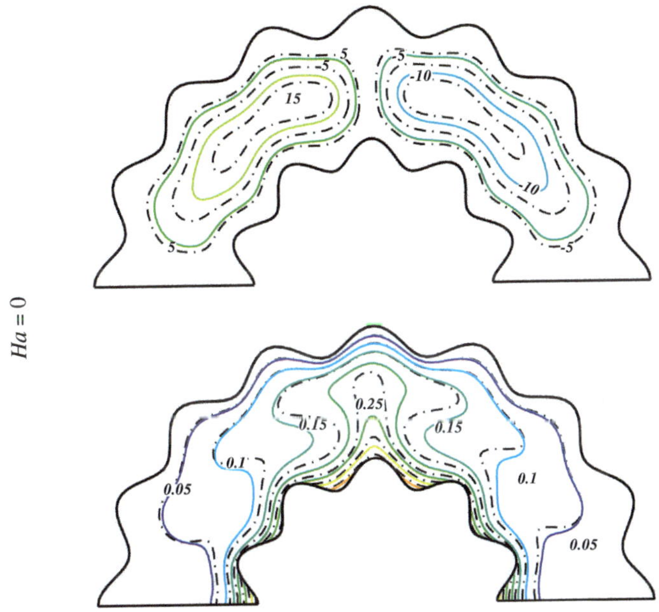

FIGURE 16.19 Impact of nanofluid volume fraction on streamlines (top) and isotherms (bottom) contours (nanofluid ($\phi = 0.04$)(—) and pure fluid ($\phi = 0$) (– · –)) when $Ra = 10^5, Da = 100, Rd = 0.8$.

with the increase of ϕ. $|\Psi_{\max}|$ increases with the addition of nanoparticles because of the increase in the solid movements. Fig. 16.20 shows the impact of the radiation parameter on velocity and temperature contours. The thermal boundary layer thickness increases with the increase of Rd. As Hartmann number increases, the impact of the radiation parameter on streamlines becomes not significant. Figs. 16.21–16.23 depict the effects of Da, Ra, Ha on isotherms and streamlines. One eddy can be seen in streamlines. By applying a magnetic field, the main eddy moves downward. The distortion of isotherms becomes less than before. As Ra increases, a thermal plume is generated near the vertical symmetric line. Augmenting Ha shifts the thermal plume to the left and decreases $|\Psi_{\max}|$. As the Darcy number increases, the convective mode becomes stronger and the primary eddy converts into two eddies. Fig. 16.24 demonstrates the influence of significant parameters on Nu_{ave}. The correlation for Nu_{ave} is:

$$\begin{aligned} Nu_{ave} = {} & 22.3 - 3.13Rd - 10.5og(Ra) - 0.43Da^* + 1.03Ha^* \\ & + 1.88Rd\log(Ra) + 0.36Rd\ Da^* - 0.54Rd\ Ha^* \\ & + 0.19Da^*\log(Ra) - 0.28Ha^*\log(Ra) - 0.52Da^*Ha^* \\ & - 2.5Rd^2 + 1.37\left(\log(Ra)\right)^2 + 0.29(Da^*)^2 - 0.17(Ha^*)^2 \end{aligned} \quad (16.73)$$

where $Da^* = 0.01Da$, $Ha^* = 0.1Ha$. Heat transfer rate augments with the rise of permeability of porous media. A similar treatment is observed for buoyancy forces. The temperature gradient is reduced with the increase of the Hartmann number.

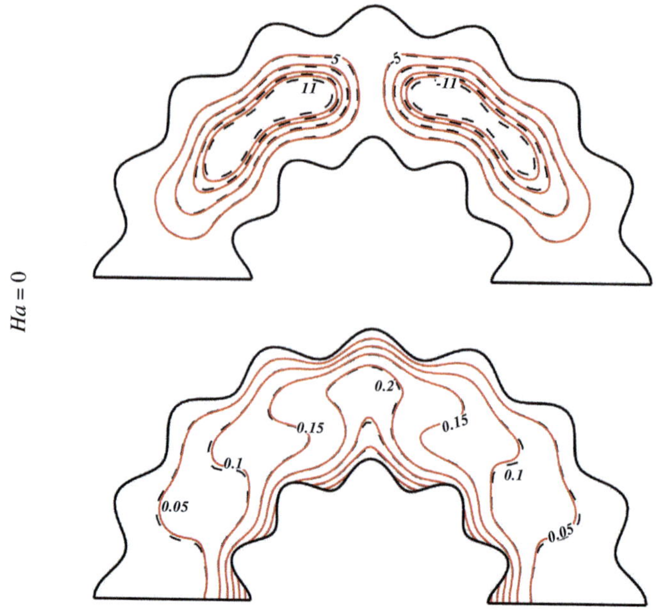

FIGURE 16.20 Impact of radiation parameter on streamlines (top) and isotherms (bottom) contours ($Rd = 0.8$(—), $Rd = 0$ (- - -)) when $Ra = 10^5, Da = 100, \phi = 0.04$.

16.5 EFFECT OF EXTERNAL MAGNETIC SOURCE ON FE_3O_4-H_2O NANOFLUID BEHAVIOR IN A PERMEABLE CAVITY CONSIDERING SHAPE EFFECT

16.5.1 Problem Definition

Fig. 16.25 shows the sample element and boundary conditions of current geometry. As shown in Fig. 16.26, the external magnetic source affects the nanofluid flow. $\overline{H}, \overline{H_x}, \overline{H_y}$ are:

$$\overline{H_y} = (\bar{a} - x)\frac{\gamma}{2\pi}\left[(\bar{b}-y)^2 + (\bar{a}-x)^2\right]^{-1}, \tag{16.74}$$

$$\overline{H_x} = (y - \bar{b})\frac{\gamma}{2\pi}\left[(\bar{b}-y)^2 + (\bar{a}-x)^2\right]^{-1}, \tag{16.75}$$

$$\overline{H} = \left(\overline{H}_y^2 + \overline{H}_x^2\right)^{0.5}. \tag{16.76}$$

16.5.2 Governing Equation

Convective nanofluid flow in existence of nonuniform magnetic field is considered using a non-Darcy model. The PDEs are:

$$\frac{\partial u}{\partial x} = -\frac{\partial v}{\partial y}, \tag{16.77}$$

$$\begin{aligned} &\left(\frac{\partial^2 u}{\partial y^2} + \frac{\partial^2 u}{\partial x^2}\right)\mu_{nf} - \frac{\partial P}{\partial x} - \mu_0^2 \sigma_{nf} \overline{H}_y^2 u + \sigma_{nf}\mu_0^2 \overline{H}_x \overline{H}_y v \\ &- \frac{\mu_{nf}}{K} u = \left(\rho_{nf}\right)\left(\frac{\partial u}{\partial x} u + \frac{\partial u}{\partial y} v\right), \end{aligned} \tag{16.78}$$

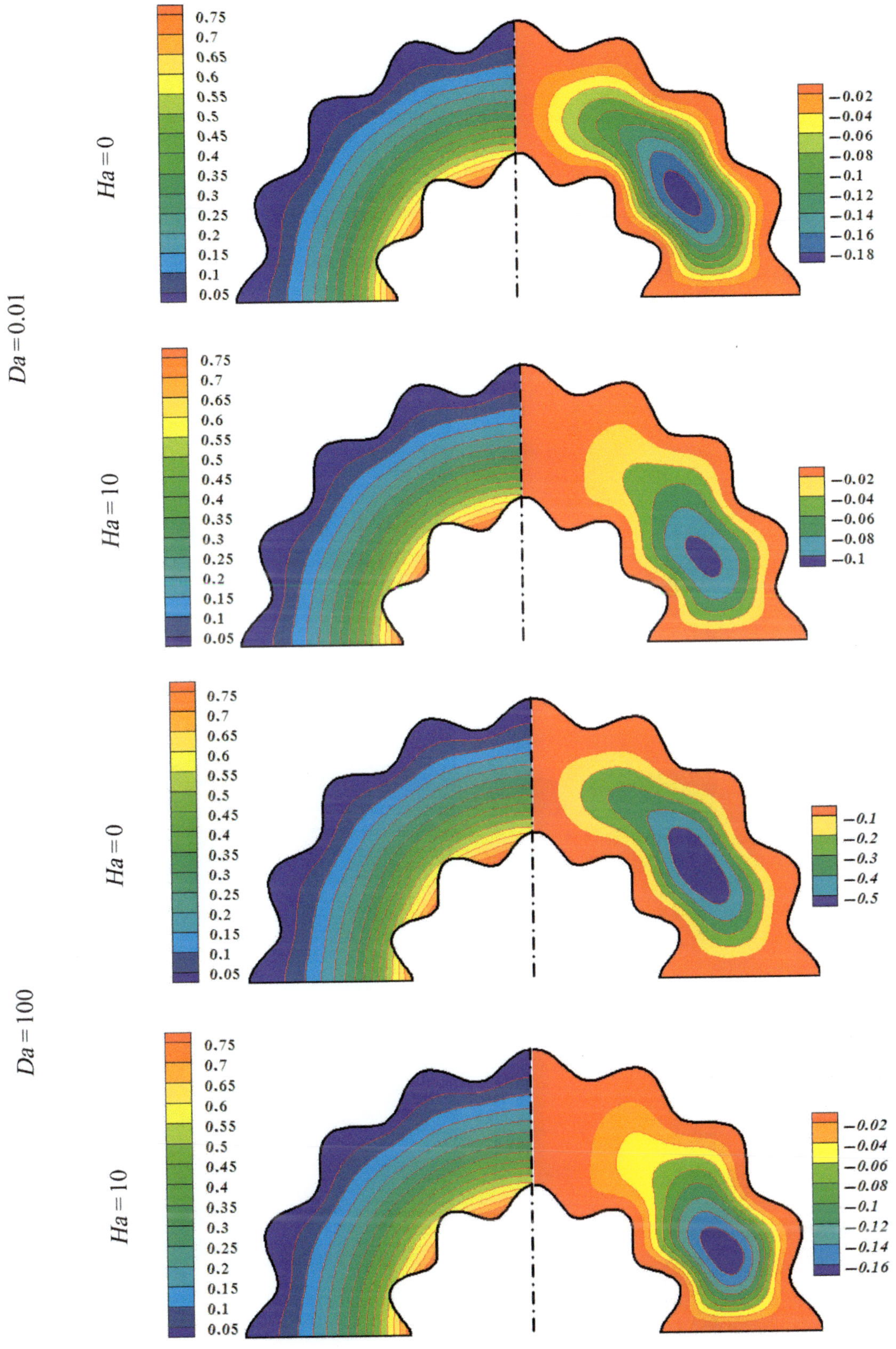

FIGURE 16.21 Influence of *Da*, *Ha* on streamline (right) and isotherm (left) contours when $\phi = 0.04$, $Ra = 10^3$.

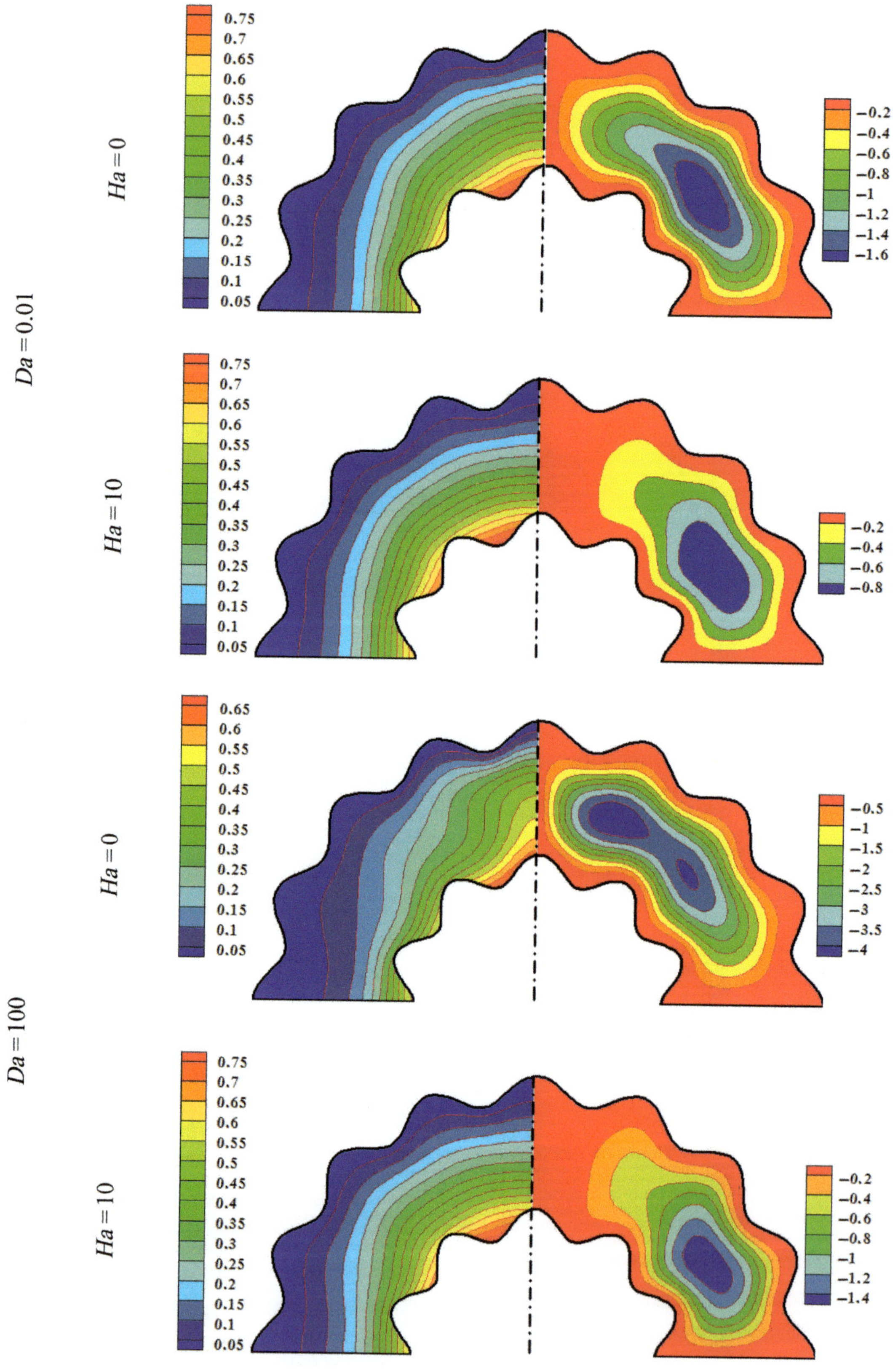

FIGURE 16.22 Influence of *Da*, *Ha* on streamline (right) and isotherm (left) contours when $\phi = 0.04$, $Ra = 10^4$.

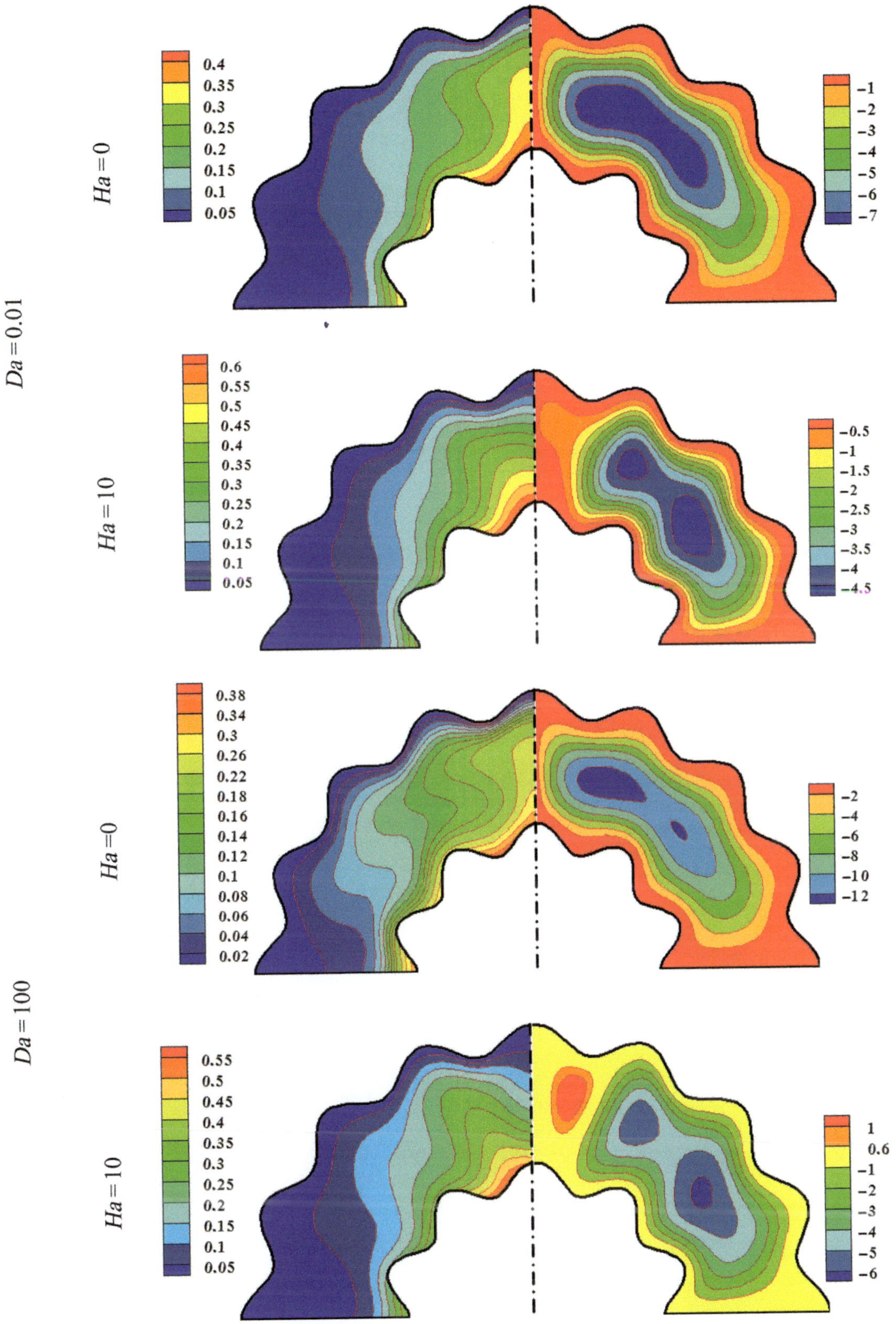

FIGURE 16.23 Influence of Da, Ha on streamline (right) and isotherm (left) contours when $\phi = 0.04$, $Ra = 10^5$.

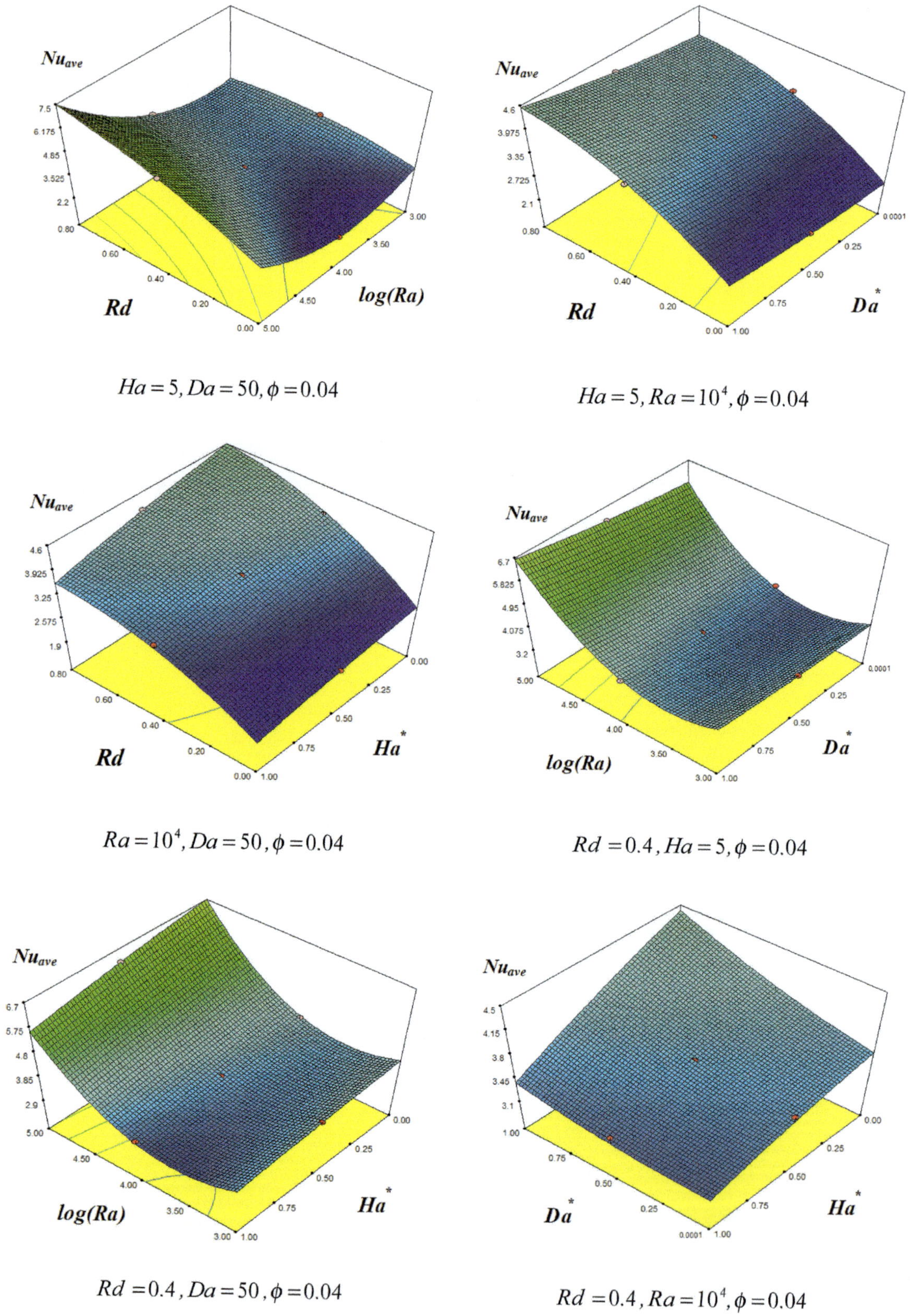

FIGURE 16.24 Effects of *Da*, *Ha*, *Rd*, and *Ra* on average Nusselt number.

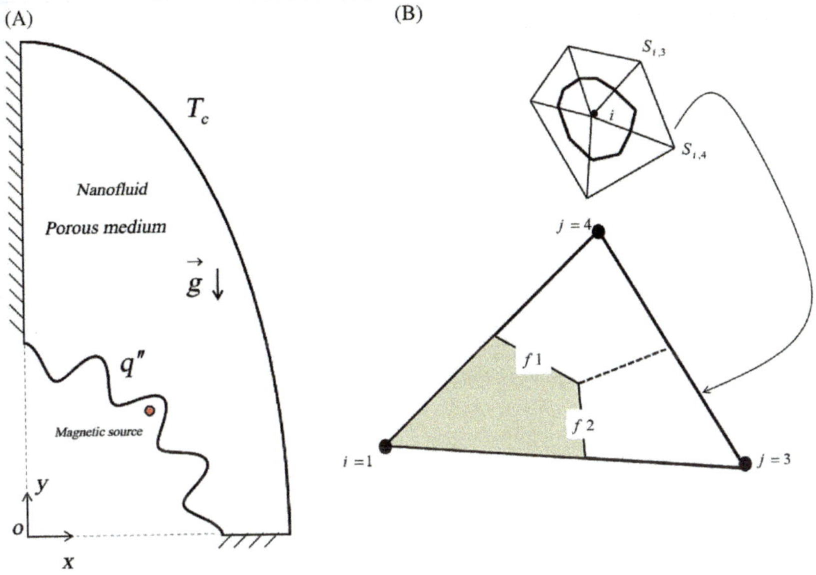

FIGURE 16.25 (A) Geometry and the boundary conditions with; (B) a sample triangular element and its corresponding control volume.

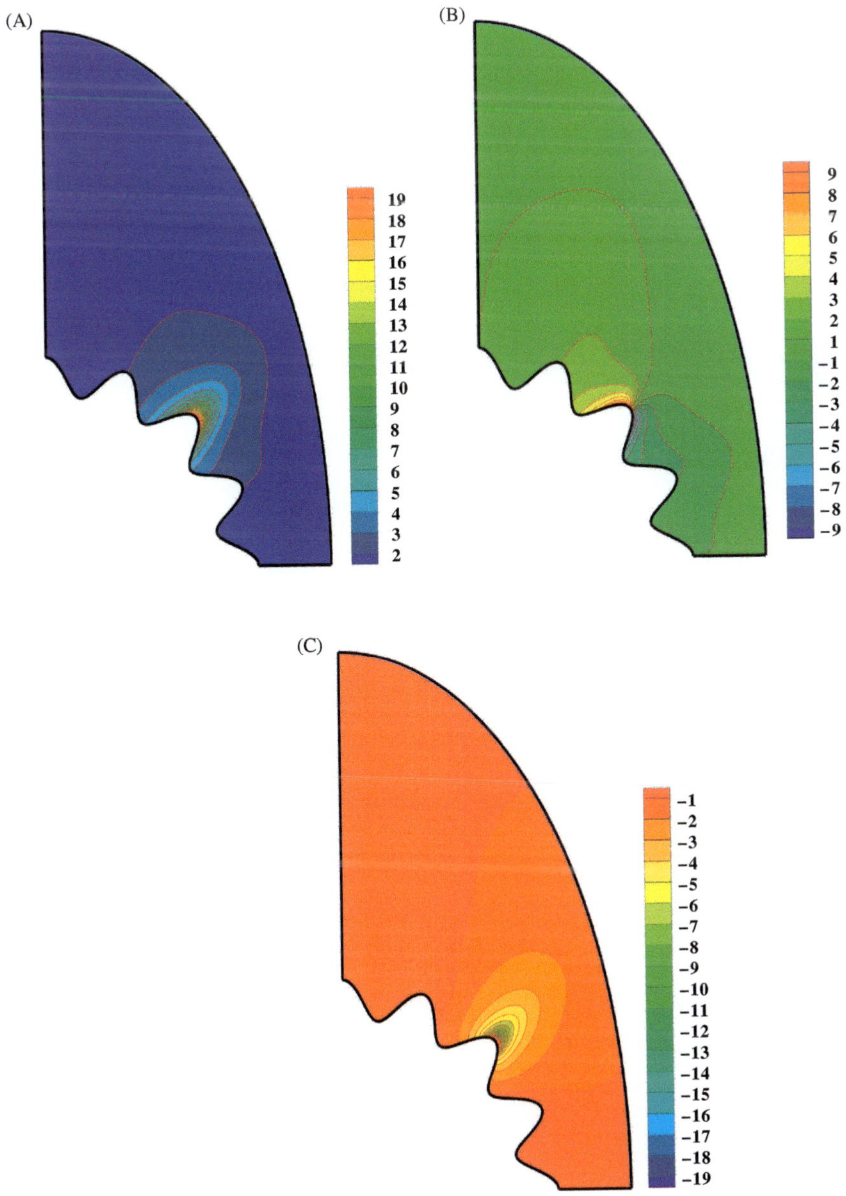

FIGURE 16.26 Contours of the (A) magnetic field strength H; (B) magnetic field intensity component in x direction Hx; (C) magnetic field intensity component in y direction Hy. (A) $H(x,y)$, (B) $Hx(x,y)$, (C) $Hy(x,y)$.

$$\mu_{nf}\left(\frac{\partial^2 v}{\partial x^2}+\frac{\partial^2 v}{\partial y^2}\right)-\frac{\partial P}{\partial y}+\mu_0^2\overline{H}_y\sigma_{nf}\overline{H}_x u-\mu_0^2\overline{H}_x\sigma_{nf}\overline{H}_x v-\frac{\mu_{nf}}{K}v$$
$$+(T-T_c)\beta_{nf}g\rho_{nf}=\rho_{nf}\left(\frac{\partial v}{\partial x}u+\frac{\partial v}{\partial y}v\right), \tag{16.79}$$

$$k_{nf}\left(\frac{\partial^2 T}{\partial y^2}+\frac{\partial^2 T}{\partial x^2}\right)-\frac{\partial q_r}{\partial y}=\left(v\frac{\partial T}{\partial y}+u\frac{\partial T}{\partial x}\right)(\rho C_p)_{nf},$$
$$\left[q_r=-\frac{4\sigma_e}{3\beta_R}\frac{\partial T^4}{\partial y},\quad T^4\cong 4T_c^3T-3T_c^4\right]. \tag{16.80}$$

$(\rho\beta)_{nf}$, $(\rho C_p)_{nf}$, ρ_{nf}, and σ_{nf} are calculated as:

$$(\rho\beta)_{nf}=(\rho\beta)_f(1-\phi)+(\rho\beta)_s\phi, \tag{16.81}$$

$$(\rho C_p)_{nf}=(\rho C_p)_f(1-\phi)+(\rho C_p)_s\phi \tag{16.82}$$

$$\rho_{nf}=\rho_f(1-\phi)+\rho_s\phi \tag{16.83}$$

$$\sigma_{nf}=\sigma_f\left[\frac{3\phi(\sigma 1-1)}{(1-\sigma 1)\phi+(2+\sigma 1)}+1\right],\qquad \sigma 1=\sigma_s/\sigma_f. \tag{16.84}$$

μ_{nf} is calculated as:

$$\mu_{nf}=\left(0.035\mu_0^2\overline{H}^2+3.1\mu_0\overline{H}-27886.4807\phi^2+4263.02\phi+316.0629\right)e^{-0.01T} \tag{16.85}$$

k_{nf} can be obtained as:

$$\frac{k_{nf}}{k_f}=\frac{-m(k_f-k_p)\phi+(k_p-k_f)\phi+mk_f+k_p+k_f}{mk_f+(k_f-k_p)\phi+k_f+k_p} \tag{16.86}$$

Tables 16.1 and 16.4 show the properties of nanofluid and shape factor values. Vorticity and stream function can be employed to eliminate pressure source terms:

$$\omega+\frac{\partial u}{\partial y}-\frac{\partial v}{\partial x}=0,\quad \frac{\partial\psi}{\partial x}=-v,\quad \frac{\partial\psi}{\partial y}=u. \tag{16.87}$$

Dimensionless parameters are defined as:

$$(H_y,H_x,H)=\frac{(\overline{H_y},\overline{H_x},\overline{H})}{\overline{H_0}},\ (b,a)=\frac{(\overline{b},\overline{a})}{L},$$
$$U=\frac{uL}{\alpha_{nf}},V=\frac{vL}{\alpha_{nf}},\ (X,Y)=\frac{(x,y)}{L},\theta=\frac{T-T_c}{\Delta T},\Delta T=q''L/k_f, \tag{16.88}$$
$$\Psi=\frac{\psi}{\alpha_{nf}},\Omega=\frac{\omega L^2}{\alpha_{nf}}.$$

So equations change to:

$$\frac{\partial^2\Psi}{\partial Y^2}+\Omega+\frac{\partial^2\Psi}{\partial X^2}=0, \tag{16.89}$$

$$\frac{\partial \Omega}{\partial X}U + V\frac{\partial \Omega}{\partial Y} = \Pr\frac{A_5}{A_1}\frac{A_2}{A_4}\left(\frac{\partial^2 \Omega}{\partial Y^2} + \frac{\partial^2 \Omega}{\partial X^2}\right)$$

$$+ \Pr Ha^2\frac{A_6}{A_1}\frac{A_2}{A_4}\left(\frac{\partial U}{\partial X}H_yH_x - \frac{\partial V}{\partial X}{H_x}^2 + \frac{\partial U}{\partial Y}H_y^2 - \frac{\partial V}{\partial Y}H_yH_x\right) \quad (16.90)$$

$$+ \Pr Ra\frac{A_3A_2^2}{A_1A_4^2}\frac{\partial \theta}{\partial X} - \frac{\Pr}{Da}\frac{A_5}{A_1}\frac{A_2}{A_4},$$

$$\frac{\partial \theta}{\partial X}U + V\frac{\partial \theta}{\partial Y} = \left(\frac{\partial^2 \theta}{\partial X^2} + \frac{\partial^2 \theta}{\partial Y^2}\right) + \frac{4}{3}\frac{1}{A_4}Rd\frac{\partial^2 \theta}{\partial Y^2}. \quad (16.91)$$

and dimensionless parameters are:

$$Ra_f = g\beta_f L^3\Delta T/(\alpha_f\upsilon_f), \Pr_f = \upsilon_f/\alpha_f, Ha = L\mu_0 H_0\sqrt{\sigma_f/\mu_f},$$

$$Da = K/L^2, Ec = \left(\mu_f\alpha_f\right)/\left[(\rho C_P)_f\Delta T\, L^2\right],$$

$$A_1 = \frac{\rho_{nf}}{\rho_f}, A_2 = \frac{(\rho C_p)_{nf}}{(\rho C_p)_f}, A_3 = \frac{(\rho\beta)_{nf}}{(\rho\beta)_f}, \quad (16.92)$$

$$A_4 = \frac{k_{nf}}{k_f}, A_5 = \frac{\mu_{nf}}{\mu_f}, A_6 = \frac{\sigma_{nf}}{\sigma_f},$$

$$Rd = 4\sigma_e T_c^3/(\beta_R k_f)$$

and boundary conditions are:

$$\begin{aligned} &\text{on inner wall} && \frac{\partial \theta}{\partial n} = 1.0 \\ &\text{on outer wall} && \theta = 0.0 \\ &\text{on other walls} && \frac{\partial \theta}{\partial n} = 0 \\ &\text{on all walls} && \Psi = 0.0 \end{aligned} \quad (16.93)$$

Nu_{loc}, Nu_{ave} over the hot wall can be calculated as:

$$Nu_{loc} = \left(\frac{k_{nf}}{k_f}\right)\left(1 + \left(\frac{k_{nf}}{k_f}\right)^{-1}\frac{4Rd}{3}\right)\frac{1}{\theta}, \quad (16.94)$$

$$Nu_{ave} = \frac{1}{S}\int_0^S Nu_{loc}\, ds. \quad (16.95)$$

16.5.3 Effects of Active Parameters

The effect of an external magnetic source on Fe_3O_4-water hydrothermal treatment in a permeable enclosure is demonstrated. Magnetic field-dependent viscosity of nanofluid is considered. The shape effect of nanoparticles on k_{nf} is taken into consideration. CVFEM is employed to find the influences of Darcy number ($Da = 0.01$ to 100), radiation parameter ($Rd = 0$ *to* 0.8), Rayleigh number ($Ra = 10^3, 10^4, 10^5$), volume fraction of Fe_3O_4 ($\phi = 0\%$ to 4%), shape factor, and Hartmann number ($Ha = 0$ to 10).

The effect of shape factor on Nu_{ave} is discussed in Table 16.6. The highest Nu_{ave} is caused by Platelet, followed by Cylinder, Brick, and Spherical. So, Platelet shape was selected for further investigation.

TABLE 16.6 Effect of Shape of Nanoparticles on Nusselt Number When $Da = 100, Ra = 10^5$, $Rd = 0.8, \phi = 0.04$

	Ha	
	0	**10**
Spherical	7.072375	5.865269
Brick	7.120273	5.901963
Cylinder	7.196075	5.960649
Platelet	7.258503	6.009695

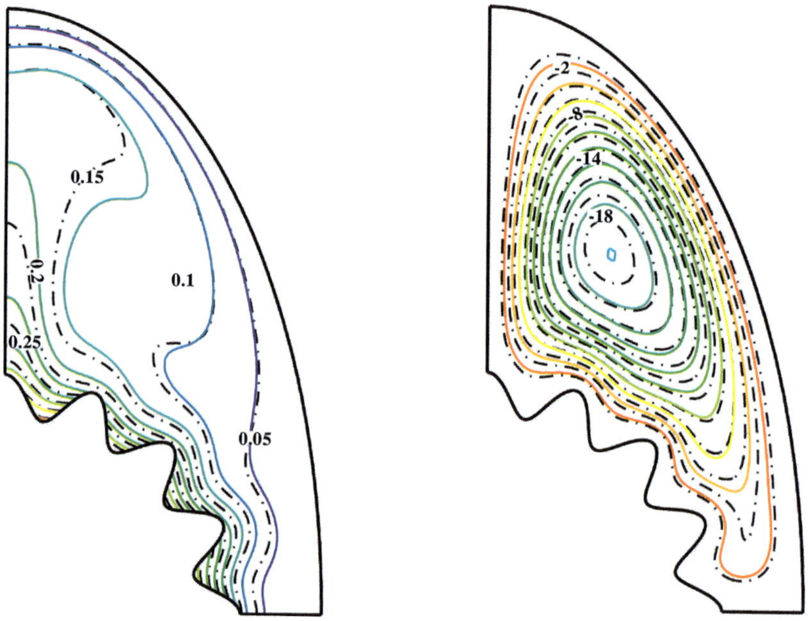

FIGURE 16.27 Impact of nanofluid volume fraction on streamline (right) and isotherm (left) contours (nanofluid ($\phi = 0.04$)(—) and pure fluid ($\phi = 0$) (– · –)) when $Ra = 10^5, Da = 100, Rd = 0.8, Ha = 0$.

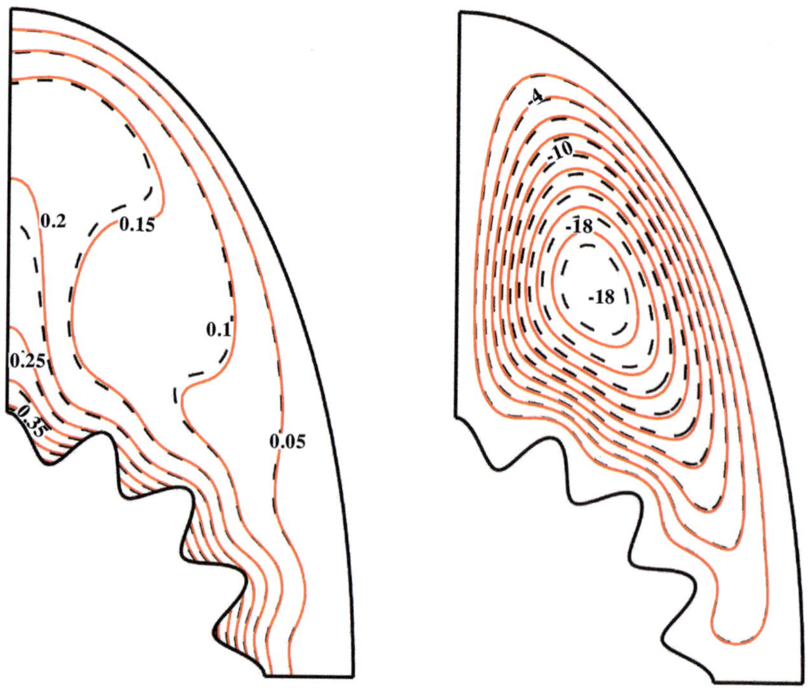

FIGURE 16.28 Impact of radiation parameter on streamline (right) and isotherm (left) contours ($Rd = 0.8$(—), $Rd = 0$ (- - -)) when $Ra = 10^5, Da = 100$, $\phi = 0.04, Ha = 0$.

Fig. 16.27 illustrates the influence of adding nanoparticles in the base fluid on velocity and temperature contours. The temperature gradient is reduced with the increase of ϕ. $|\Psi_{max}|$ increases with the addition of nanoparticles into the base fluid. Fig. 16.28 shows the impact of the radiation parameter on isotherms and streamlines.

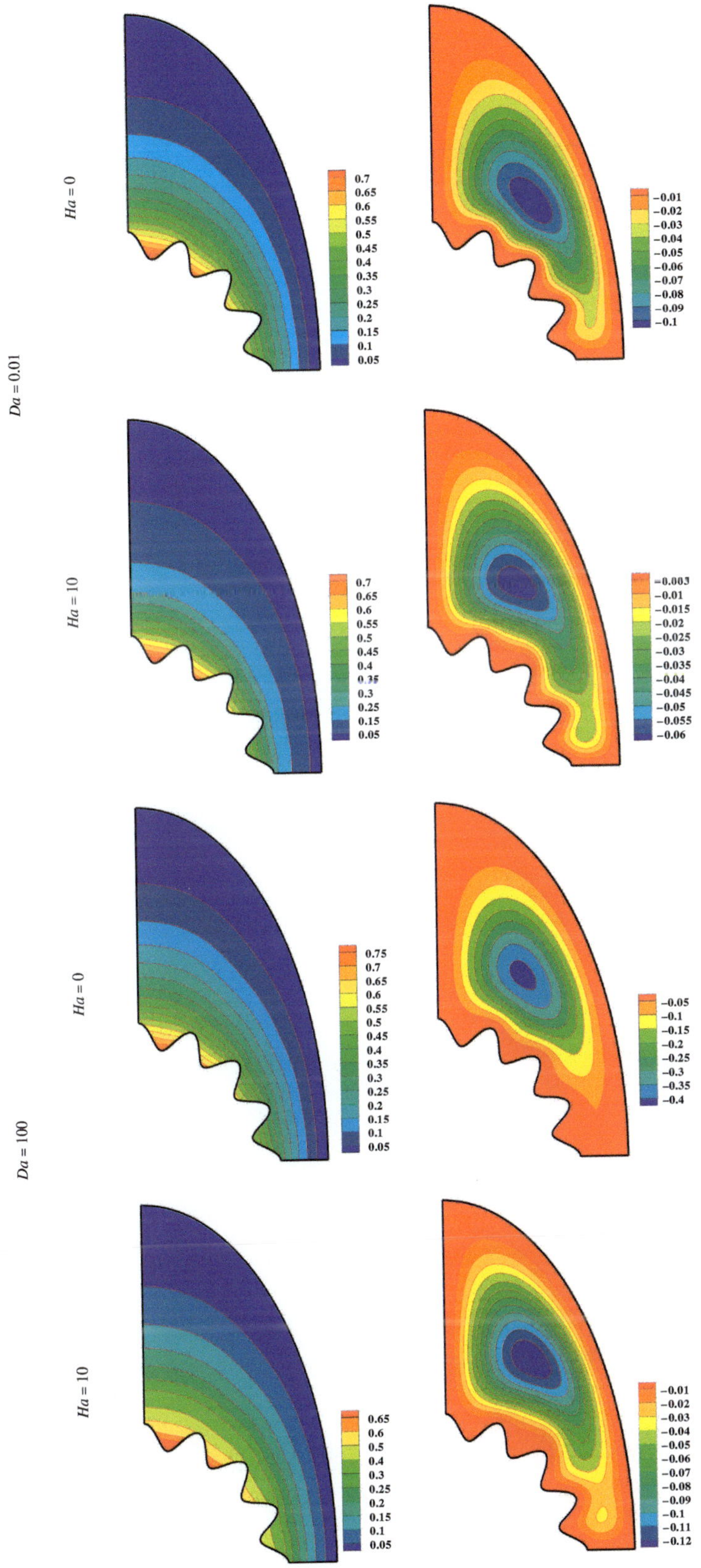

FIGURE 16.29 Influence of *Da*, *Ha* on streamline (right) and isotherm (left) contours when $\phi = 0.04$, $Ra = 10^3$.

Increasing the radiation parameter leads to an increase in the thermal boundary layer thickness. In the presence of Lorentz forces, the impact of radiation parameter on flow style becomes not sensible. Figs. 16.29–16.31 depict the influences of Da, Ra, Ha, Rd, on isotherms and streamlines. In conduction mode (low Reynolds number)

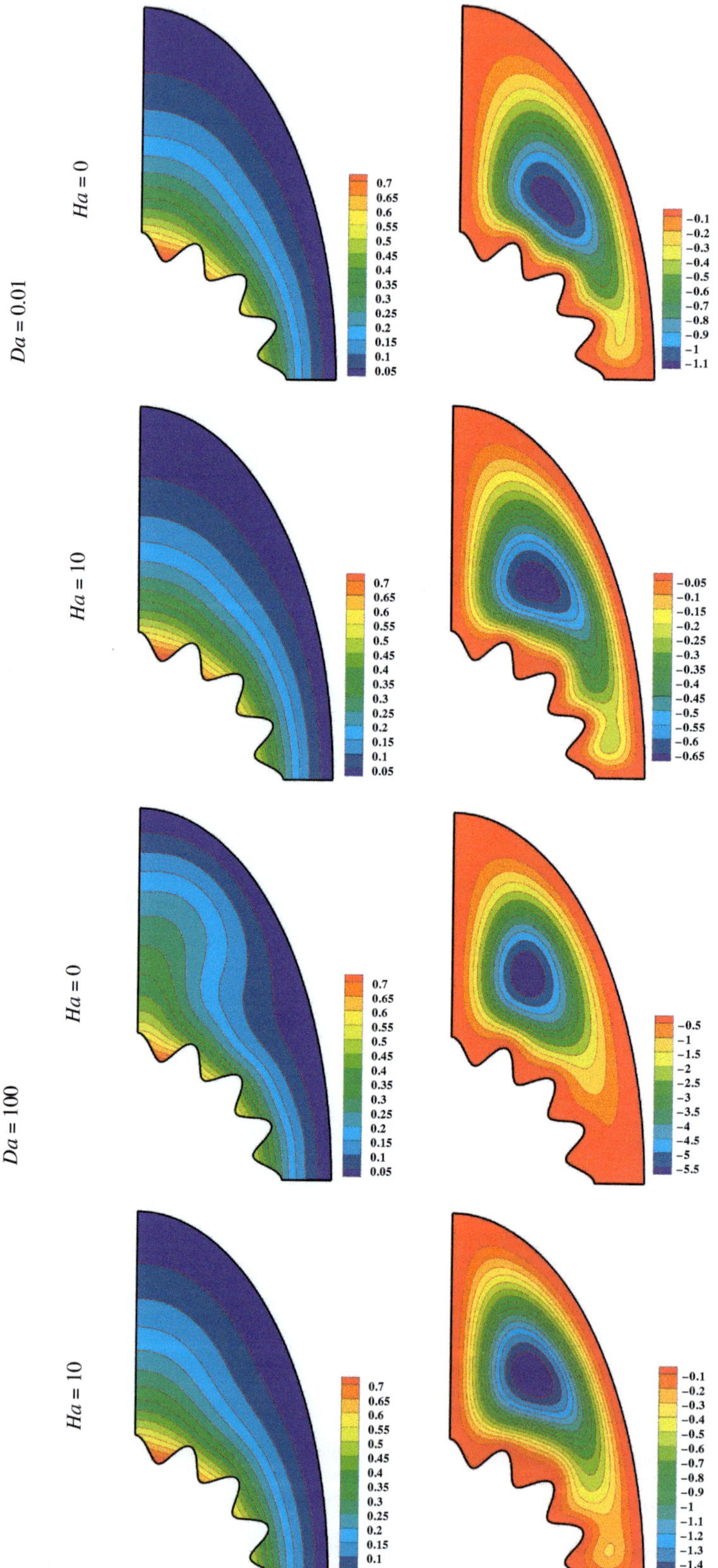

FIGURE 16.30 Influence of Da, Ha on streamline (right) and isotherm (left) contours when $\phi = 0.04$, $Ra = 10^4$.

streamlines show one rotating eddy. In the presence of a magnetic field, the main eddy moves downward. The distortion of the isotherms becomes less than before. As buoyancy force is enhanced, a thermal plume is generated. As the Darcy number is enhanced, the convective mode becomes stronger and the rate of heat transfer is enhanced.

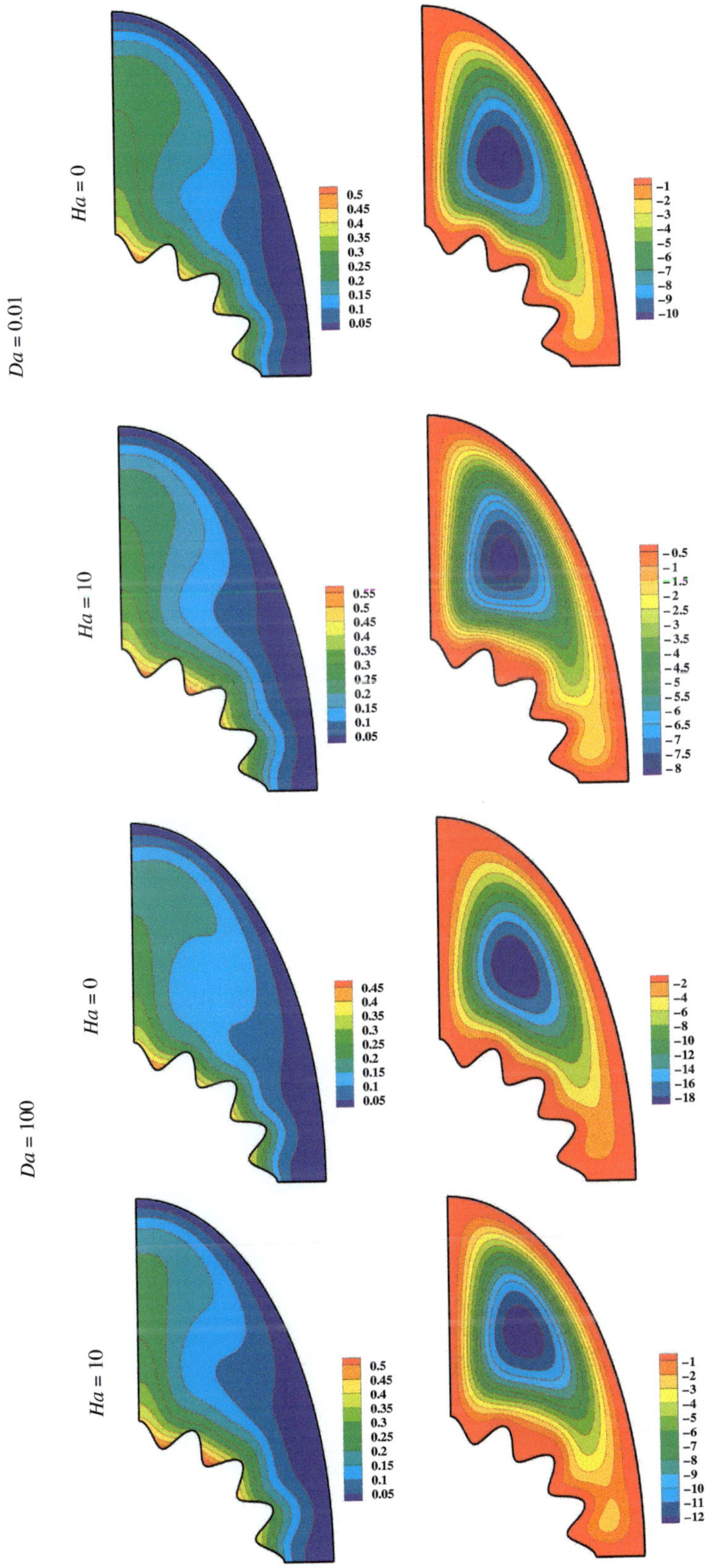

FIGURE 16.31 Influence of *Da*, *Ha* on streamline (right) and isotherm (left) contours when $\phi = 0.04$, $Ra = 10^5$.

The effect of significant parameters on Nu_{ave} is demonstrated in Fig. 16.32. The formula for Nu_{ave} is:

$$\begin{aligned} Nu_{ave} = {} & 15.28 - 1.12Rd - 6.6\log(Ra) - 0.18Da^{*} + 0.47Ha^{*} \\ & + 1.07Rd\log(Ra) + 0.32Rd\,Da^{*} - 0.44Rd\,Ha^{*} \\ & + 0.08Da^{*}\log(Ra) - 0.11Ha^{*}\log(Ra) - 0.3Da^{*}Ha^{*} \\ & - 0.9Rd^{2} + 0.85(\log(Ra))^{2} + 0.15(Da^{*})^{2} - 0.07(Ha^{*})^{2} \end{aligned} \tag{16.96}$$

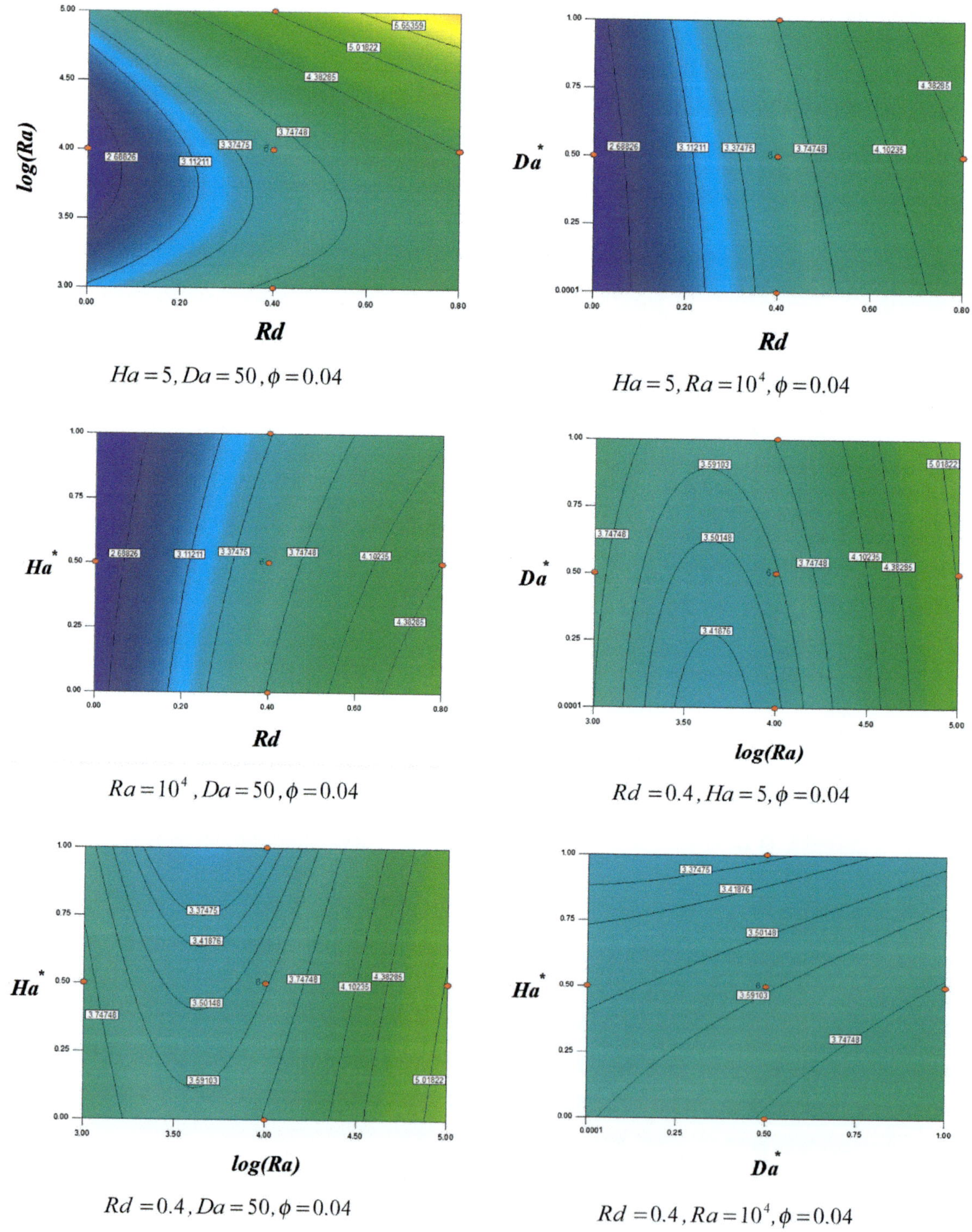

FIGURE 16.32 Effects of *Da*, *Ha*, and *Ra* on average Nusselt number.

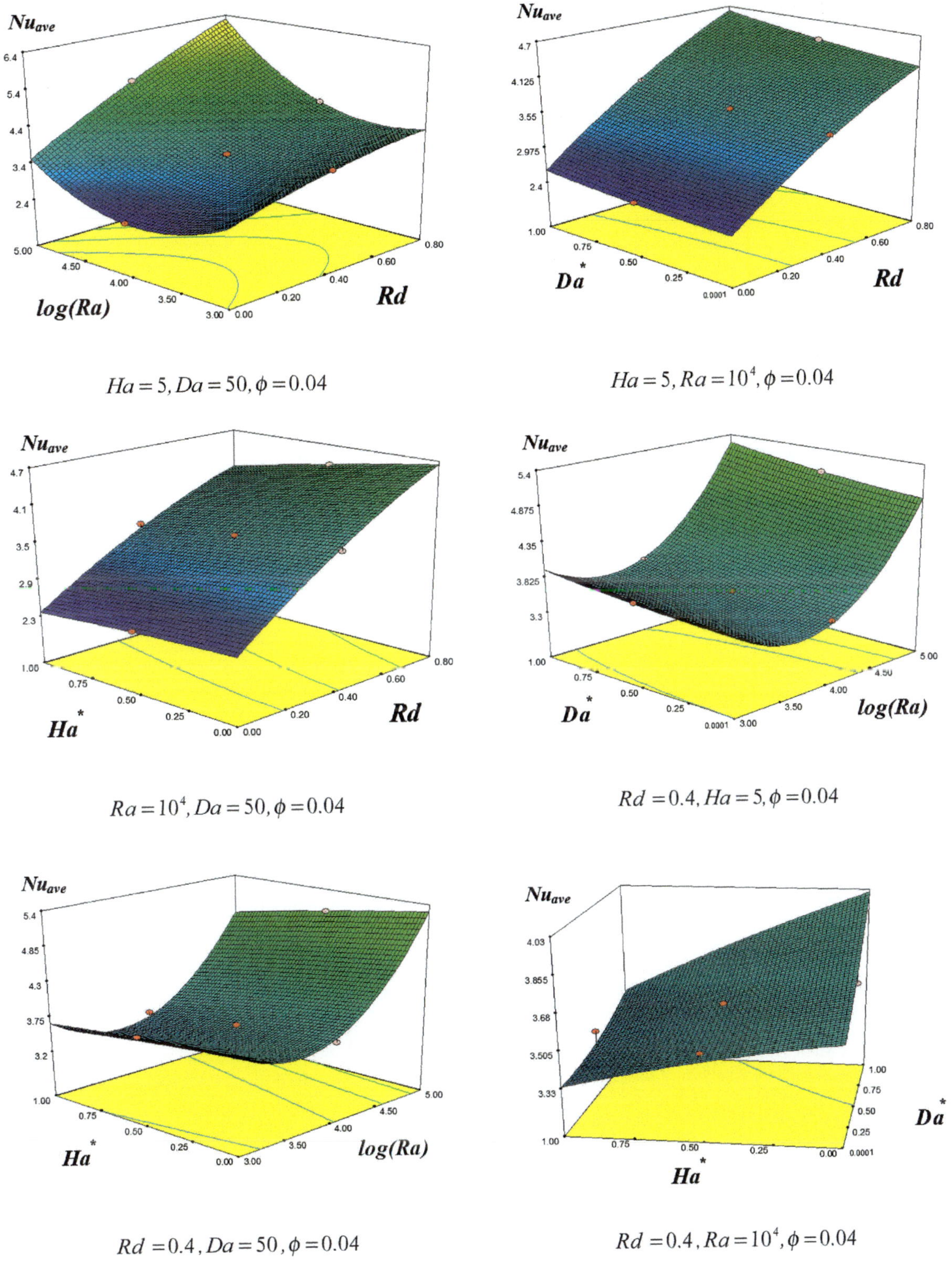

FIGURE 16.32 (*Continued*).

where $Da^* = 0.01Da$, $Ha^* = 0.1Ha$. Nusselt number is enhanced with the increase of buoyancy forces and the permeability of porous media. The temperature gradient reduces with the increase of the Hartmann number. Also, the Nusselt number is enhanced with the increase of radiation number.

References

[1] I. Pop, D.B. Ingham, Convective Heat Transfer: Mathematical and Computational Modelling of Viscous Fluids and Porous Media, Pergamon, Oxford, 2001.

[2] M. Sheikholeslami, CVFEM for magnetic nanofluid convective heat transfer in a porous curved enclosure, Eur. Phys. J. Plus 131 (2016) 413. Available from: https://doi.org/10.1140/epjp/i2016-16413-y.

[3] M. Sheikholeslami, Numerical simulation of magnetic nanofluid natural convection in porous media, Phys. Lett. A 381 (2017) 494–503.

[4] M. Sheikholeslami, D.D. Ganji, R. Moradi, Heat transfer of Fe3O4 -water nanofluid in a permeable medium with thermal radiation in existence of constant heat flux, Chem. Eng. Sci. 174 (2017) 326–336.

[5] M. Sheikholeslami, S.A. Shehzad, CVFEM for influence of external magnetic source on Fe3O4 – H2O nanofluid behavior in a permeable cavity considering shape effect, Int. J. Heat Mass Transfer 115 (2017) 180–191.

[6] M. Sheikholeslami, H.B. Rokni, Nanofluid convective heat transfer intensification in a porous circular cylinder, Chem. Eng. Proc. Process Intensif. 120 (2017) 93–104.

[7] M. Sheikholeslami, M.K. Sadoughi, Numerical modeling for Fe3O4 -water nanofluid flow in porous medium considering MFD viscosity, J. Mol. Liq. 242 (2017) 255–264.

[8] M. Sheikholeslami, M. Shamlooei, Magnetic source influence on nanofluid flow in porous medium considering shape factor effect, Phys. Lett. A 381 (2017) 3071–3078.

[9] M. Sheikholeslami, M. Shamlooei, Convective flow of nanofluid inside a lid driven porous cavity using CVFEM, Phys. B: Condensed Matter 521 (2017) 239–250.

[10] M. Sheikholeslami, K. Vajravelu, Forced convection heat transfer in Fe3O4- ethylene glycol nanofluid under the influence of Coulomb force, J. Mol. Liq. 233 (2017) 203–210.

[11] M. Sheikholeslami, H.B. Rokni, Influence of EFD viscosity on nanofluid forced convection in a cavity with sinusoidal wall, J. Mol. Liq. 232 (2017) 390–395.

[12] M. Sheikholeslami, H.B. Rokni, Magnetic nanofluid natural convection in the presence of thermal radiation considering variable viscosity, Europ. Phys. J. Plus 132 (2017) 238. Available from: https://doi.org/10.1140/epjp/i2017-11498-4.

[13] M. Sheikholeslami, H.B. Rokni, Numerical simulation for impact of Coulomb force on nanofluid heat transfer in a porous enclosure in presence of thermal radiation, Int.J. Heat Mass Transfer 118 (2018) 823–831.

[14] M. Sheikholeslami, M. Shamlooei, R. Moradi, Fe3O4- Ethylene glycol nanofluid forced convection inside a porous enclosure in existence of Coulomb force, J. Mol. Liq. 249 (2018) 429–437.

[15] M. Sheikholeslami, S.A. Shehzad, Numerical analysis of Fe_3O_4–H_2O nanofluid flow in permeable media under the effect of external magnetic source, Int. J. Heat Mass Transfer 118 (2018) 182–192.

[16] M. Sheikholeslami, S.A. Shehzad, Magnetohydrodynamic nanofluid convection in a porous enclosure considering heat flux boundary condition, Int. J. Heat Mass Transfer 106 (2017) 1261–1269.

[17] M. Sheikholeslami, S.A. Shehzad, Thermal radiation of ferrofluid in existence of Lorentz forces considering variable viscosity, Int. J. Heat Mass Transfer 109 (2017) 82–92.

[18] M. Sheikholeslami, H.B. Rokni, Numerical modeling of nanofluid natural convection in a semi annulus in existence of Lorentz force, Comput. Methods Appl. Mech. Eng. 317 (2017) 419–430.

[19] M. Sheikholeslami, K. Vajravelu, Nanofluid flow and heat transfer in a cavity with variable magnetic field, Appl. Math. Comput. 298 (2017) 272–282.

[20] M. Sheikholeslami, H.B. Rokni, Magnetohydrodynamic CuO-water nanofluid in a porous complex shaped enclosure, ASME, J. Thermal Sci. Eng. Applicat. 9 (4) (2017) 041007. Available from: https://doi.org/10.1115/1.4035973.

[21] M. Sheikholeslami, H.B. Rokni, Nanofluid two phase model analysis in existence of induced magnetic field, Int. J. Heat Mass Transfer 107 (2017) 288–299.

[22] M. Sheikholeslami, T. Hayat, A. Alsaedi, S. Abelman, Numerical analysis of EHD nanofluid force convective heat transfer considering electric field dependent viscosity, Int. J. Heat Mass Transfer 108 (2017) 2558–2565.

[23] M. Sheikholeslami, Magnetic field influence on CuO -H2O nanofluid convective flow in a permeable cavity considering various shapes for nanoparticles, Int. J. Hydrogen Energy 42 (2017) 19611–19621.

[24] M. Sheikholeslami, M.M. Bhatti, Forced convection of nanofluid in presence of constant magnetic field considering shape effects of nanoparticles, Int. J. Heat Mass Transfer 111 (2017) 1039–1049.

[25] M. Sheikholeslami, A. Zeeshan, Analysis of flow and heat transfer in water based nanofluid due to magnetic field in a porous enclosure with constant heat flux using CVFEM, Comput. Methods Appl. Mech. Eng. 320 (2017) 68–81.

[26] M. Sheikholeslami, M. Shamlooei, Fe3O4- H2O nanofluid natural convection in presence of thermal radiation, Int. J. Hydrogen Energy 42 (9) (2017) 5708–5718.

[27] M. Sheikholeslami, M.M. Bhatti, Active method for nanofluid heat transfer enhancement by means of EHD, Int. J. Heat Mass Transfer 109 (2017) 115–122.

[28] M. Sheikholeslami, T. Hayat, A. Alsaedi, Numerical study for external magnetic source influence on water based nanofluid convective heat transfer, Int. J. Heat Mass Transfer 106 (2017) 745–755.

[29] M. Sheikholeslami, A.J. Chamkha, Flow and convective heat transfer of a ferro-nanofluid in a double-sided lid-driven cavity with a wavy wall in the presence of a variable magnetic field, Numer. Heat Transfer, Part A 69 (10) (2016) 1186–1200. Available from: https://doi.org/10.1080/10407782.2015.1125709.

[30] M. Sheikholeslami, A.J. Chamkha, Electrohydrodynamic free convection heat transfer of a nanofluid in a semi-annulus enclosure with a sinusoidal wall, Numer. Heat Transfer, Part A 69 (7) (2016) 781–793. Available from: http://dx.doi.org/10.1080/10407782.2015.1090819.

[31] M. Sheikholeslami, M.K. Sadoughi, Simulation of CuO- water nanofluid heat transfer enhancement in presence of melting surface, Int. J. Heat Mass Transfer 116 (2018) 909–919.

[32] M. Sheikholeslami, H.B. Rokni, Simulation of nanofluid heat transfer in presence of magnetic field: a review, Int. J. Heat Mass Transfer 115 (2017) 1203–1233.

[33] M. Sheikholeslami, H.B. Rokni, Melting heat transfer influence on nanofluid flow inside a cavity in existence of magnetic field, Int. J. Heat Mass Transfer 114 (2017) 517–526.
[34] M. Sheikholeslami, Influence of Lorentz forces on nanofluid flow in a porous cylinder considering Darcy model, J. Mol. Liq. 225 (2017) 903–912.
[35] M. Sheikholeslami, Influence of Coulomb forces on Fe3O4-H2O nanofluid thermal improvement, Int. J. Hydrogen Energy 42 (2017) 821–829.
[36] M. Sheikholeslami, Magnetic field influence on nanofluid thermal radiation in a cavity with tilted elliptic inner cylinder, J. Mol. Liq. 229 (2017) 137–147.
[37] M. Sheikholeslami, Magnetic source impact on nanofluid heat transfer using CVFEM, Neur. Comput. Applicat. (2016). Available from: https://doi.org/10.1007/s00521-016-2740-7.
[38] M. Sheikholeslami, CuO-water nanofluid free convection in a porous cavity considering Darcy law, Europ. Phys. J. Plus 132 (2017) 55. Available from: https://doi.org/10.1140/epjp/i2017-11330-3.
[39] M. Sheikholeslami, Numerical investigation of MHD nanofluid free convective heat transfer in a porous tilted enclosure, Eng. Computat. 34 (6) (2017) 1939–1955.
[40] M.S. Kandelousi, Effect of spatially variable magnetic field on ferrofluid flow and heat transfer considering constant heat flux boundary condition, Europ. Phys. J. Plus (2014) 129–248. Available from: https://doi.org/10.1140/epjp/i2014-14248-2.
[41] M. Sheikholeslami, Numerical simulation for external magnetic field influence on Fe_3O_4-water nanofluid forced convection, Engineering Computations, In press.
[42] M. Sheikholeslami, Influence of Lorentz forces on nanofluid flow in a porous cavity by means of Non- Darcy model, Eng. Comput. 34 (8) (2017) 2651–2667. Available from: https://doi.org/10.1108/EC-01-2017-0008.
[43] M. Sheikholeslami, R. Ellahi, Three dimensional mesoscopic simulation of magnetic field effect on natural convection of nanofluid, Int. J. Heat Mass Transfer 89 (2015) 799–808.
[44] M. Sheikholeslami, S. Abelman, Two phase simulation of nanofluid flow and heat transfer in an annulus in the presence of an axial magnetic field, IEEE Trans. Nanotechnol. 14 (3) (2015) 561–569.
[45] M. Sheikholeslami, R. Ellahi, Simulation of ferrofluid flow for magnetic drug targeting using Lattice Boltzmann method, J. Zeitschrift Fur Naturforschung A 70 (2) (2015) 115–124.
[46] M. Sheikholeslami, M. Seyednezhad, Lattice Boltzmann Method simulation for CuO-water nanofluid flow in a porous enclosure with hot obstacle, J. Mol. Liq. 243 (2017) 249–256.
[47] M. Sheikholeslami, M. Sadoughi, Mesoscopic method for MHD nanofluid flow inside a porous cavity considering various shapes of nanoparticles, Int. J. Heat Mass Transfer 113 (2017) 106–114.
[48] M. Sheikholeslami, A. Zeeshan, Mesoscopic simulation of CuO-H2O nanofluid in a porous enclosure with elliptic heat source, Int. J. Hydrogen Energy 42 (22) (2017) 15393–15402.
[49] M. Sheikholeslami, M. Seyednezhad, Nanofluid heat transfer in a permeable enclosure in presence of variable magnetic field by means of CVFEM, Int. J. Heat Mass Transfer 114 (2017) 1169–1180.
[50] M. Sheikholeslami, H.B. Rokni, Free convection of CuO-H2O nanofluid in a curved porous enclosure using mesoscopic approach, Int. J. Hydrogen Energy 42 (22) (2017) 14942–14949.
[51] M. Sheikholeslami, Numerical investigation for CuO-H_2O nanofluid flow in a porous channel with magnetic field using mesoscopic method, J. Mol. Liq. 249 (2018) 739–774.
[52] M. Sheikholeslami, H.B. Rokni, Influence of melting surface on MHD nanofluid flow by means of two phase model, Chin. J. Phys. 55 (2017) 1352–1360.
[53] M. Sheikholeslami, H.B. Rokni, Effect of melting heat transfer on nanofluid flow in existence of magnetic field considering Buongiorno Model, Chin. J. Phys. 55 (2017) 1115–1126.
[54] M. Sheikholeslami, T. Hayat, A. Alsaedi, MHD free convection of Al2O3–water nanofluid considering thermal radiation: a numerical study, Int. J. Heat Mass Transfer 96 (2016) 513–524.
[55] M. Sheikholeslami, K. Vajravelu, M.M. Rashidi, Forced convection heat transfer in a semi annulus under the influence of a variable magnetic field, Int. J. Heat Mass Transfer 92 (2016) 339–348.
[56] M. Sheikholeslami, S.A. Shehzad, Magnetohydrodynamic nanofluid convective flow in a porous enclosure by means of LBM, Int. J. Heat Mass Transfer 113 (2017) 796–805.
[57] M. Sheikholeslami, Magnetohydrodynamic nanofluid forced convection in a porous lid driven cubic cavity using Lattice Boltzmann Method, J. Mol. Liq. 231 (2017) 555–565.
[58] M. Sheikholeslami Kandelousi, KKL correlation for simulation of nanofluid flow and heat transfer in a permeable channel, Phys. Lett. A 378 (45) (2014) 3331–3339.
[59] M. Sheikholeslami, Effect of uniform suction on nanofluid flow and heat transfer over a cylinder, J. Braz. Soc. Mech. Sci. Eng. 37 (2015) 1623–1633.
[60] M. Sheikholeslami, Lattice Boltzmann Method simulation of MHD non-Darcy nanofluid free convection, Physica B 516 (2017) 55–71.
[61] M. Sheikholeslami, Influence of magnetic field on nanofluid free convection in an open porous cavity by means of Lattice Boltzmann Method, J. Mol. Liq. 234 (2017) 364–374.
[62] M. Sheikholeslami, A.J. Chamkha, Influence of Lorentz forces on nanofluid forced convection considering Marangoni convection, J. Mol. Liq. 225 (2017) 750–757.
[63] M. Sheikholeslami, T. Hayat, A. Alsaedi, Numerical simulation of nanofluid forced convection heat transfer improvement in existence of magnetic field using Lattice Boltzmann Method, Int. J. Heat Mass Transfer 108 (2017) 1870–1883.
[64] M. Sheikholeslami, T. Hayat, A. Alsaedi, Numerical simulation for forced convection flow of MHD CuO-H_2O nanofluid inside a cavity by means of LBM, J. Mol. Liq. 249 (2018) 941–948.
[65] M. Sheikholeslami, H.R. Kataria, A.S. Mittal, Effect of thermal diffusion and heat-generation on MHD nanofluid flow past an oscillating vertical plate through porous medium, J. Mol. Liq. 257 (2018) 12–25.

[66] M. Sheikholeslami, M. Barzegar Gerdroodbary, S. Valiallah Mousavi, D.D. Ganji, R. Moradi, Heat transfer enhancement of ferroflui inside an 90° elbow channel by non-uniform magnetic field, J. Magn. Magn. Mater. 460 (2018) 302–311.
[67] M. Sheikholeslami, S. Soleimani, D.D. Ganji, Effect of electric field on hydrothermal behavior of nanofluid in a complex geometry, J. Mo Liq. 213 (2016) 153–161.
[68] M. Sheikholeslami, Numerical simulation for solidification in a LHTESS by means of nano-enhanced PCM, J. Taiwan Inst. Chem. Eng. 8 (2018) 25–41.
[69] M. Sheikholeslami, Numerical modeling of Nano enhanced PCM solidification in an enclosure with metallic fin, J. Mol. Liq. 259 (2018 424–438.
[70] M. Sheikholeslami, A. Ghasemi, Solidification heat transfer of nanofluid in existence of thermal radiation by means of FEM, Int. J. Hea Mass Transfer 123 (2018) 418–431.
[71] M. Sheikholeslami, S.A. Shehzad, CVFEM simulation for nanofluid migration in a porous medium using Darcy model, Int. J. Heat Mass Transfer 122 (2018) 1264–1271.
[72] M. Sheikholeslami, M. Darzi, M.K. Sadoughi, Heat transfer improvement and pressure drop during condensation of refrigerant-based nanofluid: an experimental procedure, Int. J. Heat Mass Transfer 122 (2018) 643–650.
[73] M. Sheikholeslami, H.B. Rokni, CVFEM for effect of Lorentz forces on nanofluid flow in a porous complex shaped enclosure by means of non-equilibrium model, J. Mol. Liq. 254 (2018) 446–462.
[74] M. Sheikholeslami, S.A. Shehzad, Z. Li, Water based nanofluid free convection heat transfer in a three dimensional porous cavity with hot sphere obstacle in existence of Lorenz forces, Int. J. Heat Mass Transfer 125 (2018) 375–386.
[75] M. Sheikholeslami, M. Jafaryar, D.D. Ganji, Z. Li, Exergy loss analysis for nanofluid forced convection heat transfer in a pipe with modified turbulators, J. Mol. Liq. 262 (2018) 104–110.
[76] M. Sheikholeslami, M. Jafaryar, Z. Li, Nanofluid turbulent convective flow in a circular duct with helical turbulators considering CuO nanoparticles, Int. J. Heat Mass Transfer 124 (2018) 980–989.
[77] M. Sheikholeslami, Z. Li, M. Shamlooei, Nanofluid MHD natural convection through a porous complex shaped cavity considering thermal radiation, Phys. Lett. A 382 (2018) 1615–1632.
[78] M. Sheikholeslami, H.B. Rokni, Magnetic nanofluid flow and convective heat transfer in a porous cavity considering Brownian motion effects, Phys. Fluids, Volume 30, Issue 1, https://doi.org/10.1063/1.5012517.
[79] M. Sheikholeslami, S.A. Shehzad, Simulation of water based nanofluid convective flow inside a porous enclosure via non-equilibrium model, Int. J. Heat Mass Transfer 120 (2018) 1200–1212.
[80] M. Sheikholeslami, S.A. Shehzad, Non-Darcy free convection of Fe_3O_4-water nanoliquid in a complex shaped enclosure under impact of uniform Lorentz force, Chin. J. Phys 56 (2018) 270–281.
[81] M. Sheikholeslami, M. Seyednezhad, Simulation of nanofluid flow and natural convection in a porous media under the influence of electric field using CVFEM, Int. J. Heat Mass Transfer 120 (2018) 772–781.
[82] M. Sheikholeslami, M. Shamlooei, R. Moradi, Numerical simulation for heat transfer intensification of nanofluid in a permeable curved enclosure considering shape effect of Fe_3O_4 nanoparticles, Chem.Eng. Process.: Process Intensif. 124 (2018) 71–82.
[83] M. Sheikholeslami, T. Hayat, T. Muhammad, A. Alsaedi, MHD forced convection flow of nanofluid in a porous cavity with hot elliptic obstacle by means of Lattice Boltzmann method, Int. J. Mech. Sci. 135 (2018) 532–540.
[84] M. Sheikholeslami, Numerical investigation of nanofluid free convection under the influence of electric field in a porous enclosure, J. of Mol. Liq. 249 (2018) 1212–1221.
[85] M. Sheikholeslami, CuO-water nanofluid flow due to magnetic field inside a porous media considering Brownian motion, J. Mol. Liq. 249 (2018) 921–929.
[86] M. Sheikholeslami, D.D. Ganji, Influence of electric field on Fe_3O_4-water nanofluid radiative and convective heat transfer in a permeable enclosure, J. Mol. Liq. 250 (2018) 404–412.
[87] M. Sheikholeslami, S.A. Shehzad, Z. Li, Nanofluid heat transfer intensification in a permeable channel due to magnetic field using Lattice Boltzmann method, Physica B (2018). Available from: https://doi.org/10.1016/j.physb.2018.03.036.
[88] M. Sheikholeslami, A. Zeeshan, Numerical simulation of Fe_3O_4-water nanofluid flow in a non-Darcy porous media, Int. J. Num. Methods Heat Fluid Flow Vol. 28 (Issue: 3) (2018) 641–660. Available from: https://doi.org/10.1108/HFF-04-2017-0160.
[89] M. Sheikholeslami, D.D. Ganji, Numerical approach for magnetic nanofluid flow in a porous cavity using CuO nanoparticles, Mater. Des. 120 (2017) 382–393.
[90] M. Sheikholeslami, D.D. Ganji, Numerical modeling of magnetohydrodynamic CuO-water transportation inside a porous cavity considering shape factor effect, Colloids Surf. A 529 (2017) 705–714.

C H A P T E R

17

Thermal Radiation Influence on Nanofluid Flow in a Porous Medium in the Presence of Coulomb Forces Using CVFEM

17.1 INTRODUCTION

One of the effective active techniques for heat transfer augmentation is Electrohydrodynamics. Nanofluids are widely used in coolants for computers and nuclear reactors, cancer therapy, safer surgery by cooling, lubricants, heat exchangers, microchannel heat sinks; cooling of a new class of super powerful and small computers, and other electronic devices for use in military systems; vehicle cooling, and transformer cooling; in designing the waste heat removal equipment; major process industries including materials and chemicals, oil and gas, food and drink, paper and printing etc. The thermal radiation effects are of vital importance at high absolute temperature due to basic differences between radiation and convection and conduction energy-exchange mechanisms. For space applications, some devices are designed to operate at high temperature levels in order to obtain high thermal efficiency. That is why the radiation effects are significant while determining thermal effects in the processes with high temperatures. Akber et al. [1] investigated the effect of thermal radiation on MHD convective flow of nanofluid past a shrinking surface. Natural convection and radiation heat transfer has been investigated [2–90].

17.2 COMBINED NATURAL CONVECTION AND RADIATION HEAT TRANSFER OF NANOFLUID UNDER THE IMPACT OF ELECTRIC FIELD IN A POROUS CAVITY

17.2.1 Problem Definition

Fig. 17.1 depicts the porous enclosure and its boundary conditions. Ethylene glycol-Fe_3O_4 nanofluid is utilized. All walls are stationary except for the bottom wall. The influence of Darcy and Reynolds numbers on the contour

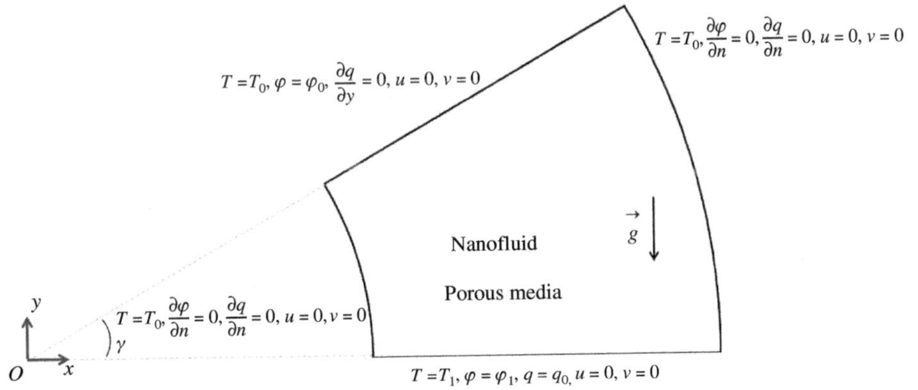

FIGURE 17.1 Geometry and the boundary conditions.

Application of Control Volume based Finite Element Method (CVFEM) for Nanofluid Flow and Heat Transfer.
DOI: https://doi.org/10.1016/B978-0-12-814152-6.00017-5

of q is demonstrated in Fig. 17.2. The effect of Re on q is less sensible than Da. As the Darcy number augments the distortion of the isoelectric density lines become greater.

17.2.2 Governing Equation

The definition of the electric field is:

$$\vec{E} = -\nabla\varphi \tag{17.1}$$

$$q = \nabla.\varepsilon\,\vec{E} \tag{17.2}$$

$$\vec{J} = q\vec{V} - D\nabla q + \sigma\vec{E} \tag{17.3}$$

$$\nabla.\vec{J} + \frac{\partial q}{\partial t} = 0 \tag{17.4}$$

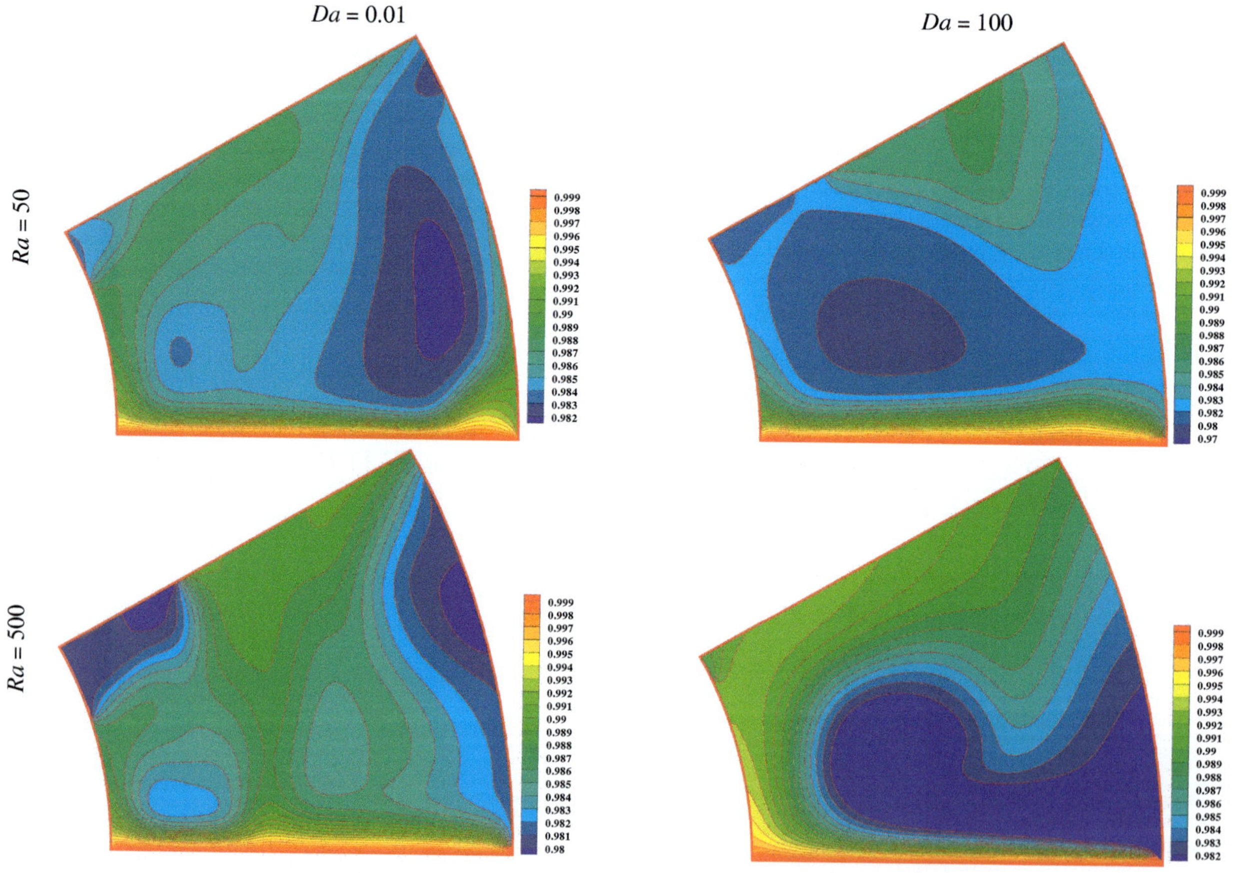

FIGURE 17.2 Electric density distribution injected by the bottom electrode when $\Delta\varphi = 6kV, \phi = 0.05, Rd = 0.8$.

The governing formulae are:

$$\begin{cases} \nabla.\vec{V}=0, \\ \rho_{nf}\left(\dfrac{\partial \vec{V}}{\partial t}+\left(\vec{V}.\nabla\right)\vec{V}\right)=-\nabla p+\mu_{nf}\nabla^2\vec{V}+q\vec{E}-(\rho\beta)_{nf}\vec{g}(T-T_0)-\dfrac{\mu_{nf}}{K}\vec{V}, \\ \left(\left(\vec{V}.\nabla\right)T+\dfrac{\partial T}{\partial t}\right)=\dfrac{k_{nf}}{(\rho C_p)_{nf}}\nabla^2 T+\dfrac{\vec{J}.\vec{E}}{(\rho C_p)_{nf}}-\dfrac{1}{(\rho C_p)_{nf}}\dfrac{\partial q_r}{\partial y}, \quad \left[q_r=-\dfrac{4\sigma_e}{3\beta_R}\dfrac{\partial T^4}{\partial y}, \quad T^4\cong 4T_c^3T-3T_c^4\right], \\ \nabla\varphi=-\vec{E}, \\ \dfrac{\partial q}{\partial t}=-\nabla.\vec{J}, \\ q=\nabla.\varepsilon\vec{E} \end{cases} \tag{17.5}$$

$(\rho C_p)_{nf}, (\beta)_{nf}, \mu_{nf}$, and ρ_{nf} can be obtained as:

$$\begin{aligned} &(\rho C_p)_{nf}=(\rho C_p)_f(1-\phi)+(\rho C_p)_s\phi, \quad (\beta)_{nf}=(\beta)_f(1-\phi)+(\beta)_s\phi, \\ &\mu_{nf}=A_1+A_2(\Delta\varphi)+A_3(\Delta\varphi)^2+A_4(\Delta\varphi)^3, \\ &\rho_{nf}=\rho_f(1-\phi)+\rho_s\phi \end{aligned} \tag{17.6}$$

Properties of Fe_3O_4 and ethylene glycol are illustrated in Table 17.1. Table 17.2 illustrates the coefficient values of this formula.

k_{nf} can be expressed as:

$$\frac{k_{nf}}{k_f}=\frac{-m(k_f-k_p)\phi+(k_p-k_f)\phi+mk_f+k_p+k_f}{mk_f+(k_f-k_p)\phi+k_f+k_p} \tag{17.7}$$

Different values of shape factors for various shapes of nanoparticles are illustrated in Table 17.3.

TABLE 17.1 Thermophysical Properties of Water and Nanoparticles

	$\rho(kg/m^3)$	$C_p(j/kgk)$	$\beta\times10^{-5}(1/k)$	$k(W/m.k)$
Ethylene glycol	1110	2400	65	0.26
Fe_3O_4	5200	670	1.3	6

TABLE 17.2 The Coefficient Values of Eq. (17.6)

Coefficient values	$\phi=0$	$\phi=0.05$
A_1	1.0603E + 001	9.5331
A_2	−2.698E-003	−3.4119E-003
A_3	2.9082E-006	5.5228E-006
A_4	−1.1876E-008	−4.1344E-008

TABLE 17.3 The Values of Shape Factor of Different Shapes of Nanoparticles

m

Spherical	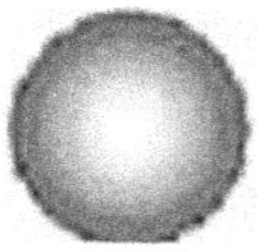	3
Platelet		5.7
Cylinder		4.8
Brick	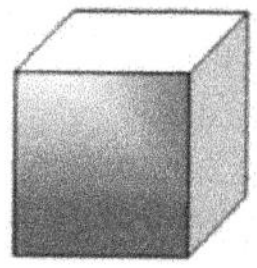	3.7

So, the final PDE in the presence of thermal radiation and an electric field in a porous medium are:

$$\begin{cases} \nabla.\vec{V}=0, \\ \left(\frac{\partial \vec{V}}{\partial t}+\left(\vec{V}.\nabla\right)\vec{V}\right)=-\nabla p+\frac{\mu_{nf}/\mu_f}{\rho_{nf}/\rho_f}\Pr\nabla^2\vec{V}+\frac{S_E}{\rho_{nf}/\rho_f}q\vec{E}-Ra\Pr\left[\frac{\beta_{nf}}{\beta_f}\right]\theta-\frac{\Pr}{Da}\frac{\mu_{nf}/\mu_f}{\rho_{nf}/\rho_f}\vec{V} \\ \left(\left(\vec{V}.\nabla\right)\theta+\frac{\partial\theta}{\partial t}\right)=\frac{k_{nf}/k_f}{(\rho C_p)_{nf}/(\rho C_p)_f}\nabla^2\theta+\frac{S_E Ec}{(\rho C_p)_{nf}/(\rho C_p)_f}\left(\vec{J}.\vec{E}\right)+\frac{4}{3}\frac{k_{nf}/k_f}{(\rho C_p)_{nf}/(\rho C_p)_f}\left(\frac{k_{nf}}{k_f}\right)^{-1}Rd\frac{\partial^2\theta}{\partial Y^2} \\ \vec{E}=-\nabla\varphi \\ q=\nabla.\varepsilon\vec{E} \\ \nabla.\vec{J}=-\frac{\partial q}{\partial t} \end{cases} \tag{17.8}$$

where

$$\bar{t}=\frac{t\alpha_f}{L^2},\bar{p}=\frac{P}{\rho(\alpha_f/L)^2},\bar{y}=\frac{y}{L},\bar{x}=\frac{x}{L},\bar{v}=\frac{v\,L}{\alpha_f},\bar{u}=\frac{u\,L}{\alpha_f},$$
$$\theta=\frac{T-T_0}{\nabla T},\nabla T=T_1-T_0,\bar{\varphi}=\frac{\varphi-\varphi_0}{\nabla\varphi},\nabla\varphi=\varphi_1-\varphi_0,\bar{q}=\frac{q}{q_0},\bar{E}=\frac{E}{E_0} \tag{17.9}$$

Vorticity and stream function should be employed in order to diminish the pressure gradient:

$$v=-\frac{\partial\psi}{\partial x},\frac{\partial\psi}{\partial y}=u,\Psi=\frac{\psi}{\alpha_f},\Omega=\frac{\omega L^2}{\alpha_f},\omega=\frac{\partial v}{\partial x}-\frac{\partial u}{\partial y} \tag{17.10}$$

Nu_{loc} and Nu_{ave} along the bottom wall are calculated as:

$$Nu_{loc}=\left(\frac{k_{nf}}{k_f}\right)\left(1+\frac{4}{3}Rd\left(\frac{k_{nf}}{k_f}\right)^{-1}\right)\frac{\partial\theta}{\partial X} \tag{17.11}$$

$$Nu_{ave}=\frac{1}{L}\int_0^L Nu_{loc}\,dY \tag{17.12}$$

17.2.3 Effects of Active Parameters

The effect of electric field on nanofluid free convection heat transfer is investigated in the presence of thermal radiation. Electric field dependent viscosity is considered for nanofluid. The porous enclosure is filled with Fe_3O_4-Ethylene glycol and has one lid wall. Roles of Darcy number ($Da = 10^{-2}$ to 10^2), Radiation parameter ($Rd = 0$ to 0.8), supplied voltage ($\Delta\varphi = 0$ to $6kV$), volume fraction of Fe_3O_4 ($\phi = 0\%$ to 5%), and Rayleigh number ($Ra = 50$ to 500) are illustrated graphically. The influence of shape factor on rate of heat transfer is reported in Table 17.4. In this table, various shapes of nanoparticles are utilized. The maximum Nu is caused by Platelet, followed by Cylinder, Brick, and Spherical. So, Platelet nanoparticle has been utilized for further investigation. Impacts of $Da, \Delta\varphi$, and Ra on isotherms and streamlines are demonstrated in Figs. 17.3–17.6. At low Rayleigh number, there are two eddies which rotate in opposite direction. In the presence of an electric field, the eddies are stretched and a thermal plume is generated. Isotherms become more disturbed in the presence of an electric field. As buoyancy

TABLE 17.4 Effect of Shape of Nanoparticles on Nusselt Number When $Rd = 0.8, Ra = 500, \Delta\varphi = 6, \phi = 0.05$

	Da	
	10^{-2}	10^2
Spherical	18.15263	21.44078
Brick	18.94481	21.67239
Cylinder	13.05472	22.01274
Platelet	19.70303	22.27031

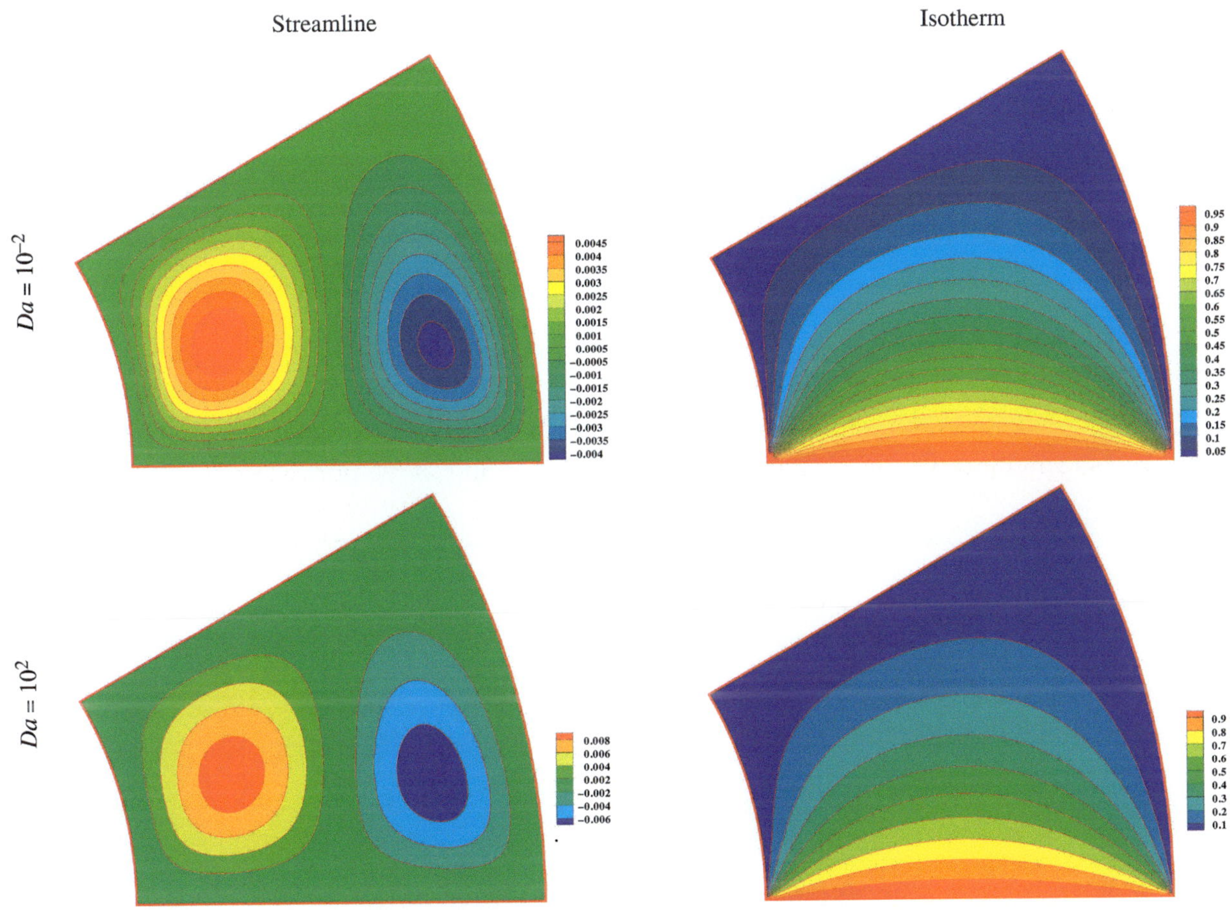

FIGURE 17.3 Effect of Darcy number on streamlines and isotherms when $Ra = 50, \Delta\varphi = 0kV, \phi = 0.05, Rd = 0.8$.

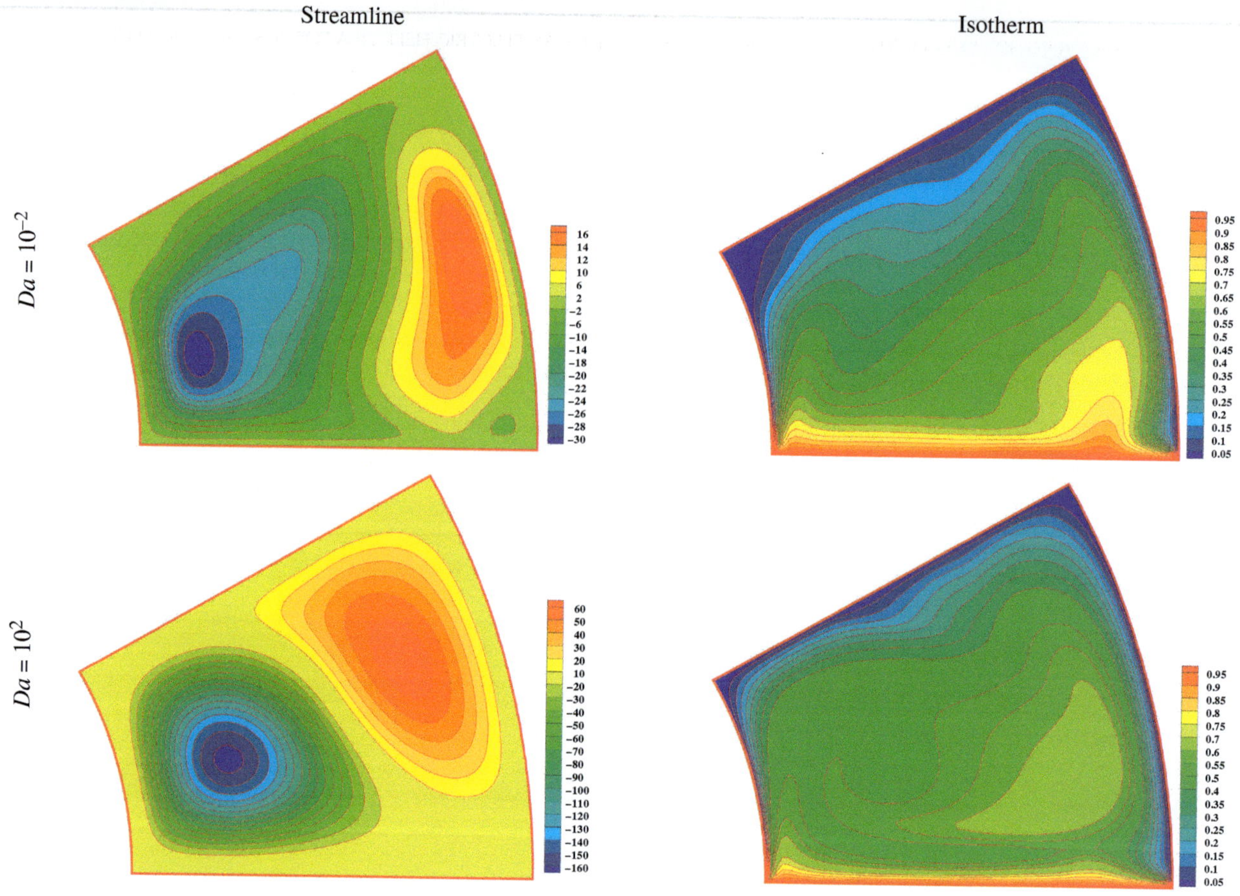

FIGURE 17.4 Effect of Darcy number on streamlines and isotherms when $Ra = 50, \Delta\varphi = 6kV, \phi = 0.05, Rd = 0.8$.

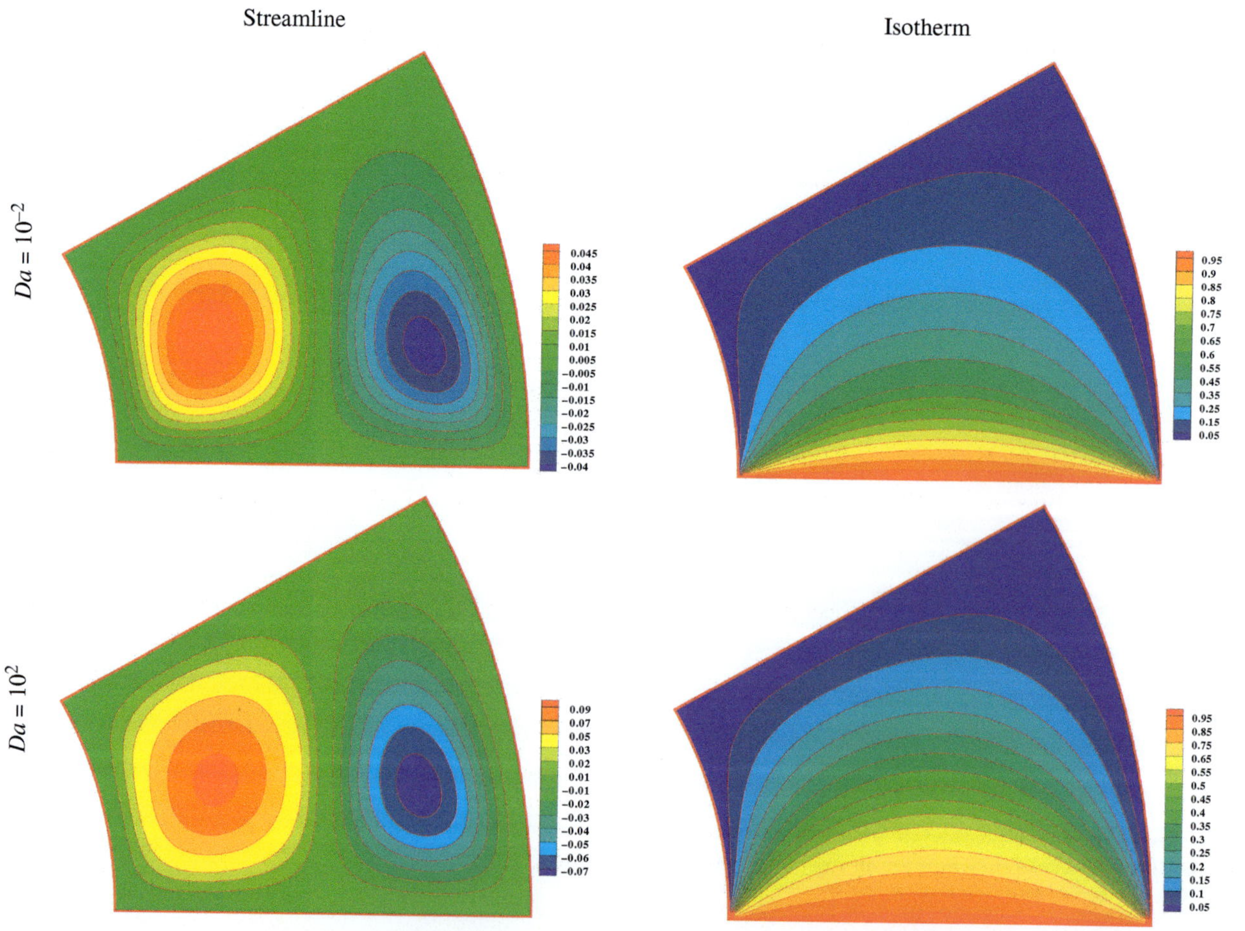

FIGURE 17.5 Effect of Darcy number on streamlines and isotherms when $Ra = 500, \Delta\varphi = 0kV, \phi = 0.05, Rd = 0.8$.

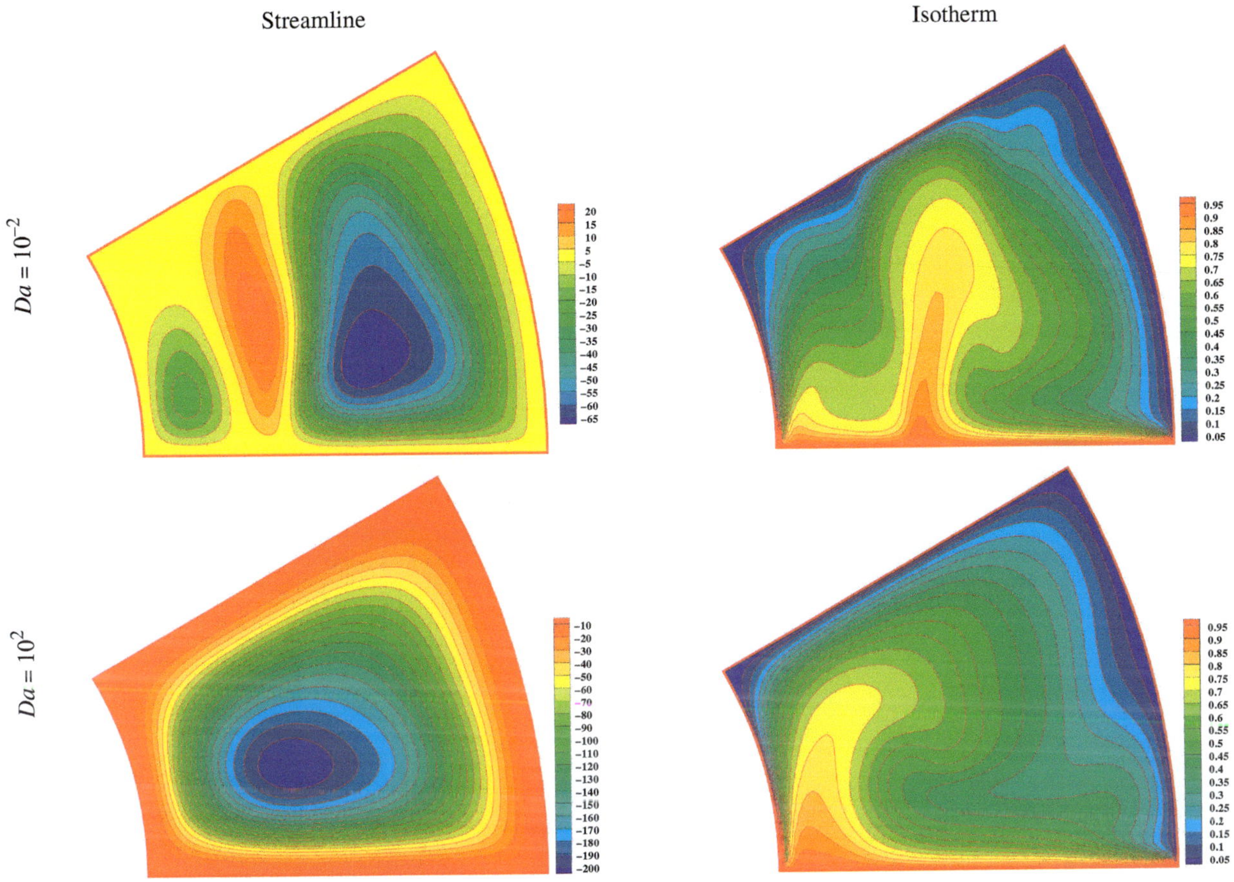

FIGURE 17.6 Effect of Darcy number on streamlines and isotherms when $Ra = 500, \Delta\varphi = 6kV, \phi = 0.05, Rd = 0.8$.

forces increase, another eddy is generated. At high values of the Rayleigh number, the increasing Darcy number leads to the conversion of all eddies to one clockwise eddy and the thermal plume shifts to the left side. Nu_{ave} versus $Ra, Da, Rd,$ and $\Delta\varphi$ is depicted in Fig. 17.7. The related formula is:

$$\begin{aligned} Nu_{ave} &= 4.07 - 0.53\Delta\varphi + 2.39Ra^* + 0.35Da^* + 0.92Rd \\ &+ 0.06\Delta\varphi Ra^* + 0.49\Delta\varphi Da^* \\ &- 0.98\Delta\varphi Rd - 0.61Ra^*Da^* \\ &+ 0.41Ra^*Rd + 3.2Da^*Rd \\ &+ 0.17\Delta\varphi^2 - 0.39(Ra^*)^2 + 0.82(Da^*)^2 + 2.5Rd^2 \end{aligned} \tag{17.13}$$

where $Ra^* = 0.01Ra, Da^* = 0.01Da,$ and $\Delta\varphi$ is voltage supply in Kilovolt. Rate of heat transfer is enhanced with the rise of Rayleigh number. The electric field helps the convective mode to enhance. So, Nu_{ave} augments with the increase of $\Delta\varphi$. Thermal radiation enhances the temperature gradient near the lid wall. The influence of Darcy number is the same as the radiation parameter. Therefore, Nu_{ave} is an increasing function of Rd, Da.

17.3 NANOFLUID FREE CONVECTION UNDER THE INFLUENCE OF AN ELECTRIC FIELD IN A POROUS WAVY ENCLOSURE

17.3.1 Problem Definition

Fig. 17.8 depicts the porous enclosure and its boundary conditions. Ethylene glycol-Fe_3O_4 nanofluid is utilized. All walls are stationary except for the bottom wall. The influence of the Darcy and Reynolds numbers on the contour of q is demonstrated in Fig. 17.9. The effect of Re on q is less sensible than Da. As the Darcy number increases the distortion of the isoelectric density lines become greater.

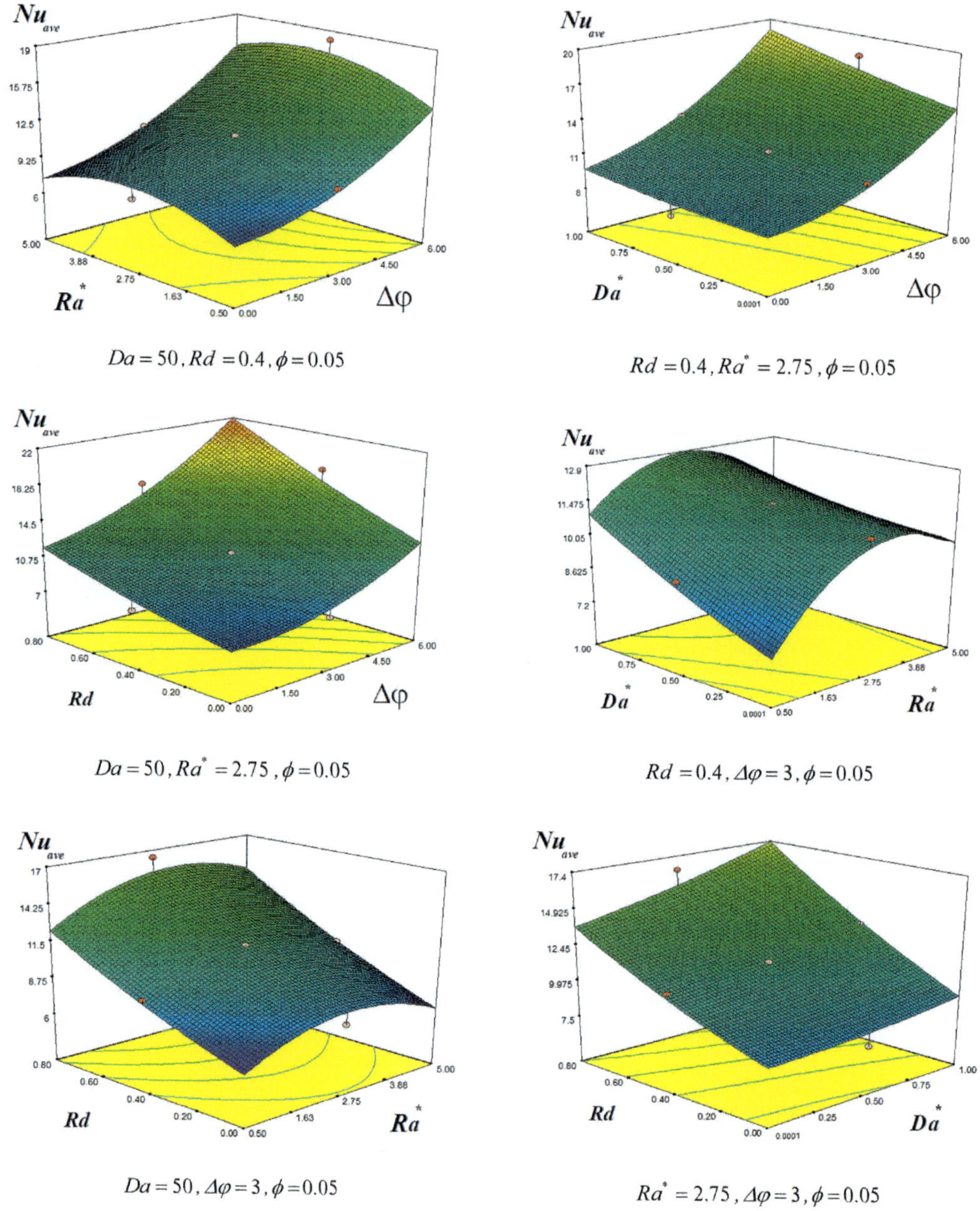

FIGURE 17.7 Effects of Da, $\Delta\varphi$, Rd, and Ra on average Nusselt number.

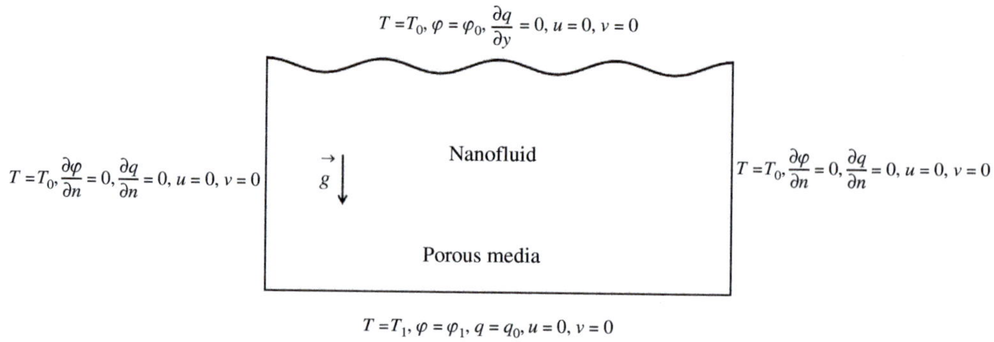

FIGURE 17.8 Geometry and the boundary conditions.

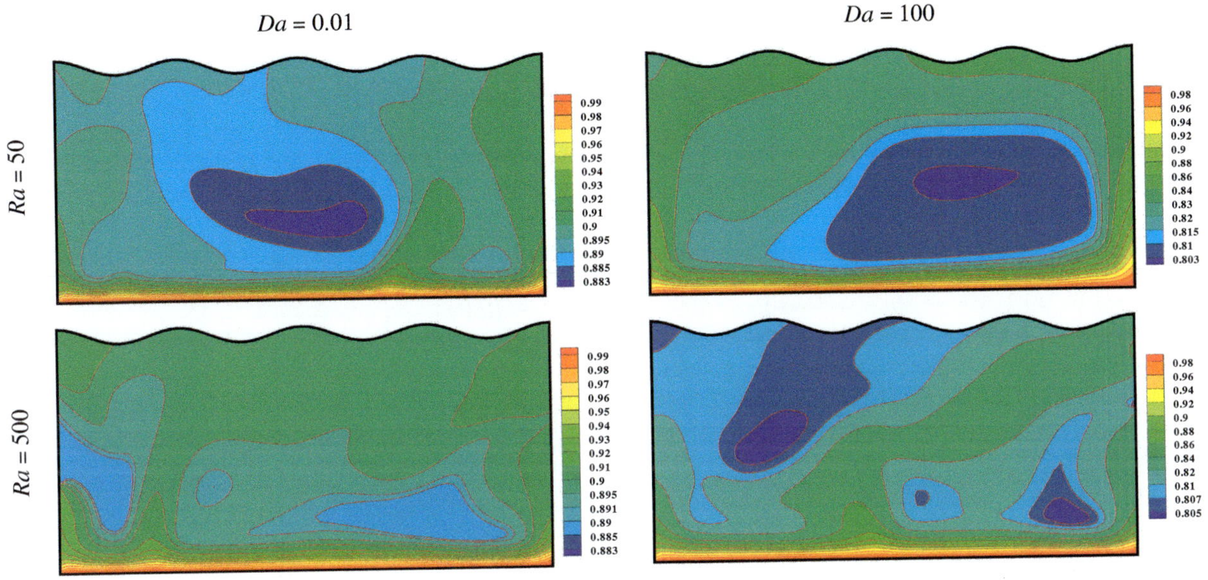

FIGURE 17.9 Electric density distribution injected by the bottom electrode when $\Delta\varphi = 6kV, \phi = 0.05, Rd = 0.8$.

17.3.2 Governing Equation

The definition of the electric field is:

$$\vec{E} = -\nabla\varphi \tag{17.14}$$

$$q = \nabla.\varepsilon\,\vec{E} \tag{17.15}$$

$$\vec{J} = q\vec{V} - D\nabla q + \sigma\,\vec{E} \tag{17.16}$$

$$\nabla.\vec{J} + \frac{\partial q}{\partial t} = 0 \tag{17.17}$$

The governing formulae are:

$$\begin{cases} \nabla.\vec{V} = 0, \\ \rho_{nf}\left(\dfrac{\partial\vec{V}}{\partial t} + \left(\vec{V}.\nabla\right)\vec{V}\right) = -\nabla p + \mu_{nf}\nabla^2\vec{V} + q\,\vec{E} - (\rho\beta)_{nf}\vec{g}(T - T_0) - \dfrac{\mu_{nf}}{K}\vec{V}, \\ \left(\left(\vec{V}.\nabla\right)T + \dfrac{\partial T}{\partial t}\right) = \dfrac{k_{nf}}{(\rho C_p)_{nf}}\nabla^2 T + \dfrac{\vec{J}.\vec{E}}{(\rho C_p)_{nf}} - \dfrac{1}{(\rho C_p)_{nf}}\dfrac{\partial q_r}{\partial y}, \quad \left[q_r = -\dfrac{4\sigma_e}{3\beta_R}\dfrac{\partial T^4}{\partial y}, \quad T^4 \cong 4T_c^3 T - 3T_c^4\right], \\ \nabla\varphi = -\vec{E}, \\ \dfrac{\partial q}{\partial t} = -\nabla.\vec{J}, \\ q = \nabla.\varepsilon\vec{E} \end{cases} \tag{17.18}$$

$(\rho C_p)_{nf}, (\beta)_{nf}, \mu_{nf}$, and ρ_{nf} can be obtained as:

$$\begin{aligned} &(\rho C_p)_{nf} = (\rho C_p)_f(1-\phi) + (\rho C_p)_s\phi, \quad (\beta)_{nf} = (\beta)_f(1-\phi) + (\beta)_s\phi, \\ &\mu_{nf} = A_1 + A_2(\Delta\varphi) + A_3(\Delta\varphi)^2 + A_4(\Delta\varphi)^3, \\ &\rho_{nf} = \rho_f(1-\phi) + \rho_s\phi \end{aligned} \tag{17.19}$$

Properties of Fe_3O_4 and ethylene glycol are illustrated in Table 17.1. Table 17.2 illustrates the coefficient values of this formula.

k_{nf} can be expressed as:

$$\frac{k_{nf}}{k_f} = \frac{-m(k_f - k_p)\phi + (k_p - k_f)\phi + mk_f + k_p + k_f}{mk_f + (k_f - k_p)\phi + k_f + k_p} \tag{17.20}$$

Different values of shape factors for various shapes of nanoparticles are illustrated in Table 17.3. So, the final PDE in the presence of thermal radiation and an electric field in a porous medium are:

$$\begin{cases} \nabla.\vec{V} = 0, \\ \left(\frac{\partial \vec{V}}{\partial t} + \left(\vec{V}.\nabla\right)\vec{V}\right) = -\nabla p + \frac{\mu_{nf}/\mu_f}{\rho_{nf}/\rho_f}\mathrm{Pr}\nabla^2\vec{V} + \frac{S_E}{\rho_{nf}/\rho_f} q\,\vec{E} - Ra\mathrm{Pr}\left[\frac{\beta_{nf}}{\beta_f}\right]\theta - \frac{\mathrm{Pr}}{Da}\frac{\mu_{nf}/\mu_f}{\rho_{nf}/\rho_f}\vec{V} \\ \left(\left(\vec{V}.\nabla\right)\theta + \frac{\partial\theta}{\partial t}\right) = \frac{k_{nf}/k_f}{(\rho C_p)_{nf}/(\rho C_p)_f}\nabla^2\theta + \frac{S_E Ec}{(\rho C_p)_{nf}/(\rho C_p)_f}\left(\vec{J}.\vec{E}\right) + \frac{4}{3}\frac{k_{nf}/k_f}{(\rho C_p)_{nf}/(\rho C_p)_f}\left(\frac{k_{nf}}{k_f}\right)^{-1} Rd\frac{\partial^2\theta}{\partial Y^2} \\ \vec{E} = -\nabla\varphi \\ q = \nabla.\varepsilon\vec{E} \\ \nabla.\vec{J} = -\frac{\partial q}{\partial t} \end{cases} \tag{17.21}$$

where

$$\begin{aligned} &\overline{t} = \frac{t\alpha_f}{L^2}, \overline{p} = \frac{P}{\rho(\alpha_f/L)^2}, \overline{y} = \frac{y}{L}, \overline{x} = \frac{x}{L}, \overline{v} = \frac{v\,L}{\alpha_f}, \overline{u} = \frac{u\,L}{\alpha_f}, \\ &\theta = \frac{T - T_0}{\nabla T}, \nabla T = T_1 - T_0, \overline{\varphi} = \frac{\varphi - \varphi_0}{\nabla\varphi}, \nabla\varphi = \varphi_1 - \varphi_0, \overline{q} = \frac{q}{q_0}, \overline{E} = \frac{E}{E_0} \end{aligned} \tag{17.22}$$

Vorticity and stream function should be employed in order to diminish the pressure gradient:

$$v = -\frac{\partial\psi}{\partial x}, \frac{\partial\psi}{\partial y} = u, \Psi = \frac{\psi}{\alpha_f}, \Omega = \frac{\omega L^2}{\alpha_f}, \omega = \frac{\partial v}{\partial x} - \frac{\partial u}{\partial y} \tag{17.23}$$

Nu_{loc} and Nu_{ave} along the bottom wall are calculated as:

$$Nu_{loc} = \left(\frac{k_{nf}}{k_f}\right)\left(1 + \frac{4}{3}Rd\left(\frac{k_{nf}}{k_f}\right)^{-1}\right)\frac{\partial\theta}{\partial X} \tag{17.24}$$

$$Nu_{ave} = \frac{1}{L}\int_0^L Nu_{loc}\,dY \tag{17.25}$$

17.3.3 Effects of Active Parameters

Electrohydrodynamic nanofluid free convection heat transfer through the porous enclosure is investigated in the presence of thermal radiation. Electric field dependent viscosity is considered for nanofluid. The porous enclosure is filled with Fe_3O_4-Ethylene glycol and has one lid wall. The roles of Darcy number ($Da = 10^{-2}$ to 10^2), Radiation parameter ($Rd = 0$ to 0.8), supplied voltage ($\Delta\varphi = 0$ to $6kV$), volume fraction of Fe_3O_4 ($\phi = 0\%$ to 5%), and Rayleigh number ($Ra = 50$ to 500) are illustrated graphically.

The influence of shape factor on the rate of heat transfer is reported in Table 17.5. In this table, various shapes of nanoparticles are utilized. The maximum Nu is caused by Platelet, followed by Cylinder, Brick, and Spherical. So, Platelet nanoparticle has been utilized for further investigation.

Impacts of $Da, \Delta\varphi$, and Ra on isotherms and streamlines are demonstrated in Figs. 17.10–17.13. At low Rayleigh number, there are two eddies which rotate in opposite directions. In the presence of an electric field, smaller eddies are generated and a thermal plume is generated. Isotherms become more disturbed in the presence of an electric field. Increasing the Darcy number makes the isotherms denser near the positive electrode. As

TABLE 17.5 Effect of Shape of Nanoparticles on Nusselt Number When $Rd = 0.8, Ra = 500, \Delta\varphi = 6, \phi = 0.05$

	Da	
	10^{-2}	10^{2}
Spherical	31.22357	48.90455
Brick	33.84407	49.91443
Cylinder	37.50945	50.10212
Platelet	40.45061	51.85191

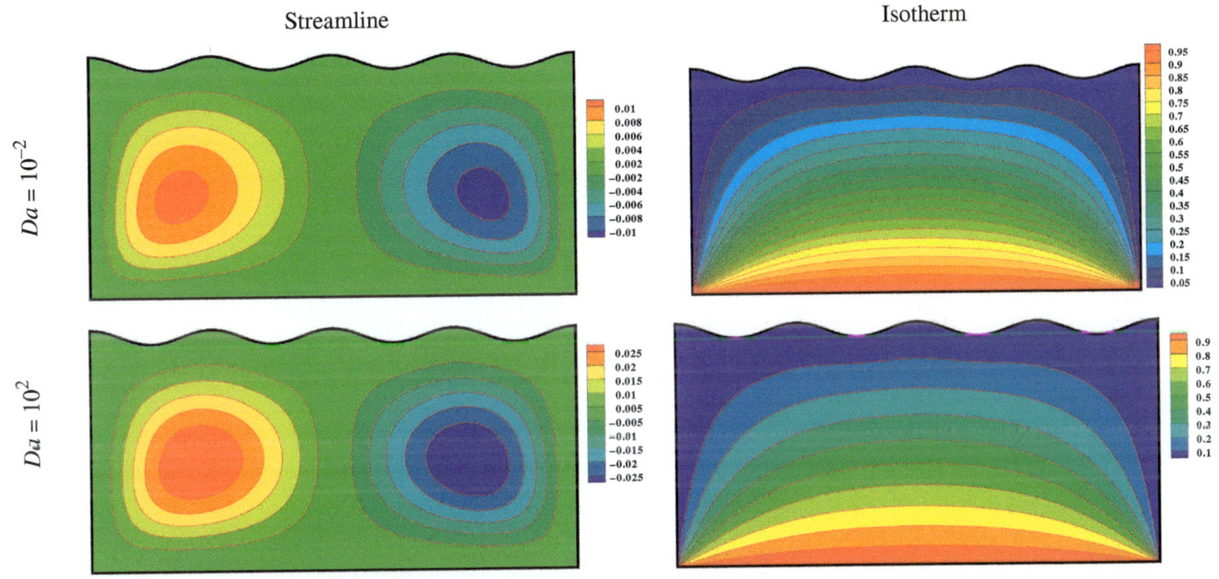

FIGURE 17.10 Effect of Darcy number on streamlines and isotherms when $Ra = 50, \Delta\varphi = 0kV, \phi = 0.05, Rd = 0.8$.

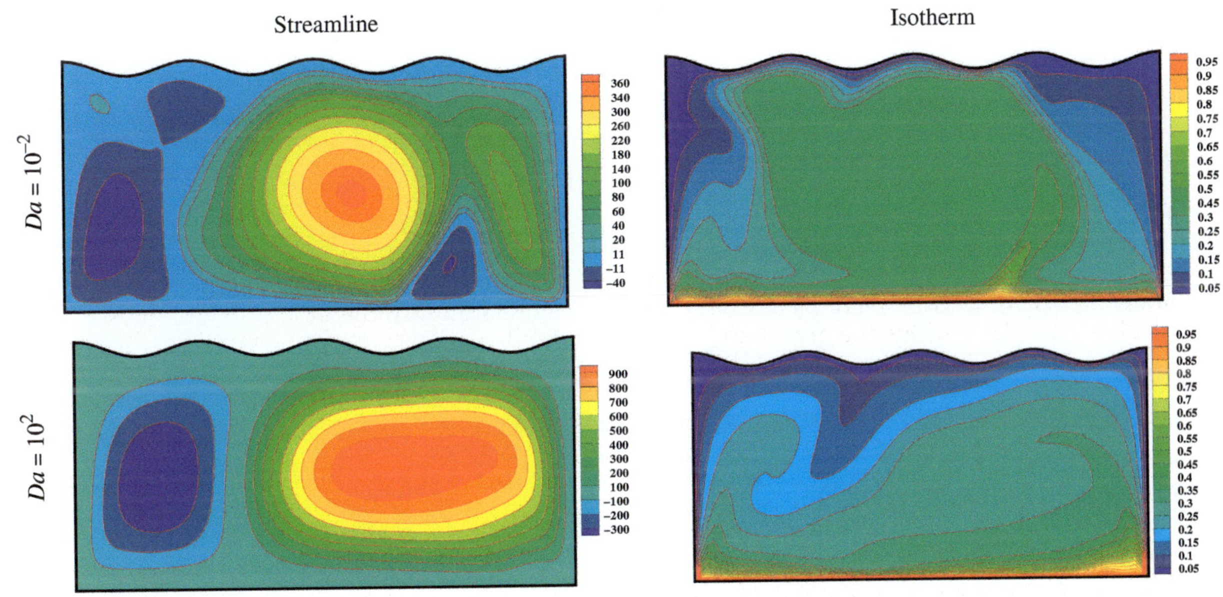

FIGURE 17.11 Effect of Darcy number on streamlines and isotherms when $Ra = 50, \Delta\varphi = 6kV, \phi = 0.05, Rd = 0.8$.

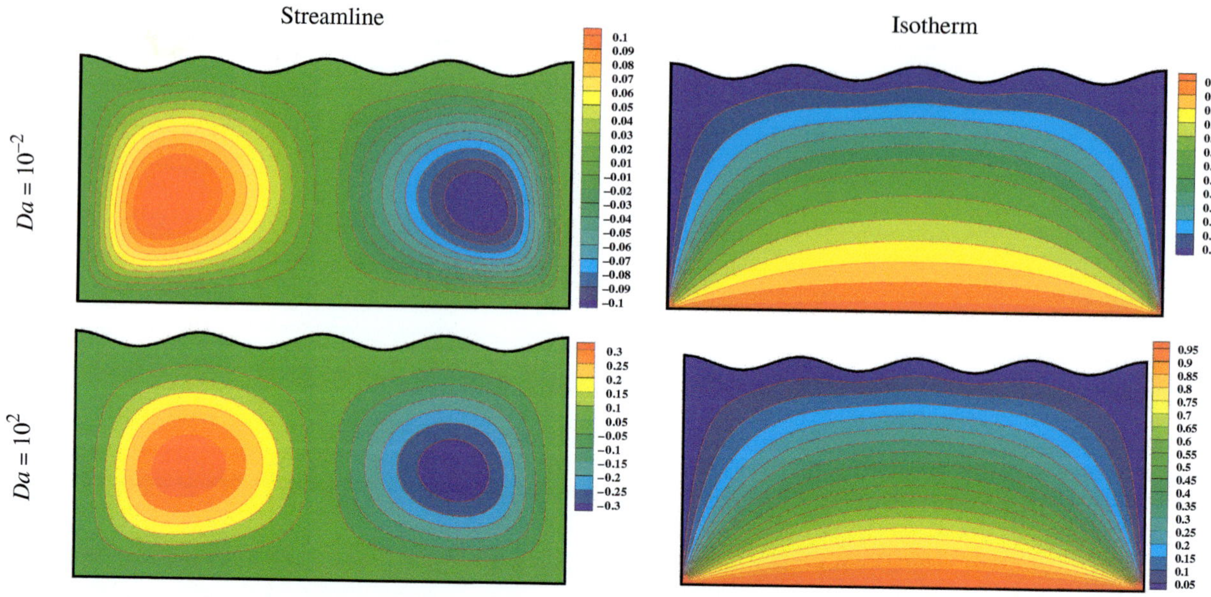

FIGURE 17.12 Effect of Darcy number on streamlines and isotherms when $Ra = 500, \Delta\varphi = 0kV, \phi = 0.05, Rd = 0.8$.

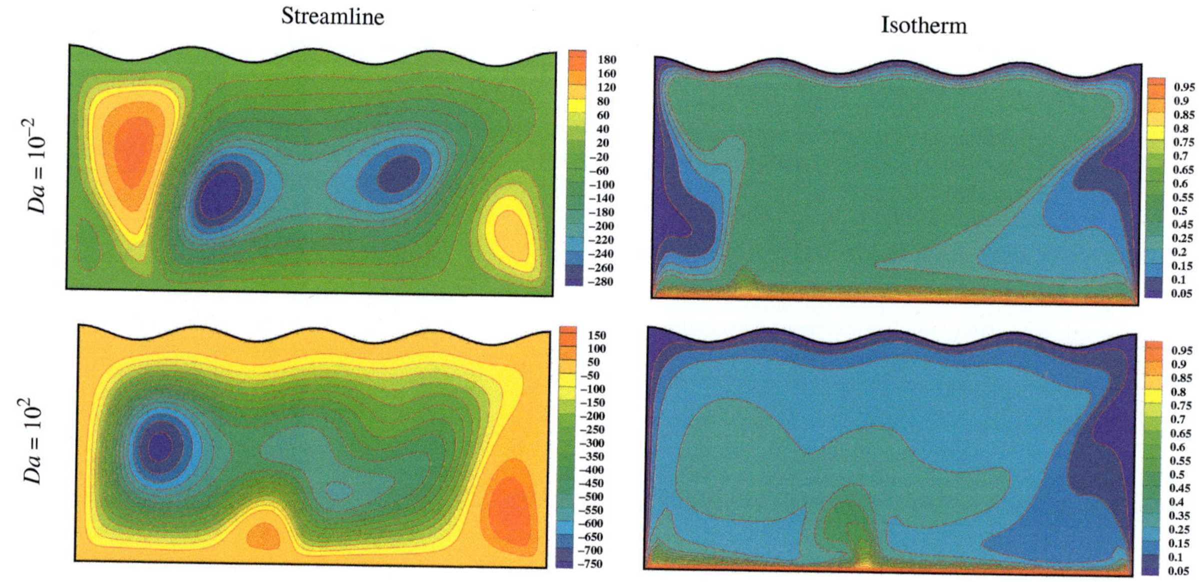

FIGURE 17.13 Effect of Darcy number on streamlines and isotherms when $Ra = 500, \Delta\varphi = 6kV, \phi = 0.05, Rd = 0.8$.

buoyancy forces increase, the shapes of the streamlines become more complex. At high values of Rayleigh number, increasing the Darcy number leads to a shift of the thermal plume to the middle of the enclosure.

Nu_{ave} versus $Ra, Da, Rd,$ and $\Delta\varphi$ is depicted in Fig. 17.14. The related formula is:

$$\begin{aligned} Nu_{ave} &= 2.04 - 0.75\Delta\varphi + 3.12Ra^* + 1.25Da^* + 0.12Rd \\ &+ 0.12\Delta\varphi Ra^* + 0.55\Delta\varphi Da^* \\ &- 1.04\times10^{-15}\Delta\varphi Rd - 0.066Ra^*Da^* \\ &+ 1.57\times10^{-15}Ra^*Rd + 6.39\times10^{-15}Da^*Rd \\ &+ 0.56\Delta\varphi^2 - 0.55(Ra^*)^2 - 1.37(Da^*)^2 - 2.15Rd^2 \end{aligned} \tag{17.26}$$

where $Ra^* = 0.01Ra, Da^* = 0.01Da,$ and $\Delta\varphi$ is voltage supply in Kilovolt. The rate of heat transfer enhances with the rise of the Rayleigh number. The electric field helps to enhance the convective mode. So, Nu_{ave} augments with the increase of $\Delta\varphi$. Thermal radiation enhances the temperature gradient near the lid wall. The influence of the Darcy number is the same as the radiation parameter. Therefore, Nu_{ave} is an increasing function of Rd, Da.

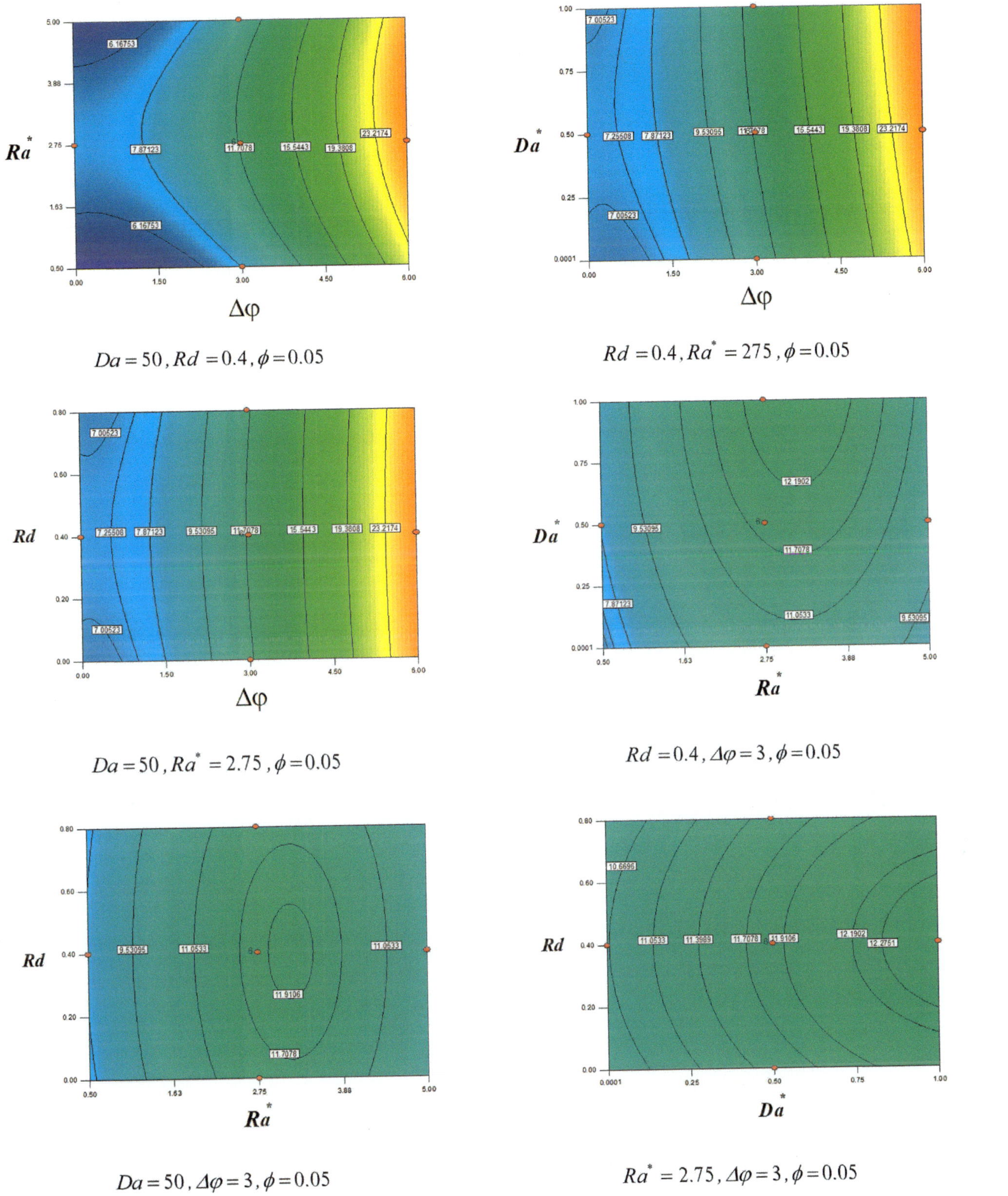

FIGURE 17.14 Effects of Da, $\Delta\varphi$, Rd, and Ra on average Nusselt number.

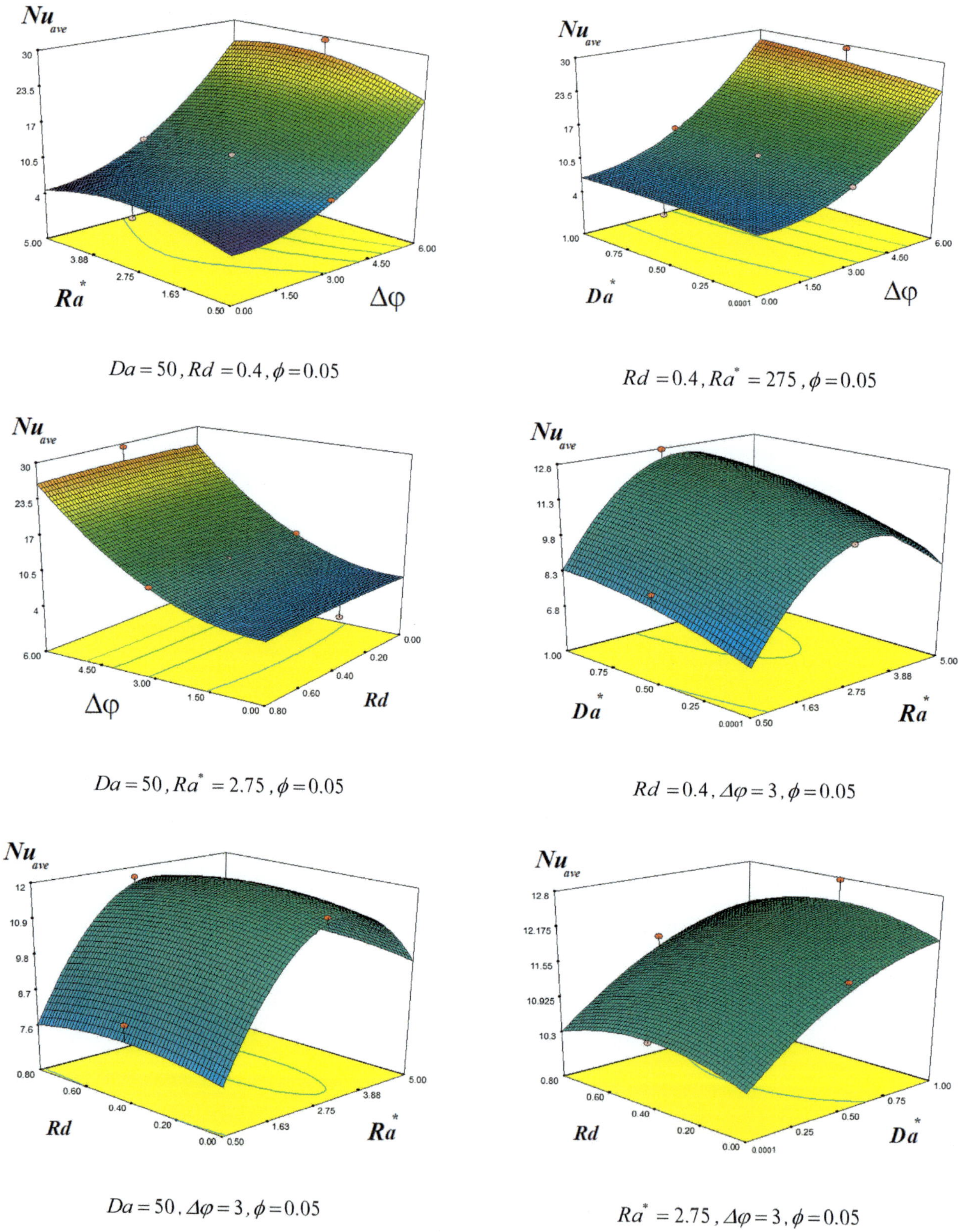

FIGURE 17.14 (Continued).

17.4 EHD NANOFLUID FLOW IN A POROUS MEDIUM CONSIDERING RADIATION PARAMETER

17.4.1 Problem Definition

Fig. 17.15 depicts the porous enclosure and its boundary conditions. Ethylene glycol-Fe_3O_4 nanofluid is utilized. All walls are stationary except for the bottom wall. The influence of Darcy and Reynolds numbers on contour of q is demonstrated in Fig. 17.16. The effect of Re on q is less sensible than Da. As the Darcy number increases, the distortion of the isoelectric density lines become greater.

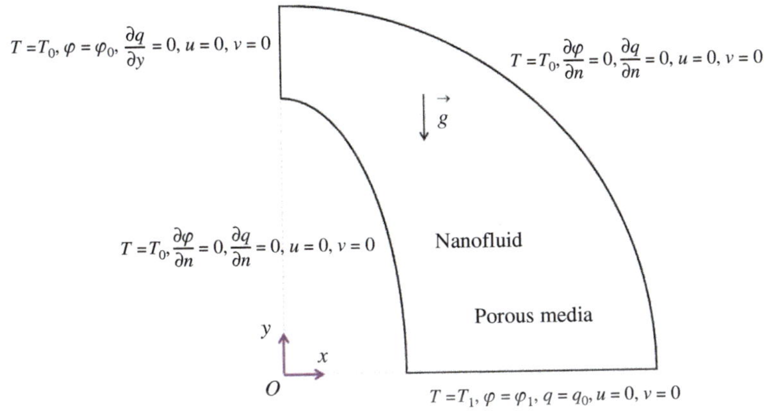

FIGURE 17.15 Geometry and the boundary conditions.

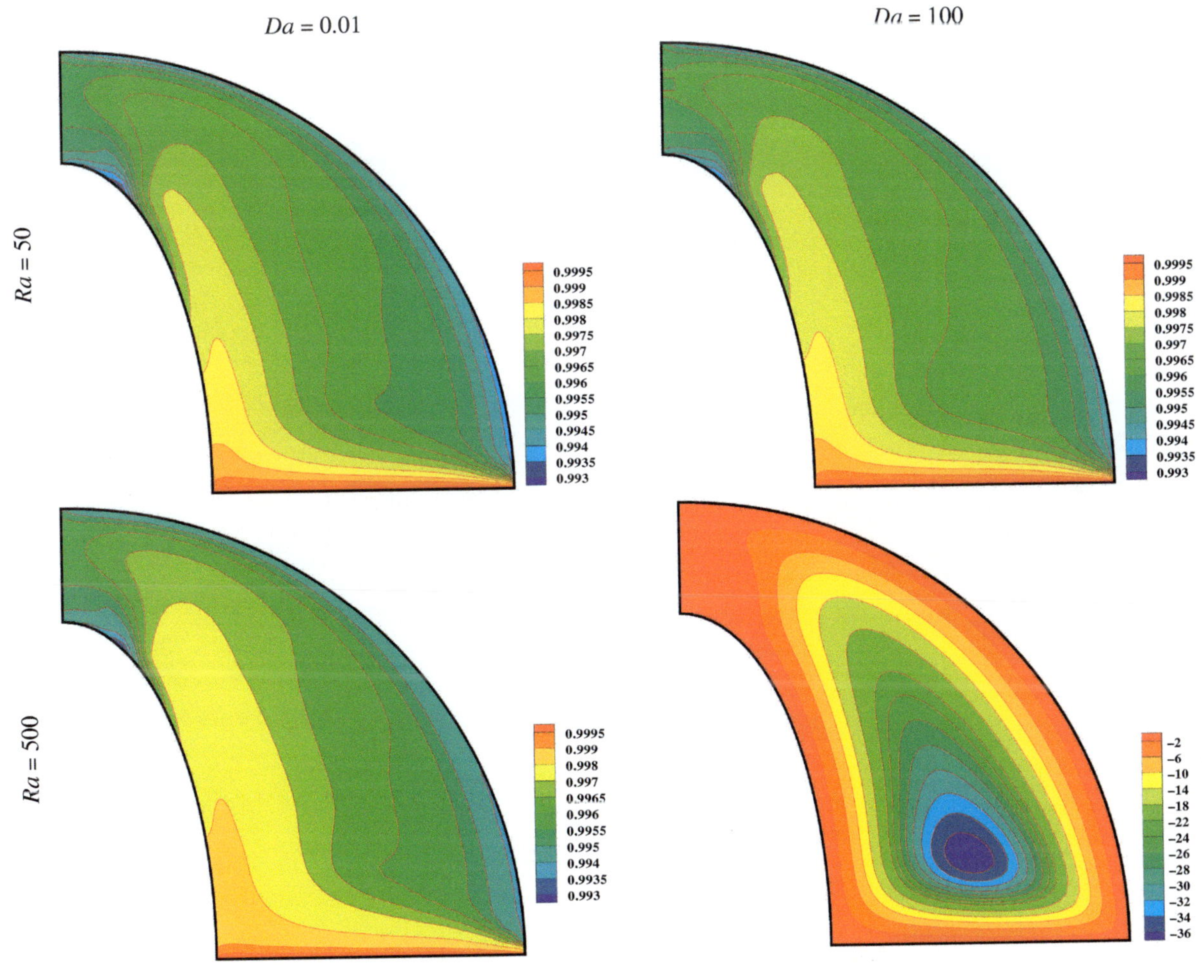

FIGURE 17.16 Electric density distribution injected by the bottom electrode when $\Delta\varphi = 6kV, \phi = 0.05, Rd = 0.8$.

17.4.2 Governing Equation

The definition of the electric field is:

$$\vec{E} = -\nabla\varphi \tag{17.27}$$

$$q = \nabla.\varepsilon\,\vec{E} \tag{17.28}$$

$$\vec{J} = q\vec{V} - D\nabla q + \sigma\vec{E} \tag{17.29}$$

TABLE 17.6 Effect of Shape of Nanoparticles on Nusselt Number When $Rd = 0.8, Ra = 500, \Delta\varphi = 6, \phi = 0.05$

	Da	
	10^{-2}	10^{2}
Spherical	17.64276	18.65795
Brick	17.83949	18.84945
Cylinder	18.12832	19.13087
Platelet	19.34793	20.34507

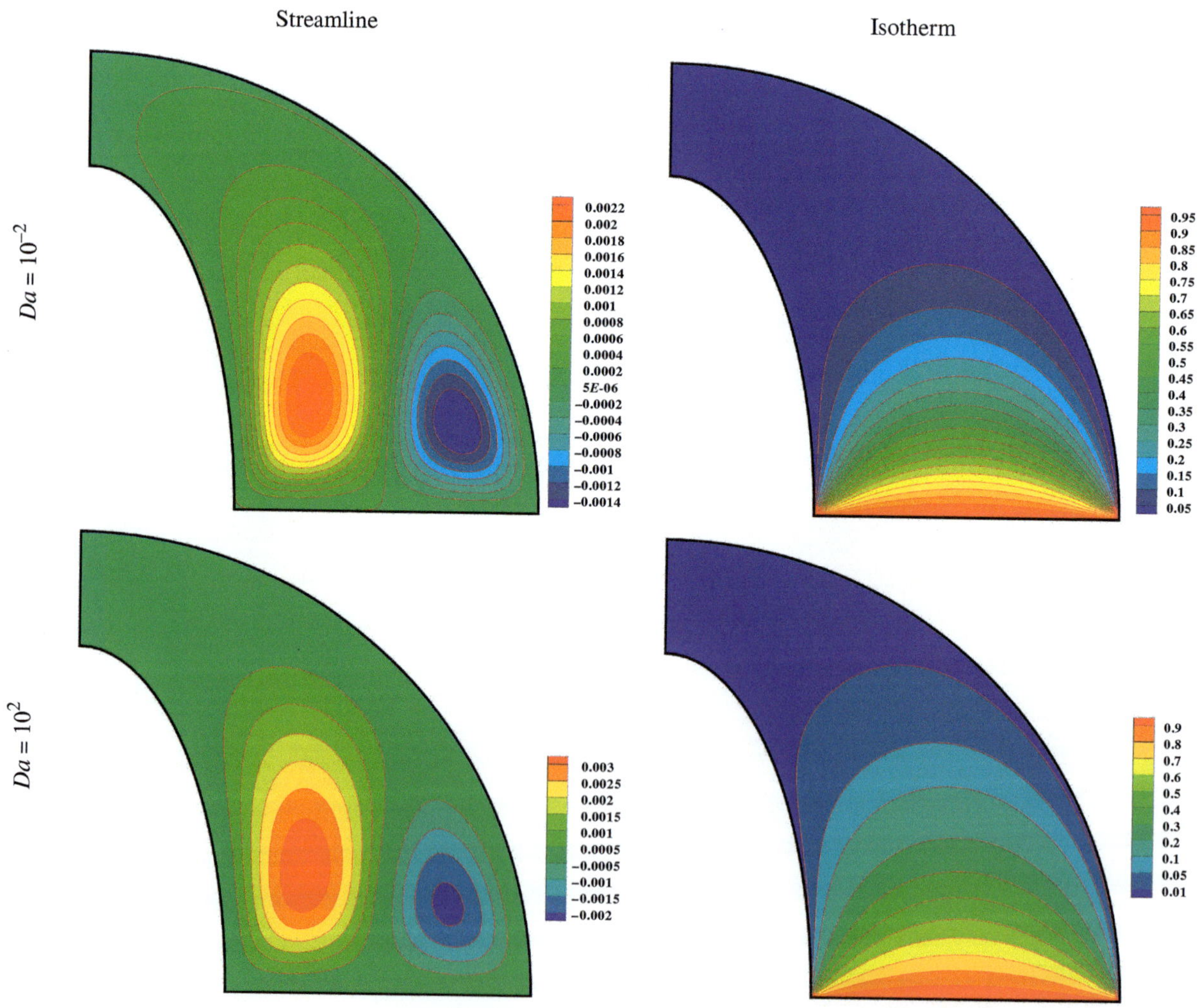

FIGURE 17.17 Effect of Darcy number on streamlines and isotherms when $Ra = 50, \Delta\varphi = 0kV, \phi = 0.05, Rd = 0.8$.

$$\nabla.\vec{J} + \frac{\partial q}{\partial t} = 0 \tag{17.30}$$

The governing formulae are:

$$\begin{cases} \nabla.\vec{V} = 0, \\ \rho_{nf}\left(\dfrac{\partial \vec{V}}{\partial t} + \left(\vec{V}.\nabla\right)\vec{V}\right) = -\nabla p + \mu_{nf}\nabla^2\vec{V} + q\,\vec{E} - (\rho\beta)_{nf}\vec{g}(T - T_0) - \dfrac{\mu_{nf}}{K}\vec{V}, \\ \left(\left(\vec{V}.\nabla\right)T + \dfrac{\partial T}{\partial t}\right) = \dfrac{k_{nf}}{(\rho C_p)_{nf}}\nabla^2 T + \dfrac{\vec{J}.\vec{E}}{(\rho C_p)_{nf}} - \dfrac{1}{(\rho C_p)_{nf}}\dfrac{\partial q_r}{\partial y}, \quad \left[q_r = -\dfrac{4\sigma_e}{3\beta_R}\dfrac{\partial T^4}{\partial y}, \quad T^4 \cong 4T_c^3 T - 3T_c^4\right], \\ \nabla\varphi = -\vec{E}, \\ \dfrac{\partial q}{\partial t} = -\nabla.\vec{J}, \\ q = \nabla.\varepsilon\vec{E} \end{cases} \tag{17.31}$$

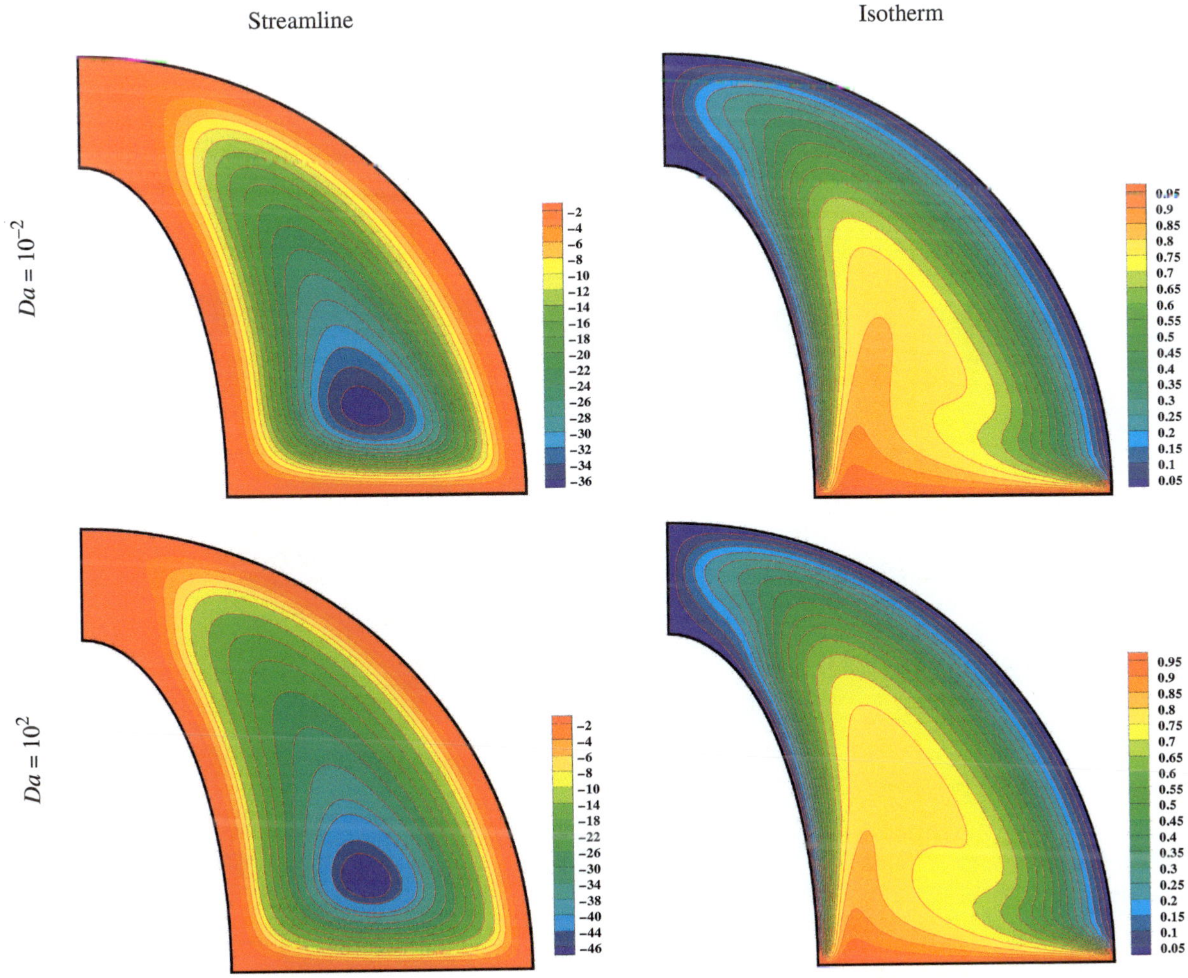

FIGURE 17.18 Effect of Darcy number on streamlines and isotherms when $Ra = 50, \Delta\varphi = 6kV, \phi = 0.05, Rd = 0.8$.

$(\rho C_p)_{nf}$, $(\beta)_{nf}$, μ_{nf}, and ρ_{nf} can be obtained as:

$$\begin{aligned}&(\rho C_p)_{nf} = (\rho C_p)_f(1-\phi) + (\rho C_p)_s\phi, \quad (\beta)_{nf} = (\beta)_f(1-\phi) + (\beta)_s\phi,\\ &\mu_{nf} = A_1 + A_2(\Delta\varphi) + A_3(\Delta\varphi)^2 + A_4(\Delta\varphi)^3,\\ &\rho_{nf} = \rho_f(1-\phi) + \rho_s\phi\end{aligned} \tag{17.32}$$

Properties of Fe_3O_4 and ethylene glycol are illustrated in Table 17.1. Table 17.2 illustrates the coefficient value of this formula.

k_{nf} can be expressed as:

$$\frac{k_{nf}}{k_f} = \frac{-m(k_f - k_p)\phi + (k_p - k_f)\phi + mk_f + k_p + k_f}{mk_f + (k_f - k_p)\phi + k_f + k_p} \tag{17.33}$$

Different values of shape factors for various shapes of nanoparticles are illustrated in Table 17.3.

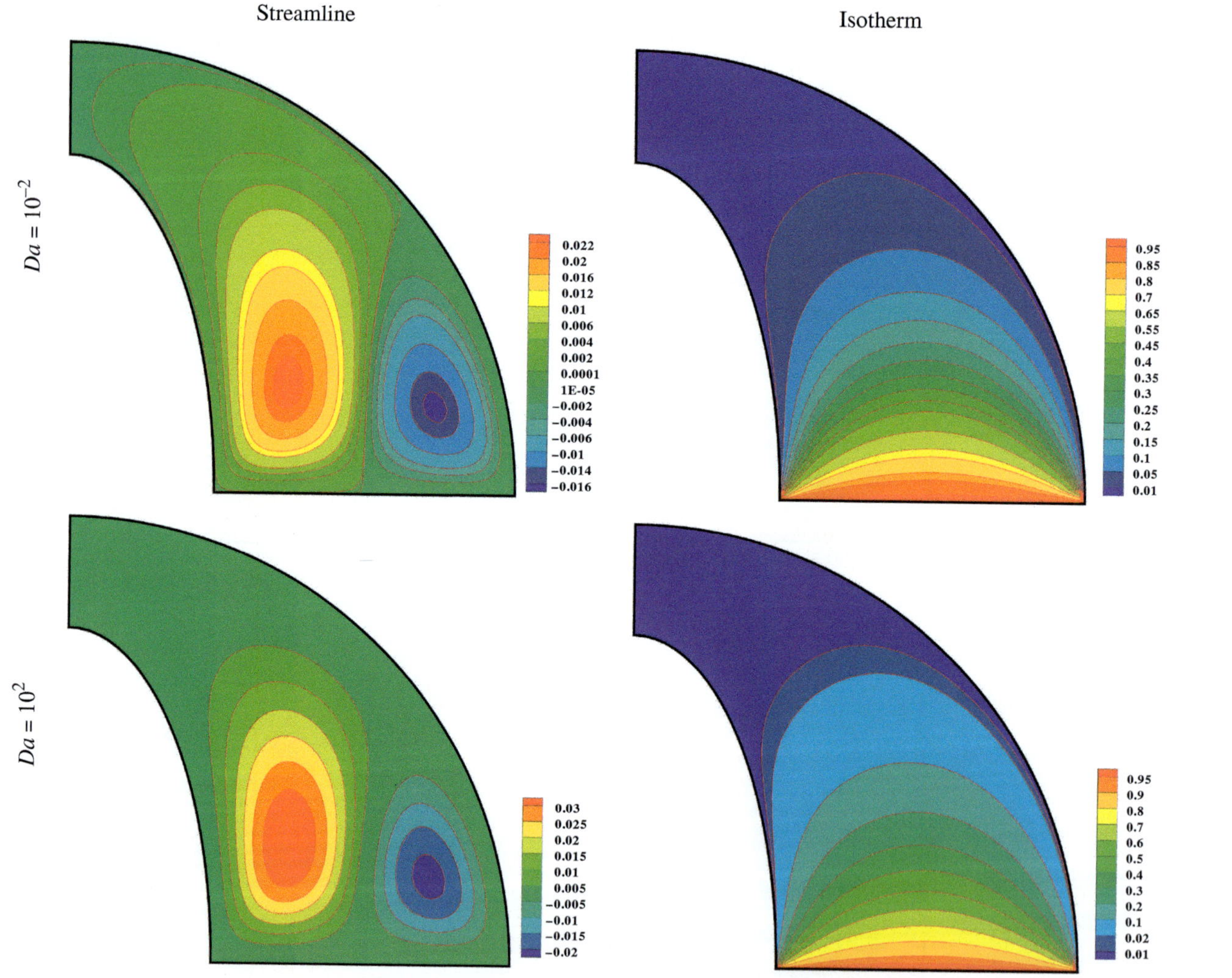

FIGURE 17.19 Effect of Darcy number on streamlines and isotherms when $Ra = 500$, $\Delta\varphi = 0kV$, $\phi = 0.05$, $Rd = 0.8$.

So, the final PDE in the presence of thermal radiation and an electric field in a porous medium are:

$$\begin{cases} \nabla.\vec{V}=0, \\ \left(\dfrac{\partial \vec{V}}{\partial t}+\left(\vec{V}.\nabla\right)\vec{V}\right)=-\nabla p+\dfrac{\mu_{nf}/\mu_f}{\rho_{nf}/\rho_f}\mathrm{Pr}\nabla^2\vec{V}+\dfrac{S_E}{\rho_{nf}/\rho_f}q\vec{E}-Ra\mathrm{Pr}\left[\dfrac{\beta_{nf}}{\beta_f}\right]\theta-\dfrac{\mathrm{Pr}}{Da}\dfrac{\mu_{nf}/\mu_f}{\rho_{nf}/\rho_f}\vec{V} \\ \left(\left(\vec{V}.\nabla\right)\theta+\dfrac{\partial\theta}{\partial t}\right)=\dfrac{k_{nf}/k_f}{(\rho C_p)_{nf}/(\rho C_p)_f}\nabla^2\theta+\dfrac{S_E Ec}{(\rho C_p)_{nf}/(\rho C_p)_f}\left(\vec{J}.\vec{E}\right)+\dfrac{4}{3}\dfrac{k_{nf}/k_f}{(\rho C_p)_{nf}/(\rho C_p)_f}\left(\dfrac{k_{nf}}{k_f}\right)^{-1}Rd\dfrac{\partial^2\theta}{\partial Y^2} \\ \vec{E}=-\nabla\varphi \\ q=\nabla.\varepsilon\vec{E} \\ \nabla.\vec{J}=-\dfrac{\partial q}{\partial t} \end{cases} \tag{17.34}$$

where

$$\begin{aligned} &\bar{t}=\frac{t\alpha_f}{L^2},\bar{p}=\frac{P}{\rho(\alpha_f/L)^2},\bar{y}=\frac{y}{L},\bar{x}=\frac{x}{L},\bar{v}=\frac{vL}{\alpha_f},\bar{u}=\frac{uL}{\alpha_f}, \\ &\theta=\frac{T-T_0}{\nabla T},\nabla T=T_1-T_0,\bar{\varphi}=\frac{\varphi-\varphi_0}{\nabla\varphi},\nabla\varphi=\varphi_1-\varphi_0,\bar{q}=\frac{q}{q_0},\bar{E}=\frac{E}{E_0} \end{aligned} \tag{17.35}$$

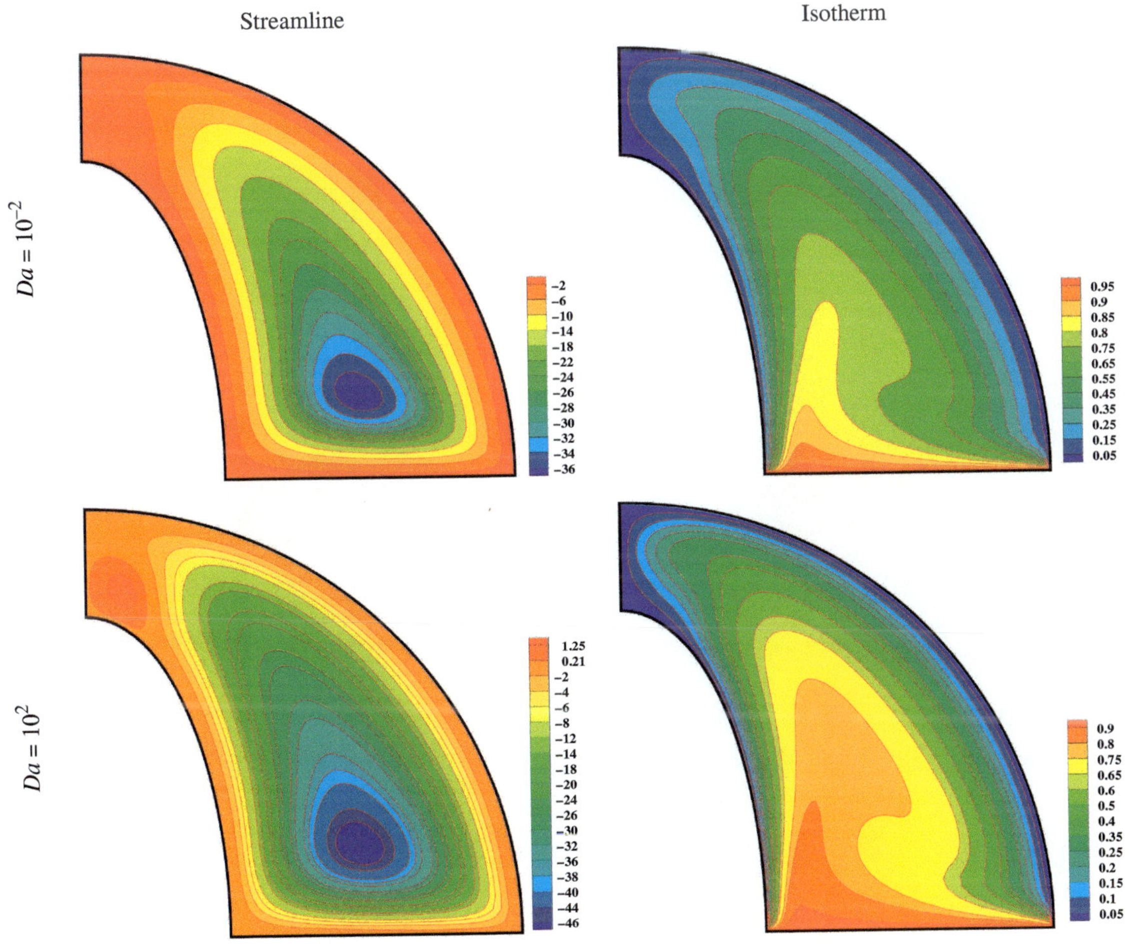

FIGURE 17.20 Effect of Darcy number on streamlines and isotherms when $Ra=500, \Delta\varphi=6kV, \phi=0.05, Rd=0.8$.

Vorticity and stream function should be employed in order to diminish the pressure gradient:

$$v = -\frac{\partial \psi}{\partial x}, \frac{\partial \psi}{\partial y} = u, \Psi = \frac{\psi}{\alpha_f}, \Omega = \frac{\omega L^2}{\alpha_f}, \omega = \frac{\partial v}{\partial x} - \frac{\partial u}{\partial y} \tag{17.36}$$

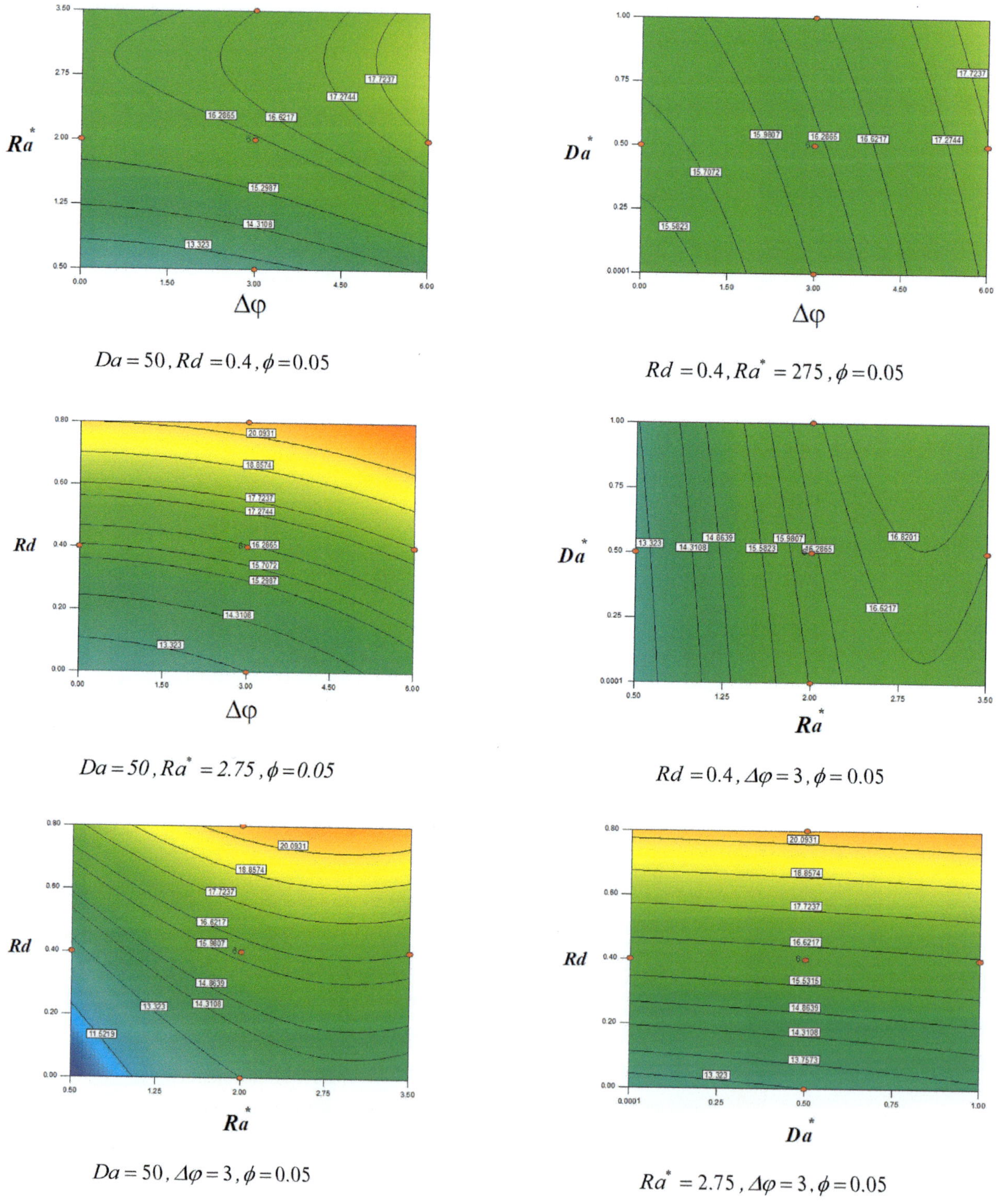

FIGURE 17.21 Effects of Da, $\Delta\varphi$, Rd, and Ra on average Nusselt number.

Nu_{loc} and Nu_{ave} along the bottom wall are calculated as:

$$Nu_{loc} = \left(\frac{k_{nf}}{k_f}\right)\left(1 + \frac{4}{3}Rd\left(\frac{k_{nf}}{k_f}\right)^{-1}\right)\frac{\partial\theta}{\partial X} \tag{17.37}$$

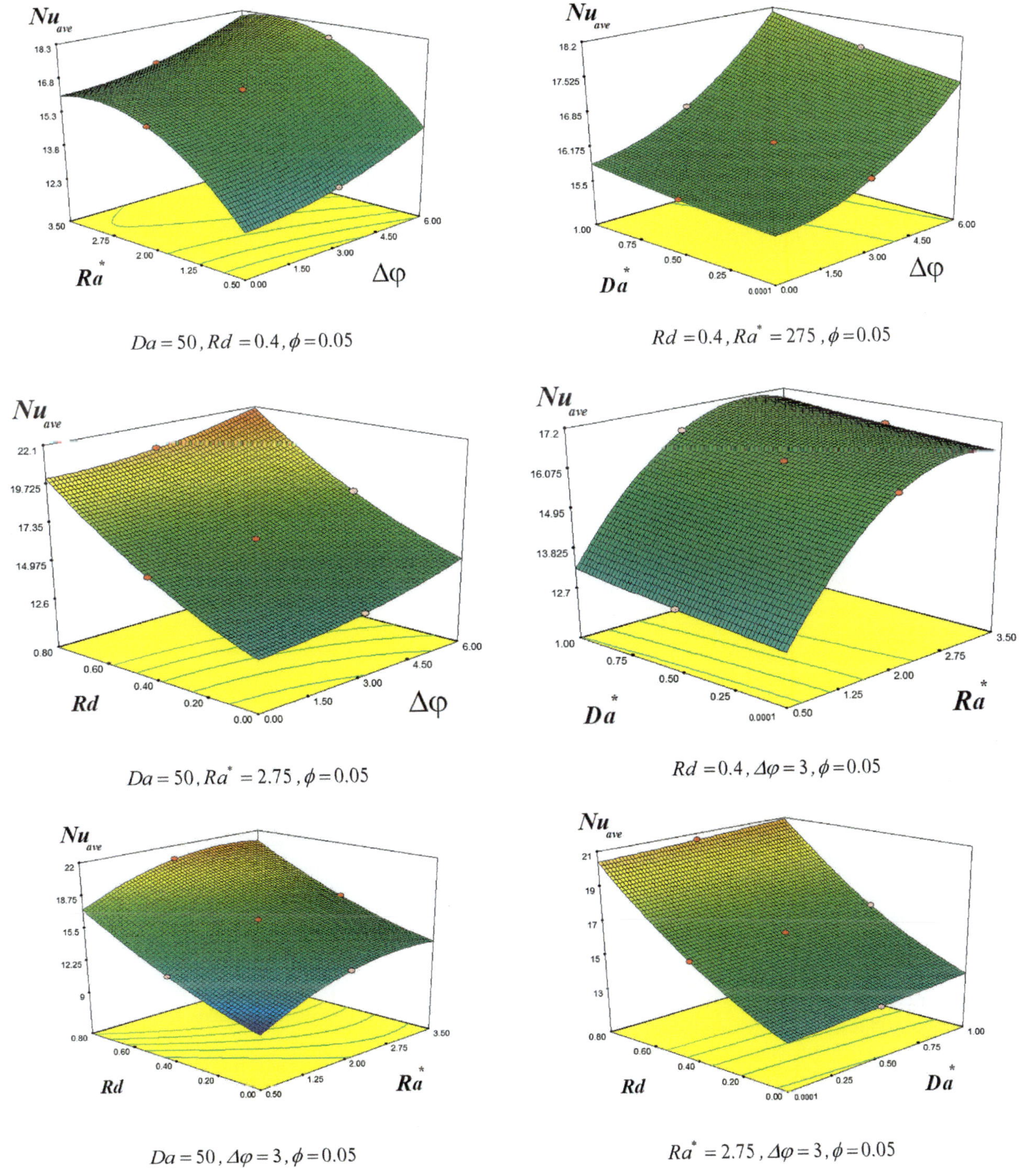

FIGURE 17.21 (Continued).

$$Nu_{ave} = \frac{1}{L}\int_0^L Nu_{loc}\, dY \tag{17.38}$$

17.4.3 Effects of Active Parameters

Electrohydrodynamic nanofluid free convection heat transfer through the porous enclosure is investigated in the presence of thermal radiation. Electric field dependent viscosity is considered for nanofluid. The porous enclosure is filled with Fe_3O_4-Ethylene glycol and has one lid wall. The roles of Darcy number ($Da = 10^{-2}$ to 10^2), Radiation parameter ($Rd = 0$ to 0.8), supplied voltage ($\Delta\varphi = 0$ to $6kV$), volume fraction of Fe_3O_4 ($\phi = 0\%$ to 5%), and Rayleigh number ($Ra = 50$ to 500) are illustrated graphically. The influence of shape factor on the rate of heat transfer is reported in Table 17.6. In this table, various shapes of nanoparticles are utilized. The maximum Nu is caused by Platelet, followed by Cylinder, Brick, and Spherical. So, Platelet nanoparticle has been utilized for further investigation. The impacts of $Da, \Delta\varphi$, and Ra on isotherms and streamlines are demonstrated in Figs. 17.17–17.20. At low Rayleigh number, there are two eddies which rotate in opposite directions. In the presence of an electric field, the two eddies convert to one clockwise eddy. The isotherms become more disturbed in the presence of an electric field and a thermal plume is generated. Increasing the Darcy number makes the isotherms denser near the positive electrode. As buoyancy forces increase, the shapes of the streamlines become more complex. At high values of Rayleigh number, increasing the Darcy number generates a secondary eddy. Nu_{ave} versus Ra, Da, Rd, and $\Delta\varphi$ is depicted in Fig. 17.21. The related formula is:

$$\begin{aligned} Nu_{ave} &= 7.36 + 0.04\Delta\varphi + 3.86Ra^* + 0.2Da^* + 5.9Rd \\ &- 6.23 \times 10^{-3}\Delta\varphi Ra^* + 0.07\Delta\varphi Da^* \\ &- 0.046\Delta\varphi Rd - 9.84 \times 10^{-3} Ra^* Da^* \\ &- 0.15Ra^* Rd - 0.073Da^* Rd \\ &+ 0.05\Delta\varphi^2 - 0.64(Ra^*)^2 + 0.16(Da^*)^2 + 4.64Rd^2 \end{aligned} \tag{17.39}$$

where $Ra^* = 0.01Ra, Da^* = 0.01Da$, and $\Delta\varphi$ is voltage supply in Kilovolt. The rate of heat transfer enhances with the increase of the Rayleigh number. The electric field helps to enhance the convective mode. So, Nu_{ave} augments with the increase of $\Delta\varphi$. Thermal radiation enhances the temperature gradient near the lid wall. The influence of the Darcy number is the same as the radiation parameter. Therefore, Nu_{ave} is an increasing function of Rd, Da.

References

[1] N.S. Akber, S. Nadeem, R. Ul Haq, Z.H. Khan, Radiation effects on MHD stagnation point flow of nanofluid towards a stretching surface with convective boundary condition, Chin. J. Aeronaut. 26 (6) (2013) 1389–1397.
[2] M. Sheikholeslami, H.B. Rokni, Nanofluid convective heat transfer intensification in a porous circular cylinder, Chem. Eng. Proc. Process Intensif. 120 (2017) 93–104.
[3] M. Sheikholeslami, M.K. Sadoughi, Numerical modeling for Fe3O4 -water nanofluid flow in porous medium considering MFD viscosity, J. Mol. Liq. 242 (2017) 255–264.
[4] M. Sheikholeslami, M. Shamlooei, Magnetic source influence on nanofluid flow in porous medium considering shape factor effect, Phys. Lett. A 381 (2017) 3071–3078.
[5] M. Sheikholeslami, M. Shamlooei, Convective flow of nanofluid inside a lid driven porous cavity using CVFEM, Phys. B: Condensed Matter 521 (2017) 239–250.
[6] M. Sheikholeslami, K. Vajravelu, Forced convection heat transfer in Fe3O4- ethylene glycol nanofluid under the influence of Coulomb force, J. Mol. Liq. 233 (2017) 203–210.
[7] M. Sheikholeslami, H.B. Rokni, Influence of EFD viscosity on nanofluid forced convection in a cavity with sinusoidal wall, J. Mol. Liq. 232 (2017) 390–395.
[8] M. Sheikholeslami, H.B. Rokni, Magnetic nanofluid natural convection in the presence of thermal radiation considering variable viscosity, Europ. Phys. J. Plus 132 (2017) 238. Available from: https://doi.org/10.1140/epjp/i2017-11498-4.
[9] M. Sheikholeslami, H.B. Rokni, Numerical simulation for impact of Coulomb force on nanofluid heat transfer in a porous enclosure in presence of thermal radiation, Int. J. Heat Mass Transfer 118 (2018) 823–831.
[10] M. Sheikholeslami, M. Shamlooei, R. Moradi, Fe3O4- Ethylene glycol nanofluid forced convection inside a porous enclosure in existence of Coulomb force, J. Mol. Liq. 249 (2018) 429–437.
[11] M. Sheikholeslami, S.A. Shehzad, Numerical analysis of Fe_3O_4–H_2O nanofluid flow in permeable media under the effect of external magnetic source, Int. J. Heat Mass Transfer 118 (2018) 182–192.
[12] M. Sheikholeslami, S.A. Shehzad, Magnetohydrodynamic nanofluid convection in a porous enclosure considering heat flux boundary condition, Int. J. Heat Mass Transfer 106 (2017) 1261–1269.

13] M. Sheikholeslami, S.A. Shehzad, Thermal radiation of ferrofluid in existence of Lorentz forces considering variable viscosity, Int. J. Heat Mass Transfer 109 (2017) 82–92.

14] M. Sheikholeslami, H.B. Rokni, Numerical modeling of nanofluid natural convection in a semi annulus in existence of Lorentz force, Comput. Methods Appl. Mech. Eng. 317 (2017) 419–430.

15] M. Sheikholeslami, K. Vajravelu, Nanofluid flow and heat transfer in a cavity with variable magnetic field, Appl. Math. Comput. 298 (2017) 272–282.

16] M. Sheikholeslami, H.B. Rokni, Magnetohydrodynamic CuO-water nanofluid in a porous complex shaped enclosure, ASME, J. Thermal Sci. Eng. Applicat. 9 (4) (2017) 041007. Available from: https://doi.org/10.1115/1.4035973.

[17] M. Sheikholeslami, H.B. Rokni, Nanofluid two phase model analysis in existence of induced magnetic field, Int. J. Heat Mass Transfer 107 (2017) 288–299.

[18] M. Sheikholeslami, T. Hayat, A. Alsaedi, S. Abelman, Numerical analysis of EHD nanofluid force convective heat transfer considering electric field dependent viscosity, Int. J. Heat Mass Transfer 108 (2017) 2558–2565.

[19] M. Sheikholeslami, Magnetic field influence on CuO -H2O nanofluid convective flow in a permeable cavity considering various shapes for nanoparticles, Int. J. Hydrogen Energy 42 (2017) 19611–19621.

[20] M. Sheikholeslami, M.M. Bhatti, Forced convection of nanofluid in presence of constant magnetic field considering shape effects of nanoparticles, Int. J. Heat Mass Transfer 111 (2017) 1039–1049.

[21] M. Sheikholeslami, A. Zeeshan, Analysis of flow and heat transfer in water based nanofluid due to magnetic field in a porous enclosure with constant heat flux using CVFEM, Comput. Methods Appl. Mech. Eng. 320 (2017) 68–81.

[22] M. Sheikholeslami, M. Shamlooei, Fe3O4- H2O nanofluid natural convection in presence of thermal radiation, Int. J. Hydrogen Energy 42 (9) (2017) 5708–5718.

[23] M. Sheikholeslami, M.M. Bhatti, Active method for nanofluid heat transfer enhancement by means of EHD, Int. J. Heat Mass Transfer 109 (2017) 115–122.

[24] M. Sheikholeslami, T. Hayat, A. Alsaedi, Numerical study for external magnetic source influence on water based nanofluid convective heat transfer, Int. J. Heat Mass Transfer 106 (2017) 745–755.

[25] M. Sheikholeslami, A.J. Chamkha, Flow and convective heat transfer of a ferro-nanofluid in a double-sided lid-driven cavity with a wavy wall in the presence of a variable magnetic field, Numer. Heat Transfer, Part A 69 (10) (2016) 1186–1200. Available from: https://doi.org/10.1080/10407782.2015.1125709.

[26] M. Sheikholeslami, A.J. Chamkha, Electrohydrodynamic free convection heat transfer of a nanofluid in a semi-annulus enclosure with a sinusoidal wall, Numer. Heat Transfer, Part A 69 (7) (2016) 781–793. Available from: http://dx.doi.org/10.1080/10407782.2015.1090819.

[27] M. Sheikholeslami, M.K. Sadoughi, Simulation of CuO- water nanofluid heat transfer enhancement in presence of melting surface, Int. J. Heat Mass Transfer 116 (2018) 909–919.

[28] M. Sheikholeslami, H.B. Rokni, Simulation of nanofluid heat transfer in presence of magnetic field: a review, Int. J. Heat Mass Transfer 115 (2017) 1203–1233.

[29] M. Sheikholeslami, H.B. Rokni, Melting heat transfer influence on nanofluid flow inside a cavity in existence of magnetic field, Int. J. Heat Mass Transfer 114 (2017) 517–526.

[30] M. Sheikholeslami, S.A. Shehzad, Magnetohydrodynamic nanofluid convective flow in a porous enclosure by means of LBM, Int. J. Heat Mass Transfer 113 (2017) 796–805.

[31] M. Sheikholeslami, Magnetohydrodynamic nanofluid forced convection in a porous lid driven cubic cavity using Lattice Boltzmann Method, J. Mol. Liq. 231 (2017) 555–565.

[32] M.S. Kandelousi, KKL correlation for simulation of nanofluid flow and heat transfer in a permeable channel, Phys. Lett. A 378 (45) (2014) 3331–3339.

[33] M. Sheikholeslami, Effect of uniform suction on nanofluid flow and heat transfer over a cylinder, J. Braz. Soc. Mech. Sci. Eng. 37 (2015) 1623–1633.

[34] M. Sheikholeslami, Lattice Boltzmann Method simulation of MHD non-Darcy nanofluid free convection, Physica B 516 (2017) 55–71.

[35] M. Sheikholeslami, Influence of magnetic field on nanofluid free convection in an open porous cavity by means of Lattice Boltzmann Method, J. Mol. Liq. 234 (2017) 364–374.

[36] M. Sheikholeslami, A.J. Chamkha, Influence of Lorentz forces on nanofluid forced convection considering Marangoni convection, J. Mol. Liq. 225 (2017) 750–757.

[37] M. Sheikholeslami, T. Hayat, A. Alsaedi, Numerical simulation of nanofluid forced convection heat transfer improvement in existence of magnetic field using Lattice Boltzmann Method, Int. J. Heat Mass Transfer 108 (2017) 1870–1883.

[38] M. Sheikholeslami, R. Ellahi, Three dimensional mesoscopic simulation of magnetic field effect on natural convection of nanofluid, Int. J. Heat Mass Transfer 89 (2015) 799–808.

[39] M. Sheikholeslami, S. Abelman, Two phase simulation of nanofluid flow and heat transfer in an annulus in the presence of an axial magnetic field, IEEE Trans. Nanotechnol. 14 (3) (2015) 561–569.

[40] M. Sheikholeslami, R. Ellahi, Simulation of ferrofluid flow for magnetic drug targeting using Lattice Boltzmann method, J. Zeitschrift Fur Naturforschung A 70 (2) (2015) 115–124.

[41] M. Sheikholeslami, M. Seyednezhad, Lattice Boltzmann Method simulation for CuO-water nanofluid flow in a porous enclosure with hot obstacle, J. Mol. Liq. 243 (2017) 249–256.

[42] M. Sheikholeslami, M. Sadoughi, Mesoscopic method for MHD nanofluid flow inside a porous cavity considering various shapes of nanoparticles, Int. J. Heat Mass Transfer 113 (2017) 106–114.

[43] M. Sheikholeslami, A. Zeeshan, Mesoscopic simulation of CuO-H2O nanofluid in a porous enclosure with elliptic heat source, Int. J. Hydrogen Energy 42 (22) (2017) 15393–15402.

[44] M. Sheikholeslami, S.A. Shehzad, CVFEM for influence of external magnetic source on Fe3O4 – H2O nanofluid behavior in a permeable cavity considering shape effect, Int. J. Heat Mass Transfer 115 (2017) 180–191.

[45] M. Sheikholeslami, M. Seyednezhad, Nanofluid heat transfer in a permeable enclosure in presence of variable magnetic field by mea of CVFEM, Int. J. Heat Mass Transfer 114 (2017) 1169–1180.
[46] M. Sheikholeslami, H.B. Rokni, Free convection of CuO-H2O nanofluid in a curved porous enclosure using mesoscopic approach, Int. Hydrogen Energy 42 (22) (2017) 14942–14949.
[47] M. Sheikholeslami, Numerical investigation for CuO-H_2O nanofluid flow in a porous channel with magnetic field using mesoscop method, J. Mol. Liq. 249 (2018) 739–746.
[48] M. Sheikholeslami, H.B. Rokni, Influence of melting surface on MHD nanofluid flow by means of two phase model, Chin. J. Phys. 5 (2017) 1352–1360.
[49] M. Sheikholeslami, H.B. Rokni, Effect of melting heat transfer on nanofluid flow in existence of magnetic field considering Buongiorn Model, Chin. J. Phys. 55 (2017) 1115–1126.
[50] M. Sheikholeslami, T. Hayat, A. Alsaedi, MHD free convection of Al2O3–water nanofluid considering thermal radiation: a numerica study, Int. J. Heat Mass Transfer 96 (2016) 513–524.
[51] M. Sheikholeslami, K. Vajravelu, M.M. Rashidi, Forced convection heat transfer in a semi annulus under the influence of a variabl magnetic field, Int. J. Heat Mass Transfer 92 (2016) 339–348.
[52] M.S. Kandelousi, Effect of spatially variable magnetic field on ferrofluid flow and heat transfer considering constant heat flux boundar condition, Eur. Phys. J. Plus (2014) 129–248. Available from: https://doi.org/10.1140/epjp/i2014-14248-2.
[53] M. Sheikholeslami, CVFEM for magnetic nanofluid convective heat transfer in a porous curved enclosure, Eur. Phys. J. Plus 131 (2016 413. Available from: https://doi.org/10.1140/epjp/i2016-16413-y.
[54] M. Sheikholeslami, Magnetic source impact on nanofluid heat transfer using CVFEM, Neural Computing and Applications (2016) Available from: https://doi.org/10.1007/s00521-016-2740-7.
[55] M. Sheikholeslami, Influence of Lorentz forces on nanofluid flow in a porous cylinder considering Darcy model, J. Mol. Liq. 225 (2017) 903–912.
[56] M. Sheikholeslami, Influence of Coulomb forces on Fe_3O_4-H_2O nanofluid thermal improvement, Int. J. Hydrogen Energy 42 (2017) 821–829.
[57] M. Sheikholeslami, Magnetic field influence on nanofluid thermal radiation in a cavity with tilted elliptic inner cylinder, J. Mol. Liq. 229 (2017) 137–147.
[58] M. Sheikholeslami, Numerical simulation of magnetic nanofluid natural convection in porous media, Phys. Lett. A 381 (2017) 494–503.
[59] M. Sheikholeslami, CuO-water nanofluid free convection in a porous cavity considering Darcy law, Eur. Phys. J. Plus 132 (2017) 55. Available from: https://doi.org/10.1140/epjp/i2017-11330-3.
[60] M. Sheikholeslami, Numerical investigation of MHD nanofluid free convective heat transfer in a porous tilted enclosure, Eng. Comput. 34 (6) (2017) 1939–1955.
[61] M. Sheikholeslami, Influence of Lorentz forces on nanofluid flow in a porous cavity by means of Non-Darcy model, Eng. Comput. 34 (8) (2017) 2651–2667. Available from: https://doi.org/10.1108/EC-01-2017-0008.
[62] M. Sheikholeslami, M. Darzi, Z. Li, Experimental investigation for entropy generation and energy loss of nano-refrigerant condensation process, Int. J. Heat Mass Transfer 125 (2018) 1087–1095.
[63] M. Sheikholeslami, Solidification of NEPCM under the effect of magnetic field in a porous thermal energy storage enclosure using CuO nanoparticles, J. Mol. Liq. 263 (2018) 303–315.
[64] M. Sheikholeslami, T. Hayat, A. Alsaedi, Numerical simulation for forced convection flow of MHD CuO-H_2O nanofluid inside a cavity by means of LBM, J. Mol. Liq. 249 (2018) 941–948.
[65] M. Sheikholeslami, Hari R. Kataria, Akhil S. Mittal, Effect of thermal diffusion and heat-generation on MHD nanofluid flow past an oscillating vertical plate through porous medium, J. Mol. Liq. 257 (2018) 12–25.
[66] M. Sheikholeslami, M.B. Gerdroodbary, S. Valiallah Mousavi, D.D. Ganji, Rasoul Moradi, Heat transfer enhancement of ferrofluid inside an 90° elbow channel by non-uniform magnetic field, J. Magn. Magn. Mater. 460 (2018) 302–311.
[67] M. Sheikholeslami, S. Soleimani, D.D. Ganji, Effect of electric field on hydrothermal behavior of nanofluid in a complex geometry, J. Mol. Liq. 213 (2016) 153–161.
[68] M. Sheikholeslami, Numerical simulation for solidification in a LHTESS by means of Nano-enhanced PCM, J. Taiwan Inst. Chem. Eng. 86 (2018) 25–41.
[69] M. Sheikholeslami, Numerical modeling of Nano enhanced PCM solidification in an enclosure with metallic fin, J. Mol. Liq. 259 (2018) 424–438.
[70] M. Sheikholeslami, A. Ghasemi, Solidification heat transfer of nanofluid in existence of thermal radiation by means of FEM, Int. J. Heat Mass Transfer 123 (2018) 418–431.
[71] M. Sheikholeslami, S.A. Shehzad, CVFEM simulation for nanofluid migration in a porous medium using Darcy model, Int. J. Heat Mass Transfer 122 (2018) 1264–1271.
[72] M. Sheikholeslami, M. Darzi, M.K. Sadoughi, Heat transfer improvement and pressure drop during condensation of refrigerant-based Nanofluid; an experimental procedure, Int. J. Heat Mass Transfer 122 (2018) 643–650.
[73] M. Sheikholeslami, H.B. Rokni, CVFEM for effect of Lorentz forces on nanofluid flow in a porous complex shaped enclosure by means of Non-equilibrium model, J. Mol. Liq. 254 (2018) 446–462.
[74] M. Sheikholeslami, S.A. Shehzad, Z. Li, Water based nanofluid free convection heat transfer in a three dimensional porous cavity with hot sphere obstacle in existence of Lorenz forces, Int. J. Heat Mass Transfer 125 (2018) 375–386.
[75] M. Sheikholeslami, M. Jafaryar, D.D. Ganji, Z. Li, Exergy loss analysis for nanofluid forced convection heat transfer in a pipe with modified turbulators, J. Mol. Liq. 262 (2018) 104–110.
[76] M. Sheikholeslami, M. Jafaryar, Z. Li, Nanofluid turbulent convective flow in a circular duct with helical turbulators considering CuO nanoparticles, Int. J. Heat Mass Transfer 124 (2018) 980–989.
[77] M. Sheikholeslami, Z. Li, M. Shamlooei, Nanofluid MHD natural convection through a porous complex shaped cavity considering thermal radiation, Phys. Lett. A 382 (2018) 1615–1632.

[78] M. Sheikholeslami, H.B. Rokni, Magnetic nanofluid flow and convective heat transfer in a porous cavity considering Brownian motion effects, Phys. Fluids 30 (1) (2018). Available from: https://doi.org/10.1063/1.5012517.
[79] M. Sheikholeslami, S.A. Shehzad, Simulation of water based nanofluid convective flow inside a porous enclosure via Non-equilibrium model, Int. J. Heat Mass Transfer 120 (2018) 1200–1212.
[80] M. Sheikholeslami, S.A. Shehzad, Non-Darcy free convection of Fe_3O_4-water nanoliquid in a complex shaped enclosure under impact of uniform Lorentz force, Chin. J. Phys. 56 (2018) 270–281.
[81] M. Sheikholeslami, M. Seyednezhad, Simulation of nanofluid flow and natural convection in a porous media under the influence of electric field using CVFEM, Int. J. Heat Mass Transfer 120 (2018) 772–781.
[82] M. Sheikholeslami, M. Shamlooei, R. Moradi, Numerical simulation for heat transfer intensification of nanofluid in a permeable curved enclosure considering shape effect of Fe_3O_4 nanoparticles, Chem. Eng. Process. Process Intensif. 124 (2018) 71–82.
[83] M. Sheikholeslami, T. Hayat, T. Muhammad, A. Alsaedi, MHD forced convection flow of nanofluid in a porous cavity with hot elliptic obstacle by means of Lattice Boltzmann method, Int. J. Mech. Sci. 135 (2018) 532–540.
[84] M. Sheikholeslami, Numerical investigation of nanofluid free convection under the influence of electric field in a porous enclosure, J. Mol. Liq. 249 (2018) 1212–1221.
[85] M. Sheikholeslami, CuO-water nanofluid flow due to magnetic field inside a porous media considering Brownian motion, J. Mol. Liq. 249 (2018) 921–929.
[86] M. Sheikholeslami, D.D. Ganji, Influence of electric field on Fe_3O_4-water nanofluid radiative and convective heat transfer in a permeable enclosure, J. Mol. Liq. 250 (2018) 404–412.
[87] M. Sheikholeslami, S.A. Shehzad, Z. Li, Nanofluid heat transfer intensification in a permeable channel due to magnetic field using Lattice Boltzmann method, Phys. B (2018). Available from: https://doi.org/10.1016/j.physb.2018.03.036.
[88] M. Sheikholeslami, A. Zeeshan, Numerical simulation of Fe_3O_4-water nanofluid flow in a non-Darcy porous media, Int. J. Numer. Methods Heat Fluid Flow 28 (3) (2018) 641–660. Available from: https://doi.org/10.1108/HFF-04-2017-0160.
[89] M. Sheikholeslami, D.D. Ganji, Numerical approach for magnetic nanofluid flow in a porous cavity using CuO nanoparticles, Mater. Des. 120 (2017) 382–393.
[90] M. Sheikholeslami, D.D. Ganji, Numerical modeling of magnetohydrodynamic CuO-water transportation inside a porous cavity considering shape factor effect, Colloids Surf. A 529 (2017) 705–714.

CHAPTER

18

Influence of Electric Field on Forced Convection of Nanofluid in a Porous Medium by Means of CVFEM

18.1 INTRODUCTION

Electrohydrodynamics (EHD), also known as electro-fluid-dynamics (EFD) or electrokinetics, is the study of the dynamics of electrically charged fluids. It is the study of the motions of ionized particles or molecules and their interactions with electric fields and the surrounding fluid. This active method can be combined with useful passive techniques such as nanofluid. Sheikholeslami [1] investigated Fe_3O_4-H_2O nanofluid thermal improvement in the presence of Coulomb forces. Sheikholeslami and Ganji [2] considered variable viscosity for simulation of nanofluid forced convection in the presence of an electric field. Sheikholeslami, and Bhatti [3] presented an active method for nanofluid heat transfer enhancement by means of EHD. Sheikholeslami and Rokni [4] presented the impact of EFD viscosity on nanofluid forced convection in a cavity with sinusoidal wall. Active and passive techniques for heat transfer enhancement can be combined [5–90].

18.2 EHD NANOFLUID FLOW IN A PERMEABLE ENCLOSURE WITH SINUSOIDAL WALL

18.2.1 Problem Definition

Fig. 18.1 depicts the porous enclosure and its boundary conditions. Ethylene glycol-Fe_3O_4 nanofluid is utilized. All walls are stationary except for the bottom wall. The influence of Darcy and Reynolds numbers on the contours of q is demonstrated in Fig. 18.2. The effect of Re on q is less sensible than Da. As the Darcy number augments, the distortion of the isoelectric density lines become greater.

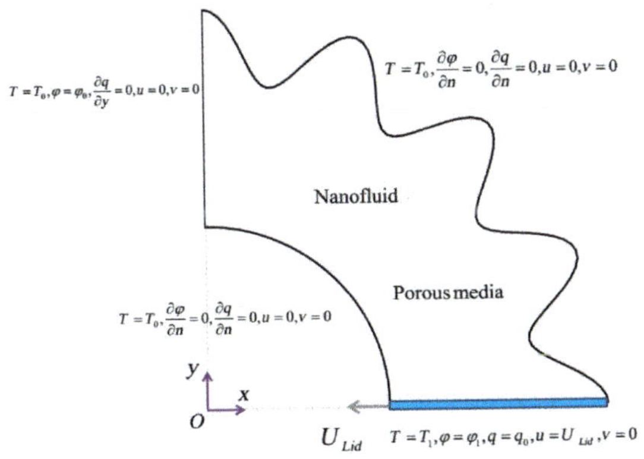

FIGURE 18.1 Geometry and the boundary conditions.

Application of Control Volume based Finite Element Method (CVFEM) for Nanofluid Flow and Heat Transfer.
DOI: https://doi.org/10.1016/B978-0-12-814152-6.00018-7

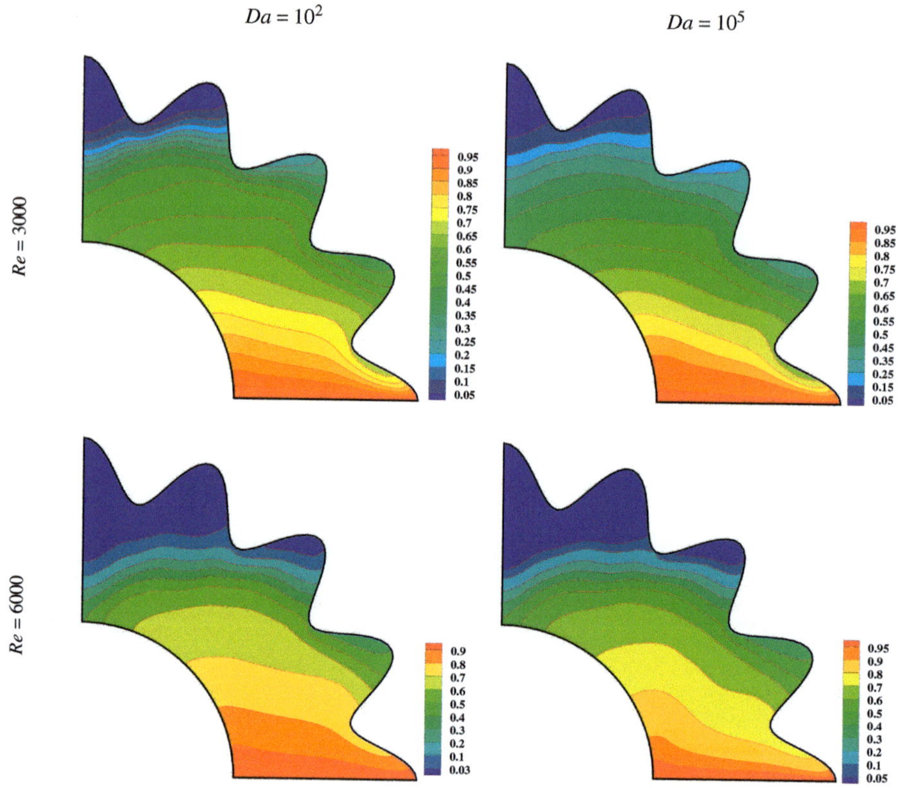

FIGURE 18.2 Electric density distribution injected by the bottom electrode when $\Delta\varphi = 10kV, \phi = 0.05, Rd = 0.8$.

18.2.2 Governing Equation

The definition of the electric field is:

$$\vec{E} = -\nabla\varphi \tag{18.1}$$

$$q = \nabla.\varepsilon\,\vec{E} \tag{18.2}$$

$$\vec{J} = q\vec{V} - D\nabla q + \sigma\,\vec{E} \tag{18.3}$$

$$\nabla.\vec{J} + \frac{\partial q}{\partial t} = 0 \tag{18.4}$$

The governing formulae are:

$$\begin{cases} \nabla.\vec{V} = 0, \\ \left(\left(\vec{V}.\nabla\right)\vec{V} + \dfrac{\partial \vec{V}}{\partial t}\right) = \dfrac{q\vec{E}}{\rho_{nf}} + \dfrac{\mu_{nf}}{\rho_{nf}}\nabla^2\vec{V} - \dfrac{\nabla p}{\rho_{nf}} - \dfrac{\mu_{nf}}{K\rho_{nf}}\vec{V}, \\ \left(\left(\vec{V}.\nabla\right)T + \dfrac{\partial T}{\partial t}\right) = \dfrac{k_{nf}}{(\rho C_p)_{nf}}\nabla^2 T + \dfrac{\vec{J}.\vec{E}}{(\rho C_p)_{nf}} - \dfrac{1}{(\rho C_p)_{nf}}\dfrac{\partial q_r}{\partial y}, \quad \left[q_r = -\dfrac{4\sigma_e}{3\beta_R}\dfrac{\partial T^4}{\partial y}, \quad T^4 \cong 4T_c^3 T - 3T_c^4\right], \\ \nabla\varphi = -\vec{E}, \\ \dfrac{\partial q}{\partial t} = -\nabla.\vec{J}, \\ q = \nabla.\varepsilon\vec{E} \end{cases} \tag{18.5}$$

$(\rho C_p)_{nf}$, μ_{nf}, and ρ_{nf} can be obtained as:

$$\begin{aligned} &(\rho C_p)_{nf} = (\rho C_p)_f(1-\phi) + (\rho C_p)_s\phi, \\ &\mu = A_1 + A_2(\Delta\varphi) + A_3(\Delta\varphi)^2 + A_4(\Delta\varphi)^3, \\ &\rho_{nf} = \rho_f(1-\phi) + \rho_s\phi \end{aligned} \tag{18.6}$$

Properties of Fe_3O_4 and ethylene glycol are illustrated in Table 18.1. Table 18.2 illustrates the coefficient values of this formula.

k_{nf} can be expressed as:

$$\frac{k_{nf}}{k_f} = \frac{-m(k_f - k_p)\phi + (k_p - k_f)\phi + mk_f + k_p + k_f}{mk_f + (k_f - k_p)\phi + k_f + k_p} \tag{18.7}$$

Different values of shape factors for various shapes of nanoparticles are illustrated in Table 18.3.

TABLE 18.1 Thermophysical Properties of Ethylene Glycol and Nanoparticles

	$\rho(kg/m^3)$	$C_p(j/kgk)$	$k(W/m.k)$
Ethylene glycol	1110	2400	0.26
Fe_3O_4	5200	670	6

TABLE 18.2 The Coefficient Values of Eq. (18.6)

Coefficient Values	$\phi = 0$	$\phi = 0.05$
A_1	1.0603E + 001	9.5331
A_2	− 2.698E-003	− 3.4119E-003
A_3	2.9082E-006	5.5228E-006
A_4	− 1.1876E-008	− 4.1344E-008

TABLE 18.3 The Values of Shape Factor of Different Shapes of Nanoparticles

	Spherical	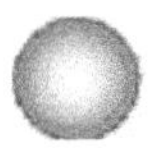	3
m	Platelet		5.7
	Cylinder		4.8
	Brick	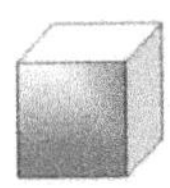	3.7

So, the final PDE in the presence of thermal radiation and an electric field in a porous medium are:

$$\begin{cases} \nabla.\vec{V}=0, \\ \left(\left(\vec{V}.\nabla\right)\vec{V}+\dfrac{\partial\vec{V}}{\partial t}\right)=\dfrac{1}{\mathrm{Re}}\dfrac{\rho_{nf}/\rho_f}{\mu_{nf}/\mu_f}\nabla^2\vec{V}-\nabla p+\dfrac{S_E}{\rho_{nf}/\rho_f}q\,\vec{E}-\dfrac{1}{\mathrm{Re}\,Da}\dfrac{\mu_{nf}}{\mu_f}\left(\dfrac{\rho_{nf}}{\rho_f}\right)^{-1}\vec{V} \\ \left(\left(\vec{V}.\nabla\right)\theta+\dfrac{\partial\theta}{\partial t}\right)=\dfrac{1}{\mathrm{PrRe}}\dfrac{k_{nf}/k_f}{(\rho C_p)_{nf}/(\rho C_p)_f}\nabla^2\theta+S_E\dfrac{1}{(\rho C_p)_{nf}/(\rho C_p)_f}Ec\left(\vec{J}\,.\,\vec{E}\right)+\dfrac{4}{3}\left(\dfrac{k_{nf}}{k_f}\right)^{-1}Rd\dfrac{\partial^2\theta}{\partial Y^2} \\ \vec{E}=-\nabla\varphi \\ q=\nabla.\varepsilon\vec{E} \\ \nabla.\vec{J}=-\dfrac{\partial q}{\partial t} \end{cases} \quad (18.8)$$

where

$$(\overline{u},\overline{v})=\frac{(u,v)}{U_{Lid}},\overline{\varphi}=\frac{\varphi-\varphi_0}{\nabla\varphi},(\overline{y},\overline{x})=\frac{(y,x)}{L},\theta=\frac{T-T_0}{\nabla T},$$
$$\overline{t}=\frac{tU_{Lid}}{L},\overline{p}=\frac{P}{\rho U_{Lid}{}^2}\overline{q}=\frac{q}{q_0},\overline{E}=\frac{E}{E_0},$$
$$\nabla T=T_1-T_0,\nabla\varphi=\varphi_1-\varphi_0, \quad (18.9)$$

Vorticity and stream function should be employed in order to diminish the pressure gradient:

$$v=-\frac{\partial\psi}{\partial x},\omega=\frac{\partial v}{\partial x}-\frac{\partial u}{\partial y},\frac{\partial\psi}{\partial y}=u,\Psi=\frac{\psi L}{U_{Lid}},\Omega=\frac{\omega}{LU_{Lid}} \quad (18.10)$$

Nu_{loc} and Nu_{ave} along the bottom wall are calculated as:

$$Nu_{loc}=\left(\frac{k_{nf}}{k_f}\right)\left(1+\frac{4}{3}Rd\left(\frac{k_{nf}}{k_f}\right)^{-1}\right)\frac{\partial\Theta}{\partial X} \quad (18.11)$$

$$Nu_{ave}=\frac{1}{L}\int_0^L Nu_{loc}\,dY \quad (18.12)$$

18.2.3 Effects of Active Parameters

Forced convection of nanofluid in the presence of Coulomb forces is presented. The radiation term is added to the energy equation. The porous enclosure is filled with Fe_3O_4-Ethylene glycol and has one lid wall. The roles of Darcy number ($Da=10^2$ to 10^5), Radiation parameter ($Rd=0$ to 0.8), supplied voltage ($\Delta\varphi=0$ to $10kV$), volume fraction of Fe_3O_4 ($\phi=0\%$ to 5%), and Reynolds number ($\mathrm{Re}=3000$ to 6000) are illustrated graphically. The impact of shape factor on the rate of heat transfer is reported in Table 18.4. In this table, various shapes of nanoparticles are utilized. The maximum Nu is caused by Platelet, followed by Cylinder, Brick, and Spherical. So, Platelet nanoparticle has been utilized for further investigation. Impacts of $Da, \Delta\varphi$, and Re on isotherms and streamlines are demonstrated in Figs. 18.3–18.6. At low Reynolds number, there is one clockwise eddy in the streamlines. The center of the main eddy is near the lid wall. Augmenting the Darcy number leads to the generation of the second eddy which rotates counterclockwise and the center of the main eddy shifts to upwards. Applying an electric field causes the strength of the main eddy to enhance and shifts the center of the eddy upwards. The isotherms become more disturbed in the presence of an electric field. A thermal plume appears when the Reynolds number is increased. Also, by increasing Re, the strengths of the rotating eddies are augmented. As the Coulomb force increases, the secondary eddy diminishes and the strength of main eddy is enhanced.

TABLE 18.4 Effect of Shape of Nanoparticles on Nusselt Number When $Rd = 0.8, \mathrm{Re} = 6000, \Delta\varphi = 10, \phi = 0.05$

	Da	
	10^2	10^5
Spherical	2.338308	2.884048
Brick	2.359831	2.913817
Cylinder	2.391644	2.957562
Platelet	2.415995	2.990852

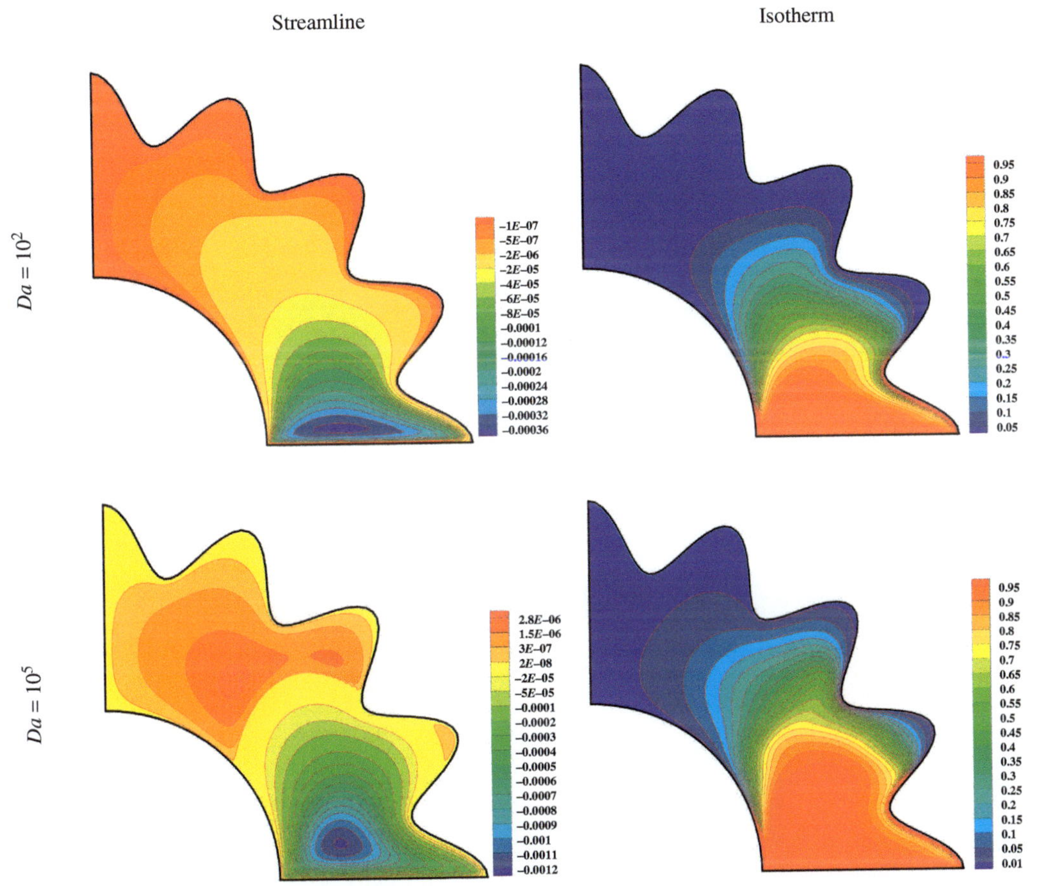

FIGURE 18.3 Effect of Darcy number on streamlines and isotherms when $\mathrm{Re} = 3000, \Delta\varphi = 0kV, \phi = 0.05, Rd = 0.8$.

Nu_{ave} versus $\mathrm{Re}, Da, Rd,$ and $\Delta\varphi$ is depicted in Fig. 18.7. The related formula is:

$$\begin{aligned} Nu_{ave} = & -2.08 - 2.65\Delta\varphi + 1.3\mathrm{Re}^* + 1.27\log(Da) + 0.8Rd \\ & - 0.3 \times 10^{-5}\Delta\varphi \mathrm{Re}^* + 0.04\Delta\varphi\log(Da) \\ & + 0.3 \times 10^{-4}\Delta\varphi Rd - 0.017\mathrm{Re}^*\log(Da) \\ & - 0.16\mathrm{Re}^* Rd + 0.16\log(Da)Rd \\ & + 0.02\Delta\varphi^2 - 0.15(\mathrm{Re}^*)^2 - 0.175\left(\log(Da)\right)^2 + 1.76Rd^2 \end{aligned} \tag{18.13}$$

where $\mathrm{Re}^* = 0.001\mathrm{Re}$ and $\Delta\varphi$ is voltage supply in Kilovolt. In the absence of Coulomb force, the Nusselt number augments with the rise of the Reynolds number but the opposite behavior is reported in the presence of such forces. The electric field helps the convective mode to enhance. So, Nu_{ave} augments with the increase of $\Delta\varphi$. Thermal radiation enhances the temperature gradient near the lid wall. The influence of the Darcy number is the same as that for the radiation parameter. Therefore, Nu_{ave} is an increasing function of Rd, Da.

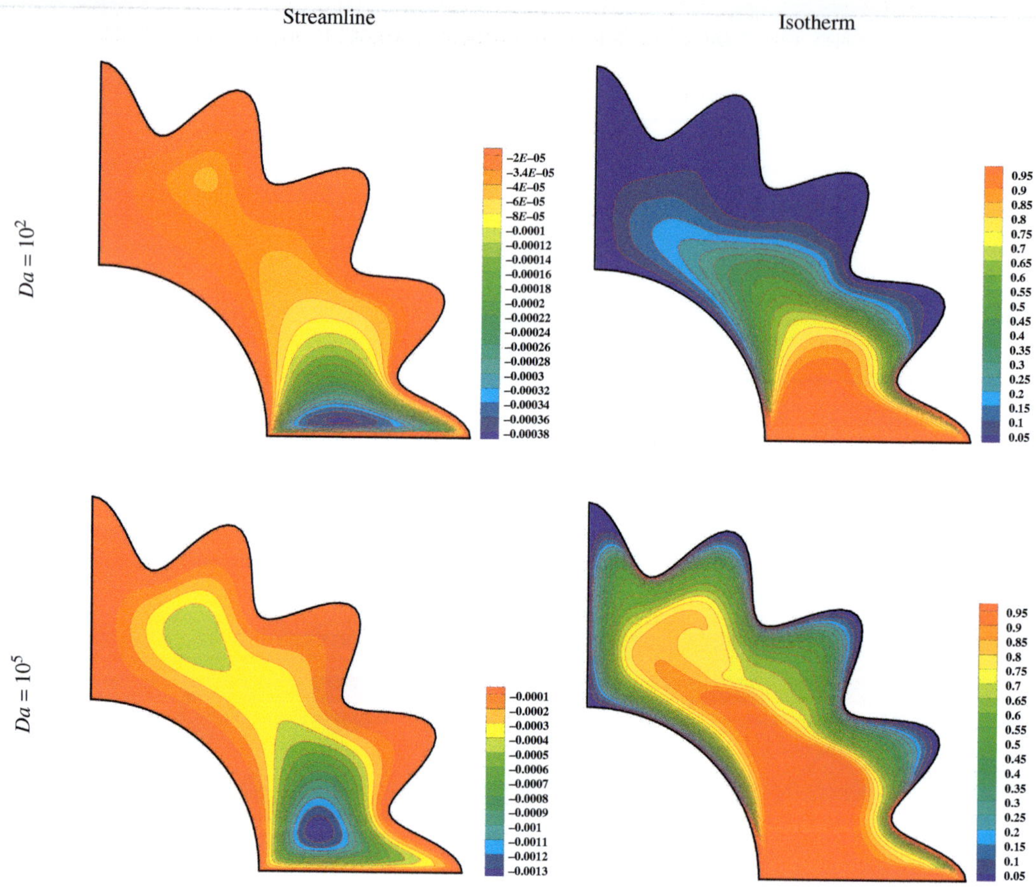

FIGURE 18.4 Effect of Darcy number on streamlines and isotherms when $\mathrm{Re} = 3000, \Delta\varphi = 10kV, \phi = 0.05, Rd = 0.8$.

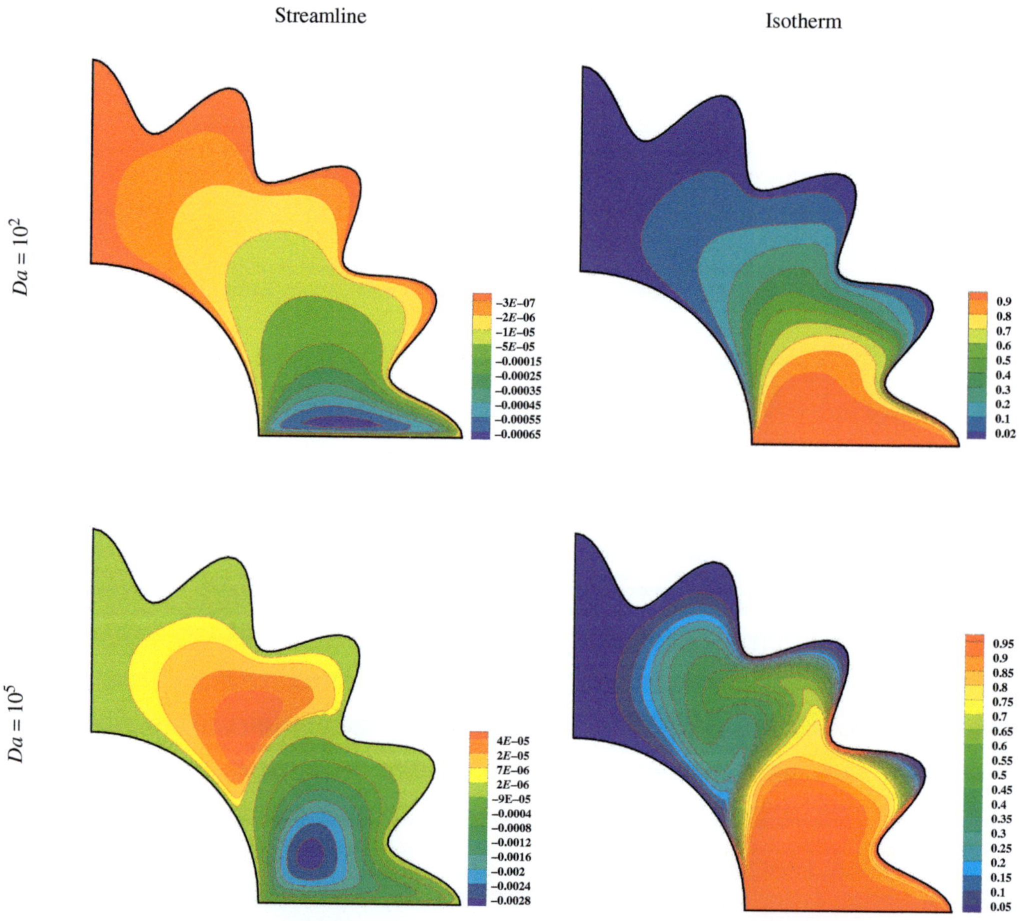

FIGURE 18.5 Effect of Darcy number on streamlines and isotherms when $\mathrm{Re} = 6000, \Delta\varphi = 0kV, \phi = 0.05, Rd = 0.8$.

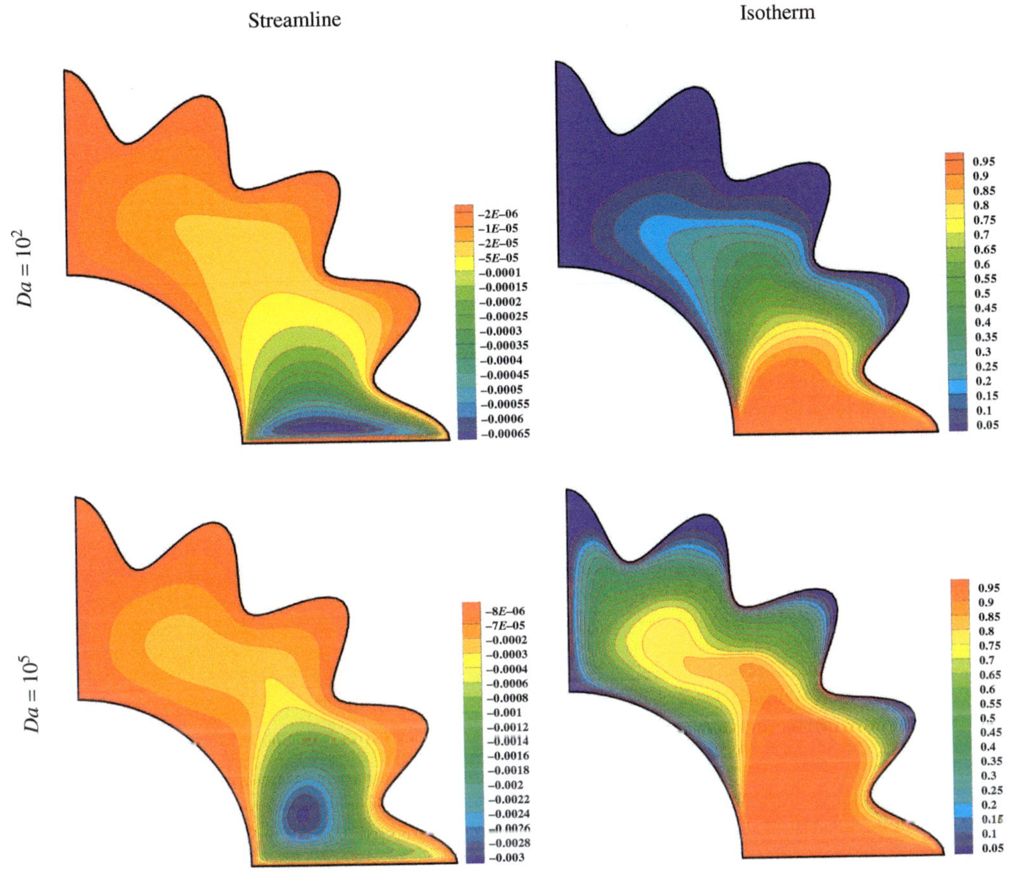

FIGURE 18.6 Effect of Darcy number on streamlines and isotherms when $\mathrm{Re} = 6000$, $\Delta\varphi = 10kV$, $\phi = 0.05$, $Rd = 0.8$.

18.3 EFFECT OF SHAPE FACTOR ON ELECTROHYDRODYNAMIC NANOFLUID FLOW IN A POROUS MEDIUM

18.3.1 Problem Definition

Fig. 18.8 depicts the porous enclosure and its boundary conditions. Ethylene glycol-Fe_3O_4 nanofluid is utilized. All walls are stationary except for the bottom wall. The influence of Darcy and Reynolds numbers on the contours of q is demonstrated in Fig. 18.9. The effect of Re on q is less sensible than forDa. As the Darcy number increases the distortion of the isoelectric density lines become greater.

18.3.2 Governing Equation

The definition of the electric field is:

$$\vec{E} = -\nabla\varphi \tag{18.14}$$

$$q = \nabla.\varepsilon\vec{E} \tag{18.15}$$

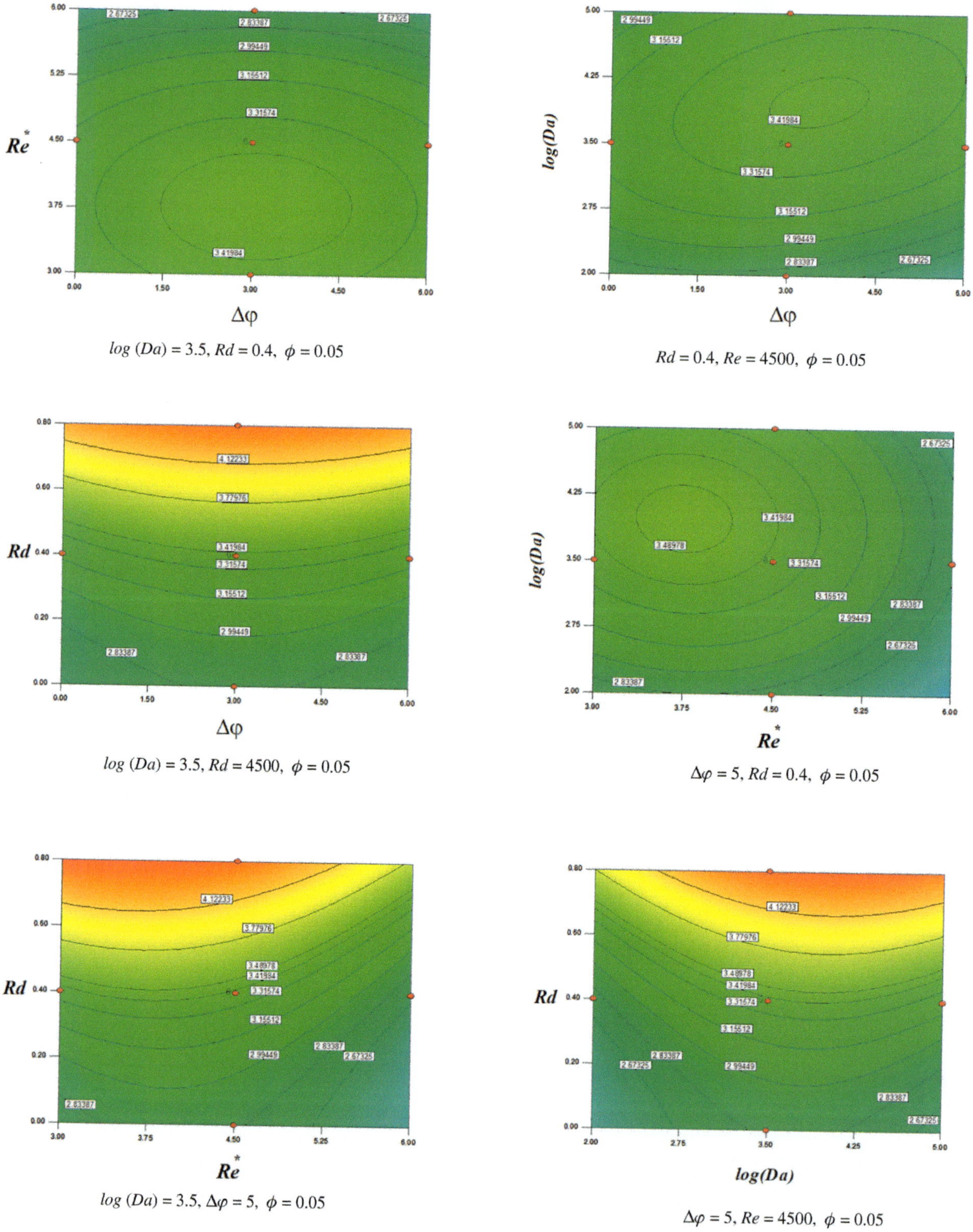

FIGURE 18.7 Effects of Da, $\Delta\varphi$, Rd, and Re on average Nusselt number.

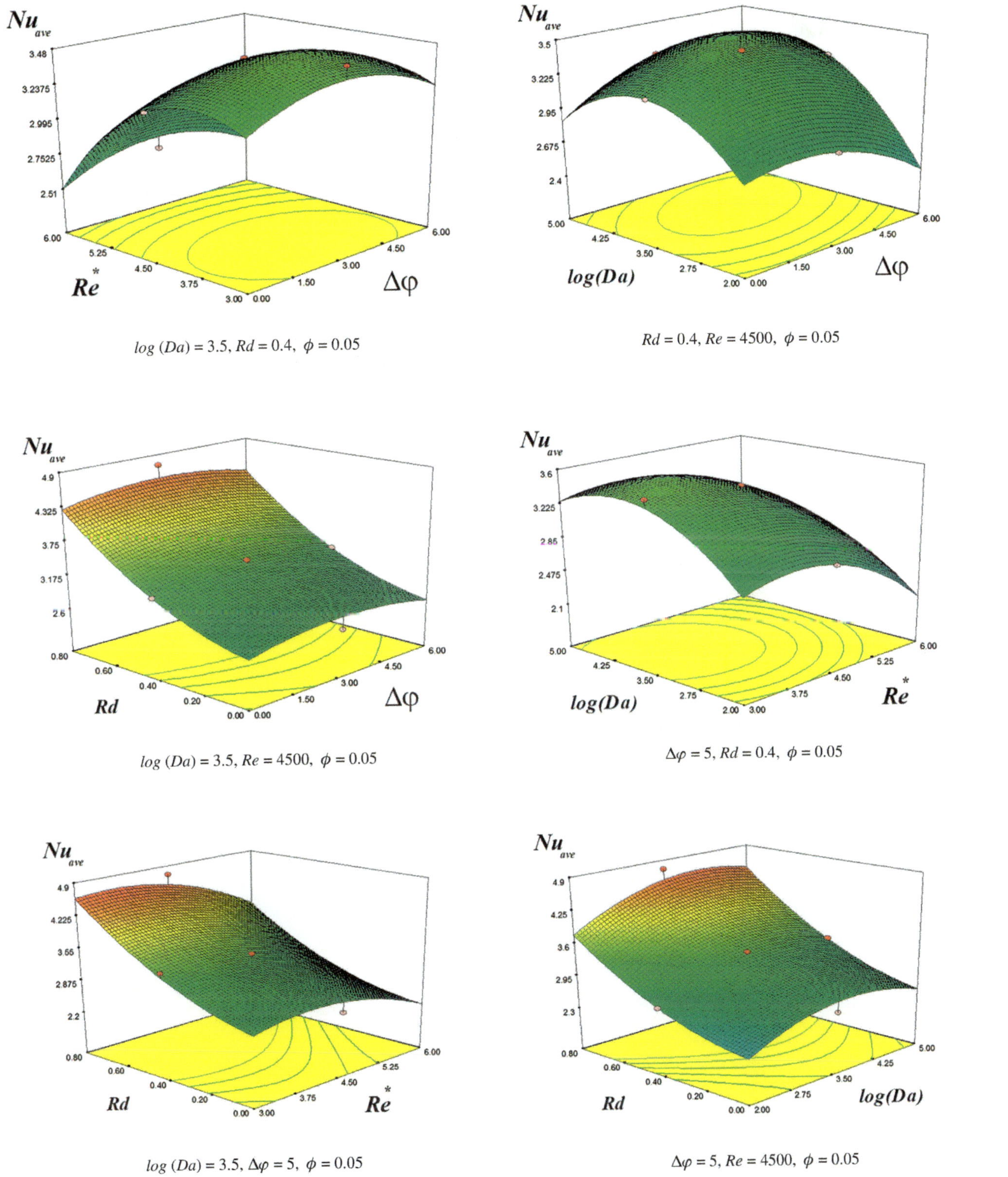

FIGURE 18.7 (Continued).

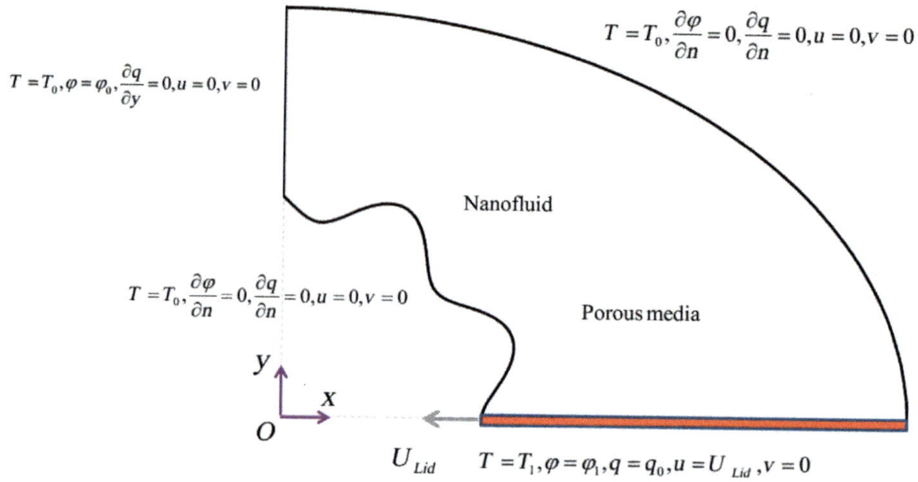

FIGURE 18.8 Geometry and the boundary conditions.

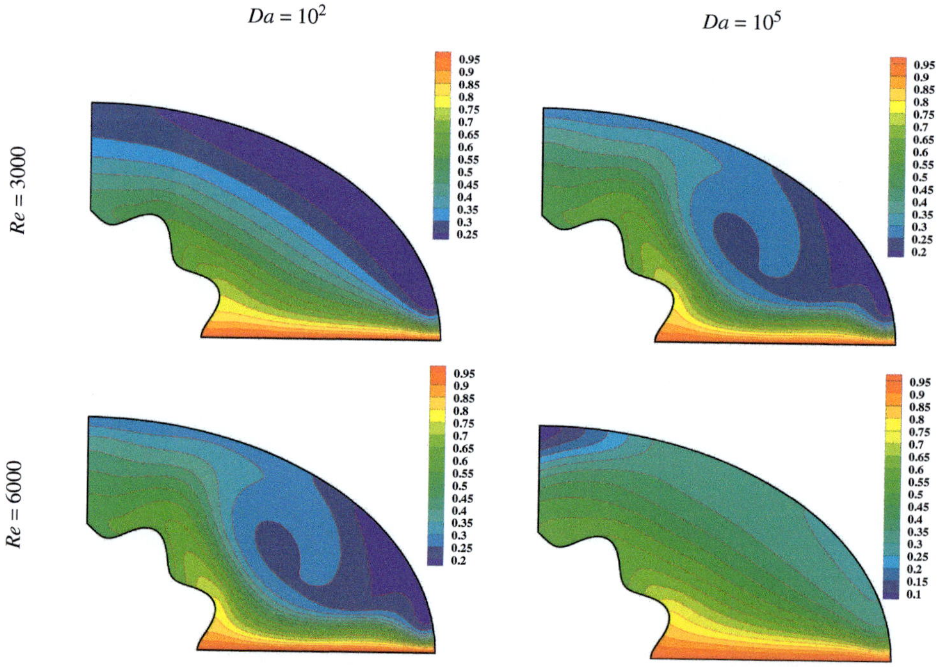

FIGURE 18.9 Electric density distribution injected by the bottom electrode when $\Delta\varphi = 10kV, \phi = 0.05, Rd = 0.8$.

$$\vec{J} = q\vec{V} - D\nabla q + \sigma\vec{E} \tag{18.16}$$

$$\nabla.\vec{J} + \frac{\partial q}{\partial t} = 0 \tag{18.17}$$

The governing formulae are:

$$\begin{cases} \nabla.\vec{V}=0, \\ \left(\left(\vec{V}.\nabla\right)\vec{V}+\dfrac{\partial \vec{V}}{\partial t}\right)=\dfrac{q\vec{E}}{\rho_{nf}}+\dfrac{\mu_{nf}}{\rho_{nf}}\nabla^2\vec{V}-\dfrac{\nabla p}{\rho_{nf}}-\dfrac{\mu_{nf}}{K\rho_{nf}}\vec{V}, \\ \left(\left(\vec{V}.\nabla\right)T+\dfrac{\partial T}{\partial t}\right)=\dfrac{k_{nf}}{(\rho C_p)_{nf}}\nabla^2 T+\dfrac{\vec{J}.\vec{E}}{(\rho C_p)_{nf}}-\dfrac{1}{(\rho C_p)_{nf}}\dfrac{\partial q_r}{\partial y}, \quad \left[q_r=-\dfrac{4\sigma_e}{3\beta_R}\dfrac{\partial T^4}{\partial y}, \quad T^4\cong 4T_c^3T-3T_c^4\right], \\ \nabla\varphi=-\vec{E}, \\ \dfrac{\partial q}{\partial t}=-\nabla.\vec{J}, \\ q=\nabla.\varepsilon\vec{E} \end{cases} \tag{18.18}$$

$(\rho C_p)_{nf}$, μ_{nf}, and ρ_{nf} can be obtained as:

$$\begin{aligned} &(\rho C_p)_{nf}=(\rho C_p)_f(1-\phi)+(\rho C_p)_s\phi, \\ &\mu=A_1+A_2(\Delta\varphi)+A_3(\Delta\varphi)^2+A_4(\Delta\varphi)^3, \\ &\rho_{nf}=\rho_f(1-\phi)+\rho_s\phi \end{aligned} \tag{18.19}$$

k_{nf} can be expressed as:

$$\frac{k_{nf}}{k_f}=\frac{-m(k_f-k_p)\phi+(k_p-k_f)\phi+mk_f+k_p+k_f}{mk_f+(k_f-k_p)\phi+k_f+k_p} \tag{18.20}$$

So, the final PDE in the presence of thermal radiation and an electric field in a porous medium are:

$$\begin{cases} \nabla.\vec{V}=0, \\ \left(\left(\vec{V}.\nabla\right)\vec{V}+\dfrac{\partial \vec{V}}{\partial t}\right)=\dfrac{1}{\text{Re}}\dfrac{\rho_{nf}/\rho_f}{\mu_{nf}/\mu_f}\nabla^2\vec{V}-\nabla p+\dfrac{S_E}{\rho_{nf}/\rho_f}q\vec{E}-\dfrac{1}{\text{Re}\,Da}\dfrac{\mu_{nf}}{\mu_f}\left(\dfrac{\rho_{nf}}{\rho_f}\right)^{-1}\vec{V} \\ \left(\left(\vec{V}.\nabla\right)\theta+\dfrac{\partial \theta}{\partial t}\right)=\dfrac{1}{Pr\text{Re}}\dfrac{k_{nf}/k_f}{(\rho C_p)_{nf}/(\rho C_p)_f}\nabla^2\theta+S_E\dfrac{1}{(\rho C_p)_{nf}/(\rho C_p)_f}Ec\left(\vec{J}.\vec{E}\right)+\dfrac{4}{3}\left(\dfrac{k_{nf}}{k_f}\right)^{-1}Rd\dfrac{\partial^2\theta}{\partial Y^2} \\ \vec{E}=-\nabla\varphi \\ q=\nabla.\varepsilon\vec{E} \\ \nabla.\vec{J}=-\dfrac{\partial q}{\partial t} \end{cases} \tag{18.21}$$

where

$$\begin{aligned} &(\bar{u},\bar{v})=\frac{(u,v)}{U_{Lid}},\bar{\varphi}=\frac{\varphi-\varphi_0}{\nabla\varphi},(\bar{y},\bar{x})=\frac{(y,x)}{L},\theta=\frac{T-T_0}{\nabla T}, \\ &\bar{t}=\frac{tU_{Lid}}{L},\bar{p}=\frac{P}{\rho U_{Lid}{}^2}\bar{q}=\frac{q}{q_0},\bar{E}=\frac{E}{E_0}, \\ &\nabla T=T_1-T_0,\nabla\varphi=\varphi_1-\varphi_0, \end{aligned} \tag{18.22}$$

Vorticity and stream function should be employed in order to diminish the pressure gradient:

$$v=-\frac{\partial\psi}{\partial x},\omega=\frac{\partial v}{\partial x}-\frac{\partial u}{\partial y},\frac{\partial\psi}{\partial y}=u,\Psi=\frac{\psi L}{U_{Lid}},\Omega=\frac{\omega}{LU_{Lid}} \tag{18.23}$$

Nu_{loc} and Nu_{ave} along the bottom wall are calculated as:

$$Nu_{loc}=\left(\frac{k_{nf}}{k_f}\right)\left(1+\frac{4}{3}Rd\left(\frac{k_{nf}}{k_f}\right)^{-1}\right)\frac{\partial\Theta}{\partial X} \tag{18.24}$$

$$Nu_{ave} = \frac{1}{L}\int_0^L Nu_{loc}\, dY \qquad (18.25)$$

18.3.3 Effects of Active Parameters

EHD nanofluid forced convection in the presence of thermal radiation is reported. The porous enclosure i filled with Fe_3O_4-Ethylene glycol and has one lid wall. The roles of Darcy number ($Da = 10^2$ to 10^5), Radiatio parameter ($Rd = 0$ to 0.8), supplied voltage ($\Delta\varphi = 0$ to $10kV$), volume fraction of Fe_3O_4 ($\phi = 0\%$ to 5%), anc Reynolds number (Re = 3000 to 6000) are illustrated graphically. The impact of shape factor on the rate of hea transfer is reported in Table 18.5. In this table, various shapes of nanoparticles are utilized. The maximum Nu i caused by Platelet, followed by Cylinder, Brick, and Spherical. So, Platelet nanoparticle has been utilized fo further investigation. Impacts of $Da, \Delta\varphi$, and Re on isotherms and streamlines are demonstrated i Figs. 18.10–18.13. At low Reynolds number, there is one clockwise eddy in the streamlines. The center of the

TABLE 18.5 Effect of Shape of Nanoparticles on Nusselt Number When $Rd = 0.8$, Re = 6000, $\Delta\varphi = 10$, $\phi = 0.05$

	Da	
	10^2	10^5
Spherical	3.697271	6.084579
Brick	3.740571	6.147401
Cylinder	3.805097	6.239713
Platelet	3.855149	6.318966

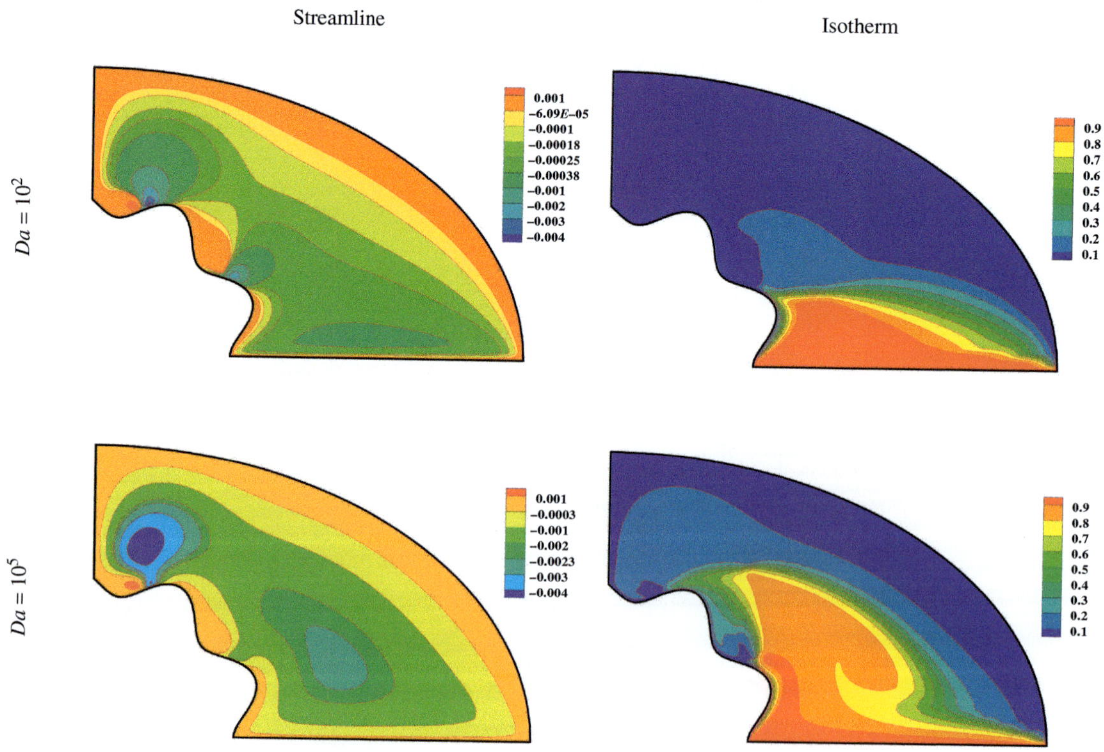

FIGURE 18.10 Effect of Darcy number on streamlines and isotherms when Re = 3000, $\Delta\varphi = 0kV$, $\phi = 0.05$, $Rd = 0.8$.

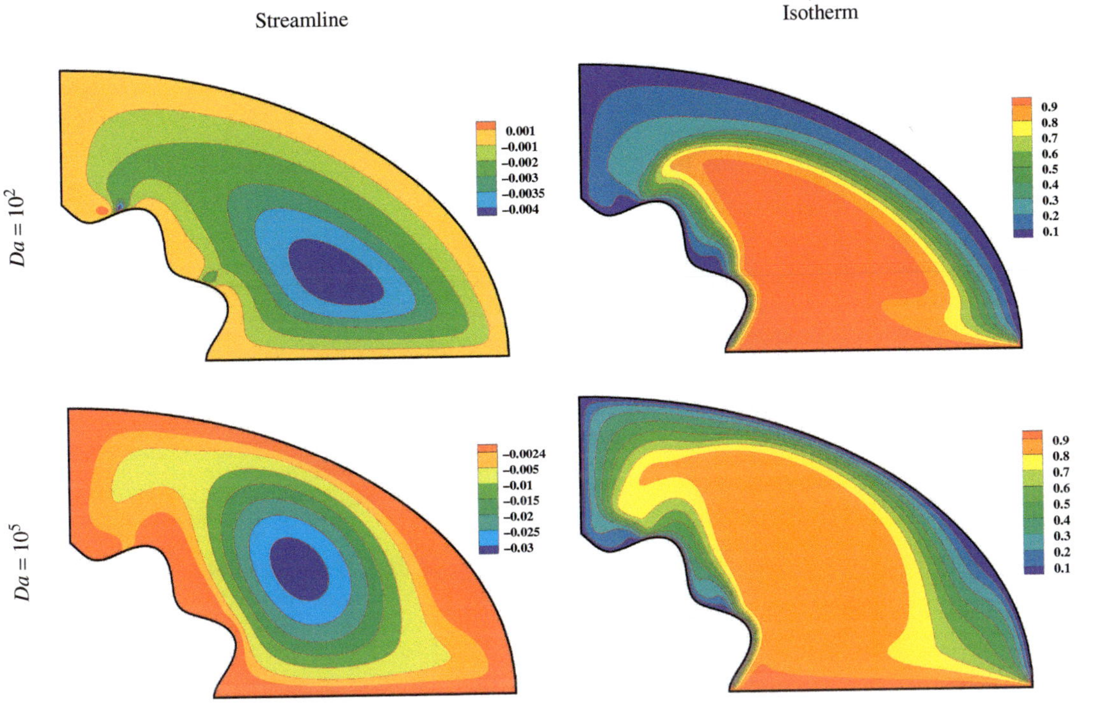

FIGURE 18.11 Effect of Darcy number on streamlines and isotherms when $Re = 3000, \Delta\varphi = 10kV, \phi = 0.05, Rd = 0.8$.

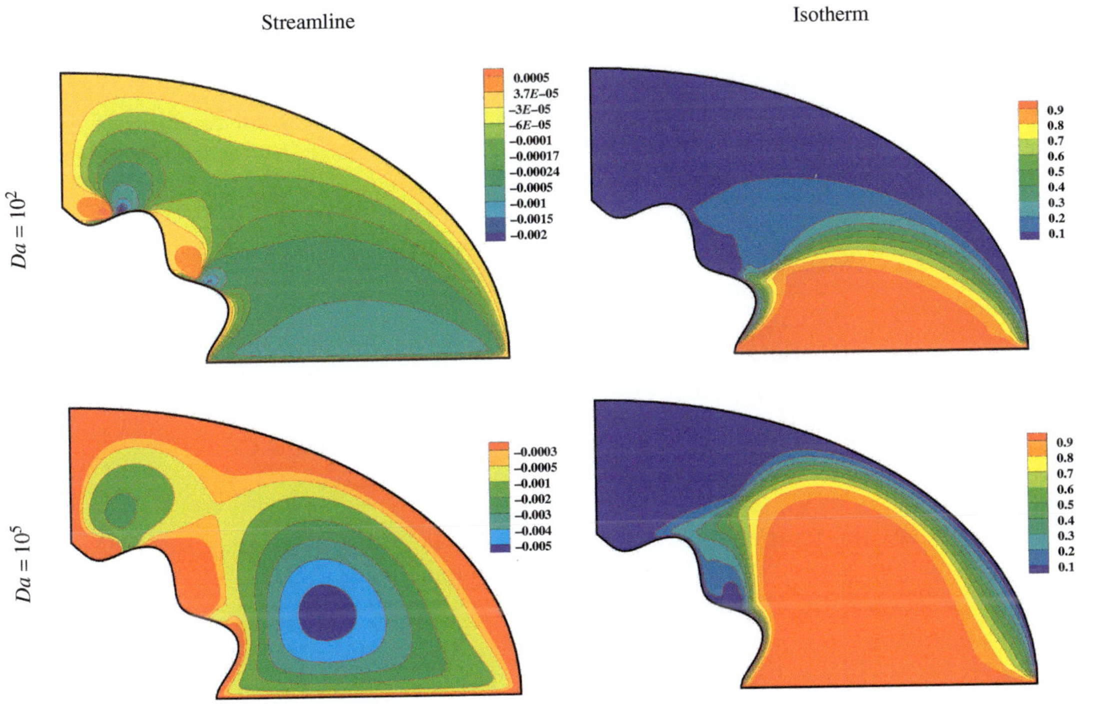

FIGURE 18.12 Effect of Darcy number on streamlines and isotherms when $Re = 6000, \Delta\varphi = 0kV, \phi = 0.05, Rd = 0.8$.

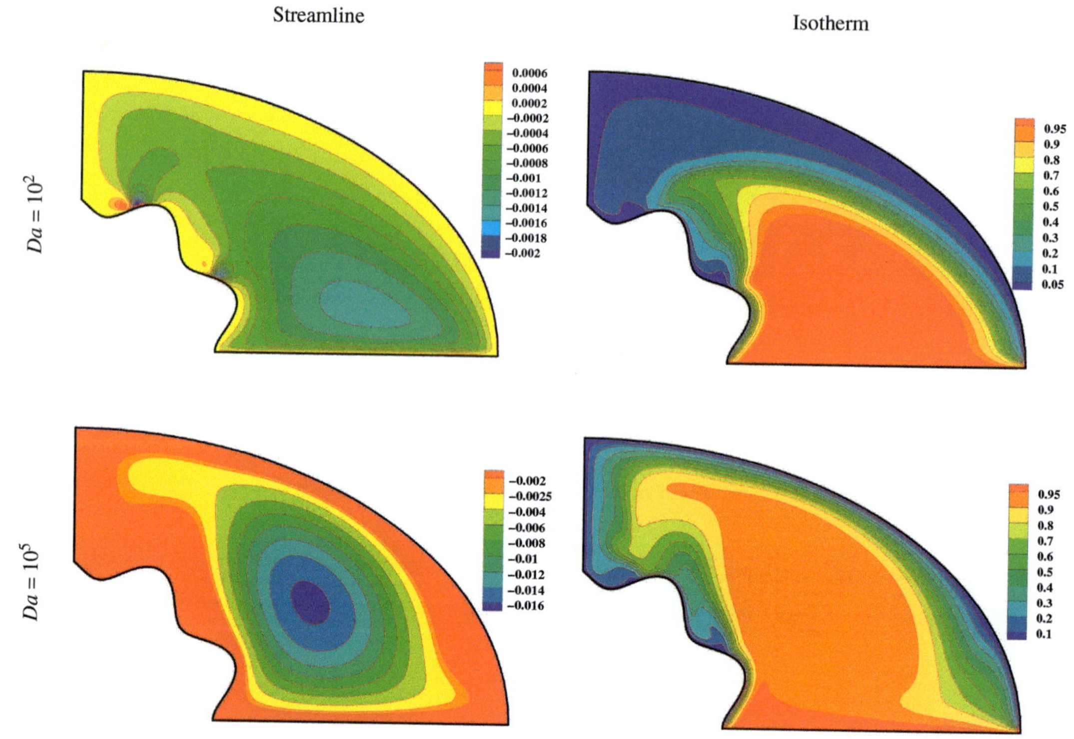

FIGURE 18.13 Effect of Darcy number on streamlines and isotherms when $\mathrm{Re} = 6000, \Delta\varphi = 10kV, \phi = 0.05, Rd = 0.8$.

main eddy is near the lid wall. Augmenting the Darcy number leads to the generation of the second eddy which rotates counterclockwise and the center of the main eddy shifts upward. Applying an electric field causes the strength of the main eddy to enhance and shifts the center of the eddy upward. The isotherms become more disturbed in the presence of an electric field. A thermal plume appears when the Reynolds number is increased. Also, by increasing Re, the strengths of the rotating eddies are augmented. As the Coulomb force increases, the secondary eddy diminishes and the strength of main eddy is enhanced.

Nu_{ave} versus Re, Da, Rd, and $\Delta\varphi$ is depicted in Fig. 18.14. The related formula is:

$$\begin{aligned} Nu_{ave} = & -2.26 + 0.08\Delta\varphi + 2.01\mathrm{Re}^* + 1.05\log(Da) + 3.25Rd \\ & - 0.036\Delta\varphi\mathrm{Re}^* + 0.023\Delta\varphi\log(Da) \\ & + 0.15\Delta\varphi Rd - 0.31\mathrm{Re}^*\log(Da) \\ & - 0.64\mathrm{Re}^* Rd + 0.41\log(Da)Rd \\ & + 0.015\Delta\varphi^2 - 0.2(\mathrm{Re}^*)^2 - 0.14(\log(Da))^2 + 1.46Rd^2 \end{aligned} \tag{18.26}$$

where $\mathrm{Re}^* = 0.001\mathrm{Re}$ and $\Delta\varphi$ is voltage supply in Kilovolt. In the absence of the Coulomb force, the Nusselt number increases with the rise of Reynolds number but the opposite behavior is reported in the presence of such forces. An electric field helps to enhance the convective mode. So, Nu_{ave} augments with the increase of $\Delta\varphi$. Thermal radiation enhances the temperature gradient near the lid wall. The influence of Darcy number is the same as that for the radiation parameter. Therefore, Nu_{ave} is an increasing function of Rd, Da.

18.4 EFFECT OF ELECTIVE FIELD ON NANOFLUID FLOW IN A POROUS LID DRIVEN CAVITY IN EXISTENCE OF ELECTRIC FIELD

18.4.1 Problem Definition

Fig. 18.15 depicts the porous enclosure and its boundary conditions. Ethylene glycol-Fe_3O_4 nanofluid is utilized. All walls are stationary except for bottom wall. The influence of Darcy and Reynolds numbers on the

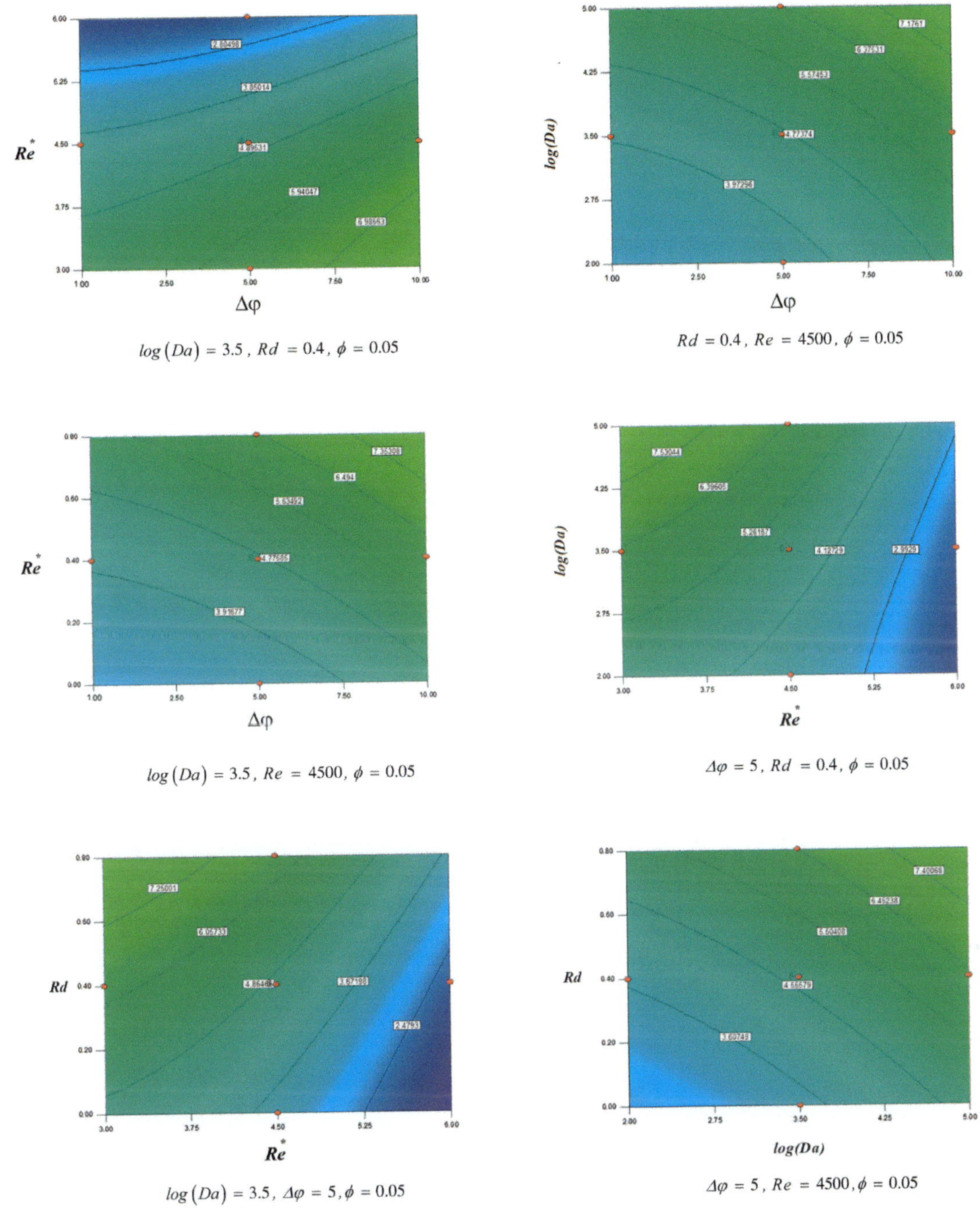

FIGURE 18.14 Effects of *Da*, $\Delta\varphi$, *Rd*, and Re on average Nusselt number.

contours of q is demonstrated in Fig. 18.16. The effect of Re on q is less sensible than for *Da*. As the Darcy number augments the distortion of the isoelectric density lines become greater.

18.4.2 Governing Equation

The definition of electric field is:

$$\vec{E} = -\nabla\varphi \tag{18.27}$$

$$q = \nabla.\varepsilon\,\vec{E} \tag{18.28}$$

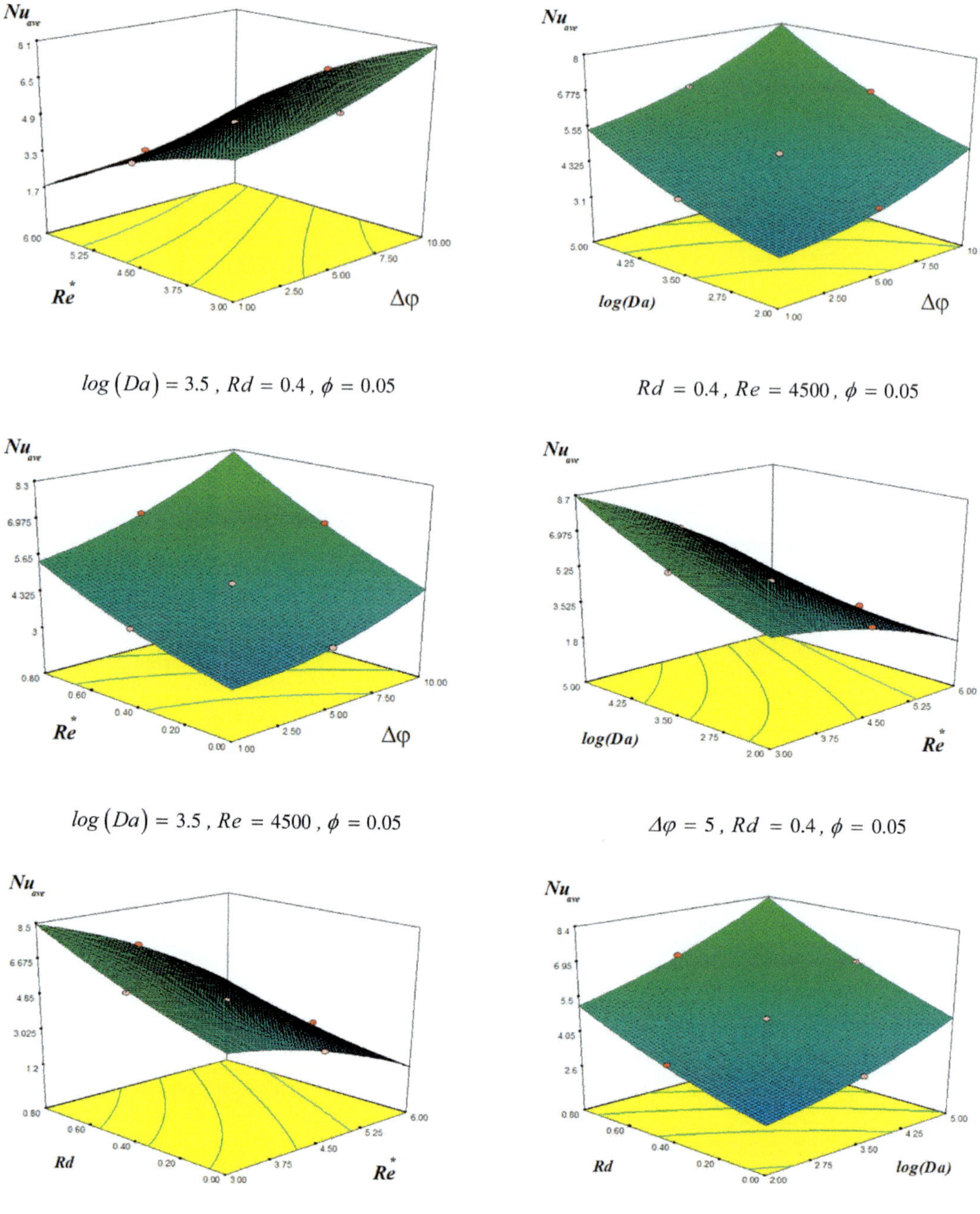

$\log(Da) = 3.5$, $Rd = 0.4$, $\phi = 0.05$

$Rd = 0.4$, $Re = 4500$, $\phi = 0.05$

$\log(Da) = 3.5$, $Re = 4500$, $\phi = 0.05$

$\Delta\varphi = 5$, $Rd = 0.4$, $\phi = 0.05$

$\log(Da) = 3.5$, $\Delta\varphi = 5$, $\phi = 0.05$

$\Delta\varphi = 5$, $Re = 4500$, $\phi = 0.05$

FIGURE 18.14 *(Continued)*

$$\vec{J} = q\vec{V} - D\nabla q + \sigma\vec{E} \tag{18.29}$$

$$\nabla . \vec{J} + \frac{\partial q}{\partial t} = 0 \tag{18.30}$$

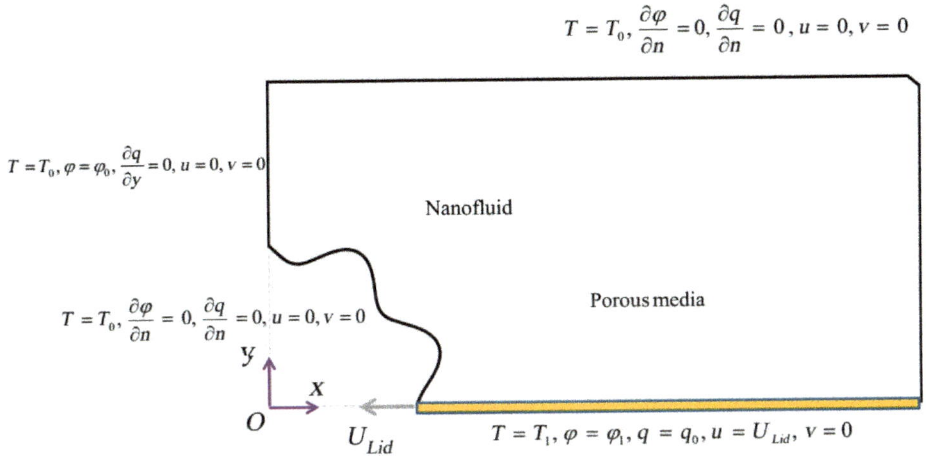

FIGURE 18.15 Geometry and the boundary conditions.

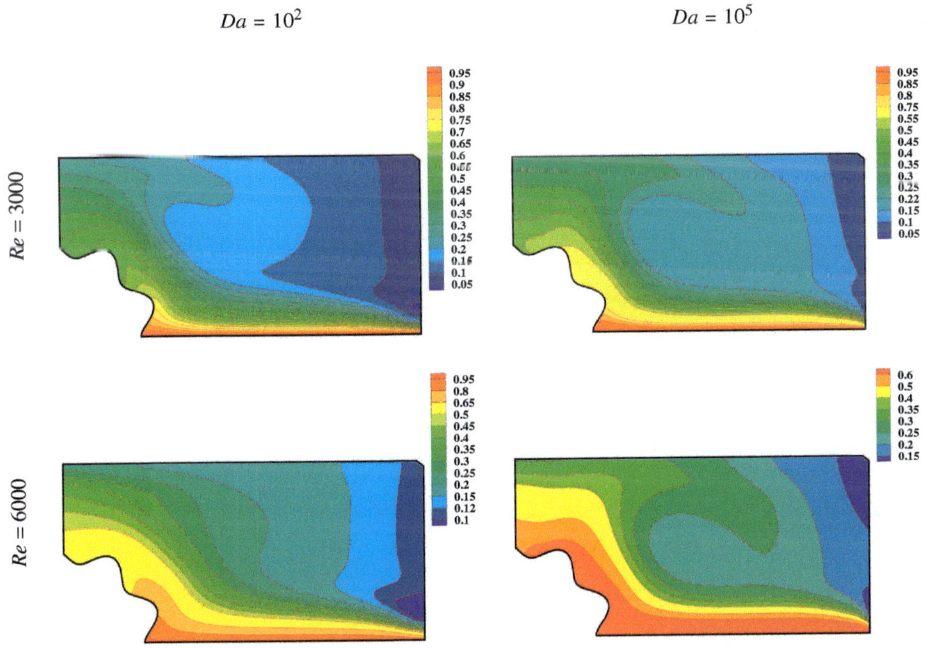

FIGURE 18.16 Electric density distribution injected by the bottom electrode when $\Delta\varphi = 10kV, \phi = 0.05, Rd = 0.8$.

The governing formulae are:

$$\begin{cases} \nabla.\vec{V} = 0, \\ \left(\left(\vec{V}.\nabla\right)\vec{V} + \dfrac{\partial \vec{V}}{\partial t}\right) = \dfrac{q\vec{E}}{\rho_{nf}} + \dfrac{\mu_{nf}}{\rho_{nf}}\nabla^2\vec{V} - \dfrac{\nabla p}{\rho_{nf}} - \dfrac{\mu_{nf}}{K\rho_{nf}}\vec{V}, \\ \left(\left(\vec{V}.\nabla\right)T + \dfrac{\partial T}{\partial t}\right) = \dfrac{k_{nf}}{(\rho C_p)_{nf}}\nabla^2 T + \dfrac{\vec{J}.\vec{E}}{(\rho C_p)_{nf}} - \dfrac{1}{(\rho C_p)_{nf}}\dfrac{\partial q_r}{\partial y}, \quad \left[q_r = -\dfrac{4\sigma_e}{3\beta_R}\dfrac{\partial T^4}{\partial y}, \quad T^4 \cong 4T_c^3 T - 3T_c^4\right], \\ \nabla\varphi = -\vec{E}, \\ \dfrac{\partial q}{\partial t} = -\nabla.\vec{J}, \\ q = \nabla.\varepsilon\vec{E} \end{cases} \tag{18.31}$$

$(\rho C_p)_{nf}$, μ_{nf}, and ρ_{nf} can be obtained as:

$$\begin{aligned}&(\rho C_p)_{nf} = (\rho C_p)_f(1-\phi) + (\rho C_p)_s\phi,\\&\mu = A_1 + A_2(\Delta\varphi) + A_3(\Delta\varphi)^2 + A_4(\Delta\varphi)^3,\\&\rho_{nf} = \rho_f(1-\phi) + \rho_s\phi\end{aligned} \tag{18.32}$$

k_{nf} can be expressed as:

$$\frac{k_{nf}}{k_f} = \frac{-m(k_f-k_p)\phi + (k_p-k_f)\phi + mk_f + k_p + k_f}{mk_f + (k_f-k_p)\phi + k_f + k_p} \tag{18.33}$$

So, the final PDE in the presence of thermal radiation and an electric field in a porous medium are:

$$\begin{cases}\nabla.\vec{V} = 0,\\ \left(\left(\vec{V}.\nabla\right)\vec{V} + \dfrac{\partial \vec{V}}{\partial t}\right) = \dfrac{1}{\mathrm{Re}}\dfrac{\rho_{nf}/\rho_f}{\mu_{nf}/\mu_f}\nabla^2\vec{V} - \nabla p + \dfrac{S_E}{\rho_{nf}/\rho_f}q\,\vec{E} - \dfrac{1}{\mathrm{Re}\,Da}\dfrac{\mu_{nf}}{\mu_f}\left(\dfrac{\rho_{nf}}{\rho_f}\right)^{-1}\vec{V}\\ \left(\left(\vec{V}.\nabla\right)\theta + \dfrac{\partial \theta}{\partial t}\right) = \dfrac{1}{Pr\mathrm{Re}}\dfrac{k_{nf}/k_f}{(\rho C_p)_{nf}/(\rho C_p)_f}\nabla^2\theta + S_E\dfrac{1}{(\rho C_p)_{nf}/(\rho C_p)_f}Ec\left(\vec{J}\,.\,\vec{E}\right) + \dfrac{4}{3}\left(\dfrac{k_{nf}}{k_f}\right)^{-1}Rd\dfrac{\partial^2\theta}{\partial Y^2}\\ \vec{E} = -\nabla\varphi\\ q = \nabla.\varepsilon\vec{E}\\ \nabla.\vec{J} = -\dfrac{\partial q}{\partial t}\end{cases} \tag{18.34}$$

where

$$\begin{aligned}&(\overline{u},\overline{v}) = \frac{(u,v)}{U_{Lid}},\overline{\varphi} = \frac{\varphi-\varphi_0}{\nabla\varphi},(\overline{y},\overline{x}) = \frac{(y,x)}{L},\theta = \frac{T-T_0}{\nabla T},\\&\overline{t} = \frac{tU_{Lid}}{L},\overline{p} = \frac{P}{\rho {U_{Lid}}^2}\overline{q} = \frac{q}{q_0},\overline{E} = \frac{E}{E_0},\\&\nabla T = T_1 - T_0, \nabla\varphi = \varphi_1 - \varphi_0,\end{aligned} \tag{18.35}$$

Vorticity and stream function should be employed in order to diminish the pressure gradient:

$$v = -\frac{\partial\psi}{\partial x},\omega = \frac{\partial v}{\partial x} - \frac{\partial u}{\partial y},\frac{\partial\psi}{\partial y} = u,\Psi = \frac{\psi L}{U_{Lid}},\Omega = \frac{\omega}{LU_{Lid}} \tag{18.36}$$

Nu_{loc} and Nu_{ave} along the bottom wall are calculated as:

$$Nu_{loc} = \left(\frac{k_{nf}}{k_f}\right)\left(1 + \frac{4}{3}Rd\left(\frac{k_{nf}}{k_f}\right)^{-1}\right)\frac{\partial\Theta}{\partial X} \tag{18.37}$$

$$Nu_{ave} = \frac{1}{L}\int_0^L Nu_{loc}\,dY \tag{18.38}$$

18.4.3 Effects of Active Parameters

Forced convection of nanofluid in the presence of Coulomb forces is presented. The radiation term is added to the energy equation. The porous enclosure is filled with Fe_3O_4-Ethylene glycol and has one lid wall. The roles of Darcy number ($Da = 10^2$ to 10^5), Radiation parameter ($Rd = 0$ to 0.8), supplied voltage ($\Delta\varphi = 0$ to $10kV$), volume fraction of Fe_3O_4 ($\phi = 0\%$ to 5%), and Reynolds number ($\mathrm{Re} = 3000$ to 6000) are illustrated graphically. The impact f shape factor on the rate of heat transfer is reported in Table 18.6. In this table, various shapes of nanoparti- les are utilized. The maximum Nu is caused by Platelet, followed by Cylinder, Brick, and Spherical. So, latelet nanoparticle has been utilized for further investigation. Impacts of $Da, \Delta\varphi$, and Re on isotherms and treamlines are demonstrated in Figs. 18.17–18.19. At low Reynolds number, there is one clockwise eddy in he streamlines. The center of the main eddy is near the lid wall. As the Darcy number increases, the center of he main eddy shifts upward and $|\Psi_{\max}|$ increases. Applying an electric field causes the strength of the main ddy to be enhanced and shifts the center of the eddy upward. The isotherms become more disturbed in the resence of an electric field. A thermal plume appears when the Reynolds number is increased. Also, by ncreasing Re, the strength of the rotating eddy is enhanced. As the Coulomb force increases, the strength of nain eddy is enhanced and a stronger thermal plume is generated. Nu_{ave} versus Re, $Da, Rd,$ and $\Delta\varphi$ is depicted n Fig. 18.20. The related formula is:

$$\begin{aligned} Nu_{ave} = & -0.99 + 0.04\Delta\varphi - 0.18\mathrm{Re}^* + 1.7\log(Da) + 1.04Rd \\ & - 0.038\Delta\varphi\mathrm{Re}^* + 0.03\Delta\varphi\log(Da) \\ & + 0.12\Delta\varphi Rd - 0.07\mathrm{Re}^*\log(Da) \\ & - 0.32\mathrm{Re}^* Rd + 0.25\log(Da)Rd \\ & + 0.013\Delta\varphi^2 - 0.05(\mathrm{Re}^*)^2 - 0.203\left(\log(Da)\right)^2 + 0.36Rd^2 \end{aligned} \tag{18.39}$$

TABLE 18.6 Effect of Shape of Nanoparticles on Nusselt Number When $Rd = 0.8$, $\mathrm{Re} = 6000$, $\Delta\varphi = 10$, $\phi = 0.05$

	Da	
	10^2	10^5
Spherical	3.225443	3.741018
Brick	3.267538	3.780537
Cylinder	3.32923	3.838521
Platelet	3.375827	3.882675

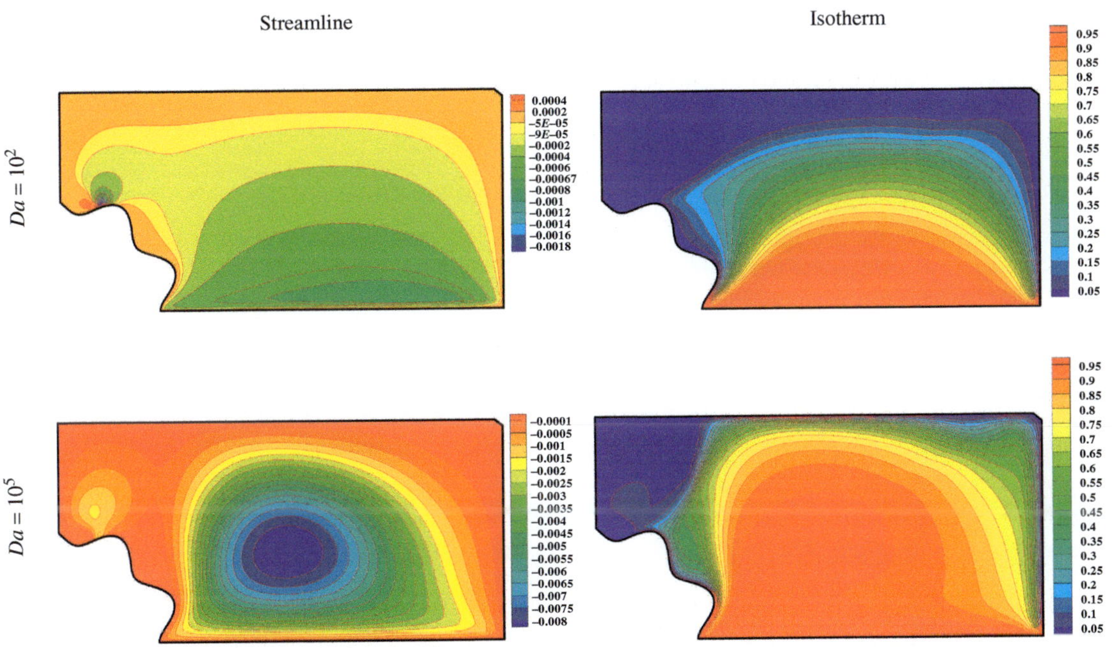

FIGURE 18.17 Effect of Darcy number on streamlines and isotherms when $\mathrm{Re} = 3000$, $\Delta\varphi = 0kV$, $\phi = 0.05$, $Rd = 0.8$.

where $\mathrm{Re}^* = 0.001\mathrm{Re}$ and $\Delta\varphi$ is voltage supply in Kilovolt. In the presence of the Coulomb force, the Nusselt number decreases with the rise of Reynolds number. The electric field helps to enhance the convective mode. So Nu_{ave} augments with the increase of $\Delta\varphi$. Thermal radiation enhances the temperature gradient near the lid wall. The influence of the Darcy number is the same as that for the radiation parameter. Therefore, Nu_{ave} is an increasing function of Rd, Da.

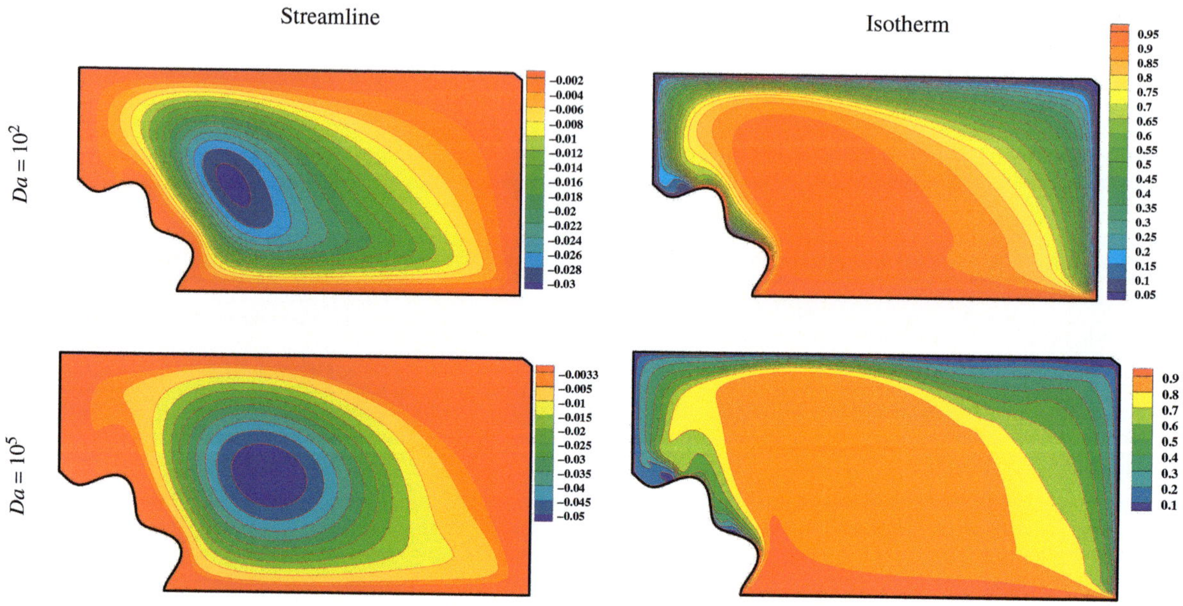

FIGURE 18.18 Effect of Darcy number on streamlines and isotherms when $\mathrm{Re} = 3000, \Delta\varphi = 10kV, \phi = 0.05, Rd = 0.8$.

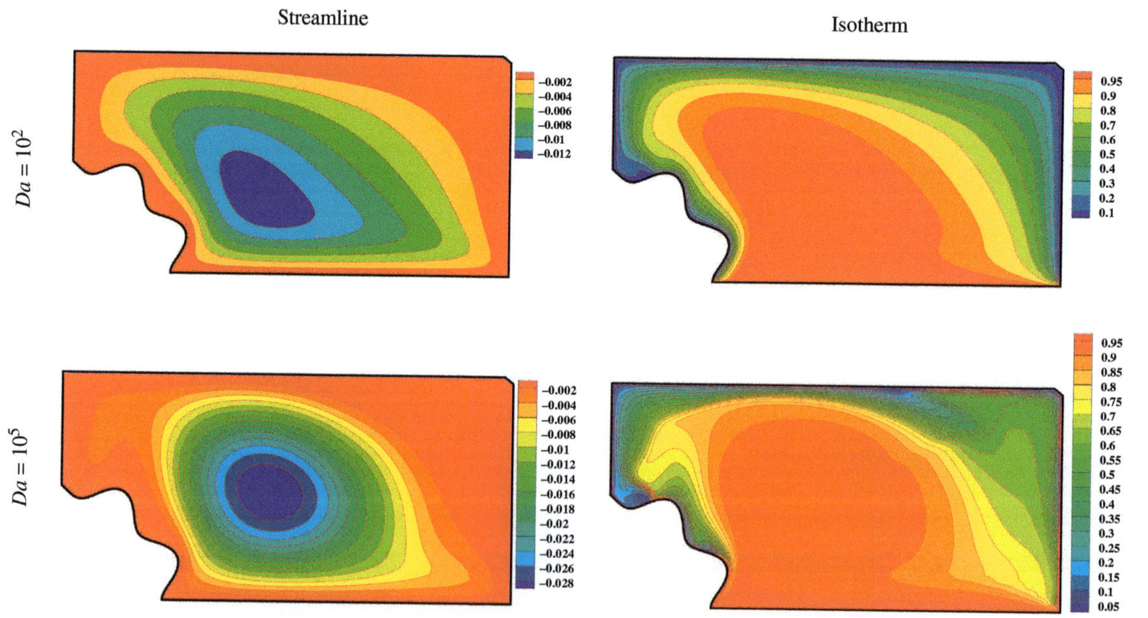

FIGURE 18.19 Effect of Darcy number on streamlines and isotherms when $\mathrm{Re} = 6000, \Delta\varphi = 10kV, \phi = 0.05, Rd = 0.8$.

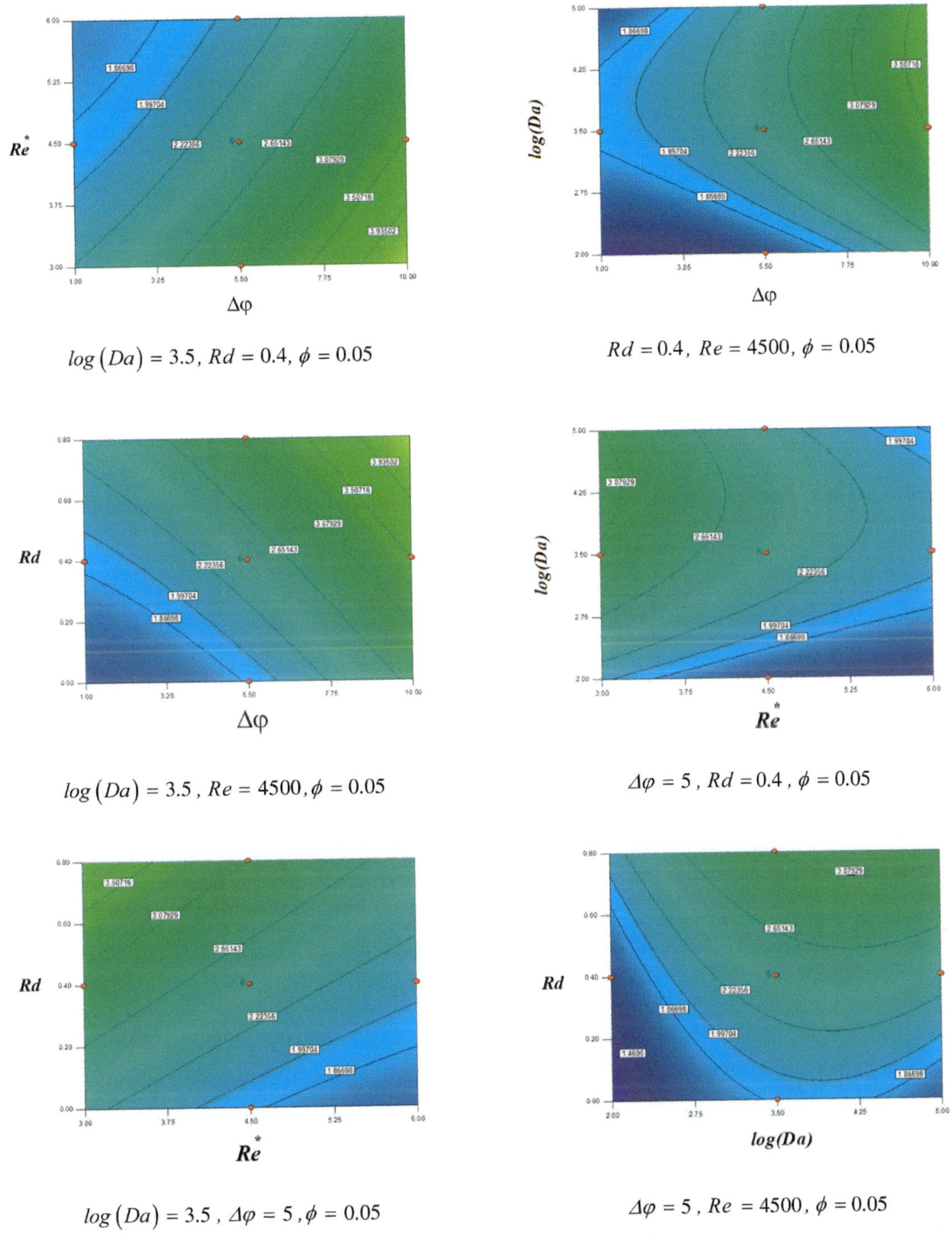

FIGURE 18.20 Effects of Da, $\Delta\varphi$, Rd, and Re on average Nusselt number.

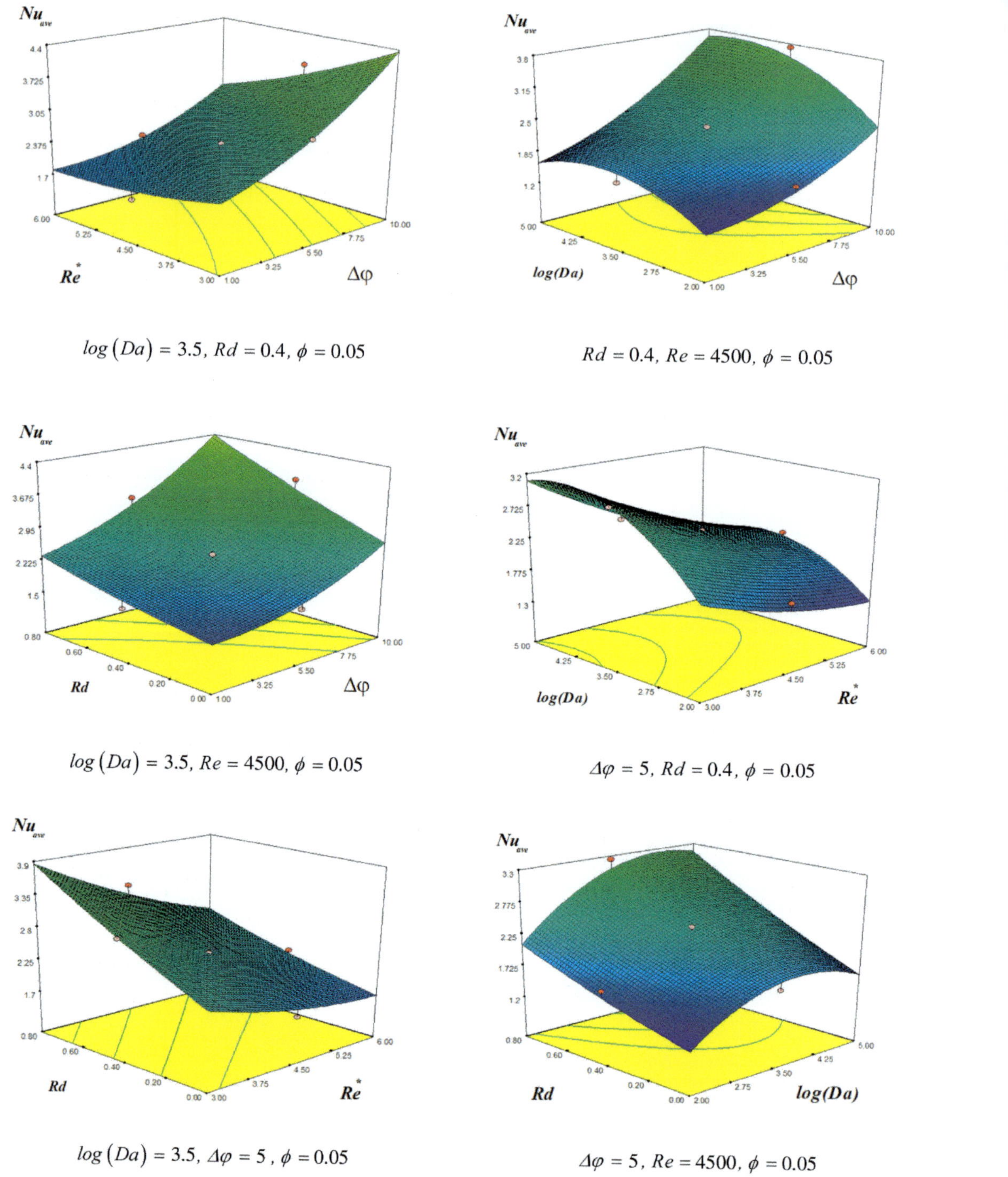

FIGURE 18.20 (*Continued*)

References

[1] M. Sheikholeslami, Influence of Coulomb forces on Fe3O4-H2O nanofluid thermal improvement, Int. J. Hydrogen Energy 42 (2017) (2016) 821–829.

[2] M. Sheikholeslami, D.D. Ganji, Impact of electric field on nanofluid forced convection heat transfer with considering variable properties, J. Mol. Liq. 229 (2017) 566–573. Available from: https://doi.org/10.1016/j.molliq.2016.12.107.

[3] M. Sheikholeslami, M.M. Bhatti, Active method for nanofluid heat transfer enhancement by means of EHD, Int. J. Heat Mass Transfer 109 (2017) 115–122.

[4] M. Sheikholeslami, H.B. Rokni, Influence of EFD viscosity on nanofluid forced convection in a cavity with sinusoidal wall, J. Mol. Liq. 232 (2017) 390–395.
[5] M. Sheikholeslami, Influence of Lorentz forces on nanofluid flow in a porous cylinder considering Darcy model, J. Mol. Liq. 225 (2017) 903–912.
[6] M. Sheikholeslami, CVFEM for magnetic nanofluid convective heat transfer in a porous curved enclosure, Eur. Phys. J. Plus 131 (2016) 413. Available from: https://doi.org/10.1140/epjp/i2016-16413-y.
[7] M. Sheikholeslami, Influence of Coulomb forces on Fe_3O_4-H_2O nanofluid thermal improvement, Int. J. Hydrogen Energy 42 (2017) 821–829.
[8] M. Sheikholeslami, Magnetic field influence on nanofluid thermal radiation in a cavity with tilted elliptic inner cylinder, J. Mol. Liq. 229 (2017) 137–147.
[9] M. Sheikholeslami, Numerical simulation of magnetic nanofluid natural convection in porous media, Phys. Lett. A 381 (2017) 494–503.
[10] M. Sheikholeslami, Magnetic source impact on nanofluid heat transfer using CVFEM, Neur. Comput. Applicat. (2016). Available from: https://doi.org/10.1007/s00521-016-2740-7.
[11] M. Sheikholeslami, CuO-water nanofluid free convection in a porous cavity considering Darcy law, Europ. Phys. J. Plus 132 (2017) 55. Available from: https://doi.org/10.1140/epjp/i2017-11330-3.
[12] M. Sheikholeslami, Numerical investigation of MHD nanofluid free convective heat transfer in a porous tilted enclosure, Eng. Comput. 34 (6) (2017) 1939–1955.
[13] M.S. Kandelousi, Effect of spatially variable magnetic field on ferrofluid flow and heat transfer considering constant heat flux boundary condition, Europ. Phys. J. Plus (2014) 129–248. Available from: https://doi.org/10.1140/epjp/i2014-14248-2.
[14] M. Sheikholeslami, Numerical simulation for external magnetic field influence on Fe_3O_4-water nanofluid forced convection, Engineering Computations, Inpress.
[15] M. Sheikholeslami, Influence of Lorentz forces on nanofluid flow in a porous cavity by means of Non- Darcy model, Eng. Computat. 34 (8) (2017) 2651–2667. Available from: https://doi.org/10.1108/EC-01-2017-0008.
[16] M. Sheikholeslami, H.B. Rokni, Nanofluid convective heat transfer intensification in a porous circular cylinder, Chem. Eng. Proc. Process Intensif. 120 (2017) 93–104.
[17] M. Sheikholeslami, M.K. Sadoughi, Numerical modeling for Fe_3O_4 -water nanofluid flow in porous medium considering MFD viscosity, J. Mol. Liq. 242 (2017) 255–264.
[18] M. Sheikholeslami, M. Shamlooei, Magnetic source influence on nanofluid flow in porous medium considering shape factor effect, Phys. Lett. A 381 (2017) 3071–3078.
[19] M. Sheikholeslami, M. Shamlooei, Convective flow of nanofluid inside a lid driven porous cavity using CVFEM, Phys. B: Condensed Matter 521 (2017) 239–250.
[20] M. Sheikholeslami, K. Vajravelu, Forced convection heat transfer in Fe_3O_4- ethylene glycol nanofluid under the influence of Coulomb force, J. Mol. Liq. 233 (2017) 203–210.
[21] M. Sheikholeslami, T. Hayat, A. Alsaedi, MHD free convection of Al_2O_3–water nanofluid considering thermal radiation: a numerical study, Int. J. Heat Mass Transfer 96 (2016) 513–524.
[22] M. Sheikholeslami, H.B. Rokni, Magnetic nanofluid natural convection in the presence of thermal radiation considering variable viscosity, Europ. Phys. J. Plus 132 (2017) 238. Available from: https://doi.org/10.1140/epjp/i2017-11498-4.
[23] M. Sheikholeslami, H.B. Rokni, Numerical simulation for impact of Coulomb force on nanofluid heat transfer in a porous enclosure in presence of thermal radiation, Int. J. Heat Mass Transfer 118 (2018) 823–831.
[24] M. Sheikholeslami, M. Shamlooei, R. Moradi, Fe_3O_4- Ethylene glycol nanofluid forced convection inside a porous enclosure in existence of Coulomb force, J. Mol. Liq. 249 (2018) 429–437.
[25] M. Sheikholeslami, S.A. Shehzad, Numerical analysis of Fe3O4 –H2O nanofluid flow in permeable media under the effect of external magnetic source, Int. J. Heat Mass Transfer 118 (2018) 182–192.
[26] M. Sheikholeslami, S.A. Shehzad, Magnetohydrodynamic nanofluid convection in a porous enclosure considering heat flux boundary condition, Int. J. Heat Mass Transfer 106 (2017) 1261–1269.
[27] M. Sheikholeslami, S.A. Shehzad, Thermal radiation of ferrofluid in existence of Lorentz forces considering variable viscosity, Int. J. Heat Mass Transfer 109 (2017) 82–92.
[28] M. Sheikholeslami, H.B. Rokni, Numerical modeling of nanofluid natural convection in a semi annulus in existence of Lorentz force, Comput. Methods Appl. Mech. Eng. 317 (2017) 419–430.
[29] M. Sheikholeslami, K. Vajravelu, Nanofluid flow and heat transfer in a cavity with variable magnetic field, Appl. Math. Comput. 298 (2017) 272–282.
[30] M. Sheikholeslami, H.B. Rokni, Magnetohydrodynamic CuO-water nanofluid in a porous complex shaped enclosure, ASME, J. Thermal Sci. Eng. Applicat. 9 (4) (2017) 041007. Available from: https://doi.org/10.1115/1.4035973.
[31] M. Sheikholeslami, H.B. Rokni, Nanofluid two phase model analysis in existence of induced magnetic field, Int. J. Heat Mass Transfer 107 (2017) 288–299.
[32] M. Sheikholeslami, T. Hayat, A. Alsaedi, S. Abelman, Numerical analysis of EHD nanofluid force convective heat transfer considering electric field dependent viscosity, Int. J. Heat Mass Transfer 108 (2017) 2558–2565.
[33] M. Sheikholeslami, Magnetic field influence on CuO -H_2O nanofluid convective flow in a permeable cavity considering various shapes for nanoparticles, Int. J. Hydrogen Energy 42 (2017) 19611–19621.
[34] M. Sheikholeslami, M.M. Bhatti, Forced convection of nanofluid in presence of constant magnetic field considering shape effects of nanoparticles, Int. J. Heat Mass Transfer 111 (2017) 1039–1049.
[35] M. Sheikholeslami, A. Zeeshan, Analysis of flow and heat transfer in water based nanofluid due to magnetic field in a porous enclosure with constant heat flux using CVFEM, Comput. Methods Appl. Mech. Eng. 320 (2017) 68–81.
[36] M. Sheikholeslami, M. Shamlooei, Fe_3O_4- H_2O nanofluid natural convection in presence of thermal radiation, Int. J. Hydrogen Energy 42 (9) (2017) 5708–5718.

[37] M. Sheikholeslami, K. Vajravelu, M.M. Rashidi, Forced convection heat transfer in a semi annulus under the influence of a variable magnetic field, Int. J. Heat Mass Transfer 92 (2016) 339–348.

[38] M. Sheikholeslami, T. Hayat, A. Alsaedi, Numerical study for external magnetic source influence on water based nanofluid convective heat transfer, Int. J. Heat Mass Transfer 106 (2017) 745–755.

[39] M. Sheikholeslami, Ali J. Chamkha, Flow and convective heat transfer of a ferro-nanofluid in a double-sided lid-driven cavity with wavy wall in the presence of a variable magnetic field, Numer. Heat Transfer, Part A 69 (10) (2016) 1186–1200. Available from: https://doi.org/10.1080/10407782.2015.1125709.

[40] M. Sheikholeslami, A.J. Chamkha, Electrohydrodynamic free convection heat transfer of a nanofluid in a semi-annulus enclosure with sinusoidal wall, Numer. Heat Transfer, Part A 69 (7) (2016) 781–793. Available from: http://dx.doi.org/10.1080/10407782.2015.1090819.

[41] M. Sheikholeslami, M.K. Sadoughi, Simulation of CuO- water nanofluid heat transfer enhancement in presence of melting surface, Int. J. Heat Mass Transfer 116 (2018) 909–919.

[42] M. Sheikholeslami, H.B. Rokni, Simulation of nanofluid heat transfer in presence of magnetic field: a review, Int. J. Heat Mass Transfer 115 (2017) 1203–1233.

[43] M. Sheikholeslami, H.B. Rokni, Melting heat transfer influence on nanofluid flow inside a cavity in existence of magnetic field, Int. J. Heat Mass Transfer 114 (2017) 517–526.

[44] M. Sheikholeslami, S.A. Shehzad, Magnetohydrodynamic nanofluid convective flow in a porous enclosure by means of LBM, Int. J. Heat Mass Transfer 113 (2017) 796–805.

[45] M. Sheikholeslami, Magnetohydrodynamic nanofluid forced convection in a porous lid driven cubic cavity using Lattice Boltzmann Method, J. Mol. Liq. 231 (2017) 555–565.

[46] M.S. Kandelousi, KKL correlation for simulation of nanofluid flow and heat transfer in a permeable channel, Phys. Lett. A 378 (45) (2014) 3331–3339.

[47] M. Sheikholeslami, Effect of uniform suction on nanofluid flow and heat transfer over a cylinder, J. Braz. Soc. Mech. Sci. Eng. 37 (2015) 1623–1633.

[48] M. Sheikholeslami, Lattice Boltzmann Method simulation of MHD non-Darcy nanofluid free convection, Physica B 516 (2017) 55–71.

[49] M. Sheikholeslami, Influence of magnetic field on nanofluid free convection in an open porous cavity by means of Lattice Boltzmann Method, J. Mol. Liq. 234 (2017) 364–374.

[50] M. Sheikholeslami, A.J. Chamkha, Influence of Lorentz forces on nanofluid forced convection considering Marangoni convection, J. Mol. Liq. 225 (2017) 750–757.

[51] M. Sheikholeslami, T. Hayat, A. Alsaedi, Numerical simulation of nanofluid forced convection heat transfer improvement in existence of magnetic field using Lattice Boltzmann Method, Int. J. Heat Mass Transfer 108 (2017) 1870–1883.

[52] M. Sheikholeslami, R. Ellahi, Three dimensional mesoscopic simulation of magnetic field effect on natural convection of nanofluid, Int. J. Heat Mass Transfer 89 (2015) 799–808.

[53] M. Sheikholeslami, S. Abelman, Two phase simulation of nanofluid flow and heat transfer in an annulus in the presence of an axial magnetic field, IEEE Trans. Nanotechnol. 14 (3) (2015) 561–569.

[54] M. Sheikholeslami, R. Ellahi, Simulation of ferrofluid flow for magnetic drug targeting using Lattice Boltzmann method, J. Zeitschrift Fur Naturforschung A 70 (2) (2015) 115–124.

[55] M. Sheikholeslami, M. Seyednezhad, Lattice Boltzmann Method simulation for CuO-water nanofluid flow in a porous enclosure with hot obstacle, J. Mol. Liq. 243 (2017) 249–256.

[56] M. Sheikholeslami, M. Sadoughi, Mesoscopic method for MHD nanofluid flow inside a porous cavity considering various shapes of nanoparticles, Int. J. Heat Mass Transfer 113 (2017) 106–114.

[57] M. Sheikholeslami, A. Zeeshan, Mesoscopic simulation of CuO-H_2O nanofluid in a porous enclosure with elliptic heat source, Int. J. Hydrogen Energy 42 (22) (2017) 15393–15402.

[58] M. Sheikholeslami, S.A. Shehzad, CVFEM for influence of external magnetic source on $Fe_3O_4 - H_2O$ nanofluid behavior in a permeable cavity considering shape effect, Int. J. Heat Mass Transfer 115 (2017) 180–191.

[59] M. Sheikholeslami, M. Seyednezhad, Nanofluid heat transfer in a permeable enclosure in presence of variable magnetic field by means of CVFEM, Int. J. Heat Mass Transfer 114 (2017) 1169–1180.

[60] M. Sheikholeslami, H.B. Rokni, Free convection of CuO-H_2O nanofluid in a curved porous enclosure using mesoscopic approach, Int. J. Hydrogen Energy 42 (22) (2017) 14942–14949.

[61] M. Sheikholeslami, Numerical investigation for CuO-H2O nanofluid flow in a porous channel with magnetic field using mesoscopic method, J. Mol. Liq. 249 (2018) 739–746.

[62] M. Sheikholeslami, H.B. Rokni, Influence of melting surface on MHD nanofluid flow by means of two phase model, Chin. J. Phys. 55 (2017) 1352–1360.

[63] M. Sheikholeslami, H.B. Rokni, Effect of melting heat transfer on nanofluid flow in existence of magnetic field considering Buongiorno Model, Chin. J. Phys. 55 (2017) 1115–1126.

[64] M. Sheikholeslami, T. Hayat, A. Alsaedi, Numerical simulation for forced convection flow of MHD CuO-H2O nanofluid inside a cavity by means of LBM, J. Mol. Liq. 249 (2018) 941–948.

[65] M. Sheikholeslami, H.R. Kataria, A.S. Mittal, Effect of thermal diffusion and heat-generation on MHD nanofluid flow past an oscillating vertical plate through porous medium, J. Mol. Liq. 257 (2018) 12–25.

[66] M. Sheikholeslami, M. Barzegar Gerdroodbary, S. Valiallah Mousavi, D.D. Ganji, R. Moradi, Heat transfer enhancement of ferrofluid inside an 90° elbow channel by non-uniform magnetic field, J. Magn. Magn. Mater. 460 (2018) 302–311.

[67] M. Sheikholeslami, S. Soleimani, D.D. Ganji, Effect of electric field on hydrothermal behavior of nanofluid in a complex geometry, J. Mol. Liq. 213 (2016) 153–161.

[68] M. Sheikholeslami, Numerical simulation for solidification in a LHTESS by means of Nano-enhanced PCM, J. Taiwan Inst. Chem. Eng. 86 (2018) 25–41.

[69] M. Sheikholeslami, Numerical modeling of Nano enhanced PCM solidification in an enclosure with metallic fin, J. Mol. Liq. 259 (2018) 424–438.
[70] M. Sheikholeslami, A. Ghasemi, Solidification heat transfer of nanofluid in existence of thermal radiation by means of FEM, Int. J. Heat Mass Transfer 123 (2018) 418–431.
[71] M. Sheikholeslami, S.A. Shehzad, CVFEM simulation for nanofluid migration in a porous medium using Darcy model, Int. J. Heat Mass Transfer 122 (2018) 1264–1271.
[72] M. Sheikholeslami, M. Darzi, M.K. Sadoughi, Heat transfer improvement and Pressure Drop during condensation of refrigerant-based Nanofluid; An Experimental Procedure, Int. J. Heat Mass Transfer 122 (2018) 643–650.
[73] M. Sheikholeslami, H.B. Rokni, CVFEM for effect of Lorentz forces on nanofluid flow in a porous complex shaped enclosure by means of Non-equilibrium model, J. Mol. Liq. 254 (2018) 446–462.
[74] M. Sheikholeslami, S.A. Shehzad, Z. Li, Water based nanofluid free convection heat transfer in a three dimensional porous cavity with hot sphere obstacle in existence of Lorenz forces, Int. J. Heat Mass Transfer 125 (2018) 375–386.
[75] M. Sheikholeslami, M. Jafaryar, D.D. Ganji, Z. Li, Exergy loss analysis for nanofluid forced convection heat transfer in a pipe with modified turbulators, J. Mol. Liq. 262 (2018) 104–110.
[76] M. Sheikholeslami, M. Jafaryar, Z. Li, Nanofluid turbulent convective flow in a circular duct with helical turbulators considering CuO nanoparticles, Int. J. Heat Mass Transfer 124 (2018) 980–989.
[77] M. Sheikholeslami, Z. Li, M. Shamlooei, Nanofluid MHD natural convection through a porous complex shaped cavity considering thermal radiation, Phys. Lett. A 382 (2018) 1615–1632.
[78] M. Sheikholeslami, H.B. Rokni, Magnetic nanofluid flow and convective heat transfer in a porous cavity considering Brownian motion effects, Phys. Fluids, 30, 1. Available from: https://doi.org/10.1063/1.5012517.
[79] M. Sheikholeslami, S.A. Shehzad, Simulation of water based nanofluid convective flow inside a porous enclosure via Non-equilibrium model, Int. J. Heat Mass Transfer 120 (2018) 1200–1212.
[80] M. Sheikholeslami, S.A. Shehzad, Non-Darcy free convection of Fe3O4-water nanoliquid in a complex shaped enclosure under impact of uniform Lorentz force, Chin. J. Phys. 56 (2018) 270–281.
[81] M. Sheikholeslami, M. Seyednezhad, Simulation of nanofluid flow and natural convection in a porous media under the influence of electric field using CVFEM, Int. J. Heat Mass Transfer 120 (2018) 772–781.
[82] M. Sheikholeslami, M. Shamlooei, R. Moradi, Numerical simulation for heat transfer intensification of nanofluid in a permeable curved enclosure considering shape effect of Fe3O4 nanoparticles, Chem. Eng. Process.: Process Intensif. 124 (2018) 71–82.
[83] M. Sheikholeslami, T. Hayat, T. Muhammad, A. Alsaedi, MHD forced convection flow of nanofluid in a porous cavity with hot elliptic obstacle by means of Lattice Boltzmann method, Int. J. Mech. Sci. 135 (2018) 532–540.
[84] M. Sheikholeslami, Numerical investigation of nanofluid free convection under the influence of electric field in a porous enclosure, J. Mol. Liq. 249 (2018) 1212–1221.
[85] M. Sheikholeslami, CuO-water nanofluid flow due to magnetic field inside a porous media considering Brownian motion, J. Mol. Liq. 249 (2018) 921–929.
[86] M. Sheikholeslami, D.D. Ganji, Influence of electric field on Fe_3O_4-water nanofluid radiative and convective heat transfer in a permeable enclosure, J. Mol. Liq. 250 (2018) 404–412.
[87] M. Sheikholeslami, S.A. Shehzad, Z. Li, Nanofluid heat transfer intensification in a permeable channel due to magnetic field using Lattice Boltzmann method, Phys. B (2018). Available from: https://doi.org/10.1016/j.physb.2018.03.036.
[88] M. Sheikholeslami, A. Zeeshan, Numerical simulation of Fe3O4 -water nanofluid flow in a non-Darcy porous media, Int. J. Numer. Methods Heat Fluid Flow 28 (3) (2018) 641–660. Available from: https://doi.org/10.1108/HFF-04-2017-0160.
[89] M. Sheikholeslami, D.D. Ganji, Numerical approach for magnetic nanofluid flow in a porous cavity using CuO nanoparticles, Mater. Des. 120 (2017) 382–393.
[90] M. Sheikholeslami, D.D. Ganji, Numerical modeling of magnetohydrodynamic CuO-water transportation inside a porous cavity considering shape factor effect, Colloids Surf., A 529 (2017) 705–714.

CHAPTER

19

Nanofluid Heat Transfer Enhancement in Presence of Melting Surface Using CVFEM

19.1 INTRODUCTION

Heat transfer generated during the melting process of a stretching sheet has various industrial applications including preparation of semiconductor materials, magma solidification, melting of permafrost, and thawing of frozen ground, etc. Epstein and Cho [1] examined the characteristics of melting heat transfer in the laminar flow over a flat plate. The work of Epstein and Cho [1] has been extended by the various researchers. Cheng and Lin [2] investigated the melting effect in mixed convective heat transfer with aiding and opposing external flows from the vertical plate in a liquid-saturated porous medium. Ishak et al. [3] studied melting heat transfer in steady laminar flow over a moving surface. Melting heat transfer in boundary layer stagnation-point flow towards a stretching/shrinking sheet in a micropolar fluid has been analyzed by Yacob et al. [4]. Recently, Ahmad and Pop [5] have studied the effect of melting parameter on the dual solutions for the problem of mixed convection boundary-layer flow past a vertical flat surface embedded in a fluid-saturated porous medium. They considered only opposing flow which shows that the effect of surface melting reduces the heat transfer rate and expedites the boundary-layer separation at the solid–liquid interface. Hayat et al. [6] analyzed the stagnation-point flow of coupled stress fluid with melting heat transfer. Sheikholeslami and Rokni [7] studied the melting heat transfer influence on nanofluid flow inside a cavity in the presence of a magnetic field. Sheikholeslami and Sadoughi [8] simulated the CuO-water nanofluid heat transfer enhancement in the presence of a melting surface. Sheikholeslami et al. [9] investigated CuO-water nanofluid magnetohydrodynamic natural convection inside a sinusoidal annulus in presence of melting heat transfer. Also, nanotechnology is utilized for other applications [10–96].

19.2 MELTING HEAT TRANSFER INFLUENCE ON NANOFLUID FLOW INSIDE A CAVITY IN THE PRESENCE OF A MAGNETIC FIELD

19.2.1 Problem Definition

Fig. 19.1 depicts the geometry, boundary condition, and sample element. The bottom wall is the hot wall ($T = T_h$) and the top one is the melting surface ($T = T_m$). Other walls are adiabatic. A horizontal magnetic field has been applied. The enclosure is filled with a nanofluid.

19.2.2 Governing Equation

2D steady convective nanofluid flow is considered in the presence of a constant magnetic field. The PDEs are:

$$\frac{\partial v}{\partial y} + \frac{\partial u}{\partial x} = 0 \tag{19.1}$$

$$\left(\rho_{nf}\right)\left(v\frac{\partial u}{\partial y} + u\frac{\partial u}{\partial x}\right) = \left[B_y \sigma_{nf} v B_x - B_y^2 \sigma_{nf} u + \left(\frac{\partial^2 u}{\partial y^2} + \frac{\partial^2 u}{\partial x^2}\right)\mu_{nf} - \frac{\partial P}{\partial x}\right] \tag{19.2}$$

Application of Control Volume based Finite Element Method (CVFEM) for Nanofluid Flow and Heat Transfer.
DOI: https://doi.org/10.1016/B978-0-12-814152-6.00019-9

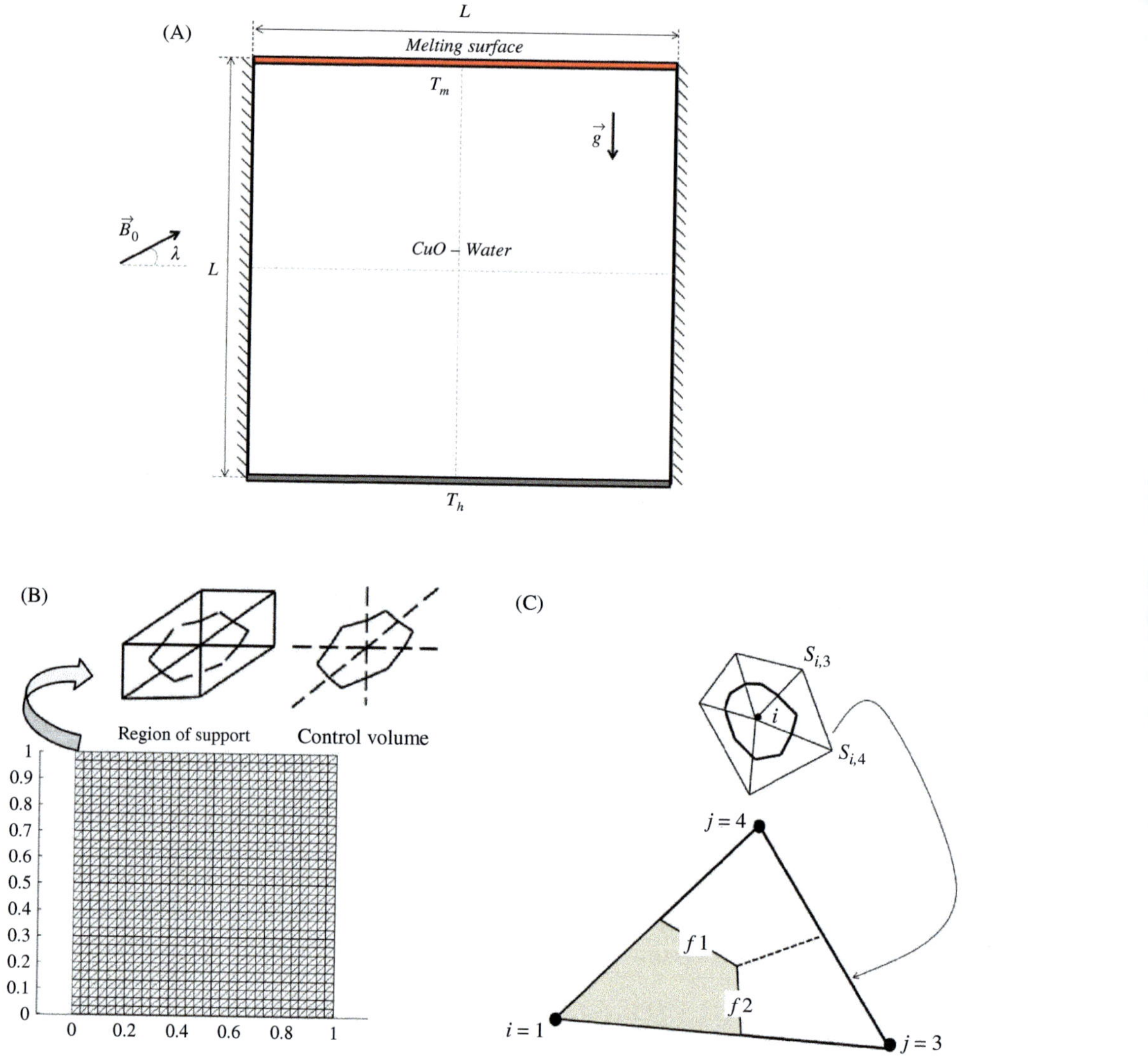

FIGURE 19.1 (A) Geometry and the boundary conditions with (B) the mesh of geometry considered in this work (C) a sample triangular element and its corresponding control volume.

$$\rho_{nf}\left(v\frac{\partial v}{\partial y}+\frac{\partial v}{\partial x}u\right)=uB_y\sigma_{nf}B_x+(T-T_c)\beta_{nf}g\rho_{nf}-vB_x\sigma_{nf}B_x$$

$$-\frac{\partial P}{\partial y}+\mu_{nf}\left(\frac{\partial^2 v}{\partial y^2}+\frac{\partial^2 v}{\partial x^2}\right), \tag{19.3}$$

$$B_x=B_o\cos\lambda,\ \ B_y=B_o\sin\lambda$$

$$(\rho C_p)_{nf}\left(v\frac{\partial T}{\partial y}+u\frac{\partial T}{\partial x}\right)=k_{nf}\left(\frac{\partial^2 T}{\partial x^2}+\frac{\partial^2 T}{\partial y^2}\right) \tag{19.4}$$

$(\rho C_p)_{nf}$, ρ_{nf}, $(\rho\beta)_{nf}$, and σ_{nf} are calculated as:

$$(\rho C_p)_{nf}=\phi(\rho C_p)_s+(1-\phi)(\rho C_p)_f \tag{19.5}$$

$$\rho_{nf}=\phi\rho_s+(1-\phi)\rho_f \tag{19.6}$$

$$(\rho\beta)_{nf} = \phi(\rho\beta)_s + (1-\phi)(\rho\beta)_f \quad (19.7)$$

$$\frac{\sigma_{nf}}{\sigma_f} = \left(\frac{+(2+\sigma_s/\sigma_f)-\phi(-1+\sigma_s/\sigma_f)}{3\phi(-1+\sigma_s/\sigma_f)}\right)^{-1} + 1 \quad (19.8)$$

k_{nf}, μ_{nf} are calculated via the Koo–Kleinstreuer–Li (KKL) model:

$$\frac{k_{nf}}{k_f} = 1 + \frac{3\left(\frac{k_p}{k_f}-1\right)\phi}{\left(\frac{k_p}{k_f}+2\right)-\left(\frac{k_p}{k_f}-1\right)\phi} + 5\times10^4 g'(\phi,T,d_p)\phi\rho_f c_{p,f}\sqrt{\frac{\kappa_b T}{d_p\rho_p}},$$
$$R_f = 4\times10^{-8} km^2/W,$$
$$R_f = -d_p(1/k_p - 1/k_{p,eff}),$$
$$g'(\phi,d_p,T) = Ln(T)\left(a_1 + a_3 Ln(\phi) + a_2 Ln(d_p) + a_5 Ln(d_p)^2 + a_4 \ln(d_p) Ln(\phi)\right)$$
$$+\left(a_6 + a_8 Ln(\phi) + a_7 Ln(d_p) + a_{10} Ln(d_p)^2 + a_9 Ln(\phi)\ln(d_p)\right) \quad (19.9)$$

$$\mu_{nf} = \frac{\mu_f}{(1-\phi)^{2.5}} + \frac{k_{Brownian}}{k_f}\times\frac{\mu_f}{\Pr} \quad (19.10)$$

Properties and needed parameters are provided in Tables 19.1 and 19.2.
Vorticity and stream function should be used to eliminate the pressure source terms:

$$\omega + \frac{\partial u}{\partial y} - \frac{\partial v}{\partial x} = 0, \quad \frac{\partial\psi}{\partial x} = -v, \quad \frac{\partial\psi}{\partial y} = u \quad (19.11)$$

Introducing dimensionless quantities:

$$P = \frac{p}{\rho_{nf}(\alpha_{nf}/L)^2}, U = \frac{uL}{\alpha_{nf}}, V = \frac{vL}{\alpha_{nf}}, \theta = \frac{T-T_m}{T_h-T_m}, (X,Y) = \frac{(x,y)}{L},$$
$$\Psi = \frac{\psi}{\alpha_{nf}}, \Omega = \frac{\omega L^2}{\alpha_{nf}} \quad (19.12)$$

TABLE 19.1 The Coefficient Values of CuO – Water Nanofluid

Coefficient values	*CuO – Water*
a_1	– 26.5933108
a_2	– 0.403818333
a_3	– 33.3516805
a_4	– 1.915825591
a_5	6.421858E-02
a_6	48.40336955
a_7	– 9.787756683
a_8	190.245610009
a_9	10.9285386565
a_{10}	– 0.72009983664

TABLE 19.2 Thermophysical Properties of Water and Nanoparticles

	$\rho(kg/m^3)$	$C_p(j/kgk)$	$k(W/m.k)$	$\beta \times 10^5(K^{-1})$	$d_p(nm)$	$\sigma(\Omega \cdot m)$
Water	997.1	4179	0.613	21	–	0.05
CuO	6500	540	18	29	45	10^{-10}

The final formulae are:

$$\Omega + \frac{\partial^2 \Psi}{\partial Y^2} + \frac{\partial^2 \Psi}{\partial X^2} = 0, \tag{19.13}$$

$$\begin{aligned} \frac{\partial \Omega}{\partial X} U + V \frac{\partial \Omega}{\partial Y} &= \Pr \frac{A_5}{A_1} \frac{A_2}{A_4} \left(\frac{\partial^2 \Omega}{\partial Y^2} + \frac{\partial^2 \Omega}{\partial X^2} \right) \\ &+ \Pr Ha^2 \frac{A_6}{A_1} \frac{A_2}{A_4} \left(-\frac{\partial V}{\partial X} B_x B_x + \frac{\partial U}{\partial X} B_x B_y - \frac{\partial V}{\partial Y} B_y B_x + \frac{\partial U}{\partial Y} B_y B_y \right) \\ &+ \Pr Ra \frac{A_3 A_2^2}{A_1 A_4^2} \frac{\partial \theta}{\partial X}, \end{aligned} \tag{19.14}$$

$$\frac{\partial \theta}{\partial X} U + \frac{\partial \theta}{\partial Y} V = \left(\frac{\partial^2 \theta}{\partial X^2} + \frac{\partial^2 \theta}{\partial Y^2} \right) \tag{19.15}$$

Boundary conditions are:

$$\begin{aligned} &\theta = 0 && \text{on top wall} \\ &\theta = 1.0 && \text{on bottom wall} \\ &\frac{\partial \theta}{\partial n} = 0.0 && \text{on other walls} \\ &\Psi = 0.0 && \text{Right, left and bottom walls} \end{aligned} \tag{19.16}$$

and on the melting surface, we have:

$$\left(\frac{\partial \theta}{\partial n} \right) \bigg|_{Y=1} = \frac{A_2}{A_1} \frac{1}{\delta} (L + c_s (T_m - T_0)) \frac{\partial \Psi}{\partial n} \bigg|_{Y=1} \tag{19.17}$$

where dimensionless and constants parameters are illustrated as:

$$\begin{aligned} &\Pr = \upsilon_f / \alpha_f, Ra = g \beta_f \Delta T L^3 / (\upsilon_f \alpha_f), \ Ha = L B_0 \sqrt{\sigma_f / \mu_f}, \ \ \delta = \frac{(\rho C_p)_f}{\rho_f} \frac{(T_h - T_m)}{1 + c_s (T_m - T_0)} \\ &A_1 = \frac{\rho_{nf}}{\rho_f}, \ A_5 = \frac{\mu_{nf}}{\mu_f}, \ A_2 = \frac{(\rho C_P)_{nf}}{(\rho C_P)_f}, A_4 = \frac{k_{nf}}{k_f}, \ A_6 = \frac{\sigma_{nf}}{\sigma_f}, \ A_3 = \frac{(\rho \beta)_{nf}}{(\rho \beta)_f} \end{aligned} \tag{19.18}$$

It should be mentioned that δ is related to Stefan numbers.
Local and average Nusselt over the hot wall can be calculated as:

$$Nu_{loc} = \left(\frac{k_{nf}}{k_f} \right) \frac{\partial \theta}{\partial y} \tag{19.19}$$

$$Nu_{ave} = \frac{1}{L} \int_0^L Nu_{loc} \, dx \tag{19.20}$$

9.2.3 Effects of Active Parameters

The impacts of melting heat transfer on nanofluid magnetohydrodynamic flow in a cavity is examined. μ_{nf}, k_{nf} f CuO-water nanofluid are estimated by means of KKL model. Graphs and tables are depicted for different mounts of CuO-water volume fraction ($\phi = 0$ to $\phi = 0.04$), melting parameter ($\delta = 0$ to 0.8), Hartmann number $Ha = 0$ to 40), and Rayleigh number ($Ra = 10^3$ to 10^4).

The impact of adding nanoparticles in the base fluid on velocity and temperature contours is depicted in ig. 19.2. Temperature gradient decreases with the increase of ϕ. $|\Psi_{max}|$ augments with the addition of nanoparti- les because of the increase in the solid movements. In the presence of melting heat transfer and magnetic field, he effect of adding nanoparticles on isotherms becomes negligible.

Fig. 19.3 depicts the impact of Rayleigh and Hartmann numbers in absence of melting heat transfer. There is nly one eddy in the streamlines. At low Rayleigh number, the conduction mechanism is dominant. As *Ra* ncreases the distortion of the isotherms is enhanced close to the hot wall. Adding a magnetic field makes iso- herms become parallel. Figs. 19.4 and 19.5 demonstrate the influence of *Ra* and *Ha* on streamlines and isotherms n the presence of the melting process. In the presence of melting heat transfer the primary eddy diminishes and he increasing Lorentz forces generate three layers for streamlines. Increasing the melting parameter increases the strength of the bottom eddy.

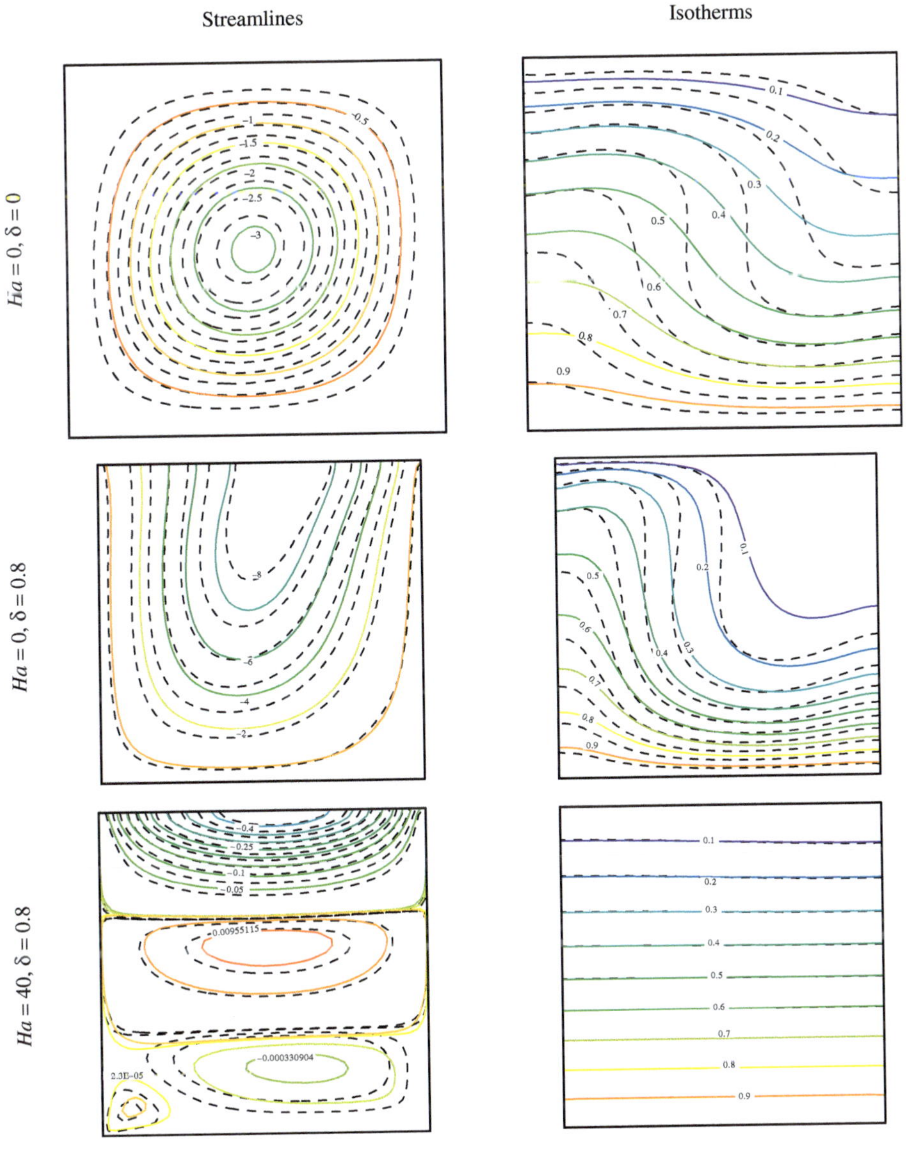

FIGURE 19.2 Influence of nanofluid volume fraction on streamline (left) and isotherm (right) contours (nanofluid ($\phi = 0.04$)(—–) and pure fluid ($\phi = 0$) (- - -)) when $Ra = 10^4$.

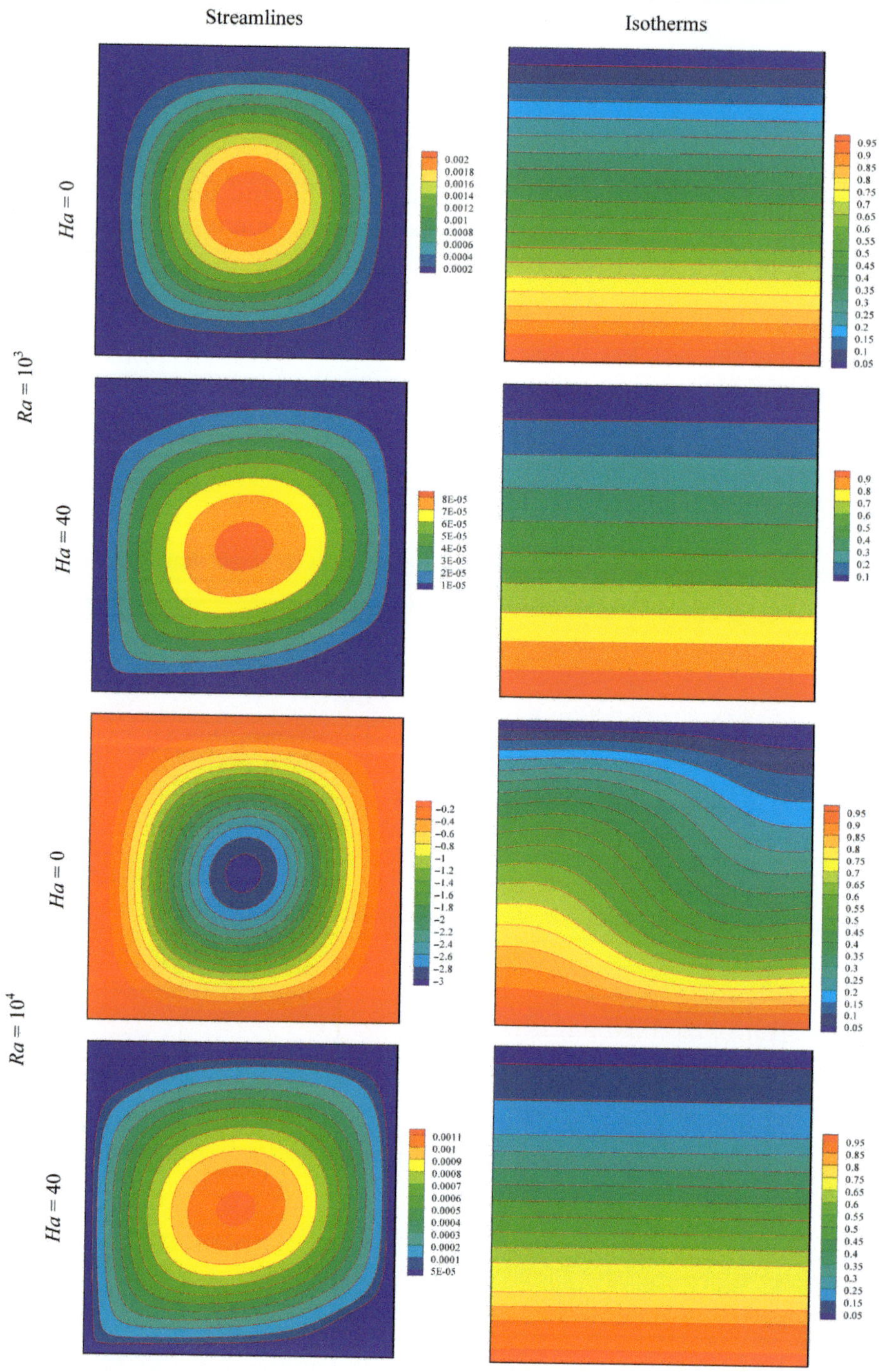

FIGURE 19.3 Streamline (left) and isotherm (right) contours for different values of Rayleigh and Hartmann numbers when $\delta = 0, \phi = 0.04$.

Fig. 19.6 illustrates the impact of δ, Ra, and Ha on Nu_{ave}. The formula for Nu_{ave} corresponding to active parameters is:

$$\begin{aligned} Nu_{ave} &= 0.425 + 0.122\delta - 1.63\log(Ra) + 0.35Ha^{*} \\ &+ 1.5\log(Ra)\delta - 0.58\delta Ha^{*} - 0.66\log(Ra)Ha^{*} \\ &+ 0.45Ha^{*2} + 0.53(\log(Ra))^{2} - 4.04\delta^{2} \end{aligned} \tag{19.21}$$

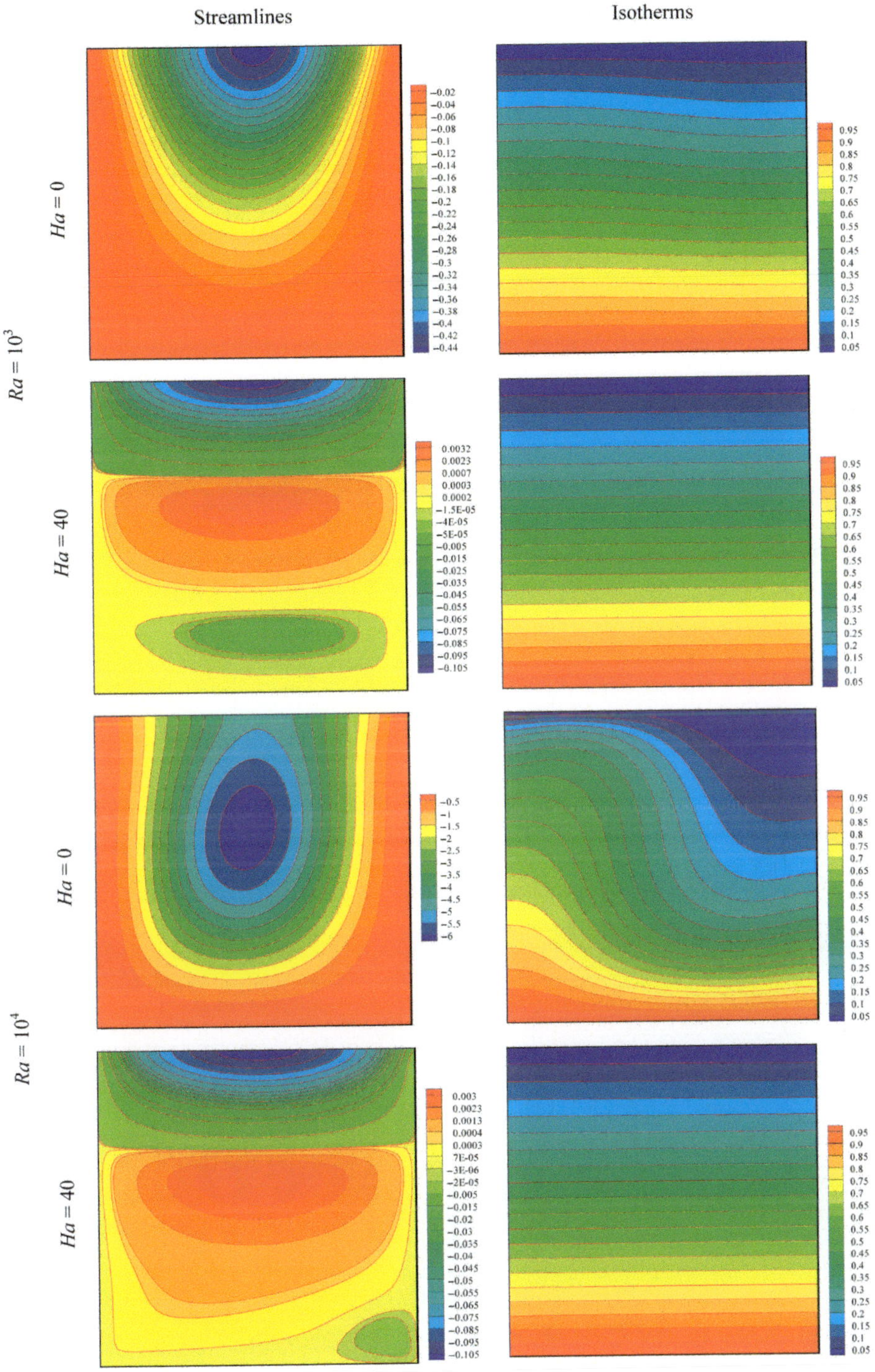

FIGURE 19.4 Streamline (left) and isotherm (right) contours for different values of Rayleigh and Hartmann numbers when $\delta = 0.2, \phi = 0.04$.

where $Ha^* = 0.1Ha$. As the melting parameter augments, the temperature gradient is enhanced, and in turn the Nusselt number increases. Increasing the bouyancy forces leads the thermal boundary layer thicknesss to reduce. So the Nusselt number increases with the enhancement of Ra. As the Hartmann number augments, the isotherms become parallel to each other. Therefore, the Nusselt number has an inverse relationship with Ha.

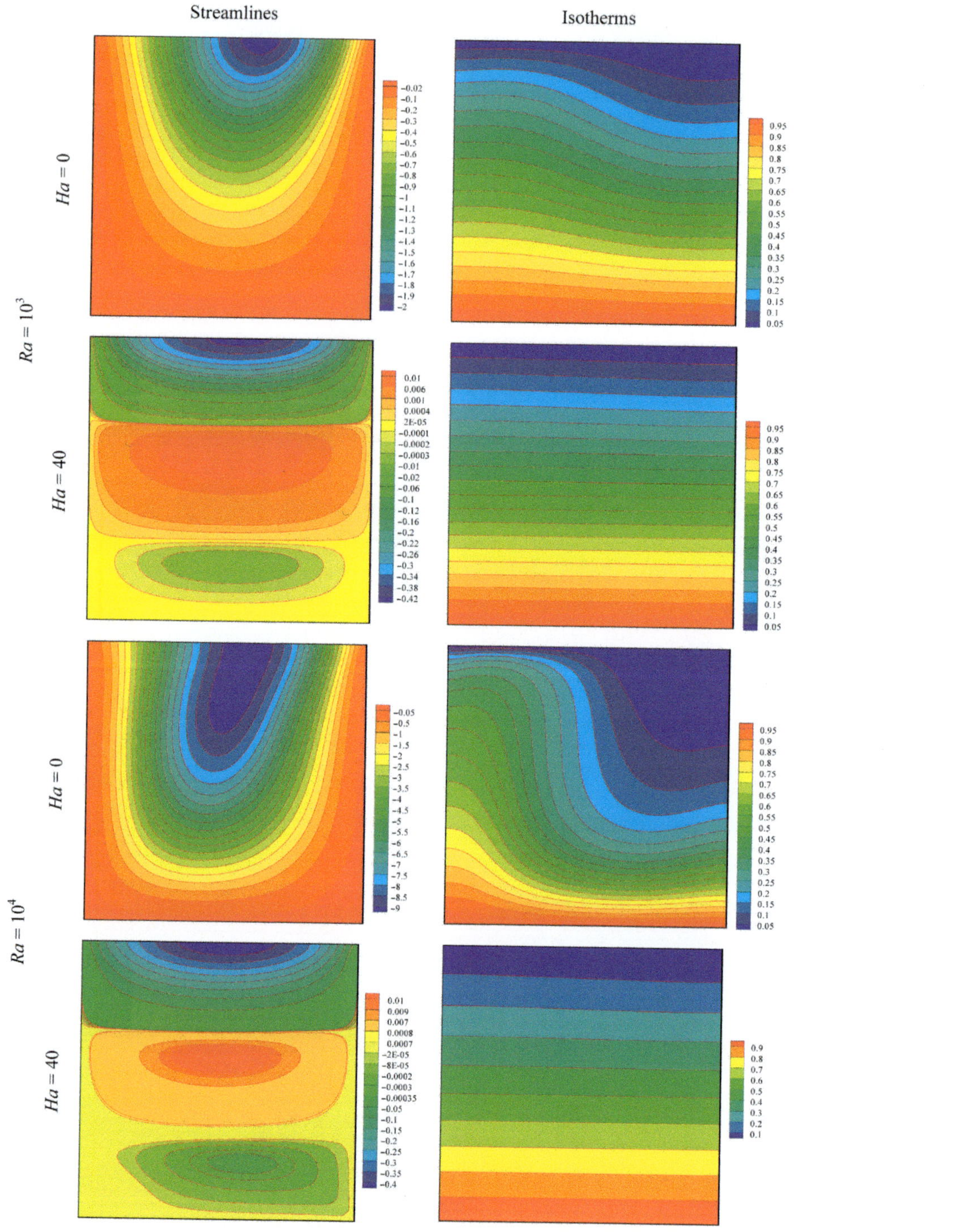

FIGURE 19.5 Streamline (left) and isotherm (right) contours for different values of Rayleigh and Hartmann numbers when $\delta = 0.8, \phi = 0.04$.

19.3 SIMULATION OF CUO-WATER NANOFLUID HEAT TRANSFER ENHANCEMENT IN THE PRESENCE OF A MELTING SURFACE

19.3.1 Problem Definition

Fig. 19.7 depicts the geometry, boundary condition, and sample element. The inner wall is the hot wall ($T = T_h$) and the outer one is the melting surface ($T = T_m$). The other walls are adiabatic. A horizontal magnetic field has been applied. The enclosure is filled with nanofluid.

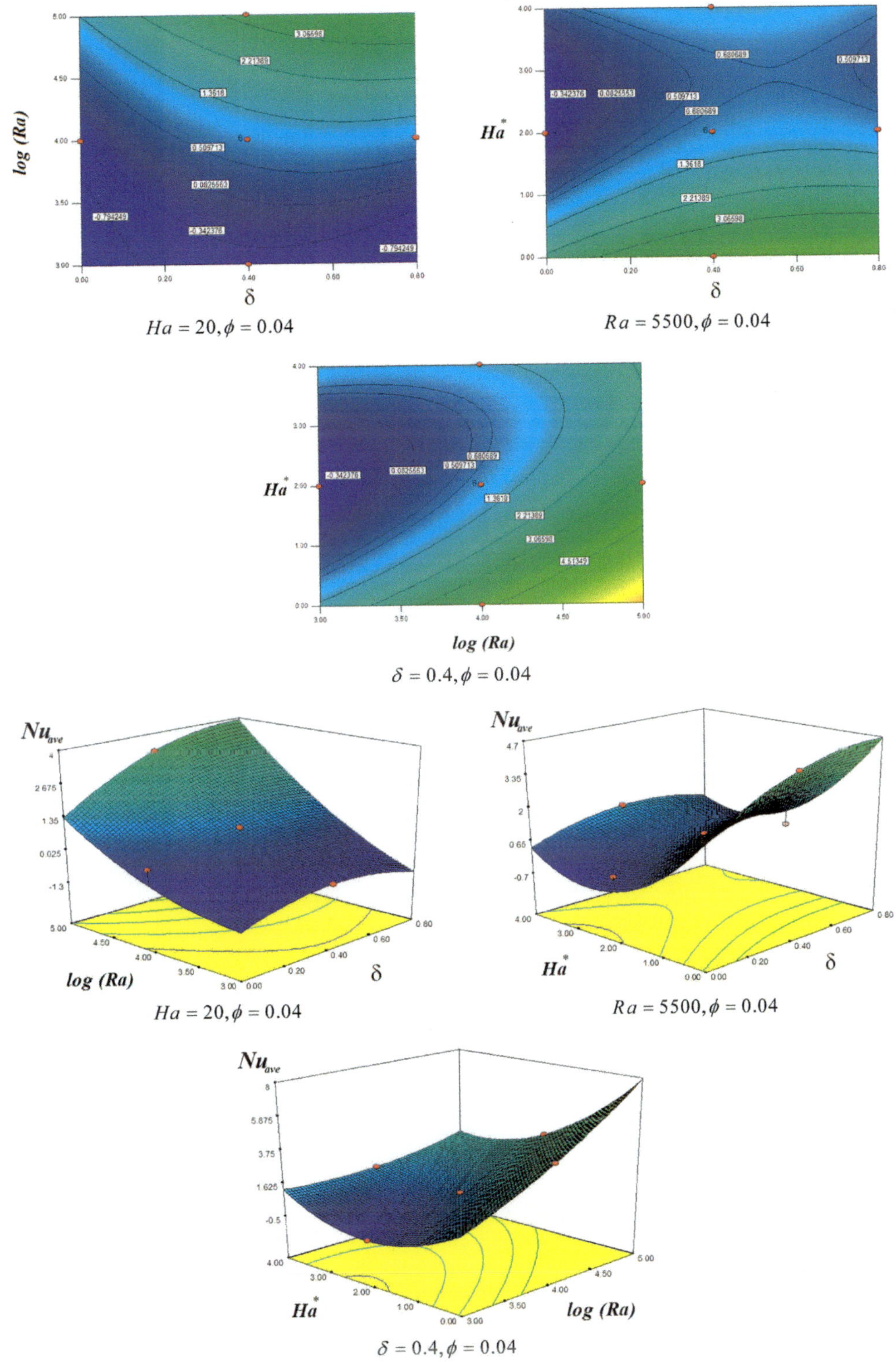

FIGURE 19.6 Influences of the melting parameter, Reynolds number, and Hartmann numbers on average Nusselt number.

19.3.2 Governing Equation

Nanofluid steady convective flow is considered in the presence of a constant magnetic field. The PDEs are:

$$\frac{\partial v}{\partial y} + \frac{\partial u}{\partial x} = 0 \tag{19.22}$$

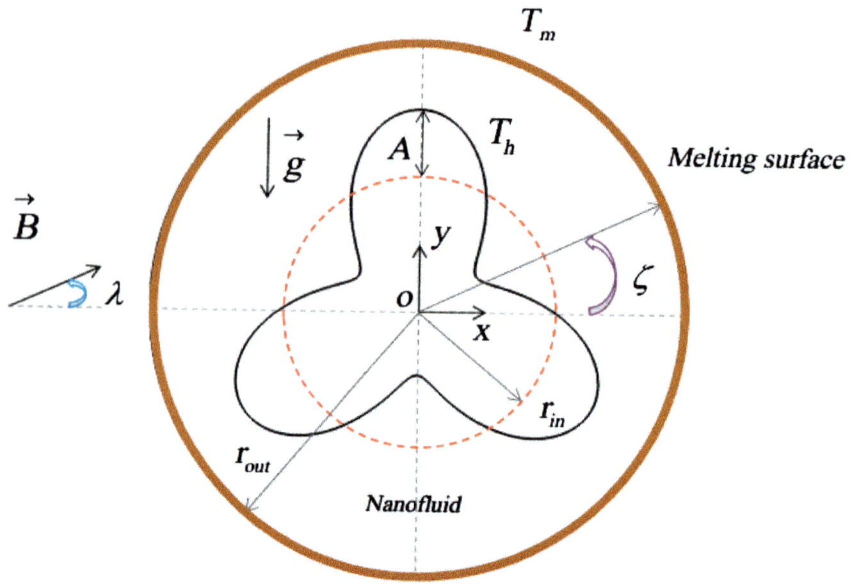

FIGURE 19.7 Geometry and the boundary conditions.

$$\left(\rho_{nf}\right)\left(v\frac{\partial u}{\partial y}+u\frac{\partial u}{\partial x}\right)=\left[B_y\sigma_{nf}vB_x-B_y^2\sigma_{nf}u+\mu_{nf}\left(\frac{\partial^2 u}{\partial y^2}+\frac{\partial^2 u}{\partial x^2}\right)-\frac{\partial P}{\partial x}\right] \tag{19.23}$$

$$\rho_{nf}\left(\frac{\partial v}{\partial y}v+\frac{\partial v}{\partial x}u\right)=uB_y\sigma_{nf}B_x+(T-T_c)\beta_{nf}g\rho_{nf}-vB_x\sigma_{nf}B_x$$
$$+\left(\frac{\partial^2 v}{\partial x^2}+\frac{\partial^2 v}{\partial y^2}\right)\mu_{nf}-\frac{\partial P}{\partial y}, \tag{19.24}$$
$$B_x=B_o\cos\lambda,\ \ B_y=B_o\sin\lambda$$

$$(\rho C_p)_{nf}\left(v\frac{\partial T}{\partial y}+u\frac{\partial T}{\partial x}\right)=k_{nf}\left(\frac{\partial^2 T}{\partial x^2}+\frac{\partial^2 T}{\partial y^2}\right) \tag{19.25}$$

$(\rho C_p)_{nf}$, ρ_{nf}, $(\rho\beta)_{nf}$, and σ_{nf} are calculated as:

$$(\rho C_p)_{nf}=\phi(\rho C_p)_s+(1-\phi)(\rho C_p)_f \tag{19.26}$$

$$\rho_{nf}=\rho_f(1-\phi)+\phi\rho_s \tag{19.27}$$

$$\frac{(\rho\beta)_{nf}}{(\rho\beta)_f}=\phi\frac{(\rho\beta)_s}{(\rho\beta)_f}+(1-\phi) \tag{19.28}$$

$$\sigma_{nf}=\left[\left(\frac{(2+\sigma_s/\sigma_f)-(\sigma_s/\sigma_f-1)\phi}{3\phi(-1+\sigma_s/\sigma_f)}\right)^{-1}+1\right]\sigma_f \tag{19.29}$$

k_{nf}, μ_{nf} are calculated via the KKL model:

$$\frac{k_{nf}}{k_f} = 1 + \frac{3\left(\frac{k_p}{k_f} - 1\right)\phi}{\left(\frac{k_p}{k_f} + 2\right) - \left(\frac{k_p}{k_f} - 1\right)\phi} + 5\times 10^4 g'(\phi, T, d_p)\phi\rho_f c_{p,f}\sqrt{\frac{\kappa_b T}{d_p \rho_p}},$$
$$R_f = 4\times 10^{-8} km^2/W,$$
$$R_f = -\,d_p(1/k_p - 1/k_{p,eff}),$$
$$g'(\phi, d_p, T) = Ln(T)\left(a_1 + a_3 Ln(\phi) + a_2 Ln(d_p) + a_5 Ln(d_p)^2 + a_4 \ln(d_p) Ln(\phi)\right)$$
$$+ \left(a_6 + a_8 Ln(\phi) + a_7 Ln(d_p) + a_{10} Ln(d_p)^2 + a_9 Ln(\phi)\ln(d_p)\right) \tag{19.30}$$

$$\mu_{nf} = \frac{\mu_f}{(1-\phi)^{2.5}} + \frac{k_{Brownian}}{k_f} \times \frac{\mu_f}{\Pr} \tag{19.31}$$

Properties and needed parameters are provided in Tables 19.1 and 19.2.
Vorticity and stream function should be used to eliminate the pressure source terms:

$$\omega + \frac{\partial u}{\partial y} - \frac{\partial v}{\partial x} = 0, \quad \frac{\partial \psi}{\partial x} = -\,v, \quad \frac{\partial \psi}{\partial y} = u \tag{19.32}$$

Introducing dimensionless quantities:

$$P = \frac{p}{\rho_{nf}(\alpha_{nf}/L)^2}, U = \frac{uL}{\alpha_{nf}}, V = \frac{vL}{\alpha_{nf}}, \theta = \frac{T - T_m}{T_h - T_m}, (X, Y) = \frac{(x, y)}{L},$$
$$\Psi = \frac{\psi}{\alpha_{nf}}, \Omega = \frac{\omega L^2}{\alpha_{nf}} \tag{19.33}$$

The final formulae are:

$$\frac{\partial^2 \Psi}{\partial Y^2} + \Omega + \frac{\partial^2 \Psi}{\partial X^2} = 0, \tag{19.34}$$

$$\frac{\partial \Omega}{\partial X} U + V \frac{\partial \Omega}{\partial Y} = \Pr \frac{A_5}{A_1} \frac{A_2}{A_4}\left(\frac{\partial^2 \Omega}{\partial Y^2} + \frac{\partial^2 \Omega}{\partial X^2}\right)$$
$$+ \Pr Ha^2 \frac{A_6}{A_1} \frac{A_2}{A_4}\left(-B_x \frac{\partial V}{\partial X} B_x + B_y \frac{\partial U}{\partial X} B_x - \frac{\partial V}{\partial Y} B_y B_x + B_y \frac{\partial U}{\partial Y} B_y\right)$$
$$+ \Pr Ra \frac{A_3 A_2^2}{A_1 A_4^2} \frac{\partial \theta}{\partial X}, \tag{19.35}$$

$$\frac{\partial \theta}{\partial X} U + \frac{\partial \theta}{\partial Y} V = \left(\frac{\partial^2 \theta}{\partial X^2} + \frac{\partial^2 \theta}{\partial Y^2}\right) \tag{19.36}$$

Boundary conditions are:

$$\begin{aligned} &\frac{\partial \theta}{\partial n} = 0.0 && \text{on other walls} \\ &\theta = 0 && \text{on outer wall} \\ &\theta = 1.0 && \text{on inner wall} \\ &\Psi = 0.0 && \text{on all walls except melting surface} \end{aligned} \tag{19.37}$$

and on the melting surface, we have:

$$\frac{\partial \theta}{\partial n} = \frac{A_2}{A_1}\frac{1}{\delta}(L + c_s(T_m - T_0))\frac{\partial \Psi}{\partial n} \tag{19.38}$$

where dimensionless and constants parameters are illustrated as:

$$\begin{aligned}&\Pr = \upsilon_f/\alpha_f, Ra = g\beta_f \Delta T L^3/(\upsilon_f \alpha_f),\ Ha = LB_0\sqrt{\sigma_f/\mu_f},\ \delta = \frac{(\rho C_p)_f}{\rho_f}\frac{(T_h - T_m)}{1 + c_s(T_m - T_0)}\\&A_1 = \frac{\rho_{nf}}{\rho_f},\ A_5 = \frac{\mu_{nf}}{\mu_f},\ A_2 = \frac{(\rho C_P)_{nf}}{(\rho C_P)_f}, A_6 = \frac{\sigma_{nf}}{\sigma_f},\ A_4 = \frac{k_{nf}}{k_f},\ A_3 = \frac{(\rho\beta)_{nf}}{(\rho\beta)_f}\end{aligned} \tag{19.39}$$

It should be mentioned that δ is related to Stefan numbers.
Nu_{loc} and Nu_{ave} over the cold wall can be calculated as:

$$Nu_{loc} = \left(\frac{k_{nf}}{k_f}\right)\frac{\partial \theta}{\partial r} \tag{19.40}$$

$$Nu_{ave} = \frac{1}{0.5\pi}\int_0^{0.5\pi} Nu_{loc}\, d\xi \tag{19.41}$$

19.3.3 Effects of Active Parameters

Nanofluid in the presence of melting heat transfer is presented under the effect of Lorentz forces. The KKL model is selected to calculate properties of the nanofluid. Outputs are demonstrated for various number of undulation ($N = 3$ to 5), volume fraction of nanofluid ($\phi = 0$ to 0.04), Hartmann number ($Ha = 0$ to 40), melting parameter ($\delta = 0$ to 0.2), and Rayleigh number ($Ra = 500$ to 5000). Fig. 19.8 presents the influence of using nanofluid instead of base fluid. Velocity increases with the use of nanofluid due to the increase in the solid movements. In the presence of a magnetic field and melting heat transfer, the impact of adding nanoparticles on isotherms becomes insignificant. The temperature gradient reduces by the use of the nanofluid. Influences of Hartmann and Rayleigh numbers without melting heat transfer are illustrated in Figs. 19.9–19.11. The streamlines have just one rotating eddy. Conduction mode is prominent in low Raleigh numbers. As the buoyancy force is enhanced, the distortion of isotherms increases close to the hot wall. Increasing the Hartmann number causes the isotherms to become parallel. In the presence of melting heat transfer the primary eddy diminishes and enhancing Lorentz forces generate three layers for streamlines. The bottom eddy becomes stronger with the increase of the melting parameter. As the number of undulations increases, the temperature gradient near the hot wall is enhanced. The effects of δ, Ra, N, and Ha on average Nusselt number are depicted in Fig. 19.12. The following correlation can be offered:

$$\begin{aligned}Nu_{ave} &= 0.068 + 0.505N - 0.15\delta + 0.08Ra^* + 0.11Ha^*\\&0.13N\delta + 0.002N\,Ra^* - 0.021NHa^* + 0.056Ra^*\delta\\&- 0.14\delta Ha^* - 0.008Ra^*Ha^* - 0.0485N^2\\&+ 0.49\delta^2 - 0.007Ra^{*2} - 0.011Ha^{*2}\end{aligned} \tag{19.42}$$

where $Ha^* = 0.1Ha$, $Ra^* = 0.001Ra$. The temperature gradient increases with the increase of melting parameter. So, the Nusselt number has a direct relationship with δ. Augmenting the Rayleigh number leads to a reduction in thermal boudary layer thicknesss. Therefore, the Nusselt number is enhanced with the increase of Ra. As the Lorentz forces increase, the conduction mechanism become stronger. Thus, Nu_{ave} reduces with the increase of the Hartmann number. In addition, the rate of heat transfer increases with the rise of N.

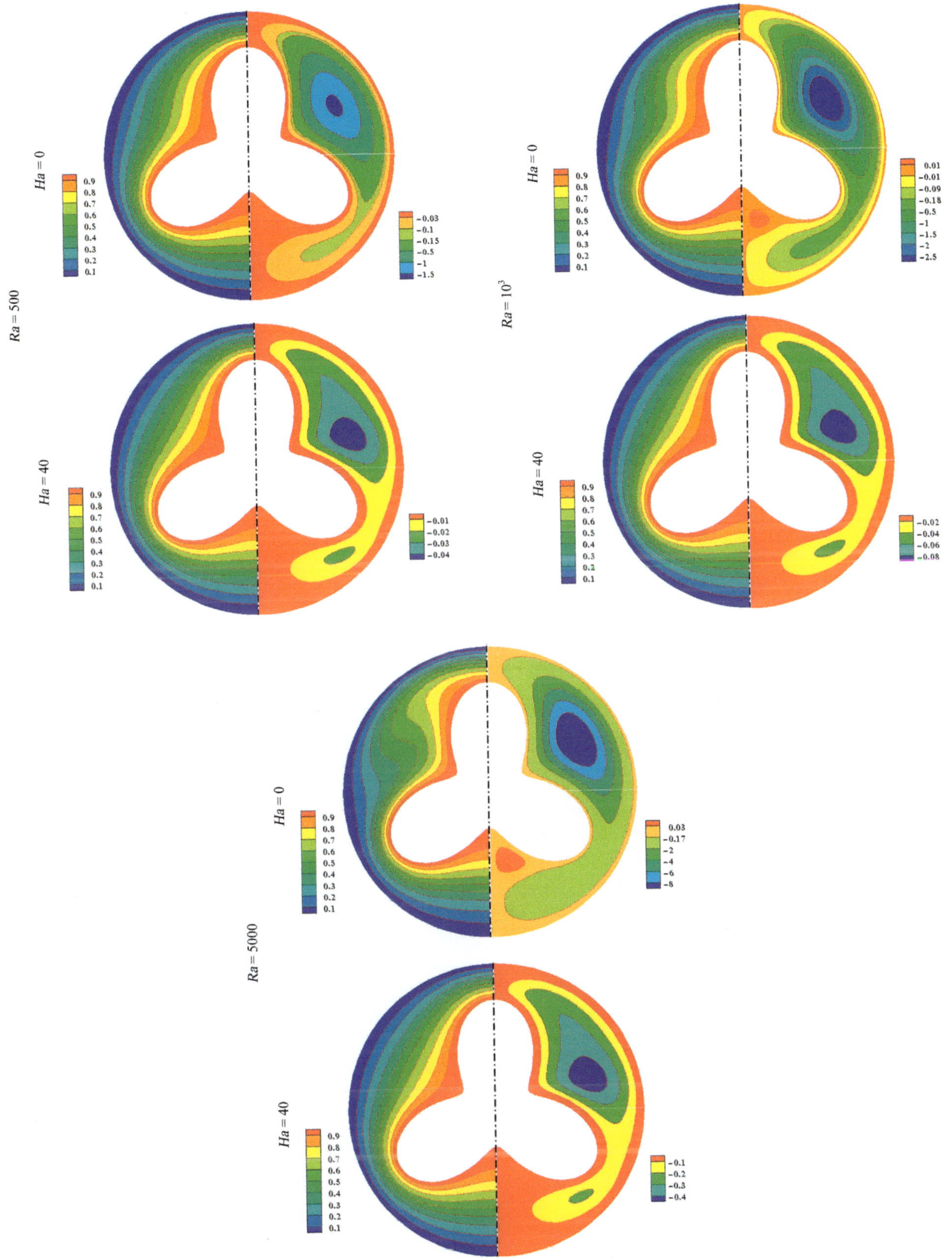

FIGURE 19.8 Isotherm (left) and streamline (right) contours for different values of Rayleigh and Hartmann numbers when $\delta = 0, \phi = 0.04, N = 3$.

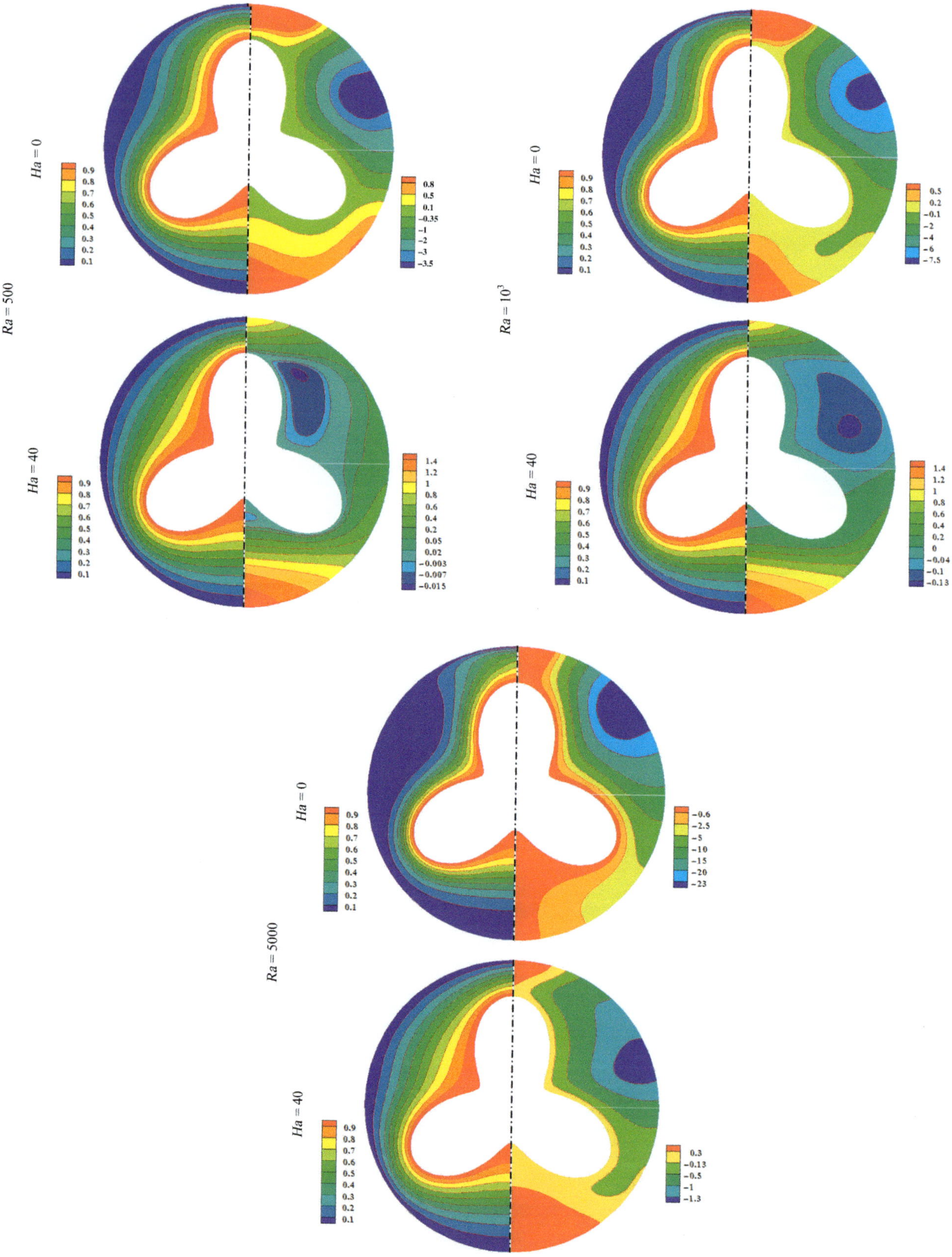

FIGURE 19.9 Isotherm (left) and streamline (right) contours for different values of Rayleigh and Hartmann numbers when $\delta = 0.2, \phi = 0.04, N = 3$.

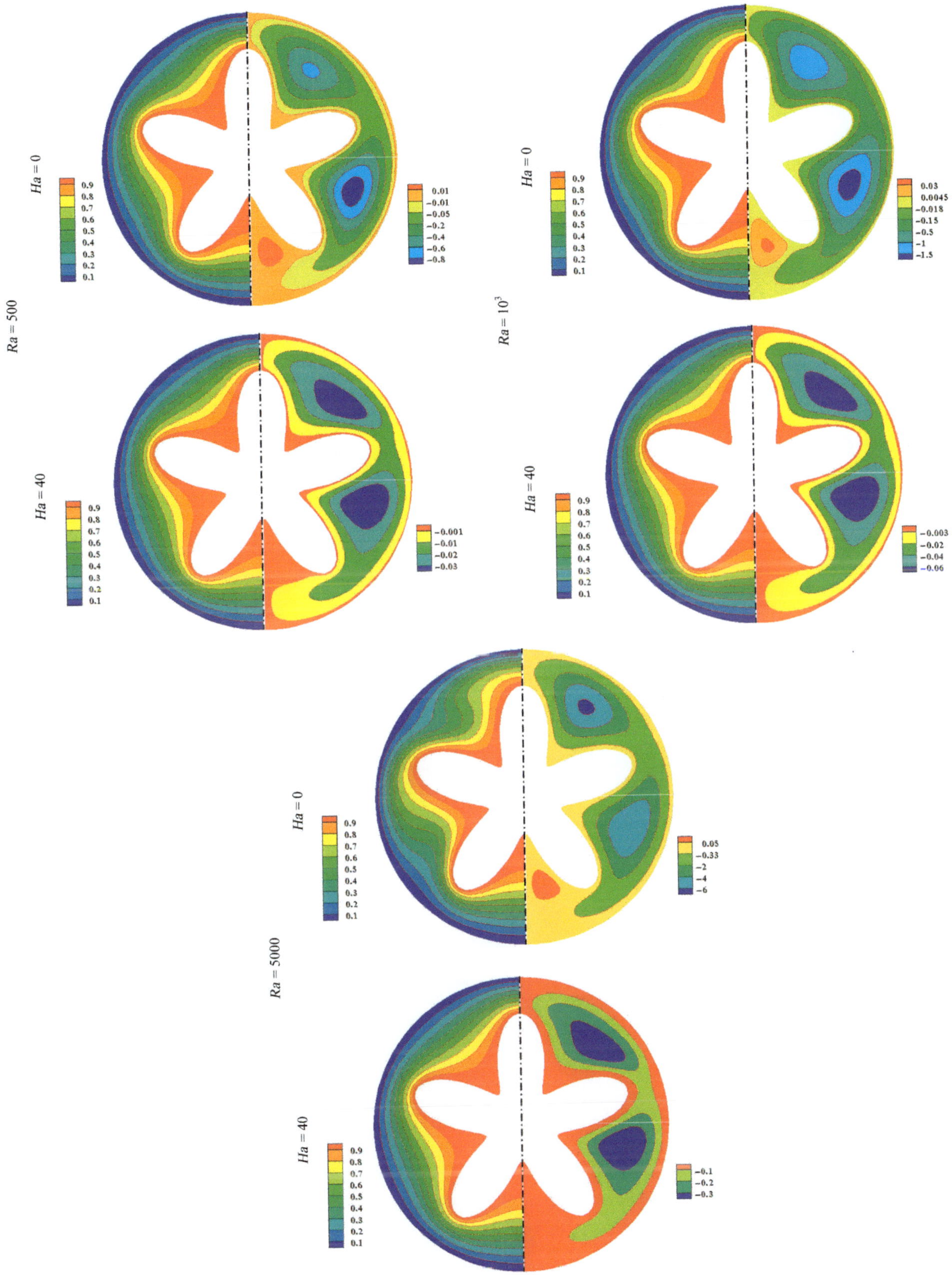

FIGURE 19.10 Isotherm (left) and streamline (right) contours for different values of Rayleigh and Hartmann numbers when $\delta = 0, \phi = 0.04, N = 5$.

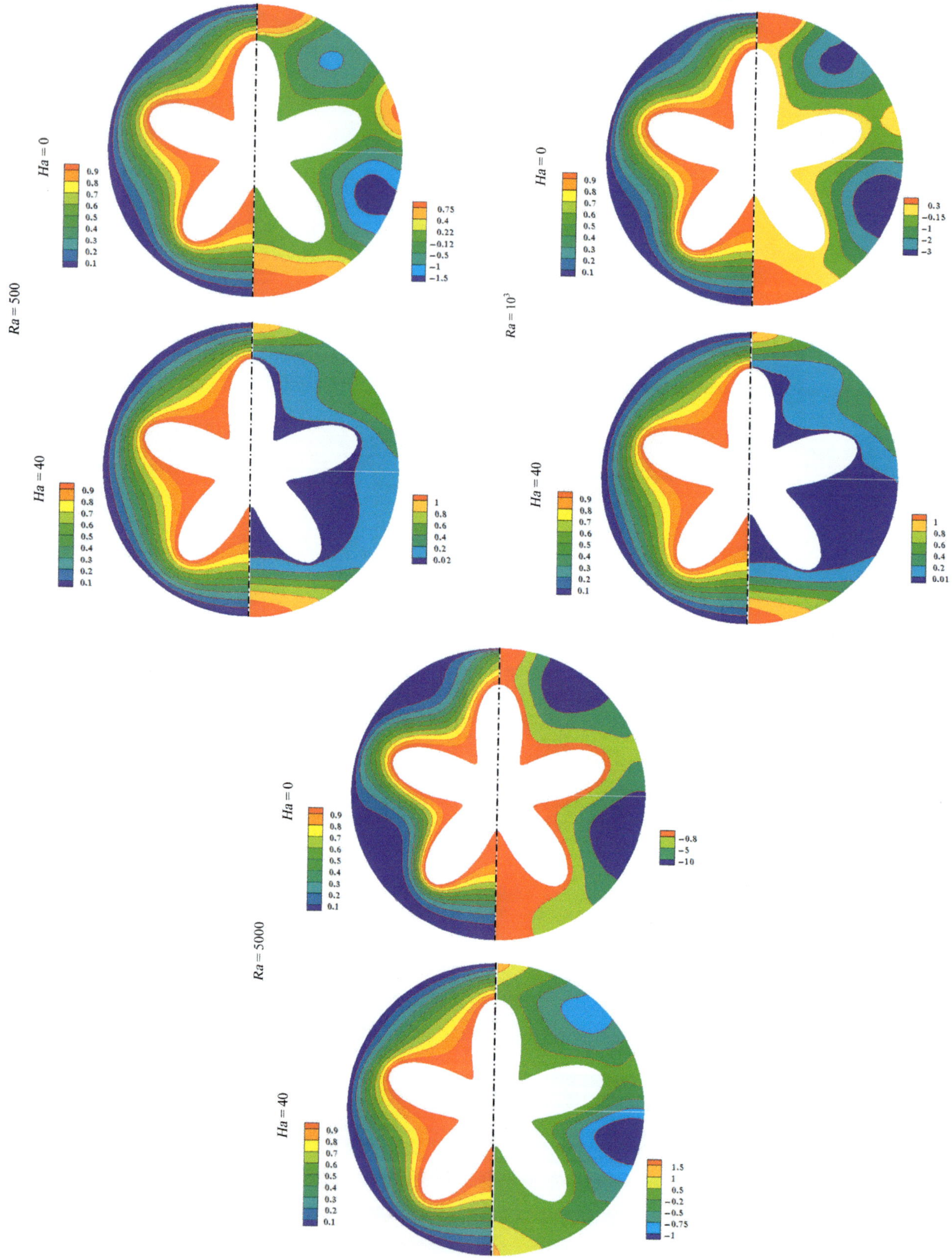

FIGURE 19.11 Isotherm (left) and streamline (right) contours for different values of Rayleigh and Hartmann numbers when $\delta = 0.2, \phi = 0.04, N = 5$.

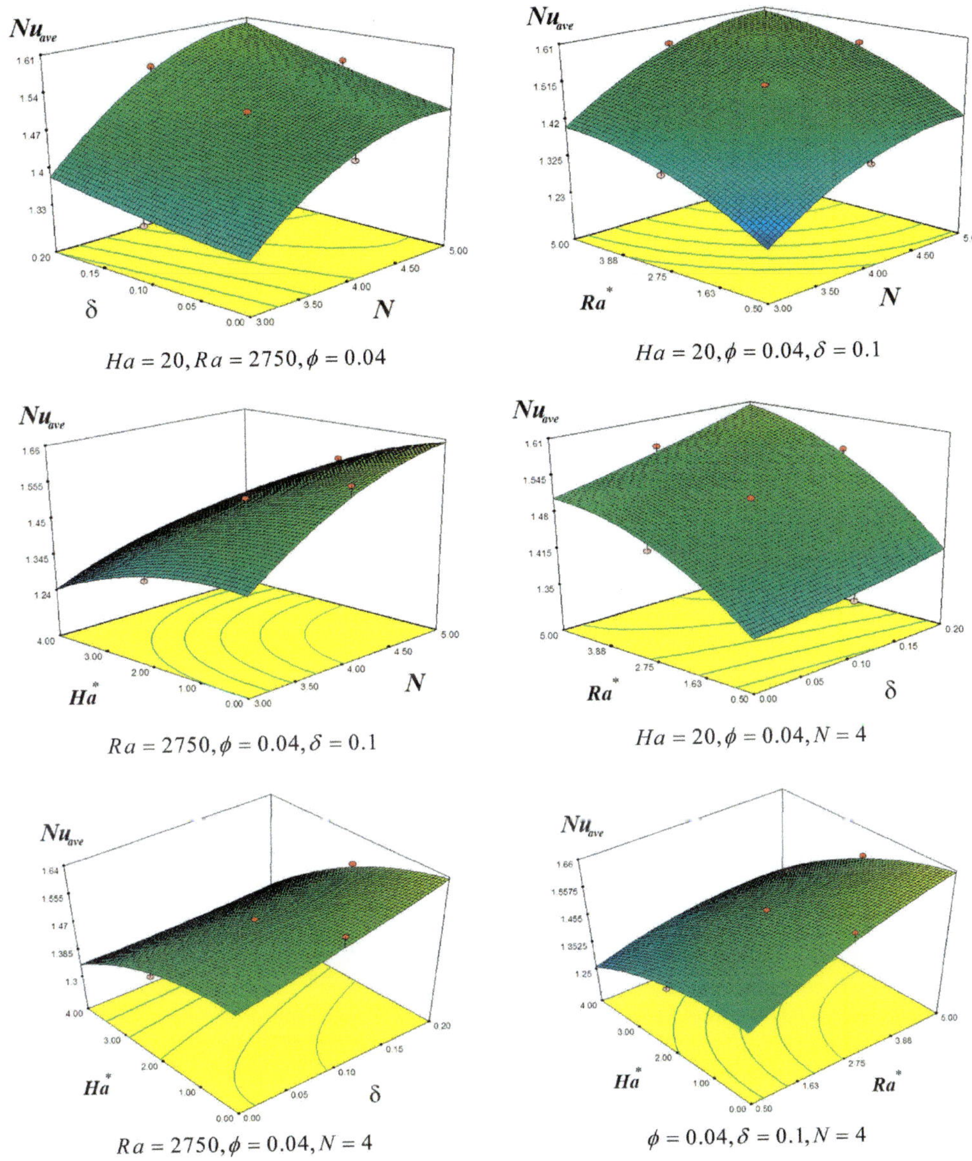

FIGURE 19.12 Influences of the number of undulation, Darcy, Reynolds, and Hartmann numbers on average Nusselt number.

19.4 CUO-WATER NANOFLUID MAGNETOHYDRODYNAMIC NATURAL CONVECTION INSIDE A SINUSOIDAL ANNULUS IN THE PRESENCE OF MELTING HEAT TRANSFER

19.4.1 Problem Definition

Fig. 19.13 depicts the geometry, boundary condition, and sample element. The inner wall is the hot wall ($T = T_h$) and the outer one is the melting surface ($T = T_m$). The other walls are adiabatic. A horizontal magnetic field has been applied. The enclosure is filled with nanofluid.

19.4.2 Governing Equation

2D steady convective nanofluid flow is considered in the presence of a constant magnetic field. The PDEs are:

$$\frac{\partial v}{\partial y} + \frac{\partial u}{\partial x} = 0 \tag{19.43}$$

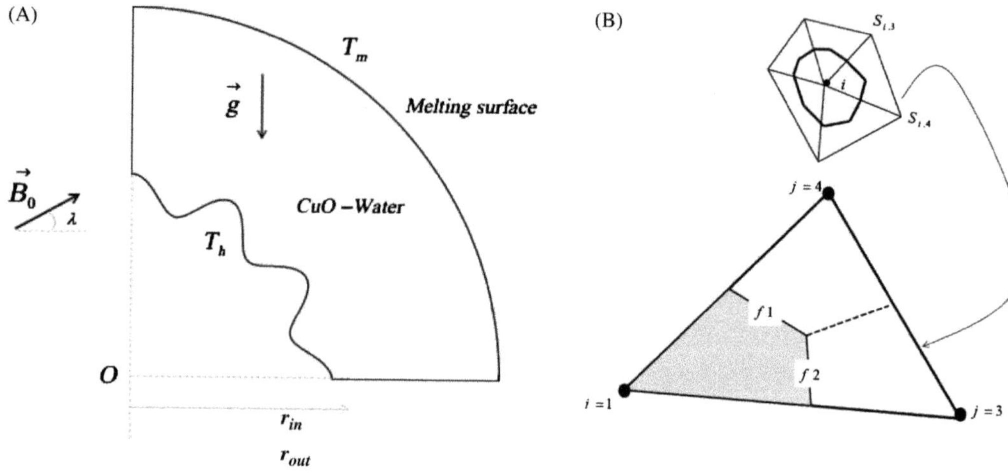

FIGURE 19.13 (A) Geometry (B) Sample element.

$$\left(\rho_{nf}\right)\left(v\frac{\partial u}{\partial y}+u\frac{\partial u}{\partial x}\right)=\left[B_y\sigma_{nf}vB_x-B_y^2\sigma_{nf}u+\mu_{nf}\left(\frac{\partial^2 u}{\partial y^2}+\frac{\partial^2 u}{\partial x^2}\right)-\frac{\partial P}{\partial x}\right] \tag{19.44}$$

$$\begin{aligned}\rho_{nf}\left(\frac{\partial v}{\partial y}v+\frac{\partial v}{\partial x}u\right)&=uB_y\sigma_{nf}B_x+(T-T_c)\beta_{nf}g\rho_{nf}-vB_x\sigma_{nf}B_x\\&+\left(\frac{\partial^2 v}{\partial x^2}+\frac{\partial^2 v}{\partial y^2}\right)\mu_{nf}-\frac{\partial P}{\partial y},\\B_x=B_o\cos\lambda,&\quad B_y=B_o\sin\lambda\end{aligned} \tag{19.45}$$

$$\left(\rho C_p\right)_{nf}\left(v\frac{\partial T}{\partial y}+u\frac{\partial T}{\partial x}\right)=k_{nf}\left(\frac{\partial^2 T}{\partial x^2}+\frac{\partial^2 T}{\partial y^2}\right) \tag{19.46}$$

$\left(\rho C_p\right)_{nf}$, ρ_{nf}, $(\rho\beta)_{nf}$, and σ_{nf} are calculated as:

$$\left(\rho C_p\right)_{nf}=\phi\left(\rho C_p\right)_s+(1-\phi)\left(\rho C_p\right)_f \tag{19.47}$$

$$\rho_{nf}=\rho_f(1-\phi)+\phi\rho_s \tag{19.48}$$

$$(\rho\beta)_{nf}=\phi(\rho\beta)_s+(1-\phi)(\rho\beta)_f \tag{19.49}$$

$$\sigma_{nf}=\sigma_f\left[\left(\frac{+(2+\sigma_s/\sigma_f)-\phi(-1+\sigma_s/\sigma_f)}{3\phi(-1+\sigma_s/\sigma_f)}\right)^{-1}+1\right] \tag{19.50}$$

k_{nf}, μ_{nf} are calculated via the KKL model:

$$\begin{aligned}\frac{k_{nf}}{k_f}&=1+\frac{3\left(\frac{k_p}{k_f}-1\right)\phi}{\left(\frac{k_p}{k_f}+2\right)-\left(\frac{k_p}{k_f}-1\right)\phi}+5\times10^4 g'(\phi,T,d_p)\phi\rho_f c_{p,f}\sqrt{\frac{\kappa_b T}{d_p\rho_p}},\\R_f&=4\times10^{-8}km^2/W,\\R_f&=-d_p(1/k_p-1/k_{p,eff}),\\g'(\phi,d_p,T)&=Ln(T)\left(a_1+a_3Ln(\phi)+a_2Ln\left(d_p\right)+a_5Ln\left(d_p\right)^2+a_4\ln\left(d_p\right)Ln(\phi)\right)\\&+\left(a_6+a_8Ln(\phi)+a_7Ln\left(d_p\right)+a_{10}Ln\left(d_p\right)^2+a_9Ln(\phi)\ln\left(d_p\right)\right)\end{aligned} \tag{19.51}$$

$$\mu_{nf} = \frac{\mu_f}{(1-\phi)^{2.5}} + \frac{k_{Brownian}}{k_f} \times \frac{\mu_f}{\Pr} \tag{19.52}$$

Vorticity and stream function should be used to eliminate the pressure source terms:

$$\omega + \frac{\partial u}{\partial y} - \frac{\partial v}{\partial x} = 0, \quad \frac{\partial \psi}{\partial x} = -v, \quad \frac{\partial \psi}{\partial y} = u \tag{19.53}$$

Introducing dimensionless quantities:

$$P = \frac{p}{\rho_{nf}\left(\alpha_{nf}/L\right)^2}, U = \frac{uL}{\alpha_{nf}}, V = \frac{vL}{\alpha_{nf}}, \ \theta = \frac{T - T_m}{T_h - T_m}, (X, Y) = \frac{(x, y)}{L},$$
$$\Psi = \frac{\psi}{\alpha_{nf}}, \Omega = \frac{\omega L^2}{\alpha_{nf}} \tag{19.54}$$

The final formulae are:

$$\Omega + \frac{\partial^2 \Psi}{\partial Y^2} + \frac{\partial^2 \Psi}{\partial X^2} = 0, \tag{19.55}$$

$$\frac{\partial \Omega}{\partial X} U + V \frac{\partial \Omega}{\partial Y} = \Pr \frac{A_5}{A_1} \frac{A_2}{A_4} \left(\frac{\partial^2 \Omega}{\partial Y^2} + \frac{\partial^2 \Omega}{\partial X^2} \right)$$
$$+ \Pr Ha^2 \frac{A_6}{A_1} \frac{A_2}{A_4} \left(-\frac{\partial V}{\partial X} B_x B_x + B_y \frac{\partial U}{\partial X} B_x - \frac{\partial V}{\partial Y} B_y B_x + B_y \frac{\partial U}{\partial Y} B_y \right) \tag{19.56}$$
$$+ \Pr Ra \frac{A_3 A_2^2}{A_1 A_4^2} \frac{\partial \theta}{\partial X},$$

$$\frac{\partial \theta}{\partial X} U + \frac{\partial \theta}{\partial Y} V = \left(\frac{\partial^2 \theta}{\partial X^2} + \frac{\partial^2 \theta}{\partial Y^2} \right) \tag{19.57}$$

Boundary conditions are:

$$\begin{aligned} &\frac{\partial \theta}{\partial n} = 0.0 && \text{on other walls} \\ &\theta = 0 && \text{on outer wall} \\ &\theta = 1.0 && \text{on inner wall} \\ &\Psi = 0.0 && \text{on all walls except melting surface} \end{aligned} \tag{19.58}$$

and on the melting surface, we have:

$$\frac{\partial \theta}{\partial n} = \frac{A_2}{A_1} \frac{1}{\delta} (L + c_s(T_m - T_0)) \frac{\partial \Psi}{\partial n} \tag{19.59}$$

where dimensionless and constants parameters are illustrated as:

$$\Pr = \upsilon_f/\alpha_f, Ra = g\beta_f \Delta T L^3/(\upsilon_f \alpha_f), \ Ha = LB_0\sqrt{\sigma_f/\mu_f}, \ \ \delta = \frac{(\rho C_p)_f}{\rho_f} \frac{(T_h - T_m)}{1 + c_s(T_m - T_0)}$$
$$A_1 = \frac{\rho_{nf}}{\rho_f}, \ \ A_5 = \frac{\mu_{nf}}{\mu_f}, \ \ A_2 = \frac{(\rho C_P)_{nf}}{(\rho C_P)_f}, A_6 = \frac{\sigma_{nf}}{\sigma_f}, \ \ A_4 = \frac{k_{nf}}{k_f}, \ \ A_3 = \frac{(\rho\beta)_{nf}}{(\rho\beta)_f} \tag{19.60}$$

It should be mentioned that δ is related to Stefan numbers.

Local and average Nusselt over the cold wall can be calculated as:

$$Nu_{loc} = \left(\frac{k_{nf}}{k_f}\right)\frac{\partial \theta}{\partial r} \tag{19.6}$$

$$Nu_{ave} = \frac{1}{0.5\pi}\int_0^{0.5\pi} Nu_{loc}\, d\xi \tag{19.6}$$

19.4.3 Effects of Active Parameters

Nanofluid flow in a half sinusoidal annulus due to magnetic field in presence of melting surface is examined. μ_{nf}, k_{nf} of CuO-water nanofluid are estimated by means of the KKL model. Graphs and tables are depicted for different amounts of CuO-H_2O volume fraction ($\phi = 0$ to 0.04), melting parameter ($\delta = 0$ to 0.2), Rayleigh number ($Ra = 500$ to 5000), and Hartmann number ($Ha = 0$ to 40). The impact of adding CuO nanoparticles in water on velocity and temperature contours is depicted in Fig. 19.14. The temperature gradient decreases with the increase of ϕ. $|\Psi_{\max}|$ augments with the addition of nanoparticles because of the increase in the solid movements. In the presence of melting heat transfer and magnetic field, the effect of the addition of nanoparticles on the isotherm becomes negligible. Figs. 19.15 and 19.16 depict the impact of Rayleigh and Hartmann numbers in the absence of melting heat transfer. There is only one eddy in the streamlines. At low Rayleigh number, the conduction mechanism is dominant. As Ra increases, the distortion of isotherms is enhanced close to the hot wall. Adding a magnetic field makes the isotherms become parallel. In the presence of melting heat transfer the primary eddy diminishes and increasing Lorentz forces generate three layers of the streamlines. Increasing the melting

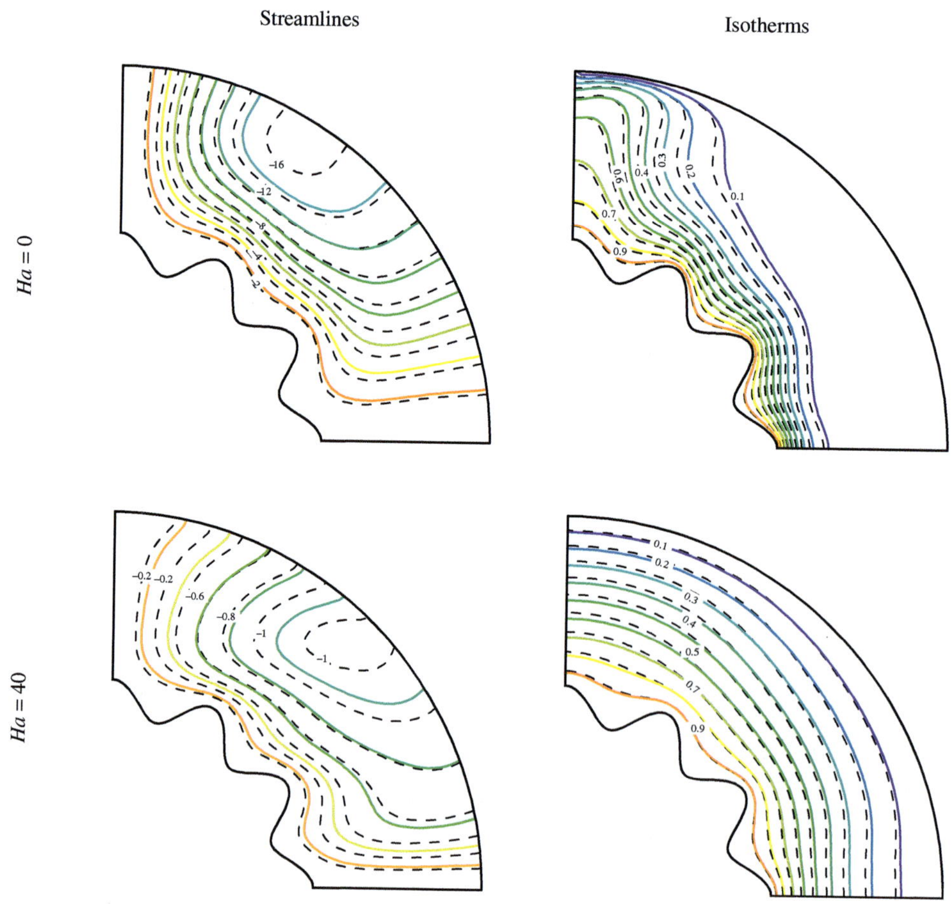

FIGURE 19.14 Impact of adding CuO in water on streamlines and isotherms (nanofluid ($\phi = 0.04$)(—) and pure fluid ($\phi = 0$) (- - -)) when $Ra = 5000$, $\delta = 0.2$.

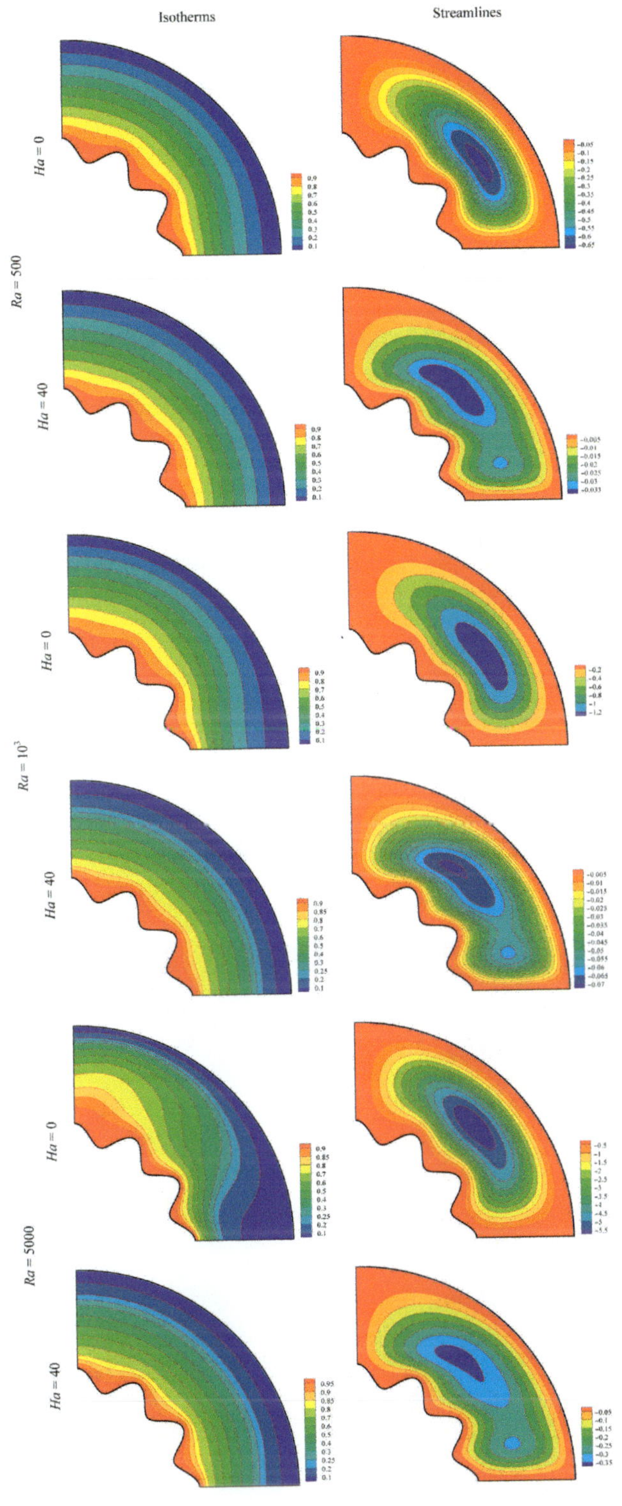

FIGURE 19.15 Isotherms and streamlines for various Ha, Ra when $\delta = 0, \phi = 0.04$.

parameter enhances the bottom eddy. Fig. 19.17 illustrates the impact of δ, Ra, and Ha on Nu_{ave}. The formula for Nu_{ave} corresponding to active parameters is:

$$
\begin{aligned}
Nu_{ave} &= 0.799 + 0.305\delta + 0.19Ra^* - 0.16Ha^* \\
&+ 0.003Ra^*\delta - 0.0039\delta Ha^* - 0.028Ra^*Ha^* \\
&+ 0.058\delta^2 - 0.012Ra^{*2} + 0.039Ha^{*2}
\end{aligned}
\tag{19.63}
$$

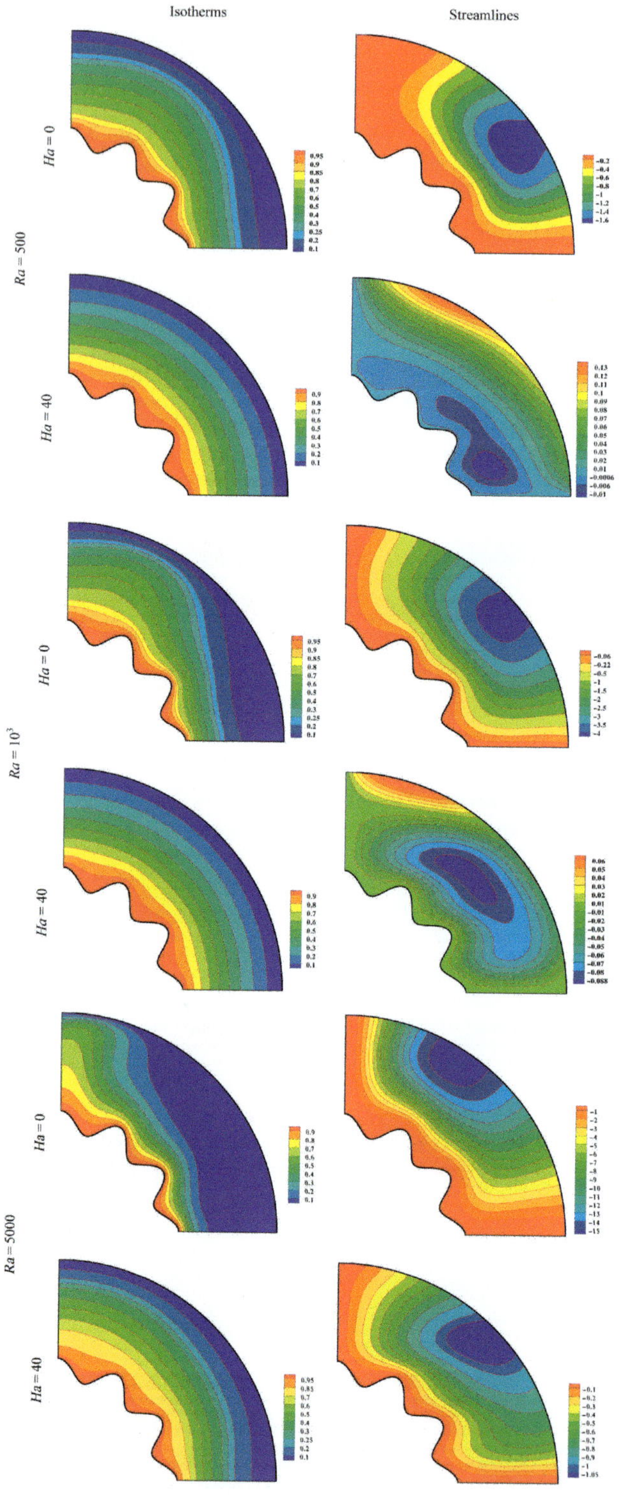

FIGURE 19.16 Isotherms and streamlines for various Ha, Ra when $\delta = 0.2, \phi = 0.04$.

where $Ha^* = 0.1Ha, Ra^* = 0.001Ra$. As the melting parameter increases, the temperature gradient is enhanced and in turn the Nusselt number is enhanced. Increasing the bouyancy forces leads to a reduction in the thickness of the thermal boundary layer . So the Nusselt number increases with the enhancement of Ra. As the Hartmann number augments, the isotherms become parallel. Therefore the Nusselt number has an inverse relationship with Ha.

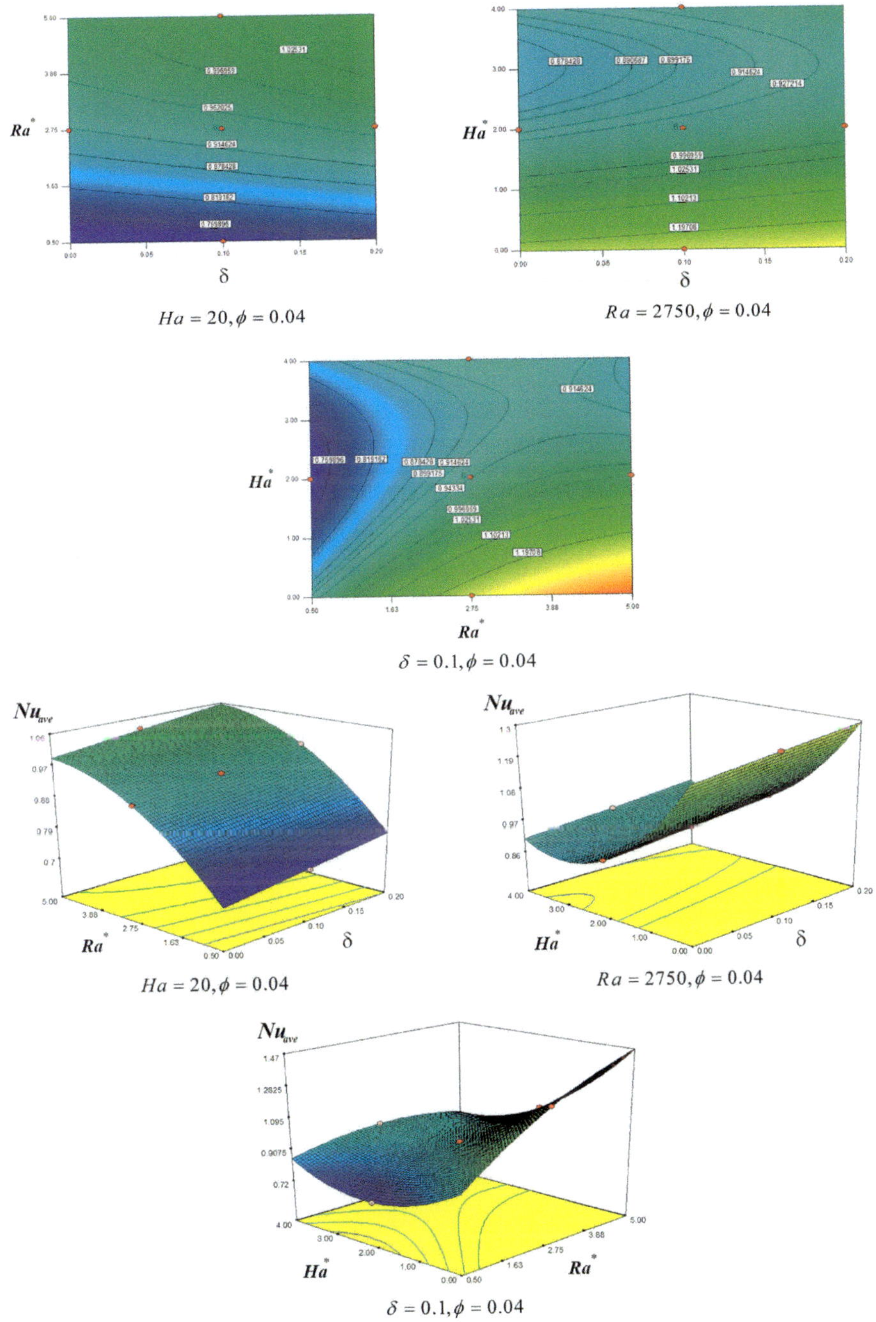

FIGURE 19.17 Impacts of the δ, Ha, Ra on Nu_{ave}.

19.5 MHD NANOFLUID NATURAL CONVECTION INSIDE A HALF ANNULUS WITH MELTING SURFACE

19.5.1 Problem Definition

Fig. 19.18 depicts the geometry, boundary condition, and sample element. The inner wall is the hot wall ($T = T_h$) and the outer one is the melting surface ($T = T_m$). The other walls are adiabatic. A horizontal magnetic field has been applied. The enclosure is filled with nanofluid.

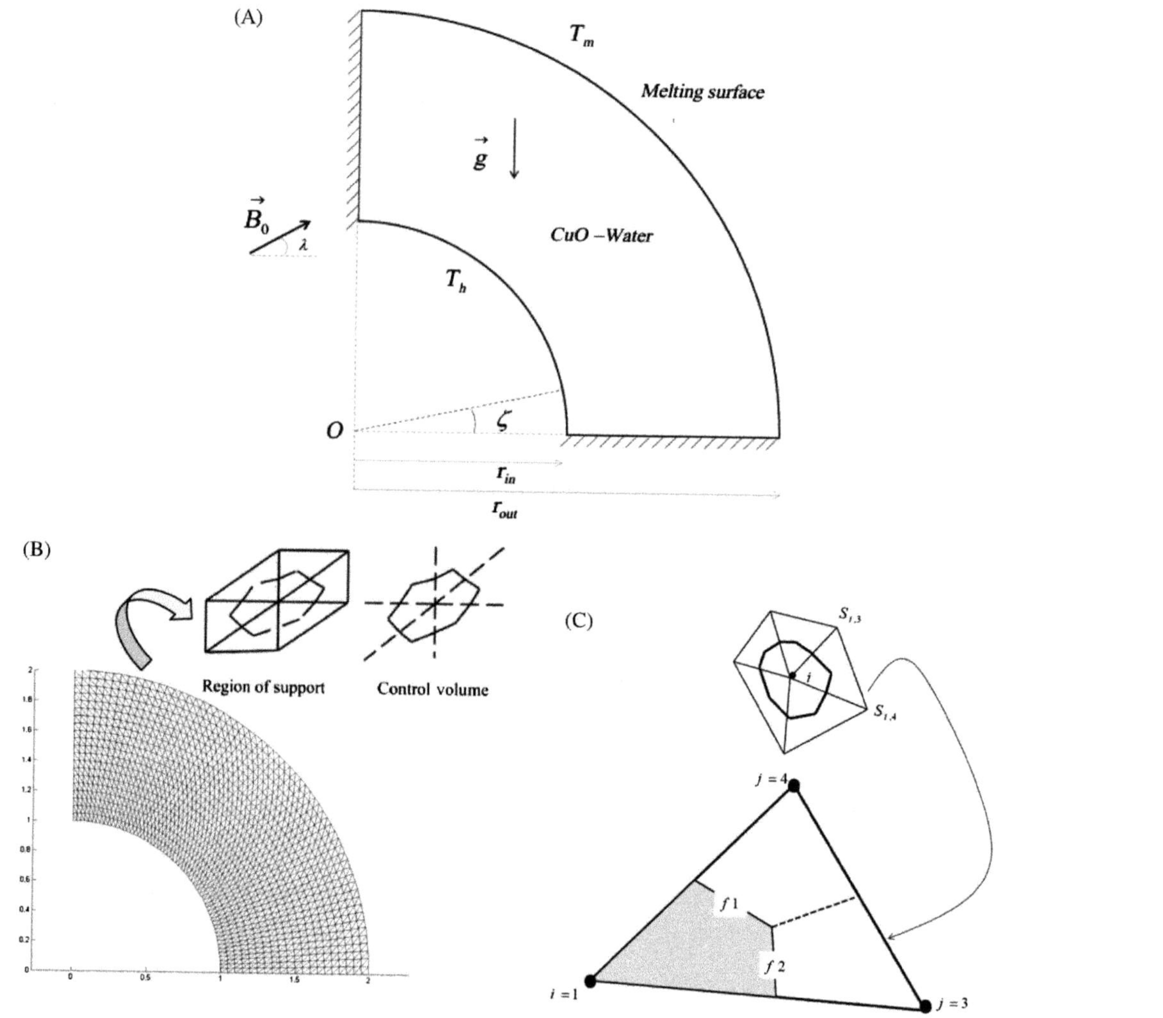

FIGURE 19.18 (A) Geometry and the boundary conditions with (B) the mesh of geometry considered in this work; (C) a sample triangular element and its corresponding control volume.

19.5.2 Governing Equation

2D steady convective nanofluid flow is considered in the presence of a constant magnetic field. The PDEs are:

$$\frac{\partial v}{\partial y}+\frac{\partial u}{\partial x}=0 \tag{19.64}$$

$$\left(\rho_{nf}\right)\left(v\frac{\partial u}{\partial y}+u\frac{\partial u}{\partial x}\right)=\left[B_y\sigma_{nf}vB_x-B_y^2\sigma_{nf}u+\left(\frac{\partial^2 u}{\partial y^2}+\frac{\partial^2 u}{\partial x^2}\right)\mu_{nf}-\frac{\partial P}{\partial x}\right] \tag{19.65}$$

$$\rho_{nf}\left(v\frac{\partial v}{\partial y}+\frac{\partial v}{\partial x}u\right)=uB_y\sigma_{nf}B_x+(T-T_c)\beta_{nf}g\rho_{nf}-vB_x\sigma_{nf}B_x$$

$$-\frac{\partial P}{\partial y}+\mu_{nf}\left(\frac{\partial^2 v}{\partial y^2}+\frac{\partial^2 v}{\partial x^2}\right), \tag{19.66}$$

$$B_x=B_o\cos\lambda,\ \ B_y=B_o\sin\lambda$$

$$\left(\rho C_p\right)_{nf}\left(v\frac{\partial T}{\partial y}+u\frac{\partial T}{\partial x}\right)=k_{nf}\left(\frac{\partial^2 T}{\partial x^2}+\frac{\partial^2 T}{\partial y^2}\right) \tag{19.67}$$

$(\rho C_p)_{nf}$, ρ_{nf}, $(\rho\beta)_{nf}$, and σ_{nf} are calculated as:

$$(\rho C_p)_{nf} = \phi(\rho C_p)_s + (1-\phi)(\rho C_p)_f \tag{19.68}$$

$$\rho_{nf} = \phi\rho_s + (1-\phi)\rho_f \tag{19.69}$$

$$(\rho\beta)_{nf} = \phi(\rho\beta)_s + (1-\phi)(\rho\beta)_f \tag{19.70}$$

$$\frac{\sigma_{nf}}{\sigma_f} = \left(\frac{+(2+\sigma_s/\sigma_f)-\phi(-1+\sigma_s/\sigma_f)}{3\phi(-1+\sigma_s/\sigma_f)}\right)^{-1} + 1 \tag{19.71}$$

$_{nf}$, μ_{nf} are calculated via the KKL model:

$$\begin{aligned}
&\frac{k_{nf}}{k_f} = 1 + \frac{3\left(\frac{k_p}{k_f}-1\right)\phi}{\left(\frac{k_p}{k_f}+2\right)-\left(\frac{k_p}{k_f}-1\right)\phi} + 5\times10^4 g'(\phi,T,d_p)\phi\rho_f c_{p,f}\sqrt{\frac{\kappa_b T}{d_p\rho_p}},\\
&R_f = 4\times10^{-8}km^2/W,\\
&R_f = -d_p(1/k_p - 1/k_{p,eff}),\\
&g'(\phi,d_p,T) = Ln(T)\left(a_1 + a_3Ln(\phi) + a_2Ln(d_p) + a_5Ln(d_p)^2 + a_4\ln(d_p)Ln(\phi)\right)\\
&+\left(a_6 + a_8Ln(\phi) + a_7Ln(d_p) + a_{10}Ln(d_p)^2 + a_9Ln(\phi)\ln(d_p)\right)
\end{aligned} \tag{19.72}$$

$$\mu_{nf} = \frac{\mu_f}{(1-\phi)^{2.5}} + \frac{k_{Brownian}}{k_f}\times\frac{\mu_f}{\Pr} \tag{19.73}$$

Vorticity and stream function should be used to eliminate the pressure source terms:

$$\omega + \frac{\partial u}{\partial y} - \frac{\partial v}{\partial x} = 0,\quad \frac{\partial\psi}{\partial x} = -v,\quad \frac{\partial\psi}{\partial y} = u \tag{19.74}$$

Introducing dimensionless quantities:

$$\begin{aligned}
&P = \frac{p}{\rho_{nf}(\alpha_{nf}/L)^2}, U = \frac{uL}{\alpha_{nf}}, V = \frac{vL}{\alpha_{nf}},\ \theta = \frac{T-T_m}{T_h-T_m}, (X,Y) = \frac{(x,y)}{L},\\
&\Psi = \frac{\psi}{\alpha_{nf}}, \Omega = \frac{\omega L^2}{\alpha_{nf}}
\end{aligned} \tag{19.75}$$

The final formulae are:

$$\Omega + \frac{\partial^2\Psi}{\partial Y^2} + \frac{\partial^2\Psi}{\partial X^2} = 0, \tag{19.76}$$

$$\begin{aligned}
&\frac{\partial\Omega}{\partial X}U + V\frac{\partial\Omega}{\partial Y} = \Pr\frac{A_5}{A_1}\frac{A_2}{A_4}\left(\frac{\partial^2\Omega}{\partial Y^2} + \frac{\partial^2\Omega}{\partial X^2}\right)\\
&+ \Pr Ha^2\frac{A_6}{A_1}\frac{A_2}{A_4}\left(-\frac{\partial V}{\partial X}B_xB_x + \frac{\partial U}{\partial X}B_xB_y - \frac{\partial V}{\partial Y}B_yB_x + \frac{\partial U}{\partial Y}B_yB_y\right)\\
&+ \Pr Ra\frac{A_3A_2^2}{A_1A_4^2}\frac{\partial\theta}{\partial X},
\end{aligned} \tag{19.77}$$

$$\frac{\partial\theta}{\partial X}U + \frac{\partial\theta}{\partial Y}V = \left(\frac{\partial^2\theta}{\partial X^2} + \frac{\partial^2\theta}{\partial Y^2}\right) \tag{19.78}$$

Boundary conditions are:

$$\begin{aligned}
\frac{\partial \theta}{\partial n} &= 0.0 \quad \text{on other walls} \\
\theta &= 0 \quad \text{on outer wall} \\
\theta &= 1.0 \quad \text{on inner wall} \\
\Psi &= 0.0 \quad \text{on all walls except melting surface}
\end{aligned} \qquad (19.7$$

and on the melting surface, we have:

$$\frac{\partial \theta}{\partial n} = \frac{A_2}{A_1}\frac{1}{\delta}(L + c_s(T_m - T_0))\frac{\partial \Psi}{\partial n} \qquad (19.8$$

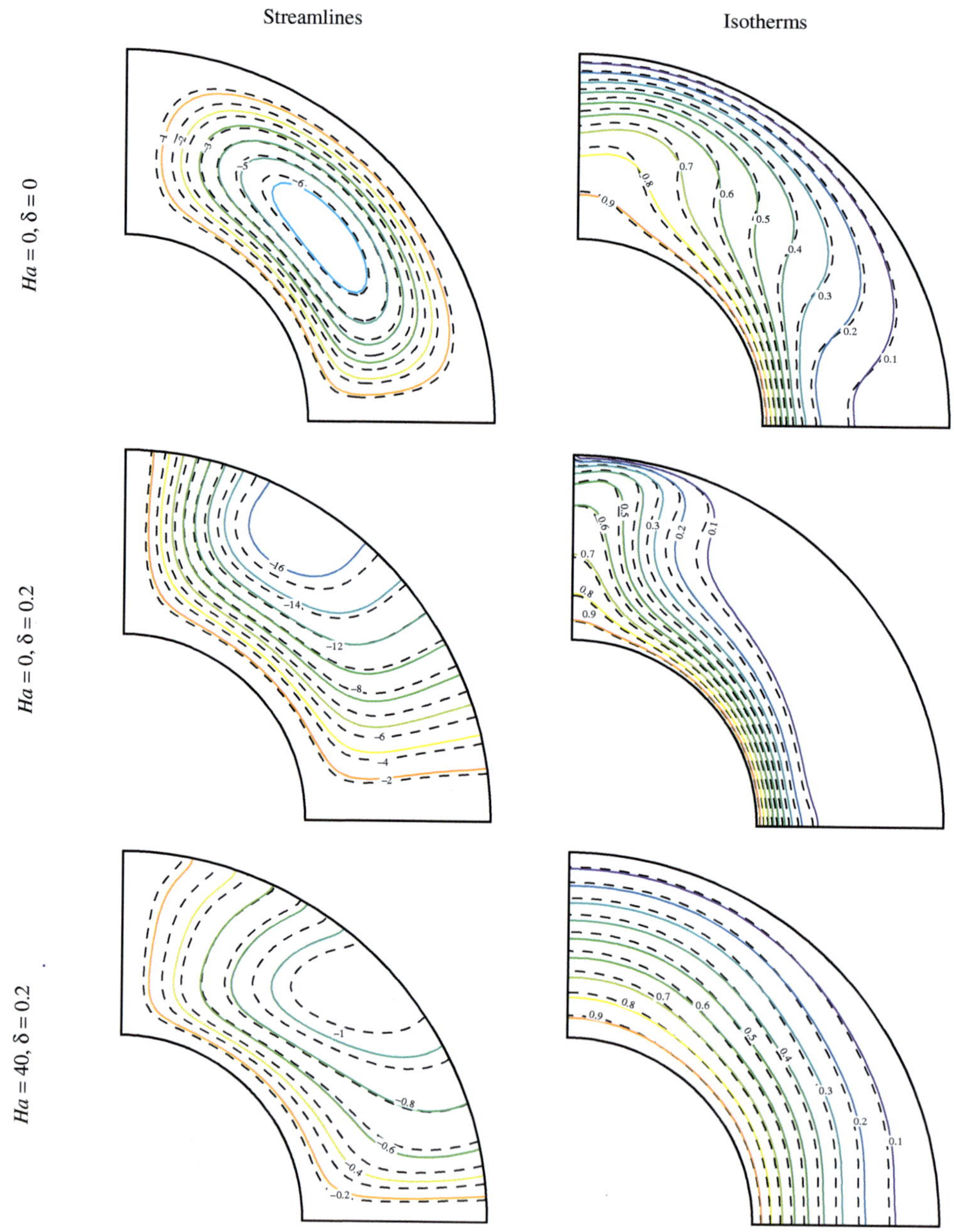

FIGURE 19.19 Influence of nanofluid volume fraction on streamlines (left) and isotherms (right) contours (nanofluid ($\phi = 0.04$)(——) and pure fluid ($\phi = 0$) (- - -)) when $Ra = 5000$

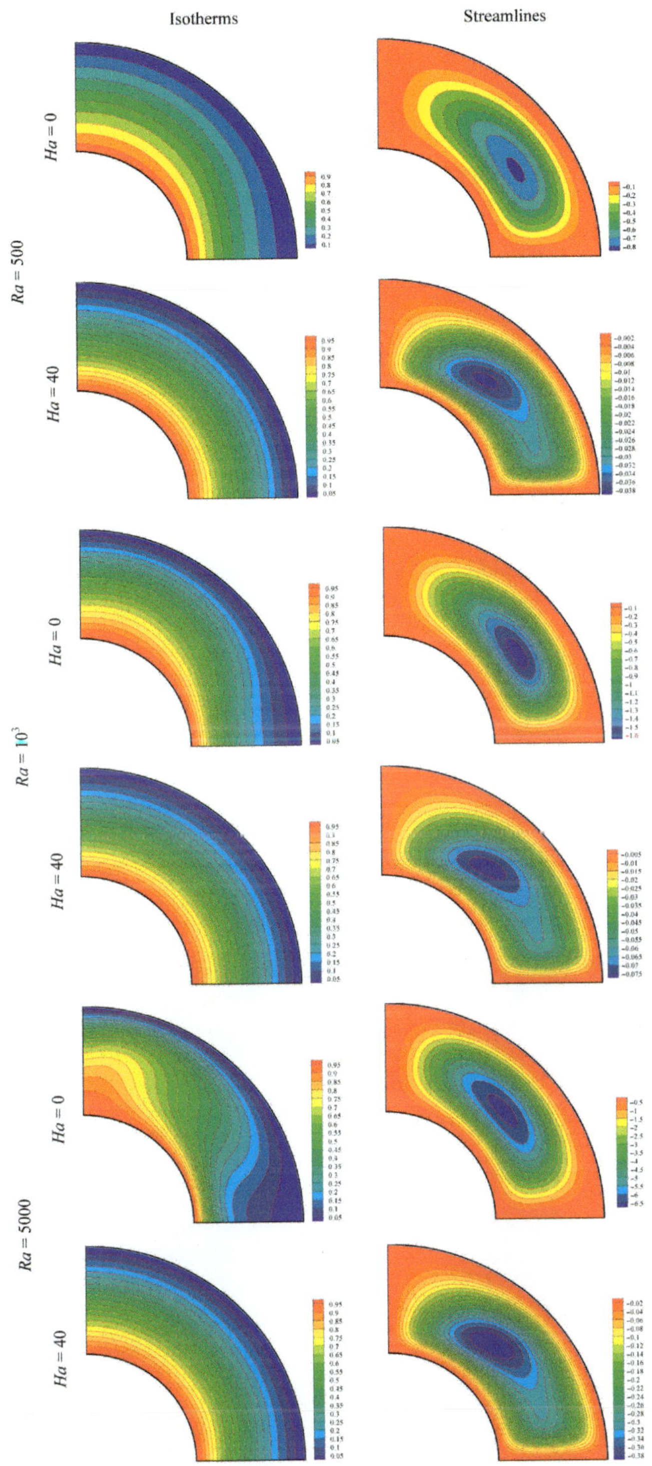

FIGURE 19.20 Isotherm (left) and streamline (right) contours for different values of Rayleigh and Hartmann numbers when $\delta = 0, \phi = 0.04$.

where dimensionless and constants parameters are illustrated as:

$$\begin{aligned} &\Pr = \upsilon_f/\alpha_f, Ra = g\beta_f \Delta T L^3/(\upsilon_f \alpha_f),\ Ha = LB_0\sqrt{\sigma_f/\mu_f},\ \ \delta = \frac{(\rho C_p)_f}{\rho_f}\frac{(T_h - T_m)}{1 + c_s(T_m - T_0)} \\ &A_1 = \frac{\rho_{nf}}{\rho_f},\ A_5 = \frac{\mu_{nf}}{\mu_f},\ \ A_2 = \frac{(\rho C_P)_{nf}}{(\rho C_P)_f}, A_4 = \frac{k_{nf}}{k_f},\ \ A_6 = \frac{\sigma_{nf}}{\sigma_f},\ A_3 = \frac{(\rho\beta)_{nf}}{(\rho\beta)_f} \end{aligned} \tag{19.81}$$

It should be mentioned that δ is related to Stefan numbers.

Local and average Nusselt over the hot wall can be calculated as:

$$Nu_{loc} = \left(\frac{k_{nf}}{k_f}\right)\frac{\partial \theta}{\partial r} \quad (19.8)$$

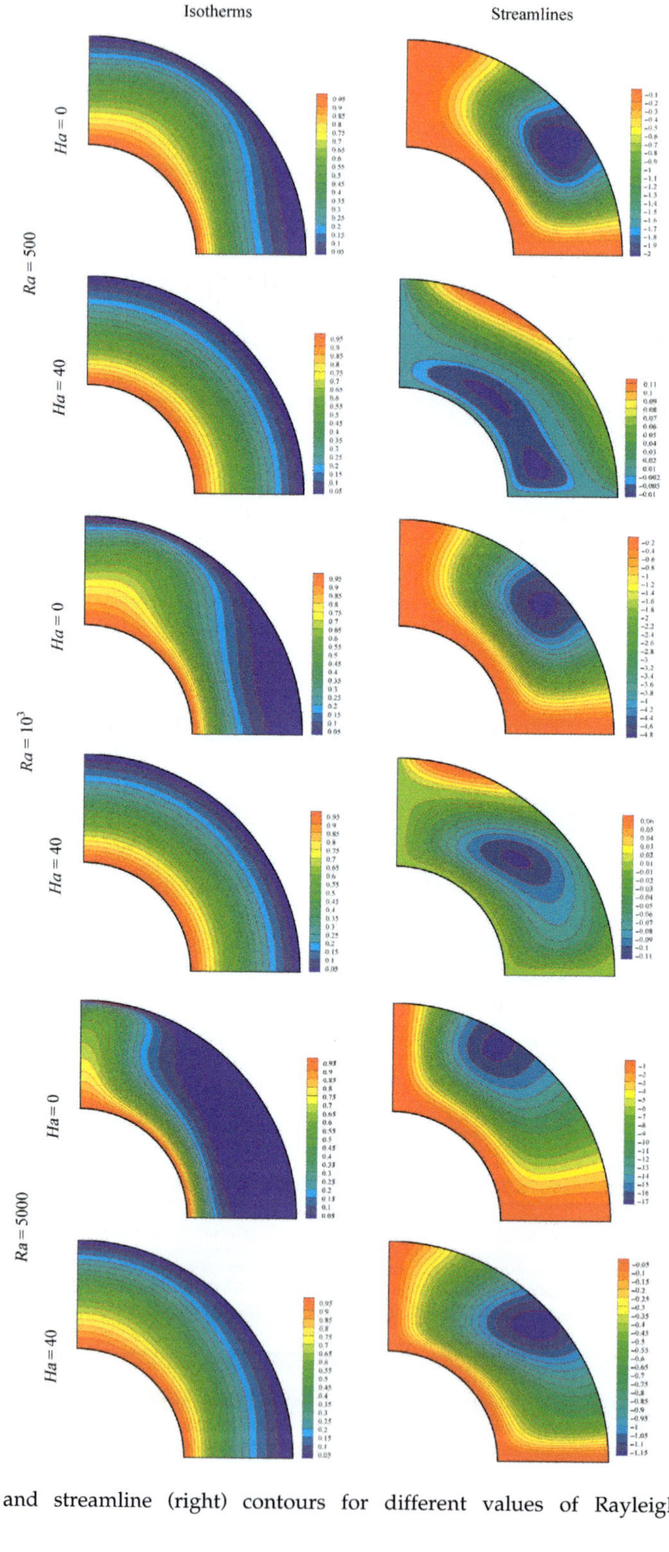

FIGURE 19.21 Isotherm (left) and streamline (right) contours for different values of Rayleigh and Hartmann numbers when $\delta = 0.2, \phi = 0.04$.

$$Nu_{ave} = \frac{1}{0.5\pi}\int_{0}^{0.5\pi} Nu_{loc}\, d\xi \tag{19.83}$$

9.5.3 Effects of Active Parameters

The effect of melting heat transfer on nanofluid MHD flow in a half annulus is examined. μ_{nf}, k_{nf} of CuO-ater nanofluid are estimated by means of the KKL model. Graphs and tables depict the different amounts of uO-H_2O volume fraction ($\phi = 0$ to $\phi = 0.04$), Melting parameter ($\delta = 0$ to 0.8), Hartmann number ($Ha = 0$ to 40), nd Rayleigh number ($Ra = 10^3$ to 10^4).

The impact of the addition of nanoparticles in the base fluid on velocity and temperature contours is depicted Fig. 19.19. The temperature gradient decreases with the increase of ϕ. $|\Psi_{\max}|$ augments with the addition of the anoparticles because of the increase in the solid movements. In the presence of melting heat transfer and a mag-etic field, the effect of adding nanoparticles on the isotherms becomes negligible.

Figs. 19.20 and 19.21 depict the impacts of Rayleigh and Hartmann numbers in the absence of melting heat trans-er. There is only one eddy in the streamlines. At low Rayleigh number, the conduction mechanism is dominant. As Ra increases the distortion of the isotherms is enhanced close to the hot wall. Adding a magnetic field makes iso-herms become parallel. In the presence of melting heat transfer the primary eddy diminishes and increasing Lorentz orces generate three layers of the streamlines. Increasing the melting parameter enhances the bottom eddy.

Fig. 19.22 illustrates the impact of δ, Ra, and Ha on Nu_{ave}. The formula for Nu_{ave} corresponding to active para-neters is:

$$\begin{aligned} Nu_{ave} &= 0.425 + 0.122\delta - 1.63\log(Ra) + 0.35Ha^* \\ &+ 1.5\log(Ra)\delta - 0.58\delta Ha^* - 0.66\log(Ra)Ha^* \\ &+ 0.45Ha^{*2} + 0.53(\log(Ra))^2 - 4.04\delta^2 \end{aligned} \tag{19.84}$$

where $Ha^* = 0.1Ha$. As the melting parameter increases, the temperature gradient is enhanced and in turn the Nusselt number is enhanced. Increasing the bouyancy forces leads to the reduction in thickness of the thermal boudary layer. So the Nusselt number increases with the enhancement of Ra. As the Hartmann number aug-ments, the isotherms become parallel. Therefore the Nusselt number has an inverse relationship with Ha.

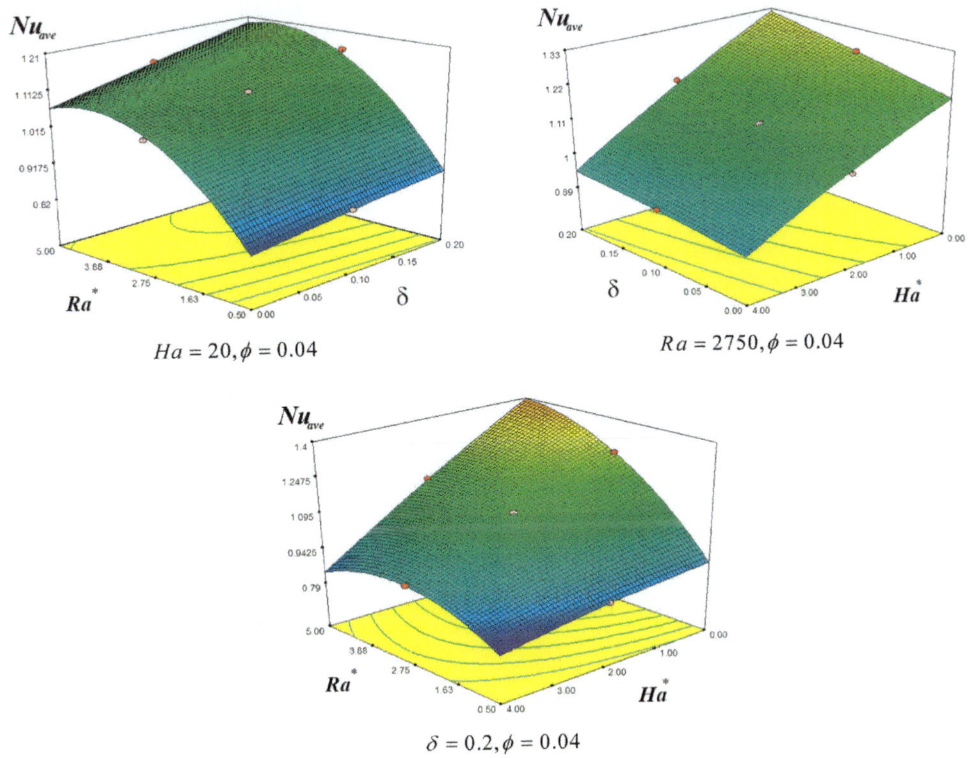

FIGURE 19.22 Influences of the Darcy, Reynolds, and Hartmann numbers on average Nusselt number.

References

[1] M. Epstein, D.H. Cho, Melting heat transfer in steady laminar flow over a flat plate, J. Heat Transfer 98 (1976) 531–533.

[2] W.T. Cheng, C.H. Lin, Melting effect on mixed convective heat transfer with aiding and opposing external flows from the vertical pla in a liquid-saturated porous medium, Int. J. Heat Mass Transfer 50 (2007) 3026–3034.

[3] A. Ishak, R. Nazar, N. Bachok, I. Pop, Melting heat transfer in steady laminar flow over a moving surface, Heat Mass Transfer 46 (201 463–468.

[4] N.A. Yacob, A. Ishak, I. Pop, Melting heat transfer in boundary layer stagnationpoint flow towards a stretching/shrinking sheet in micropolar fluid, Comput. Fluids 47 (2011) 16–21.

[5] S. Ahmad, I. Pop, Melting effect on mixed convection boundary layer flow about a vertical surface embedded in a porous mediu opposing flow case, Transp. Porous Media 102 (2014) 317–323.

[6] T. Hayat, M. Mustafa, Z. Iqbal, A. Alsaedi, Stagnation-point flow of couple stress fluid with melting heat transfer, Appl. Math. Mech. - Engl. Ed. 34 (2) (2013) 167–176.

[7] M. Sheikholeslami, H.B. Rokni, Melting heat transfer influence on nanofluid flow inside a cavity in existence of magnetic field, Int. J Heat Mass Transfer 114 (2017) 517–526.

[8] M. Sheikholeslami, M.K. Sadoughi, Simulation of CuO- water nanofluid heat transfer enhancement in presence of melting surface, Int. J Heat Mass Transfer 116 (2018) 909–919.

[9] M. Sheikholeslami, R. Ellahi, C. Fetecau, CuO-water nanofluid magnetohydrodynamic natural convection inside a sinusoidal annulus in presence of melting heat transfer, Hindawi, Math. Probl. Eng. Volume 2017 (2017). Article ID 5830279, 9 pages, https://doi.org/10.1155/2017/5830279.

[10] M. Sheikholeslami, Influence of Lorentz forces on nanofluid flow in a porous cylinder considering Darcy model, J. Mol. Liq. 225 (2017) 903–912.

[11] M. Sheikholeslami, CVFEM for magnetic nanofluid convective heat transfer in a porous curved enclosure, Eur. Phys. J. Plus 131 (2016) 413. Available from: https://doi.org/10.1140/epjp/i2016-16413-y.

[12] M. Sheikholeslami, Influence of Coulomb forces on Fe_3O_4-H_2O nanofluid thermal improvement, Int. J. Hydrogen Energy 42 (2017) 821–829.

[13] M. Sheikholeslami, Magnetic field influence on nanofluid thermal radiation in a cavity with tilted elliptic inner cylinder, J. Mol. Liq. 229 (2017) 137–147.

[14] M. Sheikholeslami, Numerical simulation of magnetic nanofluid natural convection in porous media, Phys. Lett. A 381 (2017) 494–503.

[15] M. Sheikholeslami, Magnetic source impact on nanofluid heat transfer using CVFEM, Neur. Comput. Appl. (2016). Available from: https://doi.org/10.1007/s00521-016-2740-7.

[16] M. Sheikholeslami, CuO-water nanofluid free convection in a porous cavity considering Darcy law, Europ. Phys. J. Plus 132 (2017) 55. Available from: https://doi.org/10.1140/epjp/i2017-11330-3.

[17] M. Sheikholeslami, Numerical investigation of MHD nanofluid free convective heat transfer in a porous tilted enclosure, Eng. Comput. 34 (6) (2017) 1939–1955.

[18] M.S. Kandelousi, Effect of spatially variable magnetic field on ferrofluid flow and heat transfer considering constant heat flux boundary condition, Europ. Phys. J. Plus (2014) 129–248. Available from: https://doi.org/10.1140/epjp/i2014-14248-2.

[19] M. Sheikholeslami, Numerical simulation for external magnetic field influence on Fe_3O_4-water nanofluid forced convection, Eng. Comput., Inpress.

[20] M. Sheikholeslami, Influence of Lorentz forces on nanofluid flow in a porous cavity by means of Non- Darcy model, Eng. Comput. 34 (8) (2017) 2651–2667. Available from: https://doi.org/10.1108/EC-01-2017-0008.

[21] M. Sheikholeslami, H.B. Rokni, Nanofluid convective heat transfer intensification in a porous circular cylinder, Chem. Eng. Proc. Process Intensif. 120 (2017) 93–104.

[22] M. Sheikholeslami, M.K. Sadoughi, Numerical modeling for Fe_3O_4 -water nanofluid flow in porous medium considering MFD viscosity, J. Mol. Liq. 242 (2017) 255–264.

[23] M. Sheikholeslami, M. Shamlooei, Magnetic source influence on nanofluid flow in porous medium considering shape factor effect, Phys. Lett. A 381 (2017) 3071–3078.

[24] M. Sheikholeslami, M. Shamlooei, Convective flow of nanofluid inside a lid driven porous cavity using CVFEM, Phys. B: Condensed Matter 521 (2017) 239–250.

[25] M. Sheikholeslami, K. Vajravelu, Forced convection heat transfer in Fe_3O_4- ethylene glycol nanofluid under the influence of Coulomb force, J. Mol. Liq. 233 (2017) 203–210.

[26] M. Sheikholeslami, H.B. Rokni, Influence of EFD viscosity on nanofluid forced convection in a cavity with sinusoidal wall, J. Mol. Liq. 232 (2017) 390–395.

[27] M. Sheikholeslami, H.B. Rokni, Magnetic nanofluid natural convection in the presence of thermal radiation considering variable viscosity, Europ. Phys. J. Plus 132 (2017) 238. Available from: https://doi.org/10.1140/epjp/i2017-11498-4.

[28] M. Sheikholeslami, H.B. Rokni, Numerical simulation for impact of Coulomb force on nanofluid heat transfer in a porous enclosure in presence of thermal radiation, Int. J. Heat Mass Transfer 118 (2018) 823–831.

[29] M. Sheikholeslami, M. Shamlooei, R. Moradi, Fe_3O_4- Ethylene glycol nanofluid forced convection inside a porous enclosure in existence of Coulomb force, J. Mol. Liq. 249 (2018) 429–437.

[30] M. Sheikholeslami, S.A. Shehzad, Numerical analysis of Fe_3O_4 $-H_2O$ nanofluid flow in permeable media under the effect of external magnetic source, Int. J. Heat Mass Transfer 118 (2018) 182–192.

[31] M. Sheikholeslami, S.A. Shehzad, Magnetohydrodynamic nanofluid convection in a porous enclosure considering heat flux boundary condition, Int. J. Heat Mass Transfer 106 (2017) 1261–1269.

[32] M. Sheikholeslami, S.A. Shehzad, Thermal radiation of ferrofluid in existence of Lorentz forces considering variable viscosity, Int. J. Heat Mass Transfer 109 (2017) 82–92.

3] M. Sheikholeslami, H.B. Rokni, Numerical modeling of nanofluid natural convection in a semi annulus in existence of Lorentz force, Comput. Methods Appl. Mech. Eng. 317 (2017) 419–430.

4] M. Sheikholeslami, K. Vajravelu, Nanofluid flow and heat transfer in a cavity with variable magnetic field, Appl. Math. Comput. 298 (2017) 272–282.

5] M. Sheikholeslami, H.B. Rokni, Magnetohydrodynamic CuO-water nanofluid in a porous complex shaped enclosure, ASME, J. Thermal Sci. Eng. Applicat. 9 (4) (2017) 041007. Available from: https://doi.org/10.1115/1.4035973.

6] M. Sheikholeslami, H.B. Rokni, Nanofluid two phase model analysis in existence of induced magnetic field, Int. J. Heat Mass Transfer 107 (2017) 288–299.

37] M. Sheikholeslami, T. Hayat, A. Alsaedi, S. Abelman, Numerical analysis of EHD nanofluid force convective heat transfer considering electric field dependent viscosity, Int. J. Heat Mass Transfer 108 (2017) 2558–2565.

38] M. Sheikholeslami, Magnetic field influence on CuO -H_2O nanofluid convective flow in a permeable cavity considering various shapes for nanoparticles, Int. J. Hydrogen Energy 42 (2017) 19611–19621.

39] M. Sheikholeslami, M.M. Bhatti, Forced convection of nanofluid in presence of constant magnetic field considering shape effects of nanoparticles, Int. J. Heat Mass Transfer 111 (2017) 1039–1049.

40] M. Sheikholeslami, A. Zeeshan, Analysis of flow and heat transfer in water based nanofluid due to magnetic field in a porous enclosure with constant heat flux using CVFEM, Comput. Methods Appl. Mech. Eng. 320 (2017) 68–81.

41] M. Sheikholeslami, M. Shamlooei, Fe_3O_4- H_2O nanofluid natural convection in presence of thermal radiation, Int. J. Hydrogen Energy 42 (9) (2017) 5708–5718.

42] M. Sheikholeslami, M.M. Bhatti, Active method for nanofluid heat transfer enhancement by means of EHD, Int. J. Heat Mass Transfer 109 (2017) 115–122.

43] M. Sheikholeslami, T. Hayat, A. Alsaedi, Numerical study for external magnetic source influence on water based nanofluid convective heat transfer, Int. J. Heat Mass Transfer 106 (2017) 745–755.

44] M. Sheikholeslami, A.J. Chamkha, Flow and convective heat transfer of a ferro-nanofluid in a double-sided lid-driven cavity with a wavy wall in the presence of a variable magnetic field, Numer. Heat Transfer, Part A 69 (10) (2016) 1186–1200. Available from: https://doi.org/10.1080/10407782.2015.1125709.

[45] M. Sheikholeslami, A.J. Chamkha, Electrohydrodynamic free convection heat transfer of a nanofluid in a semi-annulus enclosure with a sinusoidal wall, Numer. Heat Transfer, Part A 69 (7) (2016) 781–793. Available from: http://dx.doi.org/10.1080/10407782.2015.1090819.

[46] M. Sheikholeslami, H.B. Rokni, Simulation of nanofluid heat transfer in presence of magnetic field: a review, Int. J. Heat Mass Transfer 115 (2017) 1203–1233.

[47] M. Sheikholeslami, S.A. Shehzad, Magnetohydrodynamic nanofluid convective flow in a porous enclosure by means of LBM, Int. J. Heat Mass Transfer 113 (2017) 796–805.

[48] M. Sheikholeslami, Magnetohydrodynamic nanofluid forced convection in a porous lid driven cubic cavity using Lattice Boltzmann Method, J. Mol. Liq. 231 (2017) 555–565.

[49] M. Sheikholeslami Kandelousi, KKL correlation for simulation of nanofluid flow and heat transfer in a permeable channel, Phys. Lett. A 378 (45) (2014) 3331–3339.

[50] M. Sheikholeslami, Effect of uniform suction on nanofluid flow and heat transfer over a cylinder, J. Braz. Soc. Mech. Sci. Eng. 37 (2015) 1623–1633.

[51] M. Sheikholeslami, Lattice Boltzmann Method simulation of MHD non-Darcy nanofluid free convection, Physica B 516 (2017) 55–71.

[52] M. Sheikholeslami, Influence of magnetic field on nanofluid free convection in an open porous cavity by means of Lattice Boltzmann Method, J. Mol. Liq. 234 (2017) 364–374.

[53] M. Sheikholeslami, A.J. Chamkha, Influence of Lorentz forces on nanofluid forced convection considering Marangoni convection, J. Mol. Liq. 225 (2017) 750–757.

[54] M. Sheikholeslami, T. Hayat, A. Alsaedi, Numerical simulation of nanofluid forced convection heat transfer improvement in existence of magnetic field using Lattice Boltzmann Method, Int. J. Heat Mass Transfer 108 (2017) 1870–1883.

[55] M. Sheikholeslami, R. Ellahi, Three dimensional mesoscopic simulation of magnetic field effect on natural convection of nanofluid, Int. J. Heat Mass Transfer 89 (2015) 799–808.

[56] M. Sheikholeslami, S. Abelman, Two phase simulation of nanofluid flow and heat transfer in an annulus in the presence of an axial magnetic field, IEEE Trans. Nanotechnol. 14 (3) (2015) 561–569.

[57] M. Sheikholeslami, R. Ellahi, Simulation of ferrofluid flow for magnetic drug targeting using Lattice Boltzmann method, J. Zeitschrift Fur Naturforschung A 70 (2) (2015) 115–124.

[58] M. Sheikholeslami, M. Seyednezhad, Lattice Boltzmann Method simulation for CuO-water nanofluid flow in a porous enclosure with hot obstacle, J. Mol. Liq. 243 (2017) 249–256.

[59] M. Sheikholeslami, M. Sadoughi, Mesoscopic method for MHD nanofluid flow inside a porous cavity considering various shapes of nanoparticles, Int. J. Heat Mass Transfer 113 (2017) 106–114.

[60] M. Sheikholeslami, A. Zeeshan, Mesoscopic simulation of CuO-H_2O nanofluid in a porous enclosure with elliptic heat source, Int. J. Hydrogen Energy 42 (22) (2017) 15393–15402.

[61] M. Sheikholeslami, S.A. Shehzad, CVFEM for influence of external magnetic source on $Fe_3O_4 - H_2O$ nanofluid behavior in a permeable cavity considering shape effect, Int. J. Heat Mass Transfer 115 (2017) 180–191.

[62] M. Sheikholeslami, M. Seyednezhad, Nanofluid heat transfer in a permeable enclosure in presence of variable magnetic field by means of CVFEM, Int. J. Heat Mass Transfer 114 (2017) 1169–1180.

[63] M. Sheikholeslami, H.B. Rokni, Free convection of CuO-H_2O nanofluid in a curved porous enclosure using mesoscopic approach, Int. J. Hydrogen Energy 42 (22) (2017) 14942–14949.

[64] M. Sheikholeslami, Numerical investigation for CuO-H_2O nanofluid flow in a porous channel with magnetic field using mesoscopic method, J. Mol. Liq. 249 (2018) 739–746.

[65] M. Sheikholeslami, H.B. Rokni, Influence of melting surface on MHD nanofluid flow by means of two phase model, Chin. J. Phys. (2017) 1352–1360.
[66] M. Sheikholeslami, H.B. Rokni, Effect of melting heat transfer on nanofluid flow in existence of magnetic field considering Buongior Model, Chin. J. Phys. 55 (2017) 1115–1126.
[67] M. Sheikholeslami, T. Hayat, A. Alsaedi, MHD free convection of Al_2O_3–water nanofluid considering thermal radiation: a numeri study, Int. J. Heat Mass Transfer 96 (2016) 513–524.
[68] M. Sheikholeslami, K. Vajravelu, M.M. Rashidi, Forced convection heat transfer in a semi annulus under the influence of a variable ma netic field, Int. J. Heat Mass Transfer 92 (2016) 339–348.
[69] M. Sheikholeslami, Numerical modeling of Nano enhanced PCM solidification in an enclosure with metallic fin, J. Mol. Liq. 259 (201 424–438.
[70] M. Sheikholeslami, A. Ghasemi, Solidification heat transfer of nanofluid in existence of thermal radiation by means of FEM, Int. J. He Mass Transfer 123 (2018) 418–431.
[71] M. Sheikholeslami, S.A. Shehzad, CVFEM simulation for nanofluid migration in a porous medium using Darcy model, Int. J. Heat Ma Transfer 122 (2018) 1264–1271.
[72] M. Sheikholeslami, M. Darzi, M.K. Sadoughi, Heat transfer improvement and Pressure Drop during condensation of refrigerant-bas Nanofluid; An Experimental Procedure, Int. J. Heat Mass Transfer 122 (2018) 643–650.
[73] M. Sheikholeslami, H.B. Rokni, CVFEM for effect of Lorentz forces on nanofluid flow in a porous complex shaped enclosure by means Non-equilibrium model, J. Mol. Liq. 254 (2018) 446–462.
[74] M. Sheikholeslami, S.A. Shehzad, Z. Li, Water based nanofluid free convection heat transfer in a three dimensional porous cavity wi hot sphere obstacle in existence of Lorenz forces, Int. J. Heat Mass Transfer 125 (2018) 375–386.
[75] M. Sheikholeslami, M. Jafaryar, D.D. Ganji, Z. Li, Exergy loss analysis for nanofluid forced convection heat transfer in a pipe with mod fied turbulators, J. Mol. Liq. 262 (2018) 104–110.
[76] M. Sheikholeslami, M. Jafaryar, Z. Li, Nanofluid turbulent convective flow in a circular duct with helical turbulators considering Cu nanoparticles, Int. J. Heat Mass Transfer 124 (2018) 980–989.
[77] M. Sheikholeslami, Z. Li, M. Shamlooei, Nanofluid MHD natural convection through a porous complex shaped cavity considering the mal radiation, Phys. Lett. A 382 (2018) 1615–1632.
[78] M. Sheikholeslami, H.B. Rokni, Magnetic nanofluid flow and convective heat transfer in a porous cavity considering Brownian motio effects, Phys. Fluids 30 (1). Available from: https://doi.org/10.1063/1.5012517.
[79] M. Sheikholeslami, S.A. Shehzad, Simulation of water based nanofluid convective flow inside a porous enclosure via Non-equilibriur model, Int. J. Heat Mass Transfer 120 (2018) 1200–1212.
[80] M. Sheikholeslami, S.A. Shehzad, Non-Darcy free convection of Fe_3O_4-water nanoliquid in a complex shaped enclosure under impact o uniform Lorentz force, Chin. J. Phys. 56 (2018) 270–281.
[81] M. Sheikholeslami, M. Seyednezhad, Simulation of nanofluid flow and natural convection in a porous media under the influence of elec tric field using CVFEM, Int. J. Heat Mass Transfer 120 (2018) 772–781.
[82] M. Sheikholeslami, M. Shamlooei, R. Moradi, Numerical simulation for heat transfer intensification of nanofluid in a permeable curvec enclosure considering shape effect of Fe_3O_4 nanoparticles, Chem. Eng. Process.: Process Intensif. 124 (2018) 71–82.
[83] M. Sheikholeslami, T. Hayat, T. Muhammad, A. Alsaedi, MHD forced convection flow of nanofluid in a porous cavity with hot ellipti obstacle by means of Lattice Boltzmann method, Int. J. Mech. Sci. 135 (2018) 532–540.
[84] M. Sheikholeslami, Numerical investigation of nanofluid free convection under the influence of electric field in a porous enclosure, J. Mol. Liq. 249 (2018) 1212–1221.
[85] M. Sheikholeslami, CuO-water nanofluid flow due to magnetic field inside a porous media considering Brownian motion, J. Mol. Liq. 249 (2018) 921–929.
[86] M. Sheikholeslami, D.D. Ganji, Influence of electric field on Fe_3O_4- water nanofluid radiative and convective heat transfer in a permeable enclosure, J. Mol. Liq. 250 (2018) 404–412.
[87] M. Sheikholeslami, S.A. Shehzad, Z. Li, Nanofluid heat transfer intensification in a permeable channel due to magnetic field using Lattice Boltzmann method, Phys. B (2018). Available from: https://doi.org/10.1016/j.physb.2018.03.036.
[88] M. Sheikholeslami, A. Zeeshan, Numerical simulation of Fe_3O_4 -water nanofluid flow in a non-Darcy porous media, Int. J. Numer. Methods Heat Fluid Flow 28 (3) (2018) 641–660. Available from: https://doi.org/10.1108/HFF-04-2017-0160.
[89] M. Sheikholeslami, D.D. Ganji, Numerical approach for magnetic nanofluid flow in a porous cavity using CuO nanoparticles, Mater. Design 120 (2017) 382–393.
[90] M. Sheikholeslami, D.D. Ganji, Numerical modeling of magnetohydrodynamic CuO-water transportation inside a porous cavity considering shape factor effect, Colloids Surf., A 529 (2017) 705–714.
[91] M. Sheikholeslami, T. Hayat, A. Alsaedi, Numerical simulation for forced convection flow of MHD CuO-H_2O nanofluid inside a cavity by means of LBM, J. Mol. Liq. 249 (2018) 941–948.
[92] M. Sheikholeslami, H.R. Kataria, A.S. Mittal, Effect of thermal diffusion and heat-generation on MHD nanofluid flow past an oscillating vertical plate through porous medium, J. Mol. Liq. 257 (2018) 12–25.
[93] M. Sheikholeslami, M. Barzegar Gerdroodbary, S. Valiallah Mousavi, D.D. Ganji, R. Moradi, Heat transfer enhancement of ferrofluid inside an 90° elbow channel by non-uniform magnetic field, J. Magn. Magn. Mater. 460 (2018) 302–311.
[94] M. Sheikholeslami, S. Soleimani, D.D. Ganji, Effect of electric field on hydrothermal behavior of nanofluid in a complex geometry, J. Mol. Liq. 213 (2016) 153–161.
[95] M. Sheikholeslami, Numerical simulation for solidification in a LHTESS by means of Nano-enhanced PCM, J. Taiwan Inst. Chem. Eng. 86 (2018) 25–41.
[96] M. Sheikholeslami, M. Darzi, Z. Li, Experimental investigation for entropy generation and exergy loss of nano-refrigerant condensation process, Int. J. Heat Mass Transfer 125 (2018) 1087–1095.

CHAPTER

20

Nanofluid Convective Heat Transfer Considering Magnetic Field Dependent (MFD) Viscosity by Means of CVFEM

20.1 INTRODUCTION

Ferrohydrodynamics deals with the study of ferrofluids in the presence of a magnetic field and it is a special branch of magnetohydrodynamics. During the last decades, extensive research work has been done on these fluids since the effect of magnetization has yielded interesting information leading to diverse fascinating technological applications [1]. Ferrofluids consist of colloidal suspensions of single domain magnetic nanoparticles and it has been recognized that they have promising potential for heat transfer applications in electronics, engines, micro- and nanoelectromechanical systems (MEMS and NEMS), air-conditioning and ventilation systems [2]. Under the circumstances, the study of thermal convection in ferrofluids is gaining much importance in recent years. Moreover, many physical properties of these fluids can be tuned by varying the magnetic field. One of the well-known phenomena generated by the influence of magnetic fields on ferrofluids is the change of their viscous behavior. Realizing the importance of magnetic field dependent (MFD) viscosity on ferrofluid flows, several studies have been undertaken in the past. The effect of a homogeneous magnetic field on the viscosity of a fluid with solid particles possessing intrinsic magnetic moments has been investigated by Shliomis [3]. The effect of MFD viscosity on the onset of ferroconvection in a rotating ferrofluid layer is discussed by Vaidyanathan et al. [4]. Nanjundappa et al. [5] have investigated the effect of MFD viscosity on the onset of convection in a ferromagnetic fluid layer in the presence of a vertical magnetic field by considering the bounding surfaces are either rigid-ferromagnetic or stress- free with constant heat flux conditions. Sheikholeslami et al. [6] considered MFD viscosity effect on natural convection of ferrofluid. Sheikholeslami [7] studied the magnetic field influence on nanofluid thermal radiation in a cavity with tilted elliptic inner cylinder. Sheikholeslami and Shehzad [8] studied the thermal radiation of ferrofluid in the presence of Lorentz forces considering variable viscosity. Sheikholeslami and Rokni [9] reported the magnetic nanofluid natural convection in the presence of thermal radiation. Sheikholeslami and Abelman [10] presented the numerical analysis of the effect of magnetic field on Fe_3O_4-water ferrofluid convection with thermal radiation. Sheikholeslami et al. [11] investigated the nanofluid radiation and natural convection in an enclosure with elliptical cylinders. Sheikholeslami and Sadoughi [12] studied the Fe_3O_4-water nanofluid flow in porous medium considering MFD viscosity. The magnetic field effect on nanofluid treatment was investigated in recent years [13–98].

20.2 NATURAL CONVECTION OF MAGNETIC NANOFLUID CONSIDERING MFD VISCOSITY EFFECT

20.2.1 Problem Definition

The geometry of this problem is shown in Fig. 20.1A. The heat source is centrally located on the bottom surface and its length $L/3$. The cooling is achieved by the two vertical walls. The heat source has constant heat flux q''

Application of Control Volume based Finite Element Method (CVFEM) for Nanofluid Flow and Heat Transfer.
DOI: https://doi.org/10.1016/B978-0-12-814152-6.00020-5

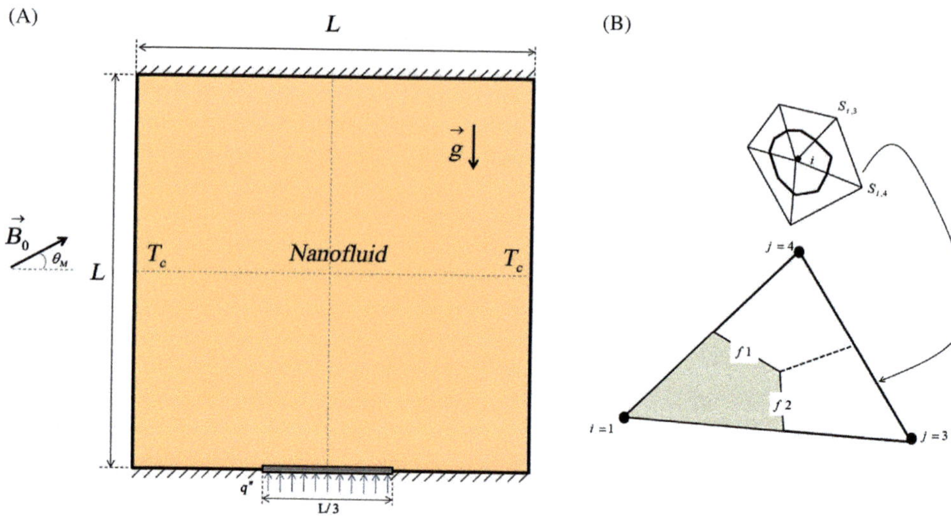

FIGURE 20.1 (A) Geometry and the boundary conditions; (B) a sample triangular element and its corresponding control volume.

while the cooling walls have a constant temperature T_c; all the other surfaces are adiabatic. Also, it is also assumed that the uniform magnetic field ($\vec{B} = B_x\vec{e_x} + B_y\vec{e_y}$) of constant magnitude $B = \sqrt{B_x^2 + B_y^2}$ is applied, where $\vec{e_x}$ and $\vec{e_y}$ are unit vectors in the Cartesian coordinate system. The orientation of the magnetic field forms an angle θ_M with horizontal axis such that $\theta_M = \cot^{-1}(B_x/B_y)$. The electric current J and the electromagnetic force F are defined by $J = \sigma\left(\vec{V} \times \vec{B}\right)$ and $F = \sigma\left(\vec{V} \times \vec{B}\right) \times \vec{B}$, respectively.

20.2.2 Governing Equation

The flow is steady, two-dimensional, laminar, and incompressible. The induced electric current and Joule heating are neglected. The magnetic Reynolds number is assumed to be small so that the induced magnetic field can be neglected compared to the applied magnetic field. Neglecting displacement currents, induced magnetic field, and using the Boussinesq approximation, the governing equations of heat transfer and fluid flow for nanofluid can be obtained as follows:

$$\frac{\partial u}{\partial x} + \frac{\partial v}{\partial y} = 0 \tag{20.1}$$

$$\begin{aligned}\rho_{nf}\left(u\frac{\partial u}{\partial x} + v\frac{\partial u}{\partial y}\right) &= -\frac{\partial P}{\partial x} + \eta\left(\frac{\partial^2 u}{\partial x^2} + \frac{\partial^2 u}{\partial y^2}\right) \\ &+ \sigma_{nf}{B_0}^2\left(v\,\sin\theta_M\,\cos\theta_M - u\,\sin^2\theta_M\right)\end{aligned} \tag{20.2}$$

$$\begin{aligned}\rho_{nf}\left(u\frac{\partial v}{\partial x} + v\frac{\partial v}{\partial y}\right) &= -\frac{\partial P}{\partial y} + \eta\left(\frac{\partial^2 v}{\partial x^2} + \frac{\partial^2 v}{\partial y^2}\right) + \rho_{nf}\beta_{nf}\,g(T - T_c) \\ &+ \sigma_{nf}{B_0}^2(u\,\sin\theta_M\,\cos\theta_M - v\,\cos^2\theta_M)\end{aligned} \tag{20.3}$$

$$u\frac{\partial T}{\partial x} + v\frac{\partial T}{\partial y} = \alpha_{nf}\left(\frac{\partial^2 T}{\partial x^2} + \frac{\partial^2 T}{\partial y^2}\right) \tag{20.4}$$

here $\eta = (1 + \vec{\delta}.\vec{B})\mu_{nf}$, the variation of MFD viscosity (δ) has been taken to be isotropic, $\delta_1 = \delta_2 = \delta_3 = \delta$. The fective density ($\rho_{nf}$), the thermal expansion coefficient (β_{nf}), heat capacitance $(\rho C_p)_{nf}$, and electrical conductivity nanofluid (σ_{nf}) of the nanofluid are defined as:

$$\rho_{nf} = \rho_f(1-\phi) + \rho_s\phi$$
$$\beta_{nf} = \beta_f(1-\phi) + \beta_s\phi \tag{20.5}$$
$$(\rho C_p)_{nf} = (\rho C_p)_f(1-\phi) + (\rho C_p)_s\phi$$

$$\frac{\sigma_{nf}}{\sigma_f} = 1 + \frac{3(\sigma_s/\sigma_f - 1)\phi}{(\sigma_s/\sigma_f + 2) - (\sigma_s/\sigma_f - 1)\phi} \tag{20.6}$$

The Koo–Kleinstreuer–Li (KKL) model is used to simulate thermal conductivity of nanofluid:

$$k_{eff} = k_{static} + k_{Brownian} \tag{20.7}$$

$$\frac{k_{static}}{k_f} = 1 + \frac{3(k_p/k_f - 1)\phi}{(k_p/k_f + 2) - (k_p/k_f - 1)\phi} \tag{20.8}$$

$$k_{Brownian} = 5 \times 10^4 \phi \rho_f c_{p,f} \sqrt{\frac{\kappa_b T}{\rho_p d_p}} g'(T, \phi, d_p) \tag{20.9}$$

$$R_f + \frac{d_p}{k_p} = \frac{d_p}{k_{p,eff}}, \; R_f = 4 \times 10^{-8} km^2/W \tag{20.10}$$

$$\begin{aligned} g'(T,\phi,d_p) = &\left(a_1 + a_2\ln(d_p) + a_3\ln(\phi) + a_4\ln(\phi)\ln(d_p) + a_5\ln(d_p)^2\right)\ln(T) \\ &+ \left(a_6 + a_7\ln(d_p) + a_8\ln(\phi) + a_9\ln(\phi)\ln(d_p) + a_{10}\ln(d_p)^2\right) \end{aligned} \tag{20.11}$$

with the coefficients a_i (i = 0.10) based on the type of nanopartices, Al_2O_3-water nanofluids has an R^2 of 98%, respectively (Table 20.1).

The effective viscosity due to micro–mixing in suspensions can be obtained as follows:

$$\mu_{eff} = \mu_{static} + \mu_{Brownian} = \mu_{static} + \frac{k_{Brownian}}{k_f} \times \frac{\mu_f}{\Pr_f} \tag{20.12}$$

where $\mu_{static} = \frac{\mu_f}{(1-\phi)^{2.5}}$ is viscosity of the nanofluid, as given originally by Brinkman.

TABLE 20.1 The Coefficient Values of Al_2O_3 – Water Nanofluid

Coefficient values	Al_2O_3 – Water
a_1	52.813488759
a_2	6.115637295
a_3	0.6955745084
a_4	4.17455552786E-02
a_5	0.176919300241
a_6	−298.19819084
a_7	−34.532716906
a_8	−3.9225289283
a_9	−0.2354329626
a_{10}	−0.999063481

The stream function and vorticity are defined as:

$$u = \frac{\partial \psi}{\partial y},\ v = -\frac{\partial \psi}{\partial x},\ \omega = \frac{\partial v}{\partial x} - \frac{\partial u}{\partial y} \tag{20.13}$$

The stream function satisfies the continuity Eq. (20.1). The vorticity equation is obtained by eliminating the pressure between the two momentum equations, i.e., by taking the y-derivative of Eq. (20.2) and subtracting from it the x-derivative of Eq. (20.3). This gives:

$$\frac{\partial \psi}{\partial y}\frac{\partial \omega}{\partial x} - \frac{\partial \psi}{\partial x}\frac{\partial \omega}{\partial y} = \upsilon_{nf}(1+\delta B_0(\cos\theta_M + \sin\theta_M))\left(\frac{\partial^2 \omega}{\partial x^2} + \frac{\partial^2 \omega}{\partial y^2}\right) + \beta_{nf} g\left(\frac{\partial T}{\partial x}\right) + \frac{\sigma_{nf} {B_0}^2}{\rho_{nf}}\left(\begin{array}{l} -\frac{\delta v}{\delta y}\sin\theta_M \cos\theta_M + \frac{\delta u}{\delta y}\sin^2\theta_M \\ +\frac{\delta u}{\delta x}\sin\theta_M\cos\theta_M - \frac{\delta v}{\delta x}\cos^2\theta_M \end{array}\right) \tag{20.14}$$

$$\frac{\partial \psi}{\partial y}\frac{\partial T}{\partial x} - \frac{\partial \psi}{\partial x}\frac{\partial T}{\partial y} = \alpha_{nf}\left(\frac{\partial^2 T}{\partial x^2} + \frac{\partial^2 T}{\partial y^2}\right) \tag{20.15}$$

$$\frac{\partial^2 \psi}{\partial x^2} + \frac{\partial^2 \psi}{\partial y^2} = -\omega \tag{20.16}$$

By introducing the following nondimensional variables:

$$X = \frac{x}{L},\ Y = \frac{y}{L},\ \Omega = \frac{\omega L^2}{\alpha_f},\ \Psi = \frac{\psi}{\alpha_f},\ \Theta = \frac{T - T_c}{(q'' L/k_f)},\ U = \frac{uL}{\alpha_f},\ V = \frac{vL}{\alpha_f} \tag{20.17}$$

Using the dimensionless parameters, the equations now become:

$$\frac{\partial \Psi}{\partial Y}\frac{\partial \Omega}{\partial X} - \frac{\partial \Psi}{\partial X}\frac{\partial \Omega}{\partial Y} = \Pr_f\left(\frac{\mu_{nf}}{\mu_f}\frac{\rho_f}{\rho_{nf}}\frac{k_f}{k_{nf}}\frac{(\rho C_p)_{nf}}{(\rho C_p)_f}\right)(1+\delta^*(\cos\theta_M + \sin\theta_M))\left(\frac{\partial^2 \Omega}{\partial X^2} + \frac{\partial^2 \Omega}{\partial Y^2}\right) + Ra_f \Pr_f \frac{\beta_{nf}}{\beta_f}\left(\frac{\partial \Theta}{\partial X}\right) + Ha^2 \Pr_f \frac{\sigma_{nf}}{\sigma_f}\frac{\rho_f}{\rho_{nf}}\left(-\frac{\delta V}{\delta Y}\tan\theta_M + \frac{\delta U}{\delta Y}\tan^2\theta_M + \frac{\delta U}{\delta X}\tan\theta_M - \frac{\delta V}{\delta X}\right) \tag{20.18}$$

$$\frac{\partial \Psi}{\partial Y}\frac{\partial \Theta}{\partial X} - \frac{\partial \Psi}{\partial X}\frac{\partial \Theta}{\partial Y} = \frac{k_{nf}}{k_f}\frac{(\rho C_p)_f}{(\rho C_p)_{nf}}\left(\frac{\partial^2 \Theta}{\partial X^2}\right) \tag{20.19}$$

$$\frac{\partial^2 \Psi}{\partial X^2} + \frac{\partial^2 \Psi}{\partial Y^2} = -\Omega \tag{20.20}$$

where $Ra_f = g\beta_f L^4 q''/(k_f\,\alpha_f \upsilon_f)$ is the Rayleigh number for the base fluid, $Ha = LB_x\sqrt{\sigma_f/\mu_f}$ is the Hartmann number, and $\Pr_f = \upsilon_f/\alpha_f$ is the Prandtl number for the base fluid. Also $\delta^* = \delta B_0$ is the viscosity parameter. The thermophysical properties of the nanofluid are given in Table 20.2. The boundary conditions as shown in Fig. 20.1 are:

$$\begin{array}{ll} \partial\Theta/\partial n = -1.0 & \text{on the heat source} \\ \Theta = 0.0 & \text{on the left and right} \\ \partial\Theta/\partial n = 0.0 & \text{on all the other adiabatic surfaces} \\ \Psi = 0.0 & \text{on all solid boundaries} \end{array} \tag{20.21}$$

TABLE 20.2 Thermophysical Properties of Water and Nanoparticles

	$\rho(kg/m^3)$	$C_p(j/kgk)$	$k(W/m.k)$	$\beta \times 10^5 (K^{-1})$	$d_p(nm)$	$\sigma(\Omega \cdot m)^{-1}$
water	997.1	4179	0.613	21	-	0.05
Al_2O_3	3970	765	25	0.85	47	1×10^{-10}

The values of vorticity on the boundary of the enclosure can be obtained using the stream function formula-on and the known velocity conditions during the iterative solution procedure. The local Nusselt number of the anofluid along the heat source can be expressed as:

$$Nu_{local} = \left(\frac{k_{nf}}{k_f}\right)\frac{1}{\theta}\bigg|_{L/3<X<2L/3,Y=0} \tag{20.22}$$

The average Nusselt number is evaluated as:

$$Nu_{ave} = \frac{1}{L/3}\int_{L/3}^{2L/3} Nu_{loc}(X)\,dX \tag{20.23}$$

The Nusselt ratio is defined as follows:

$$Nu_{ratio} = \frac{Nu_{ave}\big|_{\delta^*=1}}{Nu_{ave}\big|_{\delta^*=0}} \tag{20.24}$$

20.2.3 Effects of Active Parameters

In this study magnetohydrodynamic natural convective heat transfer in a cooling system of electronic components is investigated. MFD viscosity effect is taken into account. Brownian motion effect is considered for simulating effective thermal conductivity and viscosity. The influence of active parameters such as Rayleigh number ($Ra = 10^3, 10^4$ and 10^5), Hartmann number ($Ha = 0, 15, 30$ and 60), and viscosity parameter ($\delta^* = 0,\ 1$) on flow and heat transfer are examined when $\phi = 0.04, \Pr = 6.2$.

The effects of viscosity parameter, Hartmann number, and Rayleigh number on isotherms and streamlines are shown in Figs. 20.2–20.4. At low Rayleigh number, the streamlines take the enclosure geometry. By increasing the Rayleigh number the prominent heat transfer mechanism is switched from conduction to convection. When the magnetic field is imposed on the enclosure, the velocity field is suppressed owing to the retarding effect of the Lorenz force. So the intensity of convection weakens significantly. The decelerating effect of the magnetic field is observed from the maximum stream function value. The core vortex is shifted downwards vertically as the Hartmann number increases. Also, imposing a magnetic field leads to the omission of the thermal plume over the bottom wall. At high Hartmann number the conduction heat transfer mechanism is more pronounced.

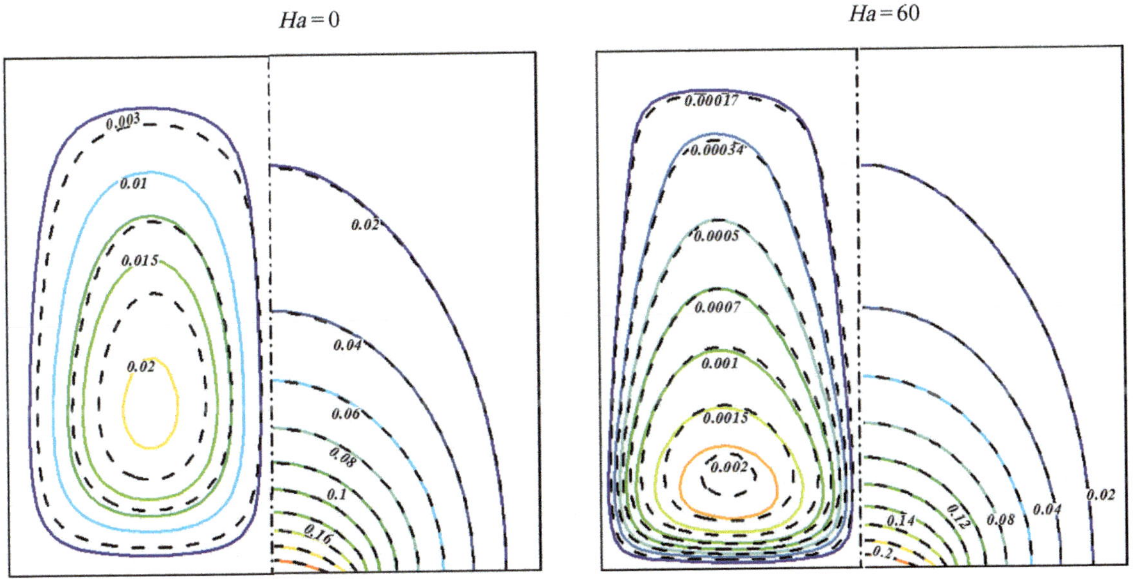

FIGUE 20.2 Effects of viscosity parameter ($\delta^* = 1$ (− − −) and $\delta^* = 0$ (—)) and Hartmann number on isotherms (right) and streamlines (left) when $\phi = 0.04, Ra = 10^3$.

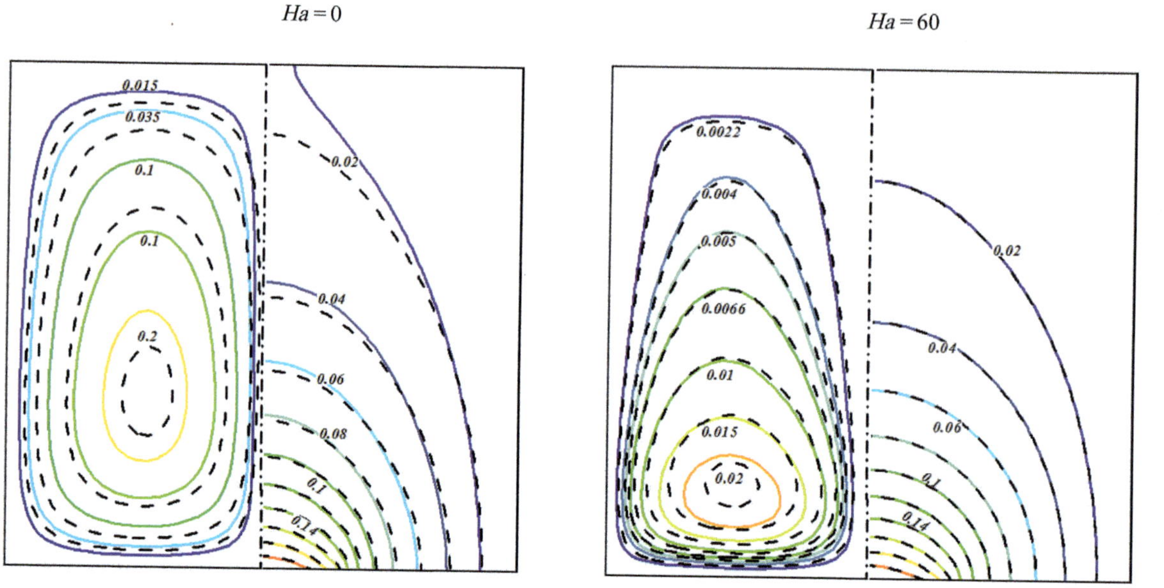

FIGURE 20.3 Effects of viscosity parameter ($\delta^* = 1$ (– – –) and $\delta^* = 0$ (–)) and Hartmann number on isotherms (right) and streamlines (left) when $\phi = 0.04$, $Ra = 10^4$.

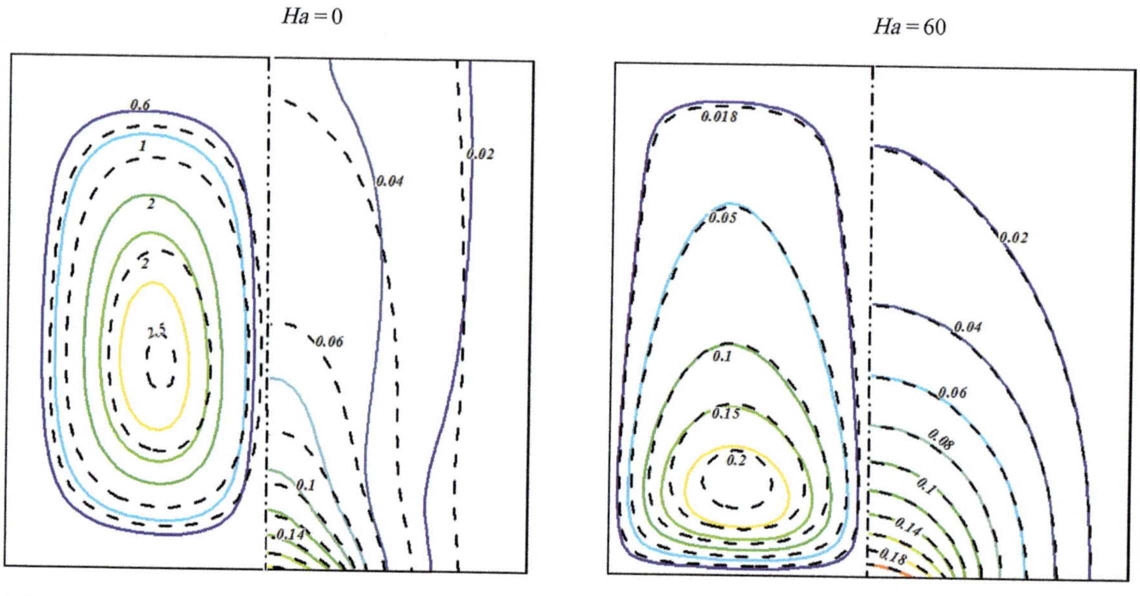

FIGURE 20.4 Effects of viscosity parameter ($\delta^* = 1$ (– – –) and $\delta^* = 0$ (–)) and Hartmann number on isotherms (right) and streamlines (left) when $\phi = 0.04$, $Ra = 10^5$.

For this reason the isotherms are parallel to each other. Considering MFD viscosity effect increases the thermal boundary layer thickness and in turn the Nusselt number decreases with the increase of the viscosity parameter. Also it can be seen that flow circulation increases with the increase of this parameter.

The effects of Hartmann number and Rayleigh number on local and average Nusselt number are shown in Figs. 20.5 and 20.6. Increasing the Rayleigh number leads to the enhancement of the Nusselt number. Increasing the Lorentz forces makes the thermal boundary layer thickness increase and in turn the Nusselt number decreases with the enhancement of the Hartmann number. Fig. 20.7 depicts the effect of the Hartmann number and the Rayleigh number on the Nusselt number ratio. As the viscosity parameter increases, the rate of heat transfer decreases. This reduction is more pronounced for higher values of Rayleigh number and lower values of Hartmann number. It is an interesting observation that increasing the Lorentz forces increases the effect of MFD viscosity on the Nusselt number. At low Rayleigh number, the MFD viscosity has no significant effect on Nusselt number.

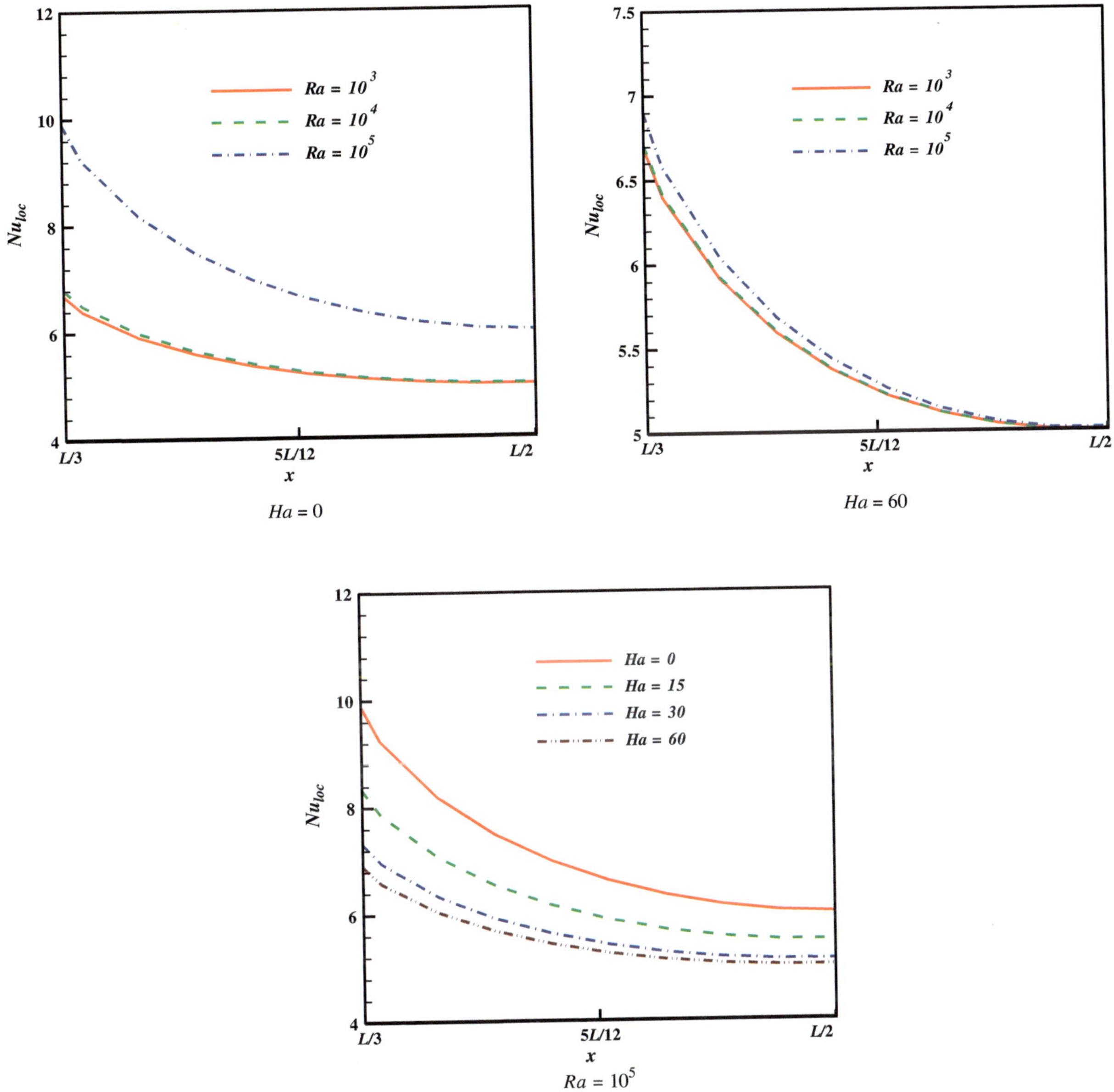

FIGURE 20.5 Effects of Hartmann number and Rayleigh number on local Nusselt number when $\phi = 0.04, \delta^* = 1, \mathrm{Pr} = 6.2$.

20.3 MAGNETIC FIELD INFLUENCE ON NANOFLUID THERMAL RADIATION IN A CAVITY WITH TILTED ELLIPTIC INNER CYLINDER

20.3.1 Problem Definition

Fig. 20.8 depicts the geometry, boundary condition, and sample element. The formula of the inner cylinder is:

$$b = \sqrt{1 - \varepsilon^2} \,.\, a \tag{20.25}$$

where a, b, ε are the major, minor axis of elliptic cylinder, and eccentricity for the inner cylinder. The inner cylinder has constant heat flux condition.

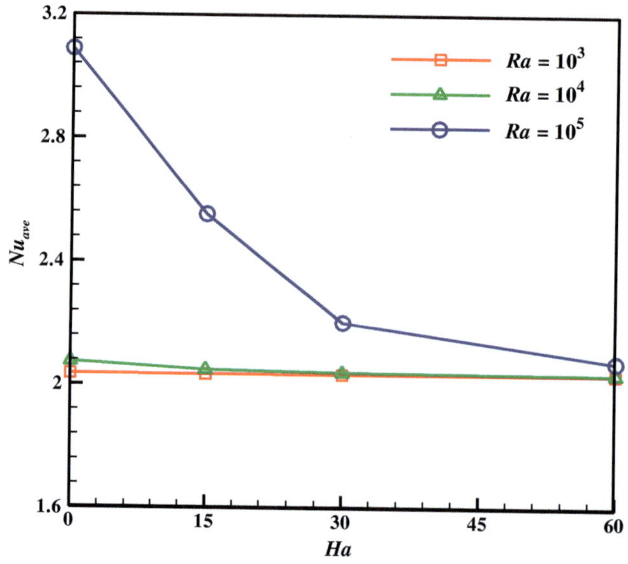

FIGURE 20.6 Effects of Hartmann number and Rayleigh number on average Nusselt number when $\phi = 0.04, \delta^* = 1, \Pr = 6.2$.

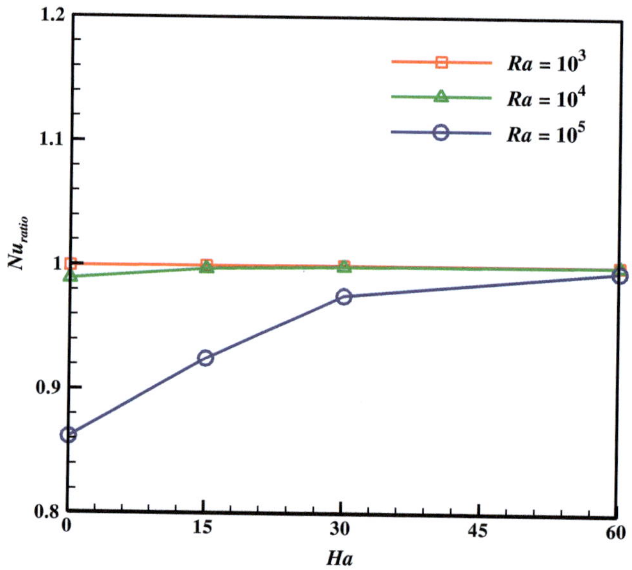

FIGURE 20.7 Effects of Hartmann number and Rayleigh number on Nusselt number ratio when $\phi = 0.04, \Pr = 6.2$.

20.3.2 Governing Equation

2D steady convective flow of nanofluid in a porous media is considered in the presence of a constant magnetic field. The PDEs equations are:

$$\frac{\partial u}{\partial x} + \frac{\partial v}{\partial y} = 0, \tag{20.26}$$

$$\left(v\frac{\partial u}{\partial y} + \frac{\partial u}{\partial x}u \right) = \left[-\sigma_{nf} B_y^2 u + \sigma_{nf} B_x B_y v + \left(\frac{\partial^2 u}{\partial y^2} + \frac{\partial^2 u}{\partial x^2} \right) \mu_{nf} - \frac{\partial P}{\partial x} \right] \left(\rho_{nf} \right)^{-1}, \tag{20.27}$$

$$\rho_{nf}\left(\frac{\partial v}{\partial x}u + \frac{\partial v}{\partial y}v \right) = +\mu_{nf}\left(\frac{\partial^2 v}{\partial x^2} + \frac{\partial^2 v}{\partial y^2} \right) - \frac{\partial P}{\partial y} + B_y \sigma_{nf} B_x u - B_x \sigma_{nf} B_x v + (T - T_c)\beta_{nf} g \rho_{nf},$$
$$B_x = B_o \cos\lambda, \quad B_y = B_o \sin\lambda, \tag{20.28}$$

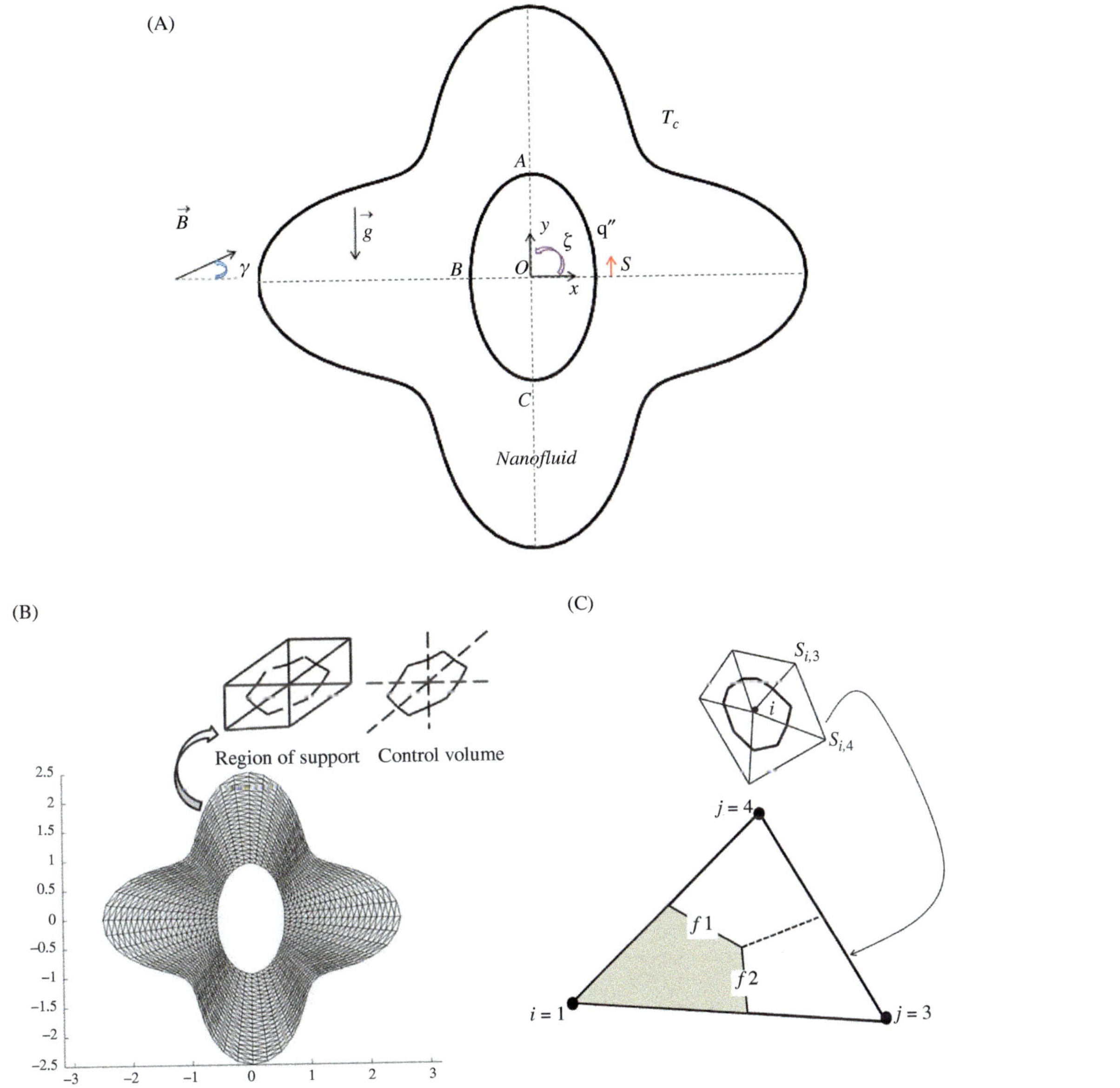

FIGURE 20.8 (A) Geometry and the boundary conditions with (B) the mesh of geometry considered in this work; (C) a sample triangular element and its corresponding control volume.

$$(\rho C_p)_{nf}\left(v\frac{\partial T}{\partial y}+u\frac{\partial T}{\partial x}\right)=k_{nf}\left(\frac{\partial^2 T}{\partial x^2}+\frac{\partial^2 T}{\partial y^2}\right)-\frac{\partial q_r}{\partial y},$$
$$\left[q_r=-\frac{4\sigma_e}{3\beta_R}\frac{\partial T^4}{\partial y},\quad T^4\cong 4T_c^3T-3T_c^4\right]. \tag{20.29}$$

$(\rho C_p)_{nf}, (\rho\beta)_{nf}, \rho_{nf}, k_{nf}$, and σ_{nf} are defined as:

$$(\rho C_p)_{nf}=(\rho C_p)_f(1-\phi)+(\rho C_p)_s\phi, \tag{20.30}$$

$$(\rho\beta)_{nf}=(\rho\beta)_f(1-\phi)+(\rho\beta)_s\phi, \tag{20.31}$$

$$\rho_{nf}=\rho_f(1-\phi)+\rho_s\phi, \tag{20.32}$$

TABLE 20.3 Thermophysical Properties of Water and Nanoparticles

	$\rho(kg/m^3)$	$C_p(j/kgk)$	$k(W/m.k)$	$d_p(nm)$	$\sigma(\Omega\cdot m)$
Pure water	997.1	4179	0.613	-	0.05
Fe_3O_4	5200	670	6	47	25000

$$k_{nf} = k_f\left(\frac{k_s + 2k_f + 2\phi(k_s - k_f)}{k_s - \phi(k_s - k_f) + 2k_f}\right), \tag{20.33}$$

$$\frac{\sigma_{nf}}{\sigma_f} = 1 + \frac{3\left(\frac{\sigma_s}{\sigma_f} - 1\right)\phi}{\left(\frac{\sigma_s}{\sigma_f} + 2\right) - \left(\frac{\sigma_s}{\sigma_f} - 1\right)\phi}. \tag{20.34}$$

μ_{nf} is obtained as follows:

$$\mu_{nf} = \left(0.035B^2 + 3.1B - 27886.4807\phi^2 + 4263.02\phi + 316.0629\right)e^{-0.01T} \tag{20.35}$$

Properties of nanofluid are shown in Table 20.3. Vorticity and stream function should be used to eliminate the pressure source terms:

$$\omega + \frac{\partial u}{\partial y} - \frac{\partial v}{\partial x} = 0,\ \frac{\partial \psi}{\partial x} = -v,\ \frac{\partial \psi}{\partial y} = u. \tag{20.36}$$

Introducing the dimensionless quantities:

$$U = \frac{uL}{\alpha_{nf}}, V = \frac{vL}{\alpha_{nf}},\ \theta = \frac{T - T_c}{\Delta T}, \Delta T = q''L/k_f,$$
$$(X, Y) = \frac{(x, y)}{L}, \Psi = \frac{\psi}{\alpha_{nf}}, \Omega = \frac{\omega L^2}{\alpha_{nf}}. \tag{20.37}$$

The final formulae are:

$$\frac{\partial^2 \Psi}{\partial Y^2} + \frac{\partial^2 \Psi}{\partial X^2} = -\Omega, \tag{20.38}$$

$$U\frac{\partial \Omega}{\partial X} + \frac{\partial \Omega}{\partial Y}V = \Pr\frac{A_5}{A_1}\frac{A_2}{A_4}\left(\frac{\partial^2 \Omega}{\partial Y^2} + \frac{\partial^2 \Omega}{\partial X^2}\right)$$
$$+ \Pr Ha^2\frac{A_6}{A_1}\frac{A_2}{A_4}\left(\frac{\partial U}{\partial X}B_xB_y - \frac{\partial V}{\partial X}{B_x}^2 + \frac{\partial U}{\partial Y}B_y^2 - \frac{\partial V}{\partial Y}B_xB_y\right) + \Pr Ra\frac{A_3A_2^2}{A_1A_4^2}\frac{\partial \theta}{\partial X}, \tag{20.39}$$

$$\frac{\partial \theta}{\partial X}U + \frac{\partial \theta}{\partial Y}V = \left(\frac{\partial^2 \theta}{\partial X^2} + \frac{\partial^2 \theta}{\partial Y^2}\right) + \frac{4}{3}\frac{1}{A_4}Rd\frac{\partial^2 \theta}{\partial Y^2}. \tag{20.40}$$

where dimensionless and constants parameters are illustrated as:

$$\Pr = \upsilon_f/\alpha_f,\ Ra = g(\rho\beta)_f\Delta T\, L^3/\left(\mu_f\alpha_f\right),\ Ha = LB_0\sqrt{\sigma_f/\mu_f},$$
$$A_1 = \frac{\rho_{nf}}{\rho_f},\ A_2 = \frac{(\rho C_p)_{nf}}{(\rho C_p)_f},\ A_3 = \frac{(\rho\beta)_{nf}}{(\rho\beta)_f},$$
$$A_4 = \frac{k_{nf}}{k_f},\ A_5 = \frac{\mu_{nf}}{\mu_f},\ A_6 = \frac{\sigma_{nf}}{\sigma_f},$$
$$Rd = 4\sigma_e T_c^3/\left(\beta_R k_f\right). \tag{20.41}$$

nd boundary conditions are:

$$\begin{aligned} &\frac{\partial \theta}{\partial n} = 1.0 \quad \text{on inner wall} \\ &\theta = 0.0 \quad \text{on outer wall} \\ &\Psi = 0.0 \quad \text{on all walls} \end{aligned} \tag{20.42}$$

Local and average Nusselt over the hot cylinder can be calculated as:

$$Nu_{loc} = \left(1 + \frac{4}{3} Rd \left(\frac{k_{nf}}{k_f}\right)^{-1}\right) \frac{1}{\theta} \left(\frac{k_{nf}}{k_f}\right), \tag{20.43}$$

$$Nu_{ave} = \frac{1}{S} \int_0^S Nu_{loc}\, ds. \tag{20.44}$$

20.3.3 Effects of Active Parameters

CVFEM is utilized to simulate the magnetic field impact on nanofluid free convection. The radiation source term is taken into account. MFD viscosity is considered. Results are reported for different levels of Radiation parameter ($Rd = 0$ to 0.8), volume fraction of Fe_3O_4-water ($\phi = 0$ to 0.04), Rayleigh number ($Ra = 10^3$, 10^4 and 10^5), inclination angle ($\xi = 0^\circ$ and 90°), and Hartmann number ($Ha = 0$ to 40).

Fig. 20.9 illustrates the impact of radiation parameter on nanofluid hydrothermal treatment. Radiation parameter can enhance the rate of heat transfer. Also augmenting Rd increases the nanofluid motion. This impact is more obvious in the presence of a magnetic field. Influences of ξ, Ha and Ra on hydrothermal characteristics are demonstrated in Figs. 20.10 and 20.11. As the nanofluid temperature increases, the nanofluid begins moving from the warm surface to the outer one and dropping along the outer cylinder. At low Rayleigh number, conduction is more signification than convection.$|\Psi_{\max}|$ augments as buoyancy force increases and it reduces as the Lorentz force increases. When $\xi = 0^\circ$, $Ha = 0$, three vortexes appear in the streamlines and these vortexes convert to two smaller ones in the presence of Lorentz forces. As ξ increases, the distortion of isotherms is enhanced. A thermal plume is generated near the vertical centerline. As the Lorentz force increases, the two upper vortexes merge together and the thermal plume vanishes.

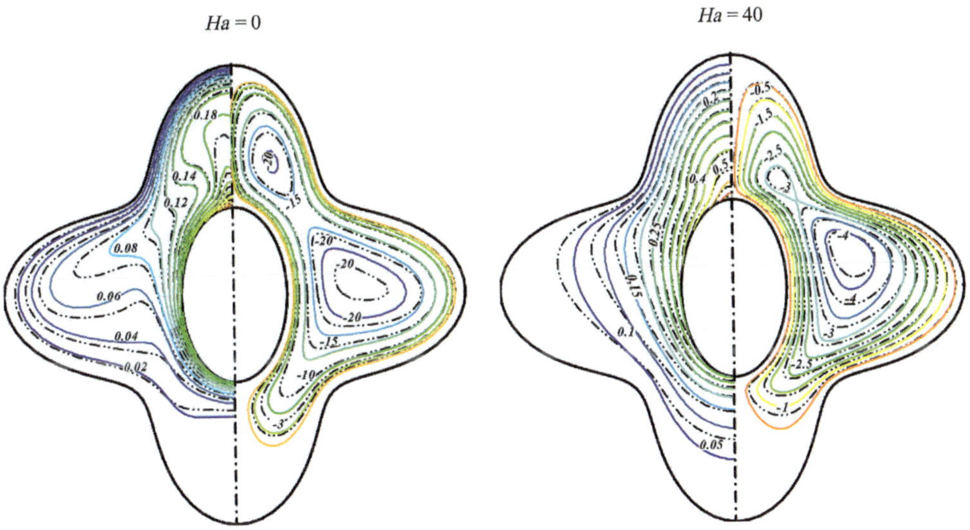

FIGURE 20.9 Effect of Hartmann number and radiation parameter on isotherm (left) and streamline (right) contours ($Rd = 0$ (– · · –), $Rd = 0.8$(–)) when $\xi = 90^\circ, \phi = 0.04, Ra = 10^5$.

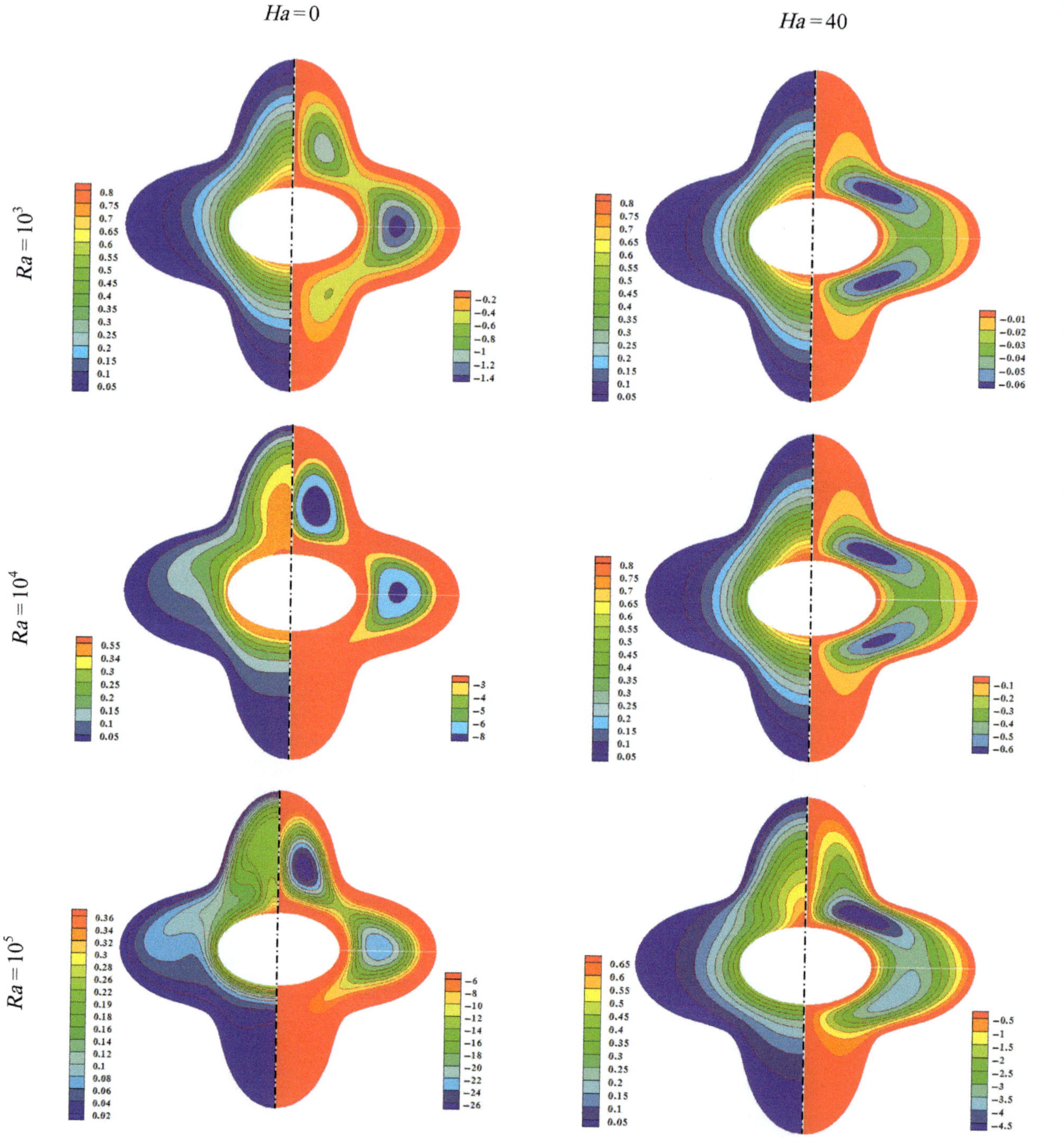

FIGURE 20.10 Effect of Rayleigh and Hartmann numbers on isotherm (left) and streamline (right) contours when $\xi = 0, \phi = 0.04, Rd = 0.8$.

Figs. 20.12–20.14 illustrate the impact of ξ, Rd, Ra, and Ha on Nu_{loc}, Nu_{ave}. The formula for Nu_{ave} corresponding to active parameters is:

$$
\begin{aligned}
Nu_{ave} = {} & 9.94 - 0.22\xi - 4.44Rd - 4.76\log(Ra) + 0.98Ha^* \\
& + 0.4\xi\, Rd - 0.06\xi\log(Ra) - 0.06\xi Ha^* + 1.69\log(Ra)Rd \\
& - 0.351\log(Ra)\, Ha^* - 0.25Rd\, Ha^* - 0.23\xi^2 + 0.72\left(\log(Ra)\right)^2 + 1.49Rd^2 \\
& - 0.06Ha^{*2}
\end{aligned}
\tag{20.45}
$$

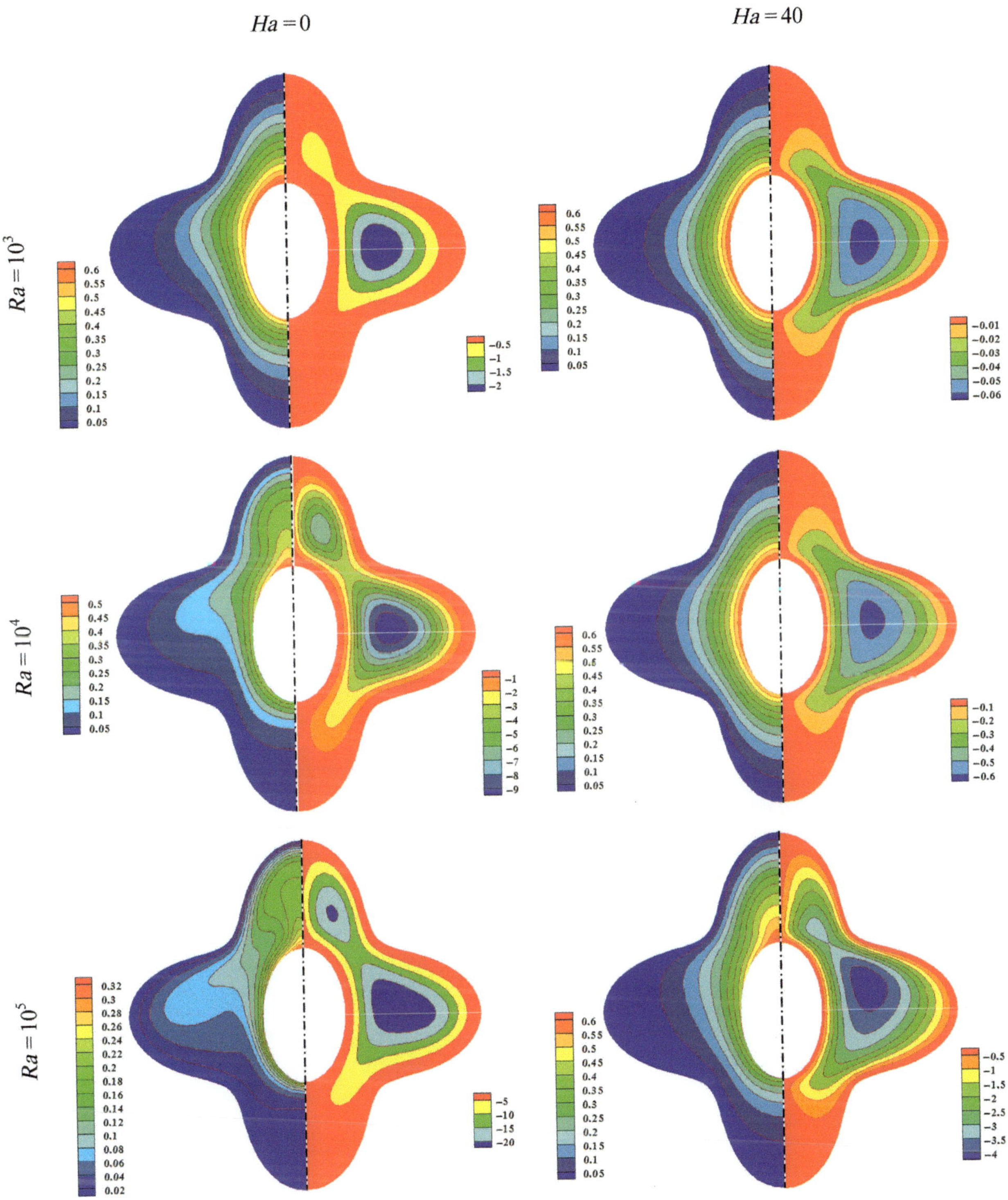

FIGURE 20.11 Effect of Rayleigh and Hartmann numbers on isotherm (left) and streamline (right) contours when $\xi = 90^{\circ}$, $\phi = 0.04, Rd = 0.8$.

Where $Ha^* = 0.1Ha$. The number of extremum in Nu_{loc} profile matches the presence of the thermal plumes and the shape of the outer cylinder. Nu_{ave} increases with the rise of Rd. As the buoyancy forces augment, the temperature decreases and in turn Nu_{ave} increases with the rise of Ra. The impact of inclination angle on Nu_{ave} is similar to that obtained for Ra. Besides, this figure depicts that the Nusselt number has a reverse relationship with the Hartmann number.

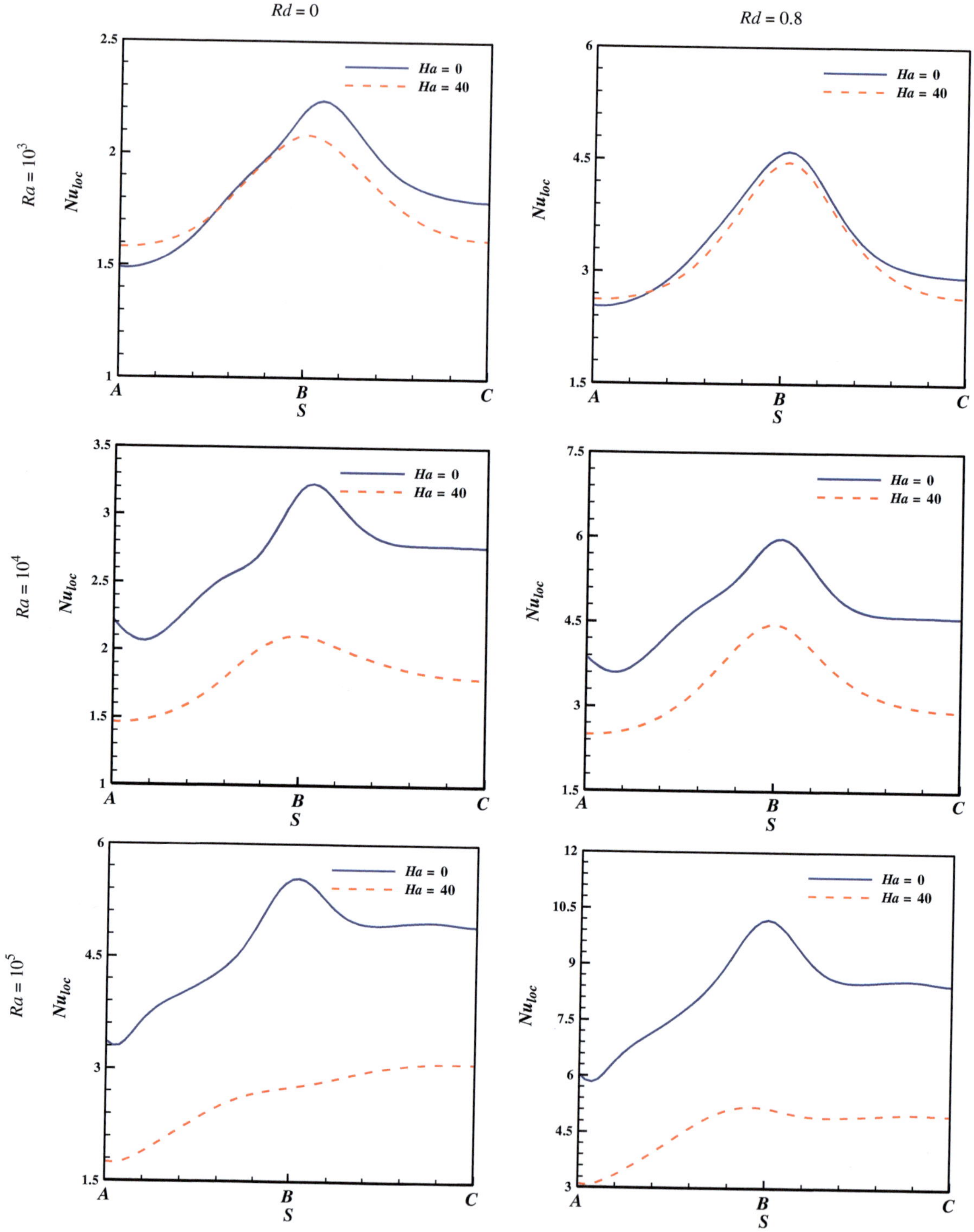

FIGURE 20.12 Effects of the radiation parameter, Hartmann number, and Rayleigh numbers on local Nusselt number when $\phi = 0.04, \xi = 0^{\circ}$

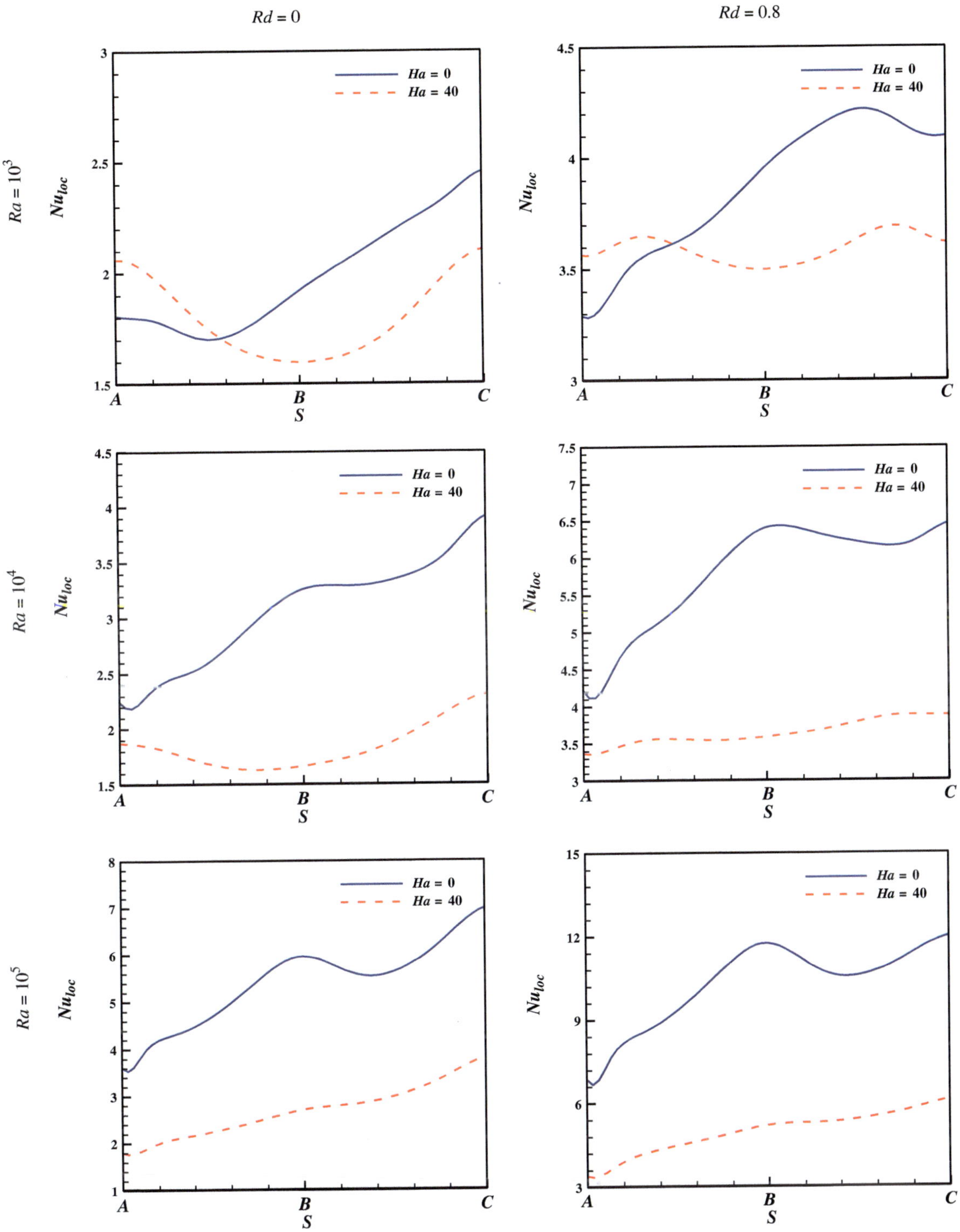

FIGURE 20.13 Effects of the radiation parameter, Hartmann number, and Rayleigh numbers on local Nusselt number when $\phi = 0.04, \xi = 90^{\circ}$.

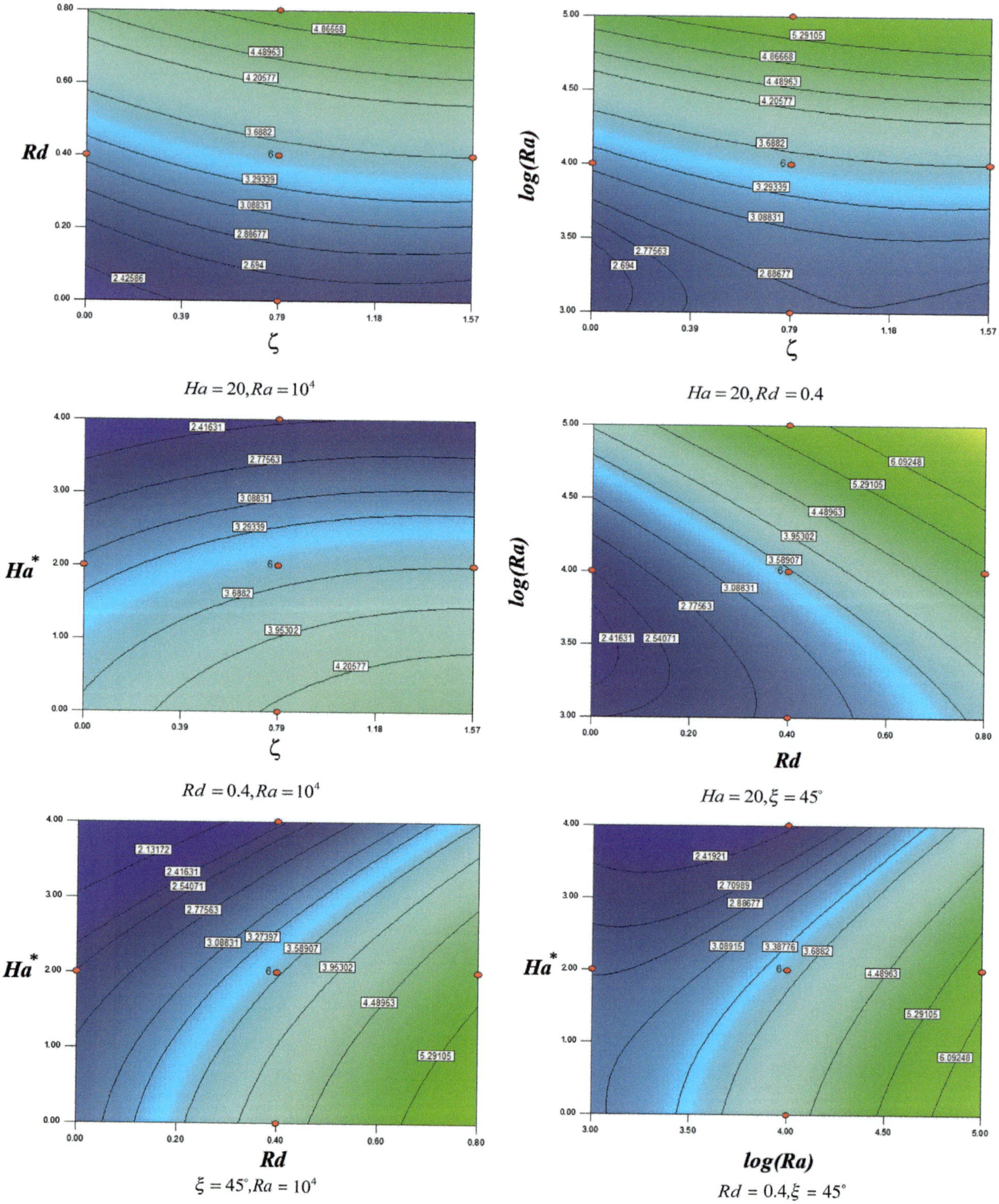

FIGURE 20.14 Effects of inclination angle, radiation parameter, Rayleigh number, and Hartmann numbers on average Nusselt number.

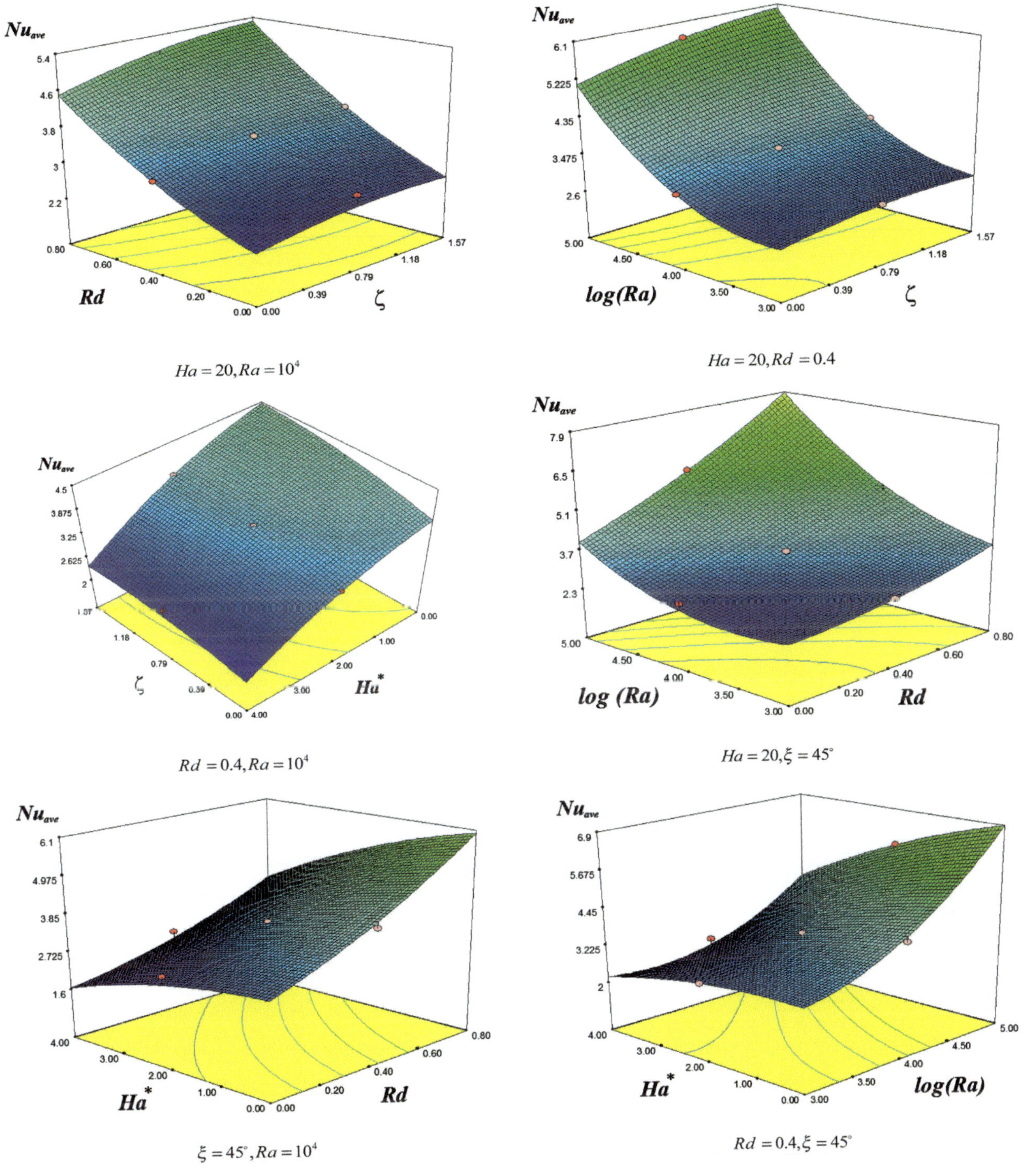

FIGURE 20.14 (Continued).

20.4 THERMAL RADIATION OF FERROFLUID IN EXISTENCE OF LORENTZ FORCES CONSIDERING VARIABLE VISCOSITY

20.4.1 Problem Definition

Fig. 20.15 depicts the geometry, boundary condition, and sample element. The formula of the inner cylinder is:

$$b = \sqrt{1 - \varepsilon^2} \, . \, a \tag{20.46}$$

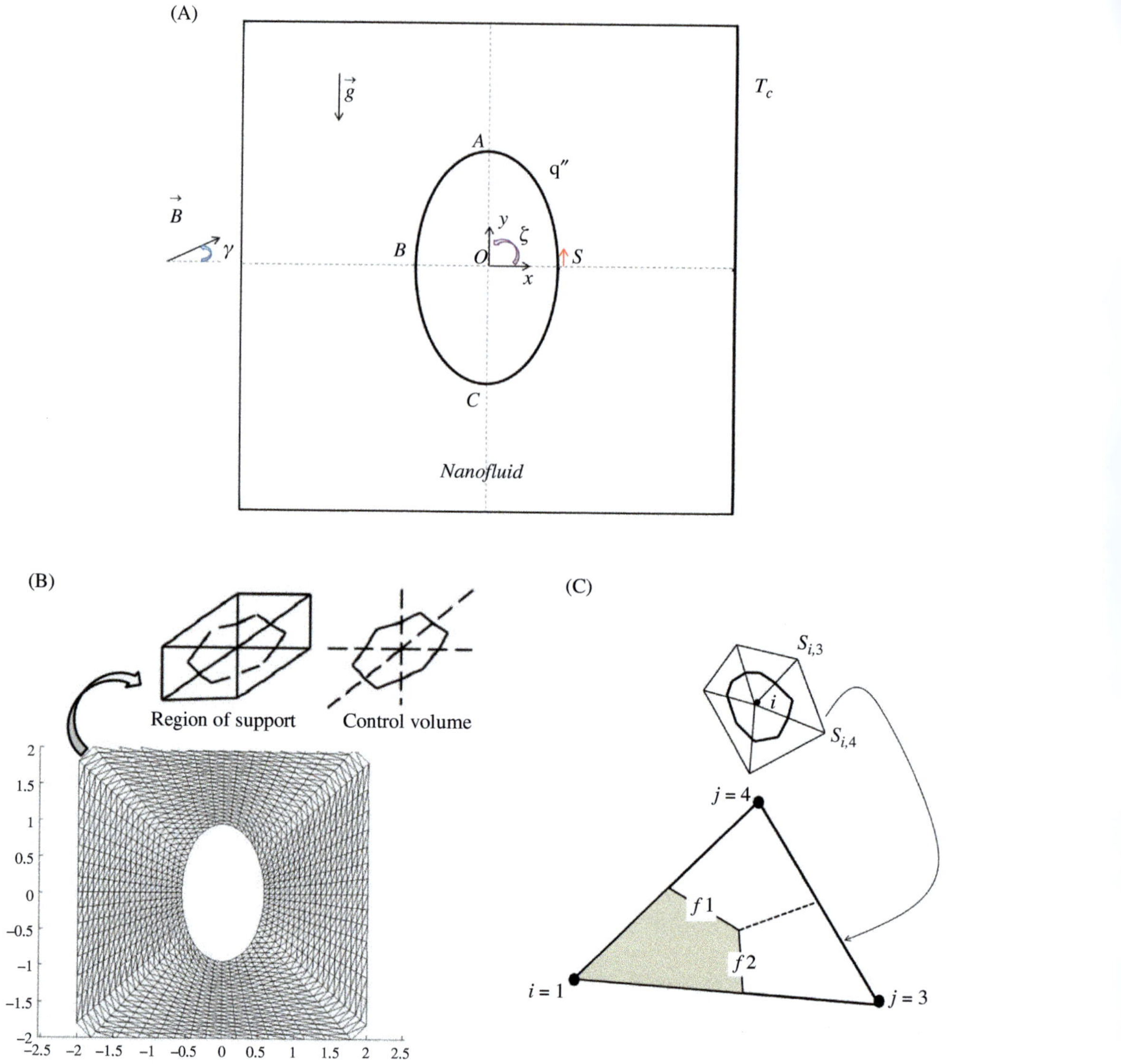

FIGURE 20.15 (A) Geometry and the boundary conditions with (B) the mesh of geometry considered in this work; (C) a sample triangular element and its corresponding control volume.

where a, b, ε are the major, minor axis of elliptic cylinder, and eccentricity for the inner cylinder. The inner cylinder has constant heat flux condition.

20.4.2 Governing Equation

2D steady convective flow of nanofluid is considered in the presence of a constant magnetic field. The PDEs equations are:

$$\frac{\partial u}{\partial x} + \frac{\partial v}{\partial y} = 0 \tag{20.47}$$

$$\left(v\frac{\partial u}{\partial y} + \frac{\partial u}{\partial x}u\right) = \left[-\sigma_{nf}B_y^2 u + \sigma_{nf}B_x B_y v\left(\frac{\partial^2 u}{\partial y^2} + \frac{\partial^2 u}{\partial x^2}\right)\mu_{nf} - \frac{\partial P}{\partial x}\right](\rho_{nf})^{-1} \tag{20.48}$$

$$\rho_{nf}\left(\frac{\partial v}{\partial x}u+\frac{\partial v}{\partial y}v\right)=+\mu_{nf}\left(\frac{\partial^2 v}{\partial x^2}+\frac{\partial^2 v}{\partial y^2}\right)-\frac{\partial P}{\partial y}+B_y\sigma_{nf}B_x u-B_x\sigma_{nf}B_x v+(T-T_c)\beta_{nf}g\rho_{nf}, \tag{20.49}$$

$$B_x=B_o\cos\lambda,\;\; B_y=B_o\sin\lambda$$

$$(\rho C_p)_{nf}\left(v\frac{\partial T}{\partial y}+u\frac{\partial T}{\partial x}\right)=k_{nf}\left(\frac{\partial^2 T}{\partial x^2}+\frac{\partial^2 T}{\partial y^2}\right)-\frac{\partial q_r}{\partial y},\quad \left[q_r=-\frac{4\sigma_e}{3\beta_R}\frac{\partial T^4}{\partial y},\quad T^4\cong 4T_c^3T-3T_c^4\right] \tag{20.50}$$

$(\rho C_p)_{nf}$, $(\beta)_{nf}$, ρ_{nf}, k_{nf}, and σ_{nf} are defined as:

$$(\rho C_p)_{nf}=(\rho C_p)_f(1-\phi)+(\rho C_p)_s\phi \tag{20.51}$$

$$(\beta)_{nf}=(\beta)_f(1-\phi)+(\beta)_s\phi \tag{20.52}$$

$$\rho_{nf}=\rho_f(1-\phi)+\rho_s\phi \tag{20.53}$$

$$k_{nf}=k_f\left(\frac{k_s+2k_f+2\phi(k_s-k_f)}{k_s-\phi(k_s-k_f)+2k_f}\right) \tag{20.54}$$

$$\frac{\sigma_{nf}}{\sigma_f}=1+\frac{3\left(\frac{\sigma_s}{\sigma_f}-1\right)\phi}{\left(\frac{\sigma_s}{\sigma_f}+2\right)-\left(\frac{\sigma_s}{\sigma_f}-1\right)\phi} \tag{20.55}$$

μ_{nf} is obtained as follows:

$$\mu_{nf}=\left(0.035B^2+3.1B-27886.4807\phi^2+4263.02\phi+316.0629\right)e^{-0.01T} \tag{20.56}$$

Vorticity and stream function should be used to eliminate pressure source terms:

$$\omega+\frac{\partial u}{\partial y}-\frac{\partial v}{\partial x}=0,\quad \frac{\partial\psi}{\partial x}=-v,\quad \frac{\partial\psi}{\partial y}=u \tag{20.57}$$

Introducing dimensionless quantities:

$$P=\frac{p}{\rho_f(\alpha_f/L)^2},\quad U=\frac{uL}{\alpha_f},\quad V=\frac{vL}{\alpha_f},\quad \Theta=\frac{T-T_c}{\Delta T},\quad \Delta T=q''L/k_f,\,(X,Y)=\frac{(x,y)}{L} \tag{20.58}$$

The final formulae are:

$$\frac{\partial V}{\partial Y}+\frac{\partial U}{\partial X}=0, \tag{20.59}$$

$$U\frac{\partial U}{\partial X}+\frac{\partial U}{\partial Y}V=\Pr\left[\frac{\mu_{nf}/\mu_f}{\rho_{nf}/\rho_f}\right]\left(\frac{\partial^2 U}{\partial Y^2}+\frac{\partial^2 U}{\partial X^2}\right)-Ha^2\Pr\left[\frac{\sigma_{nf}/\sigma_f}{\rho_{nf}/\rho_f}\right]\left(B_y^2U-B_xB_yV\right)-\frac{\partial P}{\partial X}, \tag{20.60}$$

$$V\frac{\partial V}{\partial Y}+U\frac{\partial V}{\partial X}=\Pr\left(\frac{\partial^2 V}{\partial Y^2}+\frac{\partial^2 V}{\partial X^2}\right)\left[\frac{\mu_{nf}/\mu_f}{\rho_{nf}/\rho_f}\right]-Ha^2\Pr\left[\frac{\sigma_{nf}/\sigma_f}{\rho_{nf}/\rho_f}\right]\left(B_x^2V-B_xB_yU\right)-\frac{\partial P}{\partial Y}+Ra\Pr\left[\frac{\beta_{nf}}{\beta_f}\right]\Theta, \tag{20.61}$$

$$\frac{\partial \Theta}{\partial X}U + \frac{\partial \Theta}{\partial Y}V = \left(1 + \frac{4}{3}\left(\frac{k_{nf}}{k_f}\right)^{-1} Rd\right)\left[\frac{(\rho C_P)_f}{(\rho C_P)_{nf}}\frac{k_{nf}}{k_f}\right]\frac{\partial^2 \Theta}{\partial Y^2} + \left[\frac{(\rho C_P)_f}{(\rho C_P)_{nf}}\frac{k_{nf}}{k_f}\right]\left(\frac{\partial^2 \Theta}{\partial X^2}\right)$$

where dimensionless and constants parameters are illustrated as:

$$\Pr = \upsilon_f/\alpha_f,\ \ Ra = g\beta_f q'' L^4/(k_f \upsilon_f \alpha_f),\ Ha = LB_0\sqrt{\sigma_f/\mu_f},\ Rd = 4\sigma_e T_c^3/(\beta_R k_f),$$

and boundary conditions are:

$$\begin{aligned} \frac{\partial \Theta}{\partial n} &= 1.0 \quad \text{on inner wall} \\ \Theta &= 0.0 \quad \text{on outer wall} \\ \Psi &= 0.0 \quad \text{on all walls} \end{aligned}$$

Local and average Nusselt over the hot cylinder can be calculated as:

$$Nu_{loc} = \left(1 + \frac{4}{3}Rd\left(\frac{k_{nf}}{k_f}\right)^{-1}\right)\frac{1}{\theta}\left(\frac{k_{nf}}{k_f}\right) \tag{20.65}$$

$$Nu_{ave} = \frac{1}{S}\int_0^S Nu_{loc}\, ds \tag{20.66}$$

20.4.3 Effects of Active Parameters

Fe_3O_4-H_2O nanofluid free convection in the presence of radiation source term is examined considering the magnetic field effect. MFD viscosity is taken into account. Numerical simulations are examined for various amounts of Radiation parameter ($Rd = 0$ to 0.8), volume fraction of Fe_3O_4-water ($\phi = 0$ to 0.04), Rayleigh number ($Ra = 10^3$, 10^4 and 10^5), inclination angle ($\xi = 0^\circ$ and 90°), and Hartmann number ($Ha = 0$ to 40).

The impact of radiation parameter on nanofluid hydrothermal behavior is illustrated in Fig. 20.16. Thermal boundary layer thickness increases with the increase of Rd. Also increasing Rd enhances the nanofluid motion. This influence is more evident in the presence of a magnetic field.

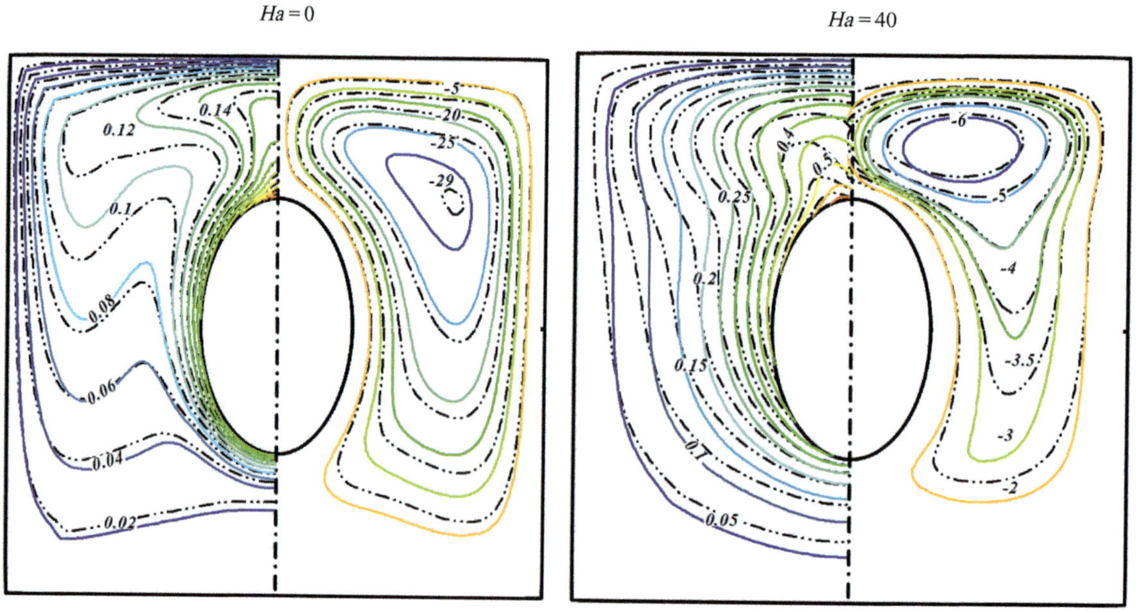

FIGURE 20.16 Effect of Hartmann number and radiation parameter on isotherm (left) and streamline (right) contours ($Rd = 0$ (− ·· −), $Rd = 0.8$(−)) when $\xi = 90^\circ, \phi = 0.04, Ra = 10^5$.

Influence of ξ, Ha and Ra on streamlines and isotherms are depicted Figs. 20.17 and 20.18. As nanofluid temrature augments, the nanofluid begins moving from the warm surface to the outer one and dropping along the ter cylinder. At low Rayleigh number, conduction is more signification than convection. $|\Psi_{\max}|$ augments as the oyancy force augments and it reduces as the Lorentz force increases. When $\xi = 0^\circ$, $Ha = 0$, as the Rayleigh umber increases, the lower vortex diminishes and the upper one becomes stronger.

As the Hartmann number increases, the centers of the vortexes move to $x = 0$. $|\Psi_{\max}|$ and the temperature graent is enhanced with the rise of Ra. As ξ increases, the distortion of the isotherms increases. Increasing the artmann number causes the thermal plume to diminish and changes the heat transfer mechanism from convecon to conduction. As the Lorentz force augments, the two upper vortexes merge together and the thermal ume vanishes.

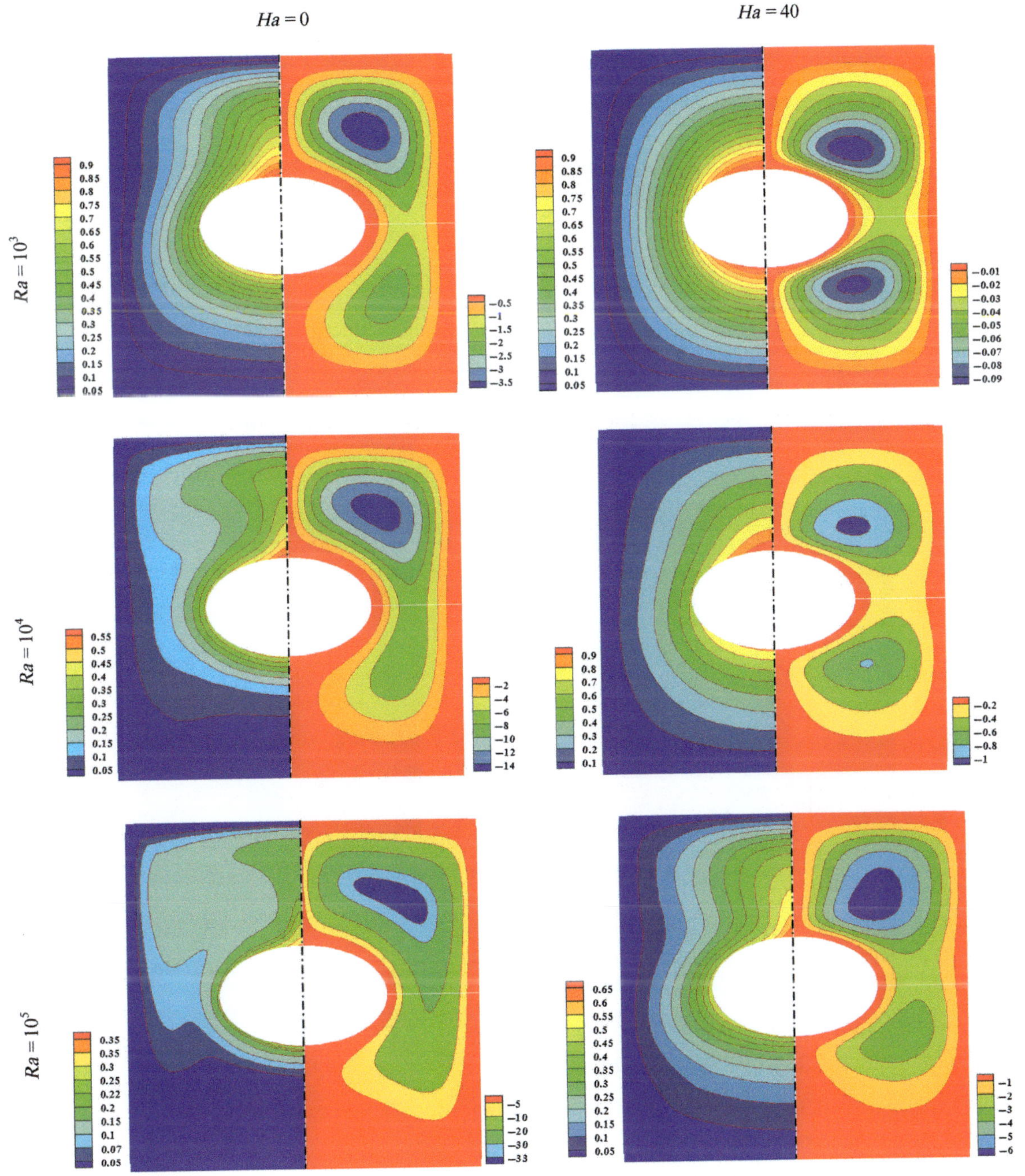

FIGURE 20.17 Effect of Rayleigh and Hartmann numbers on isotherm (left) and streamline (right) contours when $\xi = 0$, $\phi = 0.04$, $Rd = 0.8$.

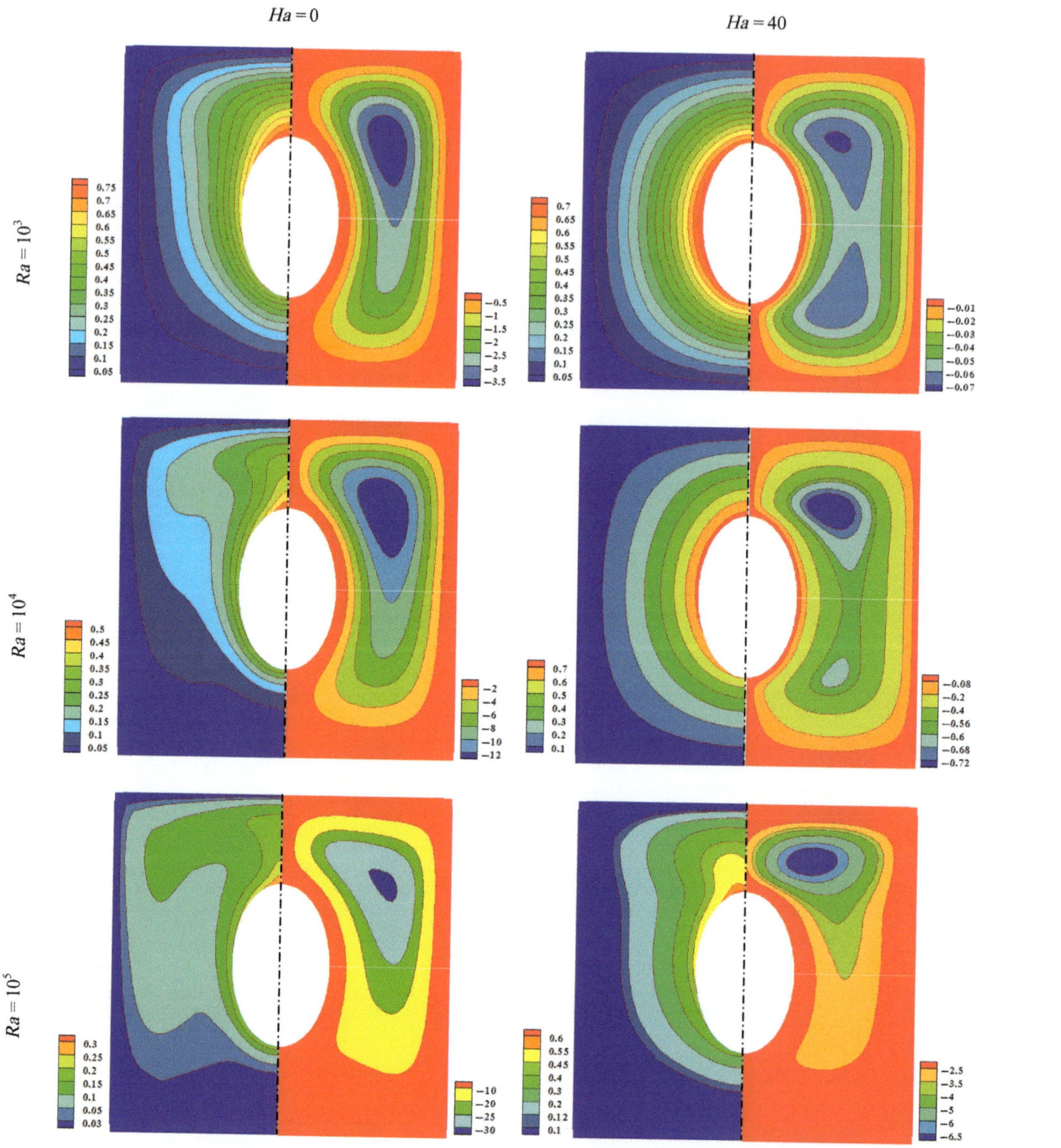

FIGURE 20.18 Effect of Rayleigh and Hartmann numbers on isotherm (left) and streamline (right) contours when $\xi = 90^{\circ}, \phi = 0.04, Rd = 0.8$.

Figs. 20.19–20.21 illustrate the impact of ξ, Rd, Ra, and Ha on Nu_{loc}, Nu_{ave}. The formula for Nu_{ave} corresponding to active parameters is:

$$\begin{aligned} Nu_{ave} &= 6.91 - 0.12\xi - 4.3Rd - 3.5\log(Ra) + 1.26Ha^{*} \\ &+ 0.36\xi\, Rd + 0.2\xi\log(Ra) - 0.07\xi Ha^{*} + 1.61\log(Ra)Rd \\ &- 0.31\log(Ra)\, Ha^{*} - 0.45Rd\, Ha^{*} - 0.2\xi^{2} + 0.62\left(\log(Ra)\right)^{2} + 1.5Rd^{2} \\ &- 0.08Ha^{*2} \end{aligned} \tag{20.67}$$

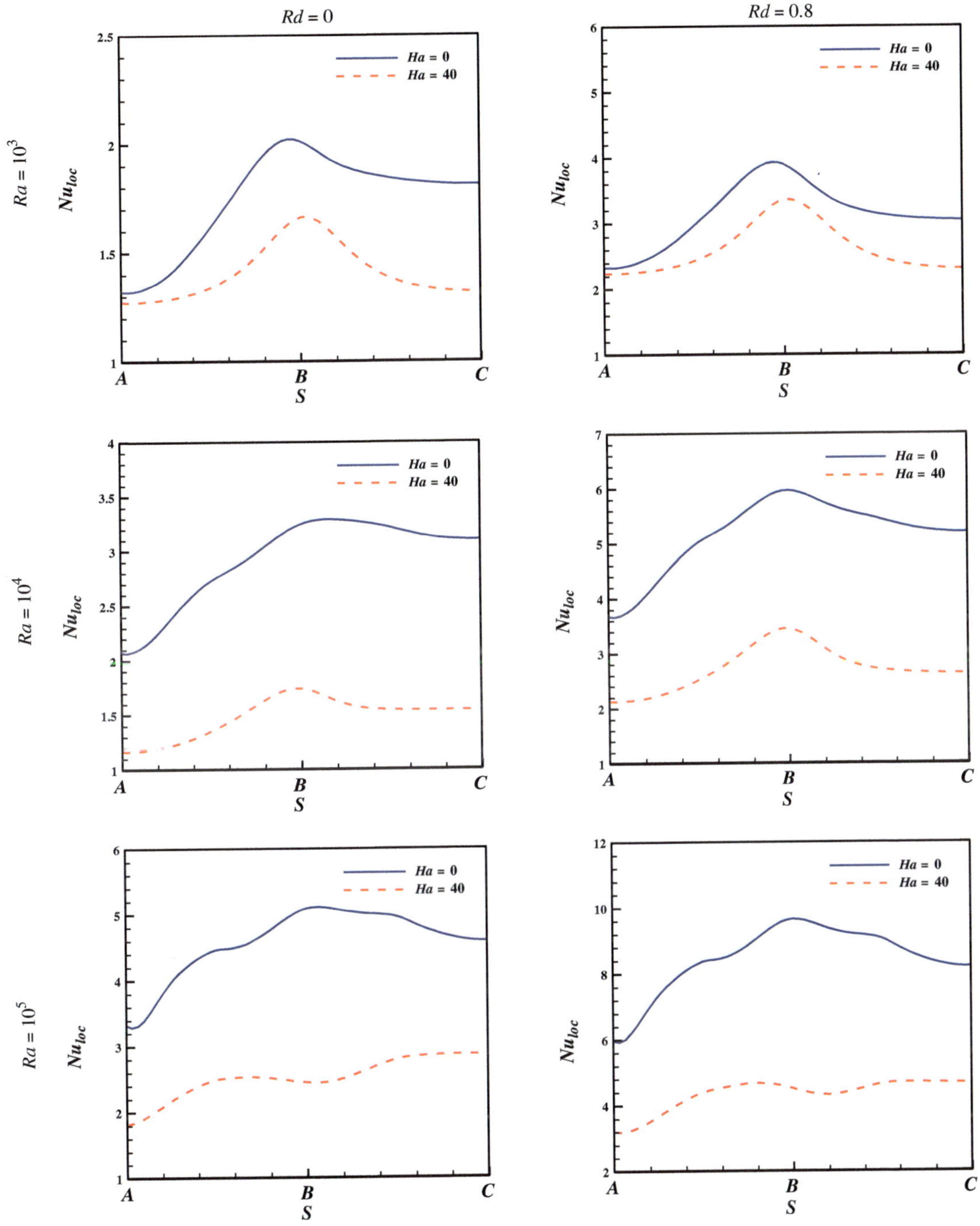

FIGURE 20.19 Effects of the radiation parameter, Hartmann number, and Rayleigh numbers on local Nusselt number when $\phi = 0.04, \xi = 0^\circ$.

Where $Ha^* = 0.1Ha$. The number of extremum in Nu_{loc} profile matches the presence of the thermal plumes. Nu_{ave} is enhanced with the rise of Rd. As the buoyancy forces increase, the temperature decreases and in turn Nu_{ave} increases with the rise of Ra. Also this figure shows that the Nusselt number has a direct relationship with inclination angle.

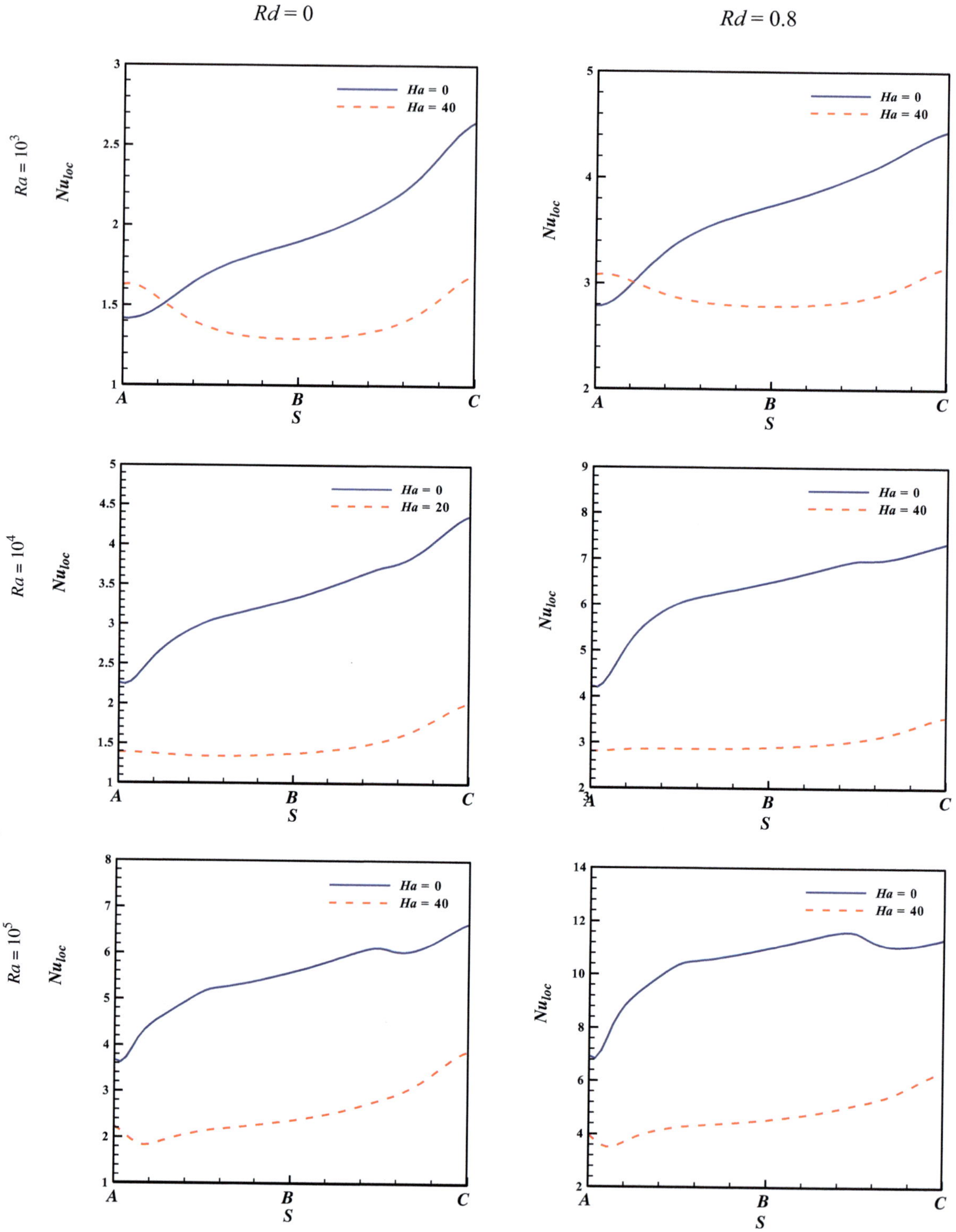

FIGURE 20.20 Effects of the radiation parameter, Hartmann number, and Rayleigh numbers on local Nusselt number when $\phi = 0.04, \xi = 90^{\circ}$.

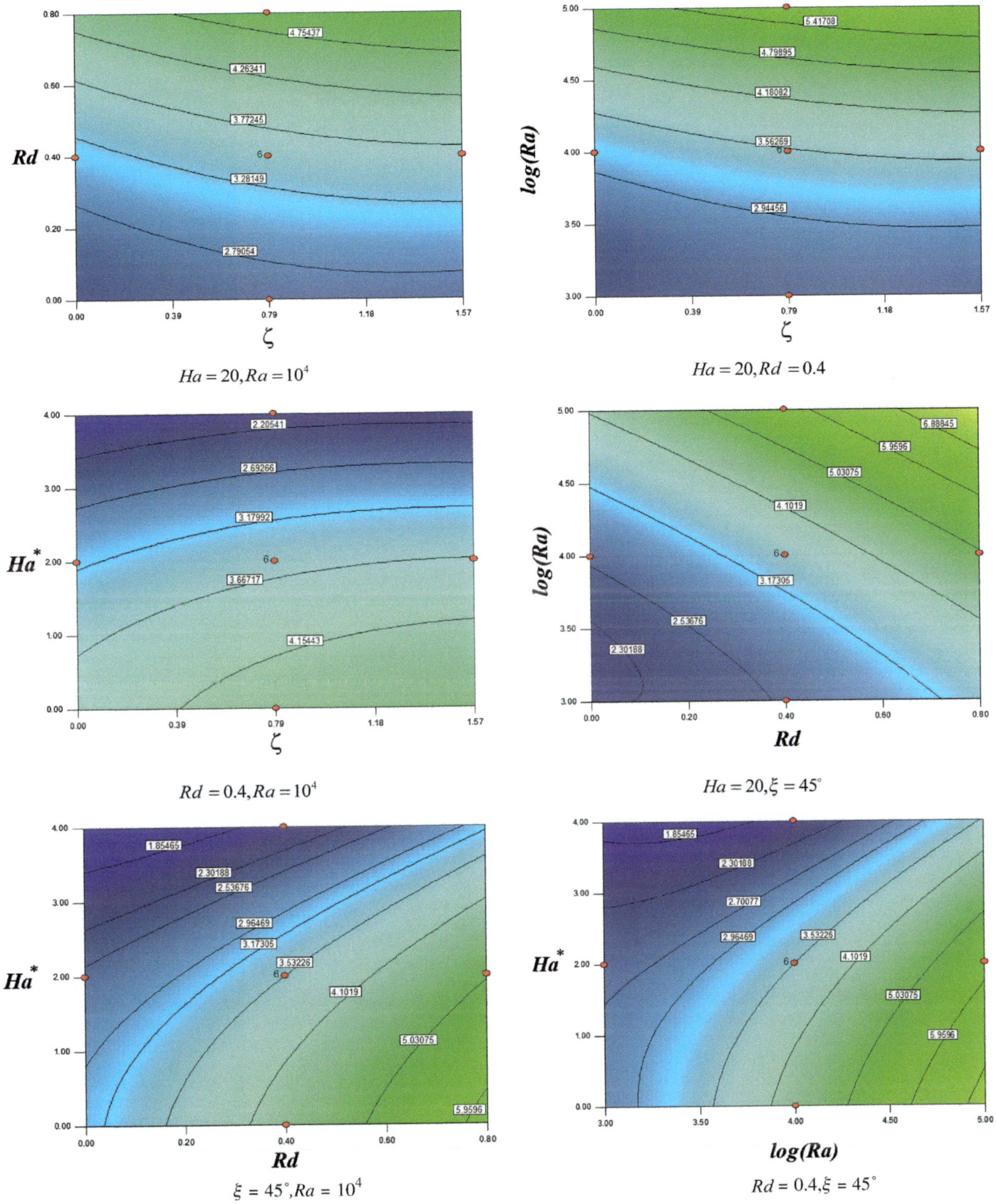

FIGURE 20.21 Effects of inclination angle, radiation parameter, Rayleigh number, and Hartmann numbers on average Nusselt number.

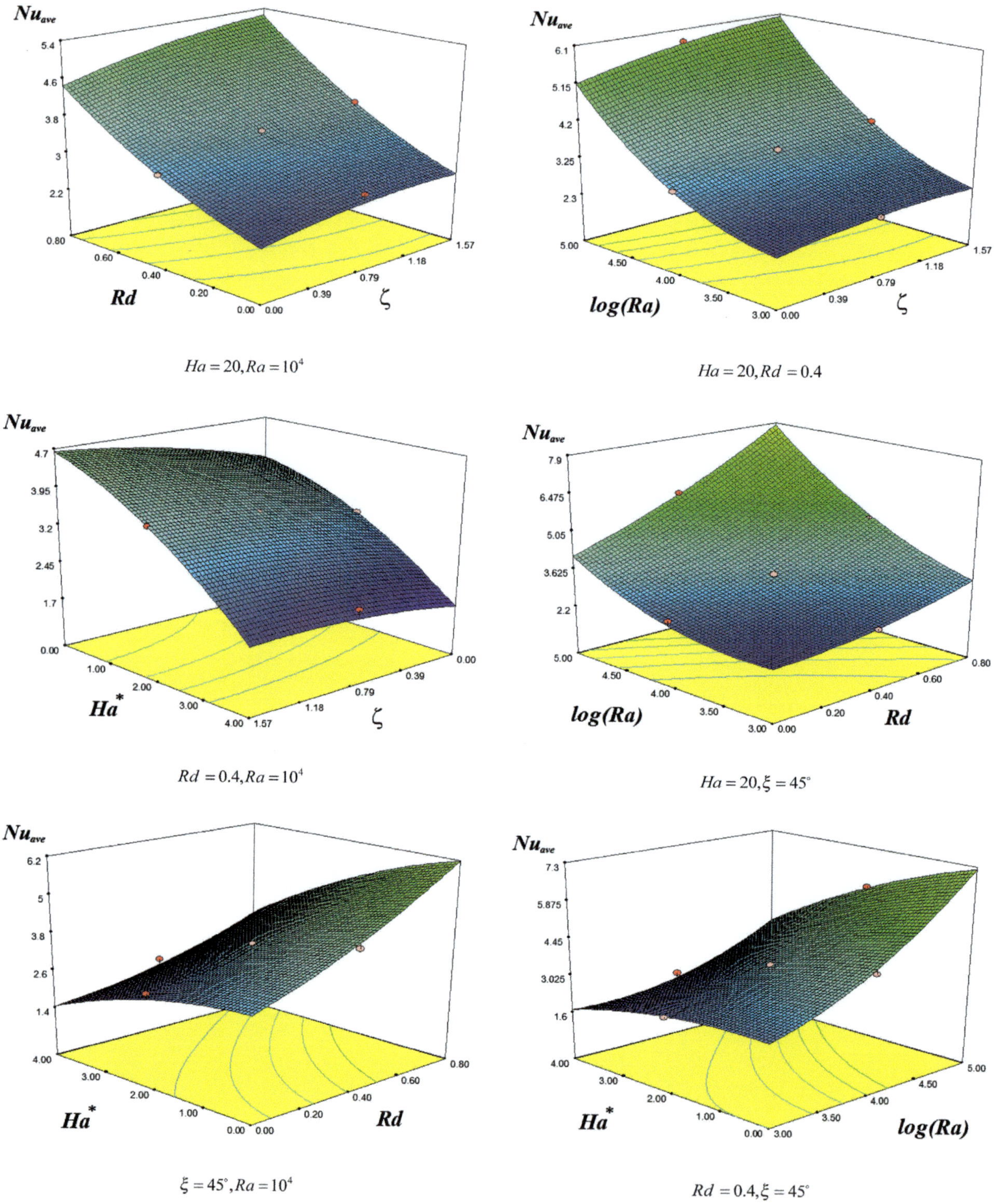

FIGURE 20.21 (Continued).

20.5 MAGNETIC NANOFLUID NATURAL CONVECTION IN PRESENCE OF THERMAL RADIATION CONSIDERING VARIABLE VISCOSITY

).5.1 Problem Definition

Fig. 20.22 shows the geometry, boundary condition, and sample element. The formula of the outer cylinder is:

$$r = r_{in} + A\cos(N(\zeta - \zeta_0)) \tag{20.68}$$

which r_{in}, r_{out} are the radii of the base circle and outer cylinder, respectively.

0.5.2 Governing Equation

2D steady convective flow of nanofluid is considered in the presence of a constant magnetic field. The PDEs quations are:

$$\frac{\partial u}{\partial x} + \frac{\partial v}{\partial y} = 0 \tag{20.69}$$

$$\left(v\frac{\partial u}{\partial y} + \frac{\partial u}{\partial x}u\right) = \left[-\sigma_{nf}B_y^2 u + \sigma_{nf}B_x B_y v\left(\frac{\partial^2 u}{\partial y^2} + \frac{\partial^2 u}{\partial x^2}\right)\mu_{nf} - \frac{\partial P}{\partial x}\right](\rho_{nf})^{-1} \tag{20.70}$$

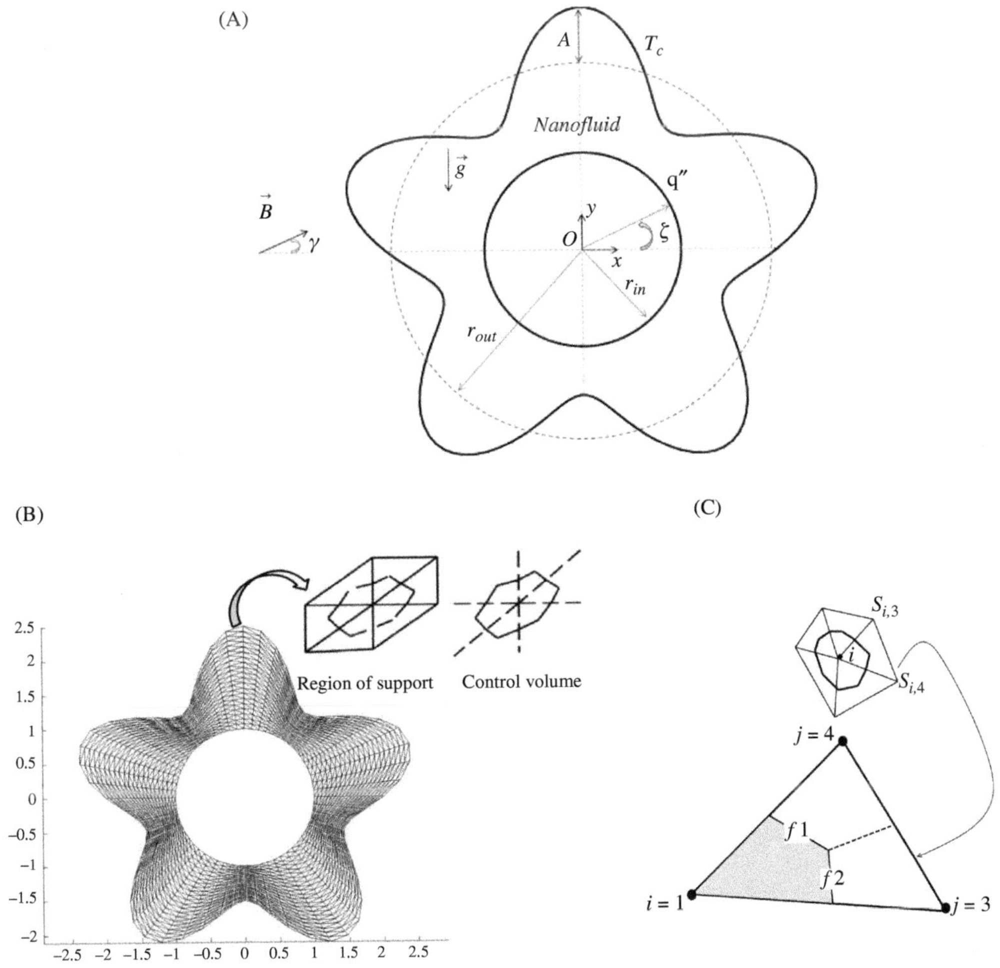

FIGURE 20.22 (A) Geometry and the boundary conditions with (B) the mesh of geometry considered in this work; (C) a sample triangular element and its corresponding control volume.

$$\rho_{nf}\left(\frac{\partial v}{\partial x}u+\frac{\partial v}{\partial y}v\right)=+\mu_{nf}\left(\frac{\partial^2 v}{\partial x^2}+\frac{\partial^2 v}{\partial y^2}\right)-\frac{\partial P}{\partial y}+B_y\sigma_{nf}B_x u-B_x\sigma_{nf}B_x v+(T-T_c)\beta_{nf}g\rho_{nf},\tag{20.71}$$

$$B_x=B_o\cos\lambda,\ \ B_y=B_o\sin\lambda$$

$$(\rho C_p)_{nf}\left(v\frac{\partial T}{\partial y}+u\frac{\partial T}{\partial x}\right)=k_{nf}\left(\frac{\partial^2 T}{\partial x^2}+\frac{\partial^2 T}{\partial y^2}\right)-\frac{\partial q_r}{\partial y},$$
$$\left[q_r=-\frac{4\sigma_e}{3\beta_R}\frac{\partial T^4}{\partial y},\quad T^4\cong 4T_c^3T-3T_c^4\right]\tag{20.72}$$

$(\rho C_p)_{nf}$, $(\beta)_{nf}$, ρ_{nf}, and σ_{nf} are defined as:

$$(\rho C_p)_{nf}=(\rho C_p)_f(1-\phi)+(\rho C_p)_s\phi\tag{20.73}$$

$$(\beta)_{nf}=(\beta)_f(1-\phi)+(\beta)_s\phi\tag{20.74}$$

$$\rho_{nf}=\rho_f(1-\phi)+\rho_s\phi\tag{20.75}$$

$$\frac{\sigma_{nf}}{\sigma_f}=1+\frac{3\left(\frac{\sigma_s}{\sigma_f}-1\right)\phi}{\left(\frac{\sigma_s}{\sigma_f}+2\right)-\left(\frac{\sigma_s}{\sigma_f}-1\right)\phi}\tag{20.76}$$

Properties of Fe_3O_4-water nanofluid are presented in Table 20.3. The Maxwell model and Hamilton–Crosser model for irregular particle geometries by introducing a shape factor can be expressed as

$$\frac{k_{nf}}{k_f}=\frac{k_p+(m+1)k_f-(m+1)\phi(k_f-k_p)}{k_p+(m+1)k_f+\phi(k_f-k_p)}\tag{20.77}$$

in which k_p and k_f are the conductivities of the particle material and the base fluid. In this equation "m" is shaper factor. Table 20.4 shows the different values of shape factors for various shapes of nanoparticles.μ_{nf} is obtained as follows [41]:

$$\mu_{nf}=\left(0.035B^2+3.1B-27886.4807\phi^2+4263.02\phi+316.0629\right)e^{-0.01T}\tag{20.78}$$

Vorticity and stream function should be used to eliminate pressure source terms:

TABLE 20.4 The Values of Shape Factor of Different Shapes of Nanoparticles

	Spherical	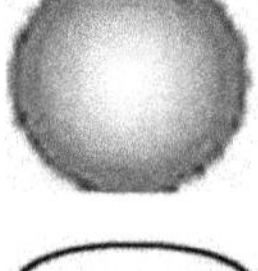	3
m	Platelet	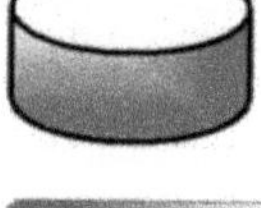	5.7
	Cylinder		4.8
	Brick	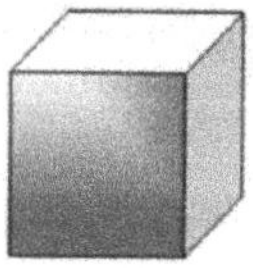	3.7

$$\omega + \frac{\partial u}{\partial y} - \frac{\partial v}{\partial x} = 0, \quad \frac{\partial \psi}{\partial x} = -v, \quad \frac{\partial \psi}{\partial y} = u \tag{20.79}$$

Introducing dimensionless quantities:

$$P = \frac{p}{\rho_f (\alpha_f / L)^2}, \quad U = \frac{uL}{\alpha_f}, \quad V = \frac{vL}{\alpha_f}, \quad \Theta = \frac{T - T_c}{\Delta T}, \quad \Delta T = q''L/k_f, \quad (X, Y) = \frac{(x, y)}{L} \tag{20.80}$$

The final formulae are:

$$\frac{\partial V}{\partial Y} + \frac{\partial U}{\partial X} = 0, \tag{20.81}$$

$$\begin{aligned} U\frac{\partial U}{\partial X} + \frac{\partial U}{\partial Y}V &= \Pr\left[\frac{\mu_{nf}/\mu_f}{\rho_{nf}/\rho_f}\right]\left(\frac{\partial^2 U}{\partial Y^2} + \frac{\partial^2 U}{\partial X^2}\right) \\ &- Ha^2\Pr\left[\frac{\sigma_{nf}/\sigma_f}{\rho_{nf}/\rho_f}\right]\left(B_y^2 U - B_x B_y V\right) - \frac{\partial P}{\partial X}, \end{aligned} \tag{20.82}$$

$$\begin{aligned} V\frac{\partial V}{\partial Y} + U\frac{\partial V}{\partial X} &= \Pr\left(\frac{\partial^2 V}{\partial Y^2} + \frac{\partial^2 V}{\partial X^2}\right)\left[\frac{\mu_{nf}/\mu_f}{\rho_{nf}/\rho_f}\right] \\ &- Ha^2\Pr\left[\frac{\sigma_{nf}/\sigma_f}{\rho_{nf}/\rho_f}\right]\left(B_x^2 V - B_x B_y U\right) - \frac{\partial P}{\partial Y} + Ra\Pr\left[\frac{\beta_{nf}}{\beta_f}\right]\Theta, \end{aligned} \tag{20.83}$$

$$\frac{\partial \Theta}{\partial X}U + \frac{\partial \Theta}{\partial Y}V = \left(1 + \frac{4}{3}\left(\frac{k_{nf}}{k_f}\right)^{-1} Rd\right)\left[\frac{(\rho C_P)_f}{(\rho C_P)_{nf}}\frac{k_{nf}}{k_f}\right]\frac{\partial^2 \Theta}{\partial Y^2} + \left[\frac{(\rho C_P)_f}{(\rho C_P)_{nf}}\frac{k_{nf}}{k_f}\right]\left(\frac{\partial^2 \Theta}{\partial X^2}\right) \tag{20.84}$$

where dimensionless and constants parameters are illustrated as:

$$\Pr = \upsilon_f/\alpha_f, Ra = g\beta_f q'' L^4/(k_f \upsilon_f \alpha_f), \; Ha = LB_0\sqrt{\sigma_f/\mu_f}, \; Rd = 4\sigma_e T_c^3/(\beta_R k_f), \tag{20.85}$$

and boundary conditions are:

$$\begin{aligned} \frac{\partial \Theta}{\partial n} &= 1.0 \quad \text{on inner wall} \\ \Theta &= 0.0 \quad \text{on outer wall} \\ \Psi &= 0.0 \quad \text{on all walls} \end{aligned} \tag{20.86}$$

Local and average Nusselt over the hot cylinder can be calculated as:

$$Nu_{loc} = \left(1 + \frac{4}{3}Rd\left(\frac{k_{nf}}{k_f}\right)^{-1}\right)\frac{1}{\theta}\left(\frac{k_{nf}}{k_f}\right) \tag{20.87}$$

$$Nu_{ave} = \frac{1}{\xi}\int_{0.5\pi}^{1.5\pi} Nu_{loc}\, d\xi \tag{20.88}$$

20.5.3 Effects of Active Parameters

Fe_3O_4-water nanofluid magnetohydrodynamic natural convection heat transfer in the presence of thermal radiation is studied. MFD viscosity has been considered. Control volume-based finite element method is selected to solve the governing equations. Numerical simulation are examined for various values of radiation parameter ($Rd = 0$ to 0.8), Rayleigh number ($Ra = 10^3$, 10^4 and 10^5), volume fraction of nanofluid ($\phi = 0$ to 0.04), shaper of nanoparticles, and Hartmann number ($Ha = 0$ to 20). Table 20.5 shows the shape of nanoparticle effect on Nusselt number. As illustrated in this table, Platelet shape has highest rate of heat transfer. So, Platelet shape is selected for further investigation.

TABLE 20.5 Effect of Shape of Nanoparticles on Nusselt Number When $Rd = 0.8, Ra = 10^5, \phi = 0.04$

	Ha	
	0	20
Spherical	8.269555	6.119096
Brick	8.336758	6.170139
Cylinder	8.442229	6.250367
Platelet	8.52838	6.316004

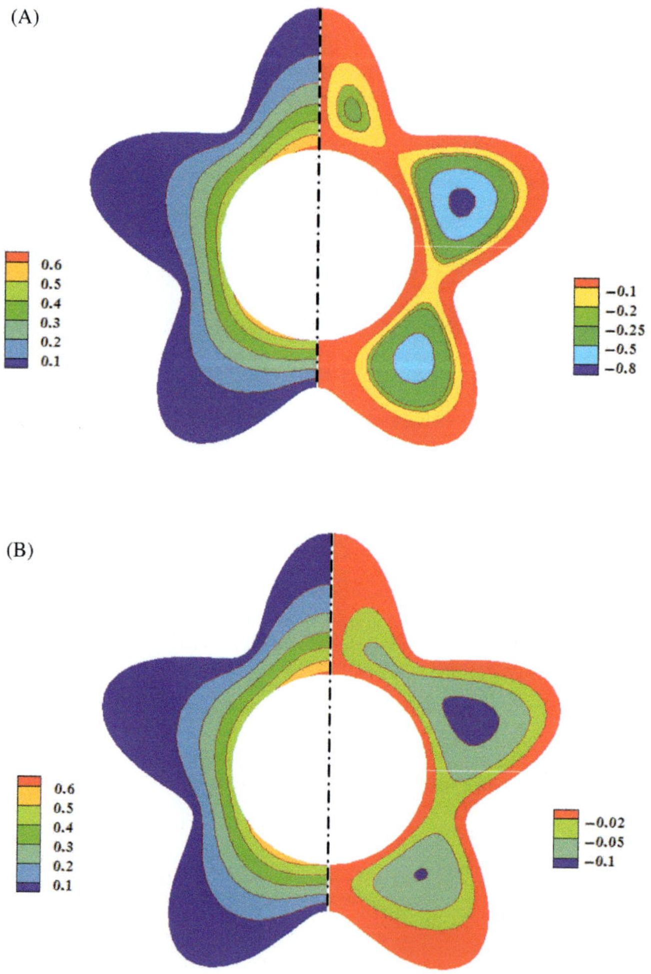

FIGURE 20.23 Effect of Hartmann number on isotherm (left) and streamline (right) contours when (A) $Ha = 0$, (B) $Ha = 20$ and $Ra = 10^3, Rd = 0.8$.

Figs. 20.23–20.25 illustrate the effect of *Ha* and *Ra* on the streamlines and isotherms. As nanofluid temperature increase, the nanofluid begins moving from the warm surface to the outer one and dropping along the outer cylinder, afterwards climbing again at the inner cylinder, generating a revolving vortex inside the cavity. At low Rayleigh number, conduction is more signification than convection.$|\Psi_{max}|$ augments as the buoyancy force augments and it reduces as the Lorentz force is enhanced. As buoyancy forces increase, the eddies become stronger and a thermal plume is generated near the centerline. Increasing the Hartmann number causes the thermal plume to diminish and changes the heat transfer mechanism from convection to conduction.

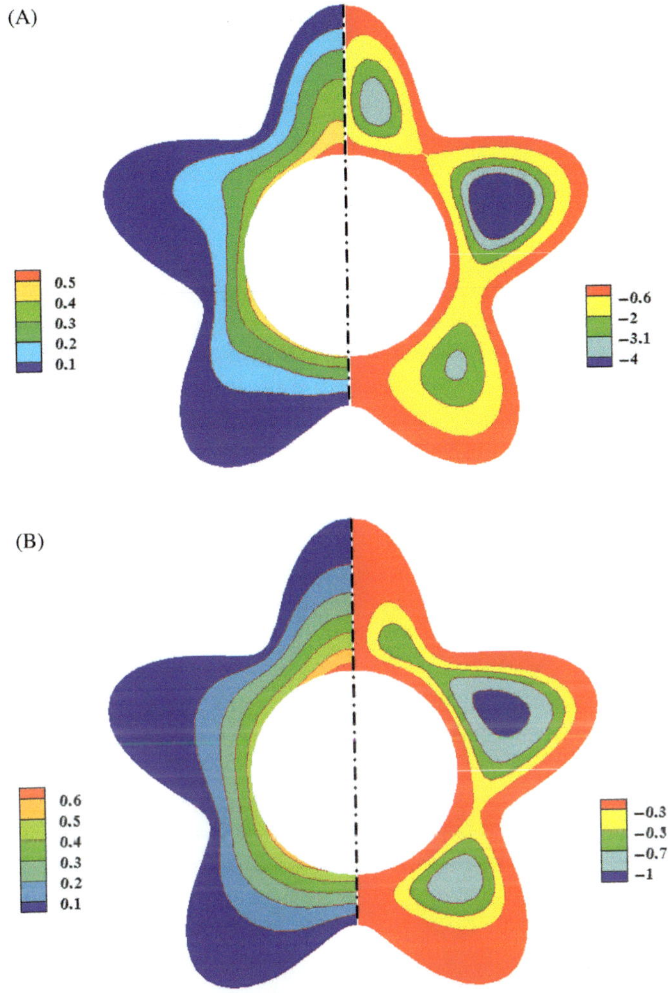

FIGURE 20.24 Effect of Hartmann number on isotherm (left) and streamline (right) contours when (A) $Ha = 0$, (B) $Ha = 20$ and $Ra = 10^4$, $Rd = 0.8$.

Figs. 20.26 and 20.27 demonstrate the influence of Rd, Ra , and Ha on Nu_{loc}, Nu_{ave}. The correlation for Nu_{ave} corresponding active parameters is:

$$\begin{aligned} Nu_{ave} &= 18.8 - 5.4Rd - 8.1\log(Ra) - 1.2Ha^* \\ &+ 1.68Rd\log(Ra) - 0.2Rd\,Ha^* - 0.32Ha^*\log(Ra) \\ &+ 1.6Rd^2 + 1.07(\log(Ra))^2 - 0.15(Ha^*)^2 \end{aligned} \tag{20.89}$$

where $Da^* = 0.01Da, Ha^* = 0.1Ha$. The number of extremum in in Nu_{loc} profile matches the presence of the thermal plumes and the number of undulations. As the buoyancy force and thermal radiation increase, Nu_{ave} is enhanced. Enhancing the Hartmann number causes the nanofluid flow to retard and the Nusselt number reduces.

20.6 NUMERICAL STUDY OF THE EFFECT OF MAGNETIC FIELD ON FE_3O_4-WATER FERROFLUID CONVECTION WITH THERMAL RADIATION

20.6.1 Problem Definition

Fig. 20.28 depicts the geometry, boundary condition, and sample element. The formula for the inner cylinder is:

$$r = r_{in} + A\cos(N(\zeta - \zeta_0)) \tag{20.90}$$

in which r_{in} is radius of the base circle. The inner cylinder has constant heat flux condition.

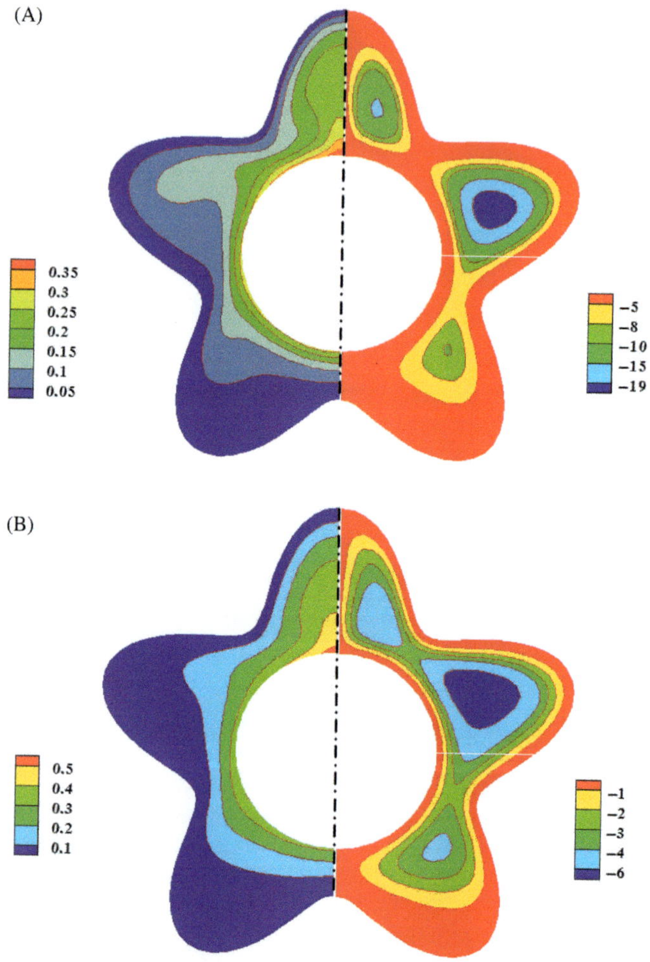

FIGURE 20.25 Effect of Hartmann number on isotherm (left) and streamline (right) contours when (A) $Ha = 0$, (B) $Ha = 20$ and $Ra = 10^5, Rd = 0.8$.

20.6.2 Governing Equation

2D steady convective flow of nanofluid in a porous medium is considered with constant magnetic field. The PDEs are:

$$\frac{\partial u}{\partial x} + \frac{\partial v}{\partial y} = 0 \tag{20.91}$$

$$\left(v\frac{\partial u}{\partial y} + \frac{\partial u}{\partial x}u\right) = \left[-\sigma_{nf}B_y^2 u + \sigma_{nf}B_x B_y v + \left(\frac{\partial^2 u}{\partial y^2} + \frac{\partial^2 u}{\partial x^2}\right)\mu_{nf} - \frac{\partial P}{\partial x}\right](\rho_{nf})^{-1} \tag{20.92}$$

$$\rho_{nf}\left(\frac{\partial v}{\partial x}u + \frac{\partial v}{\partial y}v\right) = +\mu_{nf}\left(\frac{\partial^2 v}{\partial x^2} + \frac{\partial^2 v}{\partial y^2}\right) - \frac{\partial P}{\partial y} + B_y\sigma_{nf}B_x u - B_x\sigma_{nf}B_x v + (T - T_c)\beta_{nf} g \rho_{nf},$$
$$B_x = B_o\cos\lambda,\ B_y = B_o\sin\lambda \tag{20.93}$$

$$(\rho C_p)_{nf}\left(v\frac{\partial T}{\partial y} + u\frac{\partial T}{\partial x}\right) = k_{nf}\left(\frac{\partial^2 T}{\partial x^2} + \frac{\partial^2 T}{\partial y^2}\right) - \frac{\partial q_r}{\partial y},$$
$$\left[q_r = -\frac{4\sigma_e}{3\beta_R}\frac{\partial T^4}{\partial y},\quad T^4 \cong 4T_c^3 T - 3T_c^4\right] \tag{20.94}$$

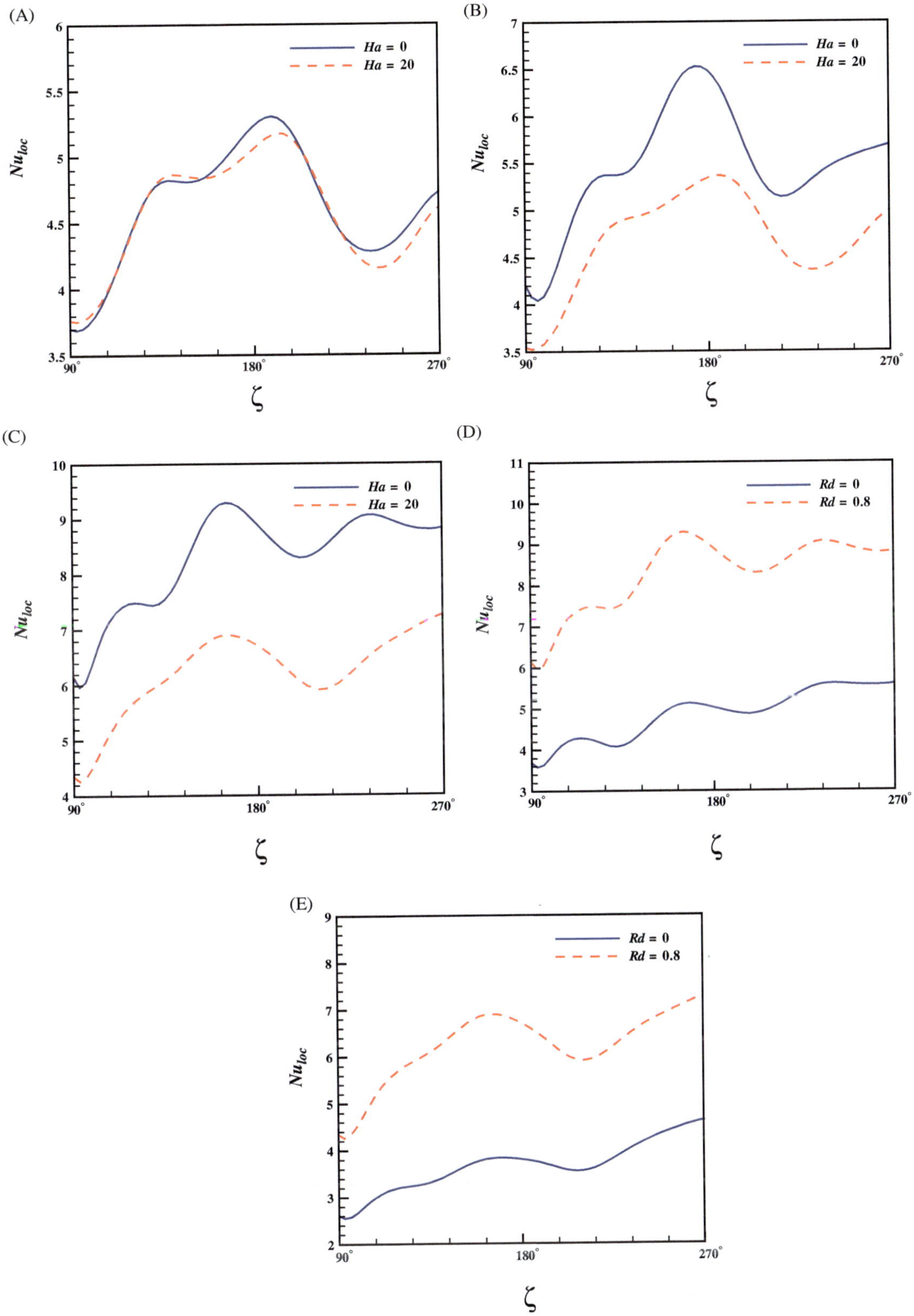

FIGURE 20.26 Effects of radiation parameter, Hartmann number, and Rayleigh number on local Nusselt number when (A) $Ra = 10^3, Rd = 0.8$, (B) $Ra = 10^4, Rd = 0.8$, (C) $Ra = 10^5, Rd = 0.8$, (D) $Ra = 10^5, Ha = 0$, (E) $Ra = 10^5, Ha = 20$ and $\phi = 0.04$.

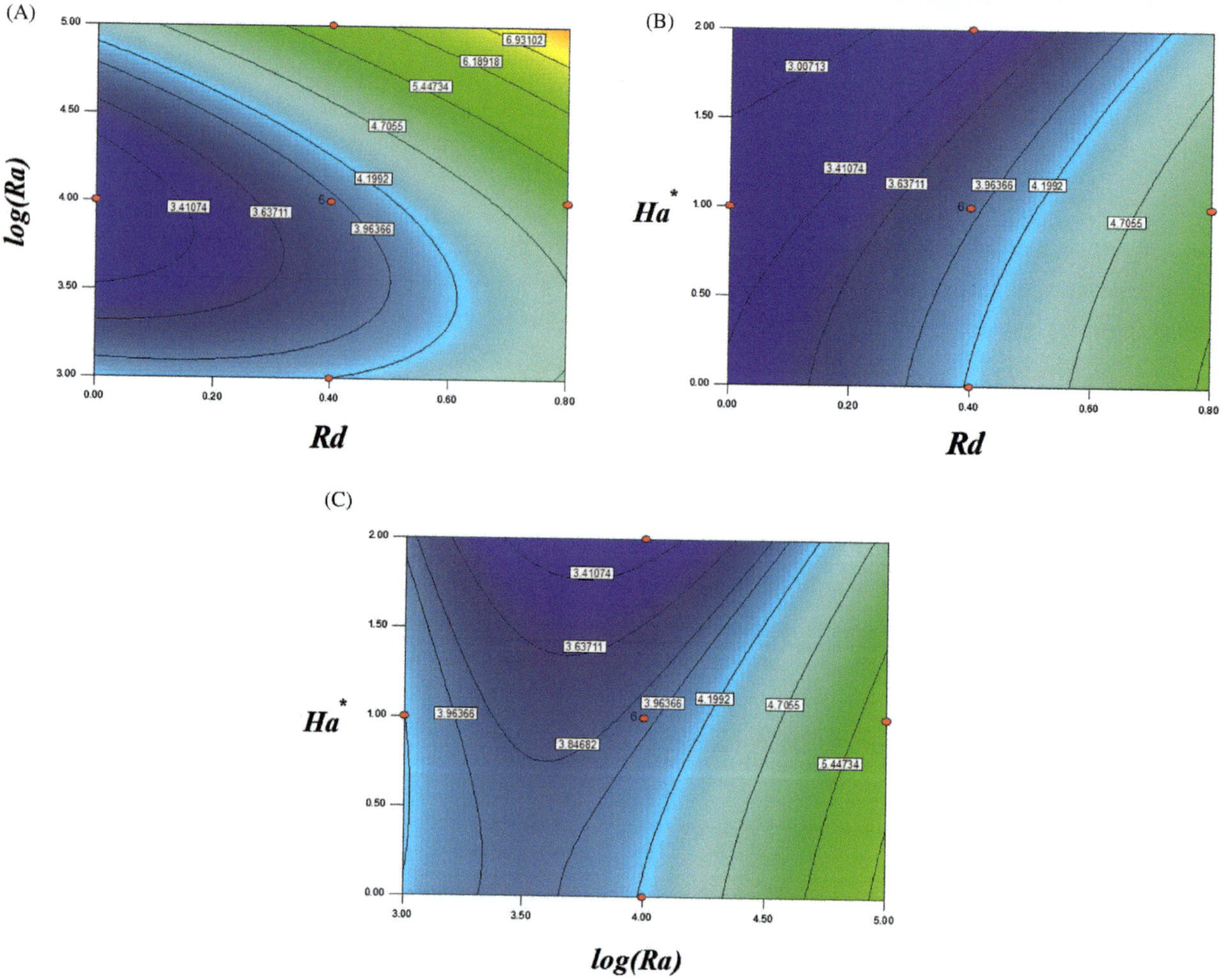

FIGURE 20.27 Effects of radiation parameter, Hartmann number, and Rayleigh number on average Nusselt number when (A) $Ha = 10, \phi = 0.04$, (B) $Ra = 10^4, \phi = 0.04$, (C) $Rd = 0.4, \phi = 0.04$, (D) $Ha = 10, \phi = 0.04$, (E) $Ra = 10^4, \phi = 0.04$, (F) $Rd = 0.4, \phi = 0.04$.

$(\rho C_p)_{nf}, (\rho\beta)_{nf}, \rho_{nf}, k_{nf}$, and σ_{nf} are defined as:

$$(\rho C_p)_{nf} = (\rho C_p)_f(1-\phi) + (\rho C_p)_s\phi \tag{20.95}$$

$$(\rho\beta)_{nf} = (\rho\beta)_f(1-\phi) + (\rho\beta)_s\phi \tag{20.96}$$

$$\rho_{nf} = \rho_f(1-\phi) + \rho_s\phi \tag{20.97}$$

$$k_{nf} = k_f\left(\frac{k_s + 2k_f + 2\phi(k_s - k_f)}{k_s - \phi(k_s - k_f) + 2k_f}\right) \tag{20.98}$$

$$\frac{\sigma_{nf}}{\sigma_f} = 1 + \frac{3\left(\dfrac{\sigma_s}{\sigma_f} - 1\right)\phi}{\left(\dfrac{\sigma_s}{\sigma_f} + 2\right) - \left(\dfrac{\sigma_s}{\sigma_f} - 1\right)\phi}. \tag{20.99}$$

μ_{nf} is obtained as follows:

$$\mu_{nf} = \left(0.035B^2 + 3.1B - 27886.4807\phi^2 + 4263.02\phi + 316.0629\right)e^{-0.01T} \tag{20.100}$$

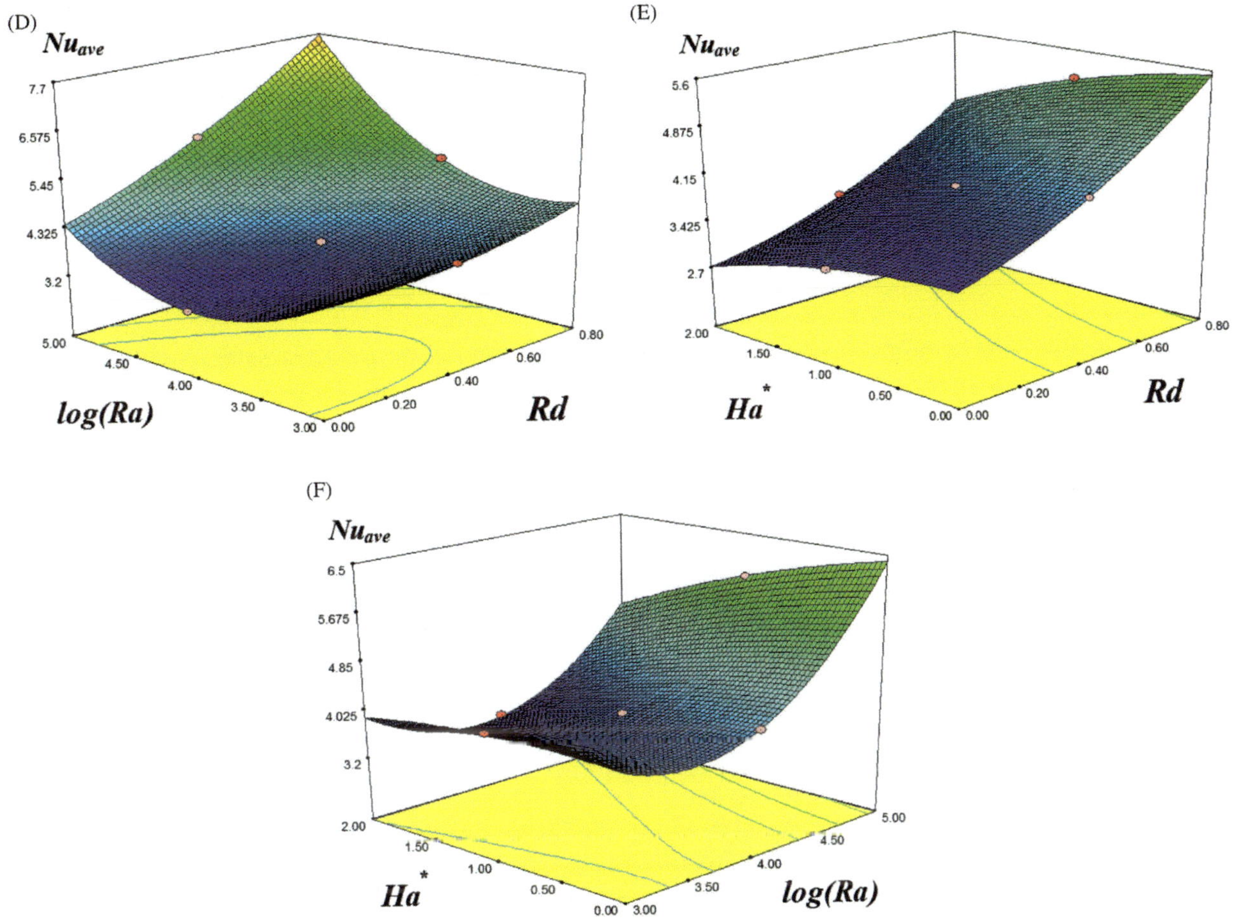

FIGURE 20.27 (Continued).

Vorticity and stream functions should be used to eliminate the pressure source terms:

$$\omega + \frac{\partial u}{\partial y} - \frac{\partial v}{\partial x} = 0, \quad \frac{\partial \psi}{\partial x} = -v, \quad \frac{\partial \psi}{\partial y} = u. \tag{20.101}$$

Introducing dimensionless quantities:

$$P = \frac{p}{\rho_f (\alpha_f / L)^2}, \quad U = \frac{uL}{\alpha_f}, \quad V = \frac{vL}{\alpha_f}, \quad \Theta = \frac{T - T_c}{\Delta T}, \quad \Delta T = q''L/k_f, \quad (X, Y) = \frac{(x, y)}{L} \tag{20.102}$$

then the final equations are:

$$\frac{\partial V}{\partial Y} + \frac{\partial U}{\partial X} = 0, \tag{20.103}$$

$$\begin{aligned} U\frac{\partial U}{\partial X} + \frac{\partial U}{\partial Y}V = \Pr\left[\frac{\mu_{nf}/\mu_f}{\rho_{nf}/\rho_f}\right]\left(\frac{\partial^2 U}{\partial Y^2} + \frac{\partial^2 U}{\partial X^2}\right) \\ - Ha^2 \Pr\left[\frac{\sigma_{nf}/\sigma_f}{\rho_{nf}/\rho_f}\right]\left(B_y^2 U - B_x B_y V\right) - \frac{\partial P}{\partial X}, \end{aligned} \tag{20.104}$$

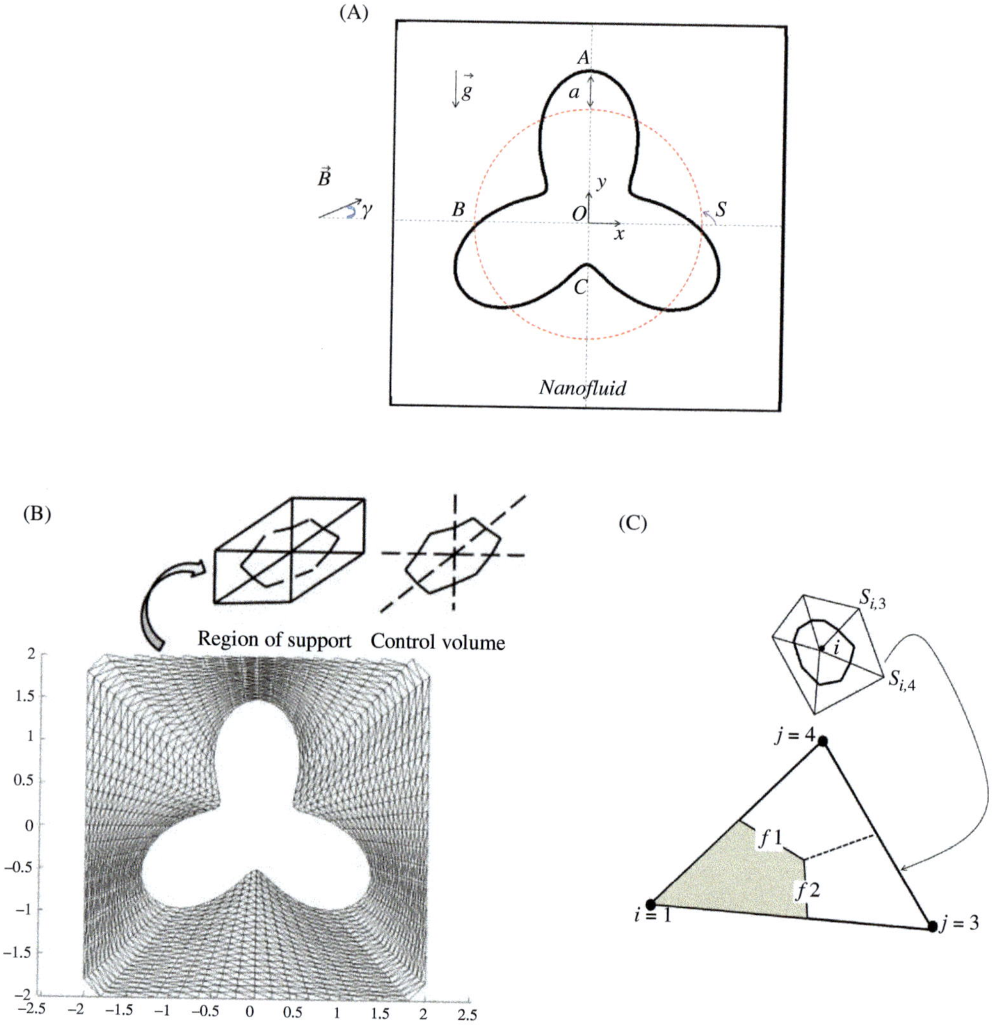

FIGURE 20.28 (A) Geometry and the boundary conditions with (B) the mesh of geometry considered in this work; (C) a sample triangular element and its corresponding control volume.

$$V\frac{\partial V}{\partial Y}+U\frac{\partial V}{\partial X}=\Pr\left(\frac{\partial^2 V}{\partial Y^2}+\frac{\partial^2 V}{\partial X^2}\right)\left[\frac{\mu_{nf}/\mu_f}{\rho_{nf}/\rho_f}\right] - Ha^2\Pr\left[\frac{\sigma_{nf}/\sigma_f}{\rho_{nf}/\rho_f}\right]\left(B_x^2V - B_xB_yU\right) - \frac{\partial P}{\partial Y} + Ra\Pr\left[\frac{\beta_{nf}}{\beta_f}\right]\Theta, \tag{20.105}$$

$$\frac{\partial \Theta}{\partial X}U+\frac{\partial \Theta}{\partial Y}V=\left(1+\frac{4}{3}\left(\frac{k_{nf}}{k_f}\right)^{-1}Rd\right)\frac{\partial^2\Theta}{\partial Y^2}+\left[\frac{(\rho C_P)_f}{(\rho C_P)_{nf}}\frac{k_{nf}}{k_f}\right]\left(\frac{\partial^2\Theta}{\partial X^2}\right) \tag{20.106}$$

where dimensionless and constants parameters are:

$$\Pr=\upsilon_f/\alpha_f,\ Ra=g\beta_f q''L^4/(k_f\upsilon_f\alpha_f),\ Ha=LB_0\sqrt{\sigma_f/\mu_f},\ Rd=4\sigma_e T_c^3/(\beta_R k_f), \tag{20.107}$$

and boundary conditions are:

$$\begin{aligned} \frac{\partial\Theta}{\partial n}&=1.0 \quad \text{on inner wall}\\ \Theta&=0.0 \quad \text{on outer wall}\\ \psi&=0.0 \quad \text{on all walls} \end{aligned} \tag{20.108}$$

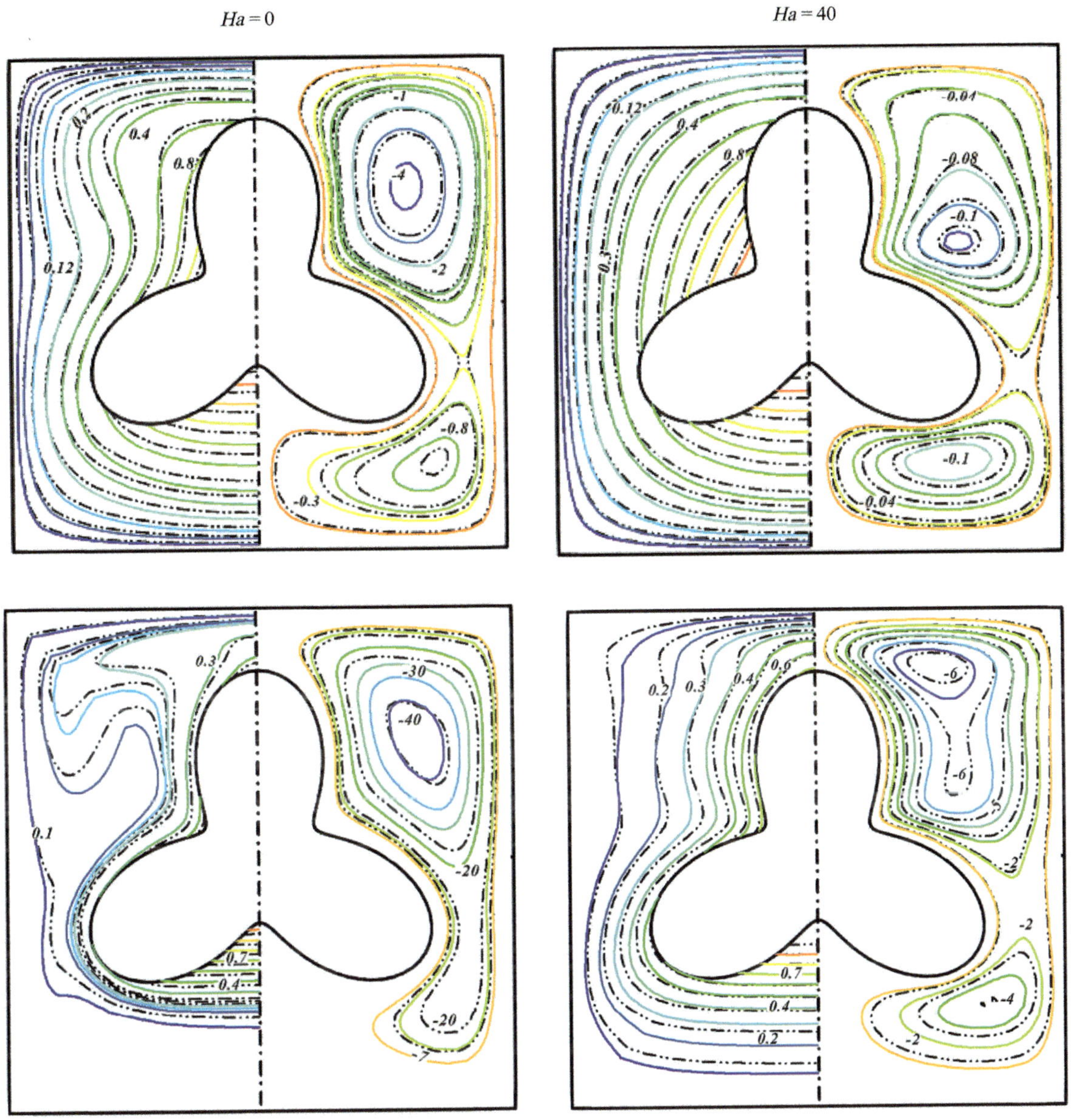

FIGURE 20.29 Effect of Rayleigh number, Hartmann number, and radiation parameter on Isotherms (left) and streamlines (right) contours ($Rd = 0$ (— · · —), $Rd = 0.8$(—)) when $N = 3, \phi = 0.04$.

Local and average Nusselt over the hot cylinder are determined from:

$$Nu_{loc} = \left(1 + \frac{4}{3}Rd\left(\frac{k_{nf}}{k_f}\right)^{-1}\right)\frac{1}{\theta}\left(\frac{k_{nf}}{k_f}\right) \tag{20.109}$$

$$Nu_{ave} = \frac{1}{S}\int_0^S Nu_{loc}\, ds \tag{20.110}$$

20.6.3 Effects of Active Parameters

MHD Fe_3O_4-water nanofluid natural convection with radiation source term is examined. MFD viscosity is taken into consideration. Numerical simulations are examined for various values of the radiation parameter ($Rd = 0$ to 0.8), volume fraction of Fe_3O_4-water ($\phi = 0$ to 0.04), Rayleigh number ($Ra = 10^3$, 10^4 and 10^5), number of undulations ($N = 3,4$ and 5), and Hartmann number ($Ha = 0$ to 40).

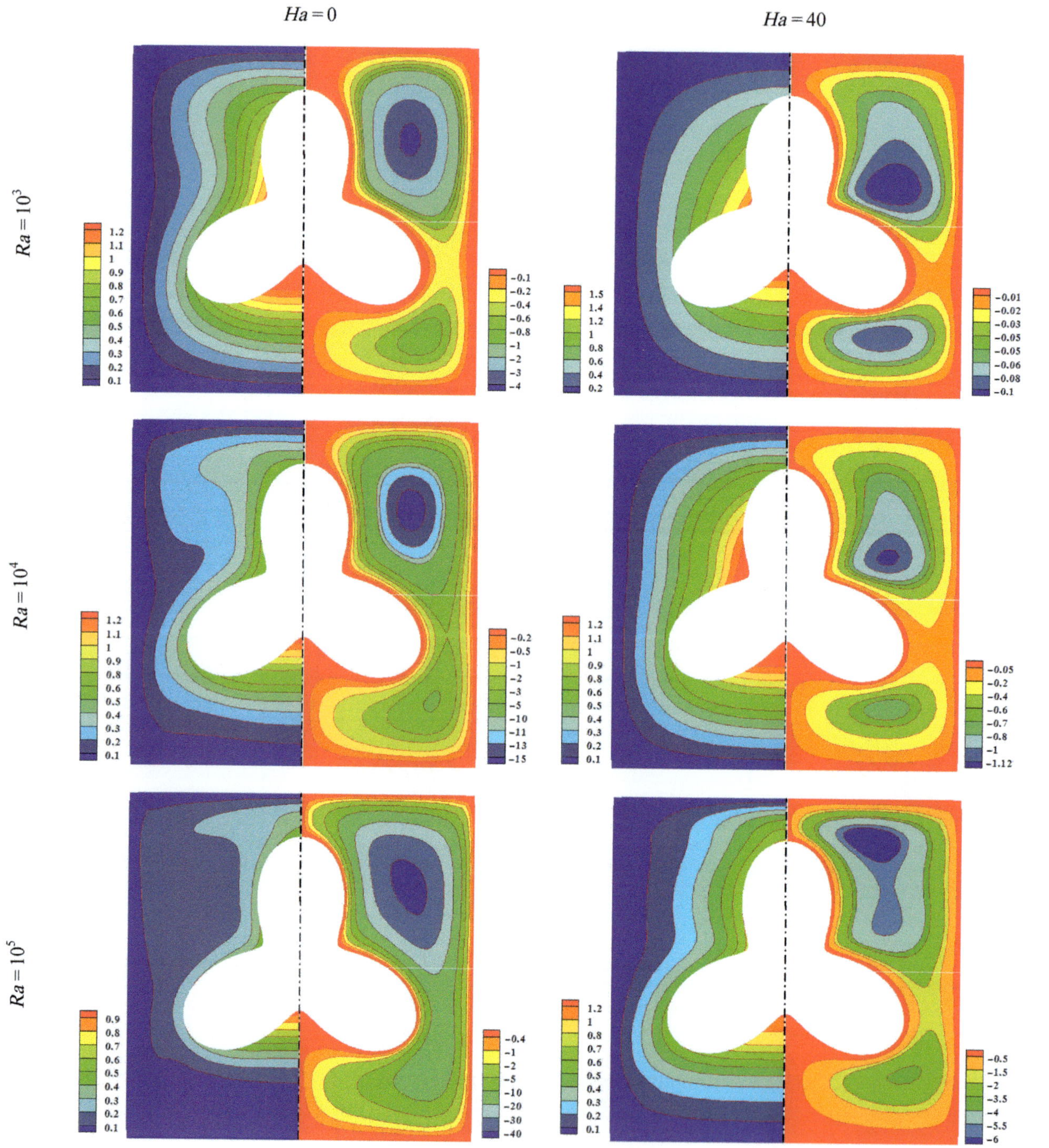

FIGURE 20.30 Effect of Rayleigh and Hartmann numbers on isotherm (left) and streamline (right) contours when $N = 3$, $\phi = 0.04$, $Rd = 0.8$.

The effect of radiation parameter on nanofluid hydrothermal behavior is illustrated in Fig. 20.29. The inner wall temperature increases with increasing Rd. Also increasing Rd results in increasing nanofluid motion. This influence is more evident when convection flow is dominant.

The influence of N, Ha, and Ra on streamlines and isotherms are depicted in Figs. 20.30–20.32. As nanofluid temperature increases, the nanofluid begins moving from the warm surface to the outer one and dropping along the circular cylinder. At low Rayleigh number, conduction is more significant than convection.$|\Psi_{max}|$ increases as

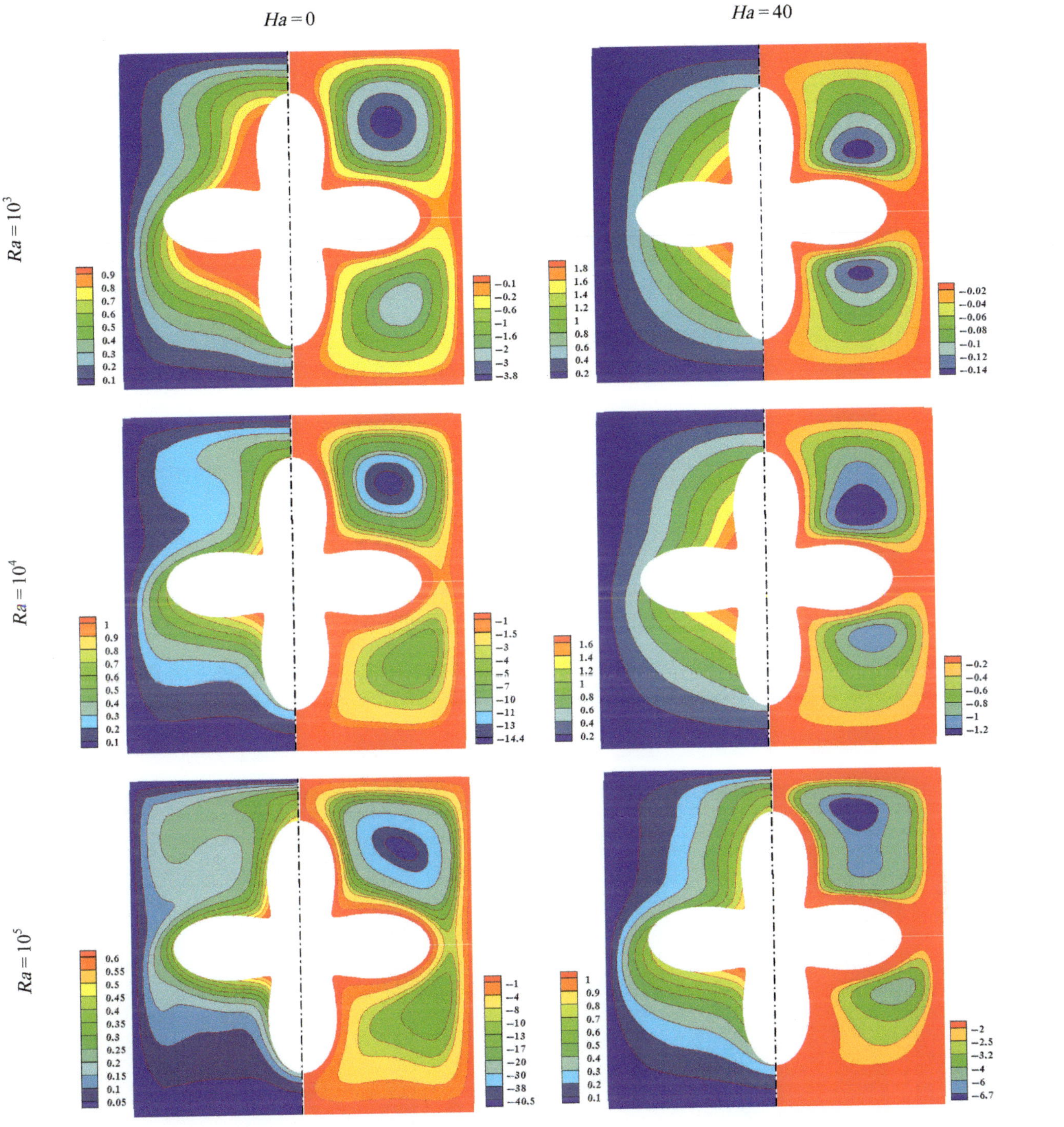

FIGURE 20.31 Effect of Rayleigh and Hartmann numbers on isotherm (left) and streamline (right) contours when $N = 4, \phi = 0.04, Rd = 0.8$.

buoyancy force increases and it decreases as the Lorentz force increases. As the Hartmann number increases, the center of the vortices moves to $x = 0$. As Ra increases, convection becomes stronger. Thus $|\Psi_{\max}|$ and the temperature gradient increase with increasing Ra. As N increases, the distortion of isotherms reduces and vortices become weaker. Increasing the Hartmann number leads to a reduction in the thermal plume and the heat transfer mechanism changes from convection to conduction. So, Nu decreases with increasing Lorentz forces. At $N = 5$, increasing the Lorentz force causes the main vortices to convert into three smaller ones. As the Lorentz force increases, the two upper vortices merge together and the thermal plume vanishes.

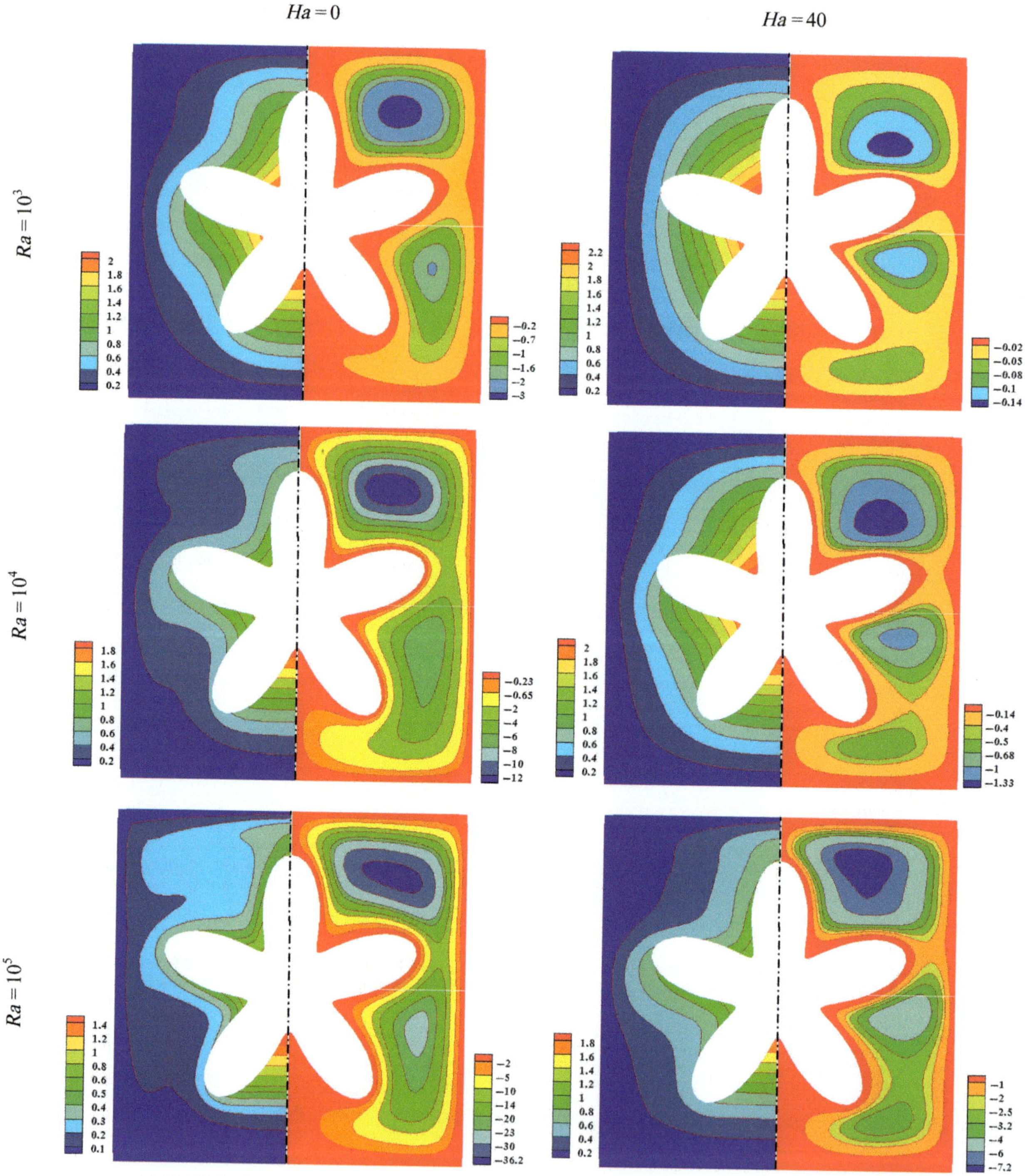

FIGURE 20.32 Effect of Rayleigh and Hartmann numbers on isotherm (left) and streamline (right) contours when $N = 5, \phi = 0.04, Rd = 0.8$.

References

[1] R.E. Rosensweig, Ferrohydrodynamics, Cambridge University Press, Cambridge, London, 1985.
[2] R. Ganguly, S. Sen, I.K. Puri, Heat transfer augmentation using a magnetic fluid under the influence of a line dipole, J. Magn. Magn. Mater. 271 (2004) 63–73.
[3] M.I. Shliomis, Effect of viscosity of magnetic suspensions, Soviet Phys. FETP 34 (1972) 1291–1294.
[4] G. Vaidyanathan, R. Sekar, A. Ramanathan, Effect of magnetic field dependent viscosity on ferroconvection in rotating medium, Ind. J. Pure Appl. Phys. 40 (2002) 159–165.
[5] C.E. Nanjundappa, I.S. Shivakumara, M. Ravisha, The onset of ferroconvection in a horizontal saturated porous layer heated from below and cooled from above with constant heat flux subject to MFD viscosity, Int. Commun. Heat Mass Transfer 37 (2010) 1246–1250.

[6] M. Sheikholeslami, M.M. Rashidi, T. Hayat, D.D. Ganji, Free convection of magnetic nanofluid considering MFD viscosity effect, J. Mol. Liq. 218 (2016) 393–399.
[7] M. Sheikholeslami, Magnetic field influence on nanofluid thermal radiation in a cavity with tilted elliptic inner cylinder, J. Mol. Liq. 229 (2017) 137–147.
[8] M. Sheikholeslami, S.A. Shehzad, Thermal radiation of ferrofluid in existence of Lorentz forces considering variable viscosity, Int. J. Heat Mass Transfer 109 (2017) 82–92.
[9] M. Sheikholeslami, H.B. Rokni, Magnetic nanofluid natural convection in presence of thermal radiation considering variable viscosity, Europ. Phys. J. Plus 132 (2017) 238. Available from: https://doi.org/10.1140/epjp/i2017-11498-4.
[10] M. Sheikholeslami, S. Abelman, Numerical study of the effect of magnetic field on Fe_3O_4-water ferrofluid convection with thermal radiation, Eng. Comput.
[11] M. Sheikholeslami, T. Hayat, A. Alsaedi, On simulation of nanofluid radiation and natural convection in an enclosure with elliptical cylinders, Int. J. Heat Mass Transfer 115 (2017) 981–991.
[12] M. Sheikholeslami, M.K. Sadoughi, Numerical modeling for Fe_3O_4 -water nanofluid flow in porous medium considering MFD viscosity, J. Mol. Liq. 242 (2017) 255–264.
[13] M. Sheikholeslami, Influence of Lorentz forces on nanofluid flow in a porous cylinder considering Darcy model, J. Mol. Liq. 225 (2017) 903–912.
[14] M. Sheikholeslami, CVFEM for magnetic nanofluid convective heat transfer in a porous curved enclosure, Eur. Phys. J. Plus 131 (2016) 413. Available from: https://doi.org/10.1140/epjp/i2016-16413-y.
[15] M. Sheikholeslami, Influence of Coulomb forces on Fe_3O_4-H_2O nanofluid thermal improvement, Int. J. Hydrogen Energy 42 (2017) 821–829.
[16] M. Sheikholeslami, Numerical simulation of magnetic nanofluid natural convection in porous media, Phys. Lett. A 381 (2017) 494–503.
[17] M. Sheikholeslami, Magnetic source impact on nanofluid heat transfer using CVFEM, Neur. Comput. Applicat. (2016). Available from: https://doi.org/10.1007/s00521-016-2740-7.
[18] M. Sheikholeslami, CuO-water nanofluid free convection in a porous cavity considering Darcy law, Europ. Phys. J. Plus 132 (2017) 55. Available from: https://doi.org/10.1140/epjp/i2017-11330-3.
[19] M. Sheikholeslami, Numerical investigation of MHD nanofluid free convective heat transfer in a porous tilted enclosure, Eng. Comput. 34 (6) (2017) 1939–1955.
[20] M.S. Kandelousi, Effect of spatially variable magnetic field on ferrofluid flow and heat transfer considering constant heat flux boundary condition, Europ. Phys. J. Plus (2014) 129–248. Available from: https://doi.org/10.1140/epjp/i2014-14248-2.
[21] M. Sheikholeslami, Numerical simulation for external magnetic field influence on Fe_3O_4-water nanofluid forced convection, Eng. Comput., Inpress.
[22] M. Sheikholeslami, Influence of Lorentz forces on nanofluid flow in a porous cavity by means of Non- Darcy model, Eng. Comput. 34 (8) (2017) 2651–2667. Available from: https://doi.org/10.1108/EC-01-2017-0008.
[23] M. Sheikholeslami, H.B. Rokni, Nanofluid convective heat transfer intensification in a porous circular cylinder, Chem. Eng. Proc. Process Intensif. 120 (2017) 93–104.
[24] M. Sheikholeslami, M. Shamlooei, Magnetic source influence on nanofluid flow in porous medium considering shape factor effect, Phys. Lett. A 381 (2017) 3071–3078.
[25] M. Sheikholeslami, M. Shamlooei, Convective flow of nanofluid inside a lid driven porous cavity using CVFEM, Phys. B: Condensed Matter 521 (2017) 239–250.
[26] M. Sheikholeslami, K. Vajravelu, Forced convection heat transfer in Fe_3O_4- ethylene glycol nanofluid under the influence of Coulomb force, J. Mol. Liq. 233 (2017) 203–210.
[27] M. Sheikholeslami, H.B. Rokni, Influence of EFD viscosity on nanofluid forced convection in a cavity with sinusoidal wall, J. Mol. Liq. 232 (2017) 390–395.
[28] M. Sheikholeslami, H.B. Rokni, Magnetic nanofluid natural convection in the presence of thermal radiation considering variable viscosity, Europ. Phys. J. Plus 132 (2017) 238. Available from: https://doi.org/10.1140/epjp/i2017-11498-4.
[29] M. Sheikholeslami, H.B. Rokni, Numerical simulation for impact of Coulomb force on nanofluid heat transfer in a porous enclosure in presence of thermal radiation, Int. J. Heat Mass Transfer 118 (2018) 823–831.
[30] M. Sheikholeslami, M. Shamlooei, R. Moradi, Fe_3O_4- Ethylene glycol nanofluid forced convection inside a porous enclosure in existence of Coulomb force, J. Mol. Liq. 249 (2018) 429–437.
[31] M. Sheikholeslami, S.A. Shehzad, Numerical analysis of Fe_3O_4 –H_2O nanofluid flow in permeable media under the effect of external magnetic source, Int. J. Heat Mass Transfer 118 (2018) 182–192.
[32] M. Sheikholeslami, S.A. Shehzad, Magnetohydrodynamic nanofluid convection in a porous enclosure considering heat flux boundary condition, Int. J. Heat Mass Transfer 106 (2017) 1261–1269.
[33] M. Sheikholeslami, H.B. Rokni, Numerical modeling of nanofluid natural convection in a semi annulus in existence of Lorentz force, Comput. Methods Appl. Mech. Eng. 317 (2017) 419–430.
[34] M. Sheikholeslami, K. Vajravelu, Nanofluid flow and heat transfer in a cavity with variable magnetic field, Appl. Math. Comput. 298 (2017) 272–282.
[35] M. Sheikholeslami, H.B. Rokni, Magnetohydrodynamic CuO-water nanofluid in a porous complex shaped enclosure, ASME, J. Thermal Sci. Eng. Applicat. 9 (4) (2017) 041007. Available from: https://doi.org/10.1115/1.4035973.
[36] M. Sheikholeslami, H.B. Rokni, Nanofluid two phase model analysis in existence of induced magnetic field, Int. J. Heat Mass Transfer 107 (2017) 288–299.
[37] M. Sheikholeslami, T. Hayat, A. Alsaedi, S. Abelman, Numerical analysis of EHD nanofluid force convective heat transfer considering electric field dependent viscosity, Int. J. Heat Mass Transfer 108 (2017) 2558–2565.
[38] M. Sheikholeslami, Magnetic field influence on CuO -H_2O nanofluid convective flow in a permeable cavity considering various shapes for nanoparticles, Int. J. Hydrogen Energy 42 (2017) 19611–19621.

[39] M. Sheikholeslami, M.M. Bhatti, Forced convection of nanofluid in presence of constant magnetic field considering shape effects of na particles, Int. J. Heat Mass Transfer 111 (2017) 1039–1049.

[40] M. Sheikholeslami, A. Zeeshan, Analysis of flow and heat transfer in water based nanofluid due to magnetic field in a porous enclos with constant heat flux using CVFEM, Comput. Methods Appl. Mech. Eng. 320 (2017) 68–81.

[41] M. Sheikholeslami, M. Shamlooei, Fe_3O_4- H_2O nanofluid natural convection in presence of thermal radiation, Int. J. Hydrogen Energy (9) (2017) 5708–5718.

[42] M. Sheikholeslami, M.M. Bhatti, Active method for nanofluid heat transfer enhancement by means of EHD, Int. J. Heat Mass Tran 109 (2017) 115–122.

[43] M. Sheikholeslami, T. Hayat, A. Alsaedi, Numerical study for external magnetic source influence on water based nanofluid convec heat transfer, Int. J. Heat Mass Transfer 106 (2017) 745–755.

[44] M. Sheikholeslami, A.J. Chamkha, Flow and convective heat transfer of a ferro-nanofluid in a double-sided lid-driven cavity wit wavy wall in the presence of a variable magnetic field, Numer. Heat Transfer, Part A 69 (10) (2016) 1186–1200. Available from: https doi.org/10.1080/10407782.2015.1125709.

[45] M. Sheikholeslami, A.J. Chamkha, Electrohydrodynamic free convection heat transfer of a nanofluid in a semi-annulus enclosure wit sinusoidal wall, Numer. Heat Transfer, Part A 69 (7) (2016) 781–793. Available from: http://dx.doi.org/10.1080/10407782.2015.109081

[46] M. Sheikholeslami, M.K. Sadoughi, Simulation of CuO- water nanofluid heat transfer enhancement in presence of melting surface, Int Heat Mass Transfer 116 (2018) 909–919.

[47] M. Sheikholeslami, H.B. Rokni, Simulation of nanofluid heat transfer in presence of magnetic field: a review, Int. J. Heat Mass Trans 115 (2017) 1203–1233.

[48] M. Sheikholeslami, H.B. Rokni, Melting heat transfer influence on nanofluid flow inside a cavity in existence of magnetic field, Int Heat Mass Transfer 114 (2017) 517–526.

[49] M. Sheikholeslami, S.A. Shehzad, Magnetohydrodynamic nanofluid convective flow in a porous enclosure by means of LBM, Int. J. He Mass Transfer 113 (2017) 796–805.

[50] M. Sheikholeslami, Magnetohydrodynamic nanofluid forced convection in a porous lid driven cubic cavity using Lattice Boltzma Method, J. Mol. Liq. 231 (2017) 555–565.

[51] M. Sheikholeslami Kandelousi, KKL correlation for simulation of nanofluid flow and heat transfer in a permeable channel, Phys. Lett. 378 (45) (2014) 3331–3339.

[52] M. Sheikholeslami, Effect of uniform suction on nanofluid flow and heat transfer over a cylinder, J. Braz. Soc. Mech. Sci. Eng. 37 (201 1623–1633.

[53] M. Sheikholeslami, Lattice Boltzmann Method simulation of MHD non-Darcy nanofluid free convection, Physica B 516 (2017) 55–71.

[54] M. Sheikholeslami, Influence of magnetic field on nanofluid free convection in an open porous cavity by means of Lattice Boltzmar Method, J. Mol. Liq. 234 (2017) 364–374.

[55] M. Sheikholeslami, A.J. Chamkha, Influence of Lorentz forces on nanofluid forced convection considering Marangoni convection, J. Mo Liq. 225 (2017) 750–757.

[56] M. Sheikholeslami, T. Hayat, A. Alsaedi, Numerical simulation of nanofluid forced convection heat transfer improvement in existence c magnetic field using Lattice Boltzmann Method, Int. J. Heat Mass Transfer 108 (2017) 1870–1883.

[57] M. Sheikholeslami, R. Ellahi, Three dimensional mesoscopic simulation of magnetic field effect on natural convection of nanofluid, Int. Heat Mass Transfer 89 (2015) 799–808.

[58] M. Sheikholeslami, Shirley Abelman, Two phase simulation of nanofluid flow and heat transfer in an annulus in the presence of an axia magnetic field, IEEE Trans. Nanotechnol. 14 (3) (2015) 561–569.

[59] M. Sheikholeslami, R. Ellahi, Simulation of ferrofluid flow for magnetic drug targeting using Lattice Boltzmann method, J. Zeitschrift Fu Naturforschung A 70 (2) (2015) 115–124.

[60] M. Sheikholeslami, M. Seyednezhad, Lattice Boltzmann Method simulation for CuO-water nanofluid flow in a porous enclosure with ho obstacle, J. Mol. Liq. 243 (2017) 249–256.

[61] M. Sheikholeslami, M. Sadoughi, Mesoscopic method for MHD nanofluid flow inside a porous cavity considering various shapes o nanoparticles, Int. J. Heat Mass Transfer 113 (2017) 106–114.

[62] M. Sheikholeslami, A. Zeeshan, Mesoscopic simulation of CuO-H_2O nanofluid in a porous enclosure with elliptic heat source, Int. J. Hydrogen Energy 42 (22) (2017) 15393–15402.

[63] M. Sheikholeslami, S.A. Shehzad, CVFEM for influence of external magnetic source on $Fe_3O_4 - H_2O$ nanofluid behavior in a permeable cavity considering shape effect, Int. J. Heat Mass Transfer 115 (2017) 180–191.

[64] M. Sheikholeslami, M. Seyednezhad, Nanofluid heat transfer in a permeable enclosure in presence of variable magnetic field by means of CVFEM, Int. J. Heat Mass Transfer 114 (2017) 1169–1180.

[65] M. Sheikholeslami, H.B. Rokni, Free convection of CuO-H_2O nanofluid in a curved porous enclosure using mesoscopic approach, Int. J. Hydrogen Energy 42 (22) (2017) 14942–14949.

[66] M. Sheikholeslami, Numerical investigation for CuO-H_2O nanofluid flow in a porous channel with magnetic field using mesoscopic method, J. Mol. Liq. 249 (2018) 739–746.

[67] M. Sheikholeslami, H.B. Rokni, Influence of melting surface on MHD nanofluid flow by means of two phase model, Chin. J. Phys. 55 (2017) 1352–1360.

[68] M. Sheikholeslami, H.B. Rokni, Effect of melting heat transfer on nanofluid flow in existence of magnetic field considering Buongiorno Model, Chin. J. Phys. 55 (2017) 1115–1126.

[69] M. Sheikholeslami, T. Hayat, A. Alsaedi, MHD free convection of Al_2O_3–water nanofluid considering thermal radiation: a numerical study, Int. J. Heat Mass Transfer 96 (2016) 513–524.

[70] M. Sheikholeslami, K. Vajravelu, M.M. Rashidi, Forced convection heat transfer in a semi annulus under the influence of a variable magnetic field, Int. J. Heat Mass Transfer 92 (2016) 339–348.

[71] M. Sheikholeslami, S.A. Shehzad, CVFEM simulation for nanofluid migration in a porous medium using Darcy model, Int. J. Heat Mass Transfer 122 (2018) 1264–1271.
[72] M. Sheikholeslami, M. Darzi, M.K. Sadoughi, Heat transfer improvement and Pressure Drop during condensation of refrigerant-based Nanofluid; An Experimental Procedure, Int. J. Heat Mass Transfer 122 (2018) 643–650.
[73] M. Sheikholeslami, H.B. Rokni, CVFEM for effect of Lorentz forces on nanofluid flow in a porous complex shaped enclosure by means of non-equilibrium model, J. Mol. Liq. 254 (2018) 446–462.
[74] M. Sheikholeslami, S.A. Shehzad, Z. Li, Water based nanofluid free convection heat transfer in a three dimensional porous cavity with hot sphere obstacle in existence of Lorenz forces, Int. J. Heat Mass Transfer 125 (2018) 375–386.
[75] M. Sheikholeslami, M. Jafaryar, D.D. Ganji, Z. Li, Exergy loss analysis for nanofluid forced convection heat transfer in a pipe with modified turbulators, J. Mol. Liq. 262 (2018) 104–110.
[76] M. Sheikholeslami, M. Jafaryar, Z. Li, Nanofluid turbulent convective flow in a circular duct with helical turbulators considering CuO nanoparticles, Int. J. Heat Mass Transfer 124 (2018) 980–989.
[77] M. Sheikholeslami, Z. Li, M. Shamlooei, Nanofluid MHD natural convection through a porous complex shaped cavity considering thermal radiation, Physics Letters A 382 (2018) 1615–1632.
[78] M. Sheikholeslami, H.B. Rokni, Magnetic nanofluid flow and convective heat transfer in a porous cavity considering Brownian motion effects, Phy. Fluids, 30 (1), Available from: https://doi.org/10.1063/1.5012517.
[79] M. Sheikholeslami, S.A. Shehzad, Simulation of water based nanofluid convective flow inside a porous enclosure via non-equilibrium model, Int. J. Heat Mass Transfer 120 (2018) 1200–1212.
[80] M. Sheikholeslami, S.A. Shehzad, Non-Darcy free convection of Fe_3O_4-water nanoliquid in a complex shaped enclosure under impact of uniform Lorentz force, Chin. J. Phys. 56 (2018) 270–281.
[81] M. Sheikholeslami, M. Seyednezhad, Simulation of nanofluid flow and natural convection in a porous media under the influence of electric field using CVFEM, Int. J. Heat Mass Transfer 120 (2018) 772–781.
[82] M. Sheikholeslami, M. Shamlooei, R. Moradi, Numerical simulation for heat transfer intensification of nanofluid in a permeable curved enclosure considering shape effect of Fe_3O_4 nanoparticles, Chem. Eng. Process.: Process Intensif. 124 (2018) 71–82.
[83] M. Sheikholeslami, T. Hayat, T. Muhammad, A. Alsaedi, MHD forced convection flow of nanofluid in a porous cavity with hot elliptic obstacle by means of Lattice Boltzmann method, Int. J. Mech. Sci. 135 (2018) 532–540.
[84] M. Sheikholeslami, Numerical investigation of nanofluid free convection under the influence of electric field in a porous enclosure, J. Mol. Liq. 249 (2018) 1212–1221.
[85] M. Sheikholeslami, CuO-water nanofluid flow due to magnetic field inside a porous media considering Brownian motion, J. Mol. Liq. 249 (2018) 921–929.
[86] M. Sheikholeslami, D.D. Ganji, Influence of electric field on Fe_3O_4-water nanofluid radiative and convective heat transfer in a permeable enclosure, J. Mol. Liq. 250 (2018) 404–412.
[87] M. Sheikholeslami, S.A. Shehzad, Z. Li, Nanofluid heat transfer intensification in a permeable channel due to magnetic field using Lattice Boltzmann method, Phys. B, 542 (2018) 51–58. Available from: https://doi.org/10.1016/j.physb.2018.03.036.
[88] M. Sheikholeslami, A. Zeeshan, Numerical simulation of Fe_3O_4 -water nanofluid flow in a non-Darcy porous media, Int. J. Numer. Methods Heat Fluid Flow 28 (3) (2018) 641–660. Available from: https://doi.org/10.1108/HFF-04-2017-0160.
[89] M. Sheikholeslami, D.D. Ganji, Numerical approach for magnetic nanofluid flow in a porous cavity using CuO nanoparticles, Mate. Des. 120 (2017) 382–393.
[90] M. Sheikholeslami, D.D. Ganji, Numerical modeling of magnetohydrodynamic CuO -water transportation inside a porous cavity considering shape factor effect, Colloids Surf., A 529 (2017) 705–714.
[91] M. Sheikholeslami, T. Hayat, A. Alsaedi, Numerical simulation for forced convection flow of MHD CuO-H_2O nanofluid inside a cavity by means of LBM, J. Mol. Liq. 249 (2018) 941–948.
[92] M. Sheikholeslami, H.R. Kataria, A.S. Mittal, Effect of thermal diffusion and heat-generation on MHD nanofluid flow past an oscillating vertical plate through porous medium, J. Mol. Liq. 257 (2018) 12–25.
[93] M. Sheikholeslami, M. Barzegar Gerdroodbary, S. Valiallah Mousavi, D.D. Ganji, R. Moradi, Heat transfer enhancement of ferrofluid inside an 90o elbow channel by non-uniform magnetic field, J. Magn. Magn. Mater. 460 (2018) 302–311.
[94] M. Sheikholeslami, S. Soleimani, D.D. Ganji, Effect of electric field on hydrothermal behavior of nanofluid in a complex geometry, J. Mol. Liq. 213 (2016) 153–161.
[95] M. Sheikholeslami, Numerical simulation for solidification in a LHTESS by means of Nano-enhanced PCM, J. Taiwan Inst. Chem. Eng. 86 (2018) 25–41.
[96] M. Sheikholeslami, M. Darzi, Z. Li, Experimental investigation for entropy generation and exergy loss of nano-refrigerant condensation process, Int. J. Heat Mass Transfer 125 (2018) 1087–1095.
[97] M. Sheikholeslami, Numerical modeling of Nano enhanced PCM solidification in an enclosure with metallic fin, J. Mol. Liq. 259 (2018) 424–438.
[98] M. Sheikholeslami, A. Ghasemi, Solidification heat transfer of nanofluid in existence of thermal radiation by means of FEM, Int. J. Heat Mass Transfer 123 (2018) 418–431.

Nomenclature

	amplitude
	magnetic induction
	specific heat at constant pressure
$_B$	Brownian diffusion coefficient
$_T$	thermophoretic diffusion coefficient
	Eckert number
n	heat transfer enhancement
	gravitational acceleration vector
$_x, H_y$	components of the magnetic field intensity
	the magnetic field strength
a	Hartmann number
	thermal conductivity
	gap between inner and outer boundary of the enclosure
e	Lewis number
An_F	magnetic number arising from FHD
	magnetization
	number of undulations
b	Brownian motion parameter
t	thermophoretic parameter
u	Nusselt number
r	Prandtl number
	nondimensional radial distance
Ra	thermal Rayleigh number
Nr	Buoyancy ratio number
T	Fluid temperature
$T_c^{'}$	curie temperature
u, v	velocity components in the x-direction and y-direction
U, V	dimensionless velocity components in the X-direction and Y-direction
$\vec{g}$	gravitational acceleration vector
q''	heat flux
x, y	space coordinates

GREEK SYMBOLS

ζ	angle measured from the isulated right plane
α	thermal diffusivity
ϕ	volume fraction
μ	dynamic viscosity
υ	kinematic viscosity
ψ & Ψ	stream function and dimensionless stream function

Θ dimensionless temperature
ρ fluid density
β thermal expansion coefficient
μ_0 magnetic permeability of vacuum

SUBSCRIPTS

c cold
h hot
loc local
ave average
nf nanofluid
f base fluid
p solid particles

Appendix: A CVFEM Code for Lid Driven Cavity

```
implicit none
!!!!!!!!!!!!================Lid Driven Cavity==========================!!!!!!!!!!!!!!!!
!!================ Mohsen Sheikholeslami and Soheil Soleimani ===============!!!!!!!!!!!
integer rows,n,k1,k2,itert,nstep
integer cols,ii,iter,loopcount
integer i,j,k,n_seg,iterv,iters
parameter (cols=51,rows=51,n=rows*cols,n_seg=4)
integer n_s(1:n),n_b(1:n_seg),B(1:n_seg,1:max(rows,cols)),S(1:n,1:12)
real*8 dx,x(1:n),y(1:n),vor(1:n),sf(1:n),t(1:n)
real*8 dy,ulid,aps(1:n),apv(1:n),as(1:n,12),av(1:n,12)
real*8 V(1:n),xL(1:3),yL(1:3),RHS,BC(1:n),BB(1:n),TAX,QC(1:n),QB(1:n)
real*8 diff(1:3),Vele,Nx(1:3),Ny(1:3),diff_f,vxf,vyf,delx,dely,qf
real*8 dely2,delx2,dely1,delx1,Nu,SAX,SAY
real*8 Re,upper,lower
nstep=10000
Re=50.
ulid=1.0
dx=1./real(rows-1)
dy=1./real(cols-1)
do k=1 , rows
do j=1 , cols
i=j+(k-1)*cols
y(i)=real(k-1)/real(rows-1)
x(i)=real(j-1)/real(cols-1)
enddo
enddo
!support
do k=1,rows
do j=1,cols
i=j+(k-1)*cols
n_s(i)=0 !Nodes in support
do ii=1,12
S(i,ii)=0.0 !Global node number of ii_th node in support i
enddo
enddo
enddo
!boundary
n_b(1)=cols
n_b(2)=rows !outter radius
n_b(3)=cols
n_b(4)=rows !inner radius
do k=1,n_seg
do j=1,n_b(k)
B(k,j)=0.0 !Global node number of j_th node on i_th boundary
enddo
enddo
!SUPPORT DATA ------------
!Assumed Morphology in Fig. A1
! Support do internal points
do k=2,rows-1
do j=2,cols-1
i=j+(k-1)*cols
```

```
n_s(i)=6
S(i,1)=i+1 !assumed morphology do support see Fig. A1
S(i,2)=i+1+cols
S(i,3)=i+cols
S(i,4)=i-1
S(i,5)=i-1-cols
S(i,6)=i-cols
S(i,7)=i+1 !last node is same as first node at internal nodes
enddo
enddo
!horizonal boundary (y=0) Non-corner points
do j=2,cols-1
i=j
n_s(i)=4
S(i,1)=i+1  ! arranged so that points are--
S(i,2)=i+1+cols  ! --contiguous in counter clockwise count
S(i,3)=i+cols
S(i,4)=i-1
S(i,5)=0  ! last point zero: a boundary node flag
enddo
!outter (rot) curve boundary Non-corner points
do k=2,rows-1
i=cols+(k-1)*cols
n_s(i)=4
S(i,1)=i+cols
S(i,2)=i-1
S(i,3)=i-1-cols
S(i,4)=i-cols
S(i,5)=0
enddo
! vertical boundary (x=0) Non-corner points
do j=2,cols-1
i=j+(rows-1)*cols
n_s(i)=4
S(i,1)=i-1
S(i,2)=i-1-cols
S(i,3)=i-cols
S(i,4)=i+1
S(i,5)=0
enddo
!inner (rin) curve boundary Non-corner points
do k=2,rows-1
i=1+(k-1)*cols
n_s(i)=4
S(i,1)=i-cols
S(i,2)=i+1
S(i,3)=i+1+cols
S(i,4)=i+cols
S(i,5)=0
enddo
!Corners
! x=0, y = rin
i=1
n_s(i)=3
S(i,1)=i+1
S(i,2)=i+1+cols
S(i,3)=i+cols
S(i,4)=0
! y = 0 x = rot
i=cols
n_s(i)=2
S(i,1)=i+cols
S(i,2)=i-1
```

```
S(i,3)=0
! x =0 y = rot
i=rows*cols
n_s(i)=3
S(i,1)=i-1
S(i,2)=i-1-cols
S(i,3)=i-cols
S(i,4)=0
! x=0 y = rin
i=1+(rows-1)*cols
n_s(i)=2
S(i,1)=i-cols
S(i,2)=i+1
S(i,3)=0
!END SUPPORT DATA-----------
! BOUNDARY DATA------------
!NOTE counter-clockwise numbering and boundary segment index
! y=0 boundary (segment 1)
do i=1,n_b(1)
B(1,i)=i
enddo
! r=rout boundary (segment 2)
do i=1,n_b(2)
B(2,i)=i*cols
enddo
! x =0 boundary (segment 3)
do i=1,n_b(3)
B(3,i)=rows*cols+1-i
enddo
! r=rin boundary (segment 4)
do i=1,n_b(4)
B(4,i)=1+rows*cols-i*cols
enddo
!==============================================
!==============initial conditions========!
!==============================================
do i=1,n
vor(i)=0.0
sf(i)=0.0
enddo
do i=1,n
aps(i)=0.0
apv(i)=0.0
do j=1,n_s(i)+1
as(i,j)=0.0
av(i,j)=0.0
enddo
enddo
do iter=1,nstep
print*,iter
!=====================================
! calculating the vorticity======
!=====================================
do i=1,n
apv(i)=0.0
V(i)=0.0
do j=1,n_s(i)+1
av(i,j)=0.0
enddo
loopcount=n_s(i)
if (S(i,n_s(i)+1)==0) then
loopcount=n_s(i)-1
endif
```

```
do j=1,loopcount
k1=S(i,j) !The numbers of the nodes in the j_th elemnt are
k2=S(i,j+1) ! i, k1=S(i,J), and k2=S(i,j+1)
xL(1) = x(i) !xL=local x-coordinate of a node within a element
xL(2) = x(k1)
xL(3) = x(k2)
yL(1) = y(i) !yL=local y-coordinate of a node within a element
yL(2) = y(k1)
yL(3) = y(k2)
! nodal diffusivity (kappa)
diff(1)=1/Re
diff(2)=1/Re
diff(3)=1/Re
! Vele= element-volume
Vele=(xL(2)*yL(3)-xL(3)*yL(2)-xL(1)*yL(3)+xL(1)*yL(2)+yL(1)*xL(3)-yL(1)*xL(2))/2.
V(i)=V(i)+Vele/3 !Contribution to control volume
Nx(1)= (yL(2)-yL(3))/(2.*Vele) !derivative of shape function wrt x
Nx(2)= (yL(3)-yL(1))/(2.*Vele)
Nx(3)= (yL(1)-yL(2))/(2.*Vele)
Ny(1)=-(xL(2)-xL(3))/(2.*Vele) !derivative of shape function wrt y
Ny(2)=-(xL(3)-xL(1))/(2.*Vele)
Ny(3)=-(xL(1)-xL(2))/(2.*Vele)
! Face 1--values of diffusivity and velocity
diff_f=(5.*diff(1)+5.*diff(2)+2.*diff(3))/12.
vxf= Ny(1)*sf(i)+Ny(2)*sf(k1)+Ny(3)*sf(k2)
vyf= -(Nx(1)*sf(i)+Nx(2)*sf(k1)+Nx(3)*sf(k2))
! Face 1--face area normal components
delx= ((xL(1)+xL(2)+xL(3))/3.)-((xL(1)+xL(2))/2.)
dely= ((yL(1)+yL(2)+yL(3))/3.)-((yL(1)+yL(2))/2.)
qf=vxf*dely-vyf*delx !Face 1 volume flux
! Face 1--Diffusion Coefficients
apv(i) = apv(i) + diff_f*(- Nx(1) * dely + Ny(1) * delx)
av(i,j) = av(i,j) + diff_f*( Nx(2) * dely - Ny(2) * delx)
av(i,j+1) = av(i,j+1) + diff_f*( Nx(3) * dely - Ny(3) * delx)
! Face 1--Advection Coefficients
apv(i)= apv(i) +max(qf,0.)
av(i,j)=av(i,j) +max(-qf,0.)
! Face 2--values of diffusivity and velocity
diff_f=(5*diff(1)+2*diff(2)+5*diff(3))/12.
vxf= Ny(1)*sf(i)+Ny(2)*sf(k1)+Ny(3)*sf(k2)    !!!!ro face
vyf= -1*(Nx(1)*sf(i)+Nx(2)*sf(k1)+Nx(3)*sf(k2))
! Face 2--face area normal components
delx= ((xL(1)+xL(3))/2)-((xL(1)+xL(2)+xL(3))/3.)
dely= ((yL(1)+yL(3))/2)-((yL(1)+yL(2)+yL(3))/3.)
qf=vxf*dely-vyf*delx ! Face 1 volume flux
! Face 2--Diffusion Coefficients
apv(i) = apv(i) + diff_f*(- Nx(1) * dely + Ny(1) * delx)
av(i,j) = av(i,j) + diff_f*( Nx(2) * dely - Ny(2) * delx)
av(i,j+1) = av(i,j+1) + diff_f*( Nx(3) * dely - Ny(3) * delx)
! Face 2--Advection Coefficients
apv(i)= apv(i) + max(qf,0.)
av(i,j+1) = av(i,j+1) + max(-qf,0.)
enddo !end loop on nodes/elements in support
av(i,1)=av(i,1)+av(i,n_s(i)+1)
av(i,n_s(i)+1)=0
enddo !end loop on nodes in domain
!!!!!!!!!!!!!!!!!!!!!!!!!!!!!!!!!!!!!!!!!!!!!!!!!!
!SETTING THE VORTICITY BOUNDARY==================
!!!!!!!!!!!!!!!!!!!!!!!!!!!!!!!!!!!!!!!!!!!!!!!!!!
do k=1,n_b(1)
RHS=0.0
do j=1,n_s(B(1,k))
RHS=RHS+as(B(1,k),j)*sf(S(B(1,k),j))
```

```
enddo
vor(B(1,k))= (1/V(B(1,k)))*(aps(B(1,k))*sf(B(1,k))- RHS - 0.0 )
enddo
do k=1,n_b(2)
RHS=0.0
do j=1,n_s(B(2,k))
RHS=RHS+as(B(2,k),j)*sf(S(B(2,k),j))
enddo
vor(B(2,k))= (1/V(B(2,k)))*(aps(B(2,k))*sf(B(2,k))- RHS - 0.0)
enddo
Do k=1,n_b(3)
RHS=0.0
Do j=1,n_s(B(3,k))
RHS=RHS+as(B(3,k),j)*sf(S(B(3,k),j));
end do
if (k==1) then
   upper=0
else
   upper=0.5*sqrt((x(B(3,k))-x(B(3,k-1)))**2+(y(B(3,k))-y(B(3,k-1)))**2)
end if
if (k==n_b(3)) then
   lower=0
else
   lower=0.5*sqrt((x(B(3,k))-x(B(3,k+1)))**2+(y(B(3,k))-y(B(3,k+1)))**2)
end if
vor(B(3,k))= (1/V(B(3,k)))*( aps(B(3,k))*sf(B(3,k))- RHS - (lower+upper)*ulid )
end do
do k=1,n_b(4)
RHS=0.0
do j=1,n_s(B(4,k))
RHS=RHS+as(B(4,k),j)*sf(S(B(4,k),j))
enddo
vor(B(4,k))= (1/V(B(4,k)))*( aps(B(4,k))*sf(B(4,k))- RHS - 0.0 )
enddo
!=====================================
do i=1,n
BC(i)=0.0
BB(i)=0.0
enddo
do k=1,n_b(1)
BC(B(1,k))=1.e18
BB(B(1,k))=vor(B(1,k))*1.e18
enddo
do k=1,n_b(2)
BC(B(2,k))=1.e18
BB(B(2,k))=vor(B(2,k))*1.e18
enddo
do k=1,n_b(3)
BC(B(3,k))=1.e18
BB(B(3,k))=vor(B(3,k))*1.e18
enddo
do k=1,n_b(4)
BC(B(4,k))=1.e18
BB(B(4,k))=vor(B(4,k))*1.e18
enddo
do i=1,n
QC(i)=0.0
if (S(i,n_s(i)+1).NE.0) then
QB(i)=0.0
else
QB(i)=0.0
endif
enddo
```

```
!================================

do iterv=1,1
do i= 1,n
RHS=BB(i)+QB(i)
do j=1,n_s(i)
RHS=RHS+av(i,j)*vor(S(i,j))
enddo
vor(i)=RHS/(apv(i)+BC(i)+QC(i))
enddo
enddo

!====================================
!====================================
!==SOLVING THE STREAMFUNCTION===
!====================================
!====================================
do i=1,n !loop on nodes

aps(i)=0.0
V(i)=0.0
do j=1,n_s(i)+1
as(i,j)=0.0
enddo

loopcount=n_s(i)
if (S(i,n_s(i)+1)==0) then
loopcount=n_s(i)-1
endif

do j=1,loopcount
k1=S(i,j)
k2=S(i,j+1)
xL(1) = x(i)
xL(2) = x(k1)
xL(3) = x(k2)
yL(1) = y(i)
yL(2) = y(k1)
yL(3) = y(k2)

diff(1)=1.0
diff(2)=1.0
diff(3)=1.0

Vele=(xL(2)*yL(3)-xL(3)*yL(2)-xL(1)*yL(3)+xL(1)*yL(2)+yL(1)*xL(3)-yL(1)*xL(2))/2.
V(i)=V(i)+Vele/3.
Nx(1)= (yL(2)-yL(3))/(2.*Vele)
Nx(2)= (yL(3)-yL(1))/(2.*Vele)
Nx(3)= (yL(1)-yL(2))/(2.*Vele)
Ny(1)=-(xL(2)-xL(3))/(2.*Vele)
Ny(2)=-(xL(3)-xL(1))/(2.*Vele)
Ny(3)=-(xL(1)-xL(2))/(2.*Vele)
! Face 1--values of diffusivity and velocity
diff_f=(5.*diff(1)+5.*diff(2)+2.*diff(3))/12.
! Face 1--face area normal components
delx= ((xL(1)+xL(2)+xL(3))/3.)-((xL(1)+xL(2))/2.)
dely= ((yL(1)+yL(2)+yL(3))/3.)-((yL(1)+yL(2))/2.)
! Face 1--Diffusion Coefficients
aps(i) = aps(i) + diff_f*(- Nx(1) * dely + Ny(1) * delx)
as(i,j) = as(i,j) + diff_f*( Nx(2) * dely - Ny(2) * delx)
as(i,j+1) = as(i,j+1) + diff_f*( Nx(3) * dely - Ny(3) * delx)
! Face 2--values of diffusivity and velocity
diff_f=(5.*diff(1)+2.*diff(2)+5*diff(3))/12.
```

```
! Face 2--face area normal components
delx= ((xL(1)+xL(3))/2.)-((xL(1)+xL(2)+xL(3))/3.)
dely= ((yL(1)+yL(3))/2.)-((yL(1)+yL(2)+yL(3))/3.)
! Face 2--Diffusion Coefficients
aps(i) = aps(i) + diff_f*(- Nx(1) * dely + Ny(1) * delx)
as(i,j) = as(i,j) + diff_f*( Nx(2) * dely - Ny(2) * delx)
as(i,j+1) = as(i,j+1) + diff_f*( Nx(3) * dely - Ny(3) * delx)
enddo !end loop on nodes/elements in support

as(i,1)=as(i,1)+as(i,n_s(i)+1)
as(i,n_s(i)+1)=0
enddo !end loop on nodes in domain
!================SET BOUNDARY
do i=1,n
BC(i)=0.0
BB(i)=0.0
enddo

do k=1,n_b(1)
BC(B(1,k))=1.e18
enddo
do k=1,n_b(2)
BC(B(2,k))=1.e18
enddo
do k=1,n_b(3)
BC(B(3,k))=1.e18
enddo
do k=1,n_b(4)
BC(B(4,k))=1.e18
enddo
!------------------------------
do i=1,n
QC(i)=0
if (S(i,n_s(i)+1).NE.0) then
QB(i)=vor(i)*V(i)
else
QB(i)=0.0
endif
enddo
!------------------------
do iters=1,20
do i= 1,n
RHS=BB(i)+QB(i)
do j=1,n_s(i)
RHS=RHS+as(i,j)*sf(S(i,j))
enddo
sf(i)=RHS/(aps(i)+BC(i)+QC(i))
enddo
enddo
enddo

end
```

Index

te: Page numbers followed by "*f*" and "*t*" refer to figures an d tables, respectively.

B

C

D

E

F

P

CPI Antony Rowe
Chippenham, UK
2018-12-12 18:09